Principles of Animal Communication

PRINCIPLES *of* ANIMAL COMMUNICATION

Jack W. Bradbury
Sandra L. Vehrencamp

Department of Biology
University of California, San Diego

Sinauer Associates, Inc., *Publishers*
Sunderland, Massachusetts

FRONT COVER
Courtship strut of male sage grouse (*Centrocercus urophasianus*). Only males in the best condition can afford to perform this display rapidly and for long periods. Females compare strut rates and strut performance of nearby males to select mates (see pages 769–770). Photo courtesy of Marc Dantzker.

BACK COVER
Hatchling squid (*Loligo edulis*). The yellow and red spots are true chromatophore pigments. It is not yet known whether these squid see the pattern in color or as shades of gray. Photo by Yasumasa Kobayashi.

PRINCIPLES OF ANIMAL COMMUNICATION

Sinauer Associates, Inc., PO Box 407
23 Plumtree Road, Sunderland, MA, 01375 U.S.A.
FAX: 413-549-1118
Internet: publish@sinauer.com; http://www.sinauer.com

Library of Congress Cataloging-in-Publication Data
Bradbury, J. W.
Principles of animal communication/Jack W. Bradbury, Sandra L. Vehrencamp.
p. cm.
Includes bibliographical references and index.
ISBN 0-87893-100-7 (hardcover)
1. Animal communication. I. Vehrencamp, Sandra Lee, 1948– .
II. Title.
QL776.B73 1998 97–44014
591.59—dc21 CIP

Printed in Canada

5 4

This book is dedicated to Bob Capranica, Don Griffin,
and Peter Marler in recognition of all they have done
to promote the study of animal communication.

Table of Contents

Preface

THIS BOOK IS THE OUTCOME OF 25 YEARS OF GOOD INTENTIONS. It is the much evolved spawn of a popular undergraduate course in animal communication first developed with Robert Capranica at Cornell University in 1970. In 1972, Bradbury moved to Rockefeller University and efforts to write a book for the Cornell course were shelved while pursuing field work in exotic locations. The two current authors began teaching a much expanded version of the original course at UCSD in 1986, and it too became a reasonable success (in one quarter attracting 220 students). Several years later, some of the former Cornell students, now established researchers and teachers, encouraged us to take up again the task of writing an appropriate text. Bob Capranica was planning to retire and fishing looked a lot better to him than writing this book, so we were stuck doing it ourselves if at all. With insufficient trepidation, we set to work. It has proved a Herculean task. So much had changed in the ensuing years that the original manuscripts were little more than springboards. The book took on a frightening scope, and it probably reflects hubris or dementia (or both) to have agreed to write it. On the other hand, it is just that scope that our UCSD undergraduates seem to like: it is the only time they ever get to use physics, chemistry, algebra, physiology, economics, and evolutionary biology all in one course. And for the same reasons, it is a fun course to teach (once you get the lectures written).

This text has more material than anyone (including us) can teach in one term. We actually teach all three parts and most topics in a 10-week quarter, but our lectures cover only a fraction of the text's depth. Our course is aimed at upper division undergraduate students; however, we have tried to include enough material (using boxes) to make the book suitable for graduate courses as well. Each of the three parts of the book could be used as the basis for a more focused and in-depth course, or one could combine any two of them into a course. Whether aimed at undergraduate or graduate students, we believe that the study of animal communication must include some algebra and mathematical concepts: this is essential to understand what communication is all about and how it evolves. Even at UCSD, undergraduate math skills have clearly declined over the last decade and it has become more difficult to cover the topics in the allocated time. To keep the math from bogging down the text, we have tried to put the more mathematical but less-essential treatments into boxes. Students can then read these at their own discretion; they are usually not critical to the remainder of the chapter. In addition, chapters differ in mathematical content. The most mathematical sections of the book are Chapters 3, 13, and 14. A course with minimal math emphasis could skip these chapters if necessary; the general take-home messages are reiterated in later chapters. Although the original Cornell course generated several legendary exam questions (e.g., "the Bulls-eye Snark" and "the Olfactory Indiscretion"), we have not included sample exam materials because faculty lectures are sure to reflect individual emphases and subsets of the material.

It is impossible when a field has so few general treatments not to be somewhat biased by one's own mentors and education. Don Griffin and Peter Marler played major roles in our introductions to behavior. Our earliest sourcebooks were Marler and Hamilton's (1967) text, Margaret Bastock's (1967) classic little volume on courtship, and Griffin's (1958) synopsis of echolocation. A challenging sabbatic in Sussex with John Maynard Smith and colleagues in 1981 clearly left its imprint on Part III. Years of work in various study sites have imprinted certain animal communication examples on us and left us ignorant of others. All of these influences tended to shape our selection of material and our emphases. Although we have tried very hard to present a balanced account, we have surely short-changed some topics, taxa, or authors inadvertently. In some areas, we intentionally avoided detailed treatments. For example, there are several new neuroethology texts out and in the works: there seemed no reason for us to try to duplicate that area, but instead to complement what those texts seek to achieve.

Animal communication is a burgeoning field at the moment. One reason is the complex interface it has with other fields (hence all the topics students get to integrate in the course). As the related fields grow, so does animal communication. The interface with cognitive science is a particularly hot area of expansion (see the 1996 book by Hauser), and evolutionary game theory has truly blossomed since Grafen and Zahavi made us all honest about signaling. There are many other interface areas that are just now getting started. It is a very exciting time to undertake research in this field. Maybe now that this book is finally done, we can get back to doing some ourselves.

JACK W. BRADBURY
SANDRA L. VEHRENCAMP
December, 1997

Acknowledgments

WE COULD NOT HAVE WRITTEN THIS TEXT without the enormous help of many colleagues and students. The numerous undergraduate and graduate students who have taken our course over the years have contributed useful ideas and feedback at every stage; in our unbiased opinion, UCSD students are the best anywhere. Early drafts were critiqued and improved by our graduate students Lisa Angeloni, Marc Dantzker, Laura Molles, Helen Neville, Cat deRivera, and Tim Wright. We are tremendously appreciative to Allison Alberts, Staffan Andersson, Kathy Cortopassi, Fred Dyer, Pascal Gagneux, Carl Gerhardt, Peter Narins, Trevor Price, Rod Suthers, Haven Wiley, and Jerry Wilkinson for reviews of early drafts and negotiations for photos. Mike Ryan, Harold Zakon, and Carl Hopkins supplied drafts of their own lecture materials to help focus our treatments, and Tom O'Neil, Howard Howland, and David Rapaport provided important consultation on light and vision topics. Oren Hasson sent us early drafts of new manuscripts at critical times. Mike Beecher, Eliot Brenowitz, and Haven Wiley provided feedback from trial runs of Part I on their students. A very large number of colleagues have contributed illustrative material; they are cited in each case and have our deepest appreciation. Our secretary, Judie Murray, oversaw the mammoth task of obtaining permissions and provided help at many other junctures. This book would never have been written without the superb mentoring and friendship of Bob Capranica, Don Griffin, and Peter Marler; they have left their mark on us and everyone in this field and this book is dedicated to them. Finally, Andy Sinauer has waited 25 years for this book; he has to be one of the world's most patient and supportive publishers. His staff, including our editor Nan Sinauer and graphics director Chris Small, did an outstanding job integrating what was a very complex operation (especially the exchange of corrected galleys while we were in Costa Rica). Nancy Haver and Abigail Rorer drew the superb line and stipple art in the book, and Precision Graphics created the many graphs and diagrams. To all of the above, our heartfelt appreciation.

Chapter 1

Introduction

THIS BOOK IS ABOUT HOW ANIMALS COMMUNICATE with each other and why they do it the way they do. The biological world is full of the smells, sounds, movements, and electric signals by which animals communicate. In fact, we could fill the entire volume just describing the diversity of animal signals. Our concern, however, is to extract the general principles that govern the evolution of animal communication systems. Is there any order in all the diversity? Can one come up with a limited set of evolutionary rules with which to predict the likely communication system of a species, given information about its life style, social system, habitat, and phylogenetic history? The study of animal communication, like the study of animal social systems to which it is closely linked, is now at a point where such rules are emerging. In this book, we try to identify as many of the major rules as possible and show how they interact to shape signal evolution.

WHAT IS ANIMAL COMMUNICATION?

Evolution hates definitions; for every tidy definition a biologist constructs, evolution will have provided some exception. In a way, this should not be surprising. The longer-term success or failure of an organism often depends upon novel combinations of traits that emancipate it from current problems with predators, prey, competition, or disease. Such new combinations favor multidimensional diversity and that diversity hinders the creation of tidy distinctions by biologists. Another problem with definitions is that they are invariably categorical. Even if the criterion trait is a quantitative one, we tend to assign all values below some threshold to one category and all others to a second. Biological definitions are useful only to the degree that they reflect underlying evolutionary processes or conditions. If the underlying processes grade into each other, a categorical definition will surely generate exceptions. The problem is compounded by our tendency to assemble suites of traits as criteria for definitional categories. When acceptable values for all traits are found in the same animal, these categorical assignments are straightforward. The problem arises when some unique animal meets all but one of the criteria. Does one then establish an entire category for this exceptional combination of trait values or should we just start over?

Our feeling is that categorical definitions are best used as rules of thumb. If properly designed, they sort out most of the cases encountered, and the exceptions can be logically explained given the underlying evolutionary rules that the definitions reflect. Under these conditions, definitions are useful tools for research and teaching. A set of definitions that generates too many exceptions is likely to be based on erroneous underlying rules; researchers then need to go back and look again. The maturity of a science is reflected in the stabilization of its definitions and underlying rules. Is the study of animal communication at such a point? It is certainly close. While there are still differences of opinion about what should and should not be considered animal communication, the various opinions usually reflect recycled versions of earlier definitions. The current differences are largely quantitative, centering around how restrictive a definition one should invoke.

Nearly all authors agree that communication involves the provision of **information** by a **sender** to a **receiver**, and the subsequent use of that information by the receiver in deciding how to respond (Figure 1.1). The vehicle that provides the information is called the **signal**. Provision of utilized information is a necessary part of any definition of communication, but few would consider it sufficient. For one thing, defining communication this way passes the problem onto the definition of information, a task we take up in Chapter 13. Accepting for the moment that we can agree on a definition of information, what other criteria are usually invoked to characterize animal communication? There are two major ones used to define "**true communication**" (Marler 1977; Markl 1985; Dusenbury 1992). The first is that the provision of information is not accidental but occurs only because it benefits the sender. Such benefits are accrued by altering the likelihood that the receiver will respond one way instead of another. Note that animals can also produce stimuli that are used by

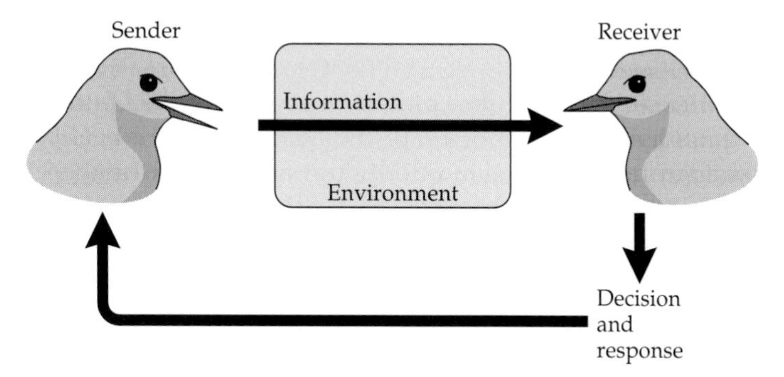

Figure 1.1 The process of communication. Communication involves two individuals, a sender and a receiver. The sender produces a signal which conveys information. The signal is transmitted through the environment and is detected by the receiver. The receiver uses the information to help make a decision about how it should respond. The receiver's response affects the fitness of the sender as well as its own. In true communication, both sender and receiver benefit (on average) from the information exchange.

other animals to the sender's detriment. For instance, owls find and attack mice as a result of the inadvertent sounds the mice produce when foraging. Stimuli whose perception by other animals is not beneficial to the emitter are usually called **cues** (Seeley 1989), and the exploitation of these by a receiver is called **eavesdropping**. The first criterion thus distinguishes between signals, in which the transfer of information benefits the sender, and cues, in which it does not. The fact that senders benefit from signal production sets the scene for evolutionary refinement and elaboration of signal form to enhance information transfer. In fact, the evidence of substantial costs for stimulus production and the absence of evidence for any function for stimulus emission other than information transfer are often used to justify identifying particular stimuli as signals. Were there not benefits to information transfer, why would the sender bear these costs? In contrast, cues either create little cost or they have clear functions other than provision of information. Hence the definitions used in this criterion reflect quite different evolutionary trajectories.

The second criterion is that the receiver must also benefit by having access to the provided information. One can certainly envision situations in which the optimal response for the receiver is not that sought by the sender. The sender may then be tempted to provide **misinformation**. Clearly, any coevolution between sender and receiver will take different paths depending upon whether the two parties have similar or different interests in **honest communication**. If the preferred receiver response is the same for sender and receiver, adaptations by either party that enhance the provision of information will tend to be favored; where they differ in interests, one expects a complicated arms race involving increasingly subtle deceit by the sender and greater discrimination or discounting by the receiver (Krebs and Dawkins 1984). Again, the relevant distinctions reflect different underlying evolutionary scenarios.

Recent definitions of communication vary depending upon how strictly they invoke these two criteria. Wiley (1994) took a very broad tack and defined communication as any alteration in a receiver induced by a signal. He then defined a signal as a stimulus released by a sender that elicits a response in a receiver but has insufficient energy to power that response. Wiley invoked this latter requirement to distinguish between sender actions that only provide information, and those that manipulate a recipient directly. Wiley's definition by itself did not directly invoke either of the two criteria. In the same review, he later argued that the receiver must benefit, on average, from attention to signals, and thus he implicitly invoked the second criterion. However, he was quite critical of some attempts to invoke the first criterion. This stance was in part a reaction to the historical invocation of "presumed intent" to distinguish between cues and signals. In this tradition, a signal reflects the intentional transfer of information, whereas cues are released without intent to provide information. The problem, of course, arises in the determination of an animal's intent, since even with humans, true intentions can be difficult to assess. One attempt to circumvent the intention problem can be seen in the more recent emphasis on whether receiver detection of animal-emitted stimuli is beneficial to the sender. Although benefit assessment is also a challenge, behavioral ecologists have gotten quite adept at measuring benefits and costs of different behavioral actions in the last decade. Wiley, in fact, invoked sender benefits as a condition of communication in an earlier review (1983).

Wilson (1975) also invoked only one criterion, but did not specify which: he accepted as communication any exchange in which at least one of the parties benefited from the information transfer. True communication, eavesdropping on cues, and provision of misinformation would all be accepted under his rubric. The most common invocation of one criterion was that represented by Otte (1974) and Burghardt (1970), who included the first criterion but omitted the second. In this definition, cues are distinguished from signals, but honest and deceitful exchanges are both considered to be communication. Finally, many researchers insist on the invocation of both criteria and thus have focused only on "true communication."

Additional distinctions have been suggested. Some authors have proposed restricting communication to intraspecific exchanges, but few modern workers would accept that limitation. Green and Marler (1979) and Hauser (1996) divided true communication into subcategories depending upon whether the signal is a "state" or an "event." States include plumage coloration, uncoverable badges, fixed body scents, and any other signal that is permanently "on." Events include sounds, electric discharges, visually detected movements, and quick touches that are only "on" for short periods. Green and Marler argued that states do not evolve as signals per se but instead are ancillary traits designed to accentuate true signals. Hauser made a similar distinction and stirred the pot by reassigning the same terms to different meanings. His reason for the distinction is that the costs for states are paid prior to their use, whereas those for events tend to be paid during their expression. We do not feel that this distinction is at all a clean one nor as important as Hauser has suggested, but we do take up the consquences of the timing of signal costs in Chapters 17 and 20.

In this book, our definition of animal communication is initially narrow and broadens with successive sections. In Part I we do not invoke an explicit definition, but focus on the mechanics of signal production, propagation, and reception for examples that nearly all readers would consider to be communication. Part II then provides an explicit definition and a formal model, given true communication. That is, throughout Part II both criteria are assumed to be met. Finally, in Part III we relax each criterion and allow senders to deceive receivers and receivers to exploit senders. How conflicts of interest might shape signal evolution is the major focus of Part III. In a sense, the entire book is our definition of animal communication. We initially outline the more general rules shaping signals, and then gradually accommodate the exceptions by refining the rules and definitions. In the process we hope to capture most of the underlying evolutionary processes and give them some hierarchical order.

WHY STUDY ANIMAL COMMUNICATION?

There are several motivations for studying animal communication, the first of which is its central role in animal societies. Anyone who examines animal behavior, whether from the perspective of classical ethology, behavioral ecology, or psychology, knows that communication is the glue that holds animal societies together. In all sexual animals (and even in vertebrate parthenogenetic ones), reproduction is not possible without communication (if not at the level of adults, then at least at the level of gametes). Highly social animals have complex interactions and communicate information above and beyond that required for mating; these transactions require their own suites of signals. One might expect communication systems in solitary or unsocial animals to be less complex. However, even in these cases, elaborate signals may be required to establish and maintain the species' dispersed spatial patterns.

Another reason to study animal communication signals is to identify the kinds and amounts of information animals convey to each other. We shall see that animals send many different types of messages: information about their identity (such as their species, sex, group membership, or individual identity), information about their status and mood (such as dominance, fear, or aggressive motivation), information about what they are likely to do next (such as approach, flee, mount, or groom), and information about relevant discoveries in the environment (such as predators or food location). Animals may also send information to their predators and even to themselves in order to navigate and obtain information about the environment. Messages vary in how much information is exchanged and how much the exchange costs either party. Such costs might include energetic outlays, losses of time, exposure to predators, and investments in sensory and brain tissues. Are some types of messages too expensive to be economically warranted except in unusual cases? How does the amount of sensory or brain tissue required for an exchange scale with the message or information task? For example, does the provision of twice as much information require twice as much brain tissue? The types and amounts of information are thus of great interest to both behavioral ecologists (who study the overall economics), and neurobiologists, cognitive scientists, and animal psychologists (who are concerned with how communication tasks are related to neural circuitry and sensory structures).

Animal signals evolve, so the study of animal communication can also be used as a tool for elucidating general evolutionary principles. For example, communication signals may be examined along with morphological traits for the taxonomic classification of species. Furthermore, signals are a component of an animal's adaptation to its environment, and selection optimizes them in the context of background noise, predator detection, and signal efficacy. Studies of signal optimization thus demonstrate the intensity and process of natural selection. Because mate attraction and selection usually involve signals, animal communication is also an important factor in studies of reproductive isolation and speciation.

Animal communication can be recruited as a partial solution to various practical issues. Signals are often very conspicuous and can be used to census endangered populations and determine species diversity. Studies of mate attraction signals have successfully been applied to the control of pests. In fact, much of the recent work on olfactory mate attraction systems in insects was motivated by biological control and reproductive suppression schemes. Finally, an understanding of animal communication can be used to improve the welfare of both wild and domestic animals and the breeding of endangered species in zoos and national parks. Current attempts to save the giant panda include detailed studies of this solitary species' chemical signals.

APPROACHES TO THE STUDY OF ANIMAL COMMUNICATION

Animal communication is a diverse topic requiring approaches ranging from basic physics through behavioral ecology to neurophysiology. Communication studies often require mastery of several disciplines; the field is a truly integrative one. Biology majors will find themselves relying heavily on their backgrounds in physiology, behavior, ecology, and evolution, and will finally discover a use for all that mathematics, physics, and chemistry they took their first two years. Students with some economics training will find numerous applications in animal communication studies. Despite the diversity of fields required, none of the material is so technical or mathematical that a person interested in animal communication cannot master it. We have tried throughout the book to focus on the principles governing signal evolution without segregating them by parental discipline. We feel that this approach provides the best way to master the topic.

The basic logic of the text is as follows. Part I examines the mechanics of communication. Not all structures, products, or behaviors are equally effective as signals. In any given environment, a song, a flash of color, a puff of odorant, and an electric pulse will not be equally detectable at a substantial distance from the sender. The physics of each potential signal will limit the propriety of that signal, and these limits will vary depending on the specific environment (e.g., air versus water, closed forest versus open savannah, day versus night). A typical example is shown in Figure 1.2. In Part I we shall look at the alternative ways senders can produce signals, the nature of signal propagation through each environment, and the types of receptor organs that receivers would need. For many scientists, study of the production, transmission, and reception of signals is an end in itself, but here it is viewed as a necessary first step in understanding the physical constraints and tradeoffs in signal evolution. There will be little behavior discussed in this part of the book, and large doses of physics, chemistry, anatomy, and physiology. We do not discuss the neurobiology of signal production and reception above peripheral organs at great length because much of this material is covered in existing texts on neuroethology.

In Part II, we take up the economics of animal communication. Animals that send and respond to signals do not do so "for free." Both the sending and receiving of signals have associated costs and benefits, and we might expect natural and sexual selection to favor certain potential signals over others because they maximize the fitness of the participants (Figure 1.3). For example, a frog that calls continuously at a very loud level may maximize its chances of attracting a female, but it also increases its chances of being located by a

Figure 1.2 Acoustic signaling by humpback whales (*Megaptera novaeangliae*). Large body size allows whales to produce sounds of high intensity and low frequency. Both properties increase the range over which they can be detected by conspecifics. Calling individuals select particular depths and bottom substrates, and channel sounds so that they are detectable hundreds of kilometers away. This demonstrates the complex matching of signal form, ambient medium, and body size which is seen during long distance communication. (Painting by Richard Ellis.)

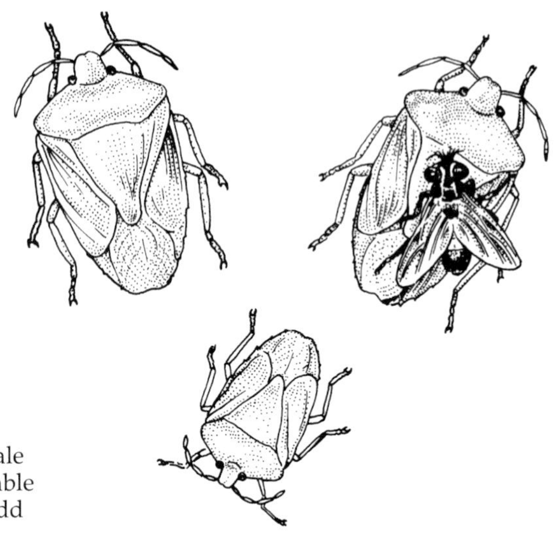

Figure 1.3 Costs and benefits of sex-attractant pheromone release by southern green stink bug (*Nezara viridula*). Male stink bugs release a pheromone to attract females for mating. The pheromone also attracts female tachinid flies (*Trichopoda pennipes*), which lay small maggots on the stink bugs. The maggots burrow into the bugs and eventually kill them. Male bugs are more likely to be parasitized than females, although both sexes may be affected. Stink bugs reduce further risk by exchanging courtship vibrations through the leaf substrate. Broadcast sounds would not solve the problem, as other tachinid flies use airborne sounds to find their prey. Calling male crickets, for example, are particularly vulnerable to this cost of signaling. (After Harris and Todd 1980; Ryan and Walter 1992.)

predator or of running out of energy before the season is over. Such tradeoffs are the rule in nature, and we expect selection to find the optimal balance of options. We thus begin this section by defining what is meant by information and we then discuss the general benefits and costs of exchanging it. We go on to outline methods to quantify the amounts of information provided by different types of signals. We then turn to the receiver's point of view and show how an ideal receiver detects and evaluates information and makes appropriate decisions about responses. Next, we examine the sender's options for encoding information in signals and the degree to which sender and receiver must share the same coding rules. The actual signals available to any given species depend both on its phylogenetic history and on the costs and benefits specific to its environment. Finally, we ask whether there are convergent patterns of signal form across species for signals that serve a similar function and thus presumably contain a similar type of information (design rules). Throughout Part II, we assume that senders and receivers share the same interest in transmitting information, i.e., that senders send the information that receivers want to have. The theme of this part of the book is thus **optimization** of signaling systems: given cooperative parties, what kinds of signals should they jointly adopt to maximize their respective fitnesses?

In Part III, we relax the presumption that sender and receiver have a common interest in effective communication. For example, must female birds comparing potential mates be on guard against false male advertising? Do fighting bull elephants have to decide how much of their opponent's trumpeting is bluff and how much is a serious commitment to win a prolonged and dangerous battle? How should male fireflies decide whether a flashed response to their own signals is that of a receptive mate or that from a predatory mimic that will eat them if they approach too closely? Conflict between sender and receiver adds a new dimension to signal evolution and requires an explicit analysis of the signal exchange as a game. The degree to which game-theoretic considerations override or modify the simple optimality models of Part II depends largely on the level of this conflict. The greater the disparity in sender and receiver interests, the more guarantees of honesty must be built into signals to justify receivers attending to them. Part III reexamines each basic category of social exchange to see whether game theoretic constraints must be added to explain signal evolution or whether the simple optimality models of Part II are sufficient (see Figure 1.4 for an example).

BALANCE AND BREADTH

Throughout the text, we have tried to maintain as even a treatment of taxa and modalities as possible. This approach is what is required for extracting general principles. However, it is difficult to maintain a completely even treatment in practice because the animal communication literature is not evenly distributed over taxa or modalities. There are obvious reasons for these biases. For one, some modalities are easier for humans to study than are others. We know most about sound communication because our own hearing is very

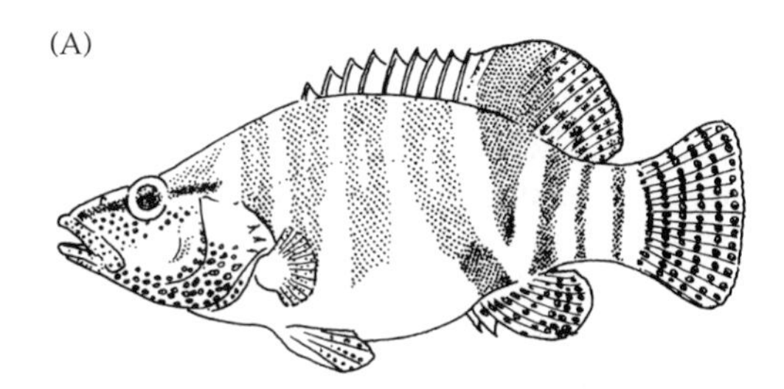

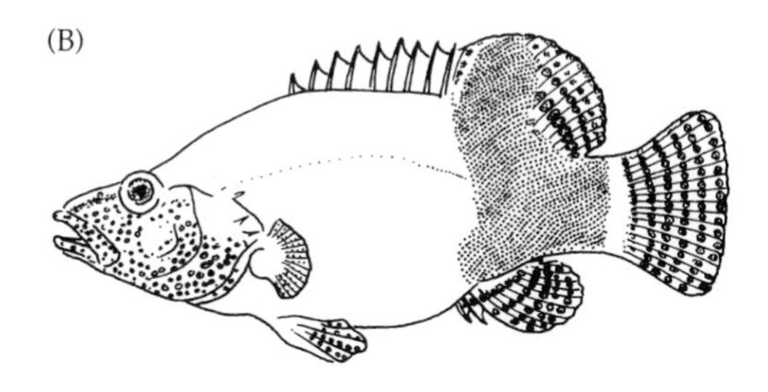

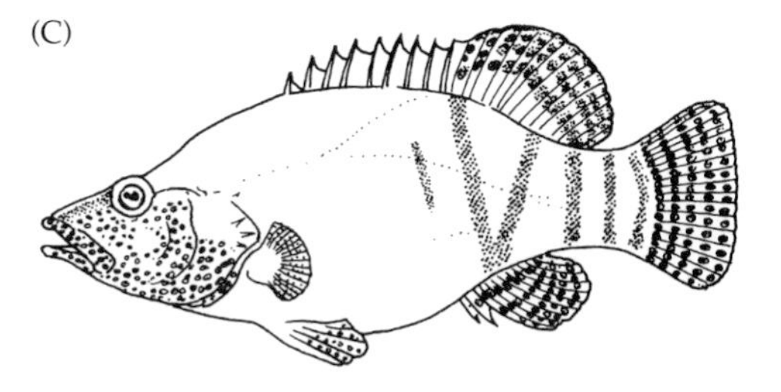

Figure 1.4 Visual signaling of sexual role during courtship in a hermaphroditic sea bass, *Serranus subligarius*. In many hermaphrodites, each individual in a spawning pair would prefer to play the male role because sperm are cheaper to produce than eggs. The conflict of interest is solved in these small marine bass by taking turns first as one sex and then as the other. This reciprocation requires careful coordination and is invariably mediated by signals. In this example, the fish currently adopting the male role shows the banded pattern in A, and that adopting the female role shows the dark pattern in B. Spawning will occur only if each party signals its intention to play a role opposite to that of the other. When this is true, the female partner changes her color pattern to that shown in C; this coloration is the signal for both to release their gametes simultaneously. (After Demski and Dulka 1986; Demski 1992.)

good, microphones can detect frequencies outside of our hearing range, and sound signals can be easily recorded and quantified. In addition, any loud sound an animal produces is bound to be a signal, so we are confident in searching for its communication function. Analysis of visual signals lags behind that for sounds. We have good visual perception and instruments to detect colors and flickering patterns beyond our visual range, but an enormous amount of visual signal processing occurs at higher levels in the brain and it is more difficult to quantify what animals actually perceive. In addition, it is often difficult to know what acts are signals and what acts are nonsignaling behavior since animals continually project a visual image. We have no perception at all of weakly electric signals, but our instruments can easily record and quantify such signals for the few animals that produce them. Chemical communication is the most difficult modality to study. Our own perception, while present, is poor relative to other animals that employ olfactory signals. It is also much more difficult to detect, identify, and quantify olfactory signals, and it is even more difficult to distinguish signals from nonsignaling chemicals exuded from the bodies of animals. As for tactile signals, we do not have a separate section in Part I but instead discuss them in various chapters in Parts II and III.

Taxonomic biases are partly due to economic considerations. It is easier to obtain government funding to study an animal model that may shed light on problems of speech development or visual processing in humans. Similarly, many more funds are available for studying chemical signals of insect pests than for studying the auditory or visual signals of commercially unimportant species. However, despite funding limitations, the unquenchable desire to undertake basic research has resulted in some studies on each likely modality in an enormous variety of taxa. The overall result is coverage that is taxonomically broad but uneven in depth.

A major goal of this book is to summarize common principles in such a way that they could be applied to any taxon or modality. Students and researchers working on a specific modality and taxon can often benefit from the greater perspective provided by knowing about prior studies on different modalities and taxa. There do seem to be basic principles governing the evolution of animal communication. Seeing how these apply in one well-studied system can often suggest new experiments and interpretations for another. It is our hope that this text will bring together enough of the basic perspectives that it will encourage more cross-talk about different taxa and modalities. What better justification for a broad and liberal education than the study of animal communication!

COMPLEMENTARY READING

Before tackling this book, some readers might want a more general introduction to animal communication. Where might they go? Several good sources are textbooks in animal behavior that have sections specifically on this topic. Examples include the volumes by Alcock (1993), Krebs and Davies (1993), Goodenough et al. (1993), Manning and Dawkins (1992), and Wilson (1975). For years, the best book available for a course on the subject was the small volume edited by Halliday and Slater (1983). The excellent chapters in this book remain a good place to start on the subject. Two older volumes discussing signal evolution are by Smith (1977) and Lewis and Gower (1980). Other treatments may prove useful to readers for more detail or a different slant on the same topics. The careful treatment by Dusenbury (1992) is an excellent complement to the topics raised in Part I of this book. Hauser (1996) takes up many of the issues raised in this volume and compares the results of animal and human studies. Readers eager to obtain more detail on communication in a particular taxon should consult the two volumes edited by Sebeok (1968, 1977).

Part I

Production, Transmission, and Reception of Signals

WE BEGIN THIS BOOK WITH A REVIEW of the physical and physiological properties that constrain the evolution of animal signals. Although animal signals are amazingly diverse, a sender does not have unbridled choice of signal form: the environment in which the signal is given and the physiological equipment that both sender and receiver can bring to the task of communication severely limit the propriety of different signal forms. To understand signal diversity we thus need to have a thorough knowledge of these constraints. We also need to understand physical and physiological properties if we want to measure and compare signals in biologically realistic ways. In the following chapters we examine each of the major signal modalities in turn:

sound, light, chemistry, and electricity. For each modality we outline the physics of signal structure and signal propagation, and the physiology of signal generation and reception. We start and spend more time with sound because it is the modality best known to most readers, and because sound provides an easier introduction to a number of topics applicable to all of the modalities.

Chapter 2

The Properties of Sound

SOUND IS WIDELY USED BY ANIMALS FOR COMMUNICATION, and is the original modality for human language. In this chapter, we review the physical properties of sound waves, how they can be described, and how they are affected during propagation through a medium. As we shall see later, a number of these principles also apply to light and electric signals. A final concept introduced in this chapter is the linearity of many signal systems. This refers to the fact that multiple signals usually combine additively and propagate without distorting each other. A receiver can then decompose the compound signal into its original parts for sensory analysis. This is a very important property that has allowed animals to evolve quite complicated types of signals.

WHAT IS SOUND?

Sound cannot propagate in a vacuum but instead requires a medium full of molecules. Left undisturbed, the molecules in a medium vibrate, and if the medium is a gas or a liquid, they will travel in random directions. The motions of a molecule in any direction are constrained by collisions with other molecules. When molecules are forced closer together than is typical for that medium, the concentrated molecules experience more collisions with each other than with molecules from outside the concentration. This results in a net movement of the concentrated molecules out of the area of concentration until they are as far apart on average as other molecules in the medium.

A sound is generated by producing a local concentration of molecules in a medium. For example, suppose a cicada is producing its loud buzzy sound while hanging from a thin twig. Cicadas are insects that produce sounds by popping the surface of a small drum called a **tymbal**. Muscles pull the semirigid tymbal membrane into a distorted position and then let it "pop" back into its original shape. As the membrane snaps back, it concentrates the air molecules in its path. These molecules collide with the adjacent layer of molecules. The collisions force the second layer of molecules outwards in turn, whereas the first layer of molecules rebounds back toward the tymbal membrane. The second layer of concentrated molecules moves out and collides with a third layer and thus the concentration of molecules moves as a sphere of increasing diameter away from the cicada. Notice that no single molecule moves along with the sphere of concentration. Instead, it is the **disturbance** that is propagated to greater distances from the buzzing insect. At each successive collision of molecules away from the tymbal, some of the disturbance energy is lost as heat (e.g., random movements of molecules). At a sufficient distance from the insect, all of the original energy released by the tymbal pop has been converted to heat, and no sound can be detected.

Transverse versus Longitudinal Waves

The cicada tymbal causes molecules near it to move away from the cicada until they are bounced back toward the sound source by collisions with the next layer of molecules. The movements imposed on surrounding molecules by the sound will be parallel to the direction in which the sound propagates. A very different kind of disturbance occurs when one ties a rope to a tree, puts some tension on the rope while holding the other end and moving it up and down. Waves will move down the rope toward the tree and away from one's hand. However, the movements imposed on any one part of the rope as these waves pass are perpendicular to the rope and thus perpendicular to the direction of propagation of the waves. Waves in which the exerted forces are perpendicular to the direction of wave propagation are called **transverse waves**. This type of wave is what a guitar string experiences when it is plucked. Light is also a transverse wave (although it does not need a molecular medium in which to propagate). Waves that move molecules back and forth along the same axis as the direction of wave propagation are called **longitudinal waves**. Sound in gases and liquids is based on longitudinal waves;

sound in solids may be based on either or both transverse and longitudinal waves. In this chapter we concentrate on longitudinal wave sounds.

Sound Pressure and Properties of Waves

Although real cicadas can generate rather complicated sounds, let us imagine a very simple species that moves its tymbal surface back and forth at a very constant rate, f times per second, and in such a way that the deviation of the membrane from its relaxed position, δ, at any time t can be predicted using a sine wave function:

$$\delta(t) = D\sin(2\pi ft + \Phi)$$

Here, D is the maximal distance away from or toward the body of the cicada that the membrane moves, Φ is the initial **phase** or position of the membrane at $t = 0$, sin (x) is the usual sine trigonometric function, and π is the usual trigonometric constant ($\pi = 3.141592654\ldots$). Such a membrane is said to be oscillating in a **sinusoidal** manner because its motion can be described by a sine wave. (Alternatively, the motion can be described by a cosine wave. Note that the two functions differ simply in that the appropriate phase, Φ, will be increased by $\pi/2$ if we use a sine function instead of a cosine function.)

Now suppose we had a sensor that could measure the force exerted on its surface by colliding molecules from the surrounding medium (see Box 2.1 for examples). The force of medium molecules per unit area of sensor is called the **pressure**. Let us place the sensor at some distance from our sinusoidally buzzing cicada. We first measure the pressure of the medium during a period when the cicada is silent. This is the **ambient pressure**. When the cicada again begins its call, our sensor will soon detect an increase in collisions as the first wave of molecular concentration passes over it and moves away. After the first concentration front moves past the sensor, there will be a shortage of molecules over the sensor (compared to ambient conditions), because many molecules have moved away to form surrounding concentrations. The sensor records this change as a lowering of the pressure below ambient levels. When the next concentration front arrives, the sensor shows a return to ambient pressure and then a rise to high values again.

For the simple cicada sound described above, the pressure measured by the sensor will rise and fall in a periodic manner. If there are no boundaries that the spreading spheres of concentration might hit, the medium is said to form a **free field**; that is, the only factor that affects the directions in which the sound disturbance spreads is the location of the cicada. In a free field, one can make several simple measurements. If we were to monitor the pressure at any fixed point, we would see a sinsuoidal rise and fall of pressure at our sensor (Figure 2.1). This temporal pattern of rise and fall in signal amplitude is called the **waveform** of the signal. In this case, the waveform is sinusoidal. The time interval between the arrival of successive sinusoidal peaks at our sensor is called the **period** of the sound wave. Since the period is the duration of each cycle of the sound wave, its reciprocal is the number of cycles per unit time or the **frequency** of the wave. The unit of frequency, cycles/sec, is called the

Box 2.1 Recording Sounds

A DEVICE WHICH CAN PROVIDE AN ELECTRICAL RECORD of varying pressure is called a **microphone**. One of the most widely used designs is called the **capacitance** (or **condenser**) **microphone**. It consists of a metal plate and a thin membrane of plastic (called the **diaphragm**) on which a fine layer of metal has been deposited:

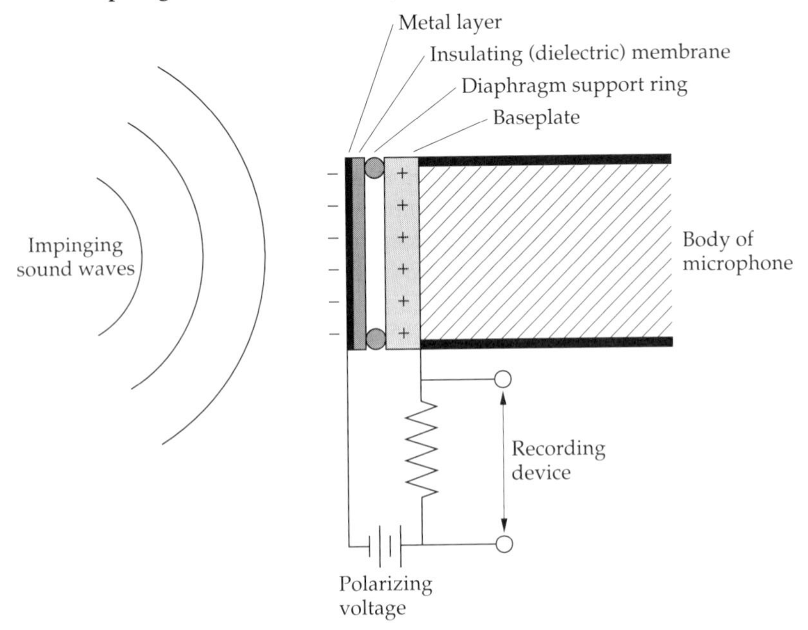

The diaphragm is placed near and parallel to the plate with the plastic side toward the plate and the metallized side facing incoming sounds. (Typically it is supported on its periphery by a supporting ring.) Any such sandwich of metal-insulator-metal is called a **capacitor**. A voltage is then applied between the metal plate and the metal layer of the diaphragm. If the metallized layer is made negative, electrons accumulate here, and their field forces electrons out of the nearby plate. The number of electrons moving into the diaphragm and out of the plate depends on the distance between the diaphragm and the plate. As long as this distance remains fixed, the number of electrons accumulated on the diaphragm remains constant. However, when sounds hit the diaphragm, the thin membrane is pushed towards and away from the plate as pressure rises and falls, respectively. When the diaphragm moves towards the plate, more electrons move out of the back plate; as the diaphragm moves away, electrons move back into the plate. This movement of electrons into and out of the plate constitutes an electric current that moves one way or the other in synchrony with the movements of the diaphragm. As long as the diaphragm moves easily and quickly, its movements, and thus the electric current, will mimic the waveform of the incident sound. This electrical model of the sound waveform can then be monitored or recorded on any number of devices.

Other types of microphones include **piezoelectric crystals**, which generate electrical signals when they are deformed by the varying pressure of impinging sounds, and dynamic and ribbon microphones, which couple the movement of a diaphragm to that of a metal conductor in a magnetic field. Moving a conductor in a magnetic field generates electrical currents in that conductor and thus creates an electrical analogue of the pressure variations.

As noted later in this chapter, most of the energy in a sound propagating in air will be reflected when it encounters a solid object because the acoustical impedances of air and solids are so different. Piezoelectrical microphones used in air may tend to produce very tiny and often ineffective signals. However, if the diaphragms of capacitor or dynamic microphones are made to be very thin and light, their average densities can be made more similar to that of air. This construction causes their acoustical impedances to be closer to that of air, and sufficient energy is transferred to them to produce sizable electrical signals. Because the acoustical impedances of water and solids are much more similar, piezoelectrical crystals in water do absorb significant amounts of incident sound energy and are often used for **hydrophones** (underwater microphones).

Hertz (abbreviated **Hz**). If T is the period of a sound in seconds and f is the frequency of a sinusoidal signal in Hertz, then, $f = 1/T$.

Suppose we now set an array of sensors at successive distances along the line of sound propagation away from the cicada. We can then take a suite of measurements at any given instant, which would also appear as a sinusoidal pattern when the pressure at each sensor is plotted versus the distance of each sensor from the insect (Figure 2.2). The distance between successive peaks at any given instant is called the **wavelength** of the sound. Let us denote this value by λ. Suppose we record the amount of time it takes any given pressure peak to travel from one sensor to the next. The distance

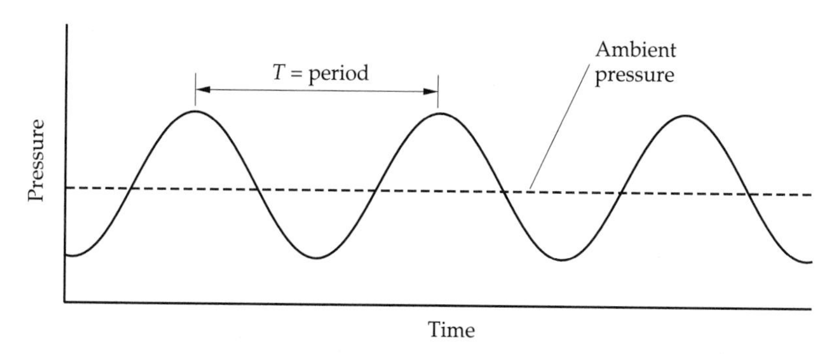

Figure 2.1 Plot of varying sound pressure at one location. Vertical axis gives variation of pressure around ambient level at sensor over time.

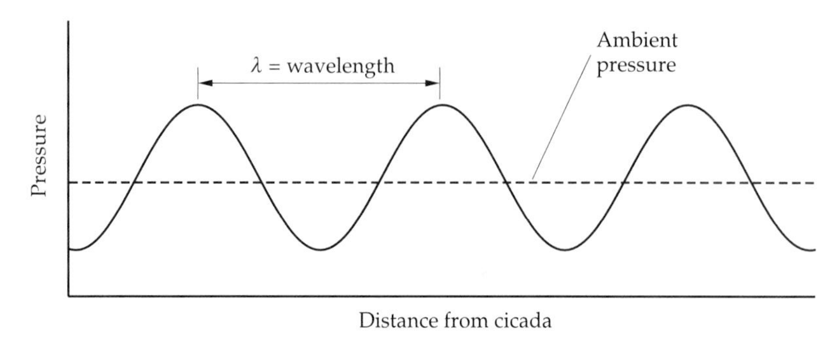

Figure 2.2 Plot of varying sound pressure at increasing distances from the source. All samples are assumed to be taken at exactly the same instant in time.

between the sensors divided by this measured delay would equal the **speed of propagation** of the sound. We shall denote the speed of propagation by c. If a single cycle of the sound takes T seconds, and each part of a given cycle of sound travels at a speed of c meters per second, a moment's reflection should convince you that the wavelength, λ, is equal to the product of period and speed of propagation: $\lambda = c\,T$. Since the frequency of the sinusoidal signal, f, is inversely related to the period, T, it follows that $\lambda = c/f$, or, rewriting, that $f\lambda = c$.

Since the speed of propagation at reasonable distances from the sound source is independent of frequency in most media, this final equation means that frequency and wavelength are inversely related for any given medium: high frequencies have short wavelengths and low frequencies have long wavelengths. It also means that this relationship will change when we shift from a medium with one speed of propagation to another with a different speed. For example, the speed of sound in air is about 344 m/sec (or on a smaller scale, about 34 cm/msec). In water, sound travels at a speed about 4.4 times faster than in air, and in solids, about 3–4 times faster than in water. If you hum a song at 1000 Hz in air, the wavelengths will be about 34 cm in length (the period of a 1000 Hz sine wave is 1 msec). If you hum the same tune while under water, you will produce waves that are about 1.5 m in length. If, once dead and buried, your last gasp is to hum the same tune, the waves propagating in the ground would have wavelengths of 5–6 m in length. As we shall discuss in detail later, the physical dimensions of sound waves set major constraints on the production, propagation, and reception of sound signals.

Note that measurements of frequency, wavelength, and period are much harder to make if the disturbance is not in a free field. Once the spreading sound hits a boundary, it is reflected and begins to propagate in a different direction. Since both the original sound and reflected components might pass simultaneously over a given sensor, it will be difficult to determine the frequency of pressure peaks, the speed of propagation, or the wavelengths of the original sound.

Near Field versus Far Field

In the immediate vicinity of the cicada, individual air molecules move back and forth substantial distances as a result of the tymbal movements. However, as a sphere of concentration moves away from the tymbal, eventually a distance is reached at which the molecules propagating the disturbance do not move much further than they would have moved without the pop. In fact, the only measurable indication that a disturbance is present is a variation in pressure. The zone immediately around the tymbal in which molecular displacement is measurably greater than normal is called the **near field** around the sound source; the outer zone in which the sound disturbance is propagated only as pressure variations is called the **far field**. The transition from near to far field occurs at one to two wavelengths from the sound source. Because it is difficult for most animals to produce wavelengths larger than themselves (see Chapter 4), this means that the near field typically extends at most several body-lengths away from the animal. The higher the frequency, the closer to the source is the transition between near and far fields.

Because the relationships between molecular movements and pressure in the near field are very complicated and highly dependent on the distance from the source, it is easier to characterize sounds in the far field. Typically, one measures variations in pressure as the sound passes a sample point. Note that pressure is a property with no directional aspects: a tiny ideal sensor should detect the same pressure no matter how it is aimed in a free field. Sensors can be made directional, however, by constructing them to distort the field around them or by sampling at several locations and mixing the resulting signals.

Doppler Shifts

Sounds emitted by a moving source or recorded by a moving detector exhibit a change in their component frequencies and durations as a result of this motion. Consider a sound source moving in the same direction as the propagation of its emitted sound. Each successive sound peak will have to propagate a shorter distance than its predecessor because the source has moved that much closer to the detector. As a result, successive sound peaks arrive at the detector with less delay between them than if the source had been static. The detector thus records a higher frequency than is seen in the vibrations at the source. Because each successive piece of the sound arrives with a smaller delay than was present during emission, the total duration of the perceived sound is also reduced. If the source is moving away from the direction of sound propagation, the reverse happens: all component frequencies are decreased, and the overall sound duration is increased. Similar changes occur if the detector is moving: trajectories toward the sound source increase perceived frequencies and decrease durations, and trajectories away from the source have the opposite effects. If both source and detector are moving in the same direction and at the same velocity, then the downward shift experienced by one party is compensated by the upward shift by the other; there is no net change. Quantitatively, the fractional change of frequency due to source or detector motion depends upon the relative velocity of sound source and detector and the velocity of sound in that medium. These changes are called **Doppler shifts**.

Note that for most animal communication problems, Doppler shifts are probably not significant. A call of 10 sec duration and 1000 Hz composition produced by a bird flying at 5 m/sec toward a static listener will be perceived by that listener as lasting 9.85 sec and having a frequency of 1015 Hz. A similar situation in water would produce only a 30 msec shortening of duration and a 3 Hz change in the perceived sound frequency. However, for certain situations, such as echolocation of prey by bats or porpoises, even these small Doppler shifts may be detectable and make a difference between dinner or no dinner. We take up this issue in Chapter 26, pages 864–866.

Interference and Beats

Consider two sources of sound that produce exactly the same frequency, but whose waves arrive at a given location with different phases. Suppose that the two waves are exactly out of phase so that one tends to concentrate molecules at the location, whereas the other tends to disperse them at the same time. If the amplitudes of the two waves are similar, the net force on any molecule is nil and the two waves simply cancel each other out. A microphone held at such a location would record no sounds. In contrast, what if the two waves arrive completely in phase? Then both will tend to create concentrations or shortages of molecules at the same time, and the result will be greater sound amplitudes than if only one wave were present.

The interaction between waves with similar frequencies and amplitudes is called **interference**. Waves that are out of phase negatively interfere; those that are in phase show positive interference. Interference in acoustic communication can arise when an animal has two sound sources (as, for example, from many birds), or when the sound produced by a single animal arrives at a location by multiple routes. A special case of interference occurs when two similar amplitude sound waves are nearly but not quite identical in frequency. In this case, the two waves cycle in and out of phase, producing a summed waveform that rises and falls in amplitude. The variations in summed amplitude are called **beats**. The **beat frequency** is the rate at which the summed amplitude rises and falls, and is equal to the difference between the frequencies of the two original sounds. Some birds produce two simultaneous sounds that are just slightly different in frequency; the overall waveform thus shows beats.

THE IMPORTANCE OF ACOUSTIC IMPEDANCE

The ease with which a disturbance can generate a propagated sound in a given environment depends upon the medium's **acoustic impedance**. Consider a small square patch of molecules one molecule thick and of area S. Suppose we exert a positive pressure P on this square for a short interval of time t. The pressure will force the small square forward where it will collide with the next layer of molecules and generate a local concentration of molecules. This local molecular concentration will press back against the square and slow down its forward motion. Depending upon how forcibly the compressed molecules counteract the forward pressure on the square, the square will move forward a distance d during the interval t. In that time, a volume of molecules equal to $S \times d$ will have been affected by the moving square. The response of the medium to a pressure P on a square patch of medium of area S is thus to move a volume of medium equal to $S \times d$ during the t seconds. Such a response is usually measured by the rate of volume movement—the **volume velocity**, $U = S\,d/t$. Alternatively, we can let $v = d/t$ be the velocity of an average particle in the square, and we then get $U = S\,v$.

Where a medium is easily compressible, there is only minor backpressure, an exerted pressure will be able to move the square a large distance d, and both the volume and the particle velocities will be high; where the medium is not compressible and backpressures are built up quickly, the response will be a much smaller d, and thus the resulting volume and particle velocities will be low. The ease with which a given environment builds up backpressure and retards the movement of the square is called the **acoustic impedance** of the environment and is usually denoted by Z. For a given environment, $U = P/Z$. (Readers familiar with electrical physics may see the similarity of this formula to Ohm's law: $i = V/Z$, where the current i is the flow of electrons in a circuit, the voltage V is the electrical force or "pressure," and Z is the electrical impedance, or for currents that are constant over time, the electrical resistance.)

In free and far field conditions, the acoustic impedance of a unit square of medium depends only on the speed of sound (c) and the density (ρ) of the propagating medium. Algebraically, $Z = \rho c$. The greater either the density of the medium or the speed of sound propagation, the higher the acoustic impedance. The speeds of sound, densities, and acoustic impedances for free and far field sounds in typical gases, liquids, and solids are given in Table 2.1. It can be seen that both the speed of sound and the density of the medium increase as one goes from gas to solid. This means that the acoustic impedance for water is about 5000 times higher, and that for solids such as rock is 13,000 to 50,000 times higher than for air. The acoustic impedances for solids are only 3–10 times higher than that of water. How might this affect sound communication within and between the various media?

Table 2.1 Acoustical properties of major media

Medium	Speed of Sound (cm/sec)	Density of Medium (g/cm^3)	Acoustic Impedance (rayls)
Air	0.3×10^5	1×10^{-3}	0.0003×10^5
Water	1.5×10^5	1	1.5×10^5
Rock	$2–5 \times 10^5$	2–3	$4–5 \times 10^5$

The intensity of a sound, I, is the amount of energy passing through a square of known area during propagation. This energy flux depends on both the sound pressure, P, and the volume velocity, U:

$$I = PU$$

Since U depends directly on the pressure and inversely on the acoustic impedance of the environment, the amount of sound energy passing through a unit area under far field and free field conditions is

$$I = \frac{P^2}{\rho c}$$

This relationship suggests that sound production costs might be different in different environments. To see whether this is so, we rewrite the relationship as follows:

$$P = \sqrt{\rho c I}$$

Clearly, the higher the acoustic impedance, the less energy that is necessary to produce a sound at a given pressure. Since the acoustic impedance of air is only 0.02% that of water, it takes only 0.02% as much energy to produce a sound at a given pressure in water as it does to produce the same pressure level in air. However, this provides no economic benefit to aquatic communicators since an aquatic sender will have to generate sounds with pressure levels 5000 times higher (e.g., 1/0.0002) than those in air just to provide the same energetic stimulation at a receiver's ears. The amount of energy put into a signal to produce a given effect is thus the same regardless of the medium.

Propagation of Sound at Boundaries

Acoustic impedance has important effects on sound propagation between media. When sound traveling in a medium with one acoustic impedance encounters a boundary of another medium with a different acoustic impedance, a fraction R^2 of the incident energy will be reflected back at the boundary into the first medium, whereas a fraction $(1 - R^2)$ will cross into the second medium. R is called the **reflection coefficient** and its value depends upon the particular boundary. The sound energy making it across the boundary will be represented in the second medium by a pressure $(1 + R)$ times that in the first medium and by a volume velocity $(1 - R)$ times the volume velocity in the first medium. If the velocity of sound in the second medium is lower than that in the first medium, the direction of propagation of the transmitted sound will be bent toward the **normal** (a line perpendicular to the boundary); if the velocity of sound in the second volume is higher than that in the first, the transmitted wave will be bent away from the normal to the surface (Figure 2.3). Put another way, sound passing from a high to a low velocity medium is bent into the second medium and away from the boundary; sound passing the opposite direction is bent back toward the boundary. This bend-

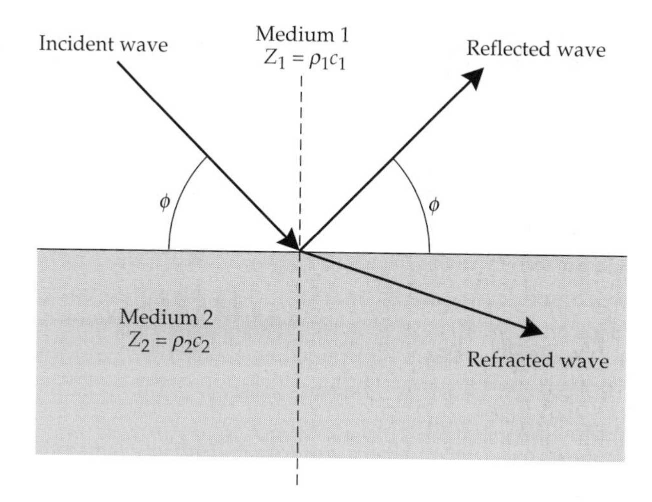

Figure 2.3 Reflection and refraction of sound wave at boundary. An incident and reflected wave have the same angle relative to surface (called the **grazing angle** and indicated here by ϕ). In the pictured situation the incident medium (e.g., air) has a lower speed of sound propagation c_1 than that, c_2, for second medium (e.g., water). Because $c_1 < c_2$, the refracted wave is bent towards the boundary and away from a line perpendicular to the boundary (called the **normal line** to the surface). If the first medium has the greater speed of sound propagation, then the refracted wave is bent towards the normal and away from the surface.

ing of transmitted waves as they cross boundaries between media is called **refraction**. The angle of the incident wave relative to the boundary, called the **grazing angle**, will be identical to that of the reflected wave relative to the boundary.

The value of the reflection coefficient, R, can be either positive or negative. When it is negative, this means that the reflected sound will leave the boundary with a different phase than had it not interacted with the second medium. Any change in phase on reflection is called a **phase shift** and can be measured in either degrees or radians (360° = 2π radians). For example, a phase shift of 180° means that the reflected wave will depart from the boundary one-half of a cycle ahead (or behind) the incident wave at the point of contact. The presence or absence of a phase shift is important because sounds often propagate between sender and receiver by several concurrent routes. If one of those routes involves reflection and a phase shift, this component may interfere negatively with other components with which it is out of phase. A receiver will then have a much more difficult time detecting the sounds.

To understand how sound is reflected at boundaries, we thus need to know the magnitude and sign of R. In addition to relative acoustic impedances, R values also vary with the grazing angle and the relative velocities of sound in the two media. The relationships can be quite complicated and are summarized for the interested reader in Box 2.2. A few patterns occur commonly, and we shall need to refer to them in later chapters. Consider a flock of sea birds. At any given time, some will be flying over the water, and some will be swimming under the surface chasing fish. The calls of flying birds propagate through the air and hit the water's surface. In this example, the impedance and velocity of sound are both much higher in the second medium, and there is a threshold grazing angle that determines the pattern of reflection. This is called the **critical angle** and equals 78° for an air-to-water

Box 2.2 *Reflection Coefficients at Different Types of Boundaries*

CONSIDER A SOUND TRAVELING in medium 1 and encountering a boundary with medium 2. The angle between the direction of propagation and the surface of the boundary is the grazing angle ϕ. The value of the reflection coefficient, R, at this boundary depends on the ratio of the acoustic impedances of the two media (Z_2/Z_1), the grazing angle ϕ, and the ratio of the velocities of sound in the two media (c_2/c_1):

$$R = \frac{\left(\frac{Z_2}{Z_1}\right)\sin\phi - \sqrt{1-\left(\frac{c_2}{c_1}\right)^2 \cos^2\phi}}{\left(\frac{Z_2}{Z_1}\right)\sin\phi + \sqrt{1-\left(\frac{c_2}{c_1}\right)^2 \cos^2\phi}}$$

Reflection coefficients vary from +1 to –1. When they are +1, all of the incident energy is reflected from the surface and the reflected wave undergoes no phase shift. Except for a change in propagation direction, it is as if the boundary were not even there. When $R = -1$, all of the energy is reflected, but the reflected wave is **phase shifted** 180° (or one-half wavelength). That is, it begins a half cycle behind that which one would expected had there been no reflection. If the incident wave were at a maximum when it hit a boundary with $R = -1$, then it would begin as a minimum in the reflected wave. The smaller the absolute value of R, the less energy that is reflected and the more that passes into the second medium.

The grazing angle of the sound, ϕ, can greatly affect the reflection coefficient at a boundary. ϕ varies between 0, when the sound is parallel to the boundary, to 90°, when the sound is traveling in a direction perpendicular to the boundary. Between these two extremes, there will be an important threshold value—for angles of incidence below the threshold, R will vary one way with ϕ, and above the threshold it will vary in another way. When the medium with the incident sound has the lower velocity, e.g., $c_1 < c_2$, the threshold value is called the **critical angle**, is denoted by ϕ_c, and is computed as $\cos\phi_c = c_1/c_2$. When the incident medium has the higher velocity, e.g., $c_1 > c_2$, then the threshold value is called the **angle of intromission**, is denoted by ϕ_i, and is computed as

$$\cos\phi_i = \sqrt{\frac{\left(\frac{Z_2}{Z_1}\right)^2 - 1}{\left(\frac{Z_2}{Z_1}\right)^2 - \left(\frac{c_2}{c_1}\right)^2}}$$

We shall invoke these two threshold grazing angles below.

We can divide the possible relationships between R and ϕ into four cases depending upon whether $Z_1 < Z_2$ or $Z_1 > Z_2$, and whether $c_1 < c_2$ or $c_1 > c_2$. Many of the situations we shall discuss in this book fit either Case I or II; however, the other two cases are not uncommon and the reader should be aware of them. In the following plots, dark gray zones are ones with a full 180° phase shift, white zones have no phase shift, and lightly shaded zones show a continuous change in phase shift with increasing grazing angle.

CASE I: If $Z_1 > Z_2$ and $c_1 > c_2$, then the value of R is always negative. This would be the case if the incident sound were in water and the sound hit the water's surface. The result is a 180° phase shift (dark gray region) regardless of incident angle. The critical angle is irrelevant in this case. As the angle of incidence increases from 0° to 90°, the-value of R increases from –1 towards its value for perpendicular incidence, R_{90}.

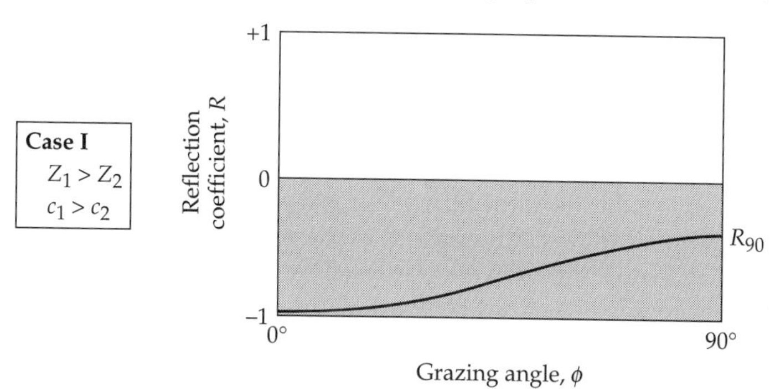

CASE II: The opposite extreme occurs when $Z_1 < Z_2$ and $c_1 < c_2$. An example would be sound in air hitting the surface of a body of water. The relevant threshold in this case is the critical angle ϕ_c: if $\phi > \phi_c$, then R (solid black line) is always positive (no phase shift, white region) and decreases the closer ϕ is to 90°. For low enough incident angles, e.g., when $\phi < \phi_c$ all energy is reflected ($|R| = 1$) but the phase shift (indicated with a dashed line) decreases from a full 180° at a 0° grazing angle down to no phase shift at a grazing angle of ϕ_c (light gray region).

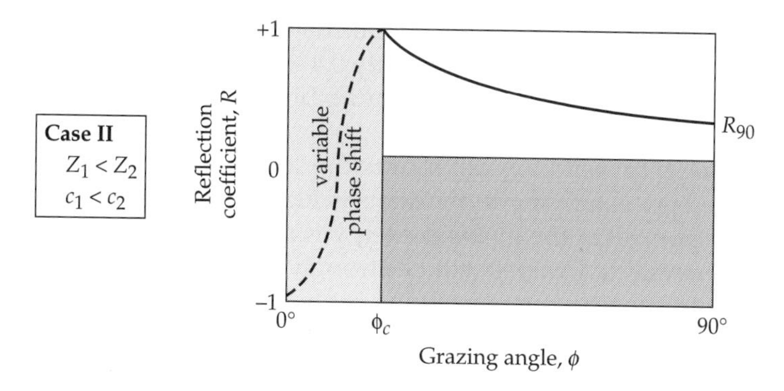

CASE III: Here, the incident medium has a lower impedance, but a higher velocity: $Z_1 < Z_2$ and $c_1 > c_2$. Examples include sound traveling in water and striking a muddy

Box 2.2 (continued)

bottom or sound traveling in air and hitting certain types of soils. The important threshold angle of incidence is here the angle of intromission, ϕ_i. For $\phi < \phi_i$, reflected waves always experience a 180° phase shift, and the value of R will increase from –1 to 0 as ϕ increases. When $\phi = \phi_i$, no energy is reflected: it all passes into the second medium! For $\phi > \phi_i$, there is no phase shift, and the fraction of energy reflected increases with the incident angle.

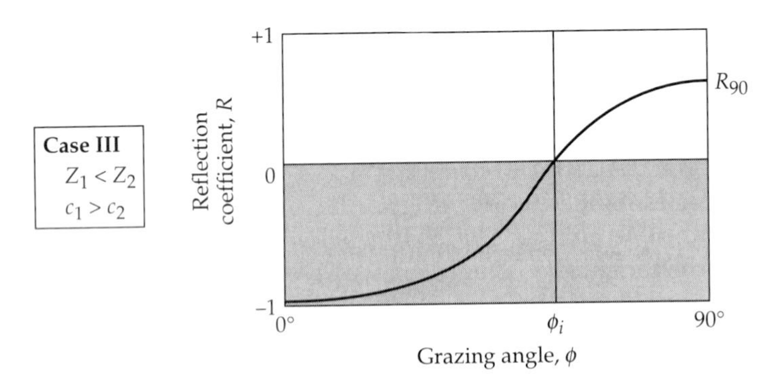

interface. If the flying birds are high above the water and not too far apart, the grazing angle between any two airborne individuals will be larger than the critical angle. R will then have a value between +0.998 to +1 and there will be no phase shift at reflection. The large value of R means that most energy will be reflected at the surface: because there is no phase shift and most of the energy remains in the air medium, flying birds will have no trouble hearing other flying birds. On the other hand, only 0.4% of the incident sound energy will make it into the water. The swimming birds are thus unlikely to hear their flockmates in the air.

If the flying birds tend to skim near to the surface (like pelicans), then the grazing angle between flying individuals is likely to be less than the critical value. In this case, all of the incident energy is reflected, none passes into the water, and the reflected waves will be strongly phase shifted. Because there are several sound paths between flying birds, (e.g., a direct route and a reflected route off of the water's surface), and these will have different phase relations, sounds traveling the two routes are likely to interfere with each other. This interface will make it hard for flying birds to communicate vocally when close to the water's surface. Because no sound energy enters the water, swimming birds at low angles relative to the flying birds will again hear nothing.

When the source of a sound is within a medium of much higher velocity and impedance, it is also the case that most of the energy will be reflected at

Case IV: This is the opposite of Case III since $Z_1 > Z_2$ and $c_1 < c_2$. It might occur when sounds propagated in a muddy or soil substrate reach the interface with the over-lying medium. Both thresholds must be invoked in this example. At incident angles less than ϕ_i, all energy is reflected but the phase shift varies from 180° when $\phi = 0°$ to none at ϕ_c. Further increases in incident angle decrease R without any phase shift until $\phi = \phi_i$ when no energy is reflected. Higher incident angles result in a 180° phase shift and variation in R from 0 to R_{90}. (After Caruthers 1977.)

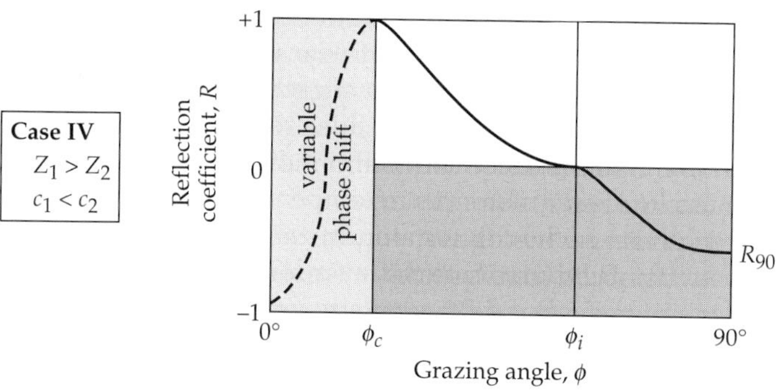

the boundary. Thus any sounds made by the swimming birds are unlikely to be heard by their flying flockmates. However, in this case, there is always a phase shift at reflection. Since there are usually multiple sound paths between swimming birds, and some of these involve reflection at the water's surface, sounds exchanged between swimming birds are likely to be severely attenuated due to sound path interactions. Swimming birds may do best just to ignore all sounds!

These effects also constrain sound production and reception options in animals. For example, airborne sounds do not enter easily into terrestrial organisms (which are largely composed of water and a few solids and thus have high R values). Unless they evolve special mechanisms to trap sounds before they are reflected away, terrestrial organisms will have a difficult time absorbing enough energy to detect sound signals. For the same reason, it is often difficult for terrestrial animals to get the sounds produced inside their bodies into the air medium. Again, special adaptations are required (see the next section). In contrast, organisms living in water have very similar acoustic impedances to the medium in which they live. Consequent R values between their bodies and the water will be close to zero and thus waterborne sounds tend to propagate into and right through aquatic organisms. As we shall see in Chapters 4–6, this creates a special set of problems for sound communication.

Spatial Variation in Acoustic Impedance

We have treated acoustic impedance so far as being constant within a given medium. In real situations, constant impedance is not the case. For example, water molecules near the surface have fewer other molecules constricting their movements (at least on the top side) than water molecules further below the surface. Such molecules can move further in response to an applied pressure than molecules at greater depths. Thus the acoustic impedance of water near the surface will be slightly less than that for water further from the surface. A similar relationship holds for the layer of water surrounding an air bubble or an air-filled sac such as the swim-bladder of a fish. By the same token, the acoustic impedance of atmospheric air near to a water surface (or air inside a bubble or air sac in water) will have a slightly higher acoustic impedance than free air because those molecules close to the water have less freedom of movement. Even in an effectively unbounded medium, acoustic impedances may vary due to local differences in temperature, altitude, or ambient pressure. As we shall see, such within-medium variations may have important consequences for propagation of sounds over long distances.

For sounds not traveling in a free field, acoustic impedance may also be affected by the geometry of the boundaries to the sound field. Consider the propagation of a sound in an air-filled tube. Because the walls of the tube and air have very different acoustic impedances, sound can only propagate along the axis of the tube and not through its walls. In addition, the movement of any volume of air molecules back and forth along the tube axis is slowed down by the friction between the air molecules and those of the tube wall. The importance of this friction depends on the radius of the tube. The number of molecules slowed down by wall friction depends on the inside perimeter of the tube and thus on the tube radius. The number of molecules the sound pressure is attempting to move depends on the cross-sectional area of the tube, which is a function of the square of the tube radius. The ratio between the perimeter and the cross-sectional area of the tube is proportional to the reciprocal of the radius. This ratio is a measure of the fraction of those molecules moved by the sound that are slowed down by wall friction. Where the radius is small, a large fraction of the molecules are slowed down by friction and the volume velocity for a given sound pressure will be substantially lower than that expected in the free field case. Thus the acoustic impedance in a thin tube will be greater than ρc. For large tubes, only a small fraction of the molecules are slowed down by friction, and the free field approximation for the tube acoustic impedance will be quite good.

When sound traveling in a tube with one radius connects to another with a substantially different radius, the differences in the acoustic impedances may cause large reflections at the boundary. Very little energy will actually pass from one tube into the other when this occurs. An extreme case of this effect occurs in animals. Many terrestrial vertebrates use their breathing "tubes" to conduct sound vibrations from inside the body to the outside. A tube that simply opens into the air may be a very poor sound radiator because the impedance of the narrow tube may be so much greater than that of the unbounded world (which will have an impedance close to ρc). Nearly all of the sound energy that makes it down the tube will be reflected back into the tube and never get out. Some of the kinds of adaptations that animals have evolved to get sounds into and out of their breathing tubes will be discussed in Chapter 4.

Finally, we noted earlier that both the pressure and the volume velocity of a sound crossing an impedance boundary will change. If the sound moves from a medium of low to one of high impedance, the transmitted pressure in the second medium will be greater than the sound pressure in the first, but the volume velocity will be lower. When sound moves from a region of high impedance into one of low impedance, the volume velocity increases at the cost of the pressure. There are times during sound communication when animals do best by increasing the volume velocity of a sound at the expense of pressure, and other times when they profit most from the reverse. Many of the organs evolved to radiate and capture sounds have also been carefully adjusted to have impedances that provide the optimal volume velocities or pressures, given the organism's environment.

SOUND INTENSITY AND THE EFFECTS OF PROPAGATION

We have envisioned a propagating sound disturbance as the surface of a sphere that expands as it moves away from the sound source. The amount of energy being propagated by that sphere is set by the sound source when it creates the disturbance. As the sphere expands, its area must increase, and the number of molecules over which that fixed amount of energy must be spread must increase (Figure 2.4). This means that the amount of energy transferred to any one molecule decreases as the sphere expands and as its surface moves away from the sound source. In the far field (and assuming free field conditions), the amount of energy per unit area decreases with the square of the distance from the sound source. If I_o is the intensity of the sound (i.e., energy passing through a unit area/unit time) at some initial distance from the source, then at a second location a distance d times farther

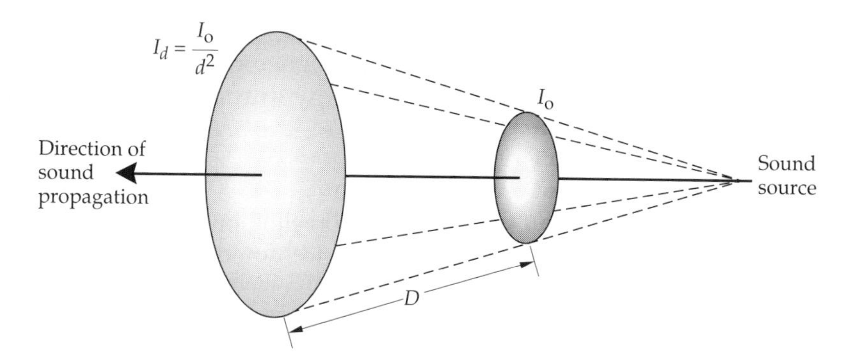

Figure 2.4 Spherical spreading loss for a propagating sound. Intensity of sound falls off with the square of the distance from the source in a far field. The variable d is the ratio of the far to near distances: $d = D_1/D_0$.

from the source than the first, the observed intensity, $I_d \leq I_o/d^2$. All other factors being constant, sound intensity should fall off at least as fast as the square of the distance from the source.

This drop in sound intensity as the sound spreads from its source is called **spreading loss**. In the near field, sound intensity tends to drop off faster than with the square of distance. Since molecules in the near field are actually moved substantially farther than they would go given normal random collisions, some of the sound energy must be used to do the work of moving these molecules back and forth. When added to the spreading loss, this loss of energy typically causes intensities in the near field to fall off with the cube of the distance traveled.

Since pressure is proportional to the square root of intensity, the drop-off of an initial sound pressure P_o over a distance d in the far field will cause $P_d \leq P_o/d$. Thus sound *pressure* falls off as the reciprocal of distance in a free field, whereas sound *intensity* falls off as the square of this reciprocal. It is important to keep in mind whether one is comparing sound intensities or sound pressures in a particular circumstance. One way to avoid problems is to use the **dB scale**, which has been designed to give the same value for both sound intensities and pressures (see Box 2.3). The dB scale is based on the logarithms of signal intensities, and so it is often a good match to vertebrate ears, which also scale sound intensities approximately logarithmically.

Medium Absorption

As sound propagates in a medium, spreading losses are only one reason that sound intensity falls off with distance from the source. We use an inequality in the expressions above because most observed intensities or pressures will be lower than that expected due to spreading losses alone. In addition to spreading loss, each collision between the molecules propagating a sound results in some loss of energy to heat, viscous properties of the medium, and absorption of energy within the molecules. The rate at which these medium losses accrue depends very much on the propagating medium, but can range from 0.01 dB/100 m of propagation to 5–6 dB/100 m. The fraction of energy lost to the medium in each collision also depends upon how rapidly the molecules are being forced to move to propagate the sound: the higher the frequency, the higher the velocity, and thus the higher the energy loss. In most situations, higher frequencies thus lose energy during propagation much faster than do lower frequencies.

Reflective Scattering

We have discussed the fact that when sound traveling in one medium encounters an object with a different acoustic impedance, some of the incident sound energy will be reflected by the object. Consider an object of acoustic impedance very different from that of the medium. When the object is much smaller than the wavelengths of the incident sound, most of the incident sound wave sweeps around the object and very little is reflected. As the size of the object is increased relative to the sound wavelengths, more and more energy is reflected and scattered in all directions. Like spreading losses and medium absorption, scattering reduces the intensity of a sound signal as it propagates away from its source.

When the ratio between the size of a reflecting object and the wavelength of an incident sound is less than about 1/6, the amount of energy scattered increases in a roughly quadratic way as the ratio is increased. The scattered sound energy is reflected equally in all directions; the process is called **Rayleigh scattering**. Note that for a given medium containing particles or objects of a given size, this relationship means that higher frequencies (smaller wavelengths) are scattered more than are low frequencies (larger wavelengths).

When objects in a sound field are similar to the size of incident wavelengths, the sound wave is broken into two components as it strikes the objects. One portion (the reflected wave) is scattered in all directions just as with Rayleigh scattering. However, because the object's size is fairly similar to the sound wavelengths, another portion (the creeping wave) is bent or **diffracted** around the object. These two waves can then be recombined in the vicinity of each scattering particle, producing what is called **diffractive** or **Mie scattering** (Figure 2.5). When objects in the sound field are of similar dimensions to the sound wavelengths, the amplitudes and frequencies of the reflected and creeping waves will be similar. The two waves are thus likely to interfere. Their relative phases depend on the acoustic impedances of the objects and the medium, and on the angle relative to the direction of propagation at which one samples. In air, reflection off of a solid object will produce no phase shift in the scattered wave. The relative phases of the reflected and creeping waves will then depend only on how many cycles pass while the creeping wave moves around the object to rejoin a reflected wave. For a larger object, it will take more cycles to move a given angle around the object than for a smaller one. Thus, as we vary the ratio between scattering object size and incident wavelength, the amount of scattered energy sampled at any location will rise and fall as the relative phases vary, and the resulting interference goes from positive to negative. The frequency dependence of diffractive scattering is thus much more complicated than for Rayleigh scattering. Diffractive scattering tends to be detectable when the ratio between objects and sound wavelengths lies between 1/6 and 6.

When objects are more than 5–10 times the size of the incident wavelengths of sound, so little energy is diffracted around the object that only the reflected wave is present. This generates a **sound shadow** behind the object. The larger the object, the greater the incident energy reflected by the object. This type of **simple scattering** is not frequency dependent. The three types of reflective scattering are summarized in Figure 2.6.

Interactive Scattering

All of the prior discussion presumed that scattering objects had significantly different acoustic impedances from that of the surrounding medium. What if the impedances are not very different? In this case, the object will absorb a significant amount of incident sound energy and begin to vibrate at frequencies characteristic of its size, shape, and composition. These vibrations of the object can then interact with both reflected waves and creeping waves. The result is a

Box 2.3 Measuring Sound Amplitudes

SINCE SUCCESSIVE CYCLES OF A PURE SINE WAVE are identical and the waveform is symmetrical above and below the value of ambient pressure, one can characterize the amplitude of such a sound by noting either the maximum or the minimum pressure recorded. The absolute value of the difference between either of these extremes and ambient pressure is called the **peak** value of the signal. Alternatively, we could measure the difference between the maximum and the minimum pressures. This is the **peak-to-peak** measure of signal amplitude; for a pure sine wave it will be twice the peak value.

For signals which are not pure sine waves, successive waves may have very different peak values, and the waveform, even within a single wave, may not be symmetrical. The obvious solution is to take a large number of measurements of pressure deviations from ambient at successive times and then generate some sort of average deviation. A simple arithmetic average would not be appropriate since positive deviations above ambient would be canceled out by succeeding negative ones, and one could easily get a mean of zero when a sound was clearly present. One solution is to compute the **RMS** (root-mean-squared) value of pressure deviations in a given sample. To do this, each measured deviation is squared, all the squared values in a given sample are summed, and the mean of these squares is computed. (Ideally, one would take an infinite number of measurements in a given sample and instead of the sum, one would compute the integral of the squared deviations. In practice, our instruments tend to use finite sums that approximate such integrals.) The square root of the mean of the squared deviations is the RMS value of the pressure deviations. When citing a pressure, you should indicate whether the value was made by peak, peak-to-peak, or RMS methods. Any comparisons between two pressures should use the same mode of measurement.

A vertebrate ear may be able to detect and respond to sounds whose pressures vary over a 100,000-fold range. The corresponding sound intensities vary over a range of 10^{10}. The **decibel scale** was developed to make characterization and comparison of sound amplitudes easier, and as we shall see later, better mimic the perceived scaling of sound amplitudes in most animals. Sounds on the decibel scale (abbreviated **dB**) are ranked logarithmically (using logs to the base 10). In addition, the decibel scale is a relative one. That is, a sound amplitude is not assigned an absolute value in decibels, but is given a value relative to some standard reference amplitude or to some other sound of comparative interest. If the intensity of the reference sound is I_R and the intensity of the sound to be characterized is I, the relative amplitude of the second sound measured as intensities is

$$\text{Relative amplitude (dB)} = 10\log_{10}\frac{I}{I_R}$$

Since sound intensity is proportional to pressure squared, we can replace each value of I in this equation with its corresponding pressure squared. Because the log of a squared value is equal to twice the log of the value, the equation for relative amplitude when the pressure of the characterized sound is P and that of the reference sound is P_R becomes

$$\text{Relative amplitude (dB)} = 20\log_{10}\frac{P}{P_R}$$

Note that for the same two sounds, the relative amplitude will be the same regardless of whether pressure or intensity is used. Where one wants to rank a sound relative to some common standard reference, most researchers set P_R equal to the human threshold at 1 kHz, which is about 2×10^{-4} dynes/cm^2. Usually, an amplitude given relative to this threshold is indicated by following the dB with **SPL** (sound pressure level). Thus a sound of 40 dB SPL is one which has a pressure of 2×10^{-2} dynes/cm^2. Where intensities are used, the reference intensity is 10^{-12} W/m^2, and the value in dB will be followed by the letters **IL**. For water-borne sounds, a different reference value is sometimes used.

There are a few simple tricks and rules to remember when using the decibel scale. As a rule of thumb, note that a sound that has twice the pressure of the reference sound will be 6 dB greater in amplitude; making that sound half of the reference pressure will make it 6 dB smaller. If you are comparing two sounds, one of amplitude 25 dB SPL and the other of amplitude 32 dB SPL, the amplitude of the second sound relative to the first is just the difference in dB levels: 32 – 25 = 7 dB. (See if you can prove this to yourself using the second formula above.) If you subtract their SPL levels to compare two sounds, do you add their SPL levels to compute the total sound produced by both? No! Suppose one cicada is producing a buzz that is 60 dB SPL by the time it reaches you, and another nearby individual produces a buzz that is 40 dB SPL at your ear. What is the amplitude of the combined sound which you hear? The answer is *not* the sum of the two SPL values. The pressure of the cicada buzz that is 60 dB SPL is 1000 times greater than the threshold of human hearing at 1 kHz, or about 0.2 dynes/cm^2. The pressure of the cicada buzz that is 40 dB SPL is 100 times greater than the same threshold or about 0.02 dynes/cm^2. Since the sum of the two signals is not a simple sine wave, we need to compute the average pressure of the combination using the RMS method:

$$\text{Combined pressures (RMS)} = \sqrt{(0.2)^2 + (0.02)^2} = 0.201 \text{ dynes/cm}^2$$

which equals 60.04 dB SPL. Clearly, the second sound, which is only 10% of the pressure of the first, contributes little to the relative amplitude of the combination. You do not add SPL levels. Finally, remember that the reference sound in the decibel formula can be set to any desired sound. If we want to compare the relative sound levels of 10 different cicadas, it is often convenient to let the buzz with the highest (or lowest) pressure be the reference sound, and then to rank the others by the number of dB by which their pressures are lower (or higher) than this extreme. If we do set a reference other than the standard one, we must be sure to omit SPL after the quoted dB values.

Below are listed some sample sounds and their measured amplitudes in dB SPL:

Rustling leaves	10 dB
Broadcast studio; soft whisper	20 dB
Purring cat; bedroom at night	30 dB
Living room	40 dB
Office; classroom; nearby bird singing	50 dB
Normal conversation	60 dB
Barking dog	70 dB
City street (no trucks nearby)	80 dB
Roaring lion; moving heavy truck	90 dB
Echolocating little brown bat	100 dB
Thunder; construction site	110 dB
Jet taking off nearby	120 dB

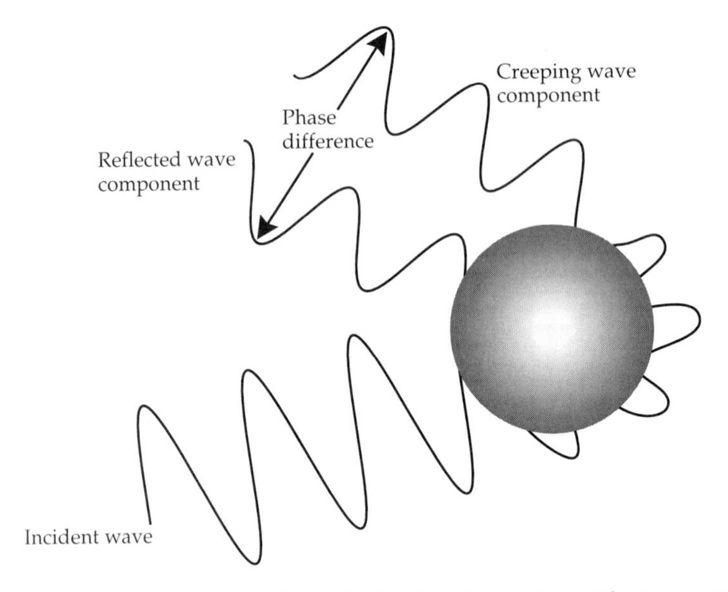

Figure 2.5 Diffractive scattering by an inelastic sphere. An incident wave traveling from lower left to upper right arrives at a sphere. Part of the incident energy is reflected directly from the surface and travels off at a 90° angle as the **reflected wave**. The remainder creeps around the sphere before some of it travels off at the same angle; this is the **creeping wave**. Depending upon how far the creeping wave had to travel before leaving the sphere, its phase may or may match that of the reflected wave. Interference will be positive or negative depending upon relative phases of these two components.

very complicated sound field around such objects in which the amplitude of sound detectable at any location depends upon the direction of sound propagation, the relative sizes of scattering objects and incident wavelengths, and the shape and composition of scattering objects. This process is called **interactive scattering** and is used by porpoises and toothed whales to discriminate between different targets using echolocation (see page 864).

LINEARITY OF SOUND

Let us return to the calling male cicada. Its tymbal oscillates and this creates temporal variations in the pressure of the nearby air. The resulting pressure waves propagate through the air until they hit the ear of a female cicada. Here they generate vibrations in the female's eardrum and eventually stimulate sensory cells in her ear. The sound thus travels through a series of media and boundaries. Each successive medium or boundary is likely to alter the sound in some way. For example, medium absorption and scattering will reduce the sound's amplitude during its propagation though the air; at the boundary between the air and the female's eardrum much of the sound energy will be reflected back, and only a portion will cross the boundary into the ear.

We can consider each successive medium and boundary as a "black box." The sound goes into each box with one set of property values and it exits the box with possibly altered ones. Suppose we put a sound whose waveform can be described by some function $x_1(t)$ into such a box and obtain an output sound whose waveform is described by the function $y_1(t)$. If we put a sound with a different waveform, $x_2(t)$, into the same box, we recover a sound with a different waveform $y_2(t)$. Suppose we then input sounds $x_1(t)$ and $x_2(t)$ to the box at the same time. The box is said to be a **linear system** if at each instant t, the output waveform is the simple sum $y_1(t) + y_2(t)$. When might a system (or box) not be linear? One possibility is that the input sounds, $x_1(t)$ and $x_2(t)$, interact with each other such that they cannot later be separated and their independent outputs no longer sum. Another possibility is that the box changes its properties when one sound is present and this affects what it does to the other sound during its transit. Finally, consider a given sound inserted into the box at successively greater amplitudes. Suppose the lowest amplitude is A, the next level is $2A$, then $3A$, and so on. Note that this is equivalent to inserting $x_1(t)$ into the box, then $x_1(t) + x_1(t) = 2x_1(t)$, then $x_1(t) + x_1(t) + x_1(t) = 3x_1(t)$, and so on. If the output for insertion of $x_1(t)$ is $y_1(t)$, then insertion of $2x_1(t)$ into a linear box should yield $2y_1(t)$, insertion of $3x_1(t)$ should yield $3y_1(t)$, and so forth. No system or box is linear over an infinite range of amplitudes; eventually, the system saturates and can no longer produce a higher amplitude output. For higher amplitude inputs, it becomes nonlinear.

Luckily for communicating animals, sound propagation in gases and liquids behaves as a linear system over most of the range of amplitudes that animals might use. This property means that separate sound waves simply add on top of each other but do not interact or combine permanently during propagation. If sound were not largely linear, we would not be able to have cocktail

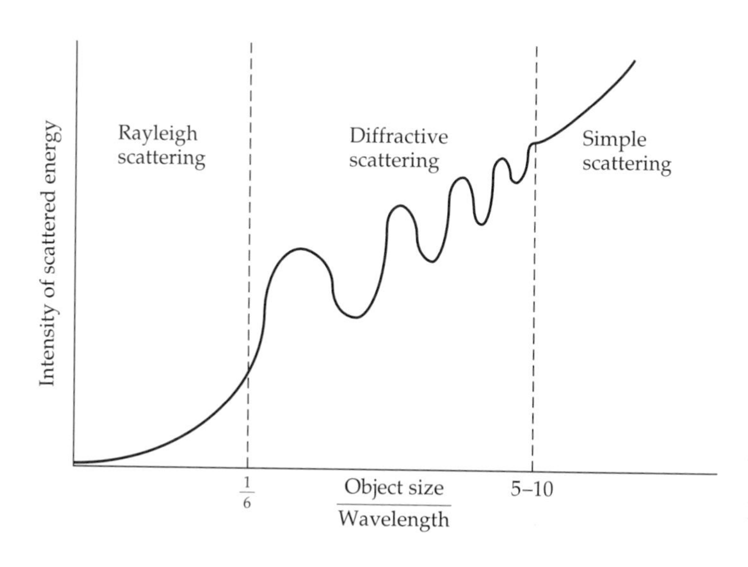

Figure 2.6 Type of scattering as function of ratio between size of scattering object and wavelength of incident sound. The plot assumes the wavelengths of incident sound are fixed and the object size is varied. For Rayleigh and diffractive scattering zones, the same effect could be achieved by holding the object size fixed and varying the wavelengths. For simple scattering, however, scattered energy is affected only by changes in object size and is unaffected by changes in wavelength.

parties—the utterances of different speakers would become inextricably mingled and we would be unable to focus on the speech of a particular person. Most natural environments are as noisy or noisier than cocktail parties—were sound not linear, the concurrent calls of singing birds, chorusing frogs, buzzing cicadas, howling monkeys, and the rustling of forest vegetation in the wind would form a loud cacophony from which no intended recipient could extract a useful signal. But to a good approximation, sound *is* linear and evolution has used this fact to shape the ability of receivers to isolate and detect specific signals amidst the noise.

The linearity of sound also facilitates the evolution of more complicated sound signals, and thus the packaging of more complicated information, than would be the case otherwise. Consider our hypothetical cicada that produces pure sinusoidal sounds. While a pure sine wave can be completely characterized by determining its frequency, amplitude, and phase relative to some reference time point, how would a cicada ear characterize a much more complicated sound in which the tymbal was moved in a nonperiodic manner and with varying amplitudes? How would such an ear compare two different sounds with different patterns? One answer is for the sender cicada to produce a tymbal movement, and thus a signal waveform, which consisted of two or more pure sine waves added together. Although this more complex waveform travels between sender and recipient as one signal, the linearity of sound means that its parts are not inextricably melded. If the ear of a recipient cicada were able to reseparate the component sine waves when it received one of these complex signals, it could then characterize the frequency, amplitude, and relative phase of each and have a complete description of the overall sound. It could then do this for several different sounds and be able to compare them quantitatively. In fact, the ears of many animals perform this process (although cicadas are not particularly good at it). It is also what researchers on animal communication do to study sounds. As we shall see in the next chapter, any complex sound can be broken down into the linear sum of a set of pure sine waves. The method for doing this, called **Fourier analysis**, gives us a very powerful tool for characterizing and comparing animal sound signals.

SUMMARY

1. Molecules in gases and fluids collide randomly with each other and with any surface inserted into them. The force of molecular collisions on a given sample surface is called the **pressure** of that medium. A sound is generated by producing a local concentration or rarefaction of molecules, which can be recorded as a local variation in pressure. In a **free field**, (a world with no boundaries), this disturbance in local molecular concentration is propagated in all directions away from the source.

2. Sounds that move molecules in directions perpendicular to the direction of sound propagation are called **transverse waves**. The oscillations of a guitar string are an example. Sounds that move molecules back and forth along the

axis of sound propagation are called **longitudinal waves**. Sounds in air and water are usually longitudinal waves; sounds in solids may be either longitudinal or transverse waves.

3. The simplest sound is one in which pressure varies sinusoidally both over space and time. A sinusoidal sound can be characterized by specifying its **frequency** in **Hertz**, its **amplitude**, and its **phase** relative to some time reference. The reciprocal of frequency is called the **period** of the sine wave. The velocity at which a sound spreads is called the **speed of propagation** and is characteristic of a given medium at a given temperature and pressure. **Wavelength** is the physical length of a single cycle of a sinusoidal sound. Wavelength and frequency are inversely related for a given medium. In water, where the speed of propagation is higher than in air, wavelengths for a given frequency are 4.4 times longer than in air. The abilities of animals to produce and detect sounds, and the amount of distortion generated during propagation of a sound, all depend critically on the wavelength of the sound.

4. If a source is moving in the same direction as its emitted sound, all component frequencies will be shifted to slightly higher values, and the sound duration reduced; the reverse occurs for sounds propagated in the direction opposite to the source's trajectory. These changes are called a **Doppler shift**, and their magnitude depends upon the velocity of the source relative to the speed of sound. When sounds of similar frequency and amplitude arrive at a location out of phase, they may interfere and cancel each other out (**negative interference**) or add positively to make a higher amplitude signal (**positive interference**). Two sounds with nearly but not identical frequencies will create a composite waveform with varying amplitude (**beats**).

5. The response of a medium to an applied force is limited by its **acoustic impedance**. In free and far field conditions, this impedance is equal to the product of density and the speed of propagation in the medium. In non-free field conditions, it may be higher or lower than this product. Where two media of very different acoustic impedances abut, sound traveling in one is mostly reflected back into that same medium at the boundary. Thus sound in air usually does not penetrate into water or into terrestrial animals (which are largely made of water). Where two media of similar acoustic impedances are in contact, sound more easily traverses the boundary. Thus sounds traveling in water may pass right through a fish or porpoise. Whenever the **grazing angle** of a sound is low, or the sound is traveling from a medium of high to low acoustic impedance, the reflected waves will be phase shifted by up to 180°. Larger grazing angles or reflections generated when a sound in one medium hits another of higher impedance usually produce no phase shift. Where sound does cross the boundary, its direction of propagation will be bent away from the boundary if the second medium has a lower sound velocity; it will be bent back toward the boundary if the second medium has a higher velocity. This is called **refraction**. Acoustic impedance is rarely uniform even within a medium. Temperature, proximity of surfaces and other media, and altitude (in air) or depth (in water) can all cause heterogeneous distributions of acoustic impedance and of sound velocity. Sound-emitting tubes and ears in air are organs in which mismatches between the acoustic impedance of the organ and that of the air require special adaptations to ensure transfer of energy across the media boundaries.

6. The **intensity** of a sound is proportional to the square of the sound's pressure. Because the same sound energy is spread over a greater sphere of disturbance as a sound propagates away from its source, both pressure and intensity decrease with distance from the sound source. In the **near field**, within a few wavelengths of the source, sound intensity and pressure can change very rapidly with distance. Outside of this range, known as the **far field**, pressure decreases as the reciprocal of distance and intensity as the square of the distance from the source. These decreases in signal magnitude during propagation are called **spreading losses**.
7. In addition to spreading losses, a propagated signal will also lose energy to **medium absorption** and to **scattering**. Both of these processes tend to be frequency dependent, with greater losses for higher frequencies. However, the amount of energy scattered can be a very complicated function of the ratio between scattering object sizes and incident sound wavelengths, relative acoustic impedances of medium and objects, and shape and composition of objects.
8. Sound propagation in liquids and gases is essentially a **linear system**. This means that many different animals can produce different sounds simultaneously and appropriate recipients can each extract the signal intended for them without distortion due to other ambient sounds. It also means that an ear (or an animal communication researcher's analytical device) can be constructed that samples sounds with complex waveforms and easily breaks them down into the sine wave components that were present in the original sound before propagation. This ability greatly extends the kinds of sounds that animals can use for communication and simplifies our task when studying these sounds.

FURTHER READING

Good introductions to the properties of sound with examples familiar to humans, such as music, can be found in the books by Rossing (1990) and Rigden (1985). Intermediate treatments of acoustics are available in Feynman et al. (Vol. I, 1989), Fletcher (1992), and Crawford (1968). The Fletcher volume integrates properties of sound with topics covered in Chapters 4–5 and is thus a good source for animal communication studies. Advanced acoustics texts include Kinsler and Frey (1962), Morse and Ingard (1968), and Beranek (1986, 1988). The complex interactions of sound at a boundary are well detailed in Caruthers (1977).

Chapter 3

Fourier Analysis

ANALYZING SOUND IS A SPECIAL CASE of the general problem of analyzing complex patterns. As a rule, we select a variable, for example local sound pressure, and we graph the measured values of this variable as a function of either location or time or both. How can we characterize any given graph that we might obtain, and how do we compare two different graphs? Suppose we record two different sounds and obtain the corresponding plots of pressure versus time (Figure 3.1). We can clearly see that these two sounds have different waveforms, but how does one go about providing a quantified comparison? If we had three such sounds, how would we decide whether two of the sounds were more similar to each other than either was to the third? If the sounds were simple sine waves, there would be no problem; we could specify the frequency, peak amplitude, and relative phases of the two sounds and they would be completely described. In this case, neither of the sounds is a pure sine wave.

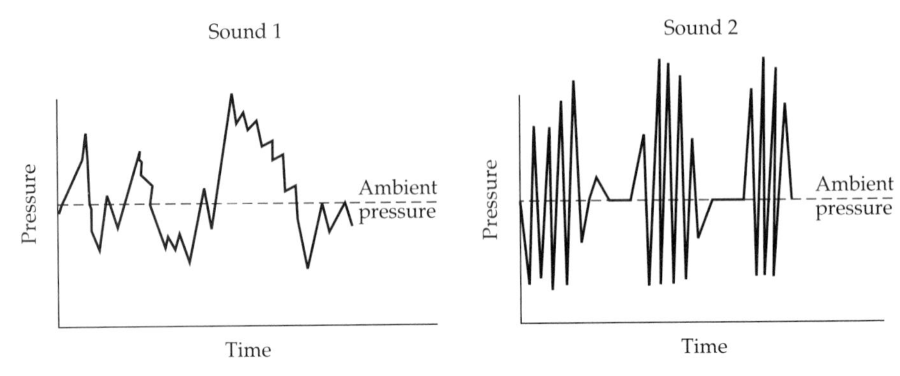

Figure 3.1 Plots of recorded pressure amplitudes versus time for two sounds.

THE LOGIC OF FOURIER ANALYSIS

Luckily, there is a technique called **Fourier analysis** that can solve these problems. The technique is based on the fact that any continuous waveform (or continuous graph line) can be broken down into a set of pure sinusoidal waves of infinite duration whose frequencies, amplitudes, and relative phases are easily quantified and compared. This list of measured components is mathematically equivalent to a complete description of the waveform. We can even synthesize the original sound by generating each of the component sinusoidal waves and adding them together with the proper phase relationships. Not only can any complex waveform be broken down and reconstituted by separating and recombining the appropriate set of sinusoidal waves, but also the pressure at any time t can be predicted exactly by summing the values of each of the component waves at that time t. The pressure $P(t)$ of a complex waveform at time t thus equals an infinite sum of cosine (or sine) waves, each of which has a specific amplitude, frequency, and relative phase. Algebraically, this is written as

$$P(t) = P_0 + \sum_{n=1}^{\infty} P_n \cos(2\pi f_n t + \Phi_n)$$

where P_0 is the mean pressure around which the waveform oscillates, P_n is the pressure amplitude of the nth cosine wave in the set, f_n is the frequency of the nth cosine wave, and Φ_n is the relative phase of the nth component. (Note that we could have used sine instead of cosine waves in this formula; since sine and cosine waves differ only by phase, using sine waves would shift the values of Φ_n). In principle, it will take an infinite number of components to completely describe any given complex waveform. In practice, animals' ears and scientists' instruments cannot measure an infinite number of components, and hence they focus on a finite number of the louder components. Usually this approximation is sufficient to characterize and compare sounds. The mean value of the waveform (P_0) is normally the ambient pressure. Where it is higher or lower, we say that the waveform has a nonzero **DC** component (from the electrical analogue of direct current).

Suppose we have decomposed a complex waveform into its Fourier cosine components. How do we summarize the necessary information to describe the sound? There are many ways to graph and tabulate the Fourier information. One method is shown in Figure 3.2. The Fourier components are arranged by their frequencies along the x-axes of two graphs. On one graph, the amplitude of each component is indicated by the height of a bar over the frequency of that component. This is called a **frequency spectrum** (or if average intensities that equal the squares of amplitudes are used, a **power spectrum**). On the second graph, the size and direction of each bar relative to the dotted line represents the phase of that Fourier component. This is called a **phase spectrum**. Since it is the relative alignment of the component waves that determines the overall waveform of the sum, one of the components is often chosen as the reference and its phase is arbitrarily set to zero. The phase spectrum then indicates the phase of each other component **relative** to this reference. If one component peaks a bit later than the reference component, then its relative phase is positive; if it peaks earlier, then its phase is negative with respect to the reference component. These two graphs combined summarize all the data required to describe or reconstitute the original signal. They are in fact mathematically equivalent to it. We shall introduce other methods for summarizing Fourier information in later sections.

In this chapter, we shall mainly focus on the Fourier analysis of sounds. You should not forget, however, that Fourier analysis can be used to characterize any continuous pattern. Most of our attention centers on graphs of sound pressure versus time recorded at a given location. We could also graph the distribution of sound pressures versus distance from the signaler, and this too could be characterized by a Fourier decomposition into pure sine or cosine waves. In later chapters, we shall use frequency decomposition to characterize visual colors and the electrical discharges of fish. Ecologists use Fourier

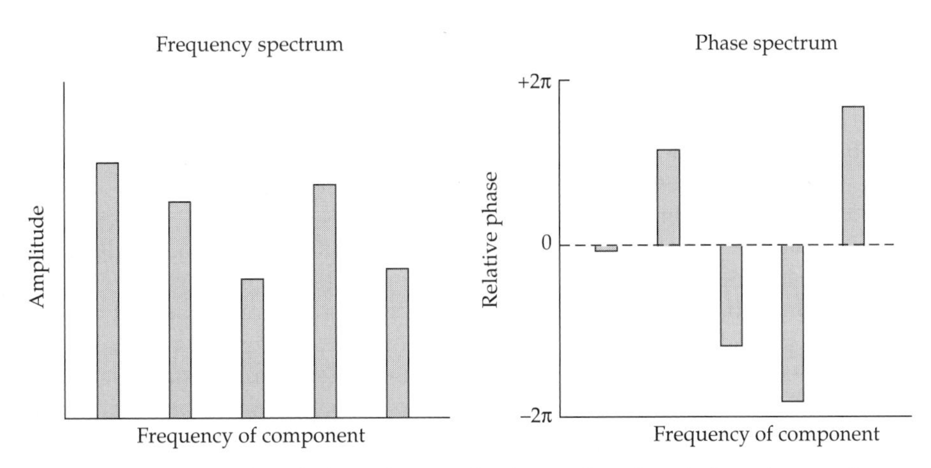

Figure 3.2 Fourier description of a sound. A complete description of a sound is provided by indicating the amplitudes of all Fourier components in a frequency spectrum and the relative phases of all components in a phase spectrum.

analysis to characterize the spatial pattern of ecological variables, such as grass height along a transect, or by doing Fourier analysis in two dimensions, they can characterize the degree of patchiness in different habitats. Clearly the method has great utility in a number of fields.

FOURIER ANALYSIS OF PERIODIC SIGNALS

When we view a record of a sound as a graph of pressure versus time, we are said to be examining the sound in the **time domain**. The time domain waveform is a complete description of the sound at the recording location. Like all signals, this waveform can be broken down into its Fourier parts. Thus a set of tables or graphs listing all the frequencies, amplitudes, and relative phases of the Fourier components of the sound is also a complete description of it. Such a description is said to be given in the **frequency domain**. Since either description is complete, it doesn't matter in principle which we use to characterize a sound. In practice, we usually have to examine both the time and frequency domain images of a sound to completely understand its structure. For this reason, it is very useful to know the basic rules by which common types of sound waveforms are translated into the frequency domain and vice versa. Most of this chapter is devoted to summarizing the more common of these rules.

The fundamental rule of Fourier analysis can be stated in the following way: the more the time domain image of a signal deviates from that of a single sinusoidal wave of infinite duration, the greater the number of other sinusoidal waves that must be added to the Fourier set and the larger must be the amplitudes of these additional components. Signals can deviate from the case of a single sinusoidal wave of infinite duration either in waveform or in repeatability. In the first case, signals have a basic pattern that repeats indefinitely (as a sine wave does), but the shape of the repeating unit is not sinusoidal. Any signal (including a sinusoidal wave) that repeats the same basic pattern indefinitely is called **periodic.** All periodic signals last forever. However, they may differ in the pattern of the repeating unit. As the shape of the repeating unit becomes more and more deviant from a sinusoidal one, the more Fourier components we can expect to see in the signal's frequency domain graphs and the greater the amplitudes these additional components will exhibit. The consequences of varying the patterns of periodic waveforms on the corresponding frequency domain representations of signals will be treated first in the next sections.

The second type of deviation away from an infinitely long single sinusoidal wave is reduced repeatability. Consider a signal which consists of 5 cycles of a pure sine wave with no pressure oscillations before or after. This signal is not periodic because it does not last forever. As signals become more and more **aperiodic**, the basic rule of Fourier analysis predicts that we shall have to add more and more additional sinusoidal components, and shift more and more energy into these components, to provide a representative frequency domain description of the sounds. We shall treat the Fourier analysis of aperiodic signals after we have discussed periodic waveforms.

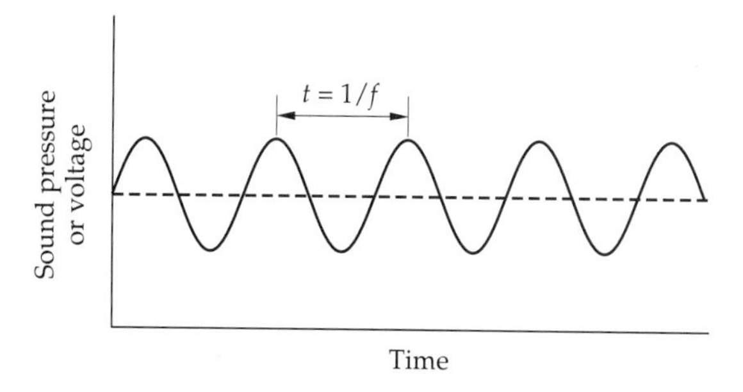

Figure 3.3 Time domain image of a pure sine wave.

Fourier Analysis of a Simple Sine Wave

Let us begin with a pure sine wave of infinite duration. Its time domain image would appear as in Figure 3.3. If there is no DC component in this sound, the dotted line will reflect ambient air pressure and the rise above and below this line will represent the increases or decreases in local air pressure as the sound disturbance passes our microphone. On most time domain instruments, early events are plotted on the left side of the graph and later events on the right. To measure the frequency of this sine wave on a time domain plot, we simply measure the period. Since period and frequency are reciprocals, a measured period of t seconds gives a frequency in Hz of $f = 1/t$. If we want to measure the amplitude, we record the peak or peak-to-peak voltage from the plot. Voltage can be converted into pressure if we know the sensitivity of our microphone.

Suppose we now perform Fourier analysis on this signal. The frequency spectrum would appear as in Figure 3.4. Notice that only a single vertical bar is shown since only one pure sine wave is present and there is no DC component (which if present would appear as a second bar at a frequency of zero). This single component should have a frequency equal to the reciprocal of the period that we measured in the time domain image, and the amplitude of the component should equal that taken off of the time domain plot. We could also plot the phase spectrum of this signal, but since there are no other components, relative phase has no real meaning in this case.

Types of Periodic Signals

There are several ways in which this single sine wave of infinite duration might be modified and still retain its periodicity. First, we could hold its shape and frequency fixed, but let its amplitude vary in some periodic way. The

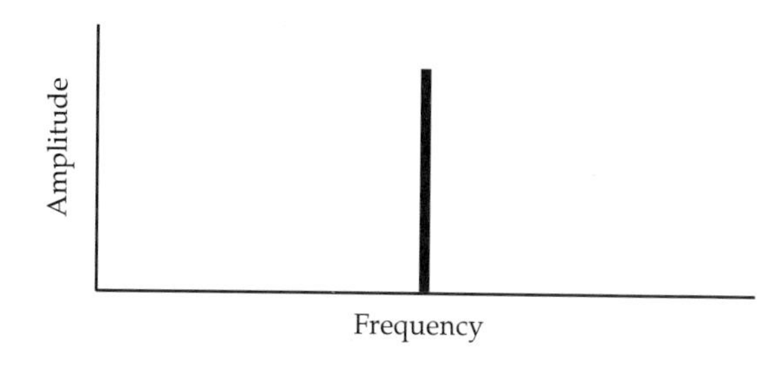

Figure 3.4 Frequency spectrum of single pure sine wave.

simplest case is to vary the amplitude sinusoidally. This is called **sinusoidal amplitude modulation**. Alternatively, we could hold its amplitude and shape fixed, but let its frequency vary in a periodic manner. Again, the simplest case would vary the frequency sinusoidally. This is called **sinusoidal frequency modulation**. A third modification would be to hold the amplitude and frequency constant, but change the repeating shape of the signal from a sine wave to something else. We shall call such a signal a **nonsinusoidal periodic wave**. It is useful to learn the frequency domain expectations for each of these three simple cases because nearly any other periodic signal can be considered as an additive combination of them. After you understand each one, we shall show you how to break down more complicated signals into pieces and apply the rules for one or more of the three basic cases to each piece.

Sinusoidal Amplitude Modulation

Suppose we take our infinitely long pure sine wave of frequency f and modulate its amplitude sinusoidally. This means that the frequency of the wave would remain unchanged, but the amplitude of successive cycles would go up and down in a sinusoidal way. We can determine whether the modulation was sinusoidal by looking at the envelope of the time domain picture (Figure 3.5). The original sine wave that is being modulated is called the **carrier** and here has frequency f; the sine wave that is imposed on the amplitude of the carrier is called the **modulating wave**. The frequency of the modulating wave can be computed by measuring the amount of time it takes to go from one maximum of the modulating wave to the next maximum. If t is the period of this modulating waveform, then the frequency of the modulating wave $w = 1/t$ Hz.

When we amplitude modulate a carrier signal, the requisite sine waves that have to be added in the frequency domain appear as **side bands** around the carrier. For a sinusoidally amplitude-modulated carrier with no DC component, there will be two side bands, one on each side of the carrier (Figure 3.6). One will have frequency $f - w$ and the other will have frequency $f + w$. Both should have amplitudes less than that of the carrier. However, the greater the amplitude of the modulating waveform, the larger the size of the

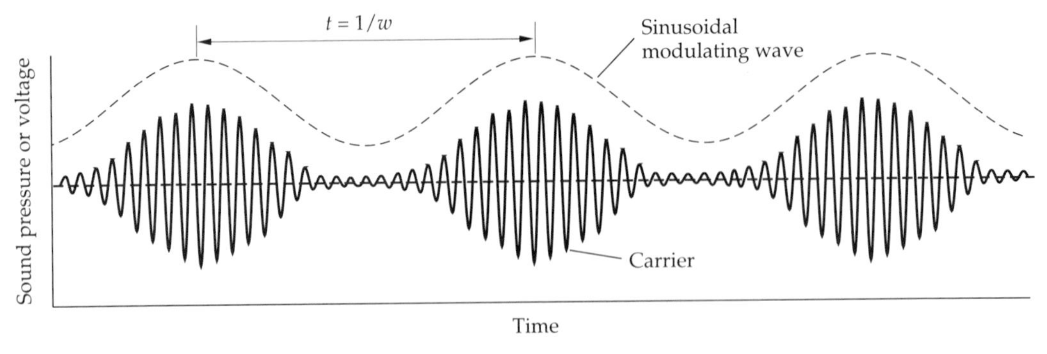

Figure 3.5 Time domain image of a sinusoidally amplitude-modulated sine wave.

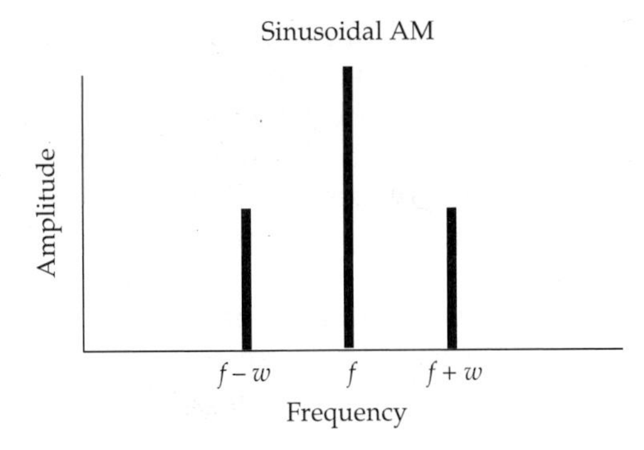

Figure 3.6 Frequency spectrum of sinusoidally amplitude-modulated sine wave. Here f is the frequency of the sine wave carrier, and w is the frequency of the modulating wave.

two side bands. We have not shown the phase spectrum for this signal, but you should realize that only one set of relative phase relationships for these three Fourier components is likely to generate the original waveform. A combination of these same three components at the amplitudes shown in Figure 3.6 that used any other set of relative phases might produce a summed waveform quite unlike sinusoidal amplitude modulation. You may have heard of amplitude modulation by its abbreviation, **AM**, since this is one of the ways in which sounds are encoded in radio waves.

Sinusoidal Frequency Modulation

Suppose we return to our original pure sine wave of frequency f, infinite duration, and no DC component. This time, let's hold the amplitude constant, but modulate the frequency. This means that the period of successive waves will change. Let us consider the simplest case in which the frequency varies above and below f in a sinusoidal fashion. The time domain plot would appear as in Figure 3.7.

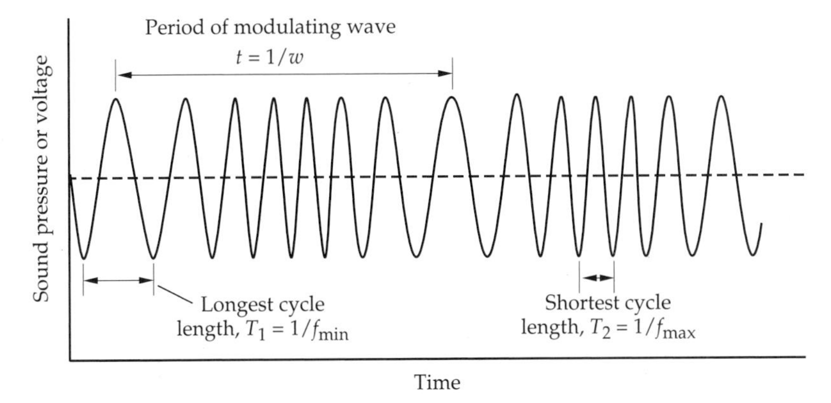

Figure 3.7 Time domain image of a sinusoidally frequency-modulated sine wave.

We can again compute how long it takes to go through one complete modulation cycle by examining the time domain plot. Let us denote this period by t and its reciprocal, the modulation frequency, by w. The frequency spectrum of a sinusoidally frequency-modulated sine wave again has a band at the carrier f and side bands around f. However, unlike sinusoidal amplitude modulation, frequency modulation (**FM**) generally produces multiple pairs of side bands around f. These side bands occur as before at $f - w$ and $f + w$, but also at $f - 2w$ and $f + 2w$, $f - 3w$ and $f + 3w$, etc. The amount of the total signal energy that gets diverted away from the carrier and into the side bands depends on the **modulation index**. This is the ratio between the range of frequencies exhibited by the modulated carrier and the modulating frequency w. To estimate the modulation index for a frequency modulated signal, compute w as usual. Then compute the time it takes for the longest cycle in the time domain plot and take its reciprocal. This gives a value of the lowest frequency, f_{min}. We next compute the time taken for the shortest cycle visible in the time domain plot. Its reciprocal is the highest frequency, f_{max}. The range of frequency is then $f_{max} - f_{min}$, and the modulation index is $(f_{max} - f_{min})/w$. The greater the modulation index, the more energy in the side bands of the frequency spectrum. These can be so large that they are in fact larger than the energy in the carrier. A frequency spectrum for a sinusoidally frequency-modulated sine wave with a low modulation index is shown in Figure 3.8; an example with a high modulation index is shown in Figure 3.9.

A thoughtful reader might have noticed that an FM signal with a sufficiently low modulation index may have only one pair of side bands visible and thus generate a spectrum identical to that of an AM signal with the same f and w. In both cases one will see a carrier band at f and side bands at frequencies $f - w$ and $f + w$. The amplitudes of carrier and side bands might even be the same for the two graphs. How can this be when we know that the time domain waveforms are quite different? The answer is that the two signals differ markedly in the relative phases of the three Fourier components. If we add together the three Fourier components with one set of phase relations, the result will be a pure AM signal. If we use another set of phase relations, the resulting waveform will be a pure FM signal. For all other sets of phase relations, the waveform will be a mixture of FM and AM.

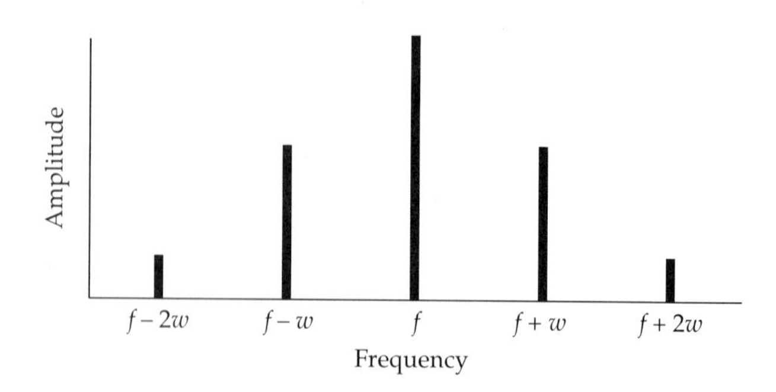

Figure 3.8 Frequency spectrum of a sinusoidally frequency-modulated sine wave with low modulation index.

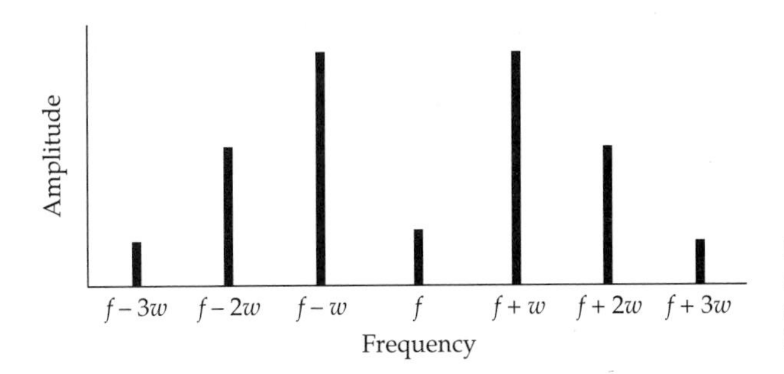

Figure 3.9 Frequency spectrum of a sinusoidally frequency-modulated sine wave with high modulation index.

Periodic Nonsinusoidal Signals

Now consider a signal which is not a sine wave but instead exhibits a repeating pattern with some nonsinusoidal shape. Examples of such signals are shown in Figure 3.10. Any signal that is periodic but is neither a sine wave nor a modulated sine wave has a frequency spectrum called a **harmonic series**. A harmonic series is a set of component frequencies that are all integer multiples of some common frequency known as the **fundamental** of the series. Figure 3.11 shows a typical frequency spectrum for a nonsinusoidal periodic signal. If the fundamental is w, then the other component frequencies $2w$, $3w$, $4w$, etc. are called **harmonics**.

Note that the differences in frequency between adjacent harmonics are equal to each other and to the fundamental w. The time domain image of such a signal will repeat the nonsinuoidal pattern w times per second. This frequency can be measured directly off of the time domain waveform by counting how many repeats occur in a fixed interval of time, or by measuring the duration of one complete repeat (the period of the waveform) and taking its reciprocal. Either method should result in a frequency that is equal to the difference between the frequencies of the corresponding Fourier components. Note that the amplitudes of higher harmonics tend to be smaller than those of

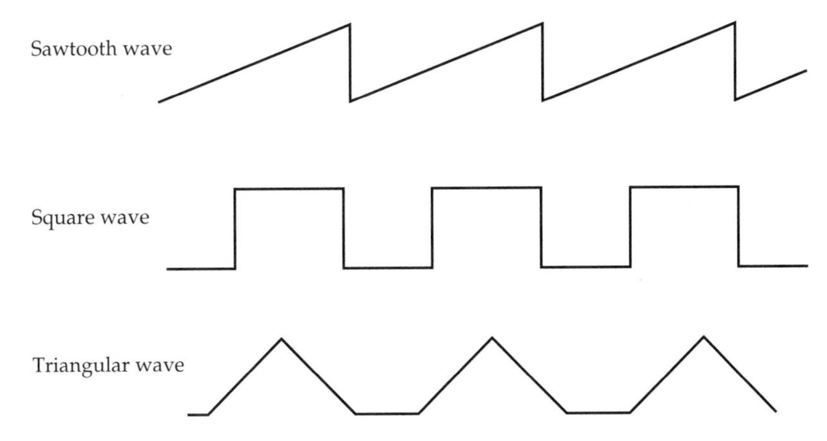

Figure 3.10 Time domain images of various periodic nonsinusoidal signals.

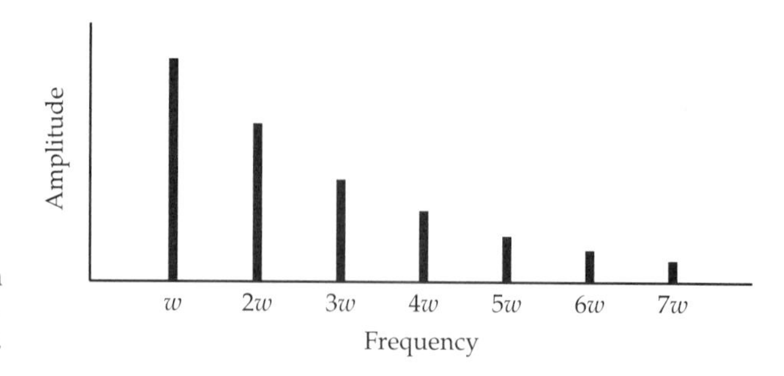

Figure 3.11 Frequency spectrum of a periodic nonsinusoidal signal. Such a spectrum is called a harmonic series built upon a fundamental frequency w.

lower harmonics. **Dirichlet's rule** states that for periodic signals with few major discontinuities in their waveforms, the energy in higher harmonics of the corresponding frequency spectrum will tend to fall off exponentially with the frequency of the harmonic.

Periodic nonsinusoidal waves that differ in shape but not in repeat rate w will all generate harmonic series built upon a component spacing of w Hz. However, the different waveforms will vary in how the energy is distributed among the successive harmonics and in the phase spectra. The most conspicuous differences between the spectra of periodic nonsinusoidal signals involve the presence or absence of specific harmonics. In the simplest case, the frequency spectrum of a nonsinusoidal periodic signal may exhibit or lack a band at a frequency of zero. As we have seen earlier, the presence or lack of such a band depends upon whether the time domain signal does or does not contain a nonzero DC component (Figure 3.12).

The remainder of the spectrum depends in part on whether the signal is **half-wave symmetric** or **half-wave asymmetric**. A half-wave symmetric signal is one whose repeating waveform can be divided into two equal halves such that the second half is identical to the first half flipped upside down

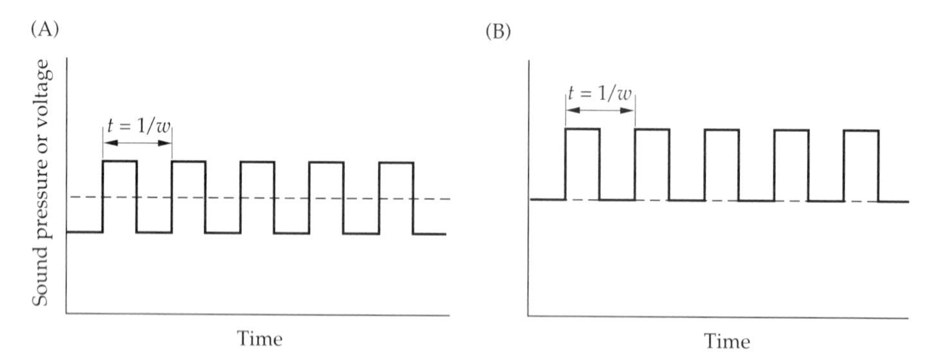

Figure 3.12 Presence or absence of a DC component in a square wave train. In both examples, square pulses occur at a rate of w/sec. In (A) the average deviation around the ambient pressure (dotted line) is zero and no DC component is present. The frequency spectrum will not include a line at a frequency of zero. (B) All deviations from ambient are greater than zero, producing a positive DC component. In this case, the frequency spectrum would include a line at a frequency of zero.

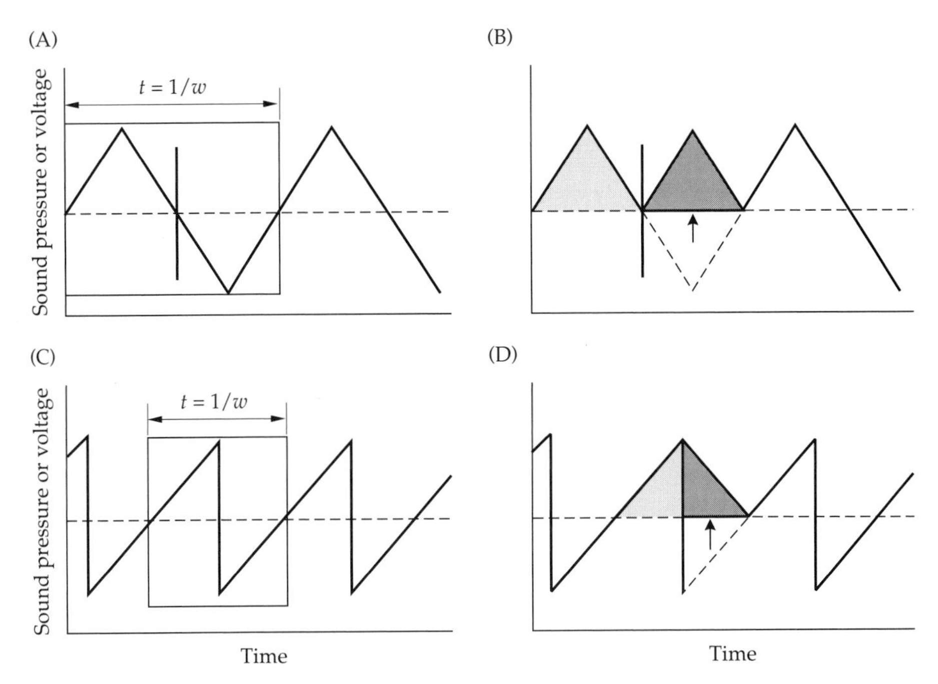

Figure 3.13 Half-wave symmetry of periodic signals with single maxima and minima. In each figure, the dotted horizontal line demarcates the average amplitude for the entire signal. (A) and (B) show a symmetrical sawtooth, and (C) and (D) show an asymmetrical one. To demonstrate symmetry in either case, draw a small box around a single period, as in A and C. Divide this period in half with a dark vertical line. Then reflect the second half of this period to the opposite side of the dotted line. In B, it is clear that the reflected second half is identical to the first half. This signal is half-wave symmetric. In D, the reflected half is not identical to the first half (it is a mirror image of it) and this wave is thus not half-wave symmetric.

(Figure 3.13). A similar operation performed on a half-wave asymmetric signal will fail to produce a match. Once you determine which type of signal you have, the frequency spectra of half-wave symmetric periodic signals are immediate: all of these signals have frequency spectra in which only the odd harmonics are present (Figure 3.14A). The spectra of half-wave asymmetric signals are a bit more complicated, but usually have all harmonics present (Figure 3.14B).

The pattern of variation in the amplitudes of the harmonics of a half-wave asymmetric signal depends upon the number and spacing of successive maxima and minima in the signal's time domain waveform. Signals that have only a single maximum and minimum per repeat period always have all harmonics present, and these tend to decrease smoothly in amplitude according to Dirichlet's rule. Signals with more than one maximum and/or minimum per period have somewhat more complicated spectral rules. The trick is to examine the time domain waveform and to find the shortest interval between either two successive maxima or two successive minima in a single period.

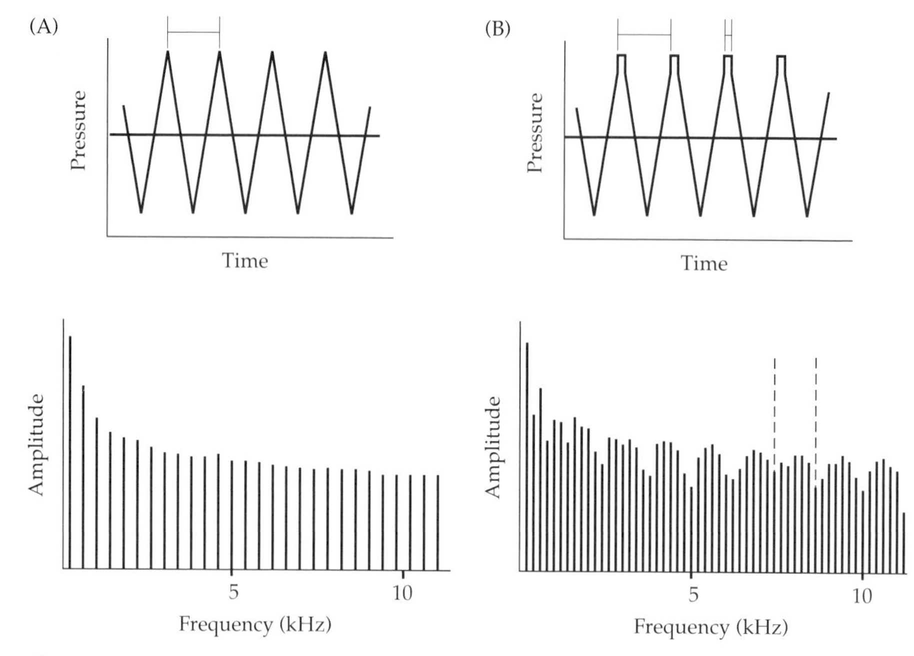

Figure 3.14 Frequency spectra of symmetric versus asymmetric periodic signals. (A) Waveform (top) and frequency spectrum (bottom) of a half-wave symmetric saw-tooth wave with a period of 5 msec and thus a fundamental frequency of 200 Hz. Note that only odd harmonics are present. (B) Waveform and frequency spectrum of the same signal as in A but with positive peaks clipped to generate a half-wave asymmetric signal. Because the signal is asymmetric, all harmonics of the 200 Hz fundamental are present. In the clipped portion, there are two successive maxima per period separated by 0.8 msec (= 1/1200 Hz). They generate a pattern of lobes and nodes across harmonics, with the width of a lobe equal to 1200 Hz (marked by vertical dashed lines).

Suppose this shortest time is τ sec and its reciprocal is the frequency z Hz. As usual, we denote the overall period of the signal by t and the corresponding repetition rate of the waveform by w. In general, all harmonics of the fundamental w will be present in the corresponding frequency spectrum. However, the closer a harmonic of w is to a harmonic of z, the lower will be its amplitude. Harmonics of w that fall closest to a harmonic of z will be maximally reduced and are called **nodes.** The adjacent clusters of more intense harmonics are called **lobes**. Nodes and lobes may not be visible among the very first harmonics of w, but they usually dominate the spectrum at higher frequencies (Figure 3.14).

Thus the frequency spectra of periodic signals with multiple maxima or minima consist of harmonic series based on the frequency of the waveform repeat but with adjacent harmonics grouped into lobes and separated by nodes. The width of lobes is inversely related to the lag between adjacent maxima or minima. When the lag between successive maxima or minima is equal to one-half the period of the waveform, then nodes will fall on the even harmonics of the waveform repeat frequency w. This pattern will cause the resulting spectrum to resemble that of a half-wave symmetric wave with the same period.

Compound Signals

Now we are ready to look at combinations of these three basic modifications of a sine wave. Because our ears and many of the instruments used to analyze sounds usually do not store or use phase spectra, we shall focus mainly on the frequency spectra. This does not mean that phase is unimportant: clearly it is very important if your analysis or the response of the receiving animal depends on the time domain waveform. As we shall see below, there is also a limit to the abilities of our instruments and any animal's ears to analyze signals in the frequency domain. Where this limit is reached, one has no choice but to examine the time domain image, including its embedded phase information. However, for the moment, we shall ignore the phase spectra of compound signals.

The simplest compound signals include (1) nonsinusoidal modulation of a simple sine carrier, (2) sinusoidal modulation of a nonsinusoidal carrier, and (3) nonsinusoidal modulation of a nonsinusoidal carrier. The first step in each case is to characterize the frequency spectrum of the carrier: What are you starting with? If the carrier is a pure sine wave, then the carrier frequency spectrum is a single line. If the carrier is a nonsinusoidal periodic signal, then it can be broken down into its component harmonics. Then we can treat each component in the compound carrier as if it were the only carrier. The second step is to examine the modulating waveform and break it down into its component sine waves. If modulation is sinusoidal, then you only have a single modulating frequency to worry about. If the modulating waveform is nonsinusoidal, then each of its component harmonic frequencies can be treated as a separate sinusoidal modulator of the carrier. The overall frequency spectrum of the compound signal will be the sum of the results of modulating each carrier component by each modulating wave component according to the simple rules. More explicitly, for every sinusoidal component in the modulating waveform, we add side bands according to the appropriate rules for AM or FM around each sinusoidal component of the carrier.

As an example, consider the case of a pure sine wave that is pulsed (turned on and off). Some frogs and insects make sounds pulsed this way. The modulating waveform is thus periodic but nonsinusoidal. Suppose that the duration of each pulse is equal to the interval between pulses. This means that the modulating waveform will be half-wave symmetric and in fact, is here a square wave. Suppose the carrier is sinusoidal. The overall waveform is shown in Figure 3.15. What does the frequency spectrum of this signal look like? We ought to be able to predict its appearance using the simple rules we have learned. The carrier will appear in the spectrum as a single line at frequency f. The modulating waveform, however, has a frequency spectrum that is a harmonic series with only odd components present because it is a half-wave symmetric square wave. The rules for AM are to add one pair of side bands around the carrier for each sine component in the modulating waveform. That means

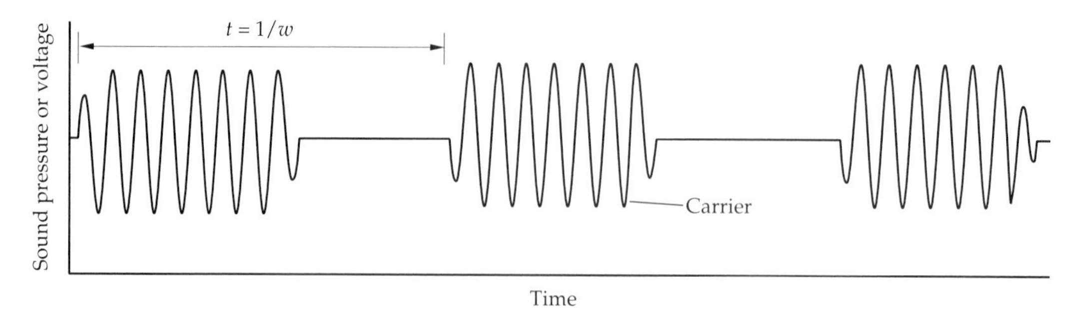

Figure 3.15 Time domain image of a pulsed sine wave. The carrier here is a pure sine wave of frequency f. The modulating waveform is a square wave with frequency w.

we expect bands at the frequency of the carrier f, and at the side bands $f + w$ and $f - w$. But this is not all. We also expect side bands at $f + 3w$ and $f - 3w$, at $f + 5w$ and $f - 5w$, etc. For each sine component in the modulating waveform, we must add two side bands around the carrier frequency in the spectrum. Because of Dirichlet's rule, the energy in these side bands ought to fall off for higher multiples of w. The frequency spectrum of the pulsed sine wave is shown in Figure 3.16A.

A special case of compound signals occurs when a carrier signal has a nonzero DC component. In addition to the component frequencies in the carrier, there is thus an additional component at a frequency of zero (the DC

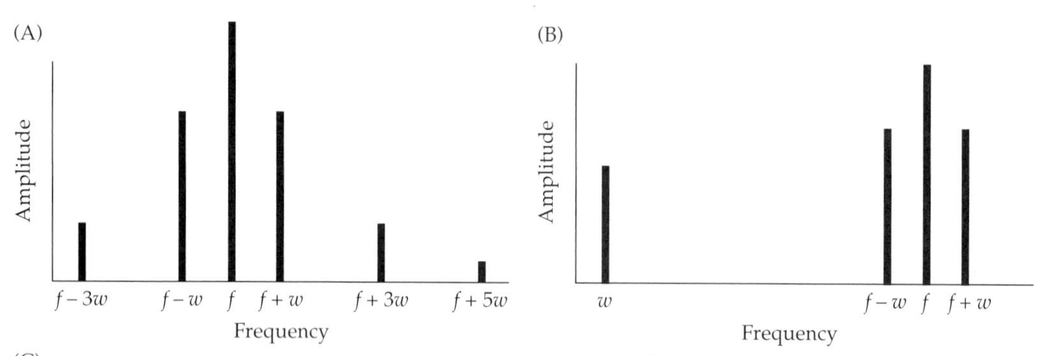

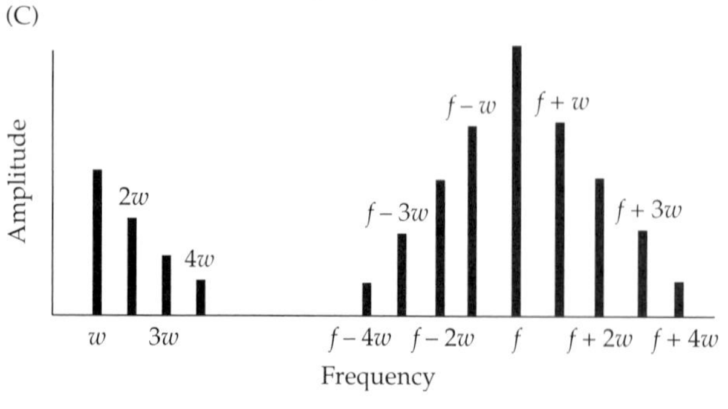

Figure 3.16 Spectra of some compound signals. (A) Frequency spectrum of a pure sine wave at frequency f pulsed at w times/sec. (B) Frequency spectrum of a sine wave of frequency f with a nonzero DC component sinusoidally modulated at rate w. (C) Frequency spectrum of a single sinusoid of frequency f with a nonzero DC component amplitude modulated by periodic nonsinusoidal waveform with repeat rate w.

component). Because all carrier components receive side bands when modulated, the DC component will also receive side bands (but just for frequencies greater than zero). Thus if a simple sine wave of frequency f has a nonzero DC component and is sinusoidally modulated at a rate w, we shall see frequency bands at w (= $0 + w$), $f - w$, f, and $f + w$ (Figure 3.16B). If the modulating waveform is periodic with repetition rate w but not sinuosidal, then we shall find a series of bands at integral multiples of w as well as bands around the carrier frequencies (Figure 3.16C).

To see if you understand the additivity of these rules, try to predict the frequency spectrum of the reverse situation—a carrier that is a square wave but that is amplitude modulated sinusoidally. Frequency modulation generates much more complicated spectra than amplitude modulation since it usually generates more than one pair of side bands per carrier component. However, the logic is no more complicated than for amplitude modulation. You may wish to try an example. Finally, try amplitude modulating a square wave carrier with a square wave modulating waveform. Being able to move freely between the time domain and frequency domain is a skill that is essential for accurate analyses of animal sounds. Box 3.1 shows how the use of both domains is often necessary for such work.

FOURIER ANALYSIS OF APERIODIC SIGNALS

All of the signals analyzed so far have been periodic. The theory presumes that these signals last forever. However, many natural signals are given only a few times, or change form in each successive rendition. Hence they may repeat only rarely, if at all. This means that we have to add even more sine waves in the frequency domain representations of the signals. The more aperiodic the signal, the more frequencies we have to add. Consider the following example. In Figures 3.15 and 3.16A, we examined the frequency spectrum of an infinite train of pulses containing a carrier frequency f. We assumed in that case that the pulse duration equaled the interval between pulses. This assumption caused the modulating waveform to be half-wave symmetric and the only side bands were based on odd harmonics of the modulating waveform repeat rate w. Let us now consider a more general case in which pulse durations do not necessarily equal inter-pulse intervals. The modulating waveform will no longer be half-wave symmetric, and thus all harmonics of the pulse rate w will appear in the spectrum as side bands around f. A piece of such a waveform is shown in the top of Figure 3.17.

The frequency domain picture of the top waveform in Figure 3.17 is a series of side bands around the carrier, f, with generally decreasing amplitudes. These side bands will be separated by a frequency interval w (equal to the pulse repeat rate). Because the modulating waveform is half-wave asymmetric and has several successive maxima or minima (here the points of onset and offset of the pulse), successive side bands will be even smaller in amplitude when they have frequencies similar to harmonics of z (the reciprocal of the pulse duration), and a bit larger in amplitude for frequencies intermediate

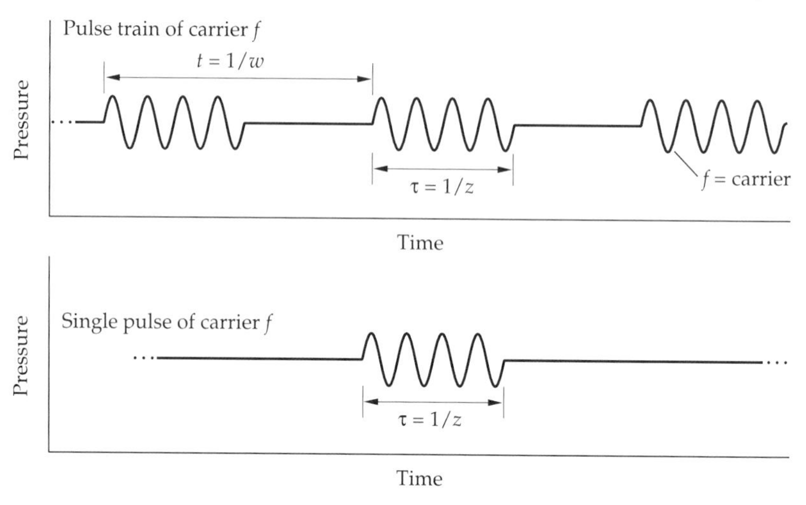

Figure 3.17 Periodic versus aperiodic pulse trains of a pure sine wave.

between the harmonics of z (nodes and lobes). The corresponding spectrum is shown in Figure 3.18A.

Now suppose we keep the carrier frequency and pulse duration fixed, but increase the interval between pulses. This change will increase the repeat period

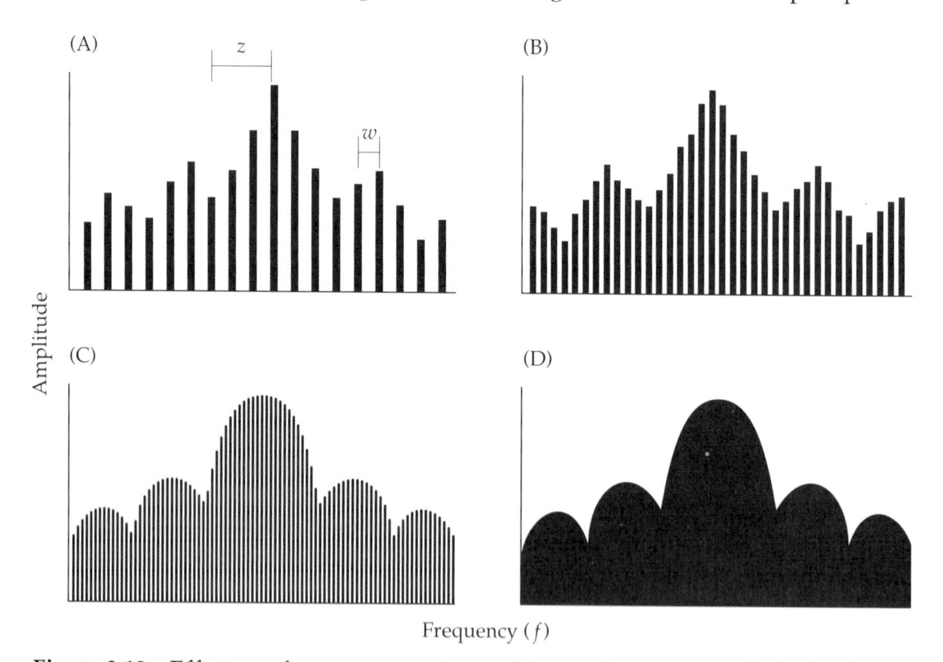

Figure 3.18 Effects on frequency spectrum of increasing repeat period of pulse. (A) through (D) represent signals with increasing intervals between repeats. f is the carrier frequency, w is the frequency of the pulse repeat, and z is the reciprocal of the pulse duration. D shows the spectrum of a single finite duration pulse.

t, decrease the pulse rate w, and thus bring adjacent side bands closer to the carrier and to each other. Figure 3.18B shows the spectrum for an infinite train of pulses with a lower pulse rate than that shown for Figure 3.18A. Note that the nodes and lobes appear in the same locations because pulse duration, and thus the internode frequency difference z, has not changed. What does change is the spacing of the many side bands around the carrier. Figure 3.18C shows the spectrum for an infinite pulse train but with an even longer interval between pulses, and thus an even smaller value of w. Finally, in the limit we make the interval between pulses infinitely long. The resulting waveform is shown in the bottom of Figure 3.17: there is a single pulse that never repeats. Because t is now infinitely long, the value of w is infinitely small, and the side bands are thus infinitely close together. The thin frequency component lines of continuously repeating signals are now replaced by continuous smears or bands of frequencies.

The previous signal has a substantial duration. It is a new type of signal only because it occurs without repetition. Now, consider what happens if we reduce the duration of that single pulse to an infinitely short period of time. The shorter the pulse duration, the larger z, and thus the wider the lobes including the main lobe around the carrier. In the limit (Figure 3.19A), we end up with a single pulse of infinitely small duration and infinite amplitude (otherwise the pulse would have no energy being infinitely short in duration). Because its duration is infinitely short, the main lobe in its spectrum is infinitely wide and fills the entire spectrum (Figure 3.19B). All frequencies in this spectrum have the same energy. This signal is as deviant from a single pure sine wave as is possible, and given our general rule for Fourier analysis, it is not surprising that we need every possible frequency at the same amplitude to synthesize it. This imaginary pulse is called a **delta function** or **delta pulse**. There is no such pulse in reality, but one can sometimes trick a machine or even an ear into thinking it was fed such a signal.

Note that another kind of signal, **white noise**, also has a frequency spectrum consisting of all frequencies at approximately the same energies. White noise has a time domain waveform that is entirely random, and the amplitude at any time t has no significant correlation with the value at subsequent times. Clearly, this is a very different time domain waveform from that of a delta

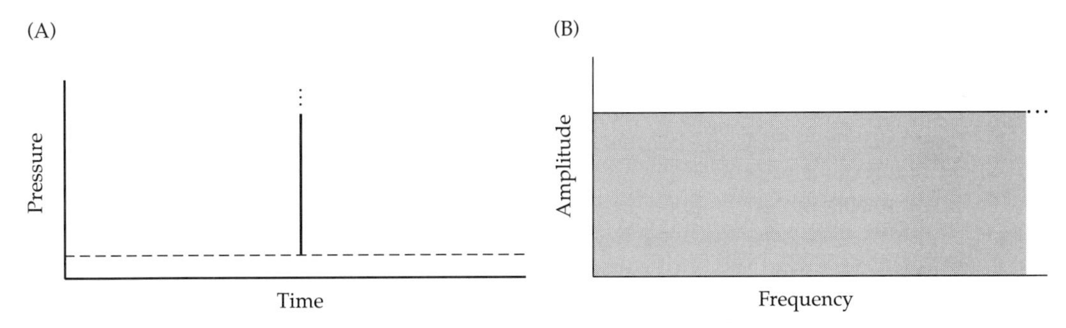

Figure 3.19 Time domain image (A) and frequency spectrum (B) of a single instantaneous pulse.

Box 3.1 *Distinguishing Between Modulations, Harmonics, and Beats*

MANY ANIMALS PRODUCE SOUNDS BY AMPLITUDE OR FREQUENCY MODULATION of a carrier signal. Other animals (and humans) produce signals that have periodic but nonsinusoidal waveforms. In both cases, the time domain waveforms of the resulting signals will be periodic, and we can measure the amount of time (the period) required for one complete cycle of each repeating pattern. The reciprocal of this period is the frequency w of the repeating pattern. If we examine the corresponding frequency spectra, both cases will exhibit numerous bands evenly spaced along the frequency axis, and the frequency difference between adjacent bands will equal the time domain measure of w. If we didn't know which animal made which sound, how could we tell which sounds were the results of modulation and which were harmonic series?

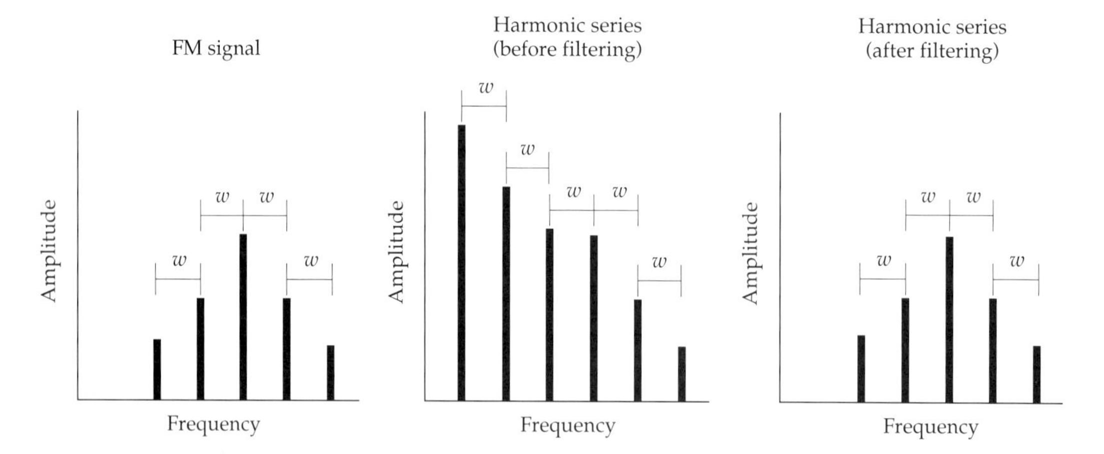

A first guess would be to see whether the spectral component with the lowest frequency was equal to equal to w or not. If it was, it is likely that the sound is a harmonic series. In addition, we would expect successive harmonics to show decreasing amplitudes because of Dirichlet conditions. However, these are not very powerful tests. What if a sound that is generated as a harmonic series suffers the loss of its fundamental and attenuation of its first few harmonics due to filtering during sound production or transmission in the medium? The frequency spectrum of this filtered signal may now look very much like that of a modulated signal, and the frequency of the lowest component will not be equal to w. How is a wise sound analyst to deal with these problems?

The best approach is as follows. First, note that the time domain periodicity of a harmonic series waveform is very robust to filtering. As long as at least two of the original components of a harmonic series remain, (and these need not be adjacent bands), the time domain image of a filtered series will still show a periodicity equal to the original w. This means that you should continue to use your time domain image to determine w. You then determine the frequencies of each of the signal components, f_i, visible in your frequency spectrum. If the original signal was a harmonic series, the quotient of f_i and w for each component should equal an integer. If these quotients are not integers, within the limits of your ability to measure w and the f_i, then it is more likely that your periodic signal was generated by modulation.

Consider a situation in which sound generation or transmission filters out alternative harmonics in a harmonic series. In this case, the value of w measured in the time domain will be much lower than the difference measured between adjacent bands in the corresponding frequency domain image. When there is a mismatch between time domain and frequency domain estimates of w, always use the time domain measure. You could still use the test outlined above to determine whether the original signal was a harmonic series or not.

Some animals, such as birds, can produce two independent sine waves at the same time. We saw in Chapter 2 that the combination of any two randomly selected sine waves yields a time domain waveform with a periodic variation in amplitude. This waxing and waning of the overall amplitude is called **beating**. If the two sine waves being added have frequencies of f_1 and f_2 with $f_1 > f_2$, then the frequency of beating $w = f_1 - f_2$. Are beats a special case of harmonics? In general, the answer is no because it will be rare that the quotients f_1/w and f_2/w both equal integers for two randomly selected frequencies f_1 and f_2 and a beating frequency w. For example, the combination of two sine waves at frequencies 519 and 530 will produce beating at a w of 11 Hz. However, neither 519/11 nor 531/11 equal integers. Thus the two frequencies are not harmonically related. Where two combined frequencies are not harmonically related, their combined waveform will not be truly periodic. If you look carefully at the sections of time domain waveform separated by beat maxima, you will see that successive sections are usually not identical. As a rule, you will have to wait for f_1 cycles of the first frequency (or the equivalent, f_2 cycles of the second), before the beating waveform really repeats itself. This corresponds to a repeat frequency of 1, which is periodic in only a trivial sense.

If it is rare to find two combined frequencies that produce truly periodic waveforms, imagine how much rarer it is to find a combination of three randomly selected frequencies that is periodic. As with pairs of random frequencies, the maximum repeat frequency is likely to be 1. From this point of view, it should be clear to you that periodic signals, including both harmonic and modulation signals, are a very small subset of the large number of possible waveforms generated by adding together randomly chosen sine waves and that most such waveforms will not be periodic.

function. Although the two types of signals show some similarities in the frequency domain, they also show major differences. Whereas the amplitudes of all frequencies in the delta function frequency spectrum are always the same, the amplitudes of different frequencies in white noise vary randomly, and it is only the average for each frequency that is constant. The relative amplitudes of each frequency in a very short segment of white noise would not be equal. Similarly, the relative phases in a delta function are precisely defined and fixed; the relative phases of different frequencies of white noise vary randomly.

SPECTROGRAMS

Visualizing Spectral Structure

Most animal vocalizations (as well as human speech) are modulated in complicated ways during the production of the signal. There is no single modulating frequency, but a complex sequence of modulations. How can we best describe these signals quantitatively? We would lose a lot of information if we pretended that the signals were infinitely long and just performed a pooled frequency spectrum on the entire signal. The alternative is to break the overall signal up into pieces or segments, perform a Fourier analysis on each segment as if it were a single unique and aperiodic signal, and then restring the set of analyses back together again to see how the frequency spectra shift through time during the course of the signal. This increase in temporal detail is achieved at some cost. As we saw in the prior section, the frequency spectrum of a single pulse has much more energy at surrounding side bands than does an infinite string of pulses. Segmenting a long signal into pieces will artifactually increase energy at side bands around each true frequency component in a segment. There are, however, ways to minimize this problem (see Box 3.2).

A series of frequency spectra extracted from successive pieces of a signal can be plotted together in a **spectrogram**. The two types of spectrograms in current use depend upon whether two or three axes are graphed. The three-dimensional spectrogram is also called a **waterfall display**. An example for part of the song of a song sparrow is shown in Figure 3.20. Each horizontal line corresponds to one temporal segment of the entire signal with the early parts of the signal in the rear and the final segments of the signal in the foreground. On each horizontal line, frequency is plotted along this axis and the amplitude of each of the frequency components on the vertical axis. Each horizontal line is thus a single frequency spectrum plotted in the same manner we have used in all of our earlier examples.

This makes for a pretty picture, but it is very difficult to see from such a plot what is happening during the song. For one thing, there is too much detail. Perhaps we don't really need to know about all the lower energy frequency components. Can we raise the floor of this plot so that only the major (more intense) components stick through? This simplification ought to allow us to see the overall patterns in the song better. If we do this to the plot in Figure 3.20, we get the new graph shown in Figure 3.21. For better temporal resolution, we then divide the song up into many shorter segments and color any component sticking out of the floor black. The resulting plot is shown in Figure 3.22. If we now get rid of the perspective and rotate the plot in Figure 3.22 90° counterclockwise, we obtain the two-dimensional spectrogram shown in Figure 3.23. (An early commercial device used to produce such plots was called a Sonagraph and the resultant plots were termed **sonagrams**. This trade name acquired wide usage in the 1960s and readers may still encounter the term sonagram used for two-dimensional spectrograms.) In this text, we shall usually mean a two-dimensional plot when we refer to spectrograms, and shall invoke the term waterfall display for three-dimensional graphs.

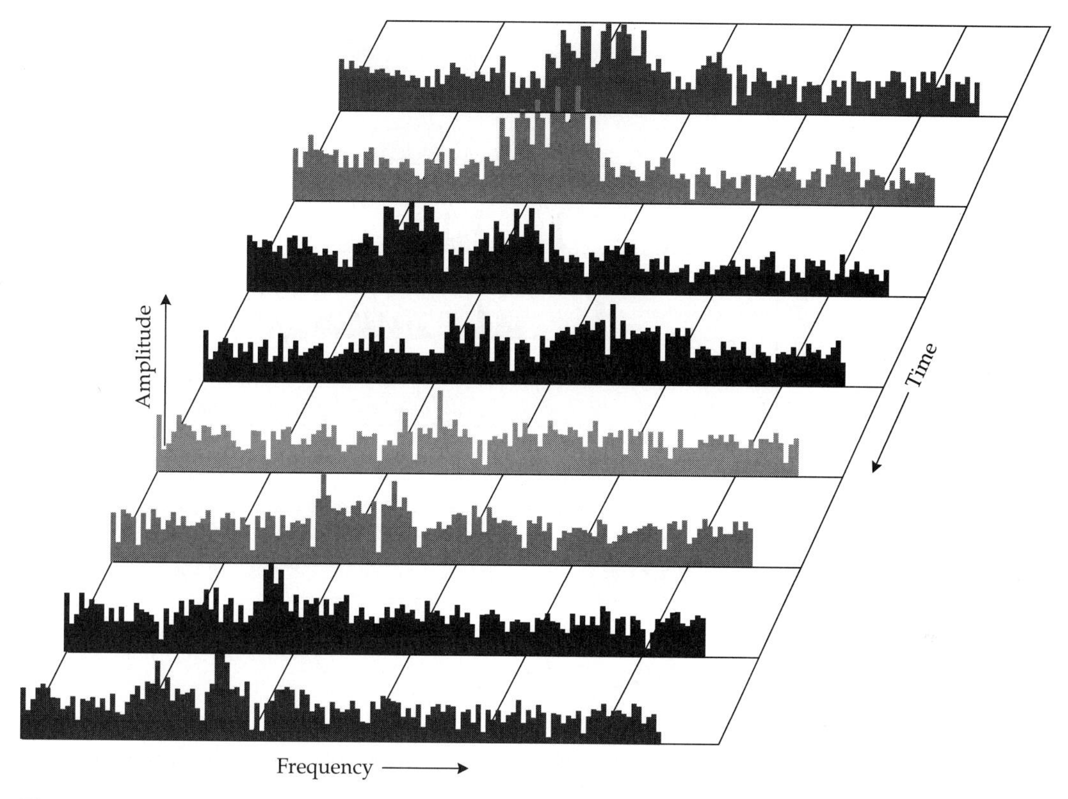

Figure 3.20 Spectrogram of a portion of a song sparrow song.

We can see more clearly the frequency structure of the sparrow song in Figure 3.23. There are several notes visible, four of which we have labeled. Note 1 (on the left top) consists of 4–5 close bands that could arise if the sparrow were (a) producing a nonsinusoidal but periodic waveform and then filtering out the lower harmonics before emitting it, (b) frequency modulating a sinusoidal carrier around 4.5 kHz, or (c) amplitude modulating such a carrier with a nonsinusoidal but periodic modulating waveform. Examination of the corresponding time domain picture shown in Figure 3.24 indicates that a fairly constant frequency carrier is being amplitude modulated in a periodic, but not quite sinusoidal, fashion at about 300 times per second.

Below this cluster of bands is note 2 which appears to be a relatively pure tone of about 2.8 kHz. Because it seems unrelated by normal modulation rules to the first note yet overlaps with note 1 in time, we presume that it was either produced by another bird, or that the song sparrow has two independent ways to make sounds at the same time (see next chapter). Note 3 is similar to note 1 with a series of parallel bands. Note 4 overlaps with the third and appears to change in frequency from 2 kHz to 3.2 kHz during its emission. If this fourth note is being frequency modulated (as it must if the frequency changes), why don't we see the side bands associated with modulation? This

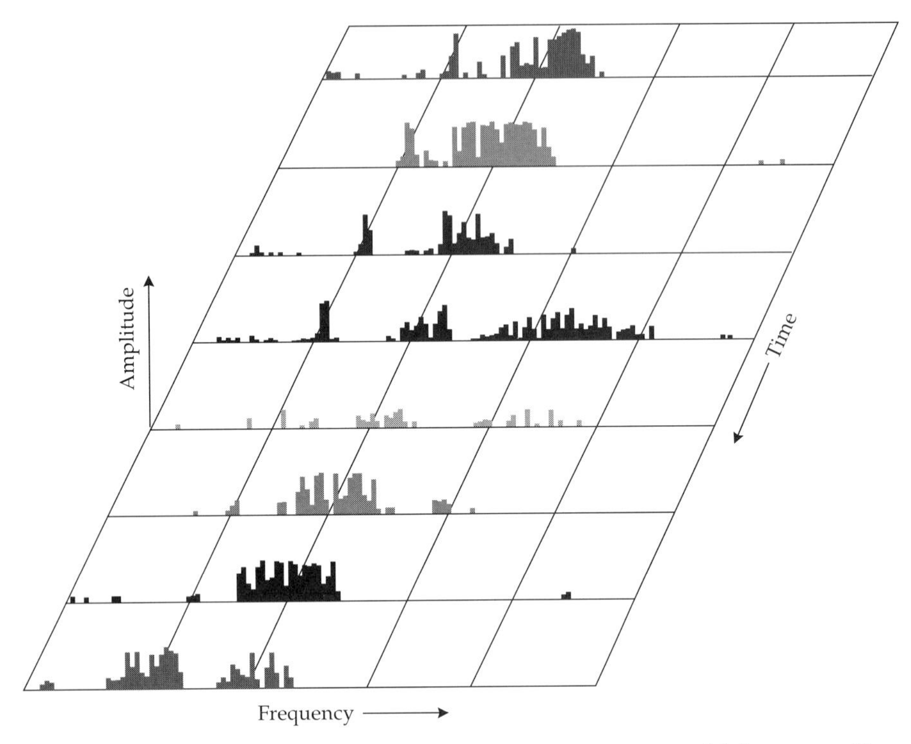

Figure 3.21 Spectrogram of a portion of a song sparrow song with lower amplitude components suppressed.

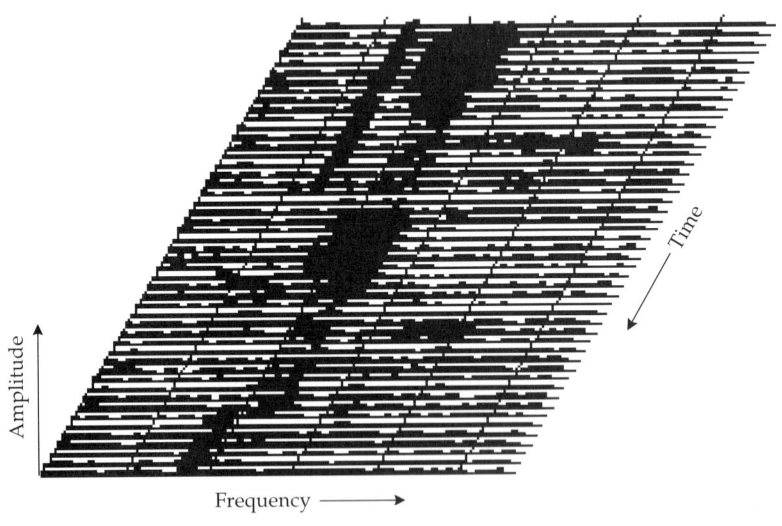

Figure 3.22 Spectrogram of a portion of a song sparrow song with all frequency components that have amplitudes greater than a critical value colored black.

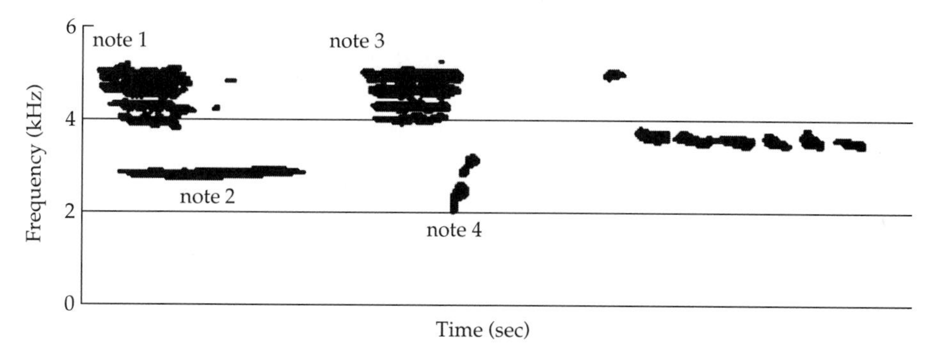

Figure 3.23 Two-dimensional spectrogram of a portion of song sparrow song. Such a plot is also called a sonagram.

raises the basic issue of bandwidth in Fourier analysis, a topic that we take up next.

Bandwidths and Analyses

Why not divide all signals into the tiniest segments possible and thus get really accurate time data? The answer is that you must have a reasonably long segment to identify the component frequencies and accurately measure their energies. Remember that as aperiodic signals get shorter and shorter, more and more energy is found at adjacent frequencies. In the extreme, a single instantaneous pulse has equal energy at all frequencies. If we cut the overall signal into long duration segments, our machine will treat the signal as if it were almost periodic and the frequency components will still appear as thin lines. As the segments become shorter, each is more like a single aperiodic event than an infinitely long periodic one. Thus a very short segment will produce a smear of energy over many adjacent frequencies and it will be difficult to know exactly where the energy lies in that smear.

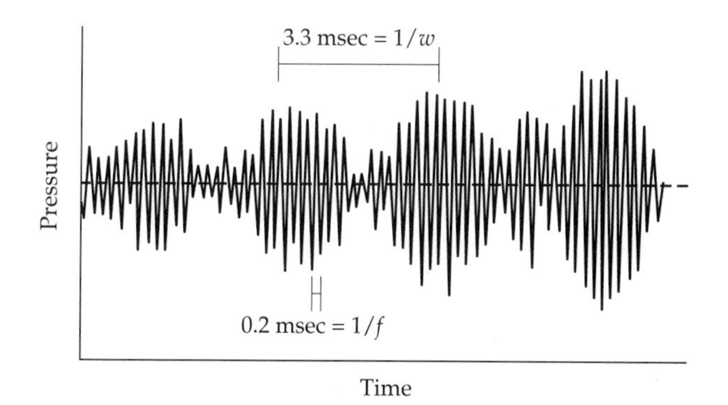

Figure 3.24 Time domain waveform of the first note of a song sparrow song. Amplitude modulations with period of 3.3 msec correspond to modulating waveform with frequency w = 300 Hz. Interval between finer peaks of 0.2 msec corresponds to carrier frequency of about 4.5 kHz.

Box 3.2 Segmenting Sounds For Spectrographic Analysis

SEGMENTING A LONG SOUND INTO PIECES, Fourier analyzing each piece as if it were an isolated sound, and then plotting the Fourier analyses together in temporal sequence provides improved temporal detail, but it can generate artifacts at the same time. Consider a long signal consisting of the sum of 1 kHz and 3 kHz sine waves. The time domain image and frequency spectrum of such a signal might appear as follows (left: waveform; right: frequency spectrum):

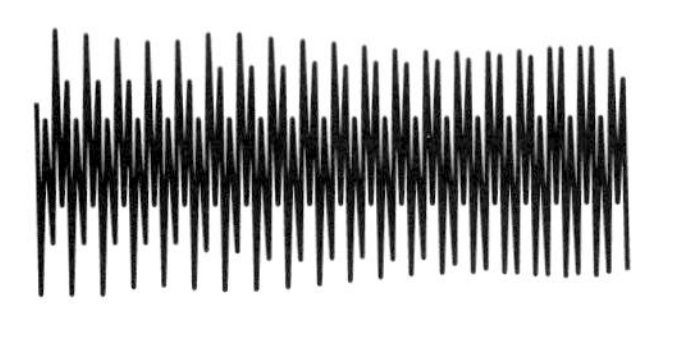

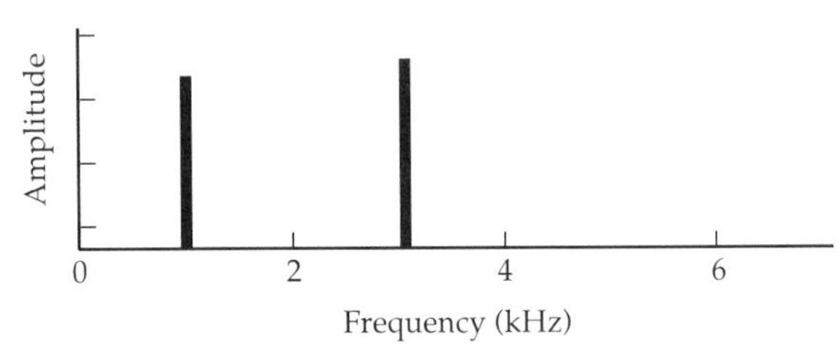

Now suppose we move a **rectangular sampling window** sequentially along the time domain waveform and only Fourier analyze that part of the signal contained within the window. The time domain image of the window's contents and the corresponding frequency spectrum for a wide sampling window would appear as follows:

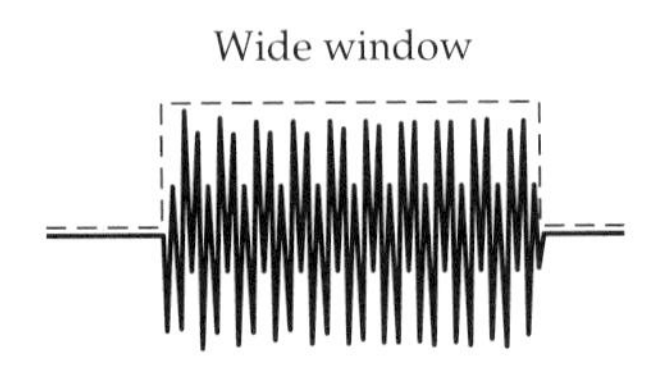

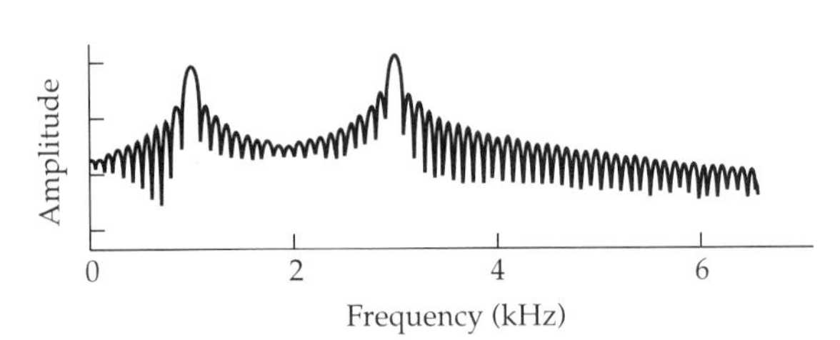

The dotted line over the waveform shows the shape and size of the sampling window. While the frequency spectrum shows the components at 1 and 3 kHz, there are now many lobes and nodes surrounding each true component. As we have seen for finite pulse trains, the lobes will be separated by a frequency equal to the reciprocal of the window width (pulse duration). Suppose we now sample the same waveform with a narrower sampling window. The waveform and spectrum will be as follows:

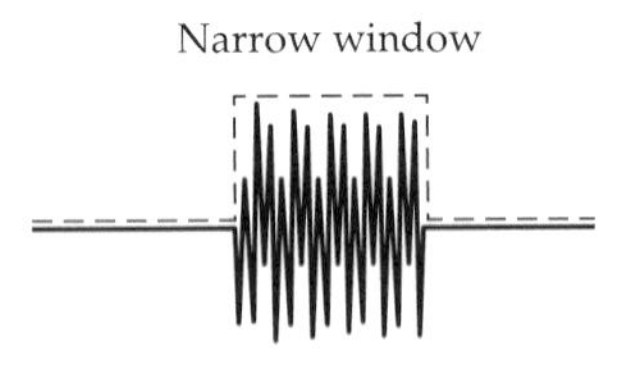

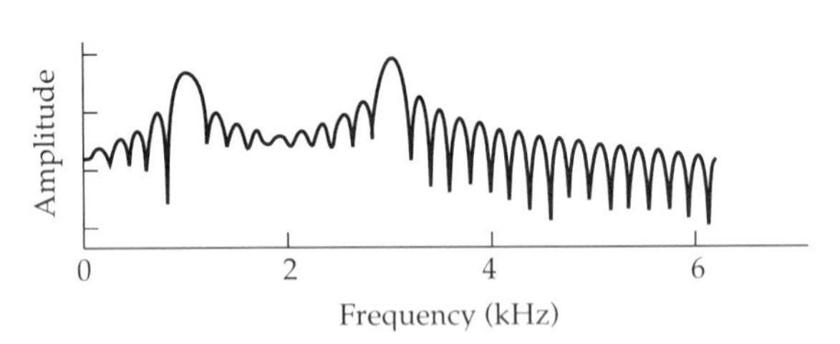

Decreasing the width of the rectangular sampling window has made the lobes wider and less numerous, but it does little to reduce the energy of the lobes relative to that of the signal components.

One way to reduce the lobe amplitudes is to make the shape of the sampling window more similar to a pure sine wave. This matching should reduce the amplitude of all but one component in the modulating waveform, and thus reduce the amplitudes of all but one pair of lobes. A wide **triangular or Bartlett window** will generate the following time domain waveform and frequency spectrum:

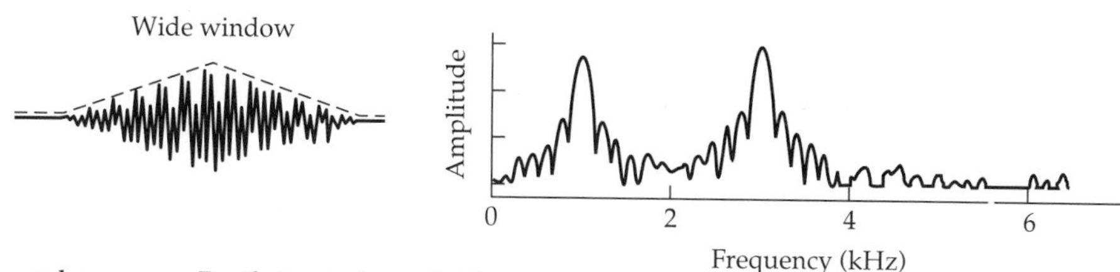

and a narrow Bartlett window yields:

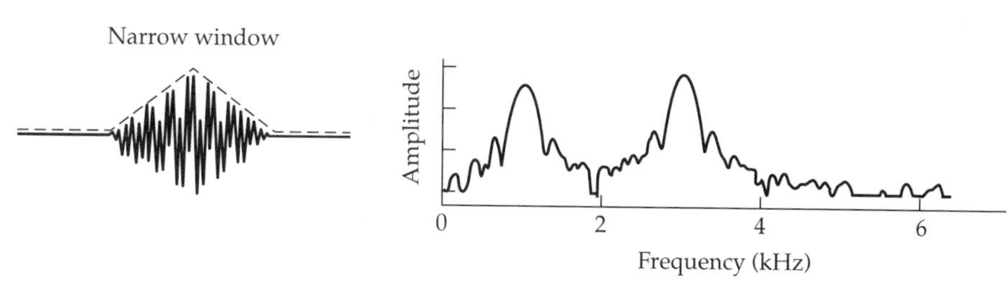

The triangular window clearly reduces the amplitudes of lobes farthest from the true components because these higher order harmonics have less energy in a triangular wave than in a square one. Narrower windows, although giving finer temporal information, again result in wider bands of energy around each true component.

Using even more sinusoidal window shapes can produce further differentiation between the magnitudes of true signal components and the artifactual side lobes. **Hamming and Hanning windows** are roughly bell-shaped in the time domain. Side lobes farthest from the true components are strongly suppressed by these windows. The two differ slightly in the rate at which the skirt of the bell shape drops to the baseline in the time domain, and in the rate at which they suppress distant lobes in the frequency domain. An example of a bell-shaped sampling window and its associated frequency spectrum appears as follows:

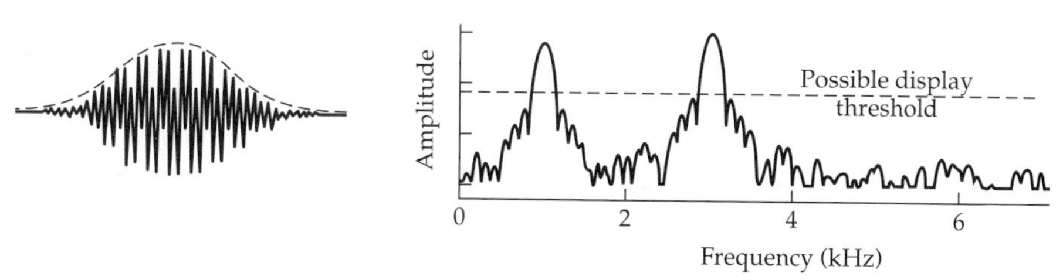

Once the amplitudes of the side lobes have been suppressed far enough below that of the true components, one can adjust the threshold above which components will be shown in the final spectra or spectrograms. This threshold will cause only the true signal components to be shown and will exclude most of the side lobes. The dashed line in the above example would be an example of such a cutoff.

When we perform a Fourier analysis on a segment of a signal, we are trying to identify each frequency component in that segment with access to only a limited number of cycles. Suppose each segment is Δt seconds in duration. The shorter Δt, the fewer the number of cycles of each component wave that will be present, and the less accurate the estimation of each component frequency. For a given sampling time, there will be a minimal difference in frequencies that can be discerned. This minimum frequency difference is called the **bandwidth** of the analysis. Let us denote the bandwidth by Δf. The smaller Δf, the more accurately we can discriminate between two adjacent frequency components. The larger Δf, the more likely we will treat two adjacent bands as part of the same component and lump them together into one very wide frequency smear. The shorter Δt, the more accurately we can distinguish between two consecutive temporal events but the fewer cycles that are available to perform an accurate frequency analysis. Clearly, there is a tradeoff between getting good temporal resolution and retaining good frequency resolution in sound analysis.

In fact, it can be shown that the product of Δf and Δt can be made no smaller than a constant. By defining Δf and Δt appropriately, the constant can be set equal to one. This relationship, (e.g., that $\Delta f \cdot \Delta t \geq 1$) is known as the **uncertainty principle** for sound analysis. It says that we can have as inaccurate measurements of frequency and time as we want, but we cannot have the contrary situation: the product of time and frequency error cannot be smaller than a certain value. As we reduce one type of error to smaller values, eventually the other has to increase to keep their product greater than one. We saw an example of this on pages 57–59 in which we showed that the width of the main lobe in the frequency spectrum of a single short pulse (one measure of signal bandwidth) is equal to the reciprocal of the pulse duration. The uncertainty principle says that once we reduce the duration of the pulse (Δt) enough, eventually the bandwidth (Δf) has to increase, and thus the pulse duration and main lobe width become reciprocally related. The uncertainty principle puts a serious constraint on signal analysis by any organism or machine. If the organism or machine opts for high frequency resolution (small Δf), it will lose temporal resolution because long segments of waveform (large Δt) will be required to achieve this accuracy. If it tries for high temporal resolution (small Δt), the cruder the frequency discrimination will be (i.e., large Δf). There is no way out of this dilemma.

The uncertainty principle means that you must be very careful in interpreting spectrograms. If you have set your bandwidth for a low Δf, you will see a spectrogram that represents most of your signal in the frequency domain. If you use a higher bandwidth, the machine will do its best to produce a spectrogram, but many components will not be separated and their sum will be plotted as time domain images. The critical factor is whether the difference between adjacent frequency components is greater or less than Δf. Suppose for example, that you set your bandwidth at 50 Hz and the signal you are examining is a 1000 Hz sine wave sinusoidally amplitude-modulated at a modulating frequency of 70 Hz. We know that the frequency domain version of this sound will consist of 3 frequency bands: a carrier at 1000 Hz, and side bands at 930 and 1070 Hz. Because your bandwidth (50 Hz) is smaller than the difference between these adjacent frequency bands (70 Hz), you will see the three bands on your spectrogram (Figure 3.25).

If however, you had selected 300 Hz for your bandwidth, your machine will be unable to separate the signal into three components. In fact, it will attempt instead to display a time domain image of the signal on the spectrogram. Since the time domain signal is a carrier of 1000 Hz being amplitude modulated, the spectrogram will show a smeared band of dark near 1000 Hz broken up into clear pulses, and if you count the number of apparent pulses per second on the spectrogram, you ought to find about 70 per second, the modulating frequency (Figure 3.26).

This explains why note 4 in the sparrows spectrogram was not broken into side bands and a carrier. The note lasted 0.017 seconds and the frequency continued climbing the entire time. A periodic modulating waveform would have taken another 0.017 seconds to go back down, giving a rough cycle length of 0.034 seconds for this frequency modulation. This cycle corresponds to a modulating frequency w of about 29 Hz, which is far below the minimum bandwidth of a typical analyzer. In short, the machine could not resolve side bands so closely spaced, and hence it plotted the frequency modulation in the time domain. In fact, with frequency modulation, it is often more useful to see the frequency patterns in the time domain so that one can measure the maximum and minimum frequencies and the rate of modulation directly from the spectrogram.

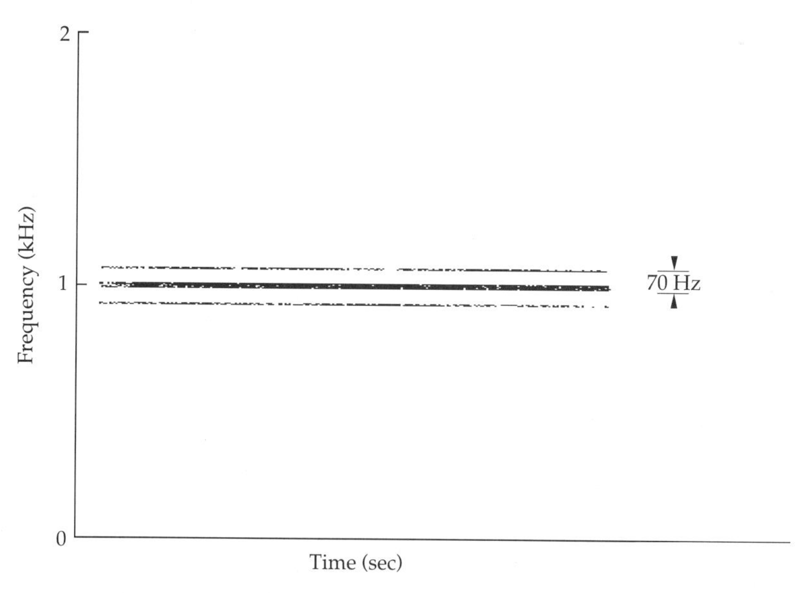

Figure 3.25 Spectrogram of 1 kHz pulses using narrow bandwidth (Δf = 50 Hz).

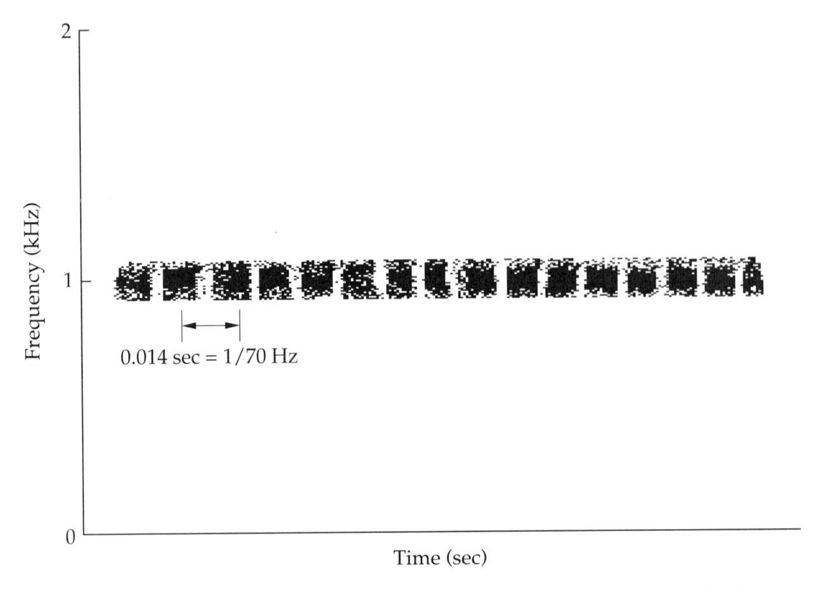

Figure 3.26 Spectrogram of 1 kHz pulses using wide bandwidth (Δf = 300 Hz).

To recap, you should expect your spectrographic analyses to be a mix of time domain and frequency domain information. The world in which you find yourself at any point in the analysis depends both on the machine and on the properties of the signal you are analyzing. If the critical parameters are different in two successive parts of a signal, you could get a frequency domain picture of the first and a time domain picture of the second. In fact, this double picture often occurs. You need to be aware of this problem, know your bandwidth, and use the time domain picture to help interpret your results. In fact, *always* examine both the time domain and frequency domain images for any signal. It is only by comparing the two that you can build a complete description and understanding of your sound (see Box 3.1). Most sound analysis is currently performed using digital techniques. These provide speed and easy data access, but carry their own set of constraints. Some of these are outlined in Box 3.3.

FOURIER ANALYSIS AND LINEAR SYSTEMS

Fourier analysis has practical uses beyond the description of animal sounds. In the next chapters, we shall repeatedly discuss how a signal becomes altered as it passes from one chunk of medium to another. For sounds, the initial vibrations generated in the animal must pass through the animal into the surrounding medium. They must then pass through an external chunk of air or water to reach the receiver. At the receiver, the sounds are collected and pass through various tissues to reach the sensory cells. Similar processes occur for light and electric signals. Passage of the signal through each chunk of medium will change some of the signal's properties. It would be very useful if we had a way to predict the changes in any given signal as it passed through a particular chunk of medium. Is this possible? As long as propagation of signals in a medium is linear, the answer is yes.

Let us again review what we mean when we say a system is linear. Consider a signal that can be described in the time domain by some waveform function, say $x(t)$. Suppose we input this signal to a zone of external medium, an organism's body, or some other chunk of material, and we record the waveform of the signal that emerges on the other side. Call the time domain waveform of this output signal $y(t)$. Let us then characterize any changes wrought in $x(t)$ during transmission through the chunk by the function F: thus $y(t) = F[x(t)]$. Suppose now we have two different signals we want to propagate through the chunk. We input $x_1(t)$ to the chunk and obtain the output signal $y_1(t)$; we later input signal $x_2(t)$ to the same chunk and get output signal $y_2(t)$. The chunk of material and the function F are both said to be **linear** if the output obtained when we input $x_1(t)$ and $x_2(t)$ simultaneously to the chunk is equal to the sum of the outputs we got when we input the two signals separately. Algebraically, $F[x_1(t) + x_2(t)] = F[x_1(t)] + F[x_2(t)]$ if the system is linear.

If a chunk of medium is linear, then all we need to characterize the way it will change a signal during propagation is the function F for that chunk. How can we measure F? One way is to describe the response in the time domain. Note that any smoothly varying input signal $x(t)$ can be mimicked by generating a series of extremely short rectangular pulses whose amplitudes rise and fall with $x(t)$. If we can characterize how the chunk would respond to one such pulse, then we should be able to extrapolate how it would respond to a series of pulses with varying amplitudes. This method is not as simple as it may seem however. Most chunks hit with a single sudden pulse will ring or oscillate a bit after the pulse hits them. This means that when we input a string of pulses to a chunk, it will still be vibrating with the effects of prior pulses when the next pulse hits. We must therefore account for this overlap in the effects of all prior pulses. The value of the signal output from the chunk at any time t, $y(t)$, will thus be the sum of the chunk's response to the current pulse and its continuing responses to all prior pulses.

We have already met the ideal pulse for modeling our input signal: the delta function. This is a pulse of infinitely short duration but sufficiently great amplitude that the area enclosed under the waveform of the pulse is greater than zero. We first examine how our system responds to a single such pulse. The chunk is hit with a delta function pulse (or a real signal as close to a delta function as possible), and we monitor the waveform of the output signal from the chunk. The plot of this response to a delta function is called the **impulse response function**, $h(t - \tau)$, where t is the time of the current sample of the response and τ is the time at which the delta pulse occurred. Because the chunk can only respond to the pulse after it has occurred, $h(t - \tau) = 0$ for all $t < \tau$. Once the pulse is input, the chunk continues to respond to the single pulse in some way which is characteristic of the material comprising the chunk. Eventually, any real chunk will dissipate the entire pulse as

Box 3.3 Aliasing and Digitized Sound

MOST FOURIER ANALYSES ARE NOW DONE ON COMPUTERS using a fast Fourier analysis algorithm. To use this method, the recorded signal must be "digitized": instead of analyzing a continuous signal, you break up the sound into many tiny samples and record the amplitude of the signal at each point. Instead of this:

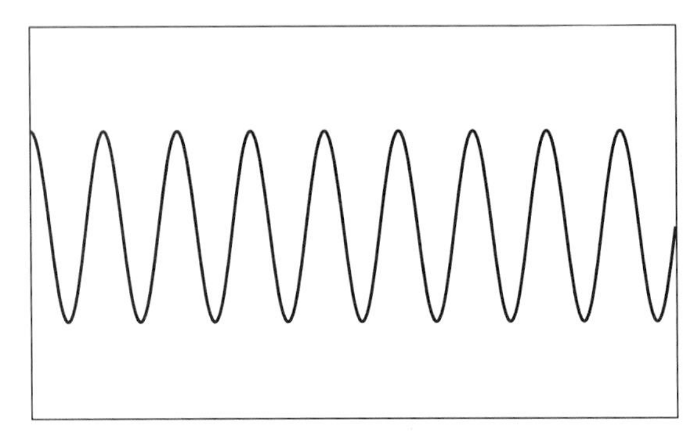

your computer stores this:

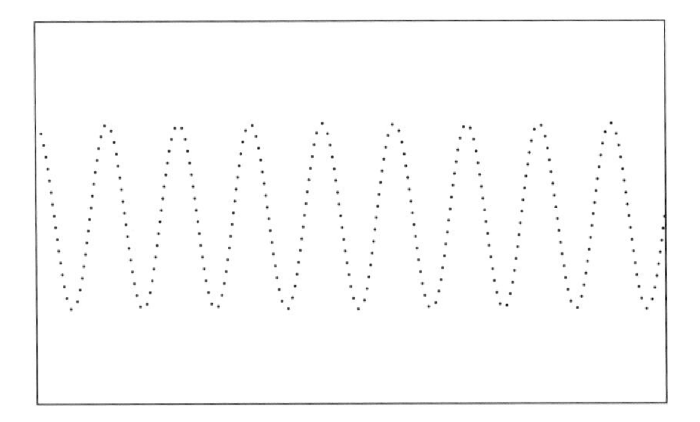

The advantage is that you now have a powerful tool for analyzing sounds quickly. The cost is that if you have too few dots per wave, the machine will have trouble identifying which frequencies are present. In fact, the sampling rate of the digitizer must be at least twice the highest frequency that you expect to find in your signal. If you cannot capture at least one maximum and one minimum of a wave, you cannot know its frequency. Were you to try to analyze a signal in which there were frequencies greater than half the sampling frequency, (called the **Nyquist frequency**), energy from these higher

heat and no further response will occur. Once we have $h(t - \tau)$, and assuming the chunk is a linear system, we can then predict the response of the chunk to any waveform by breaking the input signal, $x(t)$ into a series of scaled and successive delta functions, computing the responses of the system to all prior pulses at the appropriate time using $h(t - \tau)$, and adding these up to get each $y(t)$. Because a delta function has an infinitely short duration, we shall need

frequencies would be erroneously dumped into the lower frequency components of the frequency spectrum and produce a muddle. This artifactual spread of energy to lower frequencies is called **aliasing**. Most commercial digitizers sample at either 22 kHz (22,000 Hz) or 44 kHz. If you use the 22 kHz rate, you could analyze any signal that had component frequencies of 11 kHz or less. To prevent aliasing, you should always use a filter on your signals before digitizing them to remove any frequencies higher than your Nyquist value. For birds, which cannot hear above 10–11 kHz, no biologically relevant information is lost by filtering. For mammals, especially bats and rodents, the situation is more difficult. Many species utilize frequencies higher than 11 kHz, and if you filter these out before digitizing them, you may totally alter the important parts of the signal. The only options are to use a more expensive machine that samples at a higher rate, or to play your mammal sounds back at half or quarter speed, thus cutting the frequencies in half or in fourths respectively. Then you can filter and digitize them with the same procedures that you used on the bird sounds.

Digital Fourier analyzers, like all instruments, must obey the **uncertainty principle** of sound analysis. Thus suppose we break a signal into consecutive segments Δt seconds long and analyze each segment separately. The best frequency separation we can hope to get within each segment will be $\Delta f_{min} = 1/\Delta t$. On most digital instruments, one does not control Δt directly, but instead controls the sampling rate, F, and the number of samples assigned to each segment, n. Since $\Delta t = n/F$, the minimum bandwidth of the analysis is $\Delta f_{min} = F/n$. As the sampling frequency increases, the minimum bandwidth of the analysis increases; as the number of samples per segment increases, the lower the effective bandwidth. For example, many digital analyzers use a sampling rate (F) of 22 kHz. Suppose we set n to be 1000 samples per segment. The minimum bandwidth (Δf_{min}) of the analysis would be $22{,}000/1000 = 22$ Hz. Any sound analyzed with these settings would show a frequency spectrum with separate lines for each component separated by at least 22 Hz up to the Nyquist frequency of 11 kHz ($F/2$). The amplitude of each line would reflect the energy in the analyzed sound within that 22 Hz band.

Note that we can use smaller numbers of points per segment if we want to obtain better time resolution at the expense of frequency resolution. Commercial programs for digital sound analysis usually allow the user to make this choice by changing the segment size (sometimes called frame length); the digitization rate cannot be changed once the sound is in the computer. The programs may also improve apparent resolution by analyzing overlapping segments instead of nonoverlapping ones, invoking various smoothing algorithms, and adjusting the minimum amplitude of components that are to be visible in the resulting plots.

an infinite number of them to mimic $x(t)$. Instead of a sum, we clearly need an integral. Mathematically, the combination of an input signal, $x(t)$ and the impulse response function, $h(t - \tau)$, is called a **convolution** and $y(t)$ is thus a convolution of the form

$$y(t) = \int_{-\infty}^{\infty} x(\tau)\, h(t - \tau)\, d\tau$$

What does Fourier analysis have to do with these computations? In practice, describing the impulse response function for various chunks of material can be very difficult. There is a better way, and this involves describing the response in the frequency domain. Note that we can convert any input signal, $x(t)$ into its frequency spectrum and phase spectrum. Let us denote the input signal frequency spectrum by $X(f)$. We can also undertake Fourier analysis of the impulse response function, $h(t - \tau)$, and obtain its frequency spectrum and phase spectrum. Let us denote the frequency spectrum of the impulse response function by $H(f)$. If the frequency spectrum of the output signal, $y(t)$, is denoted by $Y(f)$, it turns out that $Y(f)$ equals a simple product (instead of a convolution) when we remain in the frequency domain:

$$Y(f) = X(f) \cdot H(f)$$

A similarly simple computation can be used to predict the phase spectrum of the output signal from the phase spectra of the input signal and the impulse response function. The frequency domain version of the impulse response function (both frequency and phase spectra) is called a **transfer function** because it characterizes the way in which a particular input signal's frequency domain components will be transferred or converted into the frequency domain image of the output signal.

This is an extremely useful result. Once we know $H(f)$ for a chunk of material in which signal propagation is linear, we can predict the frequency spectrum of the output signal for any input signal. $H(f)$ is thus the frequency domain version of the F function we sought. $H(f)$ is often called the **frequency response** of a system. One way to determine the frequency response of a system is to painstakingly input a pure sine wave of low frequency and fixed amplitude to it, measure the response, input a slightly higher frequency at the same amplitude, measure that response, and so on, until the entire frequency range has been tested. A graph of relative output amplitude versus frequency would characterize the frequency response properties of the chunk. Frequencies that were well propagated in the chunk would have high output amplitudes, whereas those which are poorly propagated would have low output amplitudes. A more efficient method would be to input one signal containing all frequencies at the same energy, perform a Fourier analysis of the output signal, and use the resulting frequency spectrum to obtain the same graph. As we saw in an earlier section, the frequency spectrum of a delta function is just such a signal: it consists of all frequencies at equal energies. The frequency spectrum of the response to a delta function is $H(f)$, and thus $H(f)$ is the function describing the frequency response of the system. Once we have the frequency response function, $H(f)$, we need only multiply the amplitude of each frequency component in the input signal by the corresponding amplitude of the same frequency in $H(f)$ to predict the frequency spectrum of the output signal.

In the ensuing chapters, we shall often refer to the frequency response of a particular medium. The medium may be the vibrating tissues of a voice box, the air-filled tube of a vertebrate trachea or mouth, volumes of external medium filled with reflecting objects, the small bones in a mammalian ear, or the dense lens of an eye. It is a rare medium indeed that propagates all input frequencies with the same efficiency. Thus nearly all media distort and change signals as they propagate them. In each case, if the medium is linear and we know the transfer function of that chunk of medium, we can predict the frequency domain image of a signal that has traversed the chunk. If we do not care about phase changes, then we need only look at the frequency response function. It is because most of the media that are relevant to animal communication are linear that it is worthwhile attempting to characterize their frequency response functions. This task will be one of the major goals in most of the remaining chapters of Part I.

SUMMARY

1. Any continuous waveform can be broken down into the sum of a set of pure sine or cosine waves. This process is called **Fourier analysis** and is usually summarized in two graphs: the **frequency spectrum** plots the amplitude of each fourier component versus its frequency, and the **phase spectrum** plots the relative phase of each Fourier component versus its frequency. A description of the original waveform is called the **time domain** image of the signal; the two Fourier graphs provide the **frequency domain** image of the signal. Mathematically, either the time domain or the frequency domain image can provide a complete description of a sound; the two images are thus totally equivalent.
2. The basic rule of Fourier analysis is as follows: **The more the time domain image of a signal deviates from that of a single sine wave of infinite duration, the greater the number of other sine waves that must be added to the frequency domain graphs and the larger must be the amplitudes of these additional components**.
3. **Periodic** signals repeat the same pattern forever. The number of times the basic waveform repeats each second is called the **repeat frequency**; it can be measured directly on a time domain image of the signal. Periodic signals include simple sine waves, sine waves that are amplitude or frequency modulated, and waveforms that repeat exactly but that are not sinusoidal in shape. The frequency spectrum of a pure sine wave with no DC component has a single band. The frequency domain image of a **modulated** sine wave includes a band at the frequency of the original wave, called the **carrier**, and **side bands** at frequencies equal to the carrier plus and minus integer multiples of the repeat frequency. Periodic nonsinusoidal signals can be broken down into frequency spectra called **harmonic series**. All bands in the frequency domain image are integer multiples of the frequency at which the waveform repeats in the time domain. If present, the band equal to this frequency is called the **fundamental** of the series, and all higher frequency bands are called **harmonics**. **Dirichlet's rule** predicts that the amplitudes of harmonics in such a series should tend to decrease with increasing frequency. Where periodic waveforms consist of several parts (defined by several successive maxima and/or minima per period), certain harmonics, called **nodes**, may be absent or show reduced amplitudes in the frequency spectrum. The difference in frequency between successive nodes is equal to the reciprocal of the duration of the shortest part in the signals' time domain images. Amplitudes of frequencies between nodes are enhanced and together form **lobes** in the spectrum. Complex modulating waveforms and carriers can be broken down into their frequency domain

components and the overall frequency spectra generated by applying the simple modulation rules across each pair of components.

4. **Aperiodic** signals do not repeat. The more a signal approaches a unique event, the wider the bands in its frequency spectrum. The extreme aperiodic signal is the **delta function**, which has infinitely short duration, infinite amplitude, and never repeats. Its frequency spectrum is one large band that includes all frequencies at equal amplitude.
5. Dividing a signal into segments, performing Fourier analysis on each segment, and plotting the resulting series of frequency spectra along a time axis generates a graph called a three-dimensional **spectrogram** or **waterfall display**. A two-dimensional **spectrogram** is a plot in which only the higher amplitude components of a spectrogram are shown, and these are graphed as a function of time on the horizontal axis and frequency on the vertical axis.
6. Any ear or instrument undertaking Fourier analysis labors under an **uncertainty principle**: the better its ability to discriminate between adjacent components in the frequency domain, the worse its ability to discriminate between successive segments in the time domain. The ability to discriminate between adjacent frequency components is called the **bandwidth** of the instrument. The smaller the bandwidth, the better the instrument's frequency resolution. Modulations and periodicities at rates less than the bandwidth of the analyzer will appear on the spectrogram in the time domain; those occuring at higher repetition rates will be shown in the frequency domain. In practice, any spectrogram is a mix of frequency and time domain images.
7. Any chunk of matter is a **linear system** if the simultaneous propagation of two signals through the chunk results in an output signal that is the sum of the outputs of the two signals propagated separately. The propagation properties of a linear chunk of medium are completely described if one knows how the chunk propagates a delta function. This rule is more easily applied in the frequency domain: the frequency spectrum of an output signal in a linear system is simply the product of the frequency spectrum of the input signal and the frequency response function of the medium. The frequency response function of a medium is the response of that system to a delta function (since the latter contains all frequencies at equal energies). Characterizing the frequency response functions of different media is fundamental to understanding why animals use the frequencies they do in specific contexts.

FURTHER READING

There are large numbers of texts discussing Fourier analysis, linear systems, and related topics. One straightforward introduction to Fourier analysis is Hsu (1970), and a concise summary of linear system analysis can be found in Faulkner (1969). The classic book by Hund (1942) discusses the Fourier compositions of modulated signals. Problems of time versus frequency domain analysis in spectrograms have been discussed by a number of authors: Watkins (1967), Greenewalt (1968), Marler (1969), Staddon et al. (1978), Hall-Craggs (1979), and Beecher (1988). Several discussions of digital sound analysis techniques are provided by Hopkins et al. (1974) and Clark et al. (1987). The journal *Bioacoustics* often has very useful articles on techniques of analysis of animal sounds and current technology.

Chapter 4

Sound Production

THE PRODUCTION OF MOST ANIMAL SOUNDS INVOLVES three successive steps: (1) the production of vibrations, (2) the modification of these vibrations to match biological functions, and (3) the coupling of the modified vibrations to the medium in which the sound is to propagate. As we shall see, none of these steps is simple and adaptations that facilitate one often complicate one of the others. As a result, only two groups of animals have evolved sound communication: the arthropods and the vertebrates. Even within these two groups, only certain taxa use sound signals. Among the arthropods, sound communication is limited to decapod crustaceans, certain insects (most orthopterans and cicadas, and some bees, flies, beetles, aphids, true bugs, and butterflies), and a few spiders and millipedes. Although sound signals are the rule among frogs, birds, and mammals, other vertebrates such as most fish, all salamanders, and most reptiles (excepting geckoes, giant tortoises, crocodiles and apparently some dinosaurs), are silent. One reason why many animal taxa do not use sound communication is surely the difficulty in producing vibrations and coupling them to the medium. Let us examine the problems.

PRODUCTION OF VIBRATIONS

All sounds begin with the generation of vibrations in some part of an animal's body. There are five basic mechanisms for producing vibrations: (1) muscular vibration of a membrane or sac, (2) stridulation (the rubbing of one part of the body against another), (3) forced flow of medium through a small orifice, (4) muscular vibration of appendages, or (5) percussion on a substrate. The waveform of generated vibrations might be sinusoidal, but it is more likely to be complex with many sinusoidal components. In addition, the shape of the radiating sound field around the vibrator may be very complex depending upon which type of vibration mechanism is used.

One mechanism of vibration used by some fish is the muscular contraction and expansion of a gas-filled sac in the middle of the fish's body. A vibrator that expands and contracts around a central point is called a **monopole** (Figure 4.1). The expansion of the sac causes the tissue fluids around it to be compressed, whereas contractions cause local rarefactions of the fluid molecules. These compressions and rarefactions radiate away from the sac through the fish's tissues and into the surrounding water. The vibrations constitute a propagating sound. Note that the sound field from a monopole is symmetrically distributed around its source. Measured sound pressures at any given distance from the fish should be identical whether one records the sound from above the fish, facing its side, or from behind its tail.

In contrast to the monopolar properties of a pulsating air sac, consider the chirping of a cricket. Cricket chirps are an example of a very common mechanism for producing sounds by arthropods—**stridulation**. Here, sound is produced by rubbing a row of small cuticular teeth (called a **file**) on one body part against a sharp ridge or blade (called a **plectrum**) on another body part (Figure 4.2). In crickets the file is on the underside of the top wing, and the plectrum is on the upper side of the bottom wing. As the wings are moved in

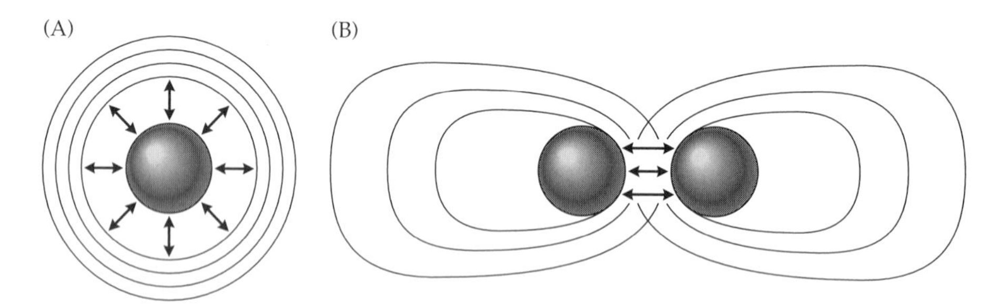

Figure 4.1 Polar sound sources. (A) A monopole sound source creates sound by alternately expanding and contracting. The resulting sound field radiates away from the source equally in all directions. (B) A dipole sound source creates sound by moving back and forth along a single axis. The resulting sound field is not uniform. Sampling at a given distance from the dipole and along the axis of movement shows higher intensities than sampling at the same distance but at a point perpendicular to the axis of movement.

(A)

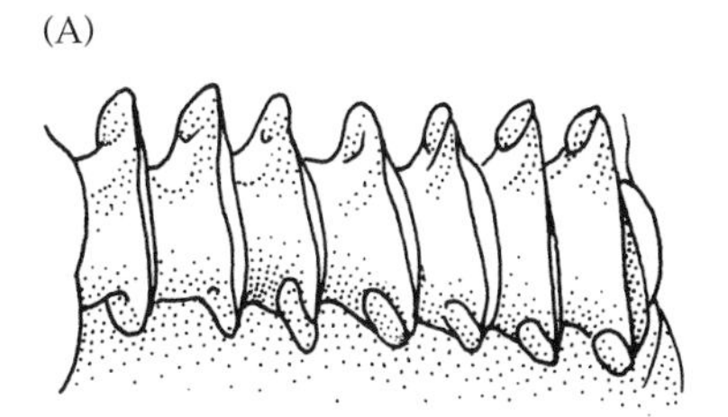

(B)

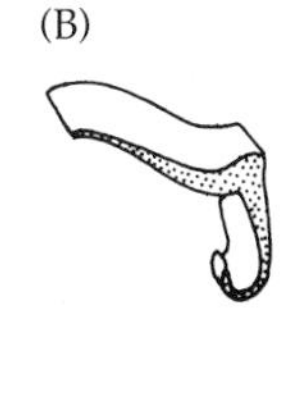

Figure 4.2 Stridulatory apparatus of a cricket. (A) File and (B) plectrum. The file is located on the underside of the top wing; the plectrum on the upper side of the bottom wing. When wings are scissored past each other, the plectrum strikes the teeth of the file. (After Lewis 1983.)

opposite directions, successive teeth on the file catch on the plectrum, are bent back slightly, and then pop forward. Each tooth then vibrates briefly back and forth along the axis in which it was bent. As it moves forward and back, it produces a series of compressions and rarefactions of air molecules depending upon the direction in which it is currently moving. These propagate away from the cricket as sound. It should be obvious that the sound field around such a source is not symmetrical. A microphone facing the teeth as they move will record the maximal sound because it is along this line that the compressions and rarefactions of air molecules are generated. A microphone placed at the side of the file and perpendicular to the tooth vibrations will pick up a signal of much lower intensity. A sound source that vibrates back and forth along a single axis is called a **dipole** (Figure 4.1). A source that vibrates along two different axes at once is called a **tetrapole**; patterns of an even higher order are known. Each type of vibrator creates a differently shaped sound field.

In addition to affecting the symmetry of the sound field, the type of vibrator also affects the amount of sound pressure the vibrator is able to radiate. As a general rule, monopoles smaller than the wavelengths of the sounds they are producing generate sound pressures proportional to the surface area of the vibrator. This result occurs because the volume velocity of the produced sound increases with the surface area of the vibrator, and the volume velocity and acoustic impedance of the medium determine the resulting pressure. Since the surface area of monopoles is proportional to the square of their linear dimensions (e.g., radius and length), this means that pressure levels of sounds with wavelengths greater than the monopole increase with the square of the vibrator's linear dimensions. Once vibrators are larger than the wavelengths they are producing, sound pressure tends to increase linearly with the radius of the vibrator.

When dipoles are smaller than the waves they are producing, the same problems apply. However, it is even harder for dipoles to produce sounds with high pressures than it is for monopoles. This difficulty arises because of **acoustical short-circuiting**. When a dipole vibrates in one direction, it creates a compression on the side toward which it is moving and a rarefaction on the side from which it is receding. The distance between a rarefaction and a condensation is set by the size of the dipole. If this is a large distance, then it is unlikely that medium molecules can leak from the condensation around the dipole to the rarefaction and cancel out the differences before the vibrator moves back and begins a new cycle. However, if the dipole is small, and if the duration between successive cycles is long (as would be the case for longer wavelengths), then short-circuiting between condensations and rarefactions will be a major problem. For a dipole smaller in length than the wavelengths it is producing, sound pressures are lower than for monopoles and the sound pressures increase with the cube of the dipole's linear dimensions (in contrast to the square for monopoles). The increase in sound pressure due to increasing vibrator surface area is augmented by a reduction in short-circuiting at larger vibrator sizes.

The significance of these distinctions for sound communication is that small animals will have a difficult time making sounds with any significant sound pressure unless they can generate higher frequencies (with their associated shorter wavelengths). Animals that use dipole or higher order vibrators must use high frequencies to produce loud signals or must try to solve the short circuiting problem in some other manner. In air, a bird 10 cm in length would face this problem for any frequency lower than about 3.4 kHz, and a 1 cm insect would have problems for frequencies below 34 kHz. In seawater, a 10 cm shrimp would have problems for frequencies below 15 kHz, and a 1 cm isopod would have trouble with sounds below 149 kHz.

How do small animals resolve this problem? One answer is that this is not a problem for all small animals. Sounds of high intensity are primarily needed for long distance communication, so if animals are signaling at short distances, the inability to produce loud signals may not be a constraint. However, for those which must communicate over some distance, solutions must be found. One solution to the problem used by some tree crickets is to create a baffle (Figure 4.3) that prevents the acoustic short-circuiting associated with dipole vibrators. The cricket chews a hole in a large leaf and stridulates while sitting in the hole. Because the path linking condensations and rarefactions now runs the considerable distance from one side of the leaf to its edge and back to the hole, short-circuiting is greatly diminished. This solution is similar to our practice of mounting hi-fi speakers in a hole in an otherwise closed cabinet. Condensations and rarefactions on the side of the speaker outside the cabinet cannot reach those on the inside and thus the cabinet prevents acoustical short-circuiting. As often happens, an animal—here, a cricket—evolved this solution before we did.

The cost of building a baffle is that the animal is forced to make sounds only at one location. For mobile animals, the only solution is to limit sound production to high frequencies. This restriction is not as easy as it sounds. The simplest vibrator would be some physical structure that was moved back and forth by a muscle. Animal muscles can twitch and relax only up to rates of about 1000 contractions per second. That means that direct muscular control would be unable to solve the problem for any of the animals listed above. However, if muscles could be coupled to a device that produced multiple vibrations per contraction, then sufficiently high frequencies might be achieved. Such a device is called a **frequency multiplier**. Stridulatory organs are excellent examples: one movement of the relevant muscles moves a succession of teeth on the file across the plectrum. Each tooth produces a click, and the result is a train of clicks at a much higher rate than that of the muscle move-

Figure 4.3 Sound baffle created by tree crickets (Gryllidae) to prevent acoustic short-circuiting. (A) *Oecanthus* chews a triangular notch into a large leaf and places its head at the notch apex with its wings and body filling remainder of notch. (B) *Neozabea* chews a hole in the middle of a leaf and extends the front half of its body on one side of the leaf, leaving the remainder on the other. In both cases, concurrent condensations and rarefactions generated by the stridulator are on opposite sides of the leaf. The closest path between condensation and rarefaction now runs from the stridulator to the leaf edge and back to the stridulator on the other side of the leaf. If path length is long enough, relative to the interval between successive condensations and rarefactions (e.g., frequency), the medium cannot move in sufficient time to short-circuit the acoustic signal at its souce. (Photos courtesy of T. G. Forrest.)

ment. The train can be envisioned as a nonsinusoidal but periodic signal whose fundamental equals the click rate, and not the muscle movement rate.

As noted earlier, stridulatory organs are very common in arthropods. This solution is facilitated by the presence of a hard exoskeleton that can be used to produce the file and plectrum and by the fact that the arthropods' many joints provide pairs of parts that can be moved in opposite directions. Not surprisingly, nearly every pair of adjacent parts has been used to make a stridulatory apparatus in some arthropod (see Figure 4.4). In some species, one organ has the file and the other the plectrum, whereas in other species the roles are reversed. Examples of stridulatory organ pairs include antennae versus antennae (walking sticks), antennae versus head (lobsters), mouth parts versus other mouth parts (orthopterans and spiders), head versus body (beetles, shrimp), adjacent thoracic parts (beetles), adjacent abdominal segments (ants), wings versus thorax (lepidopterans), body parts versus legs or claws (crabs, true bugs, aphids, grasshoppers, spiders), legs versus wings (lepidopterans,

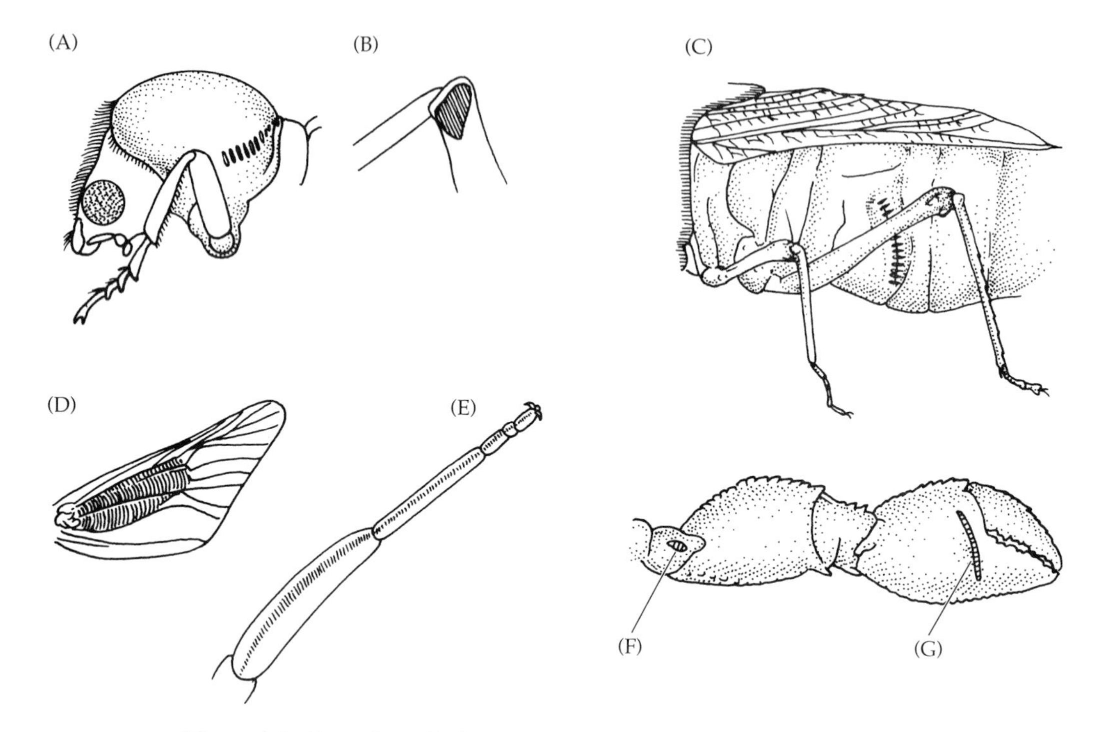

Figure 4.4 Sample stridulatory organs in arthropods. (A) Beetle with file on thorax and (B) plectrum on inner side of front leg. (C) Grasshopper with a file on its abdomen and plectrum on the inner side of its rear leg. (D) Noctuid moth with file on the underside of its wing and (E) plectrum on one side of its hind leg. (F) Crab with a file on the inner side of its claw and plectrum on the edge of the inner joint on the same appendage. (After Dumortier 1963b.)

beetles, grasshoppers), legs or claws versus legs or claws (thrips, true bugs, hermit crabs, crabs, millipedes), and wings versus wings (katydids and crickets). In the vast majority of cases, the sounds produced have dominant frequencies considerably above the maximum values for direct muscle action. Examples include lobsters (4–5 kHz), shrimp (6–8 kHz), crickets (2–20 kHz), moths (5–15 kHz), beetles (4–50 kHz), and katydids (4–90 kHz).

Stridulatory organs are only one means of producing high-frequency sounds with low-frequency muscles. Many structures will vibrate at frequencies dependent only on their own properties and not on the initial frequencies that excite them. The frequencies with which a guitar string or drum head vibrates are quite different from the rates at which the string is plucked or the head struck. Animals have made use of similar properties to excite vibrations in various organs. The vibrations in the tymbal of the cicada are a case in point. Muscles pull the tymbal membrane out of its relaxed position and let it pop back to its relaxed position. As the muscles act, the membrane briefly vibrates at frequencies determined primarily by its own physical properties (Figure 4.5).

Air-breathing vertebrates use muscles to bring air into or out of their lungs. As the air passes out, sounds may be produced in either of two ways.

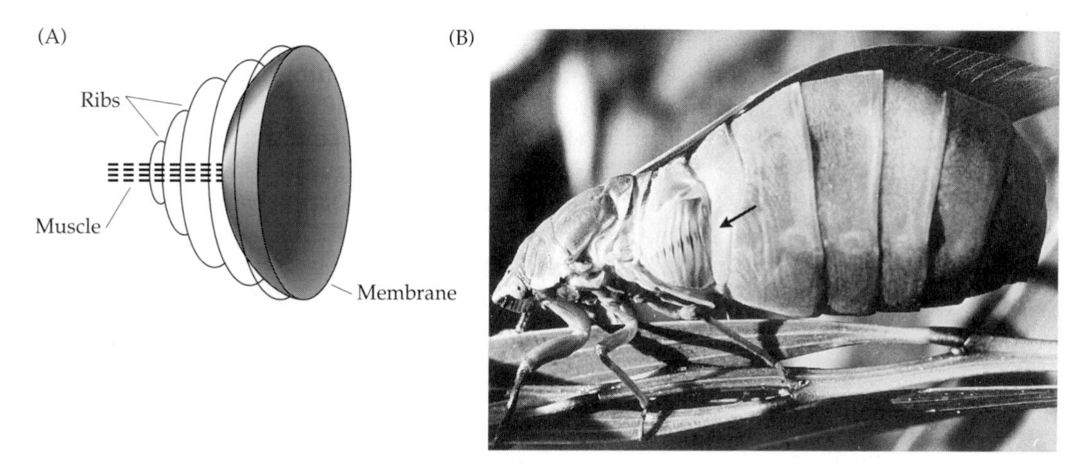

Figure 4.5 Sound production by cicada tymbal. (A) The tymbal consists of a cup supporting a stiff membrane. The cup often has ribs built into its surface. As the muscle pulls the membrane back, the cup is compressed until each rib successively buckles, giving a series of jolts, each of which vibrates the membrane. (B) Cicada with near wings removed to show the location of the tymbal (indicated by arrow). The ribs on the side of the tymbal are visible. (Photograph courtesy of David Young.)

The simplest is to place a small gate in the airflow. The gate is forced open when air pressure builds up, but then closes again as air flows through the gate and reduces the pressure buildup. The frequency at which this gate pops open and closed depends on the air pressure, the mass and elasticity of the gate, and any tension that is applied to the gate by muscles. Gated air-flows produce the sounds of anurans (frogs and toads) and most mammals; the associated sound-producing organ is called a **larynx**. Alternatively, one wall of the air-flow tube can be made very thin. As air passes over it, it will be forced to vibrate, (flutter in the breeze, one might say), and these vibrations will be added to the air flow passing out of the animal. This is the mechanism of sound generation in birds; the associated sound-producing organ is called a **syrinx**. In all of these cases, the frequency of the vibrations is not limited to the twitch-relaxation rates of individual muscles, but is instead set by the physical properties of the vibrators. The next sections describe in some detail how animals can control these frequencies.

MODIFICATION OF VIBRATIONS

Vibrations and Resonance

A vibration is the relatively cohesive movement of a group of molecules first in one direction and then in the opposite direction. All sounds begin by inducing vibrations in a fixed volume of molecules, and most last long enough for the molecules to make a number of complete cycles of vibratory motion. Sounds that are so short that only a few cycles are present are difficult to modify and difficult to couple to a propagating medium. The task for the communicating animal is to find some way to induce sustained vibrations in either a solid structure or in a fixed volume of medium.

Nearly any solid structure or closed volume of medium can be made to vibrate. However, it is not possible for such a vibrator to vibrate at any random frequency or set of frequencies for more than a few cycles. As an example, consider a string on a guitar. When we pluck the string at any point other than at the ends, the plucked portion vibrates back and forth along a line perpendicular to the string axis. This vibration is then propagated to adjacent portions of the string and eventually hits the two ends. Since the ends are fixed (i.e., have very high acoustic impedance relative to the string), the vibrations are reflected back toward the pluck point. After several vibration cycles, the entire string is experiencing the summed effects of vibrations reflected from each end and moving in opposite directions. The motion of the string at any given point depends on the local sum of these waves. At certain locations along the string, the various reflected waves will cancel each other out, and that part of the string will not move. Such locations are called **nodes**. Where the reflected waves reinforce each other, they cause the string to move further than it would have for only one reflected wave. These locations are called **antinodes**.

Although the motions of the string are unlikely to be sinusoidal, they can be broken down into a sum of simple sine waves (see Chapter 3). At any point on the string, the amplitude of a given component frequency depends upon the relative phases of the various reflected waves of that frequency. The phase difference will depend in part on any phase shift during reflection at the end and in part on how many cycles the reflected waves have experienced during their transit to the string end and back. This number of cycles will be equal to the distance traveled (which depends on the string length) divided by the size of the wavelengths of that component. For some of the component frequencies, the ratio of the string length to the component's wavelength is such that the alignment of multiple reflected waves stabilizes after a few cycles. As a result, the nodes and antinodes for these components will occur at fixed locations along the string. Such stable patterns of large and low vibration amplitude are called **standing waves**, and the frequencies that generate them are called the **natural modes** of the string (Figure 4.6). For other frequencies, the alignment of multiple reflected waves keeps changing because the string

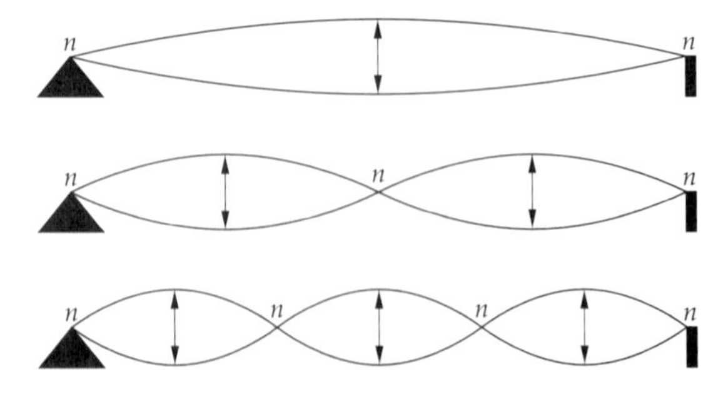

Figure 4.6 Standing waves and some natural modes of a guitar string. For certain wavelengths, string vibrations stabilize so that certain locations along the string move at maximum amplitudes. These locations are the antinodes and are indicated by arrows. At other locations one-quarter wavelength away, the string does not move at all. These are the nodes and are indicated by *n*.

length is not some simple multiple of their wavelength. These frequencies will never stabilize over time and will tend to cancel each other out. For a guitar string, the natural modes constitute a harmonic series whose fundamental has a wavelength inversely proportional to twice the length of the string. (The constant of proportionality depends on the tension of the string and on its diameter.) When the string is induced to vibrate at any of its natural modes, it is said to **resonate**.

Any vibrating system will have its own natural modes at which it tends to resonate. For one-dimensional systems such as strings and air or water vibrating in long narrow tubes, the natural modes are likely to constitute all or parts of an harmonic series. However, for two-dimensional vibrators (such as drum membranes and piano sounding-boards), or three-dimensional vibrators (such as bars, rods, or the bodies of violins and guitars), the natural modes may not be harmonically related. There are likely to be more natural modes within a given frequency range than for a one-dimensional system of similar size. Such concurrent frequencies, which need not be harmonically related, are called **overtones**. The resonant properties of a vibrator not only determine which frequencies will be found, but they also determine the relative amplitudes and phases of these frequencies. They thus determine both the frequency spectra and the phase spectra of the sounds produced.

Once excited, a vibrator does not vibrate forever. All vibrators have associated frictional and heat losses that result in a gradual decay of the vibration amplitudes back to zero. This process is called **damping**. If damping is high, then energy is quickly lost and amplitudes die down very soon after excitation. If damping is low, then there is a very slow loss of amplitude after excitation. To maintain a sustained sound, an animal must excite the vibrator at intervals to balance the losses due to damping. Alternatively, the animal might excite the vibrator at such short intervals that the energy added was greater than that lost by damping. This added energy will cause the amplitude of the vibrator to increase over time. However, because damping also increases with the amplitude of the vibrator, eventually the higher damping losses will just equal the gains from higher rates of excitation. The amplitude of the vibrator will then stabilize at a new and higher level. Thus, one way that animals can make themselves heard over longer distances is to try to increase overall vibrator amplitude by using frequent excitations of a slowly damped oscillator.

One can try to excite a vibrator with any initial waveform. If the exciting waveform is a single pure sinusoid, the vibrator will attempt to follow the exciting waveform's pattern. However, the further the exciting frequency is from the closest resonant frequency of the vibrator, the lower the amplitude of the vibrator will be. Some vibrators are so highly tuned that they only show significant amplitude vibrations for frequencies very close to resonant values. These are called **high-*Q* resonators**, where *Q* is the ratio between the frequency of a resonant peak and the **bandwidth** of that peak. By bandwidth, we here refer to the frequency range over which resonator amplitudes remain greater than some specified fraction of the amplitude at the resonant peak. In the example of Figure 4.7, we consider the range of frequencies for which amplitudes remain 5% of the peak value or greater (e.g., we allow up to a 26 dB reduction in amplitudes). If only a small change in frequency will reduce the amplitude of vibrations down to this reference amplitude, then Q is high and the resonator is narrowly tuned. If it takes a large change in frequency to produce the threshold drop in amplitude, then Q is small and the resonator is said to be broadly tuned. For the same excitation, the maximum amplitude of a high-Q resonator at its resonant frequency will be greater than that for a low-Q resonator. Thus high-Q tuning can be used by animals to increase the amplitude of the signal for a given amount of excitation.

As we saw in Chapter 3, measures of frequency domain resolution such as bandwidth are always related in an inverse way to measures of temporal resolution in the same system. The ability of a resonant vibrator to respond independently to successive excitations is a measure of its temporal resolution. This ability depends upon its damping. When damping is high, an exci-

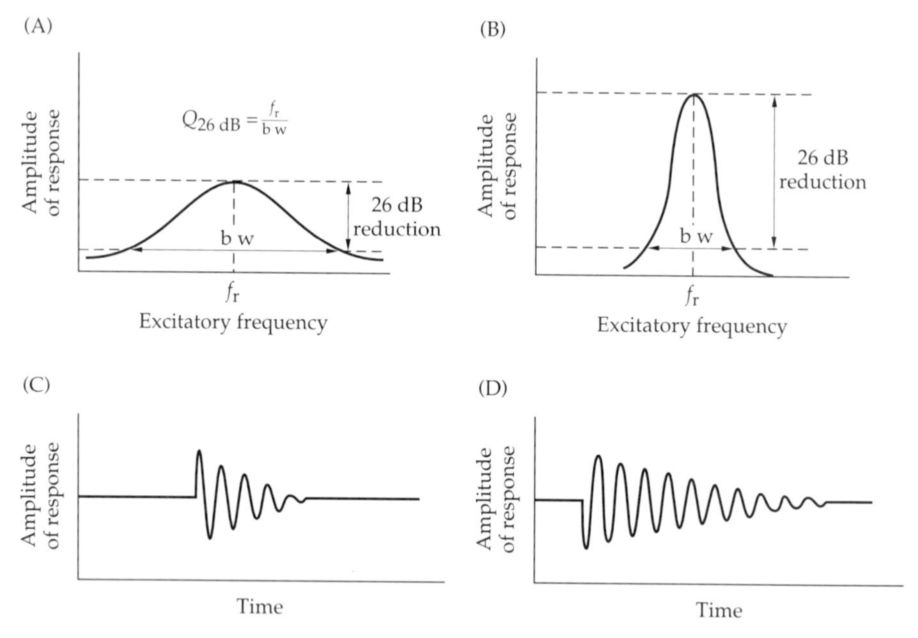

Figure 4.7 Frequency and time domain responses of resonators with different values of Q. (A) Frequency domain response of a low-Q resonator. The bandwidth (bw) is shown within which the pressure amplitude of a response is within 26 dB (5% or more) of that for the resonant frequency f_r. This resonator is broadly tuned and thus has low Q. (B) Frequency domain response of a narrowly tuned and therefore high-Q resonator. (C) Time domain response by a low-Q resonator to a short sudden stimulus. Q is the number of cycles the resonator will exhibit before pressure peaks are reduced to 5% (–26 dB) of initial values. (D) Time domain response of a high-Q resonator. Note that this resonator rings longer than does the low-Q system. The bandwidth and the duration of a response are thus inversely related.

tation will die out quickly and the system can respond independently to the next excitation. Where damping is low, the system will still be vibrating from one excitation when the next arrives, and the response will not be independent of prior events. The number of successive cycles that an excited resonator will exhibit before the response amplitude damps down to the critical threshold depends upon *Q*. Thus *Q* is an inverse measure of damping: the higher the value for *Q*, the lower the damping and the more cycles one will see before the response drops to 5% of its initial value. Systems with high *Q* values will have large amplitudes if excited at their resonant frequencies and narrow bandwidths, but they will track temporal events poorly (because they have low damping). Systems with low *Q* will have lower amplitudes at their resonant frequencies, wide bandwidths, poor frequency resolution, high damping, and better temporal tracking abilities.

This differential response of high and low *Q* systems means that animals that use vibrator resonance to produce sounds must face specific trade-offs. If they opt for a resonator with a high *Q*, there will be a strong guarantee that produced sounds will be loud and will have the preferred frequencies. However, any rapid frequency or amplitude modulation of the sound will not be tracked by the resonator. Frequency modulation will be hindered by the very limited bandwidth of the vibrator, and amplitude modulation will be impeded because damping is low and successive excitations of the vibrator will overlap and destroy any temporal pattern. Low *Q* systems will have lower amplitudes around the resonant frequency, and there will be less frequency control. However, FM and AM patterning will be more feasible.

What happens when a resonant system is excited by a waveform with many component frequencies? The plucking of a guitar string initially forces the guitar to adopt a triangular waveform with the apex at the point to which the string is stretched just before release. The Fourier composition of a triangular wave will consist of many harmonics, and initially the released string will attempt to vibrate at each of these. However, nonresonant frequency components in the excitatory waveform will cancel themselves out, and eventually the system will tend to vibrate at those frequencies in the excitatory waveform that are also its own resonant modes. This means that a wide variety of broad spectrum excitatory waveforms can generate the same final resonant sounds. Thus, although evolution tends to produce very carefully tuned vibrators in animal sound organs, it is often much less specific about how those vibrators are excited.

Post-Vibration Modification

Once a vibrator begins oscillating at its natural modes, an organism has a second chance to modify the sounds it will emit. All vibrators are attached to structures that may pick up the vibrations and be themselves excited. A guitar string is attached to the sound box through the bridge and less directly through the neck. When a string is plucked, a small amount of the vibratory energy passes through the bridge into the sound box. Depending on its own resonant properties, the sound box modifies the frequency and phase spectra supplied to it by the vibrating string and bridge. Frequencies present in the string spectrum that are not resonant properties of the box tend to be filtered out. Frequencies in the string spectrum that are also resonant frequencies of the box cause the box to begin vibrating at its surface. Successive waves of these frequencies from the string add up and cause the box to vibrate at higher and higher amplitudes. Eventually, depending on the damping of the sound box and the air it contains, the amplitudes of the resonant frequencies of the box will stabilize with a specific frequency spectrum.

Note that very little sound energy is transferred directly from the string to the air; the string has a very different acoustic impedance from air and the surface area of the string is too small to generate any significant volume velocity in the air around the string. Because the sound box is made of very light and thin wood around a large air-filled cavity, its acoustic impedance is between that of air and that of the string. This means that the sound box absorbs more energy from the string than does the air, and it releases more energy to the air than would the string. It is thus an **impedance-matching** device. If properly designed, the impedance of the box is such that it loses energy to the air more slowly than it gains energy from the string. Thus many successive cycles can build up within the box and amplify the resonant frequencies received from the string.

A guitar is a **source-driven** system. Because its acoustic impedance is so different from that of the surrounding air or the attached sound box, little energy moves between string and either air or box directly, and thus there is even less opportunity for induced vibrations in either the box or the air to leak back and affect the string's movements. In contrast, reed instruments such as the clarinet are **response-driven** systems. Although the reed has its own resonant properties, it is a very light structure with a large surface, so that its acoustic impedance is not unlike that of the surrounding air. As the clarinet reed moves, energy is quickly conveyed to the air column inside the clarinet tube and this air begins to vibrate. Because the acoustic impedance inside the tube is greater than that of the outside air (see Chapter 2), vibrations propagated inside the tube are reflected at the opening of the tube and back toward the reed. The vibrations set up standing waves in the tube, some of which fit nicely in the tube and are natural modes, and some of which do not and tend to cancel out. In short, a clarinet tube has its own resonant properties determined by its length and diameter, and also by which stop holes are open. Once resonant standing waves exist in the tube, the close match between the tube air and the reed results in an easy transfer of energy back from the air column into the reed. If the reed is moving with vibrations out of phase with the air vibrations, these tend to be canceled out; if the reed is moving in ways consistent with those in the air column, they are reinforced. The result is that the air column inside the tube forces the reed to vibrate at its own resonant frequencies. The reed is a response-driven system because its vibrations are eventually controlled by the structures which it excites. This is in marked contrast to the guitar string which essentially vibrates at frequencies independent of the action of the attached sound box or surrounding air.

In animals, sound production by a vibrator is often modified by attached resonant structures. Stridulatory organs in orthopteran insects occur close to small resonant cavities or to a high-*Q* portion of a wing (Figure 4.8). The files and plectra are sufficiently solid that their vibrations are not affected by air vibrations in the cavities or resonant vibrations in the high-*Q* part of the wing. These are probably source-driven systems, and the attached structures simply act to amplify or select those frequencies generated by the file-plectrum contacts.

In all air-breathing vertebrates, there are tubes or cavities between the sound vibration organs and the outside air. One can ask whether these resonant cavities ever drive the vibration organs in a response-driven manner or not. The issue can be tested by letting an animal vocalize in a medium of altered density (e.g., excess helium added to air) that changes the speed of sound. Although the same wavelengths fit best within the resonant cavity, the corresponding resonant frequencies will change because the speed of sound

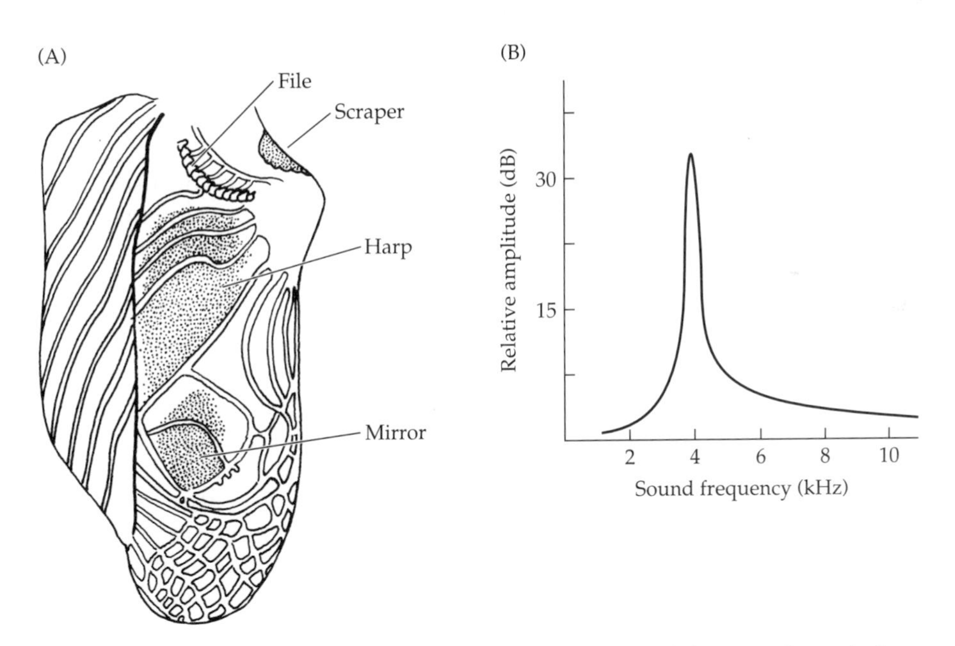

Figure 4.8 Resonant enhancement of stridulation signals. (A) Wing of a typical orthopteran. In crickets (Gryllidae), the scraper is on one side of the wing and the file is on the other. The file on one wing rubs over the scraper on the other. In katydids (Tettigoniidae), the files of two wings rub against each other. The harp is a thin part of the wing that resonates at a main frequency (in a cricket call, ~4 kHz). In katydids, the mirror is designed to resonate at species-specific frequencies. (B) The relative amplitude of harp vibrations in the cricket. Note the sharp tuning at 4 kHz; *Q* (for a bandwidth within 3 dB of peak) is 28. (After Michelsen and Nocke 1974.)

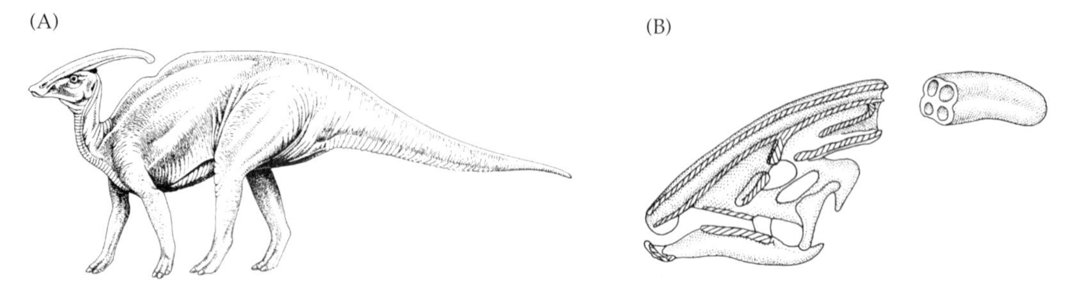

Figure 4.9 Presumed resonators in duck-billed dinosaurs. (A) Living *Parasaurolophus*. (B) Cutaway of a head showing presumed acoustic resonator. (B after Bakker 1986.)

has changed ($f = \lambda / c$). While vocalizing in a medium of higher or lower density may change which frequency components are accented, it will not change the fundamental frequency of a source-driven vibrator. It *will* change the fundamental frequency of a response-driven vibrator. This and similar tests have indicated that the larynges of mammals and frogs are essentially source-driven systems (Martin 1971; Capranica and Moffat 1983; Suthers 1988). The issue with birds is still unclear. One might presume that the fluttering membranes of the avian syrinx (see below) would be like the reed in a clarinet and thus might be particularly susceptible to acoustic feedback from standing waves in the trachea and bronchi. Based upon estimates of the relevant impedances and certain observations made on the frequency composition of several songbirds' vocalizations, Greenewalt (1968) argued that bird song was more likely to be a source-driven system. However, light gas experiments by Nowicki (1987) on sparrows and more careful contrasts between predicted tracheal resonances and observed frequencies in several species all suggest that response-driven systems may exist in a number of birds (Brackenbury 1982; Gaunt and Gaunt 1985).

Whether source- or response-driven, evolution has shaped and molded the post-vibration resonators of animals into a remarkable diversity of forms. Among birds, several species of ducks (Anatidae) have cartilaginous bullae or chambers just beyond the syrinx toward the mouth. It is presumed that these are resonators to increase signal amplitude or to control frequency spectra or both. Many cranes (Gruidae) and grouse (Tetraonidae) have long tracheae which can be stretched or enlarged during sound production, and this anatomy undoubtedly affects the nature of the emitted sounds. Some dinosaurs may have used hollow tubes in their heads as resonators (Figure 4.9). Among mammals, male howler monkeys (*Alouatta* spp.) have a very enlarged hyoid cartilage interposed between larynx and mouth (Figure 4.10), and the male hammer-headed bat (*Hypsignathus monstrosus*) has an enormous cartilaginous larynx filled with air cavities (Figure 4.11). Males of both genera produce very loud sounds with characteristic spectral properties. These hollow organs surely facilitate resonant modification of the laryngeal vibrations.

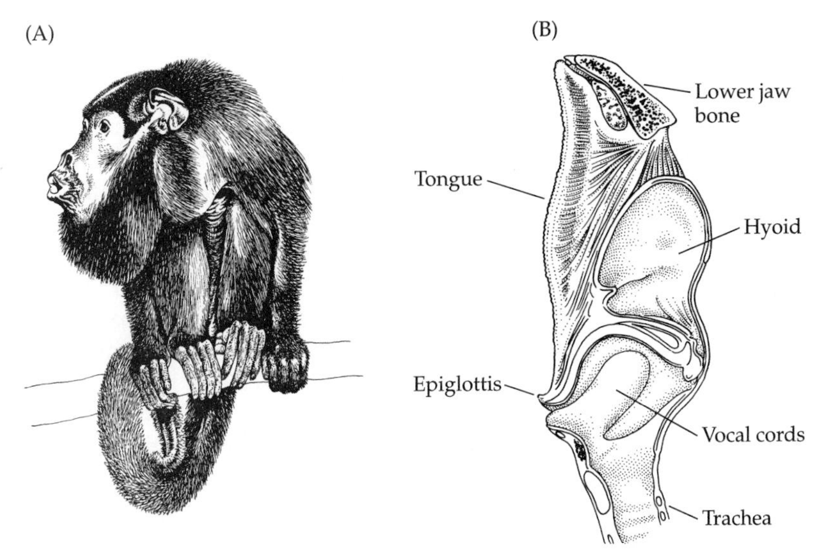

Figure 4.10 Call production organs of male howler monkey (*Alouatta palliata*). (A) Male calling. Note horn shape of pouted lips and enlargement of throat due to hollow hyoid cartilage. (B) Anatomy of upper respiratory tract. Vocal cords are long and large. Adjacent hyoid cartilage is enlarged and hollow to provide call resonance.

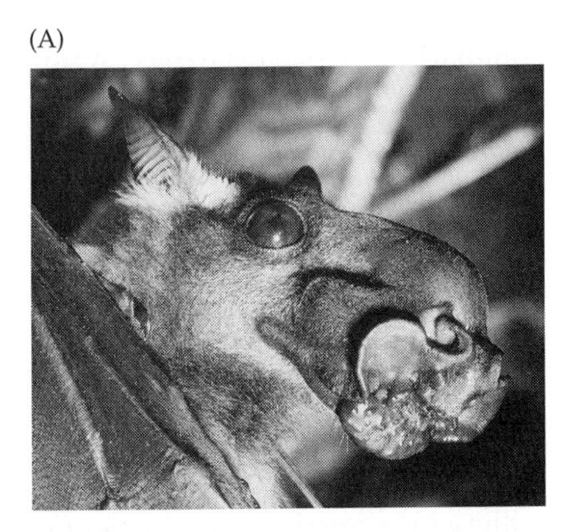

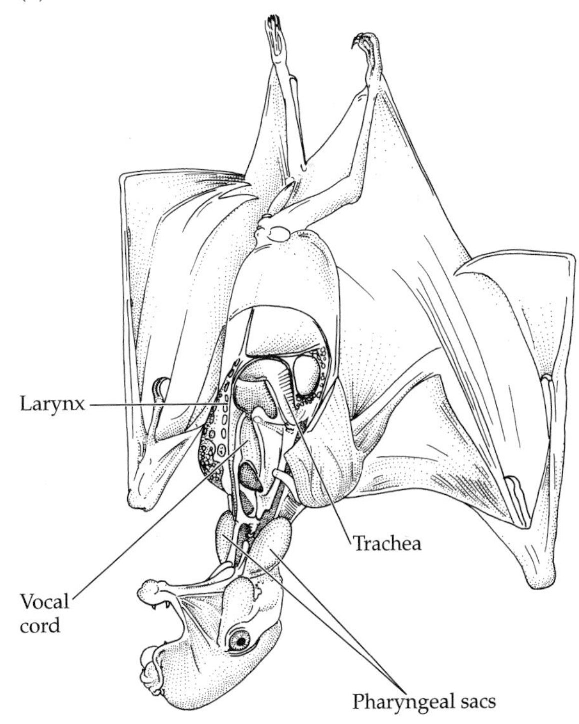

Figure 4.11 Sound producing organs in a male hammer-headed bat (*Hypsignathus monstrosus*). (A) Head of male showing large bell-shaped lips for call emission. (B) Internal anatomy of a male bat. Note enormous bony larynx that produces resonance at the sound source. Pharyngeal air sacs apparently function to increase coupling of sound to air. They may also recover air for the next vocalization, as in frogs, or allow for rapid cessation of the loud, sharp call. (Photo by A.R. Devez, ECOTROP, CNRS; B after Kingdon 1974.)

COUPLING VIBRATIONS TO THE MEDIUM

Terrestrial Animals

In terrestrial animals, coupling of sound vibrations to the surrounding medium is a difficult problem. Where the vibrating structure is organic tissue, the acoustic impedance mismatch between the vibrator and the air is so severe that less than 0.05% of the source energy ought to be transferred into the medium. Even if the vibrating structure is a column of air, the acoustic impedance of a small breathing tube or mouth is so much higher than that of the outside world that most of the sound energy will be reflected back into the animal at the tube opening. There are a number of solutions that animals have evolved to improve the coupling of sound vibrations to the propagating medium.

One solution to the vibrator/air mismatch is to couple the vibrating source to a resonator that is tuned to the most important frequencies in the sound. Ideally, the resonator should have an acoustic impedance sufficiently unlike the vibrator that it does not draw off all of the vibration energy, but sufficiently like it that it does absorb some energy. For a terrestrial animal, the resonator could thus have a greater or a smaller acoustic impedance than the vibrating organ. Using a resonator with a greater impedance would not help because the resonator would then be unable to couple the amplified sounds to the air. Using a resonator with an impedance between that of the vibrator and air would help solve the problem, and many animals do use such a device. A frog or toad's inflated throat sac is a case in point (Figure 4.12). Because the tightly stretched membrane is so thin and the interior sac is filled with air, the effective acoustic impedance of the filled air sac is only slightly greater than that of air. The laryngeal gateway adds periodic vibrations to the air flow from lungs into this sac, and these in turn induce resonant vibrations in the air and membrane of the sac. The close impedance match of the sac to the air, and the fact that the surface of the air sac is much larger relative to the wavelengths being generated than is that of the vibrator, permit the radiation of much more sound power than would be the case for the unaided larynx. This effect can be verified by puncturing the throat sac of a vocalizing frog; the radiated sound is reduced by 2–5 times (Martin 1971). Note that even with the throat sac, only a small percentage of the energy put into calling by a frog is emitted as sound (Ryan 1985b, 1988; Taigen and Wells 1985; Wells and Taigen 1984, 1986, 1989). Throat sacs occur in nearly all vocalizing anurans, in the hammerheaded bats and howler monkeys described earlier, and in gibbons and orangutans.

The problem of reflections at the opening of the breathing tube can be reduced by terminating the tube with a flared horn. The major problem with simple tube openings is that the acoustic impedances inside and outside the tube are so different. By adding a horn that increases in diameter slowly on its tube end and flares rapidly as it nears its opening, the effective impedance is changed slowly from the tube value to that of the outside world. This type of horn provides a better match between the two original impedances and allows much more energy to escape the tube. (For this reason many brass and

(A)

(B)

Figure 4.12 Filled throat sacs of calling anurans. (A) Tree frog (*Hyla ebracata*) that routinely calls from vegetation above the water. (B) Tungara frog (*Physalaemus pustulosus*) that always calls while floating in the water. Note the more lateralized sac shape in Tungara frog. (Photos courtesy of Marc Dantzker.)

woodwind instruments terminate with a flared bell.) Note that this increase in power output has its costs: the bell has its own resonant properties and its presence will change still further the frequency and phase spectra of the sound. The power increase also causes the radiated sound field to become directional. For animals that are attempting to communicate in a specific direction, this directionality is an advantage; for those that are attempting to broadcast a signal in all directions, it may be a cost. Simple horns are generated by flared mouth cavities in some birds and echolocating bats. The flares around the mouth of the male hammer-headed bat most likely function as a horn (Figure 4.11).

Finally, some animals have side-stepped the problem of the air/vibrator mismatch by switching to another medium: the substrate. Because the acoustic impedance of the substrate is only 4–10 times greater than that of an organism, whereas that of air is 4700 times the value for organisms, a number of species couple their vibrations directly to the substrate on which they live. The best known examples occur in various arachnids. Many spiders drum complex signals on their own or other spider's webs, and terrestrial species such as wolf spiders regularly drum or scratch the ground in specific ways. Fiddler crabs (*Uca* spp.) are another group that routinely produces substrate vibrations for communication. The dance of the honeybee (*Apis mellifera*) contains comb-propagated vibrations, and many ants also produce stridulations or vibrations propagated through the substrate.

Aquatic Animals

Compared to terrestrial animals, aquatic organisms have a much smaller problem with the final coupling of sounds to their medium. Stridulatory organs have an impedance sufficiently similar to water so that the sounds they produce carry long distances without the aid of resonant structures. Some fish produce sounds by grinding or clicking teeth, or by rubbing bony parts together. Snapping shrimp pop parts of their claws to make crackling noises. These sounds are all easily coupled to the medium. Porpoises and whales have abandoned the sound producing organs of their terrestrial ancestors and use entirely new parts of their respiratory systems to make sounds.

The major problem faced by aquatic animals is in the modification of sounds. An aquatic species that wishes to use resonance to limit the frequencies of its sounds or to increase signal amplitude will have a difficult time because energy is so quickly transferred to the medium that there is no time to build up resonant standing waves. For this reason, sounds with narrow bandwidths or large amplitudes are uncommon among marine invertebrates.

Some fish have solved this problem by using their swim bladders to serve as resonators for stridulatory vibrations or, more commonly, as the source vibrator. These air-filled sacs in the center of the fish regulate the buoyancy of the animal as it changes depth. In some species, stridulations between bony parts are conducted through the tissues to the swim bladder, which then amplifies those frequencies in the stridulation according to its own resonant modes. Because of the impedance mismatch between the air-filled sac and the surrounding fluid tissues, energy is lost from the vibrating sac slowly enough that resonant standing waves can be established. In a number of fish taxa, the swim bladder is used to produce loud sounds directly. Muscles in the body wall surrounding the swim bladder or even built into the bladder wall are repeatedly contracted to compress the organ, and then they are released. As the fish drives the bladder with contractions at rates similar to the fundamental of its natural resonant frequencies, signals of considerable amplitude can be built up. Since the compressed and released bladder is effectively a monopole vibrator, the higher amplitudes generated by resonance are augmented further by the absence of short-circuiting so typical of stridulatory dipoles. One cost of this direct driving of the vibrator is that the fundamental frequencies are limited by the maximal rates of the contracting muscles. However, since the waveform of each compression is periodic but not sinusoidal, harmonic series are common for such signals, and higher frequencies can be generated should they be needed.

A number of whales and porpoises use the aquatic equivalent of a horn to focus their emitted echolocation sounds (see Chapter 26) in a forward direction. Because their bodies have acoustic impedances so similar to that of water, an actual horn made of their own tissues would not confine and focus the sound waves. Their solution is an **acoustic lens** that consists of a special sac filled with an oily material. Sound waves generated in the respiratory tract behind the skull are refracted, much as light passing through a lens, by the slightly different acoustic impedances in the sac and surrounding tissue. The size and shape of the acoustic lens appear to be adapted to bend sounds of specific wavelengths into narrow beams. The melons on pilot whales and the huge heads of male sperm whales are presumed to function, at least in part, as acoustic lenses for sound emission.

VERTEBRATE SOUND PRODUCTION SYSTEMS

In the following section, we discuss the sound production mechanisms of typical mammals, anurans, and birds. These examples give us a chance to see how the preceeding adaptations might be integrated into single systems and provide necessary background on sound production in vertebrates.

Nearly all voiced sounds of vertebrates are produced by the same general mechanism. Air is pushed through a tube from one place to another. As the air flows through the tube, one or more thin membranes are forced into the flow, causing the membranes to vibrate. The resulting vibrations, which are usually periodic but not sinusoidal, are superimposed on the steady (DC) flow of air. There are four basic forces on the vibrating membrane that affect its movements (Figure 4.13). First, some force external to the tube must be used to move the membranes into the airflow. Second, the pressure difference between the two ends of the tube generates a force that tends to push the vibrating membrane downstream and back out of the air flow. A third force arises from the natural elastic properties of the membrane. This force also pulls the membrane back to its relaxed position outside of the air tube. Finally, the flow of any liquid or gas over a surface produces a suction on that surface known as a **Bernoulli force.** As the air flows down the tube it creates a Bernoulli force on the membrane, pulling it further into the tube.

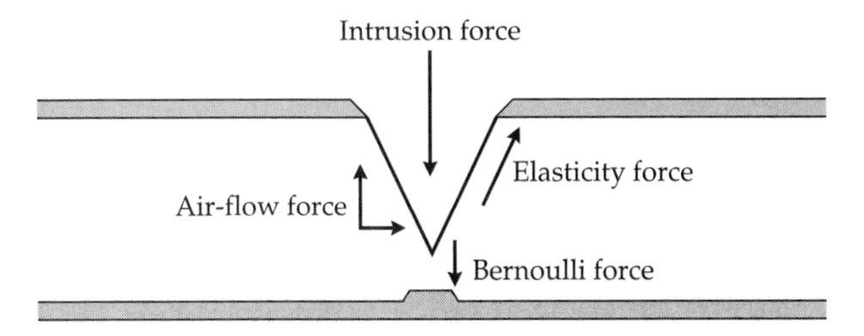

Figure 4.13 Summary of forces acting on the membrane in a typical vertebrate respiratory sound vibrator. The flow of air goes from left to right. Intrusion and Bernoulli forces extend the membrane into the air flow, whereas the air flow force and the elasticity force pull it out of the flow.

Any system in which several forces act on a common focus but with different directions and with different time delays is likely to oscillate. Begin with the membrane forced so far into the tube that it occludes the flow. Air pressure will build up on the upstream side of the membrane until finally this force and the elasticity force the membrane away from the opposite wall. As air now rushes past the membrane, the air pressure difference in the tube falls off and a Bernoulli force is established that finally sucks the membrane back toward the opposite wall and the cycle begins again. If the pressure, intrusion, and elastic forces are adjusted properly, the membrane will settle into a stable periodic motion. By changing one or more of the forces, the fundamental frequency and amplitude of the vibrations can be varied. To stop producing sounds, the intrusion force is relaxed and the elastic force will pull the membrane out of the airflow. With this basic mechanism in mind, let us first examine sound production in mammals.

Mammalian Sound Production

Mammals produce sounds with a **larynx**. Two membranes called **vocal cords** (or, jointly, the **glottis**) are forced by laryngeal muscles to totally block the respiratory air flow (Figure 4.14). Respiration in mammals utilizes a mouth or nose to connect to the open air, and a long tube called the **trachea** to connect the mouth or nose cavity at one end to two tubes called **bronchi** on the other. Each bronchus then connects to one of the two lungs. Below the lungs is a muscular layer called the **diaphragm**, which separates the lung cavity from the viscera. When the muscles of the diaphragm contract, the lung cavity is enlarged and air is drawn into the lungs through the mouth or nose via the trachea and bronchi. When the diaphragm relaxes, the elastic walls of the lung cavity return it to its relaxed state and the mammal exhales. Inhalation is thus the working step in terrestrial mammalian breathing, whereas exhaling requires no work (although forced exhalation is usually possible if needed).

A more detailed view of the mammalian larynx is seen in Figure 4.15. In addition to the vocal cords, a typical larynx consists of four small cartilages and four sets of muscles. The thin vocal cords are stretched between the **thyroid cartilage** and two **arytenoid cartilages**. When only the **cricoarytenoid muscles** contract, they pull the vocal cords back out of the air flow and the mammal exhales normally with air entering through the circular cricoid carti-

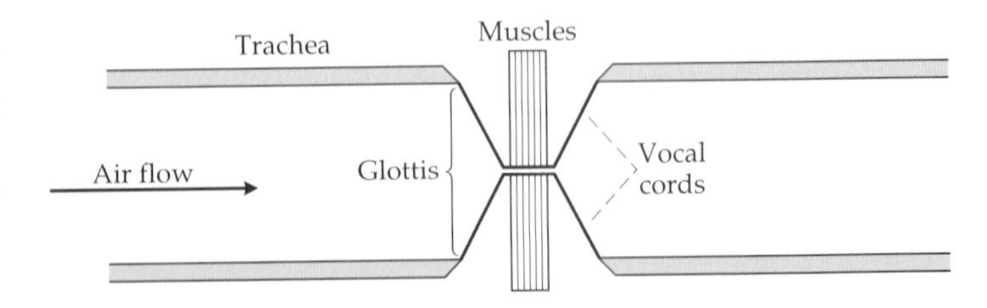

Figure 4.14 Diagrammatic map of a mammalian larynx. Vocal cords in the glottis are forced by muscles to close off the air flow. A build-up in pressure then forces the vocal cords open, until pressure drops and Bernoulli forces reclose the passage.

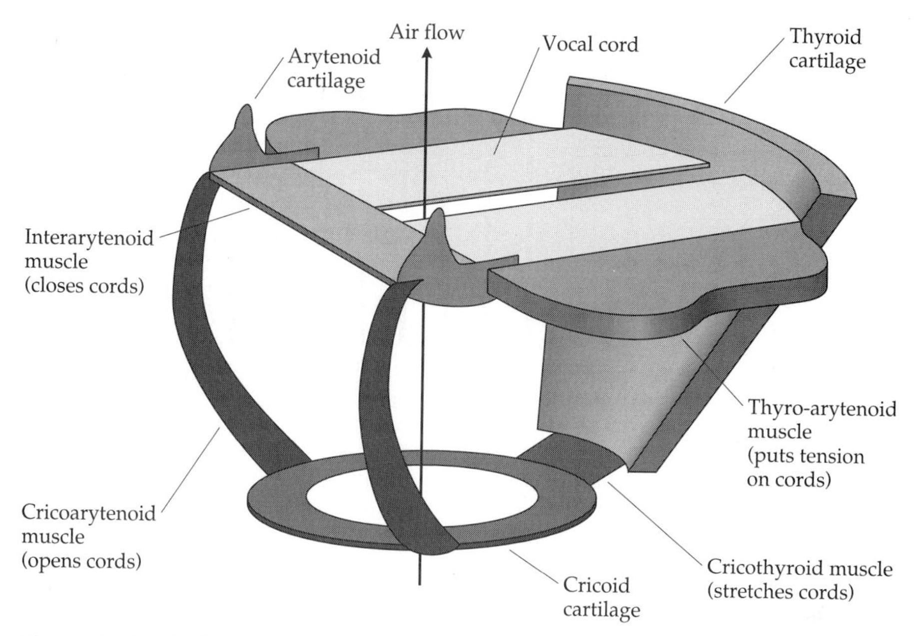

Figure 4.15 Working parts of a typical mammalian larynx.

lage and exiting through the retracted vocal cords. When the **interarytenoid muscle** contracts, it pulls the vocal cords into the laryngeal cavity until they block the air flow. Tension on the vocal cords is controlled by contraction of the **cricothyroid muscle**, which pulls the thyroid cartilage away from the air cavity, and the **thyro-arytenoid muscle**, which pulls the exterior margins of the cords back toward the laryngeal walls. Once the cords close off the passage, air pressure builds up behind them until they are forced open. The open vocal cords allow a small puff of air to pass, releasing the accumulated backpressure and generating a Bernoulli force that sucks the vocal cords shut again. As long as the vocal cords are held in the air flow and the mammal continues exhaling, a train of periodic air puffs and thus pressure signals will move out of the mammal's mouth or nose. The waveform of a typical mammalian sound is shown in Figure 4.16.

This is a periodic but nonsinusoidal signal. The Fourier composition of such a signal will consist of a harmonic series with a fundamental equal to the pulse rate. Since the waveform is not half-wave symmetric, all harmonics will be present. A typical mammalian frequency spectrum is shown in Figure 4.17. In humans, the pulses have a repetition rate of about 130 pulses/sec for men and 220 pulses/sec for women. These rates are set primarily by the thickness

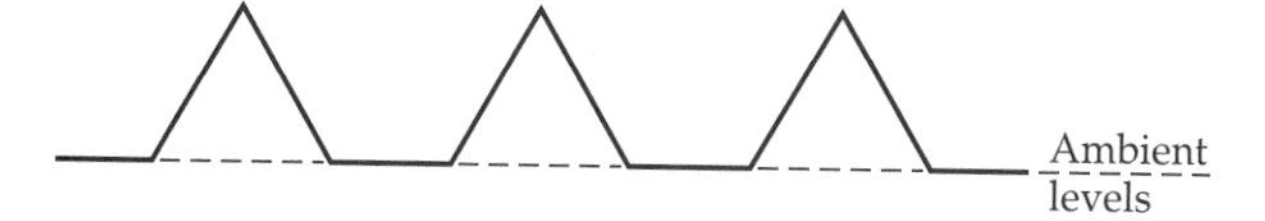

Figure 4.16 Waveform of a typical mammalian sound. Note the periodic but nonsinusoidal waveform.

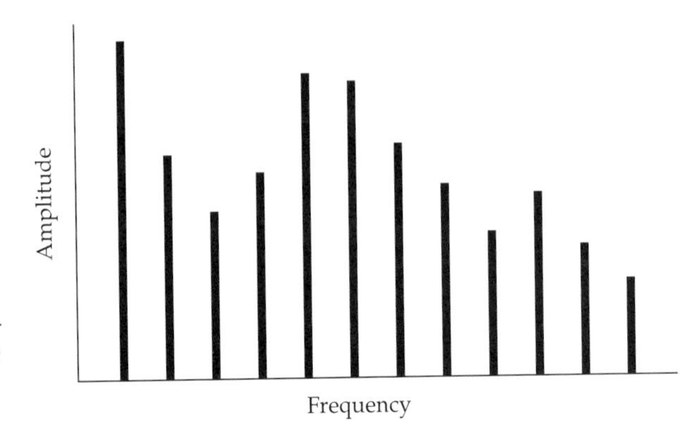

Figure 4.17 Frequency spectrum of a typical mammalian sound. The lowest frequency corresponds to the pulse rate of the sound. Both odd and even harmonics will be present.

of the membranes. When we sing, we increase the pulse rate and thus the fundamental by increasing the tension on the vocal cords.

The harmonic richness of mammalian sounds makes subsequent resonant filtering and selective amplification an obvious way to modify the final signal. In many mammalian larynges, cavities called ventricles and saccules exist in the wall of the larynx just downstream from the vocal cords. Mammals such as reindeer, a few carnivores, and many primates have enlarged these cavities and this enlargement may facilitate resonant modifications of the frequency spectra of the vocal cord sounds prior to their coupling to the medium. The enlarged cavities in the larynx of the hammer-headed bat and in the nearby hyoid cartilages of the howler monkey surely serve similar functions. The best-known example of resonant modifications of vocal cord sounds before emission occurs in humans. When we speak, the same frequency spectrum is generated by the vocal cords. However, by dramatically changing the shape of our pharyngeal cavities, we alter the distribution of energy among the constituent harmonics and thus produce our separate vowels (Figure 4.18).

The basic mechanism we have outlined here applies to the majority of mammals. However, two groups of mammals have evolved rather different means of sound production. Echolocating bats (Microchiroptera), (a group distinct from the fruit bats (Megachiroptera) to which the hammer-headed bat belongs), produce their echolocation sounds with a larynx, but do so with very thin membranes upstream from the glottis. Although the structures are quite different, echolocating bats have a mechanism of vibration production that is more like that of frogs and toads than like that of other mammals. The other odd group consists of the porpoises and whales (cetaceans). These animals have several pairs of air sacs in the front of their heads. Air trapped in the sacs and in the trachea and lungs is moved back and forth, either between lungs and sacs or between sacs to generate vibrations in small membranes in the sides of the sacs. They thus resemble frogs and toads in that trapped air is recycled back and forth between different parts of the respiratory system. The selective forces shaping sound production in cetaceans appear to be very different from those for other mammals. In addition to the effects of living their entire lives in an aquatic medium, toothed cetaceans (Odontoceti) echolocate like bats. The combination of these forces has caused them to lose the laryngeal mechanisms of their terrestrial ancestors and evolve entirely new mechanisms for producing sounds. We shall discuss the echolocation sounds of bats and cetaceans in Chapter 26.

Anuran Sound Production

Frogs and toads also use a larynx to produce sounds. As with mammals, this organ is positioned on the end of the trachea where it joins the mouth cavity. However, sound production differs in two ways from that for mammals. As in mammals, the glottis is used to occlude the air passage and the build up of air pressure by the lungs forces the glottis to pop open and closed. However, anurans have a second set of thin membranes on the sides of the larynx just upstream from the glottis (Figure 4.19). As the glottis pops open and lets air flow these membranes are forced to vibrate by a combination of Bernoulli and elastic forces. Among researchers on anuran vocalizations, it is this second set of membranes, not the glottis, that are usually called the vocal cords. Because these membranes are not directly coupled to the glottis, the frequencies at which they vibrate can be adjusted independently of the opening and closing rates of the glottis. Because they are usually thinner, they tend to vibrate at higher fundamental frequencies (e.g., 0.5–2.0 kHz) than the glottis (100–200 Hz). The waveform of a typical anuran sound thus consists of a carrier waveform (generated by the vocal cords) that is gated on and off by the periodic

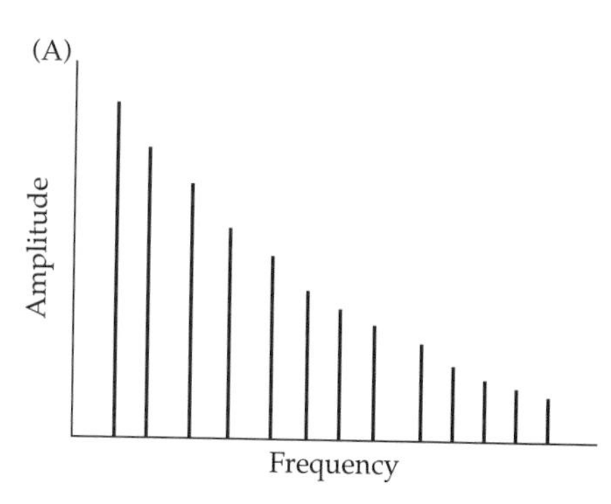

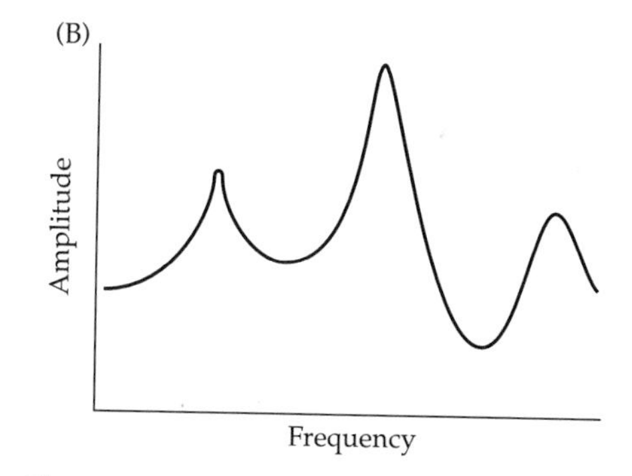

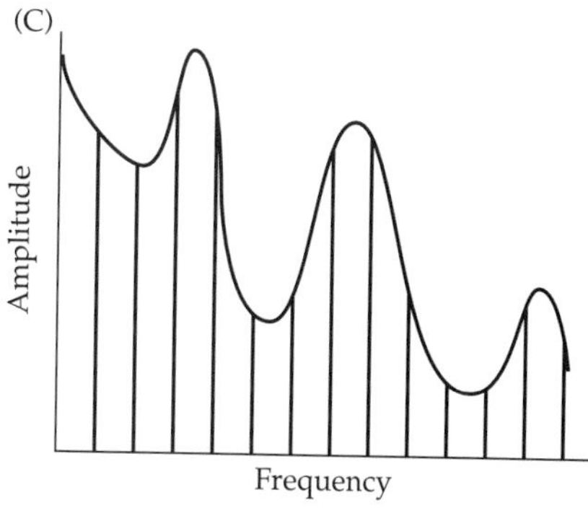

Figure 4.18 Production of human speech. (A) Frequency spectrum generated by vocal cords consisting of harmonic series. (B) Frequency response function of pharynx, mouth, and nasal cavities for one vowel. Resonant peaks are called formants. The vocal cord spectrum fed into resonant cavities produces a final spectrum shown in (C). As the human changes the shape of the mouth and pharyngeal cavity, different formants are accented and thus different vowels are produced.

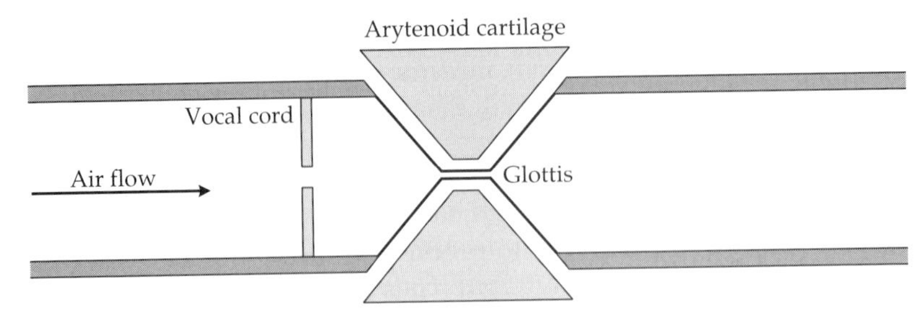

Figure 4.19 Diagrammatic map of an anuran larynx. Note the presence of separate vocal cords upstream from the glottis.

opening and closing of the glottis (Figure 4.20). Since the waveform of the glottis is also periodic but nonsinusoidal, the sound can be considered a nonsinuoidal but periodic amplitude modulation of the vocal cord carrier (which may or may not be sinusoidal). The frequency spectrum for such a signal is shown in Figure 4.21.

The second difference between anurans and most mammals is that expired sound in anurans does not escape from the mouth or nostrils. Instead, the air passing through the larynx empties into a large throat sac bounded by a very thin tissue membrane. As each call is produced, the air passing through the larynx is collected in the expandable throat sac. After the call, the frog or toad can put tension on the sac membrane and force the air back into the lungs. Given their metabolisms, anurans rarely need high ventilation rates simply to breathe and their respiratory systems cannot generate rapid movements of gas. Once air has been accumulated in the lungs, it is much more efficient to swap the same air back and forth between the lungs and the air sac when vocalizing than to try to recollect expired air. Perhaps equally important is the role of the filled air sac as a resonant coupler between the laryngeal vibrator and the outside medium. It is possible in some anurans that the backpressure of the air sac allows finer control of laryngeal vibrations. However, puncturing the sac appears to have little effect on the frequency spectra of the vibrating vocal cords and glottis (Martin 1971). It thus appears that the anuran larynx is a source-driven, as opposed to a response-driven, system.

Anurans have used this basic apparatus to produce an enormous diversity of sounds. One factor facilitating species identification is the production of sounds with specific carrier frequencies and pulse repetition rates. To pro-

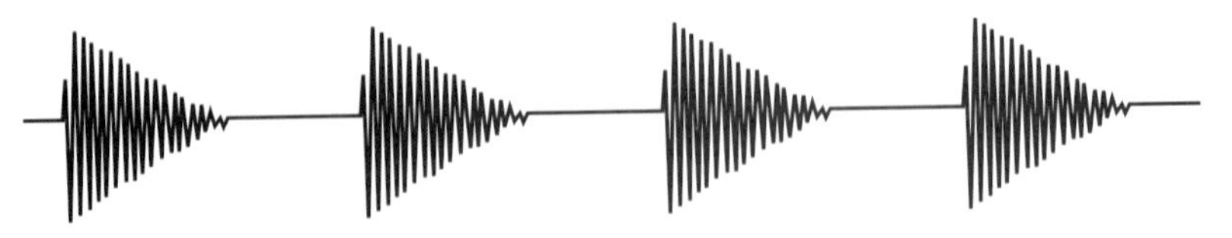

Figure 4.20 Waveform of a typical anuran vocalization. The waveform is a nonsinusoidal amplitude modulation (generated by the glottis) of a carrier frequency (generated by the vocal cords).

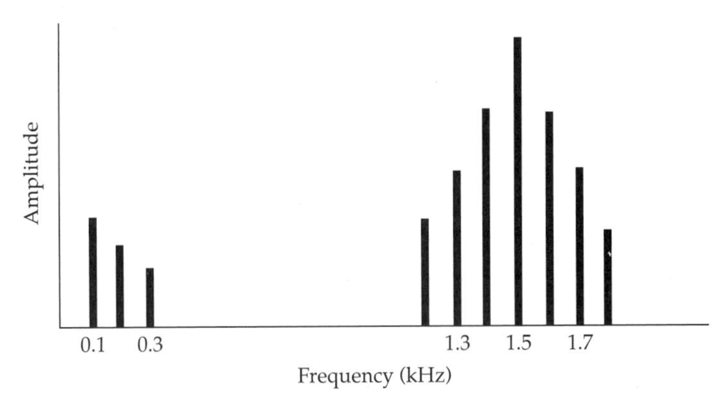

Figure 4.21 Frequency spectrum of a typical anuran call. This pattern can be viewed as a carrier frequency of 1.5 kHz (generated by the vocal cords) that is amplitude modulated by the glottis in a periodic but nonsinusoidal way at a rate of 100 Hz. Only the major harmonics are shown. Note that because the glottis is modulating a net DC flow, there may be sidebands around the baseline (a carrier frequency equal to 0 Hz) as well as around the main carrier.

duce such sounds, small masses or strings are attached to the vocal cords at critical places to force them to vibrate in very predictable ways. In some species, the waveform of the vocal cords is almost that of a pure sinusoid; in others, the waveform is periodic but not sinusoidal, and the glottal amplitude modulation produces predictable side bands around each of the harmonics in the carrier. Instead of relying on passive oscillations of the glottis in a steady air flow, some toads pulse their exhalations, generating calls with temporal properties quite different from passive mechanisms (Martin 1972). A number of anurans can also frequency modulate their signals. Toads do this by stretching the larynx to increase tension on the vocal cords and the concomitant stretching of the glottis generates coupled amplitude and frequency modulation in the waveform. In contrast, many frogs have special muscles that allow them to stretch the vocal cords without stretching the glottis. They can thus control the amplitude and frequency modulation independently. Some species even have muscles that hold the glottis open longer than the glottal physics alone would allow; they use the additional time to execute complex frequency modulations of the vocal cord vibrations. Once frequency-rich sounds are generated at the larynx, the vocal sac of the frog may further modify the final signal by filtering or accentuating particular frequency components. A sampler of spectrograms for anurans is shown in Figure 4.22.

Avian Sound Production

Birds produce sounds in quite different ways from anurans and mammals. As noted earlier, the relaxed position for respiring mammals is with the lungs largely empty; the working stroke requiring muscular action is inspiration. In birds, the relaxed position is with the air cavities partially full. Further exhalation as occurs during vocalization requires muscular action. This means that

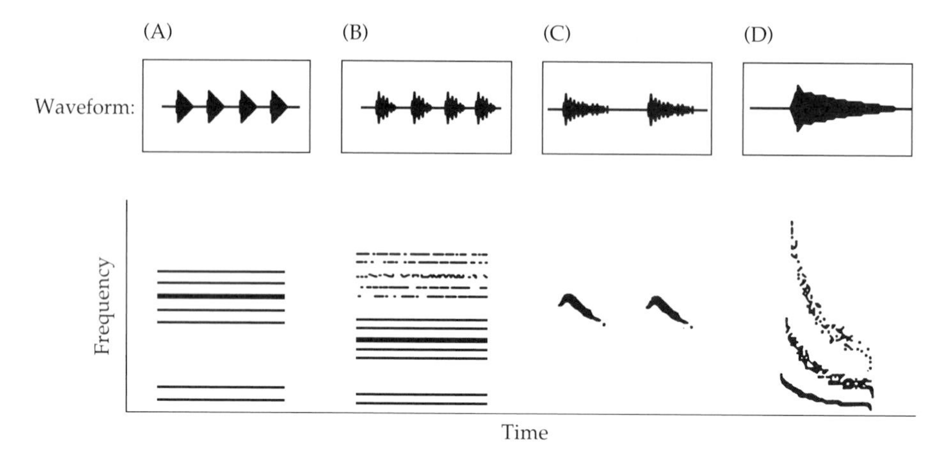

Figure 4.22 Waveforms and spectrograms of four types of anuran calls. (A) A simple system in which a single sinusoidal carrier is produced by vocal cords, and rapid nonsinusoidal glottal gating is produced primarily by the physical properties of the glottis. The nonsinusoidal modulating waveform produces sidebands around a single sinusoidal carrier. (B) Rapid nonsinusoidal glottal pulsing of a nonsinusoidal carrier. Each component of the carrier has sidebands separated by the repetition rate of glottal pulses. (C) Glottal pulses as made longer by special muscles in the larynx of some frogs, thus allowing time for frequency modulation of a single carrier frequency. Because modulations occur at a slower rate than spectrogram bandwidth, frequency modulation is seen in the time domain view. (D) Long glottal pulses facilitated by laryngeal muscles and frequency modulation of nonsinusoidal carrier waveform.

birds vocalize during the phase of respiration when they have the greatest muscular control. They use this control in sophisticated ways to produce the elaborate temporal patterns of their vocalizations. Birds do have inspiratory muscles as well: species such as canaries often take as many as 30 rapid inspirations or "mini-breaths" between song syllables (Hartley 1990). This is what allows them to produce such long songs. Birds are also unusual in their possession of a number of air sacs that pervade most parts of the body and are attached to the bronchi. When a bird breathes, it is the air sacs that expand and contract; the lungs are relatively inelastic. The air sacs facilitate gas exchange and cooling, and when under pressure, provide structural support that otherwise would require a much heavier skeleton.

Unlike anurans and mammals, birds do not produce sounds with a larynx at the point where the trachea joins the pharyngeal cavity. Instead, they have modified the junction of the two bronchi with the trachea. This modified junction is called a **syrinx**. The entire syrinx is enclosed by the **interclavicular air sac**. In all vertebrates, the bronchi and the trachea consist of cartilaginous rings (or semirings) joined by connective tissues into tubes. In the region of the avian syrinx, certain adjacent rings are incomplete and the missing cartilage is replaced by thin membranes. When **extrinsic muscles** of the syrinx contract, they change the tensions on tracheal and broncheal tubes, causing the thinned membranes in the syringeal wall to buckle. Where present, a protruding bronchial ring (as in parrots) or fleshy **labium** (as in songbirds) on the tube wall opposite to the membranes may be rotated by muscles into the air flow to reduce the tube aperture near the membranes (Chamberlain et al. 1968; Gaunt and Gaunt 1985). Air is then forced from the air sacs into the bronchi and trachea. The resulting increase in air pressure inside the syrinx tends to force the buckled membranes out of the air flow (Figure 4.23). However, by keeping the pressure in the surrounding interclavicular air sac high, the bird can balance the syringeal pressures sufficiently to let Bernoulli forces suck the membranes into the air flow, or even force the membranes into the air flow where they will begin to vibrate. It is thought that the Bernoulli forces, which are controlled by the pressure differences and the size of the aperture between labium and membranes, and the membrane tension, which is adjusted by **intrinsic muscles** of the syrinx, together regulate the frequencies of these vibrations (Brackenbury 1989; Fletcher 1988; Gaunt 1987; Goller and Suthers 1996a,b). Finally, it has been suggested that some birds such as doves may produce whistled sounds by forcing air through a highly restricted syringial space, just as humans do with their lips (Gaunt et al. 1982; Gaunt and Gaunt 1985).

The location of vibrating membranes differs for different birds. In parrots and chickens, the membranes are located on the external or lateral sides of the syrinx just downstream from the broncheal-tracheal junction (Figure 4.24A). When these two **lateral tympaniform membranes** vibrate, they jointly modulate air flow down the trachea (Brackenbury 1978, 1980, 1982; Gaunt 1987; Gaunt et al. 1976; Gaunt and Gaunt 1977; Gross 1954). Since it takes considerable air flow to vibrate the membranes, only about 2–3% of the energy invested during exhalation is converted into sound by a chicken (Brackenbury 1977, 1979a,b); this is a value very similar to that cited for frogs on page 90.

In most songbirds (Oscines), the vibrating membranes are found on the internal or medial side of each bronchus just before the junction with the trachea (Figure 4.24B). These are called **medial tympaniform membranes**, and each modulates the air flow in one bronchus. The modulation of the airflow by the membranes is enhanced by the ability of songbirds to protrude an external **labium** into the air flow just opposite the medial tympaniform membrane. This reduces the net air flow through this bronchus and makes the

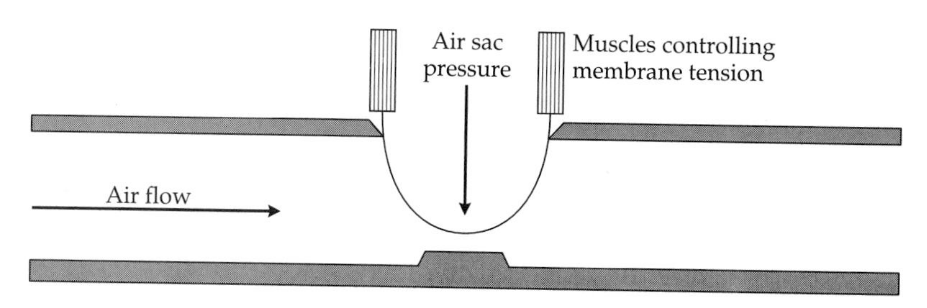

Figure 4.23 Diagrammatic map of one side of an avian syrinx. Note that there are two such structures in each individual bird.

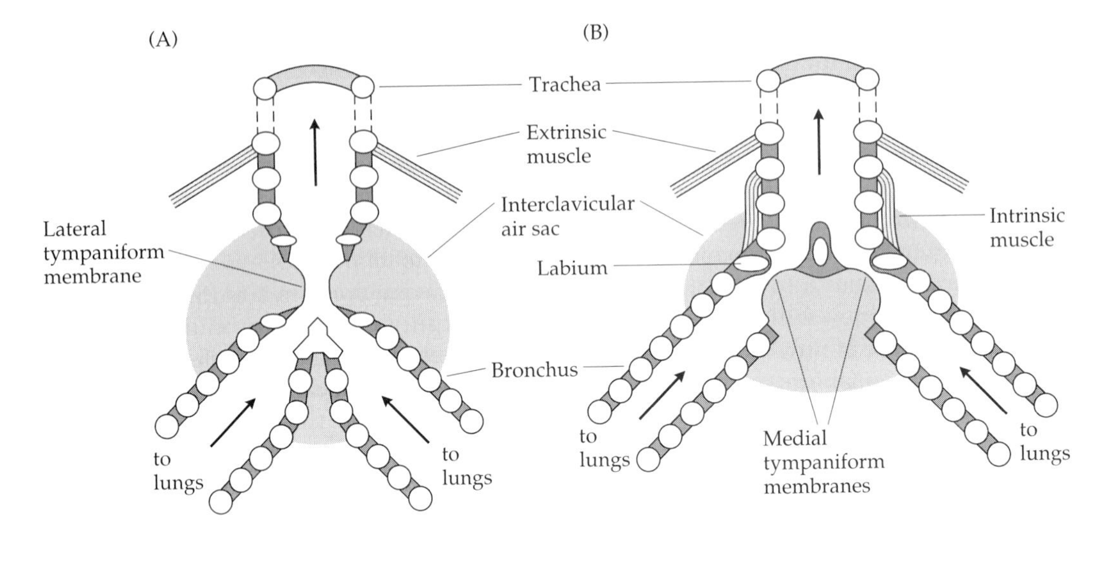

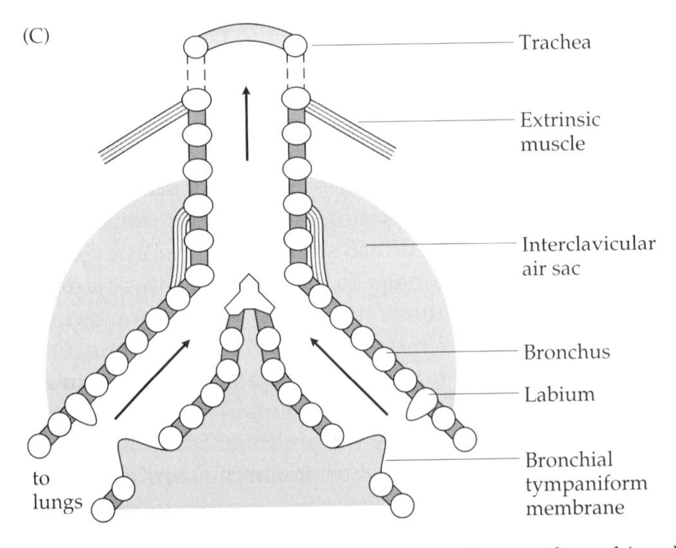

Figure 4.24 General types of avian syrinx. Bony rings forming bronchi and trachea are indicated by white ovals, membranes connecting adjacent rings by dark stippling, and the interclavicular air sac by light stippling. Incomplete rings surround patches of thin membrane called **tympaniform membranes**. Some nonsongbirds, such as chickens generate sounds with (A) **lateral tympaniform membranes** in the base of the trachea. Many other species, such as songbirds, produce sounds with (B) **medial tympaniform membranes** on the bronchi just below their junction. Some nocturnal birds, penguins and cuckoos have (C) **bronchial tympaniform membranes** on the sides of the bronchi at some distance toward the lungs from the syrinx. The extrinsic muscles of the syrinx and interclavicular sac are important in forcing tympaniform membranes into the air flow during vocalization. Intrinsic muscles modify membrane tensions; if a labium is present, these muscles move it into the air flow, making the opening between membranes and labium very small.

magnitude of the vibration, relative to the net DC flow required to generate it, much greater. The separate labia also allow songbirds to open and close each bronchial tube independently to produce rapid intercalations of notes (Allan and Suthers 1994; Suthers 1990; Suthers et al. 1994). The energy efficiency of some songbirds such as wrens is about 10–15%.

A third type of syrinx is found in a number of nocturnal birds (families Steatornithidae, Podargidae, Nyctibiidae, Strigidae, and Aegothelidae), penquins (Spheniscidae), and some cuckoos (Cuculidae). Here, each bronchus has a **bronchial tympaniform membrane** and opposing labium at some distance upstream from the junction of the two bronchi (Figure 4.24C). This type of syrinx separates the two sound sources spatially and presumably reduces the chances of direct acoustical coupling between the sources or mechanical "crosstalk" as a result of vibrations being transmitted through intermediary tissues.

One of the most important consequences of having a pair of syringeal vibrators is that a bird can produce two harmonically unrelated sounds at once. This process assumes, however, that the two sides of the syrinx can be independently controlled and that the syrinx is essentially source-driven (Greenewalt, 1968). (Were it response-driven, then the coupled resonator would force both vibrators to adopt the same waveforms.) There are numerous spectrograms of avian calls suggesting that the two sources can act independently (Figures 4.25 and 4.26; Gaunt et al. 1982; Greenwalt 1968; Marler 1969; Stein 1968). Several experiments have verified that this conjecture is true. One approach is to plug one bronchus to prevent airflow, or to cut the nerves controlling muscles on one side of the syrinx and then compare pre- and post-operative songs. Another approach is to place tiny sensors in each bronchus to monitor airflow and pressure levels during song production. Such experiments have shown that in some birds such as chaffinches (*Fringilla coelebs*), canaries (*Serinus canaria*), white-throated and white-crowned sparrows (*Zonotrichia albicollis* and *Z. leucophrys*), and Java sparrows (*Padda oryzivora*), 80–90% of the syllables in a song are produced by the left sound source in the syrinx and only 10–20% by the right side (R. S. Hartley and Suthers 1990; Lemon 1973; Nottebohm 1971, 1972; Nottebohm and Nottebohm 1976; Seller 1979). Cutting the nerves to one side eliminates the syllables produced on that side but has little effect on the remainder. In zebra finches (*Poephila guttata*) and brown-headed cowbirds (*Molothrus ater*), most of the song syllables are generated by the right side of the syrinx, although there is not as strong a lateralization as in left side dominant species (Allan and Suthers 1994; Williams et al. 1992). Finally, in catbirds (*Dumetella carolinensis*) and brown thrashers (*Toxostoma rufum*), both sides of the syrinx are used nearly equally in song production (Goller and Suthers 1995, 1996a,b; R. S. Hartley and Suthers 1990; Suthers 1990). On average, 10–23% of catbird song syllables are produced by one side of the syrinx only, 13–69% are begun by one side of the syrinx and completed by the other, and 21–67% consist of sounds produced jointly by both sides. The two sides are not completely equal, however: the right side tends to produce higher frequency notes and is more likely to be the site of rapid FM or AM patterning.

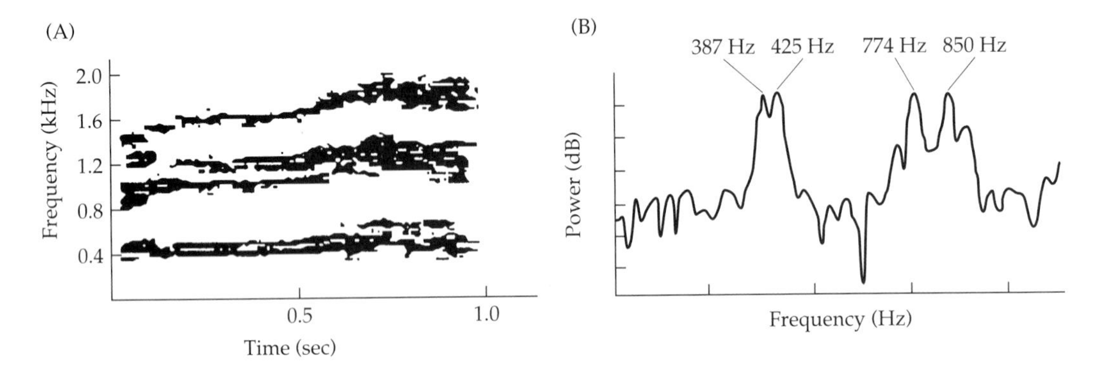
(A)
Frequency (kHz)
2.0
1.6
1.2
0.8
0.4
0.5
1.0
Time (sec)
(B)
387 Hz
425 Hz
774 Hz
850 Hz
Power (dB)
Frequency (Hz)

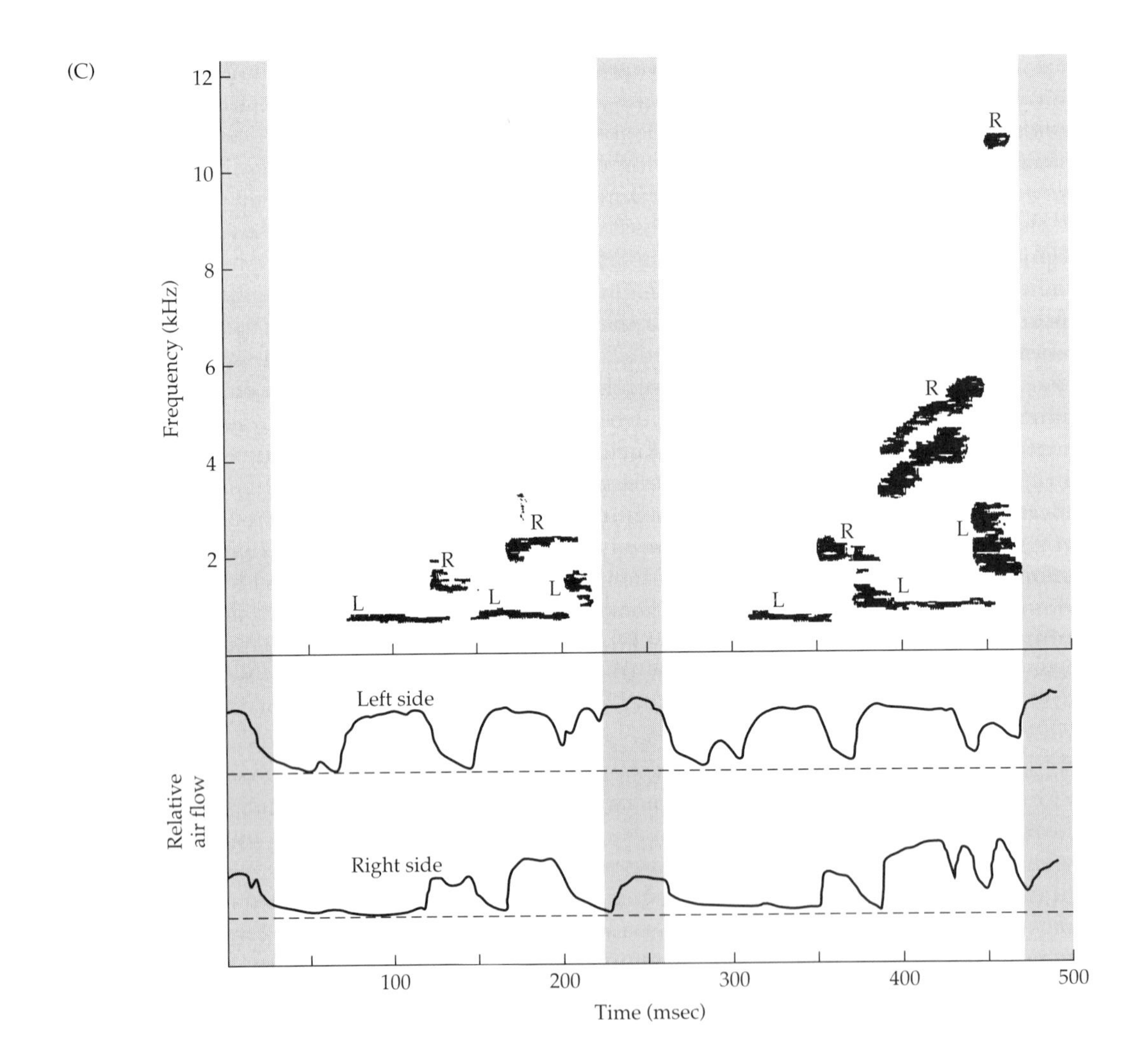
(C)
Frequency (kHz)
12
10
8
6
4
2
R
L
Relative air flow
Left side
Right side
100
200
300
400
500
Time (msec)

◀ **Figure 4.25 Evidence of two sound sources in avian vocalizations.** (A) Spectrogram of the emperor penguin (*Aptenodytes forsteri*) call. The lowest two frequencies at each point do not appear to be harmonically related, whereas higher components appear to be second and third harmonics of each member of the lower pair. (B) Power spectrum for one segment of the emperor penguin call. Note that peaks at 387 Hz and 425 Hz are not harmonically related, whereas the 774 Hz peak is the second harmonic of the 387 peak, and the 850 Hz peak is the second harmonic of the 425 Hz peak. Two sound sources are suggested here, each producing nonsinusoidal but periodic signals and at slightly different repetition rates. (C) Spectrogram (top) and contemporaneous measures of airflow through the left and right sides of the syrinx (bottom) during the song of the brown-headed cowbird (*Molothrus ater*). Components contributed by left (*L*) and right (*R*) sides of the syrinx are indicated on the spectrogram. Stippled zones in airflow graphs indicate inspiration (mini-breaths) between notes of song; remainder of airflow is expiration. Note that the bird can rapidly switch flow between sides, or it can concurrently let air flow through both sides producing harmonically unrelated notes. (B courtesy of Ann Bowles; C after Allan and Suthers 1994.)

In most cases studied to date, the two sound sources do appear to act independently and additively. Where they produce similar but unequal frequencies, the resulting signal shows amplitude modulations due to beats, with modulation rates equal to the difference between the two frequencies (Figure 4.27). This finding implies that the system is linear and the output is the simple sum of the two inputs. When the sound produced by one side of the syrinx is very loud, it may propagate to the other side via the air sac, the small space connecting the two medial tympaniform membranes, or even through the syringeal tissues (Suthers et al. 1994). However, such sounds combine additively with those produced on a given side and this "cross-talk" does not appear to alter sound production. One exception to this linearity has been reported by Nowicki and Capranica (1986) in the "dee" call of chickadees. This call does not appear to be the simple linear sum of the two sources, but instead the result of a complex interaction between them. Such nonlinearity has not been seen in any of the larger songbirds examined and may be a pecularity of smaller species in which the two sources are very close together.

Sensors and muscle recordings indicate that successive syllables in a typical bird song arise from quick movements of the labium into and out of the bronchial air passage (Goller and Suthers 1996a,b). This motion allows a very rapid juxtaposition of consecutive song elements: as one side ends a note, the other begins the next. An independent set of muscles varies the tension on the tympaniform membranes and thus controls frequency patterns. Finally, the respiratory muscles play a very complicated role during song (Hartley 1990; Vicario 1991). By rapidly varying the bronchial pressures for each syllable, these muscles are largely responsible for the overall temporal pattern in the song. Fine adjustments are even made within syllables to keep pressure constant (or intentionally to vary it) given changes in flow velocity as the labium moves in and out of the airstream. Between syllables, many songbirds take "mini-breaths," allowing them to sing continuously for minutes (Hartley and Suthers 1989; Hartley 1990).

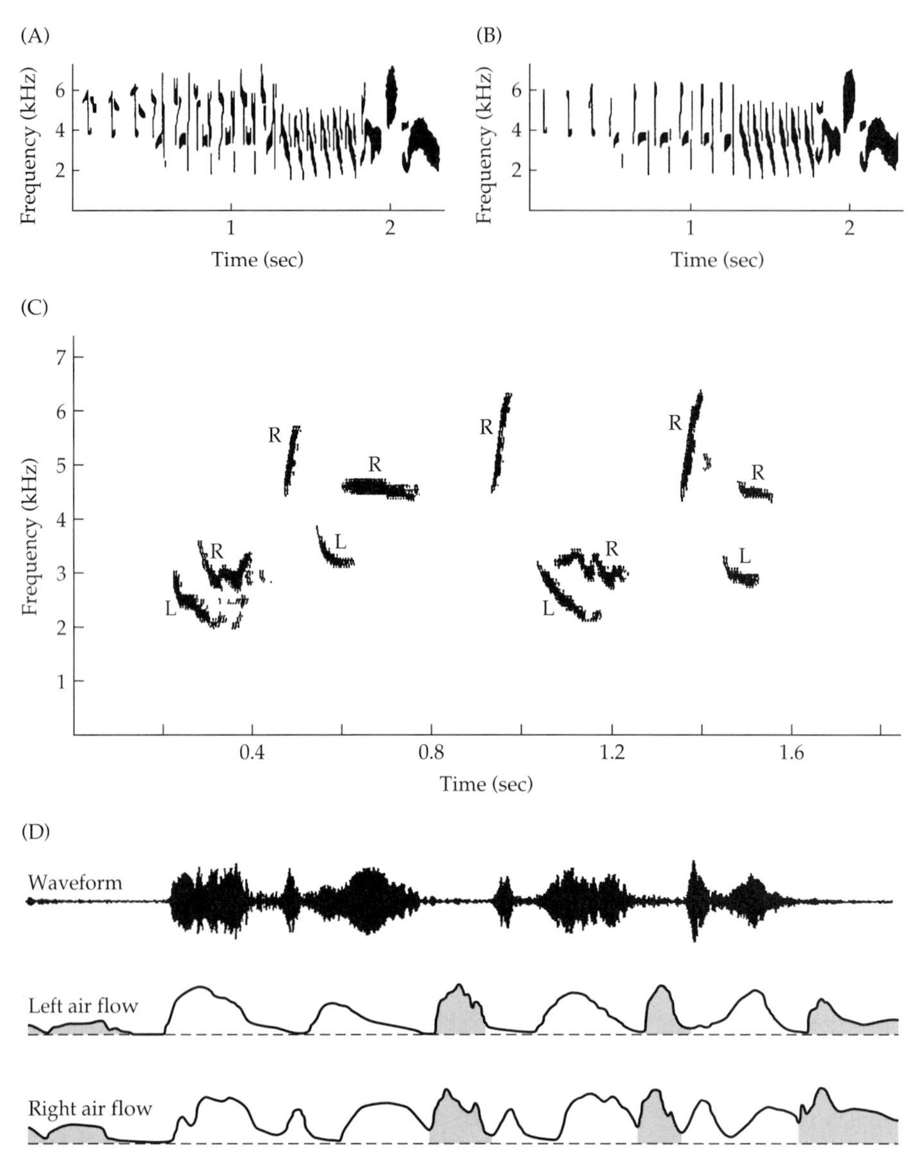

Figure 4.26 Experimental evidence of two sound sources in avian vocalizations. (A) Normal song of male chaffinch (*Fringilla coelebs*). (B) Song of same male chaffinch after cutting of the hypoglossus nerve at the right side of the syrinx. Note that much of the song is retained because this species has strong lateral dominance in song production (C) Spectrogram of vocalization by the brown thrasher (*Toxostoma rufum*). Elements generated by the left side of the syrinx are marked with *L* and those generated by the right with *R*. This species has less lateral dominance than the chaffinch and the two sides contribute about equally to the whole phrase. (D) Vocalization waveform and air flows through each side of the syrinx for the thrasher vocalization shown in C. Graphs are aligned with the spectrogram on same time axis. Inspirations (mini-breaths) are shown as stippled zones and expirations as white zones in airflow graphs. (A and B from Nottebohm 1971, © John Wiley & Sons; C and D after Suthers et al. 1994.)

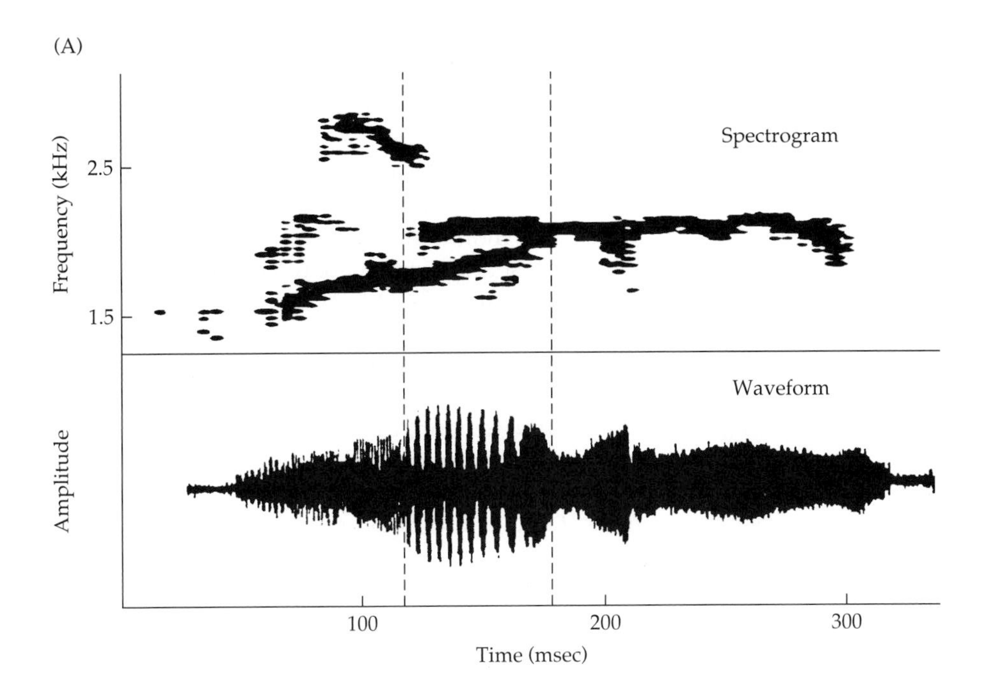

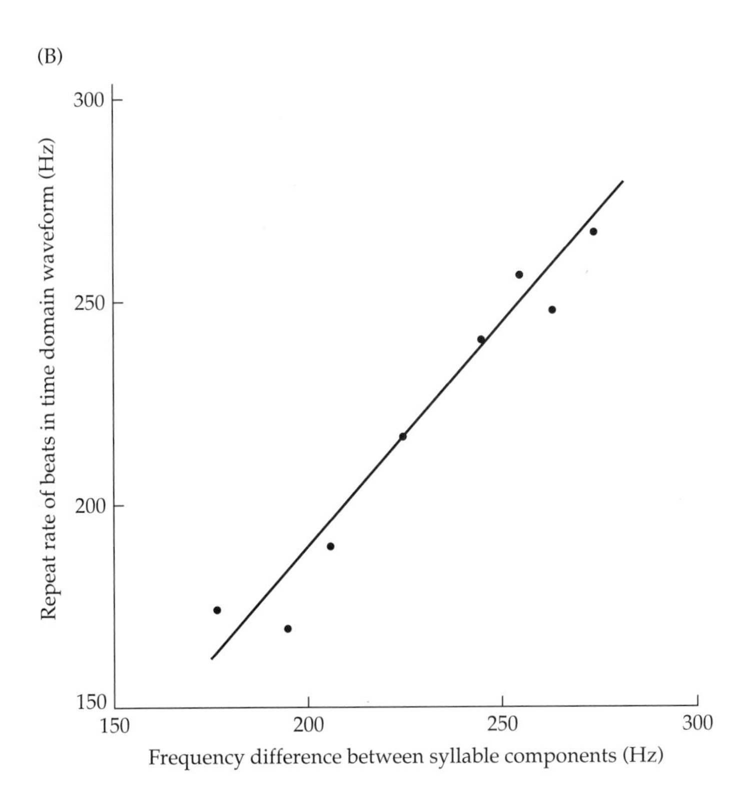

Figure 4.27 Beats in brown thrasher (*Toxostoma rufum*) song waveform due to presence of two similar but unequal and harmonically unrelated frequencies. (A) Spectrogram and waveform of a syllable whose middle section consists of two similar but harmonically unrelated frequencies, each produced by a different side of the syrinx. Note that the region of the syllable with the strongest component overlap (demarcated by two dashed lines) shows amplitude modulation in waveform due to beats between unequal frequencies. (B) Evidence that amplitude modulations in these syllables are indeed due to beats. The plot shows that the modulation rate in the time domain waveform is equal to the difference in frequency between the two components in the spectrogram. (After Suthers et al. 1994.)

The waveform of the sound produced on each side of the syrinx is thought to be controlled by the amount of tension placed upon the membranes and by the rate of air flow in the bronchi. In songbirds, it may also be affected by the degree of protrusion of the labium. According to the hypothesis first advanced by Greenwalt (1968), membrane tension affects both the frequency composition and the amplitude of the produced sounds (Figure 4.28). Sound frequency generally increases monotonically with membrane tension. Sound amplitude however is greatest at intermediate tensions. When membrane tension is low, the full movement of the membranes is constrained by proximity to the opposite side of the passage. As tension is increased, the membranes are pulled further from the opposite wall, the range of allowed membrane movement increases, and this permits greater sound amplitudes. With further increases in membrane tension, the higher tensile forces constrain membrane movements and sound amplitudes again decrease. Consequently, amplitude and frequency modulations tend to be positively correlated at low mean frequencies, (e.g., as the frequency goes up, so does the

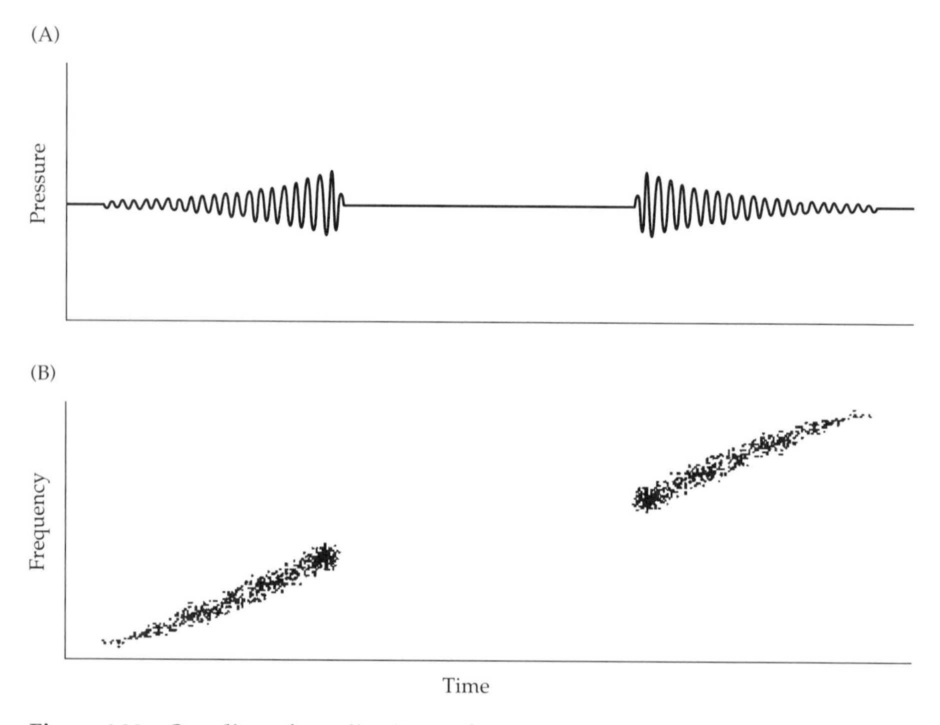

Figure 4.28 Coupling of amplitude and frequency modulations in bird vocalizations. (A) Time domain waveforms. (B) Corresponding spectrograms. When initial tympaniform membrane tension is low (left), an increase in tension increases both sound frequency and amplitude. Frequency and amplitude modulations are thus positively coupled. When initial membrane tensions are intermediate (right), increasing tension continues to increase frequency, but amplitudes decrease. Frequency and amplitude modulations are then negatively coupled.

amplitude), but negatively correlated at higher mean frequencies. (An alternative explanation for this coupling was suggested by Gaunt and Wells 1973.) Any coupling between frequency and amplitude modulation limits the combinations of amplitude and frequency that a song bird can produce. However, the remaining combinations and the ability to operate the two sides of the syrinx independently still permit a bewildering diversity of vocalizations.

The evidence for post-syringeal modification of calls and song in birds is mixed. Some studies have found no evidence for resonance or filtering despite apparent elongations of tracheae or other accessory structures (Greenewalt 1968; Gaunt et al. 1987). However, Nowicki (1987) used enhanced helium air mixtures to show that some songbirds can vary the resonance of their tracheal tubes to maximize transmission of the current fundamental frequencies. Westneat et al. (1993) showed a positive correlation between sparrow song frequencies and bill gape; a similar effect has been reported for geese (Hausberger et al. 1991). This relationship might be expected were tracheal resonances important in favoring or filtering out specific frequencies. Zebra finch males can suppress certain harmonics in their songs, and can vary the frequencies that are suppressed, depending upon the syllable type (Williams et al. 1989). Perhaps the most persuasive case arises in the oilbird (Suthers 1994). This is one of the species with bronchial tympaniform membranes (Figure 4.24C). The oilbird is interesting because the distance between the bronchial junction and the location of the tympaniform membranes is different for the two sides: the right membrane is located 10–12 bronchial rings from the junction, whereas the left membrane is 15–18 rings away. The right bronchus is also slightly smaller in diameter. The result is that the resonant frequencies for the right bronchus are higher than for the left. Both membranes generate periodic nonsinusoidal sounds with a fundamental of 500–800 Hz and up to 18 harmonics. Each bronchus and the trachea favor different harmonics in this series, thereby creating a final sound with three different bands of emphasized harmonics. Because the relative lengths of the two bronchi vary between individuals, each oilbird has its own set of emphasized frequency bands. This distinctive characteristic may facilitate individual recognition in the dark caves in which these birds roost and nest colonially.

SUMMARY

1. Sound production involves three steps: production of vibrations, modification of the vibrations, and coupling of the modified vibrations to the propagating medium.

2. A vibrator that expands and contracts identically in all directions is called a **monopole**. It produces a uniform sound field. A vibrator that moves along a single axis is called a **dipole**, and vibrations moving along two perpendicular axes are called **tetrapoles**. Dipoles and tetrapoles produce sound fields that are directional.

3. The pressure of a propagated sound depends on how fast the vibrator can move a given volume of the medium. Small animals can only move a small

volume of medium and thus have difficulty producing sounds of substantial intensity. Their small size also limits them to sounds with short wavelengths and thus high frequencies. If the animal's vibrator is a dipole or tetrapole, **acoustic short-circuiting** makes the production of sounds, especially low frequency ones, even more difficult. Thus small animals can produce sounds of substantial intensity only for high frequencies.

4. Muscles can contract at rates of 1 kHz or less. The production of high enough sound frequencies by small animals thus requires the use of **frequency multipliers**. **Stridulatory organs** are one mechanism for producing high frequencies. Another is to excite **resonant structures** by snapping them (the tymbals of cicadas and the claws of snapping shrimp) or by forcing air over fluttering membranes (the **larynges** and **syringes** of vertebrates).
5. Any vibrator excited with a waveform containing many frequencies will continue to vibrate with some of these frequencies but not with others. The frequencies retained by the vibrator are its **natural modes**; these frequencies tend to build up in amplitude with successive excitations. Such an excited vibrator is called a **resonator**. The frequencies a resonator amplifies or filters out depend on the dimensions and physical properties of the resonator. Some resonators are highly tuned to specific frequencies and will vibrate at large amplitudes for long periods after excitation. These are called **high-*Q* resonators**. Others are more broadly tuned, show rapid damping of excitations, and never produce much amplification of excitatory signals. These are **low-*Q* resonators**. High-*Q* resonators can be used to increase signal amplitude and frequency tuning; however, they hinder both AM and FM patterning of the sound. Low-Q resonators are better for AM and FM patterning, but will help less in increasing signal amplitude and frequency control.
6. Once a vibrator is excited, the sounds it produces may be modified by transmitting them to a resonator. In some animals, the initial vibrator's waveform is unaffected by attachment to a resonator. The resulting system is said to be **source-driven**. In other animals, feedback from the resonator controls the waveform of the initial vibrator. This system is **response-driven**. Most stridulatory organs and the larynges of frogs and mammals appear to be source-driven systems; some bird syringes may be response-driven. In source-driven systems, a coupled resonator may modify the frequency spectrum of the sounds before emission: some frequency components are enhanced, whereas others are removed. No additional components are added to a source-driven waveform by a resonator. The throat sacs of certain bats, birds, and monkeys are examples of resonators used to modify vibrational waveforms.
7. Coupling of sounds to air is essentially a problem of achieving a better match between the **acoustic impedances** of the vibrating organs and the medium. Thin resonant sacs are used for this purpose in frogs, whereas flared horns are used by monkeys and bats. In water, coupling between vibrators and the medium is *too* good: sound energy is leaked away before resonances can be used to amplify and modify them. To set up such resonance fish use swim bladders, which have a worse match to the impedance of the water. Many whales and porpoises use a head melon of oily material as an acoustic lens to focus their emitted echolocation sounds. This is the aquatic equivalent of a horn in air.

8. Sound production by vertebrates involves forcing air over or past thin membranes that are then set into vibration. The vibrations depend on four forces: the air pressure producing the flow, an intrusion force keeping the membranes in the air flow, elastic forces within the membranes, and the suction of **Bernoulli forces** pulling the membranes into the air flow.

9. In mammals, the **glottis** (consisting of the vocal cords) is used to block the air passage. Air pressure then forces the glottal folds open, and a subsequent drop in pressure and Bernoulli forces then pull them back together. The result is a series of small puffs of air and a sound waveform that is periodic but nonsinusoidal. This waveform consists of a harmonic series based on a fundamental equal to the puff rate. Laryngeal, hyoidal, and pharyngeal spaces are then used to modify this frequency spectrum before emission. Humans modify their pharyngeal spaces rapidly, producing different vowels from the same initial vibrational waveform.

10. Anurans (frogs and toads) also use a glottis to produce air puffs. However, they have a second pair of membranes (**anuran vocal cords**) upstream from the glottis; these vibrate at a higher frequency than the puff rate when air passes over them. Anuran sounds thus consist of a carrier frequency generated by the anuran vocal cords and then amplitude modulated in a nonsinuoidal but periodic manner by the glottis. These sounds are transmitted to the throat sac, which both recovers the spent air for subsequent vocalizations and acts as a resonant coupler to the surrounding air.

11. Birds produce sounds using thin membranes in the wall of either the trachea or the bronchi not far from the broncheal-tracheal junction (called a **syrinx**). Air passing over these membranes during expiration causes them to vibrate. In many species, the wall opposite the membranes hosts a protruberance (the **labium**) that can be rotated into the cavity to control the tube aperture and thus the onset, offset, and amplitude of sounds. Tension changes in the membranes affect both amplitude and frequency of the vibrations; amplitude and frequency modulation are thus often coupled in bird songs. Many species of birds are able to control the two membranes independently, allowing birds to produce sounds that consist of harmonically unrelated components. Birds may also have post-syringeal resonators that modify frequency spectra of sounds before emission.

FURTHER READING

Good discussions of resonance and the physical processes of making sounds can be found in Fletcher (1992) and Rossing (1990). The general types of sound production mechanisms in animals are reviewed in Michelsen (1983, 1992). Sound production in arthropods is reviewed by Bailey (1991), Bennet-Clark (1975), Haskell (1974), and Michelsen and Nocke (1974). Specific examples are provided by Bennet-Clark (1970, 1971), Elsner (1983), Hyder and Oseto (1989), Ichikawa (1976), Imafaku and Ikeda (1990), Kavanagh and Young (1989), Markl (1983),

Michelsen et al. (1982, 1986), and Pringle (1954). Sound production by fish is discussed by Harris (1964), Ladich (1989), Ladich and Kratochvil (1989), and Myrberg (1981), and in frogs and toads by Blair (1963), Capranica (1965), Gans (1973), Martin (1971, 1972), Martin and Gans (1972), Ryan (1985b, 1988), Ryan and Brenowitz (1985), Schmidt (1965), and Schneider (1988). Good reviews of the anatomy and function of the avian syrinx are provided by Brackenbury (1982, 1989), Fletcher (1988, 1989), Gaunt (1987), Gaunt and Gaunt (1985), and Warner (1972); more specialized treatments can be found in Allan and Suthers (1994), Ames (1971), Brackenbury (1977; 1978; 1979a,b; 1980; 1982; 1989), Chamberlain et al. (1968), Gaunt et al. (1973, 1976, 1982, 1987), Gaunt and Gaunt (1977, 1980), Gaunt and Wells (1973), Goller and Suthers (1995; 1996a,b), Greenwalt (1968), Gross (1954, 1964), Harris et al. (1968), R. S. Hartley and Suthers (1990), Lockner and Murrish (1975), Lockner and Youngren (1976), Mairy (1976), Miskimen (1951), Miller (1977), Nottebohm (1971, 1972), Nottebohm and Nottebohm (1976), Nowicki (1987, 1989a), Nowicki and Marler (1988), Nowicki and Capranica (1986), Rüppell (1933), Scala et al. (1990), Stein (1968), Suthers (1990, 1994), Suthers et al. (1994), Vicario (1991), Ward (1989), and Warner (1971a,b; 1972). Broad reviews of the mammalian larynx can be found in Negus (1949) and Kelemen (1963), and Tembrock (1963) provides a general review of mammalian sound signals. Specific studies include those on cats (Sissom et al. 1991), bats (D. J. Hartley and Suthers 1987, 1990; Suthers 1988; Suthers et al. 1988), cetaceans (Norris and Harvey 1974; Pilleri 1983; Reidenberg and Laitman 1988; Thompson et al. 1979), and ungulates (Kiley 1972). Reviews of the types of vocal signals produced by mammals can be found in various chapters in Sebeok (1977).

Chapter 5

Sound Propagation

WHEN AN ANIMAL PRODUCES A SOUND SIGNAL, it does so with the intention that at least one receiver will be able to detect and recognize the signal. As we shall see, there are many ways that a sound can be distorted and degraded during its propagation between the two parties. Coping with distortion during propagation is a major factor shaping and constraining the evolution of many animal sounds. The most common solution is the evolution of a signal that resists such distortions. We have already seen that producing sounds is not easy, especially for small animals, and the additional constraints imposed when communicating over any distance make the set of feasible signals even smaller. Knowing how animals cope with propagation distortion is, therefore, important in understanding the evolution of animal communication. In addition, there are practical consequences for understanding sound propagation for those of us studying animal signals in the field. We often must make recordings at some distance from our subjects. How confident can we be that what we recorded reflects what the animal emitted, and not the effects of distortion during propagation between the

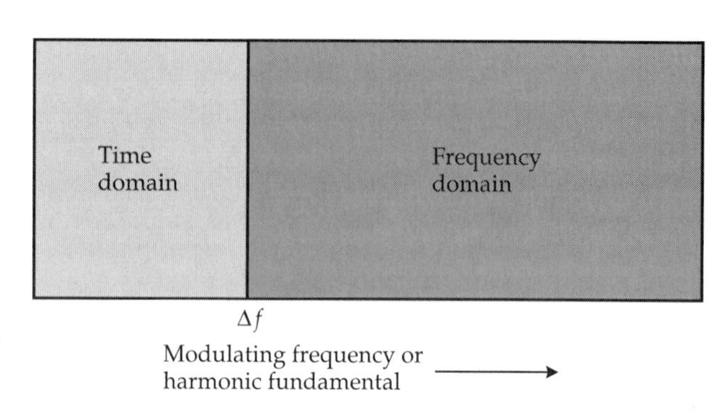

Figure 5.1 Partitioning of the perception of temporal pattern in a sound into a time domain and a frequency domain. Here Δf is the critical bandwidth in the auditory organ of the receiver.

animal and our microphone? Knowing the types of distortion that can occur, we can avoid or correct for them before undertaking studies that rely on measured sound structure. Our goal in this chapter is thus to summarize the kinds of distortion that may be imposed on a propagating sound.

It will help in our task if we classify the kinds of distortions that are possible. First note that animals analyze perceived sounds the way machines do. As outlined in Chapter 3, pages 65–68, every ear has a critical frequency bandwidth, Δf, which determines whether temporal patterns are perceived in the time domain (when frequency components are separated by intervals less than Δf), or in the frequency domain (when components are separated by intervals greater than Δf). Thus for a given ear, rapid temporal patterns are likely to be perceived in the frequency domain, whereas slower modulations and patterns will be perceived in the time domain. This partitioning of a perceived signal pattern into the two domains is illustrated in Figure 5.1. We can thus divide the various propagation distortions into those that are perceived by a receiver in the frequency domain, and those that are perceived in the time domain. Within each domain, types of distortions can be further classified according to how they perturb the signal. There are three categories: global attenuation, loss of pattern, and addition of noise. In the following discussion we treat each domain and the contributions to these three kinds of distortion in turn.

SOURCES OF DISTORTION IN THE FREQUENCY DOMAIN

For most animal ears, it is the frequency spectrum and not the phase spectrum that provides the most important information in the frequency domain. The major types of distortion in the frequency domain are therefore those that modify the frequency spectrum of the propagating sound. These sources of distortion include global attenuation, four sources of pattern loss (differential medium absorption, scattering, boundary reflections, and refraction), and the addition of noise. We take up each of these six causes of frequency domain distortion below.

Global Attenuation

As we saw in Chapter 2, spreading losses cause sound intensity in a far and free field to fall off with the square of the distance from the source (inverse square law). Since this relationship applies equally to all frequency components within a sound, the loss in intensity is not accompanied by differential filtering (Figure 5.2). Spreading losses are also identical regardless of the propagating medium. In the far field, spreading losses result in a 6 dB drop in pressure for each doubling of the distance between a measurement near the source and one further away. Note, however, that spreading losses in the near field may exhibit different rates of attenuation and may not be spatially uniform. In water, where wavelengths are 4.4 times greater than wavelengths in air for equivalent frequencies, near fields extend further from a sound source and communicating animals are more likely to be within them. The major impact of spreading loss on a receiver is the risk that intensities of some component frequencies will fall below those of the ambient noise.

Pattern Loss by Medium Absorption

All sources of pattern loss in the frequency domain operate by differentially filtering out certain frequencies during propagation. The simplest process is medium absorption. We noted in Chapter 2 that losses of sound energy to heat during propagation vary with the medium and the frequency of the sound. For a 1 kHz signal, the medium loss in water is about 0.008 dB/100 m of propagation, whereas in air the loss is 1.2 dB/100 m. Medium absorption is therefore several hundred times greater in air than water. In air, losses generally tend to be greater at higher temperatures and lower humidities, although the patterns can be complicated and dependent on the sound frequencies involved (Griffin 1971). In temperate climates where humidity and temperature rise and fall together, these effects may cancel out. Medium absorption is 1050 times greater in salt water than in fresh water, and may vary with pressure and temperature. Medium absorption for substrate-propagated sounds is generally very high; values of 6 dB/cm are typical (Markl 1968; Markl and Hölldobler 1978). Sounds propagating in the stems and leaves of plants have much lower values, but this type of communication is complicated by the establishment of traveling waves in the plant; generally, frequencies less than 1–2 kHz are used (Bell 1980; Michelsen et al. 1982). Insects on plants can ap-

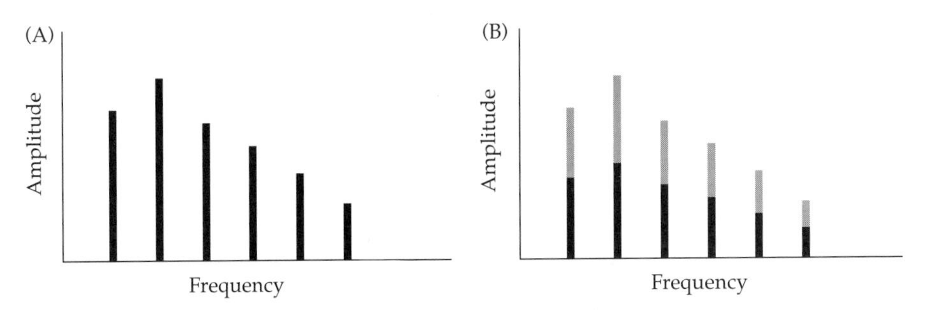

Figure 5.2 Power spectrum of signal near its source (A) as compared with that far from the source (B), showing spreading loss. The amplitude measure here is pressure and is shown by black bars; the attenuated fraction in B is shown by gray sections. Note that all frequency components are attenuated by equal proportions.

parently communicate using plant-propagated vibrations over a range of 1–2 m (Gogala et al. 1974; Ichikawa 1976).

Over the frequency ranges used by animals for communication, medium absorption increases monotonically with the frequency of the sound. For example, medium absorption for a 10 kHz sound in air is about 10 times that for a 1 kHz signal. Such losses thus alter sound frequency spectra by reducing the amplitudes of all components to some degree (the reductions being higher in air than in water), and by differentially reducing the amplitudes of high frequency components more than the amplitudes of low frequency ones (Figure 5.3).

Pattern Loss from Scattering

The second major source of frequency spectrum filtering is scattering. You may wish to refer back to the several types of scattering that were defined in Chapter 2, pages 19–22. The amount of energy scattered by objects in a medium during propagation depends on the relative sizes of the objects, the incident wavelengths, the acoustic impedances of medium and objects, and sometimes on the shape and composition of the objects. In air, a 1 kHz sound has a wavelength of about one-third of a meter. This means that any reflecting object with a diameter between 5 cm and 3 m will produce complicated diffractive scattering, and any object larger than that will produce significant simple scattering. An obvious source of scatter in terrestrial habitats is vegetation. Leaves are often small and relatively light objects with acoustic impedances only somewhat greater than that for air. They are thus potential sources for interactive scattering. Tree trunks can contribute to both diffractive and simple scattering, depending on the frequencies of the emitted sounds. The amount of propagating sound energy lost by scattering will depend on the size and density of the trees and leaves in a specific forest. Measurements of attenuation due to the

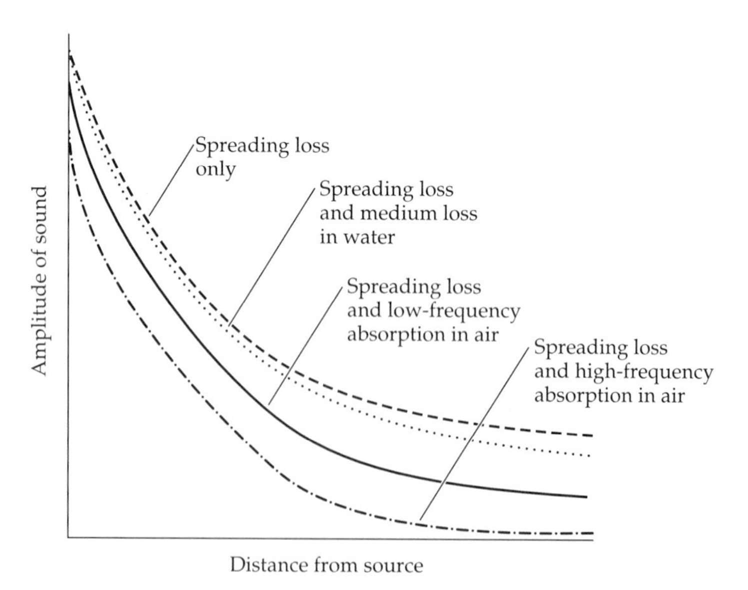

Figure 5.3 Spreading and medium losses as a function of distance from a source. Spreading losses are identical for air and water. Medium losses in water are 150–200 times lower than for air; hence sound amplitudes in air fall off more quickly with distance. In both media, the higher the frequency, the greater the medium loss for a given distance from the source.

combined effects of medium absorption and scatter in broadleaf forests (using measurements 10 m from the ground) range from 2–35 dB/100 m for frequencies between 1 and 10 kHz. Equivalent measurements in coniferous forests, which have smaller leaves and thus exhibit less scatter for a given wavelength of sound, range from 2–20 dB/100 m of propagation over the same frequency range (Marten and Marler 1977; Marten et al. 1977).

Another source of scatter arises from the generation of small pockets or bubbles of air whose acoustic impedance differs from that of the surrounding medium. When land surfaces are heated by the sun, they can produce small cells of hot and lower-density air; these air cells bubble up from the surface and are surrounded by cooler denser air as they rise. Another source of scatter can come from air flow. Wind blowing over irregular surfaces can generate local vortices within which the speed of sound and the density of the air differ from the surrounding medium. In either case, these bubbles or vortices of different acoustic impedance constitute objects that can scatter propagating sounds. These scattering objects are more likely to be found in open areas where the sun can heat the substrate or where strong winds blow over irregular surfaces. When air cells or wind vortices are absent, (e.g., when there is little sun or wind), sound attenuation over open areas will depend primarily on medium absorption (1–12 dB/100 m for frequencies of 1–10 kHz). The presence of air cells or vortices can then add an additional two to several hundred dB/100 m of attenuation (Wiley and Richards 1982).

In water, many objects have acoustic impedances similar to the medium. Such objects generate only small reflected components and thus minor scatter. However, gas bubbles and the air-filled swim bladders of fish have the required difference in impedance and these can cause significant scattering of sound. Since wavelengths of sound in water are nearly 4.4 times greater than wavelengths in air, diffractive scatter is not a problem for a 1 kHz sound unless the bubble or air sac is nearly 30 cm long. Only near a school of very large fish would scattering of this frequency be significant. Sunlight rarely heats the bottoms of bodies of water sufficiently to cause cells equivalent to terrestrial habitats; areas near hot vents might, however, produce bubbles of lower water density and thus increase scatter. Similarly, deep ocean currents are rarely rapid enough to produce significant vortices affecting sound scatter, but rough, shallow coastal waters may have such eddies. In general, we expect that losses of sound to scattering is much less of a problem in water than in terrestrial forests or open country on warm or windy days.

Scattering has different significance for different animals. For echolocating bats or porpoises that rely on scattered sound to identify and locate objects, the frequency dependence of the zones generating Rayleigh and diffractive scattering can be a useful source of information (see Chapter 26, pages 862–864). One might expect these animals to adjust the frequencies of their echolocation sounds so that wavelengths and targets were of similar size. Birds or frogs calling in a forest, however, experience scatter as a loss of emitted signal. One might expect them to select frequencies as low as possible to reduce scattering losses.

Pattern Loss from Boundary Reflections

Having discussed reflections by objects that are small relative to the wavelengths of incident sounds, we now turn to reflections of sounds from surfaces much larger than the incident wavelengths. In the air, the major reflector of large size is the ground, although pronounced temperature inversions in the air can sometimes act more like reflectors than refractors. In water, both the water's surface and the bottom act as large reflectors. As we shall see, boundaries often cause the attenuation of some frequencies in an incident sound and the accentuation of others.

The propagation of sounds near boundaries is not a simple process. However, much of the complexity can be simplified. We can describe three types of waves that can simultaneously propagate sounds: a **direct wave** that travels on a straight line between sender and receiver, a **reflected wave** that strikes the surface and rebounds toward the receiver, and a **boundary wave** that travels along the surface of the boundary (Figure 5.4). Boundary waves arise when some of the incident sound energy is absorbed by the second medium. This absorption can have two effects: the simplest is the horizontal propagation of the signal through the second medium and its eventual reradiation back into the first medium (called **substrate** or **ground wave** propagation). Alternatively, complex movements of the first medium into and out of pores in the surface of the second can arise from interactions between the incident sound field and induced vibrations in the second medium. This motion generates a **surface wave** that propagates in the first medium just above the boundary.

Whether the reflected and boundary waves have amplitudes similar to the direct wave depends upon how much farther they must travel than the direct wave and on the reflection coefficient R at the boundary (see Chapter 2, pages 24–29). When $|R|$ is high, most energy goes into the reflected wave and

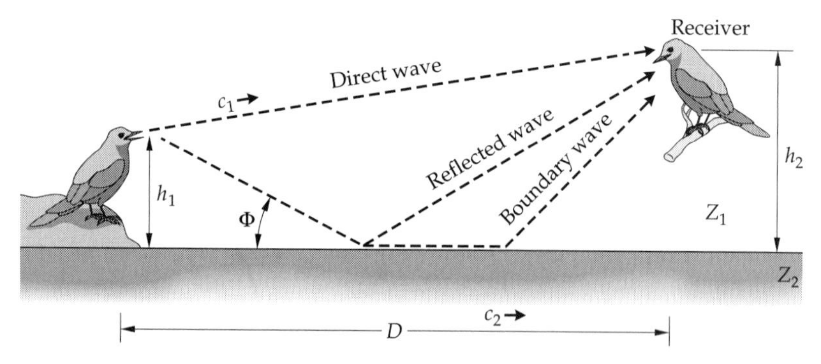

Figure 5.4 Sound propagation near a boundary surface. Sound waves arrive at a receiver through a direct path, a wave reflected from the boundary, and ground and surface waves that travel in or near the boundary. Critical parameters affecting sound intensity at the receiver include the heights above the boundary of the sender, h_1, and of the receiver, h_2; the horizontal distance D between sender and receiver; the grazing angle of the reflected sound, Φ; the relative impedances, Z_1 and Z_2 of the two media at the boundary; and the speeds of sound in the two media, c_1 and c_2.

little into the boundary wave; when $|R|$ is low, the reverse is true. As we saw in Box 2.2, the reflection coefficient R depends upon the grazing angle at which the sound strikes the boundary and bounces toward the receiver, (determined by the heights of sender and receiver off the ground and the distance between sender and receiver), the relative acoustic impedances of the two media, the speeds of sound in both media, whether the boundary is porous or elastic, and on the frequency of the sound. Of particular relevance here is the fact that little energy goes into boundary waves unless the acoustic impedance of the boundary medium is greater than that of the source medium. This means that there can be no boundary waves for sound traveling in water and striking the surface. Where boundary waves do occur, they will also be affected by a **boundary loss function** that limits how far (measured in wave cycles) the boundary wave can travel before it is fully attenuated. Boundary waves are thus only important when signal frequencies are very low (the permitted number of cycles constitutes a long distance) or, if frequencies are higher, when sender and receiver are very close together. Because watery mud and water have similar acoustic impedances, many aquatic substrates are unlikely to exhibit significant boundary waves. Some muddy bottoms may have higher acoustic impedances than the surrounding water, but propagation is complicated because they also have lower acoustic velocities. As summarized in Box 2.2, there will be a grazing angle (called the angle of intromission) over such a surface at which all incident energy is absorbed and there will be no reflected waves. Only for much steeper or much less steep incident angles will the reflected component be significant.

When they are present, direct, reflected, and boundary waves all combine additively at the receiver. Depending upon phase shifts at reflection and differences in distances traveled, the three waves for any given frequency may arrive at the receiver with different relative phases. When they arrive with similar amplitudes and phase, they add positively; that frequency appears more intense to the receiver than if the boundary were farther away. If the various waves arrive with similar amplitudes but are out of phase, they cancel each other out and the receiver will not hear this frequency. As detailed in Box 5.1, propagation near a reflecting boundary is very frequency-dependent. If there is no phase shift at reflection (as when sound in air strikes a liquid or solid surface at an angle greater than the critical one), low frequencies propagate well, intermediate frequencies show an oscillating pattern of attenuation and augmentation, and high frequencies propagate no differently than if the boundary were absent. Incident angles greater than critical are more likely when sender and receiver are close together or both are high above the boundary. Note, however, that the value of the critical angle varies markedly depending upon the relative impedances and sound velocities of the two media. For example, the critical angle for sound traveling in air and hitting a water surface is 78°; sound traveling in water experiences a critical angle of 28° over a sandy bottom and only 9° over a silty one. The chance that two fish communicating over a silty substrate will experience a phase shift of the reflected wave is thus much smaller than that for two perched birds at a similar distance.

Box 5.1 Sound Propagation and Reflection at Boundaries

THE PROXIMITY OF A BOUNDARY BETWEEN TWO MEDIA with different sound velocities and acoustic impedances can severely affect sound propagation. Consider first a situation in which there is no phase shift at reflection, such as occurs when sound in air strikes a liquid or solid boundary at a grazing angle greater than the critical angle. If c is the velocity of sound in the signal source medium, and ΔD is the difference in the travel paths of the reflected and direct waves, then there will be some minimum frequency $f_d = c/2\Delta D$ for which the direct and reflected waves cancel each other out at the receiver. This cancellation occurs because the reflected wave has to travel exactly one-half a wavelength longer to reach the receiver than does the direct wave. It will thus arrive exactly out of phase with the direct component. For the same reason, higher odd harmonics of this frequency will interfere negatively, but reflected waves will have lower amplitudes than the direct waves and will thus attenuate them less. The lower amplitudes are due in part to the greater number of cycles that must be completed by higher frequency reflected waves in order to traverse ΔD. If the surface is rough with rugosities similar in size to incident wavelengths (i.e., at sufficiently high frequencies), each rugosity will act as a separate scatter point. The result is many small amplitude reflections. In air, small heterogeneities in the medium formed by wind vortices or cells also tend to scatter both incident and reflected waves at higher frequences and thus break up the reflected wave into many smaller ones. However formed, the many smaller waves must each traverse a different ΔD and thus arrive at the receiver with a wide variety of relative phases. As a result, no general pattern of interference is established. Where the ground is soft or porous, the decrease in the absolute value of the reflection coefficient with increasing frequency also tends to decrease the amplitude of any reflected component. All of these factors favor decreasing interference effects with increasing frequencies; at high enough values, there is no detectable interference. The even harmonics of f_d will add positively to the direct wave, with the greatest enhancement for the second harmonic and decreasing effects for subsequently higher harmonics. Frequencies much less than f_d will exhibit partial positive adding since there is no phase shift at reflection and ΔD will be so small relative to a wavelength that any differences in phase between the two waves will be trivial. The transfer function for such a situation will appear as shown in Figure A.

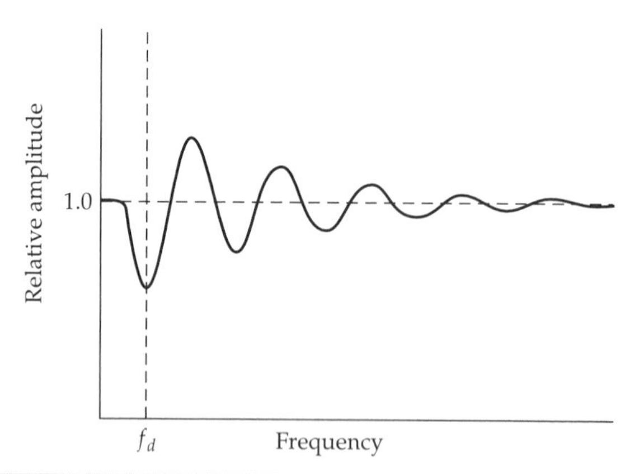

Figure A Transfer function of sound propagating in air at moderate distance above ground.

Now consider a sound produced near a boundary with a full phase shift. Two examples are (1) sound in water striking the surface and (2) sound in air hitting the ground at less than the critical angle. Because of the 180° phase shift, $f_d = c/\Delta D$. Here again, odd harmonics of f_d will tend to be filtered out, whereas even harmonics will be accentuated. The frequency of f_d is twice as high when there is a phase shift than when there is none. In water, the higher speed of sound will also increase the value of f_d by 4.4 times the value that occurs in air. All frequencies below f_d will suffer severe attenuation because the reflected wave will leave the surface with a phase opposite to that of the direct wave and the small difference in distance traveled, relative to the wavelengths, causes the two amplitudes to be similar. The transfer function when there is a phase shift at reflection is shown in Figure B.

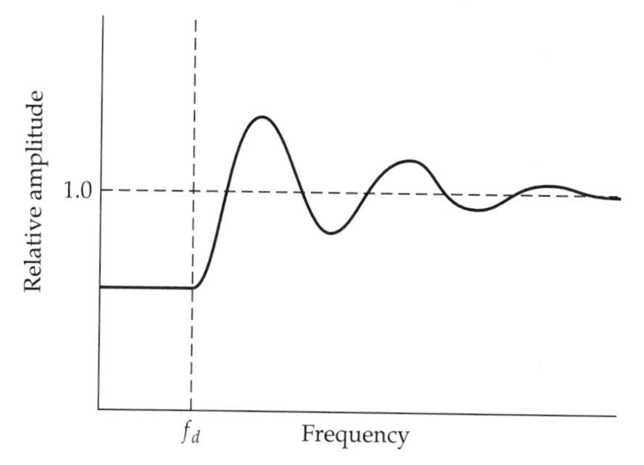

Figure B Transfer function of sound propagating in water just below the surface.

In the case of sound in air striking porous ground below the critical angle, boundary waves may be able to restore propagation of the lowest frequencies lost because of the phase shift. The notch—the zone of frequencies greater than those propagated by boundary waves and less than $2f_d$—is often avoided by ground living birds and mammals that use sound for long-distance communication. The relevant transfer function can then be pictured as in Figure C.

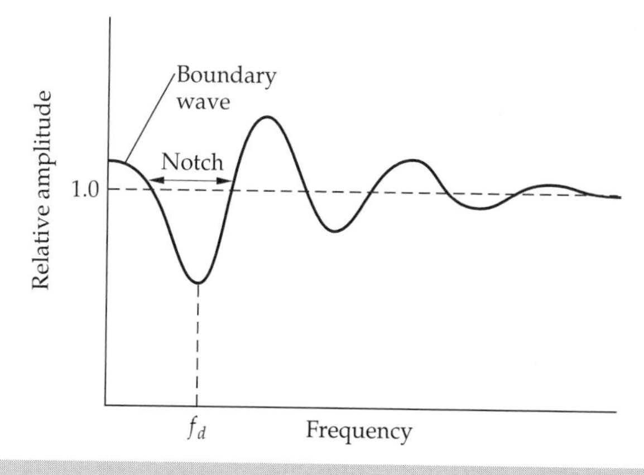

Figure C Transfer function of sound propagating in air close to porous ground.

If there is a phase shift (as when sound in water strikes the surface or sound in air or water strikes a substrate at angles below the critical value), low and some intermediate frequencies are severely attenuated, higher intermediate frequencies show an oscillation of attenuation and augmentation, and high frequency propagation is unchanged. In the case of sound in air near porous ground (but not at the water's surface), boundary waves can restore some propagation of the lowest frequencies. The remaining frequencies that are in a low to intermediate range are attenuated at low critical angles. In many terrestrial environments, this zone of attenuation (often called the **notch**) ranges from 300–800 Hz and is avoided by ground-living animals calling over long distances. Aquatic animals near the water's surface do not have the benefit of boundary waves and hence cannot use any low frequency for long distance signaling.

In shallow water, the situation can be a bit more complicated because there are two nearby boundaries: the air-water interface and the water-substrate interface. Where these are sufficiently far apart, (at least 10 complete cycles of the propagating sound), we can treat what happens at each surface as roughly independent of what happens at the other. When the water is sufficiently shallow, the reflections of sounds from the surface and the bottom can interact in complicated ways. In fact, the water then becomes rather like a large guitar box in which standing waves are set up. As with other resonant systems, there will be some frequencies or modes that build up and propagate well and others that are canceled out. Among the allowable modes, there will be a minimal frequency, f_c, below which no sound can propagate within the layer of water between the surface and the bottom. The value of this cutoff frequency is inversely related to the depth of the water. It is also inversely related to the speed of sound in the substrate. Thus f_c for a rocky bottom is lower than that for a muddy or silty one. Sandy bottoms fall in between. Values of f_c range from 400–1100 Hz for water 1 m deep, with higher values for softer bottoms, and 30–200 Hz for water 10 m deep. Clearly, as water becomes shallower, resident organisms can use sound less well for communication and use it primarily over very short distances or with very high frequencies.

A similar effect arises for insects communicating with vibrations conducted along the stems and leaves of plants. The differences in acoustic impedance between the leaves and between the air at the top of the plant and between the roots and the soil at the bottom cause vibrations to be reflected at both boundaries. This reflection sets up complicated standing waves within the plant (Michelsen et al. 1982). Which frequencies are favored or canceled depend on the size and geometry of the plant and on the location of the vibration-producing animal. Because frequency filtering of sounds propagated in solids is so unpredictable, insects that communicate in this manner, e.g., leafhoppers and cydnid bugs, tend to use very broad spectrum sounds to guarantee that at least some frequency components reach receivers.

Pattern Loss from Refraction

As we saw in Chapter 2, the sound energy that crosses a boundary between two media of different acoustic impedances is likely to change its direction of travel: if the second medium has a lower speed of sound, the direction of travel is bent away from the boundary surface and into the second medium; if the second medium has a higher speed of sound, the transmitted wave is refracted toward the boundary and away from the second medium.

In air and water, one often finds gradients of temperature or density that generate equivalent gradients of sound velocities. A gradient can be considered as a series of boundaries between layers of different sound velocities. As a propagating sound moves through each layer, it is bent successively in the same direction, producing a curved trajectory instead of a straight one. If it moves from an area of low to high sound velocity, its line of propagation will be bent back toward the layer in which it began; if it moves from high to low sound velocity, it will be bent away from the boundary and deeper into the new layer.

The velocity of sound in either air or water is determined by the ambient temperature, pressure, composition, and mass flow levels (winds in air and currents in water). Increasing temperature and pressure both increase the speed of sound. In water, the presence of more dissolved materials (i.e., higher salinity) also increases the speed of sound. Sound propagating in the same direction as a medium current has a higher speed of sound; sound propagating against a current or wind has a lower sound velocity.

Figure 5.5 shows typical effects of refraction for terrestrial animals. In general, a profile picturing sound velocity as compared with height above the ground will show one height with maximal values and decreasing sound velocities for greater or lower heights. On a sunny day, the surface of the earth heats the air nearest the ground, and the peak sound velocities are found right at the surface. Sound velocity then decreases monotonically with height above the ground. Sounds emitted by a vocalizing bird will be

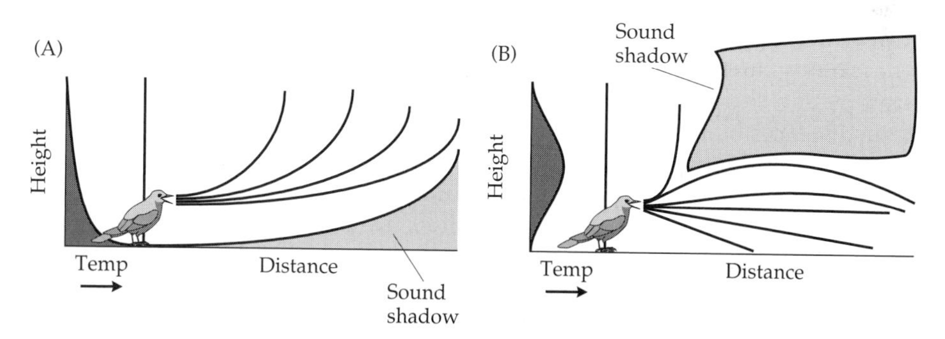

Figure 5.5 Refraction effects experienced by terrestrial animals. Sound velocity profile depends primarily on temperature of air (Temp). In (A), the ground is warm, and the air near the surface has a higher sound velocity than air at greater heights. Sounds are refracted up and away from the ground, generating a sound shadow near the ground at a moderate distance from the source. In (B), the warmest air occurs at some height above the ground. This situation is common in open areas on clear nights or at the canopy of a forest during the day. Vocalizations emitted below the warm air are refracted back down and travel long distances. The sound radiating above the warm air is bent up, generating a sound shadow above the warm layer.

refracted up into the lower sound velocity zones and thus away from the ground as they propagate. At some distance from the bird, most lines of propagation will have been bent upwards, and there will be a shadow zone near the surface at which the bird cannot be heard at all. The height of this shadow zone above the ground will be greater the farther away from the bird one moves. The closer the vocalizing bird is to the ground, the closer the shadow zone will be to it.

A similar effect arises when a strong wind is blowing. Winds have low velocities near the ground and higher velocities above it. Consider a sender calling to a receiver upwind. As the path of the propagating sound crosses from a region of low wind velocities into one of stronger headwinds (and thus lower sound velocity), it will be refracted into the region of higher wind. Not unlike a temperature gradient, a wind velocity gradient will bend the sounds upward and create a sound shadow near the ground. In air, the speed of sound is roughly 334 m/sec at 20°C. Wind velocities range from several m/sec near the ground to 100 m/sec in the upper atmosphere. Typical winds can thus change the velocity of sound by 5–10%, produce substantial refraction, and make it very difficult for animals to vocalize upwind for any appreciable distance.

Alternatively, consider a clear night. After sunset, the surface of the earth rapidly radiates its heat into the cloudless sky and cools. Eventually, the earth and the air closest to it are cooler than the air at some height above the surface. Sounds produced by a bird in the cool layer that radiate up and into the warmer air will be refracted back down toward the earth. In fact, the cool layer between the earth and the warmer air will become a **sound channel** within which sounds tend to be trapped and to propagate long distances. The area above the warm layer into which some sound would have propagated had there been no refraction thus becomes a shadow zone in which no sound can be heard. Sound channels can also exist in closed forests that are warm at the canopy surface, where sunlight is trapped, but are cool between the canopy and the ground. Birds singing beneath the canopy can be heard for long distances by receivers also beneath the canopy. However, receivers above the canopy may be in the shadow zone and hear nothing. For the reasons given earlier, sound traveling downwind propagates with lower velocities near the ground (because wind velocities are lower here) than at greater heights. The result is again a sound channel that enhances long distance propagation when both sender and receiver are close to the ground.

In water, similar effects occur but the water's surface usually takes the role that the earth's surface takes for terrestrial animals (Figure 5.6). In shallow water in summer, the surface layer of the water is warm, and both temperature and sound velocity decrease with depth. A fish producing a signal near the surface experiences a refraction of its signals away from potential listeners at the surface and at some distance, and there is thus a sound shadow at the surface. In winter, the surface water is colder than the water just below it, but the deepest waters are cold again. Sounds produced by a fish near the surface are bent back up toward the surface and channeled along it. The sig-

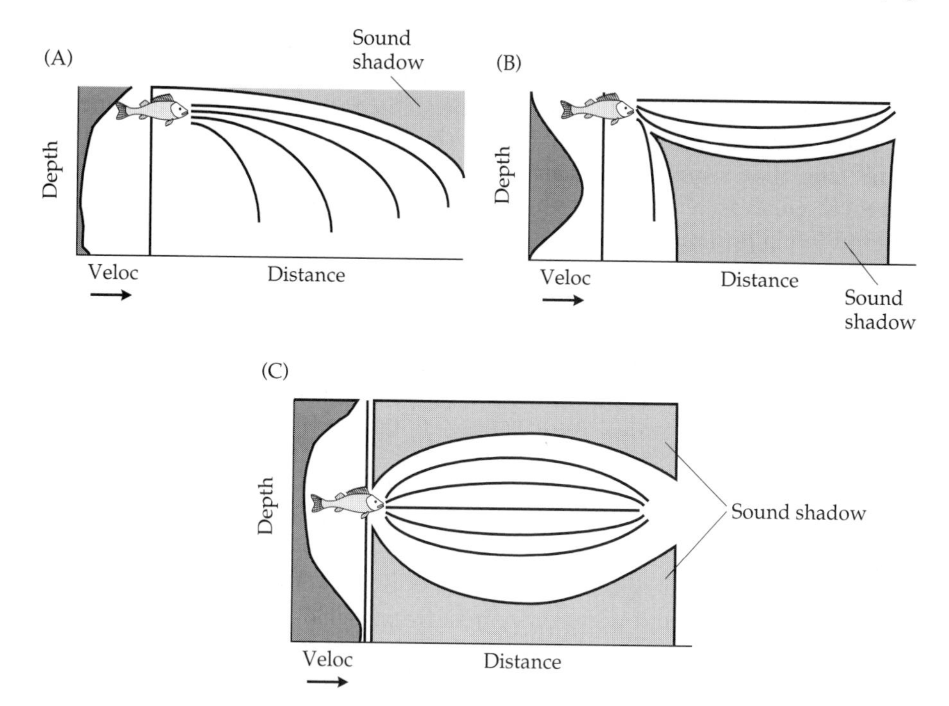

Figure 5.6 Refraction effects experienced by aquatic animals. Sound velocity (Veloc) depends on both water pressure (which increases with depth) and temperature. (A) Fish in moderate depth water in summer. The surface layer is warm and sound is refracted away from the surface. This scenario is the aquatic equivalent of a bird singing near the ground on a warm day. (B) Fish in moderate depth water in winter. The surface layer is cold, intermediate depths are warmer, and deep water is cold again. Fish that are croaking near the surface experience a sound channel effect with long-distance propagation. (C) A deep ocean band of low sound velocity at 12 km depth exists because this region is cold but has only moderate pressures. Sound emitted in this band is trapped both above and below and thus travels very long distances in the sound fixing and ranging (SOFAR) channel.

nal now has improved propagation along the resulting sound channel between the water's surface and the warm layer. Currents in oceans will rarely produce refractive effects similar to those produced by winds in air. In sea water, sound travels at speeds of about 1520 m/sec whereas most ocean currents travel at only a few meters per second. Ocean currents produce such small changes in sound velocity (less than 0.1%) that there will be little loss when the signaler is communicating upcurrent and little gain when contacting a receiver downcurrent.

Aquatic animals in the deep ocean have access to the best sound channel known. The velocity of sound is high at the water's surface where pressure is low and temperatures warm, but decreases with depth as the temperature drops in deep water. At moderate depths, water temperatures stabilize, and they generally remain constant even with increasing depth. However, pressures rise with increasing depth, and sound velocities again increase with proximity to the ocean's bottom. The sound velocity profile for deep ocean thus begins with high values at the surface (high temperatures), drops to a minimum at intermediate depths (low temperatures, moderate pressure), and then rises to high values again (low temperatures but high pressures) at great depths. The zone of low sound velocities is called the **SOFAR** (sound fixing and ranging) **channel**. At mid-latitudes, it occurs at about 1200 m. of depth, but moves closer to the surface nearer to the poles. Because sound energy which would have radiated either above or below the sender is refracted back into this narrow zone, and sound absorption and scatter in water are so low, signals can travel in the SOFAR channel for hundreds and even thousands of kilometers and still be detected. Baleen whales may use this channel to communicate over long distances (Payne and Webb, 1971; Winn and Winn, 1978; Thompson et al. 1979).

Refraction in sound channels often leads to changes in the frequency spectra of the propagated signals. For any sound channel, there is a minimum frequency, f_{min}, that can be trapped within it. In sea water, $f_{min} = 1.8 \times 10^5/d^{1.5}$ where f_{min} is in Hertz and d is the thickness of the sound channel in meters. This means that a sound with component frequencies spanning f_{min} will show greater attenuation of components below this cutoff than would a sound with component frequencies above the cutoff. In addition, propagation in sound channels is rarely linear; different frequencies may not propagate independently but may interact nonadditively. Both of these effects can result in major changes in the spectra of sounds propagated in a channel.

Noise

Noise alters the spectral composition of a propagating sound by adding new frequency components and new energy to existing components. Although we often fail to notice it, the ambient noise levels in many terrestrial environments are quite substantial. Because sound is rapidly attenuated in air, only the closest sources of noise are important to a communicating terrestrial animal. However, since many of the sources of noise are ubiquitous, the fact that most noise is locally generated does little to reduce the overall din. The major sources of low-frequency noise in air are wind and air turbulence passing over vegetation, substrate edges, and the head and body of the receiver. Regardless of habitat, the energy in wind-generated noise is greatest for low frequencies. Typical levels for frequencies under 200 Hz range from 20–30 dB (SPL) for winds of about 1 m/sec up to 60–70 dB (SPL) for winds of 8 m/sec. Wind contributes little to noise levels at frequencies above 2 kHz (usually less than 5 dB). As a general rule, there is less wind within forests than over open grasslands. One therefore expects (and finds) lower intensities of low-frequency noise within a forest and higher values over a grassland. Temporally, wind levels are usually lowest in the early morning and increase toward mid-day. Thus levels of ambient low-frequency noise tend to increase in either type of habitat during the course of the morning (Morton 1975; Waser and Waser 1977; Brenowitz 1982a,b; Ryan and Brenowitz 1985). The primary sources of continuous high-frequency noise in terrestrial habitats are chorusing insects. Major contributors include orthopterans and cicadas, which tend to produce signals with frequencies 4 kHz or higher. Measured values of noise in the 6–8 kHz range are typically 40–50 dB in both temperate and tropical forests, and 15–30 dB over grasslands and pastures. The zone of ambient noise frequencies between the bands of wind and insect sounds (e.g., 1–4 kHz) is relatively quiet, with typical intensities of only 10–20 dB (Figure 5.7).

In water, continuous noise in the range of 100 Hz to 10 kHz is largely generated by surf, wind, rain, and waves at the water's surface. Very low frequency noise may be generated by currents passing over rough substrates, and very high frequency noise arises from molecular collisions in this dense medium. Sound attenuates much more slowly in water than in air; consequently, a given noise source can be heard at much greater distances in water. Low frequencies attenuate less rapidly than high frequencies; thus the energy distribution of aquatic noise is heavily weighted toward lower frequencies.

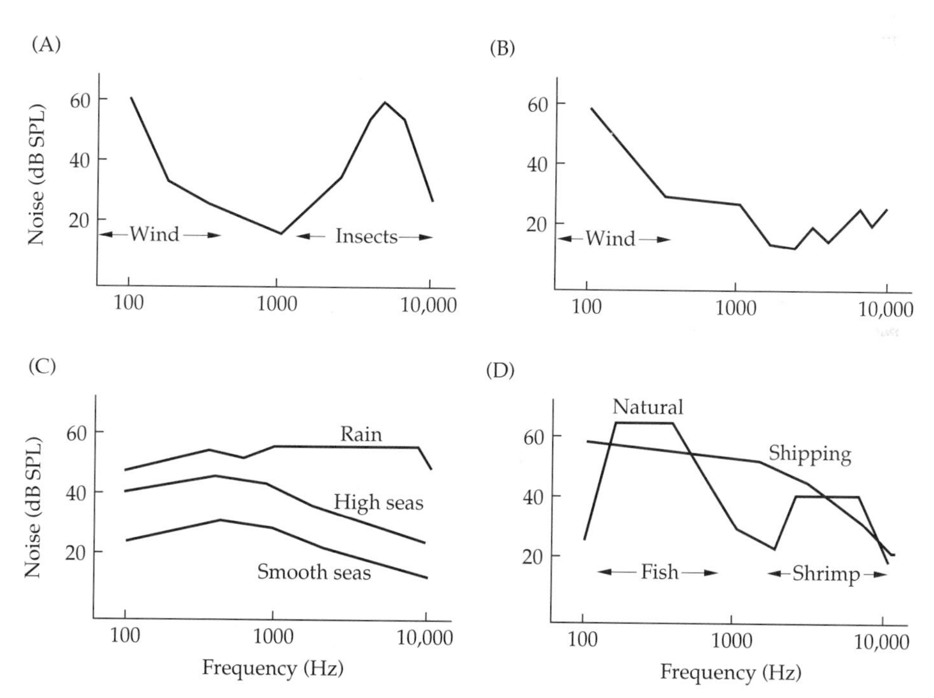

Figure 5.7 Noise levels in terrestrial and marine environments. (A) Ambient noise spectra in a typical forest. Contributions by wind and insects are marked. (B) Ambient noise levels over typical pasture or grassland. (C) Deep ocean ambient noise spectra under conditions of smooth seas, high seas, and heavy rain. (D) Shallow water noise spectra showing natural and man-made sources separately. Contributions by fish and shrimp in natural sources are marked. (A and B after Klump and Curio 1983, Klump and Shalter 1984, Morton 1975, Ryan and Brenowitz 1985, Waser and Waser 1977; C and D after Albers 1960, Rogers and Cox 1988.)

There is a certain directionality of noise sources in the ocean: most high-frequency sounds originate along a vertical line above or below the listener, whereas most low-frequency sounds come from sources at a substantial horizontal distance.

In deep ocean, noise levels for smooth seas are maximally at 20–30 dB (SPL) around 500 Hz and drop off about 5 dB for each doubling of frequency up to 10 kHz (Figure 5.7). Noise in rough seas shows a similar dependence on frequency but is 15–20 dB more intense. Heavy rain adds only 5–10 dB to ambient noise at frequencies below 1 kHz, but increases the content of nearly all higher frequencies to about 50 dB. Noise levels in deep ocean are relatively constant for depths to about 100 m; at greater depths, medium loss and scatter gradually reduce the levels of high-frequency noise.

Shallow waters are on average about 10 dB noisier than deep ocean (Tolstoy and Clay 1966). Although some of this noise is due to rough surf conditions, much of it is generated by organisms living in shallow water. Fish, particularly croakers (Sciaenidae), may produce continuous drumming sounds with levels of 50–60 dB in the range 200–600 Hz. Cod and haddock (Gadidae), gobies (Gobiidae), wrasses (Labridae), and damsel fish (Pomacentridae) are other marine fish that may produce repetitive sounds in this lower frequency range (Myrberg 1981). In tropical waters, large numbers of snapping shrimp (Alpheidae) generate 40–50 dB noise in the 4–10 kHz range. Barnacles, lobsters, mussels, squid, and sea urchins also add to the noise in this band. Although their appearance is too recent to have affected the evolution of animal signals, shipping and other human activities currently are a major source of noise in water. Typical levels in New York harbor range from 60 dB at 200 Hz to over 30 dB at 10 kHz (Albers 1960).

Overall levels of noise thus tend to be similar for both terrestrial and shallow water environments (about 45–55 dB). In both contexts, most of the noise produced by nonbiological sources has its energy in the lower frequencies: in air, this noise is largely due to wind and turbulence, whereas in water it is due to waves, wind, currents, and surf. In both water and air, arthropods (and other invertebrates) tend to produce sounds in the 4–10 kHz range. For vertebrates, this constitutes a source of ambient noise that can mask their own sounds. Where both physical and arthropod sounds are intense, there are two possible solutions for communicating vertebrates: (1) try to use the narrow window of lower noise between the physical and arthropod-generated bands; or (2) increase emitted sound output enough to be more intense than ambient noise at the receiver. As we have seen, there is a relatively quiet window between 1 and 4 kHz in many terrestrial environments, and this may be one reason why many forest bird and mammal vocalizations fall in these frequencies (Klump and Curio 1983; Klump and Shalter 1984; Morton 1975; Ryan and Brenowitz 1985; Waser and Waser 1977). Although there is a similar window in the sea, most fish produce sounds in the 200–900 Hz range where ambient noise levels are very high. One suggested explanation for why fish do not use higher frequencies is that the resonant frequencies of the swim bladders used to produce these sounds cannot be made higher than about 1 kHz. Another is that most fish use sounds only over short ranges so that it is not difficult even in noisy seas to produce nearby sound levels above that of ambient noise.

SOURCES OF DISTORTION IN THE TIME DOMAIN

As with the frequency domain, we can divide the effects of propagation on a signal perceived in the time domain into global attenuation, pattern loss, and the addition of noise. The role of global attenuation in the time domain is identical to that in the frequency domain: for a far field signal viewed in the time domain, each peak in the signal waveform decreases by 6 dB for every doubling of distance between sender and receiver. Similarly, masking by noise depends primarily on the relative amplitudes of the propagating signal and ambient noise levels. We have already summarized the levels of noise in various environments. The processes for pattern loss in the time domain are a bit more complicated. There are three major sources of pattern loss for signals perceived in the time domain: changes in phase and frequency spectra, reverberations, and added modulations.

Pattern Loss from Spectral Changes

One might naively think that if the temporal pattern of a signal is perceived in the time domain, a receiver would not have to worry about the many processes that can change the spectra of signals during propagation. Unfortunately, this expectation is not borne out. Any change in the frequency or phase spectra of a propagating signal will result in a corresponding change in the temporal pattern of the signal's time domain waveform. In prior sections, we have discussed mechanisms that might alter the frequency spectra of propagating signals. Each of these mechanisms will affect the time domain waveforms of the perceived sounds. If animals perceive the signal or parts of it in the time domain, these processes may have a double effect because many of them also alter the relative phases of the component frequencies. As we discussed in Chapter 3, page 48, some simple signals consisting only of a carrier and one pair of side bands may appear in the time domain as frequency modulation, as amplitude modulation, or as a mixture of both, depending only on the relative phases of the three components. Any shift in the relative phases of a propagating signal, even without any change in the relative amplitudes of the components, can thus have a dramatic effect on the signal's time domain waveform.

Whether phase shifts are perceived by a receiver depends very much on how its ears work. Although we shall discuss this relationship in more detail in the next chapter, it is important to recognize that some insects and most vertebrates perform at least a crude Fourier analysis of received sounds. The overall frequency spectrum of the sound is broken up into a set of consecutive bands, each of some bandwidth Δf. Usually, information about differences in phase between the bands is ignored. However, since components within a band have not been separated from each other, the band waveform depends both on the frequency amplitudes and on their relative phases. Most of the processes that affect frequency and phase spectra during propagation are smooth functions; that is, they alter the phases and amplitudes of similar frequencies by similar amounts and in similar directions. It is only disparate frequencies that show large differential shifts in either amplitude or phase.

This means that signals with very slow amplitude modulations, frequency modulations with a low modulation index, or harmonic series with very low fundamental frequencies will have components that are so similar in frequency that they suffer little differential change during propagation and thus have only minor temporal distortion. Similarly, animals whose ears have very small Δf will suffer little distortion due to propagation-induced phase shifts. Differences between different bands will exist, but will be perceived as frequency domain effects. The animals that are most likely to experience major time domain distortion due to spectral shifts are those with very crude frequency analysis (e.g., large Δf) and/or those that are attempting to detect rapidly modulated or pulsed sounds. Clearly one way to avoid temporal distortion is to use very slow modulations for signals that must be propagated over long distances.

Pattern Loss from Reverberations

When sound travels between sender and receiver by multiple paths, the waveforms traveling these different paths add up at the receiver to produce a single compound signal. Echoes from boundaries and scatter from small objects between sender and receiver both contribute to the generation of such multiple paths. Even without phase shifts at reflection, the slight differences in path lengths can result in serious distortions of any temporal patterning in the original signal. Unlike the situation with noise, the production of a louder signal by a sender does not solve the problem of reverberations; it only makes it more likely that there will be serious distortions at the receiver. The only ways to reduce reverberation distortions are to find frequencies that are poorly reflected by the ambient environment, or to emit sounds with a directional horn so as to reduce the number of objects likely to reflect sound to the receiver (Wiley and Richards 1982).

In terrestrial contexts, open grassland animals can essentially ignore the effects of reverberation. The situation is quite different in forest (Richards and Wiley 1980). Sounds arriving at a receiver in the forest are quickly followed by reverberations. Low frequencies (under 1 kHz) reflect off of the canopy and substrate and produce single intense echoes that often overlap the final parts of the sound arriving by a direct path. Higher frequencies (above 3–5 kHz) are scattered by foliage and arrive as many small overlapping echoes. Sound frequencies between 1 and 3 kHz tend to experience the least reverberation in forests. This relationship is convenient for birds since it is also the frequency band within which ambient noise tends to be lowest (see pages 126–129).

Reverberations increase the duration of the propagated waveform and modify its temporal fine structure (Figure 5.8). Because later-arriving echoes have usually traveled farther or experienced multiple reflections, the tail end

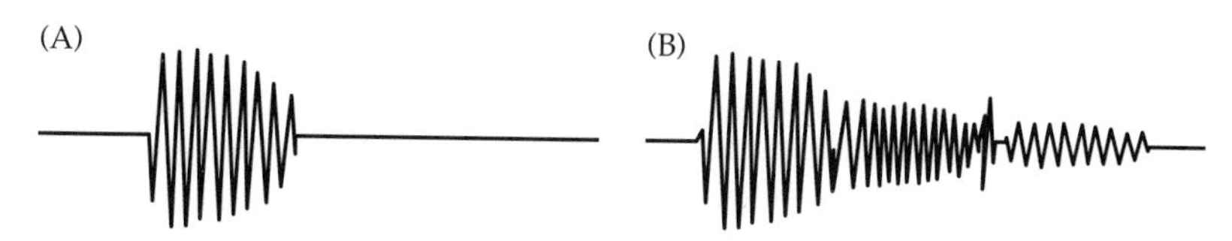

Figure 5.8 Effects of reverberations on a propagating signal. (A) Waveform of a signal emitted by sender. (B) Waveform of a signal at a receiver, consisting of the initial signal that has arrived by a direct path, followed by successively lower amplitude reverberations from scatter and boundary echoes.

of the received waveform will have an envelope of decreasing amplitude. Wiley and Richards (1982) observed that by 100 msec after initial arrival of the direct path signal, reverberation components in forest conditions had decayed to less than half of the amplitude of the direct path component. For a bird producing a signal with patterned modulations, minimal loss of temporal pattern would occur when the delay between successive modulation peaks was at least as long or longer than the time taken for a first peak plus its reverberations to die out. Given Wiley and Richard's values, the maximum modulation frequency for AM calls in a forest should thus be about 1 cycle/100 msec or 10 Hz.

Reverberations are even more likely to distort temporal waveforms in water than in air because the greater speed of sound in water makes temporal overlap between direct and reverberated components even more likely. Reverberations in water are generated as boundary echoes from above (the water's surface) and below (the substrate), and as scatter from small objects in the medium between surface and substrate (usually marine organisms). Many areas of the ocean contain large aggregations of sound-scattering organisms that live at a specific depth. These are called **deep scattering layers**, and may occur at depths ranging from 20–2000 meters. Typical contributors to oceanic scattering layers are euphasid shrimp, siphonophores, squid, lantern fish, and copepods. In shallow waters, any large school of fish with swim bladders may constitute a sound scattering mass. Densities of organisms as low as $0.05/m^3$ may create appreciable sound scattering layers (Caruthers 1977).

Reverberation amplitudes in shallow water are greatest for sound reflected from the bottom (about 25 dB below direct path signals), next greatest for surface reflections (about 40 dB below direct signals), and least for organismal scattering layers (typically 80 dB below direct signals). Reverberation amplitudes from organismal layers are highly frequency-dependent over the range 1–20 kHz. Echoes from the substrate tend to be relatively frequency-independent up to about 10 kHz; for higher frequencies, substrate details greatly affect the intensity of reverberations.

Reverberation must be a significant problem for any fish communicating by sound over a distance. Since most species produce sounds in a frequency range at which noise is considerable, amplitudes must be high to be detectable over ambient noise. This high amplitude plus the low directionality of the swim bladder (essentially a monopole) must generate high levels of reverberation at receivers. Despite this, many fish appear to rely heavily on time domain patterning of their sounds (Hawkins and Myrberg 1983). One explanation, again, is that most fish communicate only at short range thereby reducing the problems of reverberative distortion.

Pattern Loss from Added Modulations

The final source of pattern distortion in the time domain is the addition of amplitude modulations to a signal during propagation. Remember that a sound 200 msec long in air is spread out over roughly 68 m as it travels from sender to receiver. In water, such a sound would be over 300 m in length. The chances are high that as this sound passes a given location there will be some change in the density or composition of the medium, in the pattern of refractive layering, in the disposition of scattering objects, in the acoustic impedance of a nearby boundary, or in the direction and level of currents in the medium. Any of these changes can change the amplitude of one part of the signal without changing the amplitude at other parts. This alteration generates amplitude modulations in the received signal that were not present in the sound emitted by the sender.

In terrestrial environments, added modulations are very common. Except for very still periods, wind alone can modulate the amplitude of a propagating signal by at least 35 dB. Added modulations are least likely in early morning and in the interior of a forest. In contrast, open grasslands are much more susceptible to wind and thus more likely to exhibit added modulations. In addition, heat absorption by unforested land as the day proceeds generates turbulence, rising cells of hot air, and shifting gradients of air density. The importance of added modulations becomes greater as the frequency of the propagating sound increases and as the distance the sound must propagate is increased. Measurements in air show that most added modulations occur at frequencies of only a few Hz and are likely to be important to animals only up to about 20 Hz (Richards and Wiley 1980). For the majority of animals, these modulation frequencies will be less than their Δf and thus will be perceived in the time domain. To avoid masking of temporal patterns in the original signals, senders in open environments should use amplitude modulation rates greater than 10–20 Hz. Note that this is an interesting contrast to the case of animals communicating in forests, where amplitude modulation rates below 10 Hz are favored to avoid reverberations. In fact, a survey by Richards and Wiley (1980) showed that birds in forests were significantly more likely to sing songs with amplitude modulation frequencies below 10 Hz, whereas birds of open grassland were more likely to use modulation frequencies above this value.

In water, similar problems apply but to a lesser degree. Currents are less of a problem because the velocity of ocean currents is so much less than the velocity of sound. Sunlight has a much smaller effect except in very shallow water and in general, densities and sound velocities have much smaller levels of horizontal variation. Variations in scattering as schooling organisms move can be a significant modulator of propagating sounds, especially given the long distance over which a propagating sound is distributed. Whether water is better or worse than air with respect to added modulations clearly depends on the contexts.

DESIGN OF SIGNALS TO REDUCE DISTORTION

For animals communicating over short distances, most of the distortions described in this chapter are not significant. However, animals use sounds for many social functions that require communication over longer distances. What would be the preferred design for such a signal? Consider first a terrestrial species. To counter the effects of global attenuation, the sender could increase the amplitude of the emitted sound and it could attempt to emit the sound at times and from locations at which sound propagation is maximal. Temperature and wind conditions greatly affect refractive modifications of sound transmission; finding and using a sound channel, as opposed to suffering a sound shadow, will increase range, but may also result in pattern distortions. Early morning calling in open habitats is usually preferable to mid-day emission. Under these conditions wind intensities are less and air temperature profiles should not lead to strong sound shadows. The height at which a sound is emitted clearly has a major effect on its propagation: in general, greater heights are better than being close to the ground. However, remaining between the canopy of a forest and the ground during the day often produces sound channel characteristics, whereas moving above the canopy layer may lead to decreased sound propagation.

What is the most effective frequency composition of a long range sound signal? Frequency-rich sounds will suffer greater overall distortion than carefully selected bands of frequencies since the former span regions of the frequency spectrum that are affected differentially during propagation. High-frequency components will be attenuated at higher rates due to medium loss and scatter. Terrestrial callers close to the ground must avoid the notch, and those actually on the ground will be restricted to low frequencies only. Low-frequency sounds will also be less likely to pick up added modulations in open habitats. These factors all favor using narrow bands of low frequency when near the ground. However, there are other factors to be considered. As discussed in Chapter 4, it is difficult for animals to produce signals and couple them to the medium when the wavelengths of the sounds are greater than 2–5 times the body size of the sender. The larger the wavelengths relative to the animal, the lower the intensity of the emitted sound. The tradeoff between being able to produce intense sounds and avoiding losses due to attenuation and distortion leads to an intermediate range of optimum frequencies (Figure 5.9). (The general problem of **signal optimization** is taken up in Part II of this book.) For the majority of frogs, small mammals, and song birds that range in size from several inches to a foot in body size, this optimum appears to fall in the range 1–6 kHz. Conveniently for these vertebrates, this is also the range of frequencies for which ambient noise is least. Animals whose major activity is close to the ground will tend to use somewhat lower frequencies (Wiley and

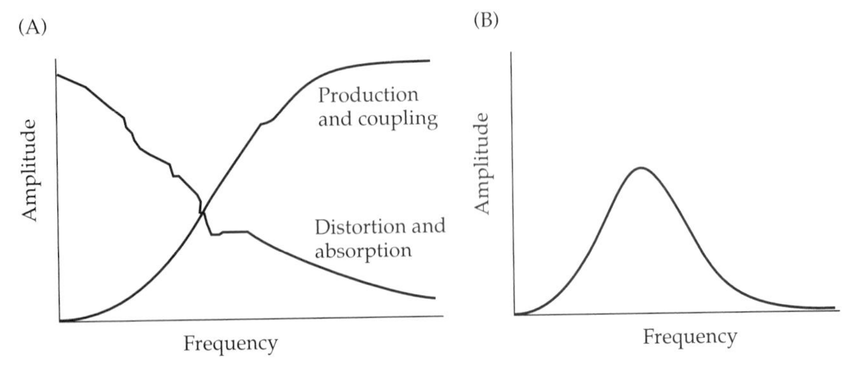

Figure 5.9 Maximization of sound signal propagation. (A) Typical curves showing frequency dependence of sound intensity given production processes versus distortion and absorption processes. (B) Composite effects of the production and distortion curves relating sound intensity to frequency. The peak in the graph occurs at the optimal frequency for communication.

Richards 1982) or to move to greater heights when producing long distance calls. For example, Lawes' Bird of Paradise (*Parotia lawesii*) builds a small court on the ground where he displays his elaborate plumage silently to visiting females. Males attract females primarily by emitting loud calls in the forest subcanopy and dashing back down to their courts as females approach (Pruett-Jones and Pruett-Jones 1990).

The small size of most insects causes their optimal frequencies to be higher, and some insect species even use ultrasonic frequencies. Very high frequency sounds have the advantage that they are less likely to be detectable by avian predators, and there will be less overlapping noise. However, the greater medium loss and scatter at these higher frequencies will certainly reduce the range of detection when compared to vertebrates. Michelsen and Nocke (1974) estimate that the maximal range of detection for a katydid (Tettigoniidae) stridulating at 50 kHz will be only 10–20 m. It is of interest that katydids that live at high densities and are thus closer together tend to use frequencies of 20–40 kHz, whereas those dispersed at greater individual distances generally rely on frequencies of 8–15 kHz (Dubrovin and Zhantiev 1970). Because the acoustic impedance of soil or vegetation is not too dissimilar to that of a stridulating insect, it is sometimes easier for such an animal to couple sounds to its substrate than to the surrounding air. The greater coupling efficiency allows these insects to use lower frequencies than would be required to produce the same intensity signal in air (Markl 1968). However, because sounds attenuate extremely rapidly in substrates such as sand or soil, the range at which such signals can be detected is usually very short (on the order of 2–10 cm). Evidence that plant-transmitted vibrations are used for communication has been obtained in species of stone flies (Plecoptera), true flies (Diptera), leafhoppers and planthoppers (Homoptera), and cydnid bugs (Hemiptera) (Chvála et al. 1974; Ichikawa 1976; Gogala et al. 1974; Michelsen et al. 1982; Rupprecht 1968). Again, the better coupling between insect and substrate allows senders to produce much lower frequency sounds than they would for airborne signals. Many insects have special receptors in their legs that may facilitate the reception of substrate-borne sounds.

If most of the animals in a given habitat elect to use the same optimal band of frequencies, species identity will depend heavily on the temporal patterning added to the signals. The optimal temporal pattern appears to be habitat specific (Richards and Wiley 1980; Wiley and Richards 1982; Wiley 1983, 1991). In species for which some sort of Fourier analysis by receivers is possible, slow modulations tend to preserve the pattern of a propagating signal better than rapid modulations do. Slow modulations are also favored for animals living in forests where reverberations are a problem. As we discussed in Chapter 4, pages 84–85, animals that use resonance to enhance signal intensity are limited to using slow modulation rates. Hence loud but slowly modulated signals are compatible goals. Animals living in open habitats, particularly diurnal ones, experience problems of added modulations. Such species should modulate their signals at rates higher than the environmentally added modulations. Producing loud but rapidly amplitude-modulated signals is quite expensive physiologically. As a result, open country species often use FM instead of AM temporal patterning because frequency modulation appears to be more resistant to distortion during propagation (Figure 5.10). Whenever AM or FM patterning is used, wide bandwidth carriers are more likely to preserve the pattern over long distances. All that is required is that some carrier frequencies and their sidebands arrive at the receiver.

For aquatic forms in shallow waters, the effects of global attenuation can be reduced by producing louder sounds and by seeking out effective sound channels. These channels are common in aquatic habitats and can propagate signals much farther than analogous channels in terrestrial contexts. They do, however, distort the spectral and temporal pattern. In water, the optimal frequencies are high frequencies, in contrast to the low frequencies that are optimal in terrestrial conditions. Sounds emitted near the water's surface experience strong negative interference unless the emitted frequency is sufficiently high. Standing wave interactions in shallow water also lead to a minimal frequency below which no sounds are propagated. If there is a sound channel, sounds will only remain within it if they are above a minimum frequency set by the vertical thickness of the channel. Although higher frequencies attenuate faster than low ones in water as well as in air, the rates of medium loss and scatter are much lower for a given high frequency in water than for equivalent conditions on land. Ambient noise levels tend to be highest at low frequencies except in tropical areas where there are large numbers of snapping shrimp. Ideally, then, communicating aquatic animals should use relatively high frequencies.

Many marine arthropods fit expectations and use frequencies similar to those of their terrestrial counterparts. Fish, however, tend to use much lower frequencies than one would have expected based on the physics of sound propagation in water. Given the high levels of reverberation in shallow

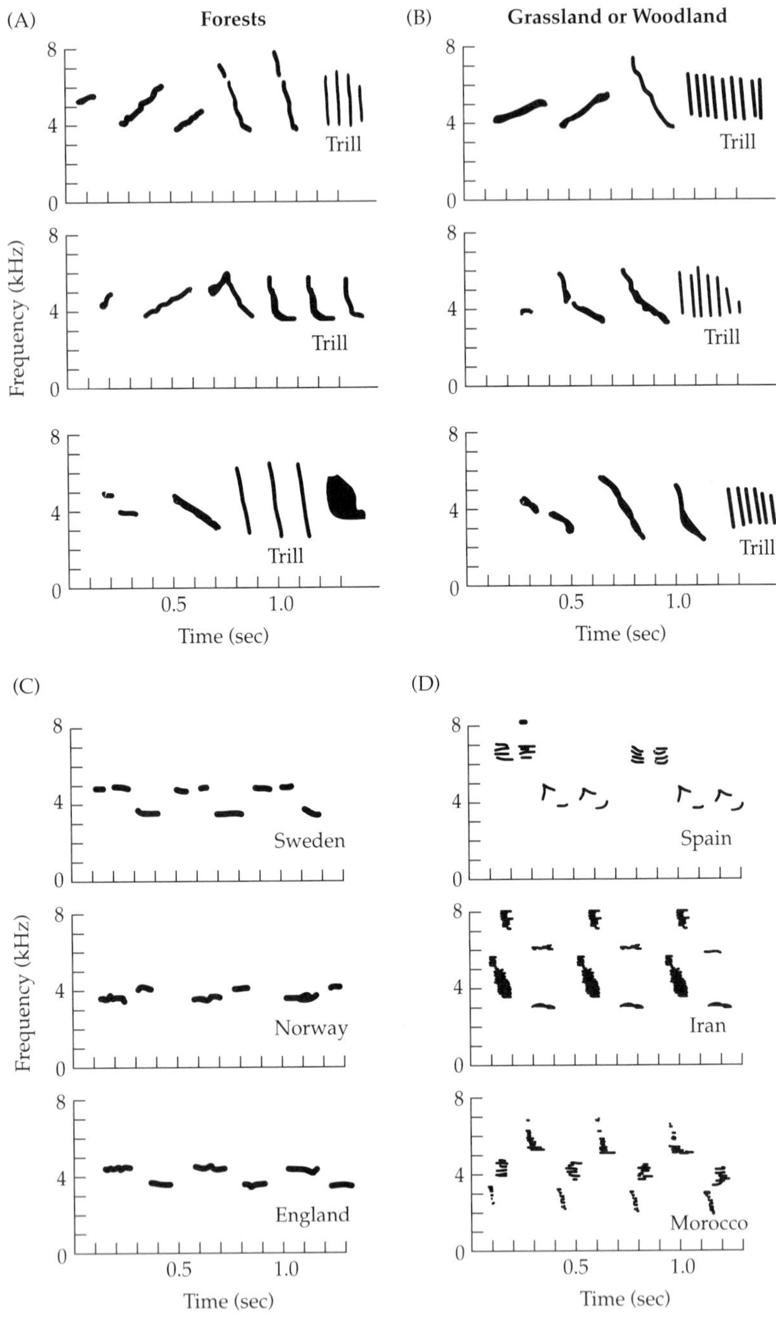

Figure 5.10 Modulation rates in forest versus open-country populations of birds. Spectrograms of male songs from (A) forest and (B) grassland populations of the chingolo sparrow (*Zonotrichia capensis*). Note slower trill rates in forest species where reverberations would distort rapid amplitude modulation. Spectrograms of songs from (C) dense forest and (D) open woodland populations of great tits (*Parus major*). Note the presence of rapid frequency modulations in open sites, but slow or no modulations in dense forest locations. (From Krebs and Davies 1993.)

waters, one would also have expected most fish to use very slow temporal modulations. In fact, they often use relatively rapid ones. It remains to be seen whether these anomalies are explained by fish having a focus on short-range communication.

Marine mammals are the aquatic group most likely to use longer-range sounds for social communication. The smaller toothed whales and porpoises produce a wide variety of social (as opposed to echolocation) sounds, including frequency modulated whistles in the 1–15 kHz range and long trains of pulses with much of their energy in the low end of this range. Many seals and sea lions also produce sounds in the 1–15 kHz range. Since most of these species live in shallower seas, this use of higher frequencies is in accord with our expectations from physical considerations. However, we must accept this fit with caution. First, the toothed cetaceans may be pre-adapted to use higher frequencies because they also echolocate with sounds of 20 kHz or higher. Second, few species have been sufficiently studied to know which sounds are used over long distances. Baleen whales (Mysticeti), in contrast, tend to produce loud low frequency sounds (20 Hz–4 kHz). These too are often frequency modulated. Since many of these species live in the deep oceans where the factors selecting against low frequencies are relaxed, this again appears to fit with expectations. The widespread use of slow frequency modulation in social signals by baleen cetaceans may well be a mechanism for reducing distortion due to reverberations.

SUMMARY

1. Depending upon the bandwidth of the receiver's ear, distortions added to a sound signal during propagation may be perceived in either the frequency or in the time domain.
2. Signals perceived in either the time or frequency domain suffer **spreading losses** as they propagate away from the sender. At a sufficient distance from the sender, the signal intensity becomes similar enough to ambient **noise** for the latter to distort and mask the signal.
3. In the frequency domain, signal propagation leads to a disproportionate loss of high-frequency energy due to **medium absorption** and **scatter**. Rates of medium loss and scatter are generally higher in air than in water, and are most severe for substrate propagated sounds.
4. When a sender and receiver are close to a **boundary** (the ground, the canopy in a forest, or the surface of the water), the existence of multiple paths for sound transmission between sender and receiver can severely modify sound propagation. When terrestrial senders and receivers are both on the ground, low frequencies are the only sounds that can propagate any distance, and they do so as **boundary waves**. When both or either are near but not on the ground, frequencies on either side of a band called the **notch** (300–800 Hz) propagate better than frequencies within the notch. Frequencies higher than the notch show some alternation of accentuation and attenuation that damps out at values greater than 4–5 kHz. Aquatic organisms near the surface suffer very severe attenuation of any

frequency less than a critical value. In shallow water, standing waves set up by reflections from the surface and the bottom prohibit propagation of any frequencies less than a critical value no matter what the depth of the sender and receiver. Sounds propagated within plants also show strong frequency dependence due to standing waves.

5. Gradients in the temperature and density of the medium or in the speed or direction of winds or currents can cause **refraction** of propagating sounds. Emitting sounds downwind or in a medium in which sound velocity increases with increasing height (air) or depth (water) favors sound propagation and can create a **sound channel**. The reverse conditions increase sound attenuation during propagation and can create a **sound shadow**. Although sound channels help ameliorate global attenuation, they can cause serious distortion to the sound pattern.

6. In both aquatic and terrestrial environments, noise distributions tend to be bimodally distributed over frequency, with a relatively quiet band in the range of 1–4 kHz. Lower-frequency noise arises from winds and currents; higher-frequency noise is primarily generated by arthropods. Overall noise levels due to nonhuman sources are similar for the two types of environment: 45–55 dB (SPL).

7. Distortion in the time domain waveforms of propagating signals arises from changes in frequency and phase spectra due to medium loss, scatter, boundary reflections, refraction, and ambient noise. It also arises from the arrival at a receiver of a directly propagated component and **reverberations** (echoes and scatter) from objects along the path. Reverberation decay times set an upper limit on modulation frequencies that can be used in signals and that can still be detected by receivers. Reverberation is most severe in forests and in aquatic situations. In open habitats, reverberation is minimal, but air turbulence and vegetational movements result in the addition of low-frequency amplitude modulations to propagating signals. The risk of **added modulations** sets a lower limit on the rate of modulation that should be used in these contexts.

8. Sound signals optimized for long-range propagation in air should be as low in frequency as the sender can efficiently produce. Animals close to the ground should avoid the notch. Since the amplitude of emitted signals decreases rapidly when the wavelengths of the sound become larger than the animal producing them, optimal frequencies depend upon body size of the sender and the rate at which higher frequencies attenuate more rapidly than low frequency ones. Ambient noise also affects the optimal choice. For many small vertebrates, optimal frequencies appear to be about 1–4 kHz; for insects, optimal frequencies will be somewhat higher and corresponding ranges of detection lower. Higher frequencies are also favored for aquatic animals in shallow waters.

FURTHER READING

More detailed treatments of the physics of scattering, sound absorption, and noise in air can be found in Aylor (1971), Evans et al. (1971), Griffin (1971), and Piercy et al. (1977). The complex interactions of sounds propagating in air over boundaries are analyzed in Chessell (1977), Daigle (1979), Donato (1976), Embleton et al. (1974, 1976), and Thomasson (1977). The biological consequences of acoustic propagation effects in air are outlined in Brenowitz (1986), Chappuis (1971), Klump and Curio (1983), Klump and Shalter (1984), Marten and Marler (1977), Marten et al. (1977), Michelsen (1978), Richards and Wiley (1980), Roberts et al. (1979), Ryan and Brenowitz (1985), Waser and Waser (1977), Wiley (1991), and Wiley and Richards (1978, 1982). Good sources for sound propagation effects in water are Albers (1960), Caruthers (1977), Clay and Medwin (1977), Tolstoy and Clay (1966), and Urick (1967). The biological implications for underwater communication are discussed in Hawkins and Myrberg (1983), Myberg (1978, 1980, 1981), Payne and Webb (1971), and Rogers and Cox (1988). Propagation of sound in plant or terrestrial substrates is discussed by Bell (1980), Gogala et al. (1974), Ichikawa (1976), Markl (1968, 1983), Markl and Hölldobler (1978), and Michelsen et al. (1982, 1986).

Chapter 6

Sound Reception

SOUND RECEPTION IS THE OPPOSITE PROCESS from sound production: sound vibrations propagating in the medium must first be coupled to the organism, modified as necessary, and then converted into useful nerve signals. Ideally, a perfect sound receiver would have a very wide range of detectable frequencies (**frequency range**), a very wide range of detectable amplitudes (**dynamic range**), very fine frequency resolution (Δf), very fine temporal resolution (Δt), very accurate measurement of signal amplitudes, very accurate determination of the angles of a sound source in the horizontal and vertical planes, very accurate determination of the distance to the source, and the ability to monitor very rapid changes in the time domain waveform of signals (pattern recognition).

Not surprisingly, it is impossible to achieve ideal conditions for all features in a single sound receiver. Evolutionary investment in one feature invariably curtails perfection in another. We have already discussed the physical tradeoff that must exist between having a low Δf and a low Δt. As we discuss later, there are other serious tradeoffs that exist between the various features. In addition, the environmental conditions under which organisms live may impose severe constraints on the perfection of specific features. Some of these constraints may be direct limits on the ways in which ears work; others may arise because the signals that are most easily produced or propagated in a given medium may not be the ones most easily detected and analyzed. Clearly, some tradeoffs between sound production, propagation, and reception will exist. In this chapter, we review the more common mechanisms for coupling sounds to organisms and converting them into neural signals. We then review the major taxa that use acoustic communication and contrast their success in designing sound receivers.

COUPLING OF PROPAGATING SOUNDS TO ORGANISMS

Ideal couplers would transfer a large fraction of incident sound energy in the propagating medium to the detector organs of the receiver organism while preserving the amplitude, phase, frequency, and directional information contained in the original signal. There are basically three kinds of couplers found in animals. Each of these varies in its ability to meet these requirements.

Particle Detectors

Particle detectors are long thin organs that move when barraged by many molecules all moving in the same direction. As we have seen, such mass movements of particles due to sound propagation occur only in the **near field**. Sensory cells attached to the base of such organs are alternately stretched and compressed as the oscillating sound field forces the organ back and forth. Most arthropods have a variety of hairs on their bodies that provide chemosensory and tactile information. In addition, many also have very fine hairs (called **trichobothria** in spiders and **trichoid sensilla** in insects); these hairs are particle detectors sensitive to air or water currents and to near field sounds (Figure 6.1). Some caterpillars use these hairs to detect the wing-induced air currents of approaching predatory wasps (Tautz and Markl 1978). Crickets and cockroaches have long cerci at the tips of their abdomens that are covered with many fine hairs. Variation in the lengths of these hairs is correlated with the particular near field sound frequencies to which each hair is sensitive (Petrovskaya 1969).

In other insects, thin antennae or mouth parts are used as particle detectors. The courtship dance of male fruit flies produces vibrations at about 200 Hz. Females within 0.5 cm of a dancing male can detect and monitor the near field signals with a fine segment of each antenna called the **arista**. The plumose antennae of male mosquitoes and midges are used to detect the species-specific wing beat frequencies of females. Water boatmen have a club

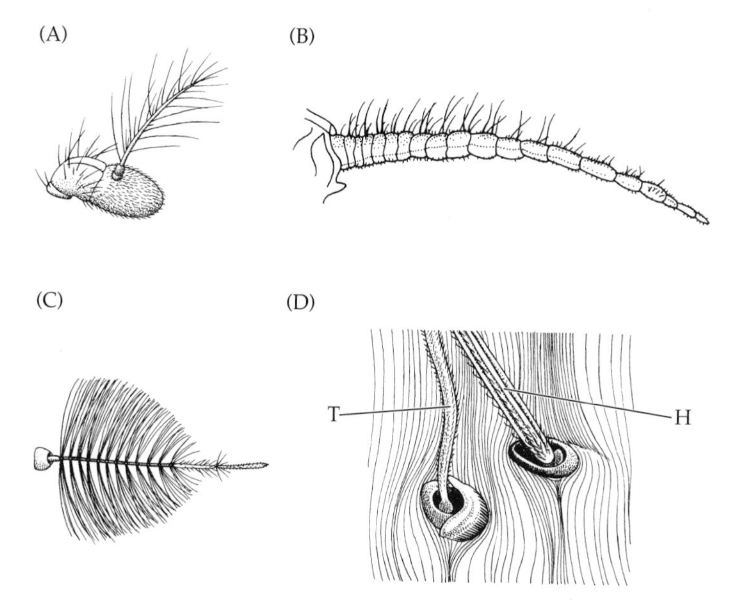

Figure 6.1 Particle detectors in arthropods. (A) Antennal arista of a fruit fly. (B) Trichoid sensilla on a cockroach cercus. (C) Antenna of a male mosquito. (D) Trichobothrium of a spider, T, and a larger sensory hair, H. (A after Burnet et al. 1971; B after Autumn 1942; C after Hutchings and Lewis 1983; D after Foelix 1982.)

that extends from their ear and responds to near field vibrations generated in the water by other individuals. Note that aquatic arthropods are more often within near fields than are terrestrial forms because of the longer wavelengths for any given frequency in water.

The frequency response of a particle detector depends upon its ability to oscillate in phase with the particles of the surrounding medium. Massive hairs will be unable to keep up with high frequencies or large amplitudes, and even the smallest particle detectors appear to be limited to frequencies well below 1 kHz. Insects such as male mosquitoes and female fruit flies use this constraint by adjusting the mass, hinging, and shape of their antennae to limit their responsiveness to the species-specific frequency band. Where they have been studied, the nerves arising from antennal particle receptors appear able to track the waveforms of conspecific sounds well as long as very high frequencies are not present. Thus they can provide considerable information to higher brain centers with only minimal distortion.

Long hairs and antennae that are attached at the cuticle of the animal provide a significant mechanical advantage to the sensory cells: a very small force distributed over the long axis of the hair or antenna is transmitted to the sensory cells as a very small displacement but one of great force. Thus very weak sound fields can be detected by such receivers. The enormous mechanical advantage of the mosquito antenna makes it one of the most sensitive auditory receptors known among the arthropods and similar in sensitivity to the far field response of the human ear (0 dB SPL).

In water, the much smaller displacement of molecules for a given sound energy makes it challenging to produce a particle receptor response. In addition, because most animal tissues have similar acoustic impedances to that of water, near field displacements cause most of the molecules in an organism to move in concert with the sound. When the tips and bases of particle detectors move together, there can be no flexing of the detector and thus no stimulation of the sensory cells. To get around this limitation, some organisms whose detectors receive vibrations through water attach the tips of sensory hairs to objects that have quite different masses and acoustic impedances from the surrounding tissues. The objects then vibrate either with a lag or with a decreased amplitude when compared to the base tissues of the hairs. The greater the differential in movement between the heavier objects and the surrounding tissues, the greater the stimulation of the sensory cells. Since the objects will have their own acoustic properties, such objects may limit the response of such a system to specific frequencies. Where heavy objects such as calcium carbonate masses are used, the response will usually be limited to lower frequencies.

Particle detectors are inherently directional (Figure 6.2). When the axis of particle movement is parallel to a sensory hair, the hair will not move; when the axis of particle movement is perpendicular to a hair, it will experience the greatest force upon it. Angles of incidence between 0° and 90° will create intermediate forces on the hair depending upon the value of the perpendicular component. Thus the amplitude of stimulation from a single hair could be used as an index of the direction of the signal source. However, there are several problems with using particle detectors to determine source direction. One is that the animal cannot know from a single hair whether variations in hair displacement are due to differing angles of incidence or to different amplitudes of the signal at the source: signal amplitude and signal direction are confounded. A second is that near fields that are dipolar or of higher-order geome-

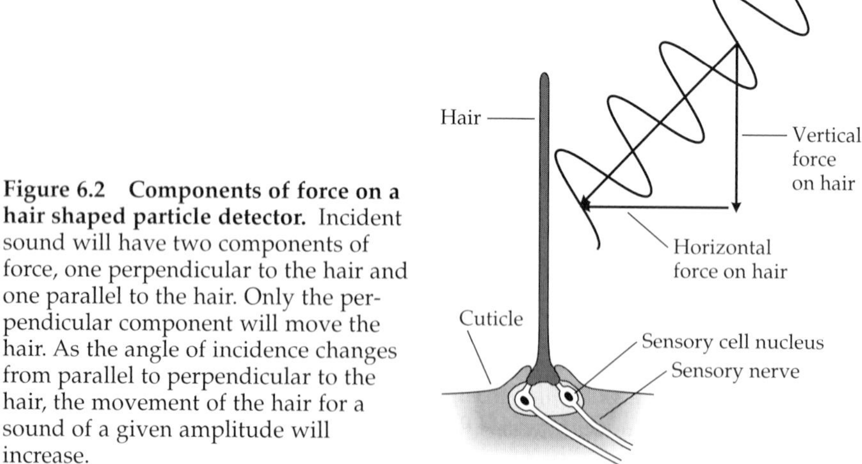

Figure 6.2 Components of force on a hair shaped particle detector. Incident sound will have two components of force, one perpendicular to the hair and one parallel to the hair. Only the perpendicular component will move the hair. As the angle of incidence changes from parallel to perpendicular to the hair, the movement of the hair for a sound of a given amplitude will increase.

try, certainly the most common source of sounds, do not necessarily produce particle movements whose axis points at the source. Even if an organism can uncouple source amplitude and directional information, it must somehow map the sound field's geometry before extrapolating the location of the source.

A common method of decoupling source angle and signal amplitude is to have multiple particle detectors. Many hairs can be spread over the body, and two or more antennae can be placed on the head. Particle detectors can be hinged or curved so that they move only along a single axis with different detectors aligned along different axes. Since the alignment and body curvature is given, a reasonable brain ought to be able to combine the information from such an array of detectors to provide separate estimates of source signal amplitude and source location.

Particle detectors thus meet some of our ideal conditions and not others. They are primarily limited to lower-frequency signals and have dynamic ranges and temporal tracking abilities limited by their mass and inertia. Within these limited frequency and amplitude ranges, they track slowly varying waveforms well and can thus provide relatively undistorted signals to sensory cells. Particle detectors are inherently directional and thus can provide information about a signal source given independent measures of signal amplitude. Their use is limited to near fields.

Pressure Detectors

Nearly all terrestrial organisms that do not utilize particle receptors for sound coupling rely on some sort of membrane (often called a **tympanum**). When the pressures on the two sides of a tympanum are unequal, the tympanum is bent away from the side of higher pressure. This bending of the tympanum can then be coupled to sensory cells to produce neural impulses. A **pressure detector** consists of a tympanum stretched over a closed cavity (Figure 6.3).

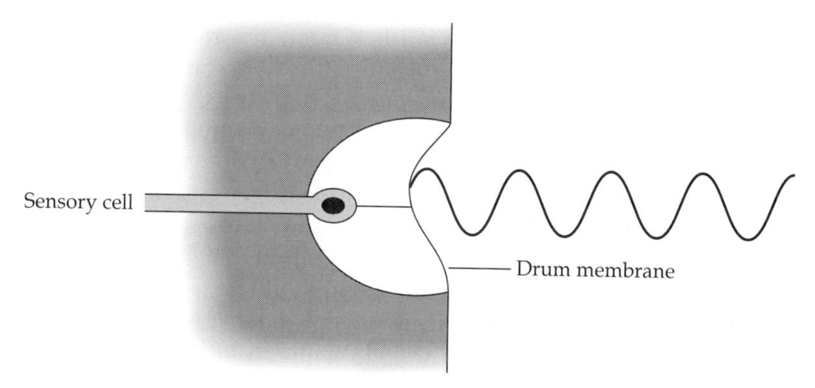

Figure 6.3 Pressure detector. A thin membrane is stretched over a closed cavity. When the incident sound pressure is greater than that inside the cavity, the membrane (called a tympanum) is forced into the cavity. When the external pressure is lower than that inside the cavity, the tympanum is forced out. Movement of the tympanum is monitored by sensory cells.

There must be no effective opening between the cavity and the outside world and the walls of the cavity must be reflective at the sound frequencies of interest. When the typanum is at rest, the pressures of the medium inside and outside the cavity are similar. When a propagating sound arrives at the tympanum, the rise in pressure outside the tympanum bends it into the cavity until the compression of the enclosed medium results in an equal but opposite pressure. Similarly, when the pressure outside of the tympanum is lower than that inside, the tympanum is bent outward until the increased cavity volume results in a compensatory lowering of cavity pressure equal to that outside. Sensory cells attached to the tympanum can then record the movements of the tympanum into and out of the cavity.

Pressure detectors are used in far fields. Because terrestrial organisms consist of water and solids, they must cope with the fact that much of the sound energy carried in air will be reflected at their ears. Tympana have several advantages as terrestrial sound couplers. They can be made very thin, so that their acoustic impedance is closer to that of air. Because the force, F, on a pressure detector tympanum equals the product of the sound pressure, P, and the membrane area, A, large thin membranes are very efficient ways to trap incident sound energy. The thinness of the membrane and the strong restorative force created by compression of the closed cavity volume also help pressure detectors to track changes in incident waveforms rapidly.

Like a string, a stretched tympanum has a number of resonant modes at which it responds strongly to incident sounds. The frequency of the lowest mode increases with the thickness of and tension on the tympanum, and decreases with the diameter of the tympanum. Unlike a string, the frequencies of successive modes are not harmonics of the lowest mode, but tend to be spaced at intervals much less than the lowest mode's frequency. Thus a very thin but large-diameter tympanum may respond to a much greater variety of frequencies within a given range than an equivalently sized string. This range makes a tympanum well-suited to respond to a wide variety of incident sound frequencies. In some organisms, however, this breadth of frequency response is undesirable. Such species have modified membrane thickness and use complex membrane shapes to constrain the resonant modes to preferred values.

Sound pressure has no inherent direction. This means that a single pressure detector lacks any intrinsic mechanism for determining the location of the sound source. One solution is to add a directional structure between the pressure detector and the ambient medium. Nearly all terrestrial mammals have a directional **pinna** that connects to the tympanum via a funnel-shaped tube (called an **auditory meatus**). Such a pinna can provide directionality only for wavelengths that are similar to or smaller than the size of its external opening. Thus the utility of a pinna is limited to higher frequencies. The simplest use of a directional pinna is to aim it in various directions and search for a maximum in the received signal amplitude. Many mammals have elaborated the structure of pinnae with numerous fine folds, ridges, and protrusions (see Figure 6.13). Each of these ridges or folds reflects sounds of a particular wavelength, or shorter. The relative phases of the separate reflections at the tympanum, for any given wavelength, will vary depending upon the vertical and horizontal angle of the source. Thus the amplitude of any given frequency component at the tympanum will vary as a function of source location. By comparing the frequency spectrum received at one tympanum with that expected for a given type of source, the animal can often reconstruct the vertical and horizontal angle of the sender. The problem with this method is that frequency spectra of sounds are easily altered during propagation and thus may never be similar to that expected. This makes it difficult to separate the effects of location and distortion when only a single ear is used. If, however, the animal uses two ears, each with a pinna, similarities in the spectra at the two ears can be used to determine the likely spectrum of the incident sound, and differences between the ears can be used to extract information about the sound source's location. Note that this type of localization mechanism requires an inner ear that can perform at least a crude Fourier analysis of the received sounds.

The use of a pair of pressure detectors also provides a number of other directional cues. If the incident sound has a sharp onset or ending, or has some other portion clearly defined by an amplitude or frequency modulation, the difference in time of arrival of that marker at the two ears can be used to estimate the location of the source in at least one plane. In the simplest case, an animal with two ears can rotate its head until the time delay is minimal. The sound source will then either be directly in front of it or directly behind it. Rotating a quarter-turn more will then resolve this final ambiguity. The problem with this method is that the sound must persist long enough to perform the head rotations. An animal could react faster by computing the angle of a sound source using the delay in arrival of a single sound at the two ears (Figure 6.4). To be useful, the time delay must be long enough to be resolvable by normal nervous systems. In humans, the maximal delay of arrival at the two ears (given a typical head diameter) is about 0.5 msec. For a small terrestrial animal with only 1 cm separating its two ears, the delay is at most 0.03 msec and for a similarly sized animal in water the delay is 22% of the terrestrial value. Although some animals can resolve very fine time delays (owls are accurate down to 0.006 msec), this detection usually takes a very large number of brain cells. Because only larger animals can afford to dedicate so many brain cells to time-delay resolution, and because larger animals will be able to separate their ears more and create longer maximum time delays, this mechanism of sound source localization is most common in larger species.

A second method for determining source location using two pressure dectectors is to compare the relative phases at the two ears. Unless an animal is facing directly toward or directly away from a sound source, the phase of an incident sound differs at the animal's two ears. Again, the animal can rotate the head until such differences are nulled out. Maximal phase differences will occur when the wavelength of the incident sound is twice as long as the distance between the two ears. At much lower frequencies, the waves are so long relative to the distance between the ears that there will be little difference in perceived pressures at the two ears. At much higher frequencies, there will be

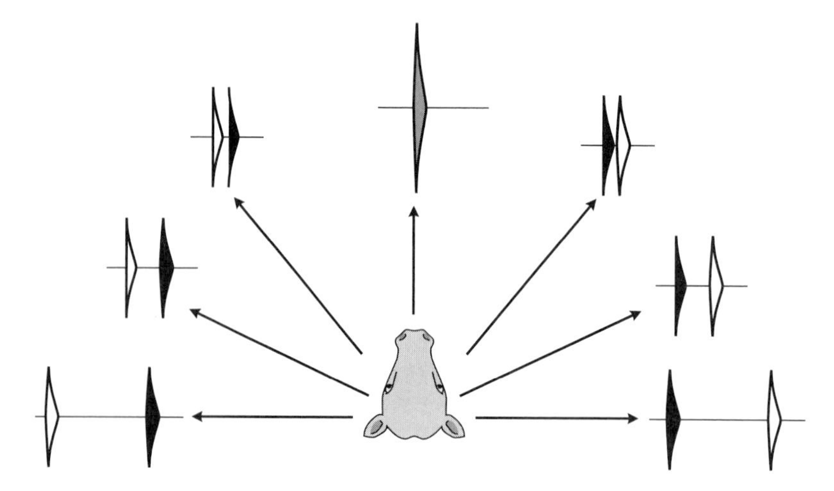

Figure 6.4 Determination of sound source using two ears and time delay. The time domain waveforms of signals from two ears are shown as perceived by a center in the brain that receives signals from both. The white waveform shows a signal from the left ear and the black waveform shows a signal from the right ear. Arrows show the angle of source in the horizontal plane of the head. When the sound source is along a line perpendicular to the line joining the two ears and midway between them, the sound arrives at the two ears simultaneously. When the sound is opposite one ear and on a line connecting the two ears, maximum delay is achieved. If the distance between the ears is known, the horizontal angle to the source can be computed from the measured delay. Ambiguity concerning whether the source is in front of or behind the animal can be resolved by rotating the head.

many different positions at which phase differences will null out. Thus a terrestrial animal with ears separated by 1 cm is likely to use phase information only if it typically responds to sounds of 12–22 kHz. An animal with ears separated by 10 cm will be able to use phase information for sound frequencies down to a few kHz.

Finally, when a far field sound encounters an object at least as large as one-tenth of a typical wavelength, diffraction causes a distortion of the sound field around the object. Sound pressure on the side of the object nearest the source will build up above free field levels, and interference between directly reflected and creeping sound waves will generate a complicated pattern of pressure levels at other angles around the object. As the ratio between the object and wavelength is increased to 5, the pressure build up on the side facing the source will approach an asymptote at about 6 dB relative to the free field values. Pressures on the far side of the object vary with location, shape of the object, and the object to wavelength ratio but are at most 8–10 dB lower than those on the near side. Further increase in the ratio above 5–10 leads to a sound shadow on the far side with differences between near and far sound pressures up to 25 dB.

The diffractive distortion of a sound field around an animal with two ears can thus be used to determine the location of a sound source. As before, a simple rotation of the head until sound pressures are similar on both sides is one way to locate a line to or from the source. Because diffractive fields around heads and bodies are highly dependent upon the wavelengths of the incident sounds as well as the sizes and shapes of the organisms, computation of a sound source's location from diffractive intensity differences is much more complicated than comparing time delays. In addition, the constraints on maximum intensity differences, particularly at intermediate object-to-wavelength ratios, limit the accuracy. One way mammals get around these limits is through the use of pinnae. As noted earlier, these organs limit the angle of acceptance of impinging sounds and, by altering the frequency spectra as a function of incident angle, provide information about source locations. Because a pinna and its attached auditory meatus are both tapered organs, they focus the sound energy captured over the large opening of the pinna onto the small opening at the end of the meatus. This process can result in a frequency-dependent amplification of the incident sound wave by as much as 30 dB. If the tuning of the pinnae and meatus are set to maximally amplify frequencies at intermediate object-to-wavelength ratios, much greater ranges of intensity difference between the two ears and thus more accurate estimates of source location can be obtained.

Pressure-Differential Detectors

When a propagating sound can reach both sides of a tympanum, the sound receiver is called a **pressure-differential** detector (also called a pressure-difference or pressure-gradient detector). Such an ear samples the sound field at two different locations, and the two samples are conducted to opposite sides of the tympanum. As long as the two samples are out of phase when they reach the tympanum, (a function of how far they have traveled and their relative phases when sampled), the tympanum will be bent toward the sample with lowest pressure (Figure 6.5).

Like particle detectors, pressure-differential detectors are intrinsically directional. Quantitatively, the magnitude of the force, F, exerted by an incident sound on a pressure-differential membrane is $F = (2\pi\, AP\, \Delta L \cos\theta)/\lambda$, where A is the surface area of the membrane, P is the incident sound pressure, λ is the wavelength of the incident sound, ΔL is the extra distance the incident

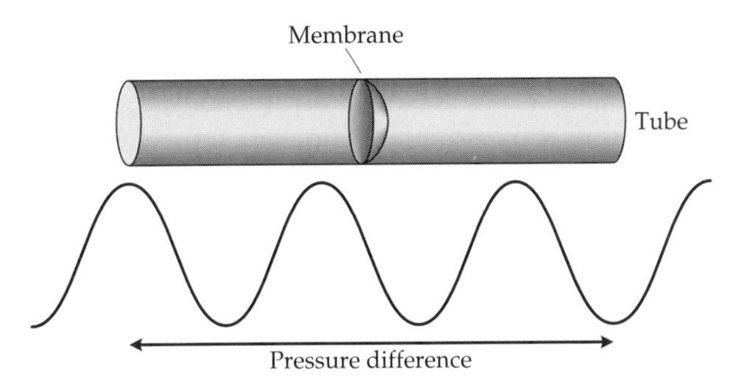

Figure 6.5 Pressure-differential detector. A closed tube samples the sound field at two different locations. The tympanic membrane can be located at any point within this tube or at either end. The pressure difference at two sampling points displaces the membrane.

waves must travel to get to one side of the membrane when compared to other side (this equation assumes that $\Delta L \leq \lambda$), and θ is the incident angle of the sound relative to its perpendicular incidence on the membrane (Michelsen and Nocke 1974). The cosine term is maximal when sound hits the tympanum perpendicularly (one sample arriving sooner at the near side of the tympanum), and is zero when the sound propagation direction is parallel to the tympanum (since both samples arrive having traveled the same distance). This term generates the directionality of the detector. Because the forces on their tympana are inversely related to wavelength, animals with pressure-differential detectors have difficulty detecting lower frequencies. One way to get around this problem is to increase ΔL sufficiently to compensate for the otherwise low force generated by low-frequency sounds. Small animals such as insects, frogs, and birds use this trick by means of a tube that connects to one side of their pressure-differential membranes and that routs the other end to a location as far from the membrane as is anatomically feasible.

In principle, the right combination of ΔL, λ, and θ could produce a net force on the tympanum twice what would be experienced by a pressure detector at the same location. However, most animals that use pressure-differential detectors do so because they are too small relative to the wavelengths of relevant sounds to use either time delays or intensity for sound-source localization. This means that the ratio $\Delta L/\lambda$ is very small, and thus the force experienced by a pressure-differential detector in a far field is often much smaller than that for an equivalent pressure detector. This cost in far field sensitivity is presumably compensated by the opportunity to have accurate directionality even when useful wavelengths are much larger than the animals detecting them. It is also the case that a pressure-differential detector, which responds most strongly when the spatial gradient of pressure is highest, will be much more sensitive in a near field than will an equivalent pressure detector. Thus losses in sensitivity of an ear that is a pressure-differential detector relative to a pressure detector at a considerable distance from a sound source become relative gains when the sender and receiver are close. In principle, the two receivers should show similar sensitivities at higher frequencies. However, friction imposed by the walls of the narrow tubes connecting the internal side of the pressure-differential tympanum to a secondary opening will attenuate the signal amplitude, and this attenuation will increase with frequency. Thus in many species with openings on the internal side of the tympanum, the ear works as a pressure-differential detector only at lower frequencies. At higher frequencies, it then acts as if the openings were absent and thus as a pressure detector. This means that the ear loses its intrinsic directionality at these higher frequencies. However, these are usually the frequencies at which other cues such as intensity differences are most available.

Pressure-differential ears can incorporate most of the refinements and external modifications used for pressure detectors. The directionality and amplification of a single ear can be improved by imposing an external cone and an auditory meatus between the tympanum and the ambient sound fields as, for example, occurs with the feathered ear funnels of birds. As long as the ear can perform some sort of Fourier analysis, directional information can be extracted by examining frequency spectra. In addition, most animals with pressure-differential detectors have paired ears that provide the same directional cues as for pressure detectors. These animals also have an extension tube connecting one tympanum to that of the opposite ear. This arrangement ensures that the interior sides of the two tympana experience the same signal. Contrasts between the two ears are thus more clearly attributable to differences in location relative to the sound source.

In summary, pressure-differential ears are less effective at low frequencies than are pressure detectors. They are also less sensitive than pressure detectors in far fields, except at higher frequencies, but respond better than pressure detectors in near fields. Pressure detectors have no single ear directionality. Animals that are large relative to the wavelengths of their sounds can get around this problem by using two pressure detectors to determine the location of a sound source. This arrangement will not work for small animals. Because pressure-differential detectors can provide quite accurate indications of the location of a source, even with only a single ear, they are widely used by small animals to guarantee directionality. Directionality and amplification in both kinds of ears can be increased by adding a pinna or funnel between the incident sounds and the tympana. Both systems provide the advantages of large membranes in capturing incident sound energy and providing good tracking of incident waveforms.

MODIFICATION OF COUPLED SOUND VIBRATIONS

As we noted on pages 22–24, the energy carried by a sound wave in a given medium equals the product of the pressure wave and the distance over which the wave moves the medium particles. Sound traveling in a low-impedance medium such as air is represented by low pressure and large particle movements of medium molecules. A sound of the same total energy traveling in a medium of high impedance, such as water, is represented by high pressure and short movements of medium molecules. When sound traveling in one medium strikes the boundary of a medium with a very different impedance, most of the energy will be reflected. Since terrestrial animals are largely made of water and solids, they have much higher acoustic impedances than the surrounding air. This makes it difficult for them to trap enough energy in incident sound waves to stimulate their ears. The compensatory adaptation that we find in the ears of organisms is a stepwise matching of the impedances of the medium to that of the organism to permit better sound-wave capture. Put another way, a good ear trades off pressure for molecular displacement with as little loss as possible as one moves from one medium to another.

One solution for a terrestrial organism is to mount a horn exterior to the tympanum. Horns are tubes with large diameters at the air interface and small diameters inside the animal's body. Just as a horn can be used to increase sound output by a signaler (see Chapter 4, pages 90–91), a horn on the ear can be use to increase sound capture by a receiver. Birds and mammals typically have funnel-shaped tubes connecting their tympana to the outside world. These tubes allow for a gradual change in acoustic impedance as the sound enters the animal. The pinnae of mammals and dish-shaped plumage of some birds further enhance this impedance matching by extending the funnel beyond the skull.

Once the sound reaches the tympanum, reflection of sound waves can be further reduced by making the tympanum thin and light, and by placing an air-filled cavity behind it. These strategies are, of course, common for pressure detectors and pressure-differential detectors. The combination of a thin, light membrane and an air cavity makes the average acoustic impedance of the tympanum intermediate between that of air and that of the fluid-filled cells of the inner ear. This arrangement thus reduces the losses due to reflection at the tympanum and provides the first body part that is made to move by the impinging sound. The final step of connecting the vibrating tympanum to the inner ear usually involves an additional impedance-matching step. In terrestrial vertebrates, the tympanum is connected by a narrow rod or chain of articulated rods to a thin membrane (**oval window**) in the wall of the inner ear. When the size of the oval window is much smaller than that of the tympanum, all of the energy collected by the large tympanum can be concentrated on this smaller window, providing greater force per unit area. This concentrated force better matches the higher pressure required to transmit sounds in the fluid-filled inner ear.

ANALYSIS OF COUPLED SOUND VIBRATIONS

Once the sound has been coupled to sensory cells of the organism, various neural configurations can be used to characterize and decompose the sound into its components. Two levels of processing generally occur. Peripheral processing occurs at the level of the sound reception organs, and is followed by central processing at higher centers such as the brain.

Peripheral Frequency Analysis

There are three common methods by which animals decompose complex waveforms into frequency spectra. The simplest is to fine-tune different groups of receptors cells to different frequency bands. If the cells tuned to a specific band vary their nerve impulse rate as a function of the intensity of stimulation, higher-order centers in the brain can compare the rates of neural firing from different groups and reconstruct the frequency spectrum. Because it is difficult and expensive to build widely different kinds of receptors into the same ear, this type of mechanism is limited to narrow frequency ranges. A second method utilizes sensory cells that are sensitive to a very wide range of frequencies. If a given cell fires whenever the signal waveform exceeds a certain threshold, it will be able to track the signal periodicity with its impulse rate. This method is sometimes called the **telephone principle** or **phase-locking**. The frequencies of simple sinusoids would thus be mapped onto nerve firing rates and identified by feeding these impulses to higher-order cells, which only trigger when stimulated at certain impulse rates. Varying the sensitivity thresholds of these higher-order cells or using cells which only respond to changes in impulse frequency would facilitate decomposition of complex waveforms. However, even with very sophisticated processing, this method is limited at the front end by the inability of the sensory cells to fire impulses at high enough rates. In practice, therefore, this method is limited to frequencies less than 1–1.4 kHz. Sensory nerves of mosquitoes, which need to track sounds of only 400 Hz, show phase-locking between sound wave peaks and nerve impulses.

The final method of frequency analysis relies on the coupling of incoming sounds to a substrate of spatially variable acoustic properties. If the coupling is properly designed, the location of maximal displacement along such a substrate varies with the frequency of the incoming sound. Sensory cells are then attached to this substrate on one side and to a less frequency-sensitive base on the other. When a sound with a given frequency is coupled to the substrate, only sensory cells in a given location are stimulated. Different frequencies displace different parts of the substrate and thus different groups of cells. This method is called **place principle analysis** and the set of spatial correlations between nerve cells and frequencies is called a **tonotopic map**.

Locusts and mole crickets trap sounds with a tympanum that has a very carefully designed shape and variable pattern of thickness (Michelsen and Nocke 1974; Michelsen 1979, 1983; Michelsen and Larson 1985). When a complex sound hits such a membrane, standing waves are set up by reflections from the membrane boundaries. Whether a particular place on the membrane is set into motion or not depends upon which, if any, of the resonant frequencies of the membrane are present in the sound. When resonant frequencies are present, the amount of movement at an antinode will depend on the amplitude of that frequency component in the signal. By attaching clusters of receptor cells at different locations on the tympanum, the presence and intensity of several discrete frequency bands within the complex signal can be determined.

More complex methods using the place principle occur in katydids, crickets, cicadas, and terrestrial vertebrates. In all of these forms, the sensory cells are not placed against the tympanum, but are located in a separate auditory organ. The motions of the tympanum are then coupled to the auditory organ by some thin and responsive structure (a ligament in cicadas, membranes in crickets and katydids, and middle ear bones in terrestrial vertebrates). This placement has the advantage that the tympanal design can be dedicated to faithful sound capture, whereas the auditory organ can be given properties that facilitate efficient frequency analysis. When the two functions must be accomplished in the same organ (as in locusts), neither can be made very efficient. Taxa that have a separate auditory organ differ in the way that sensory cells are arranged. Some organs have spatially distinct clusters of sensory cells (usually correlated with sensitivies to discrete bands of frequencies) and some have orderly rows of sensory cells (usually correlated with continuous tonotopic mappings of frequency onto location).

In both insects and vertebrates that use the place principle for frequency separation, signal intensity is coded by the firing rate of the sensory neurons. As a rule, each sensory cell has a fixed range of intensities over which it can vary its firing rate. Different cells have different minimum or **threshold intensities** that must be attained before they will fire. Increasing the sound intensities above this threshold increases the firing rate up to a limit, after which the cell impulse rate approaches an asymptote. In most organisms, the rate of sensory nerve firing tends to rise with the **logarithm** of the sound intensity or some similar power function. This means that the decibel (dB) scale is a more accurate representation of what the ear perceives than is a linear scale. It also means that each cell can hear a wider range of intensities than if impulse rates were linearly related to intensity. Typical insect ear cells have dynamic ranges of 20–30 dB; vertebrate sensory cells have dynamic ranges of 40–50 dB. Where there are enough sensory cells to have several committed to each frequency band, cells within that band are divided into several sensitivity sets, each differing from the next by 20–30 dB. This arrangement permits dynamic ranges of 100 dB or more.

Central Processing of Sound Signals

The neuronal stimulation produced in ears is usually conveyed to the receiver's central nervous system for processing. Much of the lower part of the central nervous system is used to enhance contrasts and thus improve resolution. This enhancement is often accomplished by using one kind of sensory cell to excite a higher-order neuron and other sensory cells with slightly different sensitivities to inhibit the same neuron. Maximal stimulation of the higher-order neuron will only occur when the stimulus is closest to the optimal stimulus for the excitatory cell and will drop off quickly as the stimulus is varied away from that optimum. When the criterion for similarity is frequency, such a process greatly improves the bandwidth (Δf) at the level of the central nervous system when compared to that at the level of the ear. Similar arrangements in animal central nervous systems increase the resolution of signal amplitudes and the measurement of time delays between successive events such as pulses or modulations (Δt). This improved resolution of both frequencies and amplitudes in the central nervous system facilitates the use of frequency spectra for recognition of sounds and spatial localization of sound sources. It is interesting that the tonotopic organization of ears is often preserved at successively higher levels in the brain. Although the intensity and temporal properties of the highest level brain cells may become very complex, cells with similar optimal frequency responses are still grouped together spatially. One reason for this continued segregation of cells with different optimal frequencies is the crucial role of frequency spectra in providing important information to the receiver.

In various arthropods, frogs, and reptiles, (but not birds or mammals), **peripheral tuning** of the ear is used to reduce stimulation by sound frequencies other than the bands present in species-specific signals. Within these bands, frequency-spectrum information may be used to augment species recognition or to discriminate between different conspecific signals. In bullfrogs (*Rana catesibiana*), a croak must contain energy at 200 Hz and 1400 Hz to evoke a response, but must not contain significant energy at 500–600 Hz (Capranica 1965). Adult males produce croaks with the appropriate spectra; younger males tend to have croaks with significant energy in the 500–600 Hz range and are ignored by females and adult males. Although peripheral tuning is widespread in some groups, it is often insufficient by itself to establish species identity (Capranica 1992). For example, even with strong peripheral filtering, a frog or arthropod is still likely to hear sounds of a number of other species. Additional species specificity relies on recognition of temporal patterning in the signals. Central nervous systems of frogs and arthropods invariably have phasic neurons (called **chirp** or **pulse coders**) that respond to the onset of amplitude modulations and then quickly stop firing. Higher-order cells that receive inputs from such phasic neurons can use the delay between these short bursts of stimulation to monitor the pulse or modulation rates in the received sounds and thus identify gross species-specific patterning. Additional levels of cells combining and comparing responses of lower levels are then used to improve discrimination between more complex temporal patterns.

Another major acoustic task undertaken by the central nervous system is the determination of the location of a sound. As we have seen, angular location of a sound source requires accurate resolution of time delays, signal amplitudes, and/or determination of frequency spectra. Even with particle or pressure-differential detectors, a pair of ears is required to untangle the confounding effects of signal amplitude at the source, source location, and distortion during propagation. This means that the central nervous system must attempt to improve the resolution of temporal, amplitude, and frequency information from each ear, and then combine the outputs of the two ears to determine source locations. Because the frequency and temporal properties of the sensory cells in the ear are often linked, the interpretation of time delays at the two ears must include knowledge of which frequencies were involved. This is another reason why tonotopic structure of the ear is usually preserved at higher levels in the central nervous system. It also makes the combining of information from the two ears to determine source location more complicated than if delays from all frequency stimuli could be pooled. Finally, the binaural auditory signals are often combined with information from visual, olfactory, or other modalities to maximize identification of source locations. The usual mechanism is to create a neural **map** for each modality and then to compare these maps. In a neural map, cells are stimulated only if a sound source is at a particular angular location, and adjacent cells represent adjacent angular locations. Such maps appear to be the rule in birds and mammals.

TAXONOMIC CONTRASTS OF EAR DESIGN

In the remainder of this chapter, we compare the tradeoffs that have been made as ear design has evolved for different animal groups. Arthropods are generally small and this fact has had major consequences for the appropriate ear design. Fish live in water and this leads to an entirely different set of problems when compared to terrestrial vertebrates. In addition to the ecological constraints imposed upon ear evolution, the phylogenetic history of a group sets the starting point on which subsequent selection can work. The effects of earlier historical conditions are particularly clear in the evolution of vertebrate ears.

Insect Ears

Most species of insects (except flies and beetles) have small fan-shaped clusters of sensory cells in the tibial joint of their legs called **subgenual organs**. These have no clear coupling devices other than attachment to the interior membranes of the tibia. Where they have been studied, subgenual organs appear to provide the owner with a reasonable sensitivity to substrate-borne vibrations. For some species, this ability is probably used to warn of the approach of predators. For others, such as insects stridulating on plants (see pages 115–116), fiddler crabs (genus *Uca*) tapping the mud, or water striders (Hemiptera: Gerridae) perturbing the water's surface (Wilcox 1988), substrate-borne vibrations are a major modality for conspecific communication. The fact that cells in the central nervous systems of several orthopterans receive input from both subgenual and ear sensory neurons suggests that these animals may combine or compare sound vibrations received simultaneously by the two routes (Kalmring et al. 1985).

Subgenual organs respond only to frequencies below a few kHz, although the organs in some katydids may be sensitive to airborne sounds up to 4–5 kHz. This probably reflects the better propagation of lower frequencies in solid substrates. It may also reflect limitations on the kinds of mechanisms available for generating signal vibrations. Water striders communicate by means of surface waves that they generate by pumping their forelegs (Figure 6.6). This type

Figure 6.6 Water strider sending signals on water's surface. Male *Limnoporus notabilis* producing territorial signal to repel conspecifics. Rings of surface perturbation can be seen radiating away from moving strider. (Photograph courtesy of Stim Wilcox.)

of motion limits the frequencies that they can easily produce to 25–100 Hz. The organs used to receive these signals are apparently able to discriminate between different frequencies within ± 1.5 Hz (Wilcox 1988). Tuning of subgenual organs is thought to be performed either by adjustment of the moveable ends of the sensory cells or by higher order responses to phase-locked sensory neurons. It is likely that most of the more sophisticated ears in insects have evolved from such subgenal organs.

A number of other insects have ears that are designed solely to detect the echolocation calls of bats. These calls are emitted in flight; the resulting echoes indicate to the bat the location, direction of motion, size, and shape of possible prey targets (see Chapter 26). To obtain good echoes from small targets, most bats use frequencies of from 15–125 kHz for echolocation calls. Some taxa of moths (Lepidoptera) have bat-sensitive tympana located on the thorax or first abdominal segments. In noctuids, each ear contains a single pair of sensory cells that are directly attached to the tympanum. In geometrids, there are four receptor cells for each ear. The ears of both groups respond to frequencies in the 15–125 kHz range. Although the internal sides of the tympana are connected to large tracheal air sacs that contact the corresponding sacs of the opposite ear, the very high frequencies used by the bats prohibit any pressure-differential effect and cause each ear to act as a pressure detector. Despite the small number of cells per ear, the moth's ears have very large dynamic ranges: in noctuids, signal intensities over a 40–50 dB range can be discriminated. In part, this range is achieved by one cell responding at lower intensities, and the second beginning to respond only when the first has reached saturation responses. Higher-order cells use this intensity information to determine whether the moth has time to fly out of the bat's path or instead should dive for the ground. Binaural comparisons in the moth's central nervous system determine which direction the moth should fly to evade the bat. Bat-detector ears are also known in lacewings (Neuroptera), hawk-moths (Lepidoptera), preying mantises (Mantodea), and some beetles (Coleoptera).

For insects that use airborne sounds for sexual advertisement, the auditory tasks are considerably different from those for evading predatory bats. Receivers are more likely to be able to detect the advertisements of a potential mate if the latter uses loud signals with low to moderate frequencies. The only way an insect, which is small relative to such wavelengths, can detect moderate frequencies and still extract directional information about the sender is to use a pressure-differential ear. To ensure species specificity, the sender will surely incorporate some temporal patterning in its call. Wide bandwidth signals preserve temporal patterns during propagation better than do narrow bandwidth ones (Röhmer 1992), and directional and distance information about a sender is most easily extracted from a wide bandwidth signal. Sexual advertisement thus tends to favor sounds with wide bandwidths and ears that compare frequency spectra of these wide bandwidth signals. Such ears require more sensory cells and sophisticated couplings than is the case for simple bat detectors. Note that this does not preclude a sexually receptive ear from also acting as a bat detector: ultrasonic sensitivity and bat avoidance responses have recently been found in crickets, locusts, and katydids, all of which also respond to conspecific sexual calls (Hoy 1992).

Most cicadas (Homoptera: Cicadidae) and grasshoppers (Orthoptera: Acrididae) produce sexual advertisement sounds in the range of 1–9 kHz. Both exhibit paired tympana on the anterior part of the abdomen (Figure 6.7). In male cicadas, there is a single large air sac that fills a large part of the abdomen between the two ears. In females, the sac is smaller, but apparently still provides an acoustic connection between the two ears. Acridid grasshoppers have several tracheal sacs and fat deposits connecting the two ears. For sounds below 10 kHz, these couple the two ears acoustically. Thus in both groups, acoustical connections between their ears allow each ear to act as a pressure-differential detector with intrinsic directionality and without resorting to very high frequencies. Although some grasshoppers are sensitive to frequencies above 10 kHz, the acoustic coupling between the ears becomes less effective at higher frequencies and thus the ears then become pressure detec-

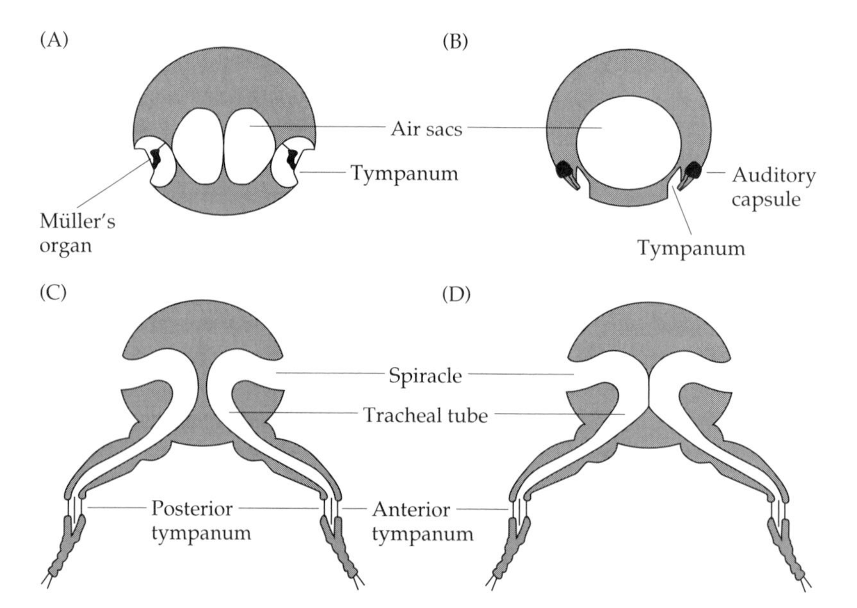

Figure 6.7 Ears and associated air spaces in insects. (A) Cross section of the body of a locust (Orthoptera: Acrididae). Central air sacs connect the internal sides of the tympanal cavities. Müller's organ contains sensory cells and is attached at several points to the internal side of the tympanum. (B) Cross section of a male cicada (Homoptera: Cicadidae). A single large air sac fills the abdominal cavity. Tympana are located inside grooves in the body. Sensory cells are located in an auditory capsule attached to the body wall and connected to the tympanum by a thin apodeme filament. (C) Cross section of a katydid (Orthoptera: Tettigoniidae). Funnel-shaped tracheal tubes connect the internal side of each tympanum (two on each leg) to a large open spiracle on the body wall. Sensory cells are located on a thin membrane between tympana inside the leg. (D) Cross section of a cricket (Orthoptera: Gryllidae). The anatomy is similar to that of the katydid except that spiracles can be closed and tracheal tubes touch in mid-body, allowing for acoustical exchange.

tors. The two taxa differ in their mechanisms of frequency analysis. Grasshoppers attach clusters of sensory cells directly to the tympanum and use the intrinsic modes of vibration by the receptor-tympanal complex to provide frequency analysis. Locusts have about 70 sensory cells organized into four clusters collectively called Müller's organ: subzones of the membrane with attached clusters have resonances of 3.5–4.1 kHz, 5.5–6.5 kHz, and 16–19 kHz (Michelsen and Nocke 1974; Stephen and Bennet-Clark 1982). Cicadas have a separate auditory organ containing up to 1000 sensory cells; this organ is attached to the tympanum by a thin rod (apodeme). Although not much is known about how these structures function, most species appear to have moderate peripheral tuning to the species-specific male advertisement frequency of their own species. Males are less narrowly tuned than females. In female cicadas, the air sac is small; it connects the two ears and thus permits pressure-differential directionality. Removal of the sac lowers female sensitivity to sounds by 10–15 dB. In males, the sac is very large because of its function as a resonant chamber to aid sound production. It also appears to resonate when exposed to nearby sounds at the species-specific frequency and thus acts as a sound reception coupling device. Removal of the sac in males reduces sensitivity to the species-specific frequency by 25 dB.

In katydids (Orthoptera: Tettigoniidae) and crickets (Orthoptera: Gryllidae), the tympana are located in the tibia of the front legs. Most species have a tympanum on each side of each foreleg, but some crickets may have tympana on one side only. Inside the leg is a tapered tracheal tube with its small opening behind the tympanum and its large opening at a spiracle on the animal's thorax (Figure 6.7). These tubes have two functions. First, they constitute amplification horns that increase the intensity of sound at the membrane by 10–30 dB. Without such amplification, the sound pressure on the back side of the tympanum would be much lower because of the longer distance the sound has traveled in a small tube. This construction generates maximal displacement of the tympanum when the exterior and interior sounds are out of phase. The second function is to make the tympanum a pressure-differential detector with concomitant directional properties, at least at lower frequencies. In crickets, the situation is even more complex because the two tracheal tubes in the thorax are acoustically coupled. This means that the sound reaching any given tympanum arrives from four sources: directly outside the tympanum, through a spiracle on the same side of the body, through a spiracle on the opposite side of the body, and through the tympanum on the other front leg. Michelsen (1979) reports that the sounds arriving from the opposite tracheal horn have only 35% of the energy directly striking a tympanum, and that arriving from the other front leg's tympanum is less than 10–20% of the direct signal.

The availability of pressure-differential detection allows these insects to use lower frequencies and still have the benefits of directionality. Perhaps because of their connections between ears across the body, crickets have particularly availed themselves of this opportunity and their sexual advertisement calls are in the range of 2–8 kHz. Crickets have a second call, called the courtship song, which consists of frequencies of 12–16 kHz. Since this high-frequency sound is only used when females are next to a male, propagation losses and the lower directionality of this sound are not of importance. Katydids tend to use much wider bandwidth calls than do crickets, with frequencies ranging from tens up to hundreds of kilohertz. These high frequencies mean that katydids' ears act only partially as pressure-differential organs, if at all. However, the wide bandwidths provide considerable information on distances between sender and receiver and result in better preservation of the temporal modulations of a particular species during the propagation of the sound. In species of katydids in which males aggregate around food resources sought by females, the advertisement calls need not propagate long distances or provide directionality because females will come to the food sites in any case. In those species in which males do advertise for females over long distances, additional directionality is achieved with each tympanum being enclosed inside the leg and sound being allowed to enter only through directional slits. Some species even have elaborate directional ridges around these slits much like the pinnae of mammals.

Both crickets and katydids have complex arrays of from 40 to 100 sensory cells that facilitate accurate frequency decomposition of sounds. In both groups, sensory cells differ in size and placement and these arrangements apparently result in different cells being attuned to different best frequencies. The sensory organs of katydids and some crickets are organized tonotopically, and the resulting tonotopic map appears to be retained at higher neural levels as well.

Most sexually advertising insects use temporal patterning of their calls for species identity. One of the most crucial temporal skills of such a species is the discrimination of its own pattern from that of sympatric species. In crickets, calls consist of repeated syllables. Auditory nerves relay the rates of syllable repetition to the cricket brain. Here selected cells act as band-pass filters by responding most strongly to the syllable repeat rate of their own species and less strongly to other rates. The European cricket has a normal pulse rate of 25 Hz. Brain cells have been identified that respond most strongly to this rate and that show lower activity with slower or faster rates. Reduction of activity to 50% of maximum requires rates as slow as 18 Hz or as high as 56 Hz. The cricket system is not highly tuned, but it does permit a reasonable amount of discrimination.

In summary, insects do surprisingly well at achieving the various goals of ideal sound reception. Despite their small size, most communicate with some degree of directionality without resorting to unreasonably high frequencies. Directional resolution appears to be most accurate for sound sources in the 60° angle in front of the animal; changing sound source locations outside of this zone usually results in little behavioral or central nervous system response. In species such as katydids that have broad bandwidths in their calls, degradation in frequency composition during sound propagation can provide cues about source distance. Where insects' ears have enough sensory cells, frequency resolution is good, but in many cases, resolution has been sacrificed because ears are tuned to the narrow spectrum used by that species for signaling. Because pressure-differential ears are intrinsically frequency dependent, use of this type of detector also reduces the range of frequencies over which the ears can respond. Frequency resolution is also limited by the small number of sensory cells per ear when compared to larger vertebrates. Regardless of the number of sensory cells, most arthropod ears require sound intensities of 40 dB (SPL) or greater; this level of sensitivity is consistently worse than one finds in vertebrates, which typically have 0 dB SPL sensitivities. Male mosquitoes (Diptera) manage 0 dB sensitivities in the near field, but only when females are within 1 m. In part, the lower sensitivity of insects is an outcome of their small size relative to the wavelengths that are most useful in their natural environments. Dynamic ranges of up to 100 dB are known in insects. Whereas locust ears are specialized to receive three to four separate frequency bands, a typical katydid breaks the 1–100 kHz range over which it is sensitive into 24 contiguous bands. Temporal resolution has been less well studied in insects. Auditory nerves of locusts can track repeated clicks up to rates of 250 Hz, whereas cicadas can track click rates up to 500 Hz. Where it has been studied, discrimination between different temporal patterns is present, although the accuracy varies widely among the taxa.

Fish Ears

Aquatic animals such as fish face rather different challenges in ear design. The first problem is the similar acoustic impedance of a fish's body and its surrounding medium. This similarity causes a fish to absorb incident sound energy well, but since its entire body vibrates in sympathy with the sound wave, no differential movement of body parts is available to stimulate sensory cells. The second problem is the 4.4 times longer wavelengths for any frequency when water is compared to air. At the sound frequencies fishes are likely to produce, most receivers will be in the near field of the sound source. Finally, the high speed of sound and longer wavelengths in water make the fishes' use of temporal and diffraction cues for the location of a sound source more challenging.

Sharks, rays, and bony fish have evolved a solution to these problems: they have three or more adjacent chambers on each side of their heads that are specifically sensitive to near field sounds. The three most common chambers are called the **utriculus**, the **sacculus**, and the **lagena** (Figure 6.8). Each chamber is lined with a tissue matrix in which are embedded a number of sensory **hair cells**, with cilia extending into the chamber. Resting on the cilia is a dense object called an **otolith**, which in bony fish is calcified (Figure 6.9). Since the otolith has a density three times that of the surrounding tissues, it responds to incident near field sounds by moving with a lag relative to the hair cells. This motion stimulates the cells. The relative movements of the otolith and the hair cells are solely due to the particle displacement of the medium in the near field; there is no way that the pressure component of the sound by itself can move these parts of the body. Because of the mass of the otolith, such a system can only respond to lower frequencies. The calcified

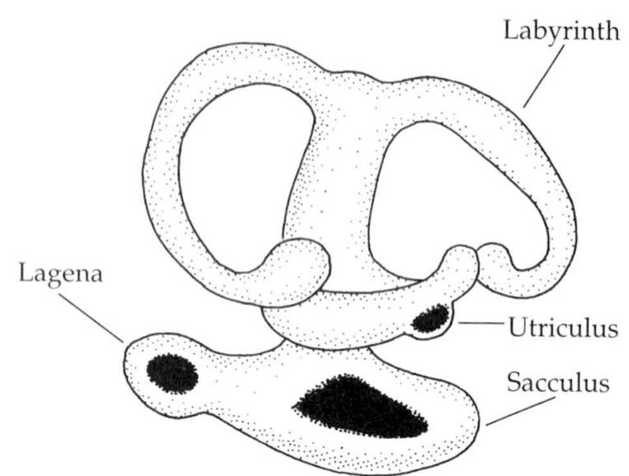

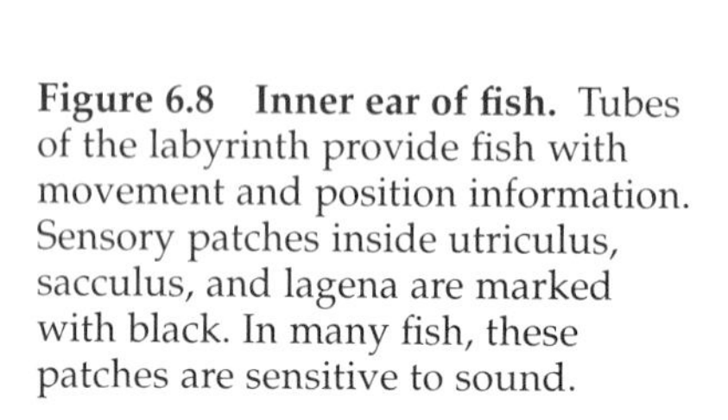

Figure 6.8 Inner ear of fish. Tubes of the labyrinth provide fish with movement and position information. Sensory patches inside utriculus, sacculus, and lagena are marked with black. In many fish, these patches are sensitive to sound.

otoliths of bony fish limit near field responses to 200 Hz or lower, whereas the smaller objects (called **otoconia**) in cartilaginous sharks and rays may respond up to 600 Hz.

In all species studied, the sensory cells in at least one of the three chambers are divided up into clusters. Within each cluster, all the cells have their cilia arranged with the same geometry. Stimulation of each cluster is maximal when particle displacement is along a particular angle relative to the cluster of cells. Many fish have different clusters oriented with their optimal angles aligned with different perpendicular axes. In addition, the three chambers are often oriented in different directions relative to each other, and the equivalent chambers on the two sides of the fish may be aligned at different angles relative to the body axis. The differences in stimulation between different clusters in the same chamber, between the different chambers on a given side of the body, and between the different sides of the body give the fish considerable potential information on the location of a sound source. Remember, however, that the patterns of particle displacement in near fields can be very complicated and even with clusters arranged along three perpendicular axes there will be ambiguities concerning where a sound source is really located. Cod

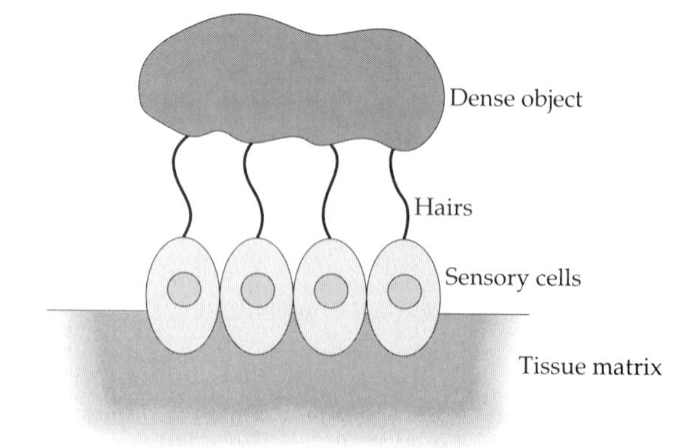

Figure 6.9 Inner ear mechanism of fish. Fish detect near field sounds or movement of adjacent tissues when a dense object (otolith; otoconium, in sharks) moves with a lag relative to the underlying hair cells. This movement bends the hairs (cilia) on the sensory cells and excites them.

(Gadidae), which have relatively typical ears, can identify the horizontal location of a sound source to within 20°, the vertical position within 16°, and the distance to the source within a few meters.

As we discussed in Chapter 5, pages 127–129, open ocean, surf zones, and rushing rivers are extremely noisy at all sound frequencies. Fish in such habitats do not need very sensitive hearing since any soft sounds made by other fish would be masked by the ambient noise. In these habitats, fish ears tend to consist only of the simple chambers, and both frequency ranges and sensitivity are limited (Schellart and Popper 1992). However, in shallow estuaries and in fresh water lakes and slow streams, ambient noise levels are much lower. This benefit is partially countered by the heavy attenuation of all frequencies below a critical cutoff (see Chapter 5, page 122). This cutoff frequency is higher the shallower the water and the softer the substrate. The low noise and the heavy attenuation of those frequencies used by fish both favor more acute auditory sensitivities in quiet shallow waters. And it is in these habitats that we find modifications of the standard fish pattern that increase both sensitivity and the upper limit on perceivable frequencies.

A mechanism evolved by many fish to increase sensitivity, localization, and frequency range is the use of the swim bladder as a pressure detector (Figure 6.10). This mechanism has several benefits. First, because the ratio between particle displacement and pressure increases as acoustic impedance of a medium decreases (see Chapter 2, pages 22–24), the lower acoustic impedance of the swim bladder converts the high pressure–low particle displacements of a sound in water into low pressure–high particle displacement in the gas and wall of the bladder. The movement imposed by a bladder on a coupled otolith is thus much greater than the otolith experiences due to movements of nearby water and tissue molecules vibrating with the near field of the sound. The result is an increase in sensitivity in the near field as a result of the coupling of the swim bladder to the ears. Second, although a swim blad-

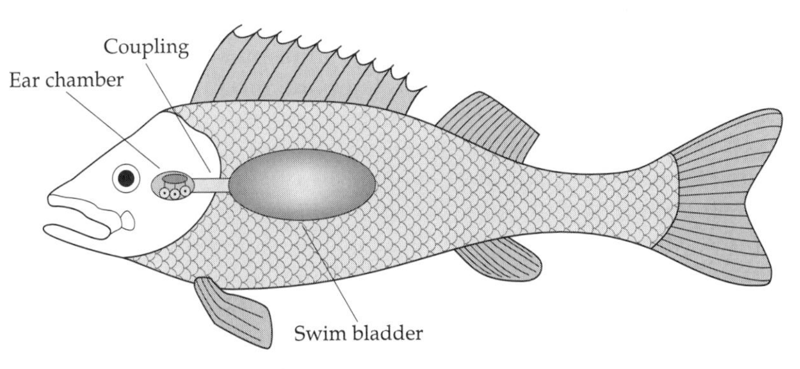

Figure 6.10 Far field sound detection by fish. The swim bladder traps pressure variations in the sound field and converts these to movements of its wall. These movements are coupled to the ear chambers by surrounding tissues, by extensions of the air bladder, and/or by small bones.

der is a pressure detector and thus provides no information about sound source angle, knowledge of the pressure amplitude can be combined with the near field information to reduce ambiguities and thus improve directional information. Third, the resonant frequencies of a typical swim bladder are usually in the kilohertz range. This range is significantly higher than that of the otoliths responding to near field forces alone. Thus coupling to a swim bladder can extend fish hearing up to 3 kHz. Finally, a swim bladder coupling allows a fish to respond to acoustic signals in the far field, and thus at greater distances.

The degree of these improvements depends on the nature of the coupling between the swim bladder and the ear chambers. In the simplest case, the swim bladder just touches the auditory portion of the skull and no special coupling is present. This is the case in some reef-dwelling squirrel fishes (Holocentridae). Auditory sensitivity is increased about 20 dB over species with no close coupling between the swim bladder and the ear, and the upper frequency limit can be extended to nearly 1 kHz. In bonefish (Albulidae), soldierfish (Holocentridae), and herrings (Clupeidae), the swim bladder has fingerlike extensions that actually insert into the ear on each side. This arrangement provides somewhat better sensitivity and increases frequency ranges in herring up to 4 kHz. The best hearing is found in goldfish, carp, and minnows (Cyprinidae); loaches (Cobitidae); neotropical electric fish (Gymnotiformes); suckers (Catastomidae); and catfish (Siluroidea), all of which have small bones connecting the swim bladder directly to each ear. These fish have upper frequency limits of several kilohertz and sensitivities up to 50 dB better than in fish whose swim bladder is not associated with the ears. Goldfish are as sensitive as humans at their respective best frequencies. A few groups of fishes, such as African electric fish (Mormyridae) and knife fish (Notopteridae), have small ear bubbles near each ear instead of a connection to the swim bladder. These air bubbles act in the same way as a bladder by trapping pressure variations and converting them into local displacements of the ear. The resulting sensitivities are as high as some species with bone couplings.

Although fish do not have tonotopic ears, the frequencies to which they are sensitive are usually below the maximal impulse rates of nerves. This means that both carrier waveforms and temporal modulations of them can be coded by phase-locking the nerve impulses to the sound wave peaks. Neural filtering is then possible at higher levels, to give frequency spectra. The result is that species with sophisticated ears like goldfish have temporal and frequency resolutions, within their frequency ranges, similar to that of higher vertebrates. Most fish, like many insects, use the temporal patterning of sounds to code information about species identity, sex, or social function. Accurate discrimination of these patterns is clearly no problem for species using swim bladder-assisted ears.

Terrestrial Vertebrate Ears

Compared with fishes, terrestrial vertebrates have opposite problems. Because they have much higher acoustic impedances than air, most incident sound energy is reflected, and sound that is absorbed is transformed from low pressure–high displacements to high pressure–low displacements. Detecting these small displacements with any kind of sensory cell is a challenge. Vertebrates have solved this challenge, as did most terrestrial insects, by evolving a large but thin tympanum that has an impedance more similar to that of air. When the sound pressures outside the tympanum differ from the reference pressure inside the middle ear cavity, the membrane moves. The movements of the tympanum as the sound pressure outside the ear rises and falls are then coupled by one or more small bones to a thin membrane (the **oval window**) set in one end of a closed fluid-filled tube (the **inner ear**). When the oval window is pushed into the tube, fluids move toward the other end, and a second small membrane (the **round window**) bulges out of the tube to accommodate the fluid movements. An opposite movement of the oval window results in fluid moving toward the oval window, and the round window bends in. At critical points within the inner ear, the fluid flows force one or more thin tectorial membranes to move. This movement stimulates the underlying hair cells. The way sound frequency is analyzed varies with the taxon. In amphibians (Figure 6.11), the hair cells are embedded in the wall of the inner ear. Like the locust tympanum, the tectorial membranes of the amphibian ear have complex modes of vibration; different signal frequencies result in maximal membrane displacements at different points on the membrane. Different underlying hair cells are thus stimulated depending upon the frequencies present.

In higher vertebrates (Figure 6.12), the hair cells are sandwiched between the tectorial membrane on one side and a **basilar membrane** on the other. This multiple layer of membranes and hair cells divides the long tube of the

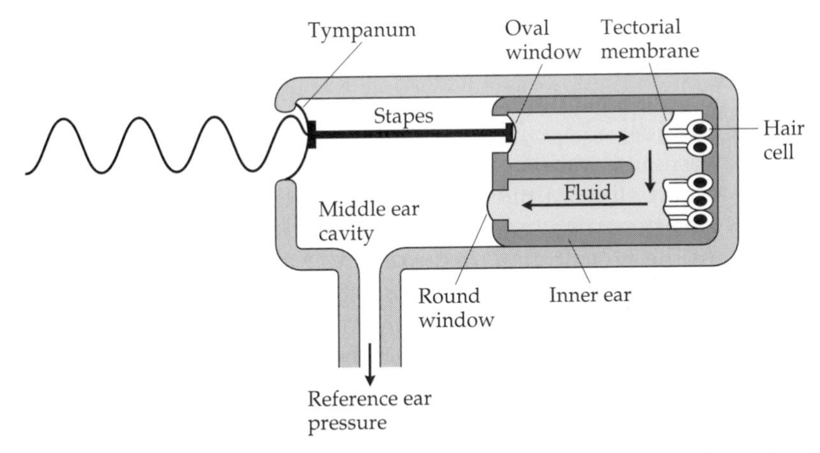

Figure 6.11 Amphibian ear. Movements of the tympanum are transferred by the stapes bone to the oval window of the inner ear. The movements of the oval window cause inner ear fluids to flow back and forth between the oval and round windows. Fluid flows induce movements of the tectorial membrane that overlies the hair cells that are embedded at various locations in the inner ear wall.

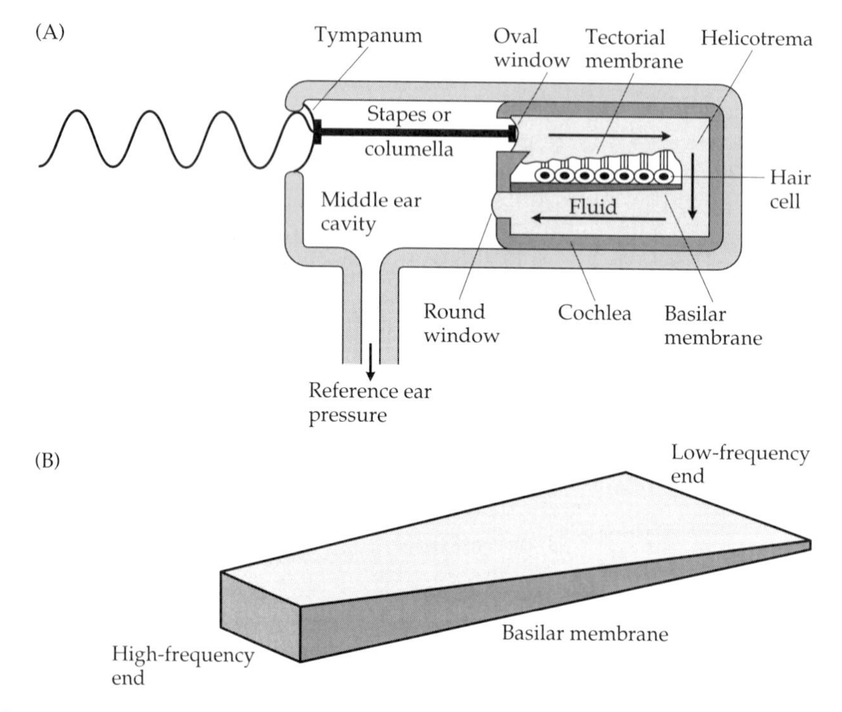

Figure 6.12 Higher vertebrate ear. (A) Basic structure. In reptiles, birds, and mammals, hair cells are attached to a basilar membrane and overlain by a tectorial membrane. Movements of fluids bend the combined membranes perpendicular to the long axis. Differential movements of tectorial and basilar membranes cause stimulation of hair cells. (B) The basilar membrane has its size and thickness adjusted so that the end near the oval window experiences maximal displacement with high frequency sounds, whereas the opposite (helicotrema) end is displaced most by low frequencies.

inner ear, here called a **cochlea**, into upper and lower chambers. At one end of the cochlea, the oval window opens into the chamber lined by the tectorial membrane, and just adjacent, the round window opens into the chamber lined by the basillar membrane. At the opposite end of the tube, a small hole (the **helicotrema**) allows fluids to move back and forth between the two chambers. When the oval window is pushed into the cochlea, a pulse of fluid moves toward the helicotrema. Because of resistance to flow through this hole, pressure builds up and bends the membrane layers into the round window chamber. When the fluids finally move through the hole, this pressure difference disappears and the membrane moves back. Because the basilar and tectorial membranes are attached to the cochlear walls at different points, their bending causes them to move relative to each other. The resulting shear stimulates the hair cells. The location along the membranes at which maximal displacement occurs varies with the frequency of the sound: high frequencies displace the membranes maximally near the oval window end of the cochlea, whereas low frequencies cause maximal displacement near the helicotrema end. This spatially variable response to frequency is produced because the basillar membrane is narrow, thick, and stiff at the oval window end, thus increasing the responsiveness of this part of the membrane to high frequencies. In contrast, the opposite end of the membrane is very wide, thin, and flexible, making it most responsive to low frequencies. The combined adaptations result in a very accurate tonotopic ear.

Frogs typically have eight separate patches of hair cells in their inner ears, four of which are responsive to sounds. Two of these correspond to the lagena and sacculus of fish, and like sharks, the hair cells of these patches are overlain by otoconia-like objects. The two additional patches are called the **amphibian** and **basilar papillae**; the hair cells for these clusters are overlain by tectorial membranes. The lagena and sacculus respond to frequencies in the range of 10–250 Hz; they appear to sense substrate vibrations and very low frequency sounds. The sacculus constitutes the largest chamber and responds to the traveling pulse of fluid movements similar to the way that bird and mammal ears operate. It is thus partly tonotopic. The amphibian papilla responds to conspecific airborne calls in the range of 100–1200 Hz, is broadly tuned, and also appears to have a tonotopic structure. The basilar papilla is sensitive to sounds in the 1–5 kHz range, is usually narrowly tuned to a species-specific set of frequencies, and is not tonotopic. This papilla is absent in some amphibians.

Frogs are generally larger than insects, but they are still small enough, relative to the wavelengths of the sounds of interest to them, that interaural time differences and interaural intensity differences due to diffraction provide poor cues for source localization. Frogs do not use an external pinna structure to increase directionality, but place the tympanum right on the head surface (Figure 6.13). Frogs typically have an opening, called a **Eustachian tube**, that joins each middle ear cavity to the mouth and thus connects the two middle ears acoustically. This connection allows each ear to act as a pressure-differential detector and thus provide some directional hearing. In addition, frogs appear to have a number of other pathways by which sounds can be captured and passed to their ears. Most amphibians have an **opercularis muscle** that connects the skeleton of each shoulder to a thin cartilaginous disk overlying the oval window of the ear on the same side of the body. When the animal is sitting up off the ground, low-frequency vibrations in the substrate as well as ground waves can propagate up through the forelimbs and through the tensed opercularis muscle to the inner ear. Low-frequency sounds can also be captured and transferred to the ears by the thin throat sacs used in call production or by the combination of body walls and adjacent lungs. Although small, many frogs are thus able to capture enough low-frequency sound energy to be very sensitive to approaching predators. Some anurans exploit this sensitivity by thumping the substrate to communicate instead of calling. The result of these adaptations is that frogs, like fish, tend to be most sensitive to frequencies of hundreds of hertz and have upper limits on hearing of about 4 kHz.

In frogs and toads that are largely aquatic, spend most of their lives underground, or live in very noisy habitats such as near to waterfalls, the tympanum and middle ear structures may be lost, and all sound and vibration

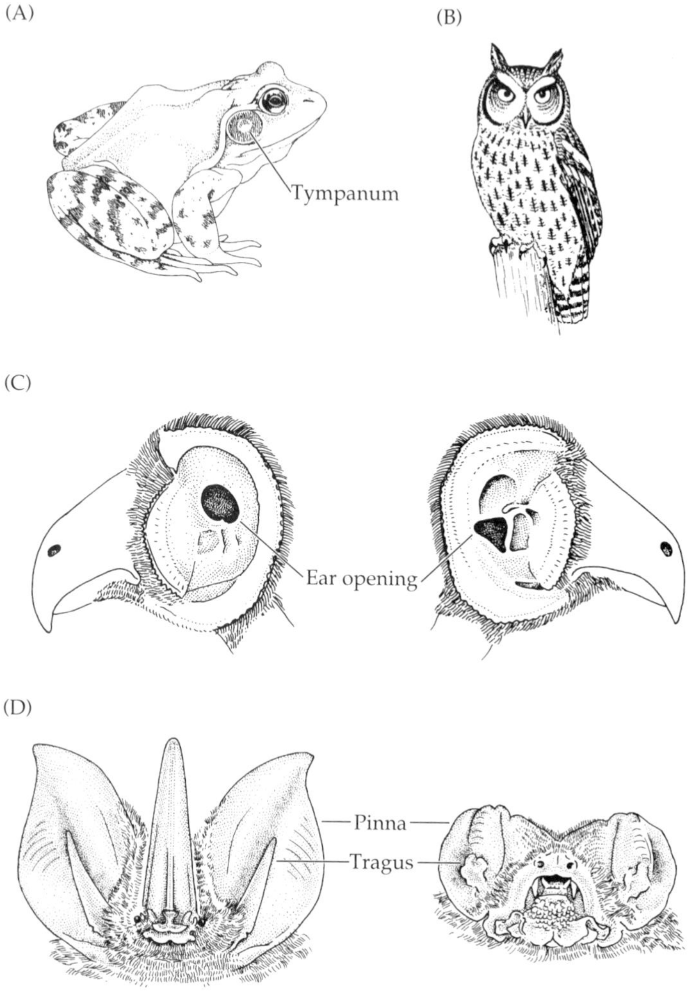

(A) Tympanum

(B)

(C) Ear opening

(D) Pinna Tragus

Figure 6.13 External ears of terrestrial vertebrates. (A) Bullfrog (*Rana catesbeiana*) with tympanum (eardrum) set in the body wall. (B) Dish-shaped faces of owls focus sound energy on ear openings. Special feather patterns provide frequency-dependent directionality. (C) Barn owl (*Tyto alba*) with feathers removed to show vertical asymmetry in placement of ear openings. This asymmetry provides directional information in the vertical plane. (D) Ears of echolocating bats have large size, complex shapes, grooved surfaces, and enlarged tragus to provide frequency-dependent directionality. The large noseleaf in bat on the left (Phyllostomatidae) is used to provide a narrow beam of outgoing echolocation pulses. (C after Welty 1962; D after Goodwin and Greenhall 1961.)

sensitivity is mediated by body reception. It is of interest that the fishlike larvae of frogs and toads use their tiny lungs like a swim bladder to trap incident sounds. Most even have special bones that couple the lung vibrations to the ear chambers exactly like some fish. These larval adaptations are lost and the adult configurations established during metamorphosis.

The ears of birds and reptiles open to the outside through a funnel-shaped tube that helps match the acoustic impedance of the middle ear to the ambient air. Although most reptiles have only a simple opening on each side of the head, birds often have special feather structures surrounding the ear openings that provide frequency-dependent directional cues. In owls, each side of the head is dish-shaped and carefully feathered to increase signal intensity and to provide forward directionality of 2–6° in the horizontal plane (Figure 6.13). Some species such as barn and great horned owls have one ear opening set above the median plane and the other opening set just below the plane. This asymmetry generates intensity differences at the two ears depending upon the vertical location of the sound source. As a result, barn owls can construct a three-dimensional map of sound sources in their brains that allows them to strike noisy prey with great accuracy in total darkness. For small birds and reptiles, interaural time and intensity differences are too small to provide good directionality. Since all birds and reptiles have cavities joining their two middle ears, each ear can act as a pressure-differential detector. This mechanism allows even small birds such as canaries to localize sounds within 20–30°.

The tympanum of birds and reptiles is attached internally to a long thin columella that transmits the tympanic movements to the oval window of the cochlea. The point of connection between the columella and the tympanum may be rigid, or it may articulate, giving a lever action. The thinness of the tympanum, the slim light columella, and the complex articulation at the tympanum allow birds and reptiles to hear higher frequencies than do amphibians. Most species are responsive to frequencies from 200 Hz to 10 kHz, with maximal sensitivities at about 2 kHz. The cochlea of higher vertebrates appears to have evolved from the basilar papilla of amphibians. From the small rounded chamber in the frog inner ear, it has become an elongated tube extending from the saccule. In some birds, the tube is slightly curved and it may be twisted along its length. The cochlea of lizards is often subdivided into separate zones, each of which appears to process different frequencies and temporal patterns. Crocodiles and birds have a continuous tonotopic cochlea in which basilar membrane width at any point is correlated inversely with the frequency stimulating that location. Low-frequency ends of the membrane are typically five times the width of high-frequency ends. Birds have much shorter and wider basilar membranes than do mammals, although they may have as many hair cells overall as an equivalently sized mammal. Thus there are more cells across the width of a typical avian cochlea when compared to that of a mammal. Size, orientation, and morphology usually vary within these cells, providing the means to improve frequency, time, and amplitude resolution at the sensory cell level.

Terrestrial mammals usually have large pinnae surrounding the external ear openings. These increase the intensity of sound at the tympanum and provide frequency-dependent directional cues. Pinnal structure varies enormously among mammals and is often highly adapted to the frequencies and likely source directions of relevant sounds. The various shapes of the pinnal membrane, the added ridges and grooves inside the pinnal surface, and the elaboration of the **tragus**, an extension of the pinnal boundary in front of the ear opening (Figure 6.13) all improve pinnal frequency dependence and directionality. Because mammals routinely hear frequencies over 5 kHz, even small mammals can achieve directionality through diffraction and interaural intensity differences. It is thus not surprising that mammals lack additional cavities linking their ears or sufficiently open Eustachian tubes to perform pressure-differential detection.

The middle ear of mammals consists of three articulated bones that form a levered chain connecting the tympanum to the oval window of the cochlea. The leverage and small size of the bones makes the system very sensitive to high frequencies. The cochlea of mammals is typically two to three times longer than that of birds. To accommodate this larger structure in the animal's head, the mammalian cochlea is coiled. The long length of the cochlea allows a much greater range of widths of the basilar membrane: the low-frequency and high-frequency ends of basilar membranes in mammals may differ in width by 10 times (compared to the fivefold range for birds). As a result, typical mammals respond to sounds over a range of several hundred hertz to 50 kHz and are most sensitive to frequencies of 5–20 kHz. Some mammals can detect sounds as low as 20 Hz, and others can detect frequencies as high as 150 kHz. As in birds, different hair cell morphologies and orientations enhance resolution within the cochlea, but with fewer cells per row, this specialization is more limited than in birds.

Marine mammals like whales and porpoises (Cetacea) face very different problems from terrestrial animals in trapping ambient sounds. Like fish, marine mammals have tissues with acoustic impedances similar to that of water. For trapping sounds, the trick is to avoid having the entire body vibrate synchronously as the sound propagates through it. All cetaceans appear to have capsules surrounding their middle and inner ears that have a density similar to that of tooth enamel. Each ear capsule is suspended within a cavity filled with a foamy mixture of air and albumin. The very high density of the capsule and the surrounding foam acoustically isolate the middle and inner ear from the rest of the animal except where specific contacts extend through the foam. Which of these contacts provide the major sound channel is currently the subject of debate. Toothed whales (which are predators on large invertebrates, fish, or other marine vertebrates) all echolocate using very high frequency sound pulses (40–200 kHz). They also make social sounds at lower frequencies. Baleen whales (which filter plankton) do not echolocate; they largely produce sounds in the tens to hundreds of hertz. In both types of cetaceans, the three middle ear bones of mammals are present within the bony capsules. Although thickened to prevent crushing by ambient pressures during deep dives, the bones are fully articulated and contact the oval window of the cochlea. They do not connect to a functional tympanum. In baleen whales, the bones are loosely connected to a membranous finger surrounded by waxy tissue. This finger extends into the closed ear canal of the whale and is surrounded by fatty tissues. In toothed whales, the bones are fused to the hard wall of the capsule at a point where fatty tissues extend through the foam to touch the capsule.

Fat has an acoustic impedance similar to that of water, and the layers of blubber on cetaceans provide a large amount of tissue that can absorb incident sounds. The fingers of fat that reach to the ear capsules thus constitute one channel by which sounds might be conducted to the ears. This channel provides the most likely route for lower-frequency social sounds. In toothed whales, the lower jaw contains a fat-filled tube that runs its length on both sides. The placement of the ear capsules in these animals suggests that the lower jaw may be the return sound channel for the high-frequency echoes used by these animals in echolocation. Such a channel would be highly directional, a characteristic critical to echolocation functions (see Chapter 26, pages 861–862).

How do higher vertebrates differ in their ability to meet the requirements for sound reception that were outlined at the beginning of the chapter? We have already seen that frogs are largely limited to sound frequencies below 4 kHz, birds respond to frequencies up to 10 kHz (12 kHz for owls), and mammals can detect frequencies as high as 150 kHz. The differences in frequency range have significant effects on the mechanisms of sound source localization that are available: low-frequency animals of small size must have pressure-differential ears to achieve any directionality at all; mammals with their high-frequency sensitivity have access to a larger variety of directional cues and have evolved very complex pinnae to exploit them.

All of the vertebrate taxa do fairly well discriminating between different frequencies within their range (Δf). Birds discriminate between frequencies differing by 1–2%; mammals can resolve frequencies differing by only 0.2%. The uncertainty principle appears again here: the higher frequency resolution of mammals is achieved at the cost of a somewhat lower temporal resolution than is found in birds. For both birds and mammals, events must be separated by gaps of 2–3 msec to be perceived as separate. Similarly, a 10–20% difference in sound durations is required before parakeets and humans can distinguish between two signals on the basis of signal length. When a signal is amplitude modulated, birds may be able to track more rapid modulation rates than can humans: parakeets can follow modulations with periods as short as 1.2 msec, whereas the shortest period a human can follow is about 5.9 msec. Birds also seem able to recover from loud masking sounds more rapidly than humans can.

Mammals are slightly more sensitive at their best frequencies than birds are. Although there are some birds that can detect signal intensities as low as –15 dB SPL, most have thresholds around 5–15 dB SPL. Typical mammals, (like humans), have threshold intensities around 0 dB SPL, although a few go as low as –20 dB SPL. Goldfish, birds, and typical mammals are all similar in their abilities to discriminate between two sounds of slightly different intensity. They require a minimal difference of 1–4 dB. Humans are on the low end, requiring a 1 dB difference; many birds require a 2–3 dB difference.

Finally, both birds and mammals show considerable variation in their abilities to localize a sound source. Much of this variation is size related: smaller animals tend to do worse (with the exception of echolocating bats, which use very high frequencies). Species that rely on sound for hunting (owls, echolocating cetaceans, echolocating bats) all tend to have very accurate localization limits at 1–2°. Similar accuracies are found in large mammals such as elephants and humans. Small animals using intermediate frequencies, such as songbirds, may only achieve localization accuracies of 20–30°.

SUMMARY

1. Sound reception requires coupling of a sound in a medium to the animal, enhancement of trapped sounds, and detection by an ear. No ear is perfect, and the various designs of ears illustrate tradeoffs between frequency range, dynamic range, frequency resolution (Δf), temporal resolution (Δt), amplitude resolution, determination of the location of the sound source, and temporal pattern. Some organisms thus emphasize one subset of these features, while others emphasize a different subset.

2. For the coupling of sound, organisms may use (a) a **particle detector** that is displaced by medium motion in a near field, (b) a **pressure detector** that consists of a membrane (**tympanum**) stretched over a closed cavity and that is bent when far field sound pressures outside the membrane differ from those inside, or (c) a **pressure-differential detector** that uses a tube to sample sound fields at two different locations, causing a tympanum inserted in the tube to bend when the samples reaching each side of it are different.

3. Animals can obtain **directionality** from coupling devices in different ways. Particle detectors and pressure-differential detectors are inherently directional. Animals using one can thus obtain an estimate of the location of a sound source. Pressure detectors are not directional; directionality can be obtained by using two such detectors and comparing sound intensities, relative phases, or times of arrival. Directionality can also be increased by adding dish-shaped external structures around the tympanum or ear opening.

4. Modification of structures most often focuses on improving the match between the acoustic impedance of the medium and that of the ear so as to increase the sound energy that is trapped instead of being reflected. Funnel-shaped horns, thin tympana, and large size differences between tympanum and inner ear windows all help reduce the impedance mismatch between air and the fluid and solid compositions of terrestrial animals. Aquatic animals have too good a match to their medium and evolve regions of their body with sufficiently different impedances so that the parts move differentially when struck by sound waves.

5. Most ears perform some frequency analysis. This analysis can be achieved for frequencies below a few kHz by **phase-locking** the nerve impulses to successive peaks of the sound waves and using higher-order nerve cells to sort out frequencies, or alternatively, by causing the Fourier components of complex sounds to be spatially separated using physical properties of the inner ear. Sensory cells placed at specific locations then indicate the presence or absence of different frequencies. This sorting out is called **place-principle analysis** and the spatial pattern of separated frequencies is called a **tonotopic map**.

6. Many insect ears are designed to detect and evade echolocating bats. These are usually tympanal organs of some kind and are most sensitive to high frequencies. They often consist of only a few sensory cells and thus do not perform complex frequency analyses. Orthopterans and cicadas

have tympanal ears largely for intraspecific sexual functions. Some types of ears may be tonotopic. Many insects have tubes interconnecting their ears, making them directional pressure-differential detectors. Insects are usually less sensitive than vertebrates to soft sounds, but some katydids are nearly as good as vertebrates at resolving temporal patterns and similar frequencies.

7. Fish with unmodified swim bladders hear primarily in the near field. Those with swim bladders modified to provide connections with the ears can respond to pressure variation in the far field as well. Fish hearing is essentially limited to a few kilohertz or less. Those fish with the best hearing live in quiet lakes, stream bottoms, and estuaries.

8. Frogs, birds, reptiles, and mammals all use an external tympanum to couple sounds in air to a fluid-filled inner ear. Frogs have several sensory patches within the ear that are mostly dedicated to detection of lower frequencies. One patch, the basilar papilla, is tuned to higher frequencies. Birds, reptiles, and mammals primarily rely on a tonotopic single patch called the **cochlea**. Frogs are largely limited to frequencies below 4 kHz, while birds and reptiles cannot hear frequencies above 10–12 kHz. Mammals have special bones connecting the tympanum to the cochlea and a specially designed cochlea that allows them to respond to sounds over 150 kHz. Although mammals are a bit better than birds at frequency resolution (Δf), some birds are slightly better at resolving rapid modulations in sounds (Δt). Mammals are also slightly more sensitive than birds at detecting low intensity sounds.

FURTHER READING

There is a voluminous literature on the physiology of hearing. Here, we only list some highlights and good places to start. Bailey (1991) provides a recent overview of sound reception in insects, and the large and excellent volume edited by Webster and colleagues (1992) provides chapter after chapter on the evolution of hearing in different taxa. A sampler of more detailed work on insect hearing includes Bailey (1990), Dubrovin and Zhantiev (1970), Hutchings and Lewis (1983), Kalmring and Elsner (1985), Lewis (1983), Michelsen (1979, 1983), Michelsen and Larsen (1985), Michelsen and Nocke (1974), Stephen and Bennet-Clark (1982), Stephen and Bailey (1982), and Towne and Kirchner (1989). The classic review by van Bergeijk (1967) on vertebrate ear evolution is good background for literature on this group. Additional reviews of vertebrate ears can be found in Smith and Takasaka (1971) and Keidel and Neff (1974). Particular chapters of interest in the latter volume are that by Shaw on external ears and by Henson on the middle ear. Details on eardrums are provided by Rabbitt (1990). Reviews of ear evolution in fish can be found in Fay (1988), Hawkins and Myrberg (1983), Kalmijn (1988b), Myberg (1981), Offutt (1974), and Popper et al. (1988). The volume edited by Fritzsch et al. (1988) has many outstanding chapters on the evolution of hearing in amphibians. Secondary routes of transmission are discussed by Ehret et al. (1990). An excellent introduction to recent work on avian and mammalian ears can be found in Chapters 25–36 of Webster et al. (1992).

Chapter 7

Properties of Light

WHAT IS LIGHT? The quick answer is that light is the radiant energy that emanates from a light bulb, a firefly, or a burning object. The sun is obviously the most important source of light on Earth. Without sunlight, life as we know it could never have evolved. Sunlight not only warms the planet, but provides the energy plants need to grow; plant growth in turn provides the primary source of food for other organisms in the ecosystem. Light is also the basis for vision. Without an external source of light, we cannot see. As with sound communication, to understand the evolution of eyes and the constraints on visual communication we need to examine the physical characteristics of the modality. In this chapter we will define light more precisely and describe its transmission properties and interactions with organic matter.

PHYSICAL CHARACTERISTICS OF LIGHT

Once we step beyond the quick definition of light, a more precise description is quite difficult. In many respects light behaves like a wave and thus it shares with sound many of the same wave-related characteristics. As with sound, waves can be described by their **frequency** (f), which is inversely proportional to the **wavelength** (λ). Algebraically,

$$f = c/\lambda$$

where c is now the speed of light. We perceive different frequencies of light as different colors. Light also varies in **intensity** and obeys the inverse square law—intensity decreases with distance from a point source as the inverse function of the distance squared. **Attenuation** of light with distance is both wavelength-specific and medium-specific, so selective filtering of certain frequencies can occur. Light is directional and travels at different speeds in different media. It thus shares with sound the wave properties of **reflection**, **refraction**, and **diffraction** (see Chapter 2).

Light waves differ from sound waves in that they require no medium for transmission and, in fact, travel fastest in a vacuum. Light is a form of energy called **electromagnetic radiation** because it generates both electrical and magnetic forces as it passes any given location. A map of the direction and magnitude of a force at each point in space is called a **field**. An **oscillating field** is one in which the direction and/or magnitude of the forces at each location varies with time in some cyclic way. An electromagnetic wave consists of both an oscillating electric field and an oscillating magnetic field. The planes of oscillation of the two fields are at right angles to each other, and both are perpendicular to the direction of travel (Figure 7.1). Electromagnetic radiation is therefore a **transverse wave**. It is generated when the motion of electrons in atoms and molecules are accelerated by an external source of energy. The vibrating, negatively charged electron generates an electric field that in turn induces the magnetic field. The coupled electric and magnetic oscillations move outward from

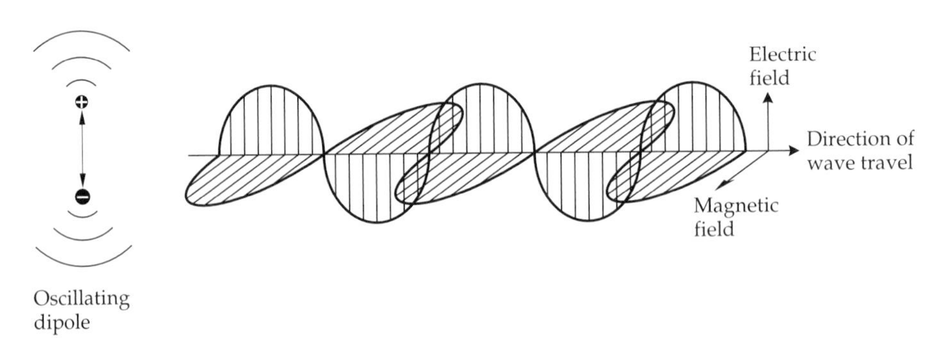

Figure 7.1 An electromagnetic wave. Waves are generated by the rapid oscillation of a negative charge relative to a positive one. They spread out from the source in all directions perpendicular to a line between the two charges. The electric force oscillates in the plane parallel to the motion of the charge vibration, and the magnetic force oscillates at right angles to the electric force.

the vibrating charge at a critical speed such that the two fields mutually induce each other indefinitely without gaining or losing energy. The traveling speed of electromagnetic radiation in a vacuum is approximately 3.0×10^8 m/sec for all frequencies. The frequency, or wavelength, of radiation depends on the composition of the excited material that produced it and the energy of that excitation.

In other respects light behaves like a stream or flux of particles. These particles of electromagnetic radiation owe their mass and their very existence to their rapid movement. The particles are viewed as packets of energy, called **quanta**, or **photons** for light in the visible range. Quanta can only contain certain discrete amounts of energy and this amount depends on the frequency of the radiation. Energy per quantum is:

$$E = hf$$

where E is energy in Joules, h is Planck's constant, and f is frequency. The higher the frequency, and thus the smaller the wavelength, the greater the flux of energy. One way to envision the combined wave and particle characteristics of electromagnetic radiation is to view it as a rapid stream of minute, elusive particles that creates a pulsating electrical disturbance in space.

Naturally occurring electromagnetic radiation spans a very wide range of frequencies, wavelengths, and energies (see Figure 7.3). The degree to which electromagnetic radiation acts like a wave or a particle depends on its wavelength and energy; long-wavelength radiation acts more like a wave, and short-wavelength radiation acts more like a stream of particles. These differences occur in part as a result of the relationship between the length of the wave relative to the size of the objects with which it interacts. Radio waves are miles long and easily bend around most objects, as do sound waves. X-rays are about 10^{-10} m long (about the size of an atom), and they either pass right through most materials or they are blocked and scattered by atomic nuclei. Gamma and cosmic rays are even smaller and contain enough energy to split atomic nuclei. Visible light is intermediate in wavelength between radio and cosmic radiation and therefore shows both wave and particle characteristics. In the following discussion we will reserve the word **light** for electromagnetic radiation in the frequency range used for vision.

The properties of any given light beam depend upon the mix of frequencies that compose it (Figure 7.2). If all of the waves have the same frequency, the light is **monochromatic**, and if all of the waves are also in phase, the light is **coherent**, as in a laser beam. Natural light from the sun or a light bulb is spectrally **complex** because it contains waves of many different frequencies and phases. When the electric (and therefore also the magnetic) vectors of the light are aligned the light is **plane-polarized**.

CONSTRAINTS ON LIGHT FREQUENCIES DETECTABLE BY BIOLOGICAL SYSTEMS

Electromagnetic waves range in wavelength from the largest radio waves (10^{-1} to 10^6 m), through the intermediate categories of microwaves, infrared, ultraviolet and visible light, to the tiny x-rays, gamma rays, and cosmic rays

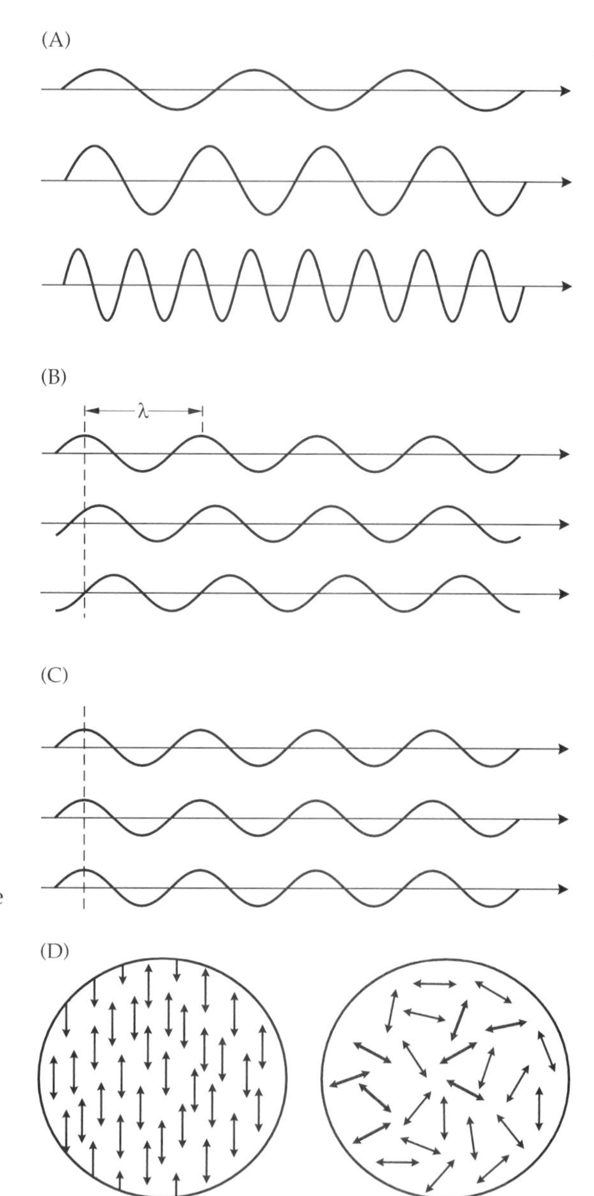

Figure 7.2 Complex, monochromatic, coherent and polarized light. A, B and C show the electric vectors in beams of light traveling from left to right, and D shows the electric vectors of a beam traveling into the page. (A) Complex light contains waves of different frequencies and therefore different wavelengths. (B) Monochromatic light contains waves of the same frequency, but they are not necessarily in phase. (C) Coherent light contains photons with the same frequency and the same phase. (D) Plane polarized light has parallel electric vectors, whereas unpolarized light has randomly oriented electric vectors.

that are similarly-sized or smaller than atoms ($\leq 10^{-10}$ m). Visible light is a narrow band of wavelengths around 10^{-6} m. Figure 7.3 summarizes the entire spectrum of electromagnetic radiation. It is customary to refer to the wavelengths of visible light in units of **nanometers** (nm). One nanometer is equal to 1×10^{-9} meters. Visible light ranges from about 300 to 800 nm.

To make use of electromagnetic waves for communication, organisms must evolve some way to detect them. For sound detection, the receptor must

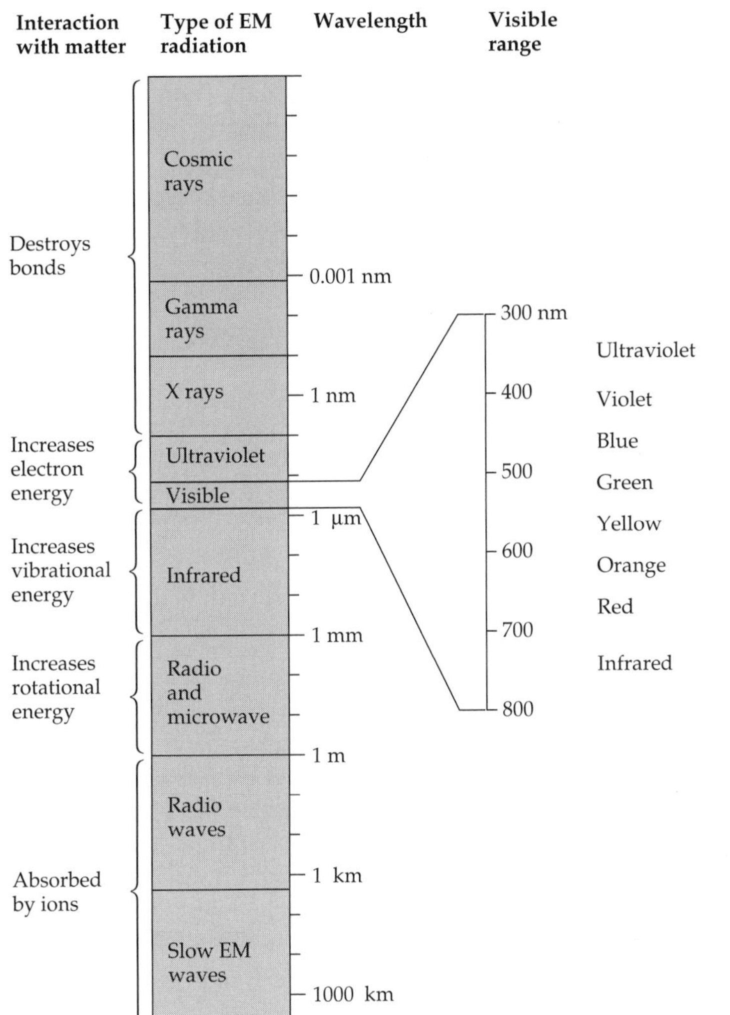

Figure 7.3 The spectrum of electromagnetic radiation. Different wavelengths of electromagnetic radiation interact with matter in different ways and are therefore used for a wide variety of functions. The visible light range has been expanded to show greater detail.

only detect the physical movement of molecules in the medium by using hair cells or membranes; this movement is then directly coupled to an excitable organelle. Detection of light is a more complex process. First, the electromagnetic energy must be trapped and absorbed by a receptor molecule. Changes wrought in the absorbing molecule must then be used to trigger an electrical response in the receptor neuron.

How might molecules trap electromagnetic radiation? Molecules can undergo several types of oscillations or vibrations (Barrow 1962). At the level of the whole molecule in gases and liquids, molecules move back and forth between collisions with each other (**translational oscillations**), and they may rotate around some axis (**rotational oscillations**). Within molecules, the component atoms vibrate towards and away from each other with the chemical bonds between them acting like springs (**bond oscillations**). Within each atom, electrons rotate in defined orbits or shells around the nucleus. Electrons can shift from their lowest-energy ground-state orbital into higher energy-level orbitals (**electron orbital oscillations**). For any given type of molecule, only certain frequencies of oscillations are allowed, and which frequency one finds in the molecule depends upon how much energy it possesses—the higher the energy, the higher the frequency for a particular type of oscillation, or for electrons, the further the orbit from the nucleus. The energy difference between oscillatory states is lowest for translational oscillations, somewhat higher for rotational ones, intermediate for bond vibrations, and highest for electron shell transitions.

A molecule can trap or absorb electromagnetic radiation only if the energy supplied by that wave is exactly equal to that needed to push this molecule into one of its higher oscillatory states. Because energy and frequency are proportional, this means that only certain frequencies of electromagnetic radiation can be absorbed by a given type of molecule. In fact, we can think of molecules as having their own vibrational **resonant frequencies** at which they absorb the corresponding frequency of light. Once energy is absorbed, it is trapped in the excited molecule until it can be lost through molecular collisions as heat (i.e., translational oscillation). It is during this time that changes in receptor molecules can be coupled to excitation of nerves. The trick to the design of a visual receptor is to use molecules that can absorb the kinds of electromagnetic radiation that will be most useful in mapping the external world.

As it turns out, there is only a narrow range of wavelengths that are suitable for vision. The long-wavelength end of the spectrum consists of radio waves that have such low frequencies and energies that they pass through and around nearly all nonmetallic and biological objects without being absorbed. Only ionized particles in gases and metallic materials absorb these frequencies. The atmosphere is currently full of high-intensity man-made radio waves that organisms cannot sense at all. In short, radio waves are not suitable as a medium for vision.

Radiation in the microwave range affects rotational states. When material with a strong natural resonance at this frequency is exposed to microwave radiation, the atoms rotate faster and collide more vigorously with their neighbors. Such collisions dissipate the energy as heat. Strongly dipolar molecules such as water absorb this type of radiation particularly well; high concentrations of microwave radiation in a microwave oven cause material containing water to heat up as a result of the rapid spin of the energy-absorbing water molecules. The absorption of microwaves by water molecules in the atmosphere makes the levels of ambient radiation at these frequencies too low to serve as a base for biological vision systems.

Infrared radiation affects bond vibrations and is also perceived by us as heat. Many types of molecules can absorb this frequency of radiation and become warmer in the process. Likewise, all objects in the environment with temperatures above absolute zero produce their own infrared radiation as a result of normal bond vibrations. This opens up the possibility of an infrared based visual system that discriminates objects in the environment radiating and reflecting different intensities of heat. Military technology uses infrared sensors to guide missiles and detect warm targets. There are two problems with this modality. The first is that the thermal world is extremely noisy and variable. The second is that infrared radiation is quickly absorbed by any biological tissues. Since even the simplest receptor organ must have a cell membrane and cytoplasm between the source of the radiation and the triggering molecules, the amount of energy reaching the latter will always be greatly reduced. Adding lenses and other structures to improve imaging would be even less useful as they would absorb all of the energy. Pit vipers have evolved an infrared sensor with which they detect the approximate location of their warm-blooded prey. The pit organ contains a thin membrane surrounded by air and loaded with an array of neurons that respond to small changes in heat. The low thermal mass of the membrane enables it to heat up quickly when the organ is aimed at an object warmer than its background. The heat receptors require five times more photons to generate a response than a visual light receptor. At best, a viper can detect a mouse at 30 cm and larger prey at somewhat longer distances (Newman and Hartline 1982). Thus an infrared detector does not come even close to an eye in its sensitivity and resolution.

Radiation in the visible and ultraviolet ranges affects electron states. Exposure to this range of frequencies causes the outer (valence) electrons of atoms to temporarily shift up to another shell or energy level. Many molecules have electron-state resonant frequencies corresponding to ultraviolet light and thus could be used as a light sensor. The problem here is that the energy absorbed and dissipated as heat after absorption of ultraviolet light is quite large and can often damage biological molecules. This problem becomes even greater with smaller wavelengths in the x-ray and cosmic ray range; these wavelengths contain so much energy that they ionize atoms, break chemical bonds, and destroy molecules. Although there are fewer molecules that have electron shell resonances in the visible light region, the heat loading and cellular damage after absorption are much more easily minimized because of the lower energy per photon. Thus radiation in the visible range is sufficiently powerful to shift electrons and produce configurational changes in molecules, but not so powerful that it breaks the chemical bonds.

The intermediate wavelength of visible light also imparts ideal transmission properties for fine-scale visual mapping of the external world. Objects are visible when they reflect light. Both larger and smaller electromagnetic waves are not reflected by organic objects but are transmitted right through them. Such objects would appear transparent or semitransparent if these wavelengths were used for vision. Water and a few other materials *are* transparent to visible light. Since life and vision evolved in the sea, this fact alone could explain the restricted wavelengths for vision. Straight-line transmission is also optimal in the visible range. Long electromagnetic waves such as radio waves are so much larger than organic objects that they bend around them much as sound waves do and thus would not be suitable for visual mapping. At the other end, shorter wavelengths are very strongly scattered by the atmosphere. Scattering impedes vision by reducing the contrast between objects and their background.

The sun is a typical **blackbody radiator** that produces electromagnetic radiation primarily in the ultraviolet to infrared region as a result of its very high temperature. In addition, most of the damaging and useless wavelengths are filtered out by Earth's atmosphere. Radio waves and microwaves (which our sun does not generate in great abundance because it is too hot) are either reflected or absorbed by the ionosphere in the outer atmosphere. Some of the infrared radiation is absorbed by atmospheric water and carbon monoxide but the remainder, mostly below 1000 nm, is able to penetrate and serves to warm Earth's surface. Most of the ultraviolet radiation below 300 nm is absorbed by the ozone layer of the atmosphere, but some reaches Earth. Finally, at the high energy end of the spectrum, x-rays and cosmic radiation are strongly scattered and gradually lose energy as they collide with particles in the atmosphere. Thus most of the electromagnetic wavelengths reaching Earth's surface fall in the region of the spectrum between 300 and 1000 nm (Figure 7.4). About 83% of this radiation falls in the visible range. Thus the

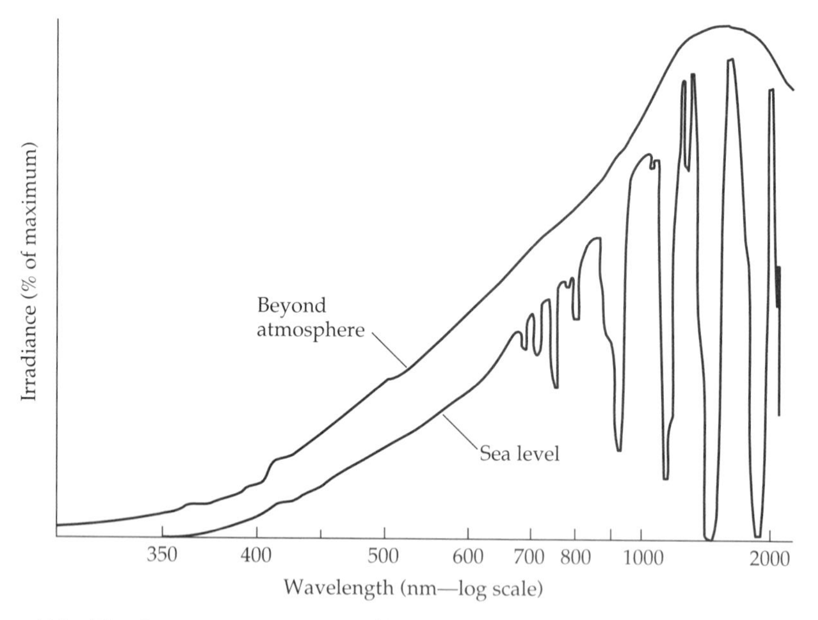

Figure 7.4 The frequency spectrum of irradiance from the sun. The smooth upper curve shows the spectral distribution of sunlight just outside Earth's atmosphere. Note that this distribution is very close to that expected from blackbody radiation. The lower curve shows the spectral distribution of sunlight at sea level. Short wavelengths are both scattered by the atmosphere and absorbed by ozone. Long wavelengths are absorbed by oxygen, carbon dioxide, and water. The sharp dips are primarily the result of absorption by water. (After Waterman 1981.)

frequency range that is biologically and physically most useful for vision is also the frequency range that is available.

TRANSMISSION OF LIGHT IN DIFFERENT MEDIA AND AT BOUNDARIES

Three crucial transmission properties of light in the visible range make vision possible: (1) reflection off of most solid objects, (2) straight-line transmission in a uniform medium without bending around objects or scattering in the medium, and (3) refraction at the boundary of two transparent media with different densities. Feature (1) ensures that light rays from the sun will bounce off of most objects in the environment in an orderly fashion rather than passing right through them. Feature (2) ensures that reflected light waves continue to travel in a straight line toward a potential receiver. This is what allows the visual receptor organ to make a spatial map using the reflected rays from different objects. The disadvantage of the straight-line-transmission property of light is, of course, that vision is blocked by larger objects or vegetation situated in front of potential objects of interest. Feature (3) facilitates focusing and image formation in the eye. Light from objects is gathered by a lens, refracted, concentrated and transmitted to sensory cells for analysis.

Medium-specific transmission speeds, reflection, refraction and diffraction are all consequences of the fact that light is transmitted as a wave. In terms of outcome, the basic principles of light propagation are similar to those we have discussed for sound waves. However, the driving force for light waves is an oscillating electric field rather than a vibrating membrane, so the reasons why light shares these principles are quite different. The interactions of light and matter are often quite different from what our intuition might suggest. Despite the illustrations in physics texts which imply that light bounces off of objects like a ball off of a wall, this is not what happens at all. Clarifying what really occurs when light strikes matter helps in understanding how colors and patterns are generated in light signals.

The Interaction of Light with Matter

Individual atoms usually consist of a balance of positive and negative charges and so are themselves electrically neutral. However, if some of the atom's electrons are only loosely bound to the nucleus, an external electrical field may pull these electrons to one side of the atom. The atom is then said to be **polarized**. Similarly, when atoms join to form molecules, the combination may not have equal numbers of electrons at all locations. Such a molecule is said to be polar because one end may be more negative than another. If an electric field were applied to such a molecule, it would tend to align itself with its positive side facing the negative side of the field and vice versa. Alternatively, two atoms in a molecule might be pulled apart or pushed together under the influence of an external electric field. Whether it is electrons, atoms, or molecules that move, the application of an electrical field to a medium will generate some separation and alignment of the charges within it.

The electric field of a light wave is just such a polarizing force. However, rather than being a static force, the polarity of the electric field oscillates with the frequency of the light wave. Molecules, atoms, and electrons in the medium are then forced by the light wave's oscillating electric field to move back and forth. Suppose the frequency of the light wave is ω, and the molecular resonant frequency of the medium closest to the light wave frequency is ω_0. As the electric field of the light wave oscillates, it forces either the entire molecule or electrically charged portions of it to move back and forth at the light frequency ω. This means that some light energy is transferred to the molecule. If $\omega = \omega_0$, successive cycles of the light wave add up to produce very large amplitude oscillations in the molecule (Weisskopf 1968). As we saw in our discussion of sound on pages 81–85, this build-up only occurs when the external oscillator has a frequency equal to a resonant frequency of the driven system. In the case of light as the external oscillator, the amplitude of the vibrations induced in the molecule build up to a critical point, at which time the molecule jumps into a higher energy state, trapping all of the incident light energy completely (and later trickling it off as heat). Because the light energy is completely absorbed, there is no electromagnetic radiation to propagate on to the next molecule. Absorption thus stops the propagation of a light wave and the medium is **opaque**.

In contrast, when $\omega \neq \omega_0$, the molecule does not resonate, and the amplitude of any induced vibrations remains at some low level. Any energy transferred to the molecule from the radiation is then reradiated by the molecule at exactly the same frequency as that absorbed. The passage of a light wave through a **transparent** medium is thus the result of a rapidly propagated series of absorption and reradiation events, and not the result of light waves merely traveling through and around the medium's atoms. A transparent medium is therefore composed of atoms and molecules whose resonant frequencies are sufficiently different from the frequencies of visible light. Water, for example, has resonant frequencies in the ultraviolet and infrared regions but not in most of the visible range; it therefore absorbs ultraviolet and infrared radiation but transmits visible light frequencies (Figure 7.5).

Transmission of Electromagnetic Radiation in Gases, Liquids, and Solids

When molecules of a medium are struck by electromagnetic radiation of frequencies other than resonant ones, the molecules acquire some of the energy of the incident waves and quickly reradiate it in all directions as light of the same frequency. The relative phases of the wavelets reradiated from adjacent molecules will depend on whether the molecules were struck by the same part of the electric wave cycle in the incident light wave or not. When molecules are far apart relative to the wavelengths of the incident light, they are likely to experience different parts of the incident light cycle and thus to reradiate out of phase with each other; where they are close together, they will all be activated by the same part of the incident light wave cycle, and thus reradiate in phase. In a gas, molecules are sufficiently far apart relative to the

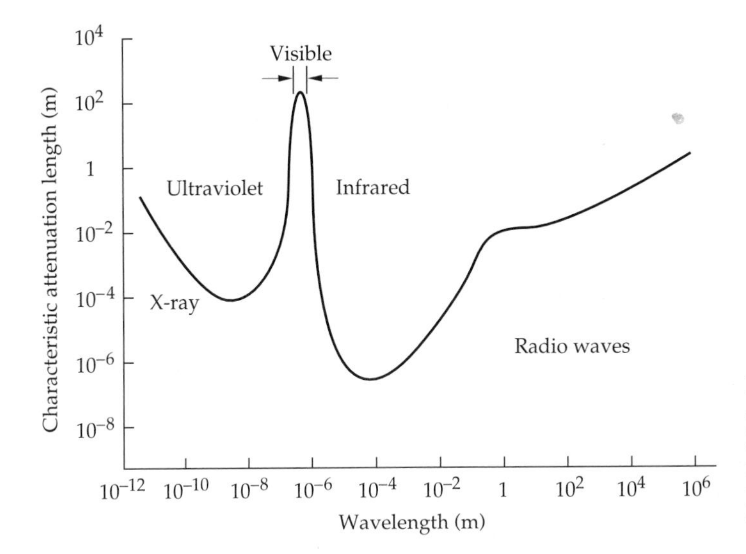

Figure 7.5 The transmission of different wavelengths of light by water. Note the narrow but strong transmission peak in the visible range. (From Williams 1970.)

wavelengths of ultraviolet, visible, and infrared light that reradiations have random phases and show no consistent pattern of interference. In liquids or solids, whole layers of molecules are sufficiently close together that they reradiate wavelets in phase with each other (coherently, as in Figure 7.2) and out of phase with the wavelets reradiated by more distant layers of molecules. Thus complex patterns of interference between the wavelets are generated.

It turns out that reradiated wavelets traveling in the forward direction through a liquid or solid (the direction of propagation of the external light wave) will add up constructively and propagate the wave in the forward direction. All wavelets reradiating in other directions (sideways and backwards) will destructively interfere with those emanating from neighboring molecules and cancel each other out. This basic process, analogous to the propagation of a sound wave, explains the forward, straight-line transmission of waves—only the forward-moving components are in phase. Thus in liquids and solids, incident electromagnetic radiation will generally be propagated in straight lines through the material unless it has a frequency equal to a resonance in the material, in which case it will be absorbed. Some light may be scattered to the sides in all materials during this propagation. The more ordered the molecules in the medium and the closer they are together relative to the wavelengths of the light, the less scatter there will be. For these reasons, gases such as air scatter light more than liquids such as water.

The Speed of Light in Different Media

When the molecules, atoms, or electrons in a medium are forced to align themselves with the electric field of an incident light wave, the positively charged parts of the molecules move closer to the negative side of the field, and the negatively charged portions closer to the positive side. This separation and alignment of the charges in the medium generates a second electric field that is opposite in polarity to that of the light wave and thus opposes it. The ability of a medium to be thus polarized and build up a counterfield is called its **permittivity** and denoted by ε. Because the electric field generated by the medium is changing with time, it also generates a perpendicular magnetic field that opposes the magnetic field of the light wave. The ability of a medium to be magnetically aligned by an external magnetic field is called its **magnetic permeability** and denoted by μ.

The speed of an electromagnetic wave is maximal when traveling through a vacuum such as outer space. In that medium the wave experiences no counterforces as it establishes its electric and magnetic fields at each point through which it passes. Even in a vacuum, however, it takes a finite amount of time to establish these fields, and thus the speed of light is not infinite. The speed of light in a vacuum is usually denoted by c. When light propagates through matter, the speed of propagation is generally reduced. This reduction occurs because both the electrical and magnetic counterfields set up in the medium work against the external fields of the light wave, and they thus hinder the process of inducing oscillations within the medium's molecules, atoms, or electrons. This in turn slows down the speed of propagation of the light wave through the medium.

The speed of light propagation in any medium is simply $v = 1/\sqrt{\varepsilon\mu}$. Although differences in magnetic permeabilities between media do play some role, the major cause of differences between media in the speed of electromagnetic propagation arises from their differing electrical permittivities. Permittivities are usually measured relative to that in a vacuum where there are no atoms or molecules to polarize. The ratio of the permittivity of a medium to that of a vacuum is called the **dielectric constant** for that medium and denoted by k. Dielectric constants range from 1 to 25 for materials that are generally considered as poor conductors of electricity, are as high as 80 for very polar molecules such as water, and are infinite for metals and other electrical conductors. Because differences in permittivity play such a large role in generating differences in the speed of electromagnetic propagation, one can usually use the rough approximation that $v \approx c/\sqrt{k}$. As with permittivities, it is useful to measure the speed of light in a medium relative to that in a vacuum. The corresponding relative measure, $n = c/v$, is called the **index of refraction**; the larger its value, the more the medium slows down the propagation of electromagnetic waves. From this definition and the fact that permittivity differences usually determine propagation speeds, it follows that $n \approx \sqrt{k}$. The degree to which this approximation holds can be seen in Table 7.1, which gives the dielectric constants, speeds of light, and refractive indices for a number of different media. The speed of light in water is very fast for its high dielectric constant. This rapid speed occurs because the highly polar molecule forms strong hydrogen bonds with neighboring molecules to create a compact lattice, like glass, that readily propagates light waves.

The dielectric constant, the speed of light in a medium, and the medium's index of refraction are all partly dependent upon the type of matter making

Table 7.1 Speeds of light and refractive indices in different media

Medium	Speed (m)	Refractive index	Square root of dielectric constant
Air	2.99×10^8	1.00028	1.000295
Water	2.25×10^8	1.33	8.9
Glass	1.99×10^8	1.5	1.4
Diamond	1.25×10^8	2.4	2.3

up the medium and partly dependent on the frequency of the light. The closer the frequency of the incident electromagnetic radiation is to a resonant frequency of the medium, the larger the amplitudes of the charge separations induced in the medium, the stronger the induced counterfields, the higher the effective dielectric constant, and thus the slower the propagation of the light. This relationship is captured in the following additional formula for the index of refraction of a medium:

$$n = \frac{c}{v} = 1 + \frac{KN}{(\omega_0^2 - \omega^2)}$$

where N is the density of atoms in the medium and K is a compound constant that incorporates the weight and charge of electrons. The density of atoms decreases the speed of propagation v, (and thus increases n), because the more molecules or atoms there are available for polarization, the larger the counterfield that can be induced in a material by an external electric force. The frequency difference $\omega_0^2 - \omega^2$ reflects the degree to which the driving frequency is similar to the resonant frequency. When the two are equal, the denominator in the above expression is zero, and thus n is infinitely large. This is equivalent to saying that the speed of propagation in the medium at this frequency is zero—everything is absorbed and nothing is transmitted. At driving frequencies close to ω_0, transmission will be poor but not zero, and there will be complicated phase effects. As the difference in frequencies increases, n drops off quickly. In fact, because the frequency terms are squared, it will not take a very large difference between ω and ω_0 before $n \approx 1$.

One medium that is an exception to these patterns is metal. Metallic atoms are electron donors, and large arrays of them form a solid medium in which the nuclei are densely packed in a matrix but the valence electrons are free to move along the matrix. The resonant frequency of these unbound electrons is effectively zero. When the driving frequency of the electromagnetic wave is much greater than the resonant frequency, the motion of the unbound electrons in the presence of a wave is 180° out of phase with the driving wave. The opposing phase and complete freedom of the moving electrons results in a counterfield in the metal that is exactly opposite to the external electric field. The result is that the electric field of an incident light wave is completely canceled by the counterfield inside the metal. Put another way, the dielectric constant of a metal is infinity, and thus the speed of light propagation inside the metal goes to zero; there is no internal propagation.

Reflection and Refraction

When a beam of light encounters a boundary between two media with different transmission speeds, it is usually partially reflected and partially refracted. If the transmission speeds in the two media are very different, most of the light will be reflected. The angle of the reflected beam is equal to the angle of the incident beam from its source to the plane of the boundary. (Both angles are measured from the normal and denoted by the symbol θ.) If the boundary's surface is very smooth, the intensity of the reflected light will be great and the object will appear shiny and highlighted (Figure 7.6C). This is called **specular reflectance**; in the animal world it is found in silvery fish scales and the elytra of many beetles. On the other hand, if the surface is rough, more of the incident light will be reflected in other directions (Figure 7.6A) and the perceived reflection will be dull and lower in intensity. This is called **diffuse reflectance** and is typical of most animal surfaces including skin, fur, and feathers. As we saw above, metals transmit none of the incident light energy; instead, all incident rays are reflected, making them look shiny.

If the transmission speeds of the two media are not very different, most of the light will cross the boundary and enter the second medium. However, it will be refracted (bent) at the boundary relative to the normal. When the second medium has a slower speed (higher index of refraction) than the first medium, the beam will be bent toward the normal (Figure 7.7; compare to Figure 2.3). When the first medium has the slower speed, the beam will be

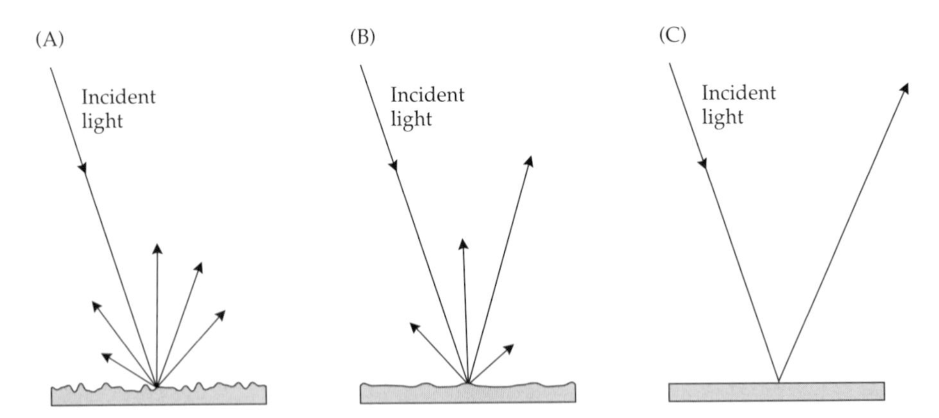

Figure 7.6 Reflection of light. The degree of optical smoothness of a surface determines the extent to which the incident radiation is reflected in all directions. A rough surface such as in (A) reflects light in all directions and the material appears matte; this is called diffuse reflection. Increasingly smooth surfaces (B) result in a more directional reflection. When the surface is perfectly smooth (C), specular reflectance occurs and all of the incident light is reflected in one direction, such that the angle of reflection equals the angle of incidence.

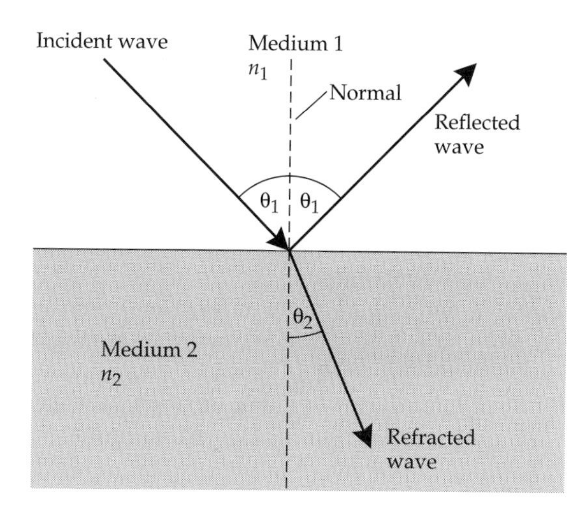

Figure 7.7 Refraction of light traveling from a medium with a faster speed such as air into a medium with a slower speed such as water. The normal is a line perpendicular to the boundary between the two media. The angle of incidence, θ_1, is measured from the normal. The incident wave is bent toward the normal in the second medium in this diagram. If the speed of the wave were faster in the second medium compared to the first medium, the wave would be bent away from the normal. Compare to Figure 2.3.

bent away from the normal. The greater the difference in the refractive indices of the two media, the more the beam is bent. The angle of refraction, θ_2, can be computed precisely from **Snell's law**:

$$n_1 \sin\theta_1 = n_2 \sin\theta_2$$

where n_1 and θ_1 are, respectively, the refractive index and angle of incidence from the normal in the first medium and n_2 and θ_2 are the equivalent measures in the second medium. The refraction of light plays a critical role in visual reception. A double convex lens collects and concentrates the incoming light waves to produce a more intense image. Light enters the lens from one side and is bent toward the center axis of the lens, i.e., toward the normal, which in this case is a line perpendicular to the surface of the lens at the point of entry. Light is bent again in the same direction (now away from the normal to the second boundary surface) as it exits the lens and enters the air. The distance of the focal point from the lens depends on the lens' curvature—the greater the curvature, the shorter the focal distance (Figure 7.8A). This occurs because light hits the surface of a more curved lens at a larger angle of incidence and is refracted more. (For example at a 10° angle of incidence a beam is bent 3.3°, whereas at a 30° angle of incidence it is bent 10.5° if $n = 1.5$.) Refraction is also responsible for our inability to accurately locate an object in water when viewed from the air (see Figure 7.8B).

Even when light encounters a boundary between two transparent media, there is always some reflection. Similarly, when light encounters a medium that completely absorbs it, there is nevertheless a strong reflection. How can we explain these phenomena? Reflection is a property of the surface molecules in a medium irradiated by electromagnetic waves. Recall in the description of wave progression in a liquid or solid that the sideways and backwards reradiations of neighboring atoms always cancel each other out. However, for the surface layer of a medium, there are no similar neighbors on the incident

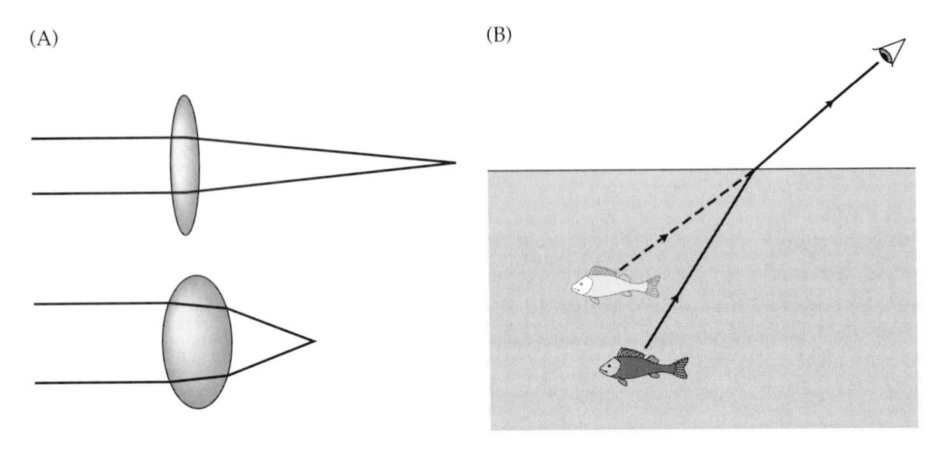

Figure 7.8 Consequences of refraction. (A) A double convex lens refracts light towards the central axis of the lens twice, once as it enters and again as it exits, and causes a parallel beam to converge at a focal point. The distance of the focal point from the lens depends on the curvature of the lens. A flat lens has a long focal length, and a round lens has a short focal length. (B) Refraction causes the deflection of an object in water when viewed from air at an angle. The object (black fish) appears both closer to the viewer and displaced at a different angle.

side, so any backwards reradiations are not subject to destructive interference. Furthermore, the surface electrons in a strongly absorbing medium do not have as many neighbors with which to collide and thus reradiate more of the energy they have absorbed as photons before losing it as heat. In all of these cases, light must travel into a material at least as far as one layer of atoms or molecules before it can be reradiated out. The distance it does travel into the second medium is called the **skin depth**, and is approximately equal to one-half wavelength of the incident light wave. (The reason why radio waves can pass through walls is because walls are much thinner than the skin depth of these long wavelengths.)

The fraction of light that is reflected versus refracted at the boundary between two transparent media depends not only on the relative indices of refraction of the media but also on the angle of incidence. The principles of reflection and refraction at such boundaries are analogous to those described for sound (pages 24–29) and are summarized in Box 7.1. For our purposes, it is sufficient to understand the conditions for **total internal reflection**. When light from a relatively dense medium (high index of refraction) hits the boundary of a medium with a lower index at a sufficiently high angle of incidence, it is completely reflected. This occurs at the point where the angle of refraction is 90° or larger and Snell's law is undefined. Internal reflection affects vision in two important ways. One is the restriction of the visual field to low angles of incidence for an animal in water looking up at the surface toward objects in the air (Figure 7.9A). A second role of internal reflection is the waveguide action of light waves trapped within single long photoreceptors (Figure 7.9B).

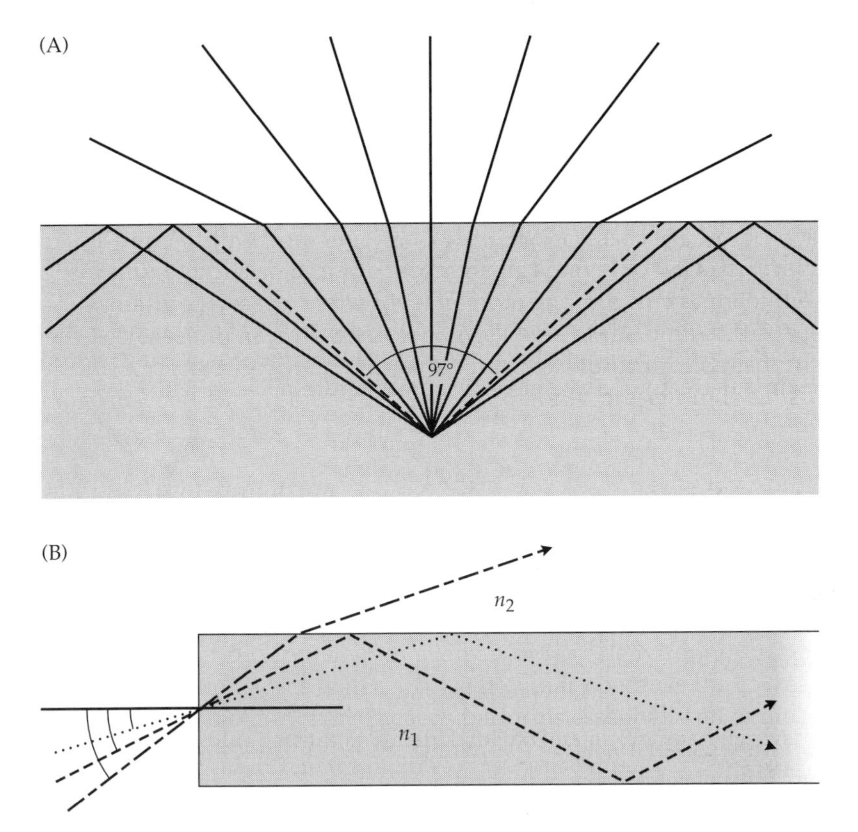

Figure 7.9 Consequences of total internal reflection. (A) Snell's window. Because of refraction, an animal in water that looks up at the surface sees the entire hemisphere above the water condensed into a solid angle of 97°. At angles of incidence greater than 48.7°, light is reflected back into the water at the surface and this region appears mirrorlike. (B) Waveguide action in a photoreceptor cell. Photocells are long tubular structures about two to three wavelengths in diameter. The refractive index within the cell is higher than that for the surrounding medium. Light that enters the cell at large angles of incidence to the longitudinal walls (i.e., at grazing angles) is reflected back into the cell. (A after Lythgoe 1979; B after Horowitz 1981.)

FREQUENCY-DEPENDENT PROCESSES AND COLOR

As mentioned earlier, we perceive different frequencies of light as different colors. Color is an important component of most visual signals. As with sound, the transmission of light through air or water is frequency-dependent, so the transmission properties of the environment affect the efficiency of signal transfer. Three frequency-dependent processes are relevant to visual signaling: scattering, frequency-dependent refraction, and differential absorption.

Scattering

As we learned in Chapter 2 (pages 19–22), waves are scattered when they intercept objects in the transmitting medium. Similar basic principles to those

Box 7.1 Polarization of Reflected and Scattered Light

WHEN LIGHT FROM A LESS DENSE MEDIUM strikes the surface of a denser dielectric medium at an angle, the reflected ray is partially polarized in the plane parallel to the surface. At a critical angle, called **Brewster's angle**, it is completely polarized. This phenomenon occurs when the incident and refracted angles add up to 90°. Brewster's angle can be computed by $\tan \theta_B = n_2/n_1$. In the illustration below for light incident in air hitting a glass boundary, Brewster's angle is about 56°. The incident ray contains electric vectors oriented in all possible directions but only two are shown, $\mathbf{E}_1$ oriented in the plane of the page and $\mathbf{E}_2$ oriented perpendicular to the page. At the critical angle, only vector $\mathbf{E}_2$ will be present in the reflected wave and $\mathbf{E}_1$ will predominate in the refracted wave.

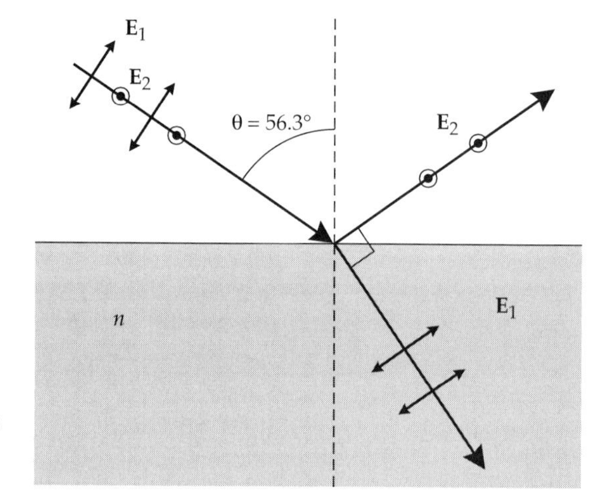

Figure A Differential reflection and transmission of the polarized components of a light wave intercepting a glass boundary at Brewster's angle of incidence.

Polaroid sunglasses that transmit only vertically polarized light are designed to cut the glare of the horizontally plane-polarized light that reflects off of flat substrates but not to reduce other light substantially. Take your glasses off and rotate them 90°. Now you will see primarily the reflected glare!

The reason why this directional effect occurs is because light polarized in different directions experiences different magnetic permeabilities when it enters the second dielectric medium. The orientation of e-vectors is expressed relative to the plane of incidence, which in Figure A is equivalent to the plane of the page. $\mathbf{E}_1$ is the **parallel e-vector** and $\mathbf{E}_2$ is the **perpendicular e-vector**. The parallel e-vector is less impeded and can pass into the medium more easily than the perpendicular e-vector, which is more likely to be reflected. This reflection property of light is analogous to the way that acoustic impedance and transmission speed affect the reflection properties of sound (described in Box 2.2). Figure B shows the equivalent plots for the reflection coefficient of light as a function of the incident angle. Recall that the reflection coefficient, R, ranges from –1 to +1 and its sign indicates whether it is phase-shifted or not. The x axis for these plots shows the incident angle θ from the normal rather than the grazing angle ϕ from the boundary surface. The case for an incident wave in a faster medium hitting a boundary with a slower medium ($v_1 > v_2$) is shown on the left, and the reverse case is shown on the right ($v_1 < v_2$).

Since a natural incident light beam contains unpolarized light with a full range of e-vector orientations, the boundary acts as a filter for waves polarized in different direc-

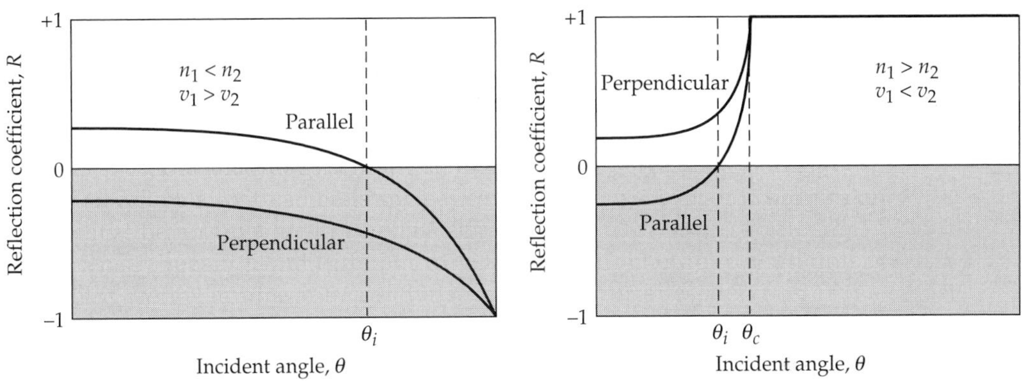

Figure B Reflection coefficient as a function of incident angle for parallel and perpendicular polarized light waves. (After Guenther 1990.)

tions. For both conditions of relative transmission speeds in medium 1 versus medium 2, there is an angle of intromission, θ_i, at which the perpendicular e-vector rays in a beam of unpolarized light are completely transmitted into the second medium ($R = 0$), whereas the parallel e-vector rays continue to be reflected ($R = 0$). This is Brewster's angle, i.e., $\theta_i = \theta_B = \tan^{-1}(n_2/n_1)$. The plot on the right also shows the critical angle θ_c at which total internal reflection occurs for a light beam incident in a slower medium hitting a boundary with a faster medium. This is the angle of total internal reflection given by $\sin \theta_c = n_2/n_1$ (defined only for $n_2 < n_1$). (The totally reflected light wave also undergoes the complex phase shifts with increasing θ as described for sound; however, this shift is not illustrated here.)

For similar reasons, scattered light is also plane polarized when viewed from a 90° angle from the incident beam. Figure C contrasts the e-vector populations of light

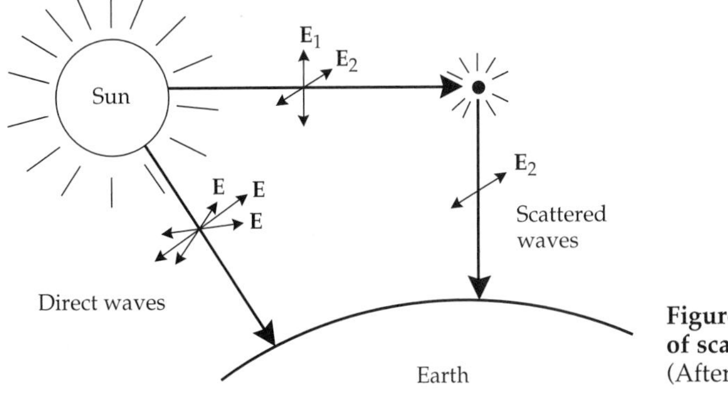

Figure C Polarization of scattered skylight. (After Camhi 1984.)

Box 7.1 (continued)

waves traveling directly from the sun to the earth with those scattered by atmospheric particles at 90° from the sun. Direct sunlight is unpolarized and contains all possible e-vector orientations perpendicular to the direction of travel. In the scattered sunlight case, the initial waves from the sun are similarly unpolarized but only two e-vectors are shown—$\mathbf{E}_1$ oriented in the plane of the page and $\mathbf{E}_2$ oriented perpendicular to the page. The particle scatters light in all directions, but since e-vectors must be perpendicular to the direction of travel, only waves with orientation $\mathbf{E}_2$ can end up in the path from the particle to the earth. At angles larger or smaller than 90° the scattered light is only partially polarized. Animals such as bees and other insects that can perceive polarized light can use the pattern of polarized light in the sky to determine the time of day (see Figure 7.11).

discussed for sound waves also apply to light waves. Rayleigh scattering occurs when the particles are much smaller than the wavelength of light, and Mie, or diffractive, scattering occurs when particles are similar in size to the wavelength. When primarily small particles are present, smaller wavelengths will be scattered more than longer wavelengths, leading to frequency-dependent transmission. However, the mechanisms by which scatter is achieved are somewhat different for light. As we learned earlier, the electrons in atoms and molecules exposed to electromagnetic radiation absorb and reradiate the waves to some degree. The reradiation is responsible for the scattering. The intensity of scattered light, I_s, is strongly frequency-dependent according to the relationship:

$$I_s = \frac{K}{\lambda^4}$$

where K is a constant. Plugging in values for the wavelengths of red and blue light, one can easily show that short, blue wavelengths are scattered about four times more intensely than long, red wavelengths (Figure 7.10). This differential scattering is ultimately a consequence of the higher energy of shorter wavelengths.

You may now be wondering why a forward-propagating wave doesn't become bluer with distance, given the stronger reradiation intensity of shorter wavelengths. The equation above expresses the radiation intensity from a single vibrating atom or molecule. When an electromagnetic wave passes through a transparent liquid or solid medium, the number of atoms or molecules that reradiate in phase and in the forward direction increases with the wavelength of the light (since the longer the wave, the more molecules that are affected by the same part of the light wave cycle). Thus a longer wavelength is reradiated at a lower intensity per molecule, but more molecules reradiate in phase than for a short wave of incident light. The two effects

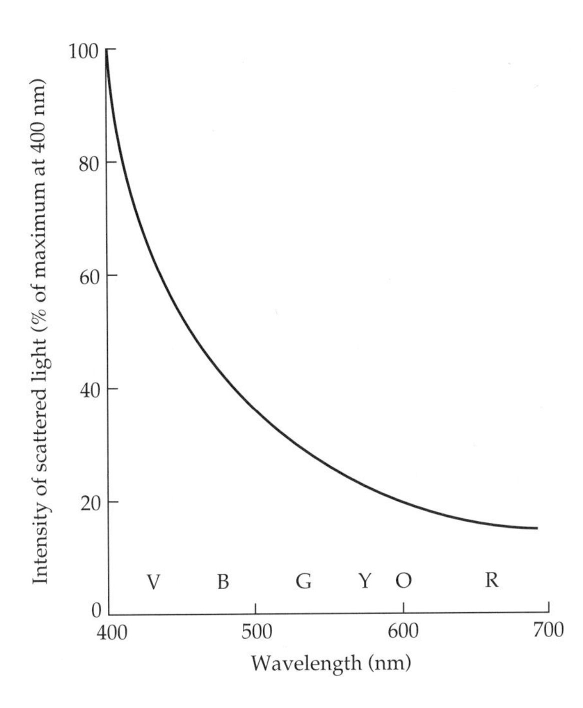

Figure 7.10 Rayleigh scattering of light. Violet (V) and blue (B) wavelengths are scattered more than the longer wavelengths of green (G), yellow (Y), orange (O), and red (R).

counteract each other, with the result that the overall intensity of forward-propagated short and long wavelengths is similar.

In a gas, however, neighboring molecules are randomly arrayed and constantly moving so sideways reradiations are not likely to be canceled out with destructive interference as in a liquid or solid medium. Each molecule is assumed to reradiate independently. Sideways scattering is strong and it favors the shorter wavelengths. Longer wavelengths in a gas are more likely to be propagated in the forward direction. This differential scattering of light by small, independent molecules in a gas medium explains why the sky is blue and why the sun appears yellow. Shorter blue wavelengths are scattered multiple times and reach a viewer via indirect paths from other parts of the sky than the straight line path between the sun and the viewer. The light traveling the straight path loses some of its blue light in transit, leaving a preponderance of longer wavelengths that makes the sun appear yellow. In fact, the intensity of longer wavelengths decreases gradually as one moves across the sky away from the sun. The intensity of Rayleigh-scattered short wavelengths is more uniform across the sky. Animals with better visual perception in both the ultraviolet and red ranges than we have can use the ratio of short to long wavelengths to evaluate the azimuth and elevation of the sun and thus estimate time of day. For such animals, the sky is probably not all blue but a gradient of color in rings around the sun (Coemans et al. 1994). We obtain a similar sensation when we view the sky at sunrise and sunset. Light traveling from the sun when it is low in the sky is being transmitted through more kilometers of atmosphere than it is at noon, and more blue light is scattered to the opposite side of the sky compared to the sunny side.

Our atmosphere also contains larger particles such as dust and water droplets that scatter all frequencies of light via the process of Mie, or diffractive, scattering (as discussed in Chapter 2). Because these particles are at least as large as the wavelengths of light and consist of denser material with a large array of regularly spaced molecules, they act like a boundary that reflects and refracts all wavelengths equally. This type of scattering tends to reduce the blueness of the sky, making it whiter in color. Skies are always bluer in dry, high altitude locations compared to dusty, humid, and coastal locations. On the moon, where there is no atmosphere, the sky appears black and the sun white because there is no atmospheric scattering.

Finally, Rayleigh scattered light is completely polarized when viewed from a 90° angle relative to the light source, and partially polarized at most other angles (see Box 7.1). If one were to look at the entire sky through a Polaroid lens, one would see a broad strip of vertically plane-polarized light at 90° from the sun (Figure 7.11). Many invertebrates, especially flying insects, use the pattern and orientation of polarized light in the sky to orient themselves and to determine the time of day.

Frequency-Dependent Refraction

A beam of complex (white) sunlight can be split into the spectrum of colors when it is passed through a triangular glass prism. The process is called **dispersion**. Dispersion occurs because the different frequencies travel at slightly different speeds in glass. The differential speed of long and short wavelengths follows directly from the equation for the refractive index, n, given earlier. The resonant frequency of electrons in glass is approximately 15.0×10^{14} Hz. Short blue wavelengths (6.7×10^{14} Hz) are closer to this resonant frequency than long red wavelengths (4.3×10^{14} Hz), so that the transmission speed for blue is slower than for red. The slower the speed in the second medium (glass), the more the waves are bent at the boundary, so blue waves are bent more at both surfaces of the prism. A similar process in raindrops produces rainbows.

Differential refraction in the lens of an eye generates chromatic aberration. Because short wavelengths are bent more, the focal point for blue and violet colors is shorter than the focal point for red. Try focusing on adjacent strips of red and blue—the boundary is fuzzy and constantly moving as the eye alternately tries to focus on each color. A red center surrounded by a blue region gives the appearance of depth because the red seems closer than the blue. In animals this red-inside-blue strategy enhances visual signals, as in the penile display of the vervet monkey and the facial color of the male mandrill.

Differential Absorption

The blue color of water is not due to Rayleigh scattering; in water the molecules are so tightly packed that wavelength-differential sideways scattering is minimal. There is substantial Mie scattering due to fine particles suspended in water. However, a more significant frequency-dependent effect is the strong absorption of longer wavelengths by water. As mentioned earlier in this chap-

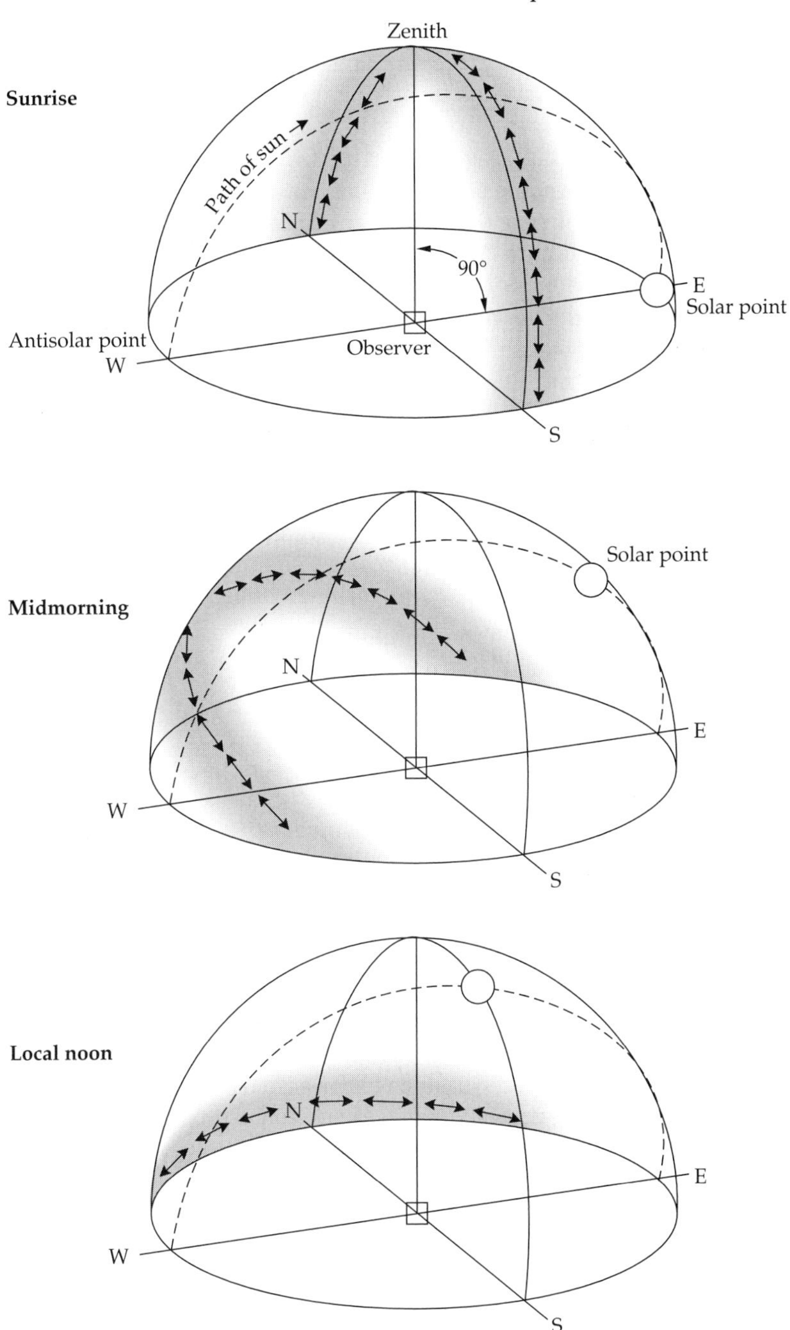

Figure 7.11 Patterns of polarized light in the sky at different times of day. Light from the sun produces a band of polarized scattered light at a viewing angle of 90°. The position of the band varies with time of day as shown. The double headed arrows show the orientation of the e-vectors within the polarized band. (After Wellington 1974; Wehner 1976.)

ter, water absorbs both infrared and ultraviolet radiation. The infrared radiation that is absorbed extends significantly into the visible red region. At about 10 meters below the water's surface, humans can no longer detect any red radiation. The deeper one descends under the water, the greater the absorption of red and eventually orange wavelengths. Dissolved oxygen in water also absorbs ultraviolet and violet wavelengths. In very clear water such as the open ocean and large clear lakes, the effect of the absorption of ultraviolet and red wavelengths result in the water becoming increasingly monochromatic blue at greater depths (Figure 7.12) (Kirk 1994).

In shallow freshwater ponds and lakes with more organic matter and phytoplankton suspended in the water, the spectral peak shifts to green and yellow. Some streams, swamps and marshes with a great deal of organic matter, tannins, and iron ore washing into the water show a reddish spectral peak. Water with a spectral peak in the near infrared is outside of our color sensitive range and therefore appears black (Lythgoe 1979).

Even in terrestrial environments, the color of the ambient light may be affected by differential absorption, reflection, and filtering caused by objects in the environment. For example, in densely vegetated habitats such as forests the ambient light is decidedly green because of the absorption of red and blue wavelengths by the chlorophyll in leaves. The color of light reflecting off soil, rocks, and, in shallow marine environments, corals, affects the spectral composition of the ambient light. Since most animals rely on reflected light to

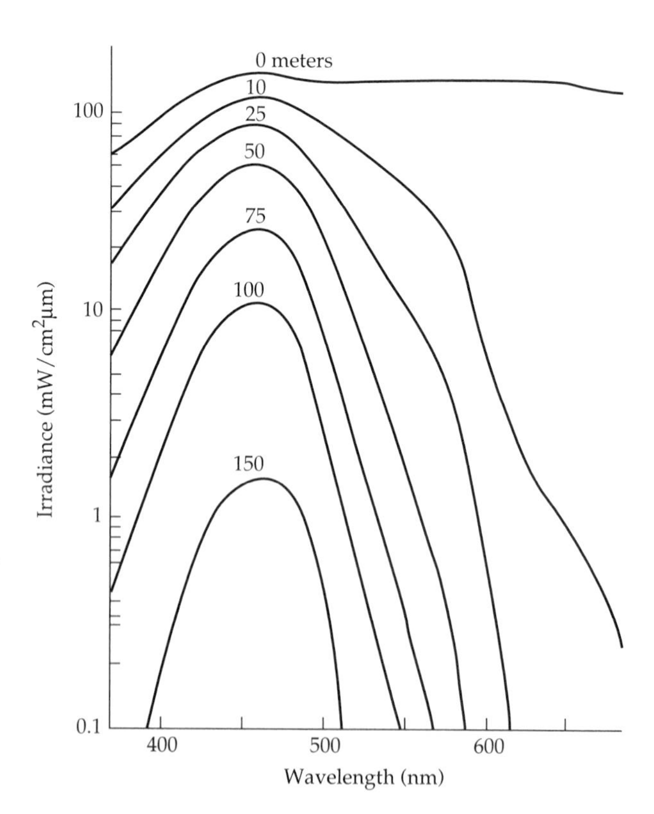

Figure 7.12 Irradiance frequency spectra in water at different depths. The intensity of longer wavelengths in the red region drops off very quickly with increasing depth because of selective absorption. Ultraviolet and violet wavelengths are also absorbed. The consequence is an increasingly monochromatic blue medium at greater depths. (From Waterman 1981, based on Lundgren and Højerslev 1971.)

send visual signals, the amount and spectral composition of the light that irradiates the environment is an important determinant of the colors and types of visual signals that animals can use (Endler 1993).

MEASURING LIGHT INTENSITY

It is important to distinguish two types of light measurements: **radiance** and **irradiance**. Both are measures of radiant flux (energy per unit time). The two differ in the acceptance angle of the sensor. Irradiance is the total amount of light incident on a surface, and includes scattered (diffuse) light as well as direct light. It is measured by an instrument that collects the light from a 180° solid angle. Radiance, on the other hand, is the flux of energy emitted from a specific radiant area such as the sun or an animal's body or signal patch. Only the light that travels directly from the source area to the receiver is measured using a tubelike or telescoping instrument that cuts out scattered light (see Box 7.2). The solid angle over which the instrument is measuring must be specified. The units for these different measures are shown in Table 7.2. Both radiance and irradiance measures can be made wavelength specific by filtering out all but a narrow range of wavelengths. Each type of sensor must correct for the angle of the primary light source relative to the orientation of the sensor. In addition, radiance measurements must take into account the orientation of the animal relative to the sensor (Endler 1990).

The photoelectric devices we use for measuring color (Box 7.2) suffer the problem that they record radiant energy or power, whereas the relevant unit for animal vision is quantum flux, the number of photons per unit time. Since a photon's energy is related to its wavelength, the conversion from energy flux to photon flux must be computed for each wavelength interval. The photon flux for a single wavelength, $Q(\lambda)$, in micromoles of photons $m^{-2}\ s^{-1}$ can be computed using $Q(\lambda) = 0.00835191E(\lambda)$ where E is energy flux in watts m^{-2} (Endler 1990).

A series of light-intensity measurements made at different wavelengths generates a radiance or irradiance spectrum equivalent to a frequency spectrum for sound. For light, however, the horizontal axis of a spectrum is usually wavelength instead of frequency. We shall use this convention in discussions of color and light intensity in subsequent chapters.

Table 7.2 Measures of light intensity

Measurement	Name	Units[a]
Flux (flow per unit time)	Radiant flux	Photons s^{-1}
Flux density at a surface	Irradiance	Photons $s^{-1}\ m^{-2}$
Flux per unit solid angle	Radiant intensity	Photons $s^{-1}\ sr^{-1}$
Flux per unit solid angle per unit area	Radiance	Photons $s^{-1}\ sr^{-1}\ m^{-2}$

[a]sr = steradian, the unit of solid angle; there are 4π steradians in the complete solid angle of a sphere.

Box 7.2 Light-Measuring Instruments

THE GENERAL TERM for light measuring instruments is **radiometer**. There are a variety of instruments designed for specific applications. They vary in the type of sensor, the range of wavelengths over which they are sensitive, and their ability to measure radiance versus irradiance (Sommer 1989).

SENSORS. There are three basic types of sensors used in radiometers. **Vacuum phototubes** and **photomultipliers** are sealed cathode tubes containing certain types of gases that generate a current of electricity in the presence of radiation. They are bulky and require a high level of thermal and voltage stability, but are very sensitive over a large range of wavelengths. **Photoconductor cells** exploit the variable conductivity of semiconductor materials such as silicon. The cell consists of a thin layer of the semiconductor material on glass. Before irradiation, the resistance of the cell is very high. Upon absorption of radiant energy, the valence electrons of the semiconductor pass into a higher energy level where they are able to conduct a current placed across the cell. The shortcomings of such detectors are the limited range of wavelength sensitivity (dependent on the specific material) and the nonlinear relationship between radiant power (watts) and resistance. **Photodiodes** exploit the photovoltaic effect at the interface between a semiconductor and a metal. Electron transitions generate either a voltage in the cell or a decrease in resistance in an electrical circuit with an external voltage source. Diodes measure accumulated radiant energy (Joules) and can be made very small. A silicon photodiode permits linear responses over 10 orders of radiant power between 100 and 1100 nm.

RADIANCE VERSUS IRRADIANCE COLLECTORS. Radiance and irradiance radiometers differ in the geometry of their light-collecting area. **Irradiance meters** are designed to collect light incident over a full 180° hemisphere. The collector is typically a flat round disc (Figure A). However, radiant flux arriving from large incident angles will result in a lower response on a perfectly flat collector compared to the same flux arriving at normal incidence because the same beam is spread over a larger area. The reduction in output is proportional to the cosine of the incident angle and is called the cosine response (recall that cos 0° = 1 and cos 90° = 0). To correct for this fall-off and provide

Figure A Cosine irradiance collector.

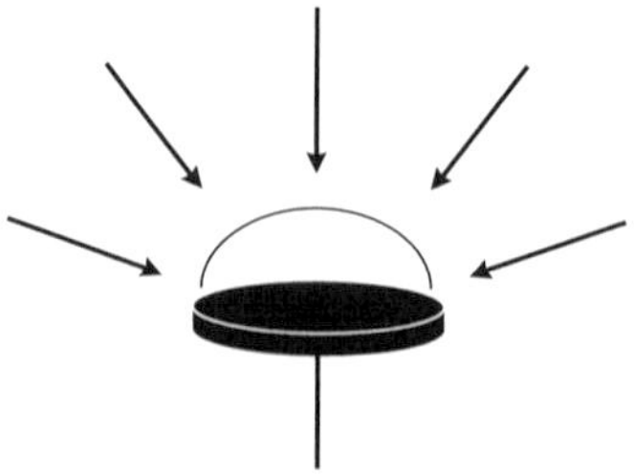

Figure B Radiance collector.

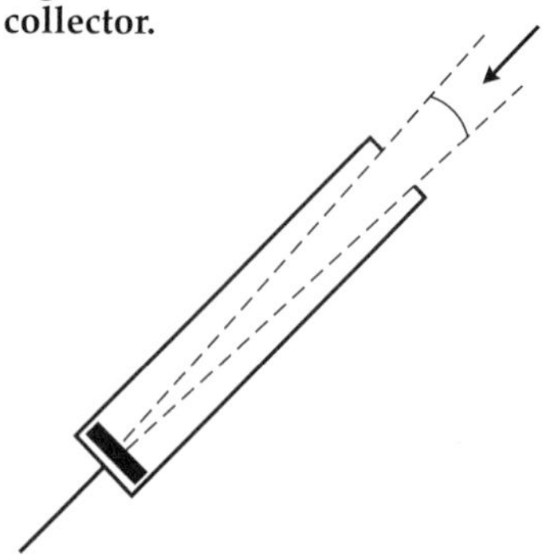

equal stimulation from all angles, the sensor disc either projects slightly above its surrounding housing or it is covered with a hemispherical diffuser. This is called a cosine-corrected collector. **Radiance meters,** on the other hand, are designed to collect light from a very narrow angle. They consist of a tubelike or telescoping probe that cuts out incident light from all but a small solid angle (Figure B). Most probes have an adjustable acceptance angle and this angle should be specified when reporting radiance measurements.

TOTAL INTENSITY RADIOMETERS. Total intensity meters measure the sum of all quanta from the range of wavelengths over which the sensor is sensitive. The range of wavelengths depends on the type of sensor, and different wavelength ranges are used for different applications. The **photometers** used in photography and commercial lighting applications are tuned to the specific sensitivity and spectral range of the human eye, e.g., no ultraviolet sensitivity, and lower sensitivity to red and blue wavelengths compared to yellow and green. Solar radiometers for agricultural purposes, on the other hand, must be sufficiently sensitive to the red and blue wavelengths absorbed by chlorophyll. They typically have a nearly flat response over wavelengths from 400 to 700 nm and sharp cutoffs above and below this range. They are often called **photosynthetically active radiation (PAR) detectors**. A **pyranometer** is a broad spectrum solar radiometer sensitive to ultraviolet and near infrared as well as visual radiation. All of these radiometers are designed to measure irradiance and thus have cosine-corrected collectors. Since this type of radiometer integrates the radiant energy of a wide range of wavelengths, it is not suitable for studies of other animal visual systems and signals.

SPECTRORADIOMETERS. To quantify the color of an animal's color signal or produce the frequency spectra shown in Figures 7.4 and 7.12, the radiometer must differentiate between wavelengths or small wavelength intervals. Such an instrument is called a **spectroradiometer** or spectrophotometer. Spectroradiometers contain a monochromator, a photosensor, analytical circuitry, and an optional external light source. Two types of monochromator and sensor systems are available, each with its advantages and disadvantages (Sommer 1989). In single-sensor instruments (Figure C), the incoming light signal is first directed through a filter wheel to narrow the range of wavelengths, then sent through a slit to a prism, diffraction grating, or holographic interference grating that disperses the remaining wavelengths. This beam is then passed through a slit to the detector. The prism or grating is then rotated so that a series of wavelengths falls on the detector. The power of each wavelength interval is measured sequentially for this range of wavelengths, then the filter wheel is turned to accept the next range of wavelengths and the process is repeated across the entire spectrum. This system has the advantage of high accuracy and the ability to vary the wavelength interval width, but it may take several minutes to scan the entire frequency range.

Diode-array instruments (Figure D) significantly shorten the time to obtain a spectrum. The light signal is passed through a slit to the prism or grating where the wavelengths are dispersed as above. An array of up to 1000 photodiodes connected with an integrating circuit simultaneously measures the accumulated charge at each wavelength interval fixed by the width of the diodes. A single scan takes a fraction of a second, and the diodes can then be discharged and prepared for a second scan. Several

Box 7.2 (continued)

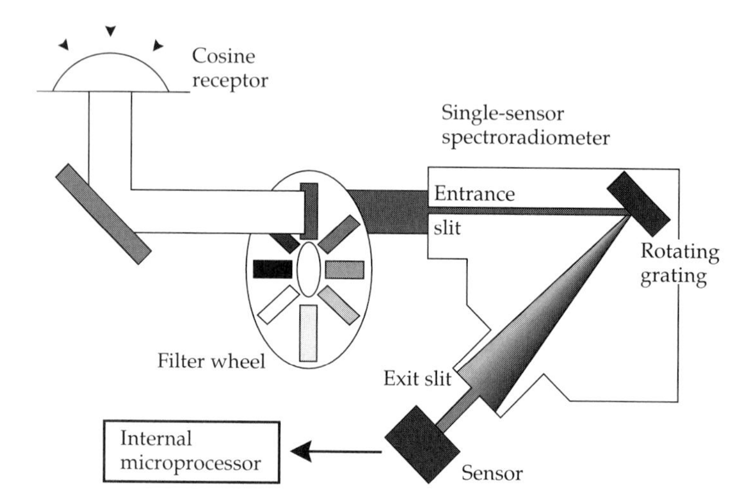

Figure C Single-sensor spectroradiometer.

scans may be made and averaged in a matter of seconds. This type of instrument does not allow the wavelength interval to be varied, but it is considerably less expensive and of comparable accuracy.

The light to be measured can be guided into the instrument by an optical fiber bundle fitted with an appropriate probe or sensor. For irradiance measurements, the fiber tip can be connected to a cosine-correcting diffuser probe. A telescope-like device can be used to narrow the acceptance angle for radiance measurements. For radiance measurements of reflected light, a controlled light source is required so that a comparison can be made between the input light intensity and the reflected light at each wavelength interval. This comparison is best accomplished by using an opaque integrating

SUMMARY

1. Light is a form of **electromagnetic energy** that, unlike sound, can travel in a vacuum. This energy consists of an oscillating electric field and magnetic field at right angles to each other and to the direction of travel. The wavelike nature of electromagnetic radiation means that it exhibits the wave properties of reflection, refraction, diffraction, spreading loss and attenuation, and its frequency is inversely proportional to its wavelength. Electromagnetic radiation can also be viewed as tiny packets of energy, or photons, that move very fast in an oscillating manner. The energy of a photon is directly related to its frequency and inversely related to its wavelength. Different frequencies of radiation behave rather differently. Low-frequency (long-wavelength) radiation in the radio range bends around most objects in the environment and seems to act more like a

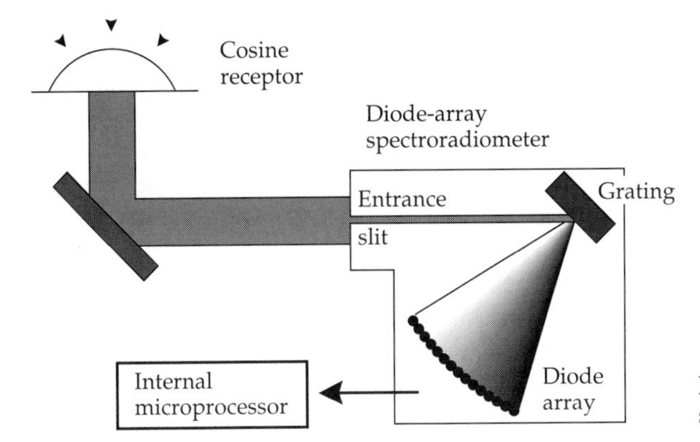

Figure D Diode-array spectroradiometer.

sphere with a diffusing inner lining. The sphere has several windows for securing the sample, probe, and light source so that the angle of the light relative to the fiber probe and the reflecting surface can be controlled.

Most spectroradiometers are connected to a computer that stores and analyzes the data collected, averages any number of sequential scans, and plots the resulting reflectance spectrum. The software provided with the instrument may or may not convert the wattage output into photon flux. Since reflectance is usually expressed as a proportion relative to a "white" standard, either measuring unit can be used as long as illumination and measuring angles are standardized. The spectral range of the white light standard is important, however, and a lamp should be chosen that possesses a relatively flat spectral response over the wavelengths perceived by the animal being studied. Spectroradiometers must be recalibrated frequently.

wave. High-frequency (short-wavelength) radiation bounces off objects or passes through them and therefore acts more like a stream of particles.

2. **Visible light** is a very narrow range of electromagnetic radiation in the middle frequency range. Only in this range can electromagnetic radiation interact constructively with organic matter. Lower frequencies possess too little energy to be detected by organisms except as heat, and higher-frequency radiation destroys molecules. Intermediate frequencies cause electron excitations that can temporarily change the shape of a molecule, and such a change can be coupled to cellular processes to generate a nerve impulse. Luckily, most of the radiation from the sun that reaches Earth's surface is in this narrow visible range because of atmospheric absorption of the higher and lower frequencies.

3. Visible light travels in **straight lines** through media such as air and water; it **reflects** off most objects in the environment rather than going around

or through them, and continues to travel in a straight line but in a new direction. Some of the reflected light enters the eye of a receiver and is **refracted** by a lens to form an image, or spatial map, of the location of all reflecting objects in the environment. These properties of light also mean that visual communication between two animals will be blocked by any large or dense opaque objects situated directly between them.

4. It is wrong to think that light just bounces off objects; it actually interacts with matter, and the nature of this interaction determines whether it will transmit through it or be reflected by it. The oscillating electric field of a light wave causes the molecules, atoms, and electrons in the medium through which it is traveling to oscillate. Molecules, atoms, and electrons have certain **natural resonant frequencies**. If the frequency of the light is very close to the frequency of one of the natural resonant frequencies of the medium, the resulting amplitude of the medium's oscillations will be very high. The highly excited molecules will collide with their neighbors and lose their energy as heat; in other words, the light energy is completely absorbed, and the propagation of the light wave is stopped. Such a medium is **opaque** to visible light. If the frequency of the light is sufficiently different from all of the resonant frequencies of the molecules in the medium, the particles in the medium will not resonate but they will oscillate slightly and reradiate electromagnetic energy of the same wavelength as the incident light wave. Although reradiation from a given molecule occurs in all directions, sideways and backwards reradiation is canceled out by destructive interference from neighboring molecules oscillating in phase. Reradiation in the forward direction, however, interferes constructively, and the additive effects promote the forward propagation of the wave. This medium is **transparent** to visible light. In a gas, the molecules are not as uniformly arrayed as in a liquid or solid. Since the sideways reradiations are not all canceled, significant scattering of light occurs.

5. The speed of light is maximal in a vacuum where it can travel unimpeded. In a medium, light causes the atoms and molecules to align their positive and negatively charged ends parallel to the direction of the light wave in a process called **polarization**. The electric field in the atoms and molecules opposes the electric field of the light wave and slows the wave's speed. The **index of refraction** is a measure of the speed of light in a given medium relative to the speed of light in a vacuum.

6. When light strikes the boundary between two transparent media with different indices of refraction, some of it is reflected and the rest enters the second medium. The direction of travel in the second medium is bent relative to the direction in the first medium, a phenomenon called **refraction**. The greater the difference in travel speeds between the two media, the greater the angle of bending. The direction of bending depends on which medium has the larger index of refraction.

7. **Reflection** of light at a smooth boundary between two media again follows the pattern expected of a wave—the angle of reflection equals the

angle of incidence. The reflected light is reradiated light from the molecules in the first few layers of the second medium, which do not have neighbors on one side with which they can collide or which can cancel all of the backward-radiated waves. Only those reradiated waves traveling in the direction of the angle of reflection are constructively in phase. The fraction of the incident light that is reflected depends on the difference in refractive indices of the two media. At certain angles, all of the light may be reflected. At some angles, the reflected light is also strongly **plane-polarized**.

8. Different frequencies of light in the visible range are perceived as **colors**. Transmission is frequency-dependent in all transparent media except a vacuum. In air, molecules of gas and tiny dust particles scatter short blue and violet wavelengths more than others. At the air/glass or air/water boundary, shorter wavelengths are refracted or bent more than longer wavelengths, and a beam of white light can be split into the spectrum of colors. Water absorbs long, red wavelengths much more than others. Scattering, selective absorption, and reflection and filtering by colored objects in the environment affect the spectral composition and total amount of ambient light in a habitat.

FURTHER READING

Le Grande (1970), van Heel and Velzel (1968) and Wolken (1995) are very readable introductory texts on light and photobiology. Good physics texts on light, waves and electromagnetic radiation include Feynman et al. (1964, vols. 1 and 2), Crawford (1965) and Reitz et al. (1980). Ditchburn (1963) and Guenther (1990) are good optics texts. Weisskopf (1968) provides an excellent layperson's description of how light interacts with matter. Barrow (1962) is the best source of information on the effects of radiation on molecular motion. Rossotti (1983) is a readable introduction to color; Nassau (1983) gives a more detailed discussion of this topic. Waterman (1981) offers the best summary of the use of polarized light by animals. The measurement of medium absorption and irradiance spectra in aquatic and terrestrial habitats is described by Kirk (1983, 1994), Williams (1970), and Endler (1990).

Chapter 8

Production and Transmission of Light Signals

AS LONG AS THERE ARE VISIBLE LIGHT RAYS BOUNCING AROUND in the environment, an organism need do nothing to make its presence and location known. All organisms, plant or animal, can therefore generate visual images passively as light reflects off their bodies and into the eyes of visual animals. To prevent other animals from seeing them, they must move to position some type of large object in the environment between themselves and the receptor animal or blend with the background. To enhance their visibility to other animals, a variety of strategies can be employed such as increasing the contrast between the animal and its background with color, pattern, and texture; moving and changing positions to attract attention; and generating one's own light signal. Most animals make use of reflected light and movement to produce visual signals. We shall examine all of these strategies in this chapter.

PROPERTIES OF LIGHT SIGNALS

Regardless of whether the signal is produced by reflected or self-generated light, there are four quantitative properties we can use to describe all light signals.

The **brightness** of a light signal is its overall intensity, measured as the total quantal flux produced by a sender in the units of radiance (photons s^{-1} sr^{-1} m^{-2}). The brightness of a reflected light signal is a function of both the range of wavelengths reflected and the surface structure of the object or signaling organism. The wider the range of wavelengths reflected, the greater the overall intensity. A smooth surface will produce a higher-intensity signal than a rough surface. If the visual receptor can discriminate between different light intensities, it can distinguish objects with subtly different reflection properties. The brightness of a self-generated signal depends on the amount of energy used to produce the signal.

Spectral composition describes the color of a light signal. Colors are defined by three parameters: **brightness**, as defined above; **hue**, which refers to the dominant wavelength or frequency of the light; and **chroma**, which refers to the saturation or purity of the dominant frequency. (The presence of other frequencies adds gray to the dominant frequency and makes the color less saturated.) Both reflected and self-generated light signals may be characterized by their color. If the visual receptor can discriminate different wavelengths of light it can enhance its ability to discriminate between objects with different reflection properties beyond what is possible with the use of only intensity differences. As we saw in the previous chapter, visible light wavelengths range from 400 nm for violet to 700 nm for red. Some organisms can detect wavelengths as short as 300 nm in the ultraviolet region and as long as 800 nm in the near infrared region.

Light signals can also be characterized by their **spatial characteristics.** These include the size, shape, surface features and color pattern of color patches and body structures, as well as the posture, position and location of the sender. The visual acuity of the receiver, or its ability to make an accurate and fine-scale spatial map of all incident light rays entering its field, determines its ability to discern both small details of shape and pattern as well as accurately locate senders in space.

Finally, **temporal variability** in intensity, color, and spatial characteristics can be used to generate a wide diversity of signals. Complex patterns of colors and surfaces, color change, flashing lights, changes in size and shape and limb movement patterns are typical visual displays exhibited by animals.

For visual receivers, it is the **contrast** between an object of interest and the object's background that determines the conspicuousness of the object. Visual backgrounds also possess the same four quantitative characteristics described above. Contrast can therefore be achieved with any one or more of these light signal characteristics. An animal sender that needs to increase its visibility to conspecifics can enhance the contrast between itself and the brightness, color, spatial pattern, or movement of its background. Conversely, an animal that needs to remain inconspicuous tries to match its brightness, color, spatial pattern, and movement to that of its background.

COLOR PRODUCTION

Animals make widespread use of color to enhance visual signals (Cott 1940; Rowland 1979; Hailman 1979; Burtt 1979). In order for a reflecting surface to appear colored it must selectively reflect certain wavelengths of light more than others. A surface that reflects all wavelengths equally appears white, and a surface that absorbs all wavelengths appears black. All other colors lie between these extremes. Figure 8.1 shows some plots of the reflectance spectra of a few colors. Curves with a high peak of reflectance over a narrow wavelength range are highly saturated colors; broader (flatter) curves are less saturated colors. The area under the curve is an approximate measure of the brightness of the color.

Humans see and categorize colors differently than most animals. Several quantitative color-naming schemes have been developed to aid people in color categorization, such as the CIE chromaticity scale (Wyszecki and Stiles

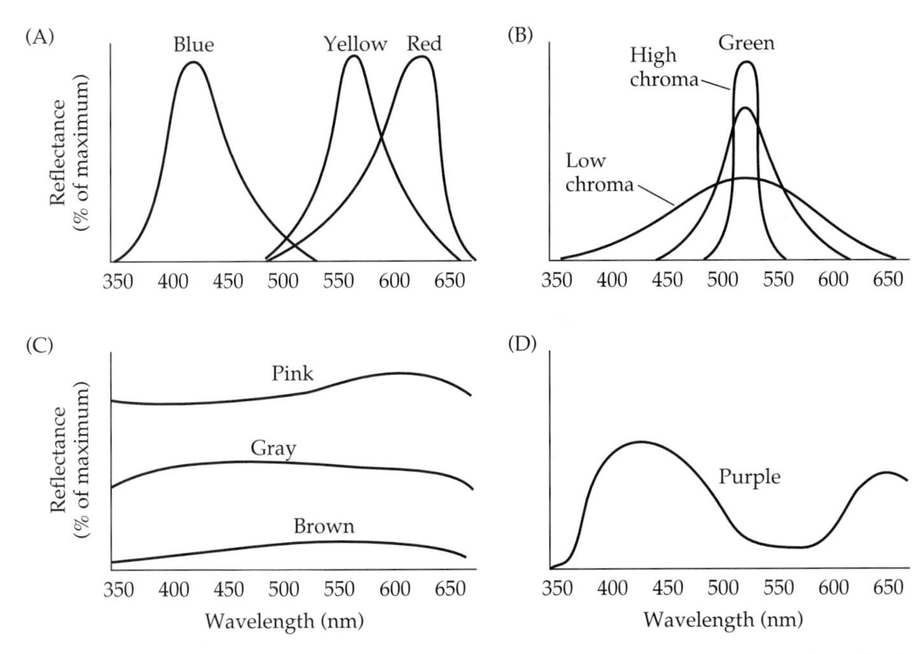

Figure 8.1 Idealized reflectance spectra for some colors. Hue is the wavelength of the peak or inflexion point of the spectral curve, chroma is the maximum slope of the curve, and brightness is approximated by the area under the curve. (A) Three different hues of similar chroma. (B) The color green at different levels of chroma or saturation. (C) Three unsaturated colors of very different brightness. (D) Some colors such as purple are not monotonic but have two hue peaks. Natural colors do not necessarily have such smooth, symmetrical curves as those shown here.

1982; Wright 1991), Munsell (1975), and Smithe (1975) systems. However, all of these systems are bounded by human color perception. A generalized color classification system is shown in Figure 8.2 that clearly differentiates the properties of hue, saturation, and brightness. The entire spectrum of hues that a species can perceive is ordered around the circle. Colors positioned around the edge of this circle are fully saturated, and the angle relative to some baseline hue (or the x, y coordinate in the horizontal plane) is a quantitative measure of the hue. Adding gray to these saturated colors leads to desaturation, so that the position of the color would move in toward the center of the circle. The degree of saturation is therefore quantified as the horizontally measured length of the radius from the central vertical line to the color. The vertical axis (z coordinate) represents brightness and contains shades of gray ranging from white at the top apex to black at the bottom apex. Any color could be represented within this three-dimensional space. To use such a system for a particular species, the range of hues in the circle would be restricted to those the species can perceive, and the spacing of the hues around the circle would reflect perceived hue differences, i.e., spacing would be irregular with respect to wavelength.

Animals that are active during the day can generally perceive a moderate-to-wide range of colors and use this ability to find food and detect predators. Many such species also employ body coloration or color patches to communicate to conspecifics. Animals use three different mechanisms to selectively reflect and absorb different wavelengths: pigments, thin-layer interference, and scattering.

Pigments

Pigments are chemical compounds whose molecules absorb certain wavelengths of light and transmit the remaining wavelengths. The receiver per-

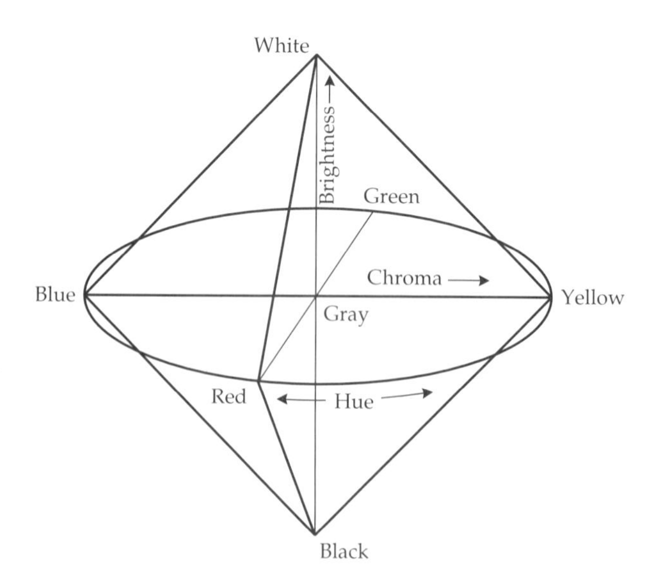

Figure 8.2 Three-dimensional representation of color space. The four compass points of the horizontal circle represent the four main colors, red, yellow, green, and blue. Colors on the edge of the circle are spectrally pure (saturated), and horizontal radii of different lengths from the central vertical line represent degrees of saturation. The vertical axis is the gray scale, with white at the top and black at the bottom.

ceives the color of the light that is not absorbed. For example, a pigment that absorbs wavelengths in the violet, blue, green, and yellow range will transmit and reflect red wavelengths. Many natural pigments absorb all wavelengths above or below a certain cutoff wavelength and pass the remaining wavelengths within the visual range. For example, the red pigment described above acts like a long-pass filter that absorbs all short wavelengths up to 600 nm. Yellow pigments are also long-pass filters with smaller cut-off values around 500 nm that pass a wider range of long wavelengths. Violet pigments, which are produced by plants but not animals, are examples of short-pass filters because they absorb all wavelengths above 450 nm. The green pigments found in plants are exceptions to this pattern, as they absorb light in two different bands, red and violet. Purple is not a spectral color, but an imaginary color produced by the combination of red and violet; such a pigment must absorb only middle (green) wavelengths.

Pigments used by animals are organic compounds that contain long chains or networks of **conjugated double bonds**. Conjugated double bonds consist of a series of carbon atoms joined by alternating single and double bonds. Such molecules have several resonance forms, as shown for the benzene ring in Figure 8.3A. The electrons that form the bonds between the carbon atoms are therefore constantly moving back and forth. These mobile electrons easily absorb light energy by shifting up to a higher energy level. The molecule is now in an excited state. When these molecules collide with other molecules, they quickly lose as heat the energy they have trapped.

Molecules differ in the wavelengths of light they can absorb. This ability is determined by the difference in energy between the molecule's ground state and its excited state; the greater the difference, the more energy that is required to excite the molecule, and the higher the frequency of light needed to excite it. Only certain elevated states are allowed, depending largely on the length of the carbon chain. Short chains require larger amounts of energy to excite electrons. Such molecules can only absorb short-wave high-energy radiation in the ultraviolet range. As the chain gets longer, lower energy levels become available to the excited electrons and lower amounts of energy are sufficient to shift electrons to higher quantum states. Thus longer-chain molecules can absorb lower energy, i.e., longer wavelength, electromagnetic radiation. The length and size of the molecule thus determine which wavelengths are absorbed and which are transmitted (Nassau 1983).

Since small molecules such as benzene shown in Figure 8.3A absorb only high energy, short wavelength radiation in the ultraviolet range, all wavelengths in the (human) visible range pass through this molecule and it appears colorless. An 18-carbon chain carotenoid such as **carotene** (Figure 8.3B), absorbs slightly lower energy (longer wavelength) radiation in the violet range. Once some wavelengths in the visible range are absorbed, selective transmission of other wavelengths occurs and color is produced. The carotene molecule transmits green, yellow, and red, which in combination appear yellow or orange. Larger carotenoids absorb blue and green and thus appear orange or red (Figure 8.3C). Carotenoids bound to proteins absorb the green wavelengths, and transmit violet and red to appear purple (Figures 9.1, 9.8, 9.9 and 9.11A).

(A) Benzene

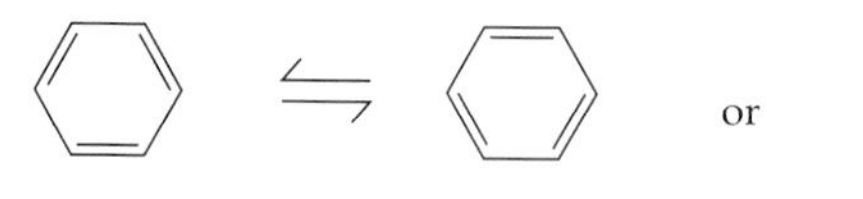

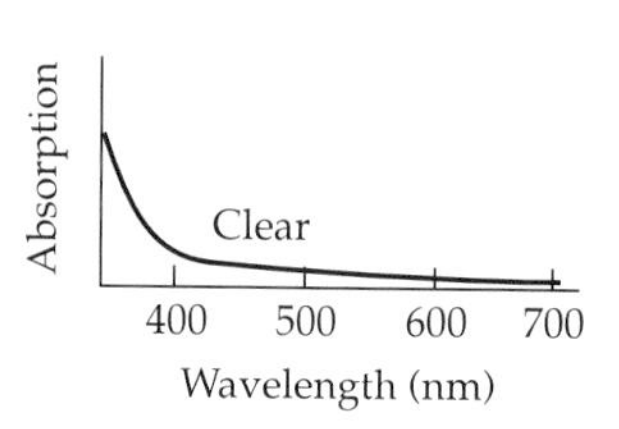

(B) Beta carotene

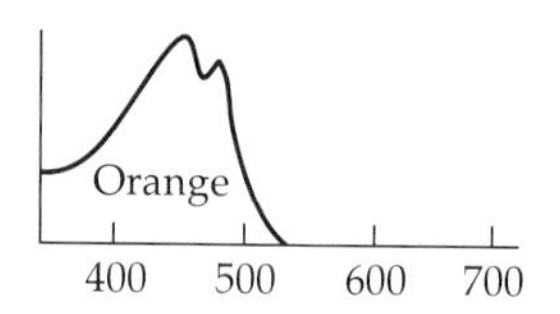

(C) Astaxanthin

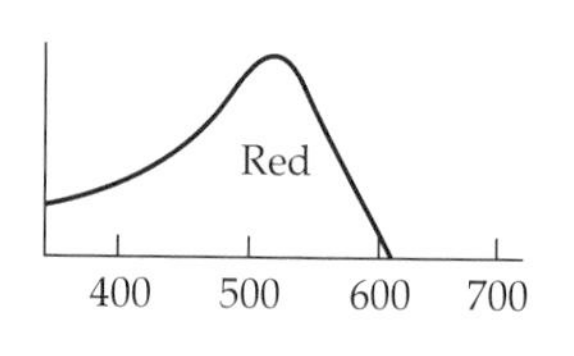

(D) Pterin

Yellow
400 500 600 700

(E) Quinone

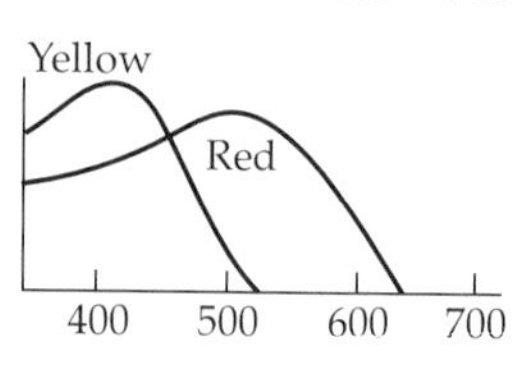

(F) Verdin

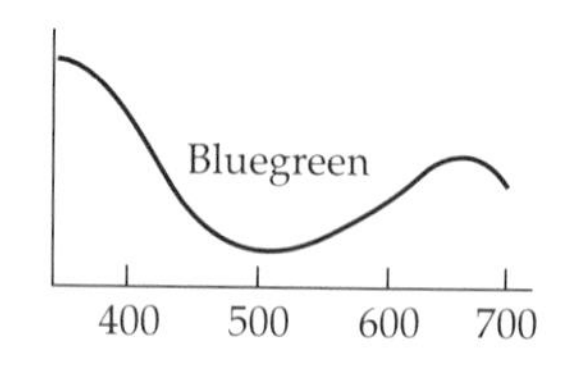

(G) Porphyrin

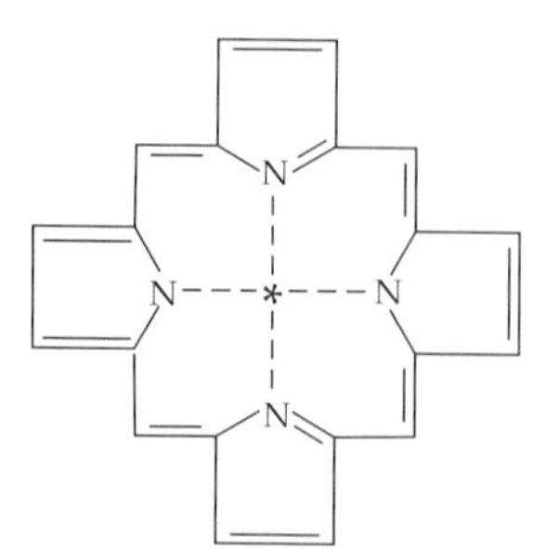

Green
Red
400 500 600 700

* = Fe, Mg, or Cu

◀ **Figure 8.3 Chemical structures and absorption spectra of natural pigments.** Note that an absorption spectrum is the inverse of a reflection spectrum (compare to Figure 8.1). (A) The resonance structures of the benzene ring. The movement of electrons shared by the carbon atoms in the ring causes a change in the position of the double bonds as shown on the left; the rapid shifting between the two forms is often indicated with a circle. The benzene molecule is too small, and the electrons too tightly bound, to absorb visible wavelengths; it absorbs only high-energy UV radiation and therefore appears colorless. (B) Carotenoids with conjugated double bonds such as beta carotene absorb blue wavelengths and reflect orange. (C) Astaxanthin has lengthened the conjugated bond chain with the addition of two ketone units; it absorbs blue and green wavelengths and reflects red. (D) Pterins contain two pyrimidine rings and produce yellow and orange pigments found in insects. (E) Quinones are a large class of pigments based on the cyclic diketone structure shown here. Addition of the two ketone groups to a benzene ring produces a conjugated bond system that absorbs violet wavelengths and reflects yellow. Fusion of the ring with more rings lengthens the conjugated bond network and allows longer wavelengths to be absorbed. (F) Verdin is based on the repetition of four pyrrole rings, creating a conjugated bond chain. (G) Porphyrins are built from a verdin molecule connected to make a ring with a central metal ion, typically magnesium, iron, or copper. Red, green, or violet pigments are produced.

Other classes of pigments are based on networks of conjugated double bonds formed by the fusion of multiple rings (Fox 1976, 1979). **Pterins** are nitrogenous rings that produce the white, yellow, and red colors often found in butterfly wings. **Quinones** are responsible for some yellow, red, and orange colors. **Verdins** are large molecules that absorb red and produce the blue-green color often found in bird's eggs. **Porphyrins** are like a verdin molecule with the ends connected into a ring and contain a central metal ion. The basic structure has been used by a variety of organisms to produce different pigments whose color depends on the specific metal ion. With iron in the center, the compound becomes hemoglobin and the color is red. Plants have placed magnesium in the center to produce chlorophyll, which absorbs both violet and red, and reflects green. Turacos, a brightly colored group of tropical birds, have placed copper in the center to produce their red, violet, and green feather pigments. Some examples of these pigments and their absorption spectra are shown in Figure 8.3D–G.

The blood of many animals is colored, typically red, as a result of the hemoglobin or other pigments used to grab and release oxygen. If small capillaries are placed near the surface of the skin, the color of the blood becomes highly visible. Common examples of the use of blood to produce red colors can be seen in the comb of the chicken and the rump of the estrous female baboon.

Melanin is a generic term for a group of dark-colored pigments found in a wide variety of vertebrates, invertebrates, and plants. It is a large protein molecule that absorbs most or all visible wavelengths to produce brown and black colors. Melanin is responsible for the brown and black color of mammalian hair, avian feathers, and the black patterning in lower vertebrates, insects, and other organisms. Variants of melanin also produce dull yellow, reddish brown, and human red hair. As we shall see shortly, it plays an important role in other color systems as well. **Guanine**, while not a differentially absorbing compound like the pigments mentioned above, is nevertheless an important source of white coloration. Guanine forms microcrystalline deposits, or platelets, that reflect all wavelengths. Very dense packing of guanine crystals causes specular reflectance and the silvery appearance of fish scales.

In order for a selectively absorbing pigment to reflect back out those wavelengths that it has not absorbed, it must be coupled with reflective structures. Pigments are located in either the outer epidermal layers of the skin of animals or in the dermal layer. Special cells called **chromatophores** produce the pigments and house them in many small granules or packets called **chromatosomes**. Incident light on the animal's surface is transmitted through the pigmented layer, where an initial round of selective absorption occurs. Nonabsorbed wavelengths continue to travel into the deeper tissue, where most will be absorbed, but some will be scattered and reflected at boundary layers. This light will pass back through the pigment layer again for a second round of selective absorption before it exits the animal's surface. Very intense colors can be produced by placing a layer of specialized chromatosomes called **iridosomes** just underneath the pigmented chromatosomes. Iridosomes contain guanine platelets that reflect all wavelengths of light like small mirrors. The platelets are responsible for reflecting all of the wavelengths that were selectively transmitted (not absorbed) by the colored pigment layer (Bagnara and Hadley 1973).

Interference

A rather different mechanism of color production is found in some birds and insects with iridescent coloration. A thin layer of a transparent material with a high index of refraction, such as wax or keratin, coats the feather, scale, or exoskeleton. White light hitting this surface at an angle is partially reflected to produce a primary reflection. Some light enters the wax and is refracted, then reflected by the boundary at the bottom of the wax layer, and refracted again as it exits the layer to produce a secondary reflection (see Figure 8.4). The primary and secondary reflections will be in phase for certain colors of light depending on the width of the layer, the refractive index of the material, and the viewing angle according to the equation

$$m\lambda = 2nx\cos\theta$$

where n is the refractive index, x is the thickness of the layer, θ is the incident light angle with respect to the normal, λ is the wavelength of light, and m is an integer (1, 2, 3, . . .). The color is very intense and shimmery. A slight change in the angle of incidence will cause the color to change slightly to blue or yellow-green, but with greater changes in angle, the color will become black. At this point, the primary and secondary reflections are out of phase and appear black as they cancel each other out. Negative interference occurs when

$$m\lambda = 2nx\cos\theta + \lambda/2$$

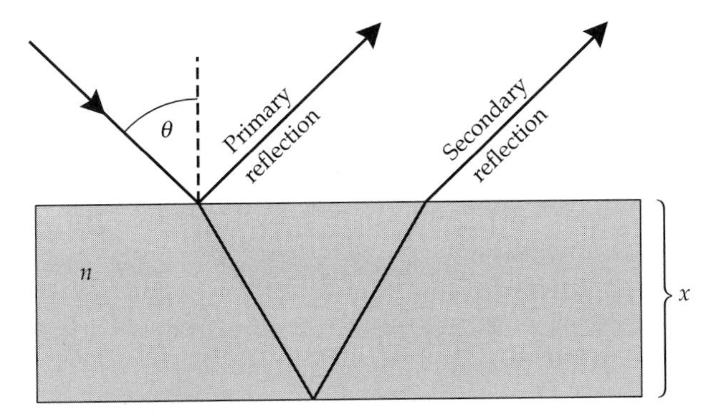

Figure 8.4 Thin layer interference. When white light is incident on a thin layer of a transparent material with a higher index of refraction than the surrounding media on top and bottom, it will produce a primary reflection (with a phase shift) off of the first boundary and a secondary reflection (no phase shift) off of the second boundary. When the thickness (x) and coefficient of refraction (n) of the layer and the angle of incidence θ meet the conditions for a maximum for a certain wavelength range, the two reflected waves will be in phase with each other and constructively interfere to produce a very intense, colored reflection. Changing the thickness of the layer by one-half a wavelength (or changing the angle of incidence or refractive index) will result in destructive interference of the two reflected waves, and black will be observed.

Most materials used to produce structural colors in animals have refractive indices of 1.5 to 2.0 and a layer thickness of one-quarter to one-half of a wavelength. A layer of melanin beneath the transparent layer intensifies the color by absorbing all nonreflecting wavelengths. Many black birds and black beetles coat the melanin layer with a thin layer of wax to produce a shiny, iridescent effect. Alternatively, several layers of refracting material such as keratin may be stacked to produce multilayered constructive reflections (Figure 8.5). Iridescent colors of red, green, blue, and violet are produced in this manner by birds such as hummingbirds and peacocks and by some butterflies (Simon 1971; Fox 1976, 1979; Nassau 1983).

Scattering

As we learned in the previous chapter, small particles with a different refractive index from the general medium can cause some wavelengths to scatter. Recall that in Rayleigh scattering, particles smaller than 300 nm in diameter cause shorter wavelengths to scatter more than longer wavelengths. The same principle is used by some animals to produce blue coloration, and with some modifications, green. The surface of the animal is coated with a transparent material that contains a matrix of tiny dense particles or air spaces considerably smaller than 300 nm in diameter. Underneath this surface is a layer of melanin. Violet, blue, and green wavelengths are scattered and the longer wavelengths that pass through this layer are absorbed by the melanin layer. This scattering and absorption generates the color blue and is the mechanism

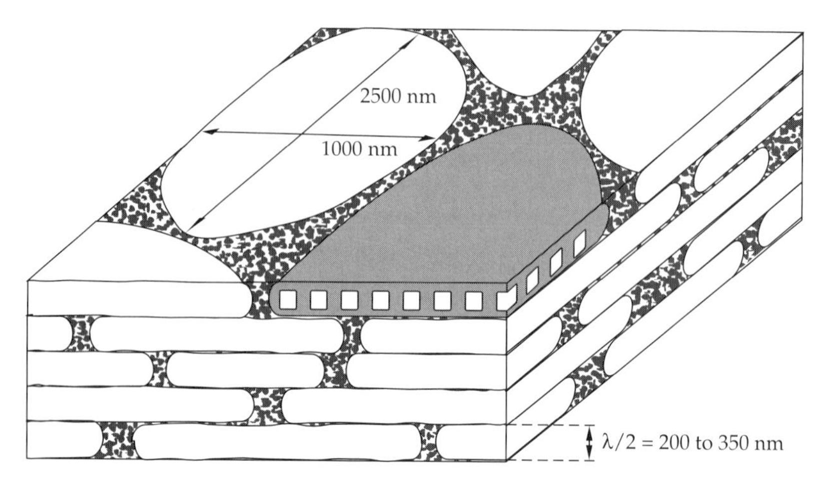

Figure 8.5 Interference coloration produced by stacks of layers. Constructive interference can be enhanced by stacking up several layers of a high refractive index material. As long as all layers are the same thickness and the equation for constructive interference is met, waves reflected from all layers will be in phase. The diagram shows a cross section through the surface structure of the barbule of a hummingbird feather. The platelets are approximately one-half wavelength thick and composed of a keratin material with a refractive index of about 2.0. Air bubbles incorporated into the platelets reduce the effective refractive index to 1.85 for red feathers and 1.5 for blue. (From Nassau 1983, © John Wiley & Sons.)

used to produce blue in birds, fish, and lizards (Figure 8.6A). Some birds use this mechanism coupled with multilayer interference to produce ultraviolet plumage coloration (which appears black to humans) (Finger et al. 1992). By placing a thin yellow carotenoid layer above this arrangement, violet and blue wavelengths are absorbed and green is the only wavelength scattered. This mechanism is used to produce green in frogs, parrots, ducks, and others (Figure 8.6B). Finally, white feathers in birds are produced by Mie scattering (Figure 8.6C). Recall that in this type of scattering the particles are considerably larger than the wavelengths of visible light, and a mixture of wavelengths is reflected to produce white. Avian feathers contain a matrix of large particles, which may be air spaces, molecules of fat, protein, keratin, or crystals, resulting in white coloration (Fox 1976; Nassau 1983).

Temporal Modulation of Color

In the case of most pigments and the two structural methods of color production, animals cannot change or vary the color on a short-term basis to modify the meaning of the color display. However, they can cover small areas of color with another body part or with feathers, and then flash the color by uncovering it. Two examples of color flashing are *Anolis* lizards, which extend a brightly colored dewlap during push-up displays to conspecifics, and the roadrunner, which pulls aside the feathers behind its eye to reveal a red and blue spot on the skin (Figure 8.7). Birds, butterflies, and other insects with folding wings can flash color patches by changing wing positions. Animals

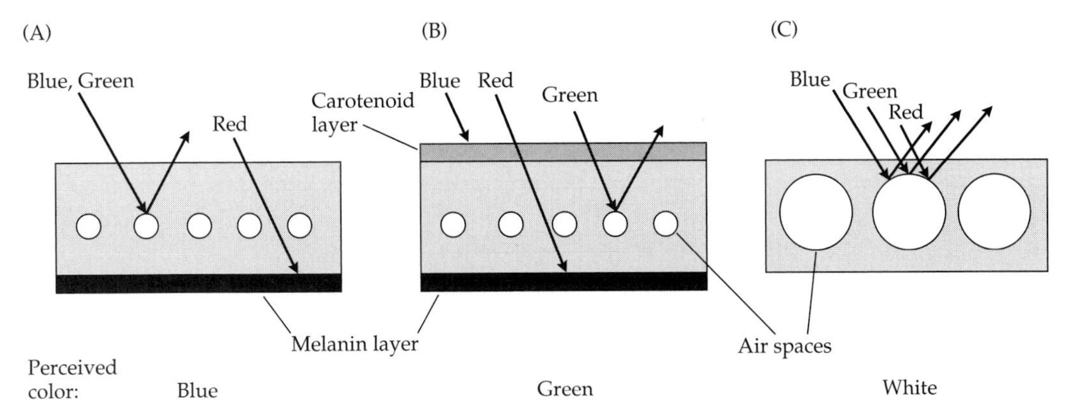

Figure 8.6 Three different colors produced by scattering. (A) Rayleigh scattering from air spaces or particles embedded in a transparent layer results in stronger reflection of short wavelengths, and a melanin layer underneath absorbs all other wavelengths, to produce blue coloration. (B) When the structure in A is overlaid with a transparent yellow filter that absorbs the violet and blue wavelengths, green is produced. (C) Larger particles or air spaces scatter all wavelengths and produce white.

that use blood to produce red coloration can in some cases vary the intensity of the color by dilating or constricting the blood vessels under the skin, as seen in the blushing response of humans.

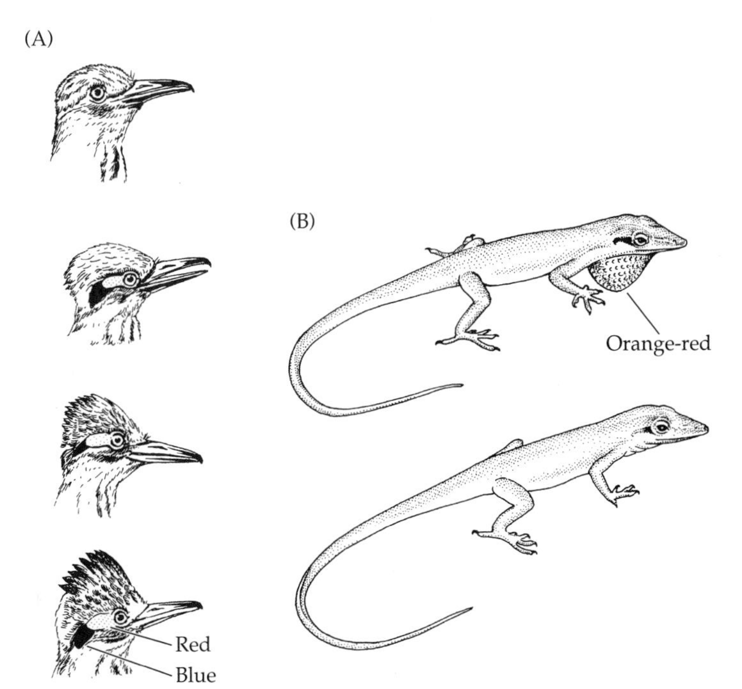

Figure 8.7 Controlled exposure of colored patches on the body resulting in rapid color change. (A) The roadrunner has patches of red and blue skin behind its eyes that are normally hidden by cryptic brown feathers (top) and exposed when the feathers are moved aside; it can also raise and lower its crest. (B) *Anolis* lizards possess a brightly colored dewlap, a flap of skin attached to a hyoid bone that is extended during push-up displays to conspecifics. (A after Whitson 1977.)

A few groups of animals, notably amphibians, some reptiles (chameleons), many fish, cephalopods, and a few insects are able to change their body color (Parker 1948; Waring 1963; Bagnara and Hadley 1973). They accomplish this color change by controlling the dispersion of pigment granules in the dermal chromatophores. The process usually involves the pigment melanin and leads to a darkening or lightening of the animal. The melanophores in such animals are large flat cells with many dendritic processes extending outward from the center. When the melanin granules are concentrated in the center of the chromatosome, the animal appears pale, and when the granules are dispersed into the dendritic arms of the cell, the animal appears dark (Figure 8.8A,B). In vertebrates the melanophores are innervated by the autonomic nervous system. Movement of the granules is under the control of a hormone that is produced by the pituitary in the brain and linked to the eyes or other light sensitive organ. The function of the color change is often to match the body color of the animal to its background for camouflage and predator avoidance. However, color change can serve a social signaling function in some fish. The intensity of the body color or the presence of a certain pattern such as stripes may reflect an individual's aggressive or reproductive state. One of the most dramatic examples is the hermaphroditic sea bass (*Serranus subligarius*), in which fish alternate sexes during a bout of courtship and mating and change from a black-and-white banding pattern in the male role to a solid black posterior in the female role (Demski and Dulka, 1986) (see Figure 1.2). Since control of the chromatophores is hormonal, it takes a few seconds to minutes to effect the color change.

The most sophisticated example of color change for signaling purposes occurs in cephalopods (squid and octopus). The chromatophores in these animals are small multicellular organs consisting of a central pigmented compartment with brown, red, or yellow pigment connected to radial muscle fibers and motor neurons (Cloney and Florey 1968). In the relaxed state the surface of the chromatophore is highly folded and the color is inconspicuous. When the radial muscles are contracted, the structure is stretched out and the color becomes visible (Figure 8.8C). The movement of the organ is under voluntary central nervous system control. By combining the expansion of its different colored chromatophores the animal can assume a variety of colors in rapid succession. These color signals are not only used to match the animal to its background, but they are also used during courtship, aggressive interactions, and alarm contexts (Moynihan 1985, Demski 1992). A particularly dramatic example of such a signal is shown in Figure 8.9.

SELF-GENERATED LIGHT

A few organisms from a wide variety of taxa can produce their own light via a process called bioluminescence. Such organisms include fireflies, cephalopods, glowworms, fishes, earthworms, snails, jellyfish, crustaceans, echinoderms, and even some fungi and bacteria. The process of light production is different in each group. Figure 8.10 shows the light-production mechanism of fireflies.

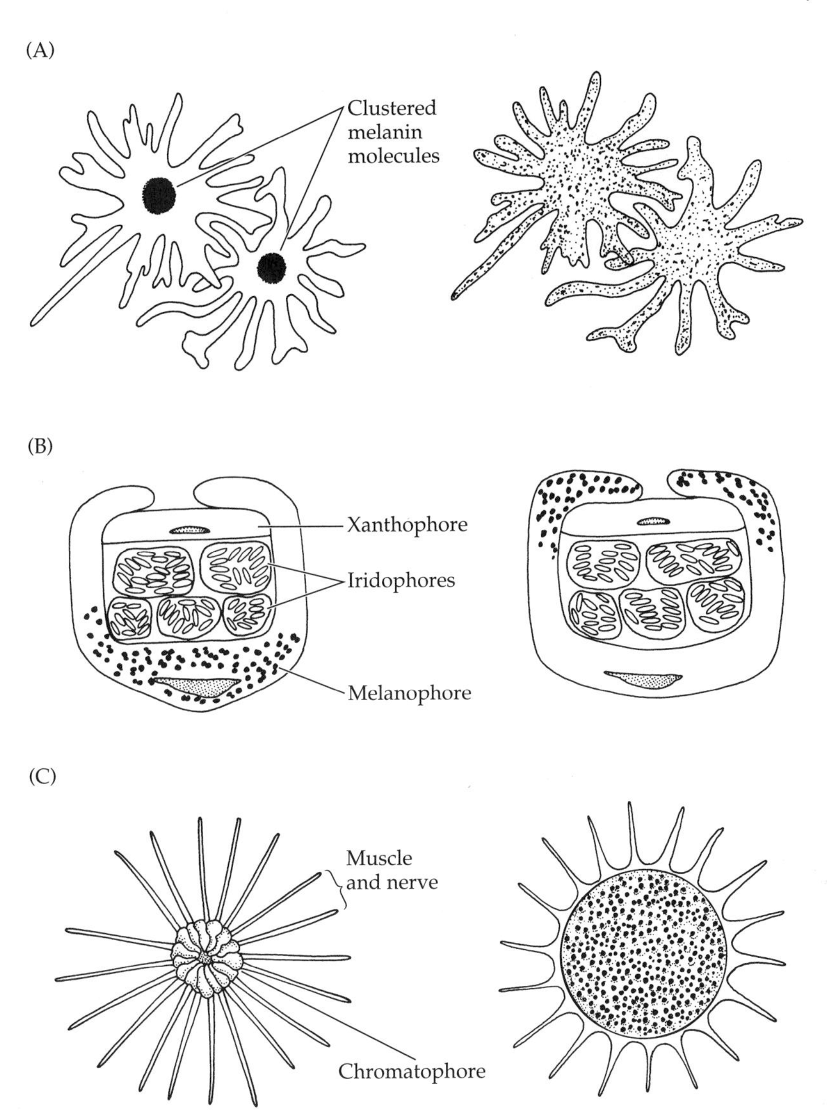

Figure 8.8 Chromatophores. (A) Typical fish melanophores consist of interdigitating cells with long dendritic arms; when the melanin granules are aggregated in the center of the cell, the animal appears lighter (left) and becomes increasingly darker when the granules are dispersed throughout the cell (right). (B) The dermal chromatophore unit in some frogs and lizards is more complex, with the dendritic processes of the melanophore wrapped around iridophores (creating small particle scattering) and xanthophores (with yellow pigment); when the melanophore granules are clustered beneath the other chromatophores, the animal appears light green (left) and when the granules are dispersed, the animal appears dark (right). (C) The cephalopod chromatophore is a multicellular organ comprised of muscles and nerves as well as the pigment-bearing cell. Brown, red, or yellow pigment granules are located in a compartment of the large folding chromarophore. In the relaxed state (left), the surface of the chromatophore is folded extensively and the pigment is hidden. When the radial muscles are contracted the chromatophore expands and flattens; the folds disappear, exposing the pigmented compartment (right). The diameter of the chromatophore increases sevenfold.

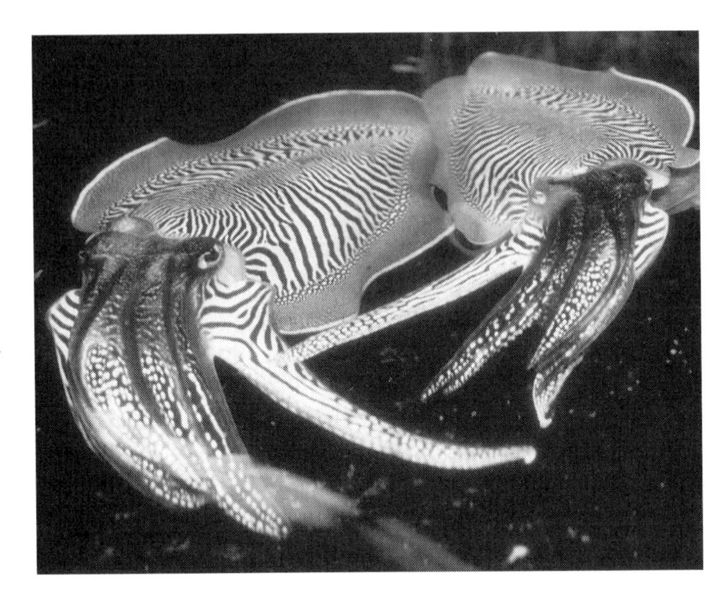

Figure 8.9 Chromatic display of squid. Two male squid, *Sepia officinalis*, display the intense zebra stripe pattern during an aggressive conflict. The normal body color pattern for this species is mottled or blotchy. (Photo courtesy of Roger T. Hanlon.)

Light is generated from a chemical reaction involving a molecule called luciferin, an enzyme luciferase, oxygen, and the energy-transporting chemical ATP. The energy from the ATP is used to push the luciferin molecule into an excited state by raising an electron into a higher orbital. When the electron falls back down to its baseline position, a photon is emitted that produces a greenish light with a wavelength of 562 nm. The process has a very high efficiency of 88% energy conversion with very little heat produced as a byproduct. However, it is a relatively expensive form of communication that is most useful for organisms active in dark environments (Nassau 1983; Lee 1989).

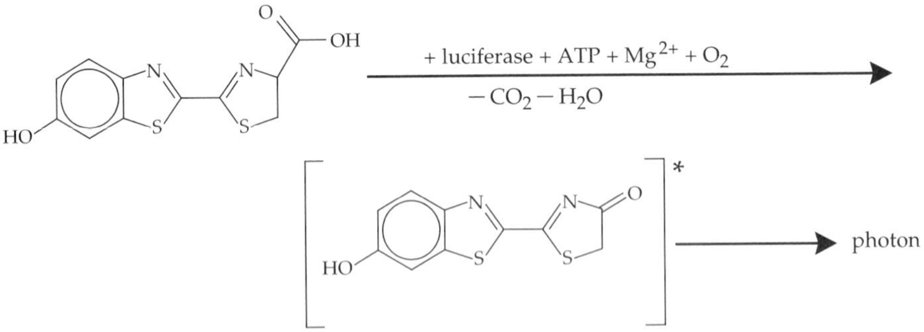

$$\text{luciferin} \xrightarrow[-CO_2 - H_2O]{+\text{luciferase} + \text{ATP} + Mg^{2+} + O_2} [\ \ ]^* \longrightarrow \text{photon}$$

Figure 8.10 The luciferin of the firefly *Photinus*. The active part of all luciferins is the COOH terminal group, which in the excited state (indicated by *) forms a double-bonded CO group allied with a system of conjugated double bonds in the rest of the molecule. One photon is released.

Almost all bioluminescent organisms are nocturnal or live in dim environments such as caves, burrows, or deep water. Self-generated light is much less intense than sunlight and would be inconspicuous if produced in the daytime. The pattern of light production depends on the communication function of the signal. Fireflies emit light in flashes with species-specific temporal patterns to attract mates (Lloyd 1983). Many cephalopods have clusters of photophores on their arm-tips or sexually dimorphic spots on other parts of the body that can be flashed on and off, suggesting a signaling function (Hanlon and Messinger 1996). Another function of bioluminescence is prey attraction, as proposed for angler fishes and glowworms. Finally, diurnally active, midwater fishes and cephalopods use bioluminescence to camouflage themselves against deeper dwelling predators with upward directed eyes that can detect their prey's silhouette against the bright background. The prey species illuminate their ventral side to match the downwelling light and thereby reduce their contrast with the background (Lythgoe 1988).

BEHAVIORAL COMPONENTS OF VISUAL SIGNALS

Behaviors may act alone or, more typically, in conjunction with colored or structured body parts, to produce a visual signal. They contribute to the temporal variability and some of the spatial properties of visual signals. All behavioral elements involve muscular movement. We have already seen how muscular control over chromatophores in cephalopods produces rapid changes in body color. Similarly, bioluminescent organisms may sometimes modulate emitted light by means of movable covers. Animals most commonly use muscular movement to change their body shapes, position themselves in space, orient with respect to a potential receiver, and make gestures and stereotyped movement displays. In the case of single-celled organisms, movement is caused by the back and forth bending motion of arrays of subcellular microtubules in cilia and flagella. In multicelled animals movement is generated by the contraction of protein microfilaments in specialized muscle cells. The anatomical details of muscular types, arrangements, and innervation vary greatly among animals and may impose certain constraints on signaling (Camhi 1984).

Insects and other arthropods typically display with limb movements. They have a large number of skeletal muscles, usually arranged in antagonistic pairs and attached to the inside of the exoskeleton. A motor neuron to a given muscle innervates all of the fibers in the muscle so that the fibers contract in unison. Several different nerves innervate each muscle, each causing the muscle to contract at a different velocity. This differential speed of contraction gives the insect very precise and rapid control over its movements, but the movements themselves are "jerky" because of the simultaneous fiber contraction. Skeletal muscles in vertebrates are innervated rather differently. Muscles are still arranged in antagonistic pairs to control extension and flexion of a limb, but there are fewer muscles overall than in a typical insect. A muscle consists of a bundle of many fibers, each of which is innervated by a single nerve. Fibers therefore contract singly, producing a graded contraction in the whole muscle as a function of the number of nerves stimulated. A muscle can contract quickly or slowly depending on the pattern of nerve stimulation to produce smooth and highly variable types of movements.

The smooth muscle system of vertebrates is designed primarily to regulate internal body processes such as breathing, vasodilation, digestion, and pupil contraction. Smooth muscle movement is generally involuntary and may be controlled by the two antagonistic nerve networks of the autonomic nervous system, by activation from the contraction of nearby fibers, by mechanical stretching, or by circulating neuroendocrine hormones. Smooth muscle contraction is therefore slow, tonic, and/or rhythmical. Some species have evolved additional neuromotor control over these processes for signaling functions. Such signals are likely to be slowly modulated.

Most species potentially have a very large number of different orientations, postures, gestures, and movements they can combine into a visual signal that symbolizes some type of information. Movement displays are the outcome of a set of neuronal pattern generators in the brain or ganglion that drives specific muscle complexes via motor neurons. (A discussion of these mechanisms is beyond the scope of this book but can be found in neuroethology texts such as Camhi 1984.) The form of movement tends to be highly conserved during evolution, so that closely related species show very similarly structured displays. Display movements are often performed repeatedly, with little variation between or among individuals (Barlow 1977). In the Part II of this book we discuss the actual type of movement employed and the way in which information is encoded in the signal.

TRANSMISSION OF VISUAL SIGNALS THROUGH THE ENVIRONMENT

The transmission of a long-distance visual signal from sender to receiver entails many of the same problems as long-distance sound transmission, such as attenuation, pattern loss, and background noise. However, the two modalities differ in several important respects. Most importantly, for reflected light signals there is an additional step in the process, since the production of a signal is dependent on the availability and quality of the ambient light. Secondly, there is always background noise for visual signals, so the contrast between the signal and the background as well as the perceptual ability of the receiver take on greater importance. Finally, light waves don't bend around objects like sound waves do, so visual signal transmission is much more strongly affected by the presence of opaque objects between sender and receiver than is sound transmission.

The major steps in visual signal transmission are illustrated in Figure 8.11. Each step in this process involves a transfer function, represented by f and an uppercase letter. The result is a wavelength-specific spectrum for that step that is indicated by the corresponding uppercase letter. For reflected light signals, we begin with the spectrum of light from the sun, S. Sunlight is scat-

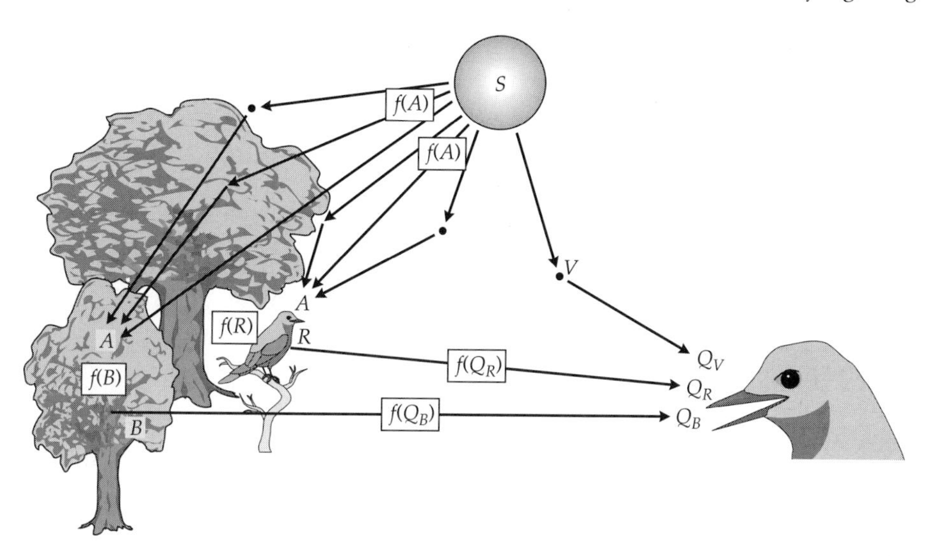

Figure 8.11 Factors affecting transmission of a reflected light signal. Each step in the transmission process involves a transfer function indicated by a box and a resulting spectrum indicated by an uppercase letter. *A* is the ambient irradiance spectrum falling on the sender or the background; the spectrum has been modified from full sunlight, *S*, via the function *f(A)*, which includes medium absorption, scattering, and filtering. *R* is the reflected radiance spectrum in the direction of the receiver after modification of *A* by the sender's surface, *f*(*R*). Similarly, *B* is the reflected radiance spectrum of the background after modification of *A* by *f(B)*. *V* is the veiling light radiance spectrum. Q_B, Q_R, and Q_V are the final radiance spectra arriving at the receiver from the background, sender and veiling light after transmission over distance *d*. (After Lythgoe 1988; Endler 1990.)

tered, filtered, and attenuated by the transfer function *f*(*A*), often in a frequency-dependent manner, depending on the medium, weather, and habitat. The available light spectrum, *A*, is produced. Senders can only reflect those wavelengths that are available. The spectrum of reflected or self-generated light from the sender, *R*, depends on the characteristics of the sender's surface, *f*(*R*). The light that reaches the receiver from the sender, Q_R, depends on the distance between them and factors such as attenuation, scatter, and total blockage from large opaque objects, $f(Q_R)$. In addition, the ability of the receiver to perceive the signal is determined by its ability to distinguish the signal from the background. Background reflectance can be quantified in the same manner as the signal by measuring the ambient light that reflects off of the background, *B*, and is transmitted to the receiver, Q_B. The conspicuousness of the signal is determined by the perceived contrast between the reflected light emanating from the sender and from the background. Finally, veiling light, Q_V, is an additional source of noise in reflected light systems that serves to reduce the contrast between signal and background.

Since scattered and reflected light can contribute significantly to the total light arriving at the sender, the available light spectrum *A* is always an irradiance measurement. The reflectance spectra R, B, Q_R, Q_B, and Q_V, on the other hand, are always angle-dependent radiance measurements because only light rays traveling directly from the sender or background to the receiver contribute to signal reception. In the first parts of this chapter we discussed the mechanisms by which animals (and the background) produce visual stimuli, $f(R)$ and $f(B)$, and the respective spectra, R and B, that these mechanisms generate under white light illumination. Now we shall examine the ways in which the perception of visual signals is affected by available light, the transmission properties of the environment, and background contrast.

Available Light

Both the amount and the spectral composition of available light can differ in different environments. A uniform reduction in ambient light across all visible frequencies, such as occurs at night and in caves, obviously reduces the total signal stimulus Q_R and makes signal perception more difficult at greater distances. In deep water below 1000 m there is insufficient light for any type of vision (Lythgoe 1979). Rather different effects occur with the differential reduction of some light frequencies caused by frequency-dependent absorption and scattering by the medium itself and by filtering and reflection from plant material and other objects in the environment. When the ambient light is itself colored, it can change the apparent color of a colored signal patch. A color patch that under white light conditions reflects strongly in a particular region of the spectrum will appear desaturated and less bright when viewed under an ambient light context that contains low levels of this spectral region. In the extreme case where the reflected frequencies of a color patch are completely missing from the ambient light spectrum, the patch will appear black. Figure 8.12 shows an example of the change in reflectance for three patches, red, green, and blue, when illuminated by a typical green forest ambient light. Although all three have equal brightness under white ambient light, the reflectance from the red and blue patches is significantly reduced relative to the green patch under ambient conditions. Thus animals can reflect colored signals only when those same frequencies are abundantly present in the ambient light.

Many environments are illuminated by colored ambient light to some degree, and certain habitats are strongly colored. As we learned in Chapter 7, clear water becomes increasingly bluer with increasing depth (see Figure 7.12). This color change occurs because the water medium strongly absorbs both ultraviolet and infrared radiation. The absence of violet and red ambient light in deep water significantly affects the appearance of color patches. Consider the reflectance spectrum R of a typical red carotenoid color patch that absorbs blue, green, and yellow and reflects orange and red (e.g., Figure 8.3C) viewed at increasing water depths. If sender and receiver are just under the surface the patch will appear red, but at depths of from 5 to 10 meters it will appear orange and increasingly darker as the long wavelengths drop out of the ambient spectrum. At 10 meters below the surface it will appear black. An

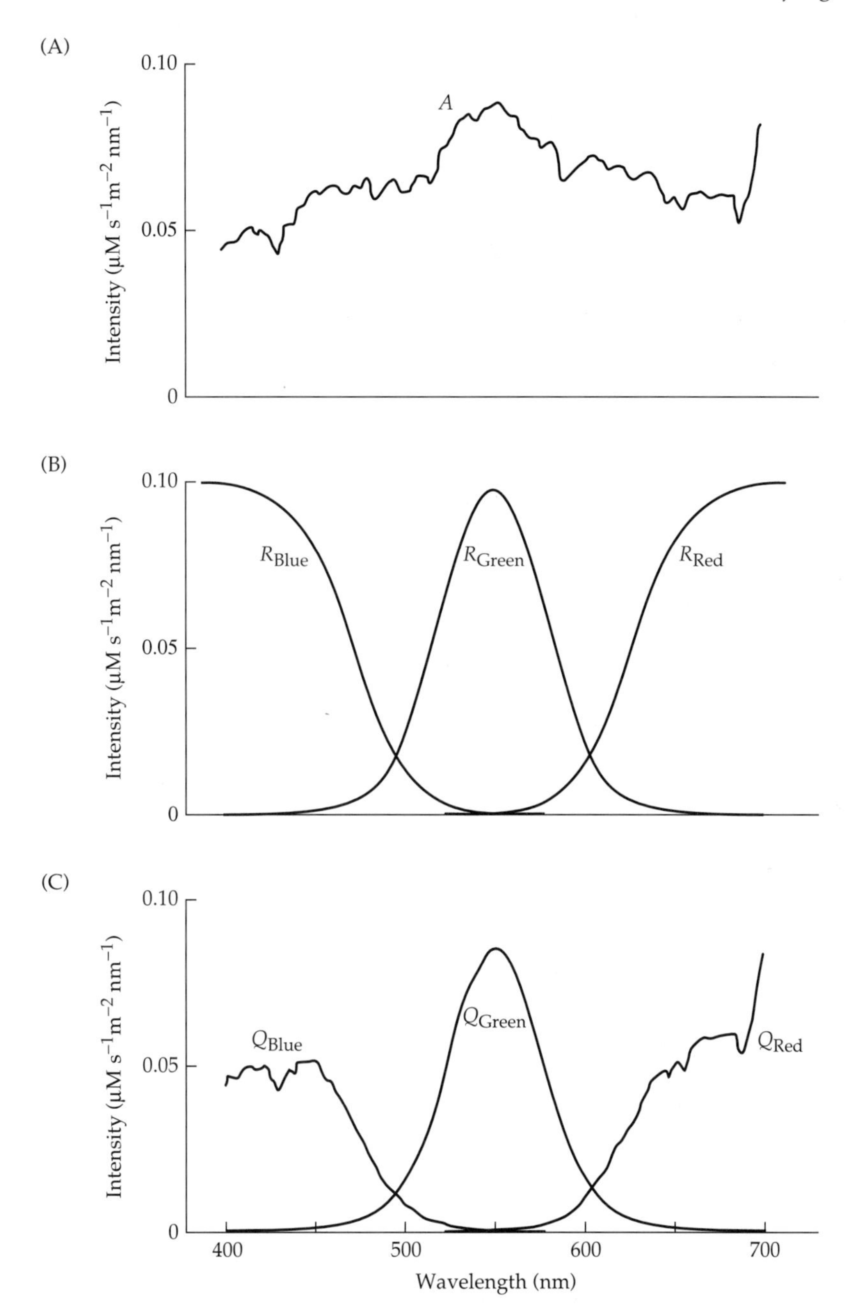

Figure 8.12 Change in color patch brightness illuminated by colored ambient light. (A) shows a typical green irradiance spectrum from a forest habitat. (B) shows the reflectance spectra of hypothetical blue, green, and red color patches under white-light illumination. (C) illustrates the change in the reflectance from the three patches when illuminated by the green ambient light in A. The blue and red patches are greatly reduced in intensity, compared to the green patch. Vertical axes in all graphs are given in units of micromoles photons $s^{-1}m^{-2}nm^{-1}$. (After Endler 1990.)

ultraviolet signal patch would similarly change to violet, dark blue, and black with increasing depth.

Blue water color is indicative of nutrient-poor water, low in phytoplankton and organic material. Freshwater lakes and ponds, on the other hand, are usually greenish as a result of the reflection of green and absorption of red and blue by phytoplankton and other plant material. Swamps and marshes often exhibit a yellowish hue from the dissolved products of decaying vegetable matter. Streams collect the runoff of minerals that reflect in the red and infrared range, making the water appear reddish, brown, and sometimes black. Scattering is also an important factor limiting vision in shallow water. Suspended particles and air bubbles increase the turbidity of the water. Turbid increases veiling light and limits light transmission to a few centimeters. Strong scattering has the additional effect of reducing one's ability to distinguish up from down, since light comes almost equally from all directions; when absorption is the more significant process, light is always strongest from the direction of the source.

In terrestrial habitats, the three principal factors influencing irradiance spectra are the sun's angle above the horizon, the weather conditions of the atmosphere, and effects of vegetation. For irradiance measurements it is important to remember that light may arrive directly from the sun, indirectly from other parts of the sky, and indirectly from vegetation and that radiance from these components may be distinctly colored. In a forest, defined by a continuous canopy, only filtered light will arrive on the ground and it is distinctly green (Figure 8.13B). In woodland shade, where the canopy is not continuous, no direct sunlight is present but there is much Rayleigh-scattered light from the blue sky, so the ambient light is bluish (Figure 8.13C). In a small gap in the forest, on the other hand, light arrives only directly from the sun and not from the blue sky, so long wavelengths predominate (Figure 8.13E). In any of these contexts, if the sky is cloudy, a large amount of Mie-scattered white light will be present and all spectra tend to look like that in Figure 8.13D. Total illumination can actually increase in shady sites when there is cloud cover. The angle of the sun determines the amount of atmosphere through which the light must travel before striking Earth, with the shortest path occurring at noon and the longest path at sunrise and sunset. The longer the path, the greater the scattering of blue wavelengths. This scattering strongly eliminates blue wavelengths from direct sunlight, but scattered blue

Figure 8.13 Effects of vegetation structure, cloud cover and sun angle on irradiance spectra of ambient light in terrestrial habitats. (A) Diagram of microhabitats showing continuous canopy forest, small gaps in the forest, broken canopy woodland, and large gaps. The gray-shaded area shows sites receiving direct sunlight while nonshaded areas receive only indirect light. Ambient light irradiance spectra for (B) forest shade, (C) woodland shade, (D) forest, woodland and gaps under cloudy conditions, (E) small gaps, (F) large gaps and (G) low sun angle. Total light intensity is shown in small boxes for adjacent microhabitats at the Santa Barbara study area in units of μmol $m^{-2}s^{-1}$. (After Endler 1993b.) ▶

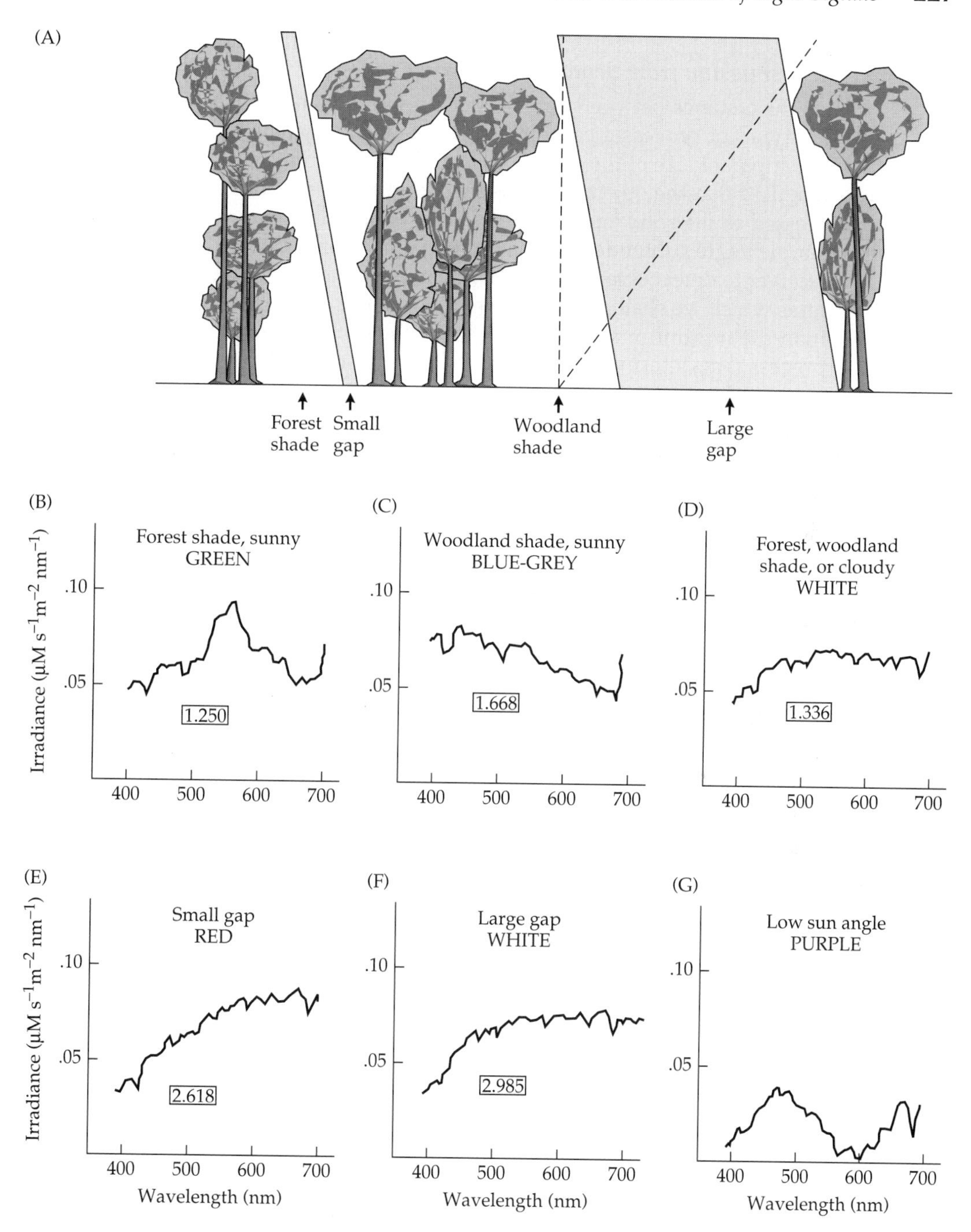

light from the rest of the sky contributes to the overall irradiance to yield purple light (Figure 8.12G). Finally, the presence of dried grass, leaf litter, woody tree trunks, and red leaves can alter irradiance spectra by reflecting yellow, orange, brown, and red wavelengths.

Transmission from Sender to Receiver

As the distance between an object and a visual receiver increases, two different types of processes combine to reduce visibility. The first is the change in the apparent size of the object. As the distance increases, the size of the receptive field that the object occupies shrinks. The apparent size of an object is expressed as the solid angle subtended by the object's area in steradians, or the simple angle subtended by the object's length or diameter. The ability of the receiver to detect objects is therefore partly a function of its resolving power, a topic which we shall discuss in the next chapter. Given that the object is larger than the minimum resolving angle for the distance in question, the second process that determines its visibility is the transmission of light waves from the object to the receiver. Factors that reduce the intensity of the light waves emanating from the object are similar to those that affect sound transmission: global attenuation or spreading loss, which is independent of frequency, and medium absorption, scattering, and filtering, which are usually both wavelength- and medium-specific.

Light spreads from a point source in all directions, so the falloff of intensity with distance, d, declines as a function of $1/d^2$ just as with sound. This decline is frequency-independent, so we refer to it as global attenuation to distinguish it from frequency-dependent types of attenuation. In principle, however, an infinitely large signal surface area experiences no falloff in intensity with distance at all. Real signal surfaces are somewhere in between these two extremes. We can only state that intensity declines approximately in proportion to $1/d$ and declines faster with smaller signal areas than with larger areas (Dusenbery 1992).

For light, the effects of global attenuation, medium absorption, scattering and filtering are all combined into a single empirically determined parameter, the **beam attenuation coefficient** a. The reciprocal of a is **attenuation length,** L_a, the maximum distance at which a large and contrasting object can just be detected. It is both medium- and wavelength-specific. Typical values for attenuation lengths are shown in Table 8.1. The beam attenuation coefficient can be measured for a specific signal patch (or the background) in a specified ambient light context. The appropriate intensity measure is radiance, and the acceptance angle of the sensor must be small enough that only the signal patch

Table 8.1 Light-beam attenuation lengths in different media.

	Wavelength (nm)					
Environment	**300**	**400**	**500**	**600**	**700**	**800**
Pure air	7.0 km	22 km	55 km	120 km	220 km	370 km
Clean air	3.8 km	5.0 km	6.0 km	6.7 km	7.4 km	7.9 km
Moderate fog	50 m	50 m	50 m	50 m	50 m	50 m
Pure water	?	23 m	28 m	5.4 m	2.0 m	0.49 m
Ocean	?	1–10 m	1–15 m	1–5 m	1–2 m	?

Source: Dusenbery 1992.

is included in the sensor's area. A radiance measure would be taken close to the patch and at varying distances from the patch. The value a is the slope of the regression line for the log-log plot of radiance versus distance. We can then use a in a transfer function of the form $f(Q) = e^{-ad}$ to calculate the falloff in radiance over any distance d. The final result is an estimate of reflected light radiance arriving at the receiver from the sender:

$$Q_R = R_0 e^{-ad}$$

where R_0 is the reflectance spectrum at the patch source. A similar type of measurement can be made for the radiance arriving at the receiver from the background, Q_B, and in the section below we will use these two quantities to estimate the contrast between the signal and its background.

Opaque objects located between sender and receiver, of course, also reduce signal visibility. Light waves cannot bend around objects like sound waves can, creating a significant problem for visual communication. In a heavily vegetated environment, signal transmission can only be improved by reducing the distance between sender and receiver. Objects in the environment also distract the receiver from the signal, so that signal objects must be different in size and shape, as well as color and brightness, to be maximally discriminated from environmental objects.

Optical Background

The transmission efficiency of a visual signal depends critically on the type of background against which the animal is seen. Background noise is always present for any signal in this modality. Most backgrounds possess four characteristics—brightness, color, pattern, and movement—which correspond to the four signal properties we identified early in this chapter. Senders can exploit any one of these, or a mixture, to develop a signal that presents a contrasting image against the backdrop. General strategies for these four types of contrast are described below, but it is important to realize that the degree to which a given species perceives these contrasts depends on its visual acuity, temporal resolution, and color perception abilities. Each of these contrast-enhancing strategies has an opposing contrast-reducing strategy for maximizing camouflage and crypticity.

STRATEGIES FOR BRIGHTNESS CONTRAST. Brightness contrast is a very common strategy for enhancing conspicuousness, especially in aquatic environments, and its principles have been well developed. In the section above, we described how one would quantify the radiance arriving at the receiver from both the signal and the background. For this analysis, Q_R and Q_B could be evaluated over a narrow range of wavelengths, or summed over all perceived wavelengths by using the areas under the spectral curves. A useful measure of the brightness contrast between signal and background as a function receiver distance is the relative difference between the two:

$$C(d) = \frac{Q_R - Q_B}{Q_R + Q_B} = \frac{R_0\, e^{-ad} - B_0\, e^{-ad}}{R_0\, e^{-ad} + B_0\, e^{-ad}}$$

C(d) is zero when brightnesses are the same, negative when the signal is darker than the background, and positive when the signal is lighter than the background. Dividing by the sum of the signal and background radiance produces a symmetrical index that varies from –1 to +1, a range that is most appropriate for questions concerning contrast maximization against any type of background. A similar index with only background radiance in the denominator is more accurate for questions concerning minimum or threshold contrast necessary for detection. Regardless of which index is used, *C(d)* can also be expressed as C_0e^{-ad} where C_0 is the inherent contrast between signal and background at a distance of zero. The farther the distance between sender and receiver, the lower the absolute contrast compared to that which would be perceived very close to the sender (Kirk 1983; Lythgoe 1979, 1988; Endler 1990; Dusenbery 1992).

The effects of veiling light on brightness contrast can easily be incorporated into this expression. Veiling light from scattering particles such as fog and particulate matter diverts more diffuse light into the receiver's eyes than if such particles were absent. Although veiling light may be colored, it is frequently white. The addition of white light to both the background and the signal therefore reduces their chroma. The intensity of veiling light increases with the distance between the sender and the receiver (because more particles intervene) according to the transfer function $f(Q_V) = 1 - e^{-ad}$, which results in $Q_V = V(1 - e^{-ad})$. Adding Q_V to both Q_R and Q_B in the equation above yields:

$$C(d) = \frac{Q_R - Q_B}{Q_R + Q_B + 2Q_V}$$

As the distance between sender and receiver increases, the veiling-light term increases and brightness contrast decreases.

Brightness contrast can be exploited best in environments with backgrounds that are either extremely dark or extremely light. With a dark background, such as occurs in forests, deep water, caves, and at night, white or bioluminescent colors contrast most, and in bright, highly illuminated contexts black or other dark colors will contrast most. An excellent example of brightness contrast is illustrated in a comparative study of eight leaf warbler species occupying a range of habitats from low montane scrub to broadleaf forest. All eight species are primarily yellow-green but vary in the number of white color patches on the wings, head, and rump (Figure 8.14). The darker the habitat occupied by a species, the more patches of white it possesses (Marchetti 1993). Some strategies of brightness contrast depend on the direction of the illuminating light. In media without a significant amount of Mie scattering, light comes from above. This means that the dorsal side of the animal is more strongly illuminated than the ventral side. **Countershading**, in which the dorsal side is more darkly pigmented than the ventral side, is a common camouflage strategy to reduce brightness contrast. **Reverse countershading,** on the other hand, serves to increase the visibility of the animal that is illuminated from above (Figure 8.15). In open habitats at low sun angles, the illumination of the background depends on the position of the viewer relative to the sun.

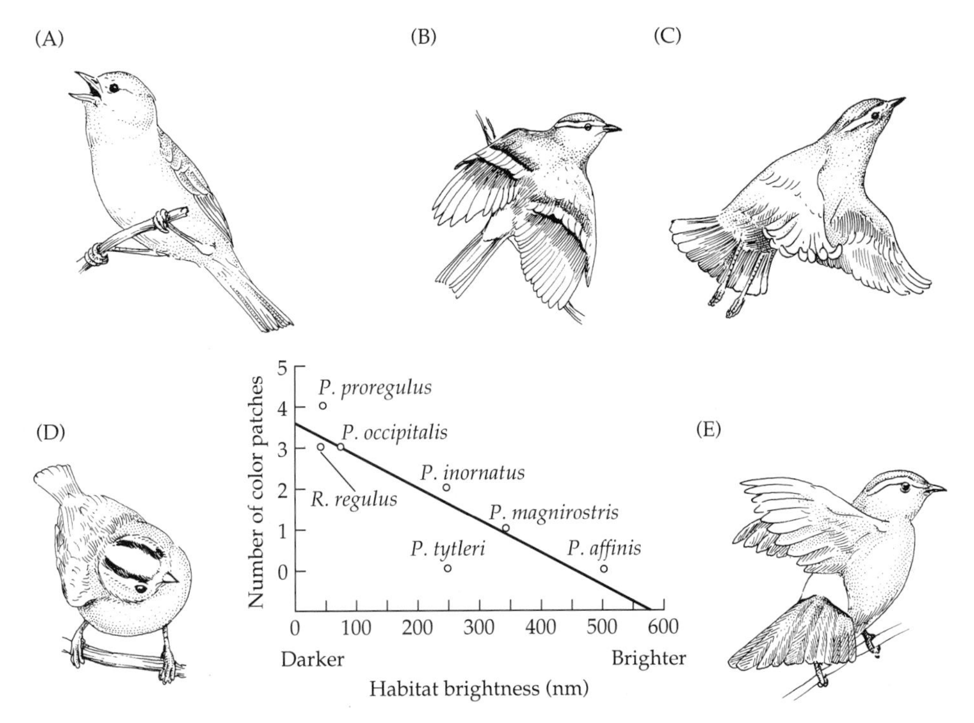

Figure 8.14 White color patches and display behaviors of *Phylloscopus* warbler species from habitats of different brightness. Species in this taxon possess yellow-green body plumage and from 0 to 5 white or pale yellow patches. Color patches are added in the following order: greater covert wing bar, median covert wing bar, crown stripe, rump patch, tail patch. These birds inhabit environments characterized by a range of available light from the brightest (open montane scrub) to intermediate (open forest) and dark (dense evergreen forest). Habitat brightness was measured with a camera meter and converted to units of food candles. Species in darker habitats have more color patches. The illustrations show how patches are featured in movement displays. (A) Species with no patches such as *P. affinis* exhibit no stereotypical display movements and rely more on vocal communication. (B) Wing bars are flashed in a wing-opening display (*P. pulcher*). (C) Tail patches are exposed when the tail is spread during flight (*P. pulcher*). (D) The white crown stripe is displayed in side-to-side tipped head movement during aggressive encounters (*Regulus regulus*). (E) *P. proregulus* turns its back to the receiver to display the rump patch. (Graph from Marchetti 1993, ©Macmillan Magazines.)

This effect is extreme in water surface environments. When the viewer faces toward the sun, it sees intensely bright, white reflectance, so black signals are more visible. When facing away from the sun the viewer sees diffuse blue light from the sky, so colored and white signals are more conspicuous. Since senders cannot control their position relative to the sun and potential receivers, they cannot maximize conspicuousness in both contexts. Most seabirds are white, probably because this minimizes their conspicuousness against the bright sky to prey in the water below and reduces thermal stress in their open environment (Hamilton 1973).

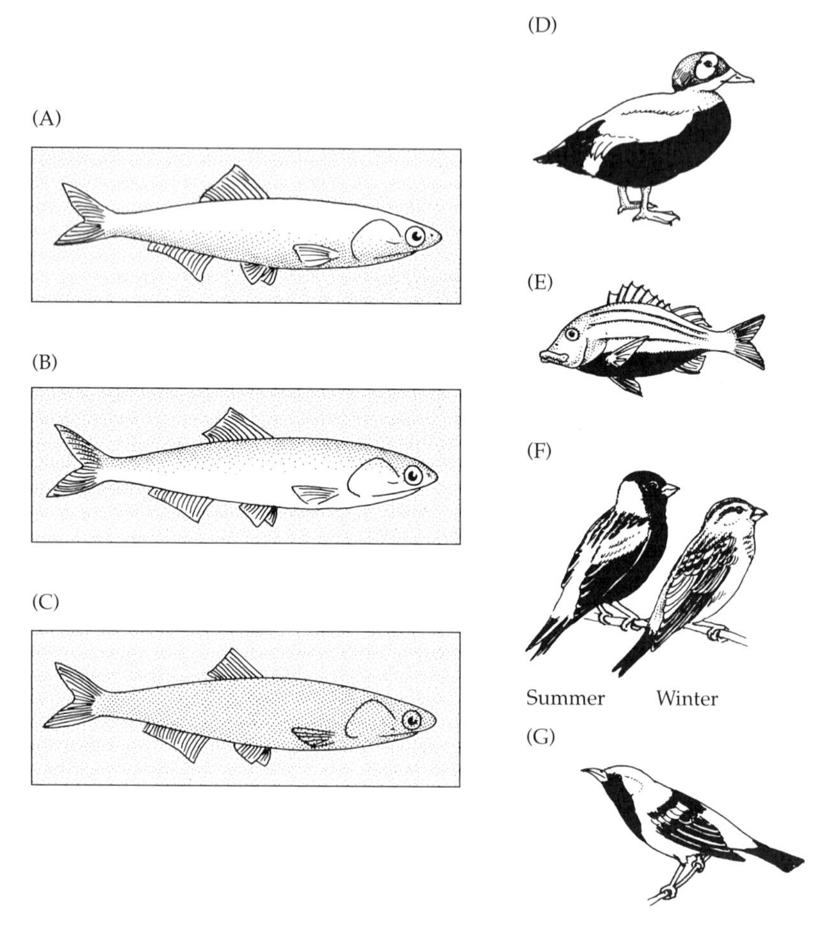

Figure 8.15 Countershading and reverse countershading. Countershading is a concealment strategy for animals illuminated from above: the dorsal side is darkened and the ventral side lightened to counteract the stronger dorsal illumination. (A) A uniformly colored animal with directional top-lighting projects a dark ventral outline. (B) A countershaded animal with diffuse or broadside illumination is also relatively conspicuous. (C) Countershading with top illumination reduces the contrast between the background and minimizes both dorsal and ventral outlines. With reverse countershading the ventral side is darkened to increase conspicuousness. Examples include: (D) spectacled eider *Somateria fischeri*, (E) Spanish grunt *Haemulon macrostomum*, (F) male bobolink *Dolichonyx oryzivorus*, and (G) male hooded oriole *Icterus cucullatus*. (A–C after Cott 1940; D–G after Hailman 1979.)

STRATEGIES FOR COLOR CONTRAST. For species with good color vision, hue and chroma contrast can be a very effective method of background contrast in addition to brightness contrast. In the absence of information on the visual pigments in the eyes of the receiver, we can make an initial evaluation of the color contrast between the signal and background spectra arriving at the receiver. A simple method is to compute the difference between the intensity of the signal and the intensity of the background at each wavelength interval, and then determine the root mean square of these differences over all wavelengths that the species can distinguish as follows (Endler 1990):

$$D(\lambda,d) = \sqrt{\sum^{\text{all }\lambda}[Q_R(\lambda,d) - Q_B(\lambda,d)]^2}$$

Any difference in either hue, chroma or brightness will increase the value of $D(\lambda,d)$. In essence, $D(\lambda,d)$ is the Euclidean distance of a line connecting the points of the signal and the background color on the three-dimensional color plot of Figure 8.2. Senders could maximize hue differences by using a saturated hue on the opposite side of the horizontal circle from the background hue, or they could maximize brightness differences along the vertical scale. In practice, this formula is only a first step in evaluating perceived color contrast by visual animals. With some knowledge of the receiver's color discriminating ability, the formula could be adjusted with weighting factors for the relative intensity sensitivity at different wavelengths, and the wavelength intervals used in the summation could also be varied to reflect regions of the spectrum in which hue discrimination varies. However, higher-level interactions in the brain may also play an important role in color contrast perception and much more work remains to be done on this topic.

Hailman (1977a, 1979) has measured the background color in a number of different terrestrial habitats. He found examples of backgrounds with a wide range of dominant frequencies: blue (sky), green (forest, marsh), yellow-green (grassland, old field), yellow (dried grass, undersides of holly leaves, broadleaf litter), orange/brown (pine needle litter, loblolly pine trunk), tan (sand) and achromatic (gray sky). An animal attempting to minimize transmission of its presence, and therefore to be cryptic, should try to match the hue of the background. An animal attempting to maximize transmission should use a contrasting color, as suggested in Table 8.2. What constitutes a contrasting color? For humans and possibly many other vertebrate species with good color vision, contrasting colors include red versus green and blue versus yellow. These color pairs appear on opposite sides of the hue circle illustrated in Figure 8.2. We shall see in the next chapter why certain color combinations are more conspicuous. If the ambient light itself is strongly biased (colored, as in Figure 8.12), animals can either use a color patch that matches the ambient light peak and surround it with colors that reflect poorly to maximize brightness contrast, or use complementary colors whose cutoff frequencies are centered on the region of greatest ambient light intensity (Endler 1992). There is some support for these color-contrast rules. Several tropical reef fish display adjacent patches of yellow and blue. Tropical forest birds, *Anolis* lizards, and stream-dwelling fish commonly use patches of red that stand out against their green backgrounds. A number of temperate zone birds that inhabit habitats with yellow-green, yellow, and orange-brown background colors have blue plumage. Birds living in open habitat, in which the bright blue sky is often their background, are often entirely black.

PATTERN CONTRAST. Habitats with lots of vegetation and/or rocks provide strongly patterned backgrounds that senders can exploit for either contrast or

Table 8.2 Optimal signal hue in different environments and backgrounds.

Habitat	Available light illumination level (hue)	Background hue	Optimal signal color
Night	Low (gray)	Black	White, biolumin
Open ocean, lake	Low to med. (blue)	Blue	Yellow
Marine reef	High (blue)	Blue	Red, yellow
Freshwater streams	Low to high (yellow-green)	Yellow-green	Blue, red
Tropical forest	Med. (green)	Green	Red
Temperate forest	Med. to high (green)	Yellow-green	Purple
Broadleaf litter	Med. (green)	Yellow	Blue
Forest tree trunk	Med. (green)	Orange	Blue-green
Grass, bush, marsh	High (white)	Yellow-green	Blue
Dried grass, old field	High (white)	Yellow	Blue
Sand dune	High (white)	Orange	Blue-green
Sky	High (white)	Blue	Black
Water surface	High (white)	Blue	Black or white
Low sun angle	Low, (purple)	Dark	White, yellow

Source: After Hailman 1979 and Lythgoe 1979.

camouflage. The basic strategy is to employ an opposing geometric pattern. For example, in vegetated habitats such as cattail marshes, grassland, eelgrass beds, and some forests, the background is a repeated vertical barred pattern. Contrasting strategies include either a horizontal barred pattern, the use of opposing geometrical shapes such as circles, or a solid (patternless) color that contrasts with the background color. Similarly, rocky habitats present a background of round or irregular shapes, and contrasting patterns include stripes, circles of very different sizes from the ambient rocks, and solid contrasting colors. In very uniform backgrounds, such as open water, any type of patterning will enhance conspicuousness. Animals can also use limbs, fins, and other appendages to exaggerate their shape. A variety of pattern-contrasting strategies are illustrated in Figure 8.16.

A well-known strategy for camouflage is **disruptive coloration**, which uses patterning to interfere with the perception of the true outline of the animal. High-contrast boundaries between adjacent color patches intercept the border of the animal at a perpendicular angle to disrupt a receiver's eye from the animal's outline. The stripes of the zebra serve such a camouflage function. The opposing, contrast-enhancing strategy, is to emphasize the outline of the animal. Figure 8.17 shows some examples of shape enhancement and outlining.

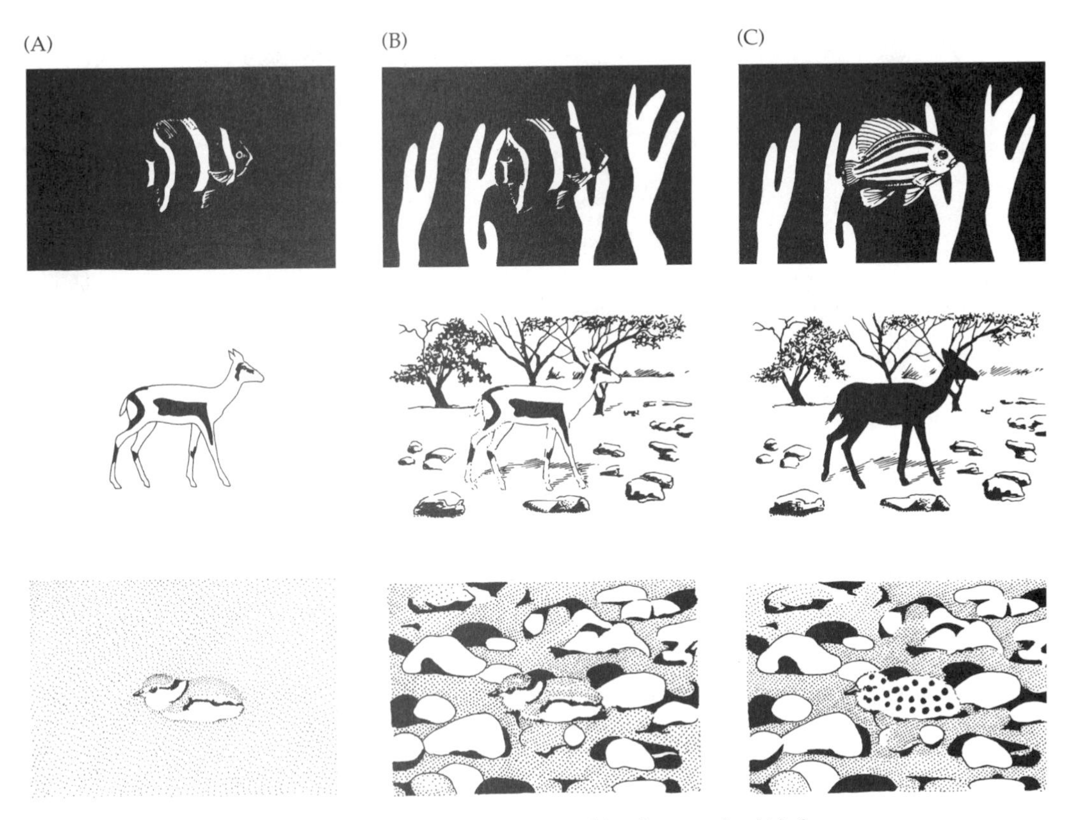

Figure 8.16 Pattern contrast. For a fish, antelope, and bird example, (A) demonstrates the conspicuous effect of a strong pattern against a uniform background, (B) illustrates the concealing effect of the same pattern when it matches the background pattern, and (C) shows the effect of a contrasting pattern in a patterned background. (After Cott 1940.)

MOVEMENT CONTRAST. Both terrestrial and shallow water environments with leaves usually present a moving background due to wind or wave action. The motion in both cases is relatively slow and sinusoidal, with a frequency around 1 to 2 Hz. Animals attempting to hide against such a background move with a similarly slow, sinusoidal pattern, e.g., vine snakes (Fleishman 1985). The most conspicuous movement pattern against such a backdrop is jerky, square-wave type motion of a higher frequency. For example, the push-up displays of *Anolis* lizards are jerky, rapidly accelerating up-and-down movements. Controlled laboratory experiments of prey detection by lizards show that changing the prey movement pattern away from that of the windy-leaf background movement to either a higher frequency or a square wave shape increased probability of detection (Fleishman 1992).

Movement and striped color patterns often operate together in creating either crypticity or conspicuousness. Striped patterns parallel to the direction of motion, such as the longitudinal stripes of some racing snakes, serve to

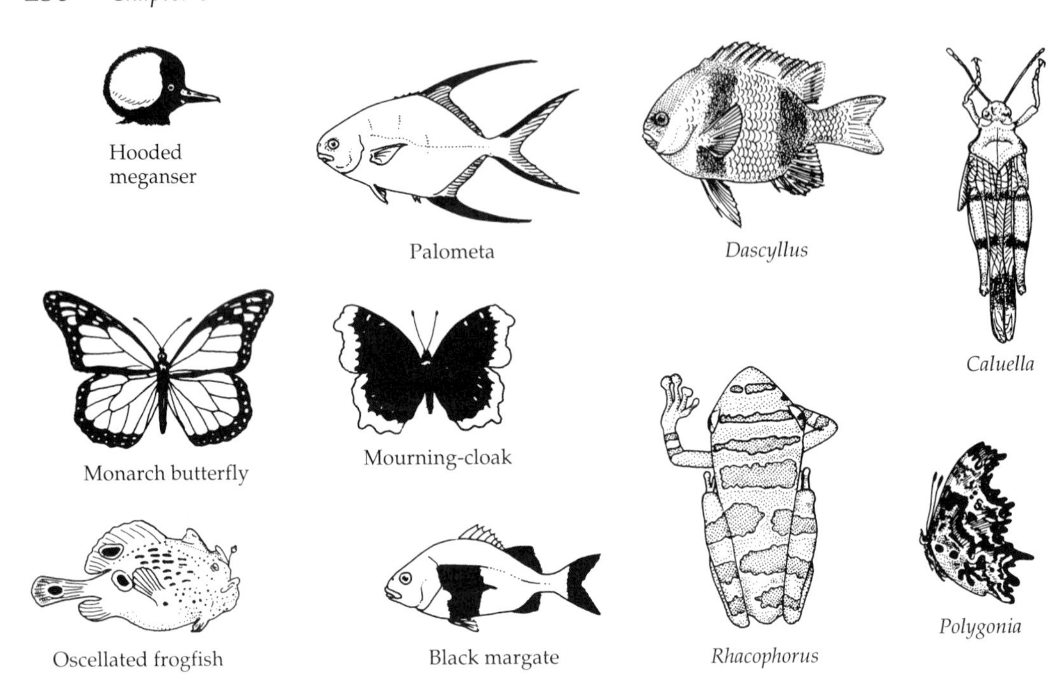

Figure 8.17 Enhancement of shape and outlining. Patterns on the left illustrate strategies that enhance the outline of the animal to make it more conspicuous, and patterns on the right illustrate concealing strategies that detract the eye from the animal's outline by intercepting the body's border. (After Cott 1940; Hailman 1979.)

conceal the motion. Conversely, movement in the opposite direction from striped patterns is highly visible. Vertical jumping displays are typically used against horizontal backgrounds, while sideways movements are often characteristic of animals living in vertically striped environments.

A serious problem with all of the strategies outlined above for maximizing contrast is that they are based on human perception. These rules can probably be extrapolated quite well to our closest relatives, the apes and old world primates. Most other mammals have much poorer-color perception abilities. Birds, some diurnal reptiles, and some fish have excellent color perception and appear to distinguish both red and violet/ultraviolet colors better than we do. Most invertebrates with color perception see well into the ultraviolet range and do not see red or orange. All of these species probably classify colors very differently and perceive different contrasting colors, compared to humans. Temporal and spatial resolution abilities also differ greatly among species. We shall describe the visual receiving organs of these other species in the next chapter, but we still do not know enough in most cases to say much about which colors, patterns, and movements provide maximum contrast and which do not. However, research in this area is proceeding rapidly (Thompson et al. 1992).

SUMMARY

1. Most animals make use of reflected light from the sun to produce visual signals, while only a few produce their own light. Regardless of the source of light, all visual signals can be characterized by the following four properties: (1) brightness or intensity of the signal, (2) spectral composition or color of the signal, (3) spatial characteristics of the signal, and (4) temporal variability in intensity, color, and spatial properties.

2. The **brightness** of a visual signal is the total amount of light emanating from it. For animals that use reflected light, brightness is largely a function of the amount of light in the environment, the range of wavelengths reflected versus absorbed by the animal's surface, and the smoothness of the surface. Animals that produce their own light can control the intensity of their signal to a much greater degree.

3. A reflecting surface appears colored if some wavelengths of light are selectively absorbed while others are reflected. Animals use three mechanisms to produce color. **Pigments** used by animals are large organic compounds with long chains or networks of conjugated double bonds. Such compounds have mobile electrons that can temporarily absorb light energy of specific frequencies. The **interference** mechanism of color production involves the construction of a thin layer of transparent material with a higher refractive index than that of the surrounding medium. The double reflection from the top and bottom of this layer produces iridescent color that changes with the incident light angle. The **scattering** mechanism makes use of the fact that small particles embedded in a transparent matrix scatter shorter wavelengths more than longer ones.

4. The **spatial properties** of a signal include the position of the sender relative to the receiver as well as the size, shape, and surface features of the sender's body or certain parts of the body. Some of these spatial characteristics, such as position and some aspects of shape, are determined by the behavior of the sender.

5. **Temporal variation** in intensity is an important feature of visual signals made by animals that generate their own light; animals using reflected light can only vary intensity by moving in and out of light patches. Temporal variation in color is achieved in several ways. Covering and uncovering of colored patches occurs in many insects and birds. The ability to change skin color is a feature of some cephalopods, frogs, salamanders, and fish, who make use of chromatophores to control the movement of pigments. Temporal variation in spatial properties of a signal are, of course, generated by movements of the body and appendages.

6. For reflected light signals, generation and transmission of a signal is a multistep process with a **transfer function** and resulting **spectrum** at each step. Sunlight is modified by attenuation, scattering, and environmental filtering to produce a spectrum of ambient light available to the sender. The sender further modifies ambient light, depending on its surface, to produce a spectrum of reflected light. Reflected light is then transmitted

a given distance through the environment to arrive at the receiver's eyes. A parallel pathway also determines the spectrum of background noise (and veiling light) arriving at the receiver's eyes at the same time.

7. The visibility of a signal declines as the receiver's distance from the sender increases as a result of two factors: the declining apparent size of the sender in the receptor field of the receiving eye, and the attenuation of the light intensity due to spreading loss, medium absorption, scattering, and filtering. All of these attenuation processes are combined into a single **beam-attenuation coefficient**, which is both medium- and wavelength-specific.
8. The **conspicuousness** of a visual signal is determined by its contrast with the background. Animals make use of four different strategies to enhance the conspicuousness of visual signals over long distances: brightness contrast, color (hue and chroma) contrast, pattern contrast, and movement contrast.

FURTHER READING

Nassau (1983) explains the physics and chemistry of color. Color production by animals is discussed in Fox (1976, 1979). Simon (1971) is an attractive coffee-table book on structural coloration in animals. Color change is described by Parker (1948), Waring (1963), Bagnara and Hadley (1973) and Demski (1992). The classic text on adaptive coloration is Cott (1940). Transmission of light signals has been discussed by Lythgoe (1979), Endler (1990) and Kirk (1983). Radiance and irradiance measurements in natural environments have been made by Jerlov (1968), Hailman (1979), Williams (1970), Endler (1993b), and Kirk (1994). Hailman (1979) and Burtt (1979) are still useful texts on visual signaling in animals.

Chapter 9

Light Signal Reception

ALMOST ALL LIVING ORGANISMS, PLANT AND ANIMAL, are sensitive to light. Plants of course use the energy from the sun to grow, and they can turn their leaves in the direction of the sun to maximize their exposure. Even primitive one-celled organisms can orient with respect to the sun and use the overhead direction of the sun to distinguish up from down. The ability to derive directional information from the sun must therefore have developed very early in the evolutionary history of life on Earth. The evolution of eyes that could form an accurate image of objects in the environment and a brain that could analyze the image required more evolutionary time. In this chapter we will describe the wide range of eyes found in animals. We cannot assume that animals see the world the way we do because their visual systems have evolved in response to the amount of light available in their environment, as well as to their needs for orientation, predator detection, and food finding. The capabilities of an animal's eyes certainly affect its use of vision for social communication.

THE VISUAL PIGMENTS

As we learned in Chapter 7, many types of organic compounds can absorb photons of light in the visible range by elevating electrons to higher energy states. These electrons then release energy in the form of heat as they slip back down to their original state. To use the energy in absorbed photons more effectively, living organisms had to devise a means of trapping the absorbed energy before it was lost as heat. The earliest type of pigment to trap light energy was chlorophyll, which can convert the energy in photons to chemical energy in the form of adenosine triphosphate (ATP). The energy stored in the phosphate bond can then be released at a later time by conversion to adenosine diphosphate (ADP). This form of chemical energy is universally used to fuel most energy-requiring metabolic processes in plants and animals. Single-celled flagellated algae such as *Euglena* have adapted several photosensitive pigments into specialized structures that detect the direction of light and orient the movement of the flagellum (Foster and Smyth 1980).

A light-sensitive pigment for true vision must trap the light energy in a photon and convert it immediately into a nerve impulse. All multicellular animals with a visual organ connected by nerves to the brain and/or locomotory muscles use a common pigment and chemical process for light entrapment. The pigment is called **rhodopsin**. Rhodopsin is found universally throughout the animal kingdom, from bacteria and primitive multicellular animals to the most advanced invertebrates and vertebrates.

Rhodopsin consists of two components: **retinal**, a conjugated double-bond molecule derived from vitamin A_1 and **opsin**, a large (40,000 atomic mass units) polypeptide protein molecule that is anchored into the plasma membrane of a visual receptor cell. The proposed structure of opsin is shown in Figure 9.1. It consists of seven helices that span the membrane and are connected and held in a circular bundle. The retinal in its relaxed state as the 11-*cis* isomer forms a bent molecule that nestles horizontally inside the bundle. It is linked to a lysine on helix 7 and forms additional weak attachments with other helices. In the presence of light in the visual range, a single photon is absorbed by the retinal. This causes the retinal component to straighten into the all-*trans* retinal isomer. The new form of the molecule is called **metarhodopsin**. Formation of metarhodopsin activates a G-protein called transferin that either directly or indirectly opens ion channels in the receptor cell's plasma membrane to generate a nerve response (Saibil 1990).

Depending on the type of eye and the amount of ambient light, metarhodopsin reverts back to rhodopsin via one of three different processes and is ready to respond again (Shichi 1983). (1) Certain forms of metarhodopsin can absorb a second photon of light and revert directly to rhodopsin. (2) The all-*trans* retinal detaches itself from the opsin and, in the presence of other as yet unknown compounds in the receptor cell, absorbs a photon of light and changes back to 11-*cis* retinal. It then reattaches to the opsin. (3) In a dark-adapted eye at very low light levels, there is insufficient ambient light energy to convert the all-*trans* retinal back to 11-*cis*. When a high concentration of all-

(A)

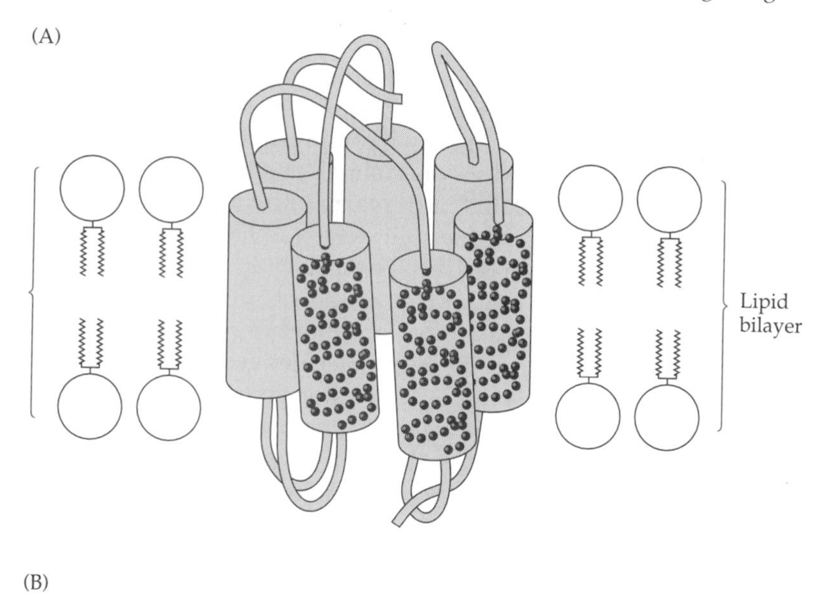

(B)

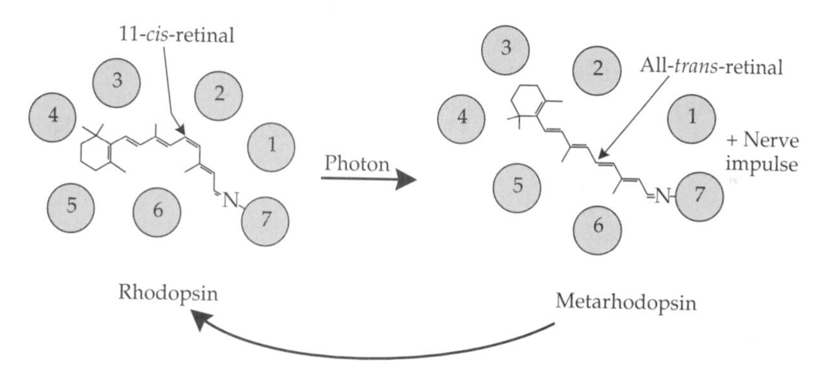

Figure 9.1 Chemical structure of rhodopsin. Rhodopsin consists of retinal bound to opsin. The opsin is anchored in the membrane of a photoreceptor cell. (A) A side view of the opsin molecule, showing the seven trans-membrane helices. (B) A top view of the molecule, showing the position of retinal in the center of the helices. When the retinal is in its 11-*cis* isomer and attached to the opsin, it can absorb a photon of light. The molecule is twisted at the point indicated by the arrows and it becomes the all-*trans* isomer called metarhodopsin. Metarhodopsin reverts to rhodopsin via one of three different processes. (After Saibil 1990; Wolken 1995.)

trans retinal builds up, it is transported to the epithelium cell layer adjacent to the receptor cells. An enzyme there converts it first to all-*trans* retinol and a second enzyme transforms it to 11-*cis* retinol (vitamin A_1). This travels back to the receptor cell and is converted to 11-*cis* retinal by a third enzyme. These conversions all require energy input in the form of ATP. The 11-*cis* retinal readily attaches to the empty opsin molecules and the rhodopsin is again ready to respond. A depleted receptor cell requires 30 to 40 minutes to recover via this process. An adequate store of vitamin A in the body has long been known to facilitate dark-adapted vision. Animals cannot synthesize vitamin A and must consume it from plants in the form of beta carotene (see Figure 8.3B). One molecule of beta carotene split exactly in half becomes two molecules of vitamin A. The vitamin is stored in the liver and transported to target organs such as the eye.

There are approximately 10 billion molecules of rhodopsin in a typical vertebrate receptor cell. Only one of these molecules needs to absorb a photon of light for the cell to produce a response. Rhodopsin operates most efficiently with light wavelengths between 400 and 600 nm. Unlike the cutoff and multi-peaked types of absorption curves described for typical pigments (see Figure 8.3), the absorption curve for rhodopsin is a smooth bell-shaped curve with a peak in the middle of the visible spectrum around 500 nm. It is called visual purple because of its violet color in the 11-*cis* form when extracted from eyes. In a later section we will discuss the ways in which pigments with different absorption peaks are produced.

The specialized nerve cells that house photopigments, called **photoreceptor cells**, all show certain elements in common. To hold the photopigment molecules in an orderly array, the cells contain specialized stacks of membranes that are derived from multiple infolding or outfolding of the cell membrane. The opsin parts of the photopigment molecules are anchored into these membranes. The stacks of pigment-laden membranes are located in the part of the cell that is most exposed to the light. The fine structure of the cells and their membrane stacks appear to fall into two general categories, called **ciliary** and **rhabdomeric** photoreceptors. These names reflect the type of cell from which the photoreceptor evolved. Vertebrates possess ciliary photoreceptors, whereas most invertebrates with good enough eyes for visual communication (arthropods and mollusks) possess rhabdomeric photoreceptors derived from microvillous cells (Eakin 1965; Westfall 1982; Cronly-Dillon 1991).

These two types of photoreceptors differ in three important ways. (1) Shape of pigment-bearing membranes. Advanced ciliary receptors have membranes shaped in flat round discs; the discs are stacked on top of each other like pancakes with their flat surfaces perpendicular to the direction of light (Figure 9.2A). Rhabdomeric cells, on the other hand, pack their photopigments on rolled microtubular membranes. The tubules are stacked parallel to each other like the bristles of a brush and oriented perpendicular to the direction of light (Figure 9.2B). As we show in greater detail later, the difference in membrane shape results in rhabdomeric cells being inherently more sensitive to polarized light. (2) Recovery process of metarhodopsin. The metarhodopsin in ciliary cells reverts to rhodopsin via processes 2 and 3, described earlier, while the metarhodopsin of rhabdomeric cells recovers primarily via process 1. This variation is a consequence of differences in the opsin structure and biochemical cascade of ciliary and rhabdomeric cells

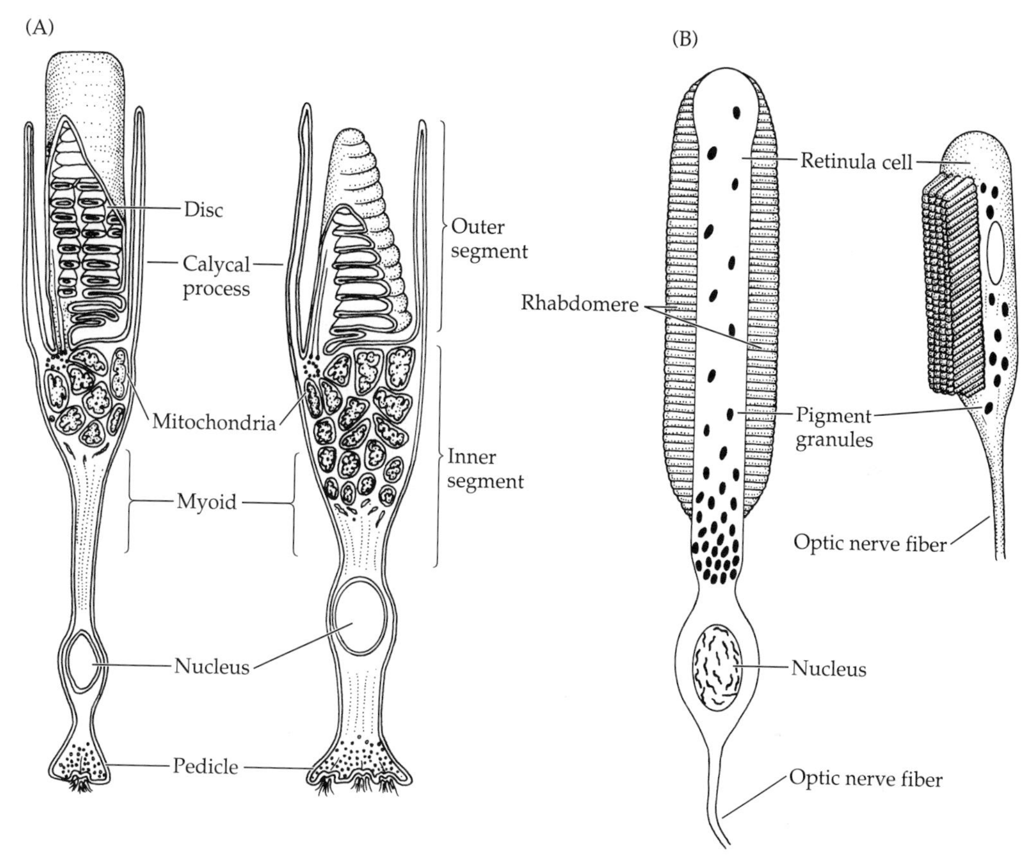

Figure 9.2 Structure of pigment-bearing membranes in ciliary and rhabdomeric photoreceptors. (A) Typical ciliary photoreceptors found in vertebrates (rod on left, cone on right). The outer segment houses the photopigment on invaginations of the cell membrane called discs. (B) The rhabdomeric photoreceptor typical of invertebrates anchors the photopigment on tubular extensions of the cell membrane called microtubules (octopus on left, insect on right). (A after Ali and Klyne 1985; B after Waterman 1981.)

(Zuker 1996). Depleted rhabdomeric cells may therefore be able to recover more rapidly and at a lower metabolic cost than ciliary cells. (3) Density of pigment molecules. The density of pigment molecules on the membrane surface is higher in ciliary cells than in rhabdomeric cells, meaning that more molecules are packed into a given length of photocell to yield a higher absorbance per mm of length (0.035 mm^{-1} versus 0.0067 mm^{-1}). Ciliary cells are therefore more sensitive to light per unit of cell length. However, rhabdomeric cells are 2 to 4 times longer and have a higher net photon capture efficiency of about 90%, compared to 20–80% for ciliary receptors (Land 1981).

The structure of photoreceptor cells is designed to maximize light absorption. The long tubular shape of both ciliary and rhabdomeric cells not only provides a tall stack of pigment-bearing membranes, but it also facilitates the trapping of light from a very narrow acceptance angle. Each cell acts like a wave guide with internal reflection propagating light waves down the length of the receptor (see Figure 7.9B). Waveguide properties operate best when the diameter of the photoreceptor is slightly more than one wavelength, or about 1 μm. In order for a photoreceptor to absorb incident light maximally, it must be aimed in the direction of the light entering the eye.

Photoreceptor cells are packed side by side to form a **retina**. The density of photoreceptors can reach 150,000 per square mm. The retina forms the photosensitive layer of all eyes, primitive and advanced. Each photoreceptor cell, or sometimes a group of photoreceptor cells, forms a synapse with one or more nerve cells, and nerve cells form simple to complex interconnections as they travel to the central nervous system. The ability of the retina to resolve fine details depends on the number of photoreceptor cells, the optical system of the eye, and the nature of the nerve connections to the brain.

THE EVOLUTION OF EYE STRUCTURE

The camera-lens eyes of vertebrates and cephalopods and the compound eyes of insects and many other arthropods are highly complex structures that could not have evolved in a single step. A brief survey of the eyes of some existing primitive organisms provides clues to the possible steps taken during evolutionary history (Wolken 1975; Land 1981; Cronly-Dillon 1991; Dusenbery 1992). The design of an animal's eyes and its resulting acuity reflect its visual needs. A sessile organism or a slow-moving herbivore does not need eyes to locate food or find its way about. However, such an organism most likely does need to perceive the direction of light so it can orient its body with respect to overhead sunlight, and it may need to detect the presence of a predator so it can retract or take appropriate evasive actions. Such eyes are very simple and are designed to detect shadows and nearby movement. Once an animal becomes mobile it needs eyes to locate landmarks in its environment, so some type of crude image formation is required. Rapidly moving animals and especially predatory animals need highly accurate visual images to move about and hunt for food, so increased acuity is essential. Intraspecific communication and social needs usually do not have a major impact on the optical design of eyes. The degree to which vision can be used for intraspecific communication is therefore dictated by the type of eye needed for foraging and predator detection.

The simplest type of eye is the **pigment cup eye** illustrated in Figure 9.3A. It consists of a cuplike depression that is lined by photoreceptor cells linked to nerves and isolated from the rest of the body by a light-absorbing pigment layer. The cup-shaped form of the eye permits detection of the direction of ambient light. Light shining directly over the eye will stimulate only the bottom of the cup, while light entering at an angle will stimulate only one side. This eye cannot form an image, but it can distinguish two sufficiently separated point sources of light. Rapid movements of a shadow-producing organism can also be distinguished from the slow movement of the sun. Animals with such eye cups show both phototaxis (directed movement toward or

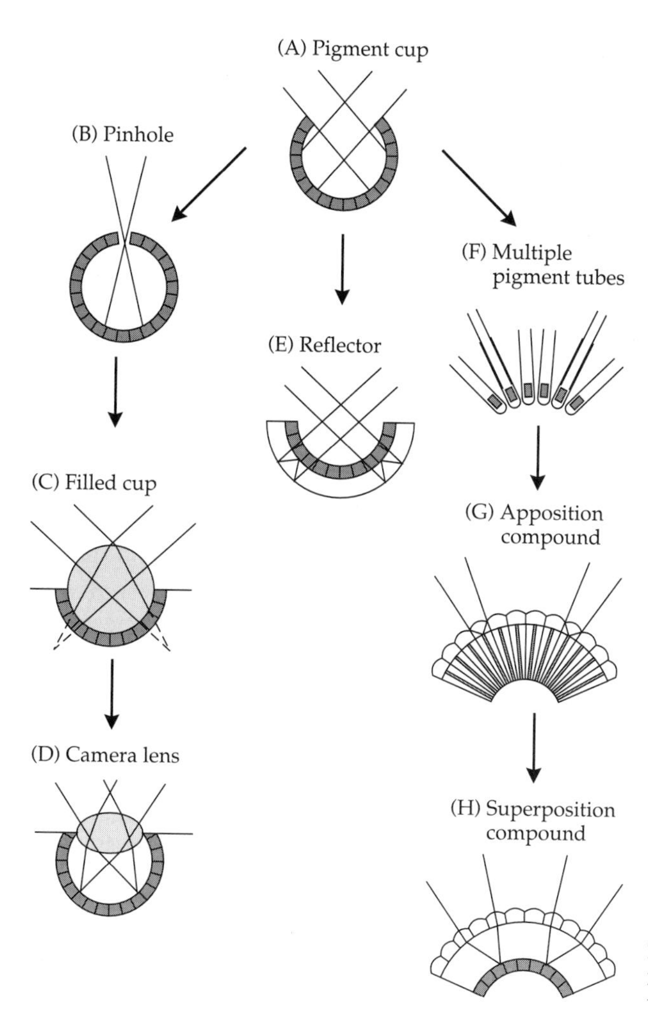

Figure 9.3 The evolution of eyes. The eye structures illustrated in A–H illustrate the possible evolutionary sequences made during eye evolution. Dark gray areas are the retina, and light gray areas are the lens. (After Land 1981.)

away from light) and rapid withdrawal or closure reflexes when light intensity decreases suddenly. Pigment-cup eyes are the only type of eye found in the most primitive animal phyla, but examples occur in virtually every invertebrate phylum. Organisms may possess two cup eyes located on the head or anterior part of the body as in many flatworms, rotifers, annelids, crustaceans, gastropods, and chaetognaths, or they may possess many eyes spread out over the entire body as in some annelids and chitons; in radially symmetrical animals such as echinoderms and cnidarians, eye cups may be located on the tips of arms.

A very simple modification can turn the cup eye into a crude imaging organ; this is the **pinhole eye**, as illustrated in Figure 9.3B. Closing the cup to a small round opening creates a pinhole pupil. Like the old pinhole cameras, the image is upside down but objects at different distances are all in focus without the need for a focusing lens. Furthermore, the size of the image varies inversely with the distance to the object. The smaller the pinhole, the better the imaging quality of the eye up to the point where diffraction effects scatter and bend the light. However, the greatest disadvantage to this eye is the very small amount of light that enters, so sensitivity and the ability to see under low light conditions is severely restricted. The only animal possessing this type of eye today is the chambered *Nautilus* and close relatives among the cephalopod mollusks. Because these animals are aquatic, the body of the eye is filled with water.

Another simple modification of the eye cup is to fill it with a transparent medium and covering as illustrated in Figure 9.3C. If the cover and filling have a higher refractive index than the external medium and the outer covering is curved, a one-sided lens is created and light entering the eye is refracted. Such a lens will have greater light-focusing power and a shorter focal length in a terrestrial animal compared to an aquatic animal because of the larger difference in the refractive indices of air and lens compared to water and lens. However, even for terrestrial eyes of this type, the calculated focal point of the eye lies behind the retina so a clearly focused image is not achieved. The lens does reduce the size of the light patterns entering the eye to produce a crude, erect (i.e., noninverted) image. This optical system provides a significant advance over the pinhole eye because the eye opening is large and the lens concentrates the light into an intensified image on the retina. However, the image quality is poorer, and image size does not vary much with distance to the object. Eyes of this type are called **ocelli** (singular: ocellus) and are found in terrestrial snails, some annelids, spiders, and many larval insect forms. These animals appear to be able to see only very short distances (a few centimeters). Insects possess ocelli as secondary eyes that seem to be used for assessment of general light levels and for orientation (horizon detection).

It is but a small step from the filled eye cup to the true **camera-lens eye** illustrated in Figure 9.3D. All that is required is a space of low refractive index between the lens and the retina of about 1.5 lens radii. This space creates a two-sided lens with two refractive surfaces; it thus shortens the focal length of the lens and provides a sufficient gap behind the lens to bring an image into focus on the retina. Camera-lens eyes produce a bright inverted image that varies inversely in size to the distance of the object. Together with a high-density array of receptor cells and complex neuronal analysis, this eye can simultaneously analyze all light entering the eye from a 180° angle and produce the most accurate spatial map of objects in the environment that animals can achieve. This type of eye has evolved independently in two major taxa, vertebrates and cephalopod mollusks. Camera-lens eyes are also found sporadically in a few species from taxa with otherwise poor eyes, such as a family of pelagic marine polychaetes, a group of marine snails, and non-web-building spiders. These organisms, along with cephalopods and the early vertebrates, are all active pursuit-type carnivores, in comparison with their less visual relatives, which suggests that a hunting lifestyle is the critical factor selecting for highly accurate visual imaging. We can expect visual communication to be most prevalent in such organisms.

A relatively rare but interesting type of imaging system is based on parabolic mirror reflection, as shown in Figure 9.3E. Here, light passes through the retinal layer as it first enters the eye, is reflected off of a curved mirrorlike cup, and the reflection is imaged on the retina. The same principle is used in mirror-type space telescopes. A moderately good image is apparently formed by such a **reflector eye**. The advantage of the system is its large light-gathering area. It is also extremely sensitive to very small movements in the environment. The scallop, a sessile bivalve mollusk, possesses about 60 small eyes of this type distributed around the edge of its mantle; it appears to be taking advantage of the motion sensitivity of the design. The only other animals with this type of eye are the giant deep sea ostracods (Crustacea), which possess two huge reflectors and appear to be taking advantage of the light-gathering potential to permit vision in the very dark environment of the oceanic depths.

The compound eye commonly found in arthropods (especially insects, crustaceans, and trilobites) evolved from the pigment cup eye through a very different route. The intermediate eye is probably the **multiple pigment tube eye** shown in Figure 9.3F and found today only in a few annelids and some starfish. This could have evolved either from the clustering and narrowing of several eye cups into a unit, or the subdivision of a single cup with the addition of tubes. Each tube accepts light only from a unique point in space. Such an eye is highly sensitive to movement, and is therefore superior to the simple cup eye in being able to detect the presence of a predator before it is close enough to cast a shadow.

With the addition of a lens system covering each tube and tight hexagonal packing, a classical **apposition compound eye** is created (Figure 9.3G). Typical eyes contain a hundred to many thousand lens/tube systems, called **ommatidia**. Each ommatidium contains several radially arranged receptor cells with their photopigment regions projecting into the center of the ommatidium. The lens system narrows the acceptance angle of each ommatidium to a small solid angle that overlaps only slightly with that of neighboring ommatidia. Although each lens produces a small inverted image at the tip of the receptor cells, the entire ommatidium responds as a unit and averages the light intensity from the point in space that the unit perceives. The result is a crude mosaic image of the world. The apposition compound eye can see distant as well as close objects with reasonable acuity and without the need to focus. This eye can also judge the distance to an object based on image size, and, like the primitive precursor, it is highly sensitive to slight movements.

A serious constraint of the apposition compound eye is its poor sensitivity in low-light conditions due to the narrow field of view of each ommatidium. The dim light environment therefore favored the evolution of the **superposition compound eye** illustrated in Figure 9.3H in some nocturnal arthropods (fireflies, moths, and nocturnal beetles and crustaceans). The tubes or baffles between the ommatidia are replaced by a clear space between the lens system and the retina, and the lenses are modified so that they can accept light from a wider field. Light from a given point is therefore received by several adjacent lenses and is concentrated at a single spot on the retina. A bright, erect image is therefore formed, and both sensitivity and acuity appear to be improved compared to the apposition eye. The compound eye in general is a marvelous lightweight solution for effective vision in a very small, highly mobile organism (Land 1980). Although fine visual patterns are lost, general shape and rapid movement are easily resolved and such an eye can be employed for visual intraspecific communication. Figure 9.4 compares the image quality of a typical compound and camera-lens eye.

NEURAL PROCESSING

The retina and the brain are not passive neural arrays that receive and project an accurate representation of the objects within view. When photons are absorbed by a photoreceptor, the photoreceptor responds with a voltage change in proportion to the amount of light received. Since there are often millions of receptors in a single eye, each able to produce a graded response over a wide range of light intensities, the total amount of information received by the retina is staggering. But the photoreceptor response is passed through a complex network of interconnecting nerves in the retina and brain, where it may be exaggerated, inhibited, or combined with the impulses from other receptors. The types of nerve connections, or "wiring", is highly species specific and determines how the animal views its world. The following discussion of neural processing will focus on the vertebrate eye, but analogous processes also occur in the image-generating eyes of invertebrates.

Vertebrate photoreceptors are tightly packed and arrayed side by side with their active light-receptor end facing the back of the eyeball. The bottom end (inner segment) of the cell faces the incoming light and performs the light-trapping function of the waveguide. The nerve cells are attached to this end and extend in several layers toward the front of the eye, where they then bend and travel to the optic nerve out of the eye and to the brain. Light must travel through the nerve layer of the retina before it strikes the pigment-bearing end of the receptor cells. This is called a **reversed retina**. It is considered to be a design flaw when compared to the compound and camera-lens eyes of invertebrates, where the active end of the photoreceptors faces the incoming light and nerves run straight back to the brain. Evolutionary history is responsible for the reversed retina in vertebrates—the eye develops as a vesicle that evaginates from the side of the brain or neural tube (Cronly-Dillon 1991).

The vertebrate eye is also characterized by a **duplex retina** with two types of photoreceptor cells: **rods** and **cones**. These form two more or less independent visual systems, rods for vision in low-light conditions and cones for vision in bright-light conditions. Rods and cones can be distinguished by their physical shape (Figure 9.2A) and by the type of nerve connections they make with neighboring cells of the same type. The main types of nerve cells in the vertebrate eye are illustrated in Figure 9.5 and their functions are discussed below (Dowling 1987; Rodeick 1988; Thompson 1991). These nerve cell types form clear layers within the retina.

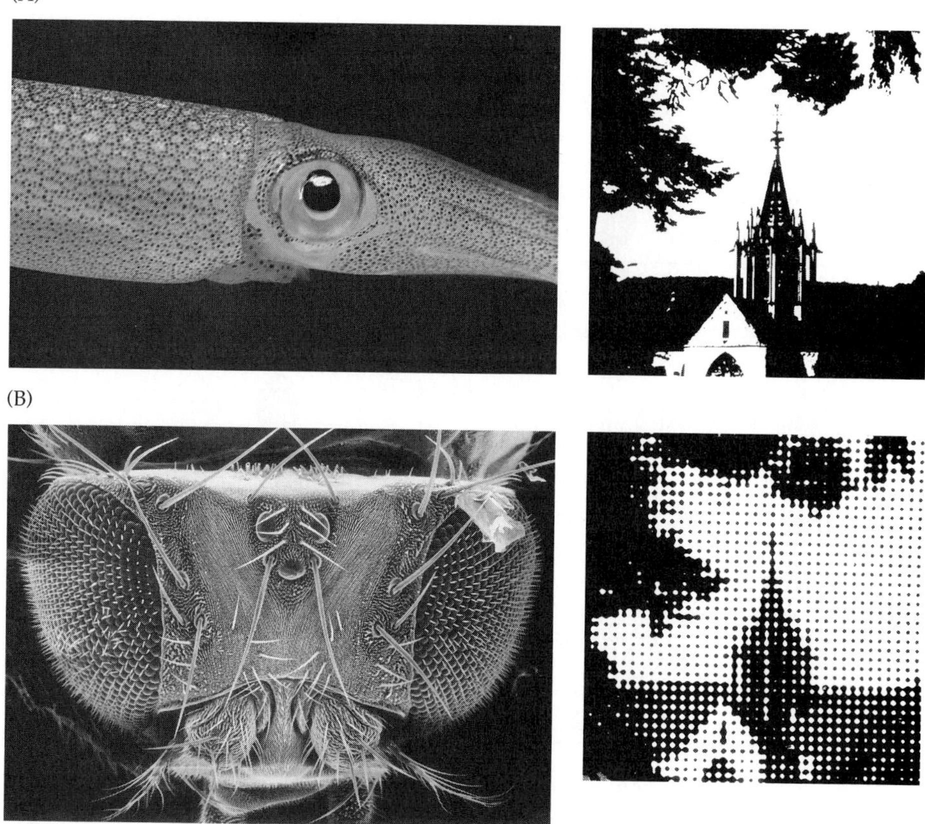

Figure 9.4 Image quality of the camera-lens versus compound eye. (A) The large camera-lens eye of cephalopods (here, the squid *Loligo pealei*) has evolved completely independently of the vertebrate camera-lens eye, but its resolution is every bit as good. It can resolve fine details such as those depicted in the scene on the right. (B) The compound eyes of the fly (here, *Drosophila*) are larger than its head but nevertheless very lightweight because they are not filled with liquid. Each facet is a single ommatidium that averages the light stimuli entering its narrow acceptance angle. The fly views the same scene shown above in a pixel-like fashion. In the central region of the fly's eye the facets are larger and the curvature of the surface is reduced. These features result in an increased overlap of the acceptance angles of adjacent ommatidia and an improvement in visual acuity in this region of the visual field. The poorer resolution of the fly eye compared to the cehalopod eye is largely a consequence of the small size of the fly rather than a function of the basic eye structure. (A courtesy of Roger T. Hanlon; B scanning electron micrograph by Eric Lai and Joshua Kavaler; visual resolution photos courtesy of Kono Kirschfeld.)

Rods are long, thin cells containing a stack of completely internal membrane discs that house the photoreceptor molecules. The base of the cell has a rounded spherule with a single invagination for synaptic connections with retinal nerves. One such nerve is the **rod bipolar cell,** which conducts the

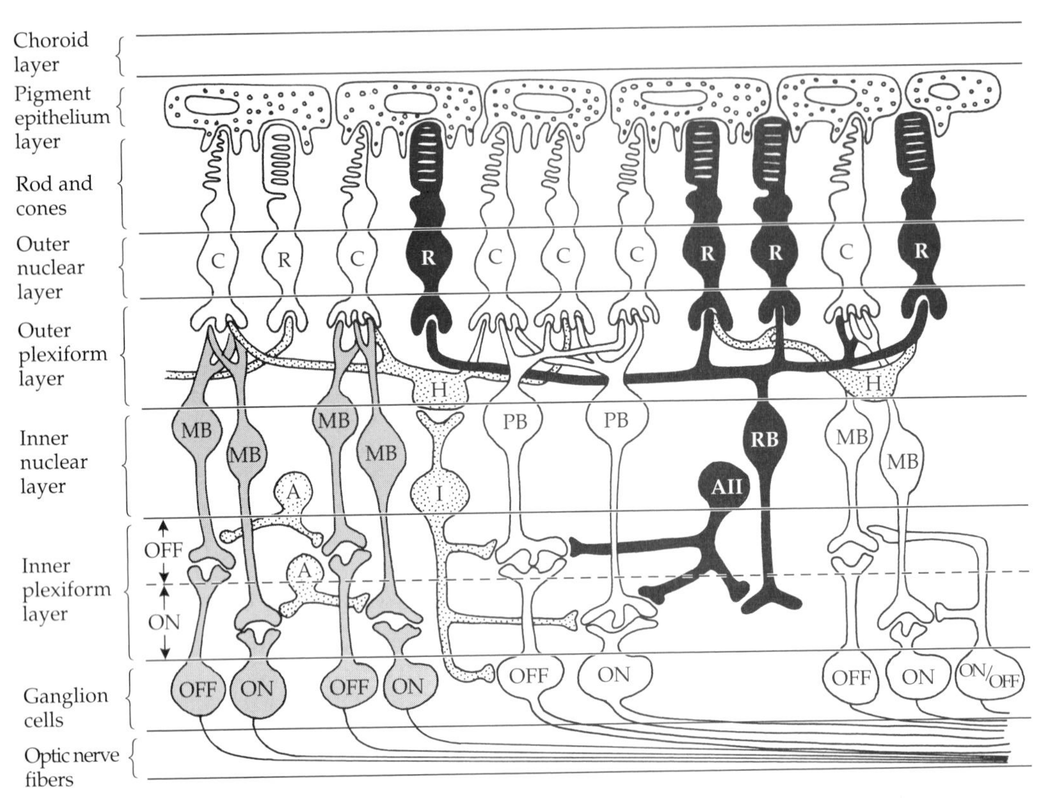

Figure 9.5 Organization of the vertebrate retina. The cross section of a vertebrate retina exhibits clear layers corresponding to functionally different cell types. The entire eyeball is encased in the choroid and fibrous sclera. The outer segments of the rods (R) and cones (C) face away from the direction of incoming light and toward the pigment epithelium cell layer, where recovery of rhodopsin takes place. Bases of the rods and cones are greatly enlarged to show the invaginations where synaptic contacts are made with retinal nerve cells. Two midget cone pathways are shown on the left in gray, with midget bipolar cells (MB) connecting to midget ganglion cells. A parasol cone pathway with parasol bipolar cells (PB) is shown in the middle in white. For both cone pathways, ON systems make their bipolar-to-ganglion connections in a different sublayer of the inner plexiform layer from the OFF systems. A rod system is shown in black, with a rod bipolar cell (RB) and its associated amacrine cell (AII) making connections to cone bipolar cells. An ON/OFF ganglion cell that responds to transient light changes is also shown. Axonless cell types are shown in stipling: horizontal cells (H) make lateral connections among cones or among rods, amacrine cells (A) make lateral connections within the inner plexiform layer, and interplexiform cells (I) make connections between horizontal cells, amacrine cells, and bipolar cells.

nerve response vertically into the retina. Between 15 and 45 neighboring rod cells connect to a single bipolar nerve, and their effects are additive. The consequence of high convergence is to increase visual sensitivity under low-light conditions—the light-gathering area on the retina for a single bipolar nerve is large, and a single photon absorbed by just one of the attached rods will cause a nerve response. However, since each rod is connected to an average of 2.5 different bipolars, there is considerable overlap in the receptive field of each bipolar and a reduction in the detail and sharpness of the eventual neural image. All of the rods contain the same photopigment. Rods are highly efficient receptors, absorbing 50–90% of the photons directed at them. Their response becomes quickly saturated when exposed to bright light, and they are then unable to respond further until some time has passed and the photopigments can be renewed.

Cones are shorter cells with a tapering point. The photopigments are located on membranous invaginations into the cell. This direct access of the pigment-bearing membranes to the extracellular environment means that renewal of the photopigments can occur more quickly and helps to explain how cones can continue to respond for long periods of time in bright-light conditions. All cones may contain the same photopigment, but more typically there are several cone types with different pigments; these, as we shall see below, permit color vision. The terminal base (pedicle) of cones is broad and contains several to many invaginations for nerve synapses. Each cone may be connected to four types of bipolar cells. **Midget bipolar cells** are attached to a single cone receptor or to a few cones with the same pigment type. **Parasol bipolar cells** receive input from six to seven neighboring cone cells comprising more than one pigment type. The number of cone cells summating on a single bipolar cell is therefore less than the typical summation for rod cells. The reduced summation of cones is responsible for the higher resolution and finer-grained image of the bright-light visual system. Midget and parasol bipolars can be further divided into two subvarieties, **ON bipolars** and **OFF bipolars**. ON bipolars generate a *positive* voltage response with an increase in the light stimulus to the photoreceptor, whereas OFF bipolars generate a *negative* (inhibitory) voltage response with an increase in the light stimulus. The bifurcation of each cone's response into a separate on and off channel is maintained in subsequent neural connections up to the visual cortex and is responsible for the analysis of brightness and color contrast (Gouras 1991a; Schiller 1995).

The responses from cone bipolars are transmitted to **ganglion cells**, which send long axons through the optic nerve and into the lateral geniculate nucleus of the brain. The ganglion cell preserves the main response characteristics of the bipolar; it is either a midget ganglion cell with a narrow-field ON or OFF response or a parasol ganglion cell with a broader-field ON or OFF response. Parasol ganglion cells have larger axons that enable the parasol system to respond faster than the midget system does. Ganglion cells respond with spike rate changes rather than voltage changes as in photoreceptors and bipolar cells. An ON ganglion therefore increases its spike rate over its baseline resting rate when stimulated, whereas an OFF ganglion cell increases its spike rate when the photoreceptor is inhibited. A third type of ganglion cell called an ON/OFF cell responds specifically to transient changes in light stimulus and receives input from both ON and OFF channels. Rod bipolars do not have their own ganglion cells. Instead, the primary connection of the rod photoreceptors to the olfactory nerve occurs via the cone bipolars. Special **AII amacrine cells** send positive signals to the ON cone bipolar cells and inhibitory signals to the OFF cone bipolar cells. The rod system therefore piggybacks on the cone system by exploiting the ON bipolars and neutralizing the negative stimulus effects of the OFF bipolars to maximize overall response sensitivity.

Other cells in the retina generate complex interactions among receptors and nerves. **Horizontal cells** are flat, axonless nerves that connect adjacent photoreceptors of the same type, e.g., rods or cones. They appear to produce inhibitory responses between neighboring receptors, a process called **lateral inhibition**. Lateral inhibition enhances the perception of a border between a light-stimulated zone and an unstimulated zone in the retina (Figure 9.6). Similarly, stimulation of a single photoreceptor with a small spot of light causes a reduction in the responses of receptors in the surrounding region. **Interplexiform cells** span the outer and inner plexiform layers and make connections between bipolar cells, ganglion cells, and horizontal cells. One of their functions appears to be turning off the inhibitory effects of the horizontal cells to maximize sensitivity during crepuscular vision (i.e., both rods and cones contribute). **Amacrine cells** are also axonless nerves that make horizontal connections among bipolar cells, ganglion cells, and other amacrine cells. There are many morphological subtypes of amacrine cells whose functions are unclear, but it is known that amacrine cells facilitate the initial phases of color analysis by making connections between midget bipolars, they mediate the input into ON/OFF ganglion cells that detect changes in light intensity, and they may also be responsible for movement detection.

Antagonistic interactions among nerve cells such as those illustrated in Figure 9.6 constitute the primary means for processing visual information at higher levels as well, including spatial, temporal, and color analysis. Animals differ in the degree to which such processing occurs in the peripheral visual organ (e.g., the eye and retina) versus in the brain. Vertebrates in general must significantly digest and reduce the information from the photoreceptors in the retina before sending it on to the brain because of the constraints imposed by the reversed retina and eyeball-penetrating optic nerve (Thompson 1991; Reichenbach and Robinson 1995). For example, the eye of the cat contains 130,000,000 receptors, but only about 170,000 fibers connect the retina to the brain. Invertebrates do not suffer this constraint and perform little retinal preprocessing. They do possess the equivalent of ON and OFF systems for contrast and hue analysis, but the bifurcation occurs at higher levels in the brain (Menzel and Backhaus 1989). Such evidence for convergent mechanisms of visual signal processing between vertebrates and invertebrates indicates strong selective pressure for optimal receiver design.

COLOR VISION

In Chapter 6 we learned that the detection and discrimination of different frequencies of sound is achieved by inherent frequency-dependent properties of the receptor cells, either due to their resonant frequency or their location

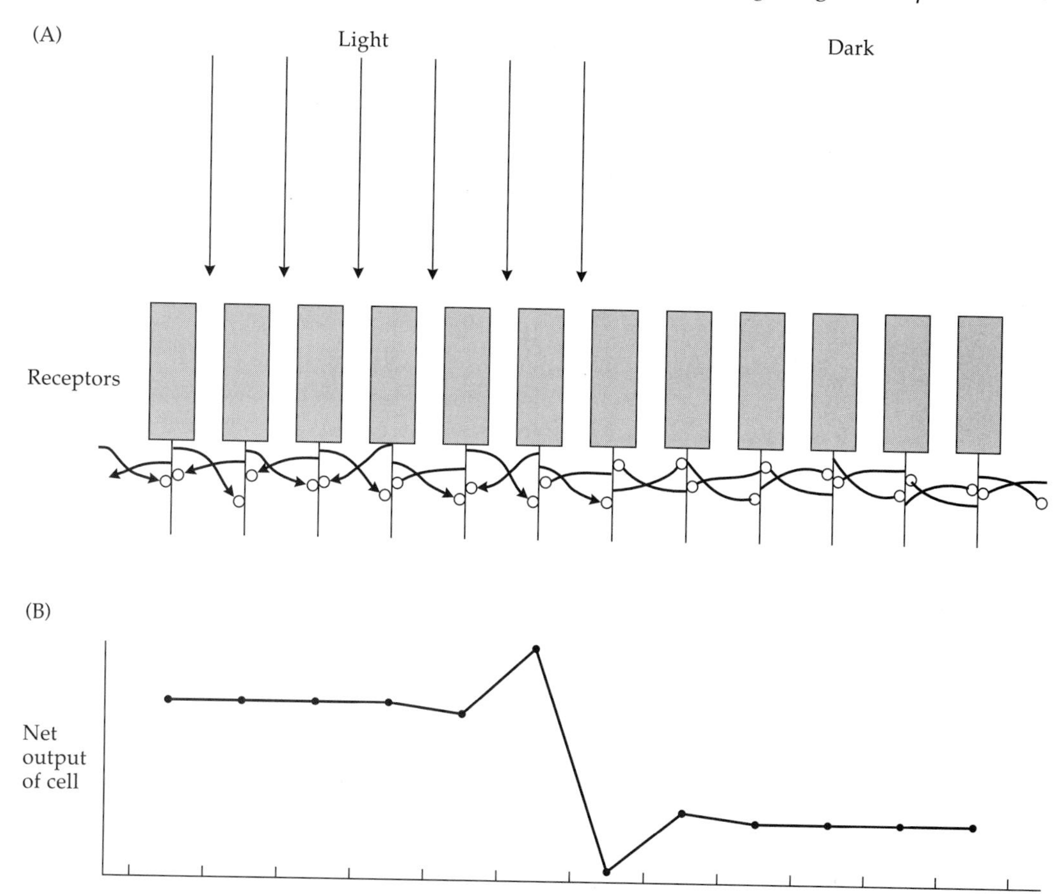

Figure 9.6 Lateral inhibition. The enhancement of boundaries in the visual field is achieved with a process called lateral inhibition that involves the horizontal cells in the retina. Adjacent photocells exert inhibitory influences on each other. (A) In the array of receptors shown here, those on the left are illuminated by bright light and those on the right by dim light. Inhibitory effects are indicated by bold arrows. (B) The net output of each receptor in A is shown in the graph. The cell on the bright edge is less inhibited than its bright neighbors and it therefore responds even more strongly. The cell on the dark edge is inhibited more than its dark neighbors and it responds even less. Perception of the edge is thereby exaggerated. (From Gould 1982, © W.W. Norton.)

along a gradient. Similarly, the detection and discrimination of different light frequencies is achieved with photoreceptors that absorb different wavelengths of light. However, the similarity ends there. Because eyes must resolve spatial information, arrays of finely tuned frequency receptors would not work. Instead, color vision employs a small number of receptor types with overlapping wavelength sensitivities dispersed throughout the retina, and the outputs of these receptors are combined and compared to produce the sensation of color differences.

Animals have found three ways to generate visual pigments with different absorption peaks (Wolken 1995). One is to use a different variant of retinal. The common form of retinal found in rhodopsin is derived from vitamin A_1 and called $retinal_1$. A second form, called $retinal_2$, is derived from vitamin A_2 and differs from $retinal_1$ only in possessing an additional double bond in the ring. When $retinal_2$ is combined with opsin, the pigment is called **porphyropsin**. The substitution of $retinal_2$ shifts the absorption peak of the pigment up about 20–25 nm. Porphyropsin is found in many aquatic vertebrates such as teleost fish, amphibians, and aquatic reptiles. The second and most widespread mechanism for altering the pigment absorption curve involves changes in the amino acid sequence of the opsin protein. These affect the way the opsin component interacts with retinal. A single substitution can shift the peak up or down, and the effects of different substitutions are additive so that pigments with absorption peaks as high as 620 nm and as low as 350 nm can be generated. The third mechanism, found in birds, amphibians, lizards, snakes, and turtles, is the addition of a **colored oil droplet** to the photoreceptor cell. Carotenoid pigments are responsible for the color in the droplets, and specific colors are associated with specific photopigments. They act as cutoff filters of the incoming light and simultaneously narrow the absorption peak, shifting it to longer wavelengths.

Animals living in dark, poorly illuminated environments must use all of their photoreceptors for spatial pattern analysis and brightness contrast; they cannot afford the luxury of color vision. Such an animal is called a **monochromat** because it sees the world in shades of gray. Nocturnal, deep sea, and cave dwelling vertebrates typically have a very high proportion of rod receptors in their eyes. Species with some cones may still be incapable of distinguishing colors if they possess only one type of cone photopigment. The cone system extends the range of dynamic vision to higher light levels than those permitted by rods alone and provides other advantages such as temporal resolving power. Even when the cone pigment has a different spectral sensitivity from the rod pigment, wavelength comparisons are unlikely to occur because of the independent operation of the rod and cone systems. Vertebrates known to have poor or no color vision include the rat, hamster, mouse, opossum, raccoon, genet, galago, owl monkey, microchiropteran bats, and deep sea fish (Jacobs 1981, 1993).

Once a species has evolved two cone pigments with different spectral sensitivities, a form of color vision becomes possible. Such an animal is called a **dichromat**. The principal mechanism of color perception is the summation of neural outputs from these two cone types; Figure 9.7 shows a schematic representation of this process. A nerve cell in the retina or brain that receives positive input from the ON system of one cone type and negative or inhibitory input from the OFF system of the other cone type gives a unique response pattern to light stimuli of different wavelengths. Such neural units that receive opposing inputs from receptor cells with two pigment types are called **chromatically opponent cells**; they are the hue detectors of the visual system. At the same time, other cells receiving same-sign input from the two cone types effectively sum the output of both receptor types. They are insensitive to hue, but highly sensitive to brightness and yield divergent responses to black versus white stimulation. Subsequent combining of the outputs of the

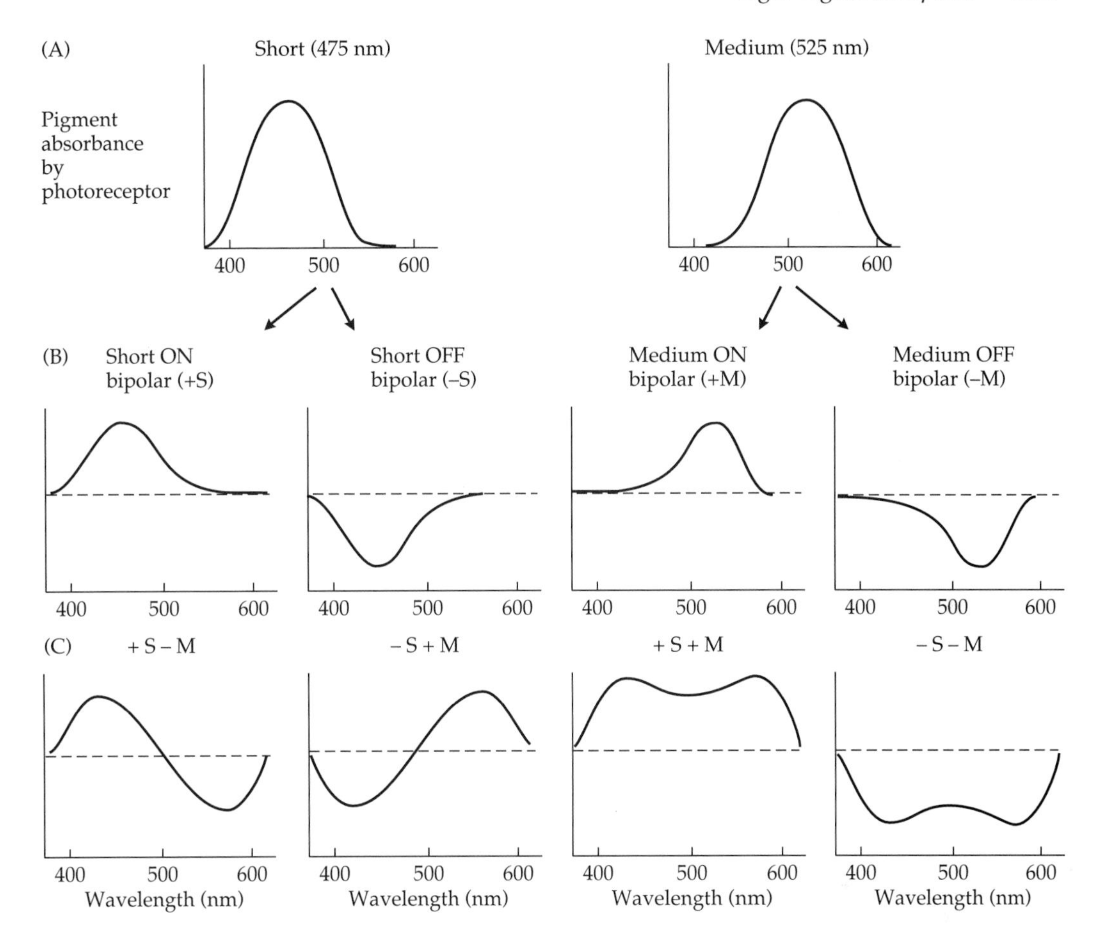

Figure 9.7 Logic of chromatically opponent cells. (A) Two cone types are required, one with a pigment absorbing maximally at a short (S) wavelength (475 nm) and another with a pigment absorbing at a medium (M) wavelength (525 nm), as shown in the top row. The absorption curves overlap each other extensively. (B) The second row of graphs shows the strength of the tonic responses of bipolar cells to light stimulation of different wavelengths. Each cone type generates a response in its associated ON midget bipolar cell that mirrors the absorption curve and a second inverse or inhibitory response in its OFF midget bipolar cell. The dashed horizontal line indicates the cell's baseline level of response. (C) The bottom row shows the strength of response by ganglion cells or other downstream nerves that receive input from different pairwise combinations of bipolar cells. There are four different response patterns. Antagonistic input from S and M bipolars leads to color-opponent cells (left two graphs in the bottom row). Note that the peaks and troughs of the neural responses occur at more extreme wavelengths than the peaks of maximum absorbance of the two photopigments. Additive input leads to brightness-sensitive cells (right two graphs in the bottom row). Converging outputs from these cells further downstream lead to the representation of hue and chroma.

two hue detectors with the black and white detectors in the visual cortex results in the perception of hue, chroma, and brightness. It should now be clear why animals living in dark environments cannot afford color vision; not only is the cone system less sensitive to light because of its reduced summation of receptor cells onto bipolar cells, but the opponent process also means that a large proportion of the cone cells stimulated by light at any one time are sending inhibitory OFF inputs to nerve cells in the brain.

What does such a dichromat actually see? Although we can't ask an animal to describe its world, we can ask them to perform discrimination tasks that reveal their visual sensitivity to monochromatic light of different wavelengths. For the dichromat described above, all objects reflecting wavelengths between about 400 nm and 600 nm (violet through orange for humans) would appear colored, and wavelengths smaller than 400 nm and larger than 600 nm would appear black. This animal would therefore not perceive red colors. The ability of the animal to distinguish different wavelengths is maximal in the zone where the two photopigments overlap, i.e., in the green region as shown in Figure 9.8A. Colors in this region would appear very bright and relatively unsaturated (as yellow is for us); violet and orange would be very dark, highly saturated colors. Finally, if a dichromat is asked to discriminate be-

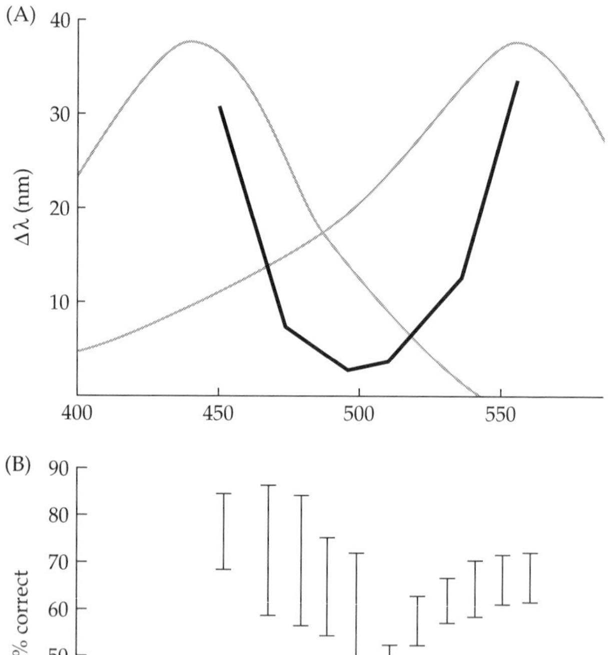

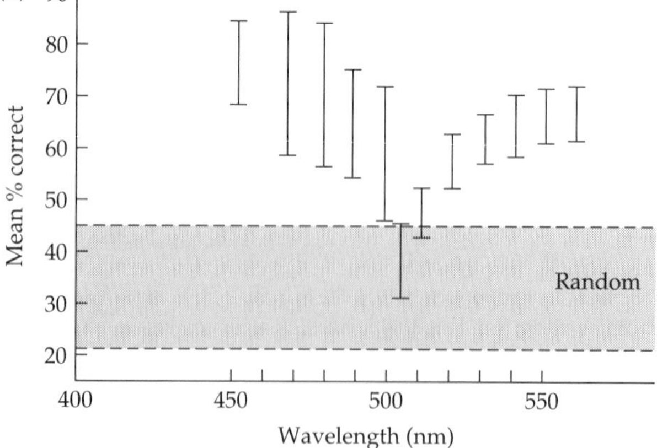

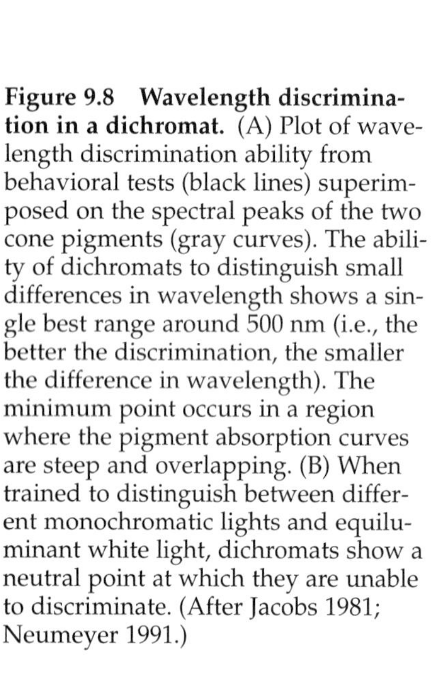

Figure 9.8 Wavelength discrimination in a dichromat. (A) Plot of wavelength discrimination ability from behavioral tests (black lines) superimposed on the spectral peaks of the two cone pigments (gray curves). The ability of dichromats to distinguish small differences in wavelength shows a single best range around 500 nm (i.e., the better the discrimination, the smaller the difference in wavelength). The minimum point occurs in a region where the pigment absorption curves are steep and overlapping. (B) When trained to distinguish between different monochromatic lights and equiluminant white light, dichromats show a neutral point at which they are unable to discriminate. (After Jacobs 1981; Neumeyer 1991.)

tween monochromatic light of different wavelengths and an equiluminant white light, there is a narrow region located between the wavelength peaks of the two photopigments in which the discrimination cannot be made (Figure 9.8B). This is called the **neutral point,** and objects reflecting only wavelengths in this region appear white. Dichromatic color vision is characteristic of the majority of mammals, including tree and ground squirrels, tree shrews, most New World monkeys, cats, dogs, ungulates, and some marine fish (Sinclair 1985; Jacobs 1993).

Animals with good color vision typically possess three photopigments and are called **trichromats**. The pigment absorption curves for humans, apes, and Old World monkeys are essentially the same, with peaks at 450 nm (blue), 530 nm (green), and 560 nm (yellow-green) (Figure 9.9). With the addition of a third pigment, a second color-opponent system becomes possible. Positive and negative inputs from the adjacent 450 and 530 pigment cones generate a yellow/blue opponent system similar to that shown for the dichromat above. Positive and negative inputs from the 530 and 560 pigment cones result in a red/green opponent system. The white/black system receives additive positive or negative input from all three cone types. Six types of ganglion cell responses have been found in the retinas of primates, as shown in Figure 9.10. Combinations of the outputs from these cells produce a rich range of hue and saturation such as that shown in Figure 8.2. Old World monkeys see colors very much as we do. In regions where two adjacent pigments overlap (440–500 blue-green, and 550–600 yellow), colors appear more saturated and wavelength discrimination is also more accurate (Figure 9.9). In trichro-

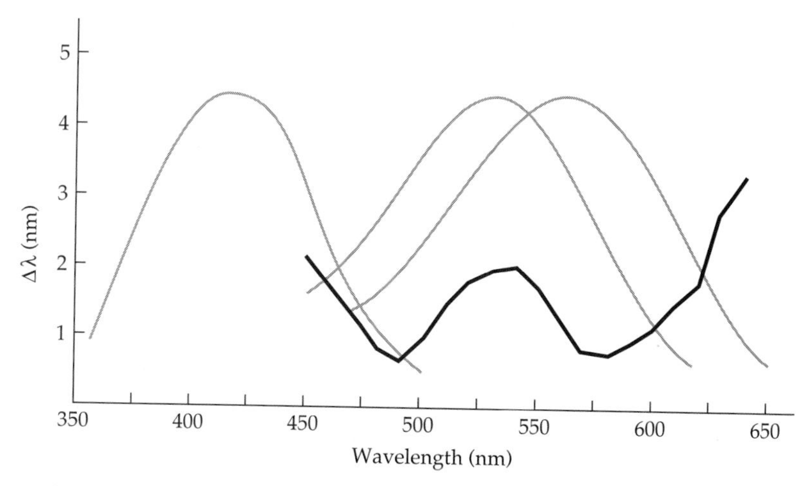

Figure 9.9 Wavelength discrimination in a mammalian trichromat. Higher primates and humans possess three pigments (shown in gray) with unequally spaced peak sensitivities at 420 nm (blue), 534 nm (green), and 564 nm (yellow). The wavelength discrimination curve (black line) shows two minima in regions where the cone absorption curves overlap; compare to Fig. 9.8A. (After Lythgoe 1979; Jacobs and Neitz 1985.)

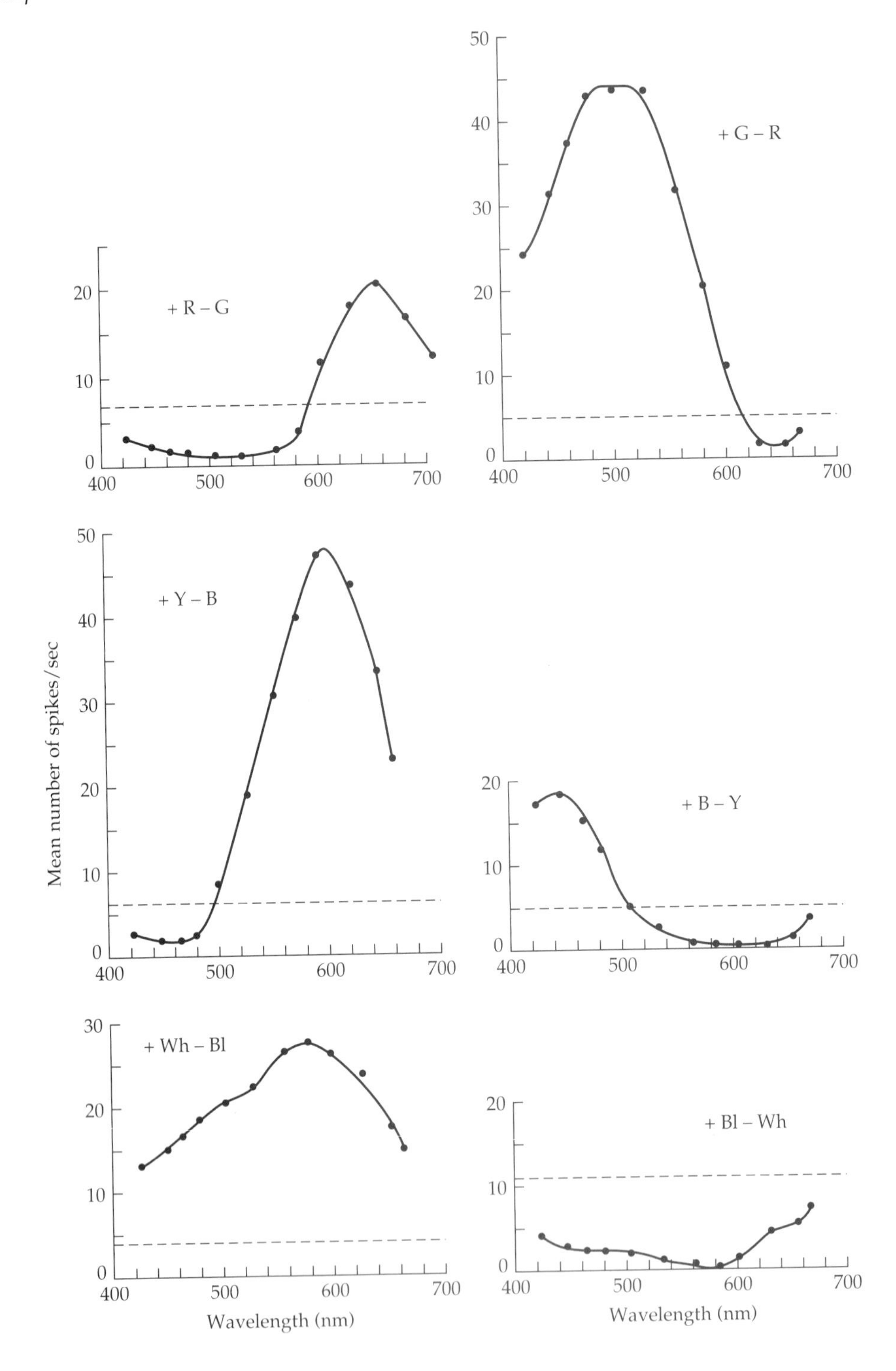
+ R – G
+ G – R
+ Y – B
+ B – Y
+ Wh – Bl
+ Bl – Wh
Mean number of spikes/sec
Wavelength (nm)
Wavelength (nm)
0
10
20
30
40
50
400
500
600
700

◀ **Figure 9.10 Six different types of opponent response cells in a typical trichromat.** The plots show the spectral response of nerve cells in the lateral geniculate nucleus of the macaque monkey. The top two plots illustrate the red/green system (R/G), the middle two plots the yellow/blue system (Y/B), and the bottom plots show the white/black system (Wh/Bl). (From Jacobs 1981, © Academic Press.)

mats there is never a neutral point or monochromatic wavelength region that is indistinguishable from white. The reason is that a monochromatic wavelength that produces a neutral balance for one of the color opponent systems will not be at the neutral balance point for the other opponent system.

In addition to monkeys and apes, other animals with trichromatic color vision can be found among freshwater fish, diurnal reptiles and amphibians, crustaceans, many insects, and some spiders. The peak absorbancies of the photoreceptors are different for these species, so the range of colors that can be seen, the categorization of hues, and the brightness of different colors are likely to be different as a result. Bees and many other color-competent insects lack red sensitivity completely but have a short-wave photoreceptor sensitive to ultraviolet wavelengths as well as blue and yellow-green receptors (Figure 9.11). Nonprimate trichromats generally show a more even spacing between the peaks of their three photopigments than do primate trichromats, i.e., the long wavelength pigment is shifted upwards. Nevertheless, all trichromats so far examined appear to possess two color-opponent systems in addition to a white/black system, and therefore perceive a rich variety of colors. New World Platyrrhine monkeys present a puzzling example of high inter-individual variation in color perception (Hunt et al. 1993; Jacobs et al. 1993; Jacobs 1993). Males are always dichromats, whereas many females are trichromats. This variation occurs because the short-wavelength opsin gene is located on an autosomal chromosome whereas the middle wavelength opsin gene is located on the X chromosome (i.e., is sex-linked) and has three alleles with different spectral peaks. Not only are heterozygous females trichromats, but there are three different trichromat phenotypes as well as three dichromat phenotypes.

Birds, turtles, some freshwater fish and butterflies possess four or occasionally five photoreceptor types (Neumeyer 1991, 1992). One function of an additional receptor type may be to extend the wavelength discrimination range into the ultraviolet and near infrared wavelengths. In species with oil droplets that narrow the spectral absorption curves, both a larger total range and improved hue discrimination may result (Bowmaker 1991b; Maier 1994). The hue space of tetrachromats must be represented in a three-dimensional tetrahedron as illustrated in Figure 9.12 rather than the two-dimensional circle of trichromats (Figure 8.2). In other cases, these extra pigments appear to be serving different functions and may not all contribute simultaneously (Goldsmith 1994). Species with more than three pigments often show a very unequal distribution of pigment types in the eye. Typically, some pigments are located in the upward-looking part of the eye while others are located in the lateral, forward, or downward-looking part. These species appear to re-

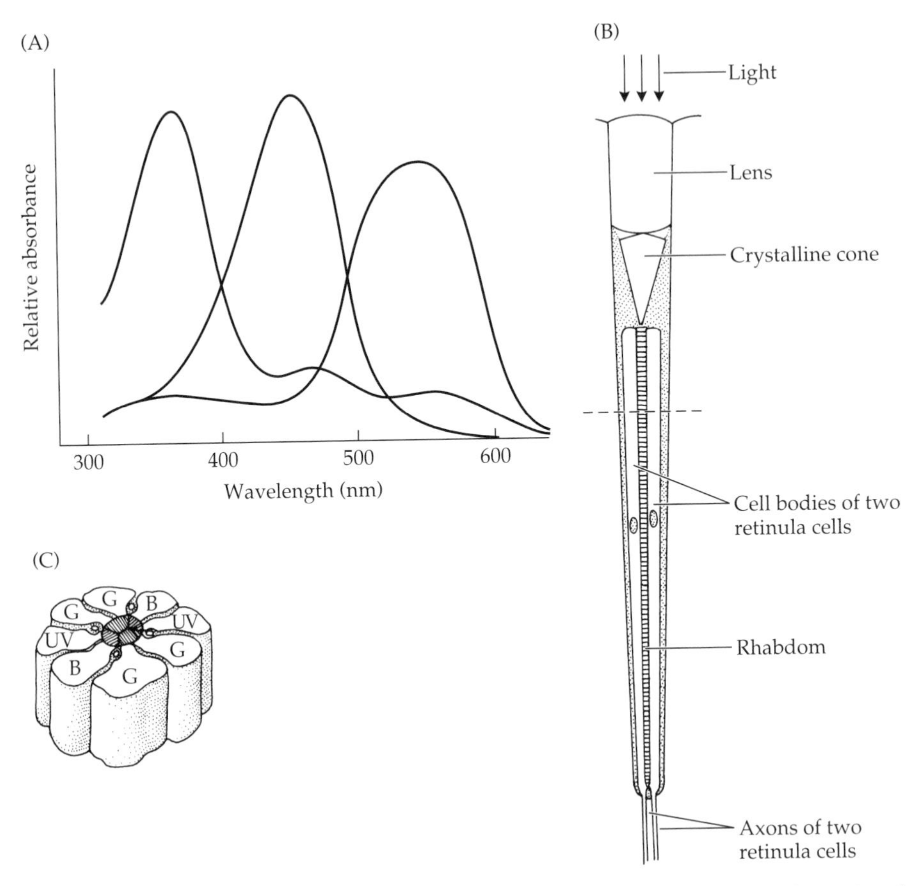

Figure 9.11 Color vision in insects. (A) Absorption curves of the three color units of the honey bee, *Apis mellifera*, based on neural responses. In human terms, the bee has receptor types that peak in the ultraviolet (344 nm), blue (436 nm), and green (556 nm) wavelength ranges. (B) A longitudinal section through a honeybee ommatidium, showing the lens system and fused rhabdom that guides light waves down the rhabdomeres. (C) A cross section of the bee's ommatidium at the point indicated by the dashed line in illustration B, showing the location of green (G), blue (B) and ultraviolet (UV) rhabdomeres. (A after Menzel 1981; B, C after Wehner 1976.)

quire different color sensitivity for different activities such as social communication versus food detection or orientation. Guppies, for example, use primarily green receptors to look upwards during foraging, and red receptors to view conspecifics from the side (Archer and Lythgoe 1990). Butterflies place those pigments necessary for food plant detection in the bottom half of the eye and pigments necessary for conspecific detection in the upper half (Bernard and Remington 1991). Bird such as pigeons may use their ultraviolet receptors in conjunction with red receptors to perceive sun-based color gradients in the sky for orientation purposes (Coemans et al. 1994; Bennett and Cuthill 1994). There is also some suggestion (in the goldfish) that pigments operate under different illumination levels, implying that color perception

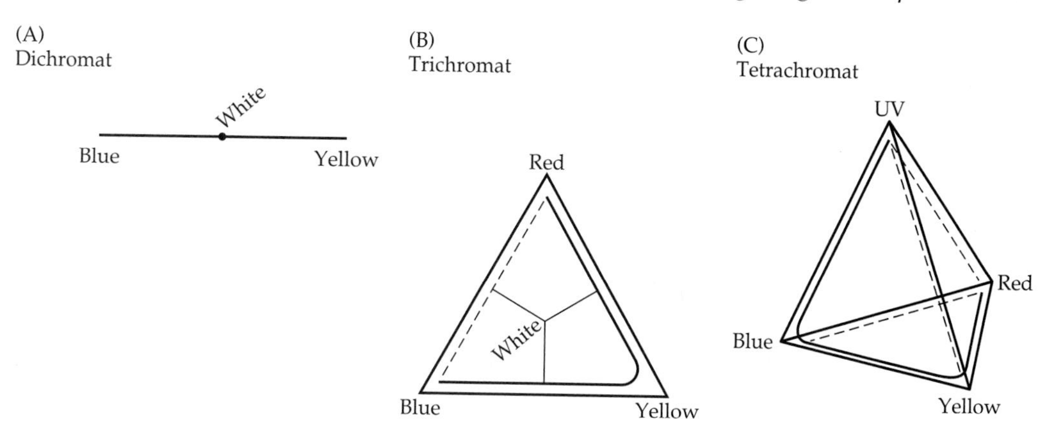

Figure 9.12 Hue space of tetrachromats compared to dichromats and trichromats. (A) The hue space of a dichromat is one-dimensional, with white occurring in the middle. (B) The hue space of trichromats such as bees and humans is a two-dimensional triangle. The heavy line within the triangle shows the range of pure spectral colors perceived and the dashed line represents nonspectral colors such as purple. White is an equal mixture of all three pigment types and occurs in the center of the triangle. (C) The hue space of a tetrachromat is a three-dimensional tetrahedron. The heavy curve snaking through the structure shows the perceived spectral colors and the dashed lines show the three nonspectral colors. Unsaturated colors occur in the interior of the structure. (After Goldsmith 1990.)

needs vary at different times of day (Neumeyer and Arnold 1989). Finally, mantis shrimp possess 10 photoreceptor types, with pairs of receptors having overlapping pigment-absorbance curves arranged in distinct bands across the eye (Cronin and Marshall 1989). This unusual system may facilitate recognition of individuals based on color patterns.

It is obvious that animals vary greatly in their ability to see color and in the quality of color perception. Why has color vision evolved? There are some clear costs of color vision even for animals living in well-illuminated environments. The color system has lower spatial and temporal resolution than does the black and white parasol system (Schiller et al. 1990). The border-enhancement effect characteristic of luminance borders does not have a color analog, so colors blend. Blue receptors in particular are so low in number that they do not contribute to the formation of chromatic borders and spatial resolution. These properties imply that color perception may make the distinction between slight brightness differences more difficult, and that reliance on color can cause two similarly hued objects to meld or be confused. Nevertheless, these disadvantages may be slight when compared to the great advantages of color vision. **Lightness constancy**, the ability to reliably order the lightnesses of objects of different spectral reflectance under different illuminance spectra, is not possible for monochromats (von Campenhausen 1986). **Object detection** in most circumstances is improved with color vision (Walls 1967). **Object recognition**, and in particular the speed with which an object can be recognized, is greatly enhanced. Color-based feature detectors have been described in retina nerve units that respond preferentially to color spots surrounded by opponent colors (Lythgoe 1979). The consequence of such units appears to be enhancement of the visual contrast between adjacent objects that reflect two colors. Detection and recognition of food and predators is undoubtedly an important driving force behind the evolution of color vision (Burkhardt 1982; Gautier-Hion et al. 1985). **Spatial relationships** among objects in the environment, especially the distance of objects from the viewer, is improved by hue changes (graying) with increasing distance. **Orientation and navigation** are additional important selective forces for color vision. Finally, **intraspecific communication** cannot be ruled out as a selective force in the refinement of color vision. The perception of color is a result of both photoreceptor characteristics and neural wiring, and we cannot presume that animals perceive colors as we do (Goldsmith 1990; Bowmaker 1991a; Neumeyer 1991; Mollon 1991).

POLARIZED LIGHT SENSITIVITY

As mentioned earlier in this chapter, rhabdomeric photoreceptors are inherently more sensitive to polarized light than ciliary receptors. Let's first examine why this is so. Visual pigment molecules themselves are sensitive to the orientation of polarized light. Rhodopsin and other pigments can only absorb a photon of light when the e-vector is parallel to the long axis of the retinal molecule. The reason why the pigment molecules are anchored into membranes in the cell and arrayed in orderly stacks is to maintain the position of the molecule for maximum photon absorption.

In a normal rod or cone cell, the long axes of the retinal molecules are maintained in a fixed horizontal position parallel to the plane of the disc membrane and perpendicular to the direction of light (Figure 9.13A). However, they are oriented at random compass angles when viewed from above. No matter what the orientation of the e-vector of polarized light is, a similar number of retinal arms will be oriented parallel to it, and the response of the cell will be the same. However, if one were to shine light transversely from the side of the rod or cone, the response of the cell would depend on the polarization angle of the light. If the light is horizontally polarized, retinal molecules oriented perpendicular to the ray's direction will absorb the light. But if the light is vertically polarized, no retinal molecules will be able to absorb it. The cell is now selectively sensitive to just one plane of polarized light.

In a typical rhabdomeric cell, recall that the membrane housing the visual pigment is rolled into a tubule (Figure 9.13B). The arms of retinal are still fixed in the same type of position as in a ciliary disc, parallel to the surface of the tubule. When light enters from the normal overhead direction, retinal arms on the top and bottom sides of the tubule are randomly oriented compass-wise and absorb all e-vectors of light, as on a disc. However, those retinal arms on the two sides of the tube that are oriented perpendicular to the light direction will preferentially absorb e-vectors of light parallel to the length of the tubule. The net output of the cell will therefore be greater for some vectors of polarized light compared to others.

In practice, specialized rhabdomeric cells in organisms that have been selected to detect polarized light have retinal arms that are all oriented parallel

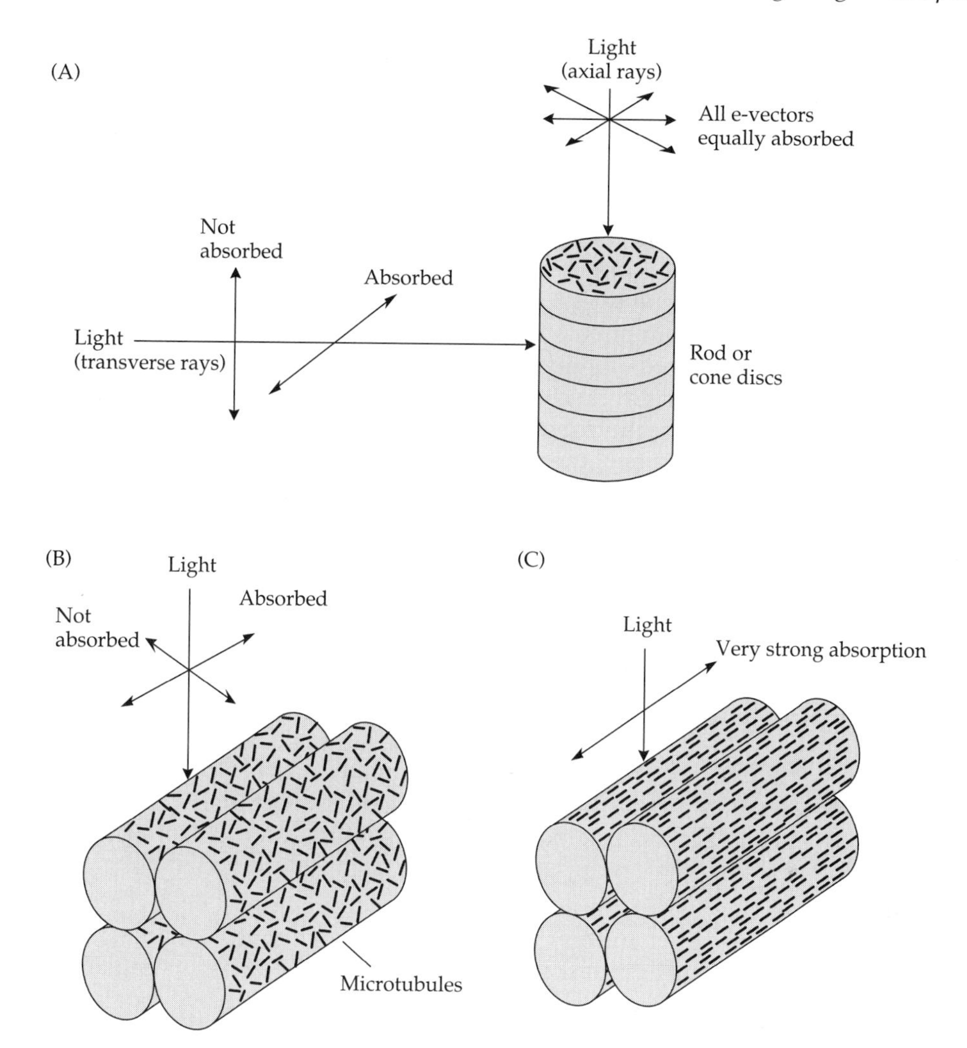

Figure 9.13 Sensitivity to polarized light in ciliary and rhabdomeric photoreceptors. (A) Ciliary receptors do not respond differently to light of different polarization orientations when shone from above (axial rays). However, if light is shone from the side (transverse rays), the photocell will respond strongly to light polarized in the horizontal direction (perpendicular to the long axis of the rod and parallel to the flat side of the discs) but weakly to light polarized in the vertical direction. (B) In rhabdomeric receptors, axial rays that are polarized parallel to the microtubules are more strongly absorbed than rays polarized perpendicular to the microtubules. (C) Alignment of retinal molecules parallel to the microtubule axis maximizes sensitivity to certain e-vectors.

to the length of the tube to maximize the absorption of polarized light in one direction (Figure 9.13C). These are then coupled with opponent systems to other cells with tubules oriented in different directions. In compound insect eyes, these specialized cells are located on the dorsal side of the two eyes (those ommatidia that look directly up at the sky). Flying insects such as bees use the pattern of Rayleigh-scattered polarized light in the sky to orient themselves (Waterman 1984; Wehner 1989).

A few vertebrates, mainly fish, are known to be sensitive to polarized light. Anchovies possess specialized cone cells in which the discs are rotated 90° so that their flat sides are parallel to the direction of light. The result is the same as shining light transversely on a normal cone cell. Different specialized cone cells have their discs oriented at different compass angles, so they selectively respond to different e-vectors. The function of polarized light sensitivity in fish is not clear, but orientation with respect to the sky is unlikely to be the reason. A possible explanation may be orientation with respect to fellow school members. The scales of schooling fish are silvery and highly reflective, and such reflected light is strongly plane-polarized. Polarization-sensitive cones may enable fish school members to reflexively maintain parallel body positions. Alternatively, fish predators may take advantage of the polarized specular reflection of prey fish to maximize prey detection (Waterman 1981).

THE IDEAL VISUAL RECEPTOR

The ideal eye should possess the following five critical attributes. First, it should have **adjustable sensitivity**, to permit vision at both low and high levels of ambient light. Second, it should have good **resolution**, to discriminate the fine details of images and small differences in the intensity and/or color of adjacent objects. Third, it should have excellent **accommodation** to bring the image into sharp focus on the retina regardless of the distance between the object viewed and the eye. Fourth, it should have good **spatial discrimination** to accurately map the location of a viewed object, both with respect to its position in the visual plane and to its distance from the eye. Fifth, it should have good **temporal resolution** to separate rapid events and to follow movements of objects regardless of their speed. There is no ideal eye, because it is not possible to maximize all of these factors. Some tradeoffs are inevitable, and the resulting compromises determine the limits within which signaling can evolve. Among vertebrates, different strategies have evolved to maximize these attributes, and we next examine the range of possibilities (see review by Charman 1991).

Optimizing Sensitivity

Animals that are active at night must have eyes that are extremely sensitive to low light levels. However, they must also cope with the bright illumination of daytime as well. A common strategy for nocturnal animals is a two-stage approach: first, to make the receptors as sensitive as possible to low-light levels, and second, to use other modifications to reduce the amount of light reaching the receptors when light intensities are too high.

The sensitivity of an eye, $\mathcal{S}$, is defined as the amount of light energy absorbed per second by a single receptor when imaging an extended object of unit intensity:

$$\mathcal{S} = \left(\frac{\pi}{4}\right)^2 \left(\frac{A}{f}\right)^2 d^2\left(1 - e^{-ax}\right)$$

where the first term is a constant, A is the aperture diameter, f is the focal length of the lens, d is the receptor diameter, x is the receptor length, and a is the absorption coefficient of the receptor per unit of receptor length (Land 1981). The ratio f/A is the ***F* number** of the lens, a notation also used with glass lenses. As the F number increases, the lens becomes flatter in shape. As the equation suggests, animals can employ a variety of strategies to increase sensitivity. One is to *increase the ratio of rods to cones* in the retina. Rods are more sensitive to low-light levels than cones because they are longer (parameter x), have a high density of photopigment molecules (parameter a), and have a large effective diameter (parameter d) because of the neural summation of many adjacent cells. A high density of rods, at the expense of cones, produces a more sensitive eye; many nocturnal animals have no cones in their eyes at all. The sensitivity of rods can be further maximized by adjusting the wavelength of peak rhodopsin absorption to the region of peak ambient light. A second strategy is to *increase the diameter of the lens and cornea* (parameter A). This increases the total amount of light entering the eye. In addition, nocturnal animals may also *decrease the focal length of the lens* (parameter f). This change is accomplished by increasing the curvature of the lens to make it rounder. The retina must also be brought forward by shortening the eye. The shorter focal length makes the image smaller but brighter (Figure 9.14A,B). In conjunction with enlarging the lens, this creates a so-called tubular eye, which can no longer be rotated in the head. Owls and deep sea fish have such an eye, with the consequence that they must turn their head to look in different directions (Figure 9.14C,D).

The final strategy for increasing sensitivity is to *install a reflecting layer* behind the photocells so that any light that passes through the retina and is not trapped by the receptor cells on the first pass is reflected back for a second pass at the receptors. A reflecting structure is called a **tapetum**. This produces the eye shine characteristic of nocturnal animals. Like fish scales and the mirrored surface of the reflector-type eye, the tapetum is constructed of layers of guanine platelets. Most invertebrate eyes do not possess tapeta, but the house spider presents an interesting example. As we shall see below, spiders possess three to four pairs of eyes and each serves a different purpose. Tapeta often occur in the secondary eyes but not in the principle eye. Eyes with tapeta also have a reversed retina, i.e., the receptor cells are turned around, with their rhodopsin-laden membranes oriented away from the direction of incoming light (Figure 9.14E). Presumably, the effectiveness of the tapetum is increased when the receptor cells are aimed at it. The common occurrence of tapeta in vertebrate eyes may be a second factor favoring the reversal of the retina in that taxon.

Once sensitivity has been raised for vision in low-light situations, animals must find ways to reduce the light reaching the retina under bright light. In

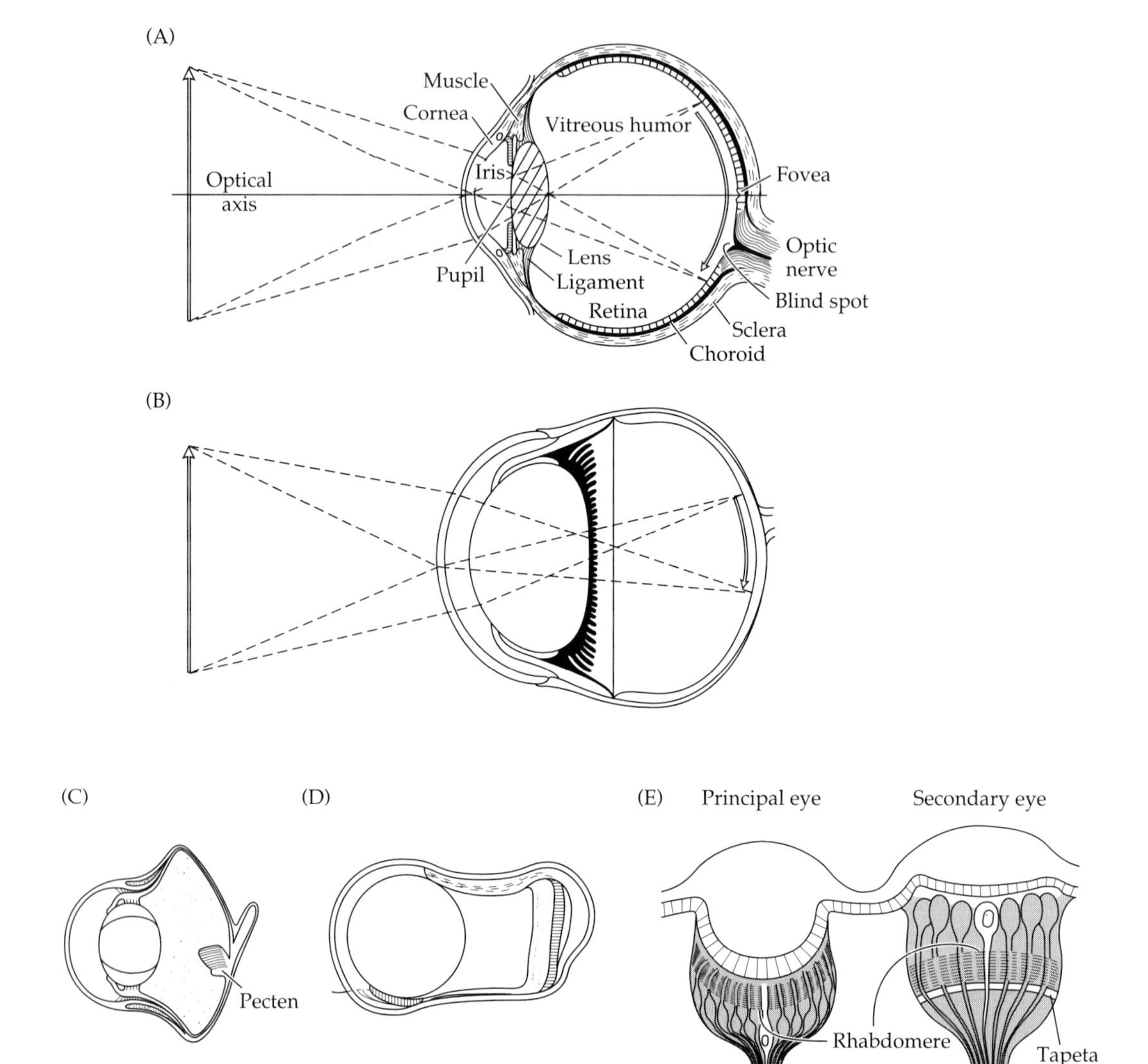

most animals the **iris** serves this purpose. In man and most diurnal vertebrates, the iris closes with a circular muscle so that the opening, or **pupil**, is round. A round pupil is ideal because it cuts out light entering the perimeter of the lens. No lens is perfect, and light passing through the edges of the lens cannot be focused at the same point as light passing through the center of the lens. This variation of focus is called **spherical aberration**. A round pupil eliminates this type of distortion. However, a round pupil can only close down so far. In nocturnal animals with large eyes, the pupil must open completely to pass the maximum amount of light at night, and must close down to a tiny opening during the day. A slit pupil is more effective in this case. The

◀ **Figure 9.14 The effect of lens shape on image brightness.** (A) The relatively large image size of a flat lens in a typical diurnal vertebrate maximizes resolution. Also shown are the key structures of the vertebrate eye, including the sclera that maintains the eyeball's shape and supports the choroid and retina, the lens and its support structures, the iris and pupil, the fovea located on the optical axis, the blind spot where the optical nerve exits the eyeball, and liquid-filled compartments. (B) A round lens, as exemplified by a the nocturnal bush baby (*Galago* spp.), produces a smaller but brighter image. To focus the image, the length of the eye must be shortened. As a result of the simultaneous lens enlargement and eye shortening in nocturnal animals, the eye becomes tubular in shape. (C) Compare the nocturnal owl eye shown here to the diurnal bird eye in Figure 9.15A. This eye can no longer be rotated in the eye socket, and for this reason, the owl must turn its head to peer in different directions. The pecten structure is also shown. (D) Deep sea fish eyes show an extreme tubular shape. The retina is several layers deep to maximize sensitivity and achieve focusing. (E) All spiders possess several pairs of ocelli-type eyes that are specialized for different visual functions. The principle or anterior medial eye (left) is used for diurnal vision and is characterized by photoreceptor cells with rhabdomeres pointing toward the light. The secondary eye (right) has a reversed retina with rhabdomeres facing away from the light coupled with reflecting tapeta structures for vision at night. (A and C after Wolken 1995; B and D after Walls 1967; E after Land 1981.)

iris muscles must only pull the halves of the iris apart, like opening a curtain that is attached to the window at the top and bottom. The slit pupil can be opened completely and can close to a tiny slit during the day. Cats, lizards, and other nocturnal animals typically have slit pupils. Whether the slit is vertical or horizontal depends on the visual field of the animal; forest dwellers must look up and down and have vertical slits, while open plains inhabitants usually have horizontal slits.

An alternative method of reducing the light falling on the receptor cells is to use masking pigments in the pigment epithelium layer that can be moved forward to cover the receptor cells during the day and retracted during the night. Usually animals with highly contractile or slit pupils do not also have masking pigments. A few birds, such as parrots, do have both, but it is the masking pigment that performs the light regulation function; the iris is freed from this function and is used for social signaling.

Optimizing Acuity

Maximizing acuity or resolving power means making a fine-grained image of the world and transmitting its details to the brain. We measure acuity as the ability to discriminate the adjacent stripes in a grating spaced a certain distance apart and placed a certain distance from the eye. At the limit of the eye's acuity, adjacent stripes appear as one. Formally, the resolving power, $\mathcal{R}$, of a single lens eye is defined as:

$$\mathcal{R} = \frac{D}{P} = \frac{f}{2p}$$

where D is the distance of the grating from the eye, P is the distance between two adjacent stripes, f is the focal length, and p is the receptor separation

(Land 1981). The 2 in the denominator incorporates the fact that two adjacent receptors are required to detect the peak and trough of a striped pattern. Since $\mathcal{R}$ is effectively the inverse of the fraction of an arc resolved by a single photoreceptor, it can be expressed in units of radians^{-1}. Alternatively, resolution may be expressed in degrees as angular resolution, $\Delta\varphi = 2 \tan^{-1}(p/2f)$; note that a smaller angle of resolution corresponds to a higher resolving power. Animals can employ several different strategies to maximize the resolving power of their eyes. Most of these strategies are diametrically opposed to the strategies previously listed for maximizing sensitivity. It is therefore not possible to maximize both sensitivity and acuity at the same time, and a compromise must be made depending on which function is the more important for the animal.

Resolution is improved by *decreasing the diameter of the photoreceptors* and packing them very tightly together to reduce receptor separation (parameter p). This permits a fine-grained analysis of the image. However, there is a lower limit on the diameter of the receptor of about one micron, which is the size of two to three wavelengths of visible light. When the receptor is smaller than this size light waves cannot be internally guided down the receptor and diffractive processes cause scattering. Light passing through one cell may be absorbed by the cell's neighbor, and this leakage means that the two cells cannot operate independently. Once this lower limit on receptor diameter has been reached, the only way to improve visual acuity is to *increase the size of the eye*. Larger animals in general have better visual acuity because of their larger eye size. Another widespread strategy for increasing acuity is to *increase the number of cones* relative to rods in the retina. Because of the reduced summation of cones on to nerve cells, the effective receptor-separation factor p is lower for cones compared to rods. Diurnal birds possess eyes with the highest cone to rod ratios known, and their acuity surpasses that of humans. However, such birds have greatly reduced ability to see in dim light. Diurnal primates, including humans, have opted for a 3:1 mix of cones to rods, which permits reasonably good acuity and moderate low-light vision. A *long focal length* also increases acuity. The lens curvature must be reduced to increase the focal length and the entire eye must be elongated. The result is a larger image on the retina, which permits greater resolution of details. However, the expanded image is less bright and sensitivity in low-light conditions is reduced. Diurnal vertebrates have a relatively flat lens with F numbers ranging from 2 to 4 to maximize acuity, whereas nocturnal vertebrates have rounder lenses with F numbers between 0.5 and 1.5 that sacrifice some acuity to maximize sensitivity (see Table 9.1).

For animals that must be active in dim light but that also require accurate resolution for foraging, several compromise strategies are possible. One compromise is an eye that has a high rod density over most of the retina but a high concentration of cones in a small area of the retina at the focal point of the lens. This cone-rich region is called an **area centralis**. Similarly, some animals possess a **visual streak** which is a horizontal band of increased cone density. With such compromises, the animal can navigate well in dim light

Table 9.1 Characteristics of vertebrate eyes.

Animal	Habit[a]	Focal length (mm)	Aperature (mm)	F-no.	$\Re$	Cone(C)/rod(R) ratio	Pupil shape	Pupil[b] control	Tapetum	Yellow filter	Area centralis	Fovea	Accommodation
Negaprion Shark	N	9.2	7.0	1.31		C <<< R	—	None (M)	Yes	No	Streak	No	Move lens
Carassius Goldfish	A	2.8	2.3	1.22	132	C = R	○	None (M)	Yes	Cornea	Yes	No	Move lens
Rana Frog	A-N	4.0	4.7	0.85	160	C = R	○	Some (M)	Yes	Cone oil	Yes	No	Move lens
Iguana Lizard	D	4.9	3.0	1.63		C >> R	○	Some	No	Cone oil	Yes	Yes	Lens shape
Gekko Gecko	N	6.5	6.0	1.08		C << R	|	Good	Yes	No	No	No	Lens shape
Aquila Eagle	D	22.6	10.0	2.26	7907	C >> R	○	Good (M)	No	Cone oil	Yes	Yes[c]	Lens shape
Columba Pigeon	D	7.9	4.0	1.98	917	C >> R	○	Good (M)	No	Cone oil	Yes	Yes[c]	Cornea shape
Strix Owl	N	17.3	13.3	1.30	860	C ≤ R	○	Good (M)	Yes	No	Yes	Yes	Lens shape
Rattus Rat	N	3.3	4.0	0.83	86	C < R	○	Good	Yes	No	No	No	Lens shape
Pteropus Megabat	N	5.7	11.4	0.50	230	C < R	○	Good	Yes	No	Streak	No	Lens shape
Oryctolagus Rabbit	N-A	10.0	10.0	1.00	195	C < R	○	Good	Yes	No	Streak	No	Lens shape
Equus Horse	A	23.7	14.8	1.60	1335	C < R	—	Good	Yes	No	Yes	No	Lens shape
Felis Cat	N	12.8	14.0	0.92	516	C < R	|	Good	Yes	No	Yes	No	Lens shape
Homo Man	D	17.0	8.0	2.13	4126	C = R	○	Good	No	Cornea	Yes	Yes	Lens shape

[a] D = diurnal, N = nocturnal, A = arhythmic [b] M = masking pigments and/or cone movements [c] Two foveas

Note: Resolving power is based on behavioral measures of grating resolution from studies with corroborating evidence from other investigators or anatomical data; computed as $\Re$ = (57.3) (cycles/degree).

Sources: Walls 1967; Hughes 1977; Northmore and Dvorak 1979; Pettigrew et al. 1988; Hall and Mitchell 1991; Pak and Cleveland 1991; Hueter 1991; Timney and Keil 1992; Aho 1997.

using its peripheral vision and can discriminate objects in line with the optical axis if some light is available. Most diurnal and some nocturnal vertebrates possess an area centralis. If even greater acuity is required, the area centralis may also contain a **fovea**, which is a rimmed depression or pit in the retina surface. Walls (1967) argued that the oblique angle of the fovea served to refract the light outward to magnify the image, but others have suggested that the fovea functions to align, fixate, and focus an image (Pumphrey 1948; Fite and Rosenfeld-Wessels 1975). Foveas are found mostly in birds and higher primates; they are absent in the majority of fish, reptiles, amphibians, and mammals. Birds of prey and a few other avian species possess two foveas, a central fovea used for high-power lateral monocular vision and a temporal fovea for binocular vision in the forward direction.

The clearer the image on the retina, the greater will be the animal's ability to distinguish fine detail. Reducing distortions generated by the cornea and the lens will improve acuity, but once again may reduce sensitivity somewhat. Spherical aberration is a serious problem for the very round lenses found in aquatic and nocturnal vertebrates. This sort of distortion can be corrected with a lens that is denser (higher refractive index) in the center than on the edge. **Chromatic aberration** is another source of fuzzy images. Recall from Chapter 7 that different frequencies of light are refracted at slightly different angles at the air/eye boundary, so the focal length is not the same for all frequencies. A common technique for reducing chromatic aberration is to filter out the most deviant violet and blue wavelengths with a yellow filter. Yellow pigments may be located in the lens or in the receptor cells themselves. The consequence is, of course, reduced sensitivity to violet and blue colors. For diurnal animals this loss may not pose a great problem, but for nocturnal animals any reduction in light reduces their sensitivity and they do not exhibit these image-clarification strategies. A final image-enhancing strategy is a structure called the **pecten**, a pleated, richly vascularized fin of tissue reaching from the retina into the vitreous humor (see Figure 9.14C). It is found in all birds and some lizards but is most highly developed in raptors and passerines. One hypothesized function of the pecten is that it provides the main source of nourishment to the eye. The avian retina has virtually no blood vessels in front of the photoreceptors to block the light path and distort the image quality.

The apposition compound eyes of invertebrates face the same types of tradeoffs between maximizing acuity and sensitivity and the same basic equations are used to express sensitivity and resolving power. However, there are some differences. In a camera-lens eye, increasing the length of receptor cells to increase sensitivity reduces acuity because the image on the retina would lie on a range of focal planes. Image focusing is not necessary in the compound eye, so receptor cells are several times longer than in the camera-lens eye. Resolving power in the apposition compound eye is expressed as $\mathcal{R} = r/2A$, where A is the aperture diameter of the facet of an ommatidium (analogous to receptor separation p) and r is the local radius of the surface of the eye (analogous to focal length f). Insect eyes often possess the equivalent of a fovea in which resolution is improved. This is partly achieved by increasing the local radius of the eye to reduce the interommatidial angle. However, to obtain the desired increase in resolution, the acceptance angle of each ommatidium must also be reduced so that the fields of view of neighboring ommatidia do not overlap. Logically, one might think that this could be done by decreasing the aperture size. Unfortunately, the aperture size is usually already minimized at the diffraction limit and a further decrease in A would not only reduce sensitivity but also blur the incoming light rays. So the aperture size of facets is actually increased in the fovea region, and the curvature of the facet lenses is decreased to narrow the acceptance angle of the ommatidia (see Figure 9.4B) (Horridge 1977). Foveas often function to improve resolution in predatory insects, but in some species they are only present in the male and are used to detect females.

Optimizing Accommodation

Accommodation is simply the ability to focus objects at different distances from the eye on the retina to produce a sharp-edged image. The compound arthropod eye has no mechanism for accommodation, but cephalopod and vertebrate eyes do have this ability. The mechanism used for accommodation depends on whether the animal lives in an air or water environment. In water, there is little refraction at the water/cornea boundary because the cornea has a similar refractive index to water. The lens performs essentially all of the refraction and therefore must be round and made of a very dense material with a high refractive index. Such a lens is hard and inflexible. To bring near and distant objects into focus on the retina, small muscles move the lens forward and backward along the optical axis (Figure 9.15B). This method of accommodation is found in cephalopods, fish, frogs, and salamanders. In terrestrial reptiles, birds, and mammals, the cornea provides the major refraction function because the refractive indices of air and corneal tissue are sufficiently different. The lens is needed only for fine focusing. The lens is therefore relatively soft, and accommodation is accomplished by changing the shape (curvature) of the lens (Figure 9.15A). For both types of lens, the relaxed state of the muscles connecting the lens to the eyeball is set at the distance most commonly needed by the animal.

Other strategies may be used for accommodation. In a pinhole eye or a camera eye with a round iris, the smaller the pupil the greater is the range of distances over which objects will be in focus. Photographers use this principle to vary the depth of focus of objects in their photographs. We often squint to aid our resolution of a distant object for the same reason. Of course, the smaller hole reduces the total amount of light entering the eye or camera. Hunting spiders with camera-lens eyes move their retina to bring objects into focus (Figure 9.15C). Another strategy is to increase the length of the receptor cells, but this option is made at the cost of poorer resolution. Finally, the once-popular idea that a few animals such as horses and skates possess asymmetries in eye shape, so that different viewing angles permit focusing on far or close objects, has fallen into disfavor (Sivak and Allen 1975).

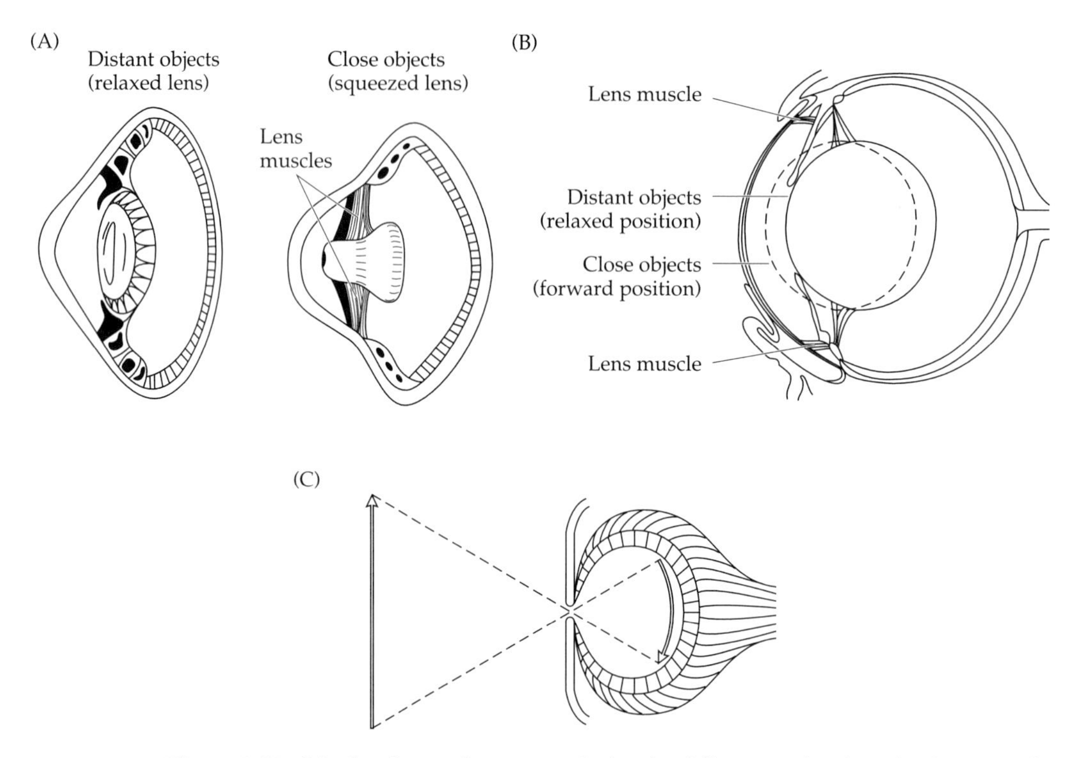

Figure 9.15 Mechanisms of accommodation in different animals. The distance of an object from an eye or lens determines the effective focal length of the image. A distant object focuses at the focal point of the lens, whereas objects at increasingly closer distances from the lens focus at increasingly longer focal lengths. (A) Terrestrial birds and mammals with a soft lens, such as the diurnal bird's shown here, focus on close objects by squeezing the lens into a rounder shape that reduces the focal length. In some birds, the cornea also changes shape. (B) Aquatic vertebrate eyes, such as the frog's eye shown here, have a very dense round lens. Instead of changing the shape of the lens, the animal uses muscles to move the lens forward, bringing close objects into focus. In many fish, the relaxed position of the eye provides good close-up focusing. The animal moves the lens backward (toward the retina) to focus on distant objects. (C) The pinhole eye of *Nautilus* needs no accommodation mechanism; objects at any distance are always in focus because there is only one path for light originating at any one point at the source. (A after Wolken 1975; B after Walls 1967; C after Land 1981.)

Optimizing Spatial Discrimination and Temporal Resolution

Good spatial discrimination and temporal resolution are necessary for mobile animals to navigate among objects in their environment. Flying animals in particular must be able to judge their position in space rapidly, and predatory animals also require especially good abilities to detect the movement of prey and judge their distance from the prey accurately.

Animals employ four methods to judge the distance to objects. The simplest method is to learn the **typical image size** of important objects. Since closer objects appear larger, the size of the object in the visual field is a measure of its distance. Another method is to use **parallax**. Moving the eye or the head causes closer objects to shift position relative to more distant objects. A few unique animals, such as chameleons and the sandlance fish *Limnichthyes fasciatus*, have evolved highly specialized lens and foveal systems for accurate monocular distance estimation using **accomodation cues** (Harkness 1977; Pettigrew and Collin 1995). These three methods are the only ones available to animals with eyes placed on the sides of the head. This type of eye placement maximizes the field of view (nearly 360°) and is found in herbivorous animals and others that do not have to hunt mobile prey (small rodents, granivorous birds, parrots, many fish). These animals have evolved so as to maximize their ability to detect predators from all directions. They have a narrow region of binocular vision of from 5 to 25°. Predatory animals such as carnivores, insectivores, and birds of prey, on the other hand, require a more accurate method of distance determination and use the fourth strategy, binocular or **stereoscopic vision**. When the visual fields of the two eyes overlap, the distance to an object can be determined precisely by triangulation of the angular deviation of the two eyes from their forward position. To achieve good distance estimation via this method, the two eyes must face forward and their fields must overlap extensively. This necessarily reduces the total field of view. At best, only close and moderately distant objects can be accurately ranged. Complex neural processing is required to integrate the images from the two eyes (Pettigrew 1991). Figure 9.16 shows the fields of view and regions of binocular vision for several different animals.

Temporal resolution is measured by determining the flicker fusion rate. This is the rate at which a rapidly blinking light is perceived as continuous. Cones can respond to light and recover much more quickly than rods, so eyes with high cone ratios have better temporal resolution. Humans have a flicker fusion rate of 16/sec, and some animals with all-cone eyes have flicker fusion rates as high as 100 to 150/sec. The rhabdomeric eyes of invertebrates have much higher flicker fusion rates than vertebrates. Invertebrate eyes, for example, can easily resolve the 60-cycle flicker of a fluorescent light and the rapid wing beat of a small flying mosquito.

Table 9.1 summarizes the features of vertebrate eyes discussed above. Several generalizations emerge. The most important factor determining the type of eye a vertebrate species possesses is whether it is diurnally or nocturnally active. A nocturnal eye has a high rod density, either a slit or highly contractile pupil, and a reflective layer behind the retina. Such an eye will be sensitive in low-light conditions but visual acuity will be poor. Spatial and temporal resolution will be limited, and the animal will be blind to colors. Visual communication involving fast, elaborate, or colorful patterns will not be possible for such an animal. The diurnal invertebrate, on the other hand, will possess an eye with a high cone density and a yellow filter, an area centralis, and/or a fovea; the iris is likely to be round in shape. This eye will have excellent acuity, spatial and temporal resolution, and color sensitivity, but will be blind under low-light conditions. The visual modality will be much more useful for communication in the diurnal animal. Table 9.2 summarizes the resolu-

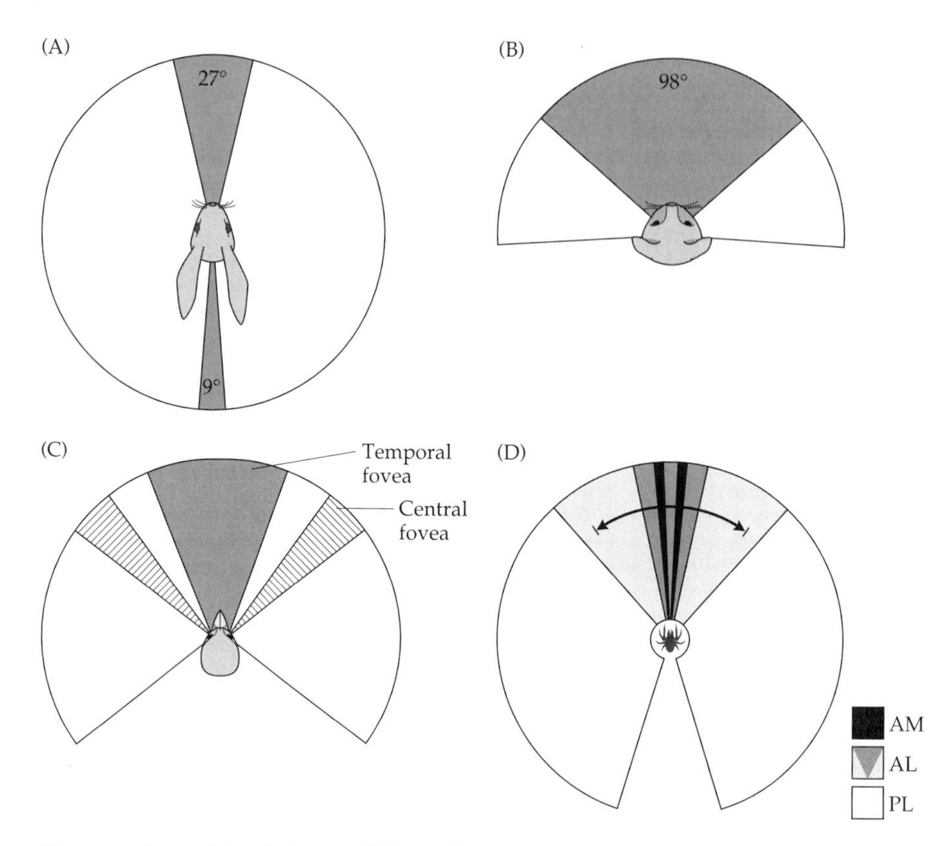

Figure 9.16 Fields of view and binocular vision. The visual field of a single eye is approximately 170° in most animals; the greater the overlap of the fields of view of the two eyes for binocular vision, the smaller the total field of view. The gray areas in the following illustrations indicate regions of binocular vision. (A) Hares and rabbits have unusually large monocular fields of view that gives them a band of binocular vision in front, behind, and above the head. (B) In cats and dogs the visual angle between the two eyes is small so the visual fields overlap extensively. This gives excellent depth perception in the forward direction but a narrow overall field of view. (C) The hawk and many other birds possess two foveas. The temporal fovea permits high resolution binocular viewing in a narrow forward area. Excellent monocular vision is enhanced by the central fovea. (D) Jumping spiders have three main pairs of eyes, each of which views a different part of the visual field. The two anterior median (AM) eyes have a small retina with only a 10° field of view that can be moved with muscles to cover a total field of 58°. The anterior lateral (AL) eyes and the posterior median (PM) eyes have fixed but wider fields of view. This gives the animal excellent resolution in the forward direction as well as moderate vision to the side. (A after Hughes 1971; B after Hughes 1976; C after Walls 1967, Symthe 1975; D after Forster 1982.)

tion and sensitivity of the eyes of selected invertebrates. The range of eye types is much greater here, and resolving power therefore varies over a very large range. As with vertebrates, sensitivity is much higher for nocturnal and deep sea species compared to diurnal species.

Table 9.2 Resolving power and sensitivity in different types of invertebrate eyes

Animal	Type of eye	Focal length f (μm)	Aperature A (μm)	Receptor separation p (d) (μm)	Receptor length x (μm)	Resolving power $\mathcal{R}$ (rad^{-1})	Angular receptor separation $\Delta\theta$ (deg)	Sensitivity $\mathcal{S}$ (μm^2)
Planaria (platyhelminth)	Simple pit	25	30	10	6	.83	35	3.2
Nautilus (cephalopod)	Pinhole	10^4	2.8×10^3 (max) 0.4×10^3 (min)	7.5	450	3.6 25	8.0 1.15	2.6 0.053
Littorina (gastropod)	Small lens in water	155	55	12 (3.5)	10	6.5	4.5	0.062
Vanadis (polycheate)	Lens in water	10^3	250	6	80	83	0.35	0.58
Octopus (cephalopod)	Large lens in water	10^4	8×10^3	3.8	200	2632	0.011	4.23
Perga (insect larva)	Small corneal lens eye in air	200	200	20 (10)	120	5	5.7	34.3
Phidippus (diurnal spider)	Corneal lens eye in air	767	380	2.0	23	192	0.15	0.087
Dinopus (nocturnal spider)	Corneal lens eye in air	771	1325	20	55	19.3	1.5	225.3
Man (fovea in daylight)	Cornea and lens in air	16.7×10^3	2×10^3	2	30	4175	0.007	0.023
Pecten (bivalve mollusk)	Reflector eye	270	450	7.5	15	18	1.6	9.3
Apis (worker) (diurnal insect)	Apposition compound eye	60	25	2 (1)	200	30	0.95	0.318
Ephestia (nocturnal insect)	Superposition compound eye	170	340	8	110	10.6	2.7	82.8
Oplophorus (deep sea crustacean)	Reflecting superposition eye	226	600	32	200	3.5	8.1	3,303

Source: After Land 1981.

SUMMARY

1. The energy of electromagnetic radiation in the visible range is captured by the visual pigment **rhodopsin**. It consists of a bent 11-*cis* carotenoid derivative (retinal) attached to a large protein (opsin). Upon absorption of a photon of light, the retinal is straightened to the all-*trans* isomer and the pigment becomes metarhodopsin. The change causes a cascade of reactions that culminates in a nerve impulse. Rhodopsin molecules are imbedded in the multiple-folded membranes of long, thin photoreceptor cells. The photoreceptor cells are packed side by side in arrays to form a retina that retains the spatial arrangement of incoming stimulus patterns.

2. Well-developed eyes are complex organs with sophisticated optical systems for collecting and concentrating light and with intricate nerve connections for spatial pattern analysis and resolution. Visual receptors evolved through a series of evolutionary steps. The first eye took the form of a simple cup lined with a retina that could detect the direction of light shining on it but that failed to form an image. One strategy for image formation involved developing a focusing mechanism such as the **pinhole eye**, which suffered from low sensitivity. The advent of the refracting lens greatly improved sensitivity and eventually led to the **camera-lens eye**, which is both sensitive and capable of producing a well-resolved image. Another strategy involved multiple cups that were grouped and lengthened into tubes to restrict the angle of light acceptance to a narrow field. This design led to the **compound eyes** of insects and crustaceans.

3. Evaluating the image falling on the retina requires a complex network of nerve connections. In most vertebrates there are two parallel systems for use under high and low illumination situations. The low-light system uses **rod** receptor cells that are very long and highly sensitive to light. The nerve output from many neighboring rods sum together to increase the probability of photon capture and nerve response. While the rod system is highly sensitive in dark environments, its resolution is poor due to the summation. The bright-light system uses **cone** receptors. Cone outputs are not extensively summated, so a fine-grain image is transmitted. They are connected to neighboring cones in complex additive and negative ways to enhance boundaries in the visual field and detect certain types of patterns. In species with several types of cone pigments with different spectral absorption peaks, interconnections among cone cells are also responsible for color perception.

4. Color perception requires at least two sets of differentially color-sensitive photoreceptor cells, and complete spectral differentiation requires three sets. **Monochromats** possess only one pigment type and therefore do not perceive color, only intensity differences. **Dichromats** possess two color-sensitive pigments with overlapping absorbance curves. Color discrimination is based on **color-opponent cells**, which are ganglion or higher-order cells that either add or subtract the neural output from neighboring photoreceptors of each color type. Cells receiving opposite-signed input from two cone types are responsible for hue differentiation, whereas cells receiv-

ing same-signed input encode brightness and saturation information. Dichromats can distinguish a wide range of colors but possess a neutral point in the middle of their spectral range. **Trichromats** have three photopigments and thus two color-opponent systems. They can perceive a wider range of colors and have no neutral point. Species with four to five receptor types frequently use certain wavelengths for special purposes.

5. **Rhabdomeric** photoreceptors are inherently sensitive to the e-vector of polarized light, while ciliary photoreceptors are not. Many invertebrates have evolved a highly developed sensitivity to polarized light that is used to detect the pattern of polarized light in the sky for orientation purposes. Aquatic predators may use polarization sensitivity to enhance the detectability of reflected light from the shiny scales of prey fish.

6. The ideal visual receptor should be sensitive to light under a wide range of ambient conditions, have good resolution of fine detail, be able to focus on far and near objects, be able to judge the distance to objects, and have good temporal resolution for following rapid events. No eye can maximize all of these features because improvements in one parameter tend to reduce one or more other parameters.

7. **Sensitivity** is maximized by increasing the eye aperture, decreasing the lens focal length, increasing the diameter and length of receptors, and increasing photopigment density. **Resolution** is maximized by increasing the lens focal length and decreasing the diameter and length of receptors. Sensitivity and resolution therefore trade off against each other. Compromises include the rod/cone system of vertebrates and a region of high resolution (area centralis plus fovea) surrounded by a region of high sensitivity. Other modifications include reflecting tapeta and tubular eyes to improve sensitivity, and lenses that correct for spherical and chromatic aberration.

8. Strategies of **accommodation** (focusing ability) in camera lens eyes depend critically on the environment. In terrestrial environments, the very different refractive indices between air and the transparent portion of the eye allows most of the refraction to occur at the cornea. Terrestrial animals therefore possess a flat, soft lens whose shape can be changed for fine focusing. In aquatic environments, little refraction can occur at the cornea because water and the eye have similar indices of refraction, so the lens is entirely responsible for focusing. It is usually very round and made of a dense material with a high refractive index. Such a lens is difficult to reshape, so focusing is accomplished by moving it toward or away from the retina.

9. **Depth perception** is important for predators and fast-moving or flying animals. Such animals evolve binocular vision in which the visual fields of the two eyes overlap. The cost is a lower overall field of view; thus herbivores and prey species usually have eyes on the sides of the head. These animals must use relative object size or parallax to judge distances. **Temporal resolution**, or flicker fusion rate, is much higher in cones than in rods, and is higher in rhabdomeric photoreceptors compared to ciliary receptors.

10. Most of the properties of visual receptors are determined by the environment, lifestyle, and diet of the animal. Nocturnal animals must maximize sensitivity and therefore usually have reduced spatial and temporal resolution. Diurnal animals have no light limitation and can therefore maximize resolution and color vision. Predators and fast-moving animals require better depth perception and spatial and temporal resolution than prey species. These constraints largely determine the optics and wiring of the visual system, and therefore the opportunity to use the visual modality for communication. One aspect of vision that does seem to evolve partly in response to communication needs is specialized regions of the eye with unique or enhanced color and spatial resolution.

FURTHER READING

An outstanding current summary of vision can be found in Wolken (1995). Lythgoe (1979) provides an excellent discussion of visual adaptation in ecological contexts, and Dusenbery (1992) expertly expands on the capabilities of visual systems to extract information from the environment. The evolution of eyes has been outlined by Wolken (1975) and Land (1981). Reviews of invertebrate photoreceptors include Horridge (1975), Autrum (1979, 1981), Ali (1984), and Stavenga and Hardie (1989). The classic book on the vertebrate eye is Walls (1967). Jacobs (1981, 1993), Hurvich (1981), Menzel (1975, 1981), and Gouras (1991b) review color perception. Additional books on vision include Ali and Klyne (1985), Leibovic (1990), Valberg and Lee (1991), and Cronly-Dillon and Gregory (1991).

Chapter 10

Chemical Signals

CHEMICAL SIGNALING IS THE OLDEST METHOD OF COMMUNICATION. From the earliest days of life in Earth's oceans, single-celled organisms possessed the ability to detect and selectively take in different classes of chemicals needed for cellular metabolism. The detection of food was and still is the primary function of most chemical reception organs. At the same time that organisms are taking in food, they are eliminating metabolic waste products. Once such organisms evolved the ability to distinguish between the chemical compounds emanating from conspecifics and the chemical components of food, a primitive social organization and communication system existed. Early metazoans relied exclusively on chemical communication to synchronize gamete release and mediate fertilization between conspecific sperm and eggs, and some colonial species even signaled alarm in the presence of a predator. More advanced organisms quickly evolved two types of chemical detection systems: smell and taste (in human terms). In this chapter we shall examine the types of chemicals and production methods used for chemical communication, the transmission properties of chemical odorants in different environments, and the evolution of chemoreception organs for the detection of chemical signals.

GENERAL FEATURES OF CHEMICAL COMMUNICATION

To transmit a chemical signal, individual molecules have to move the *entire distance* between sender and receiver. How can such movement be achieved? There are three mechanisms. (1) Senders can use the **current flow** in air or water to carry the molecule to the receiver. (2) In the absence of a current, the molecule can only move by **diffusion.** Molecules naturally move along a concentration gradient, from a point of high concentration to a point of low concentration, but this requires a certain amount of time. (3) The receiver can move toward the signal and pick up the molecule directly by **contact**, so that the molecule doesn't have to move at all.

Contrasts between the Propagation of Olfactory, Auditory, and Visual Signals

The propagation of olfactory signals differs in major ways from the propagation of auditory and visual signals.

DIRECTIONALITY. Both sound and light travel as orderly waves in a relatively straight direction away from the source of emission. Chemical signals also spread out from their source, but the movement of odorant molecules from a point of high to low concentration follows an irregular path. At any given moment, a diffusing molecule may move either away from or towards the source.

SPEED. Although both sound and diffusion require molecular motion, sound is propagated at a much higher speed than odors. In a sound far field, only the disturbance is propagated, not individual molecules; in diffusion, the individual molecules must be propagated. Typical delay times between sender and receiver are milliseconds for sound, but seconds, minutes, or even days for odors.

TEMPORAL PATTERN. Sound and light retain their temporal patterns as they propagate (although there may be some distortion over long distances). Neither diffusion nor current flow can sustain an initial pattern of modulation because molecules do not move in synchrony. Any temporal pattern imposed on an olfactory signal during emission is lost within a short distance from the source (Bossert 1968).

SPECTRUM. The spectrum of olfactory signals (i.e., the different chemical compounds) cannot be arrayed in one linear dimension as can the frequency spectra of sound and light. This means we cannot use Fourier analysis to characterize olfactory signals or to determine how they are generated, propagated, and received. We thus need a different method of analyzing olfactory signals.

Forms of Chemical Communication

Because the selective detection and uptake of chemicals is a fundamental process of all living cells, chemical communication in a broad sense occurs at many biological levels. Chemicals that operate internally and facilitate communication between the brain and organs involved in growth, digestion, and reproduction are called **hormones**. Chemicals that facilitate communication between conspecifics are called **pheromones** and are, of course, the main focus of this chapter. Chemicals that are transmitted and detected between species, such as predators and prey or sympatric competitors, are called **allomones**. For the two latter types of external communication, two different modes of detection may be employed: **olfactory reception**, which involves the detection of airborne or waterborne chemicals from a distant source (e.g., by smell), and **contact reception**, which requires direct contact of the receptors with the chemical source (e.g., by taste). Both olfactory and contact receptors may be used for the detection of food and conspecifics. As we shall see in the section on reception, many animals possess three separate chemical sensory systems, one for detection of diffusing or current-borne chemicals, another for identification of contacted food, and a third for contact reception of social signals.

PRODUCTION OF OLFACTORY SIGNALS

In this section we shall examine the range of chemicals used by animals to communicate with conspecifics, the sources of these chemical odorants, and the ways in which the chemicals are released by senders.

Types of Chemicals Used for Intraspecific Communication

The array of chemicals identified as pheromones is vast. All are, of course, organic compounds with a basic carbon skeleton. The major constraints on the chemical composition of pheromones are determined by the type of transmission, i.e., diffusion, current, or contact, and by the medium, i.e., air or water.

Airborne odorants must be volatile in air; that is, they must evaporate easily (Wilson and Bossert 1963, Wilson 1970, Wheeler 1977). Volatility is primarily a function of molecular size and weight—larger, heavier molecules have lower volatility. The upper size limit for airborne pheromones is a molecular weight (MW) of about 300. Most airborne odorants contain between 5 and 20 carbon atoms. Molecules larger than this size are both expensive to produce and too large to diffuse effectively. Pheromones with less than 5 carbons may be rare because they are too volatile and possess too few options for species-specific structural variants. Within this size range, chemical odorants show a great deal of variation in shape and type of functional group. A few examples are shown in Figure 10.1. The majority contain a single functional group with one oxygen atom, such as an alcohol, aldehyde, or ketone. Some pheromones are acids or esters with two oxygen atoms, and others are alkanes and alkenes with no oxygen. Pheromones vary greatly in the position of the functional group, the position of double and single carbon-carbon bonds, the occurrence of branches and rings in the carbon chain, and some contain other atoms such as nitrogen or sulfur. These variants have only minor effects on volatility, but they greatly affect the shape of the molecule

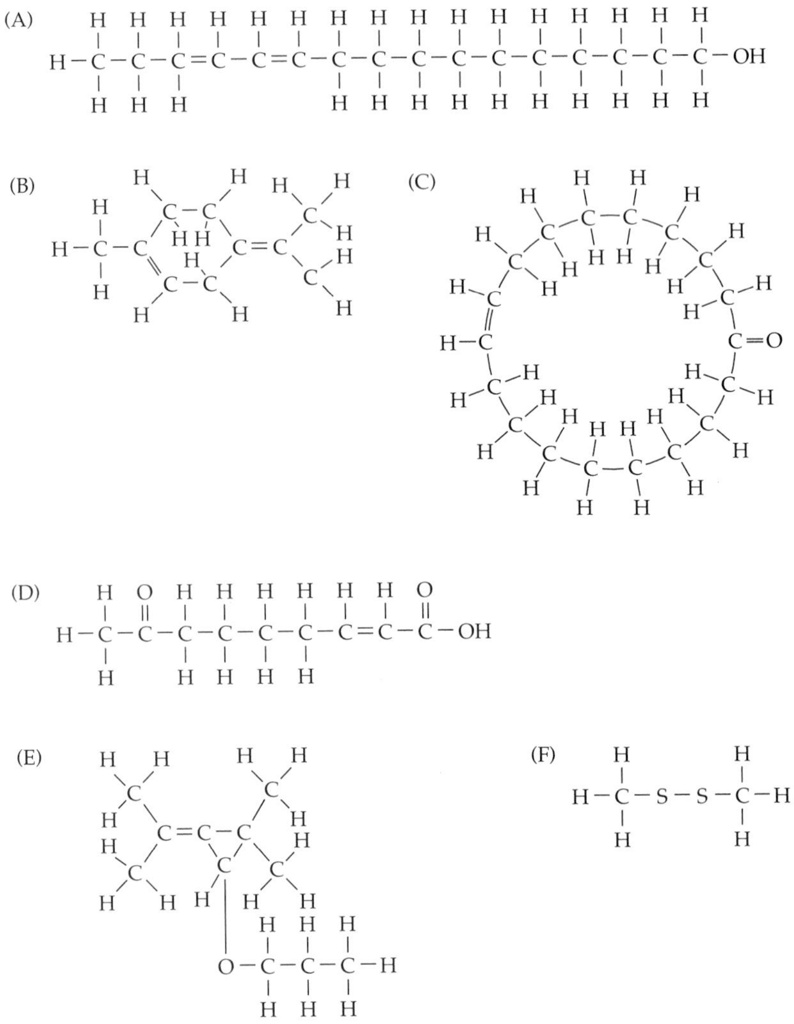

Figure 10.1 Some examples of airborne chemical odorants. (A) Silkworm moth (*Bombyx*) sex attractant, (B) a common termite alarm substance, (C) civet (*Civettictis civetta*) sex attractant, (D) honeybee (*Apis mellifera*) queen substance, (E) cockroach (*Periplaneta americana*) sex attractant, and (F) hamster (*Mesocricetus auratus*) mounting pheromone. In many cases, the pheromone is a mixture of very similar chemicals. (After Wilson 1963; Moore 1968; Johnston 1977.)

and thus the molecule's detection by receptor cells (Morse and Meighen 1986).

The size restriction does not apply for waterborne and contact pheromones. Organic compounds composed of primarily carbon (MW = 12) and hydrogen (MW = 1) can be less dense than water composed of oxygen (MW =

16) and hydrogen, and hence float regardless of their size. Large organic compounds such as lipids and proteins can therefore be used as pheromones in these circumstances. Waterborne odorants must be water soluble to disperse effectively and be detected by olfactory receptor organs. Contact pheromones in terrestrial environments are even less restricted by size constraints, and a larger variety of chemical compounds is therefore available to senders (Carr 1988).

Production Sites

Pheromones can be expelled from two fundamentally different sources: (1) well-defined secretory glands that empty their products onto the outside of an animal's body, and (2) body orifices and organs involved in digestion and reproduction such as the mouth, anus, cloaca, penis and vulva. The pheromonally active chemical components are considerably easier to identify in the case of gland secretions compared to excretory products.

The bodies of both vertebrates and invertebrates contain numerous glands composed of secretory cells that produce specific chemicals. **Endocrine** glands empty their secretory products into the blood stream; these chemicals are the hormones that regulate internal body metabolism. **Exocrine** glands, on the other hand, are those glands located either externally on the skin/integument or internally with ducts leading to the exterior of the body. Their function is to maintain the condition of the body covering and/or to produce chemical communicatory signals—pheromones and allomones. Exocrine glands can be divided into two types based on their appearance and manner of secretion (Figure 10.2). **Sebaceous** glands are flask-shaped or lobed. The basal layer of the gland's epithelium continually produces new cells, forcing the old cells into the center of the gland. As the old cells approach the lumen they become rich in lipids and then disintegrate completely to become a thick, oily product called sebum (Albone 1984). Volatile and nonvolatile pheromone chemicals are embedded in this sebum matrix, which can greatly affect the transmission properties of the pheromone. Secretion rates of sebaceous glands are always slow. **Sudoriferous** glands look like coiled tubules. The secretory cells in such glands are not layered and continuously renewed as in sebaceous glands, but form an orderly lining around the inside of the tubules. The secretory products are collected in droplets or vacuoles within the cells and then emptied into the tubular lumen, where the chemicals are stored. The secretory product is always liquid and can consist of relatively pure pheromone. Secretion rates are much faster than for sebaceous glands. The development and secretory activity of both types of glands are often controlled by endogenous hormones (Ebling 1977). Release of the secretion from the gland is much more likely to be under nervous control in sudoriferous glands, however.

Some pheromones appear to be produced over the entire external surface of some animals by many small sebaceous and/or sudoriferous glands located throughout the skin or integument. Vertebrate cutaneous (skin) glands are an excellent example (Quay 1977; Flood 1985). In mammals, each hair follicle has associated with it a pair of sebaceous glands and a single sudoriferous gland.

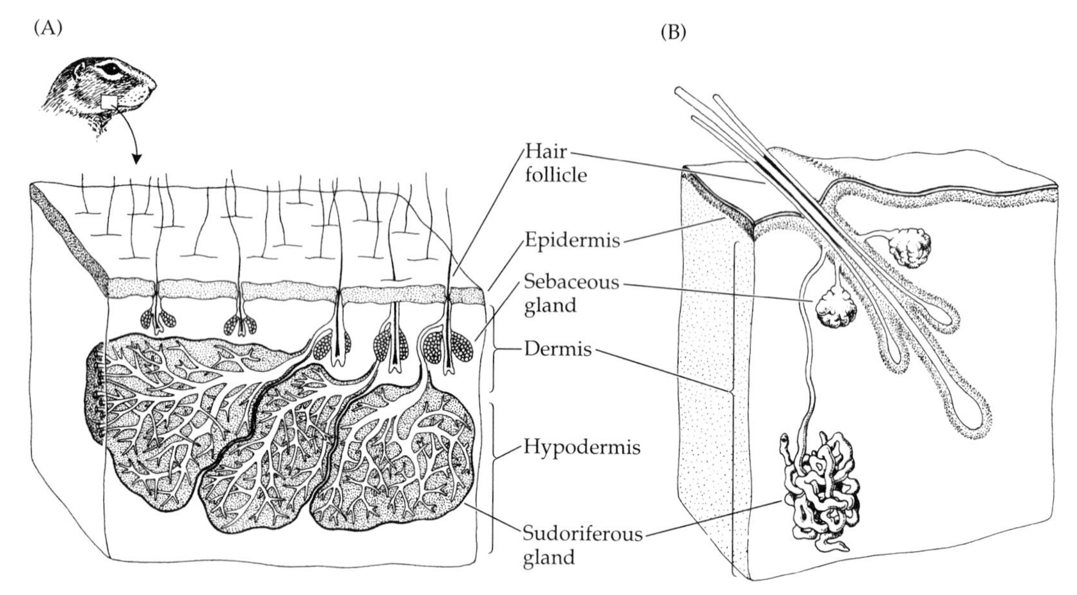

Figure 10.2 Cutaneous glands in mammals. (A) Mongolian gerbil (*Meriones unguiculatus*) cheek gland. (B) Human skin. (After Kivett 1978; Flood 1985.)

The sebaceous glands secrete sebum that maintains the condition of the skin and hair, while the sudoriferous gland produces sweat for thermoregulatory purposes. Pheromones can be added to either secretion. In the garter snake, for example, gravid females secrete a series of large, nonvolatile methyl ketones from their cutaneous glands that are deposited on the ground whenever they travel (Mason et al. 1989). Males can follow the female trail with the use of their contact chemoreceptors. Parental cichlid fish secrete a protinaceous mucus from their bodies that not only maintains contact with the brood of free-swimming fry but also provides food for the young (Barlow 1974).

Many vertebrate and invertebrate animals possess one or more major exocrine glands that produce species-specific pheromones. The chemical secretions from different glands send specific types of communication messages. The glands may be located in a variety of positions on the body where the secretion is most usefully emitted. Figure 10.3 shows some vertebrate examples, and Figure 10.4 shows the typical battery of glands present in hymenopteran insects.

Body orifices associated with digestion and reproduction are obvious locations for the leakage of chemicals out of an animal's body. Waste products such as amines and the byproducts of steroid hormones that are eliminated in urine and feces can provide important information to other individuals that can detect them. Potentially important volatile compounds are delivered to the digestive tract by the liver via the bile. Exocrine glands associated with the digestive tract whose primary function is to secrete digestive chemicals can also secondarily produce important pheromones. For example, the saliva

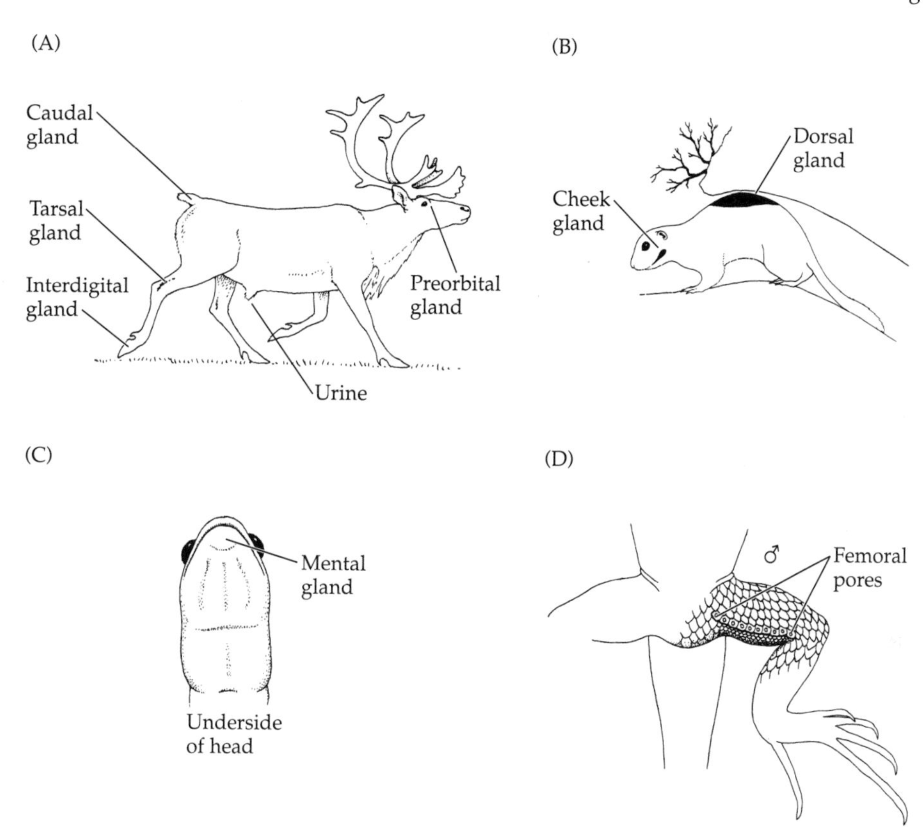

Figure 10.3 Location of glands in vertebrates. (A) Reindeer (*Rangifer tarandus*) produce airborne scents from their caudal gland, tarsal gland, and urine. The tarsal gland also marks the deer's resting sites, the interdigital gland leaves marks on the ground along the animal's path, and the preorbital gland is rubbed on upright twigs for territorial marking. (B) Ground squirrels (*Spermophilus*) mark their burrows with the dorsal gland and objects in their territories with the cheek gland. (C) Salamander males (*Aneides*) mark females during courtship with mental gland secretions. (D) Iguanid lizards mark rocks on their territories with femoral gland secretions. (After Stebbins 1966; Brown and Macdonald 1985; Gosling 1985.)

of male boars and hedgehogs is the source of pheromones used during courtship (Signoret 1970; Perry et al. 1980). The urogenital system can also provide external chemical information about internal metabolic events. The kidneys, like the liver, process cellular waste products and transfer volatile and non-volatile compounds from the blood to the urine. Steroid hormones in particular are concentrated by the urine. Chemical products from the genitals themselves, and in particular the vagina and vulva of female vertebrates, may also be expelled into the exterior environment (Michael et al. 1971; Michael and Bonsall 1977).

Microorganisms provide another source of volatile chemical odorants in some animals. Small pockets or cavities of skin that retain moisture or urine

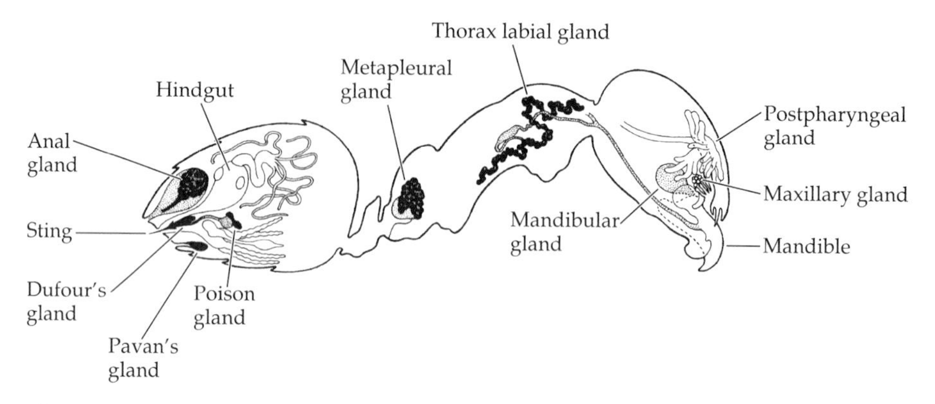

Figure 10.4 Glands of a worker ant (*Iridomyrmex humilis*). (After Wilson 1971.)

provide an ideal growth chamber for bacteria, which in turn produce a variety of small volatile metabolic products (Albone et al. 1977).

Methods of Dissemination

Animals use a variety of methods to release odorous chemicals. Some of these are illustrated in Figure 10.5. The method obviously depends on the viscosity of the secretory substance, the type and location of the gland or other source, and the target or recipient of the chemical signal. It is often difficult for human observers to know when an olfactory signal has been produced, since we may not be able to smell it ourselves. Many olfactory signals, however, are accompanied by specific behaviors, visual signals, postures, structures and even auditory signals, and in these cases we can be more certain about signal production.

Liquid secretions can be released in a brief forcible stream and therefore directed at a specific target individual or location (Figure 10.5A,B). This release requires neuromuscular control over the gland. Alarm, threat, and defensive olfactory signals are typically disseminated in this way in ants, skunks and bombadier beetles. Urination can be similarly controlled and used for marking mates, as in the South American mara, or used for marking specific locations on a territory, as in dogs and many rodents. Some arboreal mammals urinate on their feet and mark the branches in their territory as they move around (Charles-Dominique 1977). Liquid secretions can also be released slowly into a current, as in the mate-attraction pheromones of moths and other insects (Elkington and Carde 1984). Gooey sebaceous gland secretions, on the other hand, must be rubbed onto a surface, object, or target individual. Examples include marking stems or posts with anal or facial glands (Figure 10.5C), marking females with chin or facial gland secretions by many male mammals, and marking rocks with femoral gland secretions by territorial lizards (Figure 10.5J). Some animals carefully spread sebaceous gland secretions over their own bodies during grooming (Figure 10.5D). Anal, cloacal, and preputial glands add their secretions to feces and urine and are therefore spread along with the excreta.

Certain structures and behaviors are used to help transmit chemical odorants. Hairs are frequently specialized to disseminate scents (Figure 10.5G,H). They may be used like a brush to deposit secretions on objects or they may be modified with rough surfaces to form a substrate for the chemical. Hairs increase the surface area for vaporization of both liquid and oily chemical secretions. A variety of behaviors and movements enhances the transmission of chemicals by spreading the odorants or creating current flows. For example, odorant may be rubbed onto the tail and the tail waived toward the target, and secretions may be released into air or water currents generated by the sender (Figure 10.5E,F). Hairs, structures and movements also add a visual component to an olfactory signal, as in the maned rat, lizard, and butterfly examples (Figure 10.5H–J).

In most of the examples described above, there is a close association in time between the release of the odorant by the sender and its reception by the target. The behavioral responses of receivers may also be sufficiently rapid and obvious that we can ascertain the function of the signal (Figure 10.6). However, for many chemical signals such as territorial scent marks there is a significant delay between deposition by the sender and reception by the receiver. The marks are intended to last a long time and the sender may be quite far away when they are detected. This means that the signal and the sender may be dissociated in both space and time.

TRANSMISSION OF CHEMICAL SIGNALS

In this section we shall quantitatively examine the transmission properties of olfactory signals in several different environmental circumstances. The first subsections will describe simple diffusion processes in still air and water. We will then analyze how current flow affects the transmission of chemicals.

General Rules for Diffusion

Molecules move down their concentration gradients. This movement occurs not because of some form of repulsion among similar molecules, but because of unequal numbers of molecules moving in each direction. Let's set up a sampling window perpendicular to the concentration gradient and watch it. Though moving randomly, more molecules will on average cross the window from the high to low concentration sides than the reverse because there are more of them on the high concentration side available to do so. This produces a net movement on average across the window. This process is called **diffusion**.

The rate of diffusion depends upon: (a) the steepness of the concentration gradient, and (b) the ease with which a particular type of molecule can move in a particular medium. The slope of the concentration gradient can be described as the change in concentration of a molecule over a given distance. The **diffusion constant** is a measure of the ease of movement of a particular molecule type in a particular medium; it depends on the size of the molecule, how the molecule interacts with the medium, and how the medium molecules interact with each other. **Fick's first law** provides a quantitative descrip-

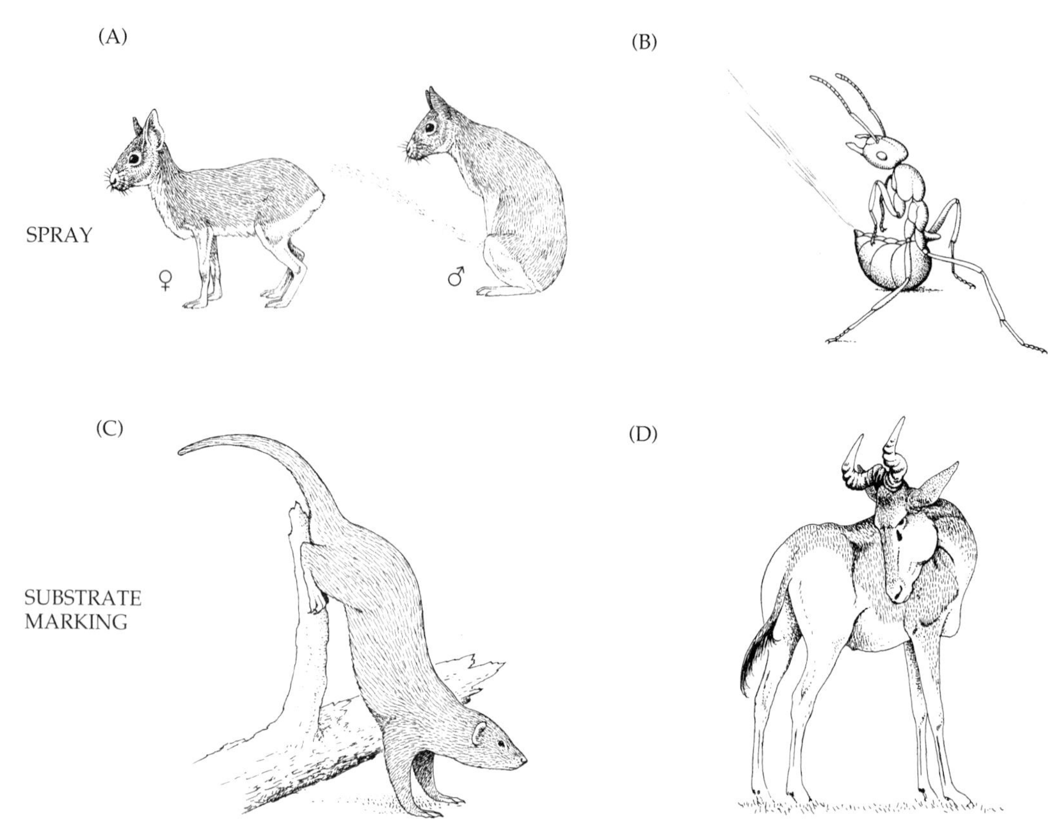

Figure 10.5 Scent dissemination methods. (A) The male mara, *Dolichotis patagonia*, marks his mate during courtship by urinating on her. She may respond by urinating backwards at the male. Similar behavior is observed in the rabbit, guinea pig, agouti, chinchilla, and porcupine. (B) Formicine ants spray formic acid from their anal gland as a defensive response to potential predators. (C) Dwarf mongooses, *Helogale parvula*, mark vertical posts around their territory by performing a handstand and rubbing anal secretions as high up on the post as possible. (D) The male hartebeest, *Alcelaphus buselaphus*, rubs the secretions from his preorbital gland to attract females during the rutting season. Similar self-annointment is found in many related antelope. (E) The female lobster, *Homarus americanus*, on the right blows her gill current containing pheromones into the male's shelter. The male draws water through his shelter and fans it out with his pleopods to advertise his mating status. (F) The male white-lined bat, *Saccopteryx bilineata*, hovers in front of a female, opens the glands in his forewings, and wafts his sweet-smelling scent toward her. (G) The male bat, *Chaerophon chapini*, has a tuft of elongated hairs on his head that disseminates a pheromone of unknown social function. (H) The male *Danio* butterfly hovers over the antennae of a flying conspecfic female and everts a brush-like structure containing a pheromone that causes a receptive female to land for mating. (I) The crested mane rat, *Lophiomys imhausi*, erects its crest to expose a lateral scent gland during extreme alarm. The olfactory signal is enhanced by the visual black and white striped pattern of its fur. (J) The desert iguana, *Dipsosaurus dorsalis*, drags its rear legs across flat rocks around its territory to deposit femoral gland secretions. The secretions strongly absorb UV radiation, and to the lizards that are sensitive to these wavelengths, the marks appear dark; they are not visible to humans. (A after Macdonald 1985a; B after Wilson 1971; C after Macdonald 1985b; D after Gosling 1985; E after Atema 1986; F after Bradbury 1972; G after Brosset 1966; H after Brower et al. 1965; I after Stoddart 1980; J after Alberts 1989.)

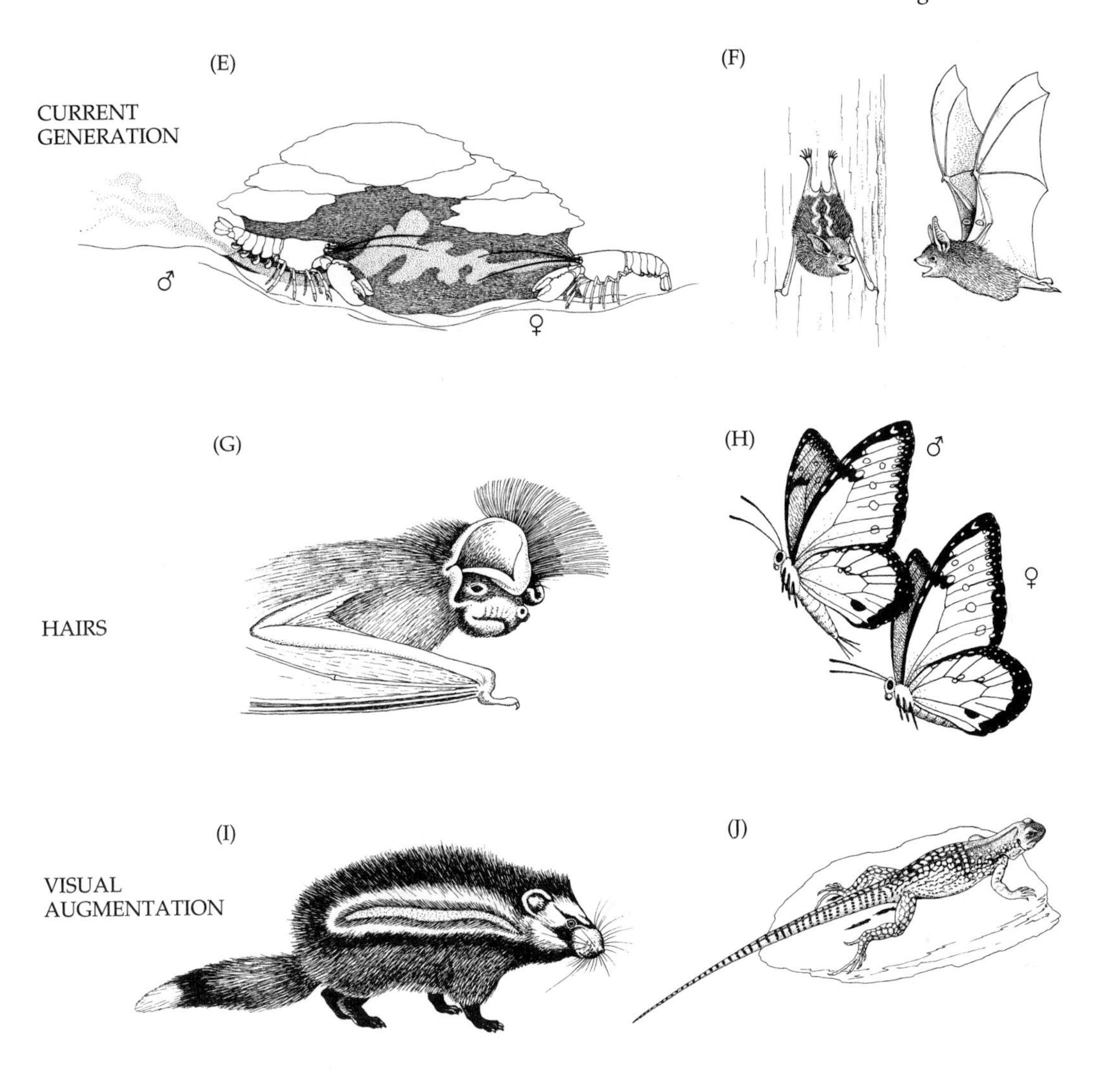

tion of the rate of movement of molecules diffusing through a small window per unit time (see Box 10.1). Armed with Fick's law, we can predict the concentration of diffusing molecules at any time and at any distance away from the source of the molecules for a variety of emission strategies and ecological conditions. Quantitative expressions for these processes are described in Box 10.1. However, we can more easily visualize what is taking place using graphical techniques. In the following discussion of chemical signal transmission strategies, we shall refer frequently to the following quantitative variables:

C = concentration of odorant molecules per unit of volume (molecules/cm^3)
K = minimum concentration of odorant molecules required for receiver detection
Q = total number of molecules released

(A)

(B)

Figure 10.6 Chemical mate attraction signals in termites. (A) The newly emerged alate (winged) queen termite (*Nasutitermes*) lifts up her abdomen in a characteristic posture and releases a pheromone that attracts males. (B) The female has attracted a male and both have dropped their wings. Unlike colonial ants, bees, and wasps, in which the queen mates, stores the sperm, and goes off on her own to found a new colony, the termite male and female remain together in a long-term monogamous relationship. The male must therefore follow the female to their new colony site. Here the female is releasing a chemical so the male can follow her, a behavior called tandem running. (Photos courtesy of Carl Rettenmeyer.)

D = medium-specific diffusion constant (cm^2/sec)
r = distance from the chemical source for circular transmission (cm)
x = distance from the chemical source for longitudinal transmission (cm)
t = time from the onset of emission (sec)

Single-Puff Case in Still Air

Suppose a single instantaneous puff of odorant is released quickly. We are interested in monitoring the concentration of the odorant at various times after the release and at various distances or radii from the source. Let the number of molecules released in the puff be Q. If the puff is emitted away from any boundaries (such as the ground), it will diffuse out in all directions. If the animal is on the ground, the same number of molecules must diffuse into half as much space, so the concentration is then twice as high for any point in space or time. Plotting the concentration at different distances r in a series of time intervals t from release of the puff, we get a set of graphs like those shown in Figure 10.7. K is the minimum threshold concentration of odorant required for detection by the receiver.

At t_0, all the odorant is concentrated at the source and there are no points other than the source at which the concentration is greater than K. Shortly thereafter, at t_1, most of the odorant is still near the source, but the peak has dropped a bit; within a distance r_A of the source, all points have a concentration of odorant above K. This distance r_A is the radius of a sphere (if the animal is near no boundaries) or a hemisphere (if the animal is on the ground) within which a receiver would be able to detect the signal. This sphere (or hemisphere) is called the **active space** of the olfactory signal. By t_2, the diameter of the active space has gotten quite large. However, we are spreading the same number of molecules over a larger and larger volume as diffusion progresses. Eventually, molecules are leaving the detection boundary of the active space faster than they are arriving from the release point. When this happens, the diameter of the active space begins to decrease. In the example, we see that r_A is smaller at t_3 than it was at t_2. By t_4 the active space has almost shrunk up to what it was just after release, and by t_5 the concentration of odorant is less than K everywhere; the puff has diffused away.

There is a maximum size to the active space. Inside the boundary of this maximum active space, an animal will detect the signal at some point in time; outside that boundary, it will not. We shall call the maximum radius of the active space r_{max}. The size of r_{max} depends on the ratio of molecules released to the detection threshold concentration raised to the one-third power, or

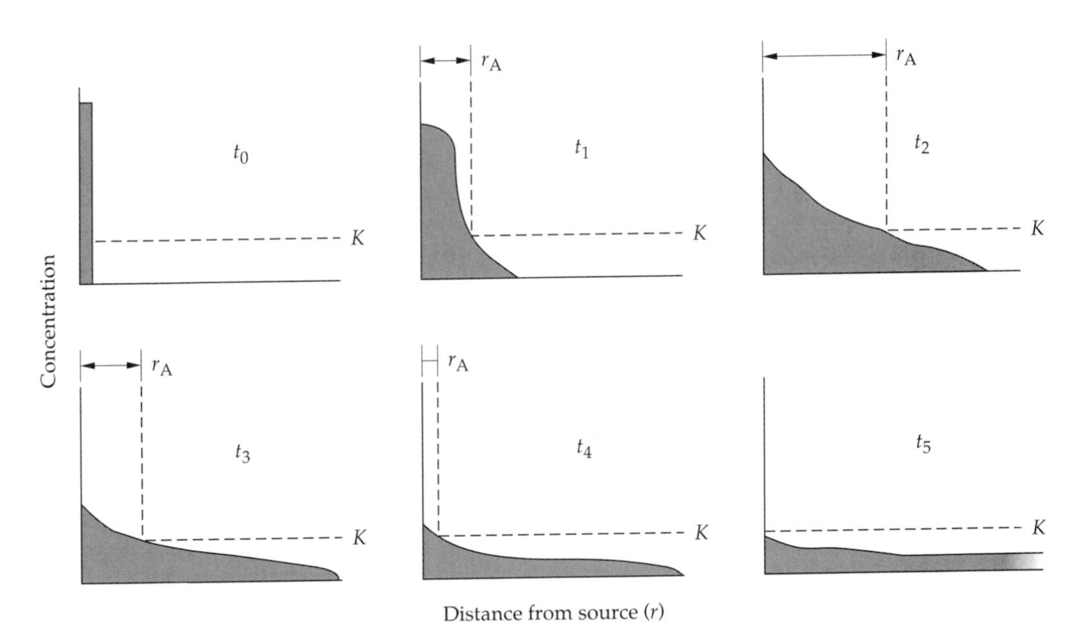

Figure 10.7 Spread of a single olfactory puff. Each graph shows a snapshot of odorant concentration (C) versus distance from the source (r) at different times since emission (t). At the instant the puff is produced (t_0), all of the odorant is highly concentrated at the point of release (r_0). At the subsequent moments in time, the odorant molecules diffuse out from the source and the local concentration (molecules per unit volume) is reduced. The active space, r_A, is the enclosed region in which the concentration of the odorant is above the threshold detection concentration, K. The active space first increases, then decreases to zero.

Box 10.1 Quantitative Expressions for the Diffusion of Chemical Signals

LAW OF DIFFUSION. Fick's first law describes the rate of movement of molecules down a concentration gradient. This rate depends on the diffusion constant D and the slope of the concentration gradient. If the concentration of the molecules at any point x is $C(x)$, then the slope of the gradient is the derivative of $C(x)$ with respect to x, or $dC(x)/dx$. The number of molecules, J, diffusing through a small window per unit time is therefore given by:

$$J = -D\frac{dC(x)}{dx}$$

where J is in units of molecules cm^{-2} sec^{-1}, D is in units of cm^2 sec^{-1}, C is in units of molecules cm^{-3}, and x is in cm. The sign of the concentration is negative since molecules are moving from a higher to a lower concentration region as x increases. Adding the minus sign to the expression therefore makes J a positive number of molecules moving down the gradient.

SINGLE PUFF IN STILL AIR. If a single instantaneous puff of odorant is released quickly from a point well away from any boundaries, the odorant will diffuse outward in all directions. Using Fick's law, it can be shown that the concentration $C\ (r,t)$ at any distance r and time t is:

$$C(r,t) = \frac{Q}{(4\pi Dt)^{3/2}} e^{-r^2/4Dt}$$

where e is the base of the natural logarithms, Q is the number of molecules released, and D is the diffusion constant. If the animal is on the ground, the same number of molecules must diffuse into half as much space, so C is then twice that predicted above for each r and t. The maximum radius of the active space when the puff is released near the ground can be calculated as (Figure 10.8A):

$$r_{max} = 0.527\left(\frac{Q}{K}\right)^{\frac{1}{3}}$$

the time required for the single puff to expand to r_{max} is (Figure 10.8B):

$$t_{r_{max}} = \frac{0.046}{D}\left(\frac{Q}{K}\right)^{\frac{2}{3}}$$

and the time to fadeout of the signal is:

$$t_{fadeout} = \frac{0.126}{D}\left(\frac{Q}{K}\right)^{\frac{2}{3}}$$

CONTINUOUS EMISSION CASE. If the sender continuously emits Q molecules/sec, the concentration is given by:

$$C(r,t) = \frac{Q}{2\pi Dr} erfc\left(\frac{r}{\sqrt{4Dt}}\right)$$

where *erfc*(x), the error function complement, is the area under the normal curve out to infinity. Here, r_{max} increases and levels off (Figure 10.10) at a value of approximately:

$$r_{max} = \frac{Q}{2\pi KD}$$

where D is measured in units of distance moved per second. The time needed to reach 95% of r_{max} is:

$$t_{95\%} = \frac{Q^2}{K^2 D^3}$$

CONTINUOUS RELEASE FROM A MOVING SOURCE. The case of an animal releasing a trail of odorant while moving can be modeled as a linear series of single puffs (Figure 10.11). If Q is the number of molecules of pheromone released per second and u is the animal's velocity, the total length of the active space, L, will be

$$L = \frac{0.160Q}{DK}$$

The location of the maximum radius of the active space occurs at a distance of 37% of the active space length from the point near the animal:

$$L_{r_{max}} = 0.37\,L$$

The maximum diameter of the active space at this location is:

$$r_{max} = \sqrt{\frac{2Q}{eK\pi u}}$$

and the time it takes before the trail at any given location has dropped below K is:

$$t_{fadeout} = \frac{L}{u}$$

CONTINUOUS EMISSION WITH LAMINAR CURRENT FLOW. When there is laminar current flow, the spread of a continuously emitted chemical can be modeled as in the case of continuous point emission, with the modification that the center of the active space moves with the flow. The concentration of the substance at any point in space is given by:

$$C(r,\theta) = \frac{Q}{4\pi Dr} e^{\left[-\frac{(1-\cos\theta)rv}{2D}\right]}$$

where v is current velocity, r is the straight-line distance from the source to the position of interest, θ is the angle between this line and the downstream direction, and D is the diffusion constant (Figure 10.13).

CONTINUOUS EMISSION WITH TURBULENT FLOW. When current flow is turbulent, precise prediction of the active space is less accurate. Using the same principle of laminar

Box 10.1 (continued)

current flow but redefining the diffusion constant as diffusivity based on the ability of molecules to be transported by the current flow, concentration is given by:

$$C(x,y,z) = \frac{2Q}{\pi D_y D_z v x^{(2-n)}} e^{\left[-x^{n-2}\left(\frac{y_2}{D_y}+\frac{z^2}{D_Z}\right)\right]}$$

where the current is moving along the x axis at a velocity v, D_y and D_z are **diffusivity constants** in the y and z planes measured at a wind speed of v, and n is a constant. The source is located on the ground at the origin ($x = y = z = 0$) and continuously emits odorant molecules at a rate of Q/sec (Figure 10.13) (Bossert and Wilson 1963; Dusenbery 1992).

$(Q/K)^{1/3}$ (Figure 10.8A). The time required for the single puff to expand to r_{max}, called $t_{r_{max}}$, depends on $(Q/K)^{2/3}$ and the reciprocal of D/r (Figure 10.8B). Since both r_{max} and $t_{r_{max}}$ depend on Q/K to a power less than 1, it takes quite a large change in Q/K to make a significant change in either the maximum radius or the time to reach that maximum. In fact, to double the diameter of the active space, an animal must increase Q/K by eight times. Since r_{max} determines the range of an olfactory signal, this relationship suggests that single puffs will require releases of very large numbers of molecules and/or very high receiver sensitivities to provide communication over any distance.

We can also calculate the time at which there is no longer any point in space with a concentration above K. This is called the **fadeout time**, $t_{fadeout}$. Like $t_{r_{max}}$, fadeout time depends on $(Q/K)^{2/3}$ and the reciprocal of D. Figure 10.9 shows how the size of the active space changes over time. Note that $t_{r_{max}}$ is always 0.37 times $t_{fadeout}$ regardless of the values of Q, K, or D.

In the case of the single puff, senders have the ability to design their signal for specific needs by independently adjusting r_{max} and the time it takes to

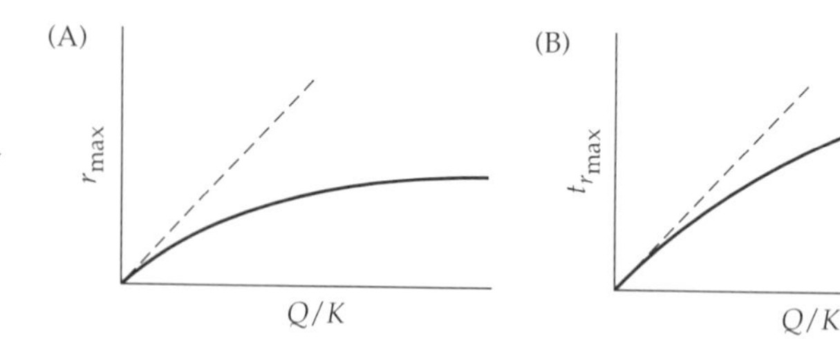

Figure 10.8 The effect of the Q/K ratio on r_{max} and $t_{r_{max}}$. Q/K is the amount of odorant released relative to the detection threshold. Both (A) the maximum radius of detection, r_{max}, and (B) the time to reach it, $t_{r_{max}}$, initially increase and then level off as the Q/K ratio is increased. The dashed line indicates a slope of 1.

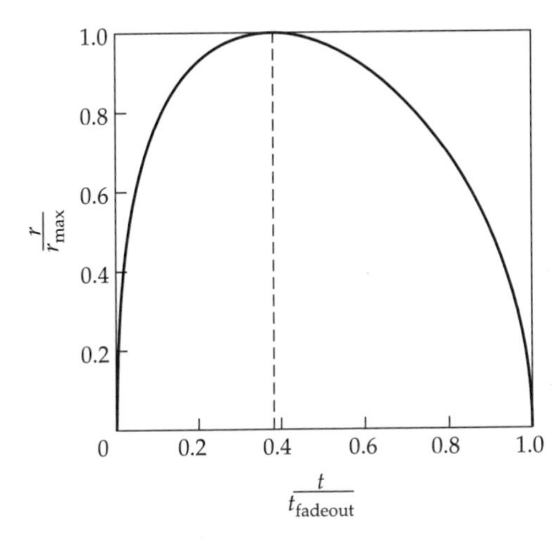

Figure 10.9 Change in the diameter of the active space over time. The active space increases, then decreases with time. The time to reach r_{max} is always 0.37 times the time to fadeout.

reach r_{max} or fadeout. Notice that r_{max} only depends on the Q/K ratio, whereas the two time parameters depend on both Q/K and D. A sender can adjust how long the signal hangs around by selecting a molecule with the appropriate D value. The maximum range over which its signal is detectable can then be finetuned by varying Q/K. For example, an ant may wish to release an alarm signal that has a very large radius (so that its nest mates are more likely to detect it and come to help), but it may not want it to hang around too long or ants will be recruited long after they are needed. Using a small molecule that diffuses quickly and thus has a high D will result in an appropriately short $t_{fadeout}$; releasing a large number of molecules will increase Q/K and thus increase r_{max}.

The effective values of r_{max} and $t_{r_{max}}$ differ in air and water. Diffusion is basically a rather slow process and the effective distances over which animals can use the diffusion process to transmit chemical signals is short, on the order of a few centimeters. Values of the diffusion constant range from .01 to 1 cm^2/sec for volatile compounds in air, but are 10,000 times slower in water. Table 10.1 shows some typical values of r_{max}, $t_{r_{max}}$ and $t_{fadeout}$ in air and water for a Q/K ratio of 1500. Diffusion transmission of chemical odorants is a viable communication channel only for small terrestrial organisms such as ants, which live at the ground/air boundary layer where there is little or no wind. The chemical can diffuse to its maximum radius in a few seconds, the transmission distance range of 1 to 10 cm is a useful scale for alarm or aggregation signals in a small colonial insect, and the signal fades in a minute or less. For most aquatic animals, however, diffusion of chemical odorants is clearly not a viable strategy. Although similar active spaces can be generated (because r_{max} is not dependent on D), the time required to reach r_{max} and $t_{fadeout}$ is on the order of many hours because both factors are dependent on the slow diffusion rates. To reduce these long rise and fadeout times, an aquatic animal would have to increase D or decrease Q/K. Its ability to increase D is limited

Table 10.1 Typical values of r_{max}, $t_{r_{max}}$ and $t_{fadeout}$ in air and water for a Q/K ratio of 1500

	Q/K	D	r_{max}	$t_{r_{max}}$	$t_{fadeout}$
Air	1500	0.5 cm^2/sec	6 cm	13 sec	35 sec
Water	1500	0.00005 cm^2/sec	6 cm	33.5 hr	92 hr

by the diffusion rate of hydrogen ions; no molecule can diffuse faster in water than hydrogen ions. It *could* reduce r_{max}, but to correct for the 10,000 times slower diffusion constant, it would have to reduce Q/K by about 464 times. This might not be too bad an option. Reducing the number of molecules may be economically beneficial. Increasing K means reducing the sensitivity of receptors and that need not be difficult. The cost is that reducing Q/K by 464 times will reduce r_{max} by nearly eight times, and the effective range of the reduced signal will be drastically curtailed.

Continuous Emission in Still Air

Now consider an animal that continuously emits Q molecules per second. Again, an active space gradually expands away from the source, but eventually it reaches a point at which diffusion across the detection boundary of the active space exactly equals the input of molecules from the source. When this happens, a steady state is achieved, and the active space will neither expand nor contract as long as the animal continues its output, as shown in Figure 10.10.

Design considerations differ from the single puff case. In the continuous emission case, the diffusion constant as well as Q/K are involved in determining r_{max}. Since the time parameters are essentially unimportant (at least in air), there is no need to independently adjust range and fadeout. However, the fact that r_{max} depends on both Q/K and D means that the same maximum range can be achieved with a wide variety of molecules. This flexibility can sometimes decrease production costs by allowing an animal to achieve a given range with a cheaply produced odorant.

For continuous emission in water, a compound with a given Q/K will generate an r_{max} 10,000 times larger than in air. This increased maximum may be beneficial for animals attempting to attract mates over long distances. However, the delay required to reach this maximum distance will be 10^{15} times longer than in air—in practice, infinitely long! That means that with continuous emission in water, the radius of the active space *could* be very large, but given the time available to most aquatic organisms, they will only be able to achieve a tiny fraction of this r_{max} in any bout of emission. One way around this problem for continuous emission is to utilize water currents, which can move contained odorant molecules at rates much faster than diffusion. In water, one expects animals to use olfactory communication only when (a) animals are at very short ranges (i.e., less than 1 cm), (b) when the emitter is sessile and has time on its "hands," or (c) when there are sufficient

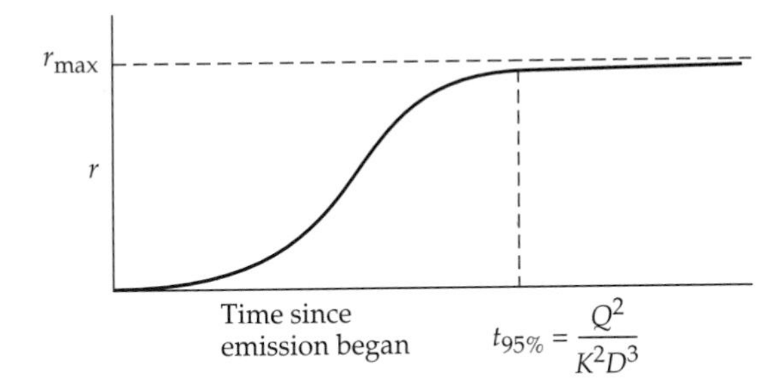

Figure 10.10 Time to achieve maximum radius with continuous emission of an odorant.

currents to increase rates of dispersion of odorant molecules above diffusion values.

Continuous Emission from a Moving Source

Animals may release odorants in more complicated ways than single puffs or continuous emission. For example, as ants forage, they often leave a thin stream of trail pheromone (odorant used as a signal) to find their way back to an earlier part of their route. Ant trails can be modeled as a series of small single puffs that are applied to successive locations along a line. Since they are close together, the active spaces generated by adjacent drops ("puffs") will overlap and merge. The overlapping active spaces will expand out to some maximum radius to the sides and above the trail and then eventually collapse back in as the odorant diffuses away. This process will produce a long hemielliptical active space at any one instant. As the ant moves on, the leading edge of the ellipse follows the ant and the trailing edge finally diffuses away enough odorant that all points near it have concentrations less than K, as shown in Figure 10.11.

Not surprisingly, the length of the active space, L, again depends on Q/K as well as D. Also not surprisingly, the location of the maximum radius of the active space occurs $0.37L$ along the active space axis from the point near the ant. The maximum diameter of the active space at this location and the time it takes before the trail at any given location has totally dropped below K can

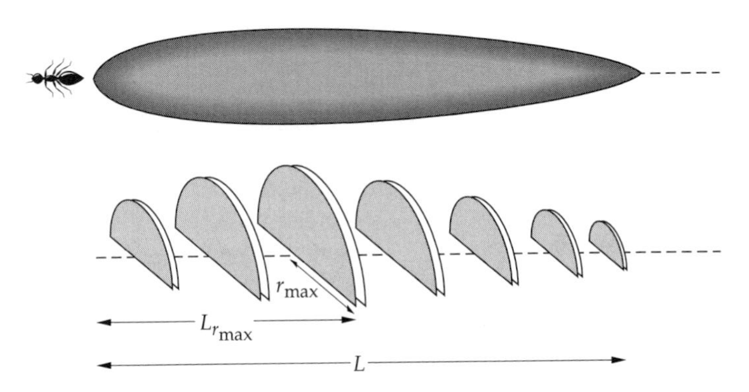

Figure 10.11 Active space of an ant trail. A trail is essentially a sequential series of small single puffs along a linear transect. L is the length of the active space, r_{max} is the width of the active space, and $L_{r_{max}}$ is the distance of the maximum radius point from the ant.

also be calculated. The length of the trail does not depend on how fast the ant walks, whereas the thickness of the active space does. Also, the thickness of the active space (like a single puff) does not depend on D. This gives an ant several degrees of freedom to control the width, length, and longevity of a trail independently.

Continuous Emission into Current Flows

An animal that emits a continuous stream of odorant for the purposes of long-distance communication is highly likely to encounter the effects of some type of current flow in the environment. Intuitively, current flow increases both the speed and maximum transmission distance of a chemical signal, but only in the down-current direction. Although diffusion processes may still be operating, they will be largely overwhelmed by the mass transport of the chemical molecules by the current. The diffusion properties of the chemical therefore become less important, and the flow dynamics of the medium dominate the transmission process. In principle, it should be possible to model the spread of molecules in a current flow if one knows the velocity of the flow. In practice, modeling this process is extremely difficult because most flows are turbulent.

When a medium flows in an orderly fashion, with the same velocity and direction of flow throughout the region of interest, it is called **laminar flow** (Figure 10.12A). When a medium flows in an irregular fashion, it is called **tur-**

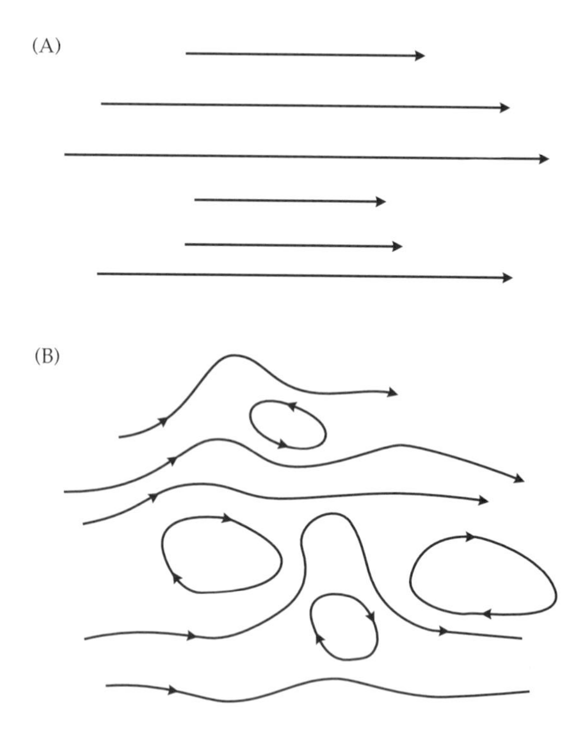

Figure 10.12 Types of current flow. (A) Laminar flow, and (B) turbulent flow. (From Okubo 1980, © Springer-Verlag.)

bulent flow (Figure 10.12B). Turbulence is created when the flow is hindered in one location relative to other adjacent locations. This differential velocity generates torque on the medium, which then generates vortices (whirls), eddies (backflow and partial whirls) and waves. The pattern of vortices and whirls is extremely complicated, with the large vortices that are initially created producing smaller vortices, which in turn produce even smaller vortices. The energy contained in the larger vortices is continually transferred to the smaller ones, and small vortices are eventually dissipated as heat. The pattern is also continually changing in time and is highly unpredictable. The most important consequence of turbulence from our perspective is the rapid mixing and irregular transport of chemical odorants in nature.

A continuous input of force, or energy, is necessary to maintain a flow. In both the atmosphere and ocean, the main physical forces that cause flows are: (a) large-scale temperature differences, which cause density differences in the medium and buoyancy differentials that result in vertical cycles of air and water as well as horizontal mass movement, and (b) Coriolis forces from the rotation of Earth that result in air and ocean circulation patterns. Normal wind speeds (with the exception of major storms) range from 1 to 10 m/sec; normal current flows in the ocean are slightly lower.

Three combined factors determine whether a flow will develop turbulence or remain laminar: the velocity of the flow, the density of the medium, and the distance scale over which one is observing the flow. The action of these factors is described more fully in Box 10.2. For our purposes, it is sufficient to understand that turbulent flow is more likely at higher velocities, in denser media (i.e., is greater in water compared to air) and over longer distances. Laminar flow is more likely with slow flow velocity, in less dense media such as air, and over shorter distances.

If the flow is laminar, then the spread of a continuously emitted chemical can be predicted with accuracy. Here we can assume that normal diffusion leads to the spherical spreading of the odorant as before, but the center of the active space moves with the flow. The active space is ovoid shaped as shown in Figure 10.13A. Along the downstream axis the concentration falls off as $1/r$ and is independent of velocity. This means that the maximum distance at which the signal can be detected directly downstream is the same as the r_{max} without any flow (Figure 10.13B). The only effects of current flow are to narrow the active space of the odorant and to increase the transmission speed.

Since most signaling situations do not meet the laminar flow condition, the above model does not have wide applicability. Early attempts to model the spread of an odorant with turbulent flow used the basic mathematics of diffusion that we have seen, but redefined D as a coefficient of eddy-diffusion (Roberts 1923; Sutton 1953; Bossert and Wilson 1963). Empirical measurements were made of the typical rates of spread of smoke along the horizontal, vertical, and cross-sectional width dimensions in different wind speeds and these **diffusivity coefficients** were incorporated into the standard diffusion equations. Plots of the active space for such an odor plume are

Box 10.2 *Laminar versus Turbulent Flow*

SEVERAL FACTORS ARE IMPORTANT in determining whether or not turbulence will develop. The most important of these is the average **velocity of the flow**. At very low velocities, flow is laminar, and at high velocities it is turbulent. A second factor is the **kinematic viscosity** of the medium. Kinematic viscosity is the ratio of viscosity to density of the medium and is similar to the diffusability of molecules in the medium. Turbulence develops more easily in a medium with a low kinematic viscosity. Because water has a lower kinematic viscosity than air, turbulence will develop at lower flow velocities compared to air. Thirdly, the **distance scale** over which one measures flow affects the appearance of turbulence. Over very small distances flow may be effectively laminar, whereas turbulence is more apparent over larger distances because there is a lower limit on the size of vortices. Below this critical size, the energy in the smallest vortices is dissipated as heat, so flow is laminar and sideways spreading of molecules only occurs via diffusion. In air flowing at 1 m/sec, the lower critical vortex size is 3 cm. This is why the smoke rising from a burning cigarette initially rises in a single thin wisp until diffusion has spread the column to the width of the critical vortex size. The figure below (From Dusenbery 1992, © W. H. Freeman), illustrates the combined effects of flow velocity, distance, and kinematic viscosity on the likelihood of turbulence development. Finally, environmental factors such as the presence of temperature gradients, the nearness of a surface boundary, and the irregularity of the surface also affect the turbulence of flow. All of these factors must be taken into account in any attempt to model the spread of a chemical odorant in a current.

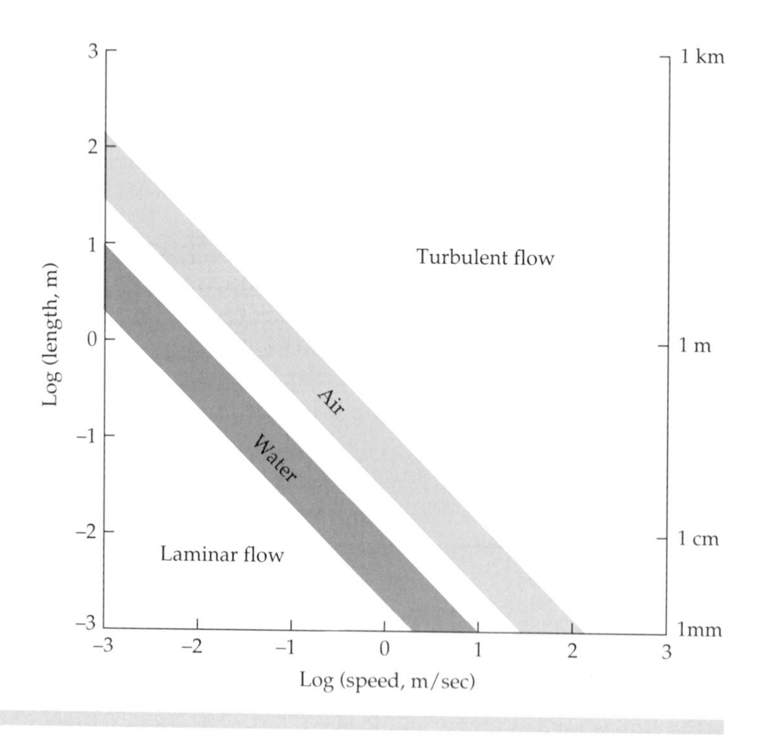

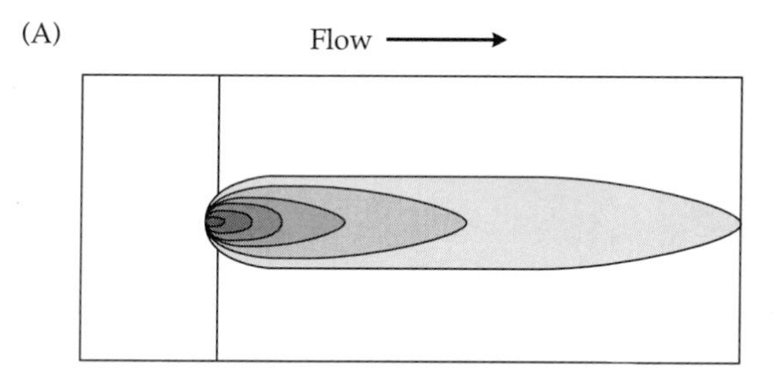

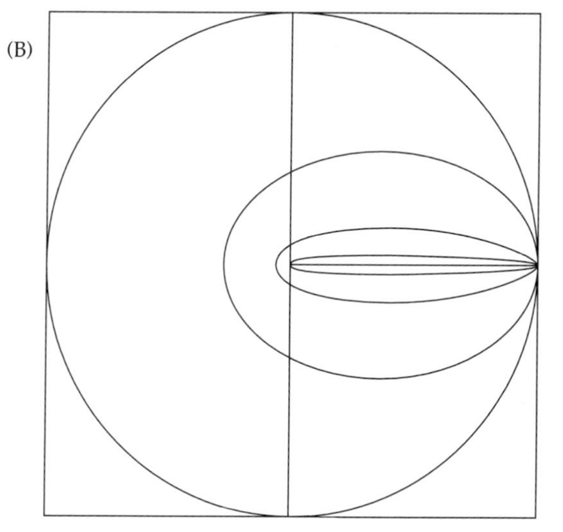

Figure 10.13 Active space for a constant source in a laminar flow field. Both graphs show a continuous source emitting at a constant rate from a point on the thin vertical line. (A) shows contours of equal intensity, with the contours differing by a factor of two and higher concentrations indicated with darker shading. (B) shows a single intensity contour for different flow speeds. The sphere represents the active space with no flow, and increasing flow narrows the space but does not increase r_{max}. (From Dusenbery 1992, © W. H. Freeman.)

shown in Figure 10.14 for three different wind speeds. In a mild wind, a large detection distance can be achieved in the downwind direction. However, if the wind is strong, it not only distorts the active space along one axis, it "whisks" molecules away faster from the surface of the active space than would simple diffusion alone. This causes the effective active space boundary to shrink in closer to the source than would be found without wind. Strong wind can thus decrease the effective range of a continuous emitter.

How well do these models estimate transmission distances of real biological odorants? We can compare the results of the models to field trials with moth sex attractants. A typical emission rate by a female is $Q = 1 \times 10^{10}$ molecules/sec and a typical male threshold detection concentration is 1×10^4 molecules/cm^3. The laminar flow model predicts a maximum range of detection of 4 km and a maximum width of 34 cm (independent of wind speed). Wilson and Bossert computed a maximum transmission distance of from 3 to 16 km. Field results suggest an actual range of 80 to 100 m. Real plumes appear to be cone-shaped, with edges about 20° to each side of the downwind axis. Field data, therefore, suggest that the active space in a turbulent current expands out to a greater width but is shorter than the models predict.

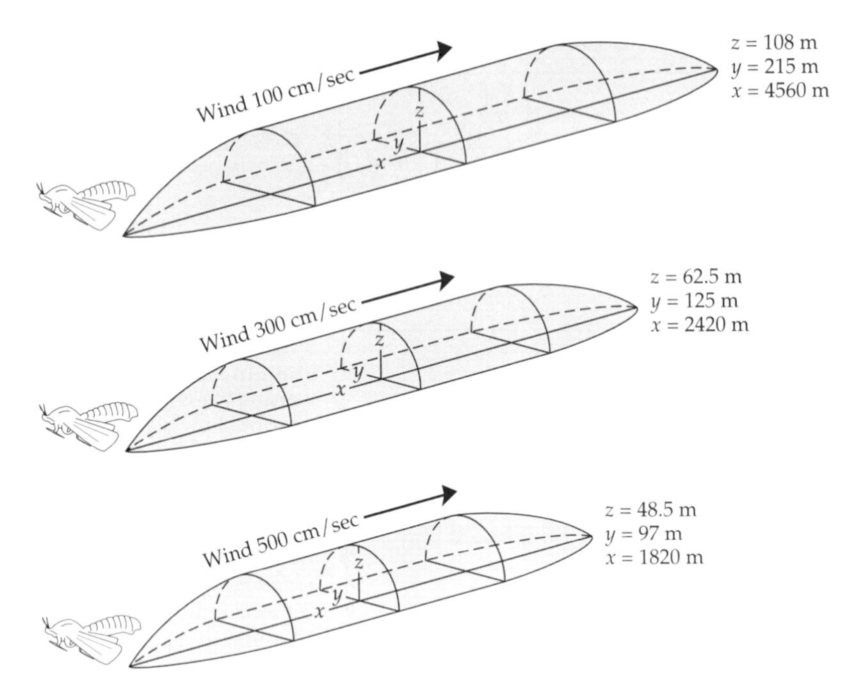

Figure 10.14 Active space of an olfactory signal in wind of different velocities. This model assumes that flow is turbulent but uses the basic diffusion equations, in which the diffusion constant is interpreted as a coefficient of eddy-diffusion, to describe the spread of the odorant. Maximum values of the dimensions are given. Transmission distances are greatest for low wind speeds and decrease with greater wind speeds. (From Bossert and Wilson 1963, © Academic Press.)

The problem with both the laminar flow model and the redefined diffusivity model described above is that they assume an idealized plume of odorant with a smooth concentration gradient as illustrated in Figure 10.13A. In fact, a more realistic vision of the distribution of odorant in the flow is an undulating filamentary plume structure (Figure 10.15). At any point in the flow, the concentration of the chemical fluctuates strongly. Measurements of real emissions in the field (using ions as tracers because they can be rapidly detected with sensors) verify the spiky, patchy nature of airborne signals (Murlis and Jones 1981). Similar results have been found in aquatic systems. In addition, ground surface roughness and variability in wind direction have enormous effects on transmission. Given this nonuniform distribution of odorant, how can animals possibly detect an odor gradient and locate the position of a long-distance olfactory signaler? We take this up later in the section on reception of chemical signals.

Transmission of Deposited Scent Marks

Scent marks that are deposited on a target individual or object in the environment are designed to last a long time in the absence of the sender. Deposited

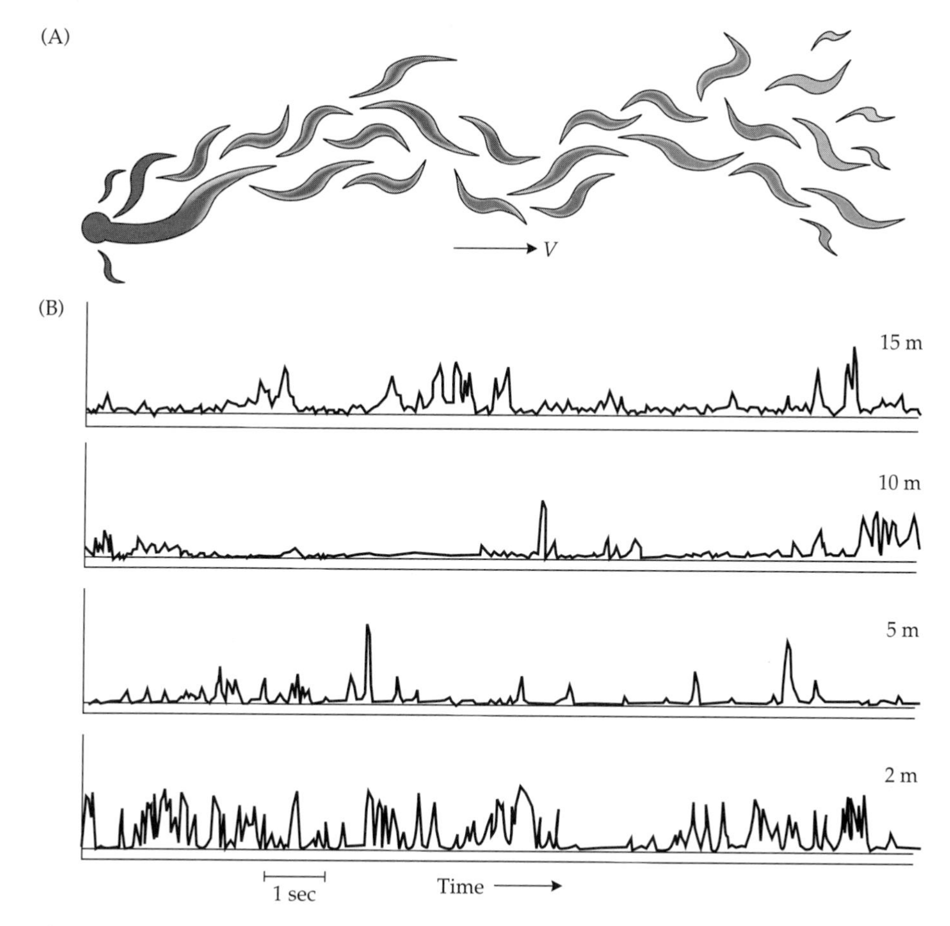

Figure 10.15 Distribution of odorant in a turbulent current flow. (A) A schematic drawing of an odor plume in a turbulent flow, showing both the filamentary nature of odor patches as well as the undulating and meandering path of the whole plume. Compare to Figure 10.13A. (B) Fluctuations in a real chemical plume. Continuous emission of an ion source was used to show fluctuations at 2, 5, 10 and 15 meters downwind from the source. Data were collected in an open field on a cloudy day at wind speeds of 5 m/sec with no vertical temperature gradient. (B from Murlis and Jones 1981.)

marks are similar to the single-puff case described above, in that a fixed amount of odorant is deposited at a point in time. The objective, however, is to maximize fadeout time rather than the active space. Recall that fadeout time is a function of $(Q/K)^{2/3}/D$. In addition to increasing the amount of odorant deposited, the sender's best strategy for maximizing fadeout time is to decrease D, the diffusion rate. Recall also that r_{max} is unaffected by D, but the increase in signal duration means that diffusion conditions no longer hold, so r_{max} is likely to be very small. If the active chemical is so large and nonvolatile that it can only be detected by contact, then the active space is effectively reduced to zero. For deposited marks, therefore, the chemical composition of the secretion, the characteristics of the surface on which it is deposited, and environmental conditions greatly affect the emissivity and longevity of the signal.

Deposited marks are often sebaceous gland products in which the active pheromone is embedded in a matrix of other compounds. The nonactive components of these secretions are anything but inert and play an important role in regulating emission. Oily, lipid-rich sebum and squalene from the glands serve as carriers for the more volatile active compounds. The more polar the carrier compound, the slower the release of the active molecules. The functional group of the pheromone affects its interaction with the carrier by determining the strength of hydrogen bonds. For example, a nonpolar hydrocarbon molecule is less likely to form hydrogen bonds with the carrier than a carboxylic acid and therefore evaporates faster than a polar hydrocarbon. Carrier compounds also buffer the active component against effects indigenous to the surface substrate and thus ensure signal emission over a prolonged period of time.

The physical and chemical characteristics of the substrate such as its surface area, porosity, and chemical polarity all affect signal emissivity and longevity (Regnier and Goodwin 1977). The surface area determines the availability of substrate binding sites for the pheromone. Porosity affects chemical release rates; porous surfaces such as clay greatly slow the release rate compared to wood surfaces. Most natural surfaces are electrically charged because they are composed of polar materials; most organic pheromone molecules are also polar, so they tend to adhere to natural surfaces, thereby reducing the emission rate.

The environmental factors affecting the longevity of a deposited signal are humidity, temperature, wind, and sun. High temperatures and wind increase evaporation and diffusion rates, causing chemicals to spread faster and fade sooner. Direct sun exposure causes chemical decomposition, so long-lasting marks need to be highly stable chemically to withstand ultraviolet radiation. Carrier compounds protect the active pheromone from degradation by exposure to wind and sun. High humidity and rain cause the most significant environmental challenge to scent mark longevity. Brief exposure of a scent mark to even intermediate levels of humidity can cause rapid release of the active pheromone. The reason for this volatility is that the strongly polar water molecules compete for the hydrogen binding sites in the carrier compound. Volatilization of odorants that are hydrogen bonded to the carrier can only occur when these bonds are disrupted. The rapid release of odorant in the presence of water may actually represent a design strategy by chemical senders to maximize mark longevity and ensure pheromone delivery only when a receiver is present. When a receiver licks, exhales, or otherwise deposits water vapor on a marked surface, a puff of active pheromone is instantly released. The rapid degradation of deposited marks by environmental humidity and rainfall may therefore be an unavoidable consequence of this otherwise useful sender strategy (Alberts 1992).

RECEPTION OF CHEMICAL SIGNALS

Chemical receptors are the easiest type of receptor to evolve because chemical stimuli participate directly in biochemical reactions without requiring a sensory transduction step. This direct link explains the universality of chemical communication and its presence in the most primitive organisms. Simple and advanced chemical receptors, whether sensitive to airborne, waterborne, or directly contacted chemicals, share many features in common. We shall first review current knowledge of the basic cellular processes that occur during chemoreception. In parallel with previous discussions of other sensory receptors, the characteristics of an ideal chemoreceptor and the inherent tradeoffs among these various features will be outlined. We will then examine the receiving organs found in vertebrates and invertebrates and discuss the strategies used to tune the receptors to the needs and capabilities of the animals. Finally, since detection of specific kinds of chemicals usually causes animals to move toward or away from the source, we shall examine the mechanisms by which animals orient in a chemical concentration gradient.

General Characteristics of Chemosensory Receptors

Regardless of whether the receptor is a single-celled organism or a cell in a sensory organ, the same basic mechanism is used to detect chemical stimuli (Vogt and Riddiford 1986; Caprio 1988; Bruch et al. 1988). Each stimulatory molecule binds to a specific protein receptor located on the cell membrane. The receptor protein either changes shape or changes chemically under the influence of the stimulatory chemical. These changes are then coupled to subsequent biochemical processes within the cell. The stimulatory molecule is usually then released from the receptor so that the receptor can bind with another stimulus molecule. The intensity of the response increases as the concentration of the chemical stimulus and the number of molecules bound to the cell increases. In more advanced **chemosensory** organs, the receptor proteins are located in the membrane of specialized sensory cells. The chemical or shape changes in the receptor protein caused by the binding of the stimulus molecule cause the cell to depolarize; secondary messenger compounds in the cell may mediate this transduction. Depolarization affects the conductance across the cell membrane and generates a nerve impulse. Buck (1996) provides a recent review of these processes.

Chemosensory cells are either nerve cells themselves or are closely bound to nerves. As with other sensory systems, they are derived from either ciliary cells or microvillous cells (Vinnikov 1975). The cilia or the microvilli extend out from the cell and make direct contact with the external environment; they provide a large surface area of membrane where binding takes place. The occurrence of either ciliary or microvillous cells in chemosensory organs follows broad taxonomic patterns—the olfactory organs of vertebrates contain mostly ciliary cells, while the taste and other contact receptors contain microvillous cells. In insects, both olfactory and taste receptors are ciliary (Moran 1987). However, many receptor organs contain both types of cells, sometimes located in distinct parts of the organ. Future research will undoubtedly show that the two types of cells respond somewhat differently and serve slightly different functions (Bruch et al. 1988).

The organization of the nerve connections to the brain is basically similar for all chemoreceptor organs. Stimuli from the primary receptors are transmitted along axons to a staging area, usually a ganglion or peripheral region of the brain. Nerve networks in the ganglion or bulb gather the combined responses of the primary receptor cells and serve to integrate the information in a fashion similar to the retina of the eye. Ganglia often contain distinct layers with different types of cells, some of which summate the input from the primary receptors and some of which provide cross connections between nerves similar to the horizontal and amacrine cells of the retina. A smaller number of ganglion nerves then transmit the partially coded information to specialized centers in the brain for further processing (Hildebrand and Montague 1986; Ache 1988; Buck 1996).

A final general characteristic of chemosensory receptor cells is their short life and continuous replacement. The life of a chemoreceptor cell ranges from a few days to a few weeks. New cells are constantly developing to replace old cells. Other types of nerve cells in the body do not show such replacement. One possible reason for the turnover is that the membranes of chemoreceptors become worn out after a period of time (Laverack 1988). Chemoreceptive cells are the only sensory receptors that are directly exposed to the environment, and the sensitive receptor membranes may become damaged or plugged up by environmental chemical noise after a period of time. Alternatively, cell turnover may provide a mechanism for changing the proportion of sensory cells sensitive to different types of chemical stimuli in different environmental contexts (Kauer 1991).

The Ideal Chemoreceptor Organ

The ideal chemoreceptor should be responsive to a broad range of different chemicals and sensitive to low concentrations of odorants. These two features, quality and quantity, are difficult to maximize simultaneously within the constraints of organ size and tend to be traded off. Let's look at how chemical range and sensitivity can each be maximized before examining what real chemoreceptors do.

Given that there are thousands of possible chemicals in the environment that an animal might need to distinguish, and that these chemicals cannot be arrayed along any single dimension, how should chemoreceptor systems be designed to maximize discriminability among a large number of chemicals? There are two basic strategies in principle (Erickson 1978). One is to make individual receptor cells narrowly responsive to single chemicals only, and generate many different classes of cells, each tuned to a different chemical. The identity of a specific chemical stimulus is made on the basis of the activation of those cells specifically sensitive to that stimulus. This is called a **labeled-line** coding system. Such a system has the advantage of high sensitivity and easy detection of a specific chemical in a noisy chemical background, but the integration of information from many such cells at higher levels may be very inefficient. The alternative is to make the receptor cells more broadly responsive to a range of chemicals, with fewer numbers of cell classes exhibiting overlapping response profiles. The identity of a particular stimulus is encoded by a distinct pattern of activity across a population of cells (Kafka 1987). This type of system is called **across-neuron pattern** coding. Such a system is less sensitive, because multiple odorant molecules must contact the different receptor types to produce the recognition pattern, but it is potentially responsive to an infinite range of chemical stimuli. Color vision, with three or four receptors sensitive to different but overlapping frequencies of light, is an example of across-neuron coding, whereas sound frequency perception is more similar to labeled-line coding.

Real chemoreceptors tend to maximize either sensitivity or chemical range, depending on the function of the receiver organ and communication task. Where extreme concentration sensitivity is required, a labeled-line type of cell specificity is used and the number of chemically-specific cell classes is reduced to a very small number of critical chemicals. This system is characteristic of sex attractant pheromones, and in some cases of specific food components and perhaps alarm and predator detection chemicals (Derby and Atema 1988). Such narrowly tuned receptors are chemically blind to all other chemicals, but a single molecule of the tuned chemical can cause a behavioral response. For most food detection systems, more broadly-tuned receptors are found and concentration sensitivity is sacrificed. Generalized chemosensory organs contain a large number of receptor types, each tuned to a different aspect of molecular structure, e.g., chain length, shape, double-bond position, or presence of a functional unit. A single receptor cell will respond to many chemical stimulants that share this molecular feature, but some stimulants may fit better than others and cause a stronger neuronal output. Depending on its suite of molecular features, a single odorant will produce responses in several receptor types. Relative responses from the different receptor types are then combined at higher levels and eventually recognized as a certain odor by the nerve population pattern. This system can not only recognize a large variety of different chemicals, but it can also perceive and learn to categorize novel chemicals not previously encountered (Sullivan et al. 1995).

Arthropod Chemosensory Systems

Most arthropods possess two distinct chemosensory organs, the antennae and the feet or mouth appendages. Contact taste receptors are located on the tips of the feet or mouth appendages, since these are the structures that manipulate the animal's food. They have little or nothing to do with social communication.

The paired antennae are often simple whip-like flagella extending forward from the head, but some species possess very complex antennae, as illustrated in Figure 10.16. Males sometimes have larger antennae than females. The antennae are primarily olfactory sensors for airborne or waterborne odorants, but they may also possess additional sensors for taste, touch, air or water current movement, sound, and temperature. Each type of function is

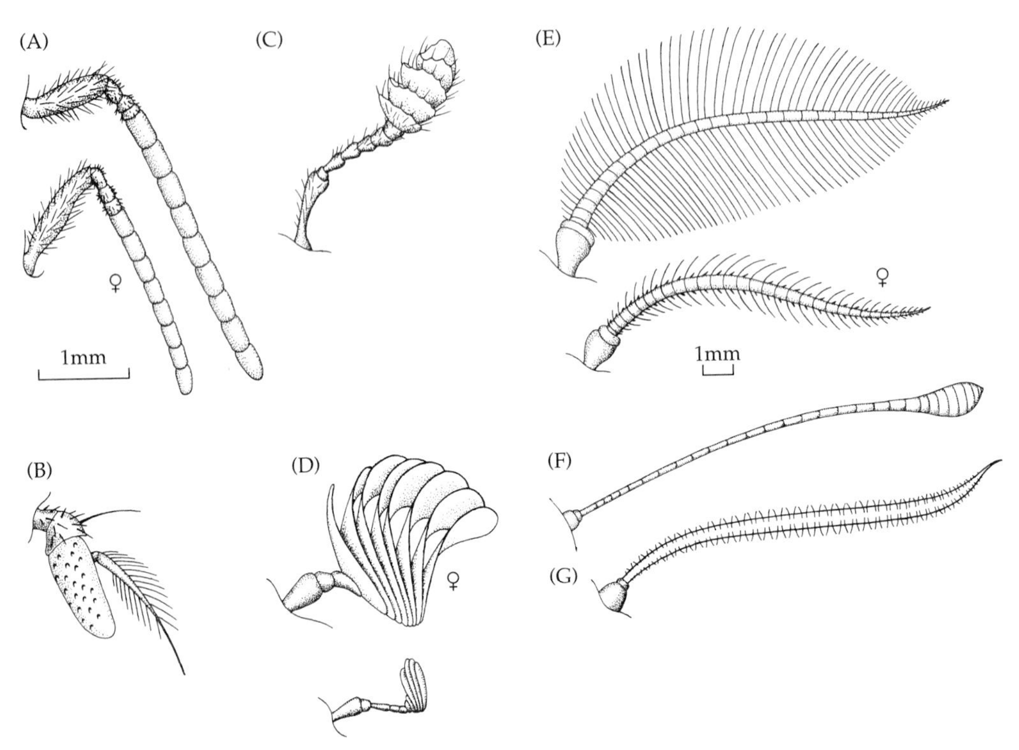

Figure 10.16 Insect antennae. (A) Male and female honey bee (*Apis mellifera*), (B) flesh fly (*Sarcophaga*), (C) carrion beetle (*Necrophorus*), (D) male and female scarabid beetle (*Rhopaea*), (E) male and female saturneed moth (*Antheraea*), (F) butterfly (*Vanessa*), and (G) sphingid hawk-moth (*Pergesa*). Common scale (1 mm) for A–D and E–G. (After Kaissling 1971.)

subserved by a distinctively structured hair, pit, or peg containing sensory cells, all of ciliary origin. The general term for these sensory structures is **sensillum** (plural sensilla). The olfactory sensilla are hairlike projections containing one to three chemosensory cells, as illustrated in Figure 10.17 (Schneider 1964; Kaissling 1971).

The large featherlike antennae of many male moths are devoted to detecting the airborne sex attractant of the female. They are estimated to intercept 80% of the molecules that pass them. The female pheromone contains between one and six chemically similar components. The pheromone-specific sensilla of the male contain identifiable cells that respond selectively to just one or a specific mixture of these components. The sensilla are highly sensitive (i.e., *K*, the threshold of detection, is very small). In some cases, a response can be produced with a single pheromone molecule. Some of the components appear to function in long distance attraction of the male while others are involved in short-range communication between the male and female (Kaissling 1979; van der Pers and Löfstedt 1986; O'Connell and Grant 1987). In species with simpler and less dimorphic antennae, contact chemore-

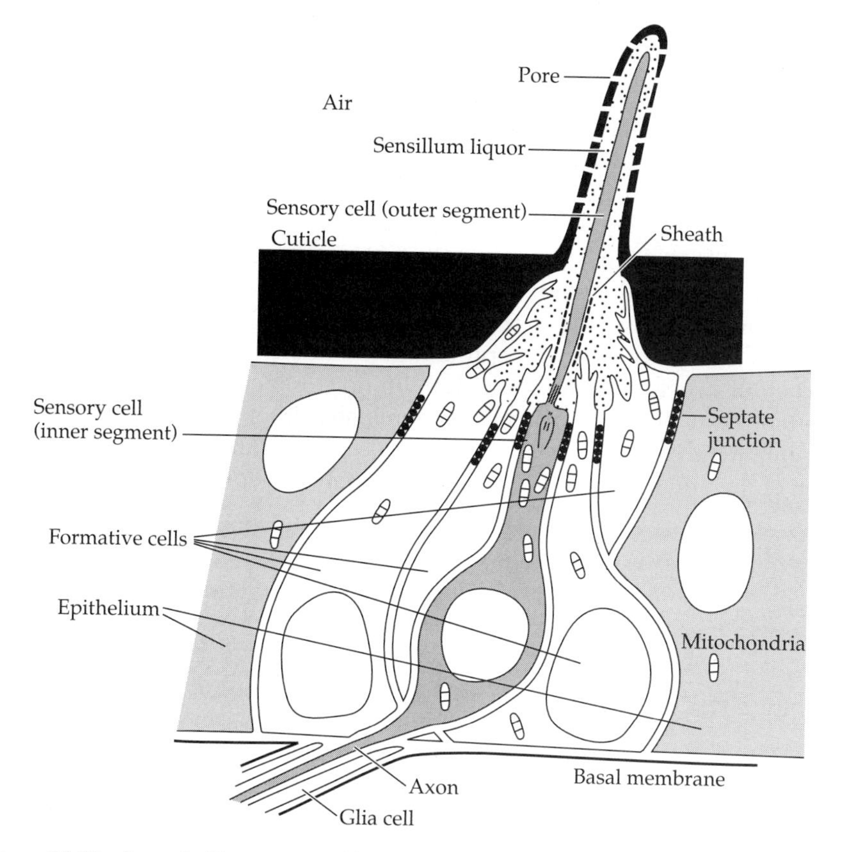

Figure 10.17 Insect olfactory sensillum. An olfactory sensillum looks like a simple hair, but the surface of the hair possesses pores to permit the passage of olfactory molecules into the interior. Inside, the outer segments of from one to three sensory cells are bathed in a liquid medium. The sensory cells are directly connected to the brain. (After Kaissling 1971.)

ceptors are an important subset of the sensilla. Animals touch each other with the antennae during encounters (called **antennation**). Information about species, sex, and individual identity may be obtained from the reception of nonvolatile compounds on the exoskeleton.

Vertebrate Chemosensory Systems

Essentially all vertebrates possess the same two basic chemosensory organs: an olfactory sensory organ for the detection of distant medium-borne signals and environmental cues, and a contact sensory organ involved in the discrimination of palatability of food items. Some vertebrates have additional chemosensory organs as well.

The olfactory organ is located at the most anterior projection of the head. In fish the organ is independent of the respiratory system. It consists of a chamber located in front of the eyes with a separate inlet (the **nares**) and outlet (Figure 10.18A). The chamber contains lamellae covered by a thin sensory epithelium to generate a large, convoluted surface area for reception of chemical odorants. In terrestrial air-breathing vertebrates (amphibians, reptiles, birds, and mammals) the organ is an integral part of the respiratory system (Figure 10.18B). The inlet is a separate duct that serves to clean and humidify the air before the sensory epithelium is subjected to it. The outlet is the lungs. Most of the air entering the nares flows straight to the lungs, but a portion is diverted into a side pocket containing the olfactory epithelium. The animal can control the fraction of air diverted to the olfactory system. The action of sniffing increases air flow to the olfactory epithelium (Stoddart 1980; Caprio 1988).

The cell structure of the epithelium lining in olfactory organs is similar in all vertebrates. Ciliary sensory cells predominate, although fish olfactory organs also contain microvillous cells. The sensory cells send their projections into the lumen of the organ (Figure 10.19). The surface of the epithelium is bathed in a

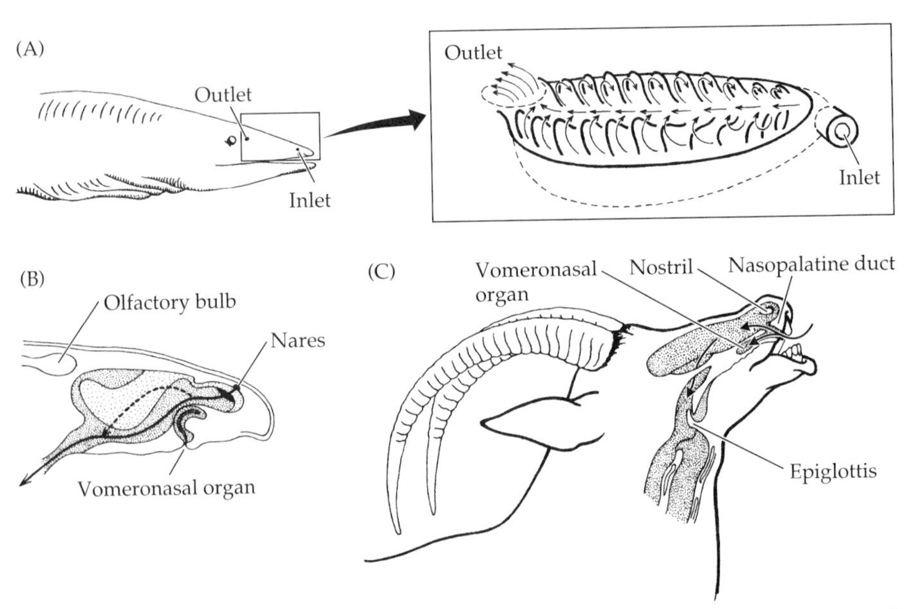

Figure 10.18 Chemosensory organs of vertebrates. All illustrations show a sagittal section through the head. (A) The eel *Anguilla anguilla* has rudimentary eyesight and relies on olfaction for navigation and communication; the inset shows the internal structure of the olfactory capsule and the water flow pattern over the lamellae. (B) The lizard *Lacerta viridis* illustrates the typical terrestrial vertebrate pattern of primary air flow from the nares to the lungs (solid line) and partial air flow over the olfactory epithelium (dashed line). Also shown is the vomeronasal organ in the roof of the mouth, which in lizards and snakes is stimulated by chemicals picked up by the forked tongue. (C) Flehmen in mammals such as this sable antelope (*Hippotragus niger*) is a conspicuous behavior that serves to stimulate the vomeronasal organ. The upper lip is curled back to close the nostrils and the head is raised to close the epiglottis so that inspired air is drawn into the vomeronasal organ. (After Estes 1972; Stoddart 1980.)

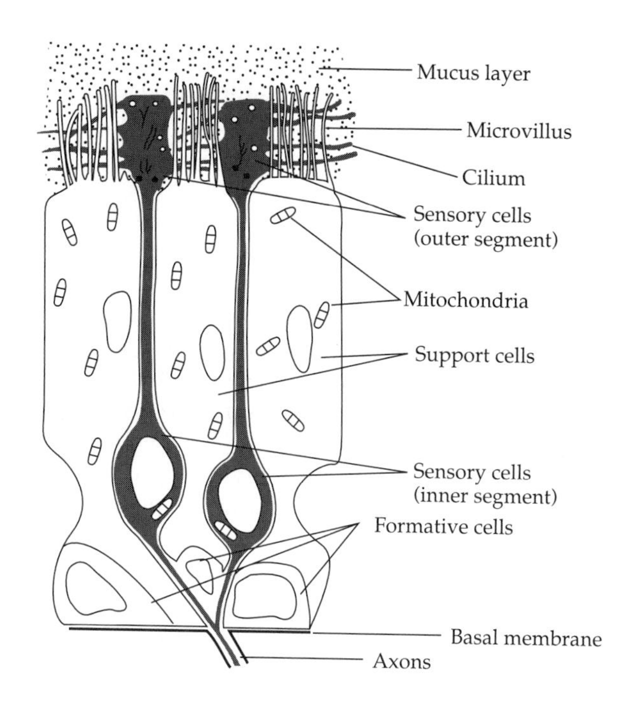

Figure 10.19 Epithelial lining of the vertebrate olfactory organ. The olfactory sensory cells, shown in dark gray, are bipolar nerve cells with cilia on the reception end and axons that project into the olfactory bulb on the other end. The receptor sites are located on the distal cilia projections. The cilia are embedded in a mucus layer. The sensory cells are sandwiched between support cells shown in white. Undifferentiated basal cells that will become new sensory cells are formed from the basal membrane. (From Stoddart 1980.)

mucus produced by secretory cells that serves to protect the sensitive membrane receptor sites on the projections and contains the ions required for sensory cell depolarization. Each sensor cell responds best to a restricted range of chemicals, but the organ as a whole responds to a wide variety of chemicals.

Contact sensory organs consists of an array of pores located around the edge of the mouth in fish and on the tongue of air-breathing vertebrates. Contact receptors such as the taste bud contain a cluster of axonless neuroepithelial cells of microvillar origin. These cells form synapses with the tips of one of the cranial nerves (glossopharyngeal, facial, or vagal nerve) that terminate in the medulla of the brain. Contact receptors in general are more narrowly tuned either to certain components of a species' diet or to its specific pheromones (Finger 1988).

Fish possess a third system called the **common chemical sense**; this system is particularly well developed in sea robins and gurnards (*Prionotus*). It is comprised of isolated chemosensory cells located throughout the body surface or on specialized fin rays and innervated by cutaneous sensory nerves (trigimenal or spinal nerves). This system is involved in predator avoidance responses. The sensors may therefore be highly specific receptor cells tuned to the olfactory products of predators or to conspecific alarm pheromones (Finger 1988).

In addition to the nose and tongue, many reptiles and mammals possess an additional chemosensory organ called the **vomeronasal organ**. It is conspicuously absent in fish, birds, Old World primates, apes, aquatic mammals, some bats, and crocodiles. The organ consists of a single-opening tube lying between the nasal cavity and the roof of the mouth (Figure 10.18B, C). The epithelial sensory cells lining the lumen of the organ are bipolar nerve cells stacked between support cells much like the olfactory epithelium. However, the receptor site projections into the mucus layer are short, stiff microvilli rather than cilia and, therefore, more similar to the contact taste receptors in the mouth than to the olfactory receptors of the nasal cavity. The organ is innervated by its own nerve tract, which projects to the accessory olfactory bulb, a structure distinct from the main olfactory bulb. The accessory system then projects to a region of the brain where connections to the hypothalamic-pituitary axis are made. The vomeronasal system is a contact chemoreceptor stimulated by nonvolatile chemicals such as proteins in the urine and secretions of conspecifics. Stimulation of the vomeronasal system affects and is strongly affected by steroid and neuroendocrine hormones involved in reproduction (Halpern 1987).

Much of the intraspecific chemical communication involved in mate attraction, courtship, copulation, aggression, and parental care in mammals is mediated via the vomeronasal organ. The mammalian organ is a blind, cigar-shaped tube connected to the nasopalatine duct or canal and thus is accessed through the mouth, nose, or both. The receiver must lick or touch its nose to the urine or secretions in order to stimulate the organ. After contacting such sources, many mammals perform a behavior called **flehmen** in which the head is raised and the upper lip retracted (Figure 10.18C). This behavior is believed to be a mechanism for delivery of nonvolatile odorants to the vomeronasal organ. Nose-touching between two individuals may also serve to transmit contact chemical signals. In reptiles the vomeronasal organ is a hemispheric structure opening into the roof of the mouth. Access of chemicals is mediated by the tongue, which is flicked in the air or touched to a substrate and then pushed into the organ. Snakes are particularly dependent on vomeronasal stimulation for both normal sexual behavior and for tracking the olfactory trail of prey.

Chemical Gradient Detection and Orientation

When an organism has detected a chemical stimulus, how does it use the information from the stimulus to guide itself either toward or away from the stimulus source? In all four of the signal modalities discussed in Part I, an intensity gradient exists that is strongest at the source and declines with increasing distance from the source. The spatial distribution of signal intensity within the stimulus field therefore provides important information about the location of the sender. Sound, light, and electric signals differ from olfactory signals in one critical respect—their stimulus fields contains additional intrinsic directional components that can potentially also be used to obtain directional information. Olfactory signals lack any additional directional components besides the intensity gradient, although as we shall see, there are sometimes special features associated with olfactory signals that help to provide spatial information.

When only intensity gradient information is available, animals can move toward (or away from) the source by employing several different mechanisms. The simplest mechanism is to couple movement and turning patterns to different concentration levels, called **kinesis**. For example, a frequent turning or circling pattern will cause an animal to remain in an area of desired concentration, while straight line movement will result in leaving a location of undesired concentration. Although the desired goal is eventually met, this indirect method of orientation is inefficient and requires little sensory sophistication. **Taxis**, on the other hand, is a more direct method of orientation involving detection of a gradient and oriented movement with respect to the gradient. Detection of the direction of a gradient can be achieved in two different ways, with simultaneous or sequential sampling, as described below. These basic principles of intensity gradient orientation apply to all modalities (Fraenkel and Gunn 1961; Kennedy 1986).

In **simultaneous sampling**, or stereolfaction, stimulus inputs from two receptors on each side of the body are simultaneously taken and compared. This gradient detection mechanism requires that paired nostrils or receptors be placed sufficiently far apart. Larger animals and those with wide heads can accomplish this easier than animals with small or narrow heads. Numerous adaptations have evolved to increase the distance between receptors or ensure that air/water is drawn from the side rather than from the front of the animal (von Bekesy 1964; Stoddart 1979). Locating receptors on long antennae provides even greater separation. In the case of attraction to a stimulus source, if the receptors are stimulated unequally the animal turns in the direction of the one more strongly stimulated. Once both receptors are receiving equal stimulation, the animal moves in a straight line and thus directly approaches the source. One way to determine whether this method is being used is to remove, deactivate or stimulate just one of the two receptors and place the animal in a uniform stimulus field. This should produce continuous turning to one side.

In **sequential sampling**, two estimates of intensity are made at adjacent locations. Since the animal must move between locations, it is essentially measuring a temporal gradient, then inferring the spatial gradient from information about how it moved between samples. Comparison of samples from two different points in time requires either a mobile neck with back and forth movement of the head or movement of the entire body. Worm-like organisms move their heads in an undulatory manner to either side of their direction of locomotion. A difference in stimulus intensity between the two sides will cause the animal to turn toward the side with the greater intensity. One way to experimentally test for the presence of this method is to pulse the signal at a rate similar to the rate of head movement, which should cause consistent turning in one direction. Comparisons of two subsequent samples requires an accurate means of detecting changes in odorant concentration. Insects possess special neurological devices to accomplish this, consisting of interneurons in the olfactory bulb that flip-flop in their response whenever there is a change in the spike rate of a receptor cell (Ache 1988). Given equal receptor sensitivity, this method of orientation requires a steeper concentration gradient than the simultaneous sampling method because sampling time is shorter.

Orientation in a chemical plume carried by a current from a stationary source usually requires an altogether different strategy because of the turbulent and intermittent nature of the stimulus (Figure 10.15). Although it was once believed that animals could average successive samples and determine the approximate direction of the concentration gradient, we now know that the discontinuities in concentration are too severe for this to be a viable strategy. Since the signal is carried by a current, the source must be upstream. If the animal can couple reception of the olfactory signal with detection of the direction of the current, then the direction of the source can be determined. An animal walking on a substrate can easily detect the direction of a current by the forces it exerts on its body and sensors, much like the particle detectors for near-field sound described in Chapter 6. A flying or swimming animal, however, also experiences the forces resulting from its own movement, and so must integrate the visual information of its flight speed and wind drift relative to the substrate or external references to estimate wind direction (David 1986). Male moths attempting to locate females dispersing odorant into the wind use this sampling method. When the male detects the odorant, he flies upwind. If he loses the scent, he flies across the wind current until he catches the scent again and turns upwind (David et al. 1983). A typical male flight trajectory is shown in Figure 10.20.

Figure 10.20 Flight path of a male moth in a turbulent odor field. Under normal field conditions, female gypsy moth (*Bombyx*) pheromone was continuously emitted from a source position marked by the star. Soap bubbles were simultaneously produced to follow the position of the plume. The tracing shows the flight path of a male moth released approximately 12 m downwind. Arrows indicate the direction of the wind. When the moth was within the odor plume (indicated by thick lines) he flew upwind with shallow zigzags. When he was outside the plume (indicated by thin lines), he flew back and forth across the wind direction until he located the plume again. (From Atema 1988, © Springer-Verlag.)

Following a chemical trail also requires special strategies. Since a trail is established by the movement of the source rather than the movement of the medium, there is no simple secondary cue that can be used to determine the directionality of the trail, as we saw for plumes in currents discussed above. Merely staying on the trail is a significant challenge. Dogs are well known for their ability to follow scent trails. They accomplish this by wandering back and forth across the trail to locate its edges. Eventually, they are able to determine the direction of movement of the sender by detecting the longitudinal intensity gradient established by the age of the trail (Gibbons 1986). Ants are able to follow substrate trails without wandering back and forth. The best hypothesis is that they sense the radial (i.e., transverse) gradient of the airborne component of the trail substance by comparing the stimulation differences between their two antennae. They move along the edge of the stimulus trail with one antenna in the high-intensity zone and the other antenna in the low-intensity zone. There is no evidence that they can determine the direction in which the trail was laid. In natural situations, many signalers contribute to trail deposition, and the direction of the nest or food source is probably evaluated with other modalities (Hölldobler and Wilson 1990). Male snakes and lizards probably exhibit the best ability to follow a trail laid by a conspecific female to its source (Ford and Low 1984; Cooper and Vitt 1986). They use their forked tongues, which can spread to twice the width of their heads, to make simultaneous spatial comparisons for locating the trail. In snakes, the direction of the female's travel can be evaluated by determining the distribution of pheromone on objects the female pushed against while traveling (Figure 10.21).

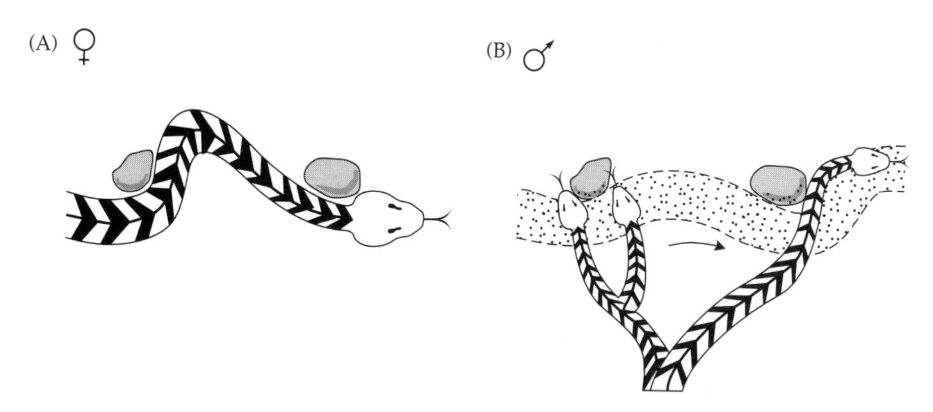

Figure 10.21 A hypothesized mechanism for snakes to determine the direction of a trail. (A) As a female deposits a pheromone trail, her body makes more contact with objects she pushes against on the side toward the direction of travel. (B) A male snake that has encountered a trail explores for chemical stimuli on both sides of nearby objects and recognizes which side the female has pushed against. The stippling indicates the locations in which the pheromone is probably deposited. (After Ford and Low 1984.)

SUMMARY

1. Olfaction is the oldest method of communication, having evolved from chemical mechanisms that primitive organisms used to identify and locate food. However, an organism's ability to control the transmission of chemical signals is limited; this limitation explains the evolution of more sophisticated signaling modalites in more advanced animals. Individual signal molecules must move the entire distance from the sender to the receiver via either current flow, diffusion, or direct contact. In comparison to light and sound transmission, olfaction is considerably slower, its stimulus field contains no directional information, temporal patterning is not possible, and it possesses no linear array of variants equivalent to the spectra of sound and light frequencies. Chemicals used for communication between conspecifics are called **pheromones**.

2. Pheromones are organic compounds differing in size, shape and composition. Airborne odorants must be sufficiently **volatile** to vaporize and waterborne pheromones must be **water soluble**. Pheromones are produced and expelled from the body in two ways. **Sebaceous** and **sudoriferous secretory glands** are well-defined structures near the body's surface that manufacture, store, and release highly specific chemical products. The deposition of these gland products is usually associated with specific behaviors and social circumstances, leaving little doubt as to the general function of the signal. **Waste products** such as urine and feces and **leakage** from body orifices and organs associated with digestion and reproduction may also provide chemical information, but here it is often difficult to distinguish between normal metabolic functions and true pheromones. When defecation and urination are associated with consistent behavioral patterns, a pheromonal function is suspected.

3. The **transmission** of chemical odorants via diffusion in still air can be explicitly modeled with great accuracy. In the case of a **single puff**, the active space is a cloud that first expands around the source, then shrinks and disappears as the chemical is entirely dissipated. If the odorant is **continuously emitted** at a constant rate, the active space expands and then reaches a maximum, steady size. Given typical values of chemical release rates, threshold sensitivities, and diffusion constants, diffusion in air can transmit a chemical signal 6 cm in 13 seconds. In water, diffusion is 10,000 times slower.

4. Long-distance transmission of chemical odorants must be coupled with environmental or sender-generated **current flows**. Quantitative models of odor transmission with flows generally fail to predict observed measurements of active spaces because they assume **laminar flow** or idealized plume structure. Most flows are **turbulent**, causing a complex, unpredictable, and variable pattern of eddies, whirls and vortices. Turbulent flow in air occurs with transmission distances over 10 cm and flow speeds greater than 1 meter per second, and at even lower speeds and distances in water. The stimulus field of an odorant in a turbulent flow is a patchwork of filaments and plumes that spreads in a wedge from the source.

5. **Reception** of chemical stimuli is based on the temporary binding of odorant molecules to specific protein-receptor molecules in a cell or chemosensory organ. The receptor protein changes in shape or in chemical composition. Either change initiates a cascade of biochemical reactions that ultimately leads to a nerve impulse. Nerves from the chemosensory cells are collected in a ganglion or **olfactory bulb**, where summation and initial processing of the sensory input occurs before it travels on to the brain. Chemosensory cells have a short life span and are continuously being replaced by new cells.

6. The ideal chemosensory organ is both highly sensitive to low concentrations and responsive to a wide range of chemicals. High sensitivity is achieved with a **labeled line coding system** responsive to a narrow range of chemicals. A broad response spectrum is achieved with an **across-neuron pattern coding system** involving many receptor types sensitive to different features of molecular structure. Since these represent an incompatible tradeoff, long-distance chemical sex attractants employ the labeled line strategy, whereas generalized food and olfactory receptors employ the across-neuron pattern strategy.

7. Arthropods have olfactory, taste, touch, and current detectors located on their paired **antennae**, and contact (taste) receptors on their mandibles, legs, and feet. The olfactory receptor in fish is a flow-through organ separated from the respiratory system, and taste receptors are located around the mouth. In terrestrial vertebrates, the olfactory system is a duct in the **respiratory system**, with the lungs functioning to draw air into the olfactory organ. Many vertebrates also possess a contact receptor, the **vomeronasal organ**, located in the roof of the mouth.

8. The only type of directional information an olfactory receiver has for orienting itself in a diffusion-based chemical stimulus field is an **intensity gradient**, where the concentration of the chemical decreases with distance form the source. Receivers can move up or down the gradient using simple kinesis mechanisms, or they can use sequential or simultaneous sampling methods to actually detect the direction of the gradient. In a current-dispersed chemical field, receivers can use the associated current flow direction to approach or avoid the sender.

FURTHER READING

Wilson (1963) provided one of the first short nontechnical overviews of olfactory signaling in animals, and Agosta (1992) is a recent layperson's review of chemical signaling. Chapters 4 and 7 of Dusenbery (1992) give the most recent description of chemical signal properties, transmission, production, and reception. Although the focus of Atema et al. (1988) is on aquatic sensory systems, the chapters dealing with chemical stimuli (1–2) and reception (11–15) provide excellent reviews relevant to communication in air as well. Summaries of gland structure may be found in Quay (1977), Brown and Macdonald (1985) and Wilson (1971). The classic diffusion models are outlined in Sutton (1953) and Wilson and Bossert (1963). Tennekes and Lumley (1990) is a good introductory text on turbulence. Invertebrate pheromones and chemical reception are described in several multiauthor books, including Birch (1974), Birch and Haynes (1982), Bell and Carde (1984), and Payne et al. (1986). Schneider (1964) and Kaissling (1971) review insect antennae. A similar set of books describes vertebrate chemical communication, including Müller-Schwarze and Mozell (1977), Duvall et al. (1986), Stoddart (1980), Vandenbergh (1983), Albone (1984), Brown and Macdonald (1985), and Bruch et al. (1988). Recent work on neurological and molecular reception in vertebrates is reviewed by Buck (1996). Good descriptions of the vomeronasal organ may be found in Halpern (1987), Hart (1983), and Wysocki et al. (1980). Orientation in a chemical field is discussed by Dusenbery (1992), Kennedy (1986), and David (1986). Finally, a multivolume series on olfaction and taste illustrates the developing knowledge of these modalities over the years, with Roper and Atema (1987) being among the more useful volumes.

Chapter 11

Electroreception

IN THIS CHAPTER, WE DISCUSS A FINAL MODALITY used in animal communication: electroreception. Whereas all of the other modalities discussed so far are available in both terrestrial and aquatic environments, electroreception can only be used in water. This limitation occurs because air is too effective an insulator to pass electric signals of biological magnitudes between senders and receivers. Even in water, the ranges over which electrical signals can be detected are quite limited. Although all of the other modalities are used by both invertebrates and vertebrates for communication, to date only aquatic vertebrates are known to use electroreception for this purpose. However, the ability to detect electrical signals is widespread among aquatic vertebrates with examples in fish, amphibians, and mammals, and the corresponding ability to produce electric signals appears to have arisen independently in both cartilaginous and bony fishes. Electrical communication exhibits certain properties that amplify and reinforce themes covered in prior chapters, and so we shall treat it in some detail below.

PROPERTIES OF ELECTRIC FIELDS

All of the matter on Earth is held together by the strong forces between electrons and protons. The total numbers of positively charged protons and negatively charged electrons are about the same and the world is largely neutral. However, within any piece of matter, there are often local regions of electron shortage that are spatially separated from regions of electron abundance. It is the attractive forces between these regions that hold them tightly together. The fact that most of the world is stable and structured is partly due to strong electrical forces.

Any piece of matter that contains a shortage or excess of electrons (relative to protons) is said to have a net **charge**: a shortage of electrons is said to confer a positive charge and an excess a negative one. The greater the excess or deficit of electrons, the greater the magnitude of the charge. Any other charged object near the first will experience a force on it due to the attraction or repulsion between the two charges. Charges of similar sign (e.g., both with electron excesses or both with electron deficits) will repel each other; opposite charges will attract. For two nonmoving objects in a vacuum with charges q_1 and q_2, respectively, the force F experienced by each along the axis joining them is given by Coulomb's law:

$$F = \frac{1}{4\pi\varepsilon_0} \cdot \frac{q_1 q_2}{d^2}$$

where ε_0 is called the **permittivity constant**, and d is the distance between the two charged entities. When more than two charged objects are present within a given volume, the forces they generate combine vectorially at any location, and the direction and magnitude of the net force is likely to vary between locations. If we were to move a small unit charge to many locations around such an array of charges and measure the direction and magnitude of the force exerted on it at each point, the resulting map would describe the **electric field** around the charge array. As one moves further and further from the array, (i.e., as d increases), the magnitude of the electric field at each point will decrease. At an infinite distance, the force drops to zero. The magnitudes of electric fields are classically measured in **newtons/coulomb** (where a newton is a unit of force and a coulomb is a unit of charge equal to 6×10^{18} electrons).

Suppose we have such a group of charged objects localized near each other. Together they will generate an electric field around themselves that differs in magnitude and direction of force depending upon the location we sample. Suppose we then take a small unit charge located an infinite distance away from these objects (at which point the net force on it is zero) and move it to some location nearer to them. If the unit charge is repelled at most locations by the electric field around the objects, we shall have to do work to move the charge to its new location. If is attracted, the electric field will help us move the unit charge and energy is released in the process (negative work). The overall work (positive or negative) expended to move the charge from infinity to a point closer to the array is called the **electric potential** at that point. As long as the charged objects are immobile, the work expended moving our unit charge to this point will be the same whatever path we take. Thus the electric potential at any point around a set of immovable charged objects is a fixed number. This is convenient, because if $\phi(a)$ is the electrical potential at point a, and $\phi(b)$ is that at point b, the work needed to move a unit charge between the points a and b is just the difference between their potentials: $\phi(a) - \phi(b)$. It is thus possible to make a plot of electric potential around our objects just as we can make a map of the electric field around them.

The electric potential and electric field maps are closely linked. If we select any point in the electric potential map, the corresponding electric field vector equals the gradient or slope in potential values at that location. Also, because the magnitude of the electric field decreases with increasing distance from the charged objects, less work must be done to move from infinity to locations distant from the charges and thus the map of potential, like the map of electric field, will show a decrease in magnitudes the farther away from the charges one samples. Electric potentials are usually measured in **volts** (a unit based on work/coulomb). The relationship between electric potentials and fields permits electric fields to be more practically measured in volts/unit distance (instead of newtons/coulomb).

Shapes of Electric Fields in a Vacuum

Let us examine the electric field and potential maps around some simple charged objects. Consider first a single charged particle with no other charged objects nearby. This is called a **monopole** situation (analogous to the acoustic monopole we discussed on page 76). The electric field and potentials surrounding a monopole charge are diagrammed in Figure 11.1. This plot may be envisioned as a topographic map in which solid lines connect locations of identical electrical potential: higher potentials constitute higher contours in the topography. Electric potential at a distance d from the charge is proportional to $1/d$; thus successively larger circles represent increasingly lower potentials. The electrical field in this plot is represented by dashed lines and indicates the trajectories down which a unit charge of the same sign would "roll" if allowed to move in this topography. Areas of high field magnitude are indicated by drawing adjacent lines closer together. For a monopole, the directions of the field at any location are pointed radially away from the point charge. Note that we have only drawn the electric field in one plane; in most situations, the field will be a three-dimensional one with lines of force radiating out of the point in all directions. Because the electric field equals the derivative of the potential with respect to location, and potential is proportional to $1/d$, the electric field at a distance d from a point charge will be proportional to $1/d^2$. In practice, we do not measure potentials or fields by moving tiny charges around, but instead by comparing the difference in electric force exerted on an instrument by potentials at two different locations. Since the locations are likely to differ in their distance from the monopole, the measured force will reflect the gradient in potential and thus the electric field. *Differences*

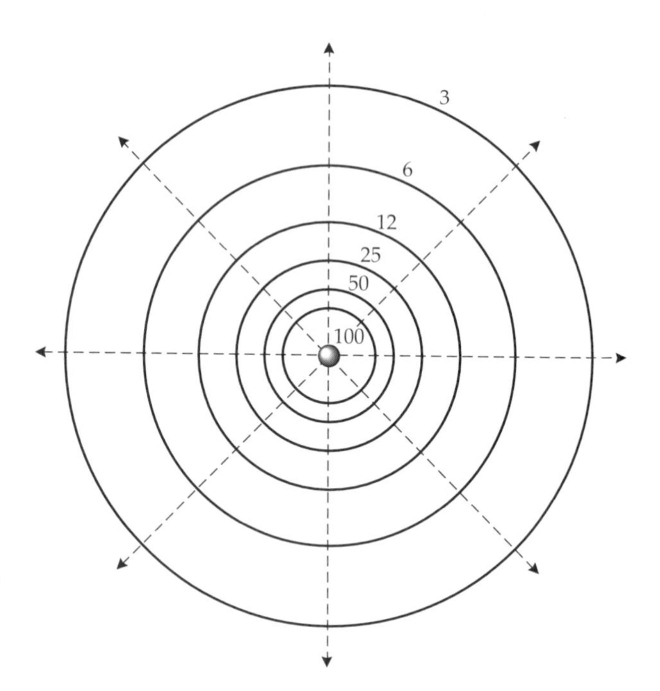

Figure 11.1 Electric field and potential around monopolar point charge. Dotted lines indicate direction of the electric field (convention is to draw arrows pointing away from a positive charge, toward a negative charge). Proximity of adjacent field lines indicates the strength of the electric field in that region (here highest strength is closest to the point charge). Solid circles indicate all locations around monopole with electric potential of a particular value. Relative values for individual isopotential circles are indicated, showing a rapid drop in potential with distance from the point charge.

in potential between two locations around a monopole thus tend to decrease with the square of the distance between them.

Now consider two small objects with equal and opposite charges. Suppose the two are separated by a distance δ, which is small relative to various distances d at which we sample the surrounding electric field and potentials. Such a charge array is called an **electric dipole**. The electric field and potentials surrounding a dipole are shown in Figure 11.2. Notice that this plot is not simply the sum of two monopoles. The electric field lines no longer radiate away from each charge in a radial fashion but are bent and distorted by the presence of the adjacent charge. This change in the electric field causes a corresponding change in the shapes of the equal potential lines around each charge: these are no longer circular, but are flattened on the side closest to the other charge. Unlike the situation with a monopole, we cannot characterize the potential around a dipole simply by specifying the distance to the sampling point d. Potential values vary with the relative angle between the line joining the two charges and that joining the sampling location to the midpoint between the charges. If we denote this angle by θ, and the absolute value of the charge on each point by q, then the potential around a dipole in a vacuum at location (d,θ) is

$$\phi(d,\theta) = \frac{1}{4\pi\varepsilon_0} \cdot \frac{q\delta\cos(\theta)}{d^2}$$

The product $q \cdot \delta$ in the numerator is often called the **dipole moment** of the array and plays a major role in the magnitude of the potential at any point;

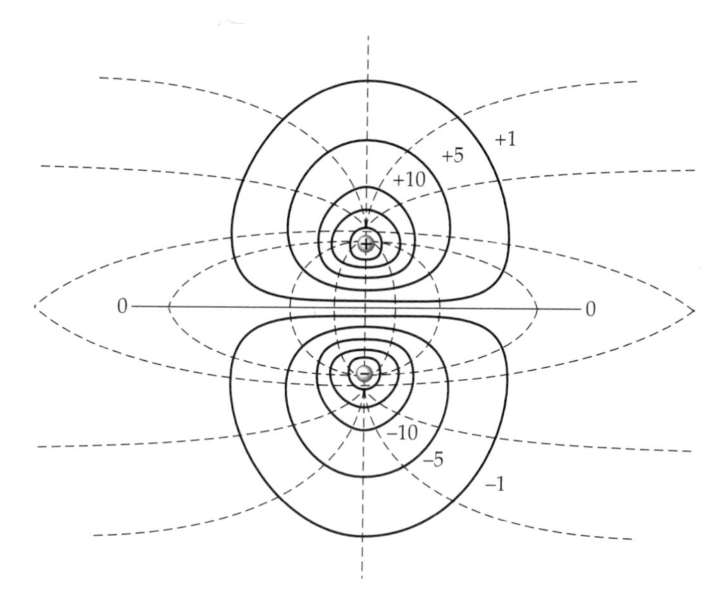

Figure 11.2 Electric field and potential around electric dipole. Conventions as in Figure 11.1. Positive charge is the top member of the pair, negative charge is the lower. Note that electric field lines are curved around the dipole, and lines connecting points of equal electric potential are distorted and no longer circular. Also, note the line midway between two points and perpendicular to the axis joining them along which potential is zero.

the farther apart the two charges are separated (δ), or the greater the charge at each point (q), the greater the potential at all d. The angle of the sampling point relative to the axis of the dipole is also important in determining the potential: $\phi(d,\theta)$ will be highest along the line joining the two charges [e.g., θ is zero and $\cos(\theta) = 1$], and zero at points perpendicular to the midpoint of the line joining the two charges [e.g., $\cos(\theta) = 0$]. As the equation indicates, the electric potential around a dipole is proportional to $1/d^2$ (and thus the electric field around a dipole, and the difference in voltage between two locations near a dipole, fall off with $1/d^3$).

The electrical fields generated by biological sources are rarely simple dipoles and are practically never monopoles. Instead, complex arrays of charges will have many axes around which the charges are distributed. The resulting field will be described by a sum of terms based on identifiable axes of charge distribution: a single axis contribution is called the dipole component, a second axis contribution is the quadrupole component, a third axis portion is the octupole component, and so on. The magnitude of each component's contribution to the overall electric field depends upon the distance between the sampling point and the charge array: the electric field of the dipolar component is proportional to $1/d^3$, that for an octupole is proportional to $1/d^4$, and that for an octupole is proportional to $1/d^5$. At very close distances to the charge array, all components of the field play significant roles in determining its shape and properties; at greater distances, all of the higher order components have such low magnitudes that only the dipole component is needed to describe the field. Just as with sound, we can thus speak of a near and far field around an array of electric charges. In the near field, all of the higher-order components of the field are required to describe the field; in the far field, only the dipole component is necessary.

Shapes of Electric Fields in Nonconducting Media

So far, we have discussed electric fields around an array of charges in a total vacuum (no surrounding matter present). Animals do not live and communicate *in vacuo* but live in some medium. How does the presence of a medium affect an electric field?

Media differ in how they respond to an applied electric field. **Conductors** (such as metals) have electrons that are relatively free to move. The moment an electric field is applied to a conductor, electrons begin moving within it to pile up on the side of the conductor facing the positive side of the field. The positive atoms that have given up their electrons are left behind on the side of the conductor facing the negative part of the field. This separation of charges inside the conductor generates an electric field exactly opposite to the external field. Eventually, this internal field balances the external one, there is no net force on any electron, and the magnitude of the net field (internal plus external) inside the conductor becomes zero. Because there is no remaining field within the conductor, it will take no work to move a point charge between any two points within this conductor. Thus the electric potential inside the conductor is the same everywhere. Note that there will still be an electric field outside of the conductor, and it will take work to bring a test charge from some point outside of the conductor to some point on or inside it.

Dielectrics are materials whose electrons have limited mobility. When placed in an electric field, dielectric electrons cannot move between atoms. However, many atoms permit their electrons to congregate on one side of the atom, or in a liquid, molecules that have one side slightly more positive than another (like water) will rotate until their positive sides face the negative side of the imposed field. This polarization of the atoms or molecules generates millions of tiny dipoles within the material with their dipole axes parallel to the lines of the electric field. Because the polarity of the dipoles is opposite to the imposed external field, the result is to reduce the net electric field inside of the dielectric. The stronger the polarization inside the material, the greater the reduction in internal field magnitude. The ability of a particular kind of matter to be polarized by an external electric field is characterized by its **dielectric constant** (k). The higher the value of k, the more easily a material is polarized and the greater the reduction of the electric field inside the material. For a vacuum (a perfect insulator), $k = 1$, whereas for a conductor, k is infinitely large. Some other sample values for k are given in Table 11.1.

The dielectric constant determines the permittivity of a medium and thus the size of an electric field at any point. Whereas for a vacuum, the permittivity constant is ε_0, the permittivity for a medium, ε, is $\varepsilon = k\varepsilon_0$. (The dielectric constant k of a medium is thus seen to be the ratio of its permittivity to that in a vacuum.) Suppose we create an electric field inside a homogeneous nonconducting medium such as a liquid. Coulomb's law then becomes:

$$F = \frac{1}{4\pi\varepsilon} \cdot \frac{q_1 q_2}{d^2} = \frac{1}{4\pi k\varepsilon_0} \cdot \frac{q_1 q_2}{d^2}$$

Table 11.1 Dielectric constants for selected media

Material	Dielectric constant (k)
Vacuum	1.0000
Air	1.0004
Oil	5.0
Glass	10.0
Water	80.0
Metal	∞

This means that the magnitude of an electric field in a homogeneous medium with a dielectric constant k is reduced by a factor $1/k$ at all points relative to that expected in a vacuum. Since the electric field is reduced by this factor at all locations, so is the electric potential. For example, the electric potential around a dipole in a homogeneous nonconducting medium is

$$\phi(d,\theta) = \frac{1}{4\pi k\varepsilon_0} \cdot \frac{q\delta\cos(\theta)}{d^2}$$

Electric Fields in Conducting Media

Suppose we place an electric dipole in a medium that is a worse conductor than a metal, but a better conductor than most dielectrics. Water is such an example. Water invariably has dissolved materials within it, and many of these, such as salts, break up in water into their component charged ions. The presence of an electric field in water will cause positive and negative ions to move in opposite directions. The ionic trajectories follow the electric field lines. This movement of ions in water (or of electrons in a metal) is called an **electric current**. The magnitude of an electric current between two points (measured in coulombs/second or **amperes**) is proportional to the potential difference or voltage between the points. The constant of proportionality between an applied voltage and a resulting current is called the **conductance** of the medium through which the current is flowing. More often, we use the reciprocal of conductance, called the **resistance**. If V is the voltage difference between two points and R is the resistance (in **ohms**), then the current I (in amperes) depends on these variables according to Ohm's law:

$$I = \frac{V}{R}$$

The convention in physics is that current flows from a region of positive voltage to one of more negative voltage. Note that this is opposite to the actual flow of electrons (from a negative to positive potential location).

Resistance in a particular context will be higher the greater the distance the current must flow, the smaller the cross-sectional area through which the current passes, and the worse the material as a conductor. The latter term is characterized by the material's intrinsic **resistivity**. Because of the resistance of the water in which we have placed our dipole, there will be a steady current of positive ions towards that part of the dipole of opposite charge to each ion. If there were no resistance, the initial current would quickly cancel the charge at each end of the dipole due to accumulations of oppositely charged ions. If the resistance is high enough, it may take some time before the dipole is fully neutralized. Alternatively, something may occur near the dipole to restore its charge. In either case, if the electric field is maintained for a sufficiently long period, we can measure the electric potential at various points around the dipole and the amount of current at each location. For a stable source of current in a conducting medium, the potential at location (d,θ) from the dipole is

$$\phi(d,\theta) = \frac{\rho_0 I}{4\pi} \cdot \frac{\delta \cos(\theta)}{d^2}$$

where the medium resistivity, ρ_0 and the current I have replaced the permittivity, ε, and the charge, q, used for nonconducting media.

Water and many other materials are both conductors and dielectrics. Some current will flow in them, but the resistance is high enough that electric fields are sustained, and their ability to be polarized and act as a dielectric permits some build-up of counterfields within the medium. For static electric fields, this may not be significant. If however, the electric field is changing in magnitude or direction, then the dielectric properties of the medium can become important. In a steady electric field, an electron in a conductor may move the entire length of the conductor. This is called a **direct current (DC)**. Now suppose we apply a sinusoidally reversing electric field to the conductor. Electrons will first move one direction and then back the other. This is an **alternating current (AC).** The faster the frequency of the alternating field, the less distance any one electron can travel before it has to turn around and go the other way. In a nonconducting dielectric, electrons or polar molecules can move a bit, but they can never move far enough to sustain a steady DC current. However, if an alternating field is applied across such a material, the distance electrons have to travel per cycle may be within the polarizing limitations of the material. The higher the dielectric constant for the material, the lower the frequency of alternation which the material can track. The effective resistances of dielectrics may thus drop if the applied electric field varies sufficiently quickly. To keep this notion of resistance distinct from classical DC resistivity, the term applied to such dielectrics is **capacitative reactance**. Capacitative reactance decreases with the dielectric constant of the material and with the frequency of the electric field oscillation. Like resistance, it is measured in ohms. Remember that even if the waveform of the electric field variations is not sinusoidal, it can be considered as the sum of a number of different sinusoids (see Chapter 3). Applying such a nonsinusoidal signal to a dielectric, we will find that the dielectric will act like a high-pass filter since it can more easily track the higher-frequency components than the lower-frequency ones. The overall **impedance** of a medium like water to a varying electrical field will thus depend on both the resistivity of the water and on the capacitative reactance of the water at the various frequencies making up the waveform of the changing field.

We have assumed so far that media are unbounded and homogeneous. The resulting electric fields can be called **free fields** (by analogy with sound). However, most media have boundaries and contain objects whose dielectric and/or resistive properties differ from those of the medium. The usual situation is thus not a free field. Boundaries and objects in the medium will distort and change the field shape from free field conditions. For example, suppose we place a monopole in a medium such as water and then place objects with resistivities greater or less than water near to the charge (Figure 11.3). Objects that have lower resistivities than the medium bend the electric field lines in the region between themselves and the charge closer together and towards the object. This region of enhanced electric field magnitude corresponds to a region of very closely spaced isopotential lines and thus a steep gradient in voltage. Objects that have higher resistivities than the medium bend the electric field lines away from themselves, lowering the field magnitude in the region between themselves and the charge, and flattening the potential gradient in this region. When many objects of differing conductivities are present, the shape of the field can become highly complex. Boundaries are also significant.

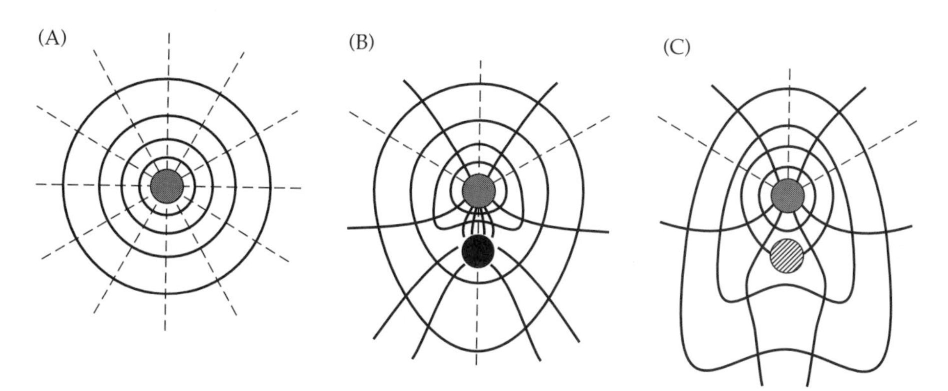

Figure 11.3 Distortion of electric field by neutral objects. (A) Simple monopole showing isopotential (solid) and electric field (dashed) lines. (B) Distortion of field and isopotential lines when an object (black) with resistivity lower than the medium is placed near the point charge (gray). Field lines between charge and object are close together, indicating a higher-magnitude field in this region, and isopotential lines are closer together, indicating a steeper potential gradient. (C) Distortion of the field due to proximity of an object (patterned) with higher resistivity than medium. Field lines are further apart in the region between charge and object, and more widely spread isopotential lines indicate a shallower gradient than elsewhere.

If we place our charge near to the air-water interface, or near a nonconducting bottom, the electric field magnitude and potential near to the charge will be twice as great as that for a charge suspended in an unbounded volume of water. This difference occurs because current can radiate in all directions in the unbounded case, but can only radiate away from the water's surface in the bounded example.

GENERATION OF BIOELECTRIC FIELDS

How might animals generate electrical fields? In fact, this is relatively straightforward since nearly all living organisms have electrical potentials across their cell membranes. Cells routinely use energy to pump potassium ions into their cytoplasm and pump sodium ions out. Both ions are missing an electron and thus are positively charged so the ion pump may not by itself generate any differences in net charge (although it does in some cases). The membranes of most cells are moderately permeable to potassium ions which tend to diffuse down their concentration gradients from the inside to the outside of the cell. The same membranes are relatively impermeable to sodium which cannot diffuse down its concentration gradient and enter the cell. Most negative charges inside the cell are large molecules (such as proteins) that cannot cross the membrane. Thus as potassium ions diffuse out, they generate a shortage of positive charges and an excess of negative ones inside the cell. This builds up an electric field across the membrane with the cell cytoplasm negative relative to the outside medium. Eventually, this electrical field becomes strong enough to draw the positively charged potassium ions back into the cell just as fast as they leave due to the concentration gradient. The result is a stable **electrochemical potential** of –60 to –80 mV across the cell membrane.

Many organisms have gone one step further by evolving excitable cells. When stimulated by a drop in the electrical potential across its membrane, an excitable cell suddenly makes its membrane much more permeable to sodium ions. These run down their gradient into the cell and generate a cytoplasmic potential with a polarity opposite to that generated by the potassium ions. In fact, the interior of the cell goes from –80 mV (relative to outside) before being excited to +50 mV at the peak of the sodium influx, a difference of 130 mV. After being depolarized by the sodium influx, the cell then changes its permeabilities back to normal and reestablishes the resting potential of –80 mV. This cyclic change in cytoplasm voltages can occur at slow or rapid speeds. Where it occurs quickly, it is called an **action potential** and is the basic mechanism of cellular stimulation in nerves and muscles.

Certain fish have evolved modified muscle cells (or in one family, the Apteronotidae, nerve cells), called **electrocytes,** which are used to generate electrical signals larger than that due to a few nerves or muscles. Tens to thousands of electrocyte cells are arranged in columns within an **electric organ**. Each column is surrounded by insulating material and each electrocyte cell is supplied with nerves. When the fish's brain sends an appropriate signal, all of the electrocytes in a column are stimulated by their respective nerves and depolarize simultaneously. In freshwater fish, the depolarization is a rapid action potential; in marine species, it is slower and only goes from resting to zero potential. Because the cells are stacked in a column and prevented from shorting each other by the surrounding insulation, their voltages add up to produce a large net voltage between the two ends of the column. The more cells present per column, the higher the voltage produced. Since electric eels have over 6000 electrocytes in series, and each cell can produce a difference of 130 mV, these fish can produce discharges up to 6000 × 130 mV = 720 volts! The amount of current that an electric organ can provide depends upon how wide each electrocyte is and how many parallel columns of electrocytes it contains—the more the number of columns, the greater the total current. A fish of a given size and a certain number of electrocytes could either organize those cells into a few long columns with many cells per column (giving high voltage but lower current capacity) or, alternatively, many columns but with fewer cells per column (giving a lower voltage but a higher current capacity).

Although electric organs are only known in fish, these structures appear to have evolved independently many times in this group. Some taxa have very large electric organs that produce high voltages and are used to stun prey and/or deter predators. Others produce much more moderate voltages and are used for intraspecific communication. Finally, some groups produce rapid trains of electric organ discharges for **electrolocation**: distortions in the induced electric fields due to nearby objects are used by the fish to detect and monitor their environment much as bats and cetaceans use sound to echolocate (see Chapter 26).

The size and shape of electric organs vary with the habitat of the fish and the function of the **electric organ discharge (EOD)**. Most taxa produce EODs of only a few volts, which are used for either social communication, electrolocation, or both. Such **weakly electric fish** would include many marine skates, the freshwater Gymnotiformes of the New World tropics, and the freshwater Mormyriformes of Africa. Organs in these species need not be very large and are usually located in the base of the tail with varying degrees of extension in the anterior (head) direction. As a general rule, species with voltages in the hundreds of volts use their organs for prey capture and self defense. Examples of such **strongly electric fish** are the freshwater electric eels of South America, the freshwater electric catfish of Africa, and the marine torpedo rays. Some torpedoes and the electric eel actually have two electric organs: one produces the high voltages for prey capture and defense, whereas the other is a low voltage organ used for social communication and/or electrolocation. The remaining electrogenic fish, the stargazer, is the only known bony marine fish with an electric organ. The function of its 5 volt EOD remains unknown. As a general rule, the higher resistivity of fresh water demands that these fish, whether weakly or strongly electric, generate somewhat higher voltages than for equivalent functions in sea water. Their electric organs thus tend to be long and thin, whereas many marine forms have shorter and fatter electric organs. The shapes and locations of the electric organs for these major taxa and EOD amplitudes are shown in Figure 11.4.

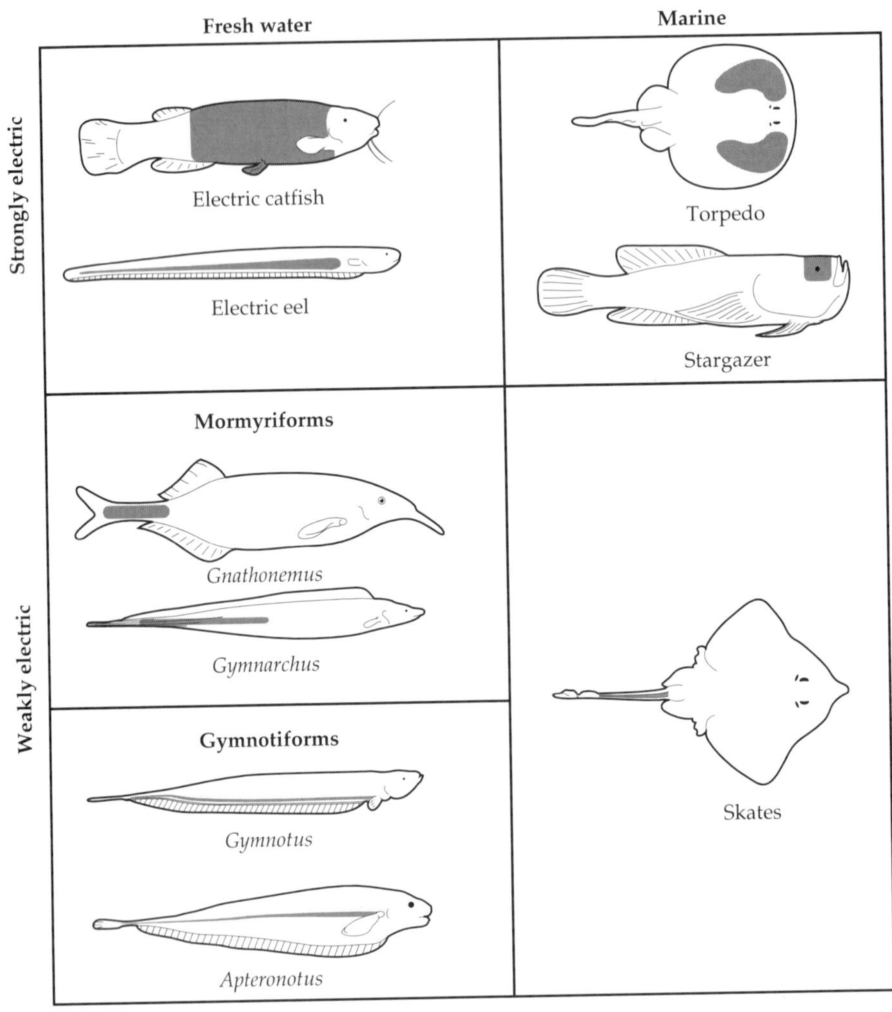

Figure 11.4 Shape and disposition of electric organs in fish. Electric organs are stippled or, if only thin rods, indicated by dotted lines. Taxa are divided according to strength of electric discharge (left axis) and habitat (top axis). Note the larger organ size in strongly electric species, and the long, thin shape of the electric organ in freshwater electric eel when compared to the short, fat shape of the organ in the marine torpedo. (After Bennett 1970.)

In addition to differences in amplitude, the waveform of the EOD also varies widely among the electrogenic taxa (Figure 11.5). These varying waveforms result from a number of factors (Bennett 1970; Bass 1986). Electrocytes are usually flattened disks with enlarged anterior and posterior sides. In skates and the high voltage organ of the electric eel, only one side of each electrocyte can be excited and undergo depolarization. The polarity of the cell during the EOD depends upon which side of the cell is active. If the anterior

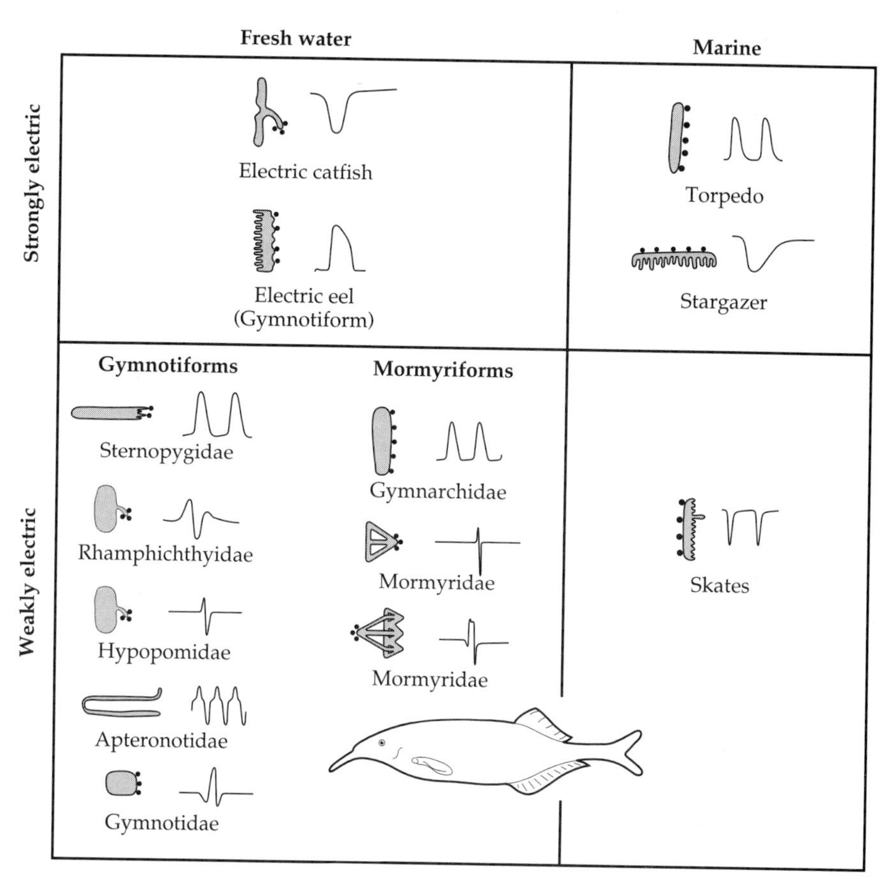

Figure 11.5 Electrocyte innervation and EOD waveform in electric fish. Each example shows disposition of electrocytes on the left and EOD waveform on the right. Electrocytes are shaded and dark dots represent contacts with stimulating nerves. Faces of electrocytes are oriented relative to position of fish shown at bottom of figure: anterior sides of cells are on the left and dorsal sides are on the top. Waveforms go upward when the anterior of the fish is positive relative to the tail, or downward when the anterior is negative. Note that electrocytes innervated on the posterior side (right side of figure) have an initial positive phase in waveform, whereas those innervated on the anterior go negative first. All electrocytes shown are derived from muscle cells, except in the gymnotiform family Apteronotidae where modified nerves are used. (After Bass 1986.)

side is the excitable one, then it is the positive membrane that is active, and current (which by definition moves from positive to negative) will flow through the animal to its posterior. Here it will emerge from the tail and flow through the water back to the head. Since the current in the external medium is flowing from tail to head, an external electrode close to the head will appear negative relative to one near the tail. Alternatively, if the electrocytes are stimulated on their posterior sides, then the external medium around the head of the animal becomes positive relative to that near the tail. In skates, innervation is on the anterior side, and the EOD waveform shows a negative potential difference when the head is compared to the tail (Figure 11.6A); the opposite innervation and waveform is seen in the electric eel (Figure 11.6B). In both cases, the waveform is monophasic (has only one maximum or minimum) because only one side of the electrocytes is excitable.

Some gymnotiform fish have electrocytes that are excitable on only one side and thus produce monophasic EODs. In others, both sides of the electrocytes are excitable but only the posterior side is innervated. When stimulated, the posterior side of each cell fires an action potential, making the head of the fish positive relative to the tail just as in the high voltage organs of electric

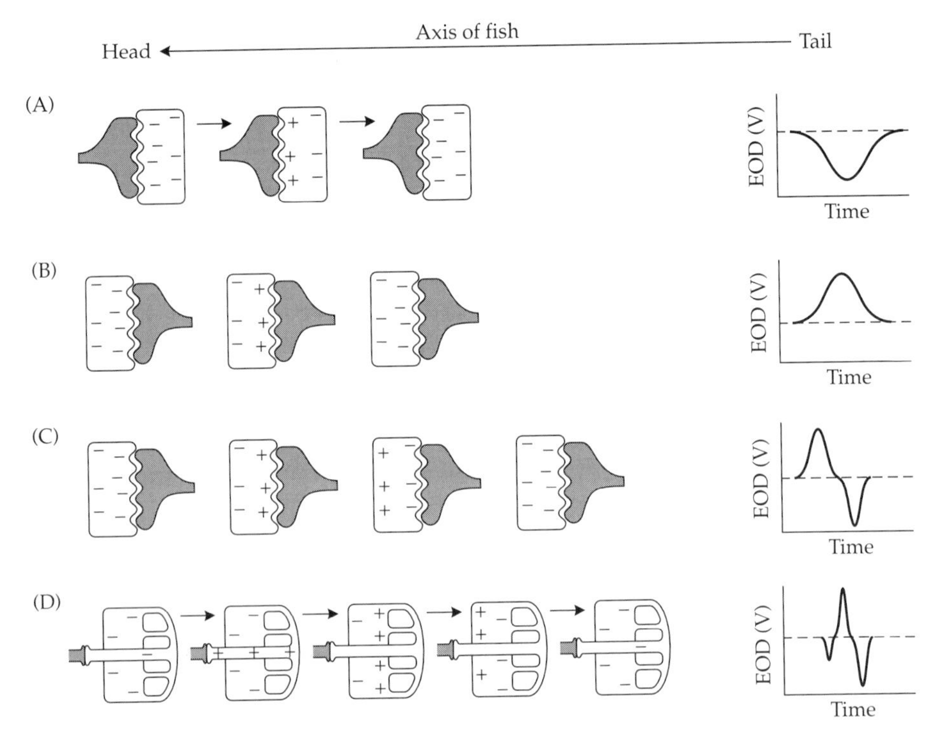

Figure 11.6 Mechanisms for generating varying EOD waveforms. The stippled cell is the innervating nerve; the empty cell is the electrocyte. (A) Skate electrocyte with innervation on the anterior face of the electrocyte and only the anterior face excitable. An EOD measured at the head relative to the tail shows a negative monophasic waveform. (B) An electric eel electrocyte with innervation on the posterior face of the electrocyte and only the posterior face excitable. The EOD near the head is a positive monophasic signal. (C) A gymnotiform electrocyte with posterior-face innervation and both faces excitable. Because the posterior face fires first, the waveform shows an initial positive phase followed by a negative phase. (D) Mormyriform electrocyte with an excitable stalk connecting to the posterior face. The stalk extends through the electrocyte towards the anterior side where it is innervated. Excitation of the anterior stalk makes the head of the fish negative; subsequent excitation of the posterior face then adds a positive phase, and eventual excitation of the anterior side of the electrocyte generates a final negative phase.

eels. However, the depolarization of the cell soon stimulates an action potential in the anterior face as well. This action potential then generates an opposite voltage, with the head of the fish negative relative to the tail as in a skate. The waveform of these gymnotiforms, EODs is thus biphasic with a positive phase at the head followed by a negative phase (Figure 11.6C). A few gymnotiforms have additional electric organs in their heads. If the fish fires the main and additional organs biphasically but slightly out of synchrony, EOD waveforms with more than two phases or with highly distinctive shapes can be generated. Some examples are shown in Figure 11.7.

Finally, in mormyriform fish, the electrocyte is divided into three regions: an anterior face, a posterior face, and a stalk (Figure 11.6D). The stalk is a tube which connects to the posterior face and then penetrates through the electrocyte towards the anterior of the fish where it connects to the stimulatory nerve. All three regions of the electrocyte are excitable: the stalk fires first and being anterior generates a head-negative phase of the EOD. The posterior face fires next, generating a head-positive phase, and finally the anterior face fires, generating a final head-negative component. Depending upon the relative sizes of stalk and cell faces, and on the geometry and complexity of the stalk, very different EOD waveforms can be generated. Examples are shown in Figure 11.7.

Most weakly electric fish emit EODs continuously. The rate and pattern of discharges is controlled by a pacemaker in the brain. Two classes of discharge pattern are known: (a) **pulse fish** produce very short duration multiphasic EODs (1–3 msec) with variable rates and low maximal rates (50–100 Hz); (b) **wave fish** produce very steady rates of monophasic EODs with slightly longer durations (3–5 msec) and very high maximal rates (300–1700 Hz). All members of the Mormyriformes except the single genus *Gymnarchus* are pulse fish; gymnotiforms in the families Apteronotidae and Sternopygidae are wave fish, whereas all other families are pulse fish. As we shall discuss later in this chapter (pages 344–346), weakly electric fish apparently use both the shapes of the EOD waveforms and the patterns of discharge to identify electrical signals from conspecifics.

COUPLING OF ELECTRICAL SIGNALS TO THE MEDIUM

Weakly electric fish use electric organ discharges either to communicate with other nearby fish or for electrolocation. In either case, producing a signal with a large range is generally advantageous. Because electric fields of animals fall off with at least the cube of the distance, it would take an eightfold increase in EOD amplitude to double the range of detection by a receiver. Clearly, this increase is not feasible for most small species. The alternative is to maximize the coupling of each EOD to the medium. As we have seen, the magnitude of the electric field at any fixed distance from the fish will depend in part on the magnitude and geometry of the charge array within it, the dielectric and resistive properties of the medium relative to the fish, the presence of nearby boundaries, the nature of any objects in the medium, and the angle relative to the fish's effective dipole axis. Some of these factors the fish can control.

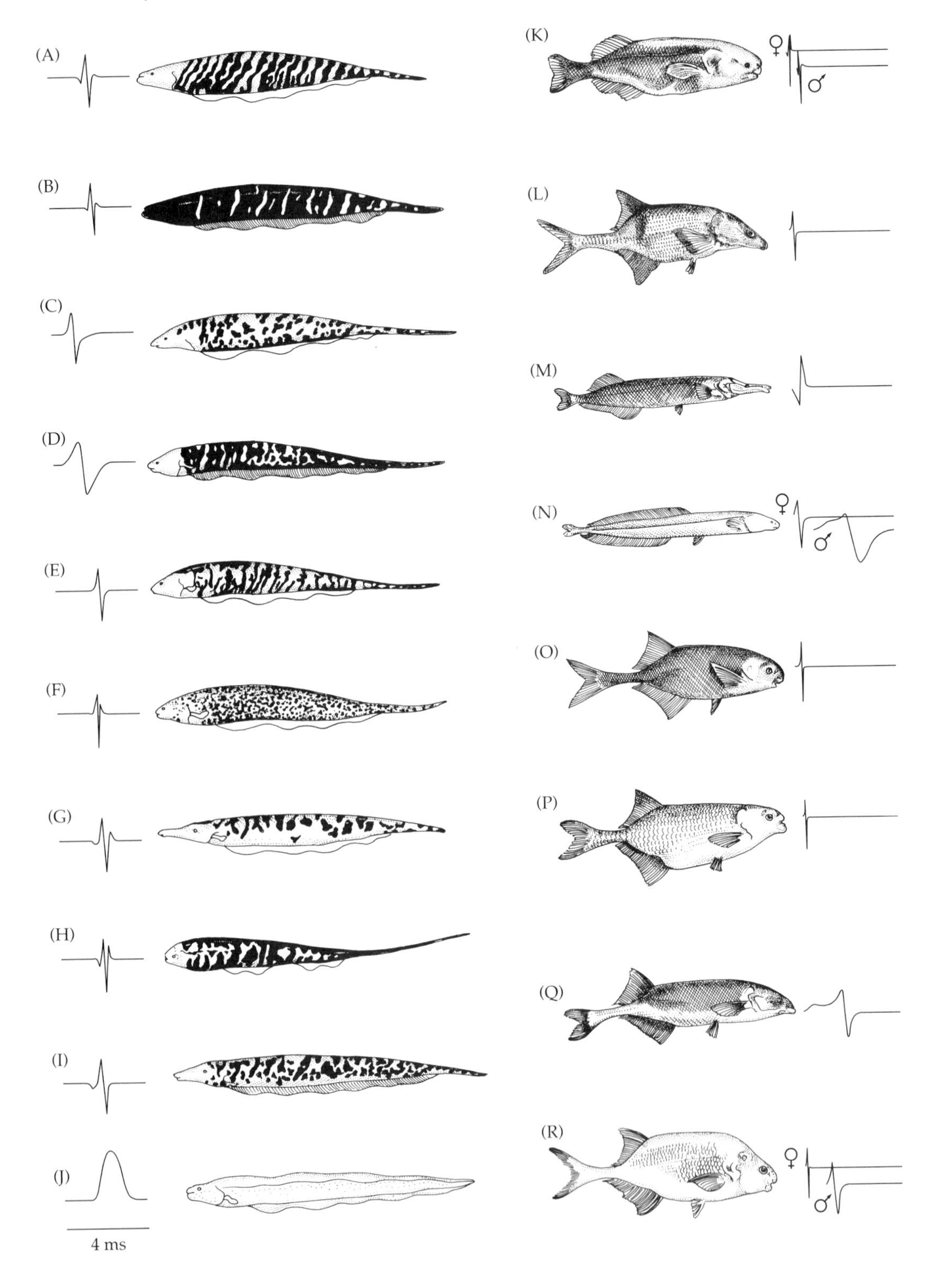
(A)
(B)
(C)
(D)
(E)
(F)
(G)
(H)
(I)
(J)
4 ms
(K)
♀
♂
(L)
(M)
(N)
♀
♂
(O)
(P)
(Q)
(R)
♀
♂

◀ **Figure 11.7 Sample EOD waveforms for selected gymnotiforms (left) and mormyriforms (right).** The species of gymnotiforms shown all occur in coastal streams of the Guianas. The selected mormyriforms all live in the Ivindo River or its feeder streams near Makokou, Gabon. It is common to find such high local species diversity of electric fish in tropical South America and Africa. Note the sexual differences in EOD for the first, fourth, and last mormyriform species. Species names: (A) *Gymnotus carapo*; (B) *Gymnotus anguillaris*; (C)*Hypopomus artedi*; (D) *Brachyhypopomus brevirostris*, (E) *Brachyhypopomus beebei*; (F) *Brachyhypopomus sp.*; (G) *Gymnorhamphichthys hypostomus*; (H) *Hypopygus lepturis*; (I) *Rhamphichthys rostratus*; (J) *Electrophorus electricus*; (K) *Stomatorhinus walkeri*; (L) *Boulengeromyrus knoepffleri*; (M) *Mormyrops zanclirostris*; (N) *Isichthys henryi*; (O) *Pollymyrus marchei*; (P) *Marcusenius paucisquamatus*; (Q) *Marcusenius conicephalus*; and (R) *Ivindomyrus opdenboschi*. (Fish based on drawings provided by Carl Hopkins.)

The electric field and corresponding isopotential lines for a typical weakly electric fish in a free field situation are shown in Figure 11.8. Although the fish roughly approximates a dipole, note the asymmetry of the isopotential lines when the two ends are compared. These plots were generated from a simplified model of such a fish by Heiligenberg (1975, 1977) and provide a good fit to observed measurements of actual fields (Knudsen 1975; Bastian 1986). The basic elements of the model are simple. An electric organ produces a voltage difference with the tail tip negative and the body tissues in contact with the anterior end of the organ positive. Ionic currents pass through the body tissues, across the skin, and then along the lines of electric field in the surrounding water. This simple circuit thus consists of a battery (the electric organ) and three resistors in series. The saline tissues of the fish are highly conductive and we can thus ignore this resistance. Because a fish's tissues are much more saline than the surrounding fresh water, water tends to move into the fish and cause it to swell. To prevent this, most freshwater fish have relatively impermeable skins to both ions and water, and thus the skin has a very high resistance. Two of the circuit resistors are thus the skin at the head and that at the tail. The third circuit resistance is that of the surrounding water which is higher than the fish's tissues, but generally lower than the skin.

By varying the geometry of model fish, Heiligenberg found that the longer the fish's tail, the farther a given isopotential line would lie from its body and the farther from such a sender a receiver would be likely to detect the signal (Figure 11.9). Increasing tail length is similar to increasing the charge separation distance, δ, in a dipole, a topic that we have discussed earlier. A quick glance at Figure 11.5 or 11.7 will show how many gymnotiform and mormyriform fishes have long tails. The model also shows that if the resistivity of the fish's skin is very high relative to that of the medium, the electric field contracts around the anterior end of the fish, making the range of detection in the anterior direction poor. If the skin resistivity is very low relative to that of the medium, the field collapses around the tail, reducing signal range in that direction. The maximal field is thus achieved when the resistivity of the fish's skin and that of the medium are similar.

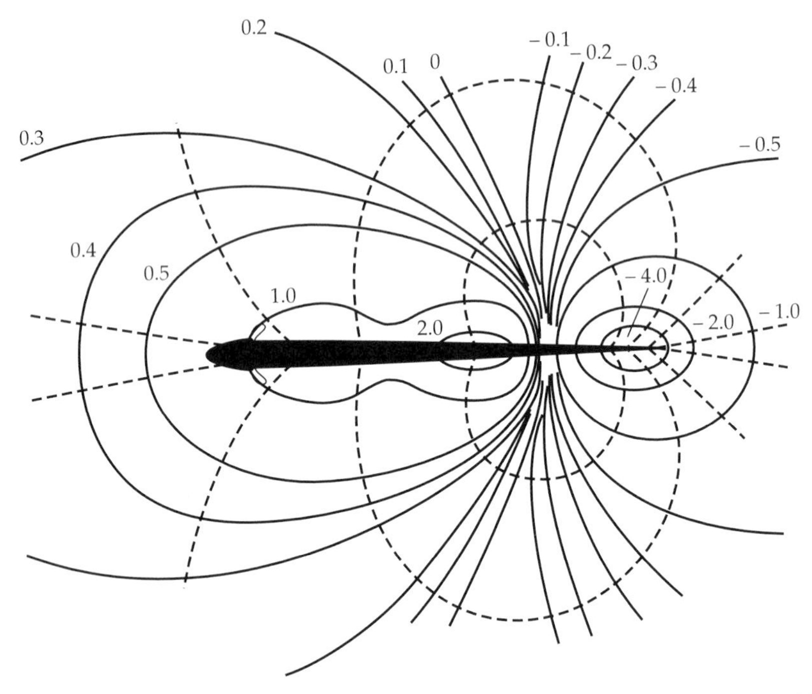

Figure 11.8 Electric field (dashed) and isopotential (solid) lines around weakly electric fish. Field shown with values of isopotential lines (V) at peak of EOD. Note the curvature of the 0 potential line away from the line perpendicular to the body that would be expected if the fish were a simple dipole. Note also the asymmetry in shapes and sizes of the isopotential lines at opposite ends of the fish's body. (From Heiligenberg 1977, © Springer-Verlag.)

Maximizing signal magnitude everywhere will help extend the range of electric signals. However, the ability of a distant receiver to identify and detect such a signal, or the ability of an electrolocating fish to interpret the distortions in its field induced by nearby objects, will depend very much on am-

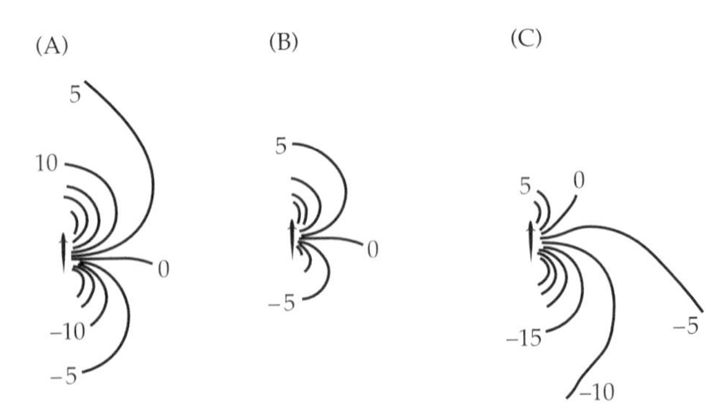

Figure 11.9 Effects of fish shape and skin resistivity on an electric field. (A) Electric field around a fish with a long tail and resistivity similar to that of the medium. Values show relative voltages. (B) Field for a fish with a shorter tail. (C) Field for a fish with a long tail but high skin resistivity. Note that the field has collapsed around the head. The equivalent plot for a fish with a long tail but much lower resistivity than the medium would show a field collapsed around the tail. (After Heiligenberg 1975.)

bient electrical noise. Throughout the tropics in which weakly electric fish live, lightning is the major source of electrical noise (Hopkins 1973). Although storms may be far away, the magnitude of the lightning discharge is large everywhere and results in a noise bandwidth of 50 Hz to 50 kHz, with peak noise levels at about 2 kHz. Not surprisingly, noise voltages in the higher-resistivity fresh water are about two orders of magnitude greater than comparable marine habitats. In addition to lightning, many tropical freshwater habitats host a variety of sympatric electric fish. These add additional noise to the environment. It should be noted, however, that compared to sound in either air or water, the number of biological sources of electrical noise is significantly fewer, allowing different species considerable latitude to vary EOD waveform and discharge patterns in ways that make conspecific signals distinctive against the background.

RECEPTION OF ELECTRIC SIGNALS

The ability to detect electrical signals is much more widespread than is the ability to generate them. In fact, there is convincing evidence that electroreception evolved very early in the vertebrate line: the earliest fish fossils from the Devonian show indications of having electroreceptors, and electroreception remains widespread in the most primitive groups of fish (Bullock and Heiligenberg 1986). Lampreys, cartilaginous fishes (rays, sharks, and skates), sturgeon, paddlefish, and coelocanths all have electroreceptors. Among the bony fishes (Teleostei), electroreception is less common being found in catfish (Siluroidea), Gymnotiformes, Mormyriformes, and the family Notopteridae. Adult aquatic salamanders and larvae of terrestial ones have electroreceptors, and the bill of the duck-billed platypus (*Ornithorhynchus anatinus*), an aquatic mammal, appears to be electroreceptive (Scheich et al. 1986). Electroreceptors appear absent in both adults and tadpoles of frogs and toads. The distribution of known electroreception over the early vertebrate tree is summarized in Figure 11.10.

The functions of this widespread electroreception appear to be varied. Sharks and rays are highly sensitive to low-intensity DC and AC electric fields generated by their prey. Most living animals have regions of the body that are at slightly different resting potentials. In water, these generate weak DC fields that are detectable at short distances from the animals. If the animal is moving, the depolarizations of its muscles generate AC fields with frequencies in the range of 0.1–8 Hz. For example, Kalmijn (1988a) reports field strengths of 10^{-7} V/cm at a distance of 10 cm from the pulsating gills of marine fish. Since sharks and rays can detect fields as low as 10^{-9} V/cm, they should be able to locate prey using electroreception.

Electric fields appear to be used by a number of marine animals to navigate. Any time a conductor is passed through a magnetic field, a perpendicular electrical field is generated in the conductor. The flow of ocean currents through Earth's magnetic field generates electric fields on the order of 10^{-9} to 10^{-7} V/cm. Rays navigate by using these induced fields to identify topographic directions and possibly locations. The magnetic field of Earth has di-

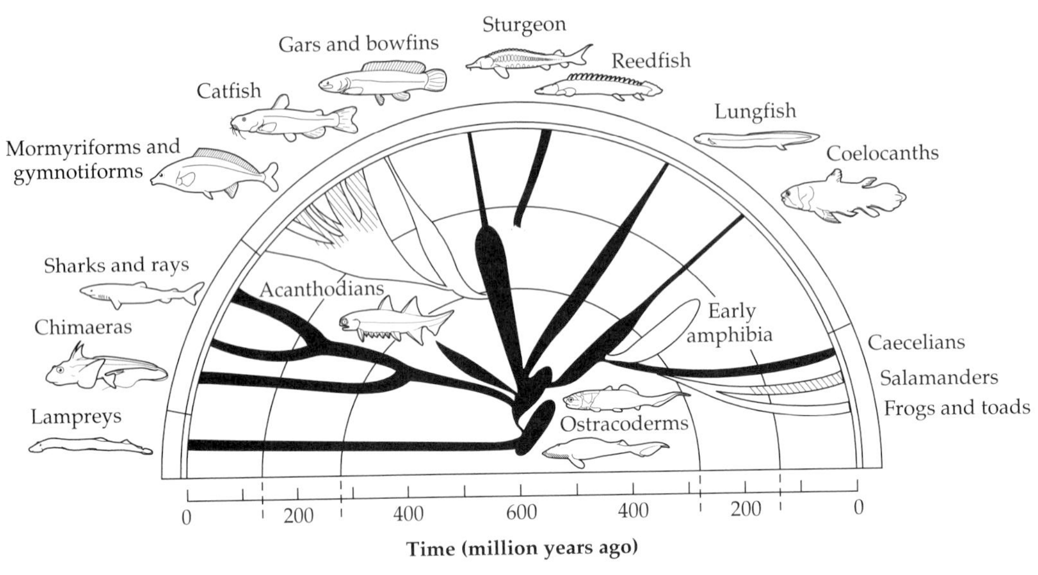

Figure 11.10 Electroreception in the early vertebrate evolutionary tree. Common names for major groups are indicated. Branches in black are those in which electroreception has been demonstrated or inferred from fossils and is the rule. Shaded branches are ones in which electroreception is present, but only in some taxa. (After Bullock and Heiligenberg 1986.)

rectional and locational properties. A shark swimming through the horizontal component of Earth's magnetic field experiences an induced electric field with an associated potential difference between its dorsal and ventral sides. The vertical component of the magnetic field generates a potential difference between its right and left sides. Both factors provide locational information that the shark might use to navigate over considerable distances. Freshwater fish do not have access to oceanic currents and thus experience 30% lower magnitudes of induced potentials as they move. However, the differential ionic compositions of the substrates over which fresh waters lie can generate electrochemical potentials that are location-specific and potentially available as navigational cues. Prey detection and navigation thus seem the likely selective forces favoring the widespread abilities of fish to detect electric fields.

Optimal Design of Electroreceptors

The sensory cells in fish electric receptor organs appear to be derived from the same kinds of sensory cells seen in the pressure receptors of the lateral lines of fish and in the equilibrium and hearing organs of all vertebrates. All electrosensory cells are stimulated when a sufficiently strong voltage difference is placed across their external and internal sides. Depending upon the voltage gradient in a particular field, the difference in voltages at any two locations may be small or large. To maximize this difference, the sensory organ needs to sample the field at points as far apart as possible (as in marine fish), or be located in that part of the field with the steepest gradient (as in freshwater fish).

In either case, it is easiest to see what is happening by treating the detecting fish as part of an electrical circuit made up of several resistors in series and connected to some voltage source that is generating the electric field. Consider a voltage V at the source applied to a chain of two resistors with resistances R_1 and R_2, respectively. Current moves from one pole of the voltage source through the first resistor, then through the second, and finally back to the opposite pole of the voltage source (Figure 11.11). Because the same amount of charge must enter the circuit as emerges from it, the value of the current I through the circuit must be equal at all points and by Ohm's law is $I = V/(R_1 + R_2)$. Given that the current through the two resistors is the same, Ohm's law also requires that $I = V_1/R_1 = V_2/R_2$, where V_1 is the voltage experienced by the first resistor and V_2 that experienced by the second. If we combine these expressions, we find that the voltages experienced by R_1 and R_2, respectively, are

$$V_1 = V\left[\frac{R_1}{R_1 + R_2}\right] \quad \text{and} \quad V_2 = V\left[\frac{R_2}{R_1 + R_2}\right]$$

Clearly, $V = V_1 + V_2$. If one imagines the trajectory of a test charge traversing this circuit, the charge will see the highest potential (e.g., be at the "top of the waterfall") as it emerges from the voltage source, will "drop" down by an amount V_1 as it emerges from the first resistor, and then will drop the remaining amount V_2 as it emerges from the second resistor and reenters the source.

Returning to the sensory cell of an electroreceptive fish, we can let R_1 be the resistance between the two sides of a sensory cell, and R_2 be all of the other resistances experienced by a current flowing from the voltage source, through the medium and the fish, and then back to the source. Clearly, the larger R_1 is relative to R_2, the bigger the fraction of the overall V that the sensory cell will see, and the more likely the cell will detect the electric field. If the electric field is not static, but varying, then the receptor cells must also deal with capacitative reactance: the higher the frequency of the field variation, the lower the effective impedance (resistance and capacitative reactance combined) across the cell, and the less likely the cell is to be stimulated. If the cell is trying to detect low-frequency signals, then a simple resistive solution

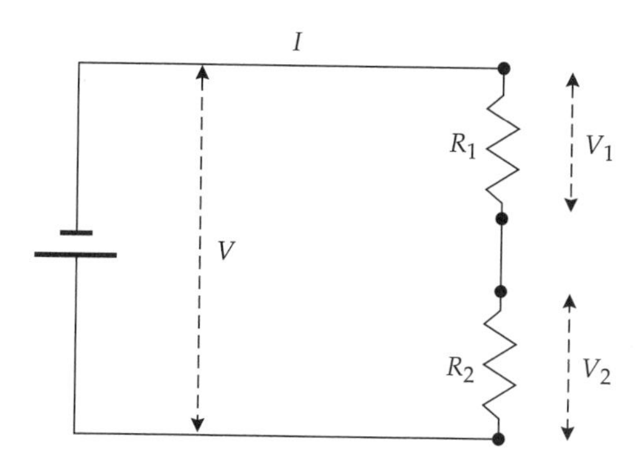

Figure 11.11 Simple circuit with two series resistances (R_1 and R_2) and voltage source (V). Ohm's law states that voltage drop across each resistor (V_1 and V_2, respectively) is the total voltage (V) times the fraction of total resistance represented by that resistor [e.g., $V_1 = V(R_1/R_1 + R_2)$].

is appropriate. If, however, it needs to detect high-frequency field variations, additional adaptations are required. In electroreceptive fish, different sensory organs have evolved to deal with the different types of signals. We describe the major types in the next sections.

Ampullary Electroreceptors

The most widespread type of electroreceptive organ among fish is the **ampullary receptor**. This organ consists of a tube opening to the outside through the fish's skin, and extending to a small cavity lined with a few (in freshwater fish) or thousands (in marine fish) of receptor cells. In freshwater forms, the resistance of the fish's skin is very high (to protect against osmotic influxes from the medium). The interior of the fish is saline and thus has very low resistance; the medium resistivity is intermediate. This means that the major drop in a source voltage occcurs across the skin, with much smaller drops occurring as current flows through the medium and the fish's body. If the fish places its sensory cells with one side facing the medium and the other facing the saline interior, e.g., as part of the skin, the cells will experience the maximal possible voltage difference across their two sides. For this reason, the tubes connecting ampullary receptors in freshwater fish are extremely short, just enough to expose the exterior side to the medium voltage and the interior side to the potential inside the skin.

In marine forms, the skin and interior of the fish are more similar in resisitivity, whereas the medium represents a much lower resistance. To obtain a detectable potential drop, the ampullary tubes are usually much longer in marine forms. The tube itself has highly insulating walls and is filled with a conductive jelly that ensures that the potential near the exterior side of the sensory cells is very close to that near the tube opening. The potential at the interior side of the cells is determined by the amount of skin and tissue that any current has had to traverse to get to that point. By making the tubes long, the amount of tissue travelled is large, and the difference in potential on the two sides of a sensory cell is increased. In sharks, ampullary organ tubes are located on the head and are up to several cm long; in rays, the tubes open all over the ventral surface and may be as long as 20 cm.

Ampullary organs are present in all electroreceptive fish and amphibians and are used to detect DC or low-frequency electric fields of low amplitude. These are the kinds of fields encountered by fish when hunting for prey or navigating. Detection of such low intensity exterior fields is complicated because the fish's body is continuously generating its own internal fields as muscles and nerves depolarize, and as DC currents move from one part of the body to another. One sensory cell exposed to a given exterior voltage might thus be stimulated because its interior side was close to an internal voltage source, whereas another sensory cell exposed to the same exterior voltage, but located more distantly from the internal one would not. As a solution, therefore, many species such as rays have large numbers of ampullary organs together in clusters, even though the tubes for these organs open at very different places on the ray's body (Figure 11.12). This placement makes the internal

Figure 11.12 Ampullary organ canals in a ray. Each canal opens separately on the ventral side of the ray where it can be used to sense the low-frequency electrical signals of potential prey. Many different canals conduct samples to the same cluster of ampullary organs (open circles). Thus the different external voltages can all be compared to the same internal one. Such organs may also play a role in local navigation abilities.

voltages seen by all cells in a cluster similar, and any differences in voltage seen by different cells in a cluster are due only to differences in the external voltages at their tube openings.

The minimum field detectable by an ampullary receptor varies with the taxon and the habitat. Marine cartilaginous fishes can detect fields as low as 5×10^{-9} V/cm, whereas freshwater species require a minimum of 5×10^{-5} V/cm. Bony fish are all much less sensitive than sympatric cartilaginous ones: marine forms require about 5×10^{-4} V/cm, whereas freshwater forms need 5×10^{-3} V/cm. Given the much higher noise levels due to lightning in fresh water, the lower sensitivities in this environment may not be surprising. The polarity of a stimulating field also varies with taxon: in cartilaginous fish, it is the exterior side of the sensory cell that is excitable: these cells are thus stimulated when the exterior medium is negative relative to the fish's body. In bony fishes, it is the interior side of the sensory cell that is excitable, and these species respond when the medium becomes positive relative to the fish. Such organs act like **rectifiers** because they only respond when the current is moving in a given direction.

Although the direction and magnitude of electric field forces vary with location, it is very difficult for a receiver to use this information to extrapolate the location of a voltage source. The reason is that the field lines rarely point directly at or away from the source; instead, they form curved trajectories. Kalmijn (1988a) has suggested that sharks find prey that generate potentials by swimming in a way that keeps their bodies oriented with a fixed angle relative to the field lines. Any deviation from a fixed angle will change the potentials experienced on the two sides of the body. One such orientation is parallel to the field lines, which would allow the shark to track a given field line to its voltage source.

Tuberous Electroreceptors

Tuberous receptors occur only in the Gymnotiformes and Mormyriformes where they are used for conspecific communication, electrolocation, or both. The electric fields that stimulate these receptors are EODs of the same fish or nearby fish. As we have seen, the EOD waveforms are highly time-variant and some modifications of the sensory organs are necessary to deal with frequency-dependent impedances.

Tuberous receptor organs also consist of a short canal opening at the skin and leading to a cavity lined with receptor cells. Unlike ampullary organs, the tuberous canal is partially filled with an epithelial plug, the walls of the canal are thickened with many cell layers, and the receptor cells are themselves encapsulated within other tissues. All of these features add to the resistances in the circuit other than the one across the sensory cell; they thus reduce that part of the voltage drop experienced by the sensory cell for DC and low-frequency signals. When the signal waveform is rapidly time-variant, then the plug and encapsulating tissues act as dielectrics, their capacitative reactances decline as the signal frequencies increase, and the contributions they make to the overall impedance of the circuit becomes small. Most of the voltage drop then occurs across the sensory cell. In short, the epithelial plug and encapsulations turn the organ into a high-pass filter so that the sensory cells can only respond to higher frequencies. The electrical properties of the sensory cell membranes are further modified to make them resonant at certain frequencies. Tuberous receptors are thus generally tuned to respond only within a fish's own specific range of electrical field frequencies.

In wave fish, (the Gymnarchidae within the Mormyriformes, and the Apteronotidae and Sternopygidae in the Gymnotiformes), the tuning of tuberous organs is very narrowly set to the specific pulse rate of the species' EOD. In some species, pulse rates differ between the two sexes, and within a sex, pulse rates of different individuals are slightly different. In all cases, the tuberous receptors of a given fish are tuned to the fish's own specific pulse rate (see review in Hopkins 1983b). In pulse fish, tuberous receptors are much more broadly tuned, but the majority of cells are best stimulated by the frequency with the largest amplitude in the power spectrum of the EOD waveform (Figure 11.13). Although the physics of tuberous receptors set an upper frequency limit of 12–15 kHz, the majority of species have EODs with major energy in the range of 200–1700 Hz.

Note that although most tuberous receptors are tuned, even in pulse fish, this does not mean that all signals are perceived only in the frequency domain. Hopkins and Bass (1981) have shown that holding the frequency spectrum of the mormyriform *Brienomyrus brachystius* EOD constant, but varying the relative phase spectrum results in quite different responses by attendant fish. Because the tuberous organs of these fish, like their ampullary organs, act like rectifiers, an organ is only stimulated when the external field is positive relative to the fish's interior. Any given field will generate opposite polarities on the two sides of a fish at any instant. When an EOD field rises to a positive value on one side, the organs on that side are stimulated, whereas those on

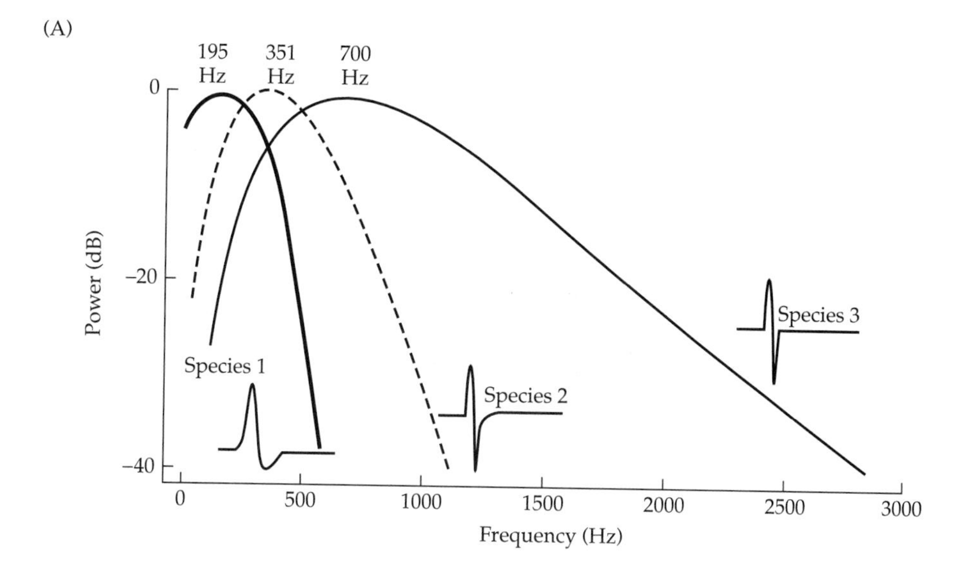

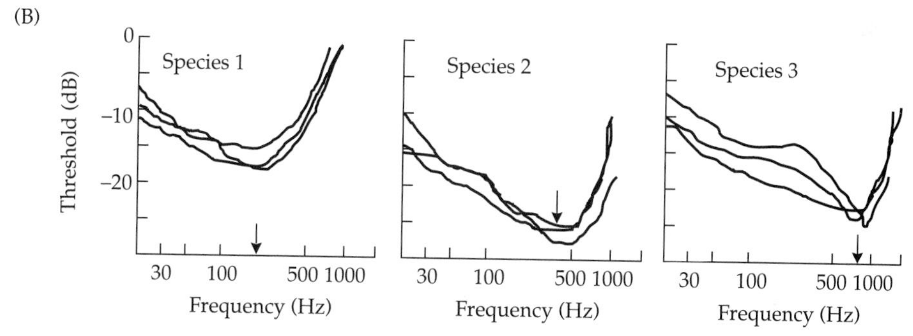

Figure 11.13 Matching of the EOD power spectrum to the best frequency of electric receptors. (A) Waveforms and power spectra of EOD waveforms for three species of the gymnotiform pulse fish, *Hypopomus*. (B) Corresponding tuning curves of receptor organs for each of the three species of fish. The reference value for the dB measure plotted on the y axis is here 90 mV/cm. In a tuning curve, the lower the threshold (y axis) for a given frequency (x axis), the more sensitive the unit is to that frequency. The best frequency is that with the lowest threshold and is marked in each plot with an arrow. In each case, best frequencies are very close to peak frequencies in power spectra. (After Heiligenberg 1977.)

the opposite side will be silent. By comparing the timing and pattern of tuberous organ stimulations on the two sides of its body, a *Brienomyrus* brain can basically reconstruct the successive maxima and minima in any received EOD waveform. It could thus perceive this signal in the time domain if both frequency and phase components were important.

In gymnotiforms, the same tuberous organs appear to mediate electrolocation and responses to the communication signals of other fish. However, there is a physiological division among organs depending upon which part of the EOD they encode in the nerves attached to them: some organs produce one nerve impulse per EOD received and thus reproduce the EOD pulse rate.

These are called **pulse markers**. Other organs produce bursts of impulses in which the number and rate of the impulses is related to the amplitude of the EOD field. These are called **amplitude coders**. In gymnotiforms, the amplitude coders are primarily used for electrolocation, whereas the pulse markers are used to identify the sex, species, or activities of nearby electric fish. Mormyriform pulse fish have separated these functions between two distinct kinds of tuberous receptors: **knollenorgans** are pulse markers devoted solely to detecting communication signals between fish; **mormyromasts** are amplitude coders dedicated to electrolocation functions.

COMMUNICATION AND ELECTRIC SIGNALS

How are these various electrogenic and receptive organs combined to effect communication? Passive electroreception of prey is not a form of communication as we define it, and will not be discussed further. However, both the exchange of electric signals between fish and electrolocation can be considered as forms of communication. We take up each of these in turn.

Social Communication

Most gymnotiforms and mormyriforms and probably many electric skates use electrical signals for social communication. Many of these species are either nocturnal or live in murky waters where visual communication is impossible or very limited. Some mormyriforms use both acoustic and electrical communication to defend territories, attract and court mates, and to determine the species, sex, or individual identity of another fish. Most other members of these groups rely only on electrical signals.

As we have noted, the shape of the EOD waveform largely depends on species and may be perceived by a receiver in either the time or frequency domain (or both). Modifications of the EOD waveform evolved by varying the durations and times of discharges of various membranes in the electric organ, or by varying the synchrony of discharges of several adjacent electric organs. Wave fish tend to produce EODs with narrow bandwidths, whereas pulse fish produce quite wide bandwidth signals. In both mormyriforms and gymnotiforms, there may be sexual differences as well, with male EODs generally being longer in duration than those of females. Playback experiments have shown that the two sexes are able to recognize each other's waveforms and males will court when played a typical female EOD.

Most social exchanges between weakly electric fish entail modulations of the EOD discharge rates. Such modulations include sudden rises in pulse rate with either abrupt returns or slow decays to normal, complete cessations of discharges, or continuous modulations up and down in wave species. The functional significance of these modulations appears to vary with the species and context. In several species, rapid increases and decreases in discharge rate are associated with aggressive interactions between fish, with subordinate individuals often responding by ceasing any discharge for some time. Dominance contests in wave fish such as *Eigenmannia virescens* result in hierarchies in which the dominant males have the lowest EOD rates, whereas the females controlling the best spawning territories have the highest EOD rates. The opposite relation holds in the wave fish *Apteronotus leptorhynchus*. Most wave fish show an ability to shift their discharge frequency slightly when a nearby fish is using exactly the same discharge rate. This response is called the **jamming avoidance reflex** and presumably serves, in part, to prevent one fish's electrolocation signals from being confounded by those of a neighbor. However evolved, dominant wave fish often alter their discharge rates to overlap those of a neighbor. Subordinate fish so threatened immediately shift their discharge frequencies.

In several wave gymnotiforms such as *Sternopygus macrurus* and *Eigenmannia virescens*, males try to attract passing females and court those that approach by adding periodic modulations to the usual discharge rate. Examples of such modulations for two gymnotiform wave fish are shown in Figure 11.14. Long strings of modulations by wave fish have been likened to "songs" of advertising male birds. Many male mormyriforms build nests and then seek to attract females who will mate with the male and lay eggs in his nest. Like some gymnotiforms, high-frequency bursts of pulses are used to advertise to nearby females and during courtship.

In fresh water, the range for electric communication is limited by the high resistivity of the medium and the size of the voltages that small fish can afford to produce repeatedly. As we have seen, it is also affected significantly by the shape of the fish: those fish with longer and thinner tails create electric fields with greater ranges. Nearly all weakly electric fish have a long thin tail. Several studies have shown that gymnotiform and mormyriform fish can communicate only over distances of about a meter.

Fish that communicate with electric organ discharges face the same problem in locating the source of an electric signal which a shark faces in localizing a prey organism producing a DC electric field: lines of force rarely point towards the voltage source but instead are curved. Recent work on the mormyriform *Brienomyrus brachystius* suggests that individuals use the same rule of thumb to locate conspecifics that Kalmijn (1988a) has suggested for sharks homing in on prey, that is, the fish basically keeps its body parallel to ambient lines of force while swimming randomly (Hopkins 1988b). The result is a trajectory that roughly tracks the current lines to the source (Figure 11.15).

Electrolocation

As we shall argue in Chapter 26, electrolocation can be considered as a form of **autocommunication**: a fish emits and receives its own signal. Differences between the outgoing and returned signal are then used to extract information about the fish's immediate environment. What kinds of differences might be detected?

The EOD of an electrolocating fish generates currents emerging from the anterior of the fish, passing through the medium, and reentering near the tail. Because the tuberous receptors are only stimulated when the medium is negative relative to the fish's interior, it is only the anterior outgoing current that can be detected. As a consequence, most of the receptors are concentrated in the head region and become much less dense as one moves towards the electric organ. In the gymnotiform, *Apteronotus albifrons*, the density of organs is

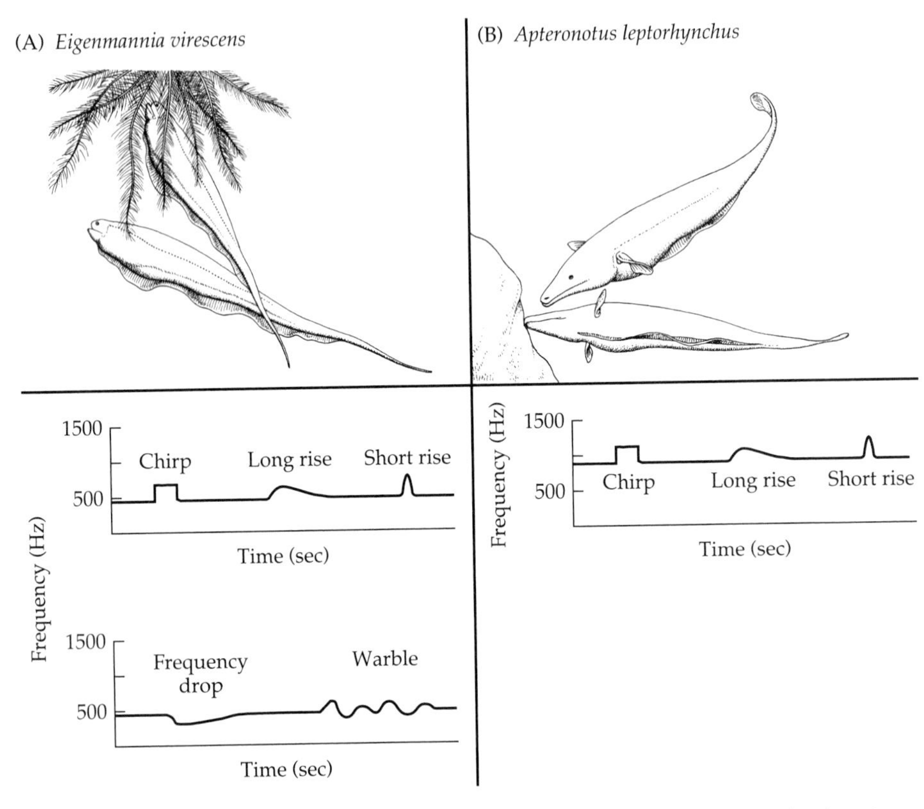

Figure 11.14 Electrical advertisement and courtship signals in a weakly electric fish. (A) Courting wave gymnotiform *Eigenmannia virescens.* The female is in the vegetation and the male below. The spectrograms show various frequency modulations in male electrical discharge rates seen during courtship. (B) Courting wave gymnotiform *Apteronotus leptorhynchus.* Male above and female below. The spectrogram shows several frequency modulations in male pulse rate during courtship. (After Hagedorn and Heiligenberg 1985.)

about $25/mm^2$ on the head and decreases to fewer than $1/mm^2$ in the region just anterior to the electric organ. We have already noted that it is the amplitude coders in gymnotiforms and the mormyromasts in mormyriforms that are the major receptors for electrolocation. It is indicative of the importance of electrolocation that in the latter group mormyromasts are generally 10 times more common than knollenorgans.

The voltage drop across the sensory cells of tuberous organs depends on their resistance relative to the total resistance through which a given current line will flow. Where objects of resistivity lower than the medium are present near to the fish, the total circuit resistance through these objects will be lower than that through medium without an object, and the fraction of the EOD voltage seen across the sensory cells will be higher; where objects of higher resistivity are present, the voltage across the sensory cells will be less. In addition, the more dielectric the objects, the lower their effective impedance to the time-variant EOD signal. The voltage across any tuberous receptor will thus vary

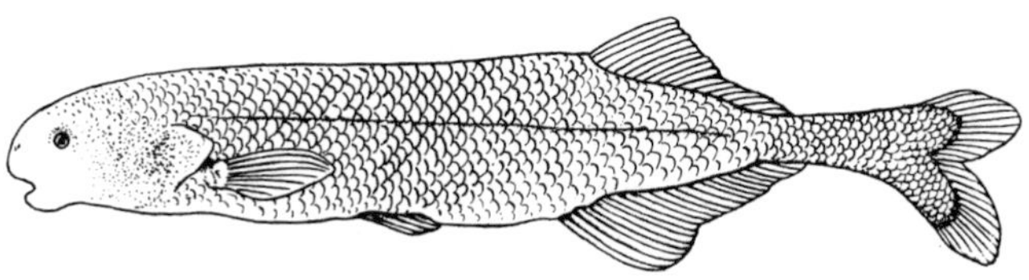

Brienomyrus brachyistius

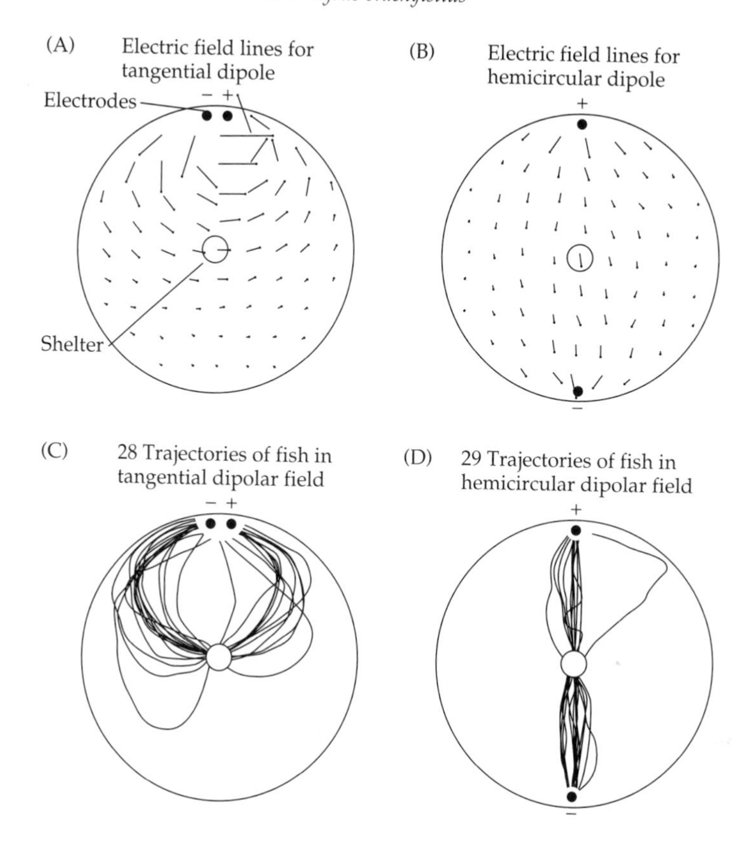

Figure 11.15 Location of the source of an electrical social signal. (A and B) Current and electrical field lines generated in two circular chambers by a pair of electrodes. Example A is a dipolar field (both electrodes close together near top of chamber) with elliptical force lines all passing through electrodes and extending into chamber. Example B is a linear field with electrodes on opposite sides of chamber. (C and D) Trajectories followed by mormyriform *Brienomyrus brachystitius* that leaves its shelter in the center of the chamber and approaches the source of the electrical field. In both cases, fish follow electric field lines to source of field. (From Hopkins 1988a, © Springer-Verlag.)

with the resistance and capacitative inductance of objects in the nearby medium. Because current flows perpendicularly across the skin and sensory cells, only those receptors near the object will experience these changes in voltage. These changes create an electrical "image" of the object on the nearest parts of the fish's body (rather like the shadow on the fish that the object would cast were a light placed somewhat beyond the object). The relative locations of receptor organs in electric fish are preserved in the fish's brain to form an electrical body map. The size and location of any electrical image can thus be monitored by the brain. As the fish moves, the location of the electrical image will move along its body and the brain can use this to infer the approximate size and location of the object. Electric fish often approach novel objects, curl their bodies around them, and then make circling movements around the objects. This curling concentrates their electric fields on the objects and presumably gives them information about object shape and composition.

The size of the voltage differences between receptors within the image of an object and those outside of that image increases as object size is increased and as the object is brought closer to the fish. However, both effects vary in ways that greatly constrain the use of electrolocation. The rise in signal amplitude as object diameter is increased is very slow: a doubling of object diameter only increases signal magnitude by about 25% and it requires a 10-fold increase in object diameter to double signal magnitude (Bastian 1986). Electroreception is thus a crude discriminator of object size. By the same token, signal magnitude falls off very quickly with distance from the fish: a doubling of distance reduces signal amplitudes to 30% of the initial value, and a 10-fold increase in distance leads to a signal 2% of the initial one. This limits the effective range of electrolocation to only 2–5 cm from the fish (in constrast to over a meter for electrical communication).

Whereas passive reception of prey electric fields favors receptor cells with high impedances, electrolocation favors impedances more similar to the medium (Heiligenberg 1975). This difference occurs because passive reception is limited by the minimal absolute voltage that a sensory cell can detect; electrolocation is limited by the minimal difference in voltages between two cells that can be detected by the brain. These are quite different goals. It is easy to show that the difference in voltages between a cell exposed to pure medium and another exposed to medium with an object in it depends on the product of two ratios: the first is the fraction of the total circuit impedance represented by the skin and sensory cells (as with passive reception), and the second is the ratio between the change in circuit impedance introduced by the object's presence relative to the total impedance in the circuit. Whereas increasing the skin impedance increases the first ratio, it simultaneously decreases the second. Because of this tradeoff, the maximal difference between regions is attained when the impedances of the medium and the skin are about the same. This result is, in fact, what is generally found (Bell et al. 1976).

SUMMARY

1. Although the ability to use electrical signals for prey detection and navigation is widespread in lower vertebrates, only certain taxa of fish have evolved specific organs for generating electrical signals for communication.

2. The basis of electrical communication is the generation of an **electrical field**. This is a map of the direction and magnitude of electrical forces at any location. Electrical fields are usually generated by the short-term separation of positive and negative **charges**. In the vicinity of these separated charges, it takes work to move any test charge against the electrical field. The amount of work it takes to move a unit test charge between two locations is called the **potential difference** between those two points and is measured in **volts**.

3. A single charge is called a **monopole** and the field around it is fairly simple in shape; this is an unlikely biological configuration. Two separated charges constitute a **dipole** surrounded by a somewhat more complicated

electrical field. Even more complicated arrays of separated charges will generate fields with dipole, **quadrupole**, **octupole**, and more complex axes and shapes. The magnitude of the electrical field around a dipole decreases with the cube of the distance from the dipole, and the potential around the dipole falls off with the square of distance. Because higher-order fields fall off much faster than dipolar fields, we can divide the electrical field around a complex array of charges into a **near field** close to the array, (where all quadrupole, octupole, and other components are still important), and a more distant **far field** (where only the dipolar component is still important).

4. A **conductor** is a material in which electrons or ions can move freely when placed in an electric field. The flux of electrons through a conductor is called **current** and its magnitude depends upon the conductor's **resistance** and the potential difference applied across it (**Ohm's law**). If the electric field is unchanging, the current is unchanging and is called a **direct current** (**DC**). A **dielectric** is a material that cannot sustain a direct current. In a field that is time-variant, however, dielectrics may be able to conduct **alternating currents** (**AC**) if the frequency of alternation of the field is high enough relative to the physical properties of the dielectric. The effective resistance of a dielectric at a certain frequency is called its **capacitative reactance**. Many substances have both conductive and dielectric properties so that one must know both their resistivity and their capacitative reactance to know how much current they will carry when placed in a particular electric field. The combination of both resistance and reactance at a given frequency is called the **impedance** of a material.

5. Organisms generate electric fields for communication by ensuring differential diffusion of ions down their concentration gradients. This diffusion creates an **electrochemical potential**. Fish produce many modified muscle or nerve cells called **electrocytes** in series to form **electric organs**. The simultaneous discharge of all the electrocytes in an organ at once creates a brief electric field which may swing both positive and negative several times before it is over. The shape of this electric organ discharge (**EOD**) waveform depends on the species and is guaranteed by particular anatomical and timing adaptations of the electric organ.

6. **Strongly electric fish** use EODs with hundreds of volts to stun prey and protect themselves. **Weakly electric fish** produce electric signals of only a few volts, which they use for social communication or for the detection of objects close to their bodies (**electrolocation**). The range over which they can detect their own or another fish's signals depends upon the relative resisitivity of the medium and the fish, and upon the levels of ambient electrical noise due to lightning. Weakly electric fish can be divided into **wave fish**, which maintain steady high rates of discharge, and **pulse fish**, which have much more erratic changes in EOD emissions.

7. **Ampullary receptor organs** are widespread in fish and the aquatic members of a few other vertebrate groups. They are very sensitive and respond only to the low frequency or DC potentials inadvertently produced by

prey and by Earth's magnetic field. Electrical communication is effected by **tuberous receptor organs**. These organs are usually tuned to respond to particular ranges of AC electric fields and only occur in fish that generate their own fields with electric organs. Some tuberous organs (**pulse markers**) respond to the rates of EODs, whereas others (**amplitude coders**) respond to the magnitudes of EOD fields. Pulse markers are usually used for social communication; amplitude coders are usually used for electrolocation.

8. **Social communication** between freshwater electric fish is limited to distances of about one meter. The waveform of the EOD and the rate of discharge, especially by a wave fish, provide important information about species identity. These same features may also vary, within species limits, according to sex, status, and individual identity. Aggressive and sexual interactions between fish are usually mediated by stereotyped modulations in the EOD emission rates.

9. Electrolocation entails emission of an EOD and careful monitoring of the consequent stimulation of sensory cells. Objects in the water near the fish change the external impedances and hence the voltage experienced by nearby tuberous receptors. Because receptor location is conserved as receptor signals travel to the fish's brain, the fish can place the location of a nearby object relative to a general body map. Different objects have different resistances, capacitative reactances, and sizes, and all of these can be used by the fish to infer information about the object. Electrolocation only occurs in the near electric field at distances of 2–5 cm (about 2% of the range for social communication by electric signals).

FURTHER READING

Good reviews of electrostatics and electric fields can be found in Feynman et al. (1964) and Paul and Sasar (1987). An early but classic review of electric signal generation and detection can be found in Bennett (1970). More recent reviews are found in Bullock and Heiligenberg (1986) and Hopkins (1974c, 1977, 1988b). An excellent comparison of electric-signal-generating organs can be found in Bass (1986), and Bell et al. (1976), Knudsen (1975), Heiligenberg (1975, 1977), and Kalmijn (1988a) analyze the shapes, ranges, and loading of fields generated by electric organs. The problems of ambient noise for both electrolocation and communication are treated by Brenowitz (1986) and Hopkins (1973). Electric reception organs are nicely reviewed and classified by Zakon (1986, 1988), and Scheich et al. (1986) describe electroreception by a primitive aquatic mammal. The process of electrolocation is discussed by Bastian (1986), Davis and Hopkins (1988), Hopkins (1983b), Heiligenberg (1977), and Meyer (1982), and electric social communication is the focus of papers by Black-Cleworth (1970), Hagedorn and Heiligenberg (1985), Hagedorn (1986, 1988), Hagedorn and Zelick (1989), Heiligenberg and Hopkins (1976), Hopkins (1972, 1974a,b, 1983a, 1986, 1988a), Hopkins and Heiligenberg (1978), Pimental-Souza and Fernandes-Souza (1987), and Westby (1975, 1988).

Part II

Optimizing Information Transfer

PART II OUTLINES THE BUILDING-BLOCKS of animal communication—the signal production organs, propagation effects, and receiver organs for each candidate modality. It also lists a variety of physical factors that constrain the signal options available to both senders and receivers. Despite the selection of a particular modality and the imposition of environmental constraints, senders still have a large number of alternative signals they might produce. Similarly, receivers might be able to respond to a wide variety of signals. Which of the allowable signals in a given situation is the best one for sender and/or receiver? What criterion do we invoke to identify the "best" signals? Since signals are presumably providing "information" to receivers, how much information should a sender seek to provide and a receiver to extract from signals? What do we mean by information when we are discussing animal communication? All of these questions are related to the **economics of communication**, our focus in Part II.

Throughout this part of the book we assume that communication is a cooperative process in which both sender and receiver benefit. The question is then how to determine which signals are optimal in any given context. In Chapter 12, we define what we mean by communication and outline a general economic model that can tell us when it pays for sender and receiver to engage in a signal exchange. Chapter 13 defines what we mean by information and how it can be measured. In Chapter 14, we use the economic model of Chapter 12 to identify decision rules by which receivers might best exploit the information provided by signals. Chapter 15 then looks at alternative coding systems through which signals provide information value. With these general economic principles in mind, we then examine the respective roles of phylogenetic history (Chapter 16), communication costs (Chapter 17), and signal function (Chapter 18) in the evolution of signal form. The joint effects of these three constraints go a long way toward explaining the observed diversity of animal signaling systems.

Chapter 12

Optimizing Communication

COMMUNICATION INVOLVES THE PROVISION OF INFORMATION by one animal to another. The exchange of information presumably benefits both parties and the acquisition of these benefits is the function of the exchange. The information exchanged will vary with the situation and so may lead to different types of signals in different contexts. We have seen how the physical environment and available physiological equipment constrain signal design. Signal exchange despite these constraints imposes costs on both senders and receivers: communication is never free. The evolutionary challenge is therefore to find the signal design that maximizes the difference between communication benefits and costs for each context. Finding the best signal is called **optimization** and we shall spend much of this chapter outlining how optimization might be achieved. To start, however, we must provide a more formal definition of animal communication.

THE COMMUNICATION TASK

What Is Communication?

Communication is the provision of information that can be utilized by a receiver to make a decision. All animals make behavioral and physiological "decisions" for given ambient conditions. If they perceive one condition to be true, they adopt one action, and when an alternative condition appears true, they adopt another. Information is important because animals are often uncertain about which alternative condition is currently the case. For example, consider cichlid fish in which mated pairs aggresively defend nest sites (Figure 12.1). Suppose a fish without a territory encounters another defending a good nesting site. If the site owner is smaller than the intruder, it may pay the intruder to attack the owner and take over the site. If the owner is larger than the intruder, then the latter may do best to leave without provocation. The uncertain condition faced by the intruding fish is the relative size of the owner. The obvious solution is to measure the owner. In many cases, **direct assessment** of the condition in question is inaccurate, impossible given available sensory modalities, or dangerous. Animals may then have to turn to **secondary sources of information**. There are four such secondary sources: (1) knowledge of prior probabilities, (2) assessment of indirect cues, (3) exploitation of amplifiers, and (4) reception of signals. Often a decision-maker will rely on several of these mechanisms at the same time.

First, consider **prior probabilities**. If the intruder has discovered during its wanderings that it is often larger than other cichlids, the prior probability that it is again larger is high and, in the absence of other information, it should attack. If it has no prior experience, then it may have to assume a chance probability of 50% that it is larger. In this case, the prior probabilities are not helpful.

Alternatively, the intruder may try to estimate properties that are correlated with the owner's body size but are also more easily, or less dangerously, measured. Assessable properties that are correlates of a condition of interest

Figure 12.1 Territorial cichlids at nest sites. Many cichlid fish mate as pairs and aggressively defend their nest sites and fry. Females tend to be more aggressive than males and here a female *Cichlasoma beani* (on left) signals by her dark coloration her willingness to attack even her own mate, whereas the male (on right) keeps his distance and exhibits a less threatening coloration. (Photo courtesy of George Barlow.)

for reasons other than the provision of information are called **cues**. [Note that we stick with Seeley's (1989) original usage of this term rather than follow Hauser's (1996) redefinition of cues as permanent signals or "states".] As the owner moves about cleaning its nest site, it will produce a local turbulence in the water that the intruder could monitor at a safe distance as an indication of the owner's body size. Turbulence is a cue; it is not produced by the owner to provide information to the intruder but is an incidental consequence of its nest cleaning activities. Put another way, the owner benefits from these movements whether an intruder is present to detect them or not.

An **amplifier** is a trait that makes direct or cue assessment easier and/or more accurate (Hasson 1989b, 1991, 1994). The amplifier itself provides no information, but its presence results in the extraction of more information than if it were absent. Many fish have constrasting lines demarcating their body margins (Figure 12.2). It has been suggested that this outlining makes measurement of body size by other fish easier and thus acts as an amplifier (Zahavi 1987). It is also possible for animals to evolve traits that make direct or cue assessment more difficult; such traits are called **attenuators**.

The final source of information is from **signals**. In Part II of this book, we shall define a signal to be any action or trait generated by one animal (the **sender**) which provides information used by another animal (the **receiver**) to select an action beneficial to both parties. To qualify as a signal, the benefit of the sender's action to both parties must depend upon this exchange of information. Suppose an owner cichlid sees a nearby intruder and performs an action that provides information about its body size. For example, it might use its mouth to lift and wave about a pebble that only a large fish could manage. Or it might approach the intruder and undulate its body to generate turbulence of a magnitude commensurate with its body size. In either case, the details of the performance are correlated with owner body size; they thus provide information to the intruder. Although both cues and signals provide information to other animals, and both are generated by actions which in some way benefit their producer, it is only in the case of signals that the provision of information to another animal is a prerequisite for obtaining that benefit. The benefits to an action generating a cue do not depend upon information transfer; in fact, the provision of information by inadvertent cues is often detrimental to their producers.

Signals are not the only actions that an animal may perform to influence the decisions of another. For example, an owner cichlid might move into a position from which it is easier to attack the intruder were it to approach further. This movement is a **tactical behavior**; it changes the relevant conditions and thus the propriety of alternative actions by the animal making the decision. By contrast, an action that is a pure signal provides information about current conditions; it does not change them. For example, the owner might adopt a darker color pattern. This is a signal indicating that it is likely to attack. In practice, actions may have both tactical and signal value (Enquist 1985; Grafen 1990a). Moving to an attack position both changes the conditions and provides information to the other fish about the likelihood of attack. Rather

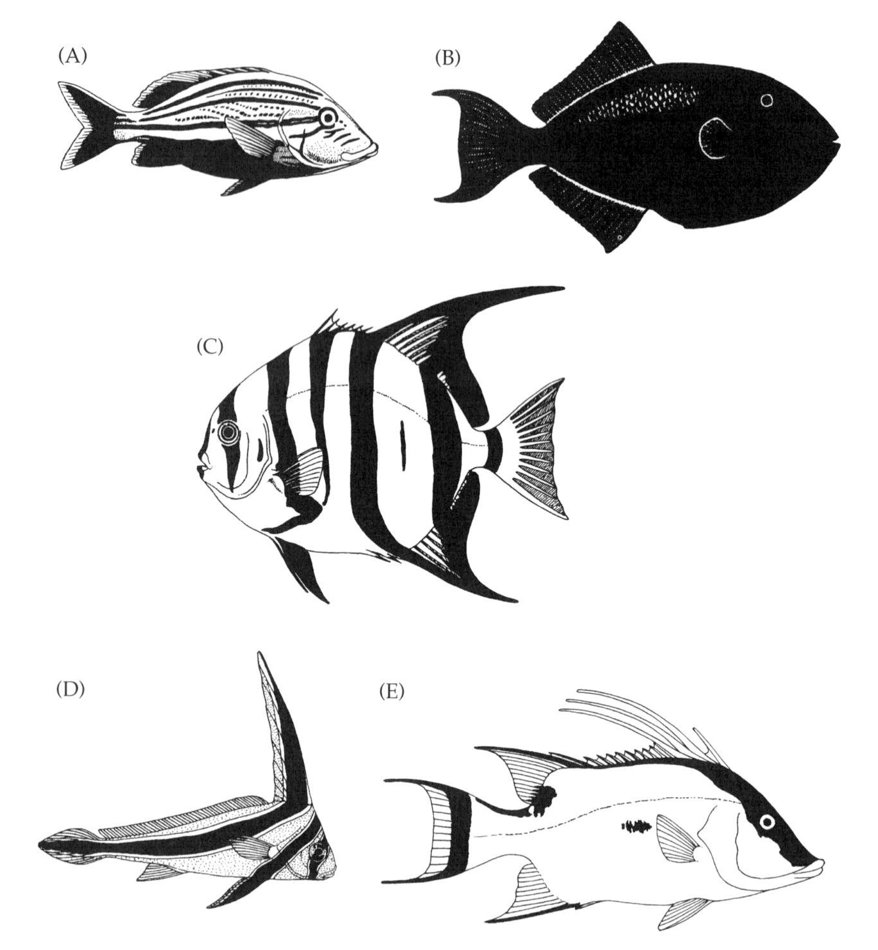

Figure 12.2 Amplifier markings on fish. Relative body size in fish is a major factor determining territorial ownership, success during conflict, and attraction of mates. The ability to measure the body size of another fish may discourage smaller fish from attacking. Many fish have evolved amplifier markings that emphasize their dimensions and thus make comparisons of body sizes easier (Zahavi 1987). In these coral reef fish examples, (A) the Spanish grunt (*Haemulon macrostomum*) and (B) the black durgon (*Melichthys niger*) accenuate their lengths with horizontal outlining of the body edges. (C) The spadefish (*Chaetodipterus faber*) emphasizes its height with vertical lines, whereas (D) the jackknife fish (*Equetus lanceolatus*) and (E) the hogfish (*Lachnolaimus maximus*) have both vertical and horizontal lines accentuating the major axis of whichever part is marked. (After Chaplin and Scott 1972.)

than ignore these hybrid actions, we shall consider any individual that performs an action with at least some signal value to be a sender. We can later partition the benefits of the action into tactical and signal components, and even rank actions by the relative importance of the signal value. This broad definition thus captures all signal situations from the viewpoint of senders.

We shall apply a similarly broad definition to receivers. A **receiver** is any animal that relies on signals, at least in part, to make a decision and that benefits as a result. We thus accept as a receiver an animal that largely relies on prior probabilities, cue assessment, or amplifiers for its decision as long as a signal plays at least some role. The quantitative issue of how large a role signals play in a decision is important, but we are only likely to consider it if we define receivers sufficiently broadly enough to include a full range of emphases. Decisions made primarily using amplifiers pose a problem: on one hand, amplifiers by themselves do not provide information; on the other, amplifiers share a number of properties with true signals, and they may be precursors of some signals (Hasson et al. 1992). In this book, we shall include consideration of decisions using amplifiers (and attenuators) because they provide some unique and important perspectives on how signals evolve.

Communication is then an exchange of a signal between a sender and a receiver to the benefit of both parties. In the parlance of behavioral ecology, we can say that the **function of sending a signal** is to increase the chances that the receiver will choose that action most beneficial to the sender; for the receiver, the **function of responding to the signal** is to increase its chances of choosing that action which is best for it. The critical element in all of this is the presence of a signal. Operationally, it is sometimes difficult to decide whether a given exchange involves a signal or not. This is especially true if the receiver relies largely on prior probabilities and cues and only marginally on received signals, and/or if the sender achieves both tactical and signal benefits with a given behavior. One factor that increases the chances that an exchange is communication is distance between sender and receiver. We have seen how distance distorts and attenuates propagated sounds, visual patterns, odorant spaces, and electric signals even if there is careful stimulus design by the emitter. Direct assessment (with or without amplifiers) will thus be more difficult the farther apart the sender and receiver. Because cues are inadvertent correlates of focal conditions, there is no reason for them to be refined or tuned to minimize distortion and attenuation when detected from a distance. Cues will thus suffer major distortions during long distance propagation. In contrast, signals function primarily to provide information. There is thus every reason for them to be refined and tuned by senders to resist attenuation and distortion. As distances between sender and receiver increase, we expect signals to play the predominant role in receiver decision-making. For similar reasons, a sender will be less able to incorporate tactical components into its actions as the distance between the parties increases. Again, signals become the primary focus of a sender when the relevant receiver is far away.

By the same token, we must be very careful when examining interactions at short ranges. Here direct assessment, cues, and amplifiers are all likely to play important roles in receiver decisions, and senders are likely to combine tactical and signal functions in their actions. It will take much more careful observation and experimentation to quantify the degree to which signals are important features of a short-range exchange. One could try to avoid this problem by invoking a more narrow definition of communication. But this would exclude many of the most interesting and illuminating cases that we wish to consider. It is also short-sighted because receivers are going to use every secondary source of information that they can find. This fact fundamentally shapes the design and role of signals, and we would be foolish to ignore it. We thus opt for our broad definition and its attendant risks in the interpretation of short-range interactions.

The Questions Answered by Animal Signals

We have posed the problem faced by the receiver as a **question** that needs to be answered. For each question, there are several alternative answers, (**conditions**) that may be true. The sender emits one of several alternative signals according to a **code** that correlates signal form with conditions. We shall call the ensemble of signals assigned to the same question a **signal set**. Different questions may be asked in different **contexts**. The entire list of signals pooled across all questions and signal sets is the **signal repertoire** for that species. These terms are summarized diagrammatically in Figure 12.3.

The size of the signal repertoire required by any species is set by the number of questions its members ask and the number of possible answers for each. Questions can be assigned to one of three broad categories. Note that the corresponding signals may not be so easily divided: a given signal might provide information dealing with questions in all three categories simultaneously. The first category of questions relates to **sender identity**: which individual is present, and to which species, social group, sex, and age class does it

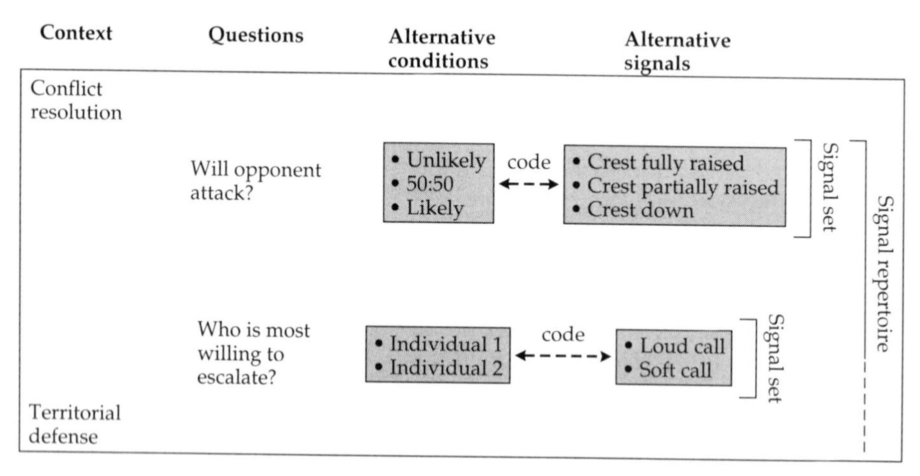

Figure 12.3 Summary of terms used to classify animal signals. Context refers to the general situation in which the signal exchange occurs. Two contexts, conflict resolution and territorial defense are listed here. Within each context, a variety of questions may be asked by receivers. For each, there are alternative conditions (answers) that may be true. Senders have access to several alternative signals that they associate with these alternative conditions according to the signal code. Those signals assigned to a particular question constitute a signal set; the entire list of signals for a species is its signal repertoire.

belong? The second category of questions relates to **sender location**: what is its distance, altitude, and compass angle relative to the receiver? Is it within or without a territorial boundary? Is it close to the nest? The final category concerns questions that are specific to particular **contexts.** Knowledge of the context in which a communication exchange occurs helps identify the likely questions that a receiver might confront, the degree to which other secondary sources of information and tactical behaviors are likely to affect a receiver's decision, the benefits and costs of communication (our concern in Part II of the book), and the likelihood of honest communication (an issue taken up in Part III). We distinguish between seven general contexts.

CONFLICT RESOLUTION. Signals produced in this context help answer questions arising during aggressive encounters between animals (Figure 12.4). They are often called **agonistic signals**. In many circumstances, escalated violence over ownership of a mate or commodity is less adaptive than an exchange of threats and submissive signals. Agonistic signals provide information about the likely intentions and levels of commitment of their senders. They may also provide information about relative fighting ability should further escalation occur. Conflict resolution usually occurs at short sender-receiver distances. Senders are thus likely to perform actions with both tactical and signal functions, and receivers are likely to make decisions based upon information pooled from all of the likely secondary sources.

TERRITORY DEFENSE. Whereas the establishment of a territory initially involves conflict resolution with its associated short-range signals, maintenance and defense of that territory raise different questions and involve a different set of longer-range signals. **Territorial signals** are thus less likely to include tactical components by senders and are more likely to be the major determinant of decisions by receivers. They indicate the presence of a territorial owner in a given

Figure 12.4 Threat signal of the great skua. Great skuas (*Stercorarius skua skua*) have at least six different displays or calls that can be followed by an attack or retreat by the sender. The calls vary subtly in the relative ratios of the two subsequent actions and thus allow skuas to communicate a wide range of agonistic intentions. The display shown here is the "oblique posture" with wing waving and loud call emission. It is more likely to be followed by attack than by retreat. The vocalization and white markings on the waving wings make this display highly conspicuous as might be expected from a more aggressive bird. (Photo courtesy of Malte Andersson.)

location, demarcate territorial boundaries, and often include concomitant information about the identity and location of the owner (Figure 12.5).

SEXUAL INTERACTIONS. **Sexual signals** mediate a two-step process: **mate attraction signals** provide information on location and availability that allows members of the two sexes to find and approach each other, whereas **courtship signals** determine whether subsequent mating will occur and effect its coordination (Figure 12.6). At one or both stages, interaction may be broken off if one party rejects the other as a potential mate. This selectivity may rely on the provision of information about species identity, age, or phenotype during the interaction. Because mate attraction signals are usually given when male and female are far apart, selectivity at this stage is most likely to rely on signal information. Selectivity during the courtship stage, when male and female are close together, will often involve information sources in addition to signals, and senders are likely to combine tactical and signal functions in the same actions. Some sexual signals are given only after mating: **postcopulatory dis-**

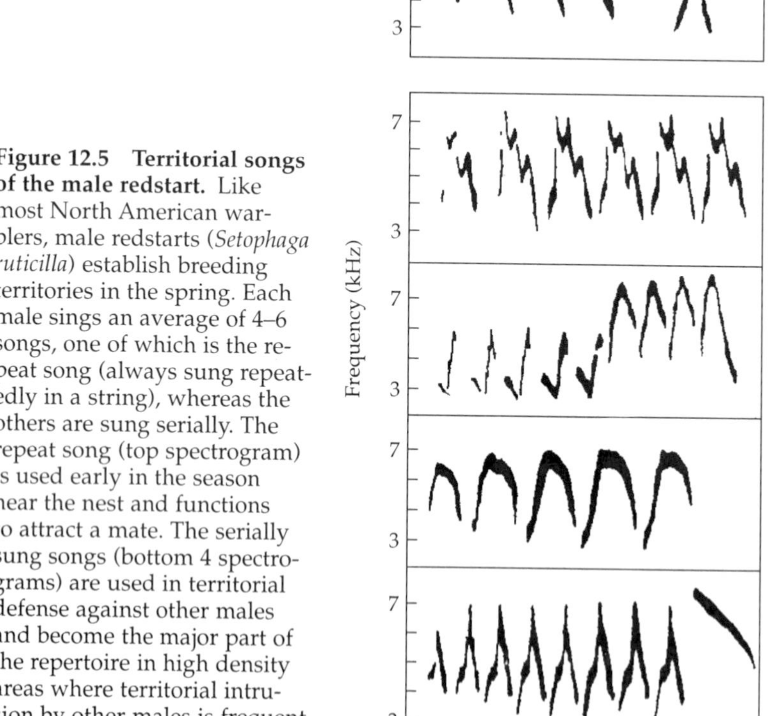

Figure 12.5 Territorial songs of the male redstart. Like most North American warblers, male redstarts (*Setophaga ruticilla*) establish breeding territories in the spring. Each male sings an average of 4–6 songs, one of which is the repeat song (always sung repeatedly in a string), whereas the others are sung serially. The repeat song (top spectrogram) is used early in the season near the nest and functions to attract a mate. The serially sung songs (bottom 4 spectrograms) are used in territorial defense against other males and become the major part of the repertoire in high density areas where territorial intrusion by other males is frequent. (From Lemon et al. 1987.)

(A)

(B)

Figure 12.6 Sexual signals produced by male Jackson's widowbird (*Euplectes jacksoni*). Males of this species set up circular display sites on the ground in high grass sites. (A) Females are attracted to males from a distance by a visual display in which males leap into the air above their display sites and emit a call. (B) Once females are attracted, the male hides behind the tall grass hub at the center of the display site and makes soft sounds until the female either solicits his approach or flies away. (Photos courtesy of Staffan Andersson.)

plays by one of the two sexes are known in ducks, bats, and some macaques, whereas both members may display after mating in geese and swans.

PARENT-OFFSPRING INTERACTIONS. There are often specific signals exchanged between parents and offspring. For example, parents announce their intention to provide food to offspring, and offspring communicate their current nurturing needs to parents. Parent-offspring signals often require concurrent information about identities and locations of senders. The range between senders and receivers in this type of signaling varies widely with the species, nesting situation, and age of the offspring (Figure 12.7).

Figure 12.7 Parent-offspring signals in passalid beetles. These temperate and tropical beetles live as monogamous pairs within rotten logs. A species from Trinidad (*Passalus* sp.) is shown here. Beetle pairs excavate tunnels in the logs and are thought to provision their larvae. Both larvae and adults maintain contact via stridulations that propagate throughout the wood substrate. (Photo courtesy of Jerry Wilkinson.)

SOCIAL INTEGRATION. Many signals are used to coordinate the activities of animals within social groups. Most of these are used at short ranges. Special signals are used to synchronize taking flight in many bird species, and inflight calls may direct movements once in the air. Many vertebrates produce low-intensity and short-range signals to maintain spacing and cohesion during foraging. Recruitment and assembly signals may be used to reduce distances between group members. Greetings and other affiliative signals are often exchanged when group members reassemble in a common location (Figure 12.8).

ENVIRONMENTAL CONTEXTS. Some signals provide information to conspecifics about conditions external to either sender or receiver, or involve exchanges with organisms in other trophic levels. **Alarm signals** indicate the presence of predators or other threats (Figure 12.9), whereas **food signals** alert conspecifics to the presence of a shareable food source. Animals may also emit signals intended to discourage attack by their predators, or to attract predators of their predators. Ranges between senders and receivers of environmental signals are highly variable.

AUTOCOMMUNICATION. Here, an animal uses the differences between the emitted and received versions of its own signal to extract information about ambient conditions. Examples include echolocation by bats and porpoises (Figure 12.10), and electrolocation by mormyrid and gymnotid fish. Given physical constraints, nearly all of these signals are short-range ones. Although the nature of this exchange seems quite different from that between different individuals, we shall see in Chapter 26 that autocommunication evolves under many of the same constraints as **heterocommunication**. That chapter

Figure 12.8 Greeting ceremony by wild dogs (*Lycaon pictus*). After sleeping for most of the afternoon, members of an African wild dog pack awaken and reaffirm relationships with oral greeting signals. Soon the pack will move off for its evening hunt. (Photo courtesy of James Malcolm.)

Figure 12.9 Female Beldings' ground squirrel (*Spermophilus beldingi*) giving alarm call to detected predator. Adult females of this species give alarms to terrestrial or distant predators only when they have close kin in the vicinity; alarms to approaching raptors are given regardless of kinship to nearby individuals. Likely reasons for these differences are discussed in Chapter 25. (Photo courtesy of George Lepp.)

sheds some interesting perspectives on the limits of information exchange by any type of signal.

In practice, animals often economize by loading information relevant to several questions into the same signal. Signals dealing with multiple questions are called **compound signals**. The answers to questions dealing with identity, location, and a specific context are often independent of each other, and compound signals frequently provide answers to all three at the same time. Because alternative contexts are often exclusive, it is less likely that a compound signal deals with questions relevant to more than one context. However, there are notable exceptions. For example, a male bird's song or a male mammal's scent mark might concurrently mediate both territorial defense against other males and the attraction of potential female mates. The questions answered thus differ depending upon the sex of the receiver.

Constraints, Contexts, and Signal Diversity

Part I of this book divided the physical and physiological constraints on signals according to modality, medium, and taxon. Using the distinctions made

Figure 12.10 Carnivorous bat, *Vampyrum spectrum*, hunting at night. This is the largest neotropical bat and feeds on sleeping birds, nocturnal rodents, and other bats. Despite the generic name, it is not a vampire. These bats hunt in dense forest using echolocation to avoid obstacles and to detect prey. (Photo by Nina Leen, Leen and Novick, 1969.)

in the prior section, we could also divide them according to the context in which they are given. Clearly, there are many possible combinations of signal modality, medium, taxon, and context. We might ask, is signal design matched to contexts at random? Or are some signal designs better than others in a specific context? Given that many different combinations will arise over successive generations, which ones will evolution retain and why? One of the most useful tools we can invoke to predict evolutionary outcomes is optimality theory. We introduce this method in the next section and then use it in subsequent chapters to explain observed patterns of convergence and divergence in signal evolution.

OPTIMALITY THEORY

We have already encountered many cases of **tradeoffs** in signal emission and detection organs. In each case, enhancement of one signal property invariably leads to detrimental effects on other properties. A resonant sac in a sender can increase signal amplitude, but it will simultaneously reduce signal bandwidth (Chapter 4, pages 83–85); increasing the frequency resolution of a receiver's ear eventually reduces temporal resolution and viceversa (Chapter 3, pages 63–68). Not all combinations of signal properties allowed by a tradeoff will be equally beneficial to the animal exhibiting them. For a given context and question, one combination is often more beneficial (or least costly) than the others. We call such a maximally beneficial (or least costly) signal **optimal**. The process of adjusting tradeoff variables to maximize or minimize some consequence is called **optimization**. How does one identify which tradeoff is optimal? To answer this question, we need to define some terms and methods used in evolutionary biology. Many of these notions were originally derived in economics, and readers who have had some economic theory will find them familiar.

Currencies

Our task is to identify that combination in a tradeoff that maximizes or minimizes some **currency**. In biological contexts, the accepted currency is **fitness**. The computation of fitness is not always straightforward. Ultimately, fitness refers to the relative abundances of specific genes over time. Actually measuring gene frequencies in natural populations is a very expensive and time-consuming process. Luckily, the reproductive output of individual animals is often a sufficient predictor of the replication and survival of the genes they contain. This predictor can be used as long as the population is fairly stable in size and age composition, and there are no major density-dependent selective processes at work. When these conditions hold true, we can determine the expected fitnesses of different tradeoffs by comparing the number of offspring produced in a lifetime by individuals exhibiting those tradeoffs. Such a measure is called **individual lifetime fitness** and equals the product of how long the animal lives (**survival**) and the average number of offspring it produces during each year of its life (**fecundity**). In some situations, one animal, the

donor, may adopt tradeoff combinations that curtail its own lifetime fitness but that concurrently increase the lifetime fitnesses of individuals, called **relatives**, who share many of the same genes by descent. If the reduced replication of donor genes is more than made up by increased replication through its relatives, the curtailed tradeoff may be favored. In such cases, an index of resulting gene frequencies can be computed by adding the final individual fitness of the donor to the increases in individual fitnesses of the helped relatives, each devalued by the likelihood that they share genes with the donor. This currency is called **inclusive fitness**. In some circumstances, the relative success of different genes depends not only on the average fitness of individuals within each social group, but also on the variation in mean values between groups. One must then compare the relative average fitnesses of different groups.

Where individual fitnesses are sufficient estimators of gene frequencies, the process by which evolution culls individuals with less optimal tradeoffs and favors individuals with optimal ones is called **individual selection**. Where inclusive fitness is a better estimator, the process of culling and favoring is called **kin selection**. In cases where differences in group fitness must be taken into account to predict subsequent evolution, the associated process is called **group selection**.

Individual lifetime fitness is the most commonly invoked currency in evolutionary optimization problems. Sometimes, alternative tradeoffs only affect one of the components of individual fitness. For example, if the only effect of a tradeoff between two signal traits is to change fecundity, then fecundity can be used as the effective currency in determining the optimal combination, and survival can be ignored. This logic can be extended still further if the only effect of varying the tradeoff is to change a single component of fecundity such as the number of matings per year. Then mating rate can be used as the relevant currency. The economy of measuring only one component is appealing. However, unless one initially measures all the components (thus giving up the economy), one never knows for sure whether reliance on a subset is a sufficient measure or not.

Strategies, Payoffs, and Optimization Criteria

Once a currency is designated, the next task is to define relevant **strategies**. To achieve a given tradeoff, an animal has to do something. For example, a territorial cichlid can either approach and display at an intruder, or remain at its territorial center and just monitor the intruder's next move. To approach and display is one strategy; to sit and watch is the alternative. The two strategies differ in the degree to which they trade off immediate risk for rapid resolution of the intrusion. In this and any optimization problem, we must first list the alternative strategies that are available (at least over evolutionary time) to the differential culling of natural selection. Not all combinations of a tradeoff are likely to be achievable as strategies. Certain combinations are more likely given the context and phylogenetic history of the animal than are others. Some may be structurally or physiologically impossible for that animal. The list must thus be one that is reasonable for the focal organism. Often, we can identify which strategies are likely by looking at related taxa.

Note that the use of the word **strategy** in evolutionary contexts is not to be taken as a presumption that animals consciously or intentionally adopt particular behaviors or structures over others (although it does not preclude this possibility either). Instead, it is used in the sense that different strategies are in competition over evolutionary time; those that result in higher fitnesses will be represented in greater relative numbers in successive generations at the expense of those with lower fitnesses. This conflict between strategies is not necessarily apparent to the animals; it may only be apparent to us as scientific observers, and it is in this sense that the alternatives are called strategies.

Once the alternative strategies are identified, the task is to ascribe to each a **payoff** in the units of the selected currency. The value of a given payoff will depend on the situation. Relevant factors include the ecological and social contexts in which the strategy will be used, and the physical, physiological, or environmental laws that generate the tradeoff. In some cases, payoffs can be predicted from knowledge of these factors. Alternatively, one can try to measure the payoffs when animals adopt different strategies under natural conditions. Measuring the payoffs of all the alternatives is difficult if (as is often the case) most of the animals in the population adopt the optimal strategy. Luckily, animals do make mistakes, and a careful observer can sometimes measure the consequences. In some systems, one can experimentally cause animals to adopt nonoptimal strategies. Finally, it is sometimes possible to estimate payoffs for alternatives by observing the use of these strategies in related or sympatric species.

Once one has listed the strategies and their payoffs, one can identify the optimal strategy by invoking an **optimization criterion**. If the currency is fitness, the criterion is one of maximization of the currency: the strategy that results in the highest fitness is the optimal one. If the currency is a component of fitness, the criterion may be different. For example, if the critical currency is mortality rate, the obvious criterion is one of minimization: the strategy that minimizes mortality risk would be the optimal strategy.

How to Compute Payoffs

The link between adopting a strategy and experiencing a payoff is often complex. Nature is rarely **deterministic** with a specific act always leading to a specific outcome. Instead, there are usually many outcomes possible, and the adoption of a given strategy at most changes the probabilities with which these different outcomes are likely to occur. We call such a situation **stochastic**. To compute the expected payoff for any strategy, we thus need some way to combine the consequences of all possible outcomes and the probabilities that each will occur. To understand how this is accomplished, readers should be familiar with the basic notation and probability principles outlined in Box 12.1.

Suppose a specific strategy can result in either of two exclusive outcomes, A or B. Each outcome will have a given payoff associated with it: event A occurs with probability $P(A)$ and when it occurs, the animal receives a fitness payoff of V_A. Similarly, the alternative outcome B occurs with probability $P(B)$ and yields a fitness of V_B. Note that both V_A and V_B must be in the same currency units. From the animal's point of view, either event A or event B will occur and thus it will obtain either V_A or V_B. However, in choosing whether to adopt this strategy, the animal (or evolution) must evaluate some overall average. The average is computed by **discounting** each payoff by the probability that it will be experienced and adding these products together. Discounting involves multiplying the value of a payoff by the probability that the payoff will be experienced. Thus the discounted payoff of outcome A is $P(A){\cdot}V_A$ and the discounted payoff of outcome B is $P(B){\cdot}V_B$. Since the events are exclusive (see Box 12.1), we can add the probabilities of the separate events to compute the overall probability of one or the other occurring. The same rule applies to computing the expected or average payoff: when events are exclusive, the expected payoff is the sum of the discounted payoffs for the separate events. In our example, the expected payoff equals $P(A){\cdot}V_A + P(B){\cdot}V_B$. One should always check to make sure that the sum of the probabilities for the entire set of exclusive events equals one. This result means that we have accounted for all possible outcomes.

As an example, suppose only large males of some cichlid species defend nest territories. A typical owner has a defended clutch of 100 eggs. When a male intruder appears, the owner threatens it with a signal. If the intruder is a small fish, it leaves immediately with no cost or benefit to the owner. If the intruder is another large fish, then it engages the owner in a prolonged exchange of threat signals that the owner invariably wins. However, the extended period during which the nest is unguarded allows other fish to eat all but 20 of the owner's eggs. Suppose that 60% of all male intruders are small fish and 40% are big fish. What is the expected payoff of performing the threat signal? When the intruder is small, the discounted payoff is $0.6 \times 100 = 60$, and that when the intruder is large is $0.4 \times 20 = 8$. The average payoff is the sum of these two discounted values or $60 + 8 = 68$ eggs. This calculation does not mean that a threatening territorial owner will end up with 68 eggs. In fact, he will have either 100 or 20, depending upon the type of intruder. However, his best guess at the value of giving a threat signal is the weighted average: 68 eggs. This is the number that should be compared to alternative strategies and their expected payoffs.

More complicated sets of alternative outcomes require a bit more effort, but follow similar logic. For example, suppose intruders can be either male or female, and either small or large fish. There will thus be four possible combinations of conditions: small male, small female, large male, and large female intruders. If sex and relative size of intruders are independent, we can compute the probability of each combination by multiplying the probability of a given sex by the probability of a given size (Box 12.1). If sex and relative size are not independent, then we must invoke conditional probabilities to compute the likelihood of each combination. In either case, we get the average payoff by summing the products of combination probability and combination payoff over all combinations.

Box 12.1 Basic Probability Logic

SUPPOSE THERE ARE TWO POSSIBLE EVENTS, A and B, that a focal animal is likely to experience. We sample the environment 10 times and tally the occurrence of the two events in the following table:

	Sample 1	2	3	4	5	6	7	8	9	10
Event A	×		×	×		×	×			×
Event B		×	×		×		×			

Assuming that the likelihood of each event remains constant during our sampling, we can estimate the probability that A tended to occur by counting the number of times that A occurred, N_A, and dividing it by the number of samples, T. Here $N_A = 6$, $T = 10$, and the probability of A, $P(A) = 6/10 = 0.6$. A similar count of the occurrences of B gives us a probability for B of $P(B) = 0.4$. We can also compute several other useful numbers. $P(A \text{ or } B)$ is the probability that at least one of the two events will occur in a typical sample; here $P(A \text{ or } B) = 8/10 = 0.8$. $P(A \text{ and } B)$ is the probability that *both* events occur on a given sample; here $P(A \text{ and } B) = 2/10 = 0.2$. A **conditional probability** is the likelihood that one event will occur given that the other has also occurred. For example, the conditional probability that A will occur given that B has occurred, denoted $P(A \mid B)$, can be computed by first counting the number of times B occurs in the T samples, N_B, and then recording on how many of those N_B occasions A also occurred (which we can call M_A). In our example, $N_B = 4$, and on $M_A = 2$ of those occasions, A also occurred. Thus, $P(A \mid B) = M_A/N_B = 2/4 = 0.5$. In words, A occurs on 50% of those occasions on which B occurs. The reciprocal computation for the likelihood that B will occur given that an A has occurred is $P(B \mid A) = M_B/N_A = 2/6 = 0.33$; on only 1/3 of the occasions that A occurred did a B also occur. Note that $P(A \mid B)$ is not necessarily equal to $P(B \mid A)$.

It is easy to show that these various probabilities have simple relationships to each other. One important relationship is the following:

$$P(A \text{ or } B) = P(A) + P(B) - P(A \text{ and } B)$$

This says that the fraction of occasions on which at least one of A or B occurred is the fraction of times that A occurred plus that on which B occurred minus the fraction of

Our calculations in these examples used the actual fitnesses of the owner as the payoffs for each possible outcome. We could have done this another way. Let the fitness of the owner before intruders appear be called the **base fitness**. We then compute the difference between the payoff for each possible outcome of owner threats and the base fitness: differences that are positive are the **benefits** of performing the threat signal, and differences that are negative are the associated **costs**. The **average relative payoff** is the sum of the various benefits and costs, each discounted by the probability that it will be experi-

times both occurred on the same sample. The latter term corrects for the fact that the sum of the number of times that A occurs and that B occurs will count the number of occasions on which both occurred twice (once in the sum of A occasions and again in the sum of B occasions). Since we should count these events only once, we need to subtract out the number of times both occurred from either the A sum or the B sum. We can see that this equation holds for our example: $P(A \text{ or } B)$ which we found to be 0.8 is indeed equal to $0.6 + 0.4 - 0.2 = 0.8$. Note that if A and B never occur on the same occasion, that is, if they are **exclusive events**, then $P(A \text{ and } B) = 0$ and in this case, $P(A \text{ or } B) = P(A) + P(B)$. This last form of the equation is very commonly used in computing payoffs because many events are in fact exclusive. We sometimes call this the **OR rule**: the probability that one or the other of several *exclusive events* will occur is simply the sum of the probabilities of the individual events.

In those cases where events can occur jointly, we may invoke a second equation:

$$P(A \text{ and } B) = P(A) \times P(B \mid A) = P(B) \times P(A \mid B)$$

This also makes intuitive sense. The fraction of times that both A and B occur, $P(A \text{ and } B)$, cannot be greater than the fraction of times, $P(A)$, that A occurs. $P(B \mid A)$ is the fraction of A occasions that also include a B event. The product of the fraction of times on which A events occur with the fraction of those events that also include B is clearly the fraction of time that both A and B occur. The same argument can be made if we start with the fraction of time B events occur, and multiply this by $P(A \mid B)$. In our example, we saw that $P(A \text{ and } B) = 0.2$. According to the equation, $P(A \text{ and } B)$ should equal $P(A) \times P(B \mid A)$; our values give 0.6×0.33 for this product, which does equal 0.2. Similarly, $P(B) \times P(A \mid B)$ should also equal 0.2. Our values for the product are $0.4 \times 0.5 = 0.2$, which is the predicted value. As with the OR rule, this equation has a simpler form in special cases. What if we find that $P(A \mid B) = P(A)$? This means that the probability that an A event will occur is the same for the full T samples as it is for the N_B samples in which B events occur (a subset of T). Put another way, what if the presence of a B event has no effect on the probability of an A event? It is easy to show that if $P(A \mid B) = P(A)$, then it also has to be true that $P(B \mid A) = P(B)$. When these conditions are met, we say that events A and B are **stochastically independent**: the occurrence of one does not alter the likelihood of the other. This condition allows us to reduce the second equation to the following **AND rule**: if two events are stochastically independent and can occur together, then $P(A \text{ and } B) = P(A) \times P(B)$.

enced. This value will be the same as the average payoff computed earlier minus the base fitness. It is thus the net gain or loss. Either method can be used in optimality analysis, but only one method should be used in any given set of comparisons.

Identifying Optimal Strategies

The set of alternative strategies available in a given evolutionary context may be **discrete** or **continuous**. In a discrete set, the different alternatives can be counted and they need have no obvious order. For example, the alternative options for a male cichlid to chase an intruder, display at it, or ignore it constitute a discrete set of strategies. Continuous sets potentially have an uncountable number of alternatives and can be ranked along some axis. The set of alternative display durations for the cichlid, say, any duration between 1 and 10 seconds, constitutes a continuous set that can be ordered by the duration times. The methods we use to identify optimal strategies differ depending upon whether the strategy set is discrete or continuous.

For a discrete strategy set, one computes the average payoff for each strategy and then invokes the optimization criterion to identify the optimal strategy. If some measure of fitness is the currency, the optimal strategy is the one that maximizes the average payoffs. Alternatively, the relevant currency might be some component of fitness. If one measures survival, the optimal strategy is that with the maximal payoff; if we measure mortality, we want the strategy creating the minimum value. An example of optimality analysis using discrete strategy sets is given in Box 12.2.

Continuous strategy sets cannot be analyzed this way because one can never list all of the possible outcomes. One solution is to derive an equation

Box 12.2 Finding Optimal Strategies in Discrete Strategy Sets

SUPPOSE THAT A TERRITORIAL MALE CICHLID can respond to an intruding fish with either a **threat** or a **sexual signal**. Intruders could be either male or female, and either larger or smaller than the owner. We assume that sex and relative size are independent of each other, and that an owner cannot discern the sex or relative body size of an intruder until after performance of its signal. The owner already has 100 fertilized eggs in its nest. The payoffs to the owner of performing the threat strategy depend upon the type of intruder: female intruders of any size and small males are scared off with no change in the owner's fitness. Males larger than the owner respond with their own threats, and during the ensuing conflict, other fish sneak into the territory and eat all but 20 of the owner's current eggs. Payoffs for the sexual signal also vary with the type of intruder. Large females mate with the male and add an average 20 eggs to the male's clutch; small females only add 10 eggs to his clutch. An owner performing a sexual signal allows an intruder to get close to its nest: as a result, a large male intruder can suddenly snatch 55 of the owner's eggs before it is chased off, whereas a smaller male can only grab 25 eggs. For each signal strategy, we have four posssible payoffs depending upon the type of intruder.

This situation creates a serious tradeoff problem for the owner. Using a sexual signal will increase the owner's clutch size if the intruder is a female, but will result in major egg loss if the intruder is a male. A threat signal avoids losses of eggs due to small male intruders, but risks substantial losses when intruders are large males and prevents additions if they are females. How can we put all this together to identify the optimal strategy? The answer is to compute average payoffs for each strategy and compare them. In this example, owners want to maximize clutch sizes: therefore the strategy with the greater average payoff is the optimal one.

relating the average payoff to the value of the strategic variable. We can then use calculus to take the first derivative of this equation with respect to the variable, set the derivative to zero, and then solve for the value of the strategic variable at which the derivative is zero. This process will give us either a minimum or maximum for the equation. We can determine which is the case by examining the sign of the second derivative. If the result matches our optimization criterion, we shall have found the optimal value of the strategic variable. It is not always possible to characterize a tradeoff relationship algebraically. Economists have thus devised a number of graphical methods for identifying optimal strategies given continuous strategy sets. One, **fitness set analysis** (Levins 1968), is widely used for evolutionary problems and is demonstrated in Box 12.3.

Sequences and Dynamic Optimization

Animals in nature often have to make a series of optimal choices before achieving some goal. A male courting a female may first have to attract a female's attention, then persuade the female that she can come close enough to investigate him without harassment, and only try to mount when he has

To compute the average payoffs, we need to know the probabilities of encountering each type of intruder. Suppose 30% of all intruders are females, and 60% of all fish are small. Because size and sex are independent, the probability of intrusion by a small female is $0.3 \times 0.6 = 0.18$, by a large female is $0.3 \times 0.4 = 0.12$, by a small male is $0.7 \times 0.6 = 0.42$, and by a large male is $0.7 \times 0.4 = 0.28$. As a check that all outcomes are accounted for, we note that $0.18 + 0.12 + 0.42 + 0.28 = 1.0$. The average payoff of the threat strategy considering the four possible types of intruders in the same order is then

$$\textbf{Threat payoff:}\quad 0.18(100) + 0.12(100) + 0.42(100) + 0.28(20) = 77.6 \text{ eggs}$$

and that for the sexual signal strategy is

$$\textbf{Sexual signal payoff:}\quad 0.18(110) + 0.12(120) + 0.42(75) + 0.28(45) = 78.3 \text{ eggs}$$

In this case, the optimal strategy is to perform the sexual signal.

Now suppose that only 20% of intruders are females (in contrast to the earlier value of 30%). This changes encounter probabilities to 0.12 for small females, 0.08 for large females, 0.48 for small males, and 0.32 for large males. The average payoff for threat is then

$$\textbf{Threat payoff:}\quad 0.12(100) + 0.08(100) + 0.48(100) + 0.32(20) = 74.4 \text{ eggs}$$

and that for the sexual signal strategy is

$$\textbf{Sexual signal payoff:}\quad 0.12(110) + 0.08(120) + 0.48(75) + 0.32(45) = 73.2 \text{ eggs}$$

The reduction in the fraction of intruders that are females now makes threats the optimal strategy. This points up the fact that the identity of the optimal strategy may be very sensitive to ambient contexts. Small changes in these contexts can therefore favor major changes in behavior.

Box 12.3 Finding Optimal Strategies Using Fitness Set Analysis

SUPPOSE THE FITNESS OF FEMALE CRICKETS depends upon their correctly recognizing sounds from males of their own species in habitats in which males of many different species are calling. Suppose that such a female uses a combination of frequency differences and temporal pattern differences to achieve this discrimination. As we saw in Part I, the female cricket's ability to discriminate different frequencies, Δf, and her ability to monitor temporal modulations, Δt, are inversely related and in fact, for most ears, $\Delta f \approx 1/\Delta t$. This relationship creates a major tradeoff that requires optimization. We can graph the allowable values of Δf and Δt as shown in Figure A.

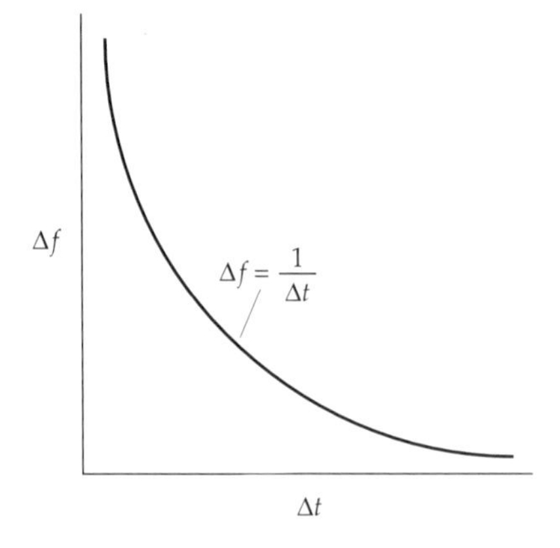

Figure A Tradeoff in values of frequency resolution, Δf, for different values of time resolution, Δt, allowed by the sound uncertainty principle. This is called the fitness set.

Suppose that the relevant currency here is the error rate in identification of males of the correct species, E, and that this rate is a simple linear function of Δf and Δt: $E = a\Delta f + b\Delta t$, where a and b are constants set by the physiology and ecology of the animals. When $a >> b$, then important errors are primarily due to mistakes in frequency discrimination; where $b >> a$, then the major errors are likely to be temporal ones. The female needs to minimize E by picking the correct combination of Δf and Δt. The optimization criterion is thus **minimization** of the currency (error rate). The evolutionary strategies that achieve the different combinations of Δf and Δt are presumably created by varying the thickness of the tympanic membranes of the ears, the pattern of tubal connections (since crickets have pressure-differential ears), and the number of sensory cells.

To combine the consequences of this second equation, $E = a\Delta f + b\Delta t$, with those of the tradeoff equation, $\Delta f \approx 1/\Delta t$, we need to rewrite the former so that Δf is the dependent variable. With a little algebra, we get that $\Delta f = \frac{E}{a} - \frac{b}{a}\Delta t$. This is an equation for a straight line with a y intercept of E/a and a slope of $-b/a$. For two sample values of E, a corresponding graph might appear as shown in Figure B.

Note that we now have two equations relating Δf and Δt: each constitutes a **constraint** on allowable values of Δf and Δt. Values of Δf and Δt that satisfied both equations for the lowest possible value of E would thus be optimal. We can thus find the answer by setting up a graph with Δf on one axis and Δt on the other, plotting the

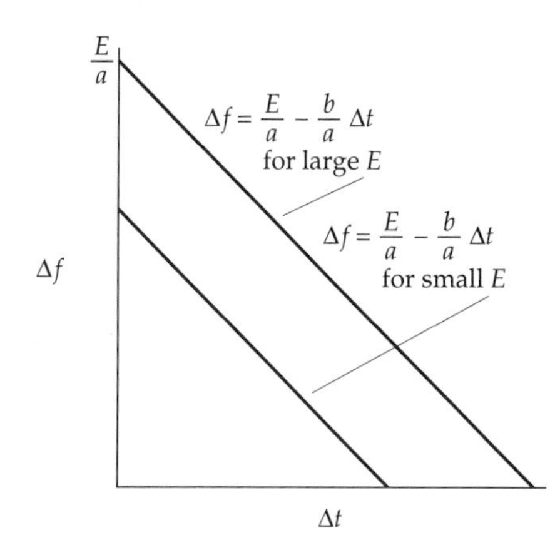

Figure B Allowable combinations of Δf and Δt for a high value of E (topline) and those for a low value of E (bottom line). Each line is called an adaptive function.

tradeoff line first, and then plotting error lines for different values of E until we find the smallest value of E that generates an intersection of the two lines. Since both lines define constraints, only points common to both, i.e., intersections, can meet both constraints. A general answer appears as shown in Figure C.

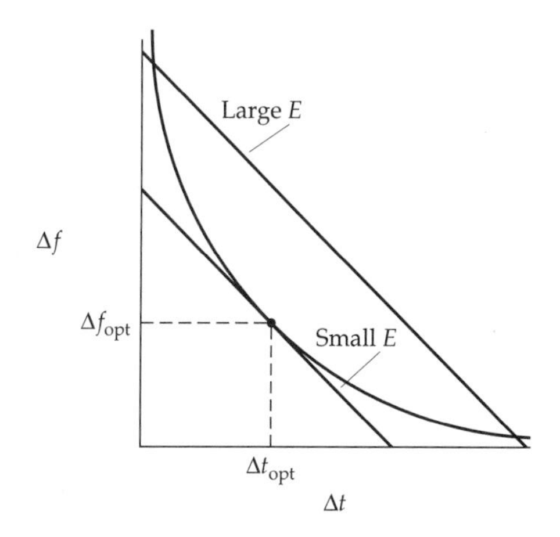

Figure C Identification of optimal allowable values of Δf and Δt by looking for intersections between the fitness set line and adaptive functions. In this situation, the intersection between the fitness set and the adaptive function for

Here the optimal value of Δt is indicated by Δt_{opt} and the corresponding optimal value of Δf is indicated by Δf_{opt}. In fitness set analysis, the tradeoff curve is known as the **fitness set**. It identifies the physically allowable combinations of the two variables involved in the tradeoff. The alternative values relating possible payoffs to variable combinations are called the **adaptive functions**. Here these are the lines of negative slope for different values of E. In our example, we want to find the adaptive function that has the smallest E but still intercepts the fitness set. In other examples, one might want to find the adaptive function that intersected the fitness set and maximized the critical currency (e.g., if the currency was fitness instead of error rate).

Box 12.3 (continued)

The utility of undertaking this analysis is more apparent if we vary the relative importance of temporal and frequency resolution to error rates by changing the values of a and b in the error equation. This variation will change the slopes of the error lines in the fitness set analyses. When $a < b$, the line will have a steeper slope than when $a > b$. This slope will change the location at which the adaptive function with lowest E value just touches the fitness set. The results for three different weightings are shown in Figure D.

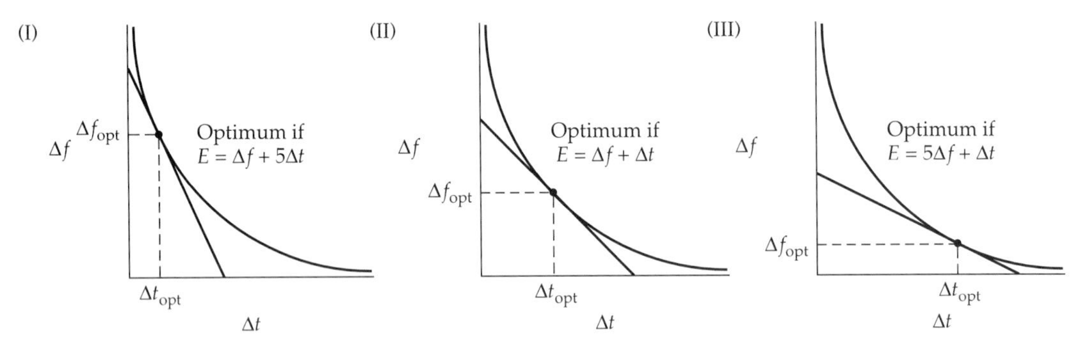

Figure D Change in the optimal combination of Δf and Δt when the relative contributions of frequency and time resolution to overall errors are varied. Relative importance of frequency errors is lowest in I and highest in III.

The optimal combination of Δf and Δt when temporal errors are either more common or more confusing (I) is a very small Δt even at the cost of a higher Δf. Equal weighting of the two sources of error results in an optimal combination in which both Δf and Δt take intermediate values (II). When frequency errors are more common or have a greater effect on E, then the optimum is for a very low Δf even at the expense of a higher temporal error rate (III). This example shows why related crickets living in different environments might evolve different optimal combinations of temporal and frequency resolution in their ears.

received convincing signals of her interest. It is tempting to break this chain of events into separate stages and apply optimality analysis to each stage independently of the others to predict what the male should do. This type of analysis is fine as long as the stages are truly independent. But consider a male that tries so hard to attract and round up females that he fails to persuade them of the safety of close approach. Similarly, a male that overinvests in the persuasion stage may miss the cues of a receptive female, causing her to become impatient and leave. Where successive stages in a chain of acts are not independent, the options that appear optimal when examining a stage in isolation may prove to be suboptimal when examining their consequences on later stages. In such cases, one needs to identify the optimal **choreography of behaviors** across the entire chain of events that maximizes fitness. **Dynamic**

optimization is the standard method for analyzing such sequences. One begins with the final state desired, and works backward through each successive decision point to find the optimal sequence of behavioral choices (Mangel and Clark, 1988).

Optimal choreographies exist in many of the contexts requiring animal communication. Conflict resolution and courtship are two contexts that often require successive chains of decisions. Optimal decisions cannot be made at just one stage without considering the consequences on later stages. The analysis of optimal choreographies has only recently been applied to the study of animal communication.

Optimization and Evolution

We have now outlined the basic methods by which we, as scientists, might try to identify optimal strategies in a particular animal system. What has this to do with evolution? The basic premise is that if we invoke the appropriate currency (e.g., fitness), the strategies that we identify as optimal ought to be those we predominantly see in nature. Research in the last two decades has shown that this premise is largely upheld. The exceptions invariably help support the rule. For example, selection can only favor a strategy over others if it is present: it may take many generations before appropriate mutants appear. A system we are examining may not yet have reached the optimum. Evolution from one strategy to another must often pass through a number of intermediate stages. If any of these intermediates are genetically or ontogenetically difficult to achieve, the apparent optimum may be unreachable. Some "traits" are, in fact, not selected themselves, but change over time because they are genetically correlated with other traits that are subject to selection. Finally, when one of the influences determining optimal strategies is the choice of action by other animals with interests different from that of the optimizer, the "optimal" strategy may be quite different from that expected were the influences passive. Once we perceive them, we can often incorporate phylogenetic, ontogenetic, and correlated response factors as costs and the resulting optimality analysis is usually more compatible with what is seen in nature. The problem of conflicts of interest is more complicated and will be dealt with in detail in Part III of this book. Despite these limitations, optimality analysis generally provides a good guide to explaining why evolution has taken the path it has, and we shall rely heavily on this approach in subsequent chapters.

THE ECONOMICS OF ANIMAL COMMUNICATION

Signals provide information that may facilitate a correct decision by a receiver. Paying attention to them can benefit the receiver and through the receiver's subsequent actions, lead to benefits for the sender. However, neither the emission nor the reception of signals is cost-free: senders require organs, time, and energy to produce signals, and receivers require organs, time, and energy to process them. There is thus a basic tradeoff between the benefits and costs of communication. For one or both parties, it may not pay to engage in the exchange. We can identify the situations in which communication is or is not favored by comparing each party's average payoffs when they engage in the exchange to those when they do not. Given that communication is favored, how much information is worth exchanging? Which of several alternative signals is optimal in a given situation? Again, we can answer these questions by computing the average payoffs for the alternatives and invoking the relevant optimization criterion. This task is a bit more challenging than it might appear. For one thing, the payoffs to senders of performing any act may depend on both tactical and signal consequences of its actions. If both effects are present, we must take each into account in computing the average payoff to the sender. Similarly, receivers will use whatever information they can to make a decision. Some of this information will be provided by signals and some by prior probabilities, direct assessment, cue assessment, or amplifier effects. The decision will depend on all of this information; the degree to which signals are favored by evolution depends only on how much they add to the final decision. We thus need some way to compare the payoffs to receivers of making decisions with and without access to signals, correcting for all other factors in receiver decisions.

The next section develops a general optimization model for assessment of sender and receiver payoffs that does just that. In the two subsequent chapters, we examine the various components of this general payoff model in greater detail. Although the algebra may initially not seem to apply to real systems, the model provides a single formulation that can fit a very large number of signal contexts. It also ties together a variety of issues that are often treated independently in the literature, but that should really be treated as part of the same economic process. In later chapters, we use the model to examine the evidence that specific animal signal systems have been optimized by evolution, and to understand why a particular level of information exchange might be optimal in a particular context.

Payoffs and the Value of Information

Consider a receiver that seeks to maximize its fitness and must now choose between two alternative actions. Suppose that Act 1 will provide the higher payoff to the receiver if condition C_1 is true, and Act 2 will yield the highest payoff if alternative condition C_2 is true. There are four possible outcomes of this receiver's decision, each of which generates a potentially different payoff. If C_1 were true, the receiver would receive a payoff of R_{11} when it correctly decides to perform Act 1; it would receive the lesser payoff R_{21} if it erroneously decided on Act 2 ($R_{11} > R_{21}$). When C_2 is true, the receiver would obtain a payoff of R_{22} if it correctly elects to perform Act 2, but receives the lesser value R_{12} if it mistakenly performs Act 1 ($R_{22} > R_{12}$). In this example, the first subscript on all payoffs will indicate what the receiver elected to do, and the second what it should have done to maximize fitness.

The problem is that the **receiver** does not know for sure which alternative condition is currently true. It may have access to the prior probabilities that C_1 and C_2 will occur in nature: we shall denote these probabilities by P_1 and P_2

respectively. [In this two-condition example, $P_2 = (1 - P_1)$.] In addition, the receiver would have attempted to integrate all other sources of information into estimates of the probabilities that each of the alternative conditions is currently true. Since we have only two alternatives in this example, the receiver's current estimate that the true condition is C_1, which we denote by $\hat{P}(C_1)$, is one minus the estimated probability that C_2 is the current state. The receiver thus only need think about one of these values.

A receiver is unlikely to have **perfect information**. That is, it is unlikely that $\hat{P}(C_1)$ is either 1.0 or 0.0. To make a decision, it will have to set some threshold value, which we can call P_c, such that if the current estimate of $\hat{P}(C_1) > P_c$, then it should perform Act 1, and if the current estimate of $\hat{P}(C_1) \le P_c$, it should perform Act 2. We shall discuss in Chapter 14 what the optimal values of P_c should be. Because $\hat{P}(C_1)$ is only an estimate, and P_c is usually less than 1.0, receivers will sometimes make mistakes and perform Act 1 when C_2 is true, and similarly, may perform Act 2 when C_1 is true. We can thus define another set of probabilities that are the chances that the receiver will make a correct decision. Suppose when the receiver relies on all sources of information except signals, the probability that it makes the correct choice when C_1 is true is ϕ_1, and the probability that it makes the correct choice when C_2 is true is ϕ_2. The corresponding probabilities of making errors are then $(1 - \phi_1)$ and $(1 - \phi_2)$.

We can now compute the average fitnesses of receivers and senders when no signals are exchanged. Using the logic outlined on pages 366–371, the average payoff to a receiver when no signals are exchanged and C_1 is true is $\phi_1 R_{11} + (1-\phi_1) R_{21}$, and when C_2 is true, the average payoff is $\phi_2 R_{22} + (1 - \phi_2)R_{12}$. The overall average payoff to a receiver given that C_1 tends to be true P_1 of the time and C_2 is true P_2 of the time is then

$$\overline{PO}\ (Receiver,\ No\ signal) = P_1[\phi_1 R_{11} + (1-\phi_1)R_{21}] + P_2[\phi_2 R_{22} + (1-\phi_2)R_{12}]$$

The first term is the average payoff when C_1 is true, discounted by the fraction of the time that C_1 occurs, and the second term is the average payoff when C_2 is true, discounted by its likelihood of occurring. Within each term, the receiver either makes the right decision (and gets the higher payoff) or it makes the wrong one and gets the lower value. These terms are each discounted by the corresponding probabilities of getting it right or getting it wrong.

Suppose there is a nearby potential sender that will benefit when the receiver adopts the correct actions. Let the payoff to the sender when the receiver correctly chooses Act 1 be T_{11}, incorrectly chooses Act 2 be T_{21}, incorrectly chooses Act 1 be T_{12}, and correctly chooses Act 2 be T_{22}. Then the average payoff to a sender when no signal is transmitted and the receiver simply decides based on prior information will be

$$\overline{PO}\ (Sender,\ No\ signal) = P_1[\phi_1 T_{11} + (1-\phi_1)T_{21}] + P_2[\phi_2 T_{22} + (1-\phi_2)T_{12}]$$

We next consider the average payoffs to both parties when the sender provides the receiver with a signal. Communication requires that there be some **code** that maps signal parameters onto conditions. The code need not be perfect and rarely is. All that is required is that different signals have different correlations with the alternative conditions, and that each party has some idea of what these correlations will be. When it detects a signal, the receiver should then combine its prior estimate of the probability that C_1 is true with its identification as to which signal was sent. It will then generate a new estimate that C_1 is true given the receipt of the signal; we can denote this new probability by $\hat{P}(C_1 \mid signal)$. We shall discuss in Chapter 13 how such updated estimates might be generated. An optimal receiver would then compare this new estimate, $\hat{P}(C_1 \mid signal)$ to the cutoff value P_c: if $\hat{P}(C_1 \mid signal) > P_c$, then it would perform Act 1, and if $\hat{P}(C_1 \mid signal) \leq P_c$, it would perform Act 2. Given the new information provided by the sender, the probability that the receiver will make a correct choice when C_1 is true will change from ϕ_1 to ϕ'_1; its probability of making a correct choice when C_2 is true will change from ϕ_2 to ϕ'_2.

The average payoff to the receiver when signals are exchanged is then

$$\overline{PO}\ (Receiver, Signal) = P_1[\phi'_1 R_{11} + (1-\phi'_1)R_{21}] + P_2[\phi'_2 R_{22} + (1-\phi'_2)R_{12}] - K_R$$

where $-K_R$ is the average cost to a receiver of attending to signals. This cost might entail time lost from other activities, increased predator risks, or the construction of receiving organs necessary to detect and decode signals. The equivalent payoff for a sender is

$$\overline{PO}\ (Sender, Signal) = P_1[\phi'_1 T_{11} + (1-\phi'_1)T_{21}] + P_2[\phi'_2 T_{22} + (1-\phi'_2)T_{12}] - K_S$$

where $-K_S$ is the average cost to the sender of producing the signals. These alternative payoffs are summarized in Figure 12.11.

When does it pay for a sender and receiver to communicate? The receiver should pay attention to signals when $\overline{PO}$(*Receiver, Signal*) > $\overline{PO}$(*Receiver, No signal*) or rewriting, when $\overline{PO}$(*Receiver, Signal*) – $\overline{PO}$(*Receiver, No signal*) > 0. This difference in average payoff with the signal versus that without the signal is called the **average value of information** to the receiver. It will be denoted by $\overline{VI}(R)$. Similarly, senders should send signals when $\overline{PO}$(*Sender, Signal*) – $\overline{PO}$(*Sender, No signal*) > 0. We shall denote this difference, the value of information for senders, by $\overline{VI}(S)$. Using the expressions derived above and simplifying, receivers should attend to signals when

$$\overline{VI}(R) = P_1(\phi'_1 - \phi_1)(R_{11} - R_{21}) + P_2(\phi'_2 - \phi_2)(R_{22} - R_{12}) - K_R > 0$$

and senders should send them when

$$\overline{VI}(S) = P_1(\phi'_1 - \phi_1)(T_{11} - T_{21}) + P_2(\phi'_2 - \phi_2)(T_{22} - T_{12}) - K_S > 0$$

Thus, each party should engage in communication only if its average value of information is positive.

This example only allows for two alternative conditions. It is easy to generalize this formulation to apply to any N discrete alternatives. The costs of communication might differ in different conditions. We can still define an average cost of communication to each party by computing the cost for each condition discounted by the likelihood it will occur. Thus

$$\overline{K}_R = \sum_{i=1}^{N} P_i K_i(R) \quad \text{and} \quad \overline{K}_S = \sum_{i=1}^{N} P_i K_i(S)$$

(A)

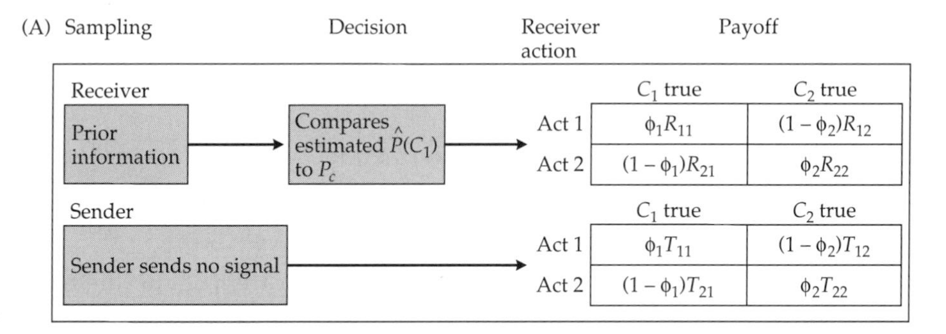

(B)

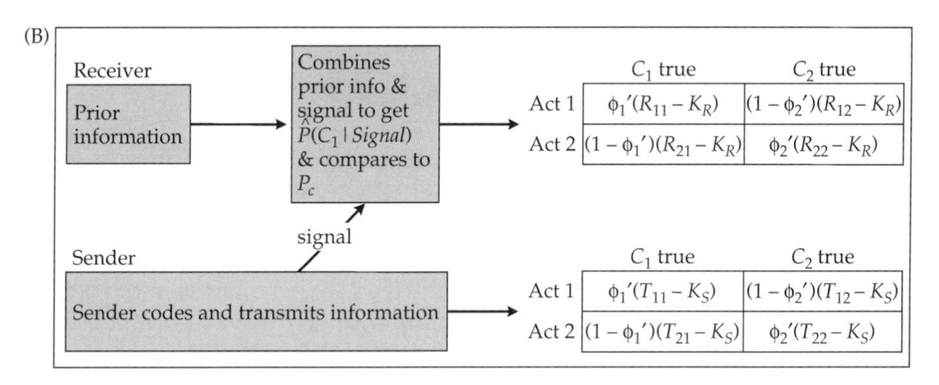

Figure 12.11 Schematic analysis of communication. A receiver must choose between performing Act 1 or Act 2. The payoffs to the receiver, R_{ij}, and those to a sender, T_{ij}, for each Act i adopted by the receiver depend upon whether a given condition, j, is C_1 or C_2. In this example, we assume that $R_{11} > R_{21}$, and that $R_{22} > R_{12}$. (A) An optimal receiver, acting on its own uses nonsignal sources of information (here denoted as "prior information") to estimate the probability $\hat{P}(C_1)$ that C_1 is currently true. It then invokes a decision rule by comparing its current estimate of $\hat{P}(C_1)$ to some cutoff value P_c. If $\hat{P}(C_1) > P_c$, the receiver performs Act 1; if $\hat{P}(C_1) \le P_c$, it performs Act 2. The probability that it will make the right choices are ϕ_1 and ϕ_2 when C_1 and C_2 are true, respectively. Thus the average payoff to the receiver is $\phi_1 R_{11} + (1 - \phi_1)R_{21}$ when C_1 is true, and $\phi_2 R_{22} + (1 - \phi_2)R_{12}$ when C_2 is true. The payoffs of alternative receiver actions to the sender when no signals are exchanged are shown in the lower table. (B) The situation is nearly the same in the bottom schematic except that here a sender provides a signal that allows the receiver to provide a better estimate, $\hat{P}(C_1 \mid Signal)$, that C_1 is true. The receiver then compares this improved estimate to its cutoff value, P_c, and makes a decision on which act to perform. The new probabilities that the receiver will make the right decision are ϕ_1' and ϕ_2' when C_1 and C_2 are true respectively. We also assume that it costs the sender $-K_S$ to send the signal, and it costs the receiver $-K_R$ to attend to the signal. The average payoff to a receiver when it attends to the signal and C_1 is true is then $\phi'_1 R_{11} + (1 - \phi'_1)R_{21} - K_R$, and $\phi'_2 R_{22} + (1 - \phi'_2)R_{12} - K_R$ when C_2 is true. Sender payoffs when communication occurs are shown in the lower table.

where P_i is the probability that condition i is currently true, $K_i(S)$ is the cost of signal production when condition i is true, and $K_i(R)$ is the average receiver cost of attending to a signal when condition i is true. Given N possible conditions, participation in communication is then favored by each party X when

$$\overline{VI}(X) = \left\{ \sum_{i=1}^{N} P_i \left[\Delta\phi_i \, \Delta W_i(X) \right] \right\} - \overline{K}_X > 0$$

where $\overline{VI}(X)$ is the average value of information for party X, $\Delta\phi_i$ is the increased chance the receiver will make a correct decision as a result of receiving a signal when condition i is true, $\Delta W_i(X)$ is the increased payoff to party X if the receiver correctly picks the right action for condition i, and $-\overline{K}_X$ is the average cost to party X of participating in communication about this set of conditions. The expression thus says that the average **benefit** of communication, $\sum_{i=1}^{N} P_i \left[\Delta\phi_i \, \Delta W_i(X) \right]$, for each party X must be greater than the average **cost**, $-\overline{K}_X$, to warrant participation in the exchange.

Note that the **amount of information** exchanged, i.e., that which allowed the receiver to upgrade its estimate that C_1 was true from $\hat{P}(C_1)$ before communication to $\hat{P}(C_1 \mid signal)$ afterwards, does not appear explicitly in the expression for the value of information. What appears instead is the change in probability that the receiver will eventually make a correct decision. It seems likely that the amount of information is somehow related to the $\Delta\phi_i$ terms in the benefit portion of the value of information computation. However, we shall see in Chapter 14 that this linkage is subtle and sensitively dependent upon the updating and decision rules invoked by the receiver after detecting a signal.

To Communicate or Not?

Given this basic model, when should a sender and receiver both enter willingly into a communication exchange? Figure 12.12 shows the four possible combinations of the average value of information for a sender and a receiver. When both parties have a positive value of information, we have called the exchange **true communication**. When the sender has a positive value of information, but the receiver a nonpositive one, the receiver is the victim of **manipulation** by the sender. Information is definitely exchanged and used by the receiver, but only the sender benefits from the resulting change in receiver decisions. When the value of information to a manipulated receiver is actually negative, (as opposed to being zero), we call the interaction **deceit** by the sender. Conversely, when the net payoff to the receiver of an exchange is positive, but the sender gains nothing, the exchange is called **eavesdropping** or

Sender value of information		Receiver value of information: Positive	Receiver value of information: Zero (or negative)
	Positive	True communication	Manipulation (deceit)
	Zero (or negative)	Eavesdropping, cueing (exploitation)	Ignoring (spite)

Figure 12.12 Possible combinations of value of information to senders and receivers and terms describing them. Terms in parentheses are those for negative values of information. (After Wiley 1983.)

cueing. The subset of these cases in which the sender is actually harmed, (e.g., experiences a negative value of information), is usually called **exploitation**. Finally, when both parties experience a nonpositive value of information from an exchange, one expects them simply to ignore each other. A more extreme case occurs when the sender inflicts a negative value of information on the receiver even though this results in a net negative value of information for itself. This exchange is called **spite**.

Do we expect any of these exchanges to evolve except true communication? The answer is yes, but only with constraints. There are two cases. First, one can envision situations in which an existing communication system is invaded by mutants that engage in deceit or exploitation sufficiently to make the other party's long-term value of information negative. There will be strong selection favoring mutant descendents of the injured party that are either more discriminating or ignore the exchange entirely. Over time, we expect evolution to restore the situation to non-negative values of information. However, were we to stumble on this situation in its early stages, we might find negative average values of information. The constraint on this scenario is thus time: negative values of information are unlikely over long time scales. In the second case, we can envision static situations in which the constraint is the frequency with which deceit, exploitation, or spite are encountered. As long as the average value of information remains positive in such situations, and the costs of greater discrimination by the injured party outweigh the benefits, this low level of deceit or exploitation might remain stable. Note that even with true communication, receivers will still make mistakes a fraction $P_1(1 - \phi'_1) + (1 - P_1)(1 - \phi'_2)$ of the time. In some situations, the errors arise because it would cost the party generating them more to improve the exchange than it would gain by more accurate receiver decisions. In other situations, the errors arise because occasional senders deceive or occasional receivers exploit. As long as these abuses are not too frequent, it may cost more to counter them than is worth it. In both situations, there is an optimal amount of error that it pays to tolerate. Such a balance could continue indefinitely, unlike the case in which the net value of information is negative.

As an example, consider fireflies. Males of the North American genus *Photinus* typically fly about at night releasing a species-specific pattern of light flashes. A receptive female responds to a male of her own species by flashing back. The male then uses the female responses to locate and approach the stationary female for mating. *Photuris* is a genus of fireflies that preys on other genera such as *Photinus*. Female *Photuris* cleverly mimic the female response flashes of their prey. Prey males are thus attracted to the female predator, which then kills and eats them after they have approached sufficiently (Figure 12.13). In this example, male *Photinus* are being deceived by sympatric *Photuris*. Why do male *Photinus* continue to respond to these signals?

In this example, we can let C_1 be the presence of a mate, and C_2 the presence of a predator. By responding to signals (flashes), the male *Photinus* increases its chances of correctly approaching mates (e.g., $\phi'_1 - \phi_1 > 0$), but it simultaneously decreases its chances of avoiding a predator (e.g., $\phi'_2 - \phi_2 < 0$).

Figure 12.13 ***Photuris* firefly eating male of another firefly genus.** This *Photuris* female imitated the response pattern of prey females to attract the victim male. She will eat all except the eyes, feet, and wing covers, and will incorporate toxic defensive compounds from the male's tissues into her own chemical defense system. (Photo courtesy of J. E. Lloyd.)

There is thus a tradeoff. For continued responsiveness to flashes to be favored, it must be the case that

$$P_1(\phi_1' - \phi_1)(R_{11} - R_{21}) + (1 - P_1)(\phi_2' - \phi_2)(R_{22} - R_{12}) - K_R > 0$$

Since $\phi'_1 - \phi_1 > 0$ but $\phi'_2 - \phi_2 < 0$, the overall inequality will only be true if

$$P_1(\phi_1' - \phi_1)(R_{11} - R_{21}) - K_R > (1 - P_1)(\phi_2 - \phi_2')(R_{22} - R_{12})$$

Is this likely? Lloyd (1986) has pointed out that competition among male fireflies to find and mate with females is extremely intense. Males may have to search for a week or more for each mate, and their short lives limit matings to only a few. If they do not find mates quickly, they do not find them at all. Responding to flashes thus increases $(\phi'_1 - \phi_1)$ significantly, but the value of $(R_{11} - R_{21})$ will not be large. On the other hand, selection has favored sufficient discrimination of flashes by male *Photinus* that *Photuris* females have a prey capture rate of only 10–15%. This makes the value of $(\phi_2 - \phi'_2)$ a small number. Since surviving males can on average expect only a few matings, the average loss of future fitness by being eaten is probably not a large value (although the experience is definitely discouraging). In fact, it seems likely that $(R_{11} - R_{21})$ and $(R_{22} - R_{12})$ are both small, and that $(R_{11} - R_{21}) \approx (R_{22} - R_{12})$. Thus if we are correct that $(\phi'_1 - \phi_1) > (\phi_2 - \phi'_2)$, as long as potential mates are generally more common than predators, [e.g., $P_1 > (1 - P_1)$], and the physiological costs of responding to flashes ($-K_R$) are not too high, male *Photinus* should continue to respond to flashes in spite of occasional deceit by *Photuris*.

Consider the opposite example of exploitation by a receiver. In Part I, we discussed the calls produced by male frogs to attract potential mates. Because suitable egg deposition sites are patchily distributed, large choruses of males vying for the attention of any visiting females are common. Selection thus favors males producing calls that are loud and frequent in an attempt to be more conspicuous to females than their nearby competitors. In the neotropics, a large bat, *Trachops cirrhosus*, exploits this behavior to find its dinner (Figure 12.14). While it can use its echolocation sounds to scan foliage and the surface

Figure 12.14 Neotropical bat, *Trachops cirrhosus*, attacking calling male frog. Although the bat is able to locate frog and lizard prey using echolocation, it can also eavesdrop on the mating calls of male frogs. (Photo courtesy of M. Tuttle, Bat Conservation International.)

of ponds and streams for suitable prey, the bat is also able to locate frogs by homing in on their mating calls. Male tungara frogs (*Physalaemus pustulosus*) produce several call variants. One, called the "chuck," is preferred by females. Unfortunately for the frogs, emitting a chuck makes a male much more vulnerable to bat predation (Ryan et al. 1982). We again find a serious tradeoff in signal design. Males can compensate a bit by not producing chucks when females are not present or there are few male competitors. However, in many cases the only way to obtain a mating is to produce a chuck. That males do produce them indicates that the overall value of information for chucks is positive. This could be the case, despite continued exploitation by bats, if (a) bat attacks were sufficiently rare (perhaps in part because frogs do not chuck all the time, making them an unpredictable source of food), (b) the lifetime of the frogs was already sufficiently short that any remaining risk did not alter fitnesses significantly, (c) the probabilities of attracting females if a male produced a chuck were sufficiently larger than if it did not, and/or (d) the ability of the bat to locate the frog with its echolocation was already sufficiently acute that the additional cues provided when the frogs called did not result in a major change in their risk of being eaten.

The outcome of these considerations is that a sender may be selected to produce a signal, even though some receivers may use these signals to exploit it, and receivers may be selected to respond to signals, even though they are sometimes deceived, as long as their respective longer-term average values of information are positive. We shall return to the issues of deceit and exploitation in Part III of this book. For the remainder of Part II, we shall focus on exchanges in which the short-term values of information, as well as the long-term averages, are positive for both senders and receivers.

SUMMARY

1. Communication **signals** are one source of information used by **receivers** to identify which of several alternative **conditions** is true before making decisions. Other sources of information include **prior probabilities** of alternative conditions, **direct assessment** of conditions, and **assessment of cues** correlated with the conditions of interest. **Amplifiers** are traits that facilitate direct or cue assessment by receivers. Sender actions may function as signals, or they may be **tactical** in that they change ambient conditions and thus the propriety of alternative receiver decisions. The relative importance of signals versus other sources of information in receiver decisions, and of signals versus tactical functions in sender actions, tends to increase the farther apart the sender and receiver are during an exchange. When a sender and receiver are at close range, observed behaviors are much more likely to combine signaling with other processes.
2. The **questions** asked by receivers that may be answered by sender signals can be classified by the **contexts** in which they arise. Thus we can distinguish between **conflict resolution signals**, **territorial signals**, **sexual signals** (for both mate attraction and courtship), **parent-offspring signals**, **social integration signals**, **environmental signals**, and **autocommunication signals**. In addition, many signals also provide information about **sender identity** and **sender location**. Different signal designs are often required, depending upon the information provided. Signals that provide information about several questions are called **compound signals**.
3. For a sender, the **function** of sending a signal is to increase the chances that the receiver will select that action most beneficial to the sender; for a receiver, the function of responding to a signal is to increase its own chances that it chooses the action best for it.
4. Evolution cannot generate perfect matches between signal design and function because **tradeoffs** lead to abridgment of one property when another is augmented. Finding that combination of properties that maximizes or minimizes some basic **currency** is called **optimization**. Whether the currency should be maximized or minimized is set by the **optimization criterion**. The basic currency in evolution is always **fitness**. This criterion is usually computed as **individual lifetime fitness**, although **inclusive fitness** is the appropriate measure if genetic relatives are involved. Components of fitness can sometimes be used as the currency, but only if there are no opposing changes in other components.
5. The alternative options between which organisms can choose are called **strategies**. A strategy set may either be **discrete** or **continuous**. The currency value resulting from the adoption of any given strategy is called the **payoff**. Where a given strategy can result in several possible outcomes, each with its own probability, we say the payoff is **stochastic**. We then compute an **average payoff** by **discounting** the fitness associated with each outcome by its probability, and combining all the discounted outcomes for that strategy. Identifying the optimal strategy when strategy sets are discrete thus entails computing their average payoffs and finding

the one that best meets the optimization criterion. When strategy sets are continuous, one can use either graphical methods or calculus to identify optimal strategy values.

6. When several successive optimizations occur in a chain, the optimal strategy at any early stage may depend not only on that stage, but on the consequences of adopting that strategy for later stages. Identifying the **optimal choreography** over a series of decision points is called **dynamic optimization**.

7. Optimality methods can be used to determine whether animals should engage in communication or not. Communication involves the transfer of information from a sender to a receiver that results in a change in the probability that the receiver will then elect an optimal action from a set of alternatives. This change in the probabilities of optimal receiver action depends on the **amount of information** transferred between the parties and the **decision rules** used by receivers. The **value of information** to each party is the increase in the average payoff with communication when compared to that without it. It is a function of the amount of information, the decision rules, the difference in payoffs to that party of alternative receiver choices, the overall probabilities of alternative conditions, and the costs of investing in sending or receiving activities. **True communication** occurs when the value of information is positive for both sender and receiver, and both parties agree on which receiver action is optimal.

8. Types of information exchange that can be distinguished from true communication, and from each other, include **eavesdropping** and **exploitation**, **deceit** and **manipulation**, and **spite**. In each of these cases, at least one party does not experience a positive value of information. Whereas individual exchanges of information between animals may involve exploitation, deceit, or spite, we expect the long term average value of information over many such exchanges to be positive for both receiver and sender.

FURTHER READING

A very readable introduction to simple and dynamic optimization in evolutionary biology can be found in Mangel and Clark (1988). Graphical identification of optima in continuous strategy sets (often called "fitness set analysis") is described in general by Ricklefs (1990, Chapter 26) and in detail by Levins (1968). Smith (1977) pioneered the classification of the types of questions answered by animal signals and the notion that information from both context and signal are required by receivers before making decisions. Wiley (1983) provides a contrasting discussion of the different kinds of information exchanges that exist among animals. Stephens (1989) outlines the distinction between amount of information and value of information and provides graphical examples.

Chapter 13

The Amount of Information

WE SAW IN THE PRIOR CHAPTER that communication provides a receiver with information that helps it decide which of several alternative conditions is true. The sender provides the information because it, as well as the receiver, will benefit if the receiver chooses correctly. Sending and receiving signals are costly activities, and neither party should engage in the exchange unless the average difference between benefits and costs is positive. This average difference is called the **value of information**. The benefits of communication depend upon the increased chances that the receiver will get it right and the difference in payoffs between getting it right and getting it wrong for each party. Nowhere in these computations did we include an explicit term for the **amount of information** provided by the signal. It seems that the amount of information exchanged between sender and receiver ought to play some role in determining whether communication is worth it or not. Is there a link, or can we just ignore the amount of information?

WHY WE NEED TO MEASURE THE AMOUNT OF INFORMATION

The amount of information affects the value of information in two ways. Increasing the amount of information in a signal generally increases the chances that the receiver will make the right decision; the relationship is complicated and we shall discuss it in detail in the next chapter. The amount of information also affects the costs of communicating. Usually the greater the amount of information, the more it costs senders to encode it into signals and receivers to extract it from signals. We thus encounter a tradeoff for both parties because increasing the amount of information increases the value of information through reduced error rates by the receiver, but it simultaneously decreases the value of information by raising costs. There will usually be some intermediate amount of exchanged information that maximizes the value of information and is thus optimal. If costs increase less rapidly than error reduction, the optimal amount of information will be large; if costs increase more rapidly than error reduction, the optimal amount of information, and thus signal sophistication, will be low. Which process increases more quickly depends on physical and physiological constraints, and on the specific function or task which the signal fulfills. To identify how function and constraints each contribute to signal propriety, we need to know how benefits and costs scale with the amount of information.

While a receiver may augment signal information with that from other sources (e.g., prior probabilities, direct assessement, or cue assessment), it cannot extract any more information from a signal than was encoded in it originally. In fact, it generally uses only a fraction of that in the emitted signal. There are several reasons for this low usage. If the signal is distorted during propagation, some of the initially encoded information may be lost. Once the signal arrives at the receiver's sensory organs, physical and physiological tradeoffs will further limit the accuracy with which the critical properties of the signal can be measured. Temporal resolution will trade off with frequency resolution; directionality will trade off with sensitivity. Some tradeoffs reflect physical laws that limit how much information *any* receiver can extract from a given signal. A receiver that can extract the maximal amount is called an **ideal receiver**; most organisms are not designed to do as well and will extract less than this maximum. If we know how to compute the maximum amount of information that even an ideal receiver could extract, we can use this as a benchmark against which we can compare what is actually exchanged. How much does it cost a given receiver species to approach ideal conditions? Why does one species come closer to this maximum than another for the same signal function? What steps in the exchange most account for the drop in efficacy below ideal expectations? To answer these questions, we need to measure the amount of information coded into the signals by the sender, the amount lost during propagation, the maximum amount extractable from the received signals by an ideal receiver, the amount the receiver already has at hand, and lastly, the amount it actually utilizes. We can only make such computations if we have a common currency for measuring information.

WHAT IS INFORMATION AND HOW IS IT MEASURED?

It is one thing to demand a common measure of the amount of information and another to derive one. Is there a currency that works as well for a unique signal exchange as it does for the average information provided by a signal set, or even by an entire repertoire? Without a common currency, we shall be unable to compare the amounts of information exchanged by different species, whether animals are close to ideal receivers or not, or how signaling costs might scale with the provision of varying amounts of information. Similarly, is there a common currency that can be used in every context? The subject matter of concern to a receiver varies depending upon whether it is attempting to resolve a conflict, attract a mate, find its own offspring in a colony, or alert others to the presence of a predator. Clearly, the answers to these questions differ markedly in content. In this sense, the information provided is as different as apples and oranges. What currency would allow contrasts between all of these different contexts?

As we shall see, there is such a currency, and the reason that it is universal is that receivers usually know beforehand what the alternative conditions are. They just do not know which one is currently true. Their concern, regardless of the specific question, is the accurate estimation of the relative probabilities of the alternatives. Improved probability estimation is both a common thread in all communication exchanges and the main point of each transaction. Might we be able to use the change in estimated probabilities after receipt of a signal as a common measure of the amount of information exchanged? The answer is yes, and we now introduce the approach with a simple example.

Binary Questions, Bits, and the Information Provided by One Signal

Consider a female bird that must decide whether to accept or reject a courting male. The female seeks a mate with sufficient health that he will be able to defend their nesting territory and provision the offspring efficiently. She brings to this decision prior knowlege about the identity of possible alternatives; she may also have prior knowledge about the a priori probabilities of each alternative. Prior information is derived from previous experience or reflects heritable biases and expectations arising from selection on the female's ancestors. If she has neither relevant experience nor heritable biases about which of M alternative conditions is most likely, then she must assume that each will occur with probability $1/M$. Let us examine a species in which females divide males into two classes: *healthy* and *sick*. The existence of these two alternative classes is part of each female's prior knowledge. A female with no prior information about the relative abundances of healthy and sick males must assume that any given male has a 50% chance of being healthy. How can she improve on this initial estimate? Since direct assessment would entail measuring antibody and parasite levels in the male, she will have to rely on her own assessment of **cues** correlated with male health or on any relevant signals that males provide to indicate their state of health. Once a female has access to such information, she updates her prior estimated probabilities that the male is healthy to either a higher or lower value. Depending upon whether this value is or is not greater than some threshold level, she then makes a decision.

At any point, a list of the alternative conditions and the female's estimates of the probabilities that each is true constitute a complete summary of the information then available to her. We suggested earlier that we might use the change in these probabilities as a measure of the **amount of information** provided by either cues or signals. One is tempted to use the difference in probabilities before and after receipt of the cue or signal for such a measure. As we shall see, the change in the probability estimates due to reception of a signal or cue varies with the prior estimate of that probability. Receipt of the same signal may result in quite different changes in the estimated probability when the prior estimate is low than when it is high. This suggests that we should use the *ratio* of the probability after signal receipt to that before as our measure of information gained. Since ratios can become infinitely large or infinitely small, one usually plots them on logarithmic scales. What logarithmic base should we use: base-10 logarithms, natural logarithms, or some other format?

Our focal female faces the simplest type of decision possible, that between two mutually exclusive alternatives. This is called a **binary question**. All uncertainty can be completely eliminated by asking a single question and obtaining an answer. In the female bird's case, she might ask, Is the male healthy or not? A "yes" or "no" answer completely resolves the question. What if there are more than two alternative answers? It can be shown that any problem with a finite set of answers can be solved by answering a smaller number of binary questions. We are particularly interested in determining the minimum number of binary questions that must be asked to solve a particular problem.

Examples in Figure 13.1 suggest that the minimum number of binary questions, H, needed to identify which of M equally likely answers is true obeys the relation $M = 2^H$. We can rewrite this as

$$H = \log_2 M$$

Figure 13.1 Number of binary questions for different numbers of equiprobable answers. (A) A male antelope detects the scent of an oestrus female when there are four nearby females. The male needs to ask only two binary questions to resolve all uncertainty about which female to court: Is the oestrus female in the northern pair? Given the answer to that question, he can then ask whether the oestrus female is the eastern member of the remaining two possibilities. (B) One of eight eggs in a nest has been laid by a cuckoo rather than by the owners of the nest. How many binary questions are needed to identify the cuckoo egg? The answer is three: (1) Is the egg among the top four eggs? (2) Is the egg among the left pair of the remaining possibilities? (3) Is the egg the left member of the remaining two possibilities? (C) Many jay species raise their crests when aggressive and lower them when submissive. Suppose members of a given species can discriminate 16 different crest positions. How may binary questions need be asked to identify the aggression level of a sender if all states are equiprobable? The answer is four (proof left to the reader). These examples suggest that when there are $M = 2^H$ equiprobable alternatives, one need ask only H binary questions to completely identify which alternative is currently true. ▶

where

$$\log_2 M = \frac{\ln M}{\ln 2} = \frac{\log_{10} M}{\log_{10} 2}$$

This relationship gives us a measure of the amount of information scaled as a logarithm with a base of 2. Because binary questions are the simplest ones we can ask, there is a certain appeal to this measure. The unit of H is called the

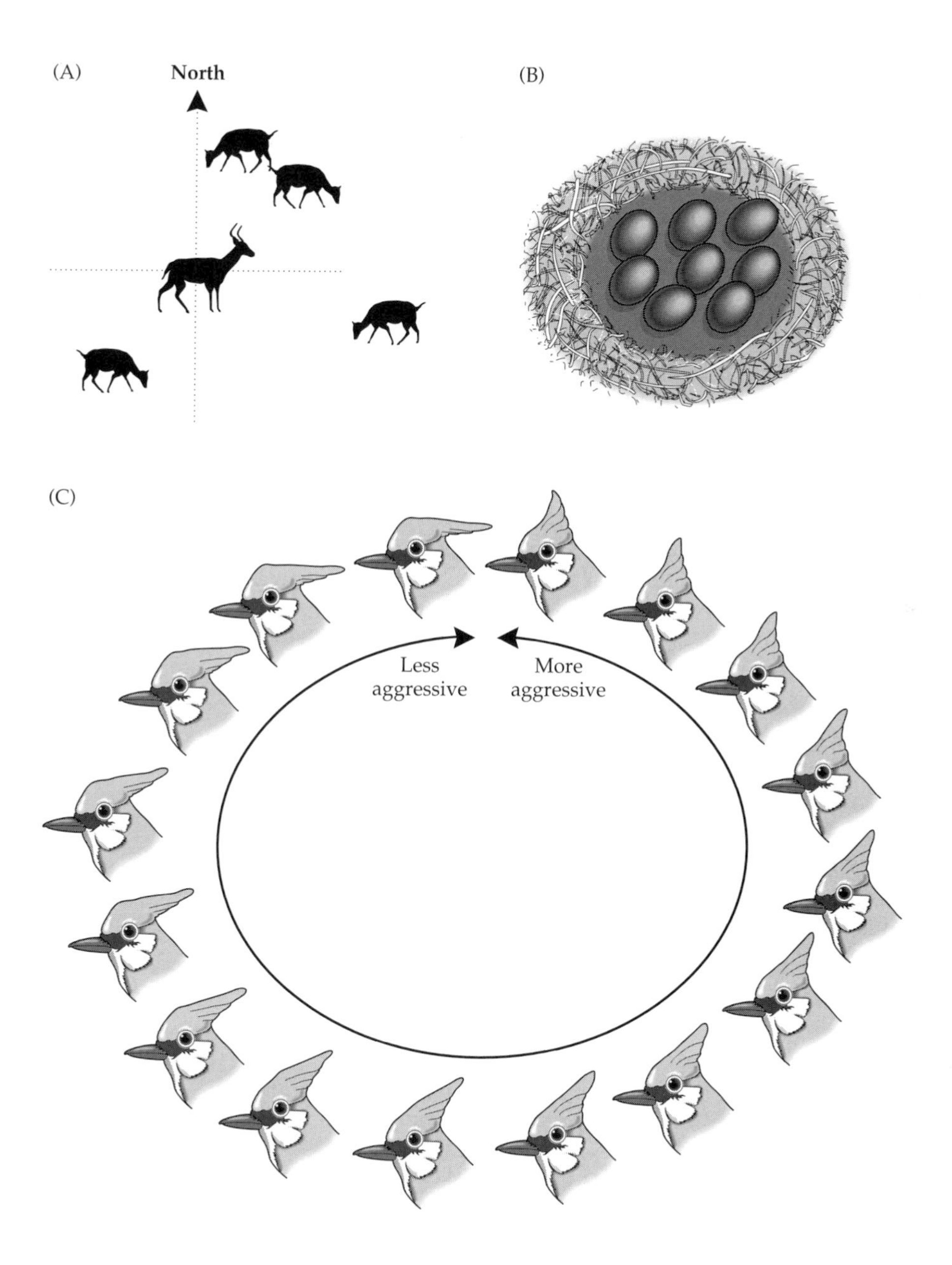

bit. Note that for most values of M, H will not be an integer. For example, if $M = 3$, then $H = 1.585$ bits. Thus if there are three equiprobable answers, it will take 1.585 binary questions to eliminate all uncertainty about which answer is true. While one cannot in practice ask 1.585 binary questions, it is intuitively reasonable that the amount of information needed to identify which of three alternatives is true ought to lie somewhere between the amount required for two (1 bit) and that required for four alternatives (2 bits). One is stuck with fractional values for H regardless of what base one elects to use for the required logarithmic scaling of information.

We now return to our female bird's problem. Suppose she begins with an a priori **probability**, P_0, that the male is healthy instead of sick. She then receives a signal from the male and updates her estimate that he is healthy to the a posteriori **probability** P_1. (How she might compute this update is discussed later in this chapter.) The amount of information she received can then be defined as

$$H_T = \log_2\left(\frac{\text{a posteriori probability}}{\text{a priori probability}}\right) = \log_2\left(\frac{P_1}{P_0}\right) \text{ bits}$$

We use the symbol H_T, instead of H, to indicate that the signal *transferred* some information, H_T, but not necessarily the amount H required to remove all uncertainty. Note that the larger the a priori probability, the larger the change in probabilities that must occur to generate a given value of H. Put another way, a given change in probabilities due to receipt of the signal results in a lower H_T the higher the a priori probability. Note also that if receipt of the signal results in an a posteriori probability smaller than the a priori one, H_T will be negative. This result is reasonable because in this case the signal has reduced the receiver's confidence that the male is healthy. Since it is possible that an a posteriori probability might be 0, the convention in information analysis is to define $\log_2(0) \equiv 0$. The only time the a priori probability is 0 is when a condition is not included on the receiver's list of alternatives; hence the expression is undefined in this case. In practice, the a priori probability can be very tiny, but it will not be 0.

Suppose a candidate male can produce either of two signals. The sender and receiver agree on a coding scheme so that males sing a fast song if healthy and a slow song if sick. All males are assumed to be honest and to know their state of health. The coding scheme is thus **perfect** in that fast songs are only given by healthy males and slow songs by sick ones. When a female hears a song, all uncertainty is removed. If the a priori probability that a male is healthy is P_0, then a fast song causes the corresponding a posteriori probability to jump to 1.0. The information provided is then

$$H_T(\textit{fast song}) = \log_2\left(\frac{1}{P_0}\right) = \log_2(1) - \log_2(P_0) = -\log_2(P_0)$$

Similarly, if the a priori probability that a male is sick is $1-P_0$, then a slow song causes that a posteriori probability to jump to 1.0 and the provided information equals

$$H_T(slow\ song) = \log_2\left(\frac{1}{(1-P_0)}\right) = \log_2(1) - \log_2(1-P_0) = -\log_2(1-P_0)$$

If $0 < P_0 < 1$, then H_T for either song will be greater than zero (since the logarithm of a fraction is a negative value); if $P_0 = 0$ or 1, then $H_T = 0$ (reflecting the fact that the female already knew his state of health and thus receipt of the song added no new information). The smaller P_0, the larger the corresponding H_T(*fast song*) and the smaller the value of H_T(*slow song*). Increasing P_0 has the reverse effect. In words, receipt of a signal coding for a rare condition and eliminating all uncertainty provides more information than receipt of a signal coding for a common one.

This relationship can be simplified further when all of the alternative signals, and thus their associated conditions, are equally likely. Suppose there are M such possible signals. Then the a priori probability that any one signal will be received is $P_0 = 1/M$. The information obtained from receipt of any one of these equally likely signals when any of them completely removes all uncertainty is then

$$H_T = -\log_2(P_0) = -\log_2\left(\frac{1}{M}\right) = \log_2(M)$$

This is precisely what we found when we considered the examples in Figure 13.1. All of these were posed as problems in which the alternative M answers, and thus signals, were equally likely, and there was complete elimination of uncertainty after receipt of the information (i.e., the a posteriori probabilities for each alternative were 1.0 or 0).

Average Information for a Signal Set

We saw in Chapter 12 that the relevant measure of evolutionary success in communication is the average value of information for a set of alternative signals. If we want to quantify the evolutionary and economic links between the amount and value of information, we thus need a corresponding measure of the average amount of information provided by such a set. This measure should take into account, as does the value of information, the fact that different conditions and thus signals will occur with different frequencies. How can we extend the rules for computation of the amount of information during a single exchange to the average amount of information provided by an entire signal set over many exchanges?

The appropriate computation follows directly from our definitions of average expectations discussed on pages 366–367. If the a priori probability that signal i will be given is P_i, and the information provided when signal i is received is H_i bits, then the average information transferred when several alternative signals are possible will be

$$\overline{H}_T = \sum_i P_i H_i$$

As we saw above, a perfect correlation between signal and condition allows us to write $H_i = -\log_2(P_i)$. We can thus compute the average information transferred per signal when perceived signals are perfectly correlated with alternative conditions as

$$\overline{H}_T = -\sum_i P_i \log_2(P_i)$$

This value will be maximized when the alternative signals are equally likely. Consider the case in which there are two alternative signals, each of which is perfectly correlated with a different condition, there is no loss of information during propagation, and the receiver makes no errors in identifying which signal has been received. Figure 13.2 shows how the average information provided by this signal set varies with the relative a priori probabilities of the two alternative conditions. When the two conditions (i and j) are equally likely, i.e., $P_i = P_j = 0.5$, then $H_i = H_j = -\log_2 0.5 = 1$ bit. $\overline{H}_T$ is here maximal and equals 0.5(1) + 0.5(1) = 1 bit. As one alternative becomes more likely than the other, the average amount of information per signal drops: even though receipt of the rare signal provides a large amount of information, it is given so rarely that the contributions of the more common and less informative signal dominate the average. When either alternative is certain, (i.e., $P_i = 1$ or 0), then $\overline{H}_T = 0$ because no received signal provides any new information.

Suppose the alternative conditions and signals are each drawn from a continuous set (instead of the discrete alternatives discussed so far). For example, sound intensity and frequency modulation rate are often used by vertebrate senders to code for a continuous range of aggressive intentions. Let us denote the value of the signal variable that codes this information by V. We can then examine the frequency with which different V values tend to appear. These can be used to estimate the **probability density function** of V, which we shall call $p(V)$. The value of this function for any small range of V values is proportional to the frequency of those values, and the entire area under the $p(V)$ distribution curve is equal to 1.0. By analogy with the earlier example, we can then define

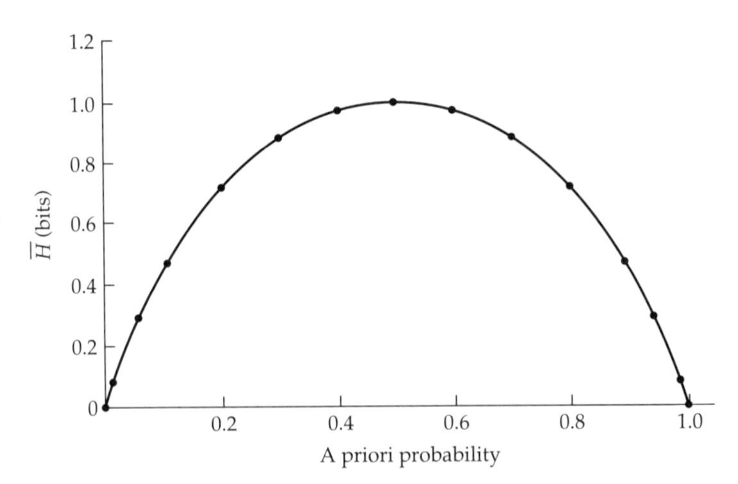

Figure 13.2 Average information as a function of a priori probabilities for a signal set with two alternative conditions and perfect coding. The horizontal axis shows the probability that one of two alternative signals will be received; the vertical axis shows the average amount of information, in bits.

$$\overline{H}_T = -\int p(V) \log_2 [p(V)]\ dV$$

The continuous case thus replaces the sum of the discrete case with an integral. For many probability distributions, Shannon and Weaver (1949) showed that this reduces to

$$\overline{H}_T = \log_2 c\sigma$$

where c is a constant with a value between 3.46 and 4.13, depending upon the shape of the probability distribution, and σ is the standard deviation of that distribution (i.e., $\sigma = \sqrt{variance}$). We shall use this result in the next chapter (Box 14.3).

Average Information for Compound Displays

Thus far we have only considered signals that help identify the alternative answers to a single question such as: Is the current male healthy or not? However, many senders produce compound signals whose components provide answers to several different questions concurrently. What is the average information transferred in this case? For example, suppose a male bird performs a display to the female that combines a bowing posture with the singing of a song: the depth of his bow indicates his willingness to mate with her, and the speed of the song is again a measure of his health. The bow and the song thus help answer two different questions that the female might ask. Let us suppose that male health and male intentions are independent variables and that males accurately advertise both health and intent. When the answers to the two questions are independent, the average information provided by the entire display is simply the sum of the average information provided by each of the display components. For example, if the average information provided by song speed is $\overline{H}_T(S)$ and that provided by bow depth is $\overline{H}_T(B)$, then the average information provided by the compound display is $\overline{H}_T(Displays) = \overline{H}_T(S) + \overline{H}_T(B)$. A similar sum can be computed for the average information provided by any finite set of independent signals or signal components.

What if male health and willingness to mate are not independent? For example, perhaps healthier males are more willing to mate. Even if males remain honest and accurate in their advertisement of health and intentions, and females remain accurate in their identification of which signal was given, the average information exchanged will be less because some combinations of health and intent are now less likely than others, and this means that some combinations of posture and song speed are less likely. How much do these correlations between questions and thus between signal sets reduce the average amount of information provided by a compound display?

Let us suppose there are only two possible song speeds, denoted by S_1 and S_2, and only two possible postures, B_1 and B_2. Let the a priori probabilities for each component be $P(S_1)$, $P(S_2)$, $P(B_1)$, and $P(B_2)$, respectively. There are four possible compound displays that might be seen. The probabilities of each of these displays can be denoted by $P(S_1 \text{ and } B_1)$, $P(S_1 \text{ and } B_2)$, $P(S_2 \text{ and } B_1)$, and $P(S_2 \text{ and } B_2)$. Note that these four probabilities must sum to 1.0 because if a display is performed, it has to be one of these four. Assuming that the performances of the different displays are independent (as opposed to the dependence of their components), the information provided by an average display is then

$$\overline{H}_T(Displays) = -\sum_i \sum_j P(S_i \text{ and } B_j) \log_2 P(S_i \text{ and } B_j)$$

As we saw in Box 12.1, if components of the displays are not independent, then $P(S_i \text{ and } B_j) = P(S_i)P(B_j \mid S_i) = P(B_j)P(S_i \mid B_j)$. Using the first of these two equalities, we can rewrite the expression for $\overline{H}_T$ (*Displays*) as

$$\overline{H}_T(Displays) = -\sum_i \sum_j P(S_i \text{ and } B_j) \log_2 \left[P(S_i) P(B_j \mid S_i)\right]$$

The logarithms of the quantity in the brackets can be expanded to give

$$\overline{H}_T(Displays) = -\sum_i \sum_j P(S_i \text{ and } B_j) \log_2 P(S_i) - \sum_i \sum_j P(S_i \text{ and } B_j) \log_2 P(B_j \mid S_i)$$

Because $P(S_1 \text{ and } B_1) + P(S_1 \text{ and } B_2) = P(S_1)$, and $P(S_2 \text{ and } B_1) + P(S_2 \text{ and } B_2) = P(S_2)$, the first term in this expression is simply

$$-\sum_i P_i(S_i) \log_2 P(S_i) = \overline{H}_T(S)$$

If we define the second term as

$$\overline{H}_T(B \mid S) \equiv -\sum_i \sum_j P(S_i \text{ and } B_j) \log_2 P(B_j \mid S_i)$$

we then get that $\overline{H}_T$ (*Displays*) $= \overline{H}_T(S) + \overline{H}_T(B \mid S)$. This says that the uncertainty resolved by the average display, (still assuming perfect coding by males), depends upon the initial uncertainty as to which song speed will be performed plus the conditional uncertainty as to which bow depth will be performed in combination with a given song speed. (Masochistic readers may wish to convince themselves that had we expanded $P(S_i \text{ and } B_j)$ using the second equality, we would have found that $\overline{H}_T$ (*Displays*) $= \overline{H}_T(B) + \overline{H}_T(S \mid B)$. Thus the average amount of information can be computed as either $\overline{H}_T(S) + \overline{H}_T(B \mid S)$ or $\overline{H}_T(B) + \overline{H}_T(S \mid B)$; the two expressions are equal.)

When health and intent were independent, we found that the average amount of information provided by a compound display was $\overline{H}(Displays) = \overline{H}_T(S) + \overline{H}_T(B)$. When they are not independent, we now find that $\overline{H}(Displays) = \overline{H}_T(S) + \overline{H}_T(B \mid S)$. The sole change is in the replacement of $\overline{H}_T(B)$ by $\overline{H}_T(B \mid S)$. Which is the larger value? Note that the terms in the sum generating $\overline{H}_T(B)$ are of the form $-P(S_i \text{ and } B_j) \log_2 P(B_j)$ whereas those used to compute $\overline{H}_T(B \mid S)$ are of the form $-P(S_i \text{ and } B_j) \log_2 P(B_j \mid S_i)$. The only differences are in the logarithmic terms. It follows that $\overline{H}_T(B) \geq \overline{H}_T(B \mid S)$ if $-\log_2 P(B_j) \geq -\log_2 P(B_j \mid S_i)$. The later expression is true only if $P(B_j \mid S_i) \geq P(B_j)$. We saw in Box 12.1 that $P(B_j \mid S_i) = 1.0$ if B and S are perfectly correlated, and $P(B_j \mid S_i) = P(B_j)$ if they are totally uncorrelated. It is thus generally true that $P(B_j \mid S_i) \geq P(B_j)$, and that $\overline{H}_T(B) \geq \overline{H}_T(B \mid S)$. $\overline{H}_T$ (*Displays*) will then be less when health and intent are not independent by an amount equal to $\overline{H}_T(B) - \overline{H}_T(B \mid S)$ bits, or $\overline{H}_T(S) - \overline{H}_T(S \mid B)$ if we are computing the average using the alternative expansion. In words, the higher the correlations between display components, the less information that is transferred by a compound display. An example of the effects of compound display component correlations on $\bar{H}_T(Displays)$ is developed in Box 13.1.

Average Information When Senders or Receivers Err

The computations in the previous sections all assume that senders unerringly match a unique signal to each underlying condition, and that receivers never make a mistake about which signal was sent. The reception of a signal thus resolves all uncertainty about which condition is true. This situation allows us to set the a posteriori probabilities to 0 or 1, and to use the negative logarithm of the a priori probability as a measure of the amount of information provided by each signal. We shall call such a situation the provision of **perfect information**. However, it is rarely true that receipt of a signal eliminates all uncertainty. Senders make errors associating signals with conditions, signals become distorted during propagation or are masked by noise, and receivers make errors in identifying signals or in associating them with conditions. Thus after receipt of a useful signal, there will often be some residual uncertainty remaining. As a result, the a posteriori probabilities do not jump to 1 or 0.

We thus need to measure the degree to which signals and conditions are correlated at each stage of the communication process. The mapping of alternative signals onto alternative conditions is based on the **coding rules** for each party. The maximum correlation is determined by the sender. Average correlations between signal and condition at later stages in the communication process can be no higher, on average, than those initially set by senders. At each successive stage of communication, the correlations become increasingly degraded. How do we compute the amount of information when correlations between signal and condition are less than 100%?

The analysis begins in a manner similar to that of the prior section. Suppose we have two possible conditions, C_1 and C_2, which occur with a priori probabilities $P(C_1)$ and $P(C_2)$, respectively. Senders provide information to receivers about which condition is true using signals S_1 and S_2. Let us suppose that S_1 signals are usually given when C_1 is true, and S_2 signals when C_2 is true. Receivers then update their estimates that either condition is true, given the signal that they received. There are four possible combinations of condition and perceived signal that can occur. The corresponding probabilities of these combinations can be denoted by $P(S_1 \text{ and } C_1)$, $P(S_2 \text{ and } C_2)$, $P(S_1 \text{ and } C_2)$, and $P(S_2 \text{ and } C_1)$. If coding were perfect, there were no propagation distortions, and reception was error free, then the last two of these probabilities would be zero: they are combinations that would never occur. However, if there are errors, then all four combinations are possible, and we must take them all into account when computing the average information provided by this pair of signals.

Using our basic definition of information (page 392), the amount of information provided when condition C_1 is true and signal S_1 received is

$$\log_2 \frac{\hat{P}(C_1 \mid S_1)}{P(C_1)}$$

Box 13.1 Computing Average Information Content of Displays

MANY ANIMALS PRODUCE DISPLAYS that provide several kinds of information concurrently. Consider two fish performing broadside displays to each other. The displays consist of both color and postural components. Suppose that the fish can expand or contract the melanophores in their skins. A fish that thinks it owns the substrate beneath it takes on a dark tone by expanding the melanophores, whereas an intruder contracts them and exhibits a light hue. Dark versus light coloration thus constitutes one signal set. At the same time, a fish with its fins raised is likely to attack; one with its fins lowered is unlikely to attack the other. Fin position constitutes the second signal set in the display. Suppose one determines the fractions of time that the possible combinations of these components are used and summarizes them in the following table:

Color	Raised	Lowered	
Dark	0.48	0.32	0.80
Light	0.12	0.08	0.20
	0.60	0.40	1.00

Fractions on the right and bottom of the table are the row and column subtotals: thus dark color is seen 0.8 of the time overall. Given these numbers, what is the average information provided by a broadside display? If 0.8 of the fish are dark (instead of light) and 0.6 of the fish have raised fins (instead of lowered ones), and if fin and color states were independent, we should see dark fish with raised fins $0.8 \times 0.6 = 0.48$ of the time, and this we do. In fact, the values in the cells of the table are all equal to the products of the corresponding row and column probabilities. This confirms that fin level and color state in these fish are independent. Since the two display components are independent, the overall $\bar{H}_T$ *(Displays)* is equal to the sum of the $\bar{H}_T$ values for each component treated alone. For the color pattern, $\bar{H}_T\,(Color) = [0.8 \log_2 (0.8) + 0.2 \log_2 (0.2)] = 0.722$ bits, and for the fin position, $\bar{H}_T\,(Fins) = [0.6 \log_2 (0.6) + 0.4 \log_2 (0.4)] = 0.971$ bits. Thus $\bar{H}_T(Displays) = 0.722 + 0.971 = 1.693$ bits.

What if we examined another population of the same fish and found the distribution table for the display components shown below:

Color	Raised	Lowered	
Dark	0.56	0.24	0.80
Light	0.04	0.16	0.20
	0.60	0.40	1.00

The row and column subtotals are the same as in the prior population, but the cell values are different. In fact, a quick glance at the cell values indicates that they are no longer equal to the products of the row and column probabilities. This means that the two components of the display are not independent. We can then compute $\bar{H}_T$(*Displays*) in several ways. The most straightforward method uses the formula

$$\bar{H}_T(Displays) = -\sum_i \sum_j P(A_i \text{ and } B_j) \log_2 P(A_i \text{ and } B_j)$$

where A_i and B_j are the alternative values of fin height and color, respectively. This is simply the negative sum of the discounted contributions from each cell: $\bar{H}_T$ (*Displays*) = –[0.56 $\log_2$ (0.56) + 0.24 $\log_2$ (0.24) + 0.04 $\log_2$ (0.04) + 0.16 $\log_2$ (0.16)] = 1.571 bits.

Sometimes, we know the conditional probabilities when signal sets are correlated but do not know the joint (cell) values. In these cases we can use the relationships $\bar{H}_T$ (*Displays*) = $\bar{H}_T(A)$ + $\bar{H}_T(B \mid A)$ or $\bar{H}_T$ (*Displays*) = $\bar{H}_T(B)$ + $\bar{H}_T(A \mid B)$. Using the second table the relevant conditional probabilities can be computed by dividing each cell value by the appropriate column or row total. For example, suppose we wish to compute $\bar{H}_T$ (*Displays*) = $\bar{H}_T$(*Fins*) + $\bar{H}_T$(*Color* | *Fins*). We use the table to determine that

$$P(Dark \mid Raised) = 0.56/0.6 = 0.933 \quad P(Light \mid Raised) = 0.04/0.6 = 0.0667$$
$$P(Dark \mid Lowered) = 0.24/0.4 = 0.600 \quad P(Light \mid Lowered) = 0.16/0.4 = 0.400$$

We can then compute

$$\bar{H}_T(Fins) = -[0.6\log_2(0.6) + 0.4\log_2(0.4)] = 0.971 \text{ bits}$$
$$\bar{H}_T(Color \mid Fins) = -[0.56\log_2(0.933) + 0.24\log_2(0.6) + 0.04\log_2(0.0667) + 0.16\log_2(0.4)]$$
$$= 0.600 \text{ bits}$$
$$\bar{H}_T(Displays) = \bar{H}_T(Fins) + \bar{H}_T(Color \mid Fins) = 0.971 + 0.600 = 1.571 \text{ bits}$$

This is the same value we obtained using the cell values directly.

Alternatively, we might analyze the second table by using the expression $\bar{H}_T$ (*Displays*) = $\bar{H}_T$(*Color*) + $\bar{H}_T$ (*Fins* | *Color*). We thus compute

$$P(Raised \mid Dark) = 0.56/0.8 = 0.7 \quad P(Lowered \mid Dark) = 0.24/0.8 = 0.3$$
$$P(Raised \mid Light) = 0.04/0.2 = 0.2 \quad P(Lowered \mid Light) = 0.16/0.2 = 0.8$$

and then find that:

$$\bar{H}_T(Color) = -[0.8 \log_2(0.8) + 0.2 \log_2(0.2)] = 0.722 \text{ bits}$$
$$\bar{H}_T(Fins \mid Color) = -[0.56 \log_2(0.7) + 0.24 \log_2(0.3) + 0.04 \log_2(0.2) + 0.16\log_2(0.8)]$$
$$= 0.849 \text{ bits}$$
$$\bar{H}_T(Displays) = \bar{H}_T(Color) + \bar{H}_T(Fins \mid Color) = 0.722 + 0.849 = 1.571 \text{ bits}$$

However calculated, the average information provided by a broadside display when color and fins are not independent is here 1.571 bits. The situation when the components were independent was 1.693 bits. Thus correlations between the components have reduced the average information provided per signal by 1.693 – 1.571 = 0.122 bits.

where $\hat{P}(C_1 \mid S_1)$ is the updated a posteriori estimate that C_1 is true after receipt of the signal. Similarly, the amount of information received when C_2 is true and signal S_2 received is

$$\log_2 \frac{\hat{P}(C_2 \mid S_2)}{P(C_2)}$$

The average information is the sum of equivalent expressions for each of the four combinations of condition and signal, each discounted by the probability it will occur:

$$\overline{H}_T = \sum_i \sum_j P(C_i \text{ and } S_j) \log_2 \frac{\hat{P}(C_i \mid S_j)}{P(C_i)}$$

We can simplify this expression by extracting the denominators and combining them in their own sum to get

$$\overline{H}_T = \sum_i \sum_j P(C_i \text{ and } S_j) \log_2 \hat{P}(C_i \mid S_j) - \sum_i \sum_j P(C_i \text{ and } S_j) \log_2 P(C_i)$$

Because

$$\sum_j P(C_i \text{ and } S_j) \log_2 P(C_i) = P(C_i) \log_2 P(C_i) \text{ and } \overline{H}(C) = -\sum_i P(C_i) \log_2 P(C_i)$$

we can replace the right-hand term in this expression with $\overline{H}(C)$. This is the amount of information which would be required to remove all initial uncertainty about which condition was true. It is the maximum amount of information the receiver could want to obtain about this situation.

Because $P(C_i$ and $S_j)$ is positive whereas $\log_2 \hat{P}(C_i \mid S_j)$ is negative, the remaining term in the expression for $\overline{H}_T$,

$$\overline{H}_T, \sum_i \sum_j P(C_i \text{ and } S_j) \log_2 \hat{P}(C_i \Big|$$

is negative. We can make it a positive value, and give it a name, by defining

$$\overline{H}(C \mid S) \equiv -\sum_i \sum_j P(C_i \text{ and } S_j) \log_2 \hat{P}(C_i \mid S_j)$$

$\overline{H}(C \mid S)$ is called the **equivocation**. It is the average uncertainty about which condition is true that remains after receipt of a signal. An example of its computation can be found in Box 13.2, page 412.

Given that $\overline{H}(C)$ is the total uncertainty before a signal is provided, we can now simplify our expression for $\overline{H}_T$ when coding is imperfect to

$$\overline{H}_T = \overline{H}(C) - \overline{H}(C \mid S)$$

This relationship says simply that the average amount of information provided per signal when there are errors is less than that expected were there no errors. The reduction is equal to the equivocation. The amount of equivocation will depend upon the accuracy of sender coding, the amount of propagation distortion, and the level of receiver error. If coding were perfect, there were no distortion, and receivers never failed to identify which signal was sent, then all the $\hat{P}(C_i \mid S_j)$ terms in the equivocation expression would be 1.0 or 0, and $\bar{H}(C \mid S) = 0$. If senders were to send signals completely at random, then each $\hat{P}(C_i \mid S_j) = P(C_i)$, and thus $\bar{H}(C \mid S) = \bar{H}(C)$. As a result, $\bar{H}_T = 0$. The overall equivocation can be partitioned into that added at each stage of the communication process: selection of signal by the sender, propagation, and intepretation by the receiver. This can be a useful way to identify which stage is the greater source of error.

The one issue we have not yet resolved is how a receiver might use receipt of a signal to update its a priori estimate $P(C_i)$ to a new a posteriori value $\hat{P}(C_i \mid S_j)$. A receiver presumably knows the code and the chances of distortion during propagation. It thus could know the value of each $P(S_j \mid C_i)$. How it might combine knowledge of the $P(S_j \mid C_i)$ and receipt of a signal into estimation of the $\hat{P}(C_i \mid S_j)$ is taken up on pages 405–409.

Redundancy

Different coding schemes result in different amounts of information being provided. The maximal amount of information that can be transferred given M alternative signals is $\bar{H}_{max} = \log_2 M$. This maximum is achieved when there is perfect coding, no propagation or receiver errors, and the M alternative conditions occur with equal probability. When conditions are equiprobable, the larger M, the larger the maximal amount of information that can be transferred. Not all signal sets provide this maximal amount of information. We have seen that if signals are assigned to conditions that do not occur with equal probabilities, then the signals will be used at unequal rates and the average amount of information transferred will be less. A signal set with six alternative signals could thus provide the same average amount of information as one with three alternative signals if conditions for the smaller set are more equiprobable than those for the larger set. Even when the probabilities of the conditions are equal, the lack of a unique mapping of signals on conditions will also result in a reduction in the average information transferred since the equivocation will not be zero. If displays consist of several components, but these do not occur independently, less information is being provided than if there were no correlations between display components. Finally, if the same signal is repeated several times, even though the context has not changed, then less information is being transferred than could be given at the same rate of signal emission.

Any coding scheme that transfers less information than the maximum possible with the same number of signals is said to be **redundant**. If the average amount of information actually transferred is $\bar{H}_T$, then the redundancy is defined as the fraction $1 - (\bar{H}_T/\bar{H}_{max})$ and usually given as a percentage. For example, in English, the redundancy of information on a per letter basis is about 75%. This high value is due to unequal use of different letters, and to the large number of combinations of letters that are obligatory, forbidden, or meaningless. The use of any letter is rarely independent of the letters preceeding and following it. These dependencies reduce the number of allowable combinations and thus the amount of information that is transferred. As we shall see in the next sections, redundancy is not always bad because it may reduce the likelihood that a receiver will make a wrong decision. In English, we can often guess which word was written even though some letters may be illegible. We are able to decode the word because the set of allowable letter combinations is limited. Optimal coding schemes may thus require redundancy; maximizing the amount of information transferred is rarely the best policy.

OPTIMAL UPDATING

We have seen that receipt of a signal is followed by a receiver's updating of the probabilities that a given condition is true. This is the whole point of communication! How can a receiver combine its prior estimates with the information in a signal to produce a new estimate? There are in fact many ways that this can be done. No method, however, can produce a more accurate estimate than Bayesian updating. We shall describe this method in detail, as it sets a benchmark against which we can compare any alternative method.

Bayesian Updating After One Signal

The best way to update probabilities in an uncertain world is specified by **Bayes' theorem**. Figure 13.3 outlines the basic logic. A receiver wishes to know whether condition C_1 or C_2 is the case. It begins with an a priori probability, $P(C_1)$, that C_1 is true and an a priori probability of $P(C_2)$ that C_2 is true. A sender then emits a signal that is assigned by the receiver into either category S_1 or S_2. Assume that if there were no errors, condition C_1 would always result in a perceived S_1 signal by the receiver, and state C_2 in an S_2 signal. Sup-

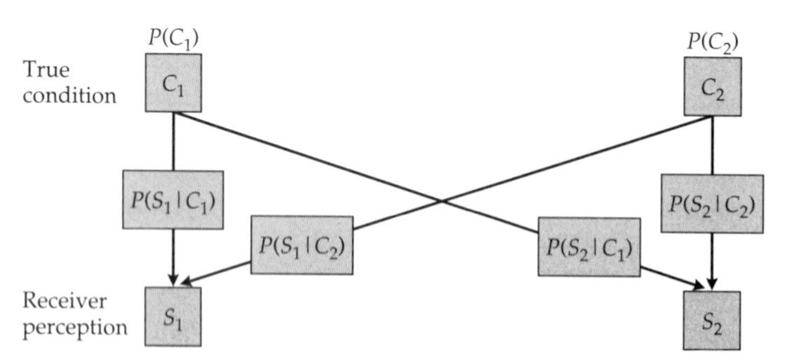

Figure 13.3 Probabilities associated with Bayesian updating. Conditions C_1 or C_2 occur with a priori probabilities $P(C_1)$ and $P(C_2)$, respectively. Sender sends correct signal S_1 when C_1 is the case with conditional probability $P(S_1 \mid C_1)$, and signal S_2 when C_2 is the case with probability $P(S_2 \mid C_2)$. The sender sends the erroneous messages S_1 when C_2 is true with probability $P(S_1 \mid C_2)$ and S_2 when C_1 is true with probability $P(S_2 \mid C_1)$. The same setup is used if propagation distortion or the receiver is the source of the errors.

pose, however, that there are coding or reception errors, and these result in some S_1 signals being perceived by the receiver when state C_2 is true, and some S_2 signals being perceived when state C_1 is true. We can assign conditional probabilities to each of the four possible outcomes of this signal exchange: $P(S_1 | C_1)$ and $P(S_2 | C_2)$ are the probabilities associated with "correct" categorizations by the receiver; $P(S_1 | C_2)$ and $P(S_2 | C_1)$ are the conditional probabilities of the two possible incorrect associations.

To perform a Bayesian update, the receiver must know the relative chances that C_1 or C_2 will occur [($P(C_1)$ and $P(C_2)$], and the average chances of the different kinds of correct and incorrect transmission occurring [$P(S_1 | C_1)$, $P(S_2 | C_1)$, $P(S_1 | C_2)$, and $P(S_2 | C_2)$]. This information is acquired through prior experience or prior evolutionary history. The task for the receiver is to use these values to estimate the a posteriori probabilities $P(C_1 | S_i)$ that C_1 is the correct state given that it has received a signal S_i, and $P(C_2 | S_i)$ that C_2 is the correct state given receipt of signal S_i. Note that these conditional probabilities look similar to those listed in Figure 13.3 and the first sentence of this paragraph, but the order of the terms is reversed. This is a critical difference! For example, $P(S_1 | C_1)$ of Figure 13.3 is the probability of perceiving a signal S_1 when C_1 is the true condition; $P(C_1 | S_1)$, on the other hand, is the probability that condition C_1 is actually the case given that the receiver has detected a signal S_1. It is this latter probability that the receiver needs to compute from prior information and the identity of the current signal. How does the receiver convert what it knows into that which it needs to know?

Bayes' theorem follows directly from the simple probability relationships outlined in Box 12.1. In our present example, it computes the best possible estimate of $P(C_1 | S_1)$ as

$$P(C_1 | S_1) = \frac{P(C_1)P(S_1 | C_1)}{P(C_1)P(S_1 | C_1) + P(C_2)P(S_1 | C_2)}$$

The numerator is the probability, $P(C_1$ and $S_1)$, that C_1 is the correct condition and that signal S_1 will be perceived. The denominator is the total probability that an S_1 signal will be perceived whatever the condition. Thus Bayes' theorem claims that the best estimator of $P(C_1 | S_1)$ is the fraction of the total time in which an S_1 is received that it is likely to be associated with condition C_1. Since either condition C_1 or C_2 is true in our example, once the receiver has $P(C_1 | S_1)$, the receiver also knows $P(C_2 | S_1) = 1.0 - P(C_1 | S_1)$. Alternatively, one could use Bayes' theorem directly to compute

$$P(C_2 | S_1) = \frac{P(C_2)P(S_1 | C_2)}{P(C_2)P(S_1 | C_2) + P(C_1)P(S_1 | C_1)}$$

What if the receiver had received the signal S_2 instead? Then we can use Bayes' theorem as well to compute

$$P(C_1 | S_2) = \frac{P(C_1)P(S_2 | C_1)}{P(C_1)P(S_2 | C_1) + P(C_2)P(S_2 | C_2)}$$

and

$$P(C_2 \mid S_2) = \frac{P(C_2)P(S_2 \mid C_2)}{P(C_2)P(S_2 \mid C_2) + P(C_1)P(S_2 \mid C_1)} = 1.0 - P(C_1 \mid S_2)$$

To see how this computation works, we return to the female bird that wants to know whether a potential mate is healthy or sick and uses the speed of male songs as an indicator of the singer's health. Suppose that healthy males sing fast songs 70% of the time and slow songs the remaining 30%; sick males sing fast songs only 40% of the time and slow songs the remaining 60%. Song rate thus does not supply perfect information; the coding rules are somewhat sloppy. Suppose further that healthy and sick males are equally common in this population. When a female spots a new male, her initial *a priori* estimate that he is healthy will thus be 50%. Suppose this male then sings a song. How should the female update her estimate that this male is healthy? An optimizing female using Bayes' theorem after detection of a fast song would compute

$$\begin{aligned} P(Healthy \mid Fast) &= \frac{P(Healthy)\, P(Fast \mid Healthy)}{P(Healthy)\, P(Fast \mid Healthy) + P(Sick)\, P(Fast \mid Sick)} \\ &= \frac{(0.50)(0.70)}{(0.50)(0.70) + (0.50)(0.40)} = 0.636 \end{aligned}$$

and, after hearing a slow song,

$$\begin{aligned} P(Healthy \mid Slow) &= \frac{P(Healthy)\, P(Slow \mid Healthy)}{P(Healthy)\, P(Slow \mid Healthy) + P(Sick)\, P(Slow \mid Sick)} \\ &= \frac{(0.50)(0.30)}{(0.50)(0.30) + (0.50)(0.60)} = 0.333 \end{aligned}$$

The female thus begins with an a priori estimate that the male is healthy of 0.50; after a fast song, her updated a posteriori estimate that he is healthy rises to 0.636, and after a slow song, the estimate drops to 0.333. Note that a fast song increases the probability the male is healthy by 0.636 – 0.5 = 0.136 whereas a slow song decreases this probability by 0.5 – 0.33 = 0.167. Fast and slow songs thus do not change the a posteriori probability by equal amounts. Put another way, alternative signals in the same signal set may provide different amounts of information.

Sequential Bayesian Updating

Suppose that a receiver is sent a sequence of signals by the sender. How should it now go about updating probabilities? Ideally, it begins before any signal is received with the a priori probabilities of condition C_1 or C_2 being the case. It receives the first signal and uses Bayes' theorem to compute the a posteriori probabilities that the current condition is C_1 or C_2. It then receives the next signal. Its most recent a posteriori probabilities are its best estimates so far of the true state, so it uses these as the new a priori probabilities and the identity of this second signal to generate the next a posteriori probabilities using Bayes' theorem. The procedure continues with each successive signal giving new a posteriori probabilities. This process can be understood most easily by way of an example.

Continuing the example of the prior section, suppose a given male sings two successive fast songs to a female. After the first song, we saw that her estimate that he is healthy rises to 0.636. To update her estimate after the second fast song, she now uses as her a priori values *P(Healthy)* = 0.636 and *P(Sick)* = 1.0 – 0.636 = 0.364 to get

$$P(Healthy \mid Fast) = \frac{(0.636)(0.70)}{(0.636)(0.70) + (0.364)(0.40)} = 0.754$$

Note that the first fast song increased the probability that the male was healthy by 0.636 – 0.50 = 0.136; the second fast song increased this probability by 0.754 – 0.636 = 0.118. This demonstrates that the same signal can result in different changes in the a posteriori probabilities depending upon the a priori values. We can see this another way: suppose the female were to receive a long sequence of fast songs. Each successive song increases the probability that the male is healthy, but this probability can never exceed 1.0. It follows that each successive fast song must change the probabilities less and hence add less information.

Sequential Bayesian updating allows the female to keep a running estimate that the male is healthy, an estimate that is modified with each successive song. A fast song pulls the estimate up, and a slow song pulls it back down again. Suppose this continues until the female has heard four consecutive songs from the same male. There are 16 possible sequences that the female might hear. For example, there are six different ways she might have heard two fast and two slow songs (using *F* for fast and *S* for slow): *FFSS*, *FSFS*, *FSSF*, *SSFF*, *SFSF*, and *SFFS*. Luckily, Bayes' theorem applied in succession will generate the same final estimate that the male is healthy regardless of the order in which the two fast and two slow songs were sung: for two fast and two slow songs, the a posteriori probability that the male is healthy is 0.434. The final estimate thus depends on the relative number of fast and slow songs, and not on their order. There are five possible histories of four successive songs by the same male. These are summarized in Figure 13.4.

Note that the possible combinations of fast and slow after listening to four consecutive songs are not equally likely nor are their probabilities the same for healthy and sick males. For example, the probability that a female will hear *FFFF* if the male is healthy is $(0.7)^4 = 0.240$, whereas the probability she will hear *FFFF* if the male is sick is only $(0.4)^4 = .026$. As a result, a female that continues sampling should eventually hear more fast songs if the male is healthy than would be the case if he were sick. Her running estimate that a healthy male *is* healthy should asymptotically approach 1.0 with continued sampling; if the male is sick, the female's estimate that he is healthy should

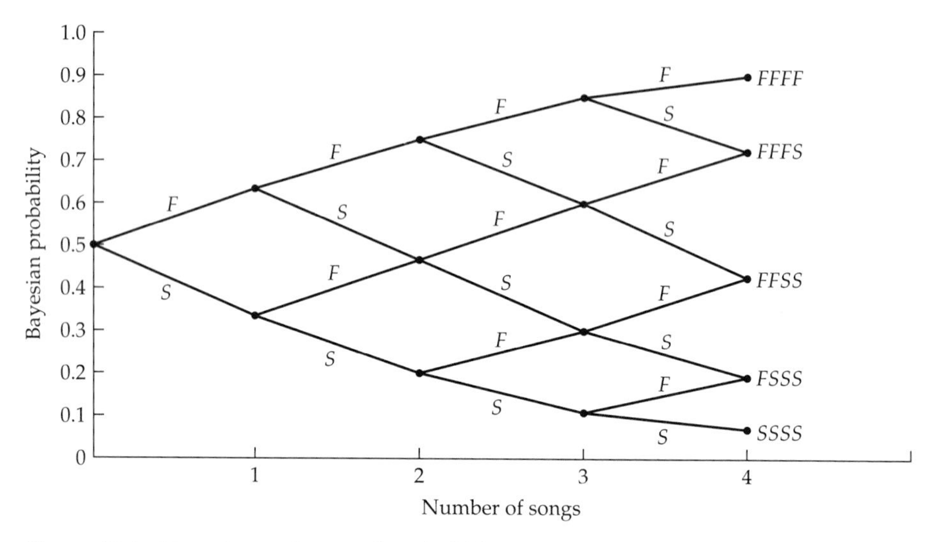

Figure 13.4 Bayesian estimate of probability male bird is healthy after four successive songs. There are five possible outcomes. Assumes healthy males sing fast songs 70% of time, and slow songs 30% of the time, whereas sick males sing fast songs only 40% of the time, and slow songs 60%. *F* = Fast song, *S* = slow song.

asymptotically approach 0. Figure 13.5 shows the cumulative estimates for healthy and sick males averaged over 1000 different Bayesian females. In both cases, successive sampling asymptotically approaches the truth about the male, and the averages do so smoothly. Note, however, that a graph of the cumulative estimate of male health for any given female's sampling would not be a smooth curve, but a jagged line erratically approaching either the top or the bottom of the graph with continued sampling.

Average Information Given Bayesian Updating

On pages 397–401, we developed expressions allowing us to compute the average information transferred by a signal set when a sender, receiver, or both err. These expressions assumed we knew how to compute terms of the form $P(C_i \mid S_j)$. Bayes' theorem provides a way to do this, given prior knowledge of the $P(C_i)$ and $P(S_j \mid C_i)$ values. (Examples are given in Box 13.2.) These expressions also allow us to compare the average information provided by each of the alternative signals in the set. In our example, this process would entail comparing the average information provided by fast versus slow songs. If *P*(*Healthy*) is the a priori probability that the male is healthy before receipt of the signal, and *P*(*Healthy* | *Fast*) is the a posteriori probability after hearing a fast song, then the amount of information provided when this combination is encountered is $H_T(Healthy, Fast) = \log_2[P(Healthy \mid Fast)/P(Healthy)]$. We saw above that for a female hearing a fast song the first time a male sang, *P*(*Healthy*) = 0.50 and *P*(*Healthy* | *Fast*) = 0.636. In this case, H_T (*Healthy, Fast*) = 0.347 bits. Note that this is a positive value, implying a reduction in uncertainty. Similarly, the information obtained after perception of a slow

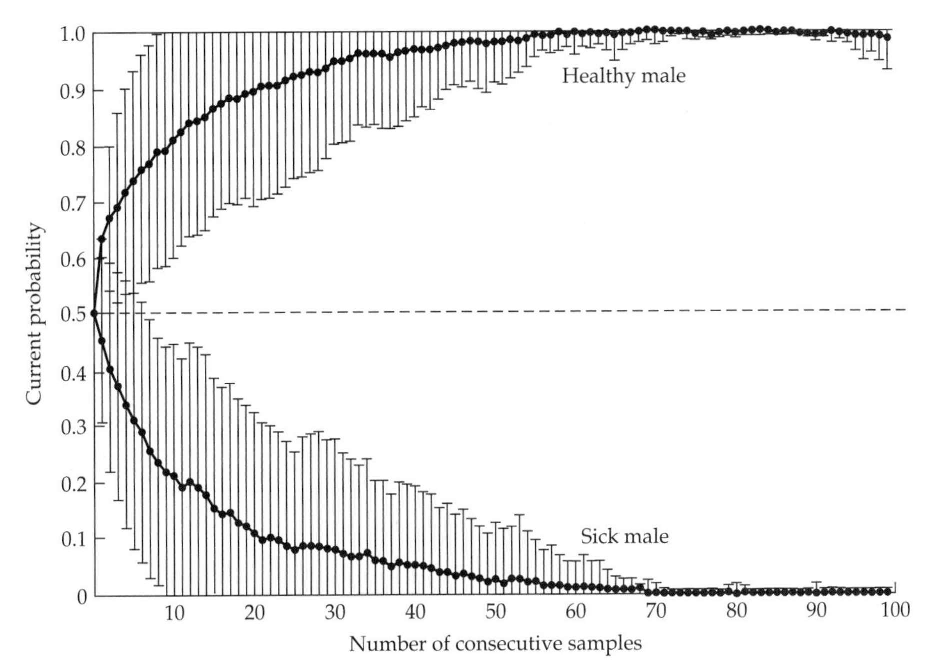

Figure 13.5 Cumulative Bayesian estimates of male health, given songs heard. Same conditions as in Figure 13.4. The results are from computer simulations with stochastic choice of signals by males. Dots show the trajectory of means for 1000 simulations and the error bars show standard deviations around each mean. Both trajectories begin with the chance expectation that a male is healthy of 0.5. The top trajectory is for a healthy male; the lower trajectory is for a sick male. Note that the average estimate that a male is healthy asymptotically approaches the truth with subsequent sampling, and that the error around any given estimate also decreases with continued sampling. Note also that a graph plotting the trajectory of any one sampling sequence would not be a smooth curve, but would jump above and below the average trajectory in an erratic way.

song when the male was in fact sick would be H_T (*Sick, Slow*) = $\log_2[P(Sick \mid Slow)/P(Sick)]$. If the first song heard by a female was slow, then *P*(*Sick*) =0.50, *P*(*Sick* | *Slow*) = 0.667, and H_T (*Sick, Slow*) = 0.416 bits. Again, this is a positive value, indicating a decrease in uncertainty. Now consider the case in which the male is sick but he sings a fast song. The information obtained will be H_T(*Sick, Fast*) = $\log_2[P(Sick \mid Fast)/P(Sick)]$ If this is the first song that this male sings to the female, then *P*(*Sick*) = 0.5, *P*(*Sick* | *Fast*) = 1.0 – 0.636 = 0.364, and thus H_T (*Sick, Fast*) = – 0.458 bits. This is a negative value because the fast song is misleading in this context. The other misleading condition occurs when a slow song is sung by a healthy male. The information transferred will be H_T(*Healthy, Slow*) = $\log_2[P(Healthy \mid Slow)/P(Healthy)]$. Again, using the first song heard, *P*(*Healthy*) = 0.5, *P*(*Healthy* | *Slow*) = 0.333, and thus H_T (*Healthy, Slow*) = – 0.586 bits. This erroneous song also generates negative information.

What is the average information provided by a fast song? Fast songs can be informative if they are produced by a healthy male [H_T(*Healthy, Fast*) > 0], but misleading if produced by a sick one [H_T(*Sick, Fast*) < 0]. The average $\overline{H}_T$ for fast songs consists of the sum of H_T(*Healthy, Fast*) and H_T(*Sick, Fast*), each discounted by its relative frequency:

$$\overline{H}_T(Fast) = \frac{P(Healthy)\,P(Fast \mid Healthy)\,H(Healthy, Fast) + P(Sick)\,P(Fast \mid Sick)\,H(Sick, Fast)}{P(Healthy)\,P(Fast \mid Healthy) + P(Sick)\,P(Fast \mid Sick)}$$

The term $P(Healthy) \cdot P(Fast \mid Healthy)$ is the probability of encountering the combination of fast songs and healthy males [e.g., *P*(*Fast* and *Healthy*)]. Similarly, the term $P(Sick) \cdot P(Fast \mid Sick)$ is the probability of encountering the combination of fast songs and sick males [e.g., *P*(*Fast* and *Sick*)]. The denominator is the overall probability of encountering a fast song by any kind of male. If we consider the first song heard by a female, the average information provided by a fast song is

$$\overline{H}_T(Fast) = \frac{(0.5)(0.7)(0.347) + (0.5)(0.4)\ (-0.364)}{(0.5)(0.7) + (0.5)(0.4)} = 0.054 \text{ bits/song}$$

Computations for succeeding songs are a little more tedious, but the logic is the same. For the second song sung by a male, the average information provided by a fast song is 0.050 bits/song, a lower value than that for the first song.

For a slow song, the average information for the first song heard is

$$\overline{H}_T(Slow) = \frac{P(Sick)\,P(Slow \mid Sick)\,H(Sick, Slow) + P(Healthy)\,P(Slow \mid Healthy)\,H(Healthy, Slow)}{P(Sick)\,P(Slow \mid Sick) + P(Healthy)\,P(Slow \mid Healthy)}$$

which in our example gives $\overline{H}_T$ (*Slow*) = 0.082 bits/song. The equivalent value for the second song drops to 0.074 bits/song.

Note again that for both first and second songs, a slow song provides a little more information than does a fast one. This difference occurs because slow songs indicate that males are sick better than fast songs indicate that they are healthy. This difference accounts for the asymmetry in the tree diagram of Figure 13.4. For example, the lowest point for each number of successive songs is closer to the bottom of the graph than the highest point for that song number is to the top of the graph. It also reinforces the more general fact that alternative signals in a signal set may provide different average amounts of information.

Note also that successive songs, whether fast or slow, tend to provide less information than earlier ones. For example, Figure 13.6 shows the average amount of information provided per song with successive sampling up to 100 songs. Plots for both healthy and sick males are shown. It is clear that the average amount of information decreases with successive samples whether the male is healthy or sick. Because the vertical axis in this plot is logarithmic,

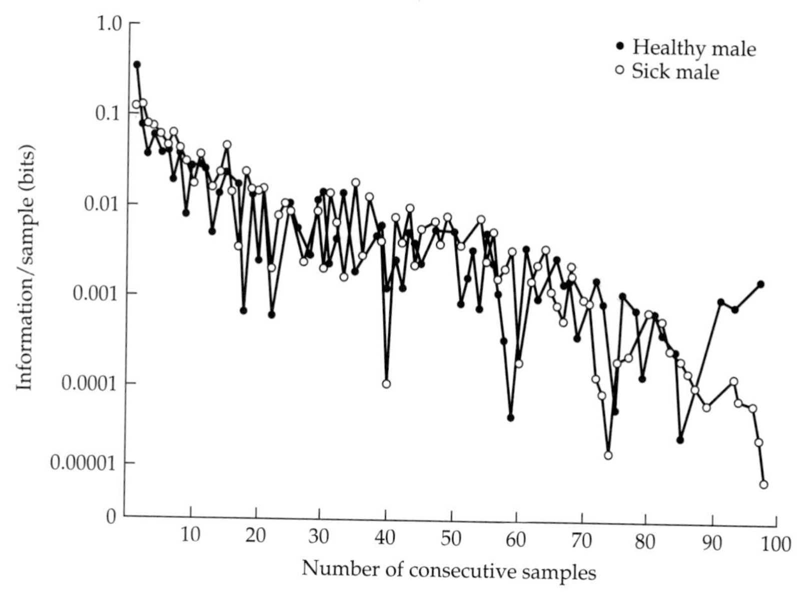

Figure 13.6 Average information provided per song with successive sampling. Averages are based on 1000 simulations using probability values outlined in the text. Note that the vertical axis is on a logarithmic scale. The plots demonstrate that successive samples generally provide less information than earlier ones.

the actual decrease in information provided is very large in the early stages of sampling and decreases with continued sampling. Although fast and slow songs may provide different amounts of information at each stage, and healthy and sick males generate different ratios of fast and slow songs, the overall amounts of information provided per song are not noticeably different for healthy and sick males.

The expressions for computing the average information provided by a signal set when errors are present do not differ if the mistakes are encoding errors by the sender or reception/decoding errors by the receiver. In our example, we assumed that senders were the major source of error. However, what if both parties err? Although the communication process is a cumulative one, it turns out that error rates are roughly additive: the more error either party contributes, the less information transmitted. In Figure 13.7, we have allowed both simulated senders and receivers to err; receiver errors are assumed to be independent of the signal that was actually sent by the sender. The graph plots the cumulative estimate that a healthy male is healthy with successive sampling. Increasing the rates of either encoding or reception errors results in a slower improvement in the cumulative estimate of male quality with successive sampling. The worst situation occurs when both parties show large error rates. Large errors by senders and small ones by receivers give a rate of increase with successive sampling similar to that with small errors by senders and large errors by receivers. One might have thought

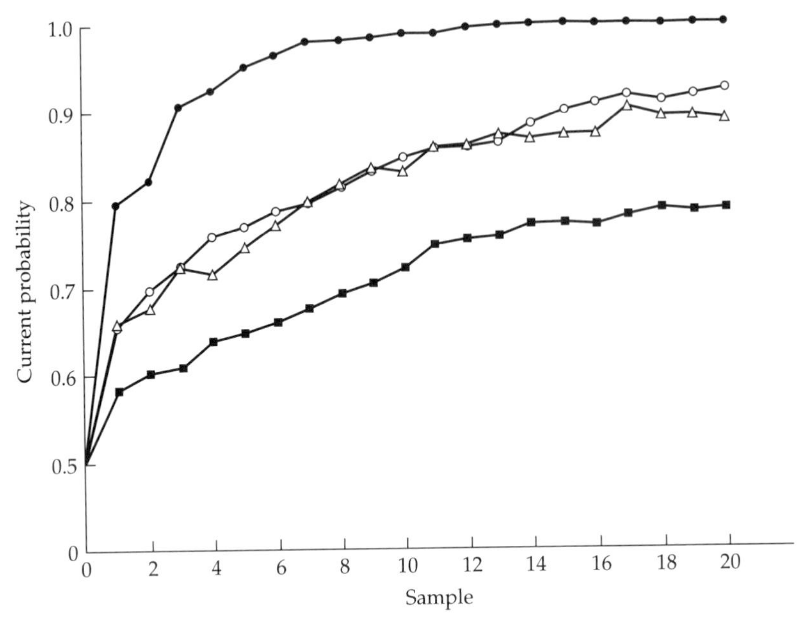

Figure 13.7 Effects of coding and reception errors on updated estimation. Each point is the mean of 1000 stochastic simulations of a receiver using Bayesian updating to estimate the probability that a singing healthy male is healthy. Dark circles: encoding and reception errors both 10%; open circles: encoding errors 30% and reception errors 10%; open triangles: encoding errors 10% and reception errors 30%; dark squares: both types of error 30%.

that high encoding errors would preempt the need for accurate reception. Clearly, this is not true; any increase in either kind of error reduces the average information provided.

Do Animals Really Use Bayesian Updating?

There is ample evidence that animals sample successive renditions of a signal before acting. Female birds and frogs typically listen to several male calls before approaching, and honeybees will observe multiple dances of a returning forager before setting out to find the advertised food source (see Chapter 25). But are animals actually using some form of Bayesian updating with these repeated samples?

Bayesian updating is the optimal strategy only if animals have reasonably accurate prior information about the likelihoods of the alternative conditions, and about the correlations between conditions and signals (e.g. the coding scheme). If this prior information is inaccurate, then Bayesian updates may be no better than random choice. Not all animals have a chance to acquire this prior information, or even if it can be acquired, have the neural equipment to store it for future use. It turns out, however, that there are a number of simple **rules of thumb** by which animals might combine new and prior information more simply than Bayesian updating, but still do relatively well (Houston et al. 1982; Houston and Sumida 1987; Stephens and Krebs 1986). The closer such approximations to actual Bayesian updating, the better these animals usually do. The cost of using approximations is that the receiver is more likely to make a mistake when exposed to unusual or unexpected conditions. As long as the unexpected does not occur too often, rules of thumb may be the most economical way to process information.

One such approximation is called a **linear operator**. A receiver using this method carries in its head a running estimate of a property of interest. The probability that a specific condition is true might be one such estimate. When the receiver perceives some signal that is associated with this condition, it weights this new information relative to its prior estimate and averages the two together. The more heavily it weights the new information, the more sensitive is the updated estimate to new events; if it weights the new information lightly, then it is being more conservative and relying more on longer-term trends. Where the probabilities for alternative conditions and the mapping of signals onto conditions are both normally distributed, it turns out that Bayesian updating and a linear operator will give equally accurate results. Even when these conditions are not met, a linear operator may perform almost as well as Bayesian updating. This type of approximation is much simpler than Bayesian updating because the receiver need only store the current estimate and a weight.

Recent studies suggest that many animals use approximations to combine prior and current information. The accuracy of these methods depends upon how closely they approximate true Bayesian updating, and in some cases, the similarity is quite striking (Green 1980; Valone 1991, 1992). Thus there appears to be a continuum of updating methods ranging from the very crude and grading into the optimal Bayesian case. Even though not all animals use Bayesian updating, it sets an upper limit on the amount of information that can be obtained from a given number of successive samples. Any approximation will provide less than this amount, and different approximations can be ranked relative to the optimal case. This ranking can be quite useful when studying the evolutionary advantages of alternative updating methods.

INFORMATION MEASURES AND ANIMAL COMMUNICATION STUDIES

Information measures have played an important but intermittant role in the study of animal communication. As early as two decades ago, animal behaviorists began using information measures derived from quantitative observations of sender and receiver interactions (e.g., Altmann 1965; Attneave 1959; Bell and Gorton 1978; Dingle 1972; Hazlett and Estabrook 1974; Hyatt and Salmon 1979; Losey 1978; Quastler 1958; Rand and Rand 1976; Rubenstein and Hazlett 1974; Steinberg and Conant 1974). The maximum information provided by senders was estimated using the number and relative frequency of use of alternative signals. This approach is still used today. In addition, a minimal measure of the amount of information transferred to receivers was

Box 13.2 Information Transfer when Senders or Receivers Err

HOW DOES ONE COMPUTE the average information transferred between sender and receiver when either sender or receiver (or both) make errors? In this context, there are two ways we can calculate $\overline{H}_T$:

$$\overline{H}_T = \sum_i \sum_j P(C_i \text{ and } S_j)\log_2 \frac{P(C_i \mid S_j)}{P(C_i)}$$

or as $$\overline{H}_T = \overline{H}(C) - \overline{H}(C \mid S), \quad \text{where } \overline{H}(C) = -\sum_i P(C_i)\log_2 P(C_i)$$

and $$\overline{H}(C \mid S) = -\sum_i \sum_j P(C_i \text{ and } S_j)\log_2 P(C_i \mid S_j)$$

In either case, we need to compute the probabilities of each combination of condition and signal occurring, and we need to compute the a posteriori probabilities, $P(C_i \mid S_j)$, that condition C_i is true given the receipt of signal S_j. We need three tables: one giving the coding scheme, $P(S_j \mid C_i)$, and a priori probabilities, $P(C_i)$, a second giving the $P(C_i$ and $S_j)$ values, and a third giving the $P(C_i \mid S_j)$ values. Once we have the tables, it is straightforward to compute $\overline{H}_T$.

Consider first a situation in which all males, regardless of health, sing fast songs 55% of the time and slow songs 45% of the time. As in the text, we shall assume that healthy and sick males are equally common. Thus the probability of being healthy is 50%. The coding scheme and a priori probabilities can be summarized as in Table (A):

(A) $P(S_j \mid C_i)$ values

		Condition	
		Healthy male	Sick male
Signal	Fast	0.55	0.55
	Slow	0.45	0.45
	$P(C_i)$	0.50	0.50

Because condition and male song are here independent, the probabilities $P(C_i$ and $S_j) = P(C_i) \cdot P(S_j)$. For example, *P(Healthy* and *Fast)*, that a male will be healthy and sing a fast song, is $0.5 \times 0.55 = 0.275$; similarly, *P(Sick* and *Slow)* $= 0.5 \times 0.45 = 0.225$. These are summarized in Table (B):

(B) $P(C_i$ and $S_j)$ values

		Condition		
		Healthy male	Sick male	$P(S_j)$
Signal	Fast	0.275	0.275	0.55
	Slow	0.225	0.225	0.45
	$P(C_i)$	0.50	0.50	1.00

(C) $P(C_i \mid S_j)$ values

		Condition		
		Healthy male	Sick male	$P(S_j)$
Signal	Fast	0.50	0.50	1.00
	Slow	0.50	0.50	1.00
	$P(C_i)$	1.00	1.00	2.00

Note that in (B), the sum of the cell values, the sum of the column subtotals, and the sum of the row subtotals should all equal 1.0. The computation of the a posteriori values in the cells in (C) relies on Bayes' theorem: the quick way to do this is to divide each cell value in the left hand table by its row subtotal, which is $P(S_j)$, and to place the result in the corresponding cell of (C). Thus the top-left cell of the right hand table is $P(Healthy \mid Fast) = 0.275/0.55 = 0.50$. The sum of the cell values, the sum of the row subtotals, and the sum of the column subtotals should all equal 2.0 in (C).

Once the two tables are assembled, it is easy to compute $\overline{H}_T$ by either method. To use the first method, we sum the products of each cell in Table (B) with the $\log_2$ of the quotient between the corresponding cell in Table (C) and the corresponding column subtotal in (B). Thus

$$\overline{H}_T = 0.275\log_2\left(\frac{.50}{.50}\right) + 0.275\log_2\left(\frac{.50}{.50}\right) + 0.225\log_2\left(\frac{.50}{.50}\right) + 0.225\log_2\left(\frac{.50}{.50}\right) = 0 \text{ bits}$$

since the $\log_2(1) = 0$. Not surprisingly, since the signals and conditions are totally independent, there is no average information exchanged given this coding scheme. The second method gives the same result: $\overline{H}(C) = -(0.5\log_2 0.5 + 0.5\log_2 0.5) = 1.0$ bit. This is the maximum information that could be coded into the signals. The average uncertainty, $\overline{H}(C \mid S)$, that a given signal will be associated with a given condition is computed as the negative of the sum of the products of each cell in the left hand table with the $\log_2$ of the corresponding cell in the right hand table. Thus $\overline{H}(C \mid S) = -(0.275\log_2 0.5 + 0.275\log_2 0.5 + 0.275\log_2 0.5 + 0.275\log_2 0.5) = 1.0$ bit. The total information encoded into the songs is again $\overline{H}_T = \overline{H}(C) - \overline{H}(C \mid S) = 1.0 - 1.0 = 0$ bits. Here the equivocation, $\overline{H}(C \mid S)$, is maximal because there is no useful correlation between signal and condition.

Now consider the opposite extreme. Suppose that males only sing fast songs when they are healthy, and only sing slow songs when they are sick. The corresponding coding table and a priori probabilities appear as shown in Table (D):

(D) $P(S_j \mid C_i)$ values

Signal	Condition: Healthy male	Condition: Sick male
Fast	1.00	0.00
Slow	0.00	1.00
$P(C_i)$	0.50	0.50

Because song and health are no longer independent, we must compute the $P(C_i$ and $S_j)$ as $P(C_i) \cdot P(S_j \mid C_i)$. The corresponding table appears as (E) below. We divide each cell value in Table (E) by the corresponding row subtotal to create the table of a posteriori values in Table (F). Then, using the first method of computing average information transfer, we get

$$\overline{H}_T = 0.5\log_2\left(\frac{1.00}{0.50}\right) + 0.0\log_2\left(\frac{0.00}{.50}\right) + 0.0\log_2\left(\frac{0.00}{.50}\right) + 0.50\log_2\left(\frac{1.00}{.50}\right)$$

Box 13.2 (continued)

(E) $P(C_i$ and $S_j)$ values

		Condition: Healthy male	Condition: Sick male	$P(S_j)$
Signal	Fast	0.50	0.00	0.50
Signal	Slow	0.00	0.50	0.50
	$P(C_i)$	0.50	0.50	1.00

(F) $P(C_i \mid S_j)$ values

		Condition: Healthy male	Condition: Sick male	$P(S_j)$
Signal	Fast	1.00	0.00	1.00
Signal	Slow	0.00	1.00	1.00
	$P(C_i)$	1.00	1.00	2.00

Using the second method, we again find that $\bar{H}(C) = -(0.5 \log_2 0.5 + 0.5 \log_2 0.5) = 1.0$ bit, but here, $\bar{H}(C \mid S) = -(0.5 \log_2 1.0 + 0.0 \log_2 0.0 + 0.0 \log_2 0.0 + 0.5 \log_2 1.0) = 0.0$ bits (since $\log_2 0.0 = 0.0$ when computing H values). This means that the information encoded is $\bar{H}_T = \bar{H}(C) - \bar{H}(C \mid S) = 1.0 - 0.0 = 1.0$ bit. Because there is a perfect correlation between condition and signal, the equivocation is 0, and thus the amount of information encoded equals the relative diversity of the conditions.

Now we turn to the situation actually described in the text: healthy males and sick males are still equally common. However, healthy males sing fast songs 70% of the time and slow songs 30% of the time, whereas sick males sing fast songs 40% of the time and slow songs 60% of the time. The coding table and a priori probabilities are shown in Table (G):

(G) $P(S_j \mid C_i)$ values

		Condition: Healthy male	Condition: Sick male
Signal	Fast	0.70	0.40
Signal	Slow	0.30	0.60
	$P(C_i)$	0.50	0.50

Health and song speed are not independent and we again need to use the fact that $P(C_i \text{ and } S_j) = P(C_i) \times P(S_j \mid C_i)$. This generates Table (H). We then divide each cell value

computed from the correlation between the signal given by the sender and the response selected by the receiver. Such a measure is a minimum value because any receiver decision will depend not only on signal reception, but also on access to prior information, the benefits and costs of using the information, and the benefits and costs of each alternative action. As we shall see in the next chapter, the link between the amount of information transferred and receiver decisions is rarely linear, and may not even be monotonic. Minimum

in that table by the corresponding row subtotal to get Table (I):

(H) $P(C_i$ and $S_j)$ values

		Condition: Healthy male	Condition: Sick male	$P(S_j)$
Signal	Fast	0.35	0.20	0.55
Signal	Slow	0.15	0.30	0.45
	$P(C_i)$	0.50	0.50	1.00

(I) $P(C_i \mid S_j)$ values

		Condition: Healthy male	Condition: Sick male	$P(S_j)$
Signal	Fast	0.636	0.364	1.00
Signal	Slow	0.333	0.667	1.00
	$P(C_i)$	0.969	1.031	2.00

Using the first method,

$$\overline{H}_T = 0.35\log_2\left(\frac{0.636}{0.50}\right) + 0.20\log_2\left(\frac{.364}{.50}\right) + 0.15\log_2\left(\frac{0.333}{.50}\right) + 0.30\log_2\left(\frac{0.667}{.50}\right) = 0.067 \text{ bits}$$

Using the second, we find again that $\overline{H}(C) = 1$ bit, and the equivocation, $\overline{H}(C \mid S) = -(0.35 \log_2 0.636 + 0.20 \log_2 0.364 + 0.15 \log_2 0.333 + 0.30 \log_2 0.667) = 0.933$ bits. Thus $\overline{H}_T = \overline{H}(C) - \overline{H}(C \mid S) = 1.0 - 0.933 = 0.0667$ bits. Note that if there were perfect coding, healthy males would always sing fast songs; in this example, they do so only 70% of the time. Similarly, with perfect coding sick males would always sing slow songs; here they do so only 60% of the time. The overall result of these **encoding errors** by both types of males is a loss of 93% (0.9333/1.0) of the potential information that could have been encoded into these signals. **Reception** and **decoding errors** by receivers can have a similar effect. In most animals, there will be both coding and reception errors and thus the equivocation, $\overline{H}(C \mid S)$, will have some value between 0 and $\overline{H}(C)$. The higher the equivocation, the less the average amount of information transferred per signal. To provide at least half of the maximum information possible with two song types, males have to be accurate an average 88% of the time or better. Small error rates thus have a big effect on the amount of information provided.

For more experience with these methods, compute the average amount of information transferred for different a priori probabilities that a male is healthy: prove to yourself that the maximal information is exchanged when alternative conditions are equiprobable, i.e., when the a priori probability a male is healthy is 50%.

measures of transferred information thus provide at best a baseline. Unfortunately, they are insufficient for the kinds of tasks we have outlined at the outset of this chapter. For this reason, they have largely been replaced by more sophisticated approaches.

One method is to examine how the correlation between sender signals and receiver responses changes as one varies the benefits and costs and/or prior probabilities. The amount of information transferred can then be estimated from the pattern of the changes in the correlations. The justification for this type of measure is provided by **signal detection theory**, which we discuss in Chapter 14. The result is a measure that is independent of payoffs and prior probabilities and thus more comparable across contexts and species. Another solution is to go directly into the sensory organs and brain of the receiver to monitor what amounts of information are received and transmitted at each stage of the decision process. Current technology permits quite sophisticated records of the responses of both sensory cells and higher-order neurons to signal input. Measures computed from such recordings can characterize the amount of information extracted from external stimuli by sensory receptors, the amount transmitted through synapses between sensory cells and second order neurons, the relationships between neural circuit design and information transfer, and the reconstruction by higher-order brain cells of input signal waveforms from trains of nerve impulses (e.g., Bialek et al. 1991; Laughlin 1990, 1994; de Ruyter van Steveninck and Laughlin 1996).

The process of information acquisition and decision-making by animals is now a topic of common interest to behavioral ecologists, psychologists, cognitive scientists, and neuroethologists. It will, in fact, take collaborative work by all of these disciplines to fully characterize what animals do and why. The existence of a single currency to characterize the transfer of information at each stage has greatly facilitated the exchange of results among these disciplines. It is one more reason why we need to measure the amount of information.

SUMMARY

1. Increasing the **amount of information** augments the value of information by increasing the chances that a receiver makes the correct decision, but it decreases the value of information by increasing the costs of communication for each party. For any situation, there is likely to be an optimal amount of information that should be exchanged and this amount is usually less than that amount (called **perfect information**) that would completely resolve a receiver's uncertainty.
2. Different signals provide information about different **questions**. Those signals relevant to the same question constitute a **signal set**. Questions may be grouped according to the general **contexts** in which they arise. The entire suite of signals used by a species is its **repertoire**. One might want to compute the amount of information provided at any level in this hierarchy: that for one signal exchange, the average information for a signal set, the average information for **compound displays** that answer multiple questions, or the average information for a repertoire.
3. The amount of information provided by one signal can be characterized by computing the ratio of a receiver's estimate of the probability that a given condition is true after signal reception to that before the signal was received. To make scaling easier, one usually uses the logarithm (to the base 2) of the ratio, instead of the ratio itself. The unit of information on this log scale is called the **bit**.

4. If a signal set consists of discrete alternatives that are perfectly and uniquely correlated with alternative conditions, then the average amount of information provided/signal in the set is

$$\bar{H}_T = -\sum_i P_i \log_2 P_i$$

where P_i is the a priori probability of condition i. When all M conditions are equally likely, the average information per signal is simply $\bar{H}_T = \log_2 M$; the more dissimilar the a priori probabilities, (i.e., the more confident the receiver is of the outcome before signal exchange), the less information that is provided by the average signal in the set.

5. Compound displays consist of multiple components, each of which provides information about a separate question. When the answers to the questions in a compound display are independent of each other, the average information for the entire display is the simple sum of the average amounts of information provided by each component. For example, if A and B are independent components of the display, $\bar{H}_T(A)$ is the average information provided by component A, and $\bar{H}_T(B)$ is that provided by component B, then the average information for the entire display is $\bar{H}_T = \bar{H}_T(A) + \bar{H}_T(B)$. If, on the other hand, observed combinations of components A and B are not independent, then the average information per display is less and is equal to $\bar{H}_T = \bar{H}_T(A) + \bar{H}_T(B \mid A) = \bar{H}_T(B) + \bar{H}_T(A \mid B)$, where $\bar{H}_T(B \mid A)$ is the average uncertainty concerning which form of B will be found with any given form of A, and $\bar{H}_T(A \mid B)$ is the corresponding uncertainty concerning which form of A will be found with any given form of B. The stronger the correlations between A and B, the smaller are the contributions by $\bar{H}_T(B \mid A)$ and $\bar{H}_T(A \mid B)$.

6. **Encoding errors** occur when senders do not match signals accurately or uniquely to contexts. **Reception** and **decoding errors** occur when receivers misidentify the transmitted signal. Both sources of error reduce the amount of information that is transmitted per signal. If $\bar{H}(C)$ is the a priori uncertainty as to which of several alternative conditions may be currently true, and $\bar{H}(C \mid S)$ is the average a posteriori uncertainty that a given condition is true after receiving a given signal (called the **equivocation**), then the average information transferred given either sender or receiver errors is $\bar{H}_T = \bar{H}(C) - \bar{H}(C \mid S)$. When there are no errors by either party, then $\bar{H}(C \mid S) = 0$; when signals are given randomly with respect to conditions, then $\bar{H}(C \mid S) = \bar{H}(C)$. Because the two sources of error are approximately additive, error reduction by either party can improve the transfer of information.

7. The maximum information is provided when the alternative signals sent by a sender, and the corresponding conditions they represent, are equally likely and each signal is only given for a single condition. If all signals within a signal set are equiprobable, a larger set thus tends to provide more information. Where alternatives are not equally likely and/or the same signal is associated with more than one condition due to coding or reception errors, then the information provided will be less than the maxi-

mum possible given that number of alternative signals. Such a system is then said to be **redundant**. A large signal set with high redundancy may thus provide less information on average than a smaller less redundant one.

8. Receivers can improve estimates as to which condition is true by sampling signals repeatedly and updating the results. The optimal method is provided by **Bayesian updating**, which requires that the receiver know the probabilities that given signals will be sent when each condition is true, and the prior probabilities of each condition being the case. The receipt of the different alternative signals usually provides different amounts of information. In addition, later samples provide less information and a smaller improvement in estimates than earlier samples. When prior probability of one alternative condition is sufficiently high, receipt of a signal makes little change in the updated value. There is thus a general pattern of diminishing returns with continued sampling or with any sampling when the likely outcome is already known. Although many animals do not use true Bayesian updating, they may use similar approximations that can be nearly as accurate as the Bayesian expectation.

FURTHER READING

Many of the best texts on information theory were written during its period of initial application and are no longer in print. One extremely good source still available in some libraries is Rosie (1973). A less technical presentation is provided by Pierce (1980). The theory as described by its originator can be found in Shannon and Weaver (1949). Recent texts written for electrical engineers cover many of the same topics but in ways which are not immediately relevant for animal communication. Some older but biologically oriented treatments can be found in Attneave (1959), Dingle (1972), Hailman (1977a), Losey (1978), Quastler (1958), and Wilson (1975). All of these citations focus on minimum measures of information transfer derived from the correlations between sender signal and receiver response. More recent measures of information transfer correct for receiver payoffs and prior knowledge using signal detection theory. Appropriate references are listed at the end of Chapter 14. A good example of the application of information theory to neurobiology can be found in Laughlin (1990, 1994). The role of Bayesian updating has largely been the province of researchers studying optimal foraging in animals, although many of the points raised in these studies are directly applicable to animal communication problems. Relevant discussions can be found in Green (1980), Houston et al. (1982), McNamara and Houston (1980, 1985, 1987), Chapter 4 of Stephens and Krebs (1986), Valone and Brown (1989), and Valone (1991, 1992).

Chapter 14

The Value of Information

THE EXCHANGE OF A GIVEN AMOUNT OF INFORMATION between a sender and a receiver is only the first step in the communication process. As we saw in Chapter 12, pages 377–380, the value of information to either party depends critically upon what the receiver decides to do after receipt of the signal. In this chapter, we examine the decision rules that receivers might invoke once they have received a signal. We first identify decisions that are optimal when using discrete signal sets, and then contrast these with decisions made using continuous signal sets. In both cases, we find that optimal decision rules rarely favor obtaining perfect information and thus the elimination of all uncertainty. We noted on pages 404–406 that continued sampling at best approaches the truth asymptotically, whereas the investment costs of continued exchanges continue to mount at the same or an increasing rate. Eventually, these costs will outstrip the benefits, and one or the other party should halt the exchange. What are the optimal amounts of information that a sender should seek to provide and a receiver to obtain? We take this up in the third section of this chapter. Finally, we conclude with some examples in which optimal decision theory has been applied to animal communication.

OPTIMAL DECISION RULES FOR DISCRETE SIGNAL SETS

Upon receipt of one of a set of discrete signals, a Bayesian receiver will update its prior information by using the apparent identity of the received signal and its knowledge about the degree of correlation between condition and signal. The updated estimate that any given condition is currently true is a fraction between zero and one. To decide upon an appropriate action, the receiver will do best if it establishes a cutoff probability P_c and lets all estimates below the cutoff favor one action, and all estimates above favor the alternative action (Raiffa 1968; Breipohl 1970). Where should an optimal receiver establish that cutoff? If the receiver is too conservative, it will miss opportunities it could have exploited; if it is too liberal, it may get itself into trouble. We first examine what a receiver should do in the absence of signal information.

Optimal Decisions with no Signal Information

We return to the female bird of Chapter 13 that seeks to decide whether to mate with a particular male or not. Candidate males may be either *healthy* or *sick*. As before, we assume that healthy males occur an average P_1 of the time and sick males the remaining fraction $(1 - P_1)$. Suppose that these a priori probabilities are known to the female. She can either *accept* or *reject* a current male as a mate. Should she accept a male when he is indeed healthy, she will receive a payoff of R_{11}; if she rejects a healthy male, she receives a payoff of R_{21}. Similarly, a female receives a payoff of R_{12} when she accepts a sick male as a mate, but receives R_{22} when she rejects a sick one. We assume that $R_{11} > R_{21}$, and $R_{22} > R_{12}$ (i.e., acceptance is the better response when a male is healthy and rejection is the better response when the male is sick). The four possible payoffs are summarized as a matrix in Figure 14.1A.

The female's current estimate that a male is healthy is denoted by $\hat{P}\,(Healthy)$. Should she accept or reject this male? To decide, she should compute the average payoff of accepting him and compare it to the average payoff of a rejection. The estimated average payoff of accepting a male when the estimated probability that he is healthy is $\hat{P}\,(Healthy)$ will be

$$\overline{PO}\,(Accept) = \hat{P}\,(Healthy)\,R_{11} + [1 - \hat{P}\,(Healthy)]\,R_{12}$$

and that for rejection will be

$$\overline{PO}\,(Reject) = \hat{P}\,(Healthy)\,R_{21} + [1 - \hat{P}\,(Healthy)]\,R_{22}$$

Acceptance is favored when $\overline{PO}(Accept) > \overline{PO}\,(Reject)$ or when

$$\hat{P}(Healthy)\,R_{11} + [1 - \hat{P}\,(Healthy)]\,R_{12} > \hat{P}\,(Healthy)\,R_{21} + [1 - \hat{P}\,(Healthy)]\,R_{22}$$

Simplifying this inequality, we find that acceptance is favored over rejection when $\hat{P}(Healthy) > P_c$, where

$$P_c = \frac{R_{22} - R_{12}}{(R_{22} - R_{12}) + (R_{11} - R_{21})}$$

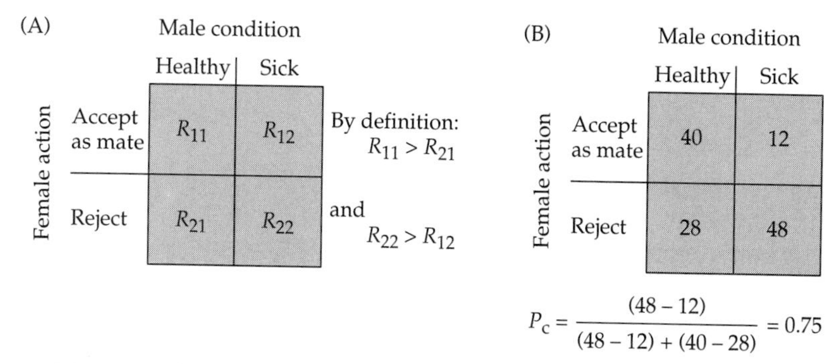

$$P_c = \frac{(48-12)}{(48-12)+(40-28)} = 0.75$$

Figure 14.1 Payoff matrix for alternative actions by a receiver with no signals exchanged. (A) General case. The female can either accept or reject the current male. If the male is healthy, acceptance has a higher payoff for the female ($R_{11} > R_{21}$), whereas rejection gives the higher payoff when the male is sick ($R_{22} > R_{12}$). (B) Quantitative example. P_c is the minimal estimated probability that the male is healthy before acceptance is favored over rejection. If the female has no information other than prior probabilities and the probability that any given male is healthy is 0.3, because $0.3 < P_c$, the female should reject all males.

Both acts give the same estimated payoff when $\hat{P}$ *(Healthy)* $= P_c$, and rejection is favored when $\hat{P}$ *(Healthy)* $< P_c$. An example of these computations is shown in Figure 14.1B. Thus the minimal probability, P_c, that is required before a female accepts a male depends only on the relative benefits of getting it right, given each type of male.

What if the receiver has **no information** except its prior experience or heritable biases. Suppose it encounters a type of male, say a yearling, and it knows from prior experience that yearlings are healthy a fraction P_1 of the time. The receiver's best current estimate that a yearling male is healthy is then $\hat{P}$ *(Healthy)* $= P_1$. If $P_1 > P_c$, the female should accept a yearling male when encountered and her average fitness will be $P_1R_{11} + (1 - P_1)R_{12}$. If $P_1 < P_c$, the female should reject all yearling males, and her average fitness will be $P_1R_{21} + (1 - P_1)R_{22}$. In the example shown in Figure 14.2, $P_1 = 0.3$, whereas $P_c = 0.75$. In this case, she should reject a yearling male and will then obtain the average fitness (0.3)(28) + (0.7)(48) = 42.0. Had she accepted him, she would have obtained only (0.3)(40) + (0.7)(12) = 20.4. Clearly, blanket rejection of yearling males is better than blanket acceptance. Perhaps a fixed cutoff is not the optimal strategy. For example, could the female do better by randomly accepting yearling males a fraction P_1 of the time and rejecting them a fraction $(1 - P_1)$ of the time? No; in this case, she would only obtain an average fitness of (0.3)(20.4) + (0.7)(42) = 35.5. The fixed cutoff is clearly the better strategy.

Optimal Decisions with Perfect Information

Now contrast this situation to one in which males provide **perfect information** to the female about their state of health. Suppose that healthy males always sing fast songs and sick males only sing slow songs. To guarantee perfect information, the medium must induce no signficant changes in the songs during propagation, and the female must never err in identifying which type

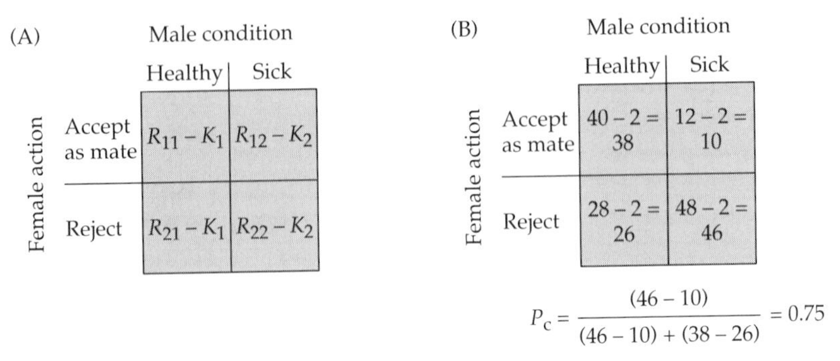

Figure 14.2 Payoff matrix for alternative actions by a receiver after signals are exchanged. (A) General case with terms as in Figure 14.1A, with incorporation of communication cost terms K_1 and K_2 for healthy and sick males, respectively. (B) Same values as in Figure 14.1B but assuming $K_1 = K_2 = 2$. Note that computation of P_c is unchanged by the incorporation of costs. (In practice, the K's will not be equal, but most subsequent algebra uses average K values, so conclusions are unchanged.)

of song was sung. Let the average cost to the female of this information be $-K_1$ if the male is healthy and $-K_2$ if he is sick. The corresponding payoff matrix is shown in Figure 14.2A.

We can again compute a minimal value of $\hat{P}$ (*Healthy*) = P_c, above which the female should switch from rejection to acceptance. Although we have included costs of communication in this payoff matrix, these are constant (at least for a given a type of male), and thus they cancel out to produce the same value of P_c computed for no communication:

$$P_c = \frac{(R_{22}-K_2)-(R_{12}-K_2)}{[(R_{22}-K_2)-(R_{12}-K_2)]+[(R_{11}-K_1)-(R_{21}-K_1)]}$$
$$= \frac{R_{22}-R_{12}}{(R_{22}-R_{12})+(R_{11}-R_{21})}$$

When a male is healthy, he will always sing fast songs and the female will use this perfect information to update her current $\hat{P}$ (*Healthy*) value to 1.0. Since $\hat{P}$ (*Healthy*) = $1.0 \geq P_c$, she will then accept this male. When a male is sick, he will only sing slow songs; the female will then get an updated $\hat{P}$ (*Healthy*) = 0, and she will then reject him. Because a fraction P_1 of the males are healthy, and a fraction $(1 - P_1)$ are sick, the average fitness of a female with access to perfect information is $P_1(R_{11} - K_1) + (1 - P_1)(R_{22} - K_2)$. The **value of perfect information** to the female is the difference between the average fitness when perfect information is provided versus that when no information is provided. Since the average cost of a signal, $\bar{K} = P_1K_1 + (1 - P_1)K_2$, it is easy to show that the female's value of perfect information when $P_1 < P_c$ reduces to $P_1(R_{11} - R_{21}) - \bar{K}$; and if $P_1 > P_c$, then perfect information is worth $(1 - P_1)(R_{22} - R_{12}) - \bar{K}$. Using the payoffs noted earlier and letting $\bar{K} = K_1 = K_2 = 2$, the average fitness obtained given perfect information is (0.3)(38) + (0.7)(46) = 43.6 (see matrix in Figure 14.2B). The value of perfect information is this fitness minus that with no information or 43.6 – 42.0 = 1.6. The alternative way to compute the value of perfect information invokes the expression $P_1(R_{11} - R_{21}) - \bar{K}$ because $P_1 < P_c$. Plugging in the values for the example, we get (0.3)(40 – 28) – 2 = 1.6, which agrees with the figure computed directly. However calculated, the value of perfect information sets an upper limit on how much can be gained through communication. In our example, the female can expect to obtain at most a 3.8% increase in her fitness by attending to male signals. Unless perfect information is provided, the value of information will be less than this. Although small, at least the female's value of perfect information is positive in our example. Were it not, then regardless of accuracy, no exchange of information would be favored.

Optimal Decisions with Imperfect Information

What if males provide less than perfect information? In this case, any male could sing either a fast or a slow song. If it produced these at random, no useful information would be provided and the female would hear fast and slow songs with an equal probability of 0.5. Suppose that males do better than this and that the conditional probability, *P*(*Fast* | *Healthy*), that a female will perceive a fast song when the male is healthy is greater than 0.5. Let us denote *P*(*Fast* | *Healthy*) by Q_1. The more accurately fast signals are linked to healthy males, the higher is Q_1. The corresponding value *P*(*Slow* | *Sick*), denoted by Q_2, is the conditional probability that a slow song will be perceived when the male is sick. Let us assume for simplicity that the two signals are given with equal accuracy. Hence $Q_1 = Q_2 = Q$. As we saw on pages 397–401, the higher Q is, (once it exceeds the chance value of 0.5), the higher the average amount of information transferred by this signal set becomes. Q can thus be used as an index of the average amount of information transferred.

A female that has detected no signals would have an a priori probability of P_1 that a given male was healthy. If she then perceives that a fast song was sung, and is a good Bayesian updater, she will revise her estimate that the male is healthy to

$$\hat{P}(Healthy \mid Fast) = \frac{P_1 Q}{P_1 Q + (1 - P_1)(1 - Q)}$$

If she thinks a slow song was sung, she will compute an updated probability that he is healthy of

$$\hat{P}(Healthy \mid Slow) = \frac{P_1(1 - Q)}{P_1(1 - Q) + (1 - P_1)Q}$$

If the female relies on only one song, her subsequent decision will then be based on the comparison of the new a posteriori $\hat{P}$ (*Healthy* | *Song*) to P_c.

Suppose that $Q = 0.7$ and that a female perceives a fast song. Given our example, her a priori expectation that the male is healthy is 0.3, whereas her updated estimate after the song would be 0.50. Given the payoff matrix, we saw that $P_c = 0.75$. The song has provided information, but because the new value of $\hat{P}$ (*Healthy* | *Fast*) is still less than P_c, she will reject this male, just as she would have had she heard no song. If Q had been 0.9, then receipt of a fast song would result in an a posteriori $\hat{P}$ (*Healthy* | *Fast*) = 0.79, a value greater than P_c, and thus she would accept this male. The example suggests that there is a minimal value of Q that is required before receipt of a single song can lead to a change in action. Put another way, there is a minimal amount of information that must be provided before receiver responses will change from the default situation. The minimal value of Q will be denoted by Q_c and is equal to that Q for which $\hat{P}$ (*Healthy* | *Fast*) = P_c. Solving for Q under these conditions gives

$$Q_c = \frac{P_c(1.0 - P_1)}{P_1 + P_c - 2P_1P_c}$$

In general, Q_c increases with P_c and decreases with P_1. When $P_1 < 0.5$, then $Q_c > P_c$, and when $P_1 > 0.5$, then $Q_c < P_c$. $P_c = Q_c$ when $P_1 = (1 - P_1) = 0.5$. Given the payoff matrix in Figure 14.2B and an a priori probability P_1 that a male is healthy of 0.3, the value of $P_c = 0.75$ and $Q_c = 0.875$. Figure 14.3 shows the relationship for this situation between different values of Q, the corresponding average amounts of information transferred for each Q, the updated value of $\hat{P}$ (*Healthy* | *Song*) after receiving either a fast or slow song, and the optimal female action for each updated value of $\hat{P}$ (*Healthy* | *Song*). Given the payoffs and the prior probabilities of healthy versus sick males, this signal system has to provide a minimum of 0.39 bits/song before females receive sufficient information to allow $\hat{P}$ (*Healthy* | *Song*) > P_c. An optimal female that is provided with less than 0.39 bits/signal should thus ignore the signals. They make no difference to her final decision and would only lead to uncompensated costs if processed. Notice that she should ignore these signals even if there are no costs involved in processing them.

We can better understand the reason why there is a minimal Q by computing the value of information for this situation. For our example, the value of information will be

$$VI = P_1\,\Delta\phi_1(R_{11} - R_{21}) + (1 - P_1)\Delta\phi_2(R_{22} - R_{12}) - \overline{K}$$

As defined on pages 376–380, the first two terms in this expression constitute the average benefit of attending to signals, and the last term is the average cost of doing so. In our example, a female without access to signals never makes the correct decision when a yearling male is healthy; she rejects all of them. However, when attending to signals, she gets it right Q of the time. Thus $\Delta\phi_1 = (Q - 0) = Q$. Since P_1, Q, and $(R_{11} - R_{21})$ are all positive, the first term contributing to the average benefit will also be positive. In contrast, a female without access to signals always makes the correct decision when males are sick by rejecting all of them. When attending to signals, she now gets it right only Q of the time. This is a decrease in the probability of making a correct decision. Thus $\Delta\phi_2 = (Q - 1)$, which is a negative number for all $Q < 1$. Because $(1 - P_1)$ and $(R_{22} - R_{12})$ are both positive, the second term contributing to the average benefit is negative. If we had begun with a larger value of P_1, so

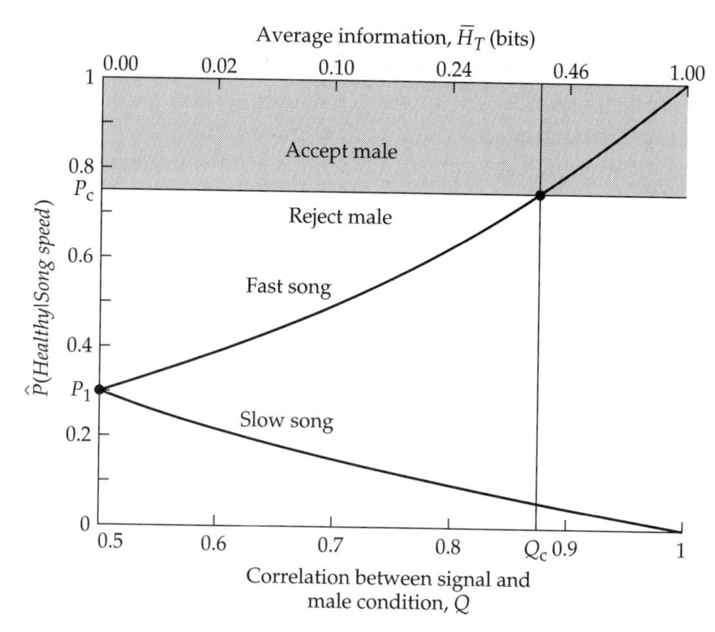

Figure 14.3 Relationship between conditional probability that female will perceive correct signal (Q) and her updated estimate of male being healthy, $\hat{P}$(*Healthy* | *Song speed*). These values are based on the example shown in Figure 14.2. P_c is the critical probability at which the receiver should switch between acceptance and rejection. P_1 is the frequency with which the average male is healthy and it is thus the a priori estimate for the receiver. Q_c is the minimal value of Q below which receiver will continue to reject males and above which she may accept them if she attends to the signals. The graph presumes that the female makes her decision after hearing a single song. The top axis shows the corresponding average amount of information (in bits) transferred by songs, given the stated prior probabilities of healthy versus sick males and various values of Q.

that the default strategy without signals was blanket acceptance of all males, we would still find that one of the two benefit terms was positive and the other negative. In fact, when either blanket acceptance or blanket rejection is the default strategy, and imperfect information is provided, the average benefit will always include both positive and negative terms.

We can now see why there is a minimal Q_c below which a receiver should ignore signals: Q_c is just that value of Q for which $P_1Q(R_{11} - R_{21}) = -(1 - P_1)(Q - 1)(R_{22} - R_{12})$ and thus the average benefit is zero. For larger Q, $P_1Q(R_{11} - R_{21}) > -(1 - P_1)(Q - 1)(R_{22} - R_{12})$ and the average benefit of attending to signals will be positive. Another way to show this is to rewrite the two terms in the average benefit expression as

$$B(Q) = Q[P_1(R_{11} - R_{21}) + (1 - P_1)(R_{22} - R_{12})] - (1 - P_1)(R_{22} - R_{12})$$

This is an equation for a straight line with slope equal to $m = [(P_1(R_{11} - R_{21}) + (1 - P_1)(R_{22} - R_{12})]$ and an ordinate intercept of $-b = -(1 - P_1)(R_{22} - R_{12})$. Given our earlier definitions of the relevant terms, the slope will be a positive number, and the intercept a negative one. This line crosses the abscissa axis at $Q = Q_c$. For $Q < Q_c$, the benefit of using signals will be negative. It will be positive and linearly increasing for higher Q values. When Q reaches 1.0, this expression simplifies to that given for perfect information.

The value of information cannot be positive unless the benefit, $B(Q)$ is positive. Thus Q_c is a minimum accuracy that must be achieved in the association of signals and conditions before communication is favored. However, it may not be sufficient if there is any cost, $-\bar{K}$, to attending to signals. When these costs are present, an even higher minimal value of Q is required to justify receiver participation in communication (Figure 14.4). It is easy to show that the minimum Q in this case is $Q_c + \bar{K}/m$ where m is the slope of the benefit line derived in the prior paragraph.

What if a receiver can use ambient cues and a priori probabilities to make correct decisions at a higher rate than one using a priori probabilities alone? Is there still a minimum correlation between signals and conditions that is required before it pays for the receiver to attend to the signals? The answer is yes, but the algebra is considerably more complicated. We can summarize the gist of the argument as follows (see also Figure 14.5). Suppose Q_0 is the probability that the receiver will view the correct cues, and Q is the probability that it will receive the correct signal. Both have to be large enough that sufficient information is provided overall to make the updated estimates greater than the requisite thresholds for changing actions. Suppose that is true. Without signals, a receiver at least gets it right Q_0 of the time. If the receiver also at-

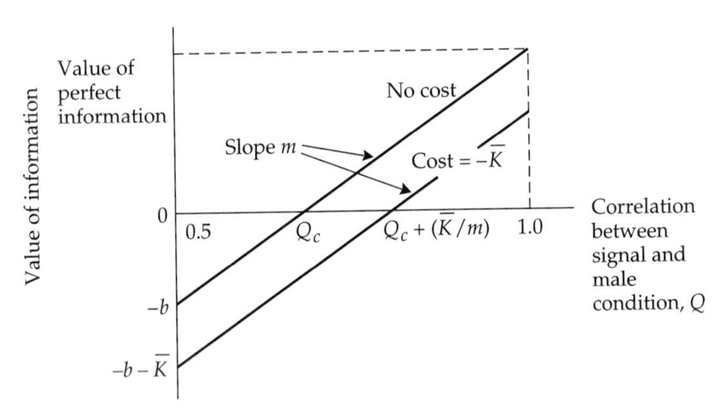

Figure 14.4 Value of information to receiver versus accuracy of correlation between discrete signals and conditions, Q. This example is based on the two-signal system described in the text. Even with no cost to receiver of attending to signals (e.g., $\bar{K} = 0$), Q must be larger than Q_c before the value of information is positive. Both the ordinate intercept, here $-b$, and the slope of the line, m, depend upon the relative payoff differences of making a correct decision and the a priori probabilities of each of the two alternative conditions occurring. If attending to signals imposes a fixed cost, $-\bar{K}$, on the receiver, the slope of the line is unchanged, but the intercept is lowered. This increases the minimal value of Q at which the value of information becomes positive to $Q_c + (\bar{K}/m)$.

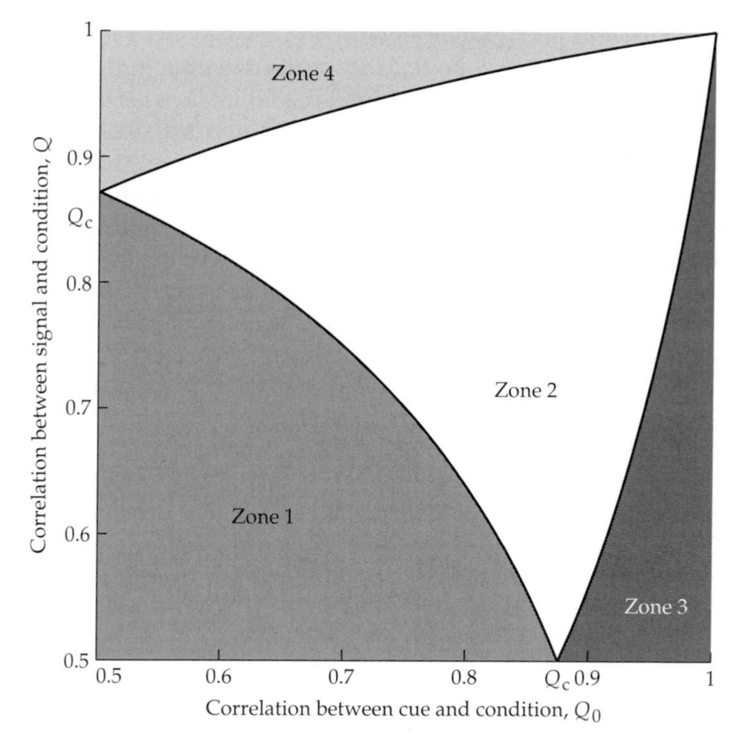

Figure 14.5 Optimal propriety of responding to signals for varying levels of signal and condition correlation, Q, versus varying levels of ambient cue and condition correlation, Q_0. Example assumes $P_1 = 0.3$, $P_c = 0\ 75$, and $Q_c = 0.875$, and that the female first generates an update using ambient cues and then incorporates the signals for a final updated estimate that the male is healthy. There are four optimal strategies depending upon the combination of Q and Q_0. In Zone 1, neither source of information is sufficiently accurate and the female does best to rely only on prior probabilities. In Zone 2, cue information is sufficiently accurate that it should be utilized, but signals are so inaccurate that reliance on them undermines the benefits of using cues. The female should use cues and ignore signals. In Zone 3, cue information is so accurate that signals neither hinder nor help. Again, if there is any cost to signal processing, cues alone should be used. In Zone 4, signals are so accurate that they correct for any errors due to cue assessment and complement cues when the latter are correct. Signals are thus favored here. Note that all combinations in Zone 4 require that signal accuracy, Q, be as great or greater than Q_c Having access to cues thus does not ameliorate the need for a minimum accuracy of coding before a receiver should attend to signals.

tends to imperfect signals, the joint probability it will receive both a correct cue and a correct signal is $Q \cdot Q_0$. This value is less than Q_0 for all $Q < 1$. As with the prior case, using signals generates some errors where before there were none. One could minimize these new errors by making Q close to chance. This would reduce the effect that including signals would have on updated estimates. The larger the Q_0 , the larger the value of Q that could be tolerated. However, the final fraction of time the receiver got it right even with low Q values and high Q_0 would only be $Q \cdot Q_0 + Q_0(1 - Q) = Q_0$. In the end, the receiver would do no better adding signals to a high Q_0 situation than if it ignored them. The only way signals could be favored is if they were sufficiently accurate that they added few errors when cue assessment was correct, and in addition, they corrected errors when it was not. It can be shown that this is only possible if $Q > Q_c$. The threshold Q will be even higher if there are costs to attending to signals. This result is just what we found when receivers only invoke a priori probabilities and ignore ambient cues.

We have focused thus far on the overall correlation between male health and the speed of songs perceived by receivers, $Q = P(Song \mid Health)$. This net Q can be broken down into the accuracy of coding by the sender, $Q_s = P(Song\ emitted \mid Health)$, the degree to which songs arrive at females unaltered by propagation, $Q_p = P(Song\ propagated \mid Song\ emitted)$, and the accuracy with which the female identifies the songs that reach her ears, $Q_r = P(Song\ perceived \mid Song\ propagated)$. An error can be made at any of these three stages in the communication process. Correct associations by the female are only possible if no errors are made at any stage, or if an error at one stage is by chance corrected by an error at a later one. Thus the overall Q for a given combination of condition and signal is equal to

$$Q = Q_s Q_p Q_r + Q_s(1 - Q_p)(1 - Q_r) + (1 - Q_s)Q_p(1 - Q_r) + (1 - Q_s)(1 - Q_p)Q_r$$

A little algebra or experimentation will show that Q cannot be larger than the smallest of Q_s, Q_p, or Q_r; in fact, unless the smallest value is 0.5 (no correlation) or 1.0 (perfect correlation), Q will always be less than the smallest of these individual stage values. It therefore follows that communication will not be worth undertaking unless the worst correlation between input and output is greater than $Q_c + (\bar{K}/m)$. In our example, and setting $\bar{K} = 0$, this means that a minimum of 0.39 bits of information must make it through the least accurate stage of information. A high error rate at one stage does not eliminate the value of low error rates at the other stages. On the contrary, the upper limit on Q set by the stage with maximal errors can be achieved only if the other stages have minimal error rates. The lower any of the stage accuracies, the lower the net Q.

Returning to our original question, what then is the optimal decision strategy for a female when the available signal set is discrete and the female has to decide after receipt of one signal? Our answer is that the female should ignore signals if $Q < Q_c + (\bar{K}/m)$; if $Q > Q_c + (\bar{K}/m)$, she should then accept a male if his song is fast, and reject him if the song is slow. The value of Q_c will depend upon the relative payoffs of getting it right when males are sick versus healthy, and on the prior probabability that a male is healthy, P_1; the lower P_1, the lower the minimum value of Q that is required. The overall Q can be no higher than that for any stage of the communication process. Thus for communication to proceed, the worst link in the chain must be sufficiently accurate. Note that we earlier assumed equal coding accuracy for fast and slow songs (e.g., that $Q_1 = Q_2$). When this coding is unequal the algebra is a bit more complicated, but the same basic conclusions apply. We have also assumed that the average cost of attending to signals, $\bar{K}$, is independent of Q. That is unlikely to be true, and we examine the effects of variable costs later in this chapter.

OPTIMAL DECISION RULES FOR CONTINUOUS SIGNAL SETS

We have assumed in our prior examples that males can produce only two kinds of songs (fast or slow) and that females can expect to receive only one or the other. There are indeed cases where animals use discrete signal sets and our prior analysis is sufficient. But what if males can vary their song speed continuously? Let us continue to assume that there are two kinds of males, healthy and sick, and on average, that healthy males sing faster songs than sick ones. However, we shall now allow any male to sing a song at any speed w between some upper and lower limits. The problem for the female is then to identify the optimal value of w, denoted by w_c, below which she will reject males and above which she will accept them. The best cutoff in this situation can be predicted by **signal detection theory**, which deals with optimal decisions in uncertain sensory contexts. The logic of signal detection theory is very similar to that outlined in the prior section for discrete signals. However, it has been formulated in a way that provides some additional insight into the communication process.

Optimal Cutoffs for Continuous Signal Sets

Suppose the probability that a given song speed w will be produced when a male is healthy, $P(w \mid Healthy)$, is normally distributed. This means that a plot of $P(w \mid Healthy)$ versus w has a symmetrical bell shape with the peak probability occurring at the **mean** w for healthy males. The width of the bell is an index of the range of w values that might be perceived when males are healthy. One measure of this range when the distribution is normal is given by the **variance** of the w values. The total area under such a curve, regardless of the location of its mean and the size of its variance is equal to 1.0; this simply means that every male has to sing a song at some w in the range under the bell-shaped curve. Let us assume that a similar normal distribution, $P(w \mid Sick)$, predicts the occurrence of w values when males are sick. For any information to be provided to a receiver, either the means or the variances or both must differ for these two distributions. Let us first consider the case in which the variances are identical, and the two means are unequal but sufficiently close together so that there is some overlap between the two distributions. This situation is shown in Figure 14.6.

We now add a cutoff value w_c at some arbitrary value of w. Females will accept all males with $w \geq w_c$ and reject all males with lower values (Figure 14.7). We saw that the total area under the healthy male curve is equal to 1.0; the fraction of that area to the right of the cutoff is the fraction of the time that females accept healthy males. In the signal detection literature, this fraction is called P_{hit}. The fraction of the area under the healthy male curve that is to the left of the cutoff line equals the fraction of the time that a female will erroneously reject a

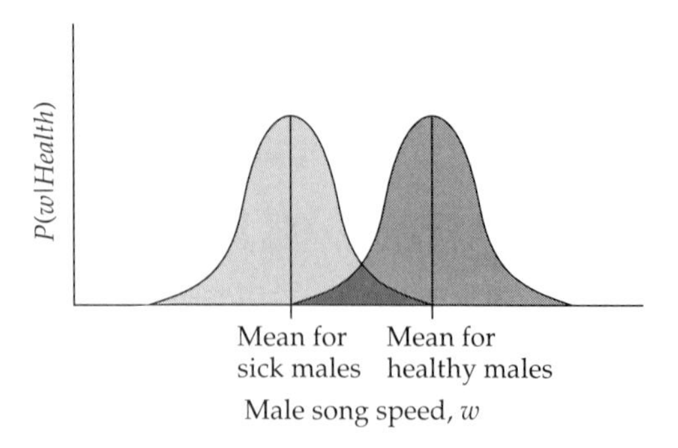

Figure 14.6 Conditional probabilities between male song speed, w, and male health (*Healthy* versus *Sick*). This example assumes that the probabilities for each type of male are normally distributed and that distributions for the two male types have equal variances but different means.

healthy male. This fraction is here called P_{miss}. The cutoff line also partitions the area under the probability curve for the sick males into two portions. The fraction of the area to the left of the line is the probability that females correctly reject sick males; this is called $P_{correct\ rejection}$. The remaining fraction to the right of the line is the probability that a female will incorrectly accept a sick male as a mate; this is called $P_{false\ alarm}$. P_{hit} and $P_{correct\ rejection}$ represent the probabilities of correct assignments by the female, whereas P_{miss} and $P_{false\ alarm}$ give the probabilities of errors. Given that P_1 of the males are healthy, and $(1 - P_1)$ are sick, the average fraction of time that a female makes a correct choice will be

$$P(Correct) = P_1 P_{hit} + (1 - P_1) P_{correct\ rejection}$$

and the average fraction of time she errs will be

$$P(Error) = P_1 P_{miss} + (1 - P_1) P_{false\ alarm}$$

Can a female maximize $P(Correct)$ or minimize $P(Error)$ by judiciously adjusting her cutoff w_c?

A quick look at Figure 14.7 shows that there is a basic tradeoff built into this system. If one moves the cutoff w_c to higher values, this will decrease false alarm errors ($P_{false\ alarm}$) and thus increase the correct rejections ($P_{correct\ rejection}$), but it will concurrently increase P_{miss} and thus reduce the correct acceptances (P_{hit}). One kind of error has simply been substituted for the other. If one moves the cutoff to lower values, one reduces P_{miss} but concurrenly increases $P_{false\ alarm}$. Thus there is no way to reduce both kinds of errors concurrently as long as the shapes and locations of the two distributions are fixed. Although

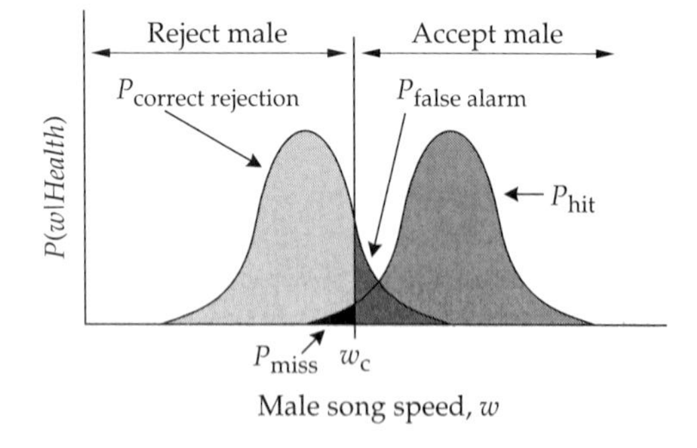

Figure 14.7 Partitioning of probability distributions with cutoff song speed, w_c. The conditional distributions are as in Figure 14.6, but with an added cutoff. Females mate with all males who have a song speed $\geq w_c$ and reject all males with lower speeds. P_{hit} is the area under the healthy male curve to the right of the cutoff, whereas P_{miss} is the area under the same curve to the left of the cutoff. $P_{false\ alarm}$ is the area of the sick male curve to the right of the cutoff, and $P_{correct\ rejection}$ is the area under that curve to the left of the cut.

the female cannot hope to find a cutoff that reduces both types of errors and maximizes correct decisisons, she still might identify an optimal cutoff by accepting the least costly errors. To find such a cutoff, we need to look at the relative payoffs for each outcome. This is how the optimal cutoff was obtained for the discrete case.

Consider a female that has just received a song of speed w. Should she accept or reject this male? Given the same payoff matrix shown in Figure 14.2, and assuming that the average cost to a female of processing any signal w is $-\overline{K}$, we can compute the expected payoff of accepting the male as

$$\overline{PO}(Accept) = P(Healthy \mid w)(R_{11} - \overline{K}) + P(Sick \mid w)(R_{12} - \overline{K})$$

and that for rejecting him as

$$\overline{PO}(Reject) = P(Healthy \mid w)(R_{21} - \overline{K}) + P(Sick \mid w)(R_{22} - \overline{K})$$

The female should accept the male when

$$P(Healthy \mid w)(R_{11} - \overline{K}) + P(Sick \mid w)(R_{12} - \overline{K}) \geq \\ P(Healthy \mid w)(R_{21} - \overline{K}) + P(Sick \mid w)(R_{22} - \overline{K})$$

Because $R_{11} > R_{21}$ by definition, this can be rewritten

$$\frac{P(Healthy \mid w)}{P(Sick \mid w)} \geq \frac{(R_{22} - R_{12})}{(R_{11} - R_{21})}$$

the cost terms canceling out.

The term on the left side of this inequality is the ratio of the a posteriori probabilites after signal reception. We can use Bayes' theorem to rewrite this term as a function of the a priori probabilities. For example, the numerator can be rewritten as

$$P(Healthy \mid w) = \frac{P_1\, P(w \mid Healthy)}{P_1\, P(w \mid Healthy) + (1 - P_1)\, P(w \mid Sick)}$$

and the denominator as

$$P(Sick \mid w) = \frac{(1 - P_1)\, P(w \mid Sick)}{P_1\, P(w \mid Healthy) + (1 - P_1)\, P(w \mid Sick)}$$

Dividing the first expression by the second, we get that

$$\frac{P(Healthy \mid w)}{P(Sick \mid w)} = \frac{P_1\, P(w \mid Healthy)}{(1 - P_1)\, P(w \mid Sick)}$$

Substituting the left hand term of the condition favoring acceptance of a male with the right term of the above expression and moving the prior probabilities to the other side, we then get that females should accept males when

$$\frac{P(w \mid Healthy)}{P(w \mid Sick)} \geq \frac{(1 - P_1)}{P_1} \cdot \frac{(R_{22} - R_{12})}{(R_{11} - R_{21})}$$

The left-hand side of this expression is called the **likelihood ratio**. It measures the relative probability that a receiver will hear a given w when the sender is healthy compared to that when he is sick (Figure 14.8). Given that the mean for healthy males is higher than that for sick males, the likelihood ratio in our example will increase as w increases. The term $(1 - P_1)/P_1$ is the ratio of the prior probabilities that a random male is sick relative to the probability that he is healthy; it is called the **prior odds**. The term $(R_{22} - R_{12})/(R_{11} - R_{21})$ is the ratio between the benefit of getting it right when a male is sick and that when he is healthy; this term is called the **payoff ratio**. The product of the prior odds and the payoff ratio is called the **operating level** and is usually denoted by β. If we let w_c be that w at which

$$\frac{P(w_c \mid Healthy)}{P(w_c \mid Sick)} = \beta = \frac{(1 - P_1)}{P_1} \cdot \frac{(R_{22} - R_{12})}{(R_{11} - R_{21})}$$

then w_c is the optimal cutoff. For all values of w less than this value, the female should reject males; for greater or equal w, she should accept them. Because the likelihood ratio increases as w_c increases, it follows that β and w_c are also positively correlated: when one goes up, so does the other. In words, the more prevalent sick males become, and the more important it is to distinguish a sick male from a healthy one, the more stringent a female should be in accepting only a high song rate in a male.

In the discrete case, we computed a cutoff probability P_c against which we compared current estimates that a male was healthy or sick. The parameter β of signal detection theory has a similar function and in fact, the two are related. It is easy to show that for either situation

$$\frac{(R_{22} - R_{12})}{(R_{11} - R_{21})} = \frac{P_c}{(1 - P_c)}$$

and thus that

$$\beta = \frac{P_c}{P_1} \cdot \frac{(1 - P_1)}{(1 - P_c)}$$

We can thus use β and P_c for either type of signal set.

ROC Curves, the Amount of Information, and the Value of Information

Choosing an optimal w_c has some interesting consequences. Note that no matter where w_c is chosen, there will still be errors. Whether misses or false alarms are the more common type of error depends on the prior odds and the payoff ratio (e.g., on β). Thus for a given pair of healthy and sick male proba-

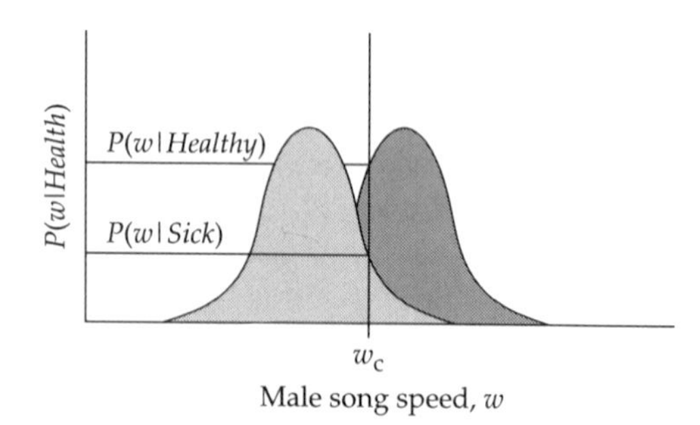

Figure 14.8 Likelihood ratio given two overlapping probability distributions. The likelihood ratio in this case is obtained by dividing $P(w \mid Healthy)$ by $P(w \mid Sick)$. Note that the ratio involves conditional probabilities at a specific value of w, not the areas under curves.

bility curves, different females might adopt different w_c values because they had experienced different prior odds or because they faced different payoff ratios or both. As w_c varies along the signal axis, the values of P_{hit}, $P_{correct\ rejection}$, $P_{false\ alarm}$, and P_{miss} all change. Note that if we know P_{hit}, we also know P_{miss} because $P_{hit} + P_{miss} = 1.0$. Similarly, if we know $P_{false\ alarm}$, then we can compute $P_{correct\ rejection}$. To describe the outcome of moving w_c to some new value, we thus only need to specify the two portions of the curves to the right of the cutoff: P_{hit} and $P_{false\ alarm}$. The sum of these two measures, each discounted by the appropriate prior probability, equals the total fraction of the time that females accept any male (healthy or sick). For a female with a high w_c, both P_{hit} and $P_{false\ alarm}$ will be small because there is only a small area under either curve to the right of the cutoff line. In contrast, females with low w_c values will exhibit high values of P_{hit} and $P_{false\ alarm}$. There is thus an inverse relationship between w_c on the one hand, and P_{hit} and $P_{false\ alarm}$ on the other.

A graph in which P_{hit} is plotted against $P_{false\ alarm}$ for females with different optimal w_c values is called an **ROC plot** (for "receiver operating characteristic" or "relative operating characteristic"). One can intepret such a plot as indicating how much a receiver has to pay (in terms of false alarms) for a given hit rate. This cost will depend upon how well signals are correlated with male health, and thus on how little overlap there is between the probability distributions of w for healthy versus sick males. Suppose the distributions for the two males have the same means and variances (i.e., they are identical). Clearly, the signals can provide no new information to a female. Decreasing w_c to increase hit rates will only result in an equal increase in false alarms. As a result, $P_{hit} = P_{false\ alarm}$ for any w_c and the corresponding ROC curve will be a straight line with a slope of 1.0 running from the lower-left corner of the graph to the upper-right corner (Figure 14.9A).

Now consider a situation in which the means for the two distributions are slightly different. A female that accepts no males will be represented on the corresponding ROC curve at a point in the lower-left corner. A female with a slightly lower w_c will achieve moderate hit rates with only small numbers of false alarms. As w_c is decreased further, hit rates will asymptote to 1.0, and further changes will only increase false alarm rates. The corresponding ROC curve is thus bent up and toward the upper left corner of the graph (Figure 14.9B). If the two male signal distributions overlap even less, the corresponding ROC curve will be bent even further toward the upper-left corner (Figure 14.9C). Put another way, the greater the bending, the lower the false alarm rate ($P_{false\ alarm}$) suffered by a receiver to achieve any given hit rate (P_{hit}). This relationship is just what one might expect. As the distributions become less overlapping, the receiver will be more accurate in its discriminations. This conclusion is reflected in the corresponding ROC curve by an increased bending toward the upper-left corner of the plot. The degree to which an ROC curve is bent away from the major diagonal thus seems useful as a measure of the amount of information provided by signals.

We discussed on pages 411–416 the problems with estimating how much information is transferred by communication when the only available data are the decisions of the receiver. For example, one female might obtain much more

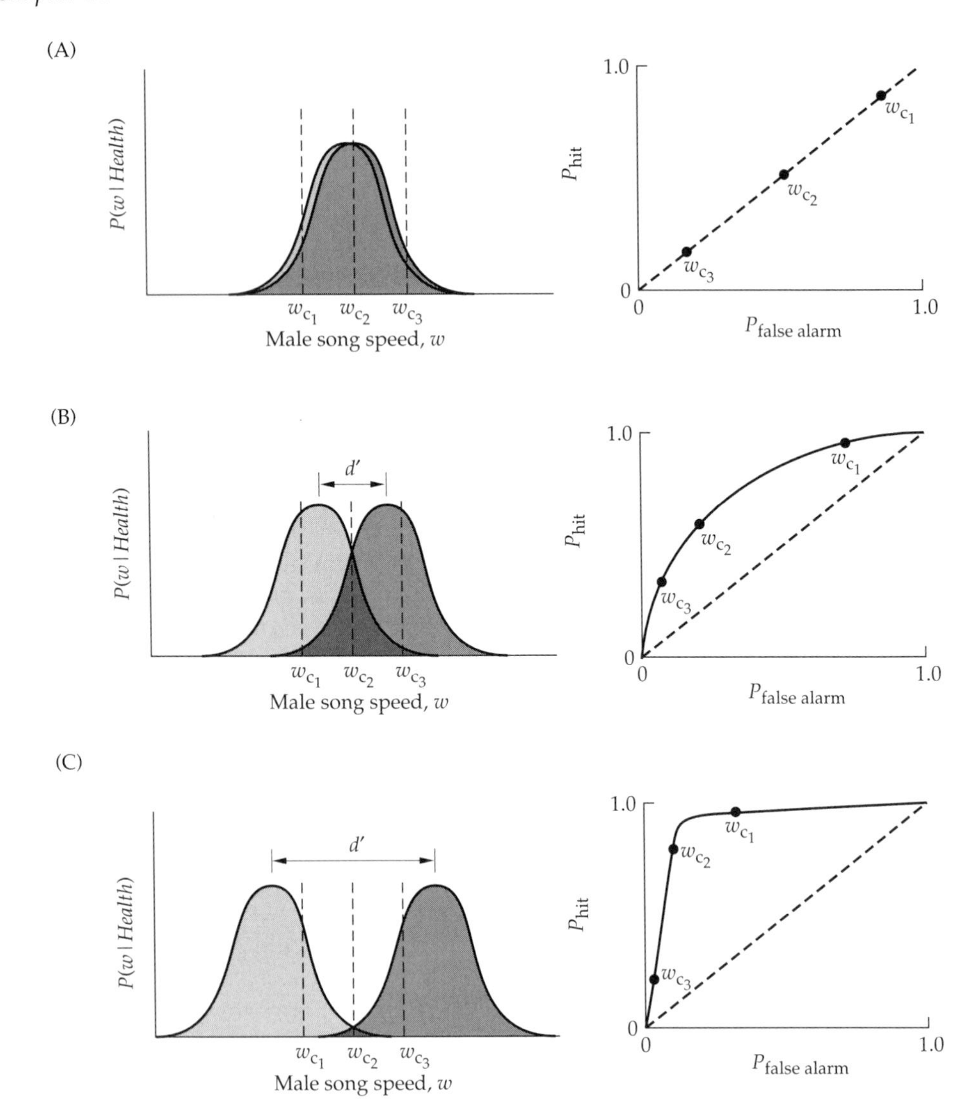

information from signals than another, and yet the two might accept the same fraction of encountered males as mates. The similarities in response despite differences in information provided are due to compensating differences in the decision criteria invoked by the females. Any decision reflects both the amount of information and the value of information. ROC curves provide a clever way to separate the effects of the amount of information from the decision process in communication. We saw above that the displacement of an ROC curve away from the major diagonal is a measure of how many false alarms must be suffered to obtain any given hit rate. It is thus a measure of the receiver's acuity of discrimination regardless of what the receiver subsequently decides to do. In signal detection theory, the degree of ROC curve bending is measured by d' and called the **sensitivity** of the receiver. The units of d' and how it can be esti-

◀ **Figure 14.9 ROC curves for distributions with differing distances between means.** Healthy (dark shading) versus sick (light shading) male conditional probability distributions are on the left; the corresponding ROC plot is on the right. d' is a measure of the differences between the shapes and dispositions of the healthy and sick male distributions. (A) Healthy and sick male distributions have same variances and same means ($d' = 0$). The corresponding ROC curve is a straight line that extends from the lower left to the upper right of the graph. Note that as w_c moves to the right, the corresponding points on the ROC plot move left and down toward the origin. (B) Moderate separation (moderate d') between the means of healthy and sick male distributions. The corresponding ROC plot bends away from diagonal. (C) Large separation between healthy and sick male distributions. The corresponding ROC plot is bent so far from diagonal that it nearly tracks the left and upper boundaries of the graph.

mated from the observed values of P_{hit} and $P_{false\ alarm}$ are outlined in Box 14.1. An increase in d' in the continuous case is thus equivalent to an increase in Q in the discrete model because both lead to an increase in the average amount of information provided by signals. (The explicit dependence of the amount of information on d' is derived in Box 14.3.) And as with Q, an overall d' can be partitioned into contributions from the sender's coding accuracy, the level of distortion during propagation, and the accuracy of the receiver's perception and analysis apparatus. Anything that decreases the overlap of the healthy and sick male w distributions will increase d'.

While d' indicates where the ROC curve will lie for a given situation, it still allows females an infinite number of combinations of hits and false alarms. Any given female in this situation has to identify which combination of hit and false alarm rates along this curve is best for her. This she does by determining β and setting w_c as described earlier. The optimal cutoff for a given receiver may be high or low. If it is high, the receiver will be conservative about her exposure to false-alarm errors; if it is low, then the receiver will be tolerant to such errors. The resistance to false alarm errors is called the receiver's **bias**, and it can also be estimated from knowledge of P_{hit} and $P_{false\ alarm}$ (see Box 14.1). Whereas d' gives us a measure of the amount of information transferred, bias is largely dependent upon the value of information to a receiver.

What is the average value of information to a female that invokes the optimal w_c? This average value of information will equal the average payoff when attending to signals minus that when signals are not present or are ignored. When we discussed discrete signals earlier in this chapter, we saw that an optimal female with no signal information will compare the prior probability that a male is healthy, P_1, to the critical threshold value, P_c. If $P_1 \geq P_c$, then the female will accept all males and obtain the average payoff $P_1 R_{11} + (1 - P_1)R_{12}$; if $P_1 < P_c$, then the female will reject all males and obtain the average payoff $P_1R_{21} + (1 - P_1) R_{22}$. The average payoff for a female exposed to continuous signals, using an optimal cutoff of w_c, and suffering a per signal cost $-\overline{K}$ is

$$\overline{PO}(w_c) = P_1P_{hit}(R_{11} - R_{21}) - (1 - P_1)P_{false\ alarm}(R_{22} - R_{12}) + P_1R_{21} + (1 - P_1)R_{22} - \overline{K}$$

Box 14.1 Estimating Parameters for Signal Detection Analysis

SIGNAL DETECTION THEORY invokes several basic parameters to explain observed receiver decisions. One of these, called **sensitivity** and denoted by ***d'***, measures the accuracy with which a receiver can discriminate between alternative conditions using signals. Where the probability of receiving a certain signal value, w, is distributed normally for each alternative condition, d' is an appropriately scaled difference between the distribution means. A scaling widely used in statistics converts all w values to **z scores** based upon the fact that the distribution is normal. The z score is computed as follows. If the mean of a normal distribution is μ, and its **standard deviation** is σ (where σ equals the square root of the variance), then the z score for w is $z(w) = (w - \mu)/\sigma$. We can thus replot any original probability distribution of w values as a probability distribution of $z(w)$ values. This distribution will have its maximum when $z(w) = 0$ (e.g., when $w = \mu$), and all $z(w)$ values to the left of this peak will be negative (e.g., $w < \mu$), and all $z(w)$ values to the right of the peak will be positive ($w > \mu$). The difference between the means of two z-scaled distributions, d', will then be given as a multiple of their common standard deviation (if it is the same for both), or as a multiple of their average standard deviation (if they are different). Because d' is measured in standard deviation units, decreasing the average standard deviation of the distributions is equivalent to increasing the distances between their means; either change reduces overlap between the distributions, and thus reduces the risk of errors.

Suppose a female receiver wishes to discriminate between healthy and sick males. There will thus be two alternative probability distributions, each with its own mean (μ_1 and μ_2 for healthy and sick males, respectively). In this example, we shall assume that the standard deviations are the same for the two distributions. The accuracy of the female's discrimination depends upon d', which is a z-score measure of the difference between μ_1 and μ_2. Unfortunately, we cannot easily determine the distributions or the means as perceived by the female, but we would like to estimate d' from the decisions she makes. How can we do this?

For each distribution we first convert the values on the w axis into $z(w)$ values. For the first distribution, $z_1(w) = (w - \mu_1)/\sigma$, and for the second distribution and the same w, $z_2(w) = (w - \mu_2)/\sigma$. We note that $z_2(w) - z_1(w) = (\mu_1 - \mu_2)/\sigma = d'$, which is the measure we seek. We can thus estimate d' if we can estimate $z_1(w)$ and $z_2(w)$ from observations of a female's decisions. An optimal female presumably has some fixed cutoff w_c such that she rejects males when $w < w_c$ and accepts them when $w \geq w_c$. This cutoff partitions each of the two distributions into two parts. The area under the healthy male distribution and to the right of the cutoff equals the fraction of time that the female accepts healthy males. This probability is called P_{hit} and is something we can measure as

The value of information when $P_1 \geq P_c$ is then found to be

$$VI(P_1 \geq P_c) = P_1 P_{hit}(R_{11} - R_{21}) - (1 - P_1) P_{false\ alarm}(R_{22} - R_{12}) - \overline{K} + P_1(R_{11} - R_{21})(1 - \beta)$$

and when $P_1 < P_c$

$$VI(P_1 < P_c) = P_1 P_{hit}(R_{11} - R_{21}) - (1 - P_1) P_{false\ alarm}(R_{22} - R_{12}) - \overline{K}$$

long as we know which males are healthy and which are sick. The area under the sick male distribution to the right of the cutoff is the fraction of sick males that the female accepts and is denoted by $P_{\text{false alarm}}$. We can also measure this quantity. Most statistics texts have tables in the back listing the area below a normal probability curve to the right or the left of some cutoff value of a z score. Usually, the reader has a z value and wants to know the corresponding probability. In our case, we know the probability, but would like to know the corresponding z score. We thus locate each of the measured probabilities P_{hit} and $P_{\text{false alarm}}$ in this table, and then find the corresponding values, z_{hit} and $z_{\text{false alarm}}$, respectively. Since these z scores are based on the same w, in this case on w_c, we can use their difference to compute $d' = z_{\text{hit}} - z_{\text{false alarm}}$. To provide a feeling for the scale of this measure, a receiver that correctly identifies both sick and healthy males 50% of the time (e.g. by chance) has a $d' = 0$, a receiver that is accurate 70% of the time has a $d' = 1.04$, a receiver that is accurate 90% of the time has a $d' = 2.56$, and a receiver that is accurate 99% of the time will have a $d' = 4.65$.

The alternative parameter of signal detection theory is **bias.** This is the degree to which a female is conservative about accepting males, and thus avoids false-alarm errors at the expense of having more miss errors. The simplest measure of bias is the **criterion** index c, which can be computed as $c = -0.5\,(z_{\text{hit}} + z_{\text{false alarm}})$. A female that has no bias accepts equal numbers of false alarms and miss errors (e.g., $P_{\text{false alarm}} = 1 - P_{\text{hit}}$), and its bias $c = 0$. When females avoid false alarms, $c > 0$, and when they avoid misses, $c < 0$. For any observed combination of P_{hit} and $P_{\text{false alarm}}$, c depends upon the distance between that point and the diagonal running from the top-left to the lower-right corner of the ROC plot (see also Box 14.2).

It is also possible to estimate the parameter β, which is equal to the ratio of the likelihood that a male is healthy to the likelihood that he is sick. Quite simply, $\ln(\beta) = cd'$, if the female is making optimal decisions. Using hit rates and false-alarm rates, we can rewrite this as $\ln(\beta) = -0.5(z_{\text{hit}} - z_{\text{false alarm}})^2$. The parameter β has some interesting graphical significance, which we take up in Box 14.2.

If we know that the distributions of w for healthy and sick males are normally distributed with equal variances, we do not have to compute an entire ROC curve to obtain estimates of d', c, and β. Instead, one pair of hit and false-alarm rates will do. However, distributions may not be normal or have equal variances. The only way to detect this case is to plot the ROC curve by giving the female different prior probabilities of male health or different payoff values. We can still compute a single d', c, and β from such a situation; however, the analysis is more complicated than that given here. See MacMillan and Creelman (1991) for details.

In either case, the value of information depends upon the payoff difference between correct and incorrect acceptance, discounted by the probability that this is what occurs, and the similarly discounted consequence of a false alarm. From this difference are also subtracted the costs of signaling and any new errors added by reliance on imperfect signals. These expressions verify one's intuition that the value of information should increase with hit rate, but decrease with false alarm rate. As with the discrete case, they are also similar to the general expressions derived on pages 378–380. Note that the sum of all terms except the last must be positive before it pays females to attend to signals. This sum is positive only if $P_{\text{hit}} \geq \beta \cdot P_{\text{false alarm}}$ when $P_1 < P_c$ and if $P_{\text{hit}} \geq \beta \cdot (P_{\text{false alarm}} + 1) - 1$ when $P_1 \geq P_c$. Since d' is a measure of the difference between P_{hit} and $P_{\text{false alarm}}$, it follows that there is a minimal value of d' that must be attained before females should ever pay attention to male songs. This is just what we found for Q in the discrete case.

Although sensitivity and bias can vary somewhat independently, in the end, bias and the value of information are both constrained by the value of d'. We saw earlier that d' measures the cost in false alarms that must be endured to achieve a given hit rate—the higher the value of d', the lower the number of false alarms for any given hit rate. If one assumes, as we did on pages 421–429, that increased accuracy comes at no additional cost to the receiver, then it follows from the expressions above that increasing d' will increase the value of information as well as the amount of information. A graphical proof of this is given in Box 14.2. Of course, it is more likely that increasing Q in the discrete case or d' in the continuous case *will* involve higher costs. We take up this complication on pages 444–448.

Signal Detection Theory in Different Contexts

Signal detection theory can be applied to a much wider variety of sensory decisions than we have discussed so far. Some decisions constitute a **detection** task in that a receiver must identify whether a signal is present or not despite ambient noise. This case can easily be modeled using signal detection theory by considering the distributions of w when only noise is present and when signals are mixed with noise. The task is to decide whether a signal is or is not present. In contrast, the female bird in our previous example was concerned with the **discrimination** between two alternative signals whose distributions had equal variances. In nature, the relevant distributions are likely to have unequal variances. The analysis must then be adjusted because unequal variances make the shape of the ROC curve asymmetric; the computation of d' and bias are still possible, but it takes a bit more work (Macmillan and Creelman 1991). Unequal variances can also generate two different values of w at

Figure 14.10 Effects on ROC curves of different variances for healthy versus sick male distributions. This layout is identical to that of Figure 14.9. (A) The variance for healthy males (dark gray) is larger than that for sick males (light gray); the means are the same. For any given value of β there will be two values of w at which the likelihood ratio equals β. Vertical dashed lines show the case for $\beta = 1$. There are thus two critical values of w_c: females should accept all males with $w < w_{c_1}$ and all males with $w > w_{c_2}$. (B) Case when healthy males have larger mean and larger variance. Dotted lines show means for each distribution. Asymmetric curve is actual ROC plot; symmetric curve shows that expected variances were equal. (C) Case when healthy males have larger mean but smaller variance than sick males. The arrows within each left hand graph indicate the distribution with the larger variance, and the arrows in the ROC plots on the right indicate consequent deviations from symmetry due to the unequal variances. ▶

which the likelihood ratio equals β (see Figure 14.10). In principle, the receiver must then have two values of w_c so that for w between the two cutoffs it should do one thing, and for values outside the two cutoffs, it should do another. Given typical prior distributions, usually only one cutoff falls within the range of likely w values. Multiple cutoff values will also be needed when the receiver must discriminate between more than two alternative conditions, all coded using the same signal parameter. A common task in animal communication that requires such multiple cutoffs is **classification**. Into which of several alternative categories should a given sender be assigned? The basic procedures still apply. First, one arrays the distributions for each of the categories along the w axis according to the magnitude of their means, and then one

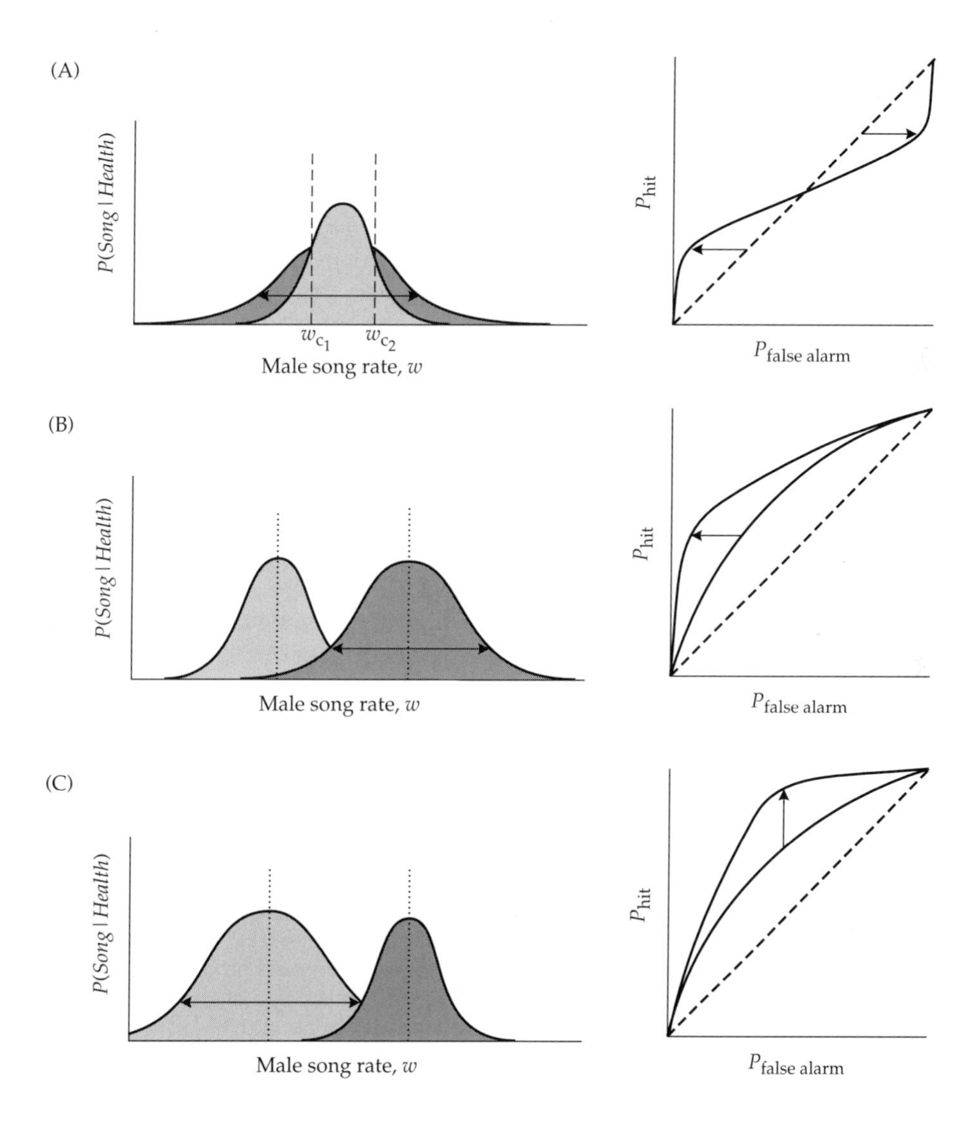

Box 14.2 Fitness Set Analysis of ROC Curves

IN BOX 12.3, we outlined a graphical method for identification of an optimal strategy when there is a tradeoff between two continuous components of fitness. This method identified a **fitness set,** which is the set of combinations of the two components that is allowed by nature, and **adaptive functions**, which are potential relationships between the two, assuming a given value of fitness. This method can be used to find the optimal bias (and hence w_c), given the observed ROC curves of a receiver. In this case, the ROC curve is the fitness set. It indicates the allowable combinations of hit rate, P_{hit}, and false alarm rate, $P_{false\ alarm}$, for a given receiver. Such a curve can be measured by varying the payoffs or prior probabilities faced by a female and observing the corresponding hit and false-alarm rates. As we saw in the text, a receiver that adjusts cutoff thresholds to increase P_{hit} (a desirable result) also increases rates of $P_{false\ alarm}$ (an undesirable result). There is thus a fitness tradeoff between P_{hit} and $P_{false\ alarm}$. What is the optimal combination that the receiver should adopt?

To compute the adaptive functions, we need to decide on a measure of fitness. We shall use the **value of information** because this incorporates all of the benefits and costs of signaling above those achieved when no signals are provided. Earlier, we derived the value of information and showed how it was affected both by hit rates and false-alarm rates and showed that its value depends upon whether the prior probability that a male was healthy, P_1, was larger or smaller than the critical threshold, P_c. The relevant adaptive functions are derived simply by rewriting the expressions for the value of information so that hit rate, P_{hit}, is a function of false-alarm rate, $P_{false\ alarm}$, and the value of information. Thus, if $P_1 \geq P_c$, the adaptive functions are of the form

$$P_{hit} = \beta P_{false\ alarm} + \alpha(VI + \overline{K}) + (1 - \beta)$$

and when $P_1 < P_c$, then

$$P_{hit} = \beta P_{false\ alarm} + \alpha(VI + \overline{K})$$

where VI is the value of information for that combination of P_{hit} and $P_{false\ alarm}$, $\alpha = 1/[P_1(R_{11} - R_{21})]$ and other variables are as defined in the text. On a graph of P_{hit} versus $P_{false\ alarm}$, these adaptive functions will be straight lines with slopes equal to β and y intercepts equal to $\alpha(VI + \overline{K}) + (1 - \beta)$ or $\alpha(VI + \overline{K})$, depending upon initial conditions. The larger the value of information, VI, the higher on the P_{hit} axis is the y intercept of the straight line.

We seek that combination of hit rate and false-alarm rate that maximizes the value of information. To find this optimum combination, we draw several adaptive functions with different possible values of information on the same graph with the fitness set (e.g., the ROC curve). The graphs are shown in Figure A. The adaptive function line that just touches the ROC curve is the maximal value of information achievable, and its point of tangency indicates the optimal combination of P_{hit} and $P_{false\ alarm}$.

In this example, adaptive functions lines are drawn for a high value of VI, a medium value and a low value. The high VI line does not touch the ROC curve: there is no way that this receiver can achieve a value of information that high. The low VI line hits the ROC curve, and thus this value of information could be achieved, but the low VI line is not the intersected line that has the maximal value of information. The medium VI line is the line with the highest value of information that also touches the

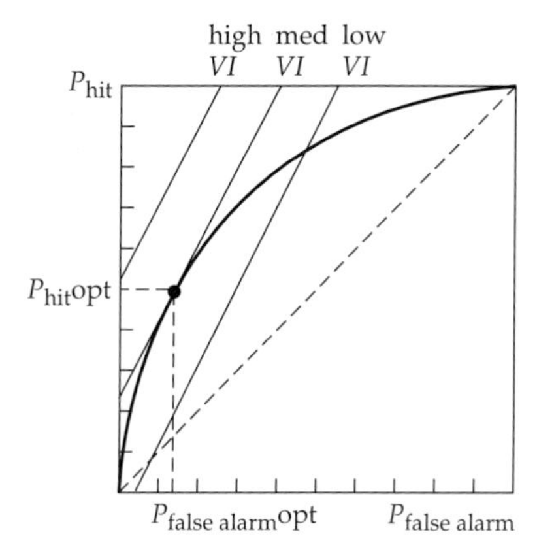

Figure A Fitness set analysis of optimal hit and false alarm rates for signal detection problem. Straight lines with positive slopes are alternative adaptive functions; the higher the line intercept, the higher the corresponding value of information (*VI*). The dark curved line is the fitness set. The intersection of the fitness set with that adaptive function with the highest *VI* defines the optimal hit rate (P_{hit} opt) and optimal false alarm rate ($P_{false\ alarm}$ opt).

ROC curve. The point on the ROC where that line touches is the optimal combination of hit and false-alarm rates. In this example, the optimal hit rate is then about 50% and the optimal false alarm rate is about 12%. An optimal female will thus adjust its w_c value to achieve this hit rate and false-alarm rate combination.

For a fixed value of β, how does the optimum bias change as d' is varied? Put another way, should a female with a lower d' be more or less conservative in her acceptance of false alarms? Suppose we compare two different females that both have the same prior experience and payoff values, (thus β and the slopes of adaptive function lines are the same), but one has defective ears (thus her d' is lower than that of the other). The corresponding ROC curves would appear as shown in Figure B.

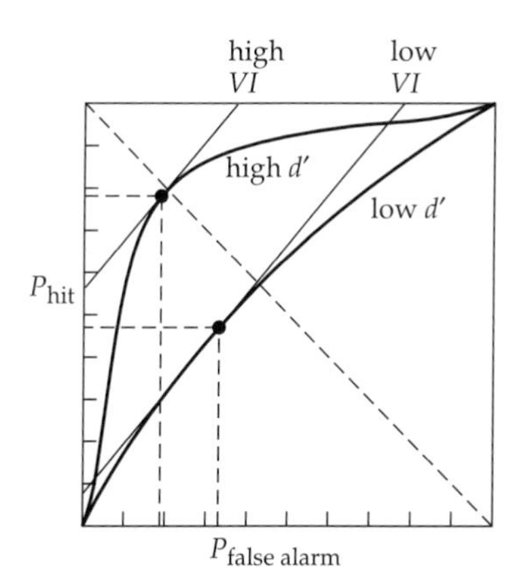

Figure B Variation in optimal bias when receivers differ in sensitivity, d'. Fitness set analysis shows that optimal hit and false alarm rates shift to more conservative bias values as d' is decreased.

The optimal combinations of P_{hit} and $P_{false\ alarm}$ are shown for each female. The female with poor ears has the lower d' and thus her ROC curve only marginally bends towards the upper-left corner. The normal female has a higher d,′ and her ROC curve

Box 14.2 (continued)

bends strongly to the upper left. The normal female will experience a higher hit rate and lower false-alarm rate at her optimum than will the defective female. She will also obtain a higher value of information. The degree of bias for each female, c, is the distance between her optimum point and the dashed line running from the upper-left to lower-right corners of the plot. This line is the locus of combinations for which $c = 0$. Both females have points to the left of this diagonal and thus both have $c > 0$. However, the defective female's optimum is further from this line than the normal female's; she thus has a larger value of c, and is thus less willing to accept false-alarm errors than is the normal female. She will be more conservative. Generally, as d' is increased, the bias, c, for a given β decreases.

finds a cutoff value between each pair of adjacent distributions based upon the relevant priors and payoff ratios.

What if the signal involves more than one parameter? In Chapter 13, we discussed the example of a compound display in which song speed and depth of a bow during singing both provided information to receivers. Suppose each parameter provides independent information about mate suitability. The appropriate plot of the relevant distributions would then be three dimensional. The relevant cutoff would be a line dividing the entire set of combinations of song speed and bow depth into two groups so that males with combinations in one group would be rejected, and those in the second group would be accepted (Figure 14.11). When different components of a compound display are not independent, one can usually extract a smaller set of independent variables, each of which is a combination of the original components weighted so as to correct for component correlations. An observer might use principal components analysis to do this (e.g., Beecher 1989); a receiver might achieve a similar end by comparing the outputs of brain circuits in which stimulations by components are summed to those in which they are subtracted.

The Ideal Receiver

There are physical limits on how much information any receiver, no matter how well-designed, can extract from a given signal. For example, suppose discrimination between healthy and sick males depends on both the frequency and the temporal details of songs. To maximize the amount of information provided by the frequency composition of songs, the female will want the most accurate frequency resolution possible. This means that the receiver bandwidth, Δf, should be small. Similarly, the female will want the most accurate temporal resolution possible: thus Δt should also be small. The uncertainty principle for sound (see Chapter 3, pages 63–68) states that the product of Δf and Δt cannot be less than 1. After both become sufficiently

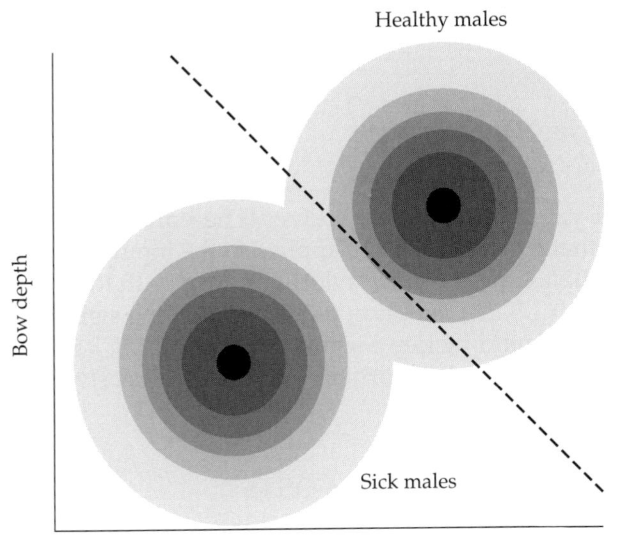

Figure 14.11 Optimal threshold cutoff rule for bivariate display. Horizontal axis indicates song speed in display; vertical axis indicates bow depth. The graph assumes that the song speed and bow depth can vary independently. The probability of a given combination of song speed and bow depth for a given male type is shown by shading: high probabilities have a darker shading. Healthy males tend to give fast songs with deep bows; sick males tend to give slower songs and shallower bows. The dashed line indicates the optimal cutoff for a female who is given one set of prior probabilities and payoffs. The female should accept all males with combinations of song speed and bow depth above and to the right of the cutoff line, and reject all males below and to the left of the line.

small, if one is decreased, the other must increase. This means that there will be a maximum amount of information, and thus a maximum d' (d'_{max}) that can be achieved for a given discrimination signal regardless of the way one designs the receiver apparatus. The same constraint applies to detection problems in which the receiver attempts to determine whether a signal is or is not present in noise. If S is the energy in the signal, and N is the level of energy for each hertz of ambient noise, it can be shown that the maximum achievable d' is

$$d'_{max} = \sqrt{\frac{2S}{N}}$$

It is physically impossible to extract more information from this situation than that specified by d'_{max}.

Note that a value of d'_{max} is specific to a particular set of signals. Even though a female cannot do better than the relevant d'_{max} when exposed to a specific type of song, she could do better if the male provided more information to the song by adding more component frequencies or singing for a longer time. There are, of course, eventual physical limits on how much information can be put into any signal; for example, a male cannot sing more than 100% of the time. However, these limits are rarely reached in nature.

A receiver that can extract the maximum possible information from a given signal, that is, one for which $d' = d'_{max}$, is called an **ideal receiver**. Most receivers are not ideal and thus do worse than is physically possible. Knowing what an ideal receiver can do provides a benchmark against which any observed receiver can be compared (see Chapter 26). The usual measure of relative receiver efficiency is

$$\eta_r = \left(\frac{d'_{\max}}{d'}\right)^2$$

where d' is the observed value for the real receiver. The value η_r can be thought of as a multiplier that would have to be applied to the input signal to bring the receiver's performance up to that of an ideal receiver with the original signal. For example, if $\eta_r = 1.5$ in a detection problem involving signal energy S, the observed receiver would require a signal energy of 1.5 S, keeping noise levels fixed, to achieve the same performance as the ideal receiver exposed to the original signal.

OPTIMAL DECISIONS WITH VARIABLE COSTS

In the prior sections, we assumed a fixed average cost to the receiver of attending to signals regardless of the value of Q or d'. This assumption is unlikely to be true. In fact, any adaptation that increases the amount of information provided to a receiver will surely impose an increased cost on the sender, the receiver, or both. Increasing Q or d' increases the amount of information exchanged by a given signal. This will surely increase costs. As we saw in Chapter 13, another way to increase the amount of provided information is to sample several consecutive signals and use sequential Bayesian or equivalent updating to obtain an improved estimate of current conditions. This mechanism, as well, will lead to increased costs for one or both parties. There is thus a tradeoff here—increased amounts of information increase the benefits of communication, but concomitant costs also tend to increase. Is there an optimal Q, d', or number of samples at which the value of information is maximized?

To find an optimum, we need to partition the value of information in each case into the component benefits and costs of communication. For our discrete example, the value of information as a function of Q, $VI(Q)$, is equal to 0 when $Q < Q_c$, (where $Q_c = \beta/(1 + \beta)$,, and β is the operating level of signal detection theory), and $VI(Q) = \bar{B}(Q) - \bar{K}(Q)$ when $Q \geq Q_c$. Here the benefit of the information is on $\bar{B}(Q) = P_1Q(R_{11} - R_{21}) - (1 - P_1)(1 - Q)(R_{22} - R_{12})$, and the average cost, now a function of Q, is $-\bar{K}(Q)$. As we saw on pages 425–426, $\bar{B}(Q)$ increases linearly with increasing Q once $Q \geq Q_c$. How might $\bar{K}$ vary with Q? The simplest case is that $\bar{K}$ also increases linearly with Q. However, it is unlikely that this is so. Remember that Q is the conditional probability that a given song will be perceived by the receiver as fast when the sender is healthy, or slow when it is sick. Whereas reduction of gross errors may be relatively inexpensive, removal of residual error is likely to be increasingly expensive. The most likely cost and benefit functions for the discrete case are shown in Figure 14.12. The value of information for any Q is the difference between these two curves, e.g., $\bar{B}(Q) - \bar{K}(Q)$. In this typical example, this differ-

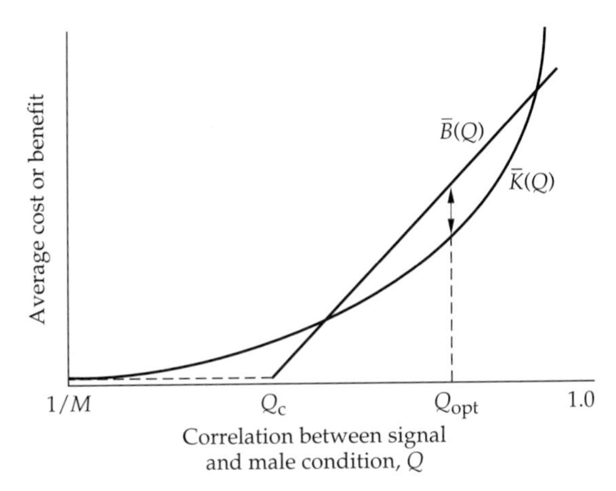

Figure 14.12 Optimal correlation between signal and condition, Q_{opt}, for discrete signal sets. The two curves indicate the magnitude of average benefit, $\bar{B}(S)$, and absolute value of average cost, $\bar{K}(S)$, for different possible correlation values Q on the horizontal axis. Q, the correlation between signal and condition, has a maximal value of 1.0 and a minimal value of $1/M$, where M is the number of alternative signals in the same signal set. The average benefit is zero for $Q < Q_c$ because it does not pay for a receiver to attend to such signals. Benefits then increase linearly with Q_c. Costs show accelerating increase with Q because it becomes increasingly expensive to remove residual errors. Because the value of information is the difference between the two curves when receivers attend to signals, this value is maximized at a $Q_{opt} < 1$, indicating that seeking perfect information is not the optimum in this case.

ence is maximal at some Q_{opt} where $1 > Q_{opt} > Q_c$. Economically, it does not pay for this receiver to seek perfect information, but instead to accept a $Q < 1$ and thus some error in its communication system.

The equivalent plot for continuous signal sets is shown in Figure 14.13. The benefit in this case is $\bar{B}(d') = P_1 P_{hit} (R_{11} - R_{21}) - (1-P_1) P_{false\ alarm} (R_{22} - R_{12})$. Because d' is a transformation of the difference, $P_{hit} - P_{false\ alarm}$, the corresponding benefit curve is not a linear function of this difference, but instead a decelerating curve. It is not well known how $\bar{K}(d')$ scales with d'. However, it is likely that costs increase at least linearly with d', and possibly at an accelerating rate. The graph in Figure 14.14 uses a linear example. Regardless of the

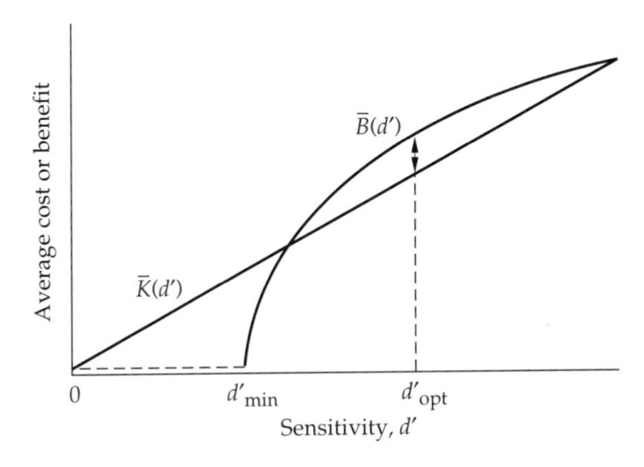

Figure 14.13 Optimal sensitivity, d'_{opt}, for continuous signal sets. The two curves indicate the magnitude of average benefit, $\bar{B}(d')$, or absolute value of average cost, $\bar{K}(d')$, for different possible sensitivities d'. The benefit is zero for $d' < d'_{min}$, where d'_{min} is the value for which $P_{hit}/P_{false\ alarm} = \beta$. The benefit then increases in a decelerating way with further increases in d'. Costs are likely to increase linearly or faster with d'. The optimal value of d' occurs where the benefit curve is maximally above the cost curve. In this case, an intermediate d' is optimum.

(A)

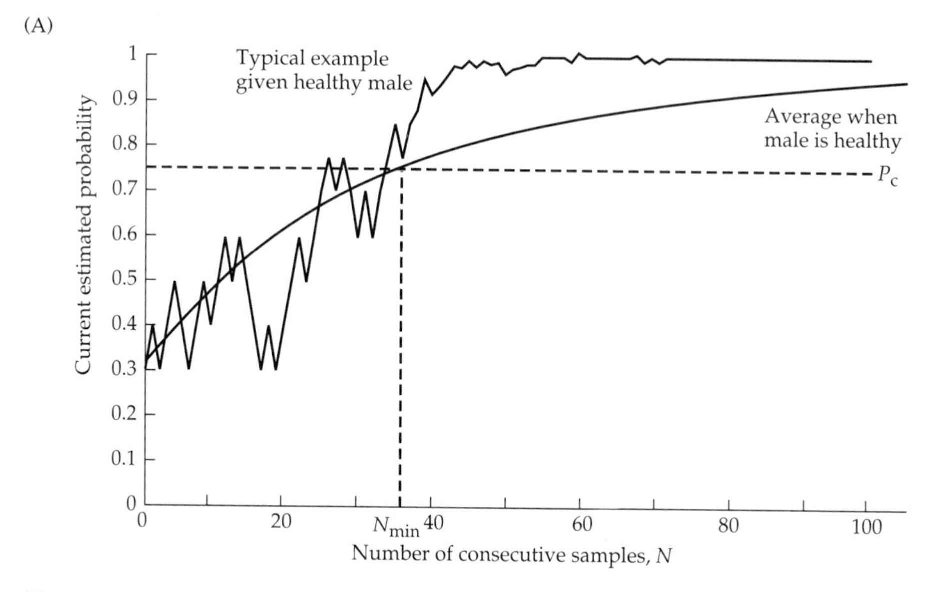

(B)

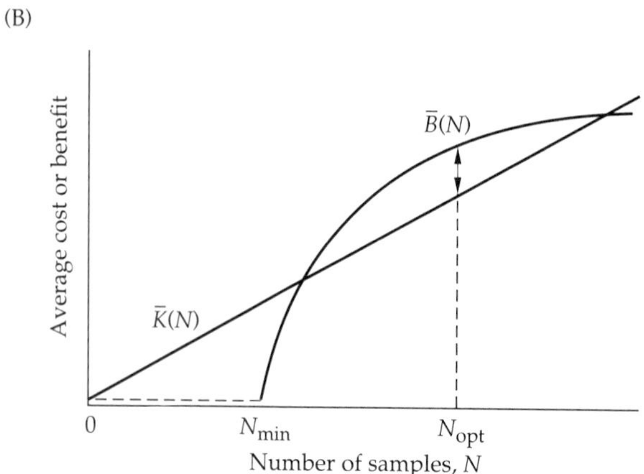

scaling of costs, it is clear that there will be an intermediate d' that is optimal. Again, it does not pay to seek perfect information.

Finally, how do the benefits and costs of multiple sampling of signals vary with the number of samples N? Figure 14.14A shows the average current estimate that a healthy male is healthy when a receiver samples different numbers of consecutive discrete signals. The actual history for one particular female is also shown. As with the other mechanisms for increasing information, successive sampling increases the benefits of communication only above a certain minimum. In this case, the critical value is the number of samples, N_{min}, at which an average female's estimate that a healthy male *is* healthy will just exceed P_c. With further sampling above N_{min} samples, the benefits rise,

◀ **Figure 14.14 Optimal number of successive samples of discrete signals for a Bayesian updater.** (A) The current probability estimates by a female that a healthy male is indeed healthy after sampling N succesive songs. The smooth curve shows the average of 1000 different sampling sequences when $P_1 = 0.3$ and $Q = 0.7$. The jagged trajectory shows the example of a single female sample history. The female accepts the male only when the current estimate that he is healthy exceeds P_c. On average, females will not exceed this threshold even if a male is healthy until they have sampled about N_{min} times. The female in this example would first reach this threshold after 26 samples. (B) Average information benefits, $\bar{B}(N)$, and costs, $\bar{K}(N)$, for different numbers of successive samples by females. $\bar{B}(N) = 0$ for $N < N_{min}$ because females ignore males giving smaller numbers of signals. Benefits increase in a decelerating way, reaching an asymptote for higher N because the probability that a male is healthy reaches a limit at 1.0. Thus successive samples provide decreasing additional benefit. If each sample costs the same, the cost curve is likely to be linear. The result is an intermediate optimal number of samples, resulting in a less-than-perfect information transfer at the moment of decision.

but in an asymptotic manner. This relationship is easy to show by substituting the current estimate that the male is healthy, P_N, for the value of P_1 in the computation of $\bar{B}(Q)$ above. The result is the benefit of one more signal given a current estimate. This value steadily decreases as N increases because the current estimate has to approach an asymptote at 1.0 eventually. If the costs of sampling are time or energy, it is likely that these costs are constant for each sample. In this case, the cost curve, $\bar{K}(N)$, increases linearly with N. Again, we find that the optimum number of samples, N_{opt}, is an intermediate value; it is rarely worth it to seek perfect information.

Most of this chapter has focused on the optimal value of information for receivers. Senders must also decide how much to invest in sending signals and the same graphical analyses can be applied to them. The relevant equations for the value of information to senders are identical to those for receivers except that payoff ratios may be different. (Using the notation of Chapter 12, we would substitute $(T_{11} - T_{21})$ for $(R_{11} - R_{21})$ and $(T_{22} - T_{12})$ for $(R_{22} - R_{12})$, respectively.) Included in the options available to senders for increasing the amount of information obtained by the receiver are the following: more careful mapping of signals onto conditions, wise selection of signal properties to compensate for distortion during propagation, and the replicated emission of the same signal while a given condition is maintained. All of these options involve some increase in the signal costs experienced by the sender. For different senders and situations, the alignment and shape of the benefit and cost curves will differ, and thus the optimal investment that the sender should put into communication will vary.

For both the sender and receiver, there is thus an optimal investment in communication. However, the optimal investments by sender and receiver are not independent of each other. The optimal investment by a receiver depends upon how much the sender has put into signal design and emission rate. Similarly, since it is the receiver's actions that determine the sender's benefits, the optimal investment for a sender depends on the current investment by the receiver. Because of these linkages, we expect the optimal investments of the two parties to coevolve. In many cases, each party in the exchange will eventually reach an investment level at which it does not pay either party to change. This stable equilibrium in investments is presumably what we observe in most animal communication systems. However, unstable systems that cycle over evolutionary time are also possible. Even if an equilibrium is reached, it is important to realize that the stable level of investment by the sender may not be that at which the fitness of the receiver is maximized, nor is that of the receiver necessarily the one that would maximize the sender's fitness. We shall return to these issues in Part III of the book.

OPTIMAL DECISION MAKING IN ANIMAL COMMUNICATION

Optimal decision theory has played an increasingly important role in studies of animal communication. Consider a general problem that has concerned many behavioral ecologists: How many displaying males should a female sample before choosing a mate and what criterion should she invoke to end the sampling? Early work (Janetos 1980; Janetos and Cole 1981) assumed that sampling imposed no costs on females and concluded that females should pick the "best" of the *N* males sampled. When costs are included, the best-of-*N* rule is not necessarily the optimal policy (Parker 1983; Real 1990, 1991). Recent studies of mate choice by females in several fish species have shown that female decisions hinge sensitively on estimates of male trait distribution acquired during prior sampling (Bakker and Milinski 1991; Downhower and Lank 1994; Miliniski and Bakker 1992). Dombrovsky and Perrin (1994) even propose an optimal rule for the proportion of mate search time that should be devoted to establishing prior distributions of male trait values versus final sampling to select a mate according to those distributions. Wiley (1994) invokes signal detection theory to explain threshold decisions in frogs about whether potential mates are conspecifics or not. His work also considers access to prior information about distributions of male display parameters. The majority of examples provide strong evidence that females combine prior sampling, relative payoffs, and decision thresholds to select mates, as predicted by the theory in this chapter.

Male pied flycatchers often mate deceptively with more than one female; usually the second female to mate with a male receives little paternal care and has lower consequent breeding success (Figure 14.15). There has been considerable debate about whether these mistakes in mate choice by females are due to intentional deception by males, or are instead the consequence of economics that do not favor the acquisition of perfect information by females (Dale and Slagsvold 1994). An analysis of this problem using signal detection theory can be found in Getty (1995, 1996).

In animals, classification of conspecifics according to kinship, group membership, or age classes is common. Reeve (1989) has used signal detec-

Figure 14.15 Errors in mate choice by female pied flycatchers. A mated pair of pied flycatchers here inspect a nesting hole. Males of this species will often sing at two different nest sites separated by a considerable distance. A female mating with the male at one nest site may not know whether the male is or is not already mated to another female elsewhere. Since the male will help only one female raise her brood, there is a substantial cost to the female who is later abandoned by the bigamous male. There is much debate over why females do not expend more effort to find out whether males are or are not already mated. Optimal decision economics have been invoked as one explanation for observed levels of bigamy in the wild. (Photo by A. R. Hamblin, FLPA-Images of Nature.)

tion theory to predict the threshold signal differences that various species might invoke to make such classifications. Although he focuses primarily on guard behavior at the entrances to the nests of social insects, his models have much wider application. Signal detection is also the basis of Beecher's (1991) analysis of the differences between swallow species in their abilities to discriminate between their own and other young (see also Box 14.3). Both studies confirm the general tenet of this chapter, that perfect classification and discrimination are rarely favored economically. Thus most receivers will make some mistakes, and the level and type of acceptable error will be closely linked to prior knowledge and the relative payoffs of getting it right versus making mistakes. How the optimal accuracies, (or put another way, the optimal error rates) are linked to the historical, environmental, and functional contexts of communication will be our focus in the remainder of Part II.

Box 14.3 Swallows and Signal Detection Theory

BARN AND ROUGH-WINGED SWALLOWS (*Hirundo rustica* and *Stelgidopteryx serripennis*) nest as single pairs, or at most in loose clusters, and young from different nests are not grouped after fledging. Cliff and bank swallows (*Hirundo pyrrhonota* and *Riparia riparia*) nest in large colonies, and after fledging, young are moved into creches consisting of hundreds of individuals. In all species, youngsters beg from parents using vocalizations. Cliff and bank swallows clearly have a need to identify their own young among the hundreds of hungry fledglings; barn and rough-winged swallows rarely need to do so. Beecher and his associates (Beecher 1989, 1990, 1991; Medvin et al. 1993) have shown that cliff and bank swallows are better able to identify their young than are barn or rough-winged swallows. What are the differences in the communication systems between the colonial versus non-colonial species?

All measurable features of swallow offspring calls vary continuously. Parents must thus establish cutoffs for each feature or for some compound feature before deciding whether to feed or ignore a given fledgling. According to signal detection theory, a parent should feed a nestling if the likelihood ratio that the call was made by its own rather than another offspring is greater than the product of the prior odds ratio and the relative payoffs. Given that there are N other nestlings present, and no alternative sources of information except the call, the prior odds ratio will be

$$\left(1-\frac{1}{N}\right)/\frac{1}{N}=N-1$$

If it costs $-C$ to feed any offspring, and the benefit of feeding your own is B, then the payoffs will be $B-C$ for a hit, $-C$ for a false alarm, and 0 for either a miss or a correct rejection. The relevant payoff ratio is then

$$\frac{0-(-C)}{(B-C)-0}=\frac{C}{B-C}$$

Beecher (1990) then argues that the likelihood ratio that the offspring is the parent's after receiving a signal will be σ_t/σ_w, where σ_t^2 is the overall variance of this feature including both between and within individual variation, and σ_w^2 is the average within-individual variance. This makes sense because the larger the ratio σ_t/σ_w, the larger the differences between individuals and the easier they will be to tell apart. Thus a parent should feed a nestling if

$$\frac{\sigma_t}{\sigma_w}\geq(N-1)\frac{C}{(B-C)}$$

If we take the $\log_2$ of each side, we then get the condition for feeding that

$$\log_2\frac{\sigma_t}{\sigma_w}\geq\log_2(N-1)+\log_2 C-\log_2(B-C)$$

As we saw on pages 394–395, the average information provided by a continuous signal feature whose probability distribution has a standard deviation σ is $\overline{H}_T=\log_2 c\sigma$, where c is a constant depending upon distribution shape. Beecher (1989) noted that if the within-individual distribution of a swallow call feature has similar shape to that for the entire population, then we can rewrite

$$\log_2 \frac{\sigma_t}{\sigma_w} = \log_2 \frac{c\sigma_t}{c\sigma_w} = \log_2 c\sigma_t - \log_2 c\sigma_w = \overline{H}_T(T) - \overline{H}_T(W) = \overline{H}_T(S)$$

$\overline{H}_T(T)$ reflects the total variation in the signal feature; $\overline{H}_T(W)$ is that portion that is due to within-fledgling variation. The difference, $\overline{H}_T(S)$, thus reflects the amount of signal variation due to differences between the calling young on which the parents must rely on for identification. The condition for feeding a fledgling then becomes

$$\overline{H}_T(S) \geq \log_2(N-1) + \log_2 C - \log_2(B-C)$$

This says that the minimum average information that must be made available by between-individual differences in fledgling calls will be higher when there are more fledglings around (N) or when costs of feeding (C) are higher; it will decrease when the net benefits of correct feeding ($B - C$) are high. If benefits and costs were similar for the four swallow species, we might expect $\overline{H}_T(S)$ to be higher for the colonial ones if only because N is higher for creches than for solitary nesters.

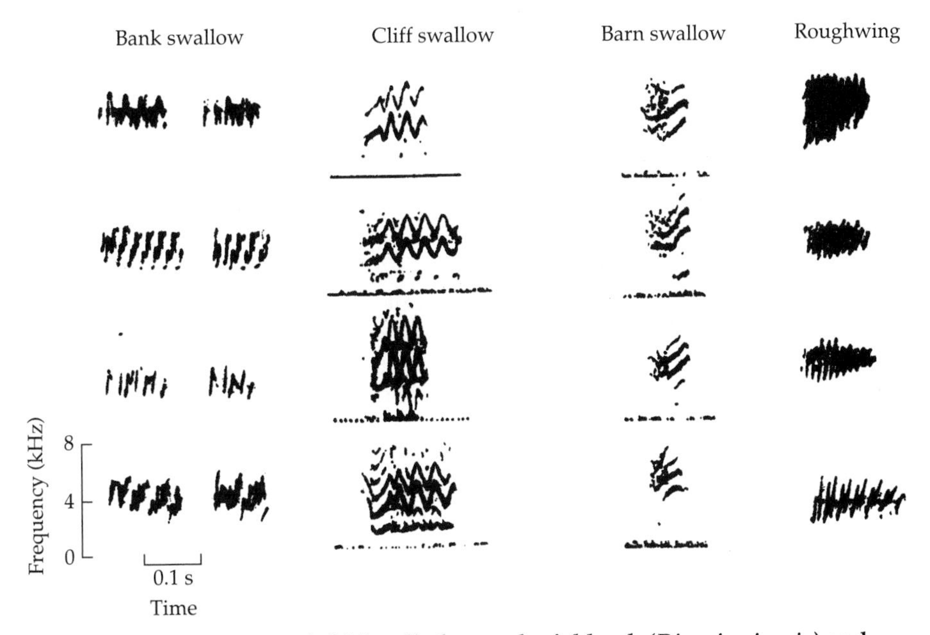

Figure A Spectrograms of chick calls from colonial bank (*Riparia riparia*) and cliff (*Hirundo pyrrhonota*) swallows versus solitary-nesting barn (*Hirundo rustica*) and rough-winged (*Stelgidopteryx serripennis*) swallows. Each cluster of spectrograms shows calls from four chicks of each species. Note higher variation between individuals in colonial nesting species. (From Beecher 1990.)

Between-individual variation in fledgling calls is higher in colonial species (see Figure A). By using values of σ_t and σ_w for a variety of fledgling call features to compute $\overline{H}_T(S)$, Beecher (1990) estimates that colonial cliff and bank swallow begging calls pro-

Box 14.3 (continued)

vide up to 8.74 and 10.2 bits of information, respectively, whereas non-colonial barn and rough-winged swallow signals provide only 4.57 and 3.83 bits. Because both barn and cliff swallows raised in the lab can learn to discriminate between cliff swallow fledgling calls better than they can calls of young barn swallows, Beecher and his associates think that selection for better discrimination in colonial nesters has favored a greater diversity of sender signals rather than refined receiver sensitivities to existing fledgling call differences.

SUMMARY

1. The value of information to either party in a communication exchange depends upon which action the receiver decides to perform. The optimal strategy for a receiver is to compare current estimates that alternative conditions are true to **threshold values**. If an estimate exceeds its threshold, the receiver should perform one action; if not, the receiver should perform the other action. The critical threshold probability depends upon the payoffs of correct actions relative to incorrect ones.

2. For discrete signal sets, the mapping of alternative signals onto alternative conditions must have a minimal accuracy before it benefits a receiver to pay attention to signals at all. The minimal accuracy is determined by the relative payoff of the receiver making a correct decision and by the prior probabilities of alternative conditions. Whether the minimum accuracy is achieved or not depends upon how well the sender matches signals to conditions, how much distortion is added to signals during propagation, and how carefully the receiver can classify perceived signals. Even with perfect accuracy at each stage in the communication sequence, there is a maximum benefit of information that can be achieved in any communication system. The maximum value of information depends upon this benefit minus the costs of signaling and reception. Increasing the accuracy of communication increases the **amount of information** provided; it may or may not increase the **value of information**. This latter value depends upon whether a receiver should pay attention to signals, and whether benefits are increased more than costs.

3. Alternative signals from a continuous set can usually be arrayed along a common signal axis. An optimal receiver will set a threshold value on this axis and respond one way for signals that are below the threshold, and another way for those above it. The optimal threshold is determined by the prior probabilities, the relative payoffs of making a correct decision, and the accuracy of the communication system. **Signal detection theory** provides methods for separating the effects of prior probabilities and payoffs (using a measure called **bias**), from the effects of accuracy (using a measure called **sensitivity** and denoted by d'). The technique used is to

plot the fraction of time that a receiver correctly identifies a given condition from signals (the **hit rate**) against the fraction of time it decides that same condition is true when it is not (the **false-alarm rate**). A high sensitivity means that a receiver makes few false alarms for a given hit rate. A graph of hit rates versus false-alarm rates for receivers with similar sensitivities but different payoffs or prior probabilities creates a curved line called an **ROC plot**. Sensitivity is measured as the degree of curvature of this plot away from the diagonal of positive slope. Receivers with high bias are conservative, set a high threshold value, and consequently have both low hit rates and low false-alarm rates. As with discrete signal sets, a receiver of signals from a continuous set must have a high enough sensitivity before it pays to attend to signals at all. Above this minimum, increased sensitivity can lead to higher amounts and values of information provided that costs do not accelerate even more quickly.

4. Signal detection theory can be applied to a wide variety of sensory decision problems including **detection** of signals in noise, **discrimination** between alternative signals, and **classification** of different signals. In each case, there is a maximum sensitivity that can be achieved for any given signal set. A receiver that achieves this maximal sensitivity is called an **ideal receiver** because no instrument or animal can do better. If follows that even at the highest achievable levels of accuracy, receivers will make errors. Errors are thus an inescapable part of even optimal decision making.
5. Accuracy and sensitivity are not free; they can be increased only at increased costs to senders, receivers, or both. Thus, even if lower errors are achievable, the extra costs may exceed the benefits. The optimum investment is usually an intermediate one in which receivers still make some errors. The same is true if more information is obtained by successive sampling. Here too, benefits of increased sampling are eventually outstripped by increasing costs. The optimum is usually an intermediate number of samples at which the receiver has still not achieved perfect information.

FURTHER READING

The classic book by Raiffa (1968) on optimal decision making is lucid, entertaining, and one of the best places to start on this topic. A quantitative general approach to the value of information in ecology and behavior can be found in Stephens (1989). Introductory sources for signal detection theory include Chapter 6 in Coombs et al. (1970), and the books by Egan (1975) and Macmillan and Creelman (1991). Examples of the invocation of signal detection theory for specific evolutionary problems can be found in Beecher (1990), Getty (1995, 1996), Getty et al. (1987), Reeve (1989), and Wiley (1994).

Chapter 15

Coding

FOR COMMUNICATION TO WORK, THERE MUST BE A **CODE** that is to some degree shared by sender and receiver. Encoding rules specify which signals senders should emit when a specific condition is true; decoding rules are invoked by a receiver to update its estimates as to which condition is true after perceiving a given signal. The rules for the two parties do not have to be identical. For example, receivers have to incorporate effects of distortion during propagation in their coding rules. But, the two sets of rules cannot be entirely independent; if they were, then no information could be exchanged.

In this chapter, we define more precisely what we mean by coding and use this definition to compare the accuracies of alternative coding schemes. We then discuss how choice of a coding scheme by one party constrains the options for the other, and how each party might acquire its coding rules. Different functions of communication often require different coding schemes, and this can lead to different communication costs for either or both parties. Finally, we discuss the evolutionary interactions between coding schemes and the sizes of signal repertoires.

CODING ACCURACY

What Is Coding?

Before we can rate the accuracies of alternative coding schemes, we need to provide a more quantitative description of coding in general. Consider a receiver that wants to know which of several alternative conditions is true. It could use prior probabilities, direct assessment, available cues, and perhaps amplifiers to make the judgment on its own. Were a sender to provide relevant signals, the receiver could also incorporate these into its assessment and decision. Suppose the sender can provide some information about N_c alternative conditions and has access to N_s alternative discrete signals to do so. If there were no correlation between the signal that a sender emits and the current condition, then the set of signals would provide no useful information to the receiver. Encoding rules generate stable correlations between each condition and each signal; the existence of such correlations is a necessary condition for communication. The relevant correlation between signal S_j and condition C_i can be characterized as a conditional probability, $P(S_j \mid C_i)$. This is the probability that signal S_j will be produced when condition C_i is true. The results of the entire set of coding rules can be summarized as an $N_s \times N_c$ matrix containing the conditional probabilities for all combinations of signal and condition (Figure 15.1). Note that the encoding procedures used by senders need not incorporate probabilities explicitly. Instead, senders may invoke simple rules of thumb or other decision protocols that generate the conditional relationships.

Decoding rules assume correlations between perceived signals and ambient conditions. Suppose that a receiver is concerned with N'_c conditions and can distinguish between N'_s different discrete signals. The ensemble of decoding rules can then be summarized as an $N'_s \times N'_c$ matrix also containing terms of the form $P(S_j \mid C_i)$. These are the conditional probabilities that an S_j signal

Signal	Condition: C_1	C_2	C_3	C_4
S_1	$P(S_1 \mid C_1)$	$P(S_1 \mid C_2)$	$P(S_1 \mid C_3)$	$P(S_1 \mid C_4)$
S_2	$P(S_2 \mid C_1)$	$P(S_2 \mid C_2)$	$P(S_2 \mid C_3)$	$P(S_2 \mid C_4)$
S_3	$P(S_3 \mid C_1)$	$P(S_3 \mid C_2)$	$P(S_3 \mid C_3)$	$P(S_3 \mid C_4)$
S_4	$P(S_4 \mid C_1)$	$P(S_4 \mid C_2)$	$P(S_4 \mid C_3)$	$P(S_4 \mid C_4)$
S_5	$P(S_5 \mid C_1)$	$P(S_5 \mid C_2)$	$P(S_5 \mid C_3)$	$P(S_5 \mid C_4)$

Figure 15.1 Typical coding matrix. Coding schemes can be summarized as matrices containing the conditional probabilities $P(S_j \mid C_i)$ that a given signal S_j will be given by the sender or perceived by the receiver when condition C_i is true. A sender's coding matrix reflects the rules by which it maps signal production onto its perception of which conditions are true. A receiver's matrix reflects the accuracy of sender coding, the degree to which signal and condition associations are altered during propagation, and the accuracy of receiver perceptions as to which signal was received and which condition it most likely represents.

will be perceived by the receiver when condition C_i is true. (These probabilities were denoted by Q_{ji} for discrete signals in Chapter 14.) Note that this overall probability depends upon the accuracy of sender coding, the amount of distortion that signals suffer during propagation, and any detection and perceptual errors made by the receiver. A receiver then combines perception of a given signal S_j with the relevant values in the matrix and any prior sources of information to improve the estimates that each alternative condition C_i is true. That is, the receiver seeks an updated value of the conditional probability $P(C_i \mid S_j)$. Note that these latter values are the results of an updating process that invokes the decoding rules; they should not be confused with the decoding scheme itself.

Perfect versus Imperfect Coding

The coding rules for both sender and receiver can thus be summarized as matrices of conditional probabilities that a given signal will be produced or perceived when a given condition is true. The distribution of values within the cells of each matrix determines how much information the associated coding scheme can provide. Consider two extreme cases in which N_s possible signals are mapped onto the same number of conditions (Figure 15.2). In the first case, the sender always gives a signal, and all signals are given with equal probability no matter which condition is true. Quantitatively, all cells in the matrix will have a probability of $1/N_s$. Clearly no information is available from such a rule. This is the **noncoding** case. At the other extreme, consider a **perfect code** in which each condition elicits one and only one signal, and no two conditions elicit the same signal. Each row and each column in the corresponding matrix will have one cell with a value of 1.0, and all other cells in the same row and column will have values of 0. Such a 1:1 mapping of signals onto conditions provides the maximum possible amount of information within the constraints set by the relative frequencies of the alternative conditions (see pages 393–394).

There are two ways in which a coding scheme might be imperfect. When each condition is mapped onto only one signal, we say that the system is **specific** (Green and Marler 1979). In the corresponding matrix, each column will have only one nonzero entry. Similarly, when each signal is elicited by only one condition, we can say that the coding system is **unique**. Each row of the corresponding matrix will have only one nonzero entry. Coding schemes can be imperfect because they have reduced specificity, reduced uniqueness, or both. The two types of deviations differ in how they affect the information provided by signals. As shown in Figure 15.2, reductions in uniqueness reduce the average information that can be provided by a set of signals. This reduction occurs because the same signal may be emitted for more than one condition. The reduction in transferred information due to reduced uniqueness is called the **equivocation** of the signal system (see pages 400–401). Reductions in specificity need not reduce the amount of information provided. Although several signals may be elicited for a given condition, the receipt of any of them indicates which condition is true. However, the redundancy of using several signals for the same condition often requires the recruitment of

Coding rule		Coding summary $\{P(S_j \mid C_i)\}$ C_1	C_2	C_3	Bayesian updates $\{P(C_i \mid S_j)\}$ C_1	C_2	C_3	Equivocation (bits)	$\bar{H}_T$ (bits)
No coding	S_1	0.33	0.33	0.33	0.33	0.33	0.33	1.585	0.000
	S_2	0.33	0.33	0.33	0.33	0.33	0.33		
	S_3	0.33	0.33	0.33	0.33	0.33	0.33		
Perfect coding	S_1	1.0	0	0	1.0	0	0	0.000	1.585
	S_2	0	1.0	0	0	1.0	0		
	S_3	0	0	1.0	0	0	1.0		
Specific but not unique coding	S_1	1.0	1.0	0	0.5	0.5	0	0.667	0.919
	S_2	0	0	1.0	0	0	1.0		
	S_3	0	0	0	0	0	0		
Unique but not specific coding	S_1	0.5	0	0	1.0	0	0	0.000	1.585
	S_2	0	1.0	0	0	1.0	0		
	S_3	0	0	1.0	0	0	1.0		
	S_4	0.5	0	0	1.0	0	0		
Neither unique nor specific coding	S_1	0.8	0.1	0	0.89	0.11	0	0.863	0.722
	S_2	0.1	0.7	0.1	0.11	0.78	0.11		
	S_3	0.1	0.2	0.9	0.08	0.17	0.75		

new signals to guarantee that all conditions can be signaled. If the number of available signals is less than or equal to the number of conditions that must be indicated, then a reduction in specificity can lead to a concomitant decrease in uniqueness. Imperfections in coding thus lead to smaller amounts of transferred information or to a compensating addition of new signals to the repertoire. An animal with an imperfect code is thus faced with three evolutionary alternatives: (1) accept the lower amount of information provided, (2) pay the costs of making the current coding scheme more perfect, or (3) pay the costs of adding new signals.

Sender versus Receiver Coding Schemes

Sender codes may be imperfect because the sender errs in the assessment of which condition is true or because the sender is sloppy in assigning signals to

◀ **Figure 15.2 Perfect versus imperfect codes.** The left column indicates various rules for encoding three equiprobable conditions [i.e. $P(C_i) = 0.33$]. The second column shows sample matrix of $P(S_j \mid C_i)$ values for that coding rule given each signal S_j and condition C_i. The third column indicates Bayesian estimates by receiver, $P(C_i \mid S_j)$, that C_i is true using this coding scheme and perceived signal S_j. The fourth column gives the equivocation (in bits) for this coding system. This is the average uncertainly remaining after signal reception and equals $-\Sigma[P(C_i)P(C_i \mid S_j)\log_2 P(S_j \mid C_i)]$ across all i and j. The fifth column gives the amount of information, $\bar{H}_T$, actually transferred by each coding rule. This can be compared to the maximum possible which is $-\Sigma[P(C_i)\log_2 P(C_i)] = 1.585$ bits. The first row shows random emission of signals regardless of condition. This **noncoding** case has maximal equivocation and provides no net information. The second row shows **perfect coding** that results in no equivocation and transfer of the maximum information possible. The third row shows a **specific** but not **unique** scheme. Note that there is significant equivocation, reduced information transfer, and failure to use one of the available signals. The fourth row shows a unique but not specific assignment. There is no equivocation and thus maximum information transfer, but to ensure that all conditions were covered, a fourth signal had to be recruited. The last row shows a more realistic example in which the coding scheme is neither unique nor specific, all signals are used and no new ones recruited, there is some equivocation, and thus less than the maximum information is transferred.

conditions. The first consequence is usually a loss of specificity because any given condition is now indicated, although perhaps infrequently, by more than one signal. As we saw earlier, this can result in a simultaneous loss of uniqueness unless new signals are added. Adding new signals is not always easy (as we discuss in the next chapter), and may be costly. The consequence is that any error usually causes the sender codes to be neither perfectly specific nor perfectly unique.

Receiver codes will nearly always deviate from perfect coding. This deviation occurs because the maximal $P(S_j \mid C_i)$ values in the relevant matrices are often reduced below 1.0 by sender errors and propagation distortions before they even reach the receiver. Receiver errors will usually exacerbate the sender and propagation errors. We noted on page 428 that the overall conditional probabilities linking perceived signals to conditions can be no greater than the smallest conditional probability in the sender-propagation-receiver chain. The high likelihood that at least one of these links is less than 1.0, particularly for long-distance signals, means that most cell values in a receiver coding matrix will be less than 1.0 and thus receiver codes will be imperfect.

It also follows that sender and receiver codes may not be identical. For example, suppose a sender achieves perfect coding in its matching of signals to conditions. If there is any signal distortion during propagation or any receiver error in identifying signals, the relevant conditional probability matrix for receivers will have to be different from that for senders. This difference arises because each term in the receiver's coding matrix depends not only on sender accuracy (which here is postulated to be perfect), but also on propagation and receiver accuracy (which is here imperfect). Should the receiver invoke the same code as a perfect sender when there is significant propagation error, it would be more likely to accept a condition as true when it is not (e.g., respond to a false alarm). If it uses a more conservative code than necessary, it will fail to respond when it should (a miss). It follows that the optimal code for receivers will usually be somewhat different from that for senders. On the other hand, the two cannot be independent. One constraint is that the probabilities within the receiver coding matrix must always depend in part on the probabilities in the sender matrix. There are other possible links between sender and receiver codes. For example, the degree of accuracy adopted by receivers may act as a selective force on senders to either upgrade or decrease the accuracy of their own codes. As we shall see below, if a sender acquires its encoding rules by learning, the receiver must also evolve an ability to learn its code. Such feedbacks reflect the fact that receiver and sender coding schemes are usually coevolved traits; we cannot consider each party in isolation.

TYPES OF CODING SCHEMES

Sources of Signal Variation

The options available for coding depend in part on how many signals are available and how much they differ from each other. Let us call any continuous emission of signal energy a **signal element**. Even the simplest elements will have several parameters that could be varied to encode information. For a sound element, the minimum variable parameters include the element's mean frequency, its amplitude, and its duration. Puffs of olfactant could vary in chemical composition, overall concentration, and relative ratios of components; electrical signals could differ in amplitude, duration, and waveform. By selectively associating different variants of any of these parameters with different conditions, a sender can generate a signal code. Additional variation can be achieved by adding temporal modulations of intrinsic variable parameters within a signal element. Such modulations are common in visual, acoustic, and electric signals, but are more difficult to add to olfactory ones (see page 280). If the parameters for a signal element can vary independently, the number of combinations, and thus the number of potential signals can be quite large. In practice, animals rarely use all combinations. Instead, certain clusters of combinations are retained, and combinations that are intermediate between the clusters are avoided. Retained clusters are often those that are most easily distinguished from each other, minimally distorted during propagation, or cheapest to produce (Barlow 1977). In some cases, such as frequency and amplitude modulations of bird songs, there may be physiological linkages between signal parameters that also limit the possible combinations.

Further diversity can be generated if a succession of signal elements is treated as a chain. In the simplest case, the rate at which successive elements are emitted in a chain can be correlated with conditions and can thus provide additional information. Alternatively, different combinations of different elements in the chain, and even different sequence orders within combinations, can each be associated with different conditions to expand the useful repertoire. Sequences of chains can in turn be linked to construct even larger units, resulting in a hierarchy of signal variants. Many bird song repertoires show this type of hierarchical structure (Figure 15.3). As with variation within signal elements, animals often have rules for forming such chains (called **syntax**) that restrict the allowable combinations and sequences. (For a bird song example, see Hailman et al. 1987.) Restrictive syntax may evolve for the same reasons that only certain signal element variations are used as signals, but also because syntactical rules allow receivers to extract information more efficiently from chains.

Rules that limit the number of salient signal variants can be imposed either by the sender or the receiver or both. If senders impose strong constraints on the production of signal variants (whether at the element or chain level), and the allowed variants are very different from each other, we say that the signals are highly **stereotyped**. That is, only a few very different signals are produced, and each is given with very little variation in signal parameters or chain structure (Figure 15.4). The highest levels of sender stereotypy may be expensive and difficult to achieve (Zahavi 1980); where high levels of stereo-

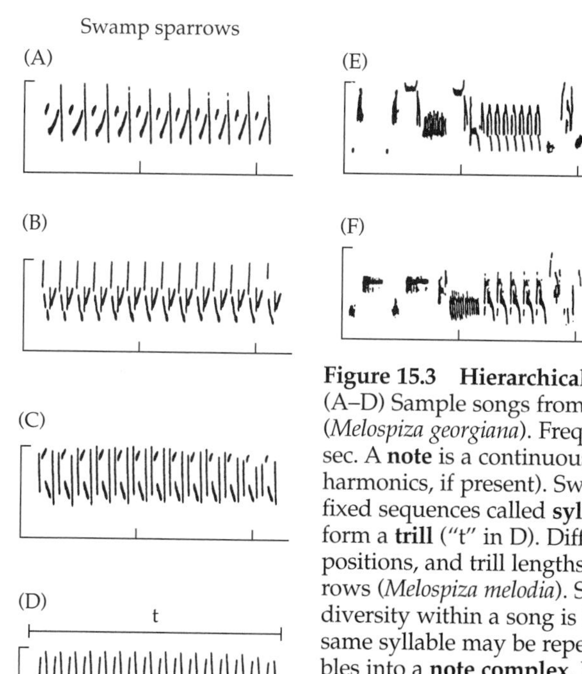

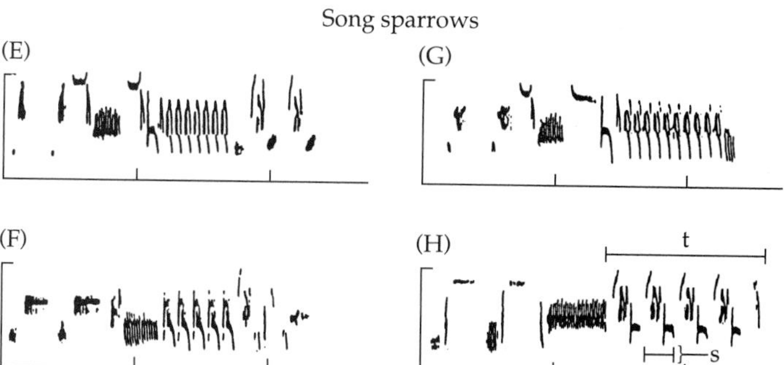

Figure 15.3 Hierarchical structure in variants of sparrow songs. (A–D) Sample songs from four different swamp sparrow males (*Melospiza georgiana*). Frequency axis is 0–8 kHz and time marker is 1 sec. A **note** is a continuous trace on the spectrogram (with associated harmonics, if present). Swamp sparrow songs use 1–4 note types in fixed sequences called **syllables**, ("s" in D), that are in turn repeated to form a **trill** ("t" in D). Different males use different notes, syllable compositions, and trill lengths. (E–H) Sample songs from two song sparrows (*Melospiza melodia*). Syllables again consist of 1–4 notes, but note diversity within a song is much higher than in swamp sparrows. The same syllable may be repeated as a trill, or combined with other syllables into a **note complex**. Each song consists of 3–4 successive trills and note complexes. Males typically sing 10–15 **songtypes** where each type has it own sequence of trills and note complexes. Songtypes are shared by neighboring males with only minor variations between individuals in note shape and number of notes per syllable. Songs E and F are different song types from one male, and G and H are the corresponding song types from a neighbor. (From Marler and Peters 1988.)

typy are observed, there are presumably benefits that compensate senders for the higher costs. In contrast, many senders show considerable variation in signal form or chain composition even when signaling the same alternative conditions. In many of these systems, it is the receivers who impose the constraining rules; they categorize signals using certain parameters, and they ignore the remaining parameters whether variable or not. If stereotypy is costly, then there is no selective force on senders to limit variation in the unused signal parameters. This fact makes it more difficult to unravel the coding schemes in species without stereotyped signals. Since not all of the variation in produced signals need be used to convey information, one must examine both sender production patterns and receiver responses to get the full picture. Some methods by which this characterization can be done are discussed in Boxes 15.1 and 15.2.

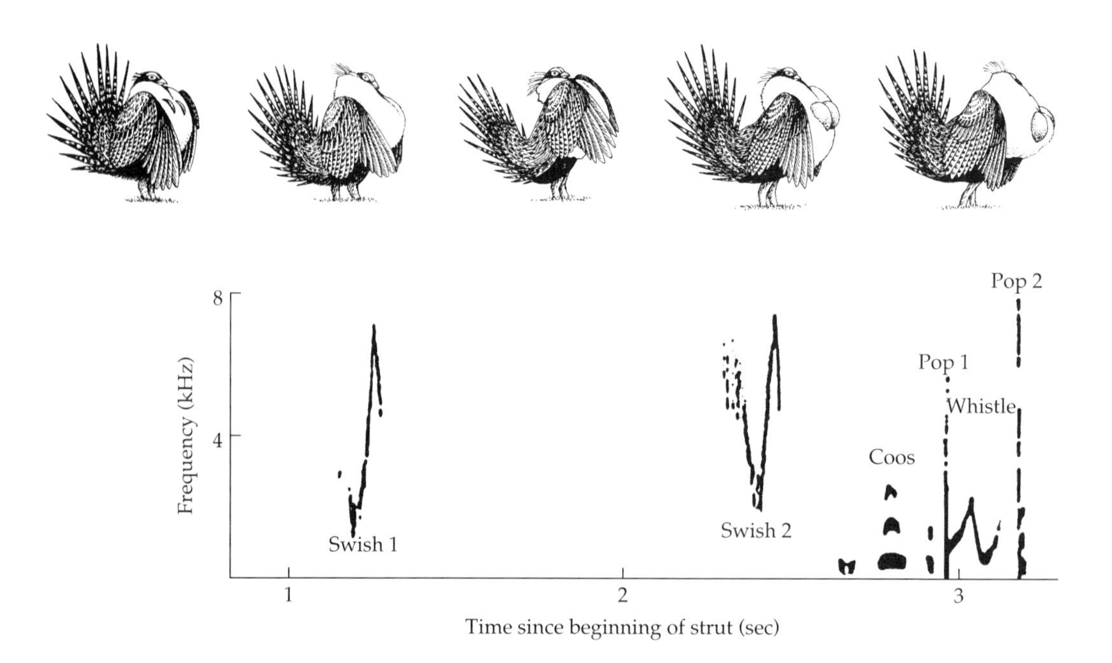

Figure 15.4 Stereotyped strut display of male sage grouse. Display is performed by males on communal arenas (leks) to which females come to select a mate. This is the only display males perform to attract and court females. (Top) Sequence of male movements during strut. Sounds are created by rubbing wings against body (swish), creating low-frequency coos, popping sacs on chest (pops), and whistling between the pops. (Bottom) Spectrogram of sounds made concurrently with strut. For a variety of display measures such as the total duration of the strut, the time interval between such events as swish 1 and pop 1, pop 1 and pop 2, pop 1 and the peak frequency of the whistle, successive struts by the same male show variation of only 1–2%, and even those by different males show only 2–4% variation (Gibson and Bradbury 1985; Wiley 1973; Young et al. 1994). The display is thus highly stereotyped. (Spectrogram after Hjorth 1970.)

The number of available signal variants in any species is thus the result of several conflicting selective pressures. These pressures presumably balance over evolutionary time and result in an overall repertoire of a given size. How might senders and receivers go about assigning these signals to alternative conditions? Receivers will be concerned with a number of different questions, each of which has several alternative answers. The entire set of signals in the repertoire must be divided among the various questions and, within each question, among the alternative answers. We call the alternative signals assigned to a given question a **signal set** (see page 358). We shall take up the options for coding within a given signal set first. Then we shall consider the coding options when dealing with several questions in a **compound signal** (see page 363).

Coding Options within a Signal Set

Senders map emitted signals within a signal set onto alternative conditions. During propagation the emitted signals are mapped onto received signals; then receivers map the received signals onto a set of perceived alternatives. At each stage (e.g., condition, emitted signal, propagation, and perceived signal), the set of alternatives may be **discrete** or **continuous**. Discrete lists may have many or a few alternatives; continuous sets may have narrow or broad ranges (Figure 15.5). It is also possible to have several discrete categories within which there is continuous variation (here called **discrete/continuous sets**). Conversely, where several discrete signals are treated as a chain, the rate at which signals in the chain are emitted can be varied continuously, thus turning a set of individually discrete alternatives into a new set of continuously variable chains. If several of these hierarchical levels are present, variation

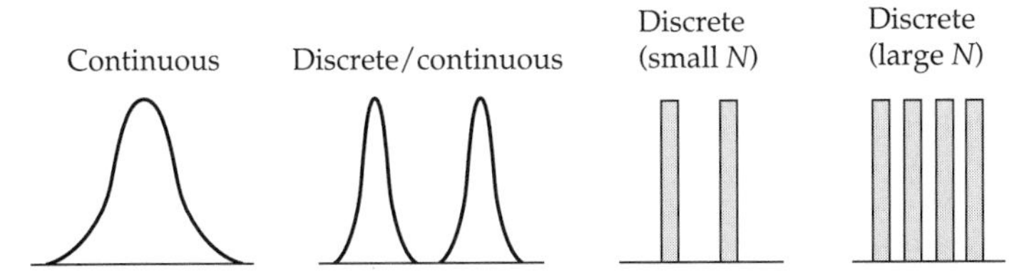

Figure 15.5 Typical sets of alternatives for conditions, emitted signals, or perceived signals. For conditions, the horizontal axis is a condition parameter, and the vertical axis is the frequency of occurrence of each value. Conditions might be distributed continuously with a modal value that is most common (e.g., for parasite loads of potential mates or for the size of an opponent); as several separate bands of encountered values with continuous variation within each band (e.g., readiness to flee, with a dominant animal having low but variable levels and a subordinate animal having high but variable levels); a few discrete alternates (e.g., male versus female); or many discrete alternatives (e.g., predator nearby is hawk, snake, leopard, or mammal). For signals, the horizontal axis shows the critical signal parameter that varies with condition, and the vertical axis represents frequency of using that signal form. In the discrete/continuous case, adjacent distributions might overlap or be completely distinct.

Box 15.1 Characterizing Sender Coding Schemes

HOW CAN WE CHARACTERIZE SENDER CODES? Where both signal variants and alternative conditions fall into discrete categories, we could sample a large number of signal emissions and create a **contingency table** summarizing the number of times each signal variant (rows) was given for each alternative condition (columns). There are standard statistical tests that can evaluate whether signal variants are significantly associated with conditions in such a table (see Everitt 1977 or Zar 1984). A significant result indicates that assignment is not random and that a code exists (e.g., Martins 1993b). We could then divide each cell in the original table by its column total to create a new table summarizing the estimated conditional probabilities that each alternative signal is given for a particular condition. This is the type of code summary described in the text.

One frequently encountered problem is that sender signals show continuous enough variation that it is not obvious how to categorize them. This problem is more likely to occur if one makes a number of different measurements on the signals and what appear to be category boundaries based on one measure conflict with boundaries suggested by another measure. Ideally, one would like to use all the measured information, but how can we do this if suggested boundaries are not congruent? The same problem occurs, although not as often, when one is categorizing alternative conditions. There are a number of statistical methods for identifying clusters of samples in a complex data set (Everitt and Dunn 1992). The most common approach is to create an **association matrix** comparing each observed signal to every other signal in the set. If one observed N signals, this would create an $N \times N$ matrix whose cell values were measures of the similarity between each pair of signal variants (a **similarity matrix**) or of the differences between each pair (a **distance matrix**). There are many alternative indices of sample similarity or distance that we could use to create association matrices (Legendre and Legendre 1983). For example, we could index the similarity of two bird songs or two pheromone samples by computing, respectively, the fraction of syllable types or chemical constituents that were common to each pair of samples. Alternatively, one could compute the linear (Euclidean) distance between two samples when each is plotted in a multidimensional space in which the axes represent the occurrence rates of each syllable type (for songs) or the concentrations of each chemical component (for pheromones). Multidimensional distances can most easily be interpreted if each measure is **standardized** before the signals are plotted. (For example, one could standardize each measure by subtracting the mean from each observed value and dividing the remainders by the standard deviation.) For sounds, there are computer algorithms available for cross-correlating different calls in the time or frequency domains and then generating indices of similarity directly (Clark et al. 1987). These indices can then be assembled into association matrices.

Instead of creating an association matrix comparing each signal variant to each other across all of the measures (an approach called **Q mode**), we might undertake the reverse operation and create an association matrix comparing each measure to each other measure across all of the signal variants (an approach called **R mode**). Because similar signal variants share similar measure values, measures will tend to be correlated if there are clusters of similar signals in the data set. Thus either an association matrix based on signal similarities or one based on measure similarities ought to preserve any categorical structure in the data.

Association matrices can be fed into a number of useful computer programs to group samples into categories. **Principle coordinate analysis** (for metric similarity or distance indices) and **multidimensional scaling** (for nonmetric indices) create two- or three-dimensional maps of the signal variants that maintain the relative proximities (or distances) of all pairs of samples. Both rely on association matrices based on signal contrasts (Q mode). Association matrices based on measure contrasts (R mode) are usually subjected to **principal components analysis**. This analysis also creates a two- or three-dimensional map of the signal variants in which clustering of signals can hopefully be discerned. Instead of creating maps, one can alternatively group samples using **cluster analysis**; this type of analysis always relies on a signal association matrix (Q mode). ***K*-means clustering** allocates the samples to *K* different groups given the similarity matrix. One can plot the variation explained by this allocation for different values of *K* and then use the clustering for the *K* that explains the greatest amount of variation (Hartigan 1975). **Hierarchical clustering** uses the similarity matrix to lump the most similar samples into a cluster, then to lump the most similar clusters into larger groups, and so forth. The resulting plot that links the signal clusters is called a **dendrogram**. An example of a multidimensional scaling plot is shown in Figure A, and a hierarchical clustering dendrogram for the same association matrix might appear as shown in Figure B (with different signal variants labeled *a, b, c,...,l*).

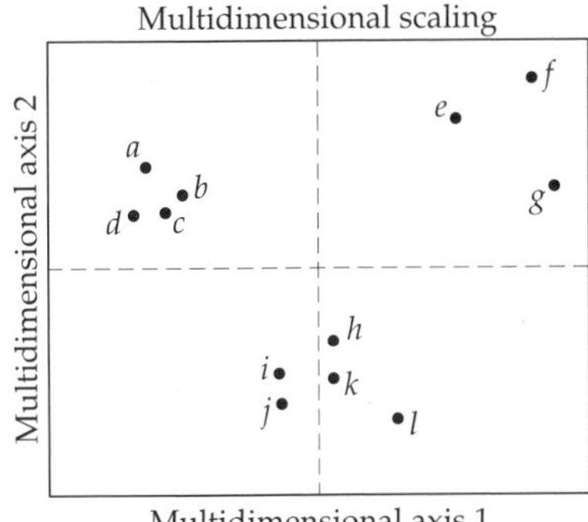

Figure A A multidimensional scaling plot showing dispersion of eight sample points along two axes. Points that are close together have similar properties.

Note that none of these methods is specifically designed to define clusters or justify them statistically. In all cases, the researcher makes a subjective classification based on the maps or cluster diagrams. Despite these limitations, these methods are widely used to define categories within an ensemble of signals (Nowicki and Nelson 1990; Podos et al. 1992; Zimmerman and Lerch 1993). It is reassuring that one usually obtains similar classifications no matter which method is used. Once signal variants are assigned to different categories, one can then go back and tally the counts of signal categories versus condition categories, to characterize the sender code.

One alternative to contingency-table analysis also relies on association matrices. We first create an association matrix for the *N* signal samples (in Q mode) using any of the

Box 15.1 (continued)

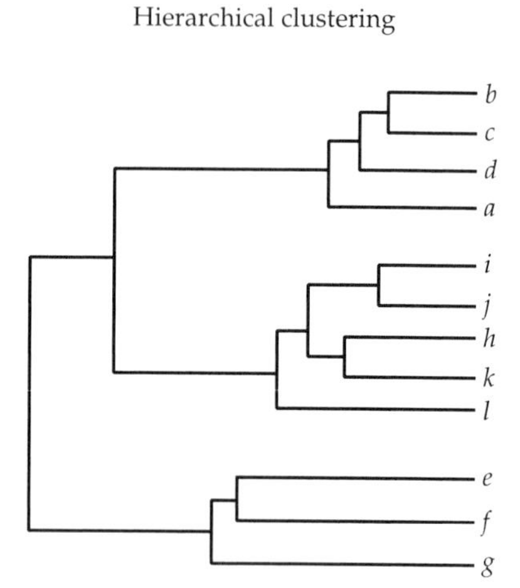

Figure B A dendrogram created by hierarchical clustering and showing the relationships of the same eight samples used in Figure A. Samples with similar properties are connected through a common link. The longer the lines, the larger the difference between samples having a common link. Note that samples clustered close together in the dendrogram are generally the same ones clustered together in the multidimensional scaling plot. The two methods thus give similar results.

possible indices noted above. This matrix characterizes the degree to which different signals share structural similarities. If alternative conditions are discrete, we can then create a second matrix in which cells contain a 1 if both signals were given in the same context and 0 if not. This matrix characterizes the degree to which different signals share contexts. The degree to which the signal-association matrix is correlated with the condition-association matrix can be examined statistically using a **Mantel test** (deVries et al. 1993; Legendre and Vaudor 1991; Mantel 1967; Schnell et al. 1985). If conditions are not discrete, one can still create an association matrix for conditions by invoking some index that uses multiple condition measures to determine similarities between conditions for each pair of samples and then one can run a Mantel test. Either approach will indicate whether there is a significant code or not; neither will generate a table from which conditional probabilities can be computed.

may shift from discrete to continuous and back again several times as one moves from lower to higher levels in the hierarchy.

As one stage is mapped onto the next, the degree of continuity and the number of alternatives in the relevant sets often changes. The ways in which sets change with mappings between stages have often been given specific names. For example, many conditions of interest to receivers are distributed continuously. Examples include the degree of hunger of a begging chick, the likelihood that a given opponent will win an escalated fight, or the state of health or parental ability of a potential mate. It would be an impossible encoding task to assign every value of a continuous range of conditions its own signal. Therefore, in many signal systems, this problem is resolved by mapping the continuous set of conditions onto a discrete set of emitted signals (Figure 15.6, scenario 1). Clearly, some information is lost by this **categoriza-**

When conditions are clearly discrete and multiple measures are taken on signals, one might alternatively use **discriminant function** or **logistic regression** analyses to relate signal variants to conditions (Everitt and Dunn 1992). The two methods do essentially the same thing but use different algorithms. In both cases, one seeks that linear combination of signal measures that best predicts the observed conditions. If the best combination predicts a significant number of cases, we can conclude that there is a code present, and we could use the relative weighting of measures in that combination to identify which properties of signals were encoding the transmitted information. If only a few signal parameters were weighted heavily, this might resolve any confusion about signal classification generated by using multiple measures. One would then use only these important signal measures to classify the signals, construct the relevant contingency table, and compute the conditional probabilities between signals and conditions (see examples in Hauser 1991; Zimmerman and Lerch 1993).

Note that a researcher's set of measures to classify signal variants may not be the same set used by senders for encoding signals. There are two ways we can err. (a) We might include measures that senders ignore, and (b) we might omit measures on which senders rely. Errors of the first kind are likely to generate more signal categories than are needed. However, there are a number of a posteriori statistical methods for contingency tables (e.g., loglinear analyses) that can be used to reduce the original table to a smaller one of significant associations. The second type of error is likely to generate a contingency table that shows no significant association of signals and conditions. If we believe that the animals are truly providing information with the signals, this type of table indicates that the researcher needs to go back and examine more signal measures or more condition measures, or both. Discriminant function and logistic regression analyses do not require a prior categorization of signals, but they may also fail to show significant correlations or they may even generate spurious relationships if the critical measures are omitted. From these considerations we can see that derivation of sender codes is never a simple process, and it often takes considerable biological insight to choose the right measures of both signals and conditions.

tion. Categorization may also occur in the mapping of detected signals onto perceived alternatives. This is known as **categorical perception** (Green and Marler 1979; Harnad 1987; Horn and Falls 1996). For long-distance communication, the distinctiveness of categorized signals may be reduced by the addition of noise or distortion during propagation. Here the receiver may do best by presuming discrete/continuous signal distributions and identifying appropriate cutoffs in signal parameters to assign signals to one or the other category (Figure 15.6, scenario 2). This is one of many situations in which the signal detection models of the last chapter would be optimal decoding schemes. Categorization by either sender or receiver tends to reduce the amount of information provided because it collapses a larger number of alternatives into fewer. It presumably reduces costs by limiting the numbers of signals that must be produced and discriminated, and it allows either party greater free-

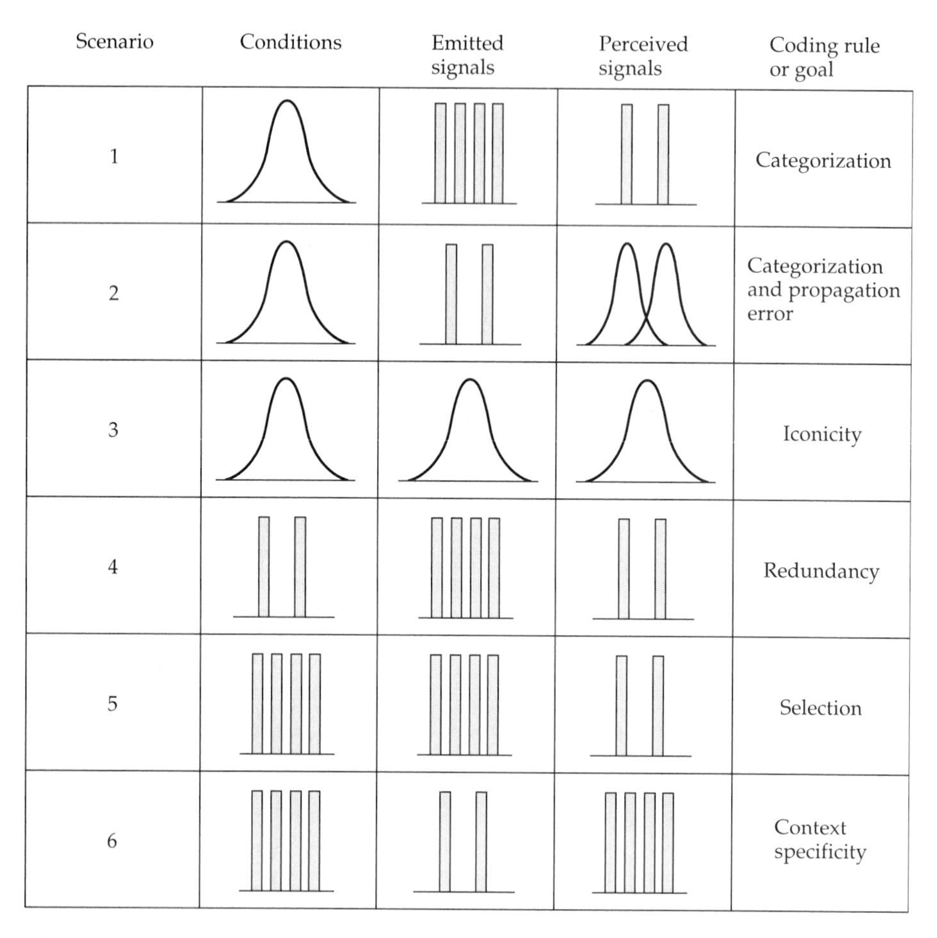

Figure 15.6 Sample coding schemes and their associated sets of alternatives. For each scenario listed in the first column, the middle column indicates set of emitted signals that senders map onto the set of alternative conditions (second column). The fourth column indicates the set of perceived signals mapped by receivers onto the signals received after propagation. (For simplicity, an intermediate column showing the mapping of emitted signals onto received signals during propagation is omitted.)

dom to select only those signals that have minimal distortion during propagation and maximal detectability.

A third method for encoding continuous conditions is the use of a general rule mapping some signal parameter onto a condition. The sender and receiver codes would then not require storage of all possible condition and signal combinations, but only storage of the rule. Suppose two opponent fish are contesting a nest site. Based on their recent fight success, the alternative actions available, and the estimate of the opponent's relative size, each fish is likely to have a different level of confidence that it would win an escalated fight. Assessing each other's confidence before fighting may be a better alternative than escalating immediately (see Chapter 21). Suppose both fish give threat displays in which the level of fin erection reflects the sender's underlying confidence of winning an escalated battle. The alternative signals thus form a continuous set that is mapped monotonically onto the continuous set of alternative confidence levels. If each fish can then measure the other's fin erection continuously and knows the coding rule, the emitted signals can be mapped onto the perceived ones. Such a communication scheme in which a general rule is used to encode and decode signals (rather than a matrix of assigned associations) is called an **iconic coding system** (Figure 15.6, scenario 3). This contrasts with the **arbitrary coding systems** we have discussed so far in which no general rules are invoked to assign signals to conditions (Green and Marler 1979). Iconic mapping could utilize either a discrete set of signals or a continuous one. Most often, iconic codes are used with continuous sets. If the iconic mapping rule causes the ranking of signals along some signal axis to be the same as the ranking of the conditions each represents along some continuous condition axis, we call the result a **graded signal system**. The example of fish fin erection is an iconic, continuous, and graded signal system.

Not all alternative conditions form continuous sets. Sex and species identity are conditions for which alternatives are clearly discrete. We have seen that propagation distortion over long distances can cause discrete emitted signals to arrive at receivers with continuous variation and overlap. If the signals are sent over short ranges or if they are carefully selected, their distortion can be minimal and discrete signals will be received and perceived as discrete. There is still room in such systems for different types of coding rules. For example, some senders may emit two different signals for the same condition, to minimize the chances of receiver error (Figure 15.6, scenario 4). This **redundancy** results in more types of emitted signals than types of conditions. Redundancy usually evolves to ensure communication despite propagation distortion, or to increase the range of possible receivers. For example, dominant male horses indicate their continued status using both vocalizations and individual odors in their feces (Rubenstein and Hack 1992). Although the two signals are largely redundant, individual identity odors provide no clues about status to intruders who will be unfamiliar with which horse produces which odor, whereas status-specific squeals can provide that information. Group members are familiar with each other's odors and can thus assess a horse's status without having to provoke the dominant male to vocalize.

Group and/or individual identity is a common condition communicated by animal signals. Although each individual sender emits only one signal, a receiver may be exposed to many possible conditions and many possible emitted signals. Determining individual identity for each encountered animal requires a rather large decoding scheme. Luckily, it is not always necessary that each individual be identified; instead, receivers often seek one animal among many and need only identify that individual. Birds or seals in a breeding colony seek their own young, and the receiver's problem is thus a binary one. (Is that individual my offspring or not?) This type of **recognition task** (Figure 15.6, scenario 5) can be associated with a large set of conditions and

Box 15.2 Characterizing Receiver Coding Schemes

THE CHARACTERIZATION OF RECEIVER CODES is as challenging as characterizing sender coding schemes (see Box 15.1). There are two steps. The first is to determine how the receiver categorizes the set of signal variants to which it is exposed (including variants generated by distortion during propagation). This determination is important because receivers may not categorize signal variants in the same way as senders. The second step is to estimate the receiver's a priori association of the signal category and alternative condition. For both steps, we are often forced to rely on receiver actions after receipt of a signal. As we saw in Chapter 14, the action that is taken by a receiver depends not only on what signal is received and the receivers coding scheme, but also on the relative payoffs of alternative actions, the prior expectations by the receiver, and the rules used to make decisions. The linkage between action and receiver code is thus a very indirect one.

The categorization of signal variants by receivers can be determined in a number of ways. One route is neurobiological; electrodes are placed at critical points along the sensory pathways, and the sensory organs are then exposed to alternative stimuli. If different variants produce exactly the same responses in the same nerves, it is unlikely that higher centers will be able to discriminate between the signals. The contrary result may not be as helpful because the ability to discriminate signals by the peripheral organs need not be used in decision making at higher levels. It is often difficult to identify brain decision centers, and this limits how useful the neurobiological approach can be by itself.

The main alternative to the use of neurobiological measures relies on differential responses by receivers to different signal variants. It really does not matter whether the test responses are natural ones (e.g., in the wild) or not (e.g., in a lab situation). The issue is whether the receiver does or does not discriminate between two signal variants. As we saw in Box 14.1, considerable manipulation of prior probabilities and relative payoffs may be required before discrimination limits can be extracted from receiver responses. The best results are usually obtained by training captive animals to respond differentially if they can distinguish signals in exchange for rewards. For example, Farabaugh and Dooling (1996) have trained budgerigars (*Melopsittacus undulatus*) to respond to paired calls of conspecifics. Where calls are perceived as similar, the birds respond strongly or with a short latency, and where they are perceived as different, response strength or latency is altered. The levels of responses are then used to create an association matrix comparing each call to each other call, and the overall categorization by the birds is displayed using **multidimensional scaling** (see Box 15.1).

A related technique can be adopted when provision of a signal elicits a receiver response. After repeated provision of the signal, most subjects' responses will gradually fade (**habituation**). If an alternative signal is then provided, subjects typically return to normal responsiveness if they consider the two signals different (**dishabituation**), but they remain at low response levels if not (Cheney and Seyfarth 1988; Eimas et al. 1971; Nelson and Marler 1989). Again, the degree to which responsiveness remains depressed can be used as an index of similarity for the two signals. A matrix comparing the habituation/dishabituation similarities of a set of signal variants can then be used to identify receiver signal categories by using any of the clustering methods outlined in Box 15.1.

Not all species can be trained or acclimatized to captivity. In some cases, one can present signals in field conditions, but they must be given in natural contexts or subjects are likely to ignore them. For many years, researchers used tape recorders for playbacks, but found it difficult to select an appropriate response call soon enough after a wild subject had provided an opening. The replacement of tape recorders by laptop computers has largely solved this problem, and interactive playback is currently one of the more powerful methods for field studies of receiver signal classification (Dabelsteen and McGregor 1996). The most difficult problem with field playbacks is ensuring that contexts are sufficiently similar across playbacks that differential responses are due to signal categorization and not to changes in the expected payoffs, prior probabilities, or decision rules invoked by the receiver. Where this condition can be met, response strength can again be invoked to generate similarity matrices and categories can be distinguished by using some clustering method.

If it is not possible to present signals in the field, one is left with recording natural exchanges between senders and receivers. Such data are even more difficult to interpret than field playbacks. One problem is that a receiver may be responding to the most recent sender signal, to the one before that, to the ordered combination of the two or to even long chains of prior signals. Extracting the correlations between signals and responses from a continuous string of exchanges requires special statistical methods. Current options include **log-linear contingency**, **lag sequential**, and **time-series** analyses. Introductions to these techniques can be found in Bradbury and Fincham (1991), Gottman (1981), and Haccou and Meelis (1992).

Once the categorization of signal variants by receivers is known, one can then try to characterize the receiver code to determine which alternative condition is usually associated with each category by the receiver. We need to be careful here to decide whether receiver codes are worth characterizing. Receivers in many species learn at least part of their codes. While we can best determine how receivers categorize signal variants by training captive animals, the very act of training may cause these animals to alter their natural codes and thus their coding may have little relevance to coding in wild populations. We thus want to be careful about *whose* codes are being characterized and what we can expect to conclude should we characterize a coding scheme.

If we assume that it is worth trying to characterize a receiver code, how might one do so? Suppose it is known that there are two alternative conditions C_1 and C_2 of relevance to the receiver, and two alternative signals S_1 and S_2. Suppose the prior probability of C_1 is P_1. As outlined on pages 421–423, we can compute a threshold probability P_c that C_1 is true, above which the optimal action of the receiver is A_1 and below which the optimal action is A_2. The value of P_c is a function of the relative payoffs of alternative actions given each condition. We can summarize the prior probabilities and relative payoffs faced by a receiver with the variable

$$k = \left(\frac{P_c}{1-P_c}\right)\left(\frac{1-P_1}{P_1}\right)$$

Let Q_1 be the conditional probability in the receiver code that signal S_1 is received when condition C_1 is true, and Q_2 be the equivalent probability that S_1 is received when C_2 is true. (The two remaining coding values we need are $(1 - Q_1)$ and $(1 - Q_2)$. We want to estimate Q_1 and Q_2 given responses of the receiver to signals S_1 and S_2. We

Box 15.2 (continued)

do this by varying k until the receiver is equally likely to respond with actions A_1 and A_2 when it is given signal S_1. We then find a second value of k, say k', at which the receiver is equally likely to respond with actions A_1 and A_2 when it is given signal S_2. If the receiver is roughly Bayesian, then we can compute

$$Q_2 = \frac{k' - 1}{k' - k}$$

and $Q_1 = kQ_2$.

It is only in the lab that one can expect to vary k values sufficiently to use the above method. The task also gets much more difficult as the numbers of alternative signals and conditions are increased. Alternative approaches involve making successive approximations to the method above. If sender codes are sufficiently perfect and signals are given over short enough distances that distortion is minimal, one can use the sender coding probabilities as a first approximation for the receiver ones. Combining some knowledge of receiver signal categorization with such sender codes improves these estimates. If we have some knowledge of the relative payoffs of different actions and the prior probabilities of different conditions, we can predict to some degree what the receiver *should* do after receiving a given signal. We can then compare observed and expected patterns of responses to see whether they are consistent with our estimates for the Q_1 values. In a two signal/two context system, we could also use sender values to predict the values of k and k' at which the receiver should equivocate over which action to take. For a Bayesian receiver, we expect

$$k \cong \frac{Q_1}{Q_2} \quad \text{and} \quad k' \cong \frac{1 - Q_1}{1 - Q_2}$$

With these values it might be easier to create or recognize a situation in which receivers are clearly ambivalent and thus to test or improve current estimates.

---

emitted signals (from the receiver's point of view), but very small discrete perceived signal sets. It is then a special case of categorization.

Finally, we have argued throughout Part II of this book that receiver decision-making invariably combines information from signals with that from nonsignal sources. Consider those situations in which both sender and receiver are likely to know the context in which signals are emitted; this is most likely for short-range signaling. A sender might then produce the same signal in a variety of contexts, but rely on the receiver to identify the relevant condition, given the context/signal combination (Figure 15.6, scenario 6). This scenario could result in considerable economy for both parties by reducing the number of different signals that have to be generated and perceived (Smith 1977, 1986). The chick-a-dee call of the black-capped chickadee (*Parus atricapillus*) and the twh-t call of the eastern phoebe (*Sayornis phoebe*) are both given in a wide variety of situations and may illustrate this context specificity (Hailman and Ficken 1996; Smith 1977). We shall call this type of scheme **contextual coding**. Its existence in nature prompted Smith

(1968, 1977) to argue that sender **messages** are best analyzed independently of their **meaning** to receivers. We take up the utility of this distinction in Box 15.3.

Note that many of these mapping rules are not exclusive. It is possible that one coding scheme might be continuous, iconic and contextual, whereas another might be redundant, involve categorical perception, and allow for selection among individuals. For most questions, there are a number of possible mixes of coding options which might be adopted. Which mix is seen will depend in part on the economics and in part on the evolutionary history of the species (see Chapter 16).

Coding for Compound Signals

There are many occasions when it is most economical for a sender to encode information about several different questions into the same signal. Thus a bird might use a compound display in which both vocal and visual components provided concurrent information. There are several ways that combinations of signal elements could be used to deal with several questions at the same time. Since even simple signals have several variant parameters, one extreme solution would be to list all combinations of parameter variants and then assign these independently to each of the alternative conditions pooled across the various questions. We can call this **combination mapping**. Combination mapping would seem most efficient when the number of alternative answers to each question is small. Thus a frog or bird that needed one compound display to attract females and repel males, and a different display during fights with intruders, might assign that combination with better long-distance propagation to the first task, and others suitable only for short ranges (whether similar in structure or not) to the latter. The method seems inefficient when each of the questions has many answers and the size of the coding matrices will be very large. However, the recruitment of different compound displays during contests (Andersson 1980) and the use of many different songs during countersinging by some territorial birds (Catchpole and Slater 1995) may be cases where coding costs are outweighed by other considerations. In both examples, it is also possible that each display does not represent an alternative condition, but instead the diversity of displays or songs per se encodes the requisite information.

As a second coding option, a sender might assign variants of one signal parameter to alternative conditions for the first question, variants of a second parameter to alternative conditions for the second question, and so forth. We can call this **parameter mapping**. In male frogs and toads, fine temporal patterns (such as pulse rates, pulse shapes, and frequency modulation patterns) function as species identifiers. The coding for body size is done through dominant call frequency, and overall call rates and call amplitudes are used by females for intraspecific mate selection (Gerhardt 1991, 1994; see Figure 18.4). For other taxa, specific signal parameters indicate group membership, whereas others indicate individual identity (Figure 15.7). In many animals that give the same signal, the choice of the repeated signal answers one ques-

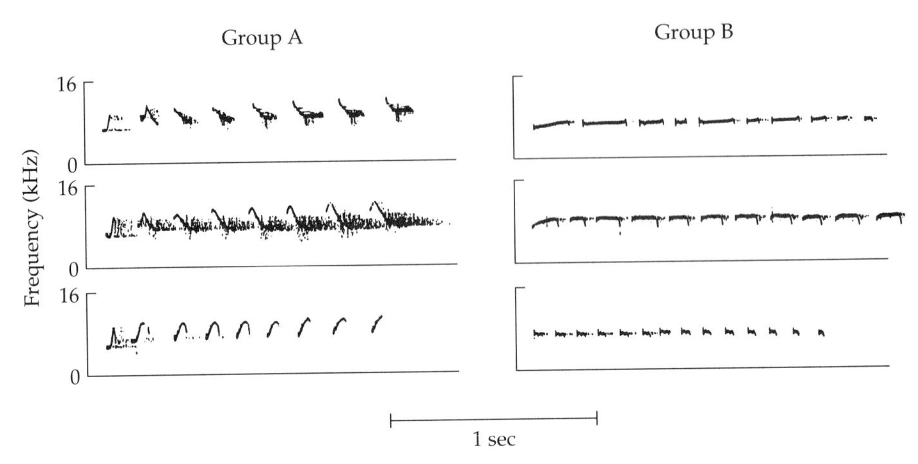

Figure 15.7 Group and individual signatures in calls of red-chested tamarins (*Saguinus labiatus*). Spectrograms of long calls of three individuals in Group A (left) and three individuals of Group B (right). Frequency axis indicates 0–8 kHz; the time marker represents 1 sec. Note similarities in the structure of long calls within each group and the slight variation around that group pattern, conferring individual signatures. Statistical analysis shows that different acoustic features are used to code for group versus individual identities. (From Maeda and Masataka 1987.)

tion (such as whether the sender is mated, ready to attack, or hungry), whereas the rate of repetition codes for the relative level of the advertised state (such as how tightly bonded, how willing to attack, or how hungry). All of these are examples of parameter mapping schemes.

Finally, if structural signals exhibit a hierarchical pattern of variation, differences between mean patterns can be used to code for one question, and different variants around that mean can be assigned to alternative answers for another question. We call this **hierarchical mapping**. Many species of song birds use the mean pattern of vocalizations to indicate species identity, whereas small deviations from the mean pattern (as in the number of repeated notes in one part of the song) indicate individual identity (Becker 1982; Falls 1982). Similar hierarchical levels of variation appear to facilitate species, sex, age, and individual distinctiveness in the discharges of electric fish (Crawford 1992; Friedman and Hopkins 1996; McGregor and Westby 1992; Figure 18.8C).

Although these three types of mapping options have been presented as alternatives, many compound signals exhibit more than one option within a given signal class. Thus territorial bird calls might indicate aggressive intent by choice of gross call type (combination mapping), species and individual identities by details of call structure (hierarchical mapping), and likelihood of attack by call-repetition rate (parameter mapping). Such a mixture is typical of most animals; failure to recognize this fact can result in tiresome arguments about how to classify observed signals.

THE ONTOGENY OF CODING SCHEMES

Heritable versus Environmental Influences on Development

How do individual animals acquire their signaling codes during development? There are several critical steps for each party. Senders must first acquire the signal variants that they will use for communication and then develop the rules that map these signal variants onto conditions of interest to receivers. Finally, they may need some rules about which animals are suitable receivers and which contexts are appropriate for each set of signals. These three steps are not totally independent, but they could be achieved by different developmental processes. Receivers must develop the receptor organs and sensitivities to detect and discriminate among alternative signals. They must then map these onto perceived signal categories and the decision rules eliciting appropriate actions. As with senders, each step for receivers might adopt a different developmental process. What alternative developmental processes are available and are there any patterns concerning which is used for each step?

Most developing traits are determined by a mix of environmental and genetically heritable influences. Heritable influences on a trait can involve many different genes, and their final expression in an adult animal may reflect not only the additive effects of each gene, but also nonlinear interactions among genes and between genes and environmental influences. Environmental influences include requisite exposure to triggering stimuli, various kinds of learning, and chance events during development. The interdigitation of the two kinds of influence to produce a trait can be very complex. For example, some animals have a sensitive period during early life when signal variants and their associations with conditions are learned (Bateson 1991; Bolhuis 1991; Marler 1984). The timing of this period and the set of acceptable signal or condition variants are usually constrained by heritable factors. However, if the codes are not successfully set up by the end of the species-typical period or no suitable models are encountered, there may be some environmental flexibility built in to extend the range (Petrinovich 1990). Once coding associations are established, subsequent trial and error learning may be required to eliminate erroneous or unprofitable ones (Galef 1995). Not all traits show such a balanced mix of influences. Some rely more heavily on heritable factors, whereas others show only the mildest of heritable constraints on what can be learned.

Although it is often challenging to quantify the relative weighting of heritable versus environmental influences in a trait's development, it is acknowledged that different traits in the same species, or the same trait in different species, often have different weightings. The adaptive benefits of different weightings have been discussed by many authors (e.g., Boyd and Richerson 1985, 1988; Galef 1988; Heyes 1993, 1994; Johnston 1982; Johnston and Turvey 1980; Laland et al. 1993; Slobodkin and Rapaport 1974; Staddon 1983; Stephens 1991, 1993). It is generally agreed that developing animals that have neither the time nor the opportunity to learn are better served by a heavy weighting on heritable acquisition of signals and codes. This is one way to en-

Box 15.3 Message-Meaning Analyses of Animal Communication

W. JOHN SMITH (1968, 1977) PIONEERED THE NOTION that the choice of signals by senders should be examined separately from the choice of responses to signals by receivers. His own research on New World flycatcher displays had persuaded him that senders often produced the same signal in multiple contexts, but that receivers somehow interpreted those signals differently. Following the lead of the linguist Cherry (1966), Smith defined the **message** of a signal to be the information that a sender encodes in a signal, and the **meaning** to be the significance attached to a signal by the receiver as indicated by its subsequent actions. The latter might depend both on the signal and on the contexts in which the signal was exchanged. He noted that the signal might also have meaning to the sender, but because this was sufficiently difficult to study, one was forced to focus on the receiver. He then defined the **function** of a signal to be the adaptive consequences of the signal exchange. These distinctions provided a useful springboard for many subsequent studies of animal communication. The terms message, meaning, and function are now widely used in the sense advocated by Smith. However, they have also been misinterpreted and applied differently by different workers. It may be useful to reexamine them in the framework of this book.

Suppose a signal has been exchanged between sender and receiver. If the sender code were perfect, the message would be that the sender perceived a given condition to be true. With an imperfect code, definition of a signal's message is more difficult. By analogy with the perfect case, we might try to determine which condition was most likely to be perceived as true by the sender given this signal. It is not sufficient to compare the probabilities that this signal will be given for each condition in the sender coding matrix. To find the most likely condition given that signal, we would have to do a Bayesian computation using the entire matrix and the prior probabilities of each condition. (Ironically, this is just what the receiver is presumably doing with the same signal.) Having undertaken this computation, it is not clear that we gain by insisting that a single condition be identified as the message. In so doing, we toss away the very data that might be vital in assessing what senders gain by sending the signals and why receivers respond the way they do. As long as we avoid that temptation, the Bayesian estimates for each condition after receipt of a signal can be considered as a quantitative summary of that signal's message. This approach assumes, as does Smith, that signals are essentially honest. The complications arising when senders are dishonest will be taken up in Part III of this book.

sure a good match between sender and receiver schemes. A similar weighting is favored when the set of alternative conditions that are indicated by signals is invariant both during the lifetime of the communicating individuals and between generations. At the other extreme, high turnover in the composition of the set of alternative conditions both within and between generations is not a situation in which communication is useful. (Human language may be an exception.) In between are intermediate levels of turnover in the set of alternative conditions, and for many of these some form of learning is favored.

Learning mechanisms can be divided into **social learning**, in which the presence of other individuals helps to learn a task, and **individual** or **trial and**

The receiver assesses contexts directly to generate the prior probabilities that each condition is true. The subsequent receipt of a signal is combined with these priors and the receiver's decoding matrix to generate new estimates that each condition is true. The receiver then applies a decision rule to this estimate and gives a response. This model is a more explicit and quantitative version of processes suggested by Smith. Because Smith never treated the process quantitatively, it is not always clear what parts of it are to be included in his sense of meaning. Even in his own publications, narrower and broader definitions are applied for different cases. Perhaps for this reason, other authors have used the term meaning to apply to various steps in the process. Some take it to refer to the receiver coding scheme, others to the updated estimates of conditions, and others to the final decisions. Although Smith carefully distinguished between the informational meaning of a signal and its fitness consequences (functions), other authors have included relative payoffs in their characterization of the significance of a signal. One can certainly argue that the signal has a different significance to the receiver at each stage of this process; however, it is not immediately clear that anything is gained by invoking the term meaning over the more quantitative version we have used in this book.

The functions of signals according to Smith are the adaptive consequences of the signal exchange. In our treatment, these are summarized as the value of information for each party. We have seen how the value of information depends upon the relative payoffs of alternative receiver responses, the degree to which signals improve estimates about which conditions are true, and the costs of communication. Signals thus function to elicit the optimal receiver response and honest signals should maximize the value of information for both parties.

Message-meaning analysis played an important role in guiding many prior studies in animal communication. In addition, W. John Smith's insights identified most of the critical terms that had to be included in formal mathematical models. We now need to move beyond the qualitative message-meaning dichotomy to more quantitative studies in which specific attention is given to the details of sender and receiver codes, computation of benefits and costs, and the optimality of alternative decision rules by receivers.

error learning in which it does not (Galef 1988, 1995; Heyes 1994). It is generally argued that social learning avoids costly errors and is faster than individual learning when conditions are static; however, social learning tends to track environmental changes more slowly than does trial and error learning. This distinction is the basis of the following predictions. When between-generation variation in condition sets is high enough to render heritable acquisition impractical, but within-generation variation in condition sets is sufficiently low, social learning can usually track the between-generation trends and is favored because it is faster than individual learning (Boyd and Richerson 1985, 1988). The type of social learning invoked can vary widely in the relative roles of model versus learner and in its utility in particular circumstances (Galef 1988). When within-generation variation is too high for tracking by heritable or social learning mechanisms, but not so high that trial and error learning cannot find solutions before conditions change, individual learning will be the favored acquisition mechanism (Stephens 1991). Individual learning also comes in a number of forms, each with its own rate of acquisition, costs, and propriety in particular contexts (Shettleworth 1993).

Acquisition of Signal Variants and Encoding Rules by Senders

Each stage of code acquisition by the sender or the receiver might draw on a different mixture of influences during development. It is thus difficult to characterize the end result of all three stages as "largely heritable" or "largely learned" (Galef 1995). For example, guards of social Hymenopteran colonies use colony-specific medleys of pheromones to decide whether entering workers should be admitted to the colony or attacked. The composite pheromones are secreted into nest substrates and then absorbed by the cuticles of colony members. Acquistion of different signal variants by different colonies can be partitioned into that due to heritable similarities among colony members (i.e., kinship), that due to the identity of the colony queen(s), and that due to the local dietary environment. The relative weightings of these different influences in creating signal variants within a population can vary with colony and with species (Breed et al. 1988; Carlin and Hölldobler 1986, 1987, 1988; Gamboa et al. 1986, 1996; Getz 1991).

The relative role of heritable and environmental influences on the acquisition of sound signals varies between and within taxa. Many insects and anurans produce normal sounds without prior exposure to conspecifics, and hybrids produce sounds with hybrid acoustic structure. This finding suggests a strong emphasis on heritable influences (Burkhardt 1977; Butlin 1989; Ewing 1989; Gerhardt 1994). The same is true for many mammals and nonoscine birds such as chickens, quail, or doves (Baptista 1996). In contrast, other mammals such as humans and porpoises (Reiss and McCowan 1993; Sayigh et al. 1990), and birds such as oscines, parrots, and some hummingbirds seem to require both individual and social learning to produce normal or appropriate vocalizations (Catchpole and Slater 1995; Kroodsma et al. 1996; Pidgeon 1981; Rowley and Chapman 1986, 1991). The duration and timing of the period during which young song birds can learn from conspecific vocalizations varies among species and is largely under genetic control. There are also genetic constraints on which signal variants a young bird is likely to copy. That is, the learning of vocalizations is inevitably biased by heritable limits and predispositions (Baptista 1996; Kroodsma 1982; Marler 1984). In some species such as tree-creepers (*Certhia brachydactyla*), individual song syllables (see Figure 15.3) are largely heritable, whereas the syntax, number of syllables, and rhythm of adult songs are acquired by exposure to conspecifics (Thielke 1970). In other species such as sparrows in the genera *Zonotrichia* and *Melospiza*, the syllables are largely learned by exposure to conspecific male singing, whereas the allowable pattern and syntax are species-specific and largely heritable (Konishi and Nottebohm 1969; Marler 1970; Marler and Peters 1982, 1988; Marler and Pickert 1984; Nowicki et al. 1992). In a wide variety of song birds, the mean pattern of the song (which confers species distinctiveness) reflects strong heritable influences, whereas small variations around this mean (which encode population, group, or individual identity) often arise through social learning, copying, and/or individual experimentation (Falls 1982; Mammen and Nowicki 1981; Mundinger 1970, 1979; Nowicki 1989b). As we note later, the likely reason male oscines learn songs is a result of competition between males to attract females. The competitive value of any male's repertoire is not fixed by the repertoire, but depends on what his neighbors are singing (see Chapters 20 and 22). The resulting arms race between males (see Chapter 20) causes the composition of optimal repertoires to change over time (at least between generations and sometimes within them), and this change favors social learning over the heritable mechanisms shown by most nonoscines.

Some birds such as parrots (Farabaugh and Dooling 1996) and hill mynahs (Bertram 1970) can learn to imitate nearly any sound (at least in captivity) and thus appear to have few genetically heritable constraints on acquisition of vocal variants. Vocalizations in these birds are used for individual and flock recognition and only rarely for competitive mate attraction. This type of function is likely to favor a very heavy emphasis on learning by both senders and receivers. Among Australian parrots, Major Mitchell cockatoos (*Cacatua leadbeateri*) may evict sympatric galahs (*Eolophus roseicapillus*) from nesting sites but fail to remove the galahs' eggs before laying their own. They then raise both their own and galah young (Figure 15.8). With the exception of alarm calls (which resemble those of their own species), the adopted galah offspring acquire their vocalizations from their foster parents (Rowley and Chapman 1986, 1991). The galah offspring do emit their own galah begging calls immediately after hatching, but quickly learn to use that of the parental species. Even with parrots, one sees some mix of heritable and environmental influences, although the weighting is clearly on the side of social learning.

The mechanisms by which senders associate signal variants with conditions and identify suitable receivers and contexts have been less well studied. For some identification signals, such as the colony-specific pheromones of the social Hymenoptera, each sender emits only one signal and does so chronically. There is thus no need to acquire an encoding rule for selective emission. For signals reflecting varying levels of confidence or motivation during agonistic contexts, as in the degree of fin erection or darkening of color pattern in fish, the rates at which pulses are given in frog calls, or the expressions on a primate's face, the codes linking signal variant to internal state are likely to be the same between and within generations, and as a consequence these signals appear to be largely heritable.

Vervet monkeys (*Cercopithecus aethiops*) produce four different alarm calls depending on whether an approaching predator is a raptor, a snake, a leopard, or some other mammalian predator (Cheney and Seyfarth 1990). Young vervets emit alarm calls to a wide variety of species, including many that

Figure 15.8 Mixed family of Major Mitchell's cockatoos (*Cacatua leadbeateri*) and adopted galah offspring (*Eolophus roseicapillus*). Major Mitchell cockatoo parents sometimes inadvertently raise offspring from galah eggs left behind when galah parents are evicted from a nest site. Galah chicks learn most of the vocal repertoire and foraging techniques of the foster species indicating that learning plays the major role in acquisition of both sender and receiver signal codes in these parrots. The photograph shows three Major Mitchell offspring on the left, and one galah offspring on the right from the same nest. Birds were removed for marking and later returned to the nest. (Photo courtesy of G. Chapman.)

never harm vervets; adults are much more discriminating and tend to emit calls only for known predators (Figure 15.9). However, from a very early age vervets do appear able to assign each of the four call types to the proper class of predator, suggesting that the gross templates that map signals on conditions are largely heritable, and it is only the fine-tuning based on local predator identities that requires experience. Vervets also produce four subtly different "grunt" vocalizations that mediate approach between animals in as many different social situations. Young vervets assign grunts to two broader situational categories and only gradually discriminate between the four categories of adults. Nestlings of altricial birds initially raise their heads and beg for food given any vibrational disturbance to the nest; with increasing age, they give the signals only when the perturbing vibrations are due to the arrival of a parent. For both the vervets and the bird nestlings, it is not clear whether these developmental changes are due to experience-dependent refinement of sender codes, or instead, to physical maturation of the sensory organs needed to discriminate between alternative conditions. If the latter were true, sender codes might be largely heritable, but not fully utilizable until conditions were accurately discriminated. In young pig-tailed macaques (*Macaca nemestrina*), learning per se is required to refine the code that links the production of agonistic screams to appropriate social contexts (Gouzoules and Gouzoules 1989). Again, there seems to be extensive variability among species in the heritability of sender code acquisition.

Figure 15.9 Vervet monkeys (*Cercopithecus aethiops*) in Kenya. Members of vervet monkey troops produce four acoustically different alarm calls evoking different responses in receivers: leopard alarm calls cause nearby monkeys to run into the trees, eagle alarms cause them to flee into dense bushes, and snake alarms cause them to stand bipedally and search around them for the serpent (as shown in the photo). Young vervets produce all three calls in approximately appropriate contexts from an early age, but also give them to organisms which pose no threat to vervets. Age and experience narrow the set of stimuli eliciting each type of call. Vervets can also distinguish between the alarm calls and respond differentially from an early age, but often make mistakes. Observation of adults plays an important role in improving response propriety. The gross patterns of sender and receiver codes thus appear to be heritable with fine-tuning dependent on learning and imitation. (Photo by Richard Wrangham/Anthro-Photo.)

Acquisition of Signal Variants and Decoding Rules by Receivers

Similar variation is seen in the heritability of receiver sensory structures and decoding schemes. Many insects, fish, and anurans have peripheral filters built into their sense organs to ensure selective responsiveness to signals of their species. Although the development of these filters is largely heritable, significant interaction with environmental factors can be seen. For example, many insect and anuran females rely on precise temporal patterns in male calls to indicate species identity. Because neither type of animal is homeothermic, the temporal patterns of emitted calls can vary slightly with ambient temperatures. At least over normal temperature ranges, female decoding schemes shift with ambient temperatures so that the call templates expected by females remain close to the signal variants produced by males (Gerhardt 1978, 1982; Stiebler and Narins 1990; Walker 1957). Although the ability to make these adjustments is itself largely heritable, the final decoding scheme of these females is clearly the product of both heritable and environmental influences. The mapping of the outputs of these organs onto perceived signals, decision thresholds, and actions shows a similar emphasis on heritable influences in these taxa.

Many more birds and mammals rely on some amount of learning to acquire proper receiver codes. As noted earlier, there is often a sensitive period during development when the exposure to suitable models establishes associations between signal variants and conditions. **Filial imprinting** occurs when young animals fixate on signals or cues that later help them to identify and follow parents; in terrestrial mammals, the relevant cues are often olfactory (Zippelius 1972), whereas in birds they are usually auditory signals (Beecher 1991). **Sexual imprinting** occurs when signal variants emitted by conspecifics are associated with future suitability as mates. This type of imprinting is often thought to ensure species recognition in mate selection, but Bateson (1978b, 1983) has argued that it facilitates subsequent avoidance of detrimental inbreeding with very close relatives or outbreeding with conspecifics not sharing local adaptations. Recognition of suitable mates may require different developmental mechanisms that depend upon whether that mate is a male or a female. For example, discrimination between males of sympatric species of dabbling ducks appears to be largely heritable, whereas that between females of different species requires sexual imprinting at an early age (Schutz 1965; Williams 1983). Female dabbling ducks do all the nest incubation, usually on the ground, and there is strong convergence in plumage to similar cryptic brown colors. Males do not participate in incubation and species differ markedly with conspicuous plumage patterns. Discriminations between males of different species are easily encoded genetically, whereas the more difficult discriminations between females of different species seem to require extensive learning.

Although male song birds acquire their song variants and encoding rules by a mechanism similar to sexual imprinting (Clayton 1989), less is known about how females of the same species acquire their decoding schemes for male songs (Baptista 1996; Baptista and Gaunt 1994; Ratcliffe and Otter 1996). Female oscines typically discriminate between conspecific and heterospecific song. Studies of song sparrows (Searcy and Marler 1981, 1984, 1987) suggest that at least some of the relevant coding is species-specific and likely heritable. The existence and acquisition of finer distinctions is the focus of current debate. A number of song bird species exhibit geographical dialects in their songs. Some studies suggest that such females are more likely to respond to natal than to alien dialects, and early learning appears to play some role in this discrimination (Baker 1983; Baker and Spitler-Nabors 1981; Baker et al. 1982, 1987; Balaban 1988); other workers have been unable to find preferences for natal dialects even in the same populations (Baptista and Morton 1982, 1988; Chilton et al. 1990). More extensive data are available for recognition of neighbors by territorial male song birds. Here, heritable influences limit which songs are considered conspecific, and within conspecific alternatives, males discriminate between neighbors by learning their song repertoires (Stoddard 1996).

The developmental mechanisms of receivers and senders cannot always be independent. For example, where a sender acquires signal variants by individual trial and error learning, there is no way in which receivers can anticipate the final signals and their associated conditions. Receivers must then acquire their code by learning as well. Individual identification requires only that there be a large number of alternative signals that each sender can make its own; this necessary signal diversity could easily be generated by mechanisms that were either highly heritable (Lenington 1994), highly dependent on environmental factors, or some mixture. If the mechanism of signal acquisition is largely heritable, receivers that are close kin might also use heritable templates to discriminate between kin and unrelated individuals. Parent-offspring recognition could thus be accomplished with high heritability of both signals and receiver templates. However, if assignment of signal variants to relevant individuals is random (whether through heritable or learned mechanisms), receivers have no choice but to acquire their decoding schemes by learning. They may learn both the relevant signal set and the coding associations. We shall consider other possible links between sender and receiver ontogenetic mechanisms in the next section.

SIGNAL FUNCTION AND CODING

Different questions faced by receivers make different demands on the codes of those engaged in communication. In general, the more alternative signals either party has to process and the closer the code to a perfect one, the harder the task and presumably the higher the costs. If similar benefits can be achieved by a simpler coding scheme with lower costs, we may expect evolution to favor the simpler scheme. However, accuracy in some tasks is sufficiently valuable, despite the costs, that a complex coding scheme may be unavoidable. In this section, we look at several communication tasks and compare the types of coding schemes that each party must evolve to deal with them. Because many questions invoked in animal signals have only two alternatives, we shall divide the tasks into those asking **binary questions** (two alternatives) and those asking **manifold questions** (more than two alternatives).

Binary Questions

There are many different ways in which binary questions can arise in animal exchanges and these differ in the burden they place on the requisite coding schemes. Two critical criteria are (1) the number of alternative signals that receivers are likely to encounter, and (2) the degree to which all receivers must share the same decoding matrices. To see why these factors are important, consider some examples.

At one extreme, suppose only two alternative conditions are encoded by senders and extracted by receivers, and the set of alternatives does not change within or between generations. As a result of the absence of variation in the condition set, all receivers should share the same decoding scheme. As an example, a sender might advertise its sex (male versus female) or mating status (mated versus unmated). This task is a **binary assignment**. Because there are only two alternative conditions, any of a wide variety of signal parameters could be varied to code for them. This allows senders to select the parameter that is cheapest, least susceptible to distortion during propagation, or more easily discriminated by receivers. Because all receivers use the same code and the condition set is unchanging, acquisition of signal variants and coding schemes by both parties can be largely heritable.

Now consider an identification task in which a receiver seeks to locate a particular individual among *N* senders. We can call this **binary recognition**. It is a problem faced by many colonial breeding birds or mammals when they must find their own offspring or mate in a crowd. Signals from successive senders are examined, and the receiver then classifies each sender as either the sought individual or not (binary alternatives). Each sender need produce only one signal; however, it must ensure that its signal is distinguishable from those of all $N - 1$ others. This requires that there be at least *N* discriminable signal variants among the senders a receiver will encounter. To achieve the requisite signal diversity, senders must either vary a number of signal parameters, or tolerate a wide range of values for each parameter. Thus even though the sender is producing only one signal, it must ensure that its signal is unique. This could be accomplished with any mixture of heritable and environmental sources of trait variation and random assignment of signal variant to each individual. Two examples relying largely on heritable differences are summarized in Figures 15.10 and 15.11. The task for the receiver is to reduce (i.e., categorize) many different received signals into two alternatives. This is certainly easier than assigning each sender to its own category (see the section on manifold questions, below). However, it requires more sophisticated machinery than that used to distinguish between only two alternative conditions, the binary assignment case above. At the minimum, each receiver must have a template specifying the acceptable values for each coded signal parameter of the sought individual. It must measure each of the parameters in each received signal and compare the results to the template. The sender's signal is then assigned to one of two categories.

The onus for matching sender and receiver codes in a binary recognition task falls on different parties in different species. In our example, the sender hits on its own signal variant, and the receiver must learn to discriminate it from alternative variants. This is apparently how mates are identified in many monogamous birds such as silver eyes (Robertson 1996) and petrels (Bretagnolle 1996). In some flocking birds, the task is to decide whether a bird belongs to the flock or not; in house finches (Mundinger 1970, 1979), parrots (Farabaugh et al. 1992), and chickadees (Ficken et al. 1978; Mammen and Nowicki 1981; Nowicki 1989b), senders from the same flock acquire a flock-specific set of call variants, and receivers concurrently modify templates to match the variants of their own flock. Thus both parties make adjustments in signal sets and codes to facilitate the discrimination. Finally, in cowbirds, males initially learn a number of alternative songs, but selective female responses cause them to eliminate all but those songs that females favor; subordinate males may later shift back to less favored songs as a result of harassment by dominant males (King and West 1983; West and King 1988). In this case, the senders must adjust their signal variants according to receivers' preexisting templates.

Figure 15.10 Individual recognition using olfactory signals in house mice (*Mus musculus*). Most chordates produce special glycoproteins which allow discrimination between own and foreign tissues. These proteins are generated by a group of as many as 50–100 genes called the major histocompatibility complex (MHC). These genes have tens to hundreds of alleles which tend to be equally common in wild populations. The number of possible genetic combinations is thus astronomical and this allows each individual to be chemically distinct from any other just by random mixing of the expressed alleles. Mammals such as the mice above (and even humans) carry breakdown products from these proteins in their sweat, urine and secretions which can be assayed by conspecifics. Although originally thought to have evolved for defense against diseases, some researchers (see Brown and Eklund 1994) have suggested that the MHC originally evolved as a signal system to avoid inbreeding with close relatives, abort pregnancies when threatened by strange males, identify kin with which to cooperate, or learn to identify specific individuals in groups. (Photo courtesy of John Nyby.)

Another common task is **binary comparison**. The receiver compares two signal variants and makes some relative judgment about them. For example, two opponents in an agonistic interaction may compare their relative fin elevations, color patterns, or call repetition rates to decide which individual has the larger value and is thus most likely to continue an escalated fight. In these examples, the receiver compares the opponent's signal parameters to its own. A related task is the comparison of another's signal variant to some stored **threshold value**. If the signal parameters exceed the threshold, the signal is assigned to one category, and if not, it is assigned to another.

Successive binary comparisons are common during mate selection. In a number of animal species, males aggregate in selected sites (leks) and display competitively to attract females for mating. Females move among the males, sample their display signals, and select one or more for mating. If a female is using a threshold comparison, she samples males until one exceeds the threshold and then she mates with him. Alternatively, the female may visit males successively and compare the signals of each to those of the best male so far encountered. After sampling a sufficient number, she then mates with the best male sampled. Among the signal parameters which appear to be used in either threshold or relative comparisons among mates are concentration and quality of emitted pheromones in insects (Löfstedt et al. 1989; Moore 1988), temporal pattern and call rate in frogs (Gerhardt 1994), brightness of color patterns in guppies (Endler 1983; Kodric-Brown 1989), and timing of display components in sage grouse (Gibson et al. 1991). These signal parameter values appear to encode information about some underlying male condition that females seek in a mate—good health, dominance of other males, foraging ability, or longevity (see Chapter 23).

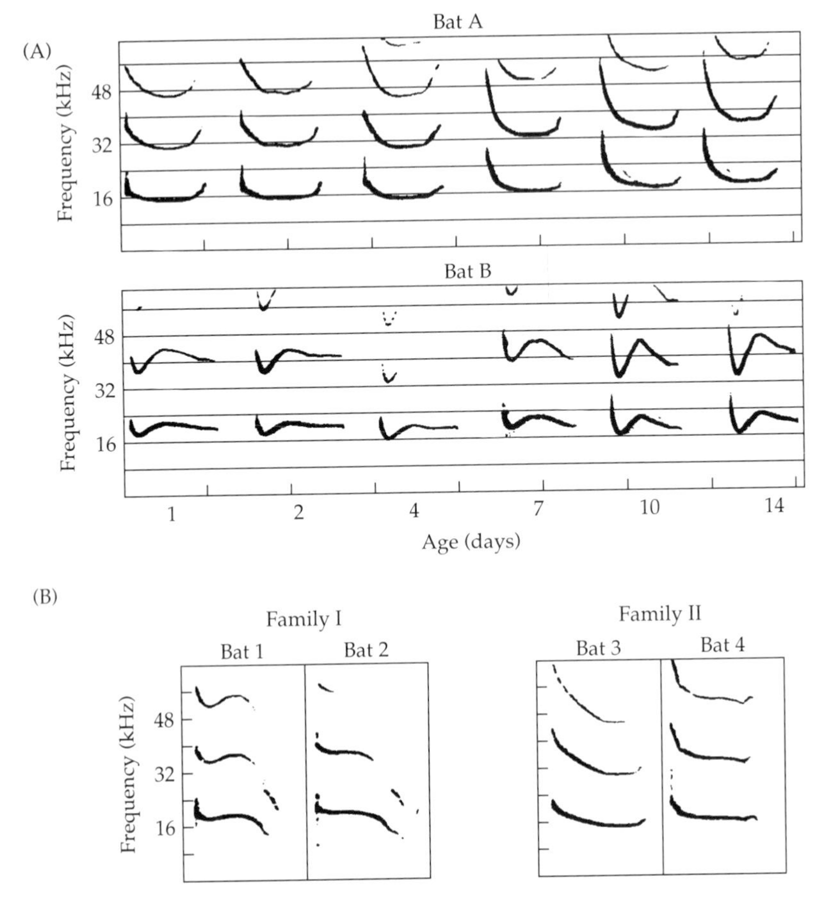

Figure 15.11 Individually distinctive isolation calls of infant evening bats (*Nycticeius humeralis*). Isolation calls are signals emitted by infant bats to attract their mothers. Evening bats have large nursery colonies, so females must be able to locate their own young among hundreds of others. Unlike some colonial birds, whose offspring mingle only after fledging, infant evening bats are immediately placed in the communal creche. They are thus under strong selective pressure to produce unique isolation calls from the minute they are born, leaving little opportunity for call differences to be learned. (A) Spectrograms of isolation calls from same two infant bats at ages of 1 to 14 days. Note that the basic structure of a call differs for two individuals from birth, and is only slightly modified with maturation. (B) Spectrograms of isolation calls from four individuals. Bats 1 and 2 are siblings with the same mother; bats 3 and 4 are siblings whose mother is unrelated to the first. Note the similarity between isolations calls within a family, but the differences between families. Both sets of evidence taken together suggest that differences between infant bats in isolation calls is largely due to heritable differences between them. (From Scherrer and Wilkinson, 1993.)

Binary comparison usually relies on sender signal sets that are continuous and coding rules that are iconic. Thus the sender signal set is no longer binary. Receivers will favor signal parameters that accurately and honestly reflect the communicated condition; whether senders are likely to be accommodating is an issue we take up in Part III. The requisite coding mechanism can be costly, but its development could rely on any mixture of heritable and environmental influences. The receiver needs only the ability to compare signal parameters from two sources. If the sender code remains unchanged across generations, the simple decoding scheme needed could easily be acquired through heritable mechanisms.

A special case of binary comparison occurs in song birds (oscines). Here males sing to attract females and to defend territories. In nearly 75% of oscine species, each male sings more than one song type. In all species, song acquisition requires some social learning. While many New World warblers use separate song types for mate attraction versus territorial defense (Kroodsma 1988; Lemon et al. 1987, 1993), most other oscines use the same set of song types in both contexts. Because (a) males of many oscines stop singing once a mate is attracted, (b) females in a variety of species have been shown to prefer males with larger or more diverse song repertoires, and (c) the immediate ancestors of songbirds such as scrub birds and lyre birds (Menurae) use large song repertoires to attract mates, it is now thought that multiple song types originally evolved as result of female mate choice (Catchpole and Slater 1995; Irwin 1988). This idea suggests a rather different coding scheme than we have discussed thus far. Instead of mapping each song type onto a different alternative condition, male oscines use the number of song types they have acquired as a measure of their current health, age, or status. The signal variants compared by females are then different repertoire sizes, not different song types. For senders, costs of encoding are higher than in other examples of binary comparisons because each male must develop more than one signal variant. For receivers, the costs of decoding are also high because females must be able to discriminate between song types to be able to assess a male's song repertoire size. Some male song birds such as marsh wrens, thrashers, mockingbirds, and nightingales have such large song repertoires that it seems unlikely that any receiver will learn them all. Instead, the receiver may simply compare the identity of the current song to a short-term memory of each of the N most recently sung ones. The amount of overlap on average would then be a measure of that male's diversity. In the end, females may evaluate successive males in a binary fashion by comparing some index of repertoire size; however, the relative ranking of any two males will surely require that the female have some ability to recognize and compare more than two signals at a time. The task is thus more like those described in the next section.

Manifold Questions

When the question facing a receiver has a large number of possible answers, the coding tasks become more complicated for both parties. If the alternative answers can be ordered along some condition axis, then an iconic rule can often be found to map a set of similarly ordered signal variants onto the alternative conditions. Where alternative conditions cannot be ordered, then pairwise associations between each signal variant and each alternative condition have to be used. Iconic rules are a more economical way to store codes than individual mappings when there are large numbers of alternative answers.

They also are easier to adjust with subsequent evolution, migration to a new context, or local adaptation.

For example, we have seen that graded signals by senders (as occurs during agonistic contests or mate selection) may not require the consideration of multiple answers by receivers if all that is needed is a relative comparison between pairs of signalers. However, there are many cases in which a receiver needs to translate the magnitude of a graded signal into the absolute value of the condition it reflects. For example, an aggressive threat by a nearby sender may need to be interpreted by the threatened receiver without comparison to its own or any other sender's state. Here is a case in which an iconic rule for translation of the sender's signal would be the most economical way for the receiver to determine which of many different answers was the case (Figure 15.12).

An example of ordered iconic coding for manifold questions involving environmental information is found in the dances of honeybees. These are performed at the hive by returning workers to indicate the direction and distance other workers should fly to find a new food source (see also pages 827–834). The main axis of the dance relative to gravity indicates to receivers the angle they should fly relative to the sun to find the food; site angle is a condition that varies continuously. The number of dance cycles per unit time indicates the distance between the food and hive, and site distance is a second continuously varying condition. The iconic rules for both the angle and distance information are largely heritable, although the rules that map dance rate on distance vary slightly with different honeybee populations. When colonies are artificially assembled with bees from populations with different rules, recruits observing dances by a bee not from their own population will fly either too far or not far enough when they search for the advertised food (von Frisch 1967).

Other environmental signals do not invoke an iconic code. Vervet monkeys do not appear to order the alternative predators they encounter, and the mapping of their alarm calls on signal variants appears to be arbitrary (Cheney and Seyfarth 1990). It is not clear whether the small number of alternative predator types makes the economic advantages of an iconic rule marginal, or whether the inability to order predator types makes iconic coding impossible, and this then limits the number of different predator classes that can be coded to four. It is more likely that other selective factors limit the number of alarm-signal variants in these monkeys and we take these up in Chapters 20 and 25.

Another receiver task that may require multiple answers is the identification of many different individuals (**manifold recognition**). This type of task puts the same burden on senders as does binary recognition, but it puts a much larger one on receivers. A receiver can no longer store a single template against which signals from possible candidates for mate or young are to be matched. Instead, the receiver must have a discriminable template for each individual to be recognized. It is unlikely that individuals are easily ordered and hence that an iconic rule can be invoked. We do not yet know how the costs of signal processing and memory storage scale with the number of individuals discriminated. Certainly, some frogs and fish can discriminate between small numbers of neighbors (Davis 1987; Myrberg and Riggio 1985;

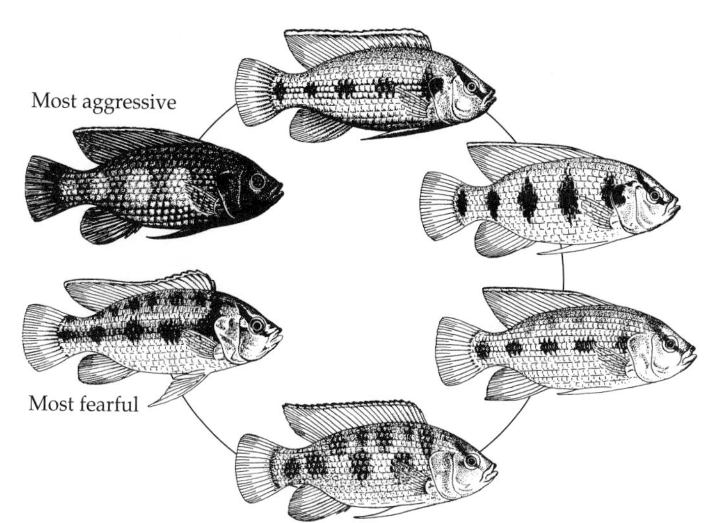

Figure 15.12 Iconic agonistic signals in cichlid fish. Examples from a continuous range of spotting patterns in the cichlid *Hemichromis fasciatus*. Fish regulate the pattern by moving pigment within melanophores in the skin. In this figure, a clockwise movement shows an increasing likelihood that the displaying fish will flee or hide; a counterclockwise movement shows an increasing likelihood that a fish will attack. (After Wickler 1964.)

Myrberg et al. 1993). A few taxa, such as echolocating bats, may be preadapted for individual recognition and thus face reduced costs. Individual variation in outgoing sonar call frequencies is known to exist in several species, presumably to help a bat recognize its own echoes in crowded environments or to reduce eavesdropping by nearby competitors (Masters et al. 1995; Obrist 1995). In addition, these bats have a superlative ability to characterize the spectra of their own echoes and this ability may make discrimination between sonar calls of other individuals much easier. The major cost would be memory storage of templates. A similar argument may apply to electric fish, which show marked individual variation in discharge patterns (Friedman and Hopkins 1996), and to porpoises, which also echolocate and use sounds to discriminate between conspecifics (Caldwell et al. 1990; McCowan and Reiss 1995). No obvious preadaptations explain the remarkable abilities of either parrots (Brown et al. 1988; Dooling et al. 1987a,b) or primates (Laska and Hudson 1995; Rendall et al. 1996) to recognize many different individuals. Why some taxa have evolved this ability and others have not remains an intriguing issue for future research.

While the origin of large song repertoires in oscines apparently evolved for female mate selection, the fact that many male oscines use the same songs for territorial defense means that males have exploited multiple song types to resolve conflicts with other males. Because individual male song repertoires are usually different, territory owners use this fact to determine whether a given singer is a neighbor or an intruder. Males who know the songs of neighbors can countersing with the same song as the neighbor when they want to escalate an encounter, or sing a song not in the neighbor's repertoire to indicate that they are present on their territory but not about to attack. These and other exploitations of multiple song types for territorial defense are discussed on pages 734–737. The cost of involving repertoire diversity in territorial defense is that each owner must learn to identify and perhaps repeat the song types of its neighbors. For receivers, this is a manifold question that will require extensive learning and memory. It may also mean that a young sender will do best by learning the song types from males in the area in which it hopes to settle. This limitation of suitable models to males in the same area is one possible reason why many oscines have evolved geographical dialects.

CODING AND REPERTOIRE SIZE

Each receiver has to deal with a number of different questions for which information from signals may be useful. The number of such questions asked is the **functional diversity** for that receiver. For each question, there will be several alternative answers and corresponding signal variants (the signal set). The average number of signals per signal set is the **variant diversity** for that receiver. The entire ensemble of signal variants that the receiver needs to distinguish, summed over all questions, is its **signal repertoire**. A signal repertoire could have a given size because it had low functional diversity but high variant diversity, or because it had high functional diversity and low variant diversity. If selection favors increasing both functional and variant diversity, then repertoire size must increase.

Several authors have argued that there is ceiling set on total repertoire size and that most vertebrate species have maximal repertoires of 30 to 40 basic categories of signals (Moynihan 1970; Smith 1977, 1986). A major reason why Smith (1968) argued for the primacy of contextual coding is his belief that there are too few distinguishable signals available to assign a separate one to each situation in which receivers seek answers. It is clear that this limitation does not widely occur. Certainly, within any taxon we can find related species, some of which have very large signal repertoires and others which have smaller ones (Figure 15.13). Unless the limits are very different for these related species, only some will have repertoires constrained by a ceiling.

Figure 15.13 Repertoire size in dabbling ducks. Dabbling ducks show great similarity in the signal functions of their calls and displays. The form of the displays, when present, is also highly similar among species. Where the species differ markedly is in the number of different courtship displays in their repertoires. Mallard males (*Anas platyrhynchos*) perform 8 major courtship displays: preen-behind-wing (A), down-up (B), headup-tailup with a burp call (C), and grunt-whistle (D) are all performed to attract females and often several males display jointly. Nod-swimming (E) followed by turning the back of the head (F) is usually performed after a female is associated with a male. Intention to mate is indicated by both sexes with head-pumping (G), and after mating, males typically perform the bridle display (H) and may nod-swim. The male Bahama pintail (*Anas bahamensis*) has a much smaller repertoire; it performs the burp call while turning the back of the head, and its major display is a down-up followed by a headup-tailup (I); both sexes head-pump only prior to mating. The other displays of the mallard male are absent. The male common shoveller (*Anas clypeata*) has the simplest courtship repertoire of all. The burp-call is given while pretending to feed (J), and males may turn the back of the head. However, the major display for male shovelers is head-pumping, which now occurs both prior to copulation and at nearly any other time males approach females. (After Lorenz 1971; Johnsgard 1965.) ▶

(A)
(B)
(C)
(D)
(E)
(F)
(G)

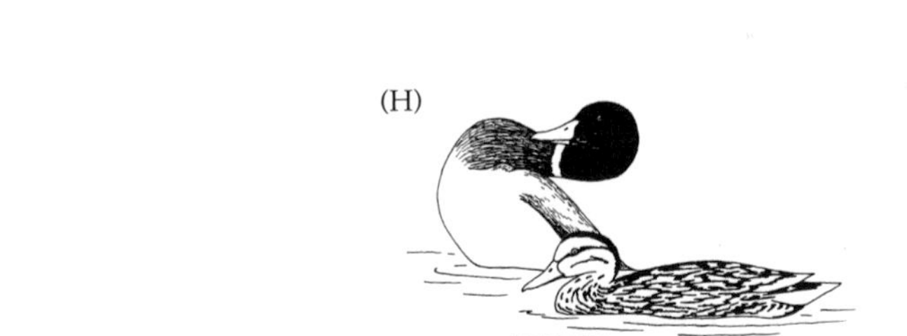
(H)

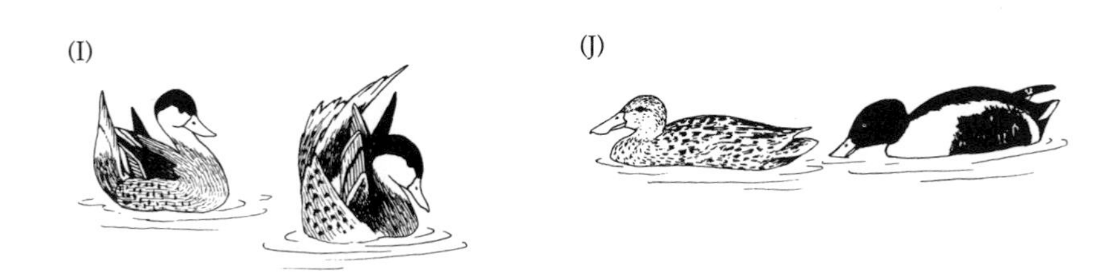
(I)
(J)

However, where large social groups are stable and interactions complex (e.g., for corvids, parrots, primates, and cetaceans), where sexual selection has favored male signal diversity (e.g., lyrebirds, bower birds, and mallard ducks), or where agonistic conflicts are prolonged and frequent (e.g., for crickets and some territorial song birds), evolution may favor larger and larger repertoires. In these situations, it seems possible that brain size and/or the size and design of sensory organs might set limits on signal number.

If there is a ceiling on repertoire size, increasing functional diversity will require a concomitant reduction in variant diversity, or vice versa. This tradeoff will have very important impacts on the kinds of coding that an animal should adopt. Thus, if the limit is low, high stereotypy of emitted signals and severe categorization of perceived ones may be necessary to ensure that all important questions are considered. Use of a complex graded signal for one task may preclude all but crude discrete signals for other functions. If there is a ceiling on repertoire size, then selection will attempt to maximize the value of information for the entire repertoire, not for each individual signal set in isolation. A repertoire ceiling would require a slightly different approach to the study of coding than we have outlined so far in this chapter.

The kinds of tradeoffs imposed by a repertoire ceiling can be envisioned graphically as in Figure 15.14. Let each of N independent signal parameters be an axis for an N-dimensional signal space. Because each parameter is limited in the values that a sender can generate or a receiver detect, this N-dimensional space is a bounded one; that is, it has a finite volume. A given signal variant will have a particular value for each parameter and can be plotted as a point in this space. We know that senders will show some variation in their production of any given signal type, that propagation will distort signals, and that receivers will make errors of assessment. Thus there will be a cloud of nearby points in the N-dimensional space for each condition which is signaled. The receiver will be able to infer different conditions if adjacent clouds in its signal space have minimal overlap. The larger any one cloud, the more likely it will overlap with adjacent clouds. Senders can reduce cloud size by increasing signal stereotypy and by selecting signal variants that are least distorted during signal propagation. Receivers can reduce cloud size by investing in more accurate sensory gear and committing more brain tissue to signal processing. The maximum repertoire size for the receiver is then set by the number of clouds that can be packed into its N-dimensional volume with less than some given level of overlap.

Repertoire size can be increased by (a) expanding the possible limits for one or more signal parameters, (b) adding more signal parameters and thus axes, or (c) reducing the size of each variant cloud. Each of these can be achieved only at some cost to one or both parties. That is, signal production or reception organs must be augmented, more brain tissue must be allocated, and if the animal is at a ceiling, new allocations will require reduced investments in other tasks. There is surely a physiological limit on both brain size and sensory acuity set by the body size and biology of any animal. Beneath that physiological limit may be an economic limit at which the costs of further

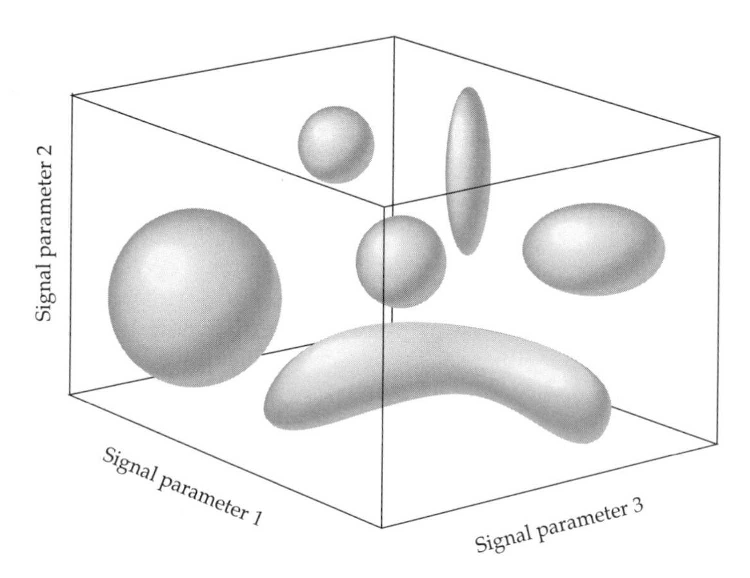

Figure 15.14 Signal space and limits on repertoire size. The axes represent different signal parameters that can be varied. The outer frame indicates maximal values that these parameters can take, thus making the possible signal space volume finite. Each signal is represented by a cloud of points that indicate typical variation as perceived by the receiver. Cloud size is larger when senders are sloppy in signal production, when there is pronounced distortion during propagation, or when receivers have only crude sensory or brain processing organs. A typical signal function requires at least two clouds to represent the alternative signals that could be given. Some functions (such as mate attraction in song birds) may require a large number of alternative signals and thus many clouds. Graded signals generate oblong clouds that extend over a large region of the space. The total number of signals that can be packed into such a space without excessive cloud overlap depends upon the size of each cloud, the range of possible values each parameter can take, and the number of axes that can be varied. Once the space is full, adding additional alternative answers to any given function will require dropping other signals, decreasing cloud sizes, or adding new axes.

increases in repertoire size outstrip the benefits. While there is good evidence that more difficult tasks require more relevant brain tissues (e.g., deVoogd et al. 1993; see Figure 17.7), we still do not know how these costs scale with task. Does requisite brain size increase linearly, in an accelerating fashion, or in a decelerating fashion as more signal variants need to be discriminated? Accelerating costs are most likely to create an economic ceiling at lower repertoire sizes than ultimate physiological limits.

At present, it is unknown whether repertoire sizes are or are not limited by any kind of ceiling. In a study of isolation calls by infant evening bats, Scherrer and Wilkinson (1993) showed that seven acoustic measures reduced statistically to four independent axes. Given natural ranges and levels of variation, they estimated that the resulting *N*-dimensional space could accommodate 1844 different calls before adjacent clouds overlapped. This number represents many more bats than is found in even the largest nursery colonies of this species. However, this study only examined the sender contributions to the space. It remains possible that mother bats cannot discriminate between such closely packed clouds, and thus many fewer signals may actually fit into the overall receiver space. Clearly, we need studies in which both sender contributions and receiver constraints are measured before we can say whether limits on overall repertoire size affect the coding schemes we see in animal communication systems.

SUMMARY

1. **Coding rules** determine the mappings of alternative signals onto alternative conditions. For either sender or receiver, the degree of association between signal S_j and condition C_i can be characterized as a conditional probability $P(S_j \mid C_i)$, and the results of a code can be expressed as a matrix of the conditional probabilities between each alternative signal and each alternative condition.

2. A code is **specific** if only one of the alternative signals is ever emitted for a given alternative condition; the code is **unique** if each condition results in the emission of only one signal. A **perfect** code is specific for every condition and unique for every signal. The more a code deviates from perfection, the smaller the average amount of information it can provide.

3. Receiver coding rules must incorporate the conditional probabilities between signals and conditions established by senders, the change in those probabilities due to signal propagation, and the further changes induced by receiver error. Final receiver probabilities can be no larger than the smallest probability in the sender-propagation-receiver sequence. As a result, receiver codes are unlikely to be perfect or identical to those of the corresponding senders.

4. Options for coding vary depending upon the number of alternative signals available. At one extreme, propagation or receiver constraints force senders to produce only a few highly **stereotyped** signals. In intermediate situations, senders can vary parameters within each emitted **signal element** to produce more signal variants. At the other extreme, senders may chain the same or different signal elements into larger units, and link small chains into larger hierarchical ones. This chaining provides a very large number of combinations and thus potential signals. In chained signals, the number of alternatives is so high that a **syntax** is usually imposed; these coding rules limit the kinds of combinations that are allowed to a more reasonable number.

5. Senders map a set of alternative signals onto a set of alternative conditions. During propagation emitted signals are mapped onto those arriving at receivers; then receivers map detected signals onto perceived categories, decisions, and actions. At each stage, the set of alternatives can be either **discrete** or **continuous**. The coding scheme at any stage may differ as to whether the degree of continuity and the number of alternatives are preserved during the mapping. For example, a sender may give the same sig-

nal in different contexts and count on the receiver interpreting the signal differently because the receiver can also determine the context. This is called **contextual coding** and puts the onus of interpretation on the receiver. On the other hand, a sender may emit more signal variants than the receiver needs; the receiver then invokes categorization to reduce these to a smaller discrete set. The decoding of **redundant signals** and **recognition** of a specific individual in a crowd are common tasks that favor such a reduction. Alternatively, it may pay both parties to exchange detailed information about a large number of alternative conditions. Coding is then easier if they share some **iconic rule** linking specific signal variants with specific conditions. An iconic rule that maps some continuous signal parameter monotonically onto a continuously varying condition is called a **graded signal system**. Codes for compound signals, which deal with more than one question, may assign different combinations of parameter and syntactical variation to different answers, regardless of question (**combination mapping**), assign variants for each signal parameter to a different question (**parameter mapping**), or may assign mean values to one question and variants around that mean to another (**hierarchical mapping**).

6. The **ontogenies** of signal variants and encoding rules by senders, and of signal categories and decoding rules by receivers, like all traits, depend upon a mix of heritable and environmental influences (the latter including individual and social learning). If developing animals are too solitary for social learning and have no time for individual learning, and if communicated conditions are relatively predictable within and between generations, ontogenies often rely most heavily on heritable factors. For example, most insects, anurans, nonoscine birds and mammals do not require learning to acquire vocalizations and codes for mate attraction and courtship. As a result of "arms races" between males, oscine birds do develop larger and more diverse song repertoires. Thus, arms races can cause frequent changes in the appropriate repertoires and favor learning over strongly heritable acquisition. Individual or flock recognition signatures are also signals that cannot be predicted between generations; receivers always have to learn which signature applies to which individual, and in some cases, senders need to learn to create the signatures.

7. The costs of encoding and decoding increase with the number of alternatives each party has to consider. Large codes take more sensory and brain tissue, and large numbers of alternative signals make difficult the optimal allocation of signals that will minimize propagation distortion, energy costs, and predator risks. Receivers can reduce costs if questions can be made binary. Examples include: Is a conspecific a male or female? (**binary assignment**), Which animal in a crowd is one's mate or offspring? (**binary recognition**), and Which male has the stronger smell (**binary comparison**). These questions can usually be handled with a minimum of templates and alternatives. **Manifold questions** are those that can have many different answers. They are best encoded and decoded by invoking an iconic rule. Examples include the advertisement of food finds by dancing honeybees, and motivational variations in territorial calls by frogs and apes.

Iconic rules are not a solution when members of a social unit must be able to recognize each individual in the group. This is the most difficult manifold task and takes considerable investment by both parties to be done effectively.

8. The total number of signal variants produced by a sender or discriminated by a receiver is its signal **repertoire size**. This depends upon the number of questions and signal sets considered (**functional diversity**) and the average number of alternative conditions and signal variants provided for each question (**variant diversity**). Some authors believe that repertoire size is limited by brain and sensory constraints. If it were, then any increase in functional diversity would result in a decrease in variant diversity and vice versa. Thus, to understand the optimality of observed signals, we could not examine signal sets in isolation, but instead would have to consider the entire repertoire and tradeoffs within it. In fact, it is not yet clear how often repertoire sizes are at an evolutionary ceiling. This issue, which has very important repercussions for how we interpret the coding systems we see in nature, awaits future research.

FURTHER READING

Many introductions to information theory include discussions of optimal coding (e.g., Pierce 1980; Rosie 1973). However, their focus is on the maximization of information transfer rates, which is rarely the optimization criterion for animal systems. Studies of human language (e.g., Cherry 1966) have been important in suggesting the types of codes that might occur in animals; representative discussions can be found in Smith (1977) and Marler (1977). To date, the most focused treatment of coding in animal communication is by Green and Marler (1979). Subsequent studies examine specific coding options. Smith (1977, 1986; Smith and Smith 1996) remains a major proponent for contextual coding. Good reviews of categorization and categorical perception can be found in Herrnstein (1991) and Harnad (1987). Marler et al. (1992) review evidence that compound signals deal with both motivational and external referent information. We have focused on the adaptive benefits of alternative ontogenies in this chapter. Readers interested in the physiological details can consult nearly any text in animal behavior or ethology. Two thoughtful papers by Stephens (1991, 1993) discuss the relative merits of learning versus nonlearning acquisition of strategies, and the classic works by Boyd and Richerson (1985, 1988) contrast the relative optimality of individual versus social learning in different contexts. Acquisition of bird song is reviewed by Catchpole and Slater (1995), and Baptista (1996) reviews the relative contributions of heritable and learning influences on acquisition of vocalizations in all bird groups.

Chapter 16

Signal Evolution

HOW DO COMMUNICATION SIGNALS ACTUALLY ARISE? There are intrinsic properties of signal exchanges that set their evolution apart from that of most other adaptations. We have seen that the communication process requires an interdigitating set of adaptations by both sender and receiver. Each party pays costs to participate, but recoups its losses through benefits provided by the other party. How can such a exchange get started? Unless both parties play their role from the start, won't an initiator have to pay costs that are not compensated? The accepted solution to this problem is that one or the other party must first evolve precursors of their eventual signaling role for reasons other than communication. Once such precursors are in place, the other party can then take advantage of them and this will be sufficient to initiate subsequent coevolution. Scenarios for signal origin thus fall into two categories: sender preadaptations and receiver preadaptations. The former were a major focus of the early days of ethology and many presumed examples can be cited; the second are of more recent concern and the current evidence is still thin. In this chapter, we first begin by examining what specific preadaptations would be required under each scenario. We then turn to evidence that particular visual, auditory, and olfactory signals arose from sender preadaptations. Finally, we discuss several models by which receiver preadaptation could lead to new signals and the degree to which these are supported by current data.

ALTERNATIVE SCENARIOS FOR SIGNAL EVOLUTION

The basic communication chain as we have described it in this book is shown diagrammatically in Figure 16.1. A sender couples certain cues or incipient signals with certain conditions according to a sender code (link 1). Once such a cue is emitted, a receiver must be able to detect it against the background (link 2), combine the detection with a decoding scheme to generate a new estimate of condition likelihoods (link 3), compare these estimates to a decision rule based on expected payoffs (link 4), and finally undertake a response (link 5). It is the response of the receiver that determines the payoffs to both parties. Over evolutionary time, these payoffs feed back on the relevant links in the chain. Links are enhanced where net payoffs are positive and diminished or broken where they are negative.

Preadaptations that install any link in this chain will make it easier for a new signal to evolve. This conclusion follows from the models developed at

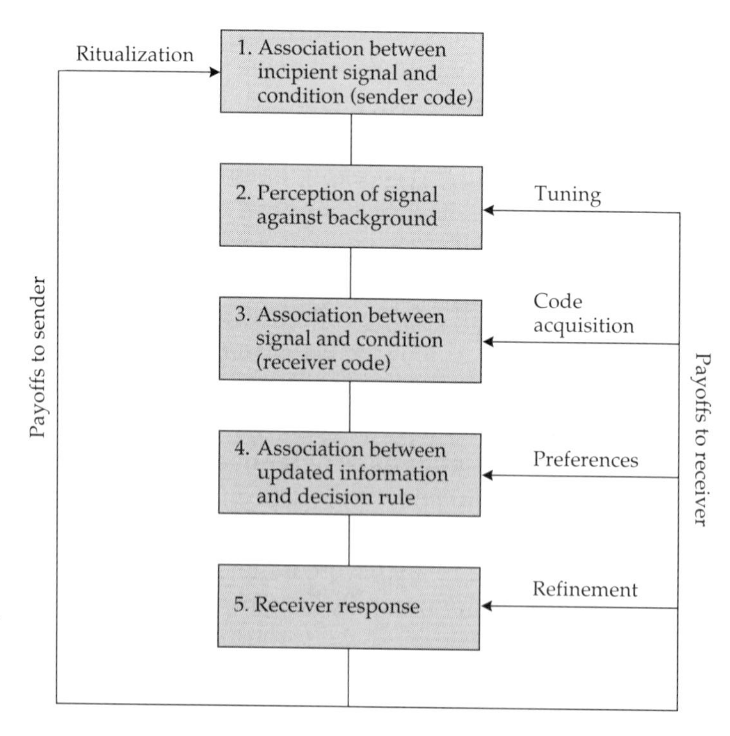

Figure 16.1 A model of the process of signal evolution. Signal evolution begins with the association between an incipient signal (such as an unintentional sender cue) and a condition. Receivers must be able to perceive the cue, and recognize its association with the condition. Receivers then incorporate the information into a decision rule and a response. If receivers benefit from their response, they will fine-tune their sensitivity, recognition code, decision rule, and response. If senders benefit from the response, the cue will be modified via ritualization to maximize information transfer and transformed into a true signal.

the end of Chapter 14 (Figures 14.13 and 14.14). The sender has to provide a minimum level of association between a signal and a context before a receiver will pay attention to it, and a receiver has to be able to make a minimum investment in reception organs before it can detect the signal. Classical ethologists focused for many years on the evidence that the relevant preadaptations occurred in senders. The basic scenario was nicely summarized by Otte (1974). The first and critical step is the establishment of an association between a specific condition and the production of a motor pattern, structure, sound, or chemical cue by the sender. This association might arise because the cue is an inadvertent or unavoidable byproduct of some activity performed in a specific context. The association may be initially imperfect, but it does constitute a potential source of information to other animals. Animals are generally designed to attend to such cues if the cues help make better decisions. One need invoke no special machinery to imagine potential receivers that might try to use these inadvertent cues in their decision making. If the information benefits the receiver, subsequent evolution may favor enhancement of the relevant links. For example, it may be worth the cost to modify the **tuning** of sensory organs to better detect and discriminate these cues. Cue reception is only useful to the degree that the receiver has an accurate a priori estimate of which conditions elicit which cues (e.g., the decoding rules). Positive feedback from cue use may favor new and improved methods of **code acquisition**. The value of cue information depends critically on how it is invoked to make decisions. Once attention to cues proves beneficial, selection may favor specific **preferences** to regulate choices. Finally, the presence of this new information may favor the receiver adding new responses or refining existing ones (**response refinement**).

To this point, the classical ethology scenario requires little that is unlikely. The bottleneck step, however, is the next one. There will be no subsequent coevolution between the parties unless the change in decisions by receivers using the cues has some fitness consequences for senders. If sender fitness is increased, there will be positive feedback on the sender to improve on the reliability, distortion-resistance, and information content of the cue. **Ritualization** is the ethological term for the refinement of an inadvertent cue into a true signal (Tinbergen 1952a). Ritualization involves one or more of the following four changes in the cue: (1) simplification, or reduction in the number of components, (2) exaggeration of the remaining components, (3) repetition of the signal, and (4) stereotypy, or reduction in the variance of signal form during repeated renditions of the signal. All signals will not undergo the same degree of ritualization; if the incipient signal from the outset is well associated with the context and provides all of the information the receiver needs, it may not be ritualized at all. A minimally ritualized signal may be difficult to distinguish from a cue.

Once selection favors sender ritualization, each change in the sender's signal design will result in corresponding selection on the receiver's perceptual tuning, code acquisition, decision preferences and alternative responses. This will in turn have effects on the sender which, if positive, will result in a new round of ritualization. The result is thus a cyclic loop of coevolution that can cause the final signals to be quite different from the original cue precursors. Finding the likely precursors of current signals has been a major preoccupation of ethology. Luckily, inadvertent cues often remain within the behavioral repertoire of a species even though they may have been the precursors of signals earlier in a species' history. Thus ethologists have attempted to trace the likely sequence of changes that must have occurred between cue and signal. The way that this is done is outlined in Box 16.1. Finally, a highly ritualized signal may become **emancipated** from the internal and external factors that originally triggered it. Emancipation thus removes the coupling between the initial cue and condition and implies that new information is provided by the signal.

The second major scenario suggested for signal evolution relies on receiver preadaptations—perceptual biases in the receiver that are not currently applied to signals. These biases will have arisen (as with sender cues) in contexts other than communication. When a mutant sender happens to produce a trait that stimulates one of these latent biases, it may trigger the entire receiver chain, including a response. For example, a mutant male might stimulate a latent bias in female receivers and be chosen as a mate even though it provided no new information of use to the female. In the case of sender preadaptations, we saw that exploitation of sender cues by receivers would lead to signal coevolution only if altered receiver responses benefited the exploited senders. In the same way, we expect exploitation of receiver preadaptations by senders to lead to sustained signal coevolution only if there is a net benefit to receivers. How this might occur is taken up later in the chapter.

It is possible that real signal systems might evolve as a consequence of preadaptations by both parties. However, these need to be compatible, or selective forces on the two parties may cancel out. Coupling of precursors could arise if the same brain center controlled both signal production by senders and signal perception by receivers. Adaptations in such a center could thus affect both parties simultaneously and in a complementary fashion. The possibility of this scenario was indicated by hybrid studies of crickets and frogs (Hoy et al. 1977; Gerhardt 1974). Male F_1 hybrids in both groups produce simple pulsed mate-attraction vocalizations that are intermediate in temporal patterning between the two parental species, and female F_1 hybrids are found to prefer these intermediate hybrid vocalizations. Although this result is suggestive of genetic coupling, a similar result would be obtained if both production pattern and receiver preference were independent but polygenic traits. Follow-up studies with F_2 hybrids and hybrid-parental backcrosses have confirmed the independent genetic basis of sender and receiver phenotypes (Butlin and Richie 1989). No clear cases of coupled precursors have been found, so either compatible mutual precursors arise by other mechanisms or signal evolution is driven primarily by preadaptations in one party.

We now turn to a review of evidence supporting sender preadaptations as the starting point in signal evolution. Examples from each of visual, auditory, and olfactory modalities are described. The breadth of examples provides strong support for the sender-precursor hypothesis, and contrasts between the modalities sheds considerable light on how the subsequent coevolution works.

SENDER PRECURSORS OF VISUAL SIGNALS

We begin our examination of the evidence for the classical scenario of signal evolution with visual signals. Visual signals consist of movements, postures, and physiological processes with visible external effects. During the process of ritualization they may become exaggerated, and color is sometimes added to enhance the conspicuousness of the signal. Because it is often possible to determine the nonsignaling source of a visual display, it is also possible to make an educated guess about the initial meaning of the cue as it evolved into a signal. The diversity of visual signals is high, but the sources of visual signals can be divided into three basic categories: intention movements, the consequences of conflicting motivational tendencies, and autonomic responses.

Intention Movements

When an animal is about to embark on some mode of action such as attacking, fleeing, or copulating (to mention only a few), it must often prepare itself by assuming a certain posture, placing its limbs in certain positions, or exposing certain parts of the body that will be used in the act. Although the subsequent behavior is always preceded by the preparatory action, the initial phases may be performed without completing the follow-up behavior. Such incomplete initial acts are called **intention movements** (Daanje 1950; Tinbergen 1952a). This term is unfortunate because it implies that the animal *intends* to perform the subsequent behavioral act. (See the discussion on sender intentions on page 6.) Although an animal may honestly indicate its intentions, so-called intention movements may be ritualized signals indicating the motivational state that the sender is in, or even be used to mislead a receiver as to what the sender will do next. Whether ritualized or not, intention movements derived from a specific type of behavioral action often provide information to the receiver about the actions the sender would like the receiver to think it will perform next.

The preparatory phases of attack behavior are intention movements that frequently become ritualized into threat signals. Some examples are illustrated in the left column of Figure 16.2. A direct stare, a tensed and forward body posture, the baring of teeth, horns, claws or other weapons, and the protection or pulling back of sensitive body parts such as ears commonly evolve into threats that are used in aggressive contexts. It is not difficult to see how such a signal can evolve. Before a dog can bite a rival or enemy, it must draw back its lips to expose its teeth. An individual that performs the preparatory movements but stops short of completing the action sends the message that it is more likely to bite now than before the signal was given. This information is likely to cause the rival to respond by retreating. Individuals that frequently prepare to bite but fail to follow through may be favored over individuals that

Box 16.1 Determining the Nonsignaling Source of a Visual Signal

IDENTIFYING THE NONSIGNALING SOURCE OF A SIGNAL is not always an easy task and depends on the signal's degree of ritualization. In many cases, the unritualized origin of a signal is clearly evident because the form of the signal is very similar to the unritualized source and both still exist in the same animal. The signaling and nonsignaling versions of the behavior are distinguished mainly by the contexts in which the behavior is performed. For example, many birds communicate with feather erection postures that are similar to the feather movements used in ordinary thermoregulation. Fluffing and sleeking of the feathers are predictable responses to changes in ambient temperature that help the bird maintain a constant internal body temperature. Fluffing and sleeking feather postures are also observed in agonistic, courtship, and allogrooming interactions with conspecifics when no change in temperature has occurred. Feather posture signals are clearly ritualized versions of thermoregulatory behaviors. In other cases, displays are highly ritualized and bear no resemblance to any nonsignaling behaviors in the species' repertoire. The only way to infer the possible source of the signal is to examine close relatives of this species that exhibit the same kind of signal in a less highly ritualized form. This technique is called the comparative approach.

The classic comparative approach, used by Darwin to build his case for evolution, is to demonstrate the existence of a logical series of steps in trait elaboration among a group of extant species (e.g., as we did for the evolution of eyes in Figure 9.4). A well-known communication signal that was examined in this way is the courtship display of the peacock and its relatives in the Phasianidae (Schenkel 1956). The fanning of the extravagant eye-spotted tail in front of the female is unlike any other nonsignaling movement of this species. However, by looking at other pheasants, one can reconstruct a plausible set of intermediates as illustrated in Figure A. These species do not constitute an evolutionary series, but the comparison suggests that ritualization from the same source behavior has proceeded to different degrees. The male bobwhite quail (*Colinus virginianus*) (A) feeds his mate and attracts her with a food call and ground pecking display called tidbitting when he has discovered food. Domestic chickens (*Gallus*) (B) and ring-necked pheasants (*Phasianus colchicus*) (C) perform a similar display with mock pecking and manipulation of food but no offering of food to attract hens when they have found it. In the impeyan pheasant (*Lophophorus impejanus*) (D) the male first rhythmically pecks at the ground, then poses with head held still and makes a low bow and tail fan. The peacock pheasant (*Polyplectron bicalcaratum*) courtship display (E) is similar to the impeyan pheasant's but is preceded by scratching and is sometimes followed by feeding the female; note the elaboration of eye spots on the tail feathers. Peacock (*Pavo cristatus*) courtship (F) contains the low bow and tail fan with extreme tail elongation; the bill is pointed to the ground but no food or pecking is involved. It therefore appears to be a highly ritualized and derived form of courtship feeding.

A second approach is to estimate phylogenetic relationships among species and then to use this estimate to reconstruct the characteristics of their ancestors (Brooks and McLennan 1991; Maddison and Maddison 1992). This method is becoming more frequently applied as phylogenies based on molecular data become readily available. Irwin (1988) used this method to study the evolution of songtype repertoires and song complexity in Emberizine sparrows from the genera *Zonotrichia*, *Junco*, *Passerella*, and *Melospiza*. In this case, the phylogeny of the taxon was estimated from allozyme data.

Figure B below shows the resulting phylogenetic tree along with the range of repertoire sizes and the average number of syllable types per song for each species. The standard way to estimate ancestors is to use **parsimony**, a procedure based on minimizing the amount of evolution required to produce the observed diversity. In the example, song-type repertoires have been coded as simple (S) or complex (C). The ancestor to the group is reconstructed as C with a single change to S associated with the origin of *Zonotrichia*. Note that if the ancestor were reconstructed as S, there would have been two originations of increased song complexity, *Junco/Passerella* and *Melospiza*, which is less parsimonious. Under this reconstruction, *Zonotrichia* sparrows reduced their song-

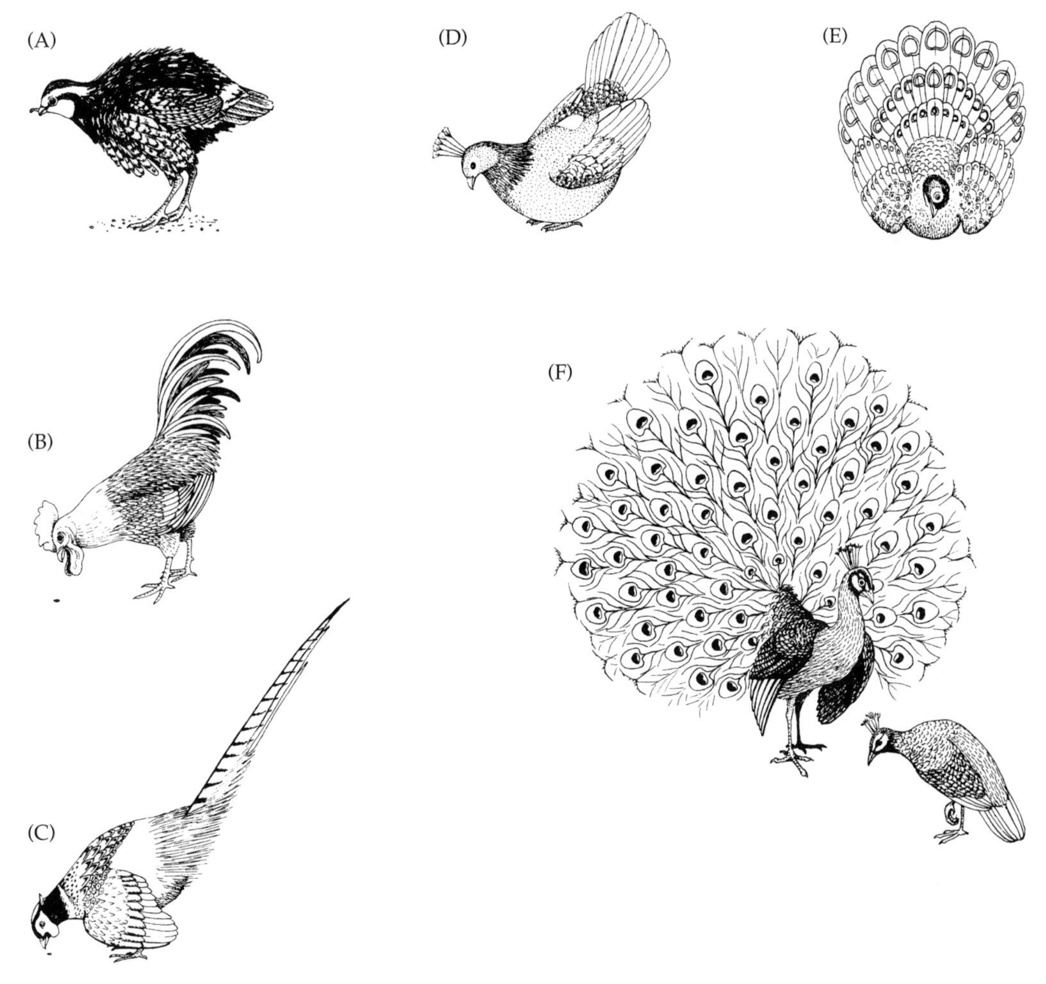

Figure A Degrees of ritualization in the courtship displays of pheasants (Phasianidae) from a food-advertising source. (After Brown 1975; Schenkel 1956.)

Box 16.1 (continued)

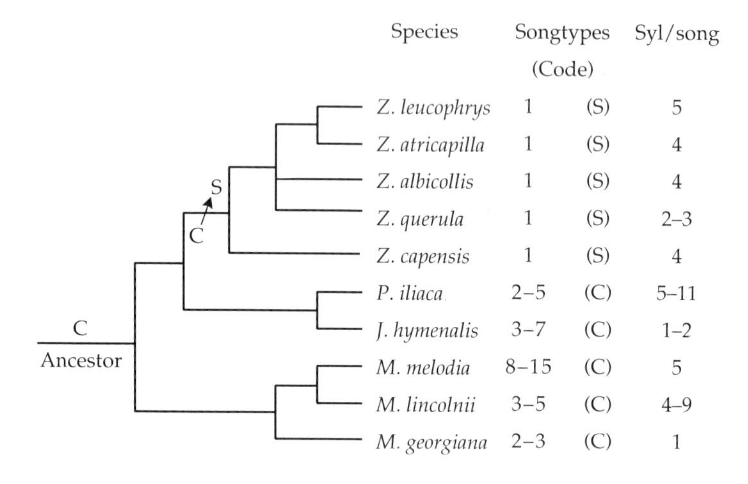

Figure B Phylogenetic tree of some Emberizine sparrow species based on allozyme data. Genera are *Zonotrichia*, *Passerella*, *Junco* and *Melospiza*. S = simple song repertoire; C = complex repertoire.

type diversity to one during evolution, whereas the song sparrow, *M. melodia*, has enlarged its repertoire and two species have increased their syllable complexity. The swamp sparrow, *M. georgiana*, has retained the intermediate ancestral repertoire size but greatly simplified the syllabic structure. An important point from this study is that signal complexity not only increases during evolution, but may also decrease.

There are several difficulties with attempts to reconstruct ancestors in this way (Frumhoff and Reeve 1994), including uncertainties in estimates of phylogenetic similarity, difficulties in coding the traits (as illustrated here), and the fact that evolution is often not parsimonious. Methods for placing confidence limits on estimates of ancestors are being developed, and these limits are often found to be large. A recent communication-related example of these difficulties involves the controversial evolution of sworded tails in *Xiphophorus* fishes (Basolo 1990; Meyer et al. 1994). Despite the problems, phylogenetically based comparative studies can tell us much more about the direction of evolution than the classic comparative method. For additional examples, see Ryan and Rand (1995), Martins (1993a), and Figure 16.16.

Figure 16.2 The principle of antithesis. For each of the species illustrated here, the ▶ primary aggressive display is shown on the left and a submissive or fearful display on the right. Notice that display features such as body posture, orientation, head position and degree of piloerection show opposite extremes in the two displays. Aggressive displays generally reflect attack preparation movements and postures, which differ for each species. A high head position is often associated with aggression as in the dog and gull, but not in the sparrow or heron. Erected head feathers or fur are also frequently associated with aggression, but the sparrow employs crest erection in its submissive display. (A after Darwin 1872; B after Meyerreicks 1960; C after Hailman 1977b; D after Moynihan 1955; E after van Hooff 1967.)

Aggressive displays

Submissive and fearful displays

(A) Domestic dog *(canis domestica)*

(B) Green heron (*Butorides virescens*)

(C) Fox sparrow (*Passerella iliaca*)

(D) Black-headed gull (*Larus ridibundus*)

(E) Macaque (*Macaca* spp.)

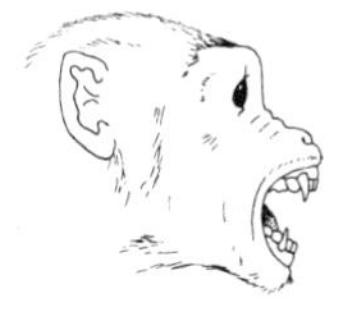

only bare the teeth when they are truly about to bite; such bluffers can win the contest without having to pay the cost of a fight. The reduced likelihood of becoming engaged in a costly fight will lead to exaggeration of the retreat-inducing intention movement, such as pulling the lips back more than is necessary. This is a clear example of the ritualization process that occurs once the sender benefits from the receiver's response to its signal.

The displays that animals use to signal nonaggression or submission are often precisely the opposite in form of those that signal threat or aggression. The right column of Figure 16.2 illustrates the opposing submissive display for each of the threat displays on the left. Looking away from the opponent, turning the body away, closing the mouth, hiding weapons, and exposing sensitive body parts are common submissive gestures. Darwin (1872) was the first to recognize the opposing nature of aggressive and submissive displays and termed the phenomenon **antithesis**. Aggressive and submissive displays may be antithetical in form because they arise independently from opposing actions (e.g., approach versus retreat). However, additional antithetical elements may be incorporated during ritualization to render them maximally divergent. A neural network model of two opposite-meaning signals by Hurd et al. (1995) demonstrates this divergence process when senders benefit from making these two intentions clear to receivers. (See Box 16.2, on pages 530–531 for a general description of neural network models.) Antithetical display systems represent an example of sender code categorization (Figure 15.6).

Intention movements provide information to receivers in a variety of other contexts (Figure 16.3). Preparatory movements for flight are frequently ritualized in flocking birds to coordinate movement (Andrew 1956; Davis 1975). When a bird on the ground is about to take off, it first crouches, raises its tail, and pulls its head in and back. It then extends its head and neck as it jumps into the air. The crouching portion of the behavior signals important information to other birds in a flock about what the individual is likely to do next. Similarly, ducks use a ritualized head jerking as a signal to synchronize taking flight within flocks. Some of the displays observed during courtship have been interpreted as ritualized reproductive intention movements (Andrew 1957). For example, in species with paternal care of offspring, male courtship displays often represent incomplete renditions of nest construction or parental care behaviors. The presentation or manipulation of nesting material is observed in many birds, and courting male fish may perform ritualized nest-fanning displays. Copulation movements and display of sex organs form the basis of ritualized courtship in spiders and some mammals (Ewer 1968; Foelix 1982).

Motivational Conflict

Ethologists view animals as possessing a complement of different motivational systems (McFarland 1985). A typical list of motivational systems might include: hunger, thirst, thermoregulation, grooming, aggression, fear, and sexual systems. For each system, a range of motivational levels is presumed to exist. The motivational state of an animal is an ever-changing dynamic vari-

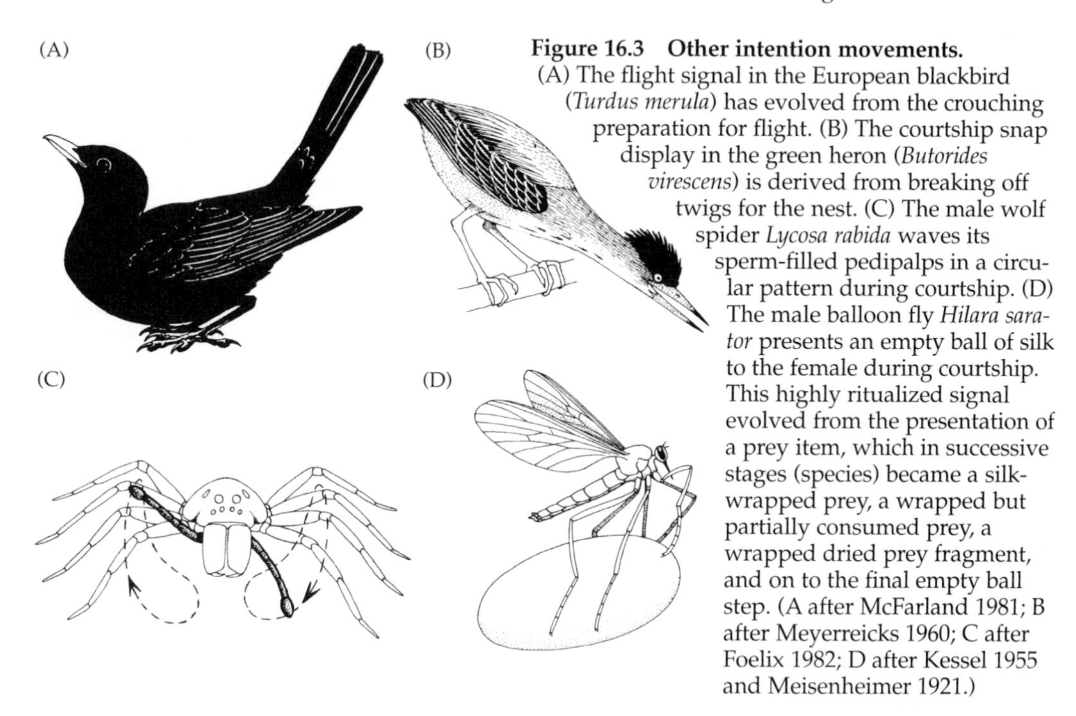

Figure 16.3 Other intention movements. (A) The flight signal in the European blackbird (*Turdus merula*) has evolved from the crouching preparation for flight. (B) The courtship snap display in the green heron (*Butorides virescens*) is derived from breaking off twigs for the nest. (C) The male wolf spider *Lycosa rabida* waves its sperm-filled pedipalps in a circular pattern during courtship. (D) The male balloon fly *Hilara sarator* presents an empty ball of silk to the female during courtship. This highly ritualized signal evolved from the presentation of a prey item, which in successive stages (species) became a silk-wrapped prey, a wrapped but partially consumed prey, a wrapped dried prey fragment, and on to the final empty ball step. (A after McFarland 1981; B after Meyerreicks 1960; C after Foelix 1982; D after Kessel 1955 and Meisenheimer 1921.)

able determined by a combination of the animal's internal (physiological) state and external conditions and stimuli relevant to each system. For example, an individual's current hunger level is a function of its current physiological need for food as well as the strength of perceived food cues in the environment. All motivational systems operate simultaneously, but the motivational level for some systems will be higher than the motivational level for other systems at any given point in time. The motivational system with the highest level will generally dictate the animal's current behavior. For example, if the animal has not fed for a while and is extremely hungry, it will be more highly motivated by hunger than by any other system and therefore will be more likely to engage in foraging behavior. Once it has fed, the hunger motivational level will drop and another motivational system will command the animal's attention.

In situations where two motivational systems are both at similar and high motivational levels, the animal is said to be in **motivational conflict**. The most common motivational conflict situation is the simultaneous activation of aggression and fear systems that invariably occurs in encounters between well-matched rivals such as two territorial males at their joint boundary. Similarly, a three-way conflict between sexual interest, fear, and aggression is believed to take place during courtship in many species. Ethologists described three types of behaviors that occur in such contexts—ambivalence, displacement, and redirected behaviors—which are thought to be the sources of some visual displays. This classical view of animal behavior leads to the conclusion that senders accurately transmit information about their current motivational state (Baerends 1975). We outline the classical view in the discussion below, but consider some alternative explanations for these behaviors as well.

Ambivalence behavior supposedly reflects the animal's indecision over which of two opposing motivational systems to attend to, and results in either: (1) the **alternation** of intention movements characteristic of the two motivational systems, or (2) the **blending** of antithetical intention movements characteristic of the two systems into an intermediate form. The classic example of a ritualized display incorporating alternation between two systems is the zigzag dance of the male stickleback during courtship. The male is simultaneously motivated to attack the female who has just entered his territory and to lead her to the nest. The result is alternated to and fro movements between the female and the nest (see Figure 18.5). An example of the blending of antithetical intention movements is the broadside threat display seen in encounters between rival males of many species. This posture is exactly intermediate between a highly aggressive opponent-facing posture and the turning away or rear-end presentation characteristic of a fearful or retreating animal. Ritualization frequently involves the "freezing" of this position into a stiffly held posture. The broadside threat not only reflects motivational conflict, but it also presents the largest possible surface area to the opponent. Since larger individuals tend to win fights against smaller ones, exaggeration of size is likely to benefit the sender if it causes the opponent to retreat. Ritualization therefore also involves exaggeration of body size with structures that enlarge the profile of the animals (Figure 16.4). More complex blended systems of the fear and aggressive motivational systems have been described in some species. Here there is a series of gradations between the pure fear and pure aggression states in which the level of each motivational system seems to be encoded (Figure 16.5).

Displacement behaviors are defined as acts that are apparently irrelevant to the motivational system in which they appear (Armstrong 1950; Tinbergen 1952a). They are recognizably similar to or derived from motor patterns normal to the species, but are usually short, incomplete, and nonfunctional in their usual sense. Displacement acts commonly appear during motivational conflict situations, where they seem to have nothing to do with either one of the conflicting motivational systems and are interspersed between relevant acts. Examples include feeding, drinking, preening, or sleeping during conflicts with rivals and during courtship (Figure 16.6). Displacement acts also appear in thwarting or frustrating contexts, where an animal is unable to achieve an expected goal.

As with other types of behaviors we have discussed, displacement acts vary in the degree to which they have been ritualized and converted into signals. In the zebra finch (*Taeniopygia guttata*), beak wiping is frequently interjected during courtship by the male (Figure 16.6F). Beak wiping normally occurs during feeding and functions to clean the bill of food particles. In the related striated and spiced finches (*Lonchura punctulata* and *L. striata*), a low bow has become an integral part of the courtship display. The body position

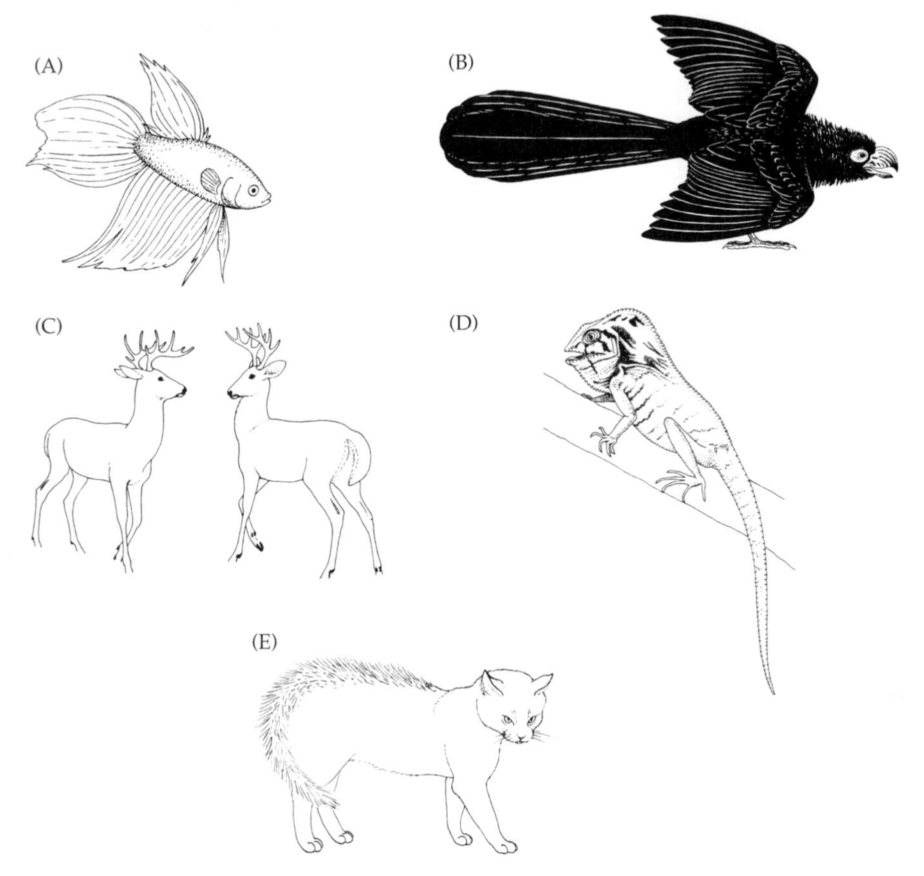

Figure 16.4 Broadside threat displays. (A) Siamese fighting fish (*Betta splendens*) with fins raised. (B) Groove-billed ani (*Crotophaga sulcirostris*) with prominant crested bill, feathers fluffed and shoulders and tail rotated. (C) Two male white-tailed deer (*Odocoileus virginianus*) with antlers held high, (D) Helmeted lizard (*Corythophanes cristatus*) with dewlap extended, opened mouth, prominant helmet, and darkened coloration. (E) Domestic cat (*Felis domestica*) with fur raised along back and tail. (A and E after Hinde 1982; C after Thomas et al. 1965; D after Carpenter 1978.)

during the bow is very similar to the position taken during beak wiping, suggesting that the bow is a ritualized form of this bill cleaning behavior (Morris 1959). Similarly, the process by which preening behaviors have become ritualized into courtship displays is quite easy to see in a series of duck displays (Figure 16.7).

There is considerable disagreement over the cause of displacement behaviors and their functional significance (McFarland 1985). Classical ethologists view them as basically nonadaptive responses to conflicting or frustrating contexts. One theory is that a tension that is built up during a conflict is released by the performance of irrelevant behaviors. Another is that the conflicting motivational tendencies cancel each other out and allow a third, lower-

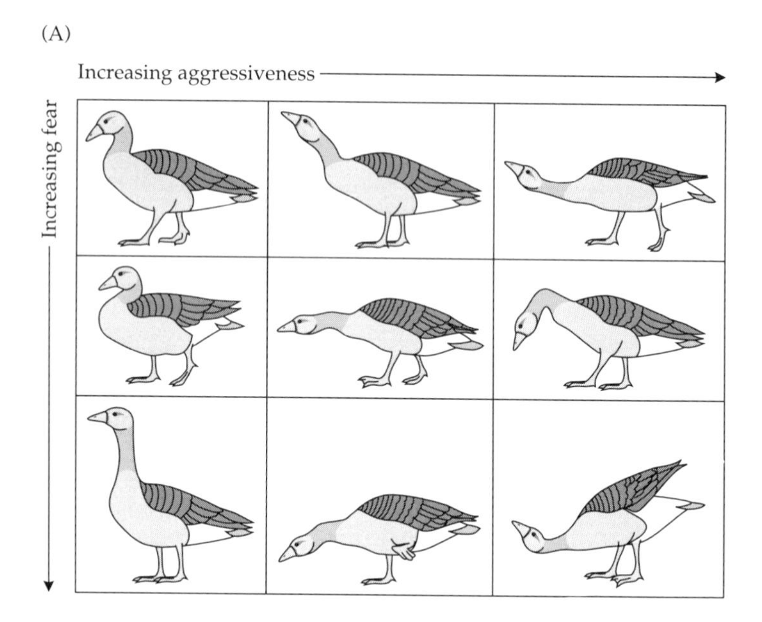

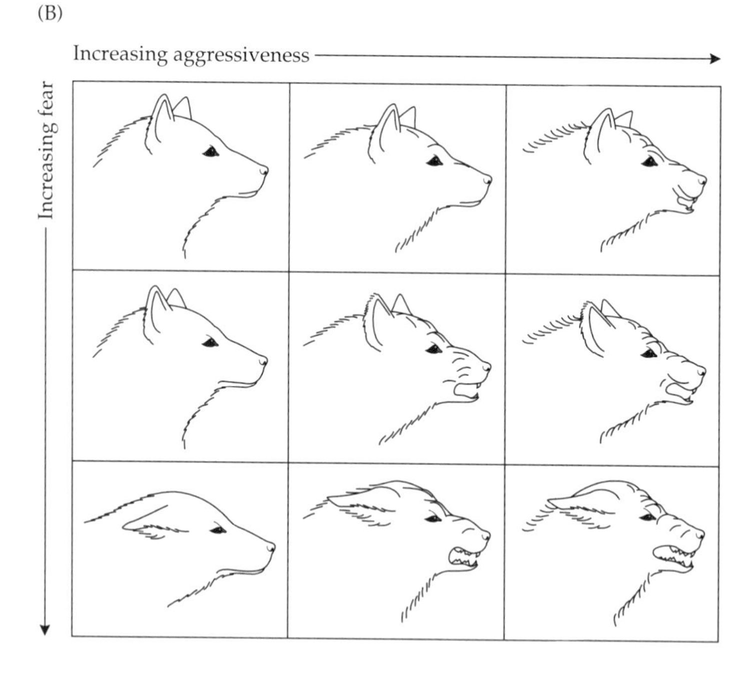

Figure 16.5 Systems of blended displays with different ratios of fear and aggression. The upper-left picture in both cases shows the relaxed posture, the lower-left shows extreme fear, the upper-right shows extreme aggression, and the lower-right shows a mixture of extreme fear and aggression. (A) The goose uses a forward-stretched neck and upturned head to express aggression, a compressed neck to express submission, and a tall head when alert or alarmed. (B) The wolf uses flattened ears to express degrees of fear and baring of the teeth plus piloerection to express aggression. (A after McFarland 1981; B after Lorenz 1953.)

level motivational system to briefly take over. If such acts reliably occur in certain conditions and result in useful information for receivers and beneficial responses for senders, the acts may become ritualized into displays. For example, in territorial disputes the territory owner is strongly aggressive toward

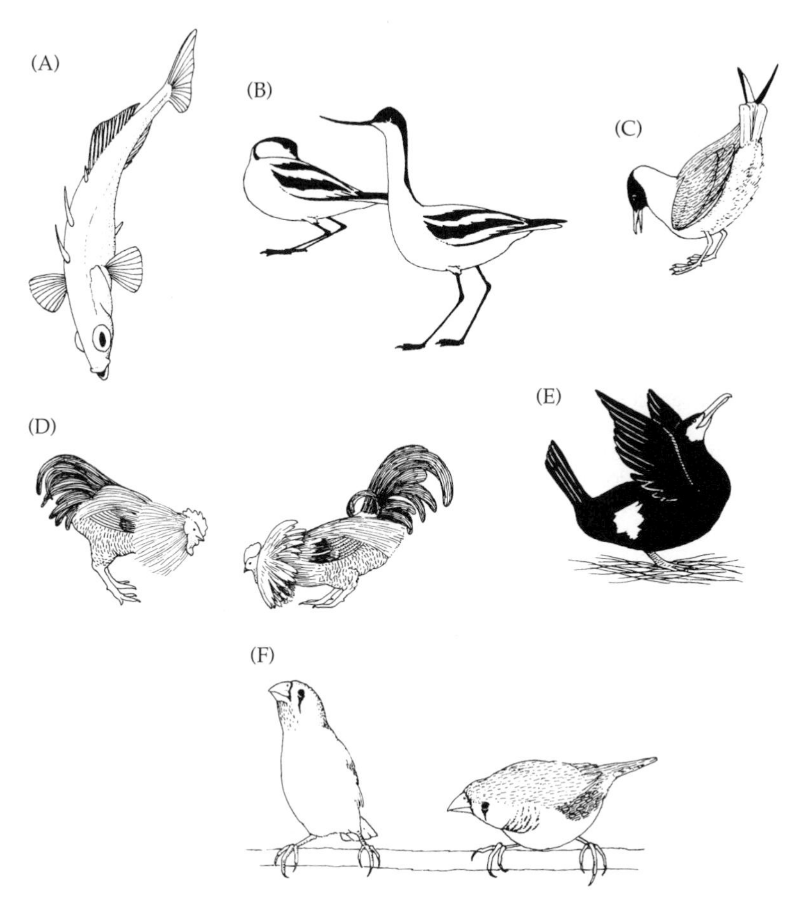

Figure 16.6 Examples of displacement acts (A) Threat display of three-spined stickleback (*Gasterosteus aculeatus*) ritualized from displacement sand-digging. (B) Displacement sleep in the European avocet (*Recurvirostra avosetta*) during a fight. (C) Choking display in the herring gull (*Larus argentatus*) observed during boundary disputes; display is similar to that between mated pairs during nest building. (D) Displacement feeding in domestic cocks (*Gallus domesticus*) during fighting. (E) Displacement sexual behavior in the European cormorant (*Phalacrocorax carbo*) during aggressive encounters. (F) Displacement bill wiping during courtship in the zebra finch (*Taeniopygia guttata*). (A–E after Tinbergen 1951; F after Morris 1959.)

any intruder it meets in the center of its territory and fearful during encounters made outside the territory. At a point near the boundary the two tendencies balance or come into conflict, and the performance of ambivalent or displacement behaviors is a good indication of how far the resident is prepared to go in defense of his territory (Sevenster 1961). Modern adaptationists are uncomfortable with this mechanistic view of behavior and argue that irrelevant acts may in fact be tactical strategies to deescalate a conflict, throw a rival off its guard, or otherwise manipulate the receiver. We shall take this issue up again in Part III.

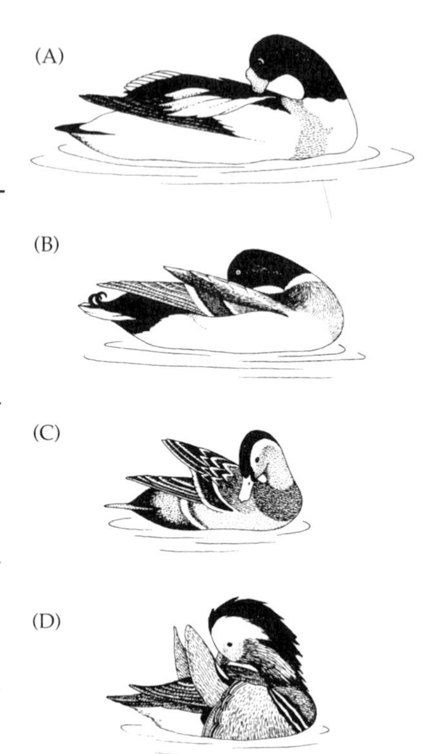

Figure 16.7 Comparative analysis of increasing degree of ritualization of displacement preening during courtship in various duck species. (A) Displacement preening in the shelduck (*Tadorna tadorna*) occurs during conflict situations but is similar to normal preening and therefore relatively unritualized. (B) In the mallard (*Anas platyrhynchos*), preening during courtship is partially ritualized, in that the preening movements are restricted in range and directed at a patch of conspicuously colored feathers on the wing. (C) In the garganey (*Anas querquedula*), ritualization has proceeded further, with incomplete movements that no longer serve any grooming function directed toward a light-blue color patch. (D) The mandarin drake (*Aix galericulata*) has the most highly ritualized preen, with the bill merely touching an enlarged conspicuous rust-red feather that becomes erected during courtship; the crest on the back of the head further emphasizes the preening movement. (After McFarland 1981, based on Lorenz 1941.)

The third type of behavior that is sometimes observed during motivational conflict situations is **redirected behavior**. These are behaviors in which the form of the act is appropriate to the context but it is directed toward an irrelevant stimulus. Examples include the attacking of an innocent bystander or an inanimate object during a fight, drinking or feeding movements directed at smooth surfaces or shiny pebbles by a thirsty or hungry animal denied access to water or food, respectively, or a male copulating with an inanimate object after being rejected by a female. Redirected behaviors can become ritualized, as in the case of grass pulling in fighting herring gulls; they appear to seize and pull on clumps of grass as if they were the wings of their opponents (Figure 16.8). As with ambivalent and displacement behaviors, redirected acts could convey information about the conflicting motivational state of the performer. However, they could also be interpreted as intention movements, tactical acts or signals that indicate the sender's fighting ability (see Chapter 21).

A final source of signals arising from conflict situations is **other displays**. Once an informative signal that elicits a specific type of response has evolved, senders can use this signal in another context in which the same type of response is desired. Such signals can appear quite manipulative. For example, senders in aggressive contexts can deflect the attention of rivals by stimulating a nonaggressive motivational system in the receiver (Chance 1962). Juvenile behaviors such as begging and grooming solicitation are commonly used

Figure 16.8 Redirected attack in fighting herring gulls. During fighting in herring gulls (*Larus argentatus*), one bird may grab a hold of a strong tuft of grass and pull on it. Some ethologists describe this behavior as a redirected version of the attack behavior in this species in which a bird grabs a hold of its opponent's wing and pulls. Others describe this as displacement nest construction behavior, since gulls use a similar motion to collect grass for nesting material. (After Tinbergen 1951, 1959.)

by adult senders to trigger a nonaggressive, parental response by the receiver. Sexual behaviors are also employed to deflect aggression. In mammals, and particularly primates, subordinates of both sexes use a form of the estrous female-mounting solicitation, presenting the rear end, against aggressive dominants (Figure 16.9). The dominant individual often acknowledges the gesture with a brief ritual mock mounting. A dominant male in particular would not normally attack a receptive female, so the mimicry of female sexual behavior by subordinate males serves a clear adaptive function. A final example of secondary display evolution is the use of nestling food-begging behavior by adult females of many bird species during courtship. The male often responds by courtship feeding the female. The exchange serves, in part, to appease the male and prevent him from attacking the female, but it also provides direct nutritional benefits to the laying female.

Autonomic Processes

The autonomic nervous system of vertebrates is an involuntary system of nerves that controls the internal organs such as the heart, blood vessels, lungs, intestines, eyes, and certain glands. The function of the autonomic system is to facilitate a rapid physiological response in stressful situations. As examples, when an individual must exert itself to run or fly, the autonomic system acts to accelerate the heartbeat rate, dilate air passages to the lungs, increase the blood supply to the muscles and reduce intestinal activity; when ambient temperature increases or decreases, the autonomic system responds by increasing or decreasing blood flow to the skin and by fluffing or sleeking feathers or fur to regulate internal body temperature; and when ambient light changes, the autonomic system dilates or constricts the pupil of the eye to regulate the amount of light entering the eye.

Any autonomic response that produces some visible change may provide the source for a ritualized visual display. Because the autonomic system is

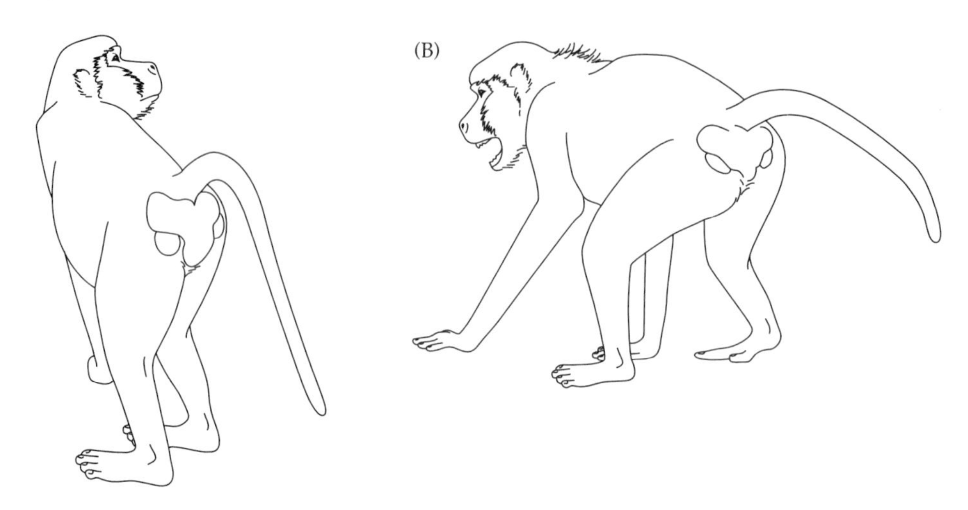

Figure 16.9 Evolution of displays from other displays. (A) A genuine sexual presentation by a female baboon (*Papio* spp.). She typically looks back over her shoulder towards the male. (B) Appeasement presentation by a subordinate male towards a threatening dominant individual. The general stance is the same but the expression on the face is fearful. (After Bolwig 1959.)

often involved in responses to stressful or fearful stimuli, visible manifestations of the response are often signals of fear. Examples include feather sleeking in birds and pallor and sweating in man. When aspects of an autonomic response become ritualized, however, they may come under some degree of voluntary control and the normal function may even be reversed.

Piloerection, or the fluffing of feathers or fur, is a common thermoregulatory response in birds and mammals with a highly visible manifestation that is frequently ritualized for visual signals. The ritualization process usually involves restricting the erected feathers or fur to a small part of the body, elongating and/or coloring the feathers or fur to exaggerate the display, and evolving voluntary control of the muscles that cause piloerection (Morris 1956). In birds, the movement of general body plumage is only slightly ritualized and follows the pattern that we would expect from thermoregulatory responses. Sleeked body feathers are associated with aggressive tendencies, a context in which individuals are preparing for action and need to minimize feather insulation to reduce overheating. Fluffed body feathers are associated with submissive tendencies and reflect the thermoregulatory needs of an inactive individual. Highly ritualized feather erection displays in birds, on the other hand, involve specialized regions of the body and show nearly the opposite relationship with motivational state (Figure 16.10A). Erected feather positions are associated with aggressive or excited tendencies, while sleeked feather positions are associated with fear or submission. For example, in the Stellar's jay illustrated in Figure 13.1C, a fully raised crest indicates maximum aggressive arousal and a fully depressed crest indicates submission. In mammals, fur erection is often highly ritualized (Figs 16.2A, 16.4E, 16.10B) and shows the same type of relationship

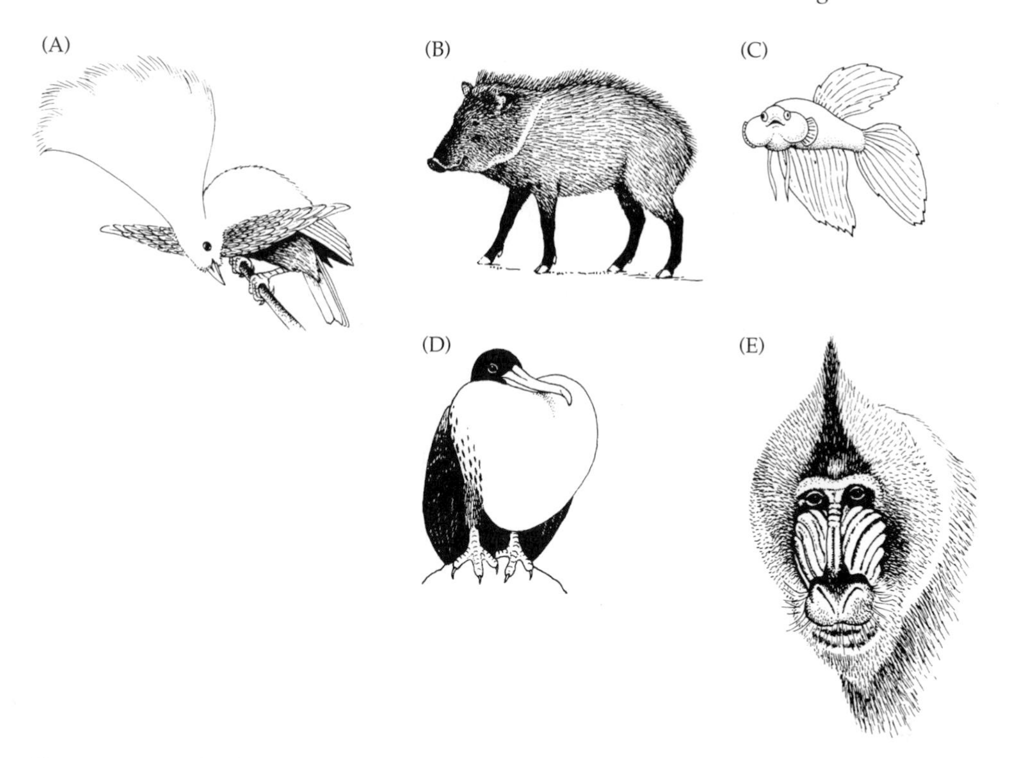

Figure 16.10 Displays derived from autonomic responses. (A) Courtship display in the superb bird-of-paradise (*Lophorina superba*). Elongated black neck feathers can be raised above the head or sleeked along the back, and shiny blue-green breast feathers can be extended to the sides. (B) Ambivalence display of the collared peccary (*Tayassu tajacu*) with fur raised along the back. (C) Erected gill covers in the Siamese fighting fish (*Betta splendens*). (D) Inflated air sacs in a frigatebird (*Fregata* spp.). (E) Colorful red, white, and blue nostril display of the male mandrill (*Papio sphinx*). (A after Cooper and Forshaw 1977; B after Hailman 1977; C and D after McFarland 1981; E after Macdonald 1984.)

with motivational state as the highly ritualized feather displays of birds. These specialized piloerection displays are thus examples of emancipation from the original function during the ritualization process.

A few other interesting examples of physiological processes that have become ritualized into visual displays are worth mentioning. In parrots, pupil dilation has been ritualized into a visual display. Recall from Chapter 9 that the pigment epithelium layer of the retina performs most of the light regulation function in birds, freeing the iris for display. Aggressive or highly excited individuals display by rapidly expanding and contracting the pupil. Parrots that use pupil dilation as a signal invariably have a brightly colored iris and a contrastingly colored ring around the eye to amplify the movement. Respiratory systems may also be ritualized for visual display (Figs. 16.10C,D,E). The vocal sacs of male prairie chickens and frigate birds used to couple sounds to the medium are brightly colored and visually conspicuous during courtship.

The proboscises of bull elephant seals, the trunks of elephants, and the flaring of nostrils in primates are all used in visual signaling. In some fishes the gill covers are enlarged and held out in frontal visual displays.

To summarize this section on visual displays, the precursors from which a given display evolved can often be identified and linked to a normal (nonsignaling) behavioral activity such as feeding, grooming, nest building, thermoregulation, or defense that directly promotes individual survival or reproduction. Ritualization modifies the behavior to varying degrees to generate a more effective signal. Slightly ritualized displays that still bear a strong resemblance to their original behavior seem to provide reasonably reliable information about what the sender is likely to do next or what type of motivational state or motivational conflict the sender is in. More highly ritualized displays often show little or no relationship to the context or motivation system of their source behavior. The type of information that these displays transmit to receivers is usually more difficult to decipher.

SENDER PRECURSORS OF AUDITORY SIGNALS

Recall from Chapter 4 that there are five basic mechanisms that animals use to produce the vibrations necessary to generate sound. These include stridulation, vibration of membranes, passage of air or liquid across an orifice, vibration of appendages, and percussion on a substrate. Sound-generating movements have evolved from a variety of different sources (Dumortier 1963a).

Visual or Tactile Courtship Displays

Existing signals can be the precursors to additional signals in another modality. For example, movements used in visual or tactile displays can be enhanced by coupling the motion to sound production. External limb movements usually lead to stridulatory or percussive types of sound-production mechanisms. The sound could be viewed as further ritualization of the visual or tactile display that renders it more exaggerated or conspicuous. Visual courtship displays are often suggested as the source of some auditory signals. The auditory component may have a larger radius of detection than the visual display and may evolve into a long-distance mate-attraction signal. In some cases the auditory component completely replaces the original visual component.

The long-horned beetles, family Cerambicidae, provide a good illustration of the transfer of modalities from tactile to auditory. In nonstridulatory species sexual attraction and species recognition are achieved by olfaction and direct contact. This is followed by the male stroking the female's abdomen, which seems to calm the female. The male then begins to tap the female's abdomen with his head in synchrony with the stroking. In one species the rhythmic head movements have been specialized to produce sounds by rubbing together the head and thoracic plates (Michelsen 1966).

Stridulation by males to attract females is characteristic of most orthopterans. The two major suborders of orthopterans show distinct differences in their method of stridulation and position of the ear, suggesting independent origins of the communication systems. In the Caelifera (short-horned grasshoppers), the femur and wing are rubbed together and the ear is located on the abdomen. In the Ensifera (crickets and long-horned grasshoppers), the two wings are rubbed together and the ears are located on the forelegs. In the latter group, at least, the stridulation motion is believed to have evolved from wing movements formerly used for visual/tactile display or pheromone release (Alexander 1962). Note that the ears of orthopterans have evolved from generalized proprioceptive organs (receptors in subcutaneous tissue that respond to stimuli produced within the body) for the explicit function of detecting social signals (Hoy 1992).

In male fiddler crabs (*Uca* spp.), the temporal pattern of waving and the coloration of the large claw form the basis of a species-specific mate-attraction and courtship signal. In some species, the male also thumps the claw on the ground during each wave to produce a percussive sound (Salmon and Atsaides 1968). Since the rhythm of the sound parallels the temporal pattern of the visual component, a second modality for species recognition becomes available for use during conditions of poor visibility and over longer distances. Similarly, many birds augment vigorous courtship displays with mechanical rattling and cracking sounds produced by modified wing and tail feathers. Aerial dives in particular are exaggerated by such sounds in some snipe, woodcock, nighthawks, bustards, hummingbirds, and manakins (Thomson 1964).

Defensive Anti-Predator Acts

Sudden bursts of sound or strong vibrations are used by a wide variety of animals when they are disturbed or grabbed by a predator. The general term for predator-startling behaviors is **protean display** (after the Greek god Proteus, noted for his ability to rapidly change his appearance). The moment of hesitation that this display often causes is sometimes sufficient to allow the victim to escape from the predator. Examples of sound-production mechanisms that have evolved in otherwise silent animals for the sole purpose of warning or escaping enemies include the click and jump of click beetles, the hiss of many salamanders and lizards, and the tail rattle of rattlesnakes. In a few notable animal groups such as the cicada, the sound production apparatus that initially evolved for this interspecific form of communication has been modified and used for intraspecific communication. These insects all feed by sucking juices from plants. Such a mode of life favors a cryptic appearance and a passive, rather than an active, defense against predators. Primitive, sound-producing relatives of the cicadas exhibit either a stridulatory or tymbal apparatus (see pages 76–81). Three factors argue in favor of a predominantly interspecific function—the tymbal apparatus is present in both sexes; sound can be elicited by touching the animal; and the ears of these species are not well-developed. In the true cicadas, only the male possesses the tymbal and both sexes have highly advanced ears with tympana and many sensory cells (Figure 4.5). Although the primary function of sounds in the cicadas is intraspecific, the male can still produce a startling vibration when grasped by a predator (Leston and Pringle 1963).

Normal Locomotory and Foraging Movements

Many animals produce characteristic sounds when moving. This is especially true of flying animals such as birds, bats, and insects. These sounds have been extensively studied in insects such as flies, mosquitoes, bees, and moths that produce a continuous tone while flying (Sotavalta 1963). The fundamental frequency of the tone is directly related to the wingbeat frequency: smaller species tend to produce a higher-pitched sound than larger species, and within a species, wingbeat frequency and pitch increase with increasing ambient temperature. However, only a few species use their flight sounds for intraspecific communication. Male mosquitoes possess sound-sensitive hairs on their antennae that are tuned to the wing-beat frequency of females of their species; they use this ability to locate and identify prospective mates. Some flies, wasps, and bees vibrate their wings in short-distance social contexts such as courtship and hive interactions, but it is not clear whether the sound or the tactile vibration is most important. Other species in the community that interact ecologically with noxious sound-producing fliers often make use of these sounds, however. The mammalian hosts of various blood-sucking and parasitic flies respond negatively to the flight sound of these pests, bats use the flight sounds of their moth prey to locate them, and innocuous insect species that visually mimic toxic butterflies or stinging Hymenoptera also mimic their flight sounds.

Another major locomotor device that has been preempted for use in communication is the swim bladder of fish. As we saw on pages 92–93 and in Figure 6.10, the swim bladder is an internal gas-filled sac that acts as a hydrostatic organ to adjust the specific gravity of the fish as it moves from one depth to another. Some fish use it as a sound transducer for auditory reception. In certain species it is used as a resonator for sound production. The sac may be set into vibration in a variety of ways, including beating the sides of the body with the fins, direct muscular contraction of the bladder, or stridulation of skeletal bones, vertebra, or teeth lying close to the bladder.

Finally, animals employ their appendages and foraging and nest building movements to produce a variety of percussive sounds. Examples include foot stomping by many rodents and lagomorphs (rabbits); tail slapping on the water's surface by beaver; bill, teeth and mandible rattling in birds and mammals; and tapping the ground with abdomen or head in several arthropod groups. The drumming of woodpeckers against hollow trees most clearly reveals the source and ritualization process of a percussive signal (McFarland 1985). The chisel-like bill of woodpeckers not only enables these birds to dig into wood to obtain wood-boring insects, but it is also employed in the excavation of nests in tree trunks. When a woodpecker forages or constructs a nest, it chops with a variable rhythm and thereby produces a sound with a variable temporal pattern and a poorly resonating quality. Males use the same type of motion to produce territorial signals, but they usually fly to known hollow or dry trees to improve the quality of the sound. The drumming in this case is more rapid and precisely patterned. A second type of drumming signal is used by paired birds to signal to their mates that they are finished with a nest excavation bout. They move to the edge of the nest entrance hole and drum with a very slow rhythm. This ritualization process of stereotyping the form and temporal pattern of the display not only distinguishes the social signal from its normal source but also serves to differentiate the two different signals. Just as we saw earlier in the example of the evolution of intraspecific stridulation from an interspecific predator defense function, the social signal exhibits a more regular temporal pattern than the normal source behavior. This characteristic of ritualized displays is called **typical intensity** (Morris 1957).

Respiration

As we saw in Chapter 4, air-breathing vertebrates use the flow of air to generate vibrations of membranes in some part of the respiratory system. Frogs and mammals co-opted the larynx for the site of membrane vibration. The larynx developed along with lungs early in the evolution of terrestrial vertebrates; its function was to prevent food and water from entering the lungs. The pre-vocal larynx was simply the enlarged terminal end of the bony trachea where it joined the pharynx. The larynx supported a muscular closure apparatus that initially was a simple sphincter muscle. With the addition of a second set of muscles to pull the sphincter open, it became the glottis. In the frog, a fold of tissue behind the glottis became the vibrating vocal cord, enabling the animal to produce a high frequency carrier that could be pulsed with the independent glottis. In mammals a new structure, the epiglottis, assumed the function of closing off the air passage against food, freeing the glottis to evolve into vocal cords. In birds, the larynx remained unmodified and instead the syrinx evolved at the junction of the two tracheal branches for sound production.

An important consequence of producing sounds with air flow across vibrating membranes is that the frequency of the sound can be rapidly modulated by changing the tension on the membrane. High tension produces a high-frequency sound, whereas low tension produces a low-frequency sound. In addition, high-tension vibrations result in a more nearly sinusoidal waveform and a purer tone or whistle-like sound, whereas low-tension vibrations result in a more nonsinusoidal waveform and a noisier, harmonically richer sound. Vertebrates, particularly birds and mammals, make use of these properties to produce a variety of different vocalizations. Aggressive senders usually produce sounds that are low in frequency and have a broad frequency band; examples include growls and harsh cries (Figure 16.11, left side). Fearful or submissive senders, on the other hand, produce sounds that are high in frequency and have a narrow frequency band; examples include high-pitched, pure tone whistles (Figure 16.11, right side). These two types of vocal signals form an antithetical signaling system equivalent to the visual displays shown in Figure 16.2.

Morton (1977) has further proposed that frequency and bandwidth variants can constitute a graded or blended system of vocal displays for the expression of different ratios of aggression and fear, analogous to the visual sys-

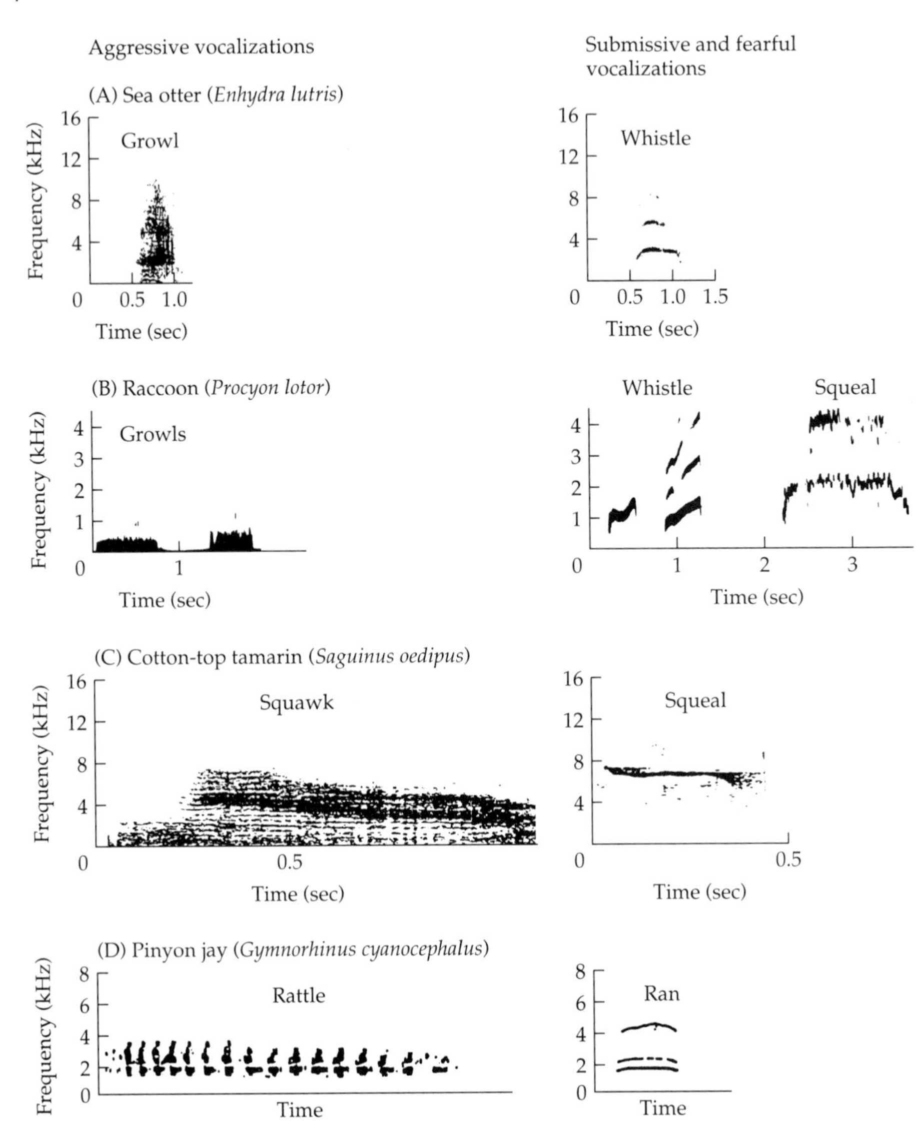

Figure 16.11 Antithetical vocal signals. For each species, an aggressive vocal signal is shown on the left, and a submissive or fearful vocal signal on the right. In the three mammals, aggressive sounds are low in frequency and very broadband (growls), whereas the submissive sounds are higher in frequency and more tonal (whistles and squeals). In the jay, females give the rattle vocalization against unwanted male approaches, and the male may then give the submissive ran vocalization in response. (A after McShane et al. 1995; B from Sieber 1984, © Brill, Leiden; C after Cleveland and Snowdon 1982, D after Marzluff and Balda 1992.)

tems depicted in Figure 16.5. His scheme of **motivation-structural rules** for the short-distance vocalizations of birds and mammals is illustrated in Figure 16.12. As suggested above, coupled frequency and bandwidth parameters express the extreme conditions of fear (high frequency, narrow bandwidth) and

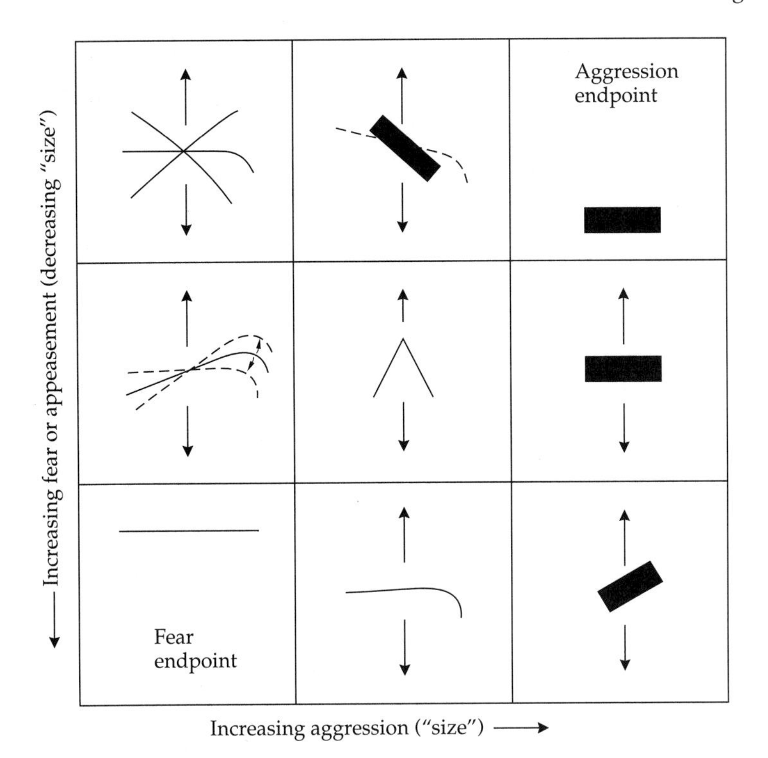

Figure 16.12 Motivation-structural rules for short-distance vocal signals in birds and mammals. Hypothesized sound structures associated with varying degrees and combinations of aggressive and fearful or submissive motivational states. The figure in each block represents a schematic sonogram. The height of the figure above the baseline indicates the frequency of the sound, and arrows suggest those blocks in which frequency may vary. The thickness of the figure represents the bandwidth or frequency range of the sound, and the dotted lines indicate that the figure's slope may change. The upper-left block shows nonaggressive or friendly vocalizations characterized by pure tones of high or low frequency. Fear is expressed with increasingly higher frequency tonal sounds. Aggression is expressed with increasingly lower frequency harsh or broadband sounds. The lower-right block shows a mixture of fear and aggression, with broadband but higher frequency sounds. The chevron in the center picture is a relatively pure tone vocalization with and up-down frequency modulation that expresses general excitement or alarm. (After Morton 1977, 1982.)

aggression (low frequency, broad bandwidth). Intermediate frequencies and bandwidths indicate conflicting motivational levels. In addition, the type of frequency modulation provides further information on the motivation state of the sender within these broad categories. A downward-modulated frequency sweep signals slightly greater aggressiveness, while an upward-modulated sweep signals less aggressiveness. In the middle of this scheme is the chirp or bark, a chevron-shaped vocalization that signals neither aggression nor fear but indicates that the sender is indecisive or attentive. Morton proposed a possible reason for the association between low frequencies for aggression and high frequencies for nonaggression. In all vocal vertebrate species, larger individuals can produce lower-frequency sounds than smaller individuals.

Since individuals with larger body size often win aggressive encounters with smaller individuals, there has been selective pressure on aggressive threat signals to lower the frequency in order to send the most intimidating type of information to an opponent. Conversely, high frequencies are associated with infant vocalizations and their tendency to attract adults. Furthermore, a fearful animal is likely to involuntarily tense all of the muscles in its body, including its vocal chords, and thus emit high-frequency sounds. There is widespread evidence supporting Morton's ideas in both birds and mammals (August and Anderson 1987; Cleveland and Snowdon 1982; Jürgens 1979; Nelson 1984; Peters 1980; Scherer 1985).

To summarize this section on auditory signal sources, sound production is primarily restricted to a handful of animal taxa that have body structures capable of producing vibrations such as the hard, articulated exoskeleton of some arthropods and the respiratory organs of frogs and higher vertebrates. Whether the sound-producing apparatus evolved initially to serve some type of survival function (e.g., defense against predators) or whether it evolved solely for intraspecific communication remains controversial in some groups, but the majority of animal sounds are used for intraspecific communication purposes. The difficulty of distinguishing signals from their nonsignaling precursors therefore does usually not arise for the auditory modality, as it does for the visual modality.

SENDER PRECURSORS OF OLFACTORY SIGNALS

Any animal's body contains a huge number of chemical compounds, many of which leak into the environment and could potentially provide specific information about the sender to conspecifics. If is often very difficult to know whether chemicals emanating from an animal that affect the behavior of recipients are truly ritualized pheromonal signals or whether they are unspecialized metabolic byproducts. Evidence for ritualization of olfactory signals includes: (1) the presence of specialized secretory structures, (2) specialized chemical products, and (3) stereotyped behaviors associated with pheromone release.

Recall from pages 283–285 that chemicals can exit the body in one of two ways: via body orifices used for digestion and reproduction such as the mouth, anus, cloaca, and vagina and via exocrine gland secretions on the exterior surface of an animal used to maintain the skin or integument. Pheromones emanating from either of these two types of locations can be derived from a variety of chemical sources. As we have seen for visual and auditory signal sources, the nature of the source of an olfactory signal in many cases provides specific types of information from the outset.

Dietary Sources

The direct use of dietary chemicals is in principle an extremely cheap way to produce a pheromone. However, most animals eat a variety of foods, and their diets change with location and season. Production of a consistent species-specific chemical or blend of chemicals would therefore be difficult to guarantee if the only source was dietary. This option, however, is available to animals that feed on a single food item. Some phytophagous insects that feed on a single host plant species are known to use plant products to generate their sexual attractant pheromones. The plant chemicals used by the insects are always toxins produced by the plant to prevent insects from eating them in the first place. An insect species that evolves the ability to detoxify a specific plant's chemical defenses often becomes completely dependent on that plant species for food. Indeed, the method of "detoxifying" the plant compound may be to pass the chemical directly through the body relatively unchanged.

Bark beetles (*Ips* spp.) are a good example. Terpenes are a common class of compounds evolved by plants to deter feeding by insects. Terpenes interfere with the metabolism of most insect species and are either lethal or prevent reproduction. When bark beetles locate their host species of tree and begin to bore into the wood, the ingested terpene is modified slightly to an alcohol and released in the feces. Depending on the species, either the male or the female first locates a suitable host and begins to signal, attracting members of both sexes. Since the host plant produces its own specific toxin and no other closely related insect has evolved the ability to deal with it, the volatile alcohol can serve as a species-specific airborne pheromone for the beetle (Figure 16.13).

Reproductive Precursors and Products

In many vertebrate species, including fish, mammals, and some reptiles, identified sexual attractant pheromones are derived from steroid hormones. These compounds are released in urine, saliva and/or sweat. Males employ pheromones derived from the male steroids androgen and testosterone to attract females and repel other males. Such pheromones clearly reveal the sex, age, and reproductive condition of the sender. The difficulty is assessing whether these are catabolic waste products or specially produced pheromones. In the boar, there is strong evidence for specialized pheromone production. The main steroid pheromone is synthesized in the testis by a different but parallel route from that leading to testosterone. It is released into the

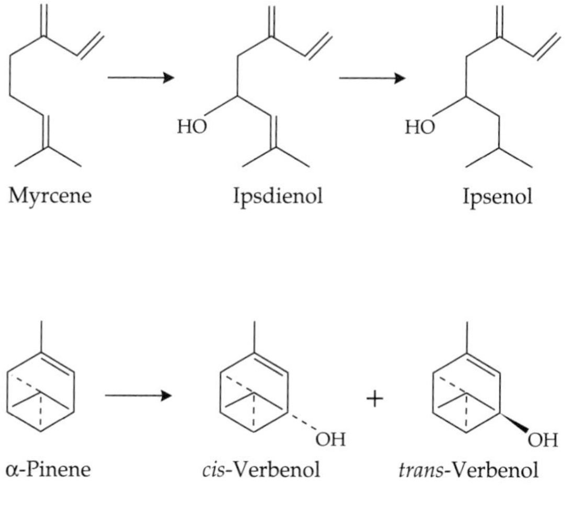

Myrcene → HO Ipsdienol → HO Ipsenol

α-Pinene → *cis*-Verbenol OH + *trans*-Verbenol OH

Figure 16.13 Pheromones derived from plant compounds. Proposed pathways for the conversion of plant monoterpenes to the aggregating pheromones of *Ips* bark beetles. Only the male produces ipsdienol and ipsenol. Once he has attracted and mated with a female, both produce verbenol, which deters further arrivals. (After Blomquist and Dillwith 1983.)

blood and transported to the salivary gland. When the male is sexually aroused he produces copious amounts of viscous, frothy drool, and the volatile chemical is released with his breath. Related compounds are also released from the boar's urine and sudoriferous sweat glands. Receptive females respond to this odor by assuming the immobile mating stance (Signoret 1970; Perry et al. 1980).

Gravid females in some fish and crustacean species release unmodified estrogen as an attractive substance (Stacey et al. 1986). The chemical thus appears to serve a second pheromonal function in addition to its primary hormonal function. Many researchers have argued that these are not true pheromones because they evolved to serve a different role. However, if senders actively release the estrogen into the medium, this clearly would qualify as a signal. This appears to be an example of a relatively unritualized signal.

The urogenital secretions of female mammals are known to play an important olfactory role in mate attraction and copulation. Identifying the critical pheromone components has been difficult, however, because a large array of chemicals enter the urogenital tract. These include: (1) bladder urine components (waste products), (2) materials arising from the upper reproductive tract and ovaries, (3) materials deriving from the vaginal walls that are highly permeable to blood plasma, and (4) secretions of the vulval glands (including large glands that produce nonvolatile substances such as sugars, amino acids, enzymes, proteins, polysaccharides and glycoproteins involved in survival and transport of sperm, as well as sebaceous and sudoriforous skin glands that may be highly specialized in some species). The volatile component in female primates, including humans, that produces behavioral responses appears to be a mixture of volatile short-chain fatty acids (Albone et al. 1977). These chemicals are produced by the breakdown of glycogen by the resident vaginal microflora (lactobacilli and facultative and strict anaerobes). The presence of both the microflora and the volatile fatty acids is affected by the female's reproductive state; the compounds increase during estrous and ovulation, and removal of the ovaries eliminates them. In female hamsters a single compound, dimethyl disulphide (see Figure 10.1F) has been found to produce a substantial part of the male response (reduced aggression and associated flank-marking behavior followed by mounting). It is not yet clear whether this chemical is metabolized in specialized glands or is a product of the resident microflora.

Defensive Chemicals

Many of the chemicals that serve as alarm pheromones and cause fleeing or hiding responses in conspecifics also serve as toxins or predator/pathogen deterrents. It has been argued that the defensive function evolved first, and that the alarm function evolved secondarily in social species where such information is useful. Even plants release volatile substances when damaged that are perceived by neighboring conspecifics and cause them to produce more defensive secondary compounds (Fowler and Lawton 1985). In these cases it is difficult to argue that the sender receives any benefit from warning conspecifics. However, the "signal" has an inherent meaning of alarm from the outset.

In some schooling fish (minnows, suckers, and catfish), an alarm substance called "shreckstoff" causes schoolmates to flee when one fish is injured. The substance is located in specialized secretory cells called club cells embedded in the middle of the epidermis. Club cells are neither close to blood vessels where they could release hormones into the blood nor do they possess pores for release of substance to the exterior. They can only release their secretions when the skin is damaged. Comparative studies have attempted to identify the source of such cells. Some species lack both the cells and the fright reaction; these tend to be species having alternative anti-predator strategies such as large size, armor, burying habits or dwelling in caves. However, many species possess club cells but lack a fright reaction. The cells may therefore have evolved to serve another purpose. In one well-studied minnow, the chemical composition of the alarm substance has been determined and judged to have possible antibiotic properties. Thus the substance may function primarily to reduce infection in the case of injury but provide a highly informative cue of dangerous conditions to nearby conspecifics (Weldon 1983; Smith 1986b).

The second major group of animals with chemical alarm substances are the Hymenoptera. Once again it has been suggested that the chemicals are the same as or derived from defensive chemicals (Wilson 1971; Hölldobler and Wilson 1990). This is clearly the case in formicine ants that spray formic acid at enemies (Figure 10.5B). In other cases, the defensive chemical is a venom or a sticky glue substance with a high molecular weight. Because this substance is not volatile enough to diffuse rapidly, a smaller and more volatile substance is secreted at the same time to signal alarm. For example, in the honey bee *Apis mellifera*, the venom consists of a complex mixture of proteins, histamine, serotonin, and acetylcholine, which together cause the inflammation experienced by the victim of a sting. Associated with the sting gland is a separate set of secretory cells that produce isoamyl acetate, the alarm pheromone. Similarly, the glands that produce the sticky substance used by termites in defense also produce volatile substances that cause the alarm reaction. This evidence makes a strong case for a ritualized alarm pheromone.

De Novo Production

Ample evidence now exists that airborne pheromones in many insect species are produced de novo from simple building blocks to produce species-specific chemicals (or blends) for mate attraction. For example, many of the known Lepidopteran sex attractants are structurally similar and consist of a 10-to-18 carbon chain with a single double bond somewhere in the middle and an acetate ester, alcohol, or aldehyde functional group at one end. The carbon chain is constructed from 2-carbon units in a stepwise fashion using ordinary fatty-acid chain elongation machinery; species specificity is incorporated during the later stages of production (Figure 16.14). The evolution of a biosynthetic pathway to produce acetate ester pheromones was probably favored as a re-

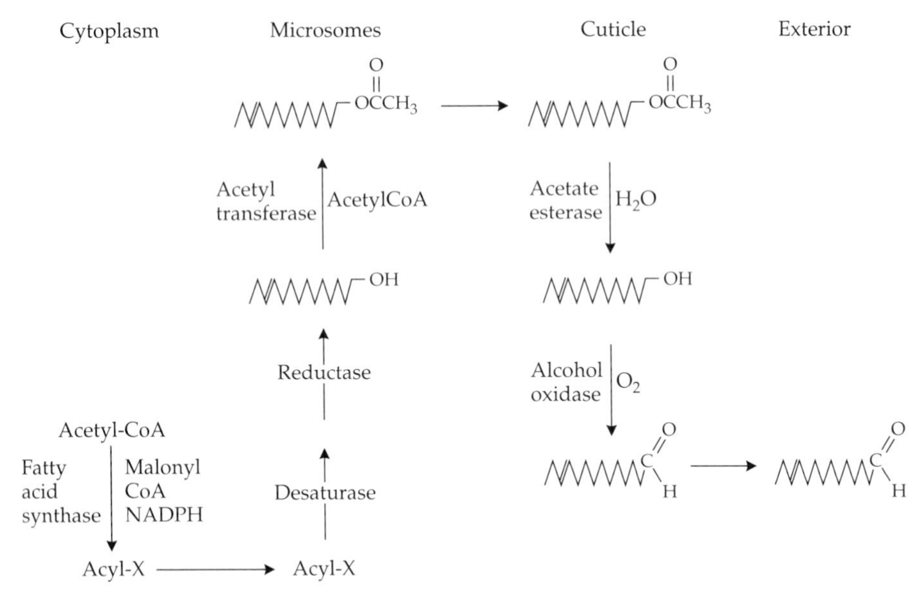

Figure 16.14 Proposed pathway for pheromone biosynthesis in the eastern spruce budworm (*Choristoneura fumiferana*). The pathway begins with fatty-acid synthesis and ends with release of the aldehyde pheromone from the insect. The first step is the construction of a long-chain fatty acid, which takes place in the cytoplasm of gland cells. Two-carbon units of malonyl-CoA are added to an acetyl-CoA base using a common fatty acid synthase enzyme. The chain is initially fully saturated. Next, a desaturase enzyme produces the double bond in the presence of NADPH and molecular oxygen in the cell's microsomes. The ester is then produced. Finally, a series of enzymes generates the alcohol or aldehyde functional group in the gland's cuticle just prior to release. (After Morse and Meighen 1986.)

sult of the high stability of esters. Hydrolytic and oxidative steps to form alcohol and aldehyde pheromones could have then evolved during species differentiation, with each species having a unique combination (see also Table 18.2) (Morse and Meighen 1986).

To sum up the olfactory sources section, elucidating the chemical sources of pheromones requires sophisticated biochemical techniques. Because this process is expensive, research has focused on humans, agriculturally relevant insects, and important domestic animals so that comparative information is incomplete. Given our existing knowledge, it is often difficult to determine which of the many chemicals exuded by animals are the signaling pheromones and which are nonsignaling byproducts of metabolic processes. Thus the olfactory modality presents us with a greater challenge in distinguishing signal from source than is the case with the visual modality.

RECEIVER PRECURSORS TO SIGNAL EVOLUTION

As we discussed in the introduction to this chapter, receiver biases might play an important role as the critical precursors to signal exchanges. The basic tenet of such **receiver-bias** models is that the receiving apparatus has undergone prior selection for other functions in addition to social communication, such as prey or predator detection, and is therefore more sensitive to certain types of stimuli. Social signals then evolve to exploit these well-developed receiver sensitivities. One model, called **sensory drive** (Endler 1992), emphasizes the role of environmental factors such as background noise, transmission properties, predators, and prey in shaping both signal form and receiver design. The model assumes that sender and receiver coevolve within severe constraints of the environment. These constraints select for senders that give only efficient signals and for receivers that respond only to those signals. If sufficiently strong, this channeling could assign signals to conditions in ways independent of historical antecedents. The model is therefore compatible with the traditional view of signal evolution outlined in Figure 16.1 but emphasizes the importance of receiver filtering at step 2, receiver tuning, and ritualization.

A second model, called **sensory exploitation** (Ryan 1990; Ryan and Rand 1990, 1993), proposes an entirely different process of signal evolution. According to this hypothesis, receivers often harbor latent preferences that can be exploited by manipulative senders to create new signals. In other words, the process *begins* at step 2 (Figure 16.1). For example, suppose female birds search preferentially for red seeds while foraging. Because the only time they encounter red is in seeds, it may prove most efficient to evolve a general preference for red objects. A mutant male that adds red to its plumage may be able to exploit a female's preference as long as the general preference for red can be expressed in a mate choice context. Thus a new male signal could evolve that had no historical link to the specific context (mate choice) but only to another currently irrelevant context (foraging). The coding that arises from such a process would thus be more arbitrary than expected from the classical scenario.

The use of the term "exploitation" here is unfortunate for two reasons. First, exploitation of receivers by senders is easily confused with the traditional usage of exploitation of senders by receivers as discussed on pages 380–381. Second, exploitation implies that the exploited party pays some type of cost. The model, however, ignores such costs and any coevolutionary responses of the receiver. As outlined here, there is no new information provided by the red males and thus no benefit to the females. There may even be costs if red sons are more likely to be spotted and killed by predators. Any costs to females will result in counter-adaptations to exploitation, such as the evolution of better discrimination abilities, or the decoupling of foraging and mating decisions. On the other hand, females may be able to find conspecific males with the red patch much more easily. Both sexes therefore benefit, and exploitation hardly characterizes the interaction. Futhermore, red males may provide additional information not provided by normal males. For example, suppose the degree of redness in males is a good indicator of their health. Thus a preference for red males over nonred ones could be a way for females to identify better mates. Whether the new signal is beneficial or costly, sensory exploitation by itself is unlikely to be stable. Instead, it will be followed by coevolution between sender and receiver. The way in which this might work when sender and receiver have different interests (the presumption behind most sensory exploitation models) is discussed in detail in Chapter 23.

Feature Detectors

Receiver-bias models of signal evolution rely on the concept of **feature detectors**. Feature detectors are refinements in the receiving apparatus that make the detection of cues and signals easier against background noise. Such receiver detection systems are more than a peripheral filter; they typically involve higher levels of neuronal integration (Ewert 1974; Martin 1994). Lateral inhibition between adjacent receptor cells plays a key role in feature detector wiring. Some examples of simple visual feature detectors are shown in Figure 16.15. Similar systems have been described for auditory and olfactory re-

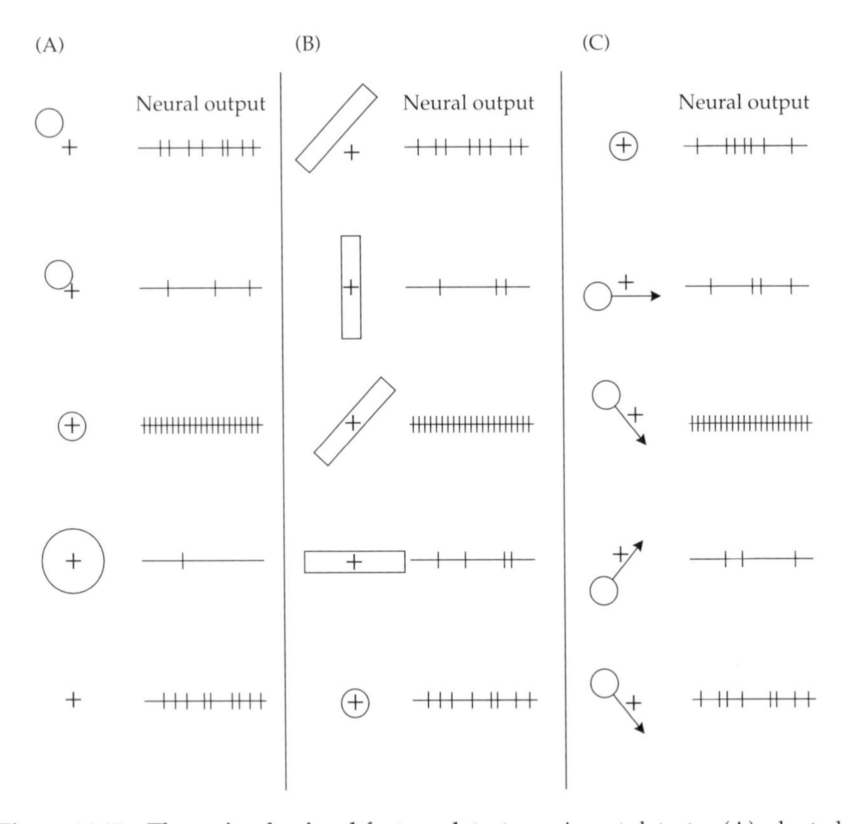

Figure 16.15 Three simple visual feature detectors. A spot detector (A), slanted line detector (B), and motion detector (C) identified in vertebrate eyes by shining specific patterns of light on the retina while recording spike activity from nerves in the retina or brain. In each example, the activity of a specific nerve is shown when test patterns of different shape, size, orientation, or movement direction are presented. The + shows the position of the photoreceptor(s) linked to the recorded nerve relative to the position of each light pattern. The optimum stimulus for each nerve is exhibited by the third test pattern from the top. (After Gould 1982, © W. W. Norton.)

ceivers. Owls, for example, have feature detectors for locating the position of prey by the sounds they make (Konishi 1993), and species-specific templates for the songs of passerine birds ensure that only conspecific notes are learned (Doupe and Konishi 1991; Doupe 1993; Lewicki and Konishi 1995). The vertebrate olfactory system also uses lateral inhibition to enhance the recognition of certain odorants (Yokoi et al. 1995).

Feature detectors are hard-wired in the sense that no learning is involved, although the neuronal pathways may need to be activated with general sensory stimulation. Specific responses upon the detection of the feature stimuli are also often hard-wired. Hard-wiring ensures that the signal or cue is recognized the first time it is encountered and that the appropriate response (approach or evasion) is made. Feature detectors can evolve as part of the receiver tuning process of signal evolution described in Figure 16.1 to enhance the detection of important social signals. Here they serve in the recognition of **sign stimuli**, or releasers for appropriate responsive behaviors. A classic example of a sign stimulus and its corresponding feature detector is the red spot on the bill of many gulls that releases begging behavior by chicks (Tinbergen and Perdeck 1950). Feature detectors can also evolve to facilitate approach toward cues from prey or avoidance of cues from enemies.

Feature detectors can be modeled with **artificial neural networks**. Neural nets represent quantitative abstractions of the initial steps of sensory input analysis by organs such as the retina, basilar papilla, or olfactory bulb. In a network model, one specifies the types and strengths of connections between nerve layers. An example of a simple network is described in Box 16.2. The network is then trained to respond to certain types of stimuli. Two important results have been obtained from these models. One is that a network trained to distinguish a certain pattern from an alternative pattern or background noise often responds even more strongly to exaggerated forms of the pattern than to the original training pattern (Enquist and Arak 1993, 1994). This outcome provides an explanation for the classical ethologist's observation of a **supernormal stimulus** (Tinbergen 1948). This is the preference sometimes shown for cues or signal features that are enlarged or exaggerated beyond the normal range. The second emergent property of neural net models is an occasional strong response to a novel pattern. No pattern recognition system performs perfectly since backgrounds vary and the pattern may appear in different orientations or at different distances. However, a variety of neural connection schemes may satisfactorily distinguish the pattern from its background on the majority of occasions. Some schemes by chance may be more sensitive to a previously unencountered pattern. This phenomenon has been termed **hidden preferences** (Arak and Enquist 1993). Supernormal stimuli and hidden preferences might drive the evolution of new signal forms as proposed by the sensory exploitation model.

Evidence for Sensory Exploitation

Several types of evidence have been invoked to support the receiver bias point of view. Comparative studies are one approach. In a major review of studies of female mate choice (Ryan and Keddy-Hector 1992), male signal-

Box 16.2 Artificial Neural Network Models

SIMPLE ARTIFICIAL NEURAL NETWORK MODELS can be used to explore some of the properties of sensory recognition systems in animals. An example of one such model is shown here from Enquist and Arak (1993, © Macmillan Magazines). It consists of three layers. The input layer is a 6×6 array of sensory receptor cells analogous to a retina (or olfactory epithelium or basilar papilla). The second layer is called the hidden layer; in this case it is comprised of 10 cells analogous to ganglion cells. Each cell in this layer is connected to every cell in the input layer above it, and thus each input layer cell is connected to every hidden layer cell. For simplicity the connections to just one ganglion cell are shown here. Each connection is associated with a quantitative weighting that regulates the strength of the signal passing between cells, analogous to a synaptic nerve connection. The total stimulus to a hidden layer cell is therefore the weighted sum of the input from all of the sensory cells in the layer above. The third layer consists of a single output cell connected to all of the hidden layer cells. Input to this output cell is also a weighted sum of the connection strengths from the hidden layer.

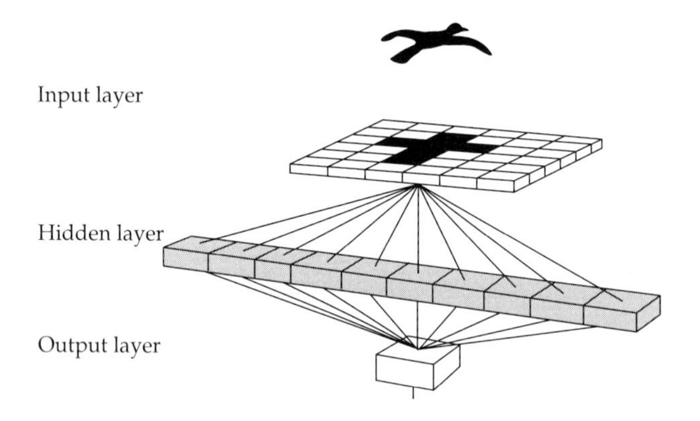

trait values that were preferred by females were compared to the mean value of the trait in the male population. In most cases where a difference existed, females preferred males with trait values of *greater magnitude* than the mean, i.e., larger body size, brighter color, higher amplitude, longer tail. For acoustic signals, females preferred higher calling rates, longer calls, more complex calls, larger repertoire size, and lower frequencies. The implication is that preference is directional for trait values that stimulate the receptor more. However, larger signals are also more costly for senders to produce, and the preference for them may have evolved because they provide important information to the female about male quality, health, or size (see Chapter 23). As a case in point, female tungara frogs (*Physalaemus pustulosus*) prefer male calls with more low-frequency energy. This preference is not merely due to the fact that lower frequencies are attenuated less and therefore louder. The female's ear exhibits maximum sensitivity at a frequency slightly lower than the average male's call (Ryan et al. 1990). Did the low tuning curve evolve first and drive the preference for lower frequency calls, as Ryan suggests, or has the

When the sensory layer is stimulated, each of the receptor cells receives an input of 0 or 1. The output from these cells is then multiplied by the weighting factors to determine the total input to the hidden layer cells. The summed output from hidden layer cells is similarly multiplied by the weighting factors associated with this layer to determine the stimulation of the third layer cell. If the stimulation of the third layer cell is above a certain threshold, the network is said to "recognize" the stimulus pattern. The weighting factors at both levels "evolve" during training sessions in a process that mimics natural selection, to achieve a certain type of discrimination, i.e., to recognize a certain pattern and reject another (or random) pattern. Weights undergo random mutational changes at each trial (generation), and the network with the better discrimination ability is retained for the next trial.

Each iterated run produces a different final set of weighting factors that enable the network to make an accurate discrimination, i.e., there are many solutions to the same problem. Novel stimulus patterns are then presented to the trained networks. Occasionally, a network is found that responds more strongly to an exaggerated version of the training pattern or to a rather different pattern. However, care must be exercised in interpreting these results. Such simple networks are not designed to perceive particular shapes or spatial patterns, but only to respond to the average level of stimulation to a given sensory cell (Cook 1995). A true feature detector that perceives spatial pattern shapes must consist of a more complex hidden layer with a two-dimensional array of cells and a third series of weighting factors among the cells in this layer. Like horizontal and amacrine cells of the retina, these connections produce patterns of reinforcement and inhibition between adjacent cells. Neural network models of this type can evolve edge, spot, and bar detectors similar to those described by neurophysiologists (Rubner and Schulten 1990).

tuning curve evolved as a consequence of a benefit to females from selecting larger males that tend to have lower frequency calls?

The need to know whether the signal trait or the preference for it evolved first has stimulated phylogenetically based comparative studies that can unravel the historical sequence of trait evolution. In the tungara frog, males may add 0 to 5 chuck notes to the end of the typical whine note. Females prefer calls with chucks. In a closely related species, *P. coloradorum*, males produce a species-specific whine but never add chucks, yet females of that species prefer whines to which chucks of the tungara frog have been added. Ryan argues that the preference for the chuck is an ancestral trait in the common ancestor of these frog species (a preexisting sensory bias) that has driven the evolution of the chuck trait in the tungara frog and one other member of this genus (Figure 16.16).

There are several clear examples of social signals that are designed to stimulate a receiver's preexisting feature detector. In some cases the selective force favoring the feature detector is also known. *Anolis* lizards are sit-and-

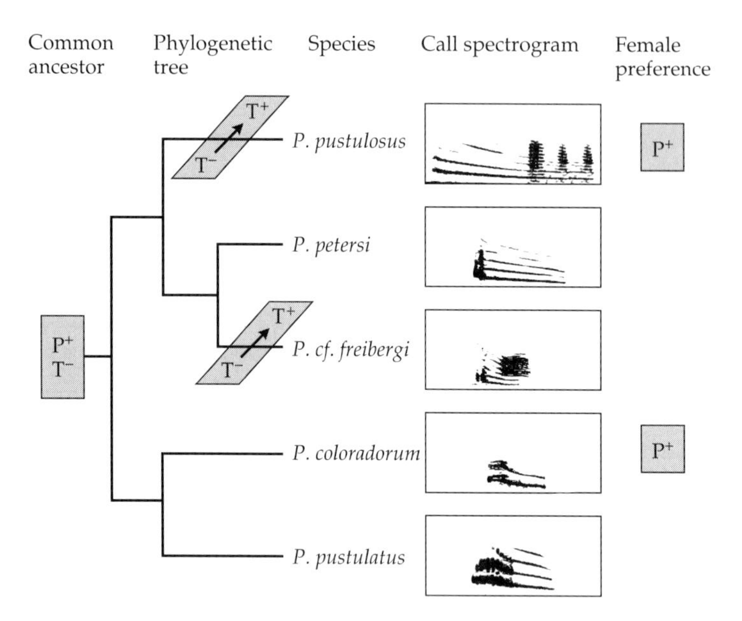

Figure 16.16 Phylogeny of male call traits and female preferences in the *Physalaemus pustulosus* species group. The tree shows estimated phylogenetic relationships among five closely related species of *Physalaemus* and their common *Physalaemus* ancestor. The spectrogram of each species' call is shown. All species produce a whine call that is necessary and sufficient for species recognition. Both *pustulosus* and *cf. freibergi* add a variable number of chucks to the whine part of the call (T^+), whereas *petersi, coloradorum,* and *pustulatus* do not (T^-). Under parsimony for this tree, the ancestor's male chuck trait is ambiguous, but no other *Physalaemus* species possess the chuck, so it is clearly T^-. The chuck trait therefore evolved independently in *pustulosus* and *cf. freibergi.* A female preference for whines with the chuck ending, symbolized by P^+, has been found in both species that have been tested, *pustulosus* and *coloradorum.* This suggests that the preference for chucks was present in the common ancestor and evolved before males added chucks to their whine. (After Ryan and Keddy-Hector 1992; Ryan and Rand 1993; spectrograms courtesy of Mike Ryan.)

wait predators on moving insects. The eyes of lizards have a large peripheral retinal region with a moderate density of receptor cells that is rich in velocity-tuned motion detectors. These cells habituate rapidly to background motion such as the 1 to 2 Hz sinusoidal movement of leaves in wind. Experiments with a prey object moved in different ways in front of typical moving vegetation show that either higher frequency movement or jerky (square-wave) movement is more readily detected. The long-distance push-up displays of territorial male lizards involve square-wave motion, high amplitude, and rapid acceleration that are maximally different from the background. The signal display pattern appears to have exploited the prey-derived motion detectors in the receptor (Fleishman 1992). Another example is the water mite (*Neumania papillator*) that locates its copepod prey by detecting water vibrations. Males mimic these vibrations with leg movements; their movements cause females to approach and clutch them as they would a prey. Proctor (1991, 1992) showed that food-deprived females were more likely to approach and clutch males than satiated females, suggesting that the female is in fact manipulated by the male display. The use or modification of dietary chemicals such as plant secondary compounds in insects (see pages 522–523) could also be interpreted as exploitation of food-finding feature detectors.

As mentioned above, Ryan and Rand (1993) argue that sensory exploitation can operate without any coevolution of receiver response. However, this argument assumes that the preexisting response of receivers to the stimulus feature benefits the sender (Gerhardt 1994). The sender thus exploits both the sensory preadaptation and its associated response. The **sensory trap** model of Christy (1995) explicitly acknowledges the role of the receiver's response during sensory exploitation as well as the importance of a benefit to receivers. A mate attraction signal, for example, should exploit a food-related feature detector that causes receivers to approach. All of the examples described above are cases of food-mimicking signals. In principle, senders could also exploit a feature detection system for predators or noxious prey and its associated retreat response as a threat signal. The vocal threat and alarm signals of many animals resemble the rattle of a rattlesnake or the conspicuous aposematic color pattern of poisonous prey, and it is tempting to suggest that this is the source of the signal. These examples of the transfer of an entire detection-recognition-response system from a predator- or prey-derived cue into a conspecific signal can occur only if the receiver benefits, which means that the response is adaptive in the social context as well as the nonsocial one.

If the receiver is truly manipulated by the exploitative sender to the detriment of the receiver, subsequent coevolution is inevitable. Receivers will improve their discrimination ability, decision rule, encoding scheme, or signal tuning to minimize their costs. However, it is still possible that a signal could initially arise from sender exploitation of receiver biases and become modified into one that benefits receivers. Cases in which the response of receivers is reversed from the original context clearly indicate the role of coevolution. The best example comes from moths that have evolved ultrasonic mate attraction signals. Several taxa of moths are characterized by the presence of ears for detection of the ultrasonic echolocation cries of bats, their major predator. The response of moths to these sounds is immediate evasion and avoidance. Some moth species have moved into bat-free niches or habitats (e.g., become diurnal) and are no longer subject to this selection pressure. Males have exploited the ultrasonic sensory capability of females and evolved ultrasonic sound production to attract mates (Fullard and Yack 1993; Fullard 1994). The initial avoidance response of female receivers has obviously been reversed to one of attraction.

In summary, sensory preadaptations can serve as important sources of signals, and the resulting signals will appear to be arbitrary in form. If the preexisting response is beneficial to receivers as well as senders, there may be little need for coevolution and receiver fine-tuning. However, it is more likely that sender and receiver interests are not the same and some amount of tuning, code refinement, and ritualization will take place and hone the communication interaction into one that benefits both parties.

SUMMARY

1. The process of signal evolution involves five initial steps: (1) **association** between a sender-produced cue and a condition, (2) **perception** of the cue by the receiver, (3) adoption of a receiver **encoding rule** relating the cue and the condition, (4) development of a receiver **decision rule**, and (5) evolution of a receiver **response.** If the response benefits the receiver, it will fine-tune its perception, encoding rule, decision rule, and response, and if the response benefits the sender, the cue will be refined into a true signal via **ritualization.** Ritualization involves simplification, exaggeration, repetition, and stereotypy to improve information transfer. Many complex changes must take place in both sender and receiver, and preadaptations in either party may serve as triggers to initiate the process.
2. The classical ethological view is that signals evolve from sender preadaptations such as existing behavioral, physiological, or morphological traits. If the presence of the trait provides useful information to receivers that can detect it, receivers will evolve to pay attention to the trait and use it to make behavioral response decisions. Signals that evolve from this type of source contain information from the outset because the sender's incipient signal is linked to the condition. Many examples of the evolution of signals from **sender preadaptations** have been described.
3. Visual signals evolve from three types of sender sources: intention movements, motivational conflict, and autonomic responses. **Intention movements** are the preparatory and incomplete initial components of normal behaviors like foraging, fleeing, attacking, or grooming. They provide information about what the sender is likely to do next. **Motivational conflict** occurs when two motivational systems such as fear and aggression are both highly stimulated. Several types of behaviors may arise from this conflict—ambivalence, displacement, and redirection—and these are often the source of ritualized displays. The **autonomic nervous system** sometimes produces visible physiological responses in a variety of stressful circumstances that serve as the source of displays. Relatively unritualized responses may provide accurate information to receivers about the fear or excitement level of the sender. As these displays become further ritualized, they become uncoupled from the source of their enervation and are used to signal very different types of information.
4. Auditory signals may arise from at least four different types of sender sources: visual or tactile displays, in which the movements are exaggerated to produce a percussive or vibrating noise; loud sounds used initially for startling predators; normal locomotor and foraging movements; and respiration. For respired-air vocal signals, a set of **motivation-structural rules** has been proposed that links the frequency and bandwidth characteristics of the signal to specific sender motivational states.

5. A large number of chemicals leak out of animals' bodies that could potentially become ritualized into olfactory signals. The presence of specialized secretory structures, specialized chemical products, and/or specialized behaviors for dissemination of scents are good indications of a ritualized chemical pheromone signal. Chemical sources for pheromones range from components of the diet to reproductive precursors and products and defensive chemicals. In other cases, pheromones are synthesized de novo from normal metabolic building blocks.
6. More recently, **receiver preadaptations** have been found to play an important role in signal evolution. Preexisting perceptual biases in the receiver can act as a filter on signal form by favoring those signal characteristics that transmit most effectively in a given environmental and social context, a concept called **sensory drive**. A more extreme hypothesis, **sensory exploitation** or **sensory trap**, proposes that signal evolution begins with receiver preadaptations (step 2). Perceptual biases or **feature detectors** can evolve for noncommunication reasons such as food-finding or predator avoidance. Senders exploit these biases for their own benefit by mimicking the feature characteristics. Sensory exploitation can lead to the evolution of new signals as long as both the receiver and sender benefit from the receiver's response to the signal. The form of signals arising from this process appears to be arbitrary because the linkage between signal and condition is established secondarily.

FURTHER READING

Otte (1974) clarifies the foundation for the traditional view of the evolution of signals. McFarland (1981) is a valuable encyclopedia of the entire field of ethology. Hailman (1977a and b) reviews the classical sources of visual displays, and McFarland (1985) provides a thorough discussion of theories for displacement behaviors. Dumortier (1963a) summarizes sound production sources in invertebrates. Morton's (1977) framework for motivation-structural rules of vocal signals remains the most provocative hypothesis for the determinants of vocal signal form in birds and mammals. Olfactory signal sources in insects are reviewed by Blomquist and Dillwith (1983). Discussions of sensory biases are more recent and include Endler (1992) and Ryan and Rand (1993).

Chapter 17

Costs and Constraints on Signal Evolution

WE HAVE SEEN THAT IT IS THE VALUE OF INFORMATION that determines whether senders should emit signals and receivers attend to them. The value of information depends both on the benefits of the exchange and on the costs. Although the benefits of sending and receiving are usually obvious, the concomitant costs can be numerous and subtle. For example, senders and receivers cannot even contemplate an exchange until each has constructed the relevant signaling organs and processing circuits. As outlined in Part I, both physical and phylogenetic constraints bracket the options available for signaling. Given the phylogenetic history of a taxon, only certain modalities are available to it, and the resulting combination of medium, modality, and propagation properties limits the range of practical signals. In addition, other species may impose costs on communication. Predators, prey, parasites, and commensals can tap the code and exploit both senders and receivers. Although a certain amount of the diversity in signals can be explained by identifying the alternative benefits they provide, an equal amount is surely due to the diversity of costs experienced by senders and receivers. In this chapter, we first classify these costs and constraints and show where they fit into our general model of animal communication. We then review the costs and constraints specifically experienced by each party and outline adaptations that animals have evolved to cope with them.

TYPES OF SIGNALING COSTS

Senders and receivers face several different types of costs when communicating. It is important that we clarify the distinctions among them because they affect the evolution of signals in different ways. Recall the basic signaling model developed in Chapter 12 (pages 375–380). Alternative conditions i occur with respective probabilities P_i. A receiver should attend to signals if the average difference in fitness between correct and incorrect decisions (ΔR_i), discounted by the change in the chances the receiver will make a correct decision if it invokes the signals ($\Delta\phi_i$), is greater than the average reception cost of communication ($\overline{K}_R$). Thus for two conditions and actions, receivers should attend to signals if

$$\sum_{i=1}^{2} P_i \Delta\phi_i \, \Delta R_i > \overline{K}_R$$

Similarly, if the improvement in sender fitness when receivers make a correct choice of action is ΔT_i and the average signal production cost is $\overline{K}_S$, then senders should send signals when

$$\sum_{i=1}^{2} P_i \Delta\phi_i \, \Delta T_i > \overline{K}_S$$

The reception and production costs expressed in the K terms on the right side of the two inequalities are always present during signaling exchanges. We shall call them **necessary costs**; they have also been called receiver-independent costs because they are paid regardless of the receiver's choice of action (Dawkins 1993). They include any investments in special structures, organs, or brain circuitry that must be undertaken prior to a signal exchange, as well as the extra time commitment, energy expenditure, or predator risk incurred while signaling or receiving. If the costs of either sending or receiving signals differ for alternative conditions, then the $\overline{K}$ terms are given by weighted averages (e.g., $\overline{K}_S = P_1 K_{S1} + P_2 K_{S2}$, where K_{S1} is the cost when C_1 is true and K_{S2} is the cost when C_2 is true). This is more likely to be the case for senders that give different signals for different conditions than it is for receivers. We shall see in Part III that condition-dependent signal production costs play an important role in maintaining signal honesty.

The left side of each inequality is the average benefit of communicating. Even when the net benefit of communicating is positive, one of the terms within that sum may be negative as an undesirable but unavoidable consequence of exchanging the signals. We shall call these negative terms the **incidental costs** of communication. They could also be called receiver-dependent costs in the sense that their magnitude is a function of the payoffs available to the party suffering the cost. The payoff that party obtains depends upon the action the receiver adopts. This definition thus includes the more specific sense of receiver-dependent costs suggested by Dawkins (1993). Note that incidental costs are specific to the contrast made in computing the value of information. In many of our examples, we compare the use of one set of signals to the use of no signals. In other situations, we might need to compare the use of one set of signals to the use of an alternative set. These two contrasts are likely to generate different incidental costs.

We have already discussed the typical source of incidental costs on pages 423–428. There we found that an optimal receiver relying only on a priori probabilities to make decisions should perform the same action regardless of condition. It thus acts as if one of the conditions, which we shall call the **default condition**, is always true. When the default condition really is present, the receiver makes correct decisions; when not, it always makes the wrong decision. Suppose such a receiver attends to imperfect signals. It will now make some errors when the default condition is true, but will hopefully make more correct decisions when it is not. The result is a negative $\Delta\phi_i$ term for the default condition when signal exchange is compared to no signal exchange. The product of this $\Delta\phi_i$ and the corresponding ΔR_i constitutes a negative term in the benefit expression for the receiver. As long as sender and receiver have common interests (i.e., all ΔR_i and ΔT_i are positive), both will then suffer an incidental cost, although the magnitude may be different for the two parties. As we saw in Chapter 14, the more accurate the coding scheme, the smaller is this reduction in average benefit for both parties.

What happens if sender and receiver do not have common interests? For example, what if the best action for the receiver when the default condition is true is not the best choice for the sender? The receiver will still suffer an incidental cost when it makes erroneous decisions during default conditions; however, each of these errors is no longer detrimental to the sender but beneficial to it. In fact, the sender will no longer suffer an incidental cost in this situation (see Box 17.1 for a numerical example). Alternatively, what if the sender and receiver agree on the proper action for default conditions, but disagree for the remaining condition? In this case, the sender suffers an incidental cost when both conditions are true. For no situation does it gain anything from communicating. It should thus avoid signal exchanges. Finally, what if the sender and receiver disagree about the optimal action for both conditions? Curiously, there is still a chance that it will pay both sender and receiver to engage in moderately accurate exchanges of signals. By making the signal accurate enough that the receiver does better in the nondefault situation, the sender induces the receiver to join in the exchange. This benefit to the receiver comes at an incidental cost to the sender. However, once the receiver begins to rely on signals, it suffers an incidental cost when the default condition occurs, and this provides

Box 17.1 Incidental Costs and Sender-Receiver Conflict

WE SAW IN CHAPTER 14, on pages 423–428, that a receiver attending to imperfect signals nearly always pays an incidental cost of attending to signals. This cost arises because the choice of an action through using signals in the default condition was always correct when only prior probabilities were used, but that choice is now only as accurate as the coding scheme. (Since the coding scheme is imperfect, it provides less than 100% accuracy.) The receiver's value of information will still be positive if the improvement in decision propriety for the alternative condition is sufficiently high. A sender that has no conflict of interest with the receiver (e.g., always agrees with the receiver about the optimal choice of action) also suffers an incidental cost of using signals when the default condition is the current one. What happens to the sender's value of information when sender and receiver do not have identical interests? In the following examples, we shall assume that necessary costs, $\bar{K}$, are negligible and thus concentrate only on the overall benefit terms in the value of information computations.

Consider first a hypothetical female choosing a mate. Suppose she encounters two males. If the first male she inspects is healthy (condition C_1), she should mate with him (action A_1); if not (condition C_2), she should wait and examine the second male (action A_2). Waiting causes the female a loss of fitness due to the delay. An encountered male wants the female to mate with him regardless of his health. Thus female and male both favor her making a correct decision when he is healthy, but they differ when he is not. Suppose the relevant payoff matrices for the two parties are as follows:

Female payoffs

Female action	Condition: Healthy male	Condition: Sick male
Mate with first male	15 offspring	6 offspring
Wait for next male	12 offspring	12 offspring

Male payoffs

Female action	Condition: Healthy male	Condition: Sick male
Mate with first male	15 offspring	6 offspring
Wait for next male	12 offspring	3 offspring

The advantage to the female of a correct decision when the male is healthy is $\Delta R_1 = 15 - 12 = 3$; that when the male is sick is $\Delta R_2 = 12 - 6 = 6$. The advantage to a healthy male of a correct female decision is $\Delta T_1 = 15 - 12 = 3$, but there is a disadvantage to a sick male of a correct female decision, namely $\Delta T_2 = 3 - 6 = -3$. Females using only a priori information should never mate with the first male encountered unless healthy males constitute more than

$$P_c = \frac{\Delta R_2}{\Delta R_1 + \Delta R_2} = 0.67$$

of the population. Suppose healthy and sick males are equally common (e.g., $P_1 = 0.5$ and $(1 - P_1) = 0.5$). The default strategy when females only use a priori information is thus to wait for the next male (A_2). They thus treat sick males as the default condition.

Now let males produce signals that provide females with some information about their relative health. Suppose these signals allow females to make a correct decision 70% of the time. What is the value of information for each party? Females not using signals always err when males are healthy; they now get it right 0.7 of the time. Thus $\Delta\phi_1 = 0.7 - 0 = 0.7$. On the other hand, $\Delta\phi_2 = 0.7 - 1.0 = -0.3$ when males are sick. The values of information are:

$$\begin{aligned} \text{Receiver:} \quad VI(R) &= P_1\Delta\phi_1\Delta R_1 + (1-P_1)\Delta\phi_2\Delta R_2 = 1.05 - 0.90 = 0.15 \\ \text{Sender:} \quad VI(S) &= P_1\Delta\phi_1\Delta T_1 + (1-P_1)\Delta\phi_2\Delta T_2 = 1.05 + 0.45 = 1.50 \end{aligned}$$

The value of information is positive for both, but only the receiver suffers the incidental cost. This is because the product of a negative $\Delta\phi_2$ and a negative ΔT_2 is positive. Thus when sender and receiver disagree about the optimal action for the default condition, both may benefit from engaging in communication, but the sender will do so without an incidental cost. Readers should convince themselves that if the opposite is true, that is, sender and receiver agree on the default condition, but disagree about the alternative condition, the value of information for senders will always be negative and senders should not produce signals.

Now consider a situation in which sender and receiver disagree about the optimal receiver action for both conditions. Two birds tend a hidden nest. One incubates and the other stands guard. Possible predators include snakes and crows. Both make similar noises during approach. The best response to a snake is to flee; the best response to a crow is to become immobile and silent. If one bird makes the wrong choice, the risk to the other is reduced. The guard bird always has less risk when snakes approach, as it is more likely to know where the snake is before fleeing. On the other hand, the guard is more likely than the nesting bird to be killed by an approaching crow if the nesting bird is immobile. Suppose the payoff matrices showing chances of survival for the two parties are as follows:

Incubator payoffs

Incubator action	Condition: Snake coming	Condition: Hawk coming
Flee	0.50	0.60
Freeze	0.30	0.90

Guard payoffs

Incubator action	Condition: Snake coming	Condition: Hawk coming
Flee	0.65	0.95
Freeze	0.80	0.60

Important parameters are then: $\Delta R_1 = 0.50 - 0.30 = 0.20$, $\Delta R_2 = 0.90 - 0.60 = 0.30$, $\Delta T_1 = 0.65 - 0.80 = -0.15$, $\Delta T_2 = 0.60 - 0.95 = -0.35$, and $P_c = 0.30/(0.30 + 0.20) = 0.60$. Suppose hawks and snakes are equally common so that $P_1 = 0.5$ and $(1 - P_1) = 0.5$. Because $P_1 < P_c$, the default strategy will be to freeze. Now suppose that guards, having a chance to

Box 17.1 (continued)

assess which type of predator is nearby, give honest, albeit imperfect, alarm calls that are different for the two predators. Suppose the type of alarm call picked is accurate 65% of the time. The corresponding values of information if receivers attend to signals are:

$$\begin{aligned}&\text{Incubator}: && VI(R) = (0.5)(0.65-0)(0.20) + (0.5)(0.65-1.0)(0.30) = 0.065 - 0.053 = 0.012\\&\text{Guard}: && VI(S) = (0.5)(0.65-0)(-0.15) + (0.5)(0.65-1.0)(-0.35) = -0.049 + 0.061 = 0.012\end{aligned}$$

Both parties have a positive value of information, and both suffer incidental costs. However, the incidental costs imposed on each party due to imperfect signals constitute benefits to the other. What is one party's gain is another's loss (although these need not be equal in value). Given the right payoffs and probabilities, it is thus possible for communication to be favored even though the two parties disagree on the optimal receiver actions for both conditions. This is a nonintuitive outcome of the existence of incidental costs.

a benefit to the sender. Box 17.1 shows an example in which the net benefits and costs generate positive values of information for both parties despite the fact that they have totally opposite interests in how the receiver should act.

A qualitatively different type of incidental cost can arise when the use of signaling causes a change in the expected payoff matrices of one or both parties. Our prior examples have assumed that payoff matrices were unaffected by whether signals were exchanged or not. Changes in payoff matrices are most likely when signals have both informative and tactical functions. Consider two animals threatening each other prior to fighting. If effective communication requires mutual approach, the greater proximity could increase the amount of injury suffered during an attack. Thus the payoff values in the matrices of both parties might change if signals are exchanged. As another example, suppose senders that are dishonest are punished by receivers that discover their deceit. The average payoff to a sender of making a mistake, T_{ij}, will thus be lower when signals are exchanged than when not. In our earlier model, we assumed that the payoff matrices for each party were constant regardless of whether signals were exchanged or not. When the payoff matrices with and without signaling are different, the expression for the value of information is a bit more complicated. We can rewrite the inequality favoring receiver attention to signals despite an altered payoff matrix as

$$\sum_i^2 P_i\left[\phi_i'\Delta R_i' - \phi_i \Delta R_i + \Delta R_{ij}\right] > \overline{K}_R$$

where $\phi_i'\Delta R_i'$ is the benefit of a correct receiver decision discounted by the probability of correct choice when signaling is used, $\phi_i \Delta R_i$ is the equivalent term when signals are not provided or are ignored, ΔR_{ij} is the difference in fitness for an incorrect choice with signaling versus that without, and other terms are as previously defined. The earlier model can be seen as a special case in which $\Delta R_i' = \Delta R_i$ and $\Delta R_{ij} = 0$. The equivalent expression for senders when matrices change is

$$\sum_i^2 P_i\left[\phi_i'\Delta T_i' - \phi_i\Delta T_i + \Delta T_{ij}\right] > \overline{K}_S$$

Let us assume that, despite changes in payoffs with signaling, the net benefits on the left sides of these inequalities are greater than the necessary costs on the right-hand side for both parties. Where might there be incidental costs? We have already noted that punishment of deceitful or erroneous senders by receivers will lower sender payoffs when a sender sends the wrong signal. Thus the term ΔT_{ij} in the sender inequality will be negative. The task of inflicting this punishment might also cause the receiver ΔR_{ij} to be negative. Either can be considered an incidental cost of signaling. It may also be the case that signaling sufficiently reduces $\Delta R_i'$ relative to ΔR_i that one entire term $[\phi_i'\Delta R_i' - \phi_i\Delta R_i + \Delta R_{ij}]$ in the summed benefit for receivers, or $[\phi_i'\Delta T_i' - \phi_i\Delta T_i + \Delta T_{ij}]$ for senders, becomes negative. The negative term can be considered an incidental cost of communication equivalent to the incidental costs described without payoff matrix changes. Note that we have so far only looked at changes in payoff matrices that reduce the average benefit of communicating. It is also possible that signaling changes payoff matrices to the benefit of one or both parties. Any such changes could then be considered **incidental benefits** of communication.

In addition to necessary and incidental costs, we also need to consider **constraints**. One could treat most physical and phylogenetic constraints as a special case of necessary costs. In fact, some evolutionary theorists argue that nearly any constraint can be overcome if the organism will pay the additional cost (Gans 1989; McKitrick 1993). For example, small body size constrains the production of low frequency sounds to trivial amplitudes (pages 76–78). A small sender could still produce a loud, low-frequency sound if it were willing to modify its sound-producing organs and expend enough energy. However, it is unlikely to do so unless the additional benefits exceed these higher costs. Constraints thus divide the list of potential signals into those in which benefits can exceed costs, and those in which they cannot. Any mutant sender or receiver that tries to adopt one of the signaling options outside of the constraint boundaries will be disfavored by selection. Thus constraints reflect costs that will have to be paid if a sender or receiver attempts to exchange a signal outside of the constraint boundaries. They are therefore potential necessary costs, but because they are rarely imposed, we shall treat them here as a special category.

In the remainder of this chapter we focus on the necessary costs and constraints faced by senders and briefly discuss the incidental costs. Well-known examples in each of the main modalities will be described and specific adaptations that ameliorate costs will be identified. We shall then examine receiver necessary costs, constraints, and incidental costs in the same way. Environmental transmission constraints have been described in detail in Part I, but they will be summarized at the end of this chapter in anticipation of the discussion of design rules in Chapter 18.

SENDER NECESSARY COSTS

Senders are probably more limited by necessary costs than receivers because they must perform the work of signaling. Often these costs affect senders by increasing predation risk, time lost, and energy expended and thereby reduce sender survival and reproductive success. The magnitude of the cost depends on the signal modality and duration of signaling (Magnhagen 1991).

Conspicuousness to Predators and Parasites

VISUAL CONSPICUOUSNESS COSTS. The bright and contrasting colors that visual senders use to communicate with conspecifics are often also visible to predators. Endler (1983, 1987, 1991b) has described a classic case of the trade-offs between attracting conspecific females versus predators in male guppies that signal with bright colors. Guppies (*Poecilia*) are small freshwater fish that live in clear tropical streams. Fertilization is internal (guppies give live birth) and the mating system is promiscuous. Females prefer to mate with the more brightly colored males. Brighter males are more likely to be taken by predators than duller males. The compromise between sexual selection and natural selection has resulted in different optimal color patterns in different populations depending on the species of predator and its color sensitivity. In general, the guppy colors and behaviors that are favored are those that are maximally conspicuous to guppy females and minimally conspicuous to the predators. The most serious fish predator of the guppies, a cichlid, has eyes that are strongly sensitive to red but lack a blue pigment (Figure 17.1). Guppy populations with this type of predator have emphasized structural blue pigment patches. Other populations suffer mainly from a prawn predator that sees well in the blue region but lacks a red visual pigment; these guppy populations show strong red and orange coloration. A third set of guppy populations coexists with a relatively low-risk killifish predator possessing color sensitivity similar to the guppy; guppy coloration here consists of diverse color patches (red, orange, blue, cream, and other colors). The evolutionary strategy of displaying colors that local predators cannot perceive is called employing **private wavelengths**. In addition, predation pressure has modified the daily timing of display activity. Predation is highest during the middle of the day when ambient light levels are highest and green and yellow ambient light prevails. Males avoid displaying during this time, and instead display during early morning or late afternoon. Ambient light in streams is strongly blue and red at these times, and this colored light further favors the evolution of blue and red color patches. Guppies in high predation populations (with the cichlid or prawn predator) lack green, yellow, and white color patches that would render them very conspicuous during midday hours.

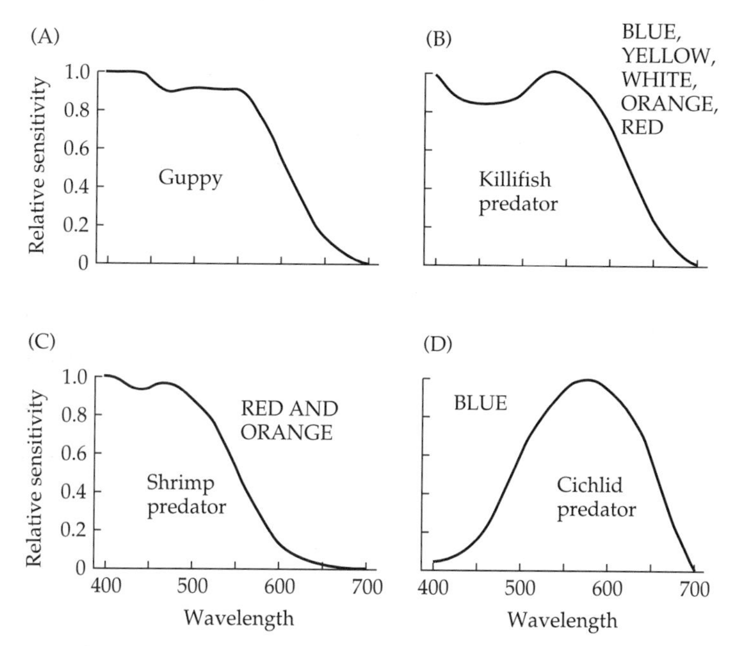

Figure 17.1 Spectral sensitivity of the guppy and three of its predators. All four curves are models based on the spectral absorbance characteristics of cones in the eyes of each species. The major hues of guppy color patches in populations sympatric with each predator are shown in uppercase letters. (A) The guppy (*Poecilia reticulata*) has three cone types and excellent sensitivity ranging from blue to red wavelengths. (B) The killifish (*Rivulus hartii*) is a nonthreatening predator with similar color sensitivity to the guppy. (C) The prawn (*Macrobrachium crenulatum*) is a moderately serious guppy predator that lacks a red cone and therefore only perceives colors ranging from blue to yellow. (D) The cichlid (*Crenicichla alta*) is a dangerous fish predator that lacks a blue cone and only perceives green through red wavelengths. (After Endler 1991b.)

McKinney (1975) attributed many of the differences in the signaling by two closely related sea ducks to differences in their vulnerability to visual predators while foraging. The common eider, *Somateria molissima,* dives to the ocean floor in relatively deep coastal water, while the smaller Steller's eider, *Polysticta stelleri,* dives in shallow water. Deep water is concealing, but shallow water leaves the ducks highly visible to aerial predators that can take them when they come up to the surface. Steller's eiders live in large flocks and dive synchronously with other birds to obtain the benefits of group detection of predators. Their communication displays and social interactions are significantly abbreviated, silent, and inconspicuous, compared to those of the common eider.

Visual signals carry a greater risk of predator attraction than other modalities because locatability of the sender is perfect (i.e., line of sight). General strategies to reduce this cost include hiding or covering the colors when signaling is not in progress, reducing the rate of display, and evolving good escape tactics. The high predation cost of visual signals may restrict the use of this modality to species that have good escape mechanisms, such as flight (in birds and insects), a nearby burrow (as in fiddler crabs), or toxic defense (as in skunks and some butterflies).

AUDITORY CONSPICUOUSNESS COSTS. Predators also cue in on the broadcast auditory signals that males use to attract females. The classic example has been described for calling frogs by Ryan et al. (1982). Male tungara frogs (*Physalaemus pustulosus*) produce repeated calls with a frequency sweep (termed the whine) plus up to five short broadband chucks (see Figure 16.16). Females show a strong preference for calls with chucks; chucks provide information on the body size of the male, since larger males produce lower frequency calls. Females prefer to mate with large males and use the frequency information in the chuck to discriminate large from small males. Bat predators (*Trachops*) home in on calling males (Figure 12.14) and are also more attracted to calls with chucks compared to calls without chucks. Because of the short, broadband nature of chucks, calls with chucks may be easier to localize than calls with only whines.

The songs of male crickets attract parasitic tachinid flies (Cade 1975, 1979). The ears of these parasites are highly specialized for locating the calling cricket. Some males avoid this cost by adopting an alternative noncalling satellite strategy. Such males go to areas where other males are singing and move about silently, attempting to intercept and copulate with females attracted to the singing males' calls. Cade found that 79% of calling males were parasitized, whereas only 14% of satellite males were parasitized. The larva that the fly lays on the host eventually kills the cricket, so the cost of calling is indeed very high. As in the tungara frog example, the tachinid fly is most strongly attracted to the same male calling characteristics that are preferred by female crickets (Figure 17.2) (Wagner 1996).

One of the most complete investigations of the costs and benefits of a signal was undertaken by Yasukawa (1989) on a short "chit" call note of the female red-winged blackbird (*Agelaius phoeniceus*). Female redwings often respond to their mate's territorial song with the chit call while sitting on their nests. However, calling from the nest may alert predators to the location of the nest. Using mock nests with and without playback of the chit call, Yasukawa found that nests with playback were more likely to be depredated than nests without the playback (Figure 17.3). Predator attraction is thus the primary cost of the vocalization. The benefit of the call is stronger defense of the nest by the male. Yasukawa placed stuffed crows near mock nests with and without playback of the chit call and showed that the male was more likely to vigorously attack the crow at the nests with playback. Females varied in their tendency to give the chit call. Those females with successful nests answered their mates with chits more often than females with unsuccessful nests, suggesting that the benefit of male defense outweighs the cost of conspicuousness to predators.

The long-distance, continuous auditory sender suffers nearly as high a risk of being localized by a predator as an equivalent visual sender. The auditory sender is not as easy to localize from a distance, especially if calling in a dark or vegetated environment. However, an auditory signal has a larger ac-

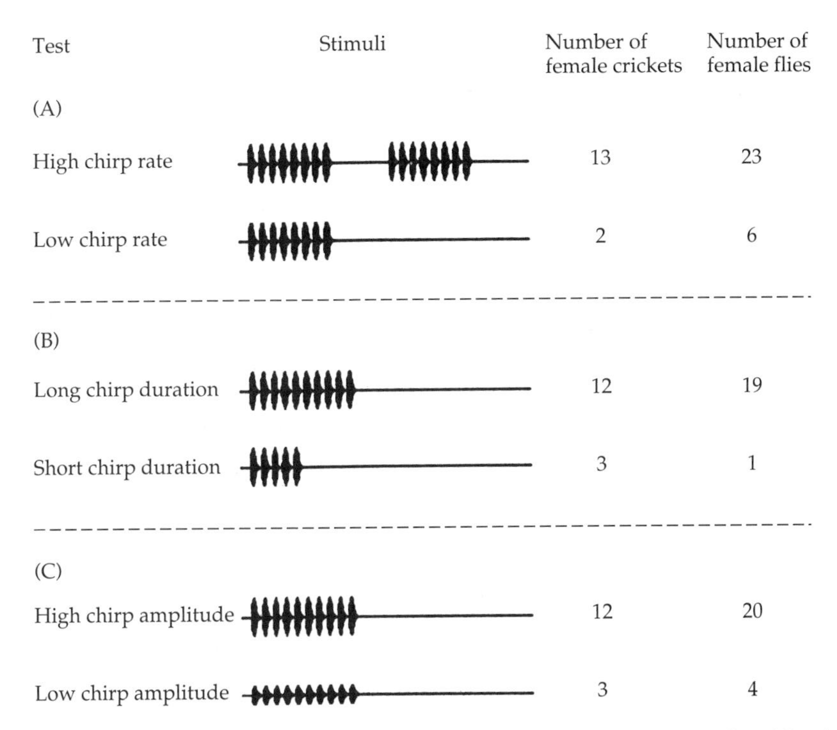

Figure 17.2 Predator attraction in calling crickets. Parasitic tachinid flies (*Ormia ochracea*) are preferentially attracted to the same types of male field cricket (*Gryllus lineaticeps*) calls that attract female field crickets. Each pair of waveforms shows the contrasting stimuli that were broadcast simultaneously from two speakers a short distance apart. The data are the number of cricket or fly receivers approaching each stimulus. Female crickets were tested in the lab whereas flies were captured as they approached speakers in the field. Both cricket and fly strongly prefer higher over lower chirp rate (A), longer over shorter chirp duration (B), and higher over lower chirp amplitude (C). (From Wagner 1996.)

tive space than a typical visual signal, and a mobile auditory predator can approach a distant sound to obtain better localization information.

OLFACTORY CONSPICUOUSNESS COSTS. Any animal that leaves an olfactory trail, whether intentional or not, has the potential to be followed by a predator that can detect the odorant. Terrestrial vertebrate predators such as snakes, weasels, and canids (dogs, wolves, and foxes) use their well-developed general olfactory sense to follow the trails of prey individuals. The long-lasting foraging trails of ants are frequently sabotaged by insect robbers that are attracted to the trails and steal the ants' food. Even airborne and water current signals can attract unwanted predators and parasites as well as conspecifics. For example, the aggregation pheromone of the female western bark beetle, *Dendroctonus brevicomis*, not only attracts conspecific males but also another beetle, *Temnochila chlorodia*, that preys on the adult and larval bark beetles (Agosta 1992). Chemical signaling probably carries the lowest risk of predator

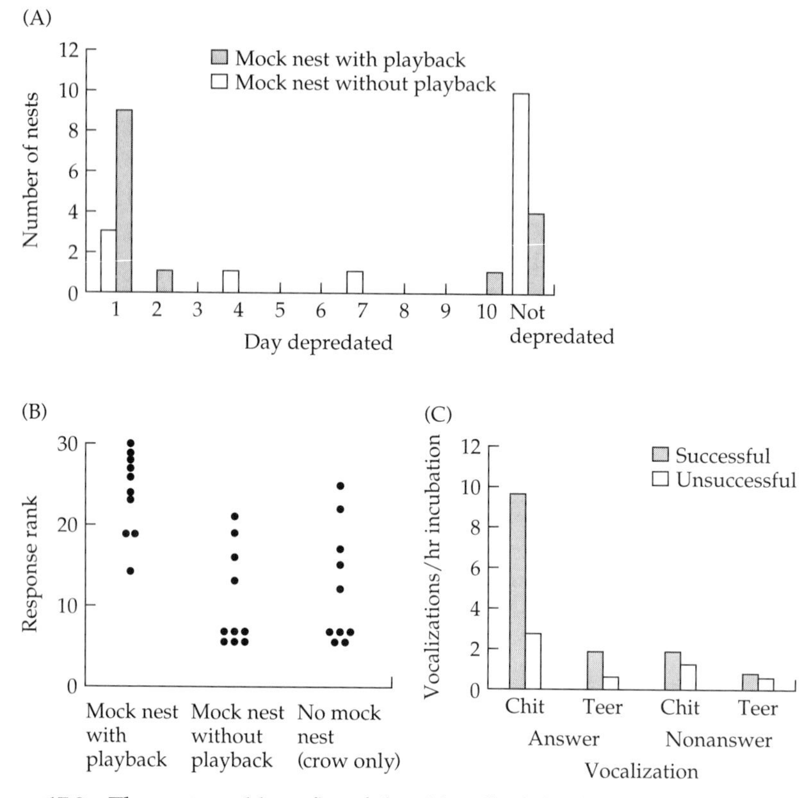

Figure 17.3 The costs and benefits of the chit call of the female red-winged blackbird. (A) The cost: Vocalizing from the nest attracts predators. Mock nests with playback of chit calls (gray bars) were more likely to be depredated within one day, whereas most mock nests without playback survived at least ten days. The experiment was performed in areas without territorial males. (B) The benefit: Males defend chit-broadcasting nests from predators. The level of male defense toward a stuffed crow model is highest at mock nests with playback, and lowest at nests with no playback and toward crow models with no nest. (C) Net benefit: Successful nesting by females (one or more fledglings) was associated with a higher rate of chit answering calls while incubating, compared to unsuccessful nesting (no fledglings). Another call, the alarm teer, showed a similar pattern. Chits and teers given in the absence of a preceeding male song were unassociated with nesting success. (After Yasukawa 1989.)

attraction of the three modalities, especially if the olfactory mark is deposited on the substrate so that the signal is not immediately associated with the sender. This may partly explain its frequent use by small terrestrial animals such as mammals and reptiles that have slower escape mechanisms and are more vulnerable to predation.

Energetic Costs of Signaling

VISUAL SIGNALS. Signals used to attract mates and defend territories must often maintain a high duty cycle, or high "on" time, to be effective. A visual movement display is by nature very short, so to achieve the requisite "on" time, the signal has to be repeated frequently. Not only is there a large time cost, but as the repetition rate increases, overall energy costs also increase. By using a variety of techniques, several studies have now demonstrated high energy costs for movement displays. Sage grouse males (*Centrocercus urophasianus*) perform a 2 second audiovisual strut display at intervals as short as 10 seconds for several hours each morning during the mating season. Using the doubly labeled water isotope technique to measure daily energy expenditure, actively displaying males were found to expend twice the amount of energy per day as nondisplaying males (Vehrencamp et al. 1989, see Figure 23.8). Weight loss, mortality, and other indicators of nutritional stress were found to be correlated with time spent in display in male Jackson's widowbirds (*Euplectes jacksonii*) that perform a jumping display to attract females and in wolf spiders (*Hygrolycosa rubrofasciata*) that drum their abdomens on dry leaf litter (S. Andersson 1994; Mappes et al. 1996).

A variety of studies have shown that males in highly sexually dimorphic species spend a great deal of time and effort attracting mates and have a higher mortality rate than females (Promislow et al. 1992; Promislow 1992; Owens and Bennett 1994; Salvador et al. 1996). The causes of mortality include not only predation, but also susceptibility to disease, injury from conspecifics, and the investment in the growth of feathers, horns, and other signaling structures. The production cost of color patches for signaling in most cases does not appear to be very expensive, but the development of red carotenoid-based coloration may be an exception. Animals cannot produce carotene and must therefore obtain the molecule from plants in their diet. There is some evidence that it is costly to accumulate a sufficient supply of carotene and that the conversion of carotene depresses immune system function (Hill 1996). The cost of producing the chemical substrate for bioluminescent signals is also high, 73 kcal/mole for the firefly. However, the efficiency of converting chemical energy into light energy is very good, at 88% efficiency, with only 1% lost as heat (Nassau 1983). Finally, investment in visual structures and large body size often results in another type of fitness cost—delayed maturation (M. Andersson 1994).

AUDITORY SIGNALS. Like visual movement displays, auditory signals are short and must be repeated to achieve a high duty cycle. Direct measurements of oxygen consumption in open-flow respirometers have been used to relate calling rate to energy expenditure in anurans, insects, and birds, as illustrated in Figure 17.4 (Bucher et al. 1982; Taigen and Wells 1985; MacNally and Young 1981; Eberhardt 1994). In all of these studies, energy expenditure was found to increase with increasing display rate. In some cases, animals were believed to be performing at their maximum aerobic capacity (Wells and Taigen 1986; Vehrencamp et al. 1989; Högland et al. 1992). The explanation for the high cost of repeated sound signals appears to be the low efficiency of sound production, as discussed in Chapter 4 and summarized by Burk (1988). Reasonable estimates for a variety of animals of the power output of sound relative to the energetic cost of sound production range from 0.5% to 4.9% efficiency (Brack-

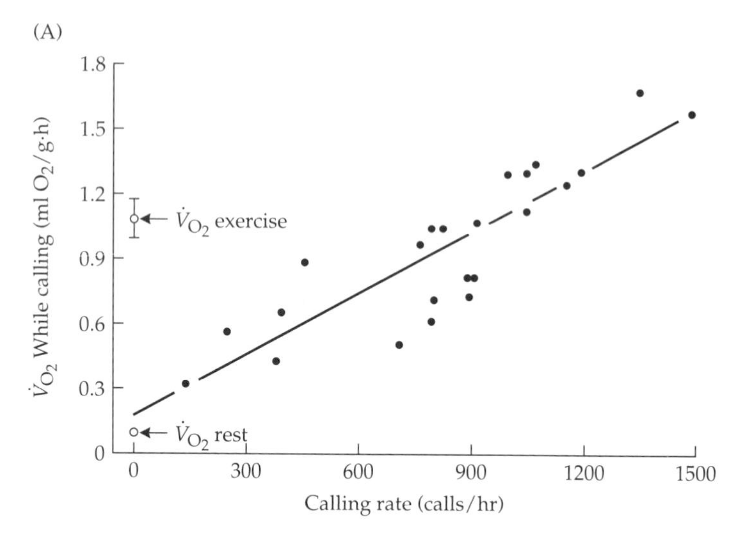

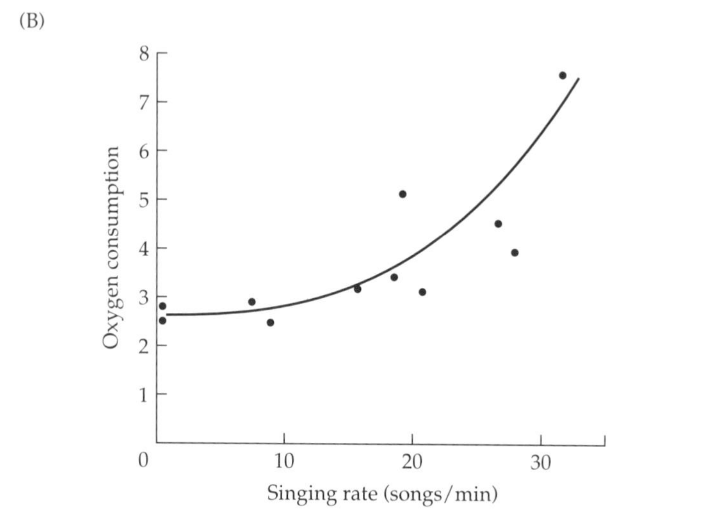

Figure 17.4 Energy expenditure increases with increasing calling rate. (A) Male gray treefrogs (*Hyla versicolor*) have a higher energy expenditure (oxygen consumption, $\dot{V}_{O_2}$) at the fastest calling rates than during forced exercise. The means for resting and forced exercise are shown in open circles. (B) Oxygen consumption of Carolina wrens (*Thryothorus carolinensis*) as a function of song rate. Oxygen consumption is given as a multiple of resting metabolic rate. (A from Taigen and Wells 1985, © Springer-Verlag; B after Eberhardt 1994.)

enbury 1979; MacNally and Young 1981; Ryan 1985b; Prestwich et al. 1989). Exceptions are some songbirds and the mole cricket (*Gryllotalpa* spp.), which achieves an efficiency of 35% by calling from its burrow with a horn-shaped entrance (Bennet-Clark 1970, 1989). A significant fraction of the energy loss is due to heat dissipated during muscular activity. The remaining loss is due to the acoustic impedance mismatch between the vibrating structures and the radiating structures that couple the sound to the environment. Long-distance acoustic signalers are faced with the tradeoff between the greater efficiency of producing high-frequency sounds (that are closer in wavelength to the size of the sound-producing and resonating structures) and the lower attenuation loss of low-frequency sounds during transmission through the environment (Ryan 1986a). Most small animals appear to opt for the production of the lowest possible frequency that body size constraints permit in order to take advantage of the transmission advantages.

CHEMICAL SIGNALS. Direct measurements of the energy cost of chemical signal production are difficult to make because the amount of chemical is usually very small (Holloway et al. 1993). In principle, the cost of building a chemical signal is the potential energy in the pheromone minus the potential energy of the source chemical or building blocks. Certainly the use of excreta and bacterial byproducts must be cheaper than the production of a complex pheromone. Glandular tissue for specialized pheromone production in mammals is highly vascularized and has a two- to threefold higher oxygen consumption rate than ordinary sweat glands (Thiessen 1977). In insects, body temperatures must be elevated by basking in the sun or by expensive muscular shivering in order to produce pheromones. On the other hand, the efficiency of chemical signal production is presumably high because metabolic (chemical) energy does not have to be converted to another form of energy. Although a single release of a chemical pheromone may not be very expensive, signals that must last a long time to be effective, or marks that must be deposited in large numbers, will certainly increase production costs. In Chapter 10 we saw how the active space of a single puff, r_{max}, increases as a function of the number of molecules released, Q, raised to the 1/3 power. This means that options for increasing Q as a mechanism for increasing signal range are limited. Nevertheless, of the three modalities, chemical signals are probably the cheapest to produce.

Time Lost

Repeated, continuous signaling as described above also entails another type of cost—time that could be spent in other activities such as foraging, parental care, and resting. Signaling is incompatible with these activities, and time spent signaling must therefore be traded off with these important functions. In some species, the time cost of signaling may be as great as, or greater than, the energy cost. Although the time cost has not been quantified to the same extent as that of energy expenditure, examples of time constraints can be found for each signaling modality. In lek-breeding species, males must spend so much time present at their arena, defending and displaying, that they have little time for foraging compared to females and nonbreeding males. Males also display during the times that females are foraging, so food resources may be depleted by the time males are able to feed. This results in significant changes and adaptations in male foraging behavior, that include major differences in diet and altered searching strategies (Bradbury 1977; Théry 1992). This cost exerts strong selection pressure on male foraging skill. In lekking antelope, males do not feed at all and rapidly lose condition while maintaining a display site (Balmford et al. 1992).

Lekking species that experience both high energy and time costs do have a few strategies available to them for ameliorating the costs. Large body size can enable males to sustain high energetic expenditure for longer periods of time (Møller 1996). Larger animals can store and carry larger fat reserves, and in homeotherms large body size reduces thermoregulatory expenditures. Changes in gut capacity and surface area can also help to get animals through periods of high energetic stress (Weiner 1992). In several lekking species, the most active and successful males are notably lean but healthy, and they seem to be able to forage and meet their daily energy needs more efficiently than less active males (McDonald 1989; Vehrencamp et al. 1989).

Territorial species that mark the periphery of their defended areas with a series of deposited chemical signals must traverse the boundary at frequent intervals to renew the marks. The time cost of olfactory territory defense is probably greater than either the travel cost or the chemical production cost. The fraction of time owners spend patrolling and marking their territories represents a substantial portion of a day, especially during territory establishment in seasonal species (Erlinge 1968; Kruuk 1972; Walther et al. 1983; Gorman and Mills 1984).

Conflict with Original Function

During the process of signal evolution, the structural source of the signal may become so exaggerated for the purpose of signal efficiency that the original function of the source is impaired. Although this type of cost is usually difficult to quantify, we know that it has occurred when another adaptation evolves to carry out the source structure's original function. For example, voluntary control of iris contraction for signaling in birds has rendered the iris much less useful for regulating the amount of light entering the eye, its original function. To compensate for the loss, pigment migration in the epithelial cells of the retina has therefore become very rapid and extensive. Analogous examples can also be found for auditory signalers. Insects that use their legs or wings for sound production may not be as well prepared for rapid escape from predators when signaling, compared to nonsignaling individuals. Terrestrial vertebrates cannot vocalize and feed at the same time. The extraordinary specialization of the glottis, larynx, and respiratory tract for vocal communication in humans results in a small but nevertheless real risk of choking on food. Olfactory examples of this type of cost are more difficult to find.

The best-documented example of costs arising from signal elaboration of functional structures is the elongated tails of birds. Tails provide lift during take-off and flight and act as a rudder for aerial turns. Elongation of the tail for display purposes increases the drag on the flying bird and greatly reduces agility, maneuverability, and flight speed. Birds that feed on the wing or migrate long distances pay a particularly large cost for tail elongation. Møller (1989; Møller and de Lope 1994; Møller et al. 1995) has quantified the cost of elongated outer tail feathers in the barn swallow, *Hirundo rustica*. This bird feeds on aerial insects and must make tight turns to pursue its prey. Larger, more profitable insects are especially difficult to capture because they are stronger fliers and make evasive movements to escape from the avian predator. Møller shortened the tails of some males and lengthened the tails of others. Elongation resulted in a smaller average size of captured insects, a reduced winter survival rate, and the growth of shorter tails the next spring (which reduced a male's mating success). Tail shortening, on the other hand, tended to improve the successful capture of large insects and survival. Female preference for long-tailed males is therefore responsible for tail elongation beyond what is aerodynamically optimal. (We shall discuss this example further in Chapter 23, see Figure 23.7.) Similar results have also been obtained for the scarlet-tufted malachite sunbird, *Nectarinia johnstoni* (Evans and Thomas 1992). The magnitude of the aerodynamic cost of tail elongation depends on the details of tail shape (Figure 17.5). Pintails (long central feathers) and deeply forked tails (long outer feathers) do not increase drag as severely as graduated tails. Several additional adaptations can ameliorate the cost of a long tail. For example, feather shape can reduce the drag of streamer feathers, increased wing length can improve overall lift, and increased muscle mass can partially compensate for the drag of a long tail (Balmford et al. 1993, 1994; Norberg 1994; Møller 1996).

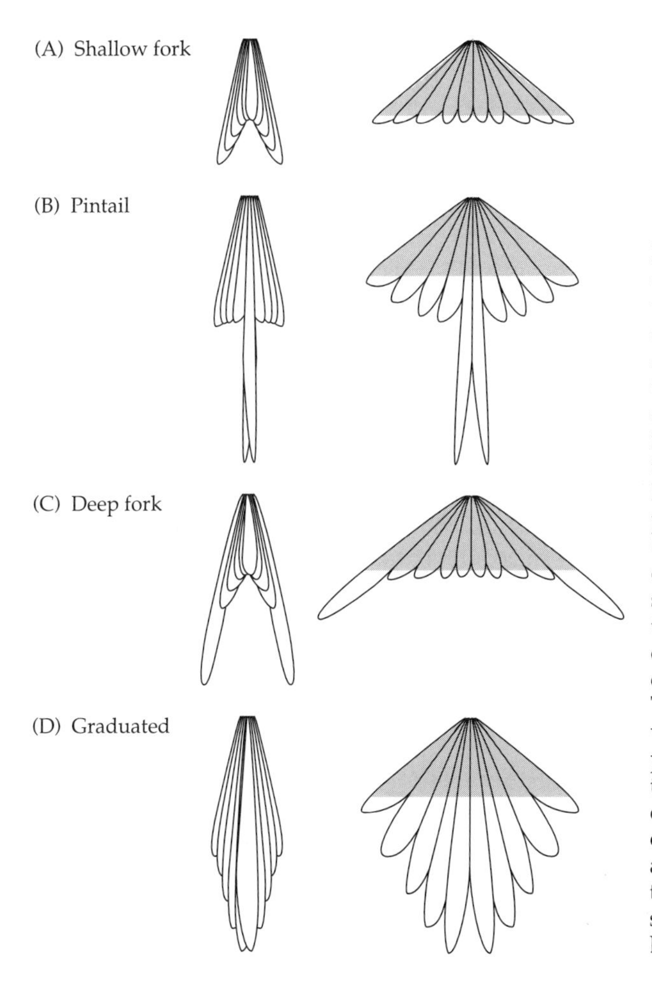

Figure 17.5 Drag caused by different tail shapes in birds. The illustration on the left shows the tail in the closed position as it would appear during signaling. The illustration on the right shows the tail spread to 120° as it would normally be during foraging flight. The shaded gray area shows the portion of the tail that generates lift, and the white portion shows the portion of the tail that generates drag. (A) A tail that is slightly forked when closed forms a perfect triangle when spread and is the optimal tail shape that maximizes lift while minimizing drag. (B) A pintail involves elongation of only the two central tail feathers. The increase in drag is relatively low while the visual effect in the closed position is high. (C) A deeply forked tail involves elongation of only the outer tail feathers and, like the pintail, only increases drag by a small amount. (D) A graduated tail produces the most drag and is the most costly strategy of tail elongation. (After Balmford et al. 1993.)

SENDER INCIDENTAL COSTS

For many types of signals such as mate attraction, social integration, begging, and alarm signals, the sender receives a benefit when the receiver responds correctly and fails to receive a benefit when the receiver does not respond. However, in other contexts the sender may experience significant costs when the receiver gives the wrong response from the point of view of the sender. This is particularly true in conflict contexts. For example, an aggressive sender may perform a threat signal or a tactical act that places it in very close proximity to its rival, in the hope that the rival will back down. However, if the rival continues its aggressive drive and is not intimidated by the threat, the sender may be attacked and injured. A threat signal can therefore be a tactical display that generates incidental costs by changing the payoff matrices. The average benefit of the signal will, of course, take into account the probability that the receiver backs down versus the probability it attacks. This benefit may be positive, but significant and immediate negative payoff terms may frequently limit the propriety of engaging in dangerous exchanges. The cost of injury during fights was shown to be a critical determinant of signaling strategies in little blue penguins (Waas 1991b; see pages 704–705). Finally, if senders are punished by receivers for emitting erroneous or false signals, there may be incidental costs to senders of participating in the exchange.

SENDER CONSTRAINTS

Physical constraints on senders (as well as receivers) can usually be understood from a thorough knowledge of the physical mechanisms involved in the particular modality, which were explained in detail in Part I of this book. Evolutionary constraints arising from a species' ancestral morphology, physiology, and behavior are more difficult to assess. A discussion of these constraints requires a broad-based phylogenetic analysis of the presence and absence of signals (or receiver mechanisms) throughout a taxonomic group. The tools for comparative analyses have only recently been developed and few studies dealing with signal evolution have been completed; this aspect of communication is bound to see major developments in the near future (Martins 1996). Acoustic senders are the most constrained by physical and phylogenetic factors, followed by visual senders, and chemical senders are the least constrained. Sender constraints are summarized in Table 17.1.

Phylogenetic Constraints

VISUAL SIGNALS. There are probably relatively few physical constraints on the production of visual signals since an animal doesn't have to do anything to generate a passive visual signal. Obviously, if a species hasn't got a specific body structure (such as tail or wings) or lacks muscles to move it, that structure can't be used for a movement display. The size of a structure or color patch determines the total intensity of the signal and the distance at which it can be detected. The body size of the sender therefore determines the maxi-

Table 17.1 Constraints on senders in each modality

Modality	Signal feature	Constrained by
Visual	Intensity/transmission distance	Small body size
	Display structures	Body form
	Movement displays	Neuromuscular preadaptations
	Carotenoid-based color	Access to dietary sources of pigment
Auditory	Low frequency	Small body size
	Intensity	Small body size
	Stridulation	Lack of hard exoskeleton or skeleton with moveable joints
	Vibration of membranes	Low-flow respiratory system, poikiothermy
	Frequency modulation	Stridulation and percussion sound production mechanisms
	Note shape and variation	Structure of vocal apparatus
Chemical	Transmission distance of airborne signals	High and low molecular weight
	Duration of deposited marks	Low molecular weight, nonpolarity
	Novel chemicals	Lack of metabolic pathways
Electric	Signal intensity/range	Body length

mum size of any visual signal it can produce. The neuronal control of muscles and the position and attachment of muscles affect the nature of movement displays, as discussed in Chapter 8, pages 221–222. Arthropods tend to produce jerky movements, whereas vertebrates can make smooth movements as a result of different strategies of enervation of muscles. The constraints on color production vary with color and taxon. Some pigments cannot be synthesized by animals and must be obtained from plants in the diet. As mentioned earlier, this is the case for most animals using red or orange carotenoid pigments. The implication is that senders that cannot synthesize the necessary pigments must acquire the chemicals from their diet.

Ritualized displays require a suite of structural, neural, and muscular adaptations that is often conserved during speciation events. Closely related species often exhibit the same form of displays even though the meaning or context may be different. Such evidence suggests that signal form is more constrained than the information content of the signal. For example, Crook (1964) describes the song bow display in weaverbirds; it has the same basic form in all species but different meanings in colonial species than in solitary species. In solitary species it is an omnidirectional display with a large active space intended for receivers at all angles relative to the sender. In colonial weaverbirds it is a highly unidirectional display with a small active space intended only for receivers directly in front of the signaler; birds located to the side of the signaler do not respond. This shift in meaning is clearly adaptive, since in colonial situations there are always many unintended receivers within visual range.

AUDITORY SIGNALS. The production of sound is considerably more constrained by sender morphology and phylogeny than visual signal production, and auditory signaling is therefore found only in a few animal taxa. Stridulation requires a hard exo- or endoskeleton and articulated appendages that can rub against other parts of the body. This requirement restricts stridulatory sound production to insects, crustaceans, and fish. Likewise, the use of the respiratory system for vocal signals is largely restricted to homothermic vertebrates (birds and mammals). Homothermy (maintenance of a constant, high body temperature) consumes about 17 times more energy than the poikilothermic (nonthermoregulating) strategy of amphibians and reptiles. High rates of oxygen consumption require an efficient, high-volume and well-controlled respiratory system. Although birds and mammals have solved this problem in very different ways (with air sacs and diaphragm, respectively), both have powerful air-flow systems with precise muscular control. Amphibians and reptiles do not have sufficiently rapid respiration rates to produce aspired air sounds. Most species found olfactory and visual communication channels sufficient for their social needs and did not bother to evolve vocal capability. Frogs, with their nocturnal habits and need for a long-distance signal to attract mates to fixed watery oviposition sites, overcame this constraint with an alternative solution—a pulse-pump breathing system that could shunt large volumes of used air back and forth between the lungs and a throat sac (Gans 1973).

Sender morphology also affects aspects of auditory signal form. The most important factor is body size, which sets a lower limit on the frequency of sounds auditory senders can produce. Figure 17.6 shows the relationship between body size and dominant frequency of vocalizations in anurans and birds. As mentioned in Chapter 4, animals can most effectively produce sounds with wavelengths approximately equal to or smaller than their body size. The power output of sounds with longer wavelengths (lower frequencies) than the size of the sound-producing apparatus is greatly reduced due to acoustic short-circuiting. Animals can only produce lower-frequency sounds than are optimal for their body size at a great increase in the cost of production.

Just as with visual displays, the form of auditory signals requires a suite of structural, neural, and muscular adaptations that tends to be conserved during evolution. The fine structure of the vocal apparatus constrains call signal form in anurans and birds. Martin (1972) showed that the type of amplitude modulation in *Bufo* is related to the fine structure of the glottis. Similarly, Tandy and Kieth (1972) showed, in the same genus, that pulse repetition rate is a conservative phylogenetic character. In hylid frogs, call characters allied with the morphological aspects of sound production such as dominant frequency and pulse shape are more similar for more closely related species, whereas behavioral and physiological characters such as call rate, duration, and pulse rate vary widely (Crocroft and Ryan 1995). In birds, the structure of the passerine syrinx enables these species to produce far more varied and complex notes than those found in nonpasserines (Raikow 1986).

CHEMICAL SIGNALS. Chemical signals are restricted to certain molecular configurations that maximize transmission or signal duration as needed for the context, but there is little evidence that chemical senders have any particular

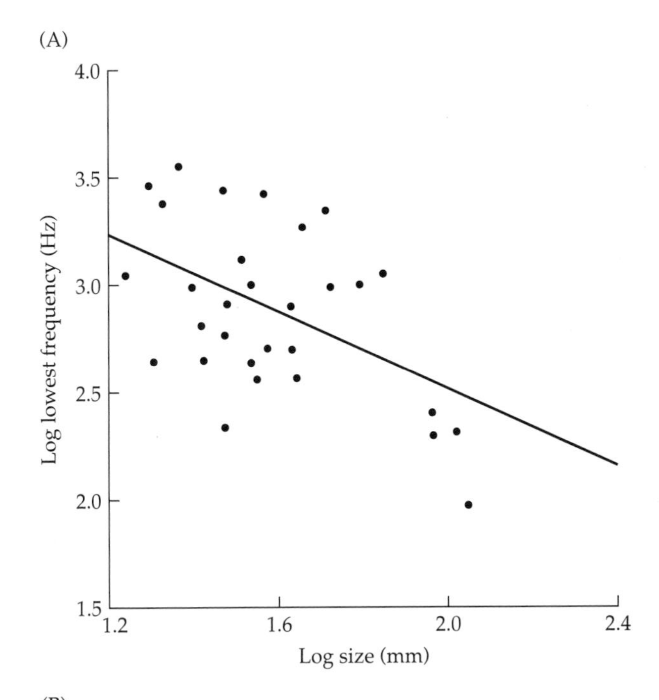

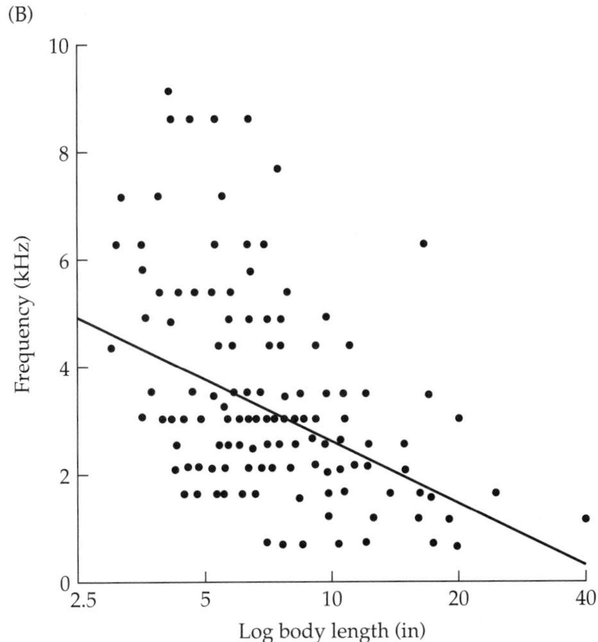

Figure 17.6 Relationship between fundamental frequency of vocalizations and body size. Frequency decreases with increasing body size. Each point represents a different species. (A) Leptodactyline frogs; (B) birds. (A after Ryan 1985a; B after Ryan and Brenowitz 1985.)

difficulty evolving adequate chemicals for signaling. This is because all organisms possess basic chemical pathways for taking in food, breaking it down, and rebuilding organic compounds for metabolic needs. Animals seem to be able to come up with novel ways of making chemicals that serve the transmission needs of the signal. The chemical alarm signal of the sea slug, *Navanax*, provides an interesting example of such solutions (Agosta 1992). Slugs normally leave a slime trail that will be followed by another individual intercepting it. The trail is composed of a long-lasting, water-insoluble mucus. If the trail-maker is molested, it adds an alarm substance to the trail that causes the follower to quickly veer off the trail. Alarm substances should dissipate quickly, since danger rarely lasts very long. To ensure the rapid degredation of the alarm substance, the chemical is a bright yellow pigment called navenone. Because of its color, the navenone absorbs light and is unstable in daylight, bleaching away within a few hours.

Other evidence suggests that phylogenetic heritage may channel the options for chemical signaling as we described for visual and auditory signals. Most moth (Lepidoptera) species use the same basic metabolic pathway to construct their mate-attraction pheromones; congeners even use the same carbon chain skeleton and differ only in the functional group (See Table 18.2). The conserved pheromone production mechanism could be construed as a phylogenetic constraint. Finally, some specific chemicals (such as carotenoids) that animals need cannot be self-manufactured and must be acquired in the diet. However, compounds derived from toxic dietary sources can be turned into species-specific pheromones, as we learned in Chapter 16.

Constraints on Sender Learning

In songbirds that learn a repertoire of songtypes for territory defense and mate attraction, several types of costs and constraints operate on the birds as a result of the learning process. One constraint is the number of songs they can learn. Recent comparative studies show that repertoire size is correlated with the size of the higher vocal center (HVC) nucleus in the brain that controls song learning (Figure 17.7). This suggests that brain size determines and constrains the capacity to learn new songs (DeVoogd et al. 1993). To show that this correlation is due to genetic constraints and not the number of songtypes young birds are exposed to, a comparative study was undertaken on two distinct populations of marsh wrens (*Cistothorus palustris*) with very different average repertoire sizes: New York birds with a repertoire of 50 and California birds with a repertoire of 150 songs. Males from both populations were exposed to 200 songs during their sensitive periods in a controlled laboratory environment. The New York males learned 40 and California males learned 100 songs (Canady et al. 1984; Kroodsma and Canady 1985). California birds also have larger HVC nuclei. Rather than construing brain size as a constraint, one could also argue that a larger repertoire is required in the California population and brain size has increased there to meet the need.

A second type of constraint concerns the duration of the sensitive period of learning in species that match their songtypes to their neighbors. Some species such as mockingbirds, starlings, red-winged blackbirds, great tits, and a few warblers can add new songtypes to their repertoire each year. Depending on the degree to which old songtypes are dropped or kept, this process can lead to a large repertoire that increases with age or a smaller repertoire

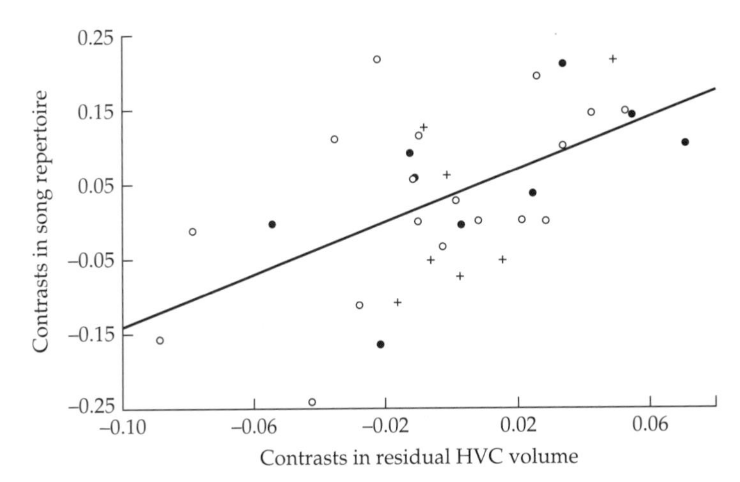

Figure 17.7 Relationship between HVC (higher vocal center) nucleus size and repertoire size in passerine birds. A larger repertoire size is associated with a larger brain area for vocal learning. Each point represents an independent contrast between two related species (•), genera (○) or families (+) with different repertoire sizes. The effect of overall brain size (telencephalon) on HVC volume has been statistically removed. (From DeVoogd et al. 1993.)

that maintains a high degree of matching with new neighbors. Other species such as sparrows and wrens exhibit a restricted period of learning during the first few months of life and do not acquire new songs after this age. Song matching also seems to be important in these species. Most males remain in the micropopulation in which they learned their songs so an extended learning period may not be necessary. However, a few males do disperse further away and appear to suffer from the restricted learning period and inability to match their new neighbors (Beecher 1996).

RECEIVER NECESSARY COSTS

Receivers are probably most strongly limited by physical and phylogenetic constraints, but they are also impacted by direct reception costs such as the time needed for evaluation, the consequences of approaching senders to assess the information in their signals, and the investment in receptive organs and neural processing. Brain tissue is very expensive metabolically, and sensory processing is therefore presumed to require substantial energy expenditure. However, no estimates of the energy cost of signal reception have been made.

Vulnerability to Predation and Harassment

Receivers that must approach senders closely in order to evaluate their signals may expose themselves to two types of costs: harassment and predation. A handful of studies have demonstrated such costs for females searching for mates. Since exposure time increases when females are more selective of mates, the magnitude of this cost is affected by receiver decisions. For example, redlip blennies, *Ophioblennius atlanticus*, defend individual feeding territories on the substrate of coral reefs. Females must leave their territories and cross the territories of other individuals to reach prospective male mates. Reynolds and Côté (1995) found that females suffered injuries from attacks by territorial damselfish during these excursions. In the guppies discussed on page 554, cryptically colored females may suffer an increased risk of predation when they school with brightly colored males (Pockington and Dill 1995). Searching female bush crickets are vulnerable to motion-detecting predators such as snakes, lizards, birds, and frogs and appear to pay a mortality cost similar to that of the advertising males (Heller 1992). In the Uganda kob (*Kobus kob*), a lekking antelope, females prefer to mate and rest on territories with low grass height because these afford higher visibility of predators (Deutsch and Weeks 1992). Several experimental studies have shown that receivers fail to exhibit mate choice preferences when they must evaluate mates in the presence of a predator or reduced cover (Berglund 1993; Forsgren 1992; Magnhagen 1991; Milinski and Bakker 1992; Hedrick and Dill 1993). The implication is that receivers reduce their assessment effort under the constraint of predation. On the other hand, Gibson and Bachman (1992) examined a range of possible costs of mate choice in the sage grouse, including travel costs, time lost from foraging, and predation risk, and found them to be extremely low. They estimated a 1% increase in daily energy expenditure from extra travel and a 0.1% increase in mortality from predation by golden eagles while on the lek.

Time Lost

Assessment by receivers takes time. Sullivan (1990, 1994) points out that the time required to extract information from a signal depends on the complexity of the signal and the type of information it encodes. Simple fixed signals such as a morphological character are quickly and cheaply assessed, and little additional benefit is obtained from repeated attention to the signal. Fixed but complex or multiple signal characters such as the tail of the peacock may require more time to assess. If a critical aspect of the signal is its degree of variability or its repetition rate, then receivers will need more time to assess the information. Finally, if the signal trait reveals information about facultative or variable aspects of the sender such as its status or condition, then assessment over an extended period of time may provide more accurate information. The degree to which receivers are under a time constraint will determine what types of traits they are likely to pay attention to, which in turn affects the nature of the signals.

Few studies have examined receiver behavior under variable time constraints, but a notable example is Backwell and Passmore's (1996) study of mate choice in the fiddler crab (*Uca annulipes*). Females must time larval release to coincide with peak nocturnal spring tides and must therefore leave sufficient time for embryonic development after mating. Early-mating females are therefore relatively unconstrained by time, whereas late-mating females are much more constrained. Since females incubate their eggs while cloistered in the male's burrow, its size, slope, depth, and stability are important for successful reproduction. Backwell and Passmore found that early females attended to both male burrow characteristics and major cheliped size. These two traits were uncorrelated among males. Early females made their choice in a two-stage process by first selecting large males and then examining their burrows. Late females, on the other hand, both checked burrows of and mated with smaller males than the early females. The implication is that late females paid less attention to male claw size.

Other studies also indicate that initial attraction of females to males is based on quickly assessed signals, whereas mating decisions are based on more complex signal characteristics. In Jackson's widowbird, long-range attraction is based on the rate of a visual jumping display; although rate evaluation requires attention to several repetitions of the display, jumping is the only way to get the female's attention because it lifts the male above the sea of tall grass. Mating decisions, on the other hand, are based on tail length, which is assessed at close range on the male's arena with a lengthy tail-quivering display (Andersson 1989). In the sage grouse, females visit the lek on two to three consecutive mornings and examine 6 to 10 males each day for about 10 minutes apiece before making a mate-choice decision. Attraction to males is based on an acoustic parameter of the strut display, the inter-pop interval, which may be correlated with signal intensity. Mating decisions are based on strut-display rate. Males increase their strut rate when females are present on their territories, and the time females spend watching a male may serve to test his stamina. Half of the females return to one or more of the males they had previously visited on a given day to reassess them, suggesting that receivers must sample extensively before making decisions (Gibson 1996). Other studies of females searching for displaying males in birds and mammals also find that females require one to three days to settle on a mate (Alatalo et al. 1988; Slagsvold et al. 1988; Bensch and Hasselquist 1992).

RECEIVER INCIDENTAL COSTS

As explained in Box 17.1, receivers are much more likely to suffer incidental costs than senders. Senders will not signal if their costs are too large, and in most cases their value of information is larger than the equivalent one for receivers, particularly in conflict situations. Errors in transmission, detection, and decoding of signals are likely to impact receivers more than senders. Receivers are also susceptible to fitness loss from manipulative or deceptive signaling. Deceptive senders can be either conspecifics or heterospecifics; the latter are often called **code-breakers** because they mimic the signals of a prey or host species. Whether conspecific or heterospecific, the sender manipulates the normally adaptive response of receivers to the conspecific signal. An example of a conspecific manipulator is the production of an alarm call in a competitive food context that causes the flockmates to flee, giving the deceptive sender sole access to the food (Møller 1988b). The many examples of heterospecific symbionts of ant colonies represent cases of olfactory code-breakers that mimic colony odor and usurp the protection and care afforded by the host ant colony (Wilson 1971). The bola spider (*Mastophora*) swings a sticky ball covered with the pheromone of the armyworm moth (*Spadoptera frugiperda*) to lure in its prey (Eberhard 1977). In Chapter 12 we described an example of visual signal deception by predatory fireflies who mimic the flash response of females to lure in the males of a prey firefly species. A quantitative illustration of this example showed that the cost of deception for a receiver is the product of the reduction in fitness from each occurrence of deceptive signaling times the probability of encountering a deceptive signal. As long as either the occurrence or fitness loss per encounter is sufficiently low to counterbalance the adaptive benefits of attending to the signal in nondeceptive circumstances, receivers will continue to be deceived on occasion. If the cost becomes higher, receivers will evolve adaptations or improved discrimination abilities to detect and avoid such manipulators.

RECEIVER CONSTRAINTS

Phylogenetic Constraints

The sensitivity of receivers certainly constrains the modalities and forms of signals that can be detected. Visual signals are probably most constrained by receiver phylogenetic history; sound signals are moderately contrained. Chemical signals seem to be the least constrained by reception. Physical and phylogenetic constraints on receivers are summarized in Table 17.2.

VISUAL RECEIVERS. The most pervasive constraint on visual reception is the effect of body size on spatial resolution or acuity. As discussed in Chapter 9, resolution increases as the distance between adjacent receptor cells decreases and as the focal length or eye radius increases. There is a lower limit to the diameter of visual receptor cells on the order of two wavelengths of visible light, or about 10^{-6} m (1 micron). Below this size, light is guided down the outside of the photoreceptor. Furthermore, a small animal cannot have a focal length or eye radius larger than its head. Small animals are therefore strongly limited in their resolving power. Figure 17.8 shows the relationship between body size and resolution in a variety of animals. The light weight of the in-

Table 17.2 Constraints on receivers in each modality

Modality	Receiver feature	Constrained by
Visual	High resolving power	Small body size
	Low-light sensitivity	Degree of summation of receptor cells
	Good temporal resolution	Speed of rhodopsin recovery
	Polarized light sensitivity	Ciliary receptor cells
	Good frequency resolution	Number of receptor pigment types
	Distance estimation	Monocular vision
	Wide field of view	Binocular vision
Auditory	Frequency range	Particle detector, pressure-differential detector
	Directionality	Body size
Chemical	Sensitivity	Number of receptor cells
	Chemical resolution	Number of receptor types
Electric	Sensitivity, directionality	Body length

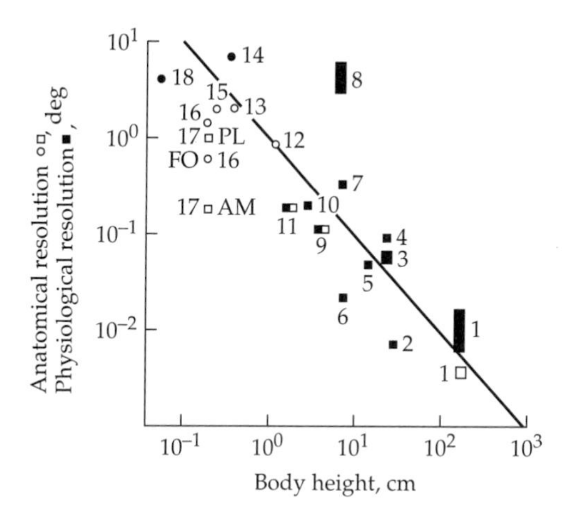

1 Man
2 Peregrine falcon
3 Hen
4 Cat
5 Pigeon
6 Chaffinch
7 Rat
8 *Myotis*
9 Frog
10 Lizard
11 Minnow
12 *Aeschna*
13 *Apis*
14 *Chlorophanus*
15 *Musca*
16 *Syritta*
17 *Methaphidippus*
18 *Drosophila*

Figure 17.8 Relationship between visual resolving power and body size. Larger animals can resolve finer spatial patterns than smaller animals. Resolution is measured in degrees and is based on either anatomical separation of individual receptors or on behavioral tests. Body height is measured at the center of the eyes above the ground. The diagonal line represents a resolution of one degree per centimeter of body height. (From Kirschfeld 1976, © Springer-Verlag.)

sect's compound eye has enabled these small, flying, vision-dependent animals to increase the size of their eyes to an extraordinary degree.

Of course, color vision is constrained by the number of cone pigments and the tradeoff with rods in sensitivity-maximizing nocturnal animals. The phylogenetic history of color vision in vertebrates has become much clearer with the recent sequencing of the rod and cone opsin genes and further studies of retina anatomy and behavioral color responses in a variety of species (Goldsmith 1990; Reichenbach and Robinson 1995; Yokoyama and Yokoyama 1996). Cones evolved first, and the most primitive living vertebrate, the lamprey, possesses an eye composed entirely of cones with two pigment types. It appears to be capable of dichromatic vision. Rods developed from cones early in the evolution of fishes to meet the needs of deep-water species. A few of these species subsequently lost all of their cones and developed entirely rod eyes, but most fish possess a duplex retina with a rod system and a two-pigment cone system. A few freshwater species then evolved three, or occasionally four, cone pigment types. Amphibians maintained the duplex retina and low number of cone pigment types as required by their mixed nocturnal and diurnal habits, but many birds and reptiles became strictly diurnal and had little need for the rod system. They reduced the number of rods and expanded their color vision capability with additional cone pigment types and colored oil droplets that enhanced their sensitivity to hue and chroma. Many lizards eliminated rods entirely. Geckos, a group of nocturnal lizards, faced a serious constraint—the need for a sensitivity-maximizing retina without the presence of rod precursors. It turns out that they reevolved a rodlike system from their blue cones. Snakes, which also evolved from lizards, may have undergone a similar process. Mammals evolved from ancient reptiles that still possessed rods. The long early history of mammals as small, nocturnal animals dominated by reptiles and dinosaurs selected for rod-biased eyes, but mammals did not lose their cones entirely and were probably capable of limited dichromatic vision. Once the dinosaurs disappeared and mammals radiated into some of the diurnal niches dinosaurs vacated, the proportion of rods decreased. However, even in animals like squirrels with strongly cone-biased eyes, the number of pigment types did not increase; these animals remained dichromats. The phylogenetic heritage of nocturnality appears to have constrained the development of a more advanced color system among mammals. Only recently, with the evolution of the primates, has a trichromatic system evolved, and this appears to have been due to the duplication of the middle-wavelength opsin gene and its subsequent slight shift to a higher peak-absorption wavelength. This brief review raises an important question. Was color vision in mammals really constrained by phylogenetic history, or did the mammals just not need a richer color sense until primates evolved? The development of a rodlike system in geckos from non-rod-possessing ancestors indicates that when the benefits are strong, phylogenetic constraints can be overcome.

Evolution also seems to have found two different alternative solutions to the constraint of the inability of vertebrate rods and cones to detect polarized light for the purpose of orientation and navigation. As illustrated in Figure 7.11, the sun generates a zone of polarized light at a viewing angle of 90°. Invertebrates use the inherent polarization sensitivity of their microvilli-based photoreceptors to estimate the sun's position. Some fish that navigate with respect to solar cues simply turned some of their blue cone receptors sideways, to a 90° angle, which makes them sensitive to the *e*-vector of polarized light (see Figure 9.13) (Waterman 1981). Turtles and birds, on the other hand, circumvented the use of the pattern of polarized light in the sky, and instead evolved a UV and red color-opponent system that can detect the position of the sun by the color gradient of scattered light in the sky (Coemans et al. 1994). Turtles have made scant use of their color-vision capabilities in signaling systems, but birds certainly have evolved a rich variety of plumage color signals as a consequence of their color reception capabilities.

AUDITORY RECEIVERS. Auditory receivers are also subject to some constraints. As discussed in Chapter 6, the type of ear determines the range of frequencies that can be detected. A particle-detector ear only picks up low-frequency near-field sound and must usually be tuned to a narrow frequency range. A pressure-differential detector cannot detect very low frequencies, but is extremely sensitive to intermediate and higher frequencies in the near field. A pressure detector can resolve a wider range of frequencies, including low frequencies, and is about as sensitive to far-field medium- to high-frequency sounds as a pressure differential ear.

Body size constrains the ability of auditory receivers to use arrival time differences, amplitude differences, phase differences, and diffraction patterns at the animal's two ears to determine the directionality of sounds. Larger animals can use these sound features much more effectively than can smaller ones. Small animals usually can get around this constraint with a pressure-differential ear that is intrinsically sensitive to sound direction. However, very small animals that need to localize sounds very accurately, such as the tachinid flies that parasitize crickets (Figure 17.2), have had to evolve yet another mechanism for localizing sounds. These very small animals possess two tympanal sensors contained within an undivided air-filled chamber. The tympana are mechanically coupled with a cuticular bridge that operates like a flexible lever to amplify small differences in arrival time and amplitude at the two ears (Robert et al. 1996). The size of a receiver's body or brain also sets an upper limit on the range of frequencies that can be detected and on the sophistication of the frequency resolution abilities of the ear. To resolve differences in frequency, a tonotopic receiving apparatus is necessary, and this type of system entails large numbers of nerves. Since nerve cells are the same size regardless of the body size of the animal, very small animals don't have room for a large, tonotopic ear. The system they develop detects a useful range of frequencies but analyzes the sound in the time domain.

CHEMICAL RECEIVERS. There appears to be little or no constraint on the ability of chemical receivers to evolve olfactory sensitivity to any chemical that is important to the animals. The more significant constraint for chemical receptors lies in the detection of small differences in intensity over a wide range of concentrations, due to the log scale of reception. Sensors must be divided into several sensitivity or threshold classes, and the most sensitive ones must be very numerous. This strategy for increasing sensitivity necessarily requires a reduction in the range of different chemicals that can be resolved. Therefore, species with extreme olfactory sensitivity can detect only a small number of chemicals, and species that can detect a broader range of chemical compounds have lower receiver sensitivity and concentration resolution (Derby and Atema 1988).

Memory Constraints

As decision makers, animals need to acquire and process information from all sources, learn to make associations between stimuli such as cues and signals and the contexts and conditions they encounter, and store and retrieve this information. All of these mental tasks require brain space. As the size and complexity of these tasks increases, brain size must also increase. Are there evolutionary constraints on brain enlargement, or encephalization? Larger brains (relative to body size) are associated with animals that have frugivorous and carnivorous diets (compared to folivorous diets), large home range size, longer lifespan, and nonfossorial habits (Harvey and Krebs 1990). Increases in total brain size are strongly associated with an increasingly larger proportion of the brain devoted to neocortex, the primary location for sensory processing and memory (Finlay and Darlington 1995). For example, a squirrel monkey (*Saimiri sciureus*) has a brain that is 10 times larger than a similarly sized mammalian insectivore (*Tenrec ecaudatus*) but a neocortex that is 60 times larger. Good evidence now exists that memory capacity for specific kinds of tasks is associated with enlargement of specific regions of the brain (Jerison 1985). In humans, specific brain regions are associated with facial memory, language learning, visual and auditory integration, and other functions. Spatial memory information, for example, is stored in the hippocampus. In birds and mammals that cache food in small amounts throughout their territories and retrieve it later, this region of the brain is significantly larger (Krebs et al. 1996). Although food storers do not have significantly larger total brain sizes than their nonstoring relatives, there is no evidence that they have had to reduce some other part of their brain to accommodate the larger hippocampus. In short, there does not seem to be any constraint on brain enlargement for special mental capacities as long as the benefit outweights the metabolic cost of the additional brain tissue.

Do animals show behavioral evidence of memory constraints? Food-storers such as the nutcracker (*Nucifraga columbiana*) scatter as many as 6000 caches of pine seeds throughout their home range and can remember the locations months later. Laboratory studies on these birds have verified the accuracy of their spatial memory (Olson et al. 1995). Earlier studies on the ability of territorial male songbirds to distinguish their neighbors from strangers concluded that species possessing only one songtype per male could accomplish the task better than species with large songtype repertoires (Falls 1982; Stoddard 1996). As the number of songtypes per male increases, the total number of songtypes a territorial owner must memorize increases rapidly. Song sparrows (*Melospiza melodia*) with an average repertoire of nine songtypes, do show neighbor-stranger discrimination, but it is more difficult to demonstrate because males in this species react very strongly to both types of intruders (Stoddard et al. 1991). Stoddard et al. (1992) then brought males into the lab to determine just how many songtypes they could remember. The study showed that male song sparrows could easily distinguish up to 64 songtypes without any evidence of confusion or memory interference and concluded that memory constraints were unimportant. McGregor (1989; McGregor and Avery 1986), on the other hand, studied memory capacity in great tits, a species in which males possess only three or four songtypes and routinely drop old songtypes and acquire new ones in response to new neighbors. He found evidence for proactive memory interference, in that the songs of new neighbors were confused with similar songtypes previously learned from older neighbors. Once again, it appears that capacity matches the typical need. Parrots are renowned for their vocal learning ability, and they possess larger brains than other birds of similar size. Although they can continue to learn new sounds throughout life, they do not produce a vast array of vocal variants as in songbirds but seem to use their ability to develop complex pair duets and group and individual-specific signature calls (Wright 1996). Parrot societies consist of populations of several hundred birds that associate in temporary flocks of variable sizes. Their large brains may therefore have evolved in part to store large numbers of individual signature variants.

TRANSMISSION CONSTRAINTS

Table 17.3 summarizes and compares the basic transmission characteristics of all five modalities. Properties of visual, auditory, chemical, and electric signals

Table 17.3 Signal transmission characteristics for each modality

Modality	Medium require-ments	Maximum range	Localiza-bility	Temporal modula-tion	Complexity	Signal duration
Visual	Ambient light	Medium	Good	Rapid	High	Variable
Auditory	Air or water	Large	Medium	Rapid	High	Short
Chemical	Current flow	Large	Variable	Slow	Low	Long
Electric	Water	Short	Good	Rapid	Low	Short
Tactile	None	Short	Good	Rapid	Medium	Short

were described in detail in Part I. We have said little about tactile signals up to this point because they are not transmitted through the environment and require little in the way of sender or receiver morphological specialization. (Tactile signals are produced by appendage movements and detected by a network of nerve endings, pressure detectors, and/or hair cells on the body surface.) Figure 17.9 shows some examples. Such stimuli are difficult to measure and manipulate, hence the paucity of studies (Markl 1983). However, together with other mechanically detected signals such as substrate-borne vibrations and near-field sounds, they represent an important signal modality in web-building spiders, social insects, species with internal fertilization, and humans (Geldard 1977). Contact signals also play a tactical role in aggressive interactions. It is therefore important to include tactile communication here in anticipation of the analysis of signal design in the next chapter.

The most important constraint for visual communication is the need for an external or internal source of light. For reflected-light signals, transmission is highly dependent on the amount and frequency range of ambient light. Reflected-light communication can occur in both air and water as long as some ambient light is available; in dark environments self-generated light is favored. Secondary constraints are blockage by objects in the environment and body-size limits on signal size, which together constrain the maximum signal range to intermediate distances. Otherwise, visual signals can encode a large and varied amount of information because of their high potential for locatability, temporal modulation, complexity, and variable signal duration.

Auditory signals are the least constrained by environmental factors. Signals can be transmitted long distances in air or water, during night or day, and in open or vegetated environments. Like visual signals, the potential for temporal modulation and complexity is high, but auditory signals are less locatable and limited to short durations unless repeated. The most significant constraint is degradation with distance due to frequency-dependent attenuation, temporal degradation, and masking environmental noise.

Long-distance chemical signals require current flow but can be used in air or water. Signal duration is long, so receivers can locate the signal source rel-

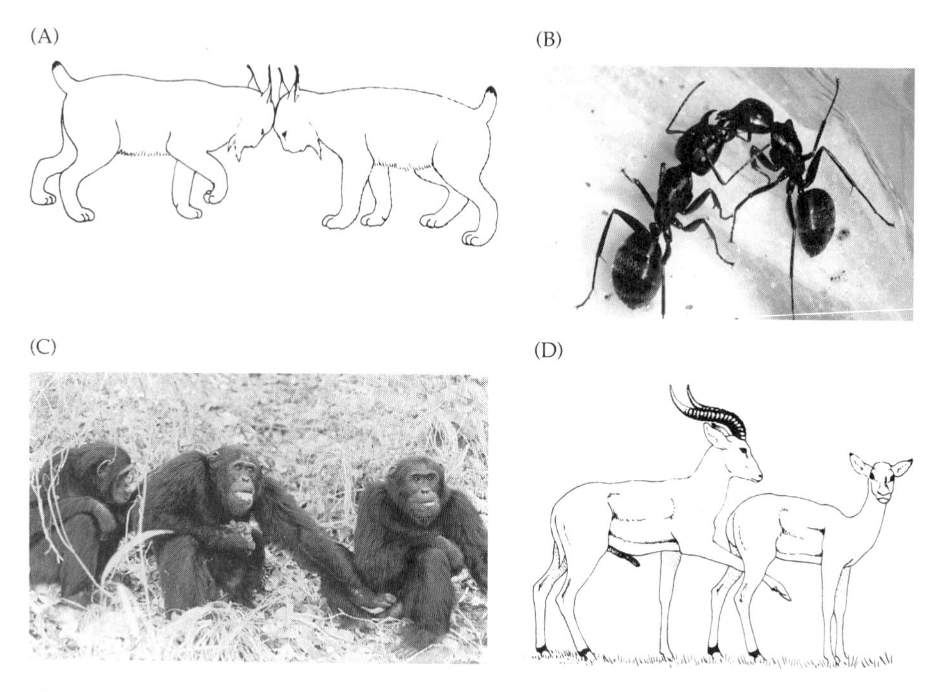

Figure 17.9 Examples of tactile communication. (A) A mutual head-offering display in the lynx (*Lynx lynx*), a common form of greeting in many cat species. (B) Food solicitation in wasps (*Polybia*) and other hymenopteran species. Using the tips of her antennae, the solicitor on the left gently strokes the antennae of a potential donor on the right until the donor regurgitates food. (C) Reassurance touch in chimpanzees (*Pan*). In fearful contexts males may touch another male's genitals as a mechanism of reassurance and alliance formation. These two males were feeding at the edge of their home range and were frightened by a snapping twig. (D) The mounting intention signal in the Uganda kob (*Kobus kob*). The male gently touches the inside of the female's hind leg with his foreleg. If she stands still and raises her tail, the male proceeds to mount. (A after Lindemann 1955; B courtesy of Carl Rettenmeyer; C courtesy of Richard Wrangham/Anthro-Photo; D after Buechner and Schloeth 1965.)

atively well with time. Such signals have poor potential for complexity and temporal modulation, however. Diffusion-transmitted signals, on the other hand, require the absence of current flow and can be designed for short-distance transmission and moderate fade-out time. Deposited marks and contact signals place the transmission burden on the receiver and can be made more complex with chemical mixes.

Electric signals are limited to aquatic environments. Signals are short in duration and can be rapidly modulated in time but have only modest potential for variant complexity. They are largely free of environmental constraints on their spectral and temporal structures but are severely attenuated by geometric spreading, so maximum range is very short. Electric signals are both accurately locatable and can travel around objects in the environment.

Tactile signals can be transmitted in dark environments and the sender is always locatable. Like electric signals, signal duration is short and can be rapidly temporally modulated. Signal complexity can range from low to moderate. A single light touch may be used during courtship to signal mounting or a more forceful touch may be used to assert dominance. The most complex tactile signals are the dances of bees that transmit information on the location of food sources, a topic we shall discuss in more detail in Part III.

SUMMARY

1. Both senders and receivers are subject to a variety of costs and constraints that can be divided into three categories. **Necessary costs** consist of immediate costs experienced while signaling or receiving and the costs of investing in signaling and receiving structures and organs. These costs may vary with the condition but are paid regardless of the outcome of decisions by the receiver. **Incidental costs** often arise in contexts of sender/receiver conflict and affect the net benefits each obtains from signaling and attending to signals. A cost occurs when either (1) the sender's code is less than 100% accurate, generating a negative $\Delta\phi$ term for both the sender and receiver, (2) the sender and receiver disagree on the optimal response for the default condition, generating occasional deceptive signals, or (3) receiver and sender payoff matrices change with the receipt of information. Finally, **constraints** are costs that are so great that they prevent a communicatory exchange from evolving or becoming further elaborated. Some constraints can be identified on the basis of physical principles, but constraints on the evolution of morphological, physiological, and behavioral traits are more difficult to pinpoint and require comparative studies and phylogenetic analysis.
2. Senders generally experience greater necessary costs than receivers because they must do the work to produce signals and become more vulnerable to predators and parasites while signaling. Receivers, on the other hand, are more likely to experience incidental costs and to be limited by constraints on their perceptual ability. However, the relative importance of these costs and constraints varies for the different signal modalities as well as for different animal taxa and different signal functions.
3. Visual communication is primarily constrained by the capability of the receiver's eye. Eyes evolved primarily for foraging, orientation, and navigation. The sensitivity of the eye is optimized for the amount and quality of ambient light in the environment, and color vision ability is determined by the types of discrimination tasks required for foraging and navigating. A major physical constraint for vision is small body size, since resolving power increases with increasing eye size. Visual signals must evolve within these receiver constraints. Reflected-light visual senders are relatively unconstrained in their ability to evolve signals because they continually emit visual information to receivers. Elaboration of signals primarily entails time and energy costs of signal production, increased conspicuousness to predators, and interference between the signaling and original function of elaborated structures. These costs are greatest for mate-attraction and territory-defense signals that require a high "on" time.

4. Looking at the entire animal kingdom, auditory communication is more constrained by signal production than by reception. Hearing organs have evolved in many taxa for the purpose of detecting predators or locating prey; quite a few animal groups possess ears but no form of auditory communication among conspecifics. Sound production requires either a hard exoskeleton and jointed appendages for stridulation, stiff tymbals, stretched vibrating chords or forceful airflow through a narrow tube. These physical requirements rule out soft-bodied animals and many others. Auditory communication is therefore largely limited to crustaceans, insects, frogs, birds, and mammals. Furthermore, body size constrains the lowest frequency a sender can produce. Since production of sounds is very inefficient, senders must expend substantial energy to make sounds. Sounds are always short, so frequent repetition to maintain a high "on" time means large time, energy, and conspicuousness costs. Environmental constraints on sound transmission are minor. They include degradation with distance and limited locatability in water. Receivers do not seem to be very constrained in their ability to evolve and tune their ears to detect, localize, and analyze the information in the sounds produced by conspecifics.
5. Chemical communication is the least costly and constraining of the three main modalities for both sender and receiver. Chemical signaling and reception for mate recognition purposes is a primeval capability for all organisms, including bacteria and plants. Senders experience some production costs that increase as the number of marks or as the active space of signals increases. Predators may also be attracted to continuous-emission senders. There does not seem to be a constraint on the ability of senders to evolve chemical signals with the transmission properties needed for the communication context. Chemical receivers do not seem to experience reception costs or constraints, but they are more vulnerable to incidental costs such as deceptive mimicry by parasites, predators, and social parasites. Environmental conditions such as wind, turbulence, rain, and aquatic media constrain chemical signal transmission.
6. Small body size not only limits visual resolution and production of low-frequency sounds, but it also determines brain size for sender and receiver functions requiring pattern generation, analysis, learning, and memory.

FURTHER READING

General treatments on constraints can be found in Goldsmith 1990 and Reichenbach and Robinson 1995 for vision and in Ryan 1986a for acoustic communication. M. Andersson 1994 and Møller 1996 discuss costs and constraints on sexually selected signals. For signal production costs, see Ryan 1988; Møller 1989; Wells and Taigen 1989; Endler 1991b; and Magnhagen 1991. Receiver costs are considered by Sullivan 1994; DeVoogd et al. 1993; and Findlay and Darlington 1995.

Chapter 18

Signal Design Rules

WE BEGAN PART II WITH THE NOTION that there is an optimal way to communicate in any given situation, and that adjustment of signal form is a critical step in achieving such an optimum. In the ensuing chapters, we have seen that the optimal form of a signal depends on the phylogenetic history of both parties, the number of alternative answers that must be coded by the relevant signal set, the number of questions dealt with by the same signal, the benefits and costs to each party of participation, and a variety of constraints set by the medium, the modality, and the distance between sender and receiver. With so many different factors shaping signal form, is it likely that there are any general patterns or rules, or is it the case that each situation is unique? For example, do signals answering the same basic question, regardless of modality or phylogenetic history, show convergent features because of their similar function? Alternatively, perhaps modality, not question, is the dominant factor limiting signal form. Do all signals used in a given modality show convergence regardless of question answered? Finally, might the specific environment in which signals are given override all other considerations in shaping signal form? The extraction of order from apparent diversity is a major task of any science. It is thus natural to ask whether some subset of the selective factors on animal signals might dominate their evolution. Then much of the variation in other factors could be ignored or included as secondary variation around general themes. The postulated links between such a subset of factors and signal form are called signal design rules. The search for design rules has been a major preoccupation of many investigators of animal communication in the last two decades. In this chapter, we shall examine the current evidence for signal design rules and some postulated rankings of the selective factors outlined in previous chapters. This will require the integration of nearly everything we have done so far in this book. Before we can discuss possible rules, however, we have to specify which properties of signals we should examine for convergence or divergence. We call these properties the design features of signals.

SIGNAL DESIGN FEATURES

A design feature is a property of a signal that is shaped by selective forces and affects the optimality of the signal form. We can refer to such features in a general sense ("The active space of the signal must be larger when attracting distant mates than when mediating conflict") or as a specific instance ("The active space of the snail's pheromone was 22 cm^3"). It will be important to remember this distinction as we proceed in our discussion. Our goal in this section is to identify the most important design features in the general sense, and in later sections, to look for design rules that might predict the value that each design feature would take in a specific context. Note that choice of a certain set of design features cannot be completely independent of any design rules present. If we pick the wrong features, we may find no evidence of design rules; if we pick other features, this may predispose us to find particular rules and not others. In practice, the process of listing design features and finding rules must be an iterative one that gradually extracts the minimal but necessary set for each.

Historical Perspective

Hockett (1963) was the first to provide a list of important signal properties. His approach was anthropocentric. He developed a list of 16 key features of the human verbal language system that included such factors as modality, sender/receiver interchangeability, semanticity (association between signals and objects/concepts), discreteness, displacement (reference to future and distant events), and learned acquisition, to mention only a few. He then determined which of these features were present in the communication systems of different animal species and found that one or more was always lacking. Altmann (1967) devised an even more complex structure based on primate communication systems. Hockett and Altmann (1968) then combined and simplified their lists by clustering related signal features into the following categories:

A. Features related to the modality (e.g., transmission properties, receiver sensitivity)
B. Features related to the social context (e.g., type, location, and number of receivers)
C. Features related to behavioral antecedents and consequences of the sig nal (Does the signal predict what the sender will do next? Is this information inherent in the signal form? Can it be deceptive?)
D. Degree of variation in signal form as a function of age, sex, or group
E. Degree to which signals in the set are graded (as opposed to discrete and stereotyped)

With this classification system, a given signal can be characterized according to a set of quantitative variables, rather than just the presence or absence of different features. Furthermore, the functional clustering of features makes it easier to search for any universal design rules related to sender and receiver strategies and social contexts.

Other attempts to categorize signal design features involve shifts in emphasis rather than fundamental differences with the Hockett/Altmann scheme. Marler (1961) and Otte (1974) stressed distinctions between the types of information contained in displays, such as information about sender identity, sender state or motivation, sender location, and environmental events. Wilson and Bossert's (1963) study of chemical communication focused on features that regulate signal range, rate of spread, and persistence for olfactory signals in different social contexts. Endler (1992, 1993a) emphasized the importance of environmental constraints and receiver biases for signal design and developed a specific list of mechanisms for maximizing signal range. Guilford and Dawkins (1991) stressed the role of receiver detectability, discriminability, and memorability in shaping the design of signals.

A Case for a Minimal Set of Design Features

An ideal design rule scheme would focus on the smallest possible set of signal design features that is common to all modalities, reflects the information content and transmission needs of all signal types, and makes explicit predictions about the values each feature would take in a given context. The design feature set proposed here represents a version of Hockett and Altmann's scheme augmented by ideas of subsequent theorists. It consists of the following six design features: (1) the range or active space of the signal, (2) the locatability of the signal, sender, intended receiver, and/or referent, (3) the duty cycle of the signal, (4) the sender identification level, (5) the modulation level, and (6) the degree of form-content linkage. These design features are defined in greater detail below. The corresponding design rules predict the relative values that the design features should take in each signaling situation. To make the translation from design rules to signal form, we shall also need to identify the **specific mechanisms** by which the design feature requirements can be achieved. Whereas the design rules might make general predictions that are modality-independent, the specific mechanisms are likely to be dependent on modality and context.

Signal range is a design feature that reflects the typical distance between sender and receiver. Although range, like the locatability design feature, encodes spatial information, range has pervasive consequences on signal design that can be quite independent of angular location (and vice versa). The mechanisms by which the range of a signal can be increased include higher amplitudes, better signal/noise ratios, and enhanced receiver sensitivity. Depending on the environment, the range of auditory and visual signals can be maximized by adjusting frequency relative to background noise. Visual movement displays can be seen from further away by increasing the speed and amplitude of motion. The r_{max} of an olfactory signal can be increased by using a moderately volatile chemical, by increasing the amount of odorant released, or by employing currents. Range for electric signals is severely limited by physical constraints but can be increased slightly with higher discharge amplitudes, choice of quiet periods to signal (e.g., times of little lightning), or choice of optimal sender locations.

The **locatability** design feature establishes the type of locational and directional information in the signal. At minimum, a sender might adjust the signal to make the determination of its location easier or harder. The sender might also include the encoding of directional information about the location of the intended receiver, a third party, or an external referent. The locatability of a sender can be enhanced or decreased by adjusting frequency, onset rise time, and repetition rate for visual movement and auditory signals and by adjusting diffusion rate for olfactory signals. With deposited olfactory signals, the signal itself can be made localizable without revealing the location of the sender. Visual senders can indicate the position of an intended receiver or referent by aiming movements in the direction of the referent or pointing at it, and auditory senders can direct attention to signaling receivers by synchronized calling or call matching. Olfactory senders can indicate specific receivers by timing and directional spraying of odorous secretions. Because of the very short range of electric signals, it is rarely necessary to increase sender locatability by signal design.

The **duty cycle** of a signal is its "on" time relative to its "off" time and is affected by the temporal pattern of signal delivery. It can be increased by lengthening the duration or increasing the repetition rate of the signal in the case of movement, auditory, electric, and tactile signals, and by decreasing volatility in the case of olfactory signals. Duty cycle can be measured over different time scales. Within a bout of signaling, the duty cycle is computed as signal duration divided by signal plus intersignal duration. At a longer time scale, it could be expressed as the fraction of total active time spent signaling.

The next two design features establish the variant diversity of the signal (see page 490). The **identification level** determines the number of different classes of individuals that are distinguished by the receiver on the basis of the signal, or the between-individual level of variation. Signals that require individual or kin group specificity must have more possible recognizable variants than signals that are designed for species or sex recognition. As the number of classes that must be distinguished within a species increases, the complexity of the signal must increase to allow for a sufficient number of variants. Auditory signals can be made more complex by adding amplitude- and frequency-modulated elements and by increasing their duration. Visual signals use color and pattern, electric signals vary in EOD waveshape, and olfactory signals employ odorant mixtures.

The **modulation potential** specifies the amount of within-individual variation possible in a given signal as measured by the position of the signal along a scale from stereotyped to graded. Variants may encode information about motivation level, aggressiveness, degree of danger, or current health. Graded signals have more variants than stereotyped ones and are encoded by varying one or more signal dimensions such as intensity, frequency, amplitude-modulation rate, frequency-modulation rate, or repetition rate, usually in an iconic fashion. There may be a tradeoff between possessing a large number of between- and within-individual variants, but in principle different signal parameters (parameter mapping, see pages 473–474) could be used to encode these two levels of variation in the same signal.

The final feature, **form-content linkage**, is the degree to which signal form is dependent upon signal content. As we saw in Chapter 15, some questions allow the assignment of any evocable signal to any alternative answer. Such signals are said to be **arbitrary** in form. However, there are cases in which signal form and content are **linked**, and signal form is thus not arbitrary. Such linkage can occur when the signal is not completely emancipated from the source from which it evolved. For example, signals derived from intention movements, redirection, or ambivalence are likely to retain a close link between the message in the signal and the actual form of the display (see pages 501–513). Similarly, the body size or motivational state of a sender may preclude it from emitting any but a particular type of signal. For example, the frequency composition of a sender's call may be obligately linked to its confidence and general muscle tension (page 522). Linkage can also occur as a result of certain types of conflict of interest between sender and receiver. Throughout Part II of this book, we have focused on situations in which sender and receiver have similar interests. Clearly, this is not always the case, and as discussed in detail in Part III of the book, receivers can invoke any of several mechanisms to ensure sender honesty. One is to insist on signals whose form is so linked to the content that sending a dishonest signal imposes an exorbitant cost on the sender. Such signals cannot be arbitrary in form. An alternative is to punish senders who are later found to have sent a dishonest signal. Signals that maintain honesty via receiver retaliation can be arbitrary (Dawkins 1993). There is thus a considerable range, even in conflict situations, in the degree of form-content linkage for signals.

Strategies in the Search for Design Rules

A number of studies have sought to identify general design rules in animal communication. One strategy has been to compare the functionally equivalent signals of different species and look for similarities and convergence. This approach was used by Marler (1959) and Klump and Shalter (1984) in describing the design of avian alarm signals, by Wilson and Bossert (1963) in examining insect chemical signals, and by Gerhardt (1991) in identifying the design features of mate-attraction calls in hylid frogs. The assumption here is that the nature of the question being asked by receivers overrides other factors in shaping signal form. This is thus a function-based approach. A complementary strategy is to compare functionally different signals and look for consistent differences in predicted directions. This is best undertaken within a modality since design mechanisms are likely to be modality-specific. An example of this approach is the comparative study by Alberts (1992) of olfactory signals in mammals. Other authors have focused on the primacy of different selective factors. Once general design rules become apparent, one can then make predictions and test them in new systems. An excellent example is the analysis of the courtship displays of male wrasses by Dawkins and Guilford (1994).

The majority of previous studies have focused, or come to focus, on the apparent primacy of signal function (or question). There are clearly convergences according to signal function that occur across modalities, and across environments. We shall follow this lead here and see how far we can get by assuming that signal function sets the broadest bounds on signal form. However, a set of design rules based only on functions/questions is not as simple as it might seem. For example, some questions are only relevant in certain contexts, and thus it is difficult to separate the independent effects of question and context. In addition, a very large number of animal signals are compound. Such signals provide information about several concurrent questions, and the final decision of the receiver may depend on the answers to all of them. It is possible that the optimal form of such a compound signal depends on the combination of questions addressed. Particular combinations of questions tend to be asked in particular signaling contexts (as defined on pages 358–363). We are confronted again with combinations of question and context. Context thus enters into signal design in two ways: at a general level where it plays some role in shaping design rules, and at the specific level by severely constraining the mechanisms by which a design rule optimum can be executed. Weaving context into the predictions based on signal function alone will be a repeated task in this chapter.

In the sections that follow, we shall examine design rules for five different signal contexts: mate attraction, courtship, territorial defense, threat, and alarm. These five have been chosen because they occur in a wide variety of animal species and because the responses to the signals are very obvious, allowing us to identify clearly the basic function and meaning of the signals in species with very different social systems. Other signaling contexts that we mentioned in Chapter 12 such as parent-offspring, social integration, and food assembly signals will be discussed in Part III. For each of the contexts discussed here, we shall develop design rules for that signaling function based on behavioral studies. The specific mechanisms employed in each modality to achieve the design rule predictions will then be outlined briefly and the relative ability of each modality to meet the signaling needs compared. Across-species comparative studies are used where available to illustrate convergences and divergences in signal design. Note that understanding optimal signal form within a modality is not a simple process. One cannot just determine the optimal value of each design feature and add them up to describe the optimal signal. This is because there are invariably tradeoffs between design features that are modality-specific, so that enhancing one design feature will automatically enhance or constrain another. The evolutionary task is to identify that combination of design feature values that maximizes the value of information, given the signal function and the modality invoked. In the final section, we shall examine the few comparative studies that have made comparisons within a modality but across signal contexts to see whether the general predictions of the design rules hold at this higher level.

MATE-ATTRACTION SIGNALS

Sexual species face the task of finding conspecific members of the opposite sex for mating. Communication plays a critical role in the mate-attraction process. Four types of questions are typically answered: species identification, sex identification, receptivity, and location. All of this information can often be provided in a single compound signal. In the simplest form of mate attraction, only one sex gives the signal, and only when it is receptive. The signal therefore need only be species-specific and localizable, since sex and receptivity are implied by the mere presence of the signal. The signaling sex is relatively stationary and the nonsignaling sex more mobile. Receivers of the nonsignaling sex are most likely to respond to the signal only when they are receptive, too. Thus the approach of an individual toward a signal tells the sender that this is probably a receptive conspecific member of the opposite sex. Subsequent responses by the approaching individual as to its true receptivity and interactions leading to mating fall into the category of courtship, a signaling function we shall discuss later. Which sex signals and which sex searches is an issue we shall take up in Chapter 23.

Design Rules for Mate Attraction

The optimal design feature values for mate-attraction signals follow fairly obviously from the types of questions the signal must answer. Unless males and females live in permanent groups, the signal must be transmitted over a long distance and thus should have a large range or active space. The sender should be locatable so that prospective mates can approach and find it. The duty cycle of the signal must be relatively high, so that it is "on" when appropriate mates might be within reception distance. The identification level is of course the species, and information about sex is implied. The signal must be complex enough to contain frequency, temporal, or structural patterns that clearly distinguish it from other similarly signaling species in the habitat. The signal should be stereotyped with little variation within and between individuals to preserve species identification. Finally, mate-attraction signals can be arbitrary in form since they primarily provide information about sender identification and location, information over whose provision sender and receiver are unlikely to be in conflict. However, in competitive mate-attraction situations and in species like frogs where the long-distance mate-attraction signal also serves as the courtship signal, conflict can arise, and senders may be forced to include components in a compound signal with high form-content linkage. The design rules and modality-specific mechanisms for mate-attraction signals are summarized in Table 18.1.

Modality-Specific Mechanisms for Mate-Attraction Signals

RANGE MAXIMIZATION. Despite major differences in maximum transmission range among the modalities, examples of mate-attraction signals can be found in all five modalities. The mechanisms for maximizing the signal range of auditory, visual, olfactory, and electric signals have been discussed in Chapters 4, 8, 10, and 11 and are summarized in the table. Auditory, visual, and chemical signals can transmit over distances of many meters. The electric modality

Table 18.1 Design rules and modality-specific mechanisms for mate attraction signals

Design feature	Rule	Visual mechanisms	Auditory mechanisms	Olfactory mechanisms
Range	Large	Brightness contrast Hue contrast Movement contrast	Lowest possible frequency Frequency notch in background noise Coupling, resonating, baffle, and sac structures	Current-borne volatile chemical Long-lasting trail
Locat-ability	Sender location	High repetition rate Rapid moves	High repetition rate Rapid onset	Concentration gradient Move up-current Directional info. in trail
Duty cycle	High	Permanent color or structure Frequent repetition	Frequent repetition Long duration signal	Continuous release of volatile chemical Deposition of nonvolatile chemical
ID level	Species	Color pattern Display pattern Structure shape	Frequency Temporal pattern Note shape Syntax	Specific chemical
Modulation level	Stereo-typed	Repetition rate	Repetition rate Call duration	Concentration
Form-content linkage	Arbitrary	Exploit preexisting visual biases	Exploit preexisting sender production mechanisms	De novo production

is limited to ranges of less than a meter, but for fish living in very muddy water it provides a viable option. The tactile modality is least likely to be used as a mate-attraction signal because its range is so short, but animals with no ears or image-forming eyes may use it to identify mates upon chance contact.

DUTY CYCLE AND SENDER LOCATABILITY. A high duty cycle is attained by different methods in each modality. Visual color and structural signals are always "on" and thus can easily meet this requirement. Current-borne olfactory signals maintain a high duty cycle by using the continuous release strategy, while deposited trail marks use a large, polar and/or nonvolatile chemical with a slow fadeout time. Auditory, electrical, and visual movement displayers must use frequent repetition to maintain a high duty cycle; they have some leeway to vary both signal duration and repetition rate to achieve an effective "on" time. The high duty cycle required of mate-attraction signals also facilitates localization of the sender so that similar mechanisms can simultaneously achieve both design feature requirements. In addition, senders employing repeated displays can improve the receiver's ability to locate the sender with rapid onset introductory notes and rapidly accelerating movement.

SPECIES IDENTIFICATION. As the number of species living sympatrically and using the same communication modality increases, their mate-attraction signals must increase in complexity to encode species distinctiveness. Species competing for a communication channel can also utilize different signaling times and microhabitat locations. All modalities except touch have the potential to encode a huge number of distinctive species-specific variants. The auditory modality can potentially use frequency range, frequency and amplitude modulation, note shape, temporal pattern, and syntax to code species identity. In songbirds (Figure 18.1A), frequency, note shape, and syntax are the most critical song characters for species identification; experimental manipulation of these parameters quickly leads to a loss in receiver response. Manipulation of temporal parameters, on the other hand, does not greatly reduce receiver responsiveness, suggesting that these parameters are less critical for species recognition (Becker 1982). As the number of songbird species in a community increases, the number of modulated elements in their songs increases, implying that note complexity facilitates species recognition. Furthermore, their songs become shorter and intersong interval increases so that the species-specific information can be inserted into brief pauses in background noise (Sorjonen 1986). In comparison, stridulators such as crickets have limited ability to modulate the frequency of stridulation sounds and rely more on temporal patterning of sound pulses (Figure 18.1B). The visual modality uses variants of color, color pattern, movement pattern, and structure shape parameters to encode species identity. Figures 18.2 and 18.3 illustrate these mechanisms in several different taxa. Olfaction mechanisms use chemical size, shape, and functional group. Lepidopteran sex attractants, for example, consist of a carbon chain backbone with one or more double bonds in a specific configuration and one functional group at the primary end. Members of a genus share the same

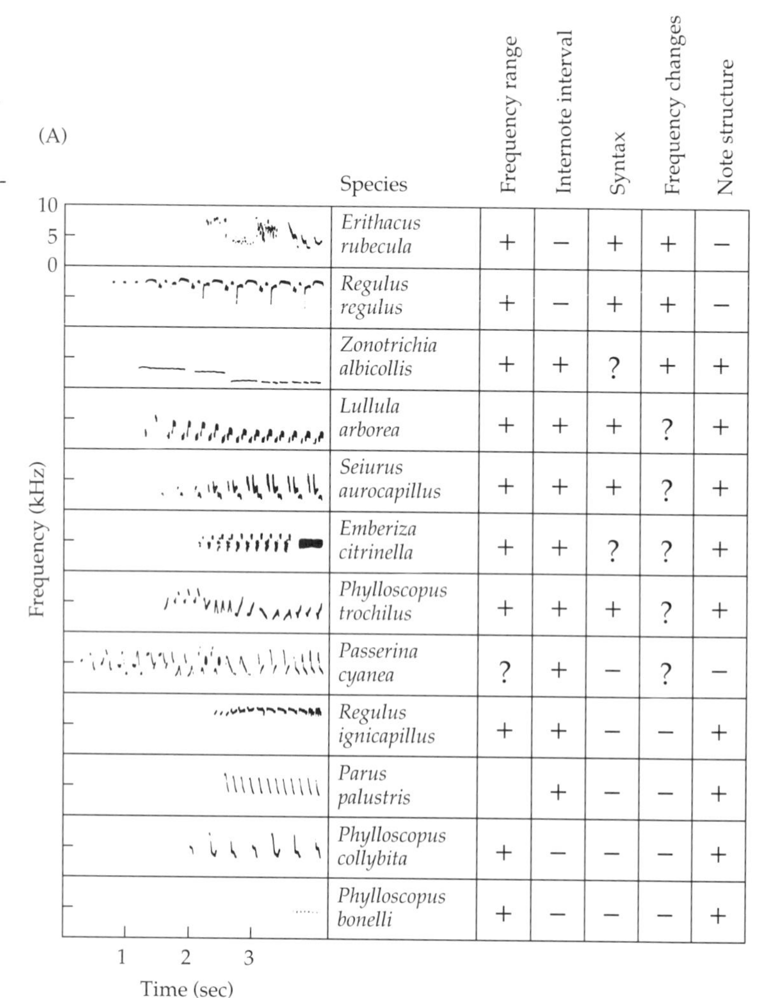

Species	Frequency range	Internote interval	Syntax	Frequency changes	Note structure
Erithacus rubecula	+	–	+	+	–
Regulus regulus	+	–	+	+	–
Zonotrichia albicollis	+	+	?	+	+
Lullula arborea	+	+	+	?	+
Seiurus aurocapillus	+	+	+	?	+
Emberiza citrinella	+	+	?	?	+
Phylloscopus trochilus	+	+	+	?	+
Passerina cyanea	?	+	–	?	–
Regulus ignicapillus	+	+	–	–	+
Parus palustris		+	–	–	+
Phylloscopus collybita	+	–	–	–	+
Phylloscopus bonelli	+	–	–	–	+

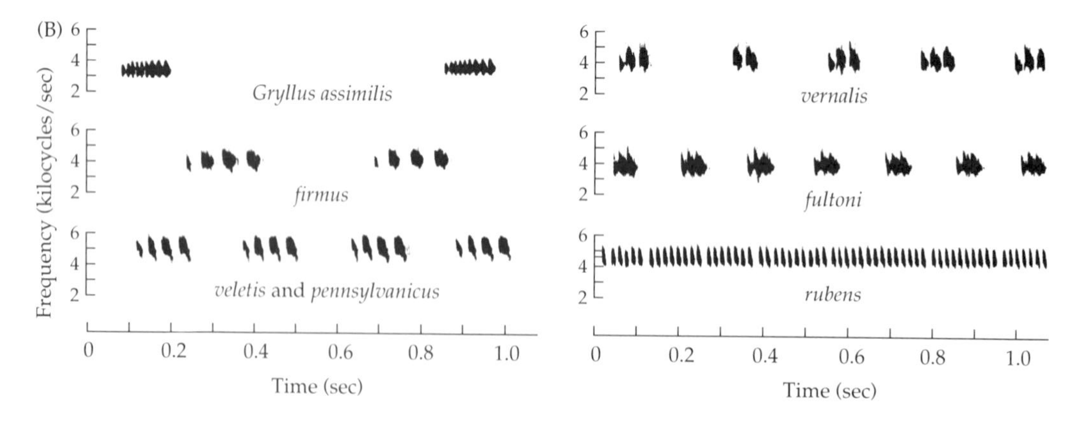

Figure 18.1 Auditory mechanisms for species identity. (A) Songbird species produce highly distinctive songs, but some parameters are more important for species recognition than others. This figure summarizes some of the experimental studies in which song parameters were systematically varied and played to territorial males to determine their level of response. Parameters indicated with a plus (+) are those that resulted in a reduced aggressive response when varied outside of the normal species range and thus indicate critical features for species recognition. Parameters indicated with a minus (–) could be varied greatly without reducing the aggressive response. (B) The stridulation mechanism of *Gryllus* crickets does not permit senders to employ frequency modulation, so note shape cannot be used to encode species-specific information as in birds and frogs. Instead, temporal patterning is the primary mechanism for species-specific coding. (A from Becker 1982, © Academic Press; B from Alexander 1968.)

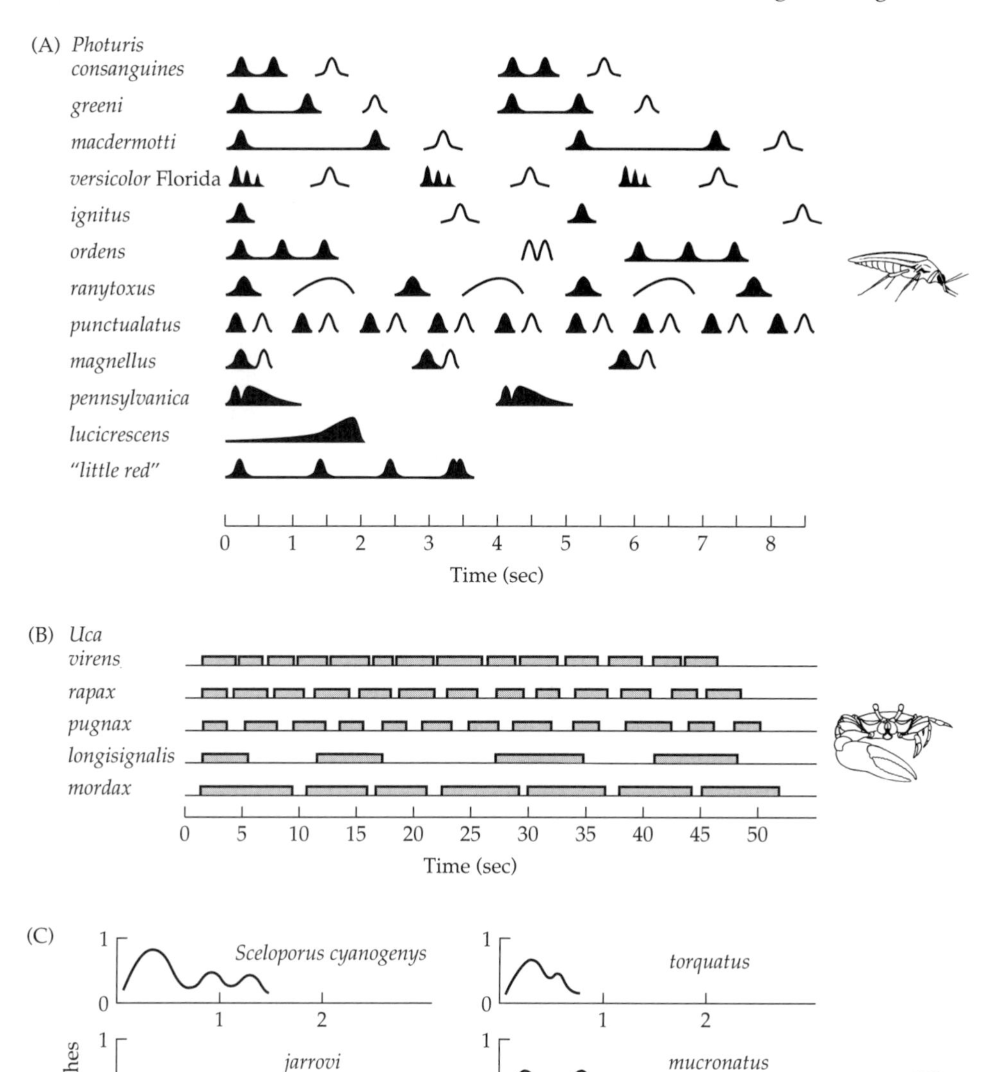

Figure 18.2 Visual mechanisms for species identity. (A) In *Photinus* fireflies, species differ most in the duration and temporal patterning of flashing. (B) In fiddler crabs (*Uca*), the primary difference among species is in the temporal pattern of the major cheliped waving display. Species also differ slightly in the arc shape of the wave, the jerkiness of the wave, and cheliped color and shape, but the significance of these components for species identification is not known. (C) *Sceloporus* lizards perform push-up displays by raising up on their front legs. The shape of the bob and its jaggedness or smoothness characterize the different species. (A after Gould 1982, based on Lloyd 1977; B after Salmon 1967; C after Hunsaker 1962.)

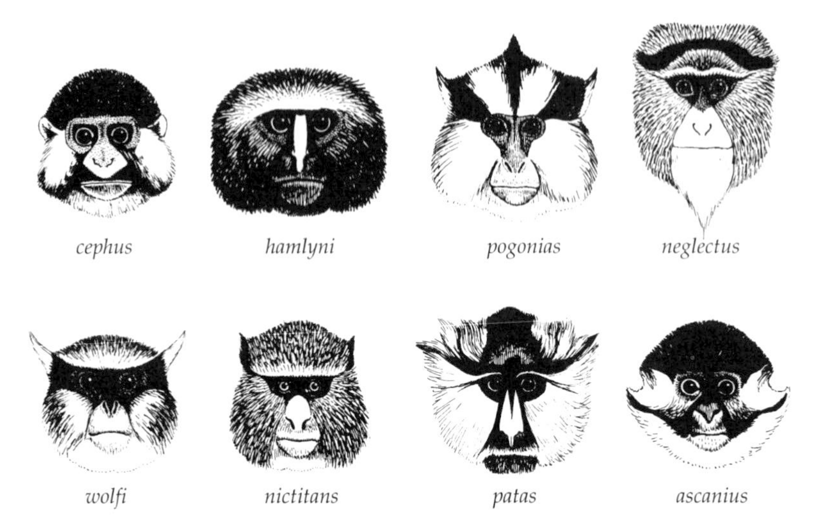

Figure 18.3 Facial color patterns in African guenons. African guenons in the genus *Cercopithecus* often live in mixed species troops and use conspicuous, colorful skin spots and hair tufts as the primary species recognition mechanism. The colors used include white, black, brown, yellow, blue, and red. (After Kingdon 1988.)

carbon chain backbone but differ in their use of functional groups (Table 18.2). Electric fish can achieve marked diversity by varying the shapes of their EOD waveforms (pulse fish) or the repetition rates of their discharges (wave fish) (Hopkins 1983a; Hopkins and Bass 1981; Friedman and Hopkins 1996).

A common feature of mate-attraction signals in all modalities is the strong correlation between signal form and receiver sensitivity. As discussed in Chapter 16, this correlation is most likely due to the coevolution of sender and receiver and the evolution of feature detectors and peripheral tuning adaptations to maximize receiver detectability. Such specificity can be built into receivers because only one variant needs to be recognized (e.g., binary receiver coding).

SIGNAL STEREOTYPY. The consequence of receiver tuning for efficient recognition is selection on senders for signal stereotypy. Mate-attraction signals are often highly stereotyped. In many species, however, the signal is complex or compound. This characteristic leaves room for some signal parameters to remain static for species recognition while other parameters can vary within and between individuals (e.g., parameter mapping). Treefrog calls show this dichotomy very clearly (Figure 18.4). Dominant frequency and pulse rate are the least variable parameters. Females prefer conspecific male calls with frequencies and pulse rates similar to those of the average males in the population. This preference pattern suggests that these call parameters are species recognition features. Call duration and repetition rate vary more within males, and females prefer calls with longer durations and higher repetition rates than those exhibited by the average males. Such parameters clearly affect finer-level choices among conspecific males (Gerhardt 1991).

Table 18.2 Coding of species specificity in the olfactory mate attractants of Lepidoptera

		Functional group				
Genus	**Species**	**Ester**	**Alcohol**	**Aldehyde**	**Length**	**Bond**
Acleris	*emargana*			+	14	*trans*-11
	paridiseana	+				
Argyrotaenia	*citrana*			+	14	*trans*/*cis*-11
	dorsalana	+	+			
	velutinana	+	–			
Autographa	*biloba*	+			12	*cis*-7
	californica	+	+			
Choristoneura	*biennis*			+	14	*trans*/*cis*-11
	fumiferana	–	–	+		
	fractivittana		+	+		
	occidentalis	+		+		
	rosaceana	+	+	+		
Croesia	*conchyloides*			+	14	*cis*-11
	askoldana	+		+		
Epiblema	*desertiana*	+			12	*cis*-8
	scudderiana	+	+			
Eurois	*occulta*	+	–		16	*cis*-11
	astricta	+				
Grapholitica	*prunivora*	+	–		12	*trans*/*cis*-8
	molesta	+		+		
Heliothus	*armiger*			+	16	*cis*-11
	virescens		+	+		
	punctiger	+		+		
Oidaematophorus	*guttalus*			+	14	*trans*-11
	mathewianus	+				
Pandemis	*limitata*	+			14	*cis*-9/*cis*-11
	heparana	+	+			
Platynota	*flavedana*	–	+		14	*trans*/*cis*-11
	sultana	+	–			
Synanthedon	*alleria*	+			18	*trans*, *cis*-3,13
	biblionipennis	+	+			
Thyris	*maculata*	+			14	*trans*-11
	usitata		+			

Note: A plus (+) indicates mate attraction, a minus (–) indicates inhibitory response. The functional group is located on the end of the carbon chain. Length is the number of carbons in the chain and Bond is the position and configuration of the double bond.

Source: Morse and Meighen 1986.

FORM-CONTENT LINKAGE. The stereotyped species recognition elements of mate-attraction displays can be arbitrary in form because the information encoded is primarily about sender species identity and location. A second factor selecting for arbitrary form is the maximization of signal range. This factor favors signal forms that exploit preexisting receiver biases. Complex note shapes, diverse song repertoires, bizarre color patterns, novel shapes, and stimulating movements are examples of arbitrary signal elements. We therefore should not expect to find convergence in these aspects of signal form, but

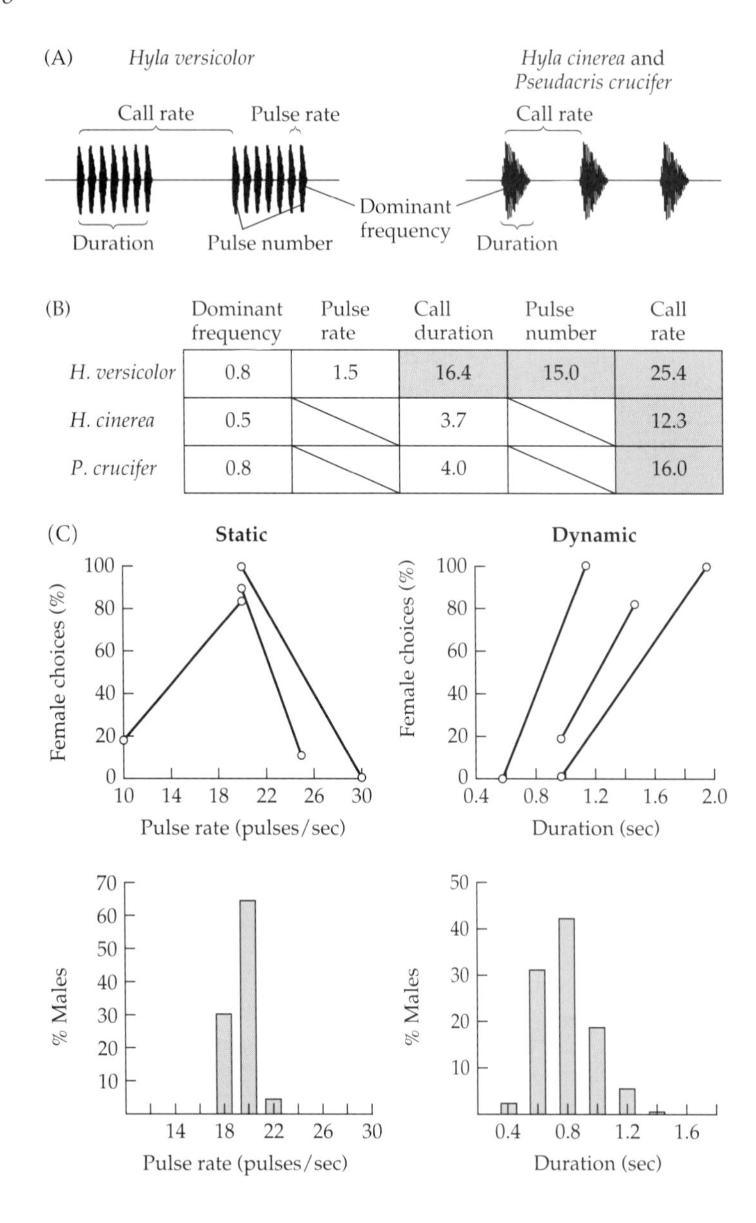

	Dominant frequency	Pulse rate	Call duration	Pulse number	Call rate
H. versicolor	0.8	1.5	16.4	15.0	25.4
H. cinerea	0.5		3.7		12.3
P. crucifer	0.8		4.0		16.0

rather species-specific divergence. However, mate-attraction signals may show form-content linkage. Competition among nearby senders for the attention of mates selects for subtle variations in signal elements that provide honest information about sender motivation, quality, or condition. As mentioned above, visual and auditory senders can vary signal duration and repetition rate. Increasing these two signal parameters greatly increases the energetic cost of signaling. Males displaying at high repetition rates may be preferred by females because this indicates good male condition. Other costly aspects of display may similarly encode information about mate quality, a topic we shall discuss at length in Chapter 23.

◀ **Figure 18.4 Static and dynamic components of treefrog calls.** (A) *Hyla versicolor* produces a pulsed call containing five potentially variable parameters whereas *Hyla cinerea* and *Pseudacris crucifer* produce nonpulsed calls with three signal parameters. (B) The table shows the coefficient of variance (CV) of within-male variation for each call parameter in the three species. Call parameters with a CV of 5 percent or less were classified as static (white boxes), and call parameters with a CV of 12% or more were classified as dynamic (gray boxes). Call rate is a highly variable parameter for all three species; the dominant frequency is a highly invariant parameter for a given individual, but can vary between males. Call duration is invariant for *H. cinerea* and *P. crucifer*, but highly variable in *H. versicolor*, which can increase call duration by adding more pulses. (C) Female preferences for artificially varied calls in *H. versicolor* relative to the mean of the male population. For static parameters such as pulse rate, females strongly prefer calls occurring at the mean of the male population, and they avoid calls with both faster and slower pulse rates. For dynamic parameters such as call duration, females strongly prefer calls with longer durations. Points connected by lines show choice alternatives offered to females. (After Gerhardt 1991.)

Evidence for Convergence in Mate-Attraction Signals

HABITAT EFFECTS. Since mate-attraction signals are selected to maximize transmission distance, we would expect to find some aspects of convergence in signal form among species living in similar habitats. Wiley (1991) examined the song characteristics of North American songbird species living in open versus forested habitats. Species signaling in vegetated habitats tend to produce pure tone whistles of lower frequency and avoid rapid modulations that would be severely degraded, whereas species signaling in open habitats use more buzzes, trills, and sidebands. Sorjonen (1986) found essentially the same structural differences in song between forested and open habitat species in a comparative study of European songbirds. In Figure 5.10 we saw how habitat structure molds song structure in similar ways within a species. These examples illustrate the coding principle of combination mapping (page 473).

Martins (1993a) examined the role of body size and habitat on the form of push-up displays in *Sceloporus* lizards. This study carefully controlled for phylogenetic effects that had confounded earlier comparative studies of the genus. Although larger species tend to be arboreal and display from branch perches and smaller species tend to display from rocks in open areas, body size and perch type themselves did not appear to constrain the form of display. Instead, certain display parameters were more strongly associated with each other and with habitat type. The displays of terrestrial species are short, fixed in length, and comprised of a few bobs separated by pauses; the bobs are jagged and complex in shape. The species-specific information is encoded in the shape of the bob. Arboreal species produce longer, variable-length displays with square-shaped headbobs. The species-specific information is encoded in the interbob temporal pattern. Martins concluded that transmission characteristics of the habitat determine which signal forms are optimal.

SIGNALING SEX AND MODALITY. There appears to be a correlation across taxa between the modality used for mate-attraction signals and which sex has the job of giving the signal. When males signal, they are far more likely to use the auditory or visual modality, whereas females are more likely to use olfaction. Among visually signaling species (birds, lizards, fiddler crabs, shallow water fish, some mammals), males sport the bright colors, develop enlarged structures, and perform elaborate displays. Females sometimes wear the same bright colors as the males of their species but almost never invest in the displays and structures. (A conspicuous exception is the swollen red bottom that females of many primate species develop to advertise their estrous period.) A similar picture emerges for auditory senders (birds, frogs, and orthopterans). Males are nearly always the calling sex. Some female birds duet with their mates after pairing, and in some owls both males and females emit solitary, long-distance calls. In these species, male and female songs show consistent sex-specific differences (Farabaugh 1982).

There are a few exceptional species in which only the female produces a mate-attraction sound. Female elephants emit ultralow-frequency sounds when they are in estrus that travel many miles and attract adult males (Poole et al. 1988). Most cat species are highly solitary, and females produce loud calls when in estrous to attract a mate. Female mosquitoes attract males with the near-field sound of their wing beat (Sotavalta 1963). This sound signal possesses more of the characteristics of a visual color-pattern signal than an auditory signal, in that it is continuous while she is flying, is very low in amplitude (short range), and does not signal whether or not she is receptive. It does provide species recognition, since different species' wings beat at a different frequencies, and the males' near-field pressure receptors are sensitive only to the narrow species-specific frequency range.

In contrast to the situation for males, females are more likely to be the signaling sex among olfactory senders, although examples of male long-distance olfactory senders can also be found (Jacobson 1974). Female olfactory signaling is especially common in mammals, snakes, spiders, and moths. The reason for these differences probably lies in the high cost of auditory and visual signals compared to olfactory signals and the differences in the sexual strategies and receptive periods of males and females. Females need to conserve energy for egg production and individual females are receptive for short periods of time. It is relatively inexpensive for them to produce an adequate supply of odorant during this period. Males are receptive for longer and therefore must compete with many other males for each receptive female. Selection is stronger on them to adopt more conspicuous, repetitive methods of signaling, and they can afford the more costly signal modalities better than can females.

COURTSHIP SIGNALS

Courtship begins once two potential conspecifics of the opposite sex are in close proximity, usually after one individual has been attracted to the other's long-distance mate-attraction signal. The functions of courtship are mate assessment and mating synchronization. A key characteristic of courtship is therefore the occurrence of several signals, some by the male and some by the female, in a specific sequence (Figure 18.5). Each signal transmits information

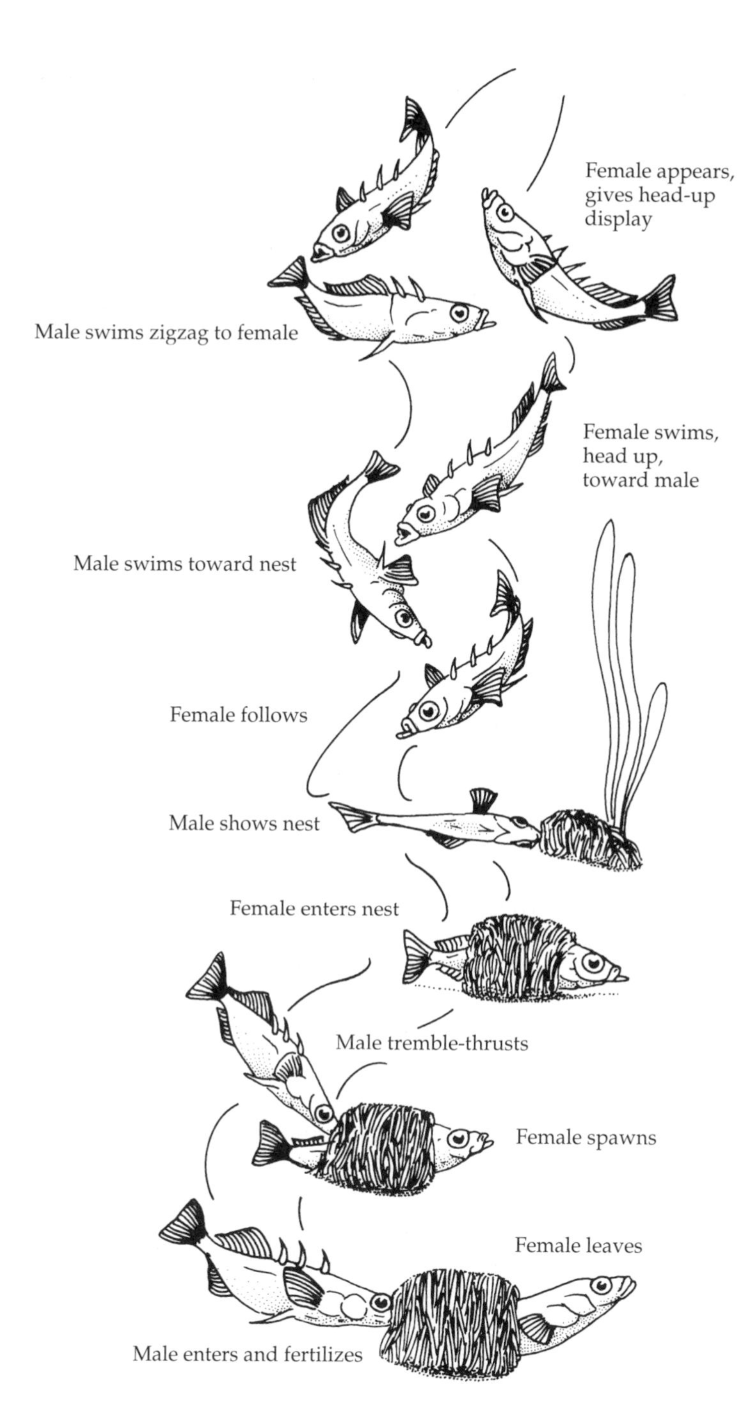

Figure 18.5 Sequence of courtship displays in the stickleback. Stickleback (*Gasterosteus*) males defend a territory that contains a nest they have made. They display their territorial status and receptivity with bright-red body coloration. A territorial male may attack a female entering his territory unless she possesses a swollen egg-filled abdomen and performs the head-up display to indicate her receptivity. The male then performs the zig-zag display consisting of alternating broadside and female-facing postures. If the female is still interested, she approaches the male, the male then swims toward his nest and she follows. The male lies on his side with his snout pointing to the nest entrance and makes a series of thrusting movements. The female enters the nest in the indicated direction. The male rhythmically prods the base of the female's tail, which induces her to release her eggs. She swims out of the nest and the male enters and fertilizes the eggs. The male then chases the female away and attempts to attract several more females. He guards the eggs and aerates them by fanning water over the nest with his fins until they hatch. (After Goodenough et al. 1991, based on Tinbergen 1952b.)

about different questions. The long-distance mate-attraction sender may need verification that the respondent is indeed a conspecific of the opposite sex, and not a predator or same-sex competitor. The approaching individual may ask questions about mate suitability. Subsequently, males initiate one or more signals designed to persuade or stimulate the female to mate. Males are less choosy and more eager to mate than females because their investment in gametes is lower, so they play the active, pushy role during courtship. Errors in species recognition and mate choice are more costly for females so they evaluate the male carefully and signal their acceptance (or rejection) of the mating. In this section, we shall focus primarily on the design of male courtship signals and only briefly touch on female signals. Other aspects of courtship signals will be deferred until we have discussed male versus female mating games in Chapter 23.

Design Rules for Courtship Signals

Courtship signals are usually sex- and species-specific. Since close approach of the two individuals has already been achieved, the range or active space of courtship signals should be short to reflect the exclusivity and intimacy of the interaction and to reduce eavesdropping by predators or mate competitors. Although the sender of the long-distance attraction signal has already been located, the sender may need to indicate which of several respondents it intends to court. A sender may also need to provide information about the location of a nest or oviposition site. Duty cycles for courtship are often higher than for mate attraction, since male courters are usually eager to lead the encounter to mating as quickly as possible. Because receivers may still be assessing species, sex, and individual identity of a sender, a complex mix of identification information is usually included in courtship signals. Similarly, both sexes may bring a mixture of motivations to the encounter, and this favors repeated exchanges of signal modulations, indicating the shifting weightings of those motivations. Courtship can be a conflict situation in which one party wants to proceed whereas the other is cautious. This is exactly the type of context in which honest signals are favored. We should thus not be surprised to find strong form-content linkages in courtship signals. Design rules and mechanisms for courtship signals are summarized in Table 18.3.

Modality-Specific Mechanisms for Courtship Signals

RANGE. The short range of courtship signals means that all modalities can easily meet this design requirement. Tactile signals are particularly effective since mating frequently involves physical contact for sperm transfer. Chemical courtship signals are typically low-volatility compounds detected by contact receptors such as the antennae of insects and the vomeronasal organ of reptiles and mammals. Auditory and visual signals can be reduced in amplitude and intensity to lower conspicuousness. Electric signals are well suited to such short-range exchanges. Modality choice is more likely to be determined by the tactics males must use to achieve successful mating. These tactics depend on the species' reproductive biology, morphology, and social system.

DIRECTIONAL INFORMATION. The visual modality is well suited for encoding directional information. Movement displays can be directed toward certain

Table 18.3 Design rules and modality-specific mechanisms for courtship signals

Design feature	Rule	Visual mechanisms	Auditory mechanisms	Olfactory mechanisms	Tactile mechanisms
Range	Short	Low color contrast Subtle movement display	Soft unstructured sounds	Contact chemical Volatile, rapid fadeout chemical	By definition
Locatability	Receiver and Nest site	Directed display Pointing	Beam sound Countercalling	Directed flow Add visual component	Herding
Duty cycle	High for short period	Flashing color High display repetition rate	High repetition rate	Contact chemical	Hold High touch rate
ID level	Species and Sex	Sexual dimorphism	Sex-specific sound pattern	Sex-specific chemicals	Sex specific pattern
Modulation level	Graded	Display rate	Repetition rate	Poor	Vary pressure
Form-content linkage	Linked:				
	Intentions	Nesting behavior	High frequency	Receptivity	Mount
	Parent skill	Courtship feed			Nuptial gift
	Stimulate		Repertoire	Hormone manip	Lick genitals
	Calm	Submissive	Soft warble	Tranquilize	Stroke

individuals or referents, and senders can approach, point toward, and look at such referents. Stripes and strategically located color patches can be used to amplify such pointing movements. Acoustic signalers can orient their bodies so that the loudest lobe of the sound field is aimed at intended receivers. Olfactants can be sprayed on receivers or referents. Tactile signals by definition can be used to indicate certain receivers, and they may be able to direct receivers to nearby sites via physical herding.

DUTY CYCLE. In species with a male-delivered visual or auditory mate-attraction signal, the courtship display is often a softer but higher duty-cycle variant of the long-distance display. In crickets, for example, when the male perceives that an approaching cricket is a conspecific female via visual and/or tactile cues, he gradually inserts soft chirps between the louder rhythmically timed chirps of the mate-attraction call. If the female indicates her interest by staying, he shifts completely into a soft arrhythmical courtship call (see Figure 18.13). Similarly, the courtship call of the male hammerheaded bat (*Hypsignathus monstrosus*), which is performed when a female hovers in front of him, is a rapid buzzy variant of the repetitive mate-attraction call (Bradbury 1977). In frogs and lek-breeding birds such as sage grouse (*Centrocercus urophasianus*), the long-distance and courtship signals are the same, but males increase their repetition rate when a female approaches them and females focus on different parameters of the display during attraction and courtship phases

(Gibson 1996). The moderate duty-cycle jumping display of the male Jackson's widowbird (Figure 12.6) switches to a high duty-cycle tail-quivering plus soft vocal display once a female has landed in his display arena (Andersson 1989). Courtship signals thus tend to be lower in amplitude but higher in duty-cycle than the preceding attraction signals.

SEX AND SPECIES INFORMATION. Olfaction is the best modality for confirmation of species, sex, and receptivity. Sex and receptivity, in particular, can be encoded in odorants derived from sexual hormones and reproductive products. Other modalities can also be recruited. For example, stickleback males recognize gravid females by their lack of red coloration, presence of a swollen belly, and a head-up display (Figure 18.5). Simultaneously hermaphroditic fish indicate which sex role they intend to play with either changeable color displays (Figure 1.4) or sex-specific sounds (Figure 24.2D). In a variety of species, respondents to mate-attraction signals must produce a species-specific reply before serious courtship is initiated. Presumably this protects long-distance senders from eavesdropping predators, or it allows them to discriminate between responsive and nonresponsive conspecifics in their proximity. For example, some orthopteran species respond to male mate-attraction calls with a female-specific answering call (Heller 1992; Dobler et al. 1994), female fireflies respond to the male's flash with a short answering flash (Lloyd 1966), and female jumping spiders may perform a version of the male's foreleg and pedipalp waving display (see Figure 18.6). Courtship does not proceed further without this response.

MODULATION LEVEL. Modulation of at least one signal element is typical of all but chemical courtship signals. Modulation level often encodes sexual arousal level and serves to synchronize the two sexes for mating. Repeated visual and auditory displays, such as the mutual head-pumping precopulatory display of ducks, can gradually increase in tempo to prepare both partners for the moment of copulation (see Figure 15.12). The male blue-headed wrasse (*Thalassoma bifasciatum*) circles around the female and gradually increases his pectoral fin-beating rate as spawning approaches; he also develops spots on the tips of the fins to amplify the fin beat and reduces the purity of his body color to decrease signal range (Dawkins and Guilford 1994). The male red rainbow fish (*Glossolepis incisus*) flashes a yellow stripe on his forehead during courtship; the rate and number of flashes increases as the female shows more interest. Tactile signals can also increase in tempo, as can rates of countercalling.

FORM-CONTENT LINKAGE. Courtship signals are often derived from reproductive intention movements or ambivalent motivations. They also can reflect considerable conflict of interest between the two parties. We should thus expect many of them to show high form-content linkage (Figure 18.6). The courtship signals of male web-building spiders are a good example. These signals are tightly linked to other behavior patterns associated with living on webs (see Figure 23.5C). Visual courtship displays that mimic nest construc-

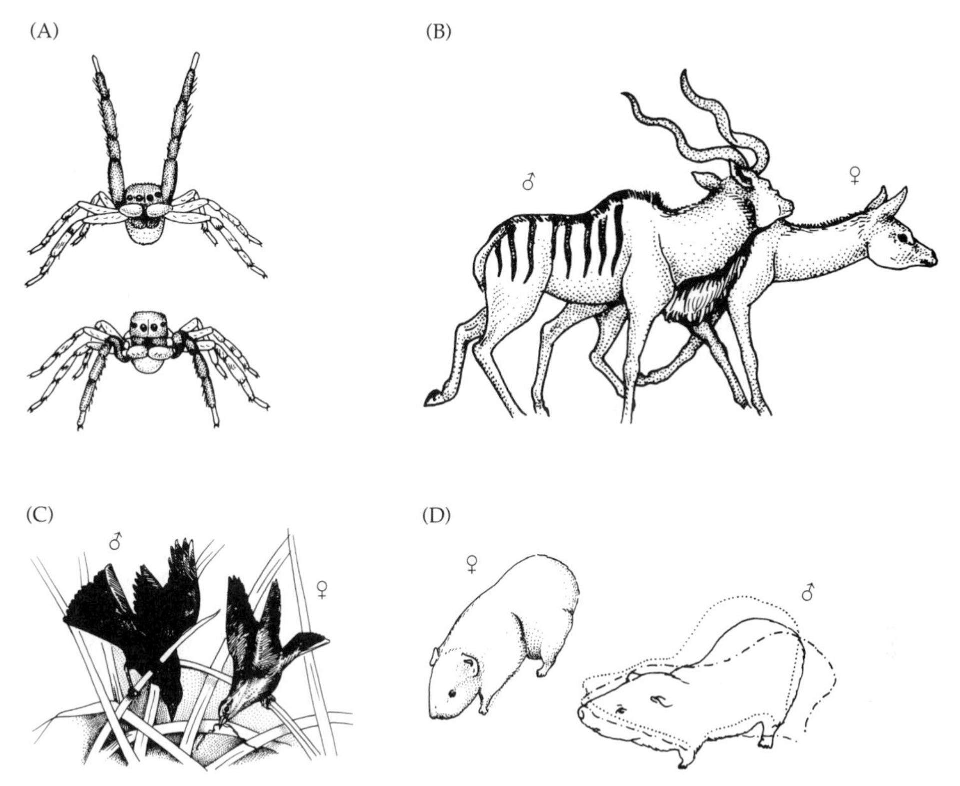

Figure 18.6 Form-content linkage in courtship displays. (A) In the jumping spider (*Euophrys frontalis*) the male approaches the female and engages in ritualized front leg raising and lowering. The front legs of spiders contain chemoreceptors, so the display indicates to the female that a male has been attracted by her pheromone. (B) The male kudu (*Tragelaphus strepsiceros*) presses his head on the female's back during mounting and uses a similar head-resting tactic during courtship. (C) The male tricolored blackbird (*Xanthocephalus xanthocephalus*) demonstrates potential nest sites on his territory with a pointing display. (D) In the precopulatory rumba display of the guinea pig (*Cavia*) the male shifts his weight from one hind leg to the other to produce a rhythmic swaying in front of the female. The motion is similar to the treading that occurs during copulation. (A after Bristowe 1958; B after Walther 1958; C after Orians and Christman 1968; D after Rood 1972.)

tion, nest material manipulation, and mate provisioning are common in birds. Soft, unstructured, and high-frequency courtship sounds may reflect nonaggressive, friendly intentions in senders. Tactile signals such as the touching of the female's hind leg by the male in some antelope species (see Figure 17.9D) are the initial stages of mounting and thus test the readiness of the female to stand still. Chemical signals of receptivity in females that are derived from reproductive hormones also are examples of strong form-content linkage. The female signal for mate acceptance, gamete release, or copulation is almost always the assumption of the same posture used during mating (see Figure 24.3). It is thus a classic intention movement signal.

Role of Reproductive Strategies in Shaping Courtship Signals

Since the purpose of courtship is to bring about mating, we would expect male and female reproductive strategies to play an important role in determining the form and modality of courtship displays. If females must oviposit in a nest, males are likely to use displays that coax females toward an appropriate site and into laying positions. The visual modality is well suited for this function. If males must provide stimulation to trigger hormonal changes leading to receptivity, the corresponding courtship displays should be highly conspicuous and distinctive. All four modalities could achieve this objective, e.g., bold visual displays, complex song repertoires, olfactory pheromones that directly stimulate the pituitary or other brain centers, and tactile stimulation of genitals. Where courtship enables females to choose mates with good paternal abilities, males could demonstrate their parental skill by courtship feeding the female or by bringing nest material and nuptial gifts. If the female is highly mobile or possesses formidable weapons, then males should engage in behaviors that calm her, such as displaying submissive postures, using soft warbling vocalizations, releasing pheromones that tranquilize or immobilize, and gentle stroking. On the other hand, if males benefit by forcing females to mate, we would expect to observe threatening visual signals, growl-like vocalizations, and physical pushing, holding, or chasing. It is interesting that many of these strategies require courtship signals with high form-content linkages; they show what a courter can or will do. Since the appropriate task often varies with species, this means that courtship signals vary idosyncratically with species' reproductive strategies. We should thus not expect large amounts of convergence in courtship signals except within a given reproductive strategy. We shall pursue the relationships between different mating strategies and optimal signals further in Chapter 23.

TERRITORIAL DEFENSE SIGNALS

A common strategy for obtaining access to critical resources such as mates or food is to stake out a piece of property containing the resource and defend it against competitors. This strategy is worthwhile only if the benefits of exclusive resource access outweigh the costs of defense. Signaling one's presence as a territorial owner is one major cost of defense; chasing out intruders is another. The signal emitted by a territorial owner is directed toward all potential intruders with an interest in the resource and conveys the message that the resource is owned and the owner is present on the site.

Design Rules for Territorial Defense Signals

Territorial defense signals share some design rules with mate-attraction signals. They must have a sufficiently large active space so that they can be perceived at or just beyond the boundaries of the territory. Either the sender or the territorial boundaries must be accurately localized by receivers. The duty cycle of the signal should be moderately high during the times that intruders are likely to encroach. Where conspecifics are the most significant competitors, a territorial defense signal should be species-specific. If one sex defends mates directly, or defends a resource to obtain mates, the signal will be directed at competitors of the same sex and thus will be sex-specific as well. Sex and species identification may be sufficient in some simple spacing systems, but in most territorial systems it is advantageous for the territorial owner also to identify itself as an individual. Territory owners thus become known to their territorial neighbors and unnecessary chases and fights with these individuals are reduced. Repeated renditions of the signal should therefore be stereotyped to preserve individual and species identification. In some cases variations in repetition rate or intensity of the signal may be used to reveal the motivation or condition of the sender. Since territorial signals primarily provide information about sender identity and location, they can be arbitrary (Table 18.4).

Modality-Specific Mechanisms for Territorial Defense Signals

RANGE, TERRITORY LOCATABILITY, AND DUTY CYCLE. The range of territorial signals increases as a function of territory size. All modalities except touch are used for territory defense and, given the constraints on transmission distance for some modalities, modality choice is affected by territory size and other environmental constraints. Modality also affects sender and territory locatabil-

Table 18.4 Design rules and modality-specific mechanisms for territory defense signals

Design feature	Rule	Visual mechanisms	Auditory mechanisms	Olfactory mechanisms
Range marks	Moderate (territory boundary)	Posture Color patch	Loud call	Deposited Low volatility chemical
Locat-ability	Territory and/or Owner	Presence on territory	Amplitude cues Degradation cues	Frequent marks along boundary Sender not locatable
Duty cycle	Variable	High when owner present	Long-duration signal Few repetitions per day	Long fadeout Embed in sebum Volatile when sniffed
ID level	Species Individual	Color pattern Presence on territory	Frequency Temporal pattern Note shape Syntax	Chemical mix
Modula-tion level	Some graded components	Coverable color patch	Repetition rate Song type variation	Add chemical related to dominance
Form-content linkage	Arbitrary	Exploit preexisting visual biases	Exploit preexisting sender production mechanisms	De novo production

ity. Electric signals are used by electric fish to defend burrows and small territories; the sender is easily localizable, but only at very short distances. Visual signals are useful for diurnal, open-habitat species with small to medium-sized territories (e.g., shallow-water fish, lizards, dragonflies, grassland birds, and ungulates). Mechanisms usually involve color patterns, postures, and more rarely, ritualized movement displays. The sender is always locatable as long as it is present and patrolling its territory, and the duty cycle is therefore high for this modality. Animals using the auditory modality can defend large territories with loud calls. The active space of auditory territory signals in birds is estimated to be about the length of two territories (Brenowitz 1982a). Locatability of the sender and its territorial boundaries is more difficult in this modality. Receivers may be able to estimate the position of the sender using amplitude, phase, or sound degradation cues. Animals that use long-distance vocalizations only for territory defense tend to produce very long calls a few times per day, e.g., tree hyraxes (*Dendrohyrax*), gibbons (*Hylobates*), howler monkeys (*Allouatta*), wolves (*Canis*), and kangaroo rats (*Dipodomys*).

The form of olfactory territorial signals is very different from that of olfactory mate-attraction signals. Dissemination of volatile diffusible or current-borne chemicals is never used; poor source locatability and turbulent currents prevent this type of release strategy from providing the boundary-defining information required of a territorial signal. Instead, substrate marking with low-volatility chemicals is the rule. The strategy of maximizing the durability of marks means that their active space is small. Most olfactory territory senders place a string of marks at intervals around the border of their territory, but since some invaders may miss the signal, additional marks are also placed along trails, around nests, and at roost sites within the territory (Figure 18.7). The territory is patrolled and marks renewed at regular intervals. The greatest advantage of olfactory signals is the high duty cycle that can be maintained even in the short-term absence of the sender; the most significant disadvantage is poor locatability of the marks. To improve locatability, many species augment olfactory marks with visual and auditory signals (Alberts 1992). Animals using feces to mark their territories produce feces piles that contrast visibly with the environment. Others deposit marks on conspicuous vertical objects, or position them close to obvious landmarks such as intersections of paths, isolated boulders, and trees. Some senders vocalize when depositing marks to draw attention to their position. Terrestrial vertebrates such as mammals and lizards are the most common practitioners of this modality for territory defense.

SPECIES AND INDIVIDUAL IDENTIFICATION. In order for territorial signals to achieve both species and individual recognition, the signals must be compound. Some components of the signal should be constant for all members of the species or population, and other components should vary between individuals. A hierarchical mapping scheme (page 474) is typically used to encode these two types of information. Visual senders employ finely variant color patterns, skin wrinkles, and other markings located on the head or face to indicate

(A)

(B)

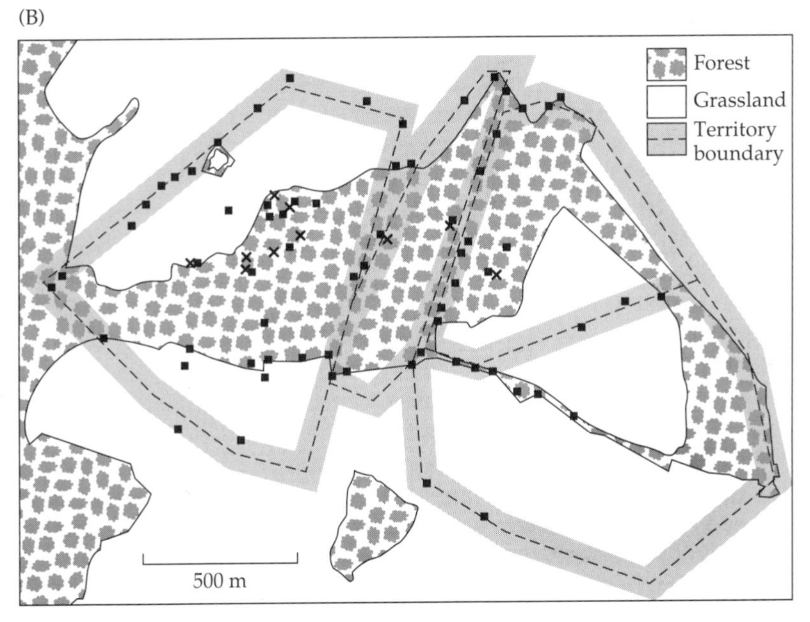

Figure 18.7 Territorial marking in the badger. (A) The Eurasian badger (*Meles meles*) lives in clans of up to 12 individuals that defend a group territory. They dig a little scrape and deposit feces along with a gelatinous material from the anal pouch. The marker squats and rubs its perineal region along the ground as shown here. (B) The location of olfactory territorial marks in relationship to territory boundaries. The boundaries of four adjacent territories are indicated by dashed lines and the 50 m zone around the boundary is shown by the wide shaded lines; the patterned area shows forested habitat. Squares show the position of scrapes and ×'s show the position of burrows. Seventy percent of the scrape sites are located within 50 m of territory boundaries. (Photo courtesy of Hans Kruuk; B after Kruuk 1978.)

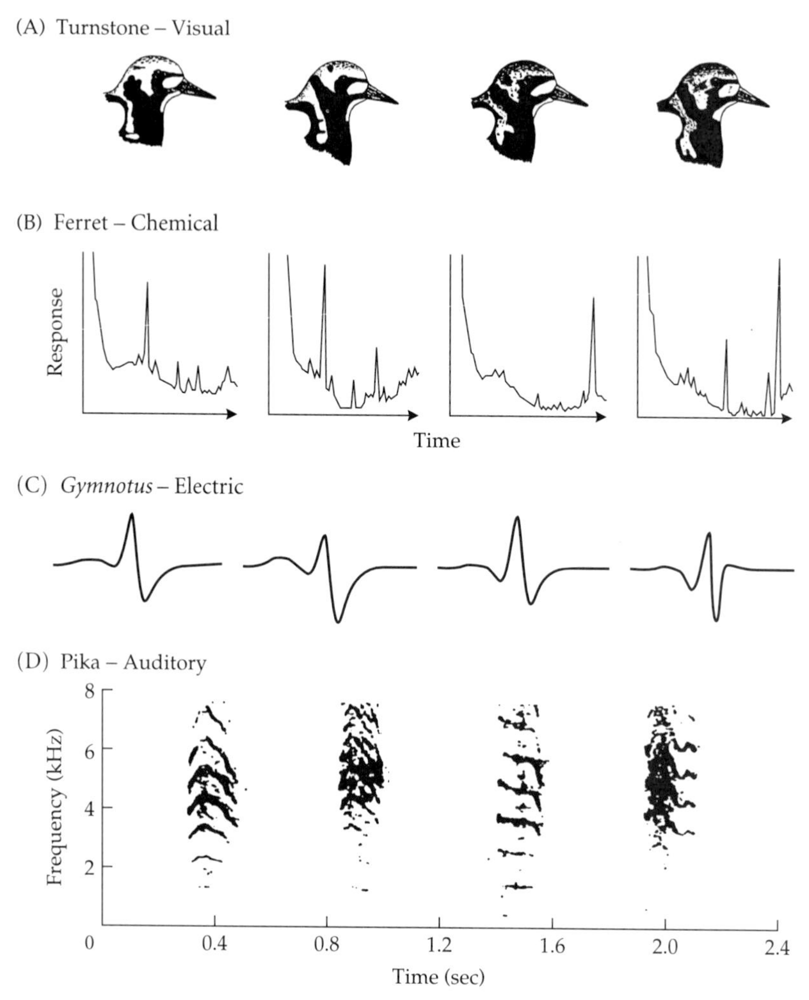

Figure 18.8 Mechanisms for achieving individual signatures in territory signals. (A) Black and white facial patterns in territorial turnstones (*Arenaria interpres*). (B) Gas chromatograph tracings illustrating differences in the chemical composition of anal gland secretions in individual ferrets (*Mustella furo*). Each numbered peak corresponds to a specific chemical component and the height of the peak reflects that component's relative concentration in the mixture. (C) Individual differences in the pulse shape of electric fish (*Gymnotus carapo*). (D) Individual differences in short territorial calls of pikas are based on frequency, frequency modulation, and harmonic structure (*Ochotona princeps*). (A after Whitfield 1986; B from Clapperton et al. 1988; C after McGregor and Westby 1992; D after Connor 1985.)

individual distinctiveness (Figure 18.8A). Visual signals may not always need to encode complex individual indentification information if the continued presence and confident posture of the owner are sufficient evidence of ownership. In the case of olfactory signals, a mix of chemicals is always used. The list of components is species-specific, whereas individuals vary in the proportions of each chemical in the mix (Figure 18.8B). In some species, individual identity is enhanced by adding small amounts of unique compounds as well. Electric fish use waveform shape for individual distinctiveness (Figure 18.8C). Auditory signals are generally long and complex to ensure the provision of both types of identification. Often the first part of the call is species-specific and highly conserved for all individuals, and subsequent parts are individually variable (Figures 18.8D and 15.3).

MODULATION LEVEL. Territorial signals must remain highly stereotyped to preserve individual distinctiveness. However, some means of modulating the signal using parameter mapping may be used to encode sender condition, status, aggressive motivation, or other aspects of defense ability. Visual color patches of territory ownership are often coverable, as in the red shoulder epaulet of red-winged blackbirds (Hansen and Rohwer 1986). Experimental removal of the signal by blackening the red feathers causes owners to lose their territories (Smith 1972). Similarly, territorial cichlid fish express black vertical stripes while present on their territories and blanche (or become completely dark) when nonterritorial or when traveling through another individual's territory (Baylis 1974) (see Figure 12.1). In *Phylloscopus* warblers, increasing the size of the white wingbars results in an increase in territory size, demonstrating clearly the territorial threat function of this signal (Marchetti 1993). Auditory senders can of course vary call rate to indicate motivation. As mentioned in previous chapters, males of many songbird species possess a repertoire of songtypes. Rates of switching between songtypes, selection of matching songtypes, and temporal ordering of songtypes can also be used to encode different levels of aggressive motivation. Group-territorial species have yet another auditory component they can vary, the number of senders. Chorus calling in lions, wolves, hyaenas (*Crocuta crocuta*), howler monkeys (*Alouatta*), and Australian magpies (*Gymnorhina tibicen*) provides information on group size (Sekulic 1982; Brown et al. 1988; Harrington 1989; East and Hofer 1991; McComb et al. 1994).

Combined Territorial and Mate-Attraction Signals

In songbirds and other territorial species in which males call to attract females, the song may also acquire the function of territory defense. The signal simultaneously attracts conspecific females and repels conspecific males. Since several of the design features for the two functions are fairly similar (i.e., they are long range, locatable, species-specific, and arbitrary), there is often no serious tradeoff imposed when the two functions are combined into a single vocalization. The two functions differ most in terms of duty cycle and individual recognition level. The repetition rate of combined territorial/mate-attraction calls is higher than the repetition rate of territorial-only calls because of the added high duty-cycle requirement of mate attraction. In songbird species that sing only until males are mated, songs are generally long, complex, and nearly continuous. In species for which the territorial function is very important and singing continues after pair-bond formation, songs are relatively short and separated by significant gaps. The argument put forward to explain this difference is selection for reception of vocal responses and countersinging in the context of male-male territorial interactions (Slater 1981; Catchpole 1982, 1987). Consistent with this notion, species that use different vocalizations for territorial defense and mate attraction have short territorial calls and long mate-attraction songs, e.g., pika (*Ocotona*), great reed warbler (*Acrocephalus arundinaceus*), and five-striped sparrow (*Amophila quinqueistriata*) (Morse 1970; Connor 1984, 1985; Baptista 1978; Catchpole 1983; Groschupf 1985). Presumably, the design requirements in these species are too different for the same vocalization to serve both functions.

THREAT SIGNALS

When two individuals compete directly and at close range over a resource such as food, a mate, a territory, or a burrow, reciprocal communication must be used to resolve the conflict. The conflict is likely to be won by the individual that is larger, stronger, healthier, more experienced, and/or more motivated. Threat signals are designed to transmit information about these questions. Once one contestant decides it would lose during further escalation, it gives a surrender signal to stop the conflict. A full understanding of the evolution of conflict-resolution signals can only be obtained with the added perspective that game theory provides (Chapter 21). However, comparisons between taxa have indicated the clear presence of design rules that we summarize below.

Design Rules for Threat Signals

Threat signals share some design features with courtship signals: range is short, displays need to be directed at specific rivals, modulations that encode gradations in motivation are required, and there is a need for signals that honestly indicate intentions. However, there are some significant differences. The duty cycle of behaviors during an encounter may be very high for a short period of time, but individual signals are usually short, forceful, and conspicuous. Identification level involves the recognition of rival classes or statuses that are largely fixed for a given individual before the contest. If there are only a few rival classes, based on sex or age, the perceptual recognition task is relatively easy. There could, however, be a continuous range of rival classes based on dominance status or body size. Since the critical question is a contestant's size or status relative to that of its rival, the perceptual recognition task may be quite difficult. Modulation of threat signals transmits information on aspects of fighting ability that vary within individuals, such as current condition and motivation. Finally, threat signals are often ritualized intention movements, ambivalent combinations of acts, or redirected behaviors, and some are physically linked to the size or condition of the animal. Threats by definition occur in situations of sender-receiver conflict. All of these factors favor a tight linkage between threat signal form and its content. These basic predictions are summarized in Table 18.5.

Table 18.5 Design rules and modality-specific mechanisms for threat signals

Design feature	Rule	Visual mechanisms	Auditory mechanisms	Olfactory mechanisms
Range	Short	Posture Movement display	Soft broadband growl Loud scream	Volatile, rapid fadeout chemical
Locat-ability	Intended receiver (rival)	Directional display or posture	Beam sound Countersinging	Directed flow Add visual component
Duty cycle	High for contest duration, Short signals	Series of single short displays	Series of single short calls Longer single growl	Single puff
ID level	Rival class: Age, body size, status	Maturational color change Broadside display Badge	Frequency shift with age or body size Intensity	Chemical derived from maturation hormone
Modula-tion level	Graded	Variable position or display form Choice of display	Frequency Intensity Repetition rate	Poor
Form-content linkage	Linked	Attack or retreat intentions Body size amplifier	Lowest possible frequency Wide bandwidth	Testosterone-derived chemical

Modality-Specific Mechanisms for Threat Signals

RANGE, DIRECTIONAL INFORMATION, AND DUTY CYCLE. Although threat signals do not need to be transmitted very far because sender and receiver are close, this does not necessarily mean that signal conspicuousness is decreased. The need to send a clearly aggressive message usually results in signals being very intense. Visual signals consist of forceful movements, assertive postures, and contrasting black or white color patches. Movements are usually of short duration, and they can be directed at specific rivals. Auditory threats can be loud, but there are examples of low-intensity grunts and growls. When two countersinging songbirds approach each other closely, they reduce their amplitude while continuing to sing (Dabelsteen and Pedersen 1990; Nielsen and Vehrencamp 1995). Olfactory threats are also effective at short range and can be directed or sprayed at specific rivals. A visual component is often added to enhance olfactory threats. The stink fights of ring-tailed lemurs are an excellent example (Figure 18.9).

RIVAL CLASS IDENTIFICATION. Visual, auditory, olfactory, and tactile signals all have some ability to encode rival class information. In the visual modality,

Figure 18.9 Ring-tailed lemur stink fight. An early stage in conflicts between male ring-tailed lemurs (*Lemur catta*) consists of an olfactory signal augmented by a visual tail display. Normally, the two animals would be several meters apart. The male on the right is shown rubbing the secretion of the wrist glands on its tail. The tail is then lifted over the head and waved in the direction of the opponent as shown by the male on the left. The olfactory display is enhanced by the conspicuously marked tail and by a staring, ears-laid-back facial expression. (Based on descriptions and photos in Jolly 1966.)

changes in color pattern with maturation are used in birds and some lizards to signal broad age-based rival classes. Status badges are color patches or black-white patterns, the size of which is correlated with an individual's aggressive tendency (see Figure 20.6). Body size can also be assessed in the visual modality, and vertical or horizontal body stripes as well as broadside displays and piloerection can be used to enhance or exaggerate the perception of this important determinant of fighting success. For auditory senders, sound frequency and intensity can encode age, sex, and body-size classes. Chemical signals derived from testosterone can provide information about age and dominance status. Finally, tactile displays involving pushing or pulling between rivals can transmit information about body size and strength.

MODULATION LEVEL. Some threat signals are graded to indicate variable levels of motivation, fighting skill, and current health. Visual mechanisms for graded threats include degree of dark coloration (Figure 15.11), degree of ear flattening in many mammals, degree of crest raising in some birds, and degree of forward leaning and proximity to the rival. Display intensity and display rate can be graded in the case of repeating displays such as branch shaking, tail twitching, and ground pawing. Auditory senders can also modulate call rate, along with call duration, sound intensity, and frequency. Domestic cats, for example, produce a nearly constant vocalization during aggressive interactions that varies in frequency and intensity. Other types of sounds used to indicate excitement level are teeth gnashing, tail rattling, and thumping, drumming, or slapping the ground, trees, or chest with the forelegs. Electric fish threaten each other by overlapping their steady discharges on top of those of their opponents. Both parties then jockey for dominance by blocking the other until one gives up and adjusts its discharge rate in a species-specific direction (up for some species, and down for others; see pages 344–345). The amplitude or intensity of an olfactory signal could be gradually varied over time to transmit information on changing arousal or motivational level, but in general olfactory signals cannot be temporally modulated to provide the subtle shifts in threat tactics and alternations of act and response that visual and auditory signals can. Again, olfactory signals are often coupled with visual or auditory signals to achieve these effects.

In contrast to these graded signals, some threat signals are short and highly stereotyped. Where threat signals are stereotyped, there are often several different stereotyped displays that encode different levels of aggressive motivation. Thus another type of modulation is choice of threat signal. Rapid switching between different visual displays characterizes conflicts in species such as penguins (discussed in Chapter 21), and switching between different songtypes characterizes territorial interactions in many songbird species (discussed in Chapter 22).

FORM-CONTENT LINKAGE. As noted earlier, the majority of threat signals show high form-content linkage and either directly reveal aspects of an individual's body size, strength, or fighting skill, or are ritualized attack or retreat intention movements. The visual modality is ideal in this regard, because each species can uniquely incorporate those body parts and movements specifically used in fighting into ritualized threat signals (Figure 18.10). For example, animals with defensive weapons such as teeth, horns, long necks, spines, quills, or powerful legs or feet display them prominently. Biting mammals bare their teeth, antelope that clash horns display them in a forward or butting posture, kangaroos and rodents that box with forelegs rear up on their hind legs, and porcupines with rear-facing quills approach their opponents with an oblique backward movement. Visual threat displays thus possess an important tactical value (in addition to information) that changes the immediate context and potentially imposes incidental costs on both sender and receiver. The postures and structures are often exaggerated to enhance the impact of the threat. Visual amplifiers can be used to make the assessment of body size easier. Surrender displays represent the exact antithetical type of posture: backing off, head turned away, mouth closed, and various crouched, lying, or belly-up postures. Auditory displays can also provide form-content linkage. The low frequency of many aggressive growls, roars, and snarls indicates the body size of contestants. Frequently repeated, high-intensity calls may use the thoracic muscles also employed in fighting and reveal the strength of contestants. As mentioned above, testosterone-derived olfactory signals can indicate age, maturation, and dominance, while tactile pushing or pulling signals can indicate strength.

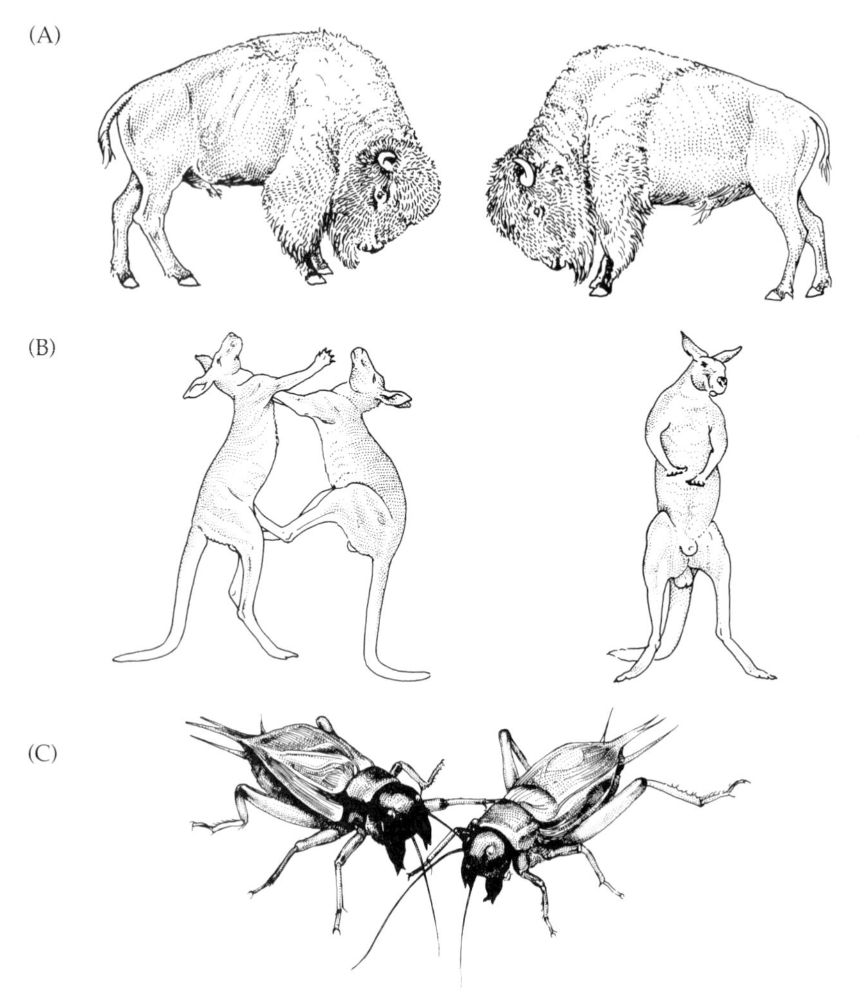

Figure 18.10 Conflict resolution signals. Visual threats often consist of intention movements and strategic behaviors related to the particular form of fighting in a species. (A) Ungulates such as the buffalo (*Bison bison*) bulls illustrated here fight by butting heads and locking horns and swing their heads from side to side in unison, a display called the nod-threat. (B) Kangaroos (here, *Macropodus rufus*) fight by locking forearms and kicking with the rear feet and threaten by standing upright on extended rear legs. (C) Animals that fight with mouth-wrestling, such as the crickets (*Acheta domesticus*) shown here, threaten with an open-mouth mandible flare display. (A and B after Macdonald 1984; C after Hack 1997b, drawn from photographs by Mace Hack.)

ALARM SIGNALS

Alarm signals are found only in relatively social species; these signals transmit information about the presence of a predator or conspecific rival. We shall distinguish three different types of alarm signals, based on the nature of the response to the signal. **Flee alarms** are given in the context of a cluster of animals in immediate danger of attack by a predator. The signal causes all receivers to rapidly disperse, run, or hide, and this response presumably serves to reduce the likelihood of a successful attack. At the other extreme are **assembly alarm** signals that cause dispersed receivers to move toward the sender. Animals might form temporary groups for a variety of reasons: to mob a predator, to rescue an endangered or lost individual, or to defend a group resource. Intermediate between these centrifugal and centripetal extremes are alarm signals that do not cause receivers to either move away from or toward the source, but to remain stationary and look around. We shall call these **alert** signals. They serve to notify receivers that some potentially important change has taken place but there is no imminent danger. They may also serve as signals to predators (see Chapter 25, pages 841–846).

Design Rules for Alarm Signals

The design features for alarm signals are very different and in some cases diametrically opposed for the two extreme types of alarms—flee versus assembly—with alert signals often intermediate (Marler 1955, 1959; Klump and Shalter 1984). The range or active space of an alarm signal depends, of course, on the distance to the intended receivers. In general, assembly signals will require a longer range than flee signals. As far as sender locatability is concerned, flee signals should be as difficult as possible to localize so that the sender does not reveal its own position to an attacking predator, while assembly signal senders should be easy to localize so group members can approach. For assembly and alert signals, the sender may need to point to the external source of alarm. The duty cycle of a flee signal is by necessity low, since the immediacy of the situation allows no time for a lengthy signal. Assembly signals require a longer "on" time. Flee signals need not be species-specific since no cost is incurred by the sender if other species hear and respond to it. On the other hand, assembly signals usually are species-specific, and in some cases group-specific as well. Flee signals do not need to express variable degrees of motivation or excitement, although some species do employ categorically different signals to code for different types of messages such as "flee up" in response to a terrestrial predator and "flee down" in response to an aerial predator. Assembly and alerting signals, on the other hand, may need to encode information about the seriousness or urgency of the situation. Finally, flee signals often show some form-content linkage since they are produced in fearful, injurious, and/or rapid flight contexts. Assembly signals are more likely to be arbitrary (Table 18.6).

Design Mechanisms for Alarm Signals

AUDITORY MODALITY. The auditory modality has the greatest versatility for adjusting the desired levels of range and locatability for alarm signals. Small mammals and birds tend to produce flee alarms that consist of a single pure-tone, high-pitched whistle with gradual onset and offset. The smooth waveform makes the signal more difficult for a predator to detect or localize using

Table 18.6 Design rules and modality-specific mechanisms for alarm signals

Design feature	Rule	Visual mechanisms	Auditory mechanisms	Olfactory mechanisms
		FLEE ALARM		
Range	Short–moderate	Color flash Appendage movement	Medium intensity call	Volatile, diffusable chemical
Locat-ability	Conceal sender location	Coverable patch	Pure tone Gradual onset	Diffusion gradient
Duty cycle	Short	Single flash Rapid movement	Single call	Single puff
ID level	None			
Modula-tion level	Stereotyped			
Form-content linkage	Linked: Fear, flight	Signal on tail or rear end	High frequency	Derive from defense chemical
		ASSEMBLY ALARM		
Range	Medium-large	Contrasting movement	Loud call	Increase Q
Locat-ability	Sender Enemy	Repeated jerky movement	Broadband note Trill	Diffusion gradient
Duty cycle	High while danger present	Regular repetition	Regular repetition	Repeated puffs
ID level	Species (Group)	Visual pattern	Note shape	Chemical mix
Modula-tion level	Graded	Repetition rate	Repetition rate	Concentration
Form-content linkage	Arbitrary	Maximize visual contrast	Maximize detection	Optimize fadeout

phase or differential arrival-time information (Figure 18.11A). The high pitch may be favored because smaller prey can produce and detect higher frequencies better than their larger predators can (e.g., small birds versus hawks), or because high vocal frequencies are linked with tense and fearful sender states. Large mammals do not seem to be as concerned about sender locatability, and instead utter a single, short broadband sound that seems to serve the same function as the small bird and mammal whistle. The snort of antelopes, clack of porpoises, tail slap on the water in beavers, and single bark of dogs and monkeys are good examples. Alerting signals, on the other hand, are often brief but broadband pulsed calls that are more easily localized by receivers. They may be repeated a few times. Assembly calls are characterized principally by the repetitive delivery at regular intervals of a short, simple note. The repetition rate is much slower than the interval between the pulses of an alert alarm. The regular repetition of the same note demands the receiver's attention, is very easy to localize, and implies that the sender is not poised to flee; repetition rate can be varied to encode information about urgency. Good examples of assembly calls include the clucking food calls of female jungle fowl with chicks, the food attraction and mobbing calls of many corvids, and the incessant barking of an isolated dog. Assembly calls can be varied in many ways to transmit information about other questions. Distress calls are a type of assembly call given by isolated individuals and designed to attract a parent or mate. They show the expected repetitive nature, but the individual notes are elongated and tonal with conspicuous frequency modulation. Figure 18.11 shows some examples of difficult-to-locate flee alarms and easy-to-locate alert and assembly calls in a variety of animals.

VISUAL MODALITY. Visual alarm signals exhibit analogous changes in form as a function of appropriate response. Visual flee signals are single, short flashes of a contrasting color, often white. They usually involve the tail, since this is the side of an animal's body that is most visible to grouped conspecifics as the sender flees. A visual alarm signal can only be perceived by a receiver that is already looking at the sender. It is also likely that the predator can see the sender as well. The option of minimizing sender locatability is therefore not possible for visual flee alarms. Alerting signals frequently involve regular, rapid-onset twitches of the tail. An added advantage of visual alert signals is that the sender can direct the gaze of receivers to the source of the alarm by looking or pointing at it. Contrasting stripes and triangles on the face or head may function as amplifiers for alerting signals. Tense, head-up postures coupled with staring serve as a generalized alerting signal in a variety of birds and mammals. Visual assembly signals take the form of up and down bobbing movements of the head, the antithesis of flee alarms involving the tail. As with vocal assembly calls, visual assembly signals are repeated regularly. Figure 18.12 illustrates some examples of visual alarm, alerting, and assembly signals.

OLFACTORY MODALITY. Patterns of timing and repetition obviously cannot be used with olfactory signals to encode flee versus assembly alarms. Several different structural mechanisms may be used to achieve the two types of responses. One is to use distinctly different odorants to signal flight and aggregation. Schooling fish use this technique. Certain chemical substances are known to cause conspecific attraction; different substances, released when a conspecific is injured, cause immediate dispersal or evasive action (Weldon 1983; Smith 1986). Another mechanism is to rely on contextual coding in which the specific context determines the meaning of the signal. In some ants, the release of alarm pheromones far from the nest elicits flight, whereas release of the same pheromones close to the nest causes aggregation and aggression. Finally, the response may be caste-specific; young worker ants flee

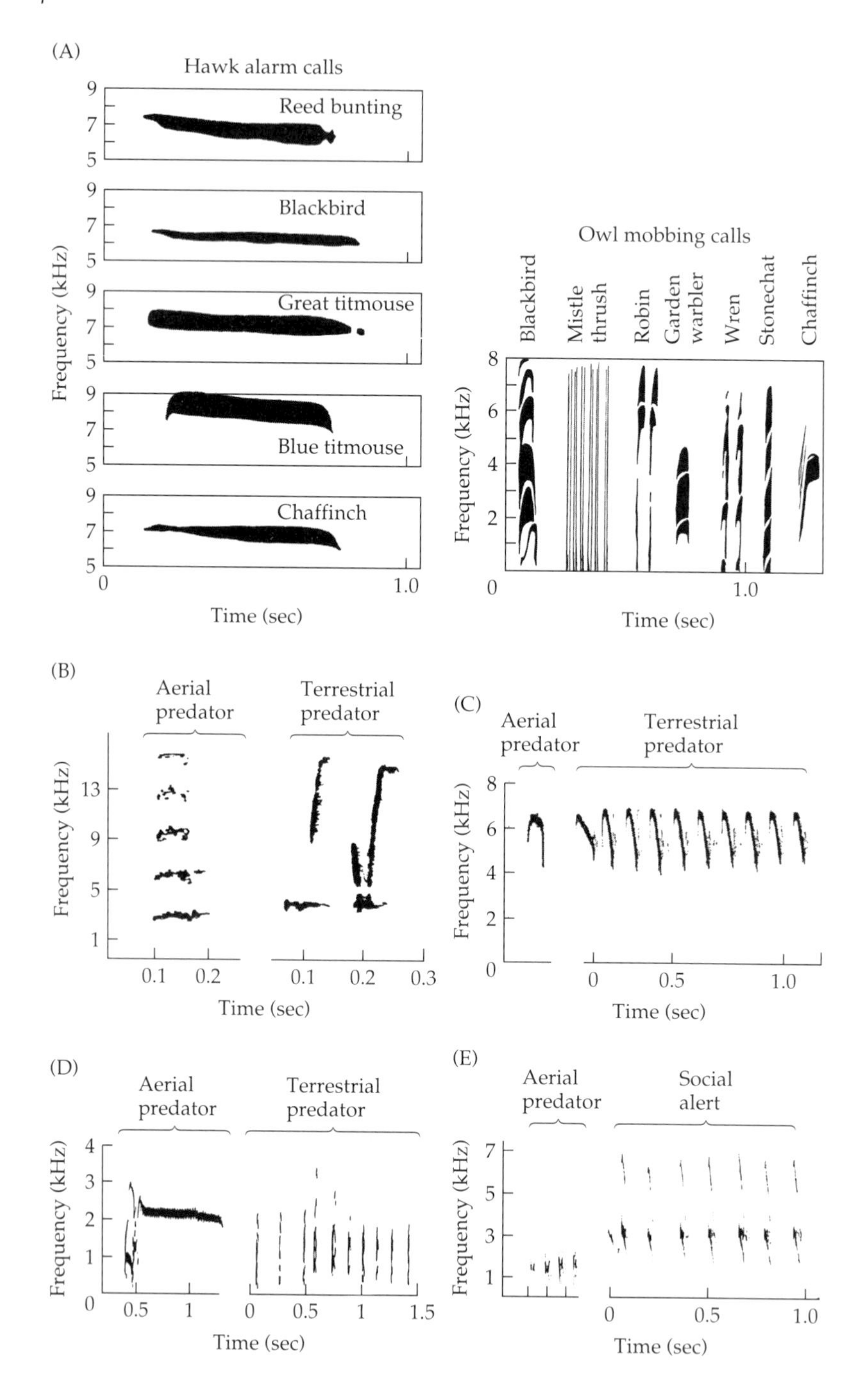

from the vicinity of alarm substances, while older workers and soldiers aggregate and attack.

Alarm signals in the social hymenoptera are designed to assemble many individuals for a group defense effort against enemies invading the nest. The

◀ **Figure 18.11 Auditory alarm calls.** (A) The classic evidence for alarm call design rules in passerine birds. Hawk alarm calls (left) are moderately loud, high-pitched whistles of pure, constant frequency tones and gradual onset and offset. Their frequency is above the sensitive hearing range of most birds of prey. They are usually given only once and are followed by immediate evasion. Owl mobbing calls (right) consist of short chirps or a burst of broadband pulses, variously described as trills, staccato calls, chutters, twitters, or scolding. Single calls may serve to alert, and repeated calls may recruit conspecifics. Other species show a similar dichotomy between flee vocalizations (left) given to dangerous aerial predators, and alerting vocalizations (right) given to terrestrial predators or conspecific intruders, including: (B) California ground squirrels (*Spermophilus beecheyi*), (C) Belding's ground squirrels (*Spermophilus beldingi*), (D) domestic chickens (*Gallus*), and (E) forest guenons (*Cercopithecus cephus*). In the guenon, the social alert vocalization sounds very much like the chirp vocalization of mobbing passerines, and similarly attracts receivers. Their raptor alarm, however, lacks the high-frequency tonal quality of the raptor alarms of the other species. Nevertheless, this very soft, broadband, and low-frequency call is difficult to hear and localize, and causes receivers to quickly hide in dense vegetation. (A from Marler 1959; B from Leger and Owings 1978, © Springer-Verlag; C from Robinson 1980; D from Gyger et al. 1987, © Brill, Leiden; E from Gautier 1978.)

signals are derived from defense secretions and, in many cases, are the same chemicals. The chemicals themselves are volatile, single-chain ketones, aldehydes, acids, or hydrocarbons with 5 to 10 carbons and molecular weights between 100 and 200. Given in single puffs, the active space reaches a maximum diameter of 6 to 10 cm in 10 to 15 seconds, and fades in 30 to 60 seconds. Individuals perceiving a low concentration of the chemical at the periphery of the active space move up the concentration gradient, causing assembly. Individuals perceiving a high concentration are presumably at the source of the danger and become aggressive. If the disturbance persists, additional puffs of the odorant by other individuals will maintain the aggressive response and enlarge the active space of attraction. Some ant species have highly elaborate olfactory alarm systems in which a mixture of substances with different diffusion rates and active spaces produces concentric rings of oriented responses (Bradshaw et al. 1979; Prestwich 1985; Hölldobler and Wilson 1990).

ACROSS-FUNCTION COMPARISONS

Across-function comparisons can only be conducted in species that produce several functionally different signals in the same modality. Although there are many taxa that fit this requirement, few such studies have been conducted. The only study undertaken with the explicit goal of elucidating and comparing signal design rules for different social functions is the comparison by Alberts (1992) of olfactory signals in terrestrial mammals. Mammals employ olfactory signals for at least four different functions: mate attraction, recognition, alarm/threat, and territorial defense (range marking). Table 18.7 shows a summary of some key chemical characteristics for signals serving these different functions. Molecular weight is lowest for mate-attraction signals and highest for range marks, in agreement with the need for a volatile, quickly diffusing signal in the former case and a nonvolatile, long-lasting signal in the

latter. The number of chemical components is highest for range marks and recognition signals, in agreement with the requirement for chemical mixtures to generate individual signatures for these two signal functions. Range marks are significantly more likely to contain aromatic rings. Rings increase the stability and durability of chemicals that must withstand degradation from wind, sun, and heat. Finally, threat/alarm and range signals are more likely to contain a carbonyl functional group. Carbonyl chemicals are more water-soluble than carboxyl or hydroxyl functional groups. This feature facilitates rapid volatilization of a range mark when sniffed by a receiver and rapid fadeout of alarm/threat signals.

◀ **Figure 18.12 Avian and mammalian visual alarm signals.** (A and B) Visual flee alarms frequently consist of white patches on the tail that are exposed in a sudden flash when the animal takes flight. The white outer tail feathers of juncos (*Junco hymenalis*) and many other ground-foraging birds are flashed when the tail is spread on takeoff. Many mammals such as the deer (*Cervus*) shown here possess white rump and under-tail patches that are exposed when the tail is raised. (C and D) Alerting signals frequently involve regular, rapid-onset twitches of the tail as illustrated here in the groove-billed ani (*Crotophaga sulcirostris*) and ground squirrel (*Spermophilus richardsonii*). (E and F) Staring postures are visual signals that direct the gaze of nearby conspecifics to distant predators. The eye and crown stripes amplify the pointing behavior of birds such as the white-crowned sparrow (*Zonotrichia leucophrys*), and a stretched neck and forward-pointing ears are typical of mammals such as the Thomson's gazelle (*Gazelle thomsoni*). (G and H) Visual assembly signals often involve head movements. Flock-forming ducks, geese, and swans use a head-bobbing movement to coordinate group departure, and female caribou (*Rangifer tarandus*) bob their heads to draw in their offspring for suckling. (Various sources.)

Several auditory examples of across-function comparisons exist in addition to the classic alarm signal studies described above. Most stridulating orthopteran species produce three distinctly different calls—a mate-attraction call, a courtship call, and an aggressive threat call (Alexander 1961; 1968). Figure 18.13 shows a comparison of these calls in two different species. The mate-attraction call is a loud, temporally patterned, and stereotyped call. The temporal pattern contains the species-specific information. Once a female has approached and shows some interest in staying with the male, he begins to shift into the courtship call, which is soft, nearly continuous, higher in frequency, and less structured. If a male approaches, the rival call is given. The rival call represents an increase in duty cycle over the mate-attraction call, and the species-specific patterning is lost; the regularly repeated chirps are longer, have a more rapid onset and offset, and the frequency is often lowered. Antiphonal calling between the two rivals is a conspicuous aspect of the exchange, and repetition rate, note duration, and amplitude of the notes vary as a function of the distance between the males and the intensity of the conflict. These differences are well predicted by the design rules for these three functions.

Read and Weary (1992) attempted to correlate song complexity in passerine birds with factors such as habitat, migratory behavior, mating system, male parental care, reproductive rate, metabolic rate, and body size. They

Table 18.7 Chemical characteristics of olfactory signals serving different social functions

	Sex attraction	Recognition	Threat/alarm	Range marks
Molecular weight	91.0	140.1	189.9	208.1
No. of components	8.4	10.5	8.3	16.1
Aromatic ring present	11.9%	21.0%	14.5%	16.1%
Carbonyl group present	0.0%	18.3%	41.9%	28.0%

Source: Alberts 1992.

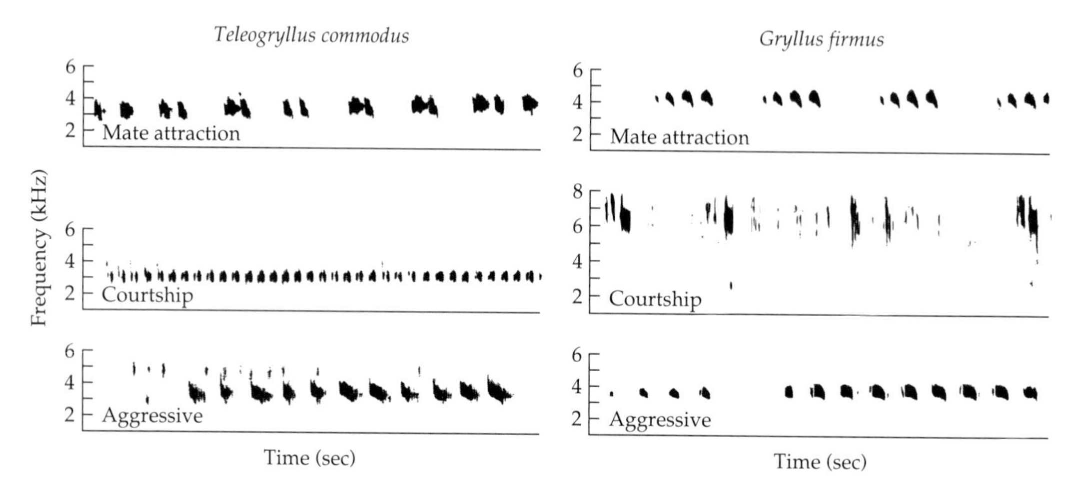

Figure 18.13 Mate attraction, courtship and aggressive calls in crickets. In *Teleogryllus commodus* and *Acheta firmus* (as well as other gryllids), the mate-attraction call, courtship call, and aggressive call exhibit characteristic differences in duty cycle, temporal patterning, and frequency that are consistent with the theoretical design rules for these functions. (From Alexander 1961, 1968.)

were particularly interested in the factors associated with songtype repertoire size (between-song complexity), syllable repertoire size (within-song complexity), and versatility (rate of switching between songtypes). The study surveyed 165 species and included corrections for phylogenetic relatedness. They found that large songtype repertoire was associated with male parental care and migratory behavior. This correlation could be caused by either a mate attraction or territory defense function for repertoires, since migratory species are under strong selection to both acquire territories and attract new mates quickly. Large syllable repertoire was associated with polygynous mating, implying a mate-attraction function. Versatility was higher in species living in forest habitat and in species with extensive male paternal care.

Plumage coloration in birds has been the subject of several reviews and comparative studies (Baker and Parker 1979; Burtt 1986; Butcher and Rohwer 1988). None of these studies specifically considers the design rules for color signals that convey functionally different messages. Such an exercise is difficult for two important reasons: (1) most color patches are always "on," so it is difficult to identify the specific context in which they are used, and (2) feather color, especially black and white colors, may be selected for important nonsignaling functions such as thermoregulation, feather durability, and prey capture (Hamilton 1973; Burtt 1986). Nevertheless, the location of conspicuous color patches on different parts of the body does seem to be associated with different social and ecological adaptations and argues for design rule patterns.

Table 18.8 shows Baker and Parker's summary of the conspicuousness of different body parts for males and females, breeders and nonbreeders, and juveniles, as well as the differences in conspicuousness between each sex, age,

Table 18.8 Average conspicuousness of different parts of the body in 516 species of western Palearctic (European, Middle Eastern and North African) bird species

Age/sex/ season class	Bill	Legs	Crown	Head	Back	Chest	Belly	Upper flash	Bottom flash
Breeding ♂	2.58	2.45	2.13	3.18	2.10	2.71	2.78	1.74	1.59
Breeding ♀	2.37	2.38	1.69	2.73	1.76	2.35	2.52	1.61	1.53
Juvenile	1.98	2.05	1.17	2.17	1.45	1.92	2.24	1.56	1.50
Winter ♂	2.53	2.40	1.96	2.99	2.01	2.54	2.71	1.70	1.56
Winter ♀	2.34	2.34	1.59	2.61	1.69	2.23	2.49	1.57	1.50
Differences between:									
Breeding ♂–♀	0.21	0.07	0.44	0.45	0.34	0.36	0.26	0.13	0.06
Breeding ♂–juvenile	0.60	0.40	0.96	1.01	0.65	0.79	0.54	0.18	0.09
Breeding ♀–juvenile	0.39	0.33	0.52	0.56	0.31	0.43	0.28	0.06	0.03
Winter ♂–♀	0.19	0.06	0.37	0.38	0.32	0.31	0.22	0.13	0.06

Source: Baker and Parker 1979.

Note: Breeding males, breeding females, juveniles, wintering males and wintering females are shown separately. Differences in mean conspicuousness between breeding males versus females, breeding males versus juveniles, breeding females versus juveniles, and wintering males versus females are shown in the lower four rows. Conspicuousness was scored on a scale of 0–5, with 0 = brown or brown grey, 1 and 2 = increased darkening or lightening of a uniform color, 3 = uniform black or white, 4 = increased color or patterning and 5 = use of red, yellow or contrasting hues. Upper flash refers to the presence of contrasting color on rump patches and wing bars, and bottom flash refers to the presence of contrasting underwing and undertail color patches.

and season classes. There are some obvious patterns: males are more conspicuous than females, adults are more conspicuous than juveniles, and breeding birds are more conspicuous than wintering nonbreeding birds. The head is the most sexually dimorphic part of the body, followed closely by crown and chest. Interestingly, the class differences in the conspicuousness of these three body parts are most marked when breeding males are compared to juveniles. This distribution of conspicuous coloration suggests that it expedites mate attraction and/or mate competition, as opposed to serving some other nonmating function.

Whether male plumage patterns are more related to mate attraction or male-male competition has proved to be a difficult question. The answer hinges on field studies concerning when coverable patches are used and on experimental covering of badges in natural contexts. Most studies so far indicate that dominance and territorial defense are more strongly affected by the addition or removal of color patches than mate attraction (Butcher and Rohwer 1988). This result may differ for different patches. The location of the most conspicuous patches on the front of the bird argues that these signals are designed to be directed at specific receivers, a feature that is most important in conflict resolution. In contrast, flashing colors on the wings and tails vary the least for the different age, sex, and season classes. Moreover, they are strongly associated with birds that spend more time on the ground or water surface, birds that forage gregariously in the nonbreeding season, and smaller bird species. These contexts all argue for a predator alarm or alerting function for this type of color patch in species that are the most vulnerable to predation.

A final outstanding taxon for the study of visual design rules is the cephalopod molluscs (Figure 18.14). They are rapid-pursuit predators with excellent vision and they communicate almost entirely in this modality, being too soft-bodied for sound production and too mobile for olfactory signals (except perhaps during mating). All species produce a large repertoire of color patterns, and these patterns can be changed in an instant owing to the muscle-controlled chromatophores (Figure 8.8C). The animals can appear all dark, very pale, mottled, barred, striped, and silvery, and body orientation and arm position can also vary. Cephalopods do not have color vision, so even though some displays show different colors, these are believed to serve a primarily cryptic function. Because of their soft bodies, the animals have little protection and rely on speed (jetting backward), crypsis, startle displays, and the inking cloud screen to escape from predators. There is no unitary antithetical meaning for dark versus light colors. Dark colors are often associated with aggressive intent, but maximum darkening is also seen during inking escapes. Pale body colors are often associated with fear, but a silvery display is used by males to ward off rivals during courtship. However, mixed black and white patterns all seem to express various types of ambivalence or mixtures of aggression and fear. Alarm signals always seem to involve a contrasting eye ring. Most species have a startle or protean display intended for predators that involves making the body very large and displaying conspicuous false eye spots. A few species produce a white, sex-specific, genital-gland color patch during courtship interactions. Moving waves of color bands seem to invite looking at the sender. Although few quantitative studies have been undertaken, the communication system is obviously as complex as that found in higher vertebrates (Hanlon et al. 1994; Hanlon and Messenger 1996).

In conclusion, design rules derived from our understanding of the social functions of signals go a long way to explain the diversity in animal signals. However, more comparative studies are greatly needed to uncover important

Figure 18.14 Design of visual signals in squid. A few of the many chromatic visual signals in the Caribbean reef squid *Sepioteuthis sepioidea*. (A) A courting pair, with the male (smaller individual in the foreground) showing the basic mottled pattern and the female showing the pied rejection pattern. (B) Lateral silver display by a courting male; female (not shown) is on the right, rival males (not shown) are on the left. (C) The startle display to predators, showing pale and flattened body, fine dark outlining around fins, and dark false eye spots. (D) The zebra stripe display with spread arms during an agonistic conflict between two males. The upper male in such conflicts is always darker, but the lower male usually wins the fight. (E) Longitudinal stripes shown during fast backward swimming. Note dark eye ring and light eyebrows. (F) Bar pattern with arms up, a cryptic pattern illustrating the disruptive coloration principle. (G) The all-dark pattern with light eye ring, often given during inking escapes. (H) The pale pattern with dark eye ring, also given in contexts of fear and alarm. (Moynihan and Rodaniche 1982; Hanlon and Messenger 1996; photos courtesy of Roger T. Hanlon.) ▶

(A)
(B)
(C)
(D)
(E)
(F)
(G)
(H)

broad trends in signal form in relation to social function and sender context. More experimental manipulation studies of color patches are needed to determine the functional context of color signals. Studies of auditory signals need to pay more attention to factors affecting duty cycle and repetition rate. In Part III we shall discover that additional aspects of signal form arise as a consequence of receiver/sender conflict and demands for signal honesty.

SUMMARY

1. The six key signal **design features** that must be optimized to achieve the information encoding and receiver recognition needs of a signal include: signal range, locatability of sender and/or receiver, duty cycle, identification level, modulation level, and form-content linkage. The combined design feature requirements for a signal serving a particular function constitute the signal's **design rules**. The **specific mechanisms** that can be employed to meet the design rules of a signal are dependent on the modality.
2. **Mate-attraction signals** require a large range, a high duty cycle, good sender locatability, and species specificity. The visual, auditory, and olfactory modalities possess adequate mechanisms to meet these requirements and are the most commonly found types of mate-attraction signal. Electric signals have limited range but can encode species specificity and provide mate recognition in muddy aquatic environments; tactile signals are very rare. Across-species comparisons show a variety of mechanisms for maximizing the differences and species distinctiveness of signals among closely related species, including frequency and temporal patterns for visual and auditory signals and chemical specificity for olfactory signals. Because mate-attraction signals provide information mainly about identity and location, they can be arbitrary in form and are designed for maximum range, transmission, efficiency, and detectability.
3. **Courtship signals** are usually given by males toward potential female mates at short sender-receiver distances. The signal range is therefore short to discourage eavesdropping, but males may increase the duty cycle to persuade females to mate. Signal form is often linked to content in order to indicate the male's amorous intentions and his suitability as a mate and/or parent. Visual signals are commonly used to direct the signal toward the intended receiver and point to a nest or oviposition site.
4. **Territorial defense signals** are designed to identify the boundaries and owner of a territory. They must contain individual signature information so that receivers can distinguish their neighbors from strange intruders. Visual and auditory signals are broadcast at long ranges to advertise the presence and location of the owner on its territory. Olfactory signals consist of long-duration, surface-deposited marks concentrated around the territory's boundary. Territorial signals can be arbitrary in form since they denote only species and individual identity.

5. The visual modality is most commonly used for **threat signals** because it can (a) provide information on body size and intentions, (b) encode information on motivational or excitational level via repetition rate, (c) allow rapid changes between different threat signals and reciprocal exchanges between opponents, (d) be made directional to specify the opponent in social group situations, and (e) provide tactical advantages. The auditory modality can (a) provide some information on body size, (b) encode the level of aggression and motivation by using frequency and temporal modulation, and (c) permit directionality of the threat via antiphonal calling. However, the variety of vocal threats in a species' repertoire is rarely as rich as its visual threat repertoire. The olfactory modality lacks the ability to encode subtle and rapid changes in threats during conflicts, but this modality can provide direct information on dominance status, general aggression level, and slow changes in arousal or excitational level.

6. All three modalities can be effectively designed to signal the broad range of **alarm signals** from flee to alert to assembly, and all have mechanisms to vary the intensity of the signal as a function of the degree of danger. However, the modalities clearly differ in the usual environmental constraints, the maximum range or active space, and the amount of information they can encode. Auditory alarm signals show the greatest complexity because they can employ repetition rate, frequency, bandwidth, and AM and FM to encode information such as the type of danger, the urgency of the desired response, and the fear, distress, or aggressive motivations of the sender. Visual signals can provide directional information about the source of danger. Olfactory signals provide simple, yet effective, alarm for animals living in very close contact.

FURTHER READING

The classic articles by Hockett and Altmann (1968), Otte (1974), Wilson and Bossert (1963), and Marler (1961) are well worth reading for their important insights. The more recent articles presented at a symposium on animal communication by Dawkins (1993) and Endler (1993a) reveal somewhat divergent views about signal design. Krebs and Davies (1995) take a signal design approach in the communication chapter of their behavioral ecology textbook. The two voluminous books by Sebeok (1968, 1977) provide endless examples of animal signals across all taxa, along with books on selected signal functions and taxa such as Ewer (1968), Peters (1980), and Macdonald (1984) on mammals; Wilson (1971) and Hölldobler and Wilson (1990) on insects; Catchpole and Slater (1995) and Kroodsma and Miller (1982, 1996) on birdsong; Bastock (1967) on courtship; and M. Andersson (1994) on mate attraction.

Part III

Game Theory and Signaling Strategies

We now turn to the dark side of animal communication: deceit, conflict, and subterfuge. In prior chapters, we assumed that senders and receivers shared a common interest in communicating; here, we make no such assumption. This complicates our prediction of optimal strategies since it is not clear which party will win a game in which each prefers that the other adopt a strategy not in its own interest. The appropriate optimality analysis for conflict is **evolutionary game theory**, which we introduce in a general way in Chapter 19. The types of problems posed for animal communication by evolutionary game theory are taken up in Chapter 20. Can receivers force senders to be honest and how might they do so? Can senders manipulate receivers? The remaining chapters revisit the basic types of communication that we discussed in Part II. However, these are now ranked in order of decreasing conflict of interest between sender and receiver. In each case, we examine the consequences on the communication system of the degree of sender-receiver conflict. What kinds of mechanisms can receivers demand to ensure honesty? In Chapter 18, we found that a number of signal design features are not easily explained by the simple optimality models of Part II. To what degree do game theoretical arguments and the demand for honesty help explain these remaining puzzles? Finally, we end Part III, and the book, by looking at what evolutionary forces are most likely to limit the amounts of information conveyed by signals. Are limits due to physiological constraints or to insufficiently common interests between senders and receivers?

As we shall see, conflict of interest is a major determinant of the accuracy and detail exchanged in animal signals, and thus of the sophistication of the system.

Chapter 19

Evolutionary Game Theory

WE SAW IN PRIOR CHAPTERS THAT WHICH ALTERNATIVE STRATEGY is evolutionarily optimal depends on the contexts. Contexts set the limits, define the tradeoffs, and determine the costs and benefits of each strategy. We have also seen that contexts do not have to be fixed for there to be an optimal strategy, and that even random variation can be taken into account in defining strategy payoffs. However, there is one type of contextual variation that is not so easily accommodated with the optimality methods we have outlined so far. This occurs when the critical contexts include other living organisms. Other living organisms are unlikely to ignore the strategy choice of a focal animal. Instead, they base their subsequent behaviors on the focal animal's choice, and this will change the context for the focal animal. The focal animal may then have to switch to a new optimal strategy, and this will evoke another round of behavioral responses by the surrounding organisms. This coevolutionary interplay between different organisms may be envisioned as a game. Whether the game stabilizes or just keeps cycling through different strategies can often be predicted by invoking **evolutionary game theory**. This method specifically examines optimization when contexts include variable responses by other organisms.

Animal communication is often such a game. A strategic choice by either sender or receiver can change the context for the other and thus alter its optimal strategy. When the optimal choice of strategy by one party is different from what the other would prefer it chose, we say that the two parties have dissimilar interests. The more dissimilar the interests of sender and receiver, the more likely that a game theoretical analysis will be required to predict communication behaviors. In this chapter, we summarize the general methods for analyzing and predicting outcomes of evolutionary games. We shall make extensive use of these methods in the remaining chapters of Part III.

BASIC PRINCIPLES

Evolutionary game theory is based in part on the game theory of modern economics, but it has been revamped to fit evolutionary situations. Modern economic treatments presume that there are at least two **players**, each of which has access to several alternative **strategies**. Each player tries to anticipate the strategy that their opponent is most likely to choose, and then selects the best response to that opponent strategy. Evolutionary game theory takes a somewhat longer term and macroscopic perspective. Instead of focusing on individual players, it deals with the strategies available to particular **roles** within a population of animals. Alternative roles might be males versus females, younger versus older animals, or smaller versus larger individuals. A certain set of alternative strategies is available to each role, and initially, different individuals are likely to adopt different strategies. The individual that adopts the best response to the actions of individuals in other roles will have the highest fitness. If strategy choice is at all heritable (whether by genetic or cultural mechanisms), this best response will become the most common strategy for that role over evolutionary time. The game will stabilize when animals in each role find the best response to the best responses of all of the other roles. When this is true, it does not pay members of any role to switch strategies. In fact, any mutants that revert to nonoptimal strategies will be selectively eliminated.

Suppose such a system can stabilize. For each role, the best response is called an **evolutionarily stable strategy** (or **ESS**). This is a strategy which, when common among members of a particular role, cannot be invaded or displaced over evolutionary time by any rare alternative. At evolutionary stability, each role will have its own ESS, and the propriety of that ESS will depend critically on the continued persistence of the ESSs of all the other roles in the game. If individuals playing the dominant strategy for any role should accidentally become rare (e.g., due to some chance catastrophe), this may reduce the suitability of the ESSs for other roles, and the system could then evolve to a different equilibrium. For any role, there may be several possible ESSs. Which one is found at a given equilibrium will depend upon which ESSs were selected by other roles. As an example, we will show in the next chapter that conflicts between sender and receiver roles may evolve to either of two equilibria—one in which the sender ESS is to cheat and lie and the receiver ESS to discount or ignore signals, and another in which the sender ESS is honest signaling and the receiver ESS is attentive responsiveness. Which equilibrium is seen may depend only on historical accident. It is also possible that some systems never reach an equilibrium. Each change in strategy by one party leads to a consequent change in strategy by the other, and the system never stabilizes. This lack of stability can arise because there are no equilibrium combinations, or because the evolutionary trajectory of the system fails to lead it to one of the potential equilibrium combinations. The latter condition is less likely with animals than with rigid mathematical models. Animals make mistakes and this causes most combinations of strategies to occur at some point over evolutionary time. In fact, some game models have no reachable equilibria *unless* the players make mistakes!

Our task is to identify the ESSs for various evolutionary games. If we have modeled a game correctly, the ESSs should be the ones predominantly observed in nature. Because achievement of an equilibrium is such a complex and interactive process, the outcomes of these games are not always intuitively obvious. Not surprisingly, evolutionary game theory often makes predictions that are quite different from those of the noninteractive optimality methods of Part II. When animals are at all in conflict, game theory is usually the better approach. The invocation of game theory is complicated because there is no single method to find the ESSs for all evolutionary games. To help sort out methods and applications, we first classify games according to a few basic criteria, and then use the category of game to select the method of analysis. Game classification has a second benefit. It clarifies the simplifying assumptions that went into the game model and thus the propriety of the game for any real animal example. It does not make sense to expect a particular ESS in nature when the assumptions on which the game model is based are totally inappropriate.

CLASSIFYING EVOLUTIONARY GAMES

There are four independent criteria that we can invoke to classify any given evolutionary game. Different combinations of values for the four criteria define different game categories. Each category will have its own methods for identifying ESSs, its own set of assumptions, and its own types of ESSs.

Type of Strategy Set

The first criterion for classifying games involves the nature of the strategy set. The strategy set includes a list of alternative behaviors or anatomical structures which each individual in that role could adopt. As we discussed on pages 369–371, these alternative strategies can either be **discrete** (e.g., produce a red tail or a green tail), or **continuous** (e.g., modulate a sine wave at some rate between 50 Hz and 453 Hz). In either case, the strategy set may also include mixtures of the alternative behaviors or structures (e.g., sing a 60 Hz song 20% of the time and a 130 Hz song the remaining 80% of the time). A strategy that consists of a single behavior or structure is called a **pure strategy**, and one that consists of some mixture of several behaviors or structures is called a **mixed strategy**. There are two ways that a particular role might achieve a mixed strategy (also called a **polymorphism**). A **genetic polymorphism** occurs when each member of the role performs only one strategy, but different subsets of the role perform different strategies. For example, suppose 40% of male moths in a given species emit organic acid A as a pheromone, whereas the remaining 60% emit organic acid B. No male can emit both. Females, which represent the opposing role, perceive the male strategy as a 2:3 mixture of the two pheromones. The second way to achieve a mixed strategy is in the form of a **behavioral polymorphism**. Here each individual performs a mixture of alternative strategies. Continuing our moth example, suppose all males emit pheromone A 40% of the time and pheromone B 60% of the time. Females still encounter male pheromones with a 2:3 ratio, but the mechanism by which the mixture is achieved is different. While it usually does not matter to opponents whether a mixed strategy is generated through a genetic or behavioral polymorphism, evolutionary trajectories often differ for the two mechanisms, and this affects the likelihood that a given role will ever achieve a particular mixture. For example, inheritance patterns might allow behavioral polymorphisms but prohibit genetic ones. Strategy sets will thus differ in the degree to which mixtures are options.

Role Symmetry

The second criterion for game classification is role symmetry. The opponent players in a **symmetrical game** have access to identical strategy sets, have equal chances of winning when playing the same strategy, and have identical payoffs from winning a particular exchange. The players are thus entirely interchangeable; or put another way, there is really only one role. Two identically aged male baboons exchanging vocalizations over access to a single sleeping site are likely involved in a symmetrical game. In an **asymmetrical game**, there is more than one role, and each may have access to different alternative strategies, different probabilities of winning with a given strategy, different payoffs when they win with a given strategy, or some combination of these conditions. Asymmetric games are typical of interactions between male versus female, young versus old animals, or big versus small players. A vocalizing contest between a large adult male baboon and a small juvenile over a sleeping site is unlikely to be a symmetrical game as the probabilities of winning a fight are surely unequal, each individual is likely to be able to adopt strategies that the other cannot, and the payoffs of winning may have different fitness consequences for each party. A single exchange of animal communication is always an asymmetrical game because sender and receiver do not have identical strategies. If there are multiple exchanges, and the parties alternate roles as sender and receiver, then the overall sequence can be symmetric.

Payoff Frequency Dependence

The third criterion depends upon the number of relevant opponents that a player must face at a time. If there is only one relevant opponent at any time, the game is called a **contest**; if there are more than one, the game is a **scramble** (also called an ***n*-person game** or **playing against the field**). The critical word here is relevant. An animal may meet multiple opponents over time, but if the chances of winning and the average payoffs of alternative actions during any one exchange only depend on what the current opponent elects to do, the game is a contest. As a result, the possible payoff values for a contest are fixed. They are unaffected by the frequencies with which opponents adopt particular strategies. In contrast, imagine a situation in which the payoffs of using particular strategies against an opponent's action depend upon how many potential opponents have adopted the same strategy. Here, the possible payoff values are not fixed but depend on the current frequencies with which alternatives are being used in the population. Frequency-dependent payoffs imply scramble games. The distinction between contests and scrambles is easier to see with examples. Consider two similarly sized chicks that are fed by their parents. If neither begs, each has an equal chance of receiving the worm. The average payoff is thus a half a worm. If one begs and the other does not, the beggar gets the entire worm and the silent chick gets none. If both beg, each again has an average payoff of half a worm. As long as the parents exclusively feed their own chicks, the payoffs to each chick in this game depend only on what the other chick in the same nest elects to do. Average payoffs are fixed at 0, 0.5, and 1.0 worms and this is thus a contest. Now consider a different species in which nests routinely contain T chicks where $T > 2$. When the parent arrives, N of these decide to compete for the worm, and $T - N$ elect not to. If the begging competitors are equally matched, the average payoff of begging is $1/N$ of a worm. Clearly, this payoff depends upon N, the number of nestlings that elect to beg. Payoffs here are frequency-dependent and the game is thus a scramble.

Sequential Dependence

The final criterion has to do with the degree to which successive interactions should be treated as independent games or not. All animals face a sequence of decisions during their lives. Even during a single fight, courtship, or parent-offspring interaction, there may be several successive points at which each player must make a strategic choice. It is critical to know whether the outcome of an earlier choice of strategy does or does not affect the suitability of alternative strategies at a later decision point. If the earlier decision has no bearing on the later one, then we can treat each decision in the sequence as if it were an independent game. If the outcomes of early decisions do constrain later choices, then we must think of the entire sequence as the game, and each decision as a **bout** within that game. Such sequences are called **dynamic games**. An ESS is then an optimal choreography of successive strategy choices. This choreography will be fixed if the opponent's choreography can be anticipated, or conditional if not. For example, if the opponent does A in the first bout, the best response in the second bout is strategy 1; if the opponent adopts B, then do strategy 2. In some dynamic games, one can identify an **evolutionarily stable policy** (Houston and McNamara, 1987; Mangel 1990), a guiding rule of thumb to find the optimal strategy at each point given current conditions. Consider two male crickets that periodically try to take over the same mating burrow. If they have no signals that allow for individual recognition, neither knows on successive occasions whether its opponent is a new individual or one it defeated (or was defeated by) on a previous occasion. Each contest may then be treated as an independent game, and the optimal strategy for both crickets might be to threaten the other at the same level on each occasion. If, however, they can recognize and remember each other, then a prior winner would benefit from escalated aggression in subsequent interactions whereas a prior loser would do best to avoid confrontation. The latter series of interactions can only be analyzed as a dynamic game. An appropriate ESS policy would adjust the aggressive levels of signals in successive bouts according to the fraction of prior wins (or losses) with this opponent.

Combinations of Criteria

Games are thus classified according to their values for each of these four criteria. One game might best be modeled as a discrete symmetric contest in which successive interactions are independent; another might be seen as a continuous asymmetric scramble with long-term sequential dependence. For each combination of criterion values, there are specific types of ESSs possible and specific methods for finding these ESSs.

In the remainder of this chapter, we shall examine examples of games from several of the possible categories. For the simpler cases, we shall outline how one finds the ESSs and what kinds of ESS solutions one might expect. In all cases, we shall show how the ESS analysis provides insights into the evolution of animal behavior and frequently clarifies otherwise inexplicable observations from nature. Not all of the classic examples reviewed in this chapter involve animal communication: however, most do, and those that do not will be easily adapted to signaling situations in later chapters.

SOME EVOLUTIONARY GAMES AND THEIR ESSs

General Analysis of Discrete Symmetric Contests

Discrete symmetric contests that are not sequentially dependent are the most easily analyzed games. There is only one role, and the question is which strategy should be invoked when members of this role play the game against each other. The simplest case involves two players, each of which has access to only two strategies. Because the game is symmetric, the two available strategies are the same for the two players. In such a 2×2 discrete symmetric game, there are four possible combinations of plays and thus four possible payoffs to each player. These can be tabulated in a **payoff matrix** (Figure 19.1). One player is arbitrarily assigned the focal status, and the matrix then lists the payoffs it would receive given each strategy it can play and each strategy its opponent can play. For a symmetric game, it does not matter which player is selected as the focal individual.

To find the ESSs in this game, we can use the **dot method**. First, we look at the left column in this matrix. In this column, the opponent plays strategy 1.

		Opponent plays	
		Strategy 1	Strategy 2
Focal player plays	Strategy 1	PO_{11}	PO_{12}
	Strategy 2	PO_{21}	PO_{22}

Figure 19.1 Payoff matrix for 2 × 2 discrete symmetric contest. Because the game is symmetric, the two players are interchangeable and we only need to list one payoff for each pair of strategies played. The convention is to list the playoffs to the player on the left. Payoffs are given in the same units for all cells in the matrix and denoted here by PO_{ij}. The first subscript (*i*) is the strategy played by the focal player and the second (*j*) is that of its opponent.

What is the best response of the focal player to this move? Assuming that the payoffs (denoted PO_{ij} where i is the focal player's strategy and j the opponent's) are in units of fitness and that natural selection will tend to maximize fitness, the best response by the focal role is strategy 1 if $PO_{11} > PO_{21}$, and strategy 2 if $PO_{11} < PO_{21}$. We put a dot where the value is larger in this column. If the two payoffs are equal (e.g., there is a tie), then it does not matter which strategy is used as a response, and we put dots in neither cell. We then look at the second column in which the opponent will play strategy 2. In this case, we compare PO_{12} to PO_{22} and place a dot in that cell with the larger value. If there are no ties in either column, the outcome has to be one of the four cases shown in Figure 19.2.

In Case I, the best response is always strategy 1 regardless of which strategy is employed by the opponent. Strategy 1 is then the ESS, and because it consists of a single strategy (as opposed to a mixture of strategies), we call it a **pure ESS**. In Case II, the best response is always strategy 2. We can then call strategy 2 the pure ESS. If there is a tie in one of the two columns, then that column exerts no selective force on the choice of strategy by the focal player. The row with a dot in the untied column is then the ESS. If both columns are tied, then there are no selective forces on strategies and hence no ESSs. When a strategy is a pure ESS, we expect this strategy to be the only one commonly seen in the population.

Case III is a bit more interesting. In this case, the best response is the strategy not used by the opponent. Imagine a population in which all players only performed strategy 1. A mutant animal that hit on strategy 2 would have higher fitness than the majority playing strategy 1, and strategy 2 would increase in frequency with successive generations. Similarly, if we started with a population in which all played strategy 2, a mutant performing strategy 1 would be able to invade, and this strategy would become increasingly common over time. If strategy 2 is better when strategy 1 is common, but strategy

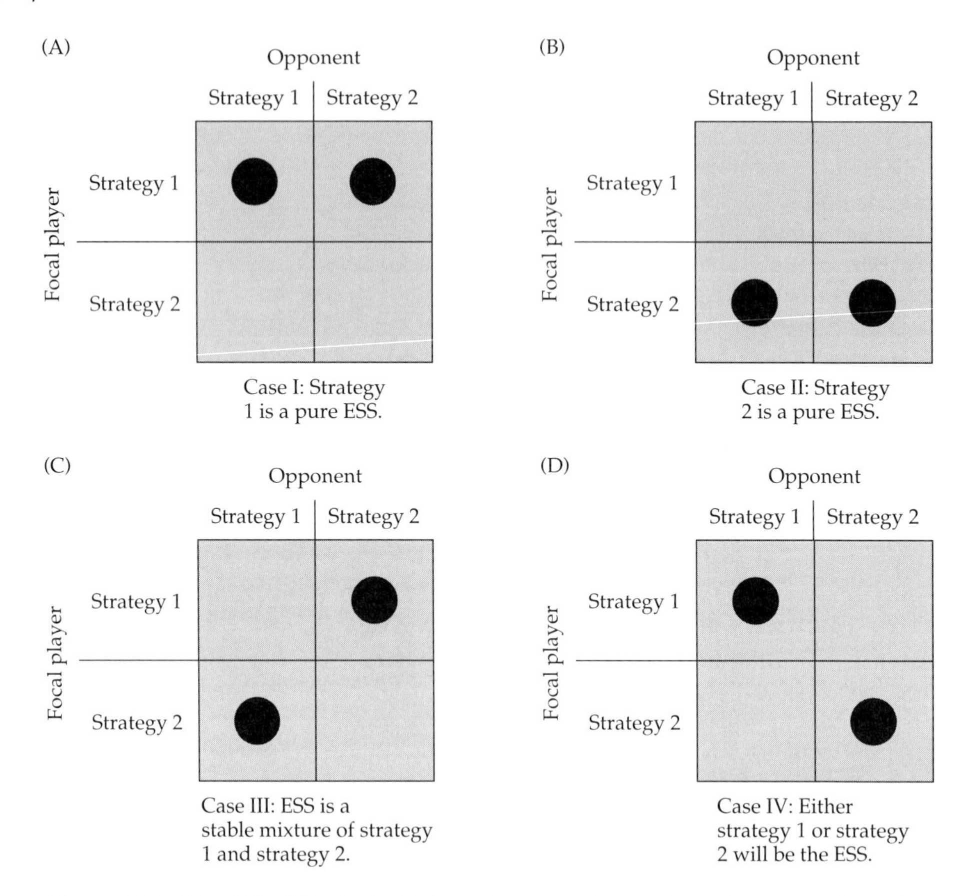

Figure 19.2 Possible ESSs in 2 × 2 discrete symmetric contest. For each column in the payoff matrix, a dot is placed in that cell conferring the best payoff to the focal player. There are four possible outcomes. (A) Case I: Strategy 1 is the best response regardless of what the opponent plays. Strategy 1 is thus a pure ESS. (B) Case II: Strategy 2 is the best response regardless of the opponent's play. It is then a pure ESS. (C) Case III: The best response is always the strategy opposite to that played by the opponent. If mixtures are allowed, the ESS is a fixed mixture of strategy 1 and strategy 2. (D) Case IV: The best response is to match the play by the opponent. There are two ESSs. Which strategy is the ESS depends upon the initial frequencies of the two strategies in the population.

1 is better when strategy 2 is common, it seems obvious that there ought to be some intermediate mixture of strategy 1 and strategy 2 at which neither has an advantage over the other. Any time the population drifts away from the optimal mixture, the rarer strategy will have an advantage and will increase to bring the population back to the ESS. The system is thus stable even in the face of some drift in frequencies.

What would the ESS mixture be? Suppose we let f be the fraction of the population exhibiting strategy 1 at any time. A focal player adopting strategy 1 will encounter opponents playing strategy 1 a fraction f of the time and will receive a payoff of PO_{11}. It will also encounter opponents playing strategy 2 a fraction $(1 - f)$ of the time and will receive a payoff equal to PO_{12}. Recalling the methods for calculating average payoffs detailed in Chapter 12, we can compute the average payoff to a focal player playing strategy 1 as

$$\overline{PO}_1 = f \cdot PO_{11} + (1 - f) \cdot PO_{12}$$

Similarly, if the focal player were to play strategy 2, its average payoff would be

$$\overline{PO}_2 = f \cdot PO_{21} + (1 - f) \cdot PO_{22}$$

We have argued that there must be some value of f, denoted by $\hat{f}$, at which equilibrium occurs and thus it does not matter which strategy you play. This means that $\overline{PO}_1 = \overline{PO}_2$ when $f = \hat{f}$. Setting the above equations equal to each other and solving for $\hat{f}$, we get that

$$\hat{f} = \frac{PO_{12} - PO_{22}}{(PO_{12} - PO_{22}) + (PO_{21} - PO_{11})}$$

In words, the fraction of the time we should see strategy 1 at equilibrium is equal to the absolute difference between the values in the right-hand column of the matrix (the one occurring with frequency $1 - f$), divided by the sum of the absolute differences between the values in each column. Clearly, this must be some fraction between 0 and 1.

As we noted earlier, an ESS mixture might be achieved by a genetic polymorphism, in which $\hat{f}$ of the players always play strategy 1 and $(1 - \hat{f})$ of the players always play strategy 2, or by a behavioral polymorphism in which each player performs strategy 1 a fraction $\hat{f}$ of the time and performs strategy 2 the remaining $(1 - \hat{f})$ of the time. As far as the analysis is concerned, the two solutions are equivalent. In real life, not all genetic systems will allow the establishment of a genetic polymorphism at any given frequency $\hat{f}$. Genetic impediments are less likely for behavioral polymorphisms.

Case IV is the reverse of Case III. The optimal choice is to match your opponent's choice of strategy. If we imagine an initial population all performing strategy 1, we can see that strategy 2 is always at a disadvantage and thus cannot invade. If we begin with a population all playing strategy 2, strategy 1 is now the less favored option. Again, it seems likely that there is some intermediate mixture of the two strategies at which payoffs of playing either are equal. In fact, the calculation of $\hat{f}$ for Case IV by setting $\overline{PO}_1 = \overline{PO}_2$ gives exactly the same result as for Case III. However, in this case the equilibrium point is not stable. Suppose one begins with a population at this equilibrium mixture. Should the frequency of players adopting strategy 1 drift upward so that $f > \hat{f}$, strategy 1 is now the better strategy and subsequent evolution will drive the system toward 100% of the animals using strategy 1 (e.g., to $f = 1$). If f had drifted lower so that $f < \hat{f}$, strategy 2 would be favored and strategy 1 would disappear (e.g., $f = 0$).

Case IV thus has two ESSs—either pure strategy 1 or pure strategy 2. Which is observed in the short term depends upon whether the initial frequency of strategy 1 is greater or less than $\hat{f}$. Over the long term, most population frequencies drift even under moderate selection pressures. Although one strategy may be most common and thus the ESS at the start, drift may eventually push f past $\hat{f}$, causing the system to flip to the other ESS. The ESS most likely to be found over the long term is the strategy that is least common at the equilibrium point. The logic is as follows. Suppose $\hat{f}$ is less than 0.5; that is, the frequency of strategy 1 at the equilibrium is less than that of strategy 2 [e.g., $\hat{f} < (1 - \hat{f})$]. If we begin with a population at the strategy 2 ESS ($f \approx 0$), strategy 1 players will appear only as rare mutants (genetic heritability) or innovators (cultural heritability). If $\hat{f}$ is a small value, only a few such mutants are needed before f exceeds $\hat{f}$. Once this occurs, strategy 1 becomes the favored strategy, and the system switches to a strategy 1 ESS. By the same token, if we begin with a strategy 1 ESS ($f \approx 1$), it will take a very large number of strategy 2 mutants, since $\hat{f}$ is small, before f can drift below $\hat{f}$ and cause a switch to strategy 2. The strategy that is least common at the equilibrium is thus the one which is most likely to invade by chance if the other is the ESS, and the one least likely to suffer invasion when it is the ESS. As we saw above, the value of $\hat{f}$ is determined entirely by the relative payoffs in the matrix.

A 2×2 discrete symmetric game with no ties always has at least one ESS. Discrete symmetric contests with more than two alternative strategies may have no ESS, one ESS, or several. Solving for the ESSs in such games is more difficult than in the 2×2 case and is best undertaken using matrix algebra (see Haigh 1975; Hines 1980, 1987). However, one can sometimes solve games with small matrices by inspection.

A Discrete Symmetric Contest: Hawk versus Dove

One of the classical evolutionary games is hawk versus dove (Maynard Smith 1982a). The biological context is a conflict between two individuals over some indivisible commodity such as a single food item, a mate, or a roost. Winning the commodity increases the fitness of the winner. The naive expectation might be that the two animals should always fight over the commodity. The observation in nature is that animals often use relatively low-risk display signals to decide who should get the commodity; one animal then leaves without a fight. It is not immediately clear why the two animals do not always fight over the commodity. Hawk versus dove is a simplified game invoked to explain these observations. It assumes that all contestants are equal (making the game symmetric) and allows for only two discrete strategies: fight (hawk) and display peaceably (dove). Since only two players are allowed to encounter the same commodity at a time, the game is a contest instead of a scramble. Players have no memory and may never meet again; thus this is not a dynamic game.

Given these assumptions, the payoffs are straightforward. When two hawks meet, they immediately fight over the commodity. Whereas the victor gets V fitness units (based upon the value of the commodity), the loser suffers fight injuries and other costs of defeat equal to $-D$ fitness units. Since the game is assumed to be symmetric, each hawk wins half of its battles with other hawks. The average payoff to each party when two hawks meet is thus $1/2V + 1/2(-D) = 1/2(V - D)$. When a hawk meets a dove, the hawk becomes aggressive and the dove runs away. The hawk thus gets V and the dove gets

0. When two doves meet, they use some costless exchange of displays to decide who gets the commodity and who leaves peacefully. It is assumed that the signal exchange leads to a random winner. The average payoff when two doves meet is thus $1/2V + 1/2(0) = 1/2V$. The payoff matrix is shown in Figure 19.3.

First examine the right-hand column of the matrix in part A of Figure 19.3. Because any $V > 0$ is always greater than $1/2$ V, we can put a dot in the upper cell of the right-hand column. Where the dot is placed in the left-hand column depends upon whether V or D is the greater number. If the benefit of winning control of the commodity, V, is greater than the absolute value of the cost of losing a fight, D, then the animals should always be hawks and always escalate to full fighting. Hawk is then a pure ESS when $V > D$. If instead the commodity is worth less than the cost of losing a fight (e.g., if $V < D$), then the left-hand column dot goes in the lower cell and this becomes a Class III situation. As we have seen, such situations lead to stable mixtures of strategies (if allowed given the heritability mechanism). If f is the frequency of hawks in the population at any given time, then an equilibrium occurs when

$$\hat{f} = \frac{V - \frac{1}{2}V}{\left(V - \frac{1}{2}V\right) + \left[0 - \frac{1}{2}(V - D)\right]} = \frac{V}{D}$$

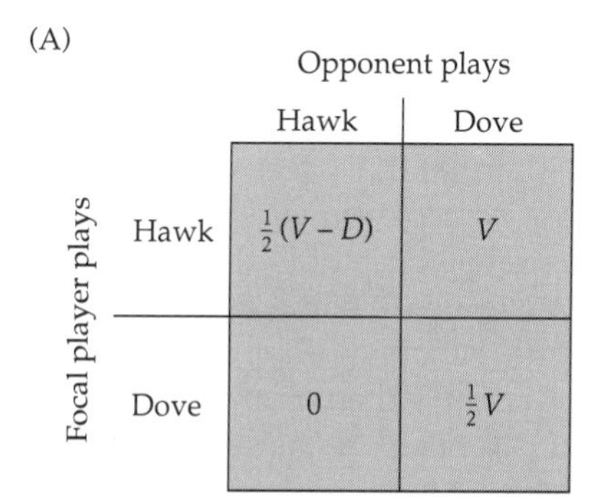

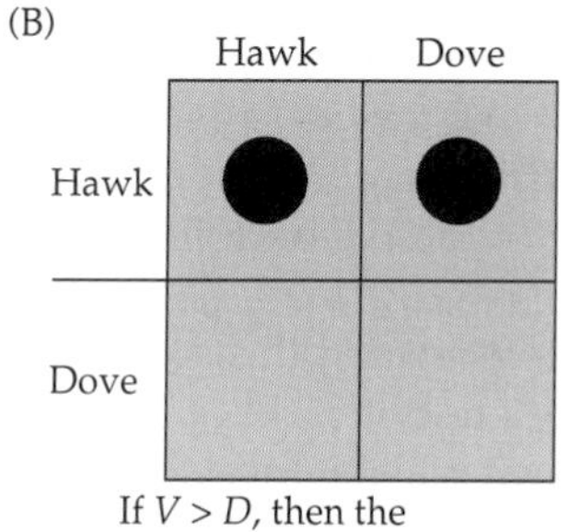

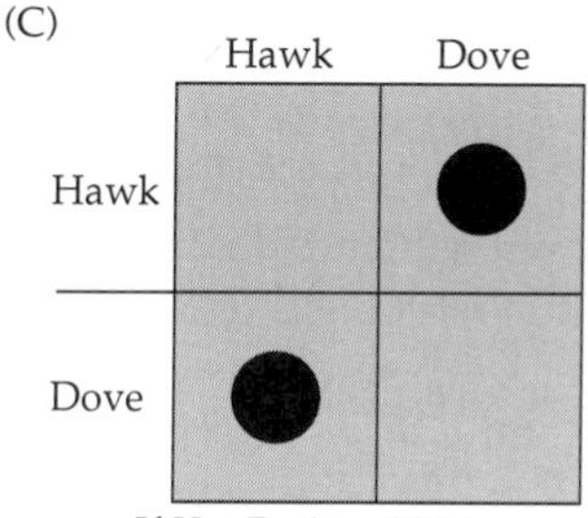

Figure 19.3 Hawk versus dove game. (A) Payoff matrix for hawk-dove game. V is the fitness benefit of winning a contest over some some commodity. $-D$ is the fitness cost of losing a prolonged fight with a hawk. The game is assumed to be a discrete symmetric contest with no sequential dependence. (B) Dot method identification of ESS when $V > D$. Hawk is a pure ESS (Case I). (C) Dot-method identification of ESS when $V < D$. ESS is a stable mixture (Case III) in which V/D of the players should play hawk and $(1 - V/D)$ play dove.

This makes sense because if $V < D$, then V/D is some fraction between 0 and 1. At the ESS, V/D of the animals should play hawk and $(1 - V/D)$ should play dove. As V decreases relative to D, more animals should settle conflicts with peaceable displays instead of a risky fight.

This is a very simplified game model. However, it shows the power of evolutionary game theory by explaining an observation that might otherwise seem counterintuitive. One can make this game much more complex (as we shall do in subsequent chapters), but the same basic conclusion emerges, namely that it does not always pay to escalate a fight. The use of signals or some other conventional way to resolve a conflict may well be favored at least part of the time.

More Discrete Symmetric Contests: Take Games

One of the major revelations of modern behavioral biology is the recognition that most animals are selfish. Even apparent cooperation often turns out to be selfish in ways that we did not realize immediately. This is in contrast to the expectation in one popular school of modern economics in which it is expected that civilized humans will act cooperatively for the common good. Recently, economists have begun to question whether humans are that different from animals, and several have turned to evolutionary game theory to better explain human behavior (e.g., Crawford 1989, 1990, 1995; Friedman 1991).

Whether evolutionary game theory has new perspectives to offer modern economists or not, it is abundantly clear that most animals act selfishly (Dawkins 1989). We can see why by invoking some simple discrete symmetrical contest games. The first is called a **take game**, and it is one which turns out to be applicable to many natural situations including animal communication. There are again two strategies: **passive** and **cheat**. A passive animal minds its own business. A cheat, on the other hand, increases its own fitness at the expense of the fitness of others. For example, consider gulls and terns fishing at the seashore. Some birds concentrate on catching their own fish (here the passive strategy). Others (the cheats) give up some of their own fishing time to monitor the success of other birds. When another catches a fish, cheats chase it until the prey is dropped and can be stolen. Suppose a passive bird catches P fish each day and that a cheat can take B of these fish away from the fisher. To chase and steal these fish, the cheat must reduce its own time spent fishing. As a result, instead of catching its usual P fish daily, it will only catch $P - C$ fish on its own. When two passive birds meet, they each mind their own business and each gets a daily payoff of P. When a cheat and a passive bird meet, the passive bird loses B fish to the cheat. Its payoff is then $P - B$. The cheat catches $P - C$ of its own fish and steals B fish from the passive bird. Its payoff in this situation is thus $P - C + B$. Because the game is symmetric, half of the time that two cheats meet, one steals B fish from the other, and the remaining half of the time this same bird has B fish stolen from it. Both still only catch $P - C$ fish by themselves. The average payoff to a cheat when it meets another cheat is then $P - C + 1/2\ B - 1/2\ B = P - C$. The payoff matrix is shown in Figure 19.4.

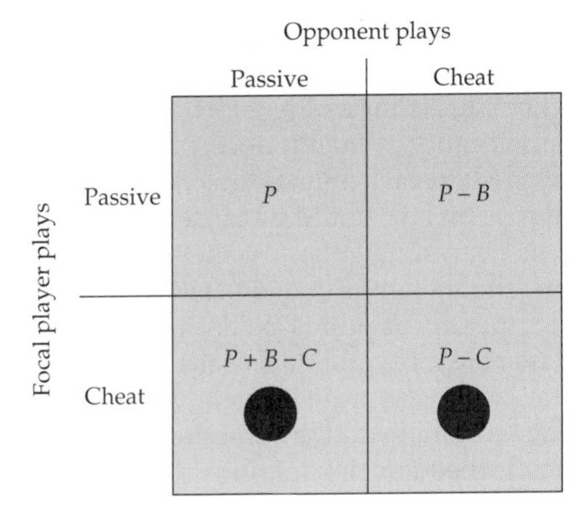

Figure 19.4 Payoff matrix for a take game. Passive animals mind their own business. P is the payoff when both players mind their own business. Cheats steal an average of B fitness units from other players at a cost to themselves of $-C$ fitness units. Dot analysis indicates that cheat is a pure ESS as long as $B > C$.

The ESS in this game depends on whether a cheat can steal more fish (B) than it loses by not hunting itself (C). If it can, i.e., if $B > C$, then $P + B - C > P$ and the appropriate dot in the left column goes in the row for cheating. Similarly, when $B > C$, then $P - C > P - B$ and thus the dot in the right column also goes in the row for cheating. The result is that cheating is a pure ESS when $B > C$. When cheating costs more than it pays (i.e., $B < C$), then passive is the pure ESS.

There are some serious lessons to be taken from this example. Note that were all birds to remain passive, they would each receive a payoff P, which is clearly better for the average bird than the ESS payoff of $P - C$. Despite this, once any cheats appear in a population for which $B > C$, the population will evolve to the ESS in which everyone cheats. Evolution will thus have lowered the average fitness of the population. This is a nonintuitive outcome that may surprise readers who assume that evolution generally improves the average fitness of populations. Where evolution can be modeled by passive optimality (the assumption of Part II in this book), improvements in mean fitness are common. In contrast, many evolutionary game models lead to lower average fitness, and this simply reflects the costs of competition. Examples abound in nature. Arms races both within and between species are take games. Players have to invest in costly defenses or aggressive displays ($-C$) just to protect themselves from competitors or predators. Were the aggressors absent, the defenders would be able to redirect those costs into reproduction and enjoy higher average fitness. Sexual selection is a similar take game. Males that invest in costly displays to attract females or in aggressive weapons to chase off competitor males will have a mating advantage over those that do not. The result is an ESS in which all males have to invest if they want to keep up with the competition, but the average male gets no more matings than before. Mean male fitness thus goes down because of the costs of the display or aggressor traits; unless the results of male competition enhance female fitness, the overall fitness of the population will have decreased.

A final example of a take game is the **selfish-herd effect** (Hamilton 1971). Imagine a scattered herd of gazelles foraging on a savannah. Somewhere within this savannah lies a hidden cheetah. As far as the gazelles are concerned, the cheetah could appear at any location. When it does, it will chase and kill the nearest gazelle. The relative risk to each animal thus depends on the size of the area around itself within which it would be closer to a suddenly appearing cheetah than would its nearest neighbors. Gazelles close to the edge of the herd have no nearby neighbors on the outside, and thus they will be the closest to any cheetah that appears at points far from the herd. Gazelles inside the herd have many close neighbors and thus a smaller area of risk. Passive animals ignore these facts and concentrate on foraging. Cheats consider them, interrupt their foraging, and move to the center of the herd. This eventually places passive gazelles at the high-risk margins of the herd. The ESS is again for all animals to adopt the cheat strategy. The cost will be a reduction in foraging efficiency due to overcrowding and continual interruptions of feeding to move to the center. Since the cheetah will still appear and take one of the herd, the average risk remains the same; however, there is now a cost of reduced foraging efficiency. Take games are inexorable once cheats appear; they never lead to an improvement in average fitness unless other factors come into play.

A combination of strategies in a game for which no player can improve its payoffs without decreasing the payoff to other players is called a **pareto optimum**. Because a cheat can only do better at a cost to other players in a take game, the adoption of the passive strategy by both players is a pareto optimum. Pareto optima are of great interest to economists because they represent the combinations of strategies that yield the greatest common good. A whole branch of modern economics has spent decades studying pareto optima and how they might be achieved in real societies. In animal societies, we see little evidence that pareto optima are maintained. Instead, animals cheat, arm, and court in selfish ways, and this invariably pushes the system to the ESS instead of the pareto optimum. Part II of this book assumed that sender and receiver shared similar interests in honest and accurate information transfer. The optimal strategies discussed in those chapters were thus pareto optima. What if sender and receiver do not share common interests? What if one party cheats? The resulting take game is likely to favor both parties switching to ESSs with lower levels of information exchange even at a cost to fitnesses. How much are game theoretic predictions likely to differ from the passive optimality predictions of Part II? Does the level of conflict of interest affect the degree of difference between the two kinds of models? We shall try to answer these questions in succeeding chapters.

A Final Discrete Symmetric Contest: Give Games

A **give game** is the opposite situation to a take game. It also involves two strategies: **passive** players, as before, mind their own business, whereas **donors** give B fitness units to other players at a cost of $-C$ fitness units to themselves. Clearly, a persistent donor that always gives and never receives a donation will have lower fitness than passive animals. However, if both parties are willing to reciprocate by playing the donor strategy one day and accepting donations from other donors the next, and if $B > C$, then donors might gain more than they lose on the average. Note that if donors perform their giving indiscriminately, passive animals will benefit from donations at no cost, whereas donors giving to passives never recoup their costs. For there to be any chance of a donor strategy ESS, we must also invoke some selectivity by donors that enables them to limit unrepaid losses to passives. The payoff matrix when these assumptions are invoked is shown in Figure 19.5. We first

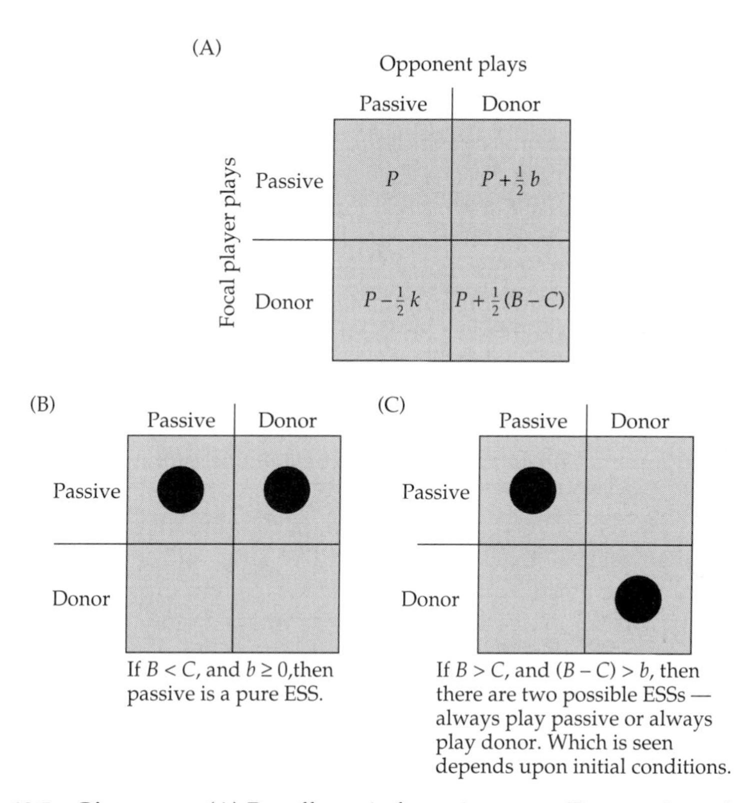

Figure 19.5 Give game. (A) Payoff matrix for a give game. Two passive animals mind their own business and obtain payoff P. Two donors alternately give the other B units at a cost $-C$ or receive B units from the other donor. When donors meet passives, on half of the occasions the donor gives b units to the passive player at a cost of $-k$ to the donor; the passive player then receives the b units at no cost to itself. It is assumed that $B > b \geq 0$. On the other half of the occasions, the donor waits to be given a donation but when nothing is forthcoming, moves on at neither benefit nor cost to itself. (B) Dot analysis when $B < C$. The costs of donating always exceed the benefits, and the passive strategy is thus a pure ESS. (C) Dot analysis when $B > C$ and $(B - C) > b$. There are two possible ESSs here (Case IV)—always play passive or always play donor. Which is seen in the short term depends upon initial conditions, and in the long term on which strategy is most rare at the equilibrium frequency $\hat{f}$.

assume that $B > C$. As before, the payoff to each party when two passive animals meet is P. When two donors meet, half of the time the focal animal receives B fitness units from the other player. The remaining half of the time, the focal animal donates B fitness units to the other at the cost of $-C$ to itself. The average payoff to a focal animal when two donors meet is thus $P + 1/2\,B - 1/2\,C = P + 1/2(B - C)$. When a focal donor meets a passive player, half of the time it awaits donation from the other party (and then neither receives nor loses anything), and half of the time it is prepared to donate. However, being selective it either recognizes that the intended recipient is not a donor, and thus reduces its donation to $b < B$ at reduced cost $k < C$, or it gives the full donation of B at a cost C, but only to a fraction q of nondonor players ($0 < q < 1$). In the latter case, we can let $b = qB$ and $k = qC$ and thus the payoffs are the same regardless of the form of the discrimination. In fact, it is not important how the discrimination works as long as donors are able to give other donors more on average than they give to nondonors.

If $B - C < b$, then the passive strategy is a pure ESS and giving is not favored. If $B - C > b$, then we have a Case IV situation in which there are two ESSs. We expect to see either a pure ESS for the passive strategy, or a pure ESS for the donor strategy. In the short term, the strategy that is initially common will become the ESS. It seems more likely that a population would begin dominated by passives and donor mutants would then have to accumulate to some critical frequency before the donor strategy became the ESS. Let f be the frequency of the passive strategy in the population. We need to find the value of the critical frequency $\hat{f}$. Using the method described on pages 626–627, it is easy to show that

$$\hat{f} = \frac{(B - C - b)}{(B - C - b) + k}$$

When k is small relative to $(B - C - b)$, $\hat{f}$ will be large and the donor strategy will be the rare one at the equilibrium. This means that only a small number of donor mutants must accumulate before f drops below $\hat{f}$ and the donor strategy becomes the ESS. The larger k is relative to $(B - C - b)$, the harder it will be for a donor strategy to invade a population of passive animals. If we think of $(B - C - b)$ as the average benefit of being a discriminating donor and k as the average cost, we can say that the higher the benefit to cost ratio, the more likely the donor strategy will be the observed ESS.

Give games model reciprocal cooperation in which one animal helps another now, and is repaid at a later time. It is telling that at best, reciprocal cooperation is only one of two likely ESSs and then only if $(B - C)$ is sufficiently large (Trivers 1971). Since we have defined the game in a specific way, one might wonder whether other formulations might allow the donor strategy to be a pure ESS. For example, perhaps the proper model is a dynamic game in which a policy of preferential sharing by donors over many successive interactions is pitted against a passive policy. This game has been examined and the result is the same. There are at best two alternative ESSs: cooperation and noncooperation (Axelrod and Hamilton 1981). Both take and give games can also be more generally modeled as scrambles instead of contests, but we again get the same conclusions. In fact, no matter how we formulate reciprocal cooperation, we never find that cooperation is a single pure ESS when opposed to passive or selfish opponents. This is an important lesson. While selfish behavior easily invades a population of passive players, reciprocally cooperative behavior is much harder to evolve. As a result, most apparently cooperative acts in animals are either due to mutually immediate benefits (both parties benefit now with no delay in repayment), or to kin selection (donor costs are more than made up by the benefits to a recipient carrying many of the same genes). There are examples in nature of true reciprocity (see Figure 19.6), but they remain the exception and this game indicates why. We might expect the relationship between communicating animals to share some of the properties of give games including two alternative ESSs. This is in fact what is found (see Chapter 20).

General Analysis of Discrete Asymmetric Contests

We now consider discrete games in which there are two or more roles. The appropriate payoff matrix will need to show two payoffs for each combination of strategies (one for each player). The convention is to divide the matrix cell for each pair of strategies diagonally in half and to enter the payoff to the player on the left in the lower-left corner of the cell, and that to the player at the top in the upper-right-hand corner (Figure 19.7). A 2×2 asymmetric discrete contest may have no ESS, one ESS, or several ESSs. It will not have mixed strategies similar to those seen for symmetric contests if players are fixed in a given role and cannot, by definition, perform one of their own

Figure 19.6 Reciprocal exchange of blood meals by vampire bats. Vampire bat females form affiliations with "partners" that are reinforced by shared grooming and huddling during the day (Wilkinson 1984). The bats must feed every second night or die. Individuals that fail to secure a blood meal beg from successful partners, and the donor regurgitates a fraction of its meal. Benefits of receiving a meal exceed costs, (i.e., $B > C$), because the same volume of blood improves the survival of a starving recipient more than it reduces the survival of a satiated donor. Cheaters are controlled by only donating to known partners. (Photo by Gunter Ziesler, Peter Arnold Inc.)

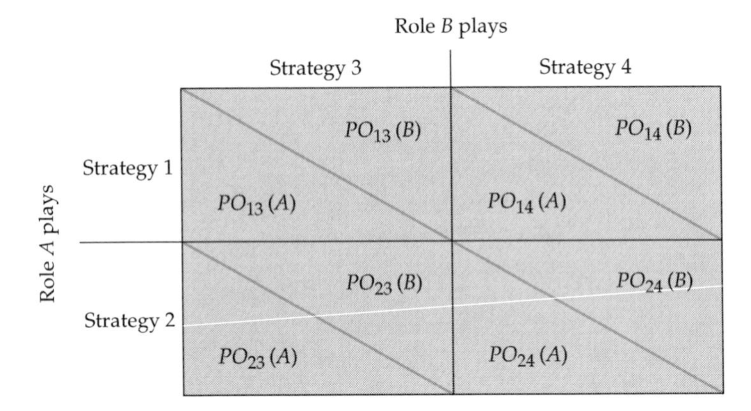

Figure 19.7 Payoff matrix for a discrete asymmetric contest. In this example, there are two players who represent different roles (*A* and *B*) and each role has access to two alternative strategies. Because the strategies available to the roles in an asymmetric game are usually different, we assign strategies 1 and 2 to role *A* and strategies 3 and 4 to role *B*. Any combination of strategies by the two players may result in different payoffs for the two players. As before, we use subscripts to indicate the strategy used by player 1 followed by the strategy used by player 2. In parentheses, we indicate to whom that particular payoff is given.

strategies part of the time and one of the strategies of another role the rest of the time. If animals are not fixed in a given role, as might occur if they make mistakes about their relative roles, then mixed asymmetric ESSs are possible.

To identify the pure ESSs (if any) in a 2×2 discrete asymmetric contest, we use the **arrow method**. This identifies the best response of each player to a given strategy of its opponent. We indicate the best responses by placing arrows parallel to each of the four sides of the payoff matrix. For example, we might first consider what the player in role *A* of Figure 19.7 should do if its role *B* opponent plays strategy 3. If $PO_{13}(A) > PO_{23}(A)$, then the role *A* player should adopt strategy 1 over strategy 2. We would then place an arrow parallel to the left side of the payoff matrix pointing up. If $PO_{13}(A) < PO_{23}(A)$, then we make that arrow point downwards. We then consider what the role *A* player should do when the role *B* player opts for strategy 4. If $PO_{14}(A) > PO_{24}(A)$, then we place an arrow parallel to the right side of the matrix pointing up; if the converse is true, then we point the arrow down. We next consider the optimal strategies for the role *B* player. If player A elects to use strategy 1, what should the role *B* player do? If $PO_{13}(B) > PO_{14}(B)$, then the role *B* player should adopt strategy 3. We then place an arrow parallel to the top of the matrix pointing to the left. If the converse were true, we would make that arrow point to the right. Finally, we identify the best response for the role *B* player when its opponent adopts strategy 2. This results in an arrow parallel to the bottom of the matrix pointing either to the right or to the left.

Any point at which two arrowheads meet is an ESS for that game. For a simple 2×2 discrete asymmetric contest, there are 16 possible outcomes (Figure 19.8). Of these outcomes, two will result in no ESSs, 12 will result in a single ESS pair of strategies, and two will result in two alternative ESS pairs. In the latter case, which of the two possible ESSs is seen depends on the initial frequencies within each role of the alternative strategies. These complex outcomes are best understood with an example.

A Discrete Asymmetric Contest: Dominance versus Subordination

Let us consider an asymmetric version of the hawk-dove game (Hammerstein 1981). As in the earlier hawk-dove game, we assume that the contest is over some indivisible commodity worth V fitness units. If two animals both esca-

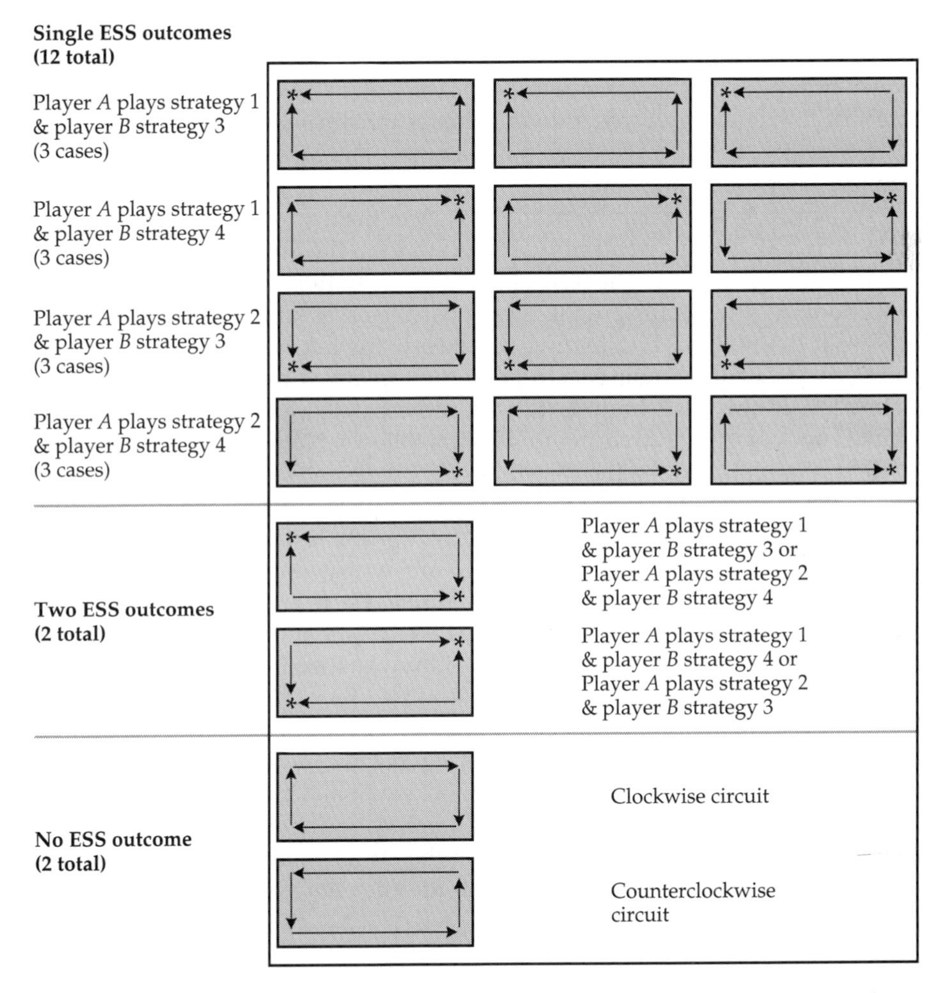

Figure 19.8 All possible outcomes of a 2 × 2 discrete asymmetric contest. Analysis using the arrow method. Each corner at which arrowheads meet (marked by asterisks) constitutes an ESS. There are 16 possible outcomes of such a table: 12 have a single ESS each, two have alternative ESSs, and two have no ESS.

late (play hawk), one will lose and suffer a cost of $-D$ fitness units. When two doves meet, they display to each other and on average, each wins the commodity half of the time. The only difference with this version of the game is that we do not assume that the two opponents have an equal 0.5 probability of winning a fight when they both escalate (play hawk). Instead we identify one animal, filling the **dominant role**, which has a probability $P_d > 0.5$ of winning an escalated fight with its opponent. The opponent, representing the **subordinate role**, thus has a probability $(1 - P_d) = P_s < 0.5$ of winning such an escalated fight. We assume that the two opponents know their relative likelihoods of winning a fight. This knowledge might have arisen from prior experience between them, or they may be able to assess each other's fighting potential based on body size, age, or some other anatomical or behavioral feature. A priori, one might expect an opponent who knew ahead of time that it would lose a fight to play a nonescalated strategy during a confrontation. It would thus act in a subordinate manner relative to the likely winner. In nature, one finds that prior losers do not always remain subordinate. This asymmetric version of the hawk-dove game shows why and when a subordinate should escalate. The relevant payoff matrix is shown in Figure 19.9A.

We notice in this matrix that if $V > 0$, then the arrow on the right side of the matrix will always point upward and that on the bottom side of the matrix will always point to the left. That leaves only two arrows that can vary conditionally. The results are the three possible cases shown in Figure 19.9B. In Case I, $P_dV - P_sD > 0$ (allowing us to draw the left-hand arrow upward), and $P_sV - P_dD > 0$ (allowing us to point the top arrow to the left). There will be one ESS in this case: both parties should escalate and play hawk. We can simplify the conditions favoring this ESS as follows. Note that because $P_d > P_s$ by definition, it is always the case that $P_dV - P_sD > P_sV - P_dD$. Case I is thus true when $P_dV - P_sD > P_sV - P_dD > 0$. With a little algebra, we can rewrite these conditions for Case I as $P_d > P_s > D/(V + D)$. In Case II, $P_dV - P_sD > 0$ (again allowing us to draw the left-hand arrow upward), but $P_sV - P_dD < 0$ (leading us to point the top arrow to the right). There is again a single ESS in this situation: the dominant should play hawk and the subordinate should play dove. The two conditions can be combined into the inequality string $P_dV - P_sD > 0 > P_sV - P_dD$, which simplifies algebraically to $P_d > D/(V + D) > P_s$. Finally, Case III occurs when both $P_dV - P_sD < 0$ and $P_sV - P_dD < 0$. The top arrow thus points to the right, and the left-hand arrow downward. Case III has two alternative ESSs: either the dominant plays hawk and the subordinate plays dove, or the subordinate plays hawk and the dominant plays dove. Either combination can be an ESS depending on the initial situation. The two conditions favoring Case III can be combined into the expression $0 > P_dV - P_sD > P_sV - P_dD$ which simplifies to $D/(V + D) > P_d > P_s$.

Let us see how these results fit together. As with the simple hawk-dove game, we have found that the relative sizes of V and D affect the ESSs. In this case, we focus on the ratio $D/(V + D)$. This value is small when the cost of losing a fight is minimal, and large when the cost of losing a fight exceeds the value of the commodity being fought over. When $D/(V + D) = 0.5$, then $V =$

(A)

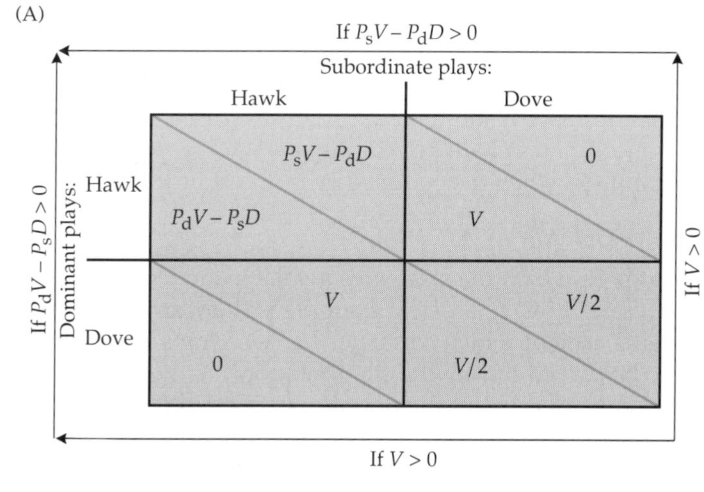

(B)

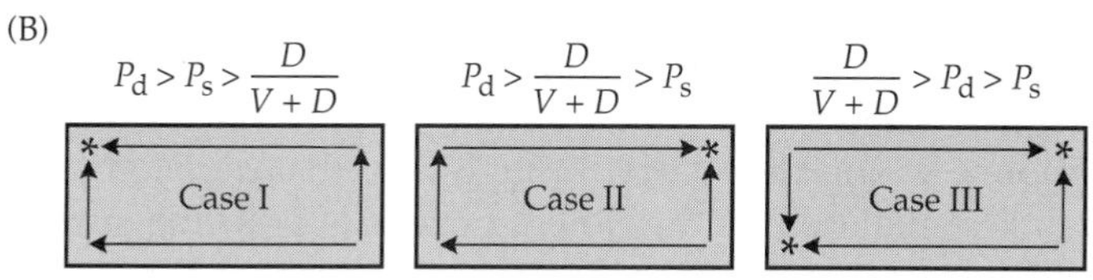

(C)

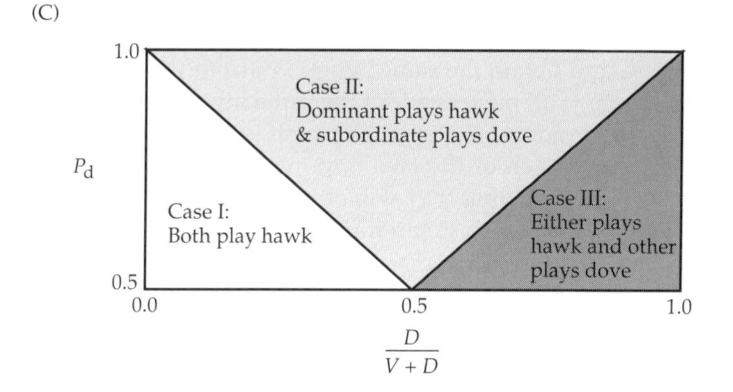

Figure 19.9 Dominance versus subordination: An asymmetric discrete contest. (A) Payoff matrix for the game. Both roles have access to the same strategies (hawk or dove), but differ in their probabilities of winning an escalated fight. Dominants are likely to win a fight with probability $P_d > 0.5$, whereas subordinates are likely to win such a fight with probability $P_s < 0.5$. Since someone has to win, $P_d + P_s = 1.0$. The commodity being fought over has fitness value V, and the cost of losing an escalated fight is $-D$. If $V > 0$, hawk is the best response by either player to the playing of dove by its opponent. The arrow on the right of the matrix always points up, and that on the bottom always points left. The remaining two arrows depend on the relative sizes of P_d, P_s, V, and D. (B) The three possible outcomes for this game and the algebraic conditions that favor them. (C) A plot of possible values of P_d versus $D/(V + D)$ divided into three regions. Within any given region, all combinations of P_d and $D/(V + D)$ lead to the same ESS outcomes. (After Hammerstein 1981.)

D. For which values of $D/(V + D)$ does it pay for a subordinate to challenge a dominant by playing hawk despite the fact that the dominant always has a better chance of winning such a fight? In Figure 19.9C we see that both subordinates and dominants should escalate when $D/(V + D) < 0.5$ and P_d is not too large. The smaller $D/(V + D)$, the higher P_d can be and still justify escalation by the subordinate. If the cost of losing is small enough, or the benefit of winning high enough, then subordinates are likely to escalate. For intermediate values of $D/(V + D)$, we are more likely to find the dominant playing hawk and the subordinate playing dove. These are the expected behaviors for dominant versus subordinate roles. At higher values of $D/(V + D)$, even the dominant should not escalate when the subordinate does unless the dominant has a very high chance of winning. The costs of losing are simply too high relative to the benefit of winning. In this case, the ESS is for one party to play hawk and the other to play dove. Which plays hawk may depend on initial conditions. This is a less intuitive result than the other two cases, but field observations in which subordinates challenge dominants who respond nonaggressively are not uncommon when the commodity is one of little value.

A Continuous Contest: The War of Attrition

We now examine a game in which the strategies available are drawn from a continuous range of possibilities. A classical example is the **war of attrition**. Consider two opponents each of which competes by selecting an amount of strategic investment to be played during the confrontation. Neither knows before the confrontation what investment the opponent has chosen. During the confrontation, the opponent that selected the larger investment wins the interaction. The investment might be the amount of time each is prepared to display to the other (hence the name war of attrition: whoever lasts longest wins). Alternatively, it might be how much energy the players put into a display. Whatever the nature of the investment, each player must decide ahead of time on a given amount, x, and then carry through with that decision during the confrontation. We shall denote the fitness benefit of winning such a contest by V, and the total cost of each investment by kx where k is the rate at which costs accrue per unit investment.

We first consider a **symmetric war of attrition** in which all players suffer the same cost of display, k, and obtain the same benefit, V, from winning. Are there pure ESSs in this game? If all players adopt the same investment x, winners can only be assigned by random. The average payoff to any player in this population is $V/2 - kx$ (since each of the two opponents wins 50% of the time), but each has a cost of $-kx$ whether they win or not. It should be obvious that any mutant strategy that invests a greater amount x' will be able to invade such a population as long as $V/2k > x' > x$; once $x' > V/2k'$, then the average payoff will be negative, and any new mutant with $x = 0$ will be able to invade. The result is that this game cannot settle into any given pure ESS. Each time a fixed value is established, it is invadable by a mutant using a higher value or no investment, depending upon the payoffs.

There is, however, a mixed ESS for this game in which each player plays all possible investments but with varying probabilities. Since no player can know ahead of time what investment its opponent will select, this mixed strategy is not invadable in the way any pure strategy would be. Because there is an infinite number of possible investments, the ESS must be described as a probability density function that is calculated using differential calculus (see Parker 1984a). For the symmetric war of attrition, the ESS is for each player to choose an investment time x randomly with probability $P(x)$ where

$$P(x) = \frac{k}{V} e^{-\frac{kx}{V}}$$

This means that smaller values of x will be chosen by chance more often than larger ones (Figure 19.10A).

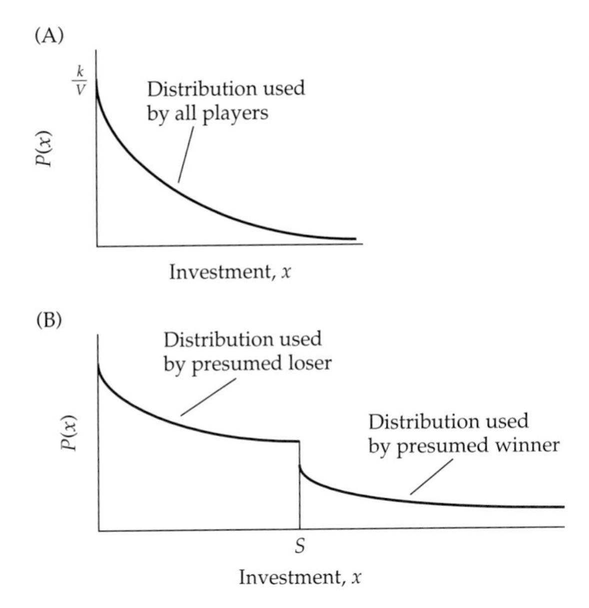

Figure 19.10 War-of-attrition ESS solutions. (A) Probability density function defining mixed ESS for symmetric war of attrition. Each player should select an investment, x, at random with the relative probability specified by this function. Thus shorter investments will be selected more often than longer ones. (B) Probability density functions for ESS of asymmetric war of attrition when players are fairly accurate in assessing each other's V/k ratio, where V is the benefit of winning, and k is the rate of cost accrual to that player. A player that thinks it has the lower ratio should select a random investment from the left-hand distribution (with possible investments between 0 and S). A player that thinks it has the higher ratio should randomly select an investment from the right-hand distribution (with values between S and infinity).

It is unlikely that many animals meet the conditions for the symmetric war of attrition. Usually, the rate at which costs accrue, k, will not be the same for any two players. It is also likely that the same commodity may have different fitness values, V, for each player. Suppose two players with different k and V values meet. The largest investment player i should choose is that x_i at which it would just break even if it won; that is, where $V_i - k_i x_i = 0$. For individual i, the maximum investment is thus $x_i = V_i/k_i$. If the opponent, player j, has a larger maximum investment, $x_j > x_i$, then it would clearly win a prolonged confrontation. The critical issue is thus which player has the larger value of the ratio V/k. If the two players know for sure at the outset which has the larger V/k value, then there should be no confrontation (and thus no game): the player with the smaller value should retire immediately. However, animals rarely have perfect information and there is always a chance that a player that suspects it might have the lower V/k really does not. In this case, it should play the game, and that game is called the **asymmetric war of attrition**.

As with the symmetric war of attrition, there is no pure ESS that is optimal for either party in the asymmetric case, and again, the mixed strategy which is the ESS for each party is found using differential calculus. The ESSs are shown graphically in Figure 19.10B. A player that suspects that it represents the lower V/k role should randomly choose an investment between 0 and a critical value S according to a probability density function that is highest at $x = 0$ and decreases for x closer to S. A player that thinks it has the winner role should randomly select an investment between S and infinity, again with lower choices of x being more likely than higher ones. Note that it is possible for both players to decide they occupy the same role. This reflects the fact that uncertainty is inherent in this game. If both think they merit the winner role, both will select large investments and the confrontation may last a long time or be very vigorous. If both presume they would lose a prolonged battle, they will both select small investments and the confrontation will be over quickly or of low vigor. This game appears to fit a number of animal examples quite well, and we shall take it up in more detail in Chapters 22 and 23.

Asymmetric Scramble Games: Mating Competition

In the prior example, the chances of winning and the average payoffs to a focal animal depended only on the focal animal's choice of investment, the value of the resource, and the choice of investment by the opponent. While the abundances of opponents with different V/k ratios affect the rate at which each type of opponent is encountered, they do not affect the outcome of any given contest. That outcome is set only by the V/k ratios of the two contestants and chance. In contrast, let us now consider situations in which the distribution of different types of opponents affects not only the frequency with which each type is encountered, but also affects the chances of winning and/or the payoffs of encounters. Such a game is a scramble. Many situations involving competition among members of one sex for matings with the opposite sex fit these conditions. Here the fraction of available matings obtained by any player depends not on the absolute value of its investment, but on the relative value of this investment when compared to all other competitors. Let us look at some examples.

We first examine a discrete asymmetric scramble. Males adopt either of two alternative strategies. **Displayers** exclude other males from their territories and display to attract receptive females; **roamers** move around as a group in those areas not occupied by displayer males and mate communally with any receptive females. This situation is found in some reef fish such as wrasses (Figure 19.11). We assume that each male in a group of N individuals adopting the roamer strategy will obtain $1/N$ of the matings (or paternity) available to that group. Among displayer males, we assume that relative mating success is an increasing function of some continuous phenotypic trait, z, such as age, body size, or experience. The larger a male's z, the larger the fraction of matings it obtains by being territorial. The distribution of z in the population can have any form. How large a value of z should a male have before it is better for it to adopt the displayer rather than the roamer strategy? Clearly, the available payoffs will depend upon how many other males have adopted each strategy, making this a scramble game.

The ESS is again found using calculus. It is defined by identifying a critical value of the trait, z^*. All males with z values less than z^* should adopt the roamer strategy and all those with larger z values should become displayers. The value of z^* depends upon (a) the relative numbers of mates accessible through each of the two strategies, (b) the shape of the distribution of the trait z in the male population, and (c) the shape of the function relating relative mating success of displayer males to their z values. If this latter function rises in a decelerating manner (Figure 19.12A), z^* will be small; if it rises in an accelerating manner, z^* will be large (Figure 19.12B). The ESS for a situation in which z values are normally distributed is shown in Figure 19.12C. In the case

(A)

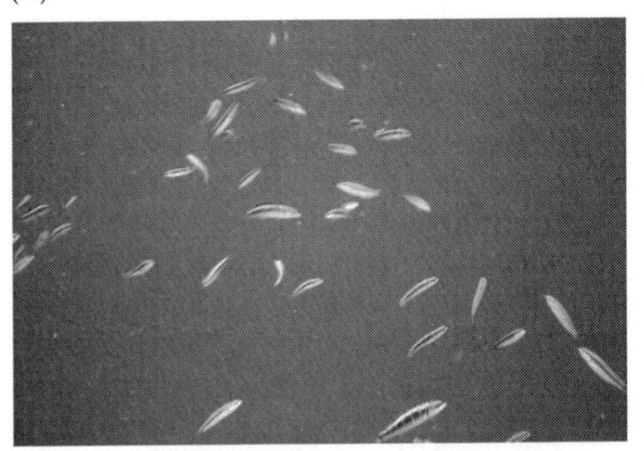

(B)

Figure 19.11 Multiple male strategies in the blue-headed wrasse (*Thalassoma bifasciatum*). In this species, primary males and females are produced from pelagic eggs and settle onto reefs. (A) Primary males form roaming schools that surround spawning females and all males release sperm onto the female's eggs concurrently. On average, each male in a school of N individuals fertilizes $1/N$ of the eggs. Growth eventually permits larger primary males and primary females to turn into terminal males. (B) A terminal male that has set up a defended territory on the down-current side of the reef where most females prefer to spawn. It will then display to attract potential mates. Except on very large reefs, terminal males are uninterrupted during spawning by primary males and thus father all of the eggs released in their territories. Ratios of primary to terminal males vary depending upon reef size, and this affects the degree to which secondary males can defend territories and thus the shape of the curve relating reproductive success to body size. (Warner 1984; Warner and Hoffman 1980; photos courtesy of Ken Clifton.)

of wrasses, the shapes of the mating success versus z curves vary for different reef sizes, and this is reflected in different ratios of males adopting the two alternative strategies. The observations generally fit this model (Warner and Hoffman 1980).

In many mating competition contexts, the strategic options available to males are not discrete but continuous. Consider mating on a **lek**, an aggregation of males on small territories used solely for mating. Males display competitively to attract females onto their own territories and persuade them to mate. How much effort should males put into display? Since a male could perform any effort between zero and some upper physiological limit, the strategy set here is a continuous one. Because the costs of any given level of display are likely to differ according to the physical condition of the male performing the display, the relevant game is most likely asymmetric. This game is thus best modeled as an asymmetric continuous scramble. It might seem at first glance that there is no stable ESS in such a game. Would not the invasion of increasingly higher effort strategies just keep pushing the average display level up and up? In fact, there are stable ESSs for this game that vary depending upon male condition. Males in worse condition should display at lower levels than those in better condition, even though this will result in their obtaining fewer matings. We shall review this model in some detail in Chapter 23.

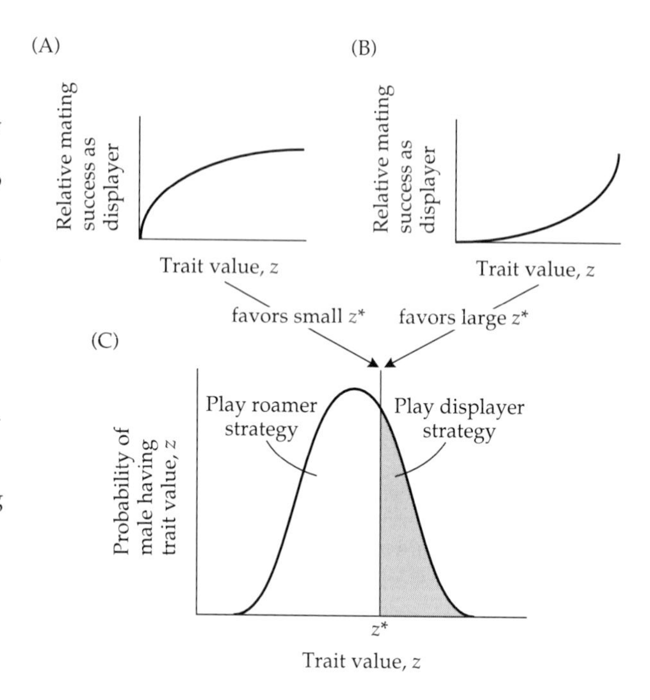

Figure 19.12 ESS for asymmetric discrete scramble game involving male mating strategies. Males can either defend territories in which they display to females (**displayer**), or move in nonterritorial groups with no display (**roamers**). The relative success of the displayer strategy increases with the value of some continuously varying phenotypic trait such as age, size, or experience (denoted by z). The function relating mating success to z could be decelerating (A) or accelerating (B). Among the N males adopting the roamer strategy, individual mating success equals $1/N$ of the matings. (C) ESS analysis assuming normal distribution of z values among males. All males with $z > z^*$ should become displayers and those with $z < z^*$ should become roamers. The value of z^* is lower if the function relating mating success for displayers increases in an decelerating manner (A) and higher if that function is accelerating (B). (After Parker 1982.)

MATCHING BIOLOGY TO GAME MODELS

Evolutionary game models are always simplified versions of what animals really do. Matching a particular kind of model to a particular biological situation thus requires assumptions about which contexts need to be included and which can be ignored. Although many theorists warn that one should never expect animals to act the way models predict, the goal of most researchers is to extract general rules from the noisy data of nature. If a model has the power to predict, even partially, what we see, most biologists are ecstatic. Evolutionary game theory models have resolved many paradoxes in animal behavior. This success has spawned increasing numbers of game models in the literature, and the process of matching real biological situations with the appropriate model can be intimidating. Luckily, reasonably alternative models applied to the same situation often yield the same general result. The use of ritual displays instead of aggressive fighting can be modeled as a discrete symmetric contest with the hawk-dove game, as an asymmetric contest with dominance or subordination, or by a number of other games that we discuss in later chapters. Although the model assumptions differ markedly, the basic conclusions turn out to be quite similar. Many ESS solutions are thus contextually robust. If your model focuses on the essential biological question, the type of game you use may not be crucial to your conclusions.

Our task in the ensuing chapters is to match existing game models with the different contexts for animal communication. One criterion for matching suggested by the discussion on pages 619–620 is the degree to which sender and receiver have commensurate interests in accurate communication. At one extreme the interests of players are so disparate that mortal risk is associated with interactions. At the other, players have such overlapping interests that it would be erroneous even to use the word "conflict" to describe their interactions. Somewhere between these two extremes, players shift from selfish take games to cooperative give games. Game propriety will thus depend critically on similarity of interests. ESSs are more likely to coincide with pareto optima at the cooperative end of this spectrum, and it is here that we expect to see animals optimizing information transfer according to the methods of Part II. Examples from the other end of the spectrum are more likely to favor signaling systems that do not follow simply from physical constraints and contextual information needs. Without careful consideration of the competitive interplay between sender and receiver, one cannot hope to make sense of how these animals communicate. Here a game-theoretical perspective will be essential.

SUMMARY

1. **Evolutionary game models** differ from **simple optimality models** in that the contexts determining which strategies are optimal are not passive and immutable environmental conditions, but instead active and responsive **opponents**. As one subset (**role**) of a population adopts some focal strategy favored by evolution, this can change the selective pressures on opponent roles. The evolutionary responses of these other roles can in turn alter the selective pressures favoring the use of the focal strategy in the first role. This results in a complex coevolution of strategies that can be most easily analyzed using **evolutionary game theory**.

2. An **evolutionarily stable strategy** (**ESS**) is a strategy which, when common, cannot be invaded and displaced by another strategy from the same set of alternatives. Note that this allows several strategies within a set to be potential ESSs. Which is the current ESS may depend on which was more common historically. Some sets may have no ESS because the coevolutionary process never settles down to an equilibrium combination. Other sets may have such fixed evolutionary trajectories that ESSs are reachable only if players make mistakes. Luckily, mistakes are a common feature of animal behavior.

3. Evolutionary games may be based on sets of **pure strategies** that are either **discrete** (the alternatives are finite) or **continuous** (there is an infinite number of alternatives differing infinitesimally from each other in some variable). In either case, the set may also allow for **mixed strategies** in which a player randomly performs several pure strategies according to specific probabilities (**behavioral polymorphism**). A mixed strategy may also be achieved if each player performs a single pure strategy, but fixed fractions of the population perform different strategies (**genetic polymorphism**).

4. Evolutionary games in which opponents have access to the same strategy sets, have equal chances of winning given a particular combination of strategies, and obtain the same payoffs when they win a particular

exchange are called **symmetric** games. When one or more of these conditions is not met, the game is **asymmetric**.

5. When the payoffs of a particular combination of strategies do not depend upon how many other individuals in the population are adopting particular strategies, the relevant evolutionary game is called a **contest**. The outcome of a contest does not depend on what players other than the two opponents are doing. When payoff values vary depending upon the number of players adopting a given strategy, the game is called a **scramble**. In this case, each player can be envisioned as playing against more than one opponent at a time. Hence the alternative names **playing against the field** or ***n*-person games**.
6. In many cases, a focal player has to play the same game on several consecutive occasions. If the available strategies, chances of winning, and payoffs of winning on later occasions are not dependent on what happened on prior occasions, then each occasion can be treated as a **single-bout game**. If successive occasions are not independent, then the entire sequence must be treated as a **dynamic game** and the ESS is the **optimal choreography of strategies** through the sequence. For some dynamic games, a optimal rule of thumb or **evolutionarily stable policy** can be invoked to ensure adherence to the ESS choreography.
7. **Discrete symmetric single-bout contests** have the most easily identified ESSs. If there are only two possible pure strategies, there will always be at least one ESS (which may be pure or mixed), and there may be two alternative ESSs. When there are more than two alternative strategies, there also may be no ESS or even more than two. **Hawk-dove** is a classic two-strategy discrete symmetric contest that helps explain why some animals resolve conflicts using ritual signals. A **take game** is one in which a population of animals at high fitness is invaded by cheaters who force all members to adopt the cheater strategy even though the result is lower average fitness for all. The original state is called the **pareto optimum**, which though better on average, is not stable against the cheater strategy. Arms races are examples of a take game. **Give games** involve reciprocal sharing by cooperators. At best they have two ESSs—one in which everyone is a reciprocal cooperator and one in which no one shares. Many communication exchanges can be modeled as give games.
8. **Discrete asymmetric single-bout contests** have only pure ESSs unless players make mistakes about their roles. Even when both roles only have two strategies each, there can be no ESS, one ESS, or two ESSs. An asymmetric version of hawk-dove explains why subordinate animals do not frequently escalate against dominant animals.
9. The **war of attrition** is a continuous contest that can be analyzed either as a symmetric or an asymmetric game. It explains why the duration of display contests between animals is often highly variable and why no individual should use a fixed display time.
10. **Mating competition** is best analyzed as an asymmetric scramble because the mating success of competitors depends on their relative performance

and not their absolute one. It can be analyzed either as a discrete or a continuous game and explains why not all animals in a population display to attract females or why not all display at the same high level.

11. Matching evolutionary game models to biological situations requires assumptions about which factors in the biological situation are most important. One major factor for analysis of animal communication games is the degree of conflict of interest between sender and receiver. At one extreme, conflict is high and games such as hawk-dove or take games are relevant. At the other, interests of sender and receiver are highly overlapping and give games and other cooperative exchanges are appropriate. We expect game theory predictions to be most similar to those based on passive optimization (Part II) when conflicts of interest are minimal; when they are not, the two methods may predict quite different optimal strategies.

FURTHER READING

Two excellent introductory reviews of evolutionary game theory are Maynard Smith (1982a) and Parker (1984a). Readers may also want to consult Hofbauer and Sigmund (1988). For those with economics backgrounds, the links between evolutionary game theory and modern economics are made clear in Thomas (1986). Matrix methods for finding ESSs in discrete games are outlined in Haigh (1975) and Hines (1980, 1987). Methods for solving dynamic games are discussed in Mangel and Clark (1988). Many references to specific evolutionary game theory models can be found in the following chapters.

Chapter 20

Signal Honesty

THE PRIOR CHAPTER BROACHED THE POSSIBILITY that sender and receiver may have conflicting interests in the accurate exchange of information. This is rather different from the happy collaboration assumed in Part II. When might we find discordant expectations between communicating animals? Consider the opponents in an aggressive encounter over some resource. Each performs threat displays to persuade the other to retreat without engaging in an escalated battle. Successful persuasion avoids the risk of injury, but whoever retreats fails to gain the resource. There is thus a conflict of interest. Alternatively, consider a courting pair in which the male benefits by mating regardless of his suitability, whereas the female prefers to mate only if the male meets certain criteria. Suppose the major information available to females is that signaled by the body ornamentation or displays of male suitors. Males and females are unlikely to have similar enthusiasm for reliable signaling. Finally, consider a hungry cheetah stalking a gazelle. If the gazelle can perform a display indicating that it is too agile to be worth chasing, perhaps the cheetah will give up the hunt. But should a less agile gazelle be honest about its greater vulnerability? In each of these cases, at least some senders will be tempted to provide false information. Receivers subject to such false signals will do best to discount or ignore them. How the corresponding evolutionary game will play out has been a major preoccupation of researchers in animal communication. The history of views on this topic is itself instructive and we shall briefly review it below. First, however, we need to refine our definition of honest communication.

HONESTY VERSUS CHEATING IN COMMUNICATION

The opposite of honesty is cheating. In a communication system, senders might cheat on receivers, and receivers might cheat on senders. In Chapter 12, we called the first form of cheating **deceit**. There are several ways in which a sender might deceive a receiver. One type of deceit occurs when the condition that is important to the receiver has only a few discrete and unordered alternatives, and the sender emits a signal that falsely identifies which alternative is true. For example, young males in lek mating species may mimic the patterning or behaviors of females to avoid harassment by older males and sneak closer to receptive females. Another example is a bird that gives a false alarm call, causing competing foragers to flee and leave any food to the caller. Hasson (1994) calls such false categorical signals **lies**. A special case occurs when a sender **witholds** production of a signal that could benefit a nearby receiver. A bird in a foraging flock might refuse to give an alarm when a hawk or snake is spotted, or a lone monkey might fail to produce recruitment calls when it discovers a rich patch of fallen fruit. By witholding the signal, the sender is falsely implying that one of two alternative conditions is the case. Deceit can also occur when the alternative conditions that receivers seek to identify are from a continuous set and reflect the value of some single property. Typically, the communication code maps each possible signal on a subset of the ranked values of the property. A sender may cheat by exaggerating or bluffing about which value of the property is currently true. One of two opponent dogs might growl more loudly at another than is justified by its willingness to fight. Its signal is thus a **bluff** or **exaggeration**. Finally, we discussed the possible role of amplifiers in Chapter 12. These are traits that make the direct assessment by a receiver of some property in a sender simpler or more accurate. The cheating version of an amplifier is an **attenuator**. This is a trait that makes direct receiver assessment of the sender property more difficult (Hasson 1994). An example is broken and patchy body marking on a fish that hinders the estimation of its body size by conspecifics.

In each case, naive acceptance of the dishonest signal (or absence of a relevant signal) can cause the receiver to select a subsequent action that is worse for it than the alternatives. Although this might enhance the value of information for the sender, it will reduce the value of information for the receiver. From its own point of view, the receiver has been induced to err as a result of reliance on the cheating signal. Deceit thus implies receiver error. It is important to note that the reverse is not necessarily true and receiver error is not a useful indication of deceit (Dawkins 1993; Dawkins and Guilford 1991; Wiley 1994). We saw in Chapter 14 that imperfect signals always lead to errors that a receiver without reliance on signals would not make. These errors are one of the incidental costs of communication. We also saw that it is rarely optimal for a sender to transmit and a receiver to demand perfect information; the costs usually outweigh the benefits. As a result, receiver error is nearly always a part of optimal communication strategies; how much error is tolerated will depend on phylogenetic and physiological limits on signal design, the costs of encoding and decoding signals, effects of the propagation medium, the relative benefits to both parties of reliable signaling, and the decision rules utilized by receivers. Whereas receiver error occurs in both honest and dishonest communication, both parties enjoy a positive value of information only when the signal is honest.

The alternative form of cheating is **exploitation of a sender** by a receiver. A sender sends honest information, but the subsequent decision made by the receiver is one that benefits itself at the expense of the sender. Here the value of information is negative for the sender, but positive for the receiver. Again, the best indicator of cheating is a difference in the sign of the values of information for the two parties.

A BRIEF HISTORY OF HONEST SIGNALING

Early ethological studies of communication focused on the evolutionary origins of signals. As we discussed in Chapter 16, many signals appear to have evolved through the ritualization of behaviors that are or were functionally appropriate to the contexts in which signals are now used. Erection of feathers or fur during agonistic conflicts can have tactical advantages if an animal is likely to be pecked or bitten. Not surprisingly, fur and feather fluffing has been incorporated into many agonistic signals. A bird uncertain whether to attack or flee may vacillate, and the resulting mixture of motor patterns can be ritualized to become a display. If this is the general mechanism by which signals evolve, it is logical to conclude that the actions performed in displays will be accurate indicators of underlying motivations. Put simply, many early ethological studies presumed that most signals were honest because the sources of signals were physiologically or anatomically linked to the motivations of the sender (Cullen 1966; Moynihan 1970; Smith 1977).

With the rise of evolutionary game theory in the 1970s, students of animal communication became more skeptical and even cynical about the honesty of animal signals. If a sender wears its heart on its coatsleeve, what prevents a clever receiver from using this information to exploit the sender? Is it really optimal to announce the limits of one's true willingness to fight at the onset of an agonistic encounter? Should a gazelle try to perform an agility display to an approaching cheetah when it is not agile enough to escape an actual chase? If two animals are engaged in a war-of-attrition contest, should either party announce how long it intends to display? Such a declaration would surely allow the opponent to select a longer display time and thus win.

Dawkins and Krebs (1978; Krebs and Dawkins 1984) went even further. They suggested that senders are likely to be deceitful manipulators, trying to mask their true intentions and trick receivers into actions benefiting the sender; receivers should be mind-readers trying to discount false signals, anticipate the true intent of the sender, and thus identify their own best countermove. The result is a never-ending arms race in which increased deceit and concealment of true intentions by senders is parried by increased discrimination and more efficient exploitation by receivers. Except where sender and receiver have common interests, senders should never reveal their true intentions, and so signals would largely be deceitful or uninformative.

This view was countered by Zahavi (1977a, 1980, 1987, 1993, 1997). He argued that receivers ought not respond to signals unless they were honest. If a receiver does not respond to a signal, then there is no selection on senders to provide one. Receivers should thus have the upper hand in any arms race. Given this control, the optimal strategy for a receiver would be to respond only to those signals that carry some guarantee of honesty. One way to do this is to require that signals impose a cost such that the sender could not afford to produce the signal, or would produce it in an ineffective manner, were the provided information untrue. Zahavi called such costly signals **handicaps**. For example, suppose two dogs threaten each other. In principle, threats could be symbolized by any doggy sound. However, dogs, like most vertebrates, favor harsh, low-frequency sounds as threats and higher pure tonal frequencies to indicate submission (Morton 1977). As we have seen, body size limits the lowest frequencies that a sender can emit at a reasonable amplitude. Thus, insisting on low-frequency threat sounds allows for honest assessment of relative body sizes and assessment of whether further escalation is a good idea or not. Zahavi (1987) added that it may be more difficult to produce a low-frequency sound than a high-frequency one when the vocalizer is tense and prepared to flee. A dog that is confident it would win an escalated fight can afford to relax enough to produce a low frequency, whereas the production of a low-frequency sound by a tense and unsure animal might either be physically impossible or require the relaxation of muscles that should be kept tense in preparation for flight. By insisting on a signal that is reliably linked to the true motivation of the sender, the receiver is thus able to guarantee honest information. As a second example, consider a female bird that wants to mate with the most proficient forager she can locate. One way to compare males is to require each to bring her multiple samples of favored prey. Many birds do just this. Finally, consider a cheetah approaching a gazelle. A display of agility by the gazelle will surely attract the attention of the cheetah to it rather than to alternative prey. Only a truly agile gazelle can hope that the risk of increased conspicuousness will be outweighed by the persuasiveness of its performance. Some other possible handicap signals are shown in Figure 20.1.

What about displays that are highly stereotyped? Doesn't such stereotypy preclude the extraction of honest information about the sender? The classical explanation for stereotypy is that there is an optimal signal form that minimizes distortion during propagation between sender and receiver (pages 461–462). Zahavi proposed an alternative explanation. He argued that stereo-

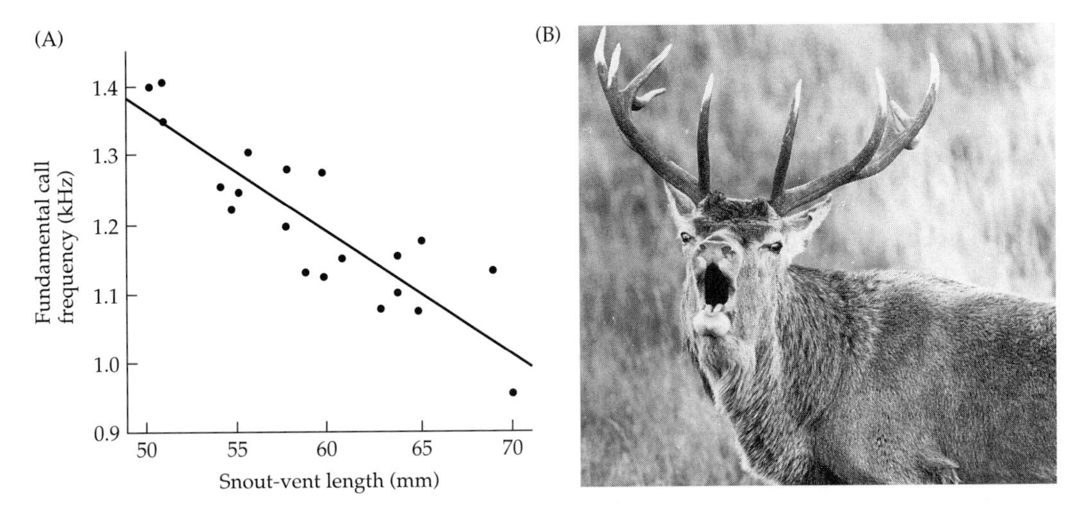

Figure 20.1 Some possible examples of handicap signaling as suggested by Zahavi. (A) Deep croaks in toads. Some workers have suggested that producing a deep frequency call is an uncheatable and honest indicator of male body size. This plot shows the relationship between male body size (as measured by snout to vent length), and the fundamental frequency of 20 male toads' calls. Only larger toads can produce lower fundamental frequencies in their calls. (B) Roaring by male red deer. Stags compete vigorously for females each fall. Before fighting, competing males will often roar at each other. This is an expensive behavior that mimics the use of many muscles and behaviors actually invoked during fighting. It is thought that only those males in good enough condition to fight can produce roaring at a winning level (Clutton-Brock and Albon, 1979). (A after Davies and Halliday 1978; B courtesy of Tim Clutton-Brock.)

typed displays are used by receivers to compare important qualities in senders that are not easily measured by direct assessment. A clever receiver would require that each sender perform a display according to some difficult standard—only those of the highest quality could perform the display with a close fit to the ideal protocol. The result would be a high degree of stereotypy, but just enough remaining variation to assay sender qualities.

Zahavi's suggestions were initially received with considerable scepticism. One reason is that he did not specify formal evolutionary models for how handicaps might evolve, but instead gave many examples that could be plausibly interpreted as handicaps. Different examples on these lists relied on different handicap mechanisms. The earliest attempts to model handicap evolution focused on mechanisms that turn out to be difficult to establish evolutionarily (Maynard Smith 1976b). Some concurrent surveys of displays used in agonistic contexts concluded that signals were rarely honest indicators of the sender's next actions (Caryl 1979; Paton and Caryl 1986). These initial doubts have been succeeded by alternative game-theoretical models that demonstrate the evolutionary plausibility of handicap signaling and verify the need for a costly guarantee if signals are to be honest. The result has been a wider acceptance of Zahavi's principle. In addition, refinements in the methods for assessing sender honesty have resulted in increasing numbers of field and lab studies supporting moderately honest signaling by senders. In the following section, we shall examine some of these more recent models briefly, identify the kinds of traits that might constitute suitable handicaps for each context, and review evidence for and against honesty in animal communication.

GAME MODELS OF SIGNALING INVOKING HANDICAPS

It is unlikely that there is a single communication game for all situations. In Chapter 12, we classified communication exchanges according to the contexts in which the exchanges occur. As we demonstrate in subsequent chapters, these contextual categories differ markedly in the degree to which sender and receiver share common interests, the degree to which a sender might be tempted to deceive the receiver, and the selective forces on the receiver to demand guarantees of honesty. This means that different game-theoretical models may be needed for different contextual categories of signaling.

Within each contextual category, the relevant game may also depend upon the degree to which the information sought by the receiver is provided by signals. As discussed in Chapter 12, a receiver has potential access to as many as four sources of information about any given set of contigencies: prior probabilities of occurrence, direct assessment by the receiver unaided by a sender, direct assessment by a receiver enhanced by amplifier traits, and coded information in sender signals. In any given situation, the relative emphasis on each source of information will vary. The degree to which information sources other than signals are available to verify or test sender veracity would seem to be important in the type of evolutionary game considered. Similarly, the sender has the option of trying to force a receiver to act a certain way (e.g., through tactical behaviors, page 355), combining some elements of force with signals, or just providing signals. The relative weighting of tactical versus signaling components in an exchange may also affect the kind of game modeled.

The nature of the handicap costs will also affect game propriety. We distinguished between two kinds of costs on pages 538–544. **Necessary costs** are intrinsic to communication and are independent of receiver responses. Both parties suffer necessary costs, but it is the necessary costs inflicted on senders that affect signal honesty. **Incidental costs** arise because imperfect signals generate errors in receiver decisions, or because payoff matrices change when signaling is adopted. Only the latter changes are likely to affect senders. The two types of costs guarantee honesty in somewhat different ways. The former are usually experienced before or during signal production. To be handicaps, they thus must be physically linked in some way to the information sought by the receiver. For example, we have noted many times that a small animal will have difficulty using a dipole source to produce low-frequency sounds. In some cases, it might be able to do so, but the requisite energy outlay would surely exceed the benefits. If low-frequency codes for body size, honesty is guaranteed by physical laws and the necessary costs of signal production. The most common incidental cost enforcing honesty is punishment. Consider a solitary monkey that discovers a patch of fruit. It will get more food if no other monkey elects to feed with it. However, if discoverers are expected to advertise their food finds with signals, the discoverer may have a higher payoff by sharing than by being punished when found not sharing. The payoff matrix for the sender is thus different depending upon whether advertisement signals are expected by dominant receivers or not. If the average amount of food consumed by a subordinate monkey declines when signals are produced, there is thus an incidental cost to their giving these signals. Punishment may be a powerful and widespread deterrent to deceit in animal societies (Clutton-Brock and Parker 1995a).

As Dawkins (1993) and Guilford and Dawkins (1995) point out, a reliance on necessary costs to ensure honesty will limit suitable signal forms. This in turn constrains signal optimization for efficient coding, minimal distortion during transmission, and maximal detectability and discriminability by receivers. Where honesty is largely ensured by incidental costs such as punishment, a much wider range of signal forms can be used, and this will affect both the amount of information provided and the values of information for both sender and receiver. The type of honesty-enforcing cost demanded by receivers thus has major repercussions on subsequent evolution of the signal exchange at both the game-theoretical and simple optimality levels.

We now turn to some of the more instructive game-theoretical treatments of honest signaling. These vary in the contextual categories of information provided, the availability of alternative sources of information, and whether costs are necessary or incidental. As a guide, the games are summarized in Table 20.2 (page 669). Despite the variation in game structure and assumptions, we shall see that the basic take-home message of the models is remarkably similar.

Honesty and Agonistic Signals

Consider two equally matched opponents that both seek the same resource. They differ in the value that they place on this commodity. For example, the same item of food might be much more important to a starving animal than to a sated one. Should they end up fighting, it is likely that the animal that values the resource the most will fight harder and/or longer and thus win. It might seem optimal for each party to signal their perceived value of the resource at the outset. This signal would lead to the same outcome as if they had the fight but without all the risks and costs. However, this peaceful strategy could be easily invaded by a cheat that always signaled a very high valuation. Are there any conditions when it would still pay to provide an honest signal?

One version of this game has been examined by Enquist and colleagues (Enquist et al. 1985). In this model, information about valuation is provided by the sender through its choice of alternative actions. These could be pure displays or they could be a mixture of display and tactical actions. Relevant sender costs could be necessary, incidental, or some mixture of the two. Available sender options will vary in the degree to which they inflict such costs. For example, a display close to an opponent is much more risky than one performed at a distance. At the same time, displays are likely to vary in their ef-

Table 20.1 Cost, effectiveness, and honesty of agonistic displays of American goldfinches (*Carduelis tristis*) during conflicts over access to seeds at a bird feeder

Sender display[a]	Receiver responses (%)[b]				
	LF	HHF	WF	Attack	Retreat
LHF	15.5	24.1	60.3	0	0
HHF	3.8	7.6	74.7	2.5	11.4
WF	0	0.2	48.8	31.1	19.9

Sender display[a]	Sender's next act (%)[c]					
	LF	HHF	WF	Attack	Retreat	Win
LHF	0	12.1	67.2	5.2	15.5	0
HHF	0	2.5	62.0	13.9	10.1	11.4
WF	0	0	11.1	19.7	40.3	28.9

[a]There are three main displays: low-intensity head forward (LHF) in which displayer faces opponent with neck partially extended; high-intensity head forward (HHF) with the next further extended, head lowered, and partially open bill pointed at opponent; and wing flap (WF), similar to the HHF with the addition of raised and spread wings.

[b]The percentage of receiver responses to each sender display. Using the probability that a receiver will subsequently retreat as an index of display effectiveness, the three displays can be ranked as LHF < HHF < WF. Similarly, if the cost of each display is the chance that it will elicit an attack by the receiver, the displays can again be ranked in the order LHF < HHF < WF. If the receiver responds by displaying, it usually selects a more effective display and thus increases the risk to the sender. Thus, as predicted by Enquist et al. (1985), display effectiveness and cost are positively correlated.

[c]The next act performed by senders after giving each type of display. Senders giving a higher-intensity display were more likely to follow the display with attack; if the next act of the sender was a display, it was never a less-effective display and, where possible, usually a more effective one. Both suggest that the sender's choice of display is an honest indicator of its intentions. The probability of subsequent sender retreat increases only for WF displays, an effect often due to loss of the contest by the sender (Popp 1987).

fectiveness. For example, a loud growl accompanied by a swipe with bared claws may partly disable an opponent, whereas a snort from a distance will do little to alter the outcome of a fight. Let $-C_i$ be the average cost to the sender of display i (including both the costs of performance and any increases in the risk of subsequent injury), P_i the probability of winning the interaction if this display is used (e.g. the effectiveness), and V the perceived value of winning the contest from the point of view of the animal choosing which display to perform. The average payoff of performing any given display is thus $\overline{PO}(i) = P_iV - C_i$. If we plot the average payoff for any display i against various values of V, we shall obtain a straight line with a slope equal to P_i and an intercept on the ordinate of $-C_i$. All such display lines will have a positive slope; however, displays with higher effectiveness will have steeper slopes, and those with higher costs will have a lower intercept.

It seems most likely that the higher the effectiveness of a display (hence the steeper its slope), the higher the cost of that display to the sender. In Figure 20.2, Displays 3 and 4 show this type of positive correlation between effectiveness and cost. Display 3 is more effective, but its higher cost causes it to begin lower down on the ordinate axis. The result is that the two lines cross at some point. For V values greater than that at the crossing point, the sender gets a higher payoff by choosing to perform Display 3; for lower V values, it should perform Display 4. Its choice of a display will thus be an honest indicator of how much it values the resource and hence how much it would be willing to fight an escalated battle. Put another way, a positive correlation between signal effectiveness and the absolute value of costs leads to honest signals. What about the converse? Does honest signaling require a positive correlation between signal costs and effectiveness? Figure 20.2 shows lines for two other displays for which effectiveness and costs are negatively correlated. Display 1 has higher effectiveness than Display 2, but it has a lower cost. In this

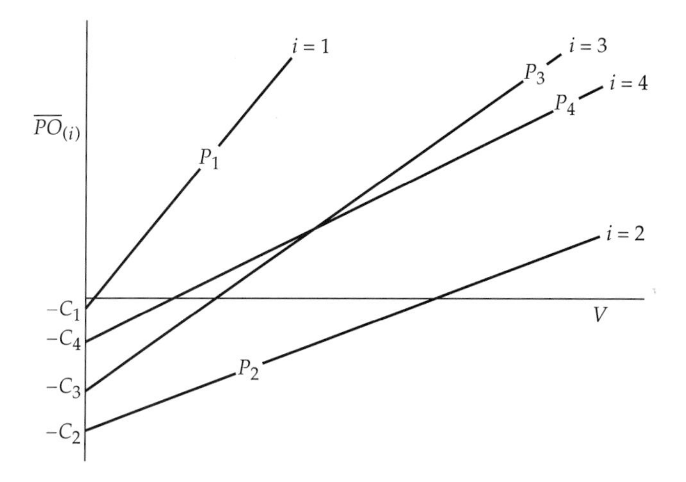

Figure 20.2 Enquist and colleagues' (1985) model for honest signaling by choice of display in agonistic encounters. This model assumes that each player can choose any of several displays or actions. These displays differ in the cost ($-C_i$) to the sender of producing display i, and in the display's effectiveness, characterized here by the probability (P_i) that using that its performance will lead to winning the contest. For any perceived value (V) of the prize to be won, the average payoff of using display i is $\overline{PO}(i) = P_iV - C_i$. The lines in this example show the average payoff values as V varies for four different displays. If Display 1 is an option, it is clearly the best, as it yields the highest average payoff for any V. This is because it has a very high value of P_i (which determines the slope of the payoff line) and a very small value of C_i (leading to a high intercept of the payoff line with the ordinate axis). Similarly, the animal should never use Display 2, as it always has the lowest payoff as a result of a small P_i and large C_i. As a rule, one expects higher effectiveness to require higher costs. This type of pattern is shown with Displays 3 and 4. Although the higher cost suffered by performing Display 3 instead of Display 4 leads to lower payoffs at low values of V, the lines cross as V is increased, and at higher V values, the steeper slope of the Display 3 line eventually makes it the better option. The optimal animal will use Display 4 if it does not value the prize highly, but will choose Display 3 if it has a high valuation of the prize. Where display effectiveness and cost are positively correlated (as with Displays 3 and 4), an animal's choice of behavior will be an honest indication to its opponent of its motivation and subsequent willingness to fight.

case, the optimal choice is to use the same option (Display 1) regardless of perceived V value. Signaling here provides no information about sender motivation. At least in this model of agonistic contests, honest signaling is likely only if costs and effectiveness are positively correlated. A more formal analysis of this evolutionary game is given by Enquist (1985).

This model thus supports Zahavi's handicap notion. Sender costs are necessary for honest signaling, and if these costs have the right relationship to benefits to make cheating uneconomical, honest signaling is the ESS outcome. What evidence is there that display costs and effectiveness are positively correlated in real agonistic contests? Popp (1987) focused on precisely this question using aggressive interactions between goldfinches at bird feeders during the winter. The results show a clear correlation between the risks of subsequent attack and the effectiveness of different displays (Table 20.2). In addition, as the value of the resource increased (either because of decreased food or increased needs due to low temperatures), birds were more likely to adopt the riskier displays. These results fit the Enquist et al. (1985) model nicely. Several other studies also show support for honest agonistic signaling when costs and effectiveness are positively correlated (Andersson 1976; Enquist et al. 1985; Hansen 1986; Nelson 1984; Waas 1991a,b). One reason these studies contradict the earlier reviews of Caryl (1979) and Paton and Caryl (1986) may be the better controls for complex sequential interactions in the more recent papers (see pages 704–706).

Honesty and Courtship

Mate choice was the original context that spurred Zahavi (1975) to develop his handicap principle. In the typical scenario, females seek to choose mates based on some male quality that is difficult to assess directly. Females might seek males that are good foragers, those with superior immune systems, those most able to protect a female against harassment, and so on. This choice could lead to direct benefits for the female's survival and/or fecundity, or to indirect benefits through the acquisition of good genes for her offspring. Because the quality of interest is hard for females to assess directly, males that honestly signaled their quality value would greatly facilitate female choice. In species where females have the upper hand in mate selection, there would thus be strong selection favoring males that produced honest signals. However, the usual problem arises: Why should lower-quality males produce honest signals?

Grafen (1990a,b) modeled this communication game as a continuous, asymmetric scramble. In his model, males vary continuously in the quality females wish to assess. Males can also differ continuously in at least one parameter of the courtship display that they perform to females. For example, the variable parameter might be display intensity. Alternative male strategies differ according to the function by which the variable display parameter codes for sender quality. Males adopting an uninformative strategy display at the same intensity regardless of their quality values; males adopting an informative and honest strategy adjust the intensity of their display to reflect their relative quality. Females in this model observe courting males and use their perceived differences in male display to infer relative male qualities. Female strategies differ in the function by which they translate these perceived display differences into estimates of male quality. Thus females adopting a sceptical strategy might treat all display intensities the same. That is, they would ignore any signal information. Others might use a strategy in which the ranks of perceived display intensities were used to estimate the quality ranks of males. Male fitness in this model depends upon (A) the true quality value of that male, (B) the necessary costs it pays to perform its display, and (C) the degree to which females infer from its signals that one male has a higher quality than other males. Note that this last emphasis on relative inferred quality makes this game a scramble because the payoff for any one male depends on what other males are doing and how they are perceived. Female fitness in the model is assumed to be maximal when female estimates of male quality are closest to true male quality. The more accurately females assess males, the higher their fitness. Clearly this game is asymmetric with different available strategies and different payoffs for the two roles (here, the two sexes).

This is a difficult game to solve. With some reasonable assumptions and simplifications, Grafen showed that there are two ESSs: (a) males do not display honestly, and females ignore all signals; and (b) all or most males are honest, and all or most females attend to signals. The latter ESS can occur if and only if four specific conditions are met. These are illustrated in Figure 20.3. One condition follows directly from the assumptions of the model. It must be the case that the higher the perceived quality of a male, the more likely females are to mate with him. That is the whole point of female assessment in the first place. A second condition for the ESS, and one that supports Zahavi's general principle, is that signaling must be costly to male senders. If we ignore subsequent reproductive benefits, a male's fitness must decrease as a result of performing the signal. The third condition is the one that actually guarantees honesty: a given investment in display must cost a high-quality male less than that same display investment would cost a poor-quality male. How does this ensure honesty? Male mating success depends upon which male displays at the highest intensity. This could lead to an arms race in which successive males increased their display effort until they were just ahead of the competition. The resulting escalation would lead to higher and higher mean costs of display. Note that the number of females waiting to mate remains constant. Thus the benefits of being top male would not change, whereas the costs keep going up. Eventually, costs would exceed benefits and no further escalation would be favored. The third condition in Grafen's model thus states that the break-even point halting escalation will occur at a lower total cost for low-quality males than for high-quality ones. The fourth and final condition requires that low- and high-quality males who are perceived to have the same quality (either due to female error or to higher display investment by the low-quality male) are treated the same as potential mates by females. This ensures that the higher costs that low-quality males pay to display at a given level are not more than made up by special treatment from females later on.

Mate choice is one driving force in **sexual selection**. This is the process Darwin invoked to explain why members of the sex in greatest competition for mates (usually males) should exhibit traits that so often reduce their via-

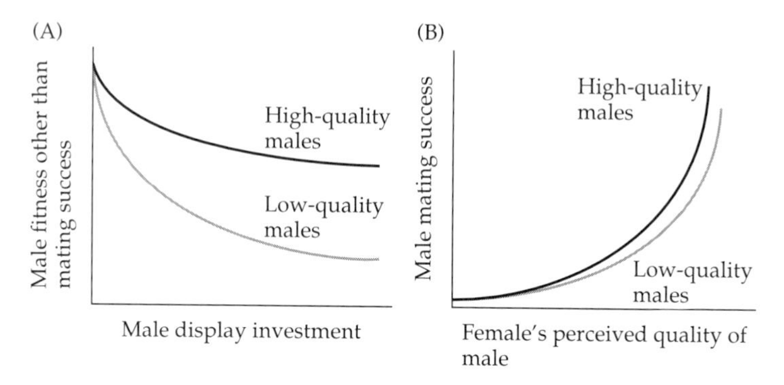

Figure 20.3 Four basic conditions required for honest courtship signaling to be an ESS. (A) Male fitness, holding the effects of mating success fixed, must decrease as male display investment increases. Thus both lines in this plot decrease as display investment increases. This simply means that signaling has to impose a fitness cost on the senders. The second condition is that, for any given investment in display, the fitness costs must be less for high-quality males than for low-quality ones. Thus the fitness curve in the plot for low-quality males must always lie below that of high-quality ones. (B) A third condition is that the higher in quality that females perceive a male to be, the more likely he is to mate. This is the starting point for the entire game, namely that females prefer to mate with high-quality males. Finally, the rate at which male mating success increases with female perception of their quality must be no higher for low-quality males than for high-quality ones. Put another way, low-quality males are treated the same or worse by females when perceived as having the same quality (albeit erroneously) as high-quality males. (After Grafen 1990a,b.)

bility. As we shall discuss in detail in Chapter 23, sexual selection must be, at least in part, a genetic process. Grafen's models of courtship signal evolution explicitly exclude some of these genetic processes. However, his goal was to show that even without special genetic processes, courtship signals could evolve as long as they were handicaps. A variety of genetic models can also explain the evolution of such signals, and it is reassuring that these predict the evolution of costly courtship signals only if low- and high-quality males differ in their ability to produce the signals (Andersson 1994; Iwasa and Pomiankowski 1991). Despite very different assumptions, the game-theoretical and genetic models have largely led to similar conclusions about the necessity for handicap costs in signal evolution.

Testing the handicap principle in male advertisement signals has proved challenging. Where the benefits to females of choosing a male affect their fecundity or survival directly, the qualities that females seek can often be identified and predictions tested. Where the benefits are paternal traits inherited by offspring, the task is more difficult. Despite the difficulties, there are increasing numbers of studies supporting honest handicap signaling during courtship. For example, Knapp and Kovach (1991) showed that the display rates of male damselfish were an accurate indicator of the abilities of males to tend the eggs laid for them by females (Figure 20.4). Females would benefit directly by selecting male mates with high rates of display, and this is in fact

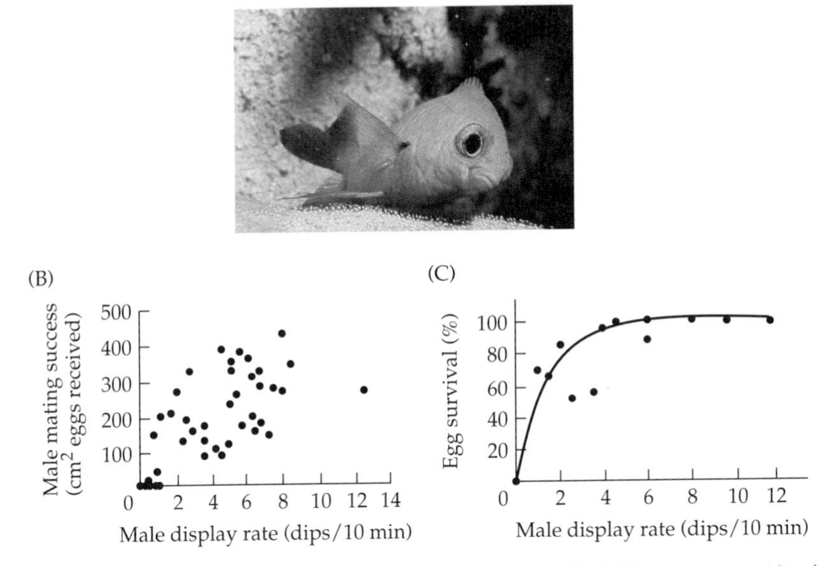

Figure 20.4 Honest courtship displays by male damselfish (*Stegastes partitus*). (A) Male damselfish tending the mass of eggs laid for it by the females with which it mated. (B) Male damselfish mating success, measured as number of eggs laid in the territory, versus male display rate measured as dips/min. Males with higher rates of display attract more females and thus obtain more fertilized eggs to tend ($r = 0.85$; $p < 0.0001$, $N = 48$). (C) Egg survival as a function of male display rate. Males with higher display rates are better parents. (A courtesy of Ken Clifton; B and C after Knapp and Kovach 1991.)

what is observed. Display rate is an honest but costly indicator of male paternal ability that is used by female receivers to select mates. Another example is the use by displaying males of red or orange carotenoid pigments in their plumage, skin, or scales (see page 549). No vertebrate can synthesize carotenoid pigments; they can only be acquired by foraging, and different levels of coloration should be honest indicators of male foraging abilities. In guppies and house finches, this appears to be the case; the better the male as a forager of sources of carotenoid pigments, the redder his coloration (Endler 1980, 1983; Hill 1991, 1994; Kodric-Brown, 1989). Females in these species appear to rely, at least in part, on male coloration in the selection of mates (Figure 20.5; Milinski and Bakker 1990). We shall take up other examples of honest courtship signaling in Chapter 23.

One final comment about Grafen's model. Because he explicitly excluded genetic processes peculiar to sexual selection, his model can be applied to any kind of signaling game in which a sceptical receiver must deal with a possibly devious sender. Grafen has thus provided the closest thing to a general communication game that we shall discuss in this chapter.

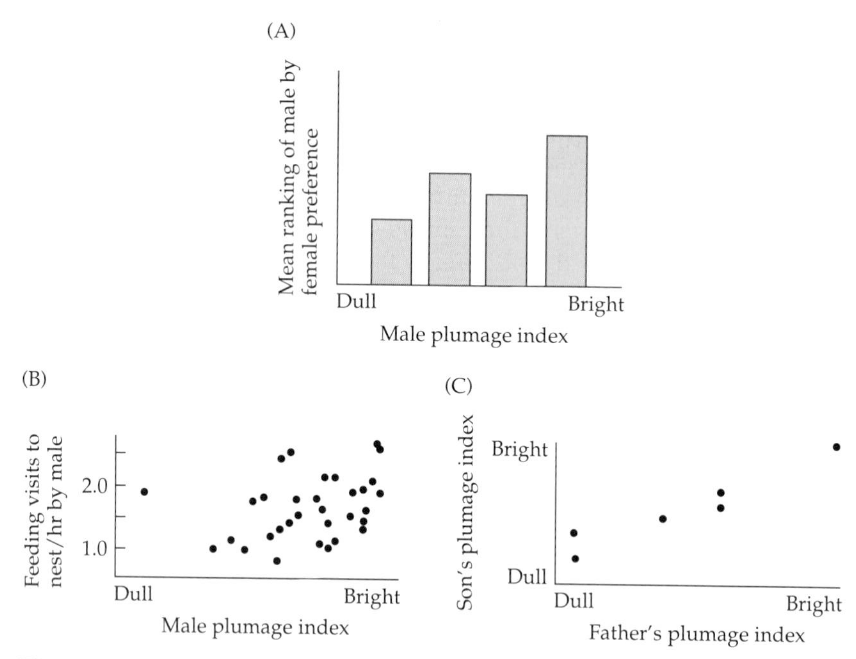

Figure 20.5 Female choice of mate and carotenoid colors of male house finches (*Carpodacus mexicanus*). House finch males vary in levels of carotenoid pigments in their feathers. Some are yellow and have little coloration (dull plumage index), whereas others have lots of conspicuous red coloration (bright plumage index). (A) Captive females offered proximity to four males of different plumage indices show significant preferences for brighter males ($p < 0.05$; $N = 14$). (B) The number of feeding trips made to the nest by wild male house finches is higher for brighter plumaged males ($r_s = 0.63$; $p < 0.05$; $N = 13$). Male coloration may thus be an honest indicator of male paternal qualities. (C) Females may also benefit by mating with bright males because they are likely to father sons who will themselves have bright plumage, be preferred by females, and thus be successful. Plot shows correlation between father plumage index and that of sons ($r_s = 0.91$; $p < 0.01$; $N = 6$; only one son from each nest was included to preserve independence of points). (After Hill 1990, 1991.)

Honesty and Badges of Status

Once animals that are capable of individual recognition have fought each other, it may pay to avoid escalation in future encounters because the outcome would be largely predictable. Even without individual recognition, if individuals that frequently won escalated contests exhibited some **badge** reflecting their currently dominant status, this might save all parties from unnecessary risk and injury (Rohwer 1975, 1982). One apparent example is shown in Figure 20.6. As discussed on pages 637–640, it often pays animals to adopt submissive behaviors when their chances of winning an escalated fight or the ratio of benefits to costs is low. Wearing or not wearing a badge makes observance of such status roles easier. But as with agonistic and courtship signals, one cannot help wondering what would prevent a sender from cheating by sporting a badge exaggerating its prior contest success or its willingness to fight if challenged.

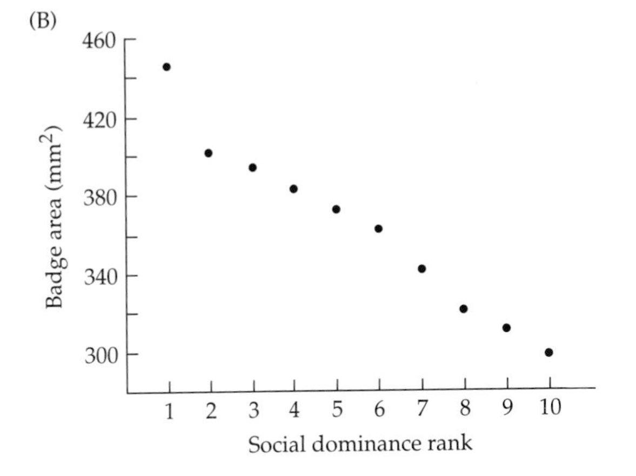

Figure 20.6 Badges of status in male house sparrows (*Passer domesticus*). (A) Variation in badge size among adult males. (B) Plot of male badge area versus social dominance rank in one flock of house sparrows captured in Denmark. Smaller numbers on the abscissa mean higher dominance rank. (After Møller 1987a,b.)

Maynard Smith and Harper (1988) modeled one badge-of-status game. They noted that badges in many species are better correlated with aggressive motivation than with fighting ability (although the two may be linked to some degree). They therefore considered a continuous symmetric contest in which players select both level of aggressiveness and size of status badges. Honest signaling would consist of a stable mixture of aggressive animals with large badges and nonaggressive ones with small badges. They found that such an honest mixture was an ESS only if the costs of escalated fights, relative to the benefits of winning, increased rapidly with increasing badge size, and animals with large badges were often challenged by others with large badges. Cheating would be disadvantageous because sporting large badges but lacking the corresponding motivation to win would be very dangerous. Again, we find that honesty exists only if there are signaling costs that are most felt by deceitful senders. As with the dominance game discussed in the prior chapter, honest signaling and observance of status ranks is favored only when the benefit to cost ratios or the chances of winning are sufficiently low; when either or both are high, even less aggressive players should ignore the signals and escalate.

The situation is a bit more complicated than this model implies because it is possible to identify additional mutant strategies that can invade a mixture of honest badges. For example, several studies have considered a cryptic aggressor strategy in which a very aggressive animal exhibits a small badge. In Owens and Hartley's (1991) version, this "Trojan sparrow" is nonaggressive when contested resources are abundant, but exhibits its true aggressiveness when they are scarce. Johnstone and Norris (1993) examined a small-badged mutant that is never challenged and attacked by animals with large badges, but itself attacks and wins fights with less aggressive small-badged individuals. Either of these mutants would invade and destabilize the honest ESS mixture described by Maynard Smith and Harper. However, Johnstone and Norris have also shown that if there is a common cost to being aggressive, whether one engages in escalated fights or not, cryptic aggressors cannot invade and the honest mixture remains an ESS. This turns out to be true whether the costs of aggression are the same for all players, or instead individually variable. If the latter is true, the expected outcome is for animals that can best sustain the costs of aggression to sport large badges and be most aggressive; those with less resilience to these costs will have small badges and be subordinate. They note that the badges would then constitute honest indicators of physical condition, and be useful for functions above and beyond social status. For example, females might use badge size as an indicator of suitor health even though the original function of the badges was to signal aggressive status. There is evidence in several bird species that badges both determine male status and attract female mates (Møller 1988a; Norris 1990a,b).

Johnstone and Norris thus argue that two different costs must exist to ensure honest badges of status: a risk that cheats will be challenged and suffer the full costs of fights, and a general cost of being aggressive that is independent of badge size or escalation rates. What might impose the latter costs? They and other authors note that testosterone is frequently a mechanism for modulating aggression in vertebrate animals. Testosterone levels are also known to impact immune systems adversely (Alexander and Stimson 1988; Folstad and Karter 1992; Grossman 1985; Wedekind 1992; Zuk 1990). The high testosterone required to be aggressive may thus impose a cost on the animal's immune system. High testosterone levels may also reduce survivorship by increasing metabolic or time costs or by increasing predation risks (Hogstad 1987; Røskaft et al. 1986). This appears to be the case in at least one lizard species (Marler and Moore 1988, 1989).

Badges clearly play a role in settling priority of access to resources or mates in some species; not surprisingly given the models, they appear to be honored only when benefit to cost ratios or the chances of subordinates winning a contest are low (Evans 1991; Evans and Hatchwell 1991; Fugle et al. 1984; Hansen and Rohwer 1986; Jarvi and Bakken 1984; Marchetti 1993; Møller 1987a,b, 1988a; Norris 1990a; Parsons and Baptista 1980; Petrie 1988; Rohwer 1977, 1985; Rohwer and Ewald 1981; Røskaft and Rohwer 1987; Searcy 1979; Studd and Robertson 1985; Whitfield 1987). In house sparrows, both types of costs required by the Johnstone and Norris (1993) models have been described. The bearers of deceitfully large badges are severely punished by other large-badged birds (Møller 1987b), and large-badged birds bear a contest-independent cost of increased autumn mortality (Møller 1989). Not all studies have shown the presence of both kinds of costs. In fact, in several other species of sparrows, experimental enlargement of badge size allowed cheats to prosper without serious punishment (Fugle and Rothstein 1987; Rohwer and Rowher 1978). In at least one case, the badges seemed not to be indicators of status but instead related to differential foraging roles within the group (Rohwer and Ewald 1981).

Honesty and Begging

In mate choice and agonistic situations, a receiver seeks information to identify which of several responses will best benefit itself. The fact that the receiver's choice may benefit the sender is often incidental to the receiver's interest in the interaction. Begging is somewhat different. Here the sender tries to persuade the receiver to perform an action whose sole purpose is to benefit the sender at a direct cost to the receiver. Presumably, receivers would not perform this action unless there were some compensatory deferred or indirect benefits. For example, the sender and receiver might be genetic relatives (kin selection). Alternatively, perhaps helping the sender now will ensure help to the receiver should the roles be reversed at a later time (reciprocity). An honest sender would only beg when it truly was in need, and it would adjust its level of begging to match its need. But why should a beggar be honest? What can receivers do to prevent being deceived about sender needs?

This situation has been modeled by Maynard Smith (1991, 1994) as the Sir Philip Sidney game. The name refers to a story about a wounded British officer who had to choose between donating his last water to a begging wounded soldier or keeping it for himself. Maynard Smith's version of the game was a 3×3 discrete asymmetric contest. The possible sender strategies were: (a) only beg when in need, (b) always beg regardless of need, and (c) never beg. Receiver strategies were: (a) only give to senders when they beg, (b) always give to senders regardless of their actions, and (c) never give to senders even if they beg. As always, the solution of this game depends on the relative values of benefits and costs accruing to each party given each combination of strategies played. Maynard Smith assumed that the two parties were at least partially related genetically, and thus allowed for a kin selection repayment to the donor. When critical parameters are favorable for communication despite a conflict of interest, there are two ESSs: (a) the sender never begs and the receiver ignores all signals, and (b) the sender only begs when in need, and the receiver only gives when the sender begs. The latter ESS, which is basically honest signaling, turns out to be possible only if begging imposes some nonzero cost on senders. This cost could be a loss of energy or time during begging (necessary costs), a risk of later punishment if found cheating (incidental costs), or both. We thus find again that honest signaling requires imposition of some handicap cost on the sender. One other outcome of Maynard Smith's analysis is that if sender and receiver do not have a conflict of interest, honest signaling can then evolve without the requisite handicap costs; note that only a similar ranking of alternative outcomes, and not a quantitative identity of interests, is required in this case.

The Sir Philip Sidney game has also been modeled by Godfray (1991) and by Johnstone and Grafen (1992a) with both sender need and sender begging level as continuous variables (a continuous asymmetric contest). In the latter study, the two parties are relatives, but the coefficient of genetic relatedness between them is allowed to vary continuously. Not surprisingly, receivers never give unless the sender is sufficiently closely related. Thus it does not pay the sender to beg much from distant relatives. By the same token, receivers give readily to very close relatives; the latter need not beg much to elicit the response. As a result, the ESS is for senders to adjust begging effort, and thus costs, as a function of genetic relatedness to the receiver; maximal costs will be expended by a sender begging from relatives of intermediate relatedness. And it is again these costs that guarantee honesty.

Begging, broadly defined, occurs in many animal species. Perhaps the best studied system is begging by altricial nestling birds (Figure 20.7). It is widely recognized that parents and nestlings are likely to differ in the amount of investment a parent should make in any given offspring (Bengtsson and Ryden 1983; Godfray 1991, 1995b; Gottlander 1987; Harper 1986; Henderson 1975; Hussell 1988; McGillvray and Levenson 1986; Mondloch 1995; Redondo and Castro 1992a; Stamps et al. 1989; Trivers 1974). For nests with a single offspring, the parents must optimally allocate their efforts and risks between this current individual and any future ones; the current nestling will usually want more care than is optimal for the parents to give and may thus be tempted to beg dishonestly. When nests contain multiple offspring, parents must choose between equal allocations to all nestlings, or favored investments in selected individuals. Should nestlings beg according to need or should even those without need beg vigorously? The single nestling case has been modeled by Godfray (1991). An ESS for honest begging by a nestling is assured only if begging is costly, and the benefits of the begging are higher for those truly in need. Note the difference from other functional categories where senders differ in the costs but not in the benefits of signaling. Sender costs are still required for an ESS. What might they be? In addition to the energy expended during vigorous begging, several authors have pointed out that begging makes nestlings more conspicuous to nest predators (Haskell 1994; Redondo and Castro 1992b). We discuss these models further in Chapter 24.

Begging occurs in many other contexts but has received less theoretical attention. Juvenile primates use special expressions and sounds to solicit food tidbits from adults, and adult chimpanzees will beg from other adults that have just killed a monkey (Boesch and Boesch 1983, 1989). Female magpie jays beg food from visiting males to avoid having to leave the nest (Langen 1996a). Adult vampire bats that have failed to feed return to their day roost at dawn and use specific behaviors to beg for regurgitated blood from roostmates (Wilkinson 1984). Social insects such as bees, ants, wasps, and termites beg for food collected by returning foragers (Michelsen et al. 1986; Wilson 1971). Individuals in at least the insect colonies are highly related genetically; the Johnstone and Grafen model would predict that begging among the insects would require only minor sender costs to guarantee honesty. Distress calls in which a frightened or threatened animal signals for assistance also may be considered begging (but see discussion on pages 846–847). In few of these examples have both the costs of signaling and the degree of honesty been assessed.

Figure 20.7 Begging signals by altricial nestling birds. Typical nestlings wag their heads, show brightly patterned and gaping mouths, and emit sounds as they beg for food from parents. Game-theoretical models suggest that nestling begging will only be honest if signal production is costly and the benefits of begging are higher for those that most need feeding (Godfray 1991, 1995b). Although there seems to be little energetic cost to nestling begging (McCarty 1996), increased attraction of predators when nestlings beg may provide sufficient incentive to insure signal honesty (Haskell 1994; Redondo and Castro 1992b). (Photo courtesy of Marc Dantzker.)

Honesty, Amplifiers, and Attenuators

Amplifiers are a special case of a handicap whose very function penalizes those individuals most tempted to cheat. The aim of such traits is to make it easier for direct assessment of sender qualities by a receiver. Clearly, animals of low quality would do better to hinder such assessment, not promote it. Exhibiting an amplifier thus has a higher cost (and probably less advantage) to a lower-quality animal than to a higher-quality one. Attenuators are the opposite type of signal; they make direct assessment of some trait by a receiver more difficult.

Can amplifiers evolve despite the costs to low-quality individuals? Amplifier evolution has only been examined with genetic models, but the results are similar to those one would obtain using game theory (Hasson 1989, 1990; Hasson et al. 1992). Regardless of genetic assumptions, amplifiers can evolve as long as high-quality individuals are sufficiently common, the benefits to high-quality individuals of displaying an amplifier exceed the costs, and low- and high-quality individuals differ sufficiently in fitness that average fitness benefits of the trait to high-quality individuals are greater than the average costs to fitness suffered by low-quality ones. Note that if low-quality individuals respond to the costs by failing to exhibit the amplifier, the simple presence or absence of the amplifier trait becomes an honest indicator of sender quality (e.g., the amplifier evolves into a signal). Note also that the evolution of amplifiers does not require an additional cost to guarantee honesty; honesty is the whole point of amplifier function.

Attenuators can evolve in a population if low-quality individuals are sufficiently common relative to high-quality ones (Hasson et al. 1992). However, the dynamics are different from those of amplifiers. Amplifier expression leads selectively to higher fitnesses for high-quality individuals. The latter become relatively more abundant over time, and this eventually favors amplifiers over attenuators. Attenuators, if successful, result in random receiver choice and thus do not selectively favor low-quality individuals in particular. This makes the evolution of attenuators slower and less stable than amplifiers.

Honesty and Predator Notification

A final type of signal we shall consider here is that sent by prey to dissuade predators from chasing them. There are two classes of such signals (see detailed treatment in Chapter 25). First, the prey may perform some display of agility, speed, or stamina that indicates its ability to escape if chased. One example is stotting by gazelles to a nearby cheetah or wild dog (Fitzgibbon and Fanshawe 1988). The second class of predator signal is some action (e.g., a snort, whistle, or tail flash) that all prey perform the same way to let the predator know it has been spotted and therefore attack is unlikely to be successful. Both kinds of predator signal can evolve into honest indicators of relative prey vulnerability. How well the first class of signals is performed can be an honest measure of prey condition. The proximity that prey allow a predator to achieve before giving the second class of signal can be an honest measure of the prey's confidence that it would escape any attack. Vega-Redondo and Hasson (1993) have shown that honest indications of vulnerability are only evolutionarily stable if there is a cost to the sender that is greatest for low-condition animals. The energetic costs and the drawing of predator attention would seem to meet these conditions for the first class of signal. Being too close to a predator before giving a signal of the second class would also constitute such a cost. As with most other cases we have examined, differential costs appear to be a necessary condition for honest signaling of prey vulnerability.

ERRORS, SIGNAL EVOLUTION, AND HONESTY

The prior models all imply that deceit will be rare or absent in animals. Yet, we know that this is not the case. Mantis shrimp that have recently moulted and are thus vulnerable in an escalated fight will bluff and produce threat displays as if there were no problem (Figure 20.8; Adams and Caldwell 1990; Steger and Caldwell 1983). Foraging birds will falsely emit alarm calls to scare competitors away from food finds (Møller 1988b). A number of primates are known to practice quite complicated patterns of deceit (Byrne and Whiten 1988; de Waal 1986). Although deceit is less common than honesty, it does seem widespread in nature. In fact, mixtures of honesty and deceit may be the rule (Bond 1989; Dawkins 1993; Gardner and Morris 1989). There are at least three explanations for why some deceit may be present even when most signals are honest: (a) perceptual error by receivers allows some cheaters to escape detection, (b) evolving signaling systems have yet to reach the ESS, and (c) a single type of receiver may have to deal with multiple senders. We take up each of these in turn below.

Table 20.2 Summary of evolutionary games on honest signaling

Function	Sender strategies	Game type	Honesty conditions	References
Agonistic contests	Select display according to effectiveness and cost to sender. Honest sender selects display indicating true motivation; selection of dishonest sender exaggerates true motivation.	Discrete asymmetric contest	Requires a positive correlation between display effectiveness and sender cost.	Enquist 1985; Enquist et al. 1985
Courtship	Display with intensity higher than competitors, to attract mates. Honest senders adjust intensity to match relative quality assayed by females; dishonest senders exaggerate by giving higher intensity than justified by quality.	Continuous asymmetric scramble	Display must be costly to senders, with greater costs at given intensity for lower-quality males.	Grafen 1990a,b
Badges of status	Display badge with size indicating dominance rank. Honest senders adjust badge size to reflect true status; dishonest senders sport badge with either too large or too small a size.	Continuous symmetric contest	Cost of escalated fights must increase with badge size, and large-badged animals must be challenged often by other large-badged animals.	Maynard Smith and Harper 1988
		Discrete symmetric contest	Same as above plus there must be a contest-independent cost of being aggressive.	Owens and Hartley 1991; Johnstone and Norris 1993
Begging (Sir Philip Sidney game)	Sender signals demand for help to receiver. Honest senders only signal when in need; dishonest senders always signal. Receiver benefits only indirectly from giving to sender.	Discrete asymmetric contest	Begging must be costly to senders.	Maynard Smith 1991
		Continuous asymmetric contest	Sender costs are highest when an intermediate level of relatedness between sender and receiver exists.	Godfray 1991; Johnstone and Grafen 1992
Amplifiers	Display trait facilitating accurate direct assessment of sender qualities. Honest sender sports amplifier; dishonest shows attenuator.	Genetic models	Amplifiers can evolve if average benefits to high-quality senders are greater than average costs to low-quality ones.	Hasson 1989b, 1990; Hasson et al. 1992; Michod and Hasson 1990
Predator notification	Prey display to predator that they are not worth chasing. Honest sender shows true agility or condition; dishonest sender uses noninformative display.	Continuous asymmetric contest	Display must be costly, with lower-quality senders paying higher cost for a given display level.	Vega-Redondo and Hasson 1993

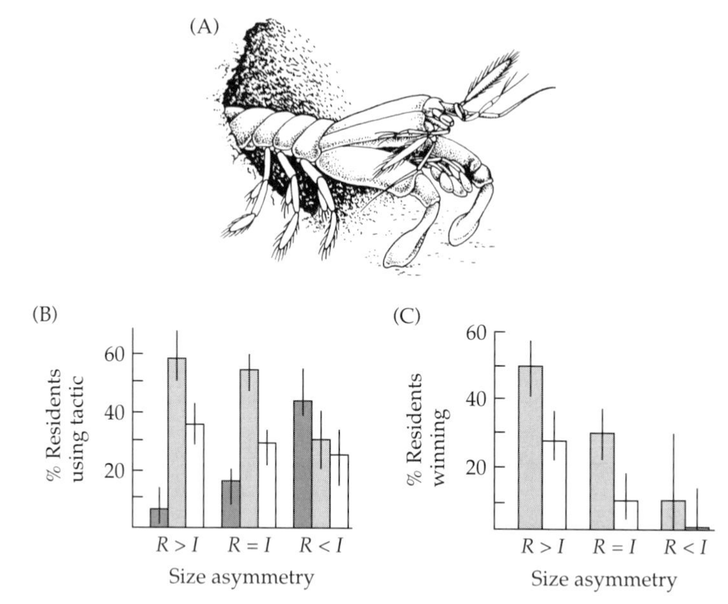

Figure 20.8 Bluff in the threat display of resident mantis shrimp (*Gonodactylus bredini*). Tropical mantis shrimp live solitarily in burrows in coral reefs. Suitable holes are in short supply, and fights between residents and intruders are vigorous and even lethal. (A) One of several displays used by a resident to threaten approaching intruders that are not too much larger than the resident. When intruders are not larger, threat display is sufficient to ward off attack in more than half of intrusions. Residents often flee without display if intruders are sufficiently larger. Adult mantis shrimp moult every two months. For three days postmolt, their exoskeletons are soft and they are easily injured or killed in an escalated fight. At most, 20% of animals on a reef are in this vulnerable condition. Despite their soft exoskeletons, many recently moulted residents bluff by giving the threat display even though they will be unable to back it up with attack. (B) Fractions of newly moulted residents that flee (dark bars), threaten (hatched bars), or do nothing (open bars) as intruders of different relative sizes approach. *R* and *I* refer to relative sizes of resident and intruder respectively. Threats are more common than fleeing or inaction when intruders are no larger than residents, but flight becomes the dominant response when intruders are larger ($G = 39.89$; $df = 4$; $p < 0.001$). (C) Fractions of intrusions won by newly moulted resident if it threatens the intruder (hatched bars) versus does nothing (open bars) for different relative sizes of intruder. Chances of retaining residence are significantly enhanced ($p < 0.05$ or smaller) if newly moulted resident threatens intruders of similar or smaller size. Thus these animals show a low but persistent level of dishonest signaling that is significantly effective to the detriment of receivers (Adams and Caldwell 1990; Steger and Caldwell 1983). (A after Trivers 1985; B, C after Adams and Caldwell, 1990.)

Signaling Games When Receivers Have Perceptual Errors

All of the models in the prior section ignore receiver error due to causes other than dishonesty. However, we saw in Part II that few signaling systems will favor (for economic reasons) or allow (for physical ones) the provision of perfect information and that some receiver error is the rule. Dawkins and Guilford (1991) have argued that it may be too costly for receivers to process all the information that would be required for honest signaling, even were the sender to provide it. Does the incorporation of receiver error not due to cheating alter the ESS outcomes of the various games?

The answer is that noise and error-prone reception do not change the ESS outcomes qualitatively, but they may lead to quantitative changes (Grafen and Johnstone 1993; Johnstone 1994; Johnstone and Grafen 1992b). It is still the case in error-prone systems that there are two possible ESSs: (a) senders are not necessarily honest, and receivers do not use signal information; and (b) on average signals are honest, this honesty is guaranteed by imposing significant and differential costs on senders, and receivers attend to signals. The term "on average" acknowledges the fact that not every signal emitted is going to be correctly interpreted by a receiver; however, reliable information must be provided often enough to justify receiver attention. These results allow generally honest signaling to be compatible with simple optimality constraints on the sending and processing of information and thus resolve the problem raised by Dawkins and Guilford. On the other hand, they raise the question of what determines the relative roles that honesty guarantees and optimality constraints play in different signals; surely the ratio varies, and how it is related to function and context is critical to understanding signal diversity (Dawkins 1993).

Quantitative effects of adding error to the ESS analyses are interesting. One common property of complex games is that each combination of possible strategies must be likely to occur or else it may be impossible to move evolutionarily from certain initial states to the ESS. In game theory parlance, all relevant strategies must be tested at least occasionally. Adding player error to such games resolves this problem because now, by chance, all possible combinations of roles and strategies are possible, and nearly any evolutionary trajectory is likely to occur sometime. This same principle applies to signaling games. Adding receiver error makes the predicted ESSs more likely and thus more globally stable.

The second change of adding receiver error is that it may no longer be optimal for senders to display with exact honesty. If two males differ in quality by 10%, and receiver error is at least that high, why should the higher-quality male advertise 10% more energetically than its competitor? The female is likely to perceive them as the same and thus the extra output by the higher quality male will be wasted. In error-free situations, the signaling ESS is a smooth monotonic function relating display investment to sender quality (Figure 20.3). The corresponding ESS functions when there is perceptual error are step-shaped; the more receivers err, the fewer the number of steps and the wider each step (Figure 20.9). Since a given animal is just as likely to be confused with its next higher ranking competitor as with its next lower one, where the steps occur is arbitrary, and there are thus many alternative ESS functions. When error is very high, the ESS is for most low-quality males to display at one low level or not at all, and higher-quality ones to display at another higher level (Grafen and Johnstone 1993; Johnstone 1994).

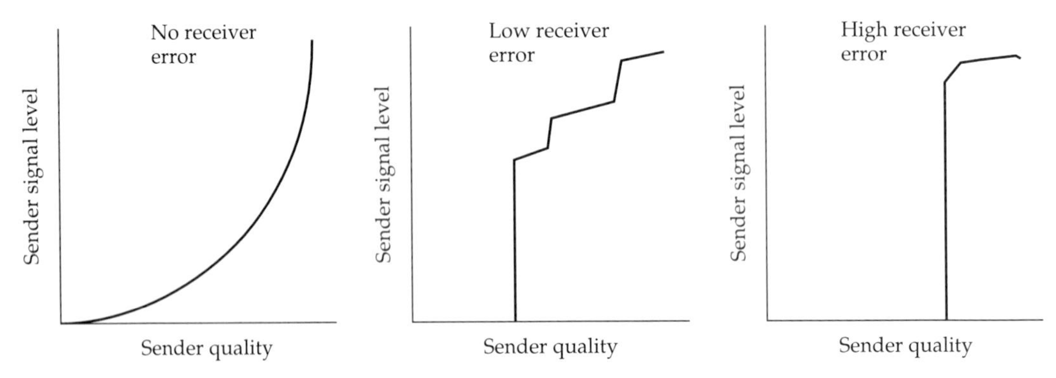

Figure 20.9 Effect of receiver error on ESS sender strategies for handicap models of signaling. When receivers have no perceptual error, the ESS is for senders to adjust the display level according to a monotonic function relating sender quality to display (left). When there is even some receiver error, low-quality senders should produce no display or a minimal one; higher-quality senders should display according to a stepwise function in which there is rough matching of quality rank and signal rank, but within clusters of similar senders, all should display at the same level (middle). For high receiver error, senders are effectively divided into two groups: low-quality individuals, who do not display, and high-quality ones, all of which display at nearly the same level. (After Johnstone 1994.)

This model thus provides an alternative explanation for the widespread use of stereotyped all-or-nothing signals given at typical intensities (as described on page 519). Stereotypy is simply the honest ESS when receiver perceptual error is high. This contrasts with simple optimality arguments that stereotypy evolves to ensure accurate signal transmission and detection. A third explanation is that posed by Zahavi (1980, 1987, 1993) and discussed on pages 652–653. He argues that receivers force senders to perform displays according to some standard. A good match to the standard indicates a high value of the quality receivers seek to measure. It is not easy to come up with tests that discriminate conclusively among these alternatives. Certainly, none of the explanations are incompatible with the others, and all may play some role in real systems.

Evolutionary Equilibria versus Systems in Continuous Flux

Is it reasonable to treat all signal systems as if they were at an evolutionary equilibrium? Is there any stage in the evolution of new signals when senders have the upper hand and cheating is common (Dawkins and Krebs 1978)? Andersson (1980) argued that animals have many more threat displays than they appear to need because many signals are the now uninformative relics of prior arms races between senders and receivers; new signals that recapture the attention of receivers would always be favored.

We discussed the mechanisms by which new signals might evolve in Chapter 16. Not all of these would seem to favor honest signaling at least at the outset. We take up the special case of Fisherian sexual selection in Chapter

23. Sensory exploitation of receivers was reviewed on pages 526–534. Here, senders produce signals that trigger latent receiver preferences to the detriment of the receivers (Basolo 1990; Leimar et al. 1986; Ryan 1990; Ryan et al. 1990; Staddon 1975). Such signals are initially dishonest. Can we expect them to evolve to honest ESSs or is such dishonesty stable?

Krakauer and Johnstone (1995) have examined a neural net simulation in which there are two players: a sender and a receiver. The model population was similar to that studied by Enquist and Arak (1993) and outlined in Box 16.2, but here allowed both senders and receivers to evolve. Senders varied in some quality that receivers wished to assess accurately. They produced multidimensional signals that were the only source of information to receivers. Each party used a three-layer neural net to deal with signal exchanges. The sender's net generated signals indicating its quality (honestly or falsely) and the receiver's net evaluated the signals to find the sender with the highest quality. Receivers locating the highest quality senders had the highest "fitness" and were thus made more common in the next generation. Sender representation in the next generation depended on the number of times each was chosen by a receiver and any costs they had to pay for signaling. Each generation, mutations occurred that varied the linkages in the neural nets allowing for new sender and receiver strategies. This world was allowed to coevolve until strategies stabilized or until some maximum number of generations had passed. The model thus examined the coevolution of sender and receiver using simple neural processing. Two different worlds were considered: one in which senders could produce signals without any costs to themselves, and one in which signals were differentially costly as a function of sender quality and thus potential handicaps.

The results of these simulations are illuminating. When signaling is cost-free, mean fitness of senders tends to increase steadily with successive generations, whereas receiver fitnesses remain low at levels similar to that when no information is being provided. The increase in sender fitness arises largely from sensory exploitation of latent receiver preferences. When signaling is differentially costly, sender fitness remains near its initial levels, but receiver fitness rises and plateaus at a higher value. The more signal dimensions that are used by the receiver, the higher this eventual receiver fitness. Thus, if costly signaling is allowed, signals are on average honest both in the final equilibrium and during much of the history. However, the simulations with signal costs clearly show periods in which mutant senders hit on signals exploiting latent biases of receivers. For a period, sender fitness increases and receiver fitness drops. But this is invariably followed by the appearance of mutant receivers that devalue the exploited biases and focus on other more honest signal dimensions. Receiver fitness then rebounds and sender fitness drops back to lower levels. These episodes recur over time because each change in sender signals causes a change in optimal neural net linkages in the receiver. This invariably generates new latent biases in receivers that future sender mutants can exploit. The process is thus inherently never-ending. This coevolutionary dynamic is exactly the kind of process envisioned by Dawkins and Krebs

(1978) and Andersson (1980). However, it is also compatible with the handicap principle because costly signals are honest at least on average and over long time scales.

Multiple Senders and Single Receivers

All of the game models on pages 654–668 assume a single class of senders, and so the ESS is a single rule for mapping information onto signal form or frequency. What if there are several classes of senders with different stakes and potential costs? This situation has been examined by Johnstone and Grafen (1993) using the Sir Philip Sidney game as an example. They allow there to be two classes of begging senders: one that begs honestly, and one that always begs regardless of need. The latter are thus cheats some of the time. Receivers cannot tell the two types apart. The ESS is for receivers to respond to all begging as long as honest beggars are more closely related to receivers, and/or honest beggars pay a higher cost for begging than do constant beggars. It must also be the case that honest beggars are sufficiently common relative to constant beggars. How common they must be depends upon the cost to receivers of responding to a constant beggar that is cheating. The higher this cost, the higher the fraction of honest beggars required to justify receiver response.

Johnstone and Grafen argue that variation in sender economics is likely to be common and so should be the existence of multiple sender classes. What is critical to this ESS mixture of honesty and deceit is that receivers cannot discriminate between the multiple classes of senders. Were they able to, then they would use different interpretive rules for signals from each class and honesty would be assured. This may be what is occuring in mantis shrimp where vulnerable and invulnerable senders cannot be distinguished without a risky close approach, but vulnerable senders are clearly a minority of the population (Figure 20.8). In a way, deceit is here seen again as a consequence of imperfect receiver assessment. However, it is precisely because receivers cannot assess everything directly that they have recourse to signals in the first place. Thus it may not be surprising that they are sometimes also unable to assess different sender classes. If all of this is so, it provides a very widespread reason for why most signaling systems should be largely honest, but exhibit some persistent low levels of deceit.

SUMMARY

1. **Deceit** is the provision of inaccurate information by a sender to a receiver. It is associated with a positive value of information for the sender, but a negative value for the receiver. Types of deceit include **lies** (use of the wrong categorical signal when alternatives are few and discrete), **witholding information** (not giving a signal when appropriate), **exaggeration** or **bluff** (using a signal whose rank among ordered alternatives is different from that for the corresponding condition values), and **attenuators** (traits that make it more difficult for a receiver to directly assess some trait). The opposite of deceit is **honest signaling**.

2. Early ethologists focused on the evolutionary sources of signals. They presumed that most signals were honest because they were obligately linked to the motivations prompting them. Behavioral ecologists focused on the strategic importance of signals, and many concluded that animal communication was an arms race between deceitful senders and sceptical receivers and thus rarely honest. Zahavi proposed the **handicap principle** that says that receivers should only attend to signals sufficiently costly to senders that it does not pay to cheat. Put another way, signals should impose **handicaps** on senders to ensure honesty.

3. The outcomes of game-theoretical treatments of honest signaling might be expected to differ depending upon the nature of the information provided during communication, the degree to which a receiver can compare signal information to direct assessment or prior expectations, and the nature of the costs of sending signals.

4. Nearly all models of honest signaling between parties without similar interests conclude that honesty can be guaranteed only if receivers demand signals that impose a higher cost on deceitful than on honest senders. **Agonistic signals** can be honest indicators of sender motivation as long as the more costly signals are the more effective ones. Females choosing mates should favor **courtship signals** that are costly to displaying males, and costly in a way that tests mate suitability. **Badges of status** will be honest only if large badges are challenged by other large-badge individuals, and there is a contest-independent cost of being aggressive and seeking high rank. **Amplifiers** are inherently more costly to potential cheaters than to those that most benefit from honesty. **Begging** and **predator notification** signals are both unlikely to be honest unless they impose costs on senders that are higher for potential deceivers. In most of these models, differential handicap costs lead to honest signals, and honest signals are possible only if there exist differential handicap costs. Many studies of animal communication support the notion that signals are usually honest and costly to senders.

5. Although the simple game models predict that animal signals will always be honest, most animal communication systems exhibit a mixture of honesty and deceit, with deceit being the more rare moiety. This outcome is in fact predicted by modified game models in which receivers are allowed to err in their identification or interpretation of signals, communication systems are allowed to continue evolving over time, and it is recognized that any given receiver may encounter multiple classes of senders each with its own costs and benefits of signaling. Part II of this book argues that all of these are likely conditions in nature.

FURTHER READING

Concise and very readable introductions to these issues can be found in Harper (1991) and Johnstone (1997). Wiley (1994) reviews the interface between deceit on the one hand, and other sources of receiver error on the other. Hasson (1994) provides a useful classification of the different ways in which senders may deceive receivers. Zahavi's own presentations (1980, 1987, 1993, Zahavi and Zahavi 1997) of the handicap principle are full of ideas and examples that will stimulate a reader's thinking about these issues. Grafen's (1990a,b) pivotal papers on signal honesty are difficult but entertaining reading. Advanced readers are encouraged to work through them, as Grafen is very good at identifying the critical steps and consequences of his models. Dawkins (1993), Dawkins and Guilford (1991), and Guilford and Dawkins (1995) provide important perspectives on the interface between simple optimality (Part II) and game-theoretical (Part III) approaches to animal communication. This interface is bound to be a major focus of future research. Finally, neural net simulations have provided many new insights into signal evolution. Studies by Arak and Enquist (1993), Enquist and Arak (1993, 1994), and Krakauer and Johnstone (1995) are surely only the beginning of this approach.

Chapter 21

Conflict Resolution

IN DISAGREEMENTS BETWEEN TWO ANIMALS over limited resources such as food, mates, or breeding sites, the interests of sender and receiver conflict to the maximum degree. Physical fighting sometimes results, but in many cases conflicts are settled with displays and ritualized methods of combat. How can communication resolve such disagreements? We have discussed the notion that threat displays inform the receiver of the aggressive intention or motivation of the sender. However, given the conflict of interest, we would also expect senders to bluff or exaggerate as a way of improving their chances of winning. Both true signals and tactical acts are likely to be used by opponents because the physical distance between the interactants is short. The number of threatening behaviors used in conflict resolution also varies considerably. Some species seem to have only one threat display, others have a few distinctive displays, and yet other species have a larger number of displays and acts they employ during conflicts. The goal of this chapter is to explore the variety of threat-display repertoires found in different species. Game theory has been used extensively over the past 15 years to describe the process of conflict resolution in animals. The available models differ substantially in their assumptions about how a contest is settled, the number of displays used, and the type of information conveyed in displays. We shall use these different models as a basis for categorizing conflict-resolution strategies. The game-theory approach leads us to a better understanding of the forms of threat displays than the design-features approach taken in Chapter 18.

WHAT HAPPENS DURING A CONTEST?

Whenever two individuals simultaneously attempt to gain access to the same valuable resource, and that resource cannot be shared without a loss in fitness, a conflict situation arises. Examples of conflict situations include two males attempting to acquire the same territory or female mate, two hungry individuals desiring the same nonsharable piece of food, or two females or pairs interested in the same nest site. Conflicts usually arise between two more or less equal individuals that need the same types of resources to increase their fitness. Both would prefer to win the resource without paying the cost of an escalated fight, so both would like the other to back down and retreat. In practice, however, the two opponents are rarely equal in fighting ability, or **resource-holding potential**. Each contestant would like to convey to its competitor that it is the superior fighter. At the same time, each is trying to decide whether it could win a fight with this competitor or whether it is likely to lose, in which case it is best to cut one's loss and retreat. It is important to note that each player must assess its opponent's fighting ability *relative* to its own. Both individuals are therefore simultaneously senders and receivers and an exchange of signals takes place between them. The way in which the conflict is resolved must have something to do with the number of signals and tactical acts they possess and the degree to which truthful information about relative fighting ability can be obtained.

How does game theory shed light on the dynamics of conflicts? The games we shall discuss in this chapter are primarily contests, that is, games between pairs of individuals, as opposed to scrambles among many individuals. This type of game model is relatively easy to construct because the payoffs to the opponents during each contest do not depend on the number of other individuals in the population playing each of the alternative behavioral strategies. The net fitness of individuals does, of course, depend on the rate of encounter with opponents playing different strategies. Existing game models differ in their assumptions about the way decisions to end a contest are made and the number and nature of alternative behavioral strategies available to contestants. Since alternative behavioral strategies can include a range of aggressive acts from mild threats to all-out fighting, these models make explicit assumptions about the parameters we are interested in: the number of threat displays and the information content of displays. We shall begin with simple games involving one type of threat display, and move up to more complex games with more behavioral options.

SIMPLE THREAT-DISPLAY CONTESTS

In some species a single type of threat display is given to a rival in conflict contexts. The display can sometimes settle the conflict and cause one individual to retreat, but if displaying is not successful, the conflict escalates to fighting. Examples of simple, yet highly effective, threats include agonistic vocalizations in many frogs, visual broadside displays in many fish, and color patches in some lizards and birds. How can such a display resolve a conflict? If the display indicates the intention to attack, then individuals that display with no intention of attacking can invade, and the signal gradually becomes meaningless (Maynard Smith 1979a).

The Hawk-Dove-Assessor Game

The hawk-dove game introduced on pages 628-630 provides a first step toward understanding species with a simple conflict-resolution system involving two distinct levels of engagement: display versus fight. Although the game model does not explicitly specify the number of threat displays, it is nevertheless the most appropriate model for single-display systems. The game is a discrete symmetric contest in which contestants have two alternative strategies: playing dove (i.e., display first, then retreat if attacked) or playing hawk (i.e., immediately escalate to fighting) (see Figure 19.3). The display is presumed to be a ritualized threat display derived from an intention-to-attack behavior. The loser pays a fixed cost of injury D and receives no benefit, while the winner pays no cost and obtains the entire benefit V. If the cost of losing a fight is higher than the value of winning ($V < D$), the ESS is a mixed strategy of sometimes choosing to display and sometimes choosing to fight. Otherwise, if $V > D$, then hawk is a pure ESS. This game seems far too crude to realistically describe animal contests, but it does tell us one thing: in species with dangerous weapons that can inflict serious injury costs on combatants, some form of conventional or ritualized interaction will be favored, whereas in species with lower fighting costs immediate escalation is more likely.

The hawk-dove model can be made a bit more realistic by including iterated dynamic strategies with different choreographies. For example, **bully** is a strategy that escalates on the first move and continues to escalate if the opponent displays, but retreats if the opponent escalates. Another possibility is a **retaliator** that initially displays, but escalates if its opponent escalates. A third strategy is **prober-retaliator** that displays most of the time, but occasionally escalates and will retaliate if its opponent also escalates. Bully is basically a bluffing strategy that never does well against the other strategies. Retaliator and prober-retaliator, on the other hand, can be quite successful strategies, even when the cost is less than the benefit. This suggests that some responsiveness to the opponent is better than all-or-nothing strategies such as hawk and dove or a random mixture of these two behaviors (Maynard Smith 1976).

In real fights between two rivals, the relative size or strength of the contestants plays an important role in determining which individual will win. Intuitively, the idea of a small individual playing hawk against a much larger individual seems quite foolish. Another alternative strategy of first evaluating one's chances of winning against a particular opponent before deciding whether to retreat or escalate seems much more prudent. **Assessor** is a strategy that displays first, evaluates whether it is the larger or smaller contestant, and then retreats if it assesses itself to be smaller or escalates if it assesses itself to be larger. In setting up the payoff matrix, it is assumed that a given contestant has a 50% probability of being the larger individual and that assessors can accurately determine their relative size. Over a wide range of reasonable values for fighting costs, benefits of winning and contestant asymmetries, assessor is a pure ESS over hawk and dove (Figure 21.1). Assessor does well because it avoids escalating and paying fight costs in contests it is likely to lose.

Characteristics of Assessor Threat Displays

With the assessor strategy, display takes on a new significance. Instead of signaling the intention to attack, it is a graded signal that reveals information about body size, strength, or other aspects of fighting ability and resource-holding power. The display must be inexpensive to execute, in the sense that (1) it does not put the displayer so close to its opponent that it would risk injury, (2) it does not require a great deal of performance energy, and (3) it does not take too much time or need to be repeated extensively. On the other hand, it must be a reasonably accurate and honest reflection of fighting ability. Recall from the discussion of honesty in the prior chapter that to guarantee honesty, either signals must be more costly for senders to produce in an exaggerated form or receivers must impose an incidental cost on cheaters upon detection.

How accurate does assessment have to be? Maynard Smith and Parker (1976) developed an asymmetric version of the assessor game to explore the effects of imperfect information. Interactants were assumed to use features of their opponents to evaluate their fighting ability. We shall call these features displays, even if they are largely cues, since the players will presumably at

	Hawk	Dove	Assessor
Hawk	$\frac{1}{2}(V-D)$	V	$\frac{1}{2}(V-D)$
Dove	0	$\frac{1}{2}V$	$\frac{1}{4}V$
Assessor	$\frac{1}{2}V$	$\frac{3}{4}V$	$\frac{1}{2}V$

Figure 21.1 The payoff matrix for the hawk-dove-assessor game. V is the benefit of winning, D is the cost of losing an escalated fight. The assessor evaluates its body size relative to its opponent and retreats if it is smaller (which it is on 50% of occasions). The assessor strategy is a pure ESS whenever D and V are greater than 0; D can be greater or less than V. If the cost is 0, hawk and assessor are equally good strategies. (After Maynard Smith 1982a.)

least pose so that the cues are revealed to their opponents. Two sources of uncertainty were identified: uncertainty in receiver estimation of the magnitude of the display, and uncertainty in the degree to which display determines the outcome of an escalated contest. If the magnitude of the display can be accurately estimated by receivers, then it will be used as a means of settling contests, even if it is a mediocre predictor of the outcome; it only needs to predict the outcome slightly above the chance level. Escalated contests in this case will be rare except when fighting is not very costly. If display magnitude cannot be estimated reliably, receivers will make more errors and escalation will occur more frequently.

When there is a cheap and low-risk display that provides a reasonably accurate, unbluffable, estimate of fighting ability, this is likely to be the only display individuals need to resolve conflicts without resorting to fighting. In order to be relatively risk-free, the display must be perceivable at a distance, so vocal and visual displays are likely to predominate. Different kinds of cues and signals vary in the accuracy with which they can be estimated and in the strength of their correlation with fighting ability. Some may be easier to bluff and exaggerate than others, so we need to consider the factors that constrain the signal to be honest as well.

Relative body size is the most commonly observed determinant of winning or losing a fight (reviewed by Archer 1988). This observation is especially true for species such as fish, amphibians, reptiles, and some arthropods with indeterminate or very slow growth that results in a large variance in body sizes among adult competitors. The classic example of a low-risk body-size-revealing signal is the vocal threat of frogs and toads (Davies and Halliday 1978; Robertson 1986a). Honesty is promoted by the body-size constraint on vocal signal frequency that we have discussed in Chapters 4 and 17 and in Figure 20.1. A small individual would have to expend a great deal of energy to produce a low-frequency threat, whereas a larger individual can do this more easily. The frequency is therefore inversely related to body size and lower-frequency calls convey a stronger threat (Figure 21.2). The correlation between body size and call frequency is extremely strong in some species and less so in others (r ranges from 0.6–0.8). Highly accurate (and therefore honest) body-size encoding may be favored as a consequence of punishment costs or other fitness losses of deceptive signaling (Robertson 1990). Some species exhibit a weaker correlation because they employ variation in call frequency to signal other types of information, a point we shall take up later (Wagner 1989, 1992).

Visual displays can also reveal information about body size at some distance from the opponent. Broadside postures are an obvious example. Terrestrial animals can compare their height to that of opponents during a parallel walk display (see Figure 16.4). Vertical or horizontal lines that stretch across the entire body may act as amplifiers to make this assessment easier (see Figure 12.2). Broadside threats can therefore provide reasonably accurate information about relative body size in addition to information about motivational ambivalence, as suggested on pages 508–509. Other visual displays may re-

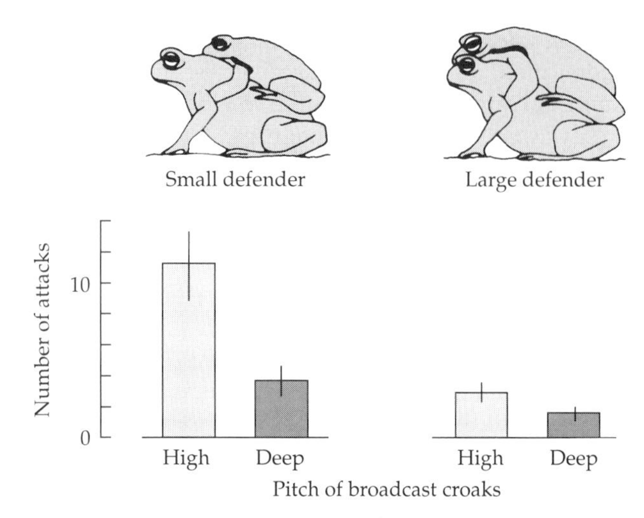

Figure 21.2 Low-frequency vocalizations are stronger threats. The frequency of threat vocalizations is often inversely related to body size and appears to be physically constrained to reveal honest information. This is clearly the case for *Bufo bufo* (see Figure 20.1). Playback experiments show that the frequency of the call is used to assess rivals as predicted by the assessor model. Males fight over access to gravid females. In this experiment, either a large or a small male was allowed to amplex with a female and was temporarily muted with a rubber band around his jaw. A medium-sized intruder male (the receiver) was then introduced. A high or low frequency threat vocalization was played as if eminating from the defender (the sender). The intruder was much more likely to attack the defender if he heard a high-frequency call and less likely to attack if he heard a low-frequency call, indicating that a lower-frequency call is a more effective threat. The difference between high- and low-frequency calls was considerably less when the defender was larger than the intruder, suggesting that other cues, perhaps visual, were also being used. Fighting only occurred when males were very similar in size. (From Davies and Halliday 1978, © Macmillan Magazines.)

veal the size of weapons. In mountain sheep, horn size is more variable than body size and is used for dominance assessment in males (Geist 1966). The meral-spread display of many crustaceans not only reveals the general body size of an individual but also the size of its weapons. As shown in Figure 20.9, the display may signal attack intention as well but becomes a bluff when employed by a recently molted individual.

Color patches, or badges, are another example of visual threat (Figure 21.3). The size, hue, and chroma of the patch has been found to affect the probability of winning a fight in several species of birds (see page 663 and Figure 20.6) and lizards (Madsen and Loman 1987; Thompson and Moore 1991; Olsson 1994). It is more difficult to see how such signals can remain honest, however, since the color patch is not expensive to produce. As discussed in Chapter 20, the cost of cheating by producing too large a badge for one's fighting ability is a punishment cost imposed by receivers. Escalated fights rarely occur between individuals possessing very different badge sizes because the smaller-badged individual always retreats, but individuals with

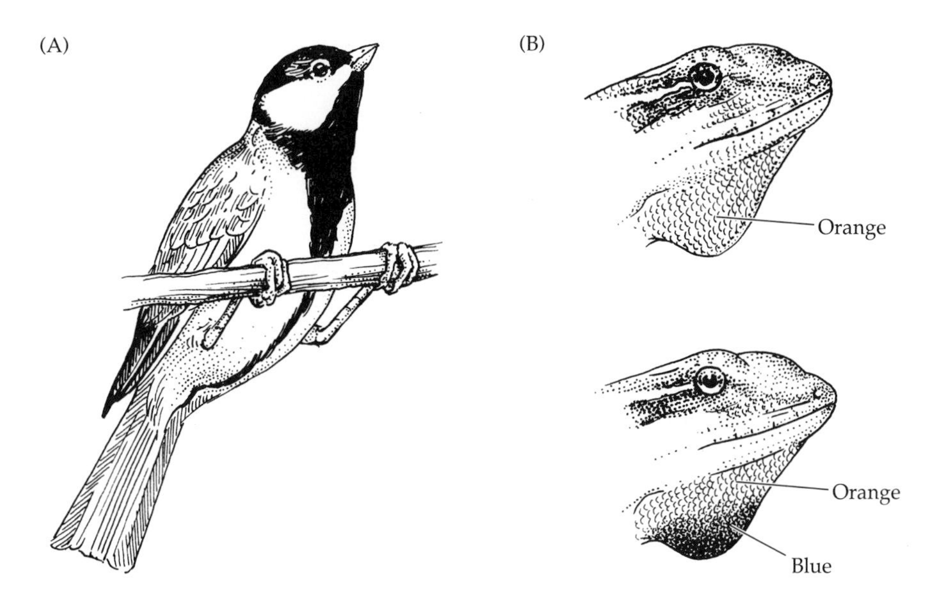

Figure 21.3 Color patch threat displays. (A) The black stripe on the white chest of the great tit (*Parus major*) is larger in males than in females, and in males the width of the stripe is correlated with their dominant status. The chest stripe is prominently displayed in the upright threat posture. (B) Male tree lizards (*Urosaurus ornatus*) display the color on their throats when they extend their dewlap. The color ranges from solid yellow or orange (top) to bicolored combinations of an orange or yellow background with a blue or green spot that varies in size (bottom). Males with larger blue areas are more likely to win contests against males with smaller blue areas. (A from Hinde 1982; B drawn from photos courtesy of Michael Moore.)

similar badge sizes frequently engage each other in contests and therefore test each other. A cheater will always lose in such encounters. Furthermore, the information such badges convey is often closer to a true estimate of fighting ability than mere body size since the development and appearance of the color patch evolve to be dependent on age and/or body condition. For example, the area of the green lateral body patch of the sand lizard *Lacerta agilis* is more strongly correlated with male body condition, measured as the weight to length ratio ($r = 0.82$), than it is with body mass alone ($r = 0.37$), length alone ($r = 0.8$), or age ($r = 0.18$). Color saturation, on the other hand, is more strongly correlated with mass ($r = 0.49$) than with condition ($r = 0.35$) (Olsson 1994). Such color patches, like other body-size-revealing displays, can only transmit information about static long-term aspects of fighting ability. Short-term motivational information, on the other hand, is better transmitted with visual displays involving movements, postures, and coverable color patches.

A final example of fighting-ability assessment with a single threat display can be found in Crespi's (1986) study of fighting in thrips (Figure 21.4). These insects have neither the visual acuity nor the sound-producing mechanism needed to assess body size with these modalities, so they use the tactile

(A)

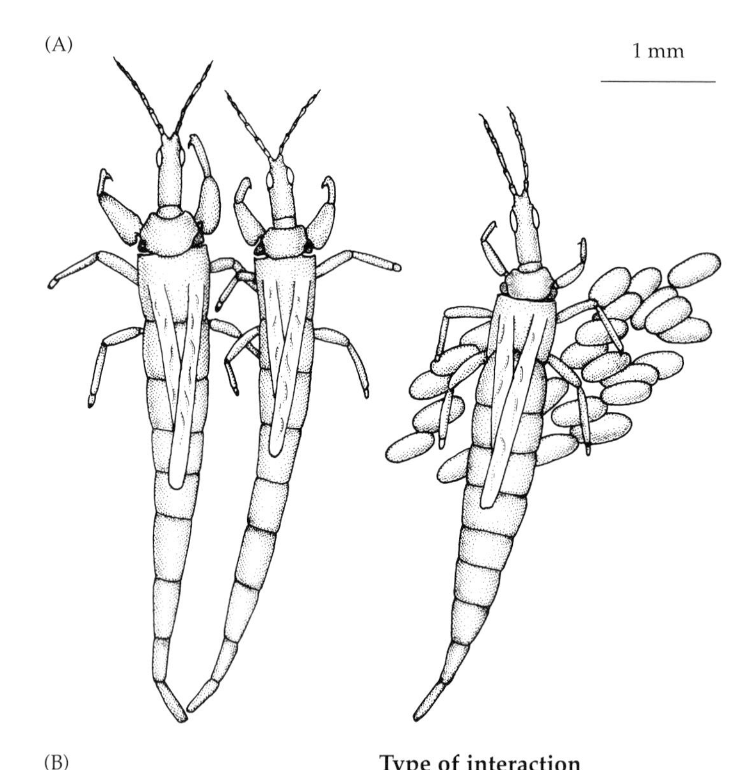

(B)

Male size	Type of interaction: Withdrawal	One parallel bout	Multiple parallel bouts	No. of interactions with stabbing (%)
Small versus small	5	2	14	11 (52%)
Large versus large	19	14	14	7 (19%)
Small versus large	57	13	2	0

Figure 21.4 Display and escalation in thrips. (A) Two male thrips (*Elaphrothrips tuberculatus*) on the left are fighting for access to the female. The assessment display by the two males is tactile and consists of aligning their bodies side to side and hitting at each other with their abdomens (called a parallel bout). The escalation behavior involves one individual (usually the larger) climbing onto the opponent's back, grasping its body with armed forelegs and squeezing (called stabbing). The smaller individual may attempt a defensive flipping maneuver to prevent stabbing. (B) The table shows the type of interaction between males and the likelihood of stabbing in relation to absolute male size. (After Crespi 1986.)

modality. Rival males align their bodies in parallel and bat at each other with their abdomens. Although no injury results from this display and the size of the opponent is clearly revealed, this is not a low-risk display because the contestants are in close proximity. The abdomen-hitting display is often followed by stabbing attacks. If the males are similar in size, the duration of displaying and the likelihood of attack depends on their absolute size. When both are small, contests are longer and more likely to escalate. When both are large, contests are shorter and less likely to escalate. This observation agrees well with the predictions of the hawk-dove-assessor game. Large males can injure each other severely, whereas two small males cannot. The cost of escalation is therefore low for small males, so they play hawk more readily. The cost is high for large males, so they continue to display in the hope that one will retreat. However, the assessor game model does not help us understand how long such displaying should continue.

VARIABLE-LENGTH CONTESTS

The thrips example above illustrates some of the constraining assumptions of discrete strategy games such as hawk-dove-assessor. Contestants cannot vary their time and energy investment in displaying and cannot control contest costs. Furthermore, such models do not specify how contestants assess each other and how they use the information acquired to make the decision to quit or continue. In a number of species, opponents engage in variable-length interactions involving a single repeated behavior in which the contestant that persists the longest wins. The contest ends when the first contestant decides to quit. The behavior is typically some form of ritualized but moderately costly fighting such as wrestling or grappling. Good examples include newts and the bowl and doily spider (see Figures 21.6A and 21.7A). The type of game model needed to encapsulate such a conflict-resolution strategy must use a continuous alternative strategy set and must specify the mechanism by which the decision to quit is made. Two continuous-strategy game models have been developed, the war of attrition game and the sequential assessment game. The two models differ in their assumptions about the decision-making process.

The Asymmetric War-of-Attrition Game

The **war-of-attrition game** was introduced in Chapter 19, on pages 640–641. Reviewing briefly, the symmetric version of the game assumes that there is a continuous range of fighting durations from which a contestant can choose and that animals can fine-tune their costs as a function of the length of time they are prepared to persist in the contest. Before the fight, both opponents make a sealed bid of how long they will persist; the animal with the longer persistance time wins, and the duration of the contest is determined by the contestant choosing the shorter persistence time. The game assumes that the individuals don't know the bid of their opponent. The solution to this game is a mixed ESS in the form of a distribution from which animals should randomly choose their persistence times (see Figure 19.10).

In the asymmetric version of the game, contestants differ in the rate at which they accumulate costs with time, k, and perhaps also in their perceived value of winning ,V (Parker and Rubenstein 1981; Hammerstein and Parker 1982). The animal with the higher V/k ratio should be willing or able to persist longer. Before the contest begins, the opponents make an assessment of their relative V/k ratios and evaluate whether they would win or lose. A contestant that estimates itself as having the higher ratio (winner role) chooses a longer persistence time above a critical value S, whereas a contestant estimating it will lose chooses a shorter persistence time below S (Figure 21.5). The persistence time distributions, $P(t)$ are determined by:

$$P_w(t)=\frac{\bar{k}}{\bar{V}}e^{-\bar{k}(t-S)/\bar{V}} \quad \text{and} \quad P_l(t)=\frac{(1-\phi_w k_w)+(\phi_l k_l)}{\phi_l(1-\phi_w)(V_w+V_l)}e^{-\bar{k}t/\bar{V}}$$

where w and l subscripts denote winner and loser role, respectively, $\bar{V}$ and $\bar{k}$ are the averages of the resource values and costs for the two contestants, and ϕ is the probability that a contestant estimates its role correctly given the relative V/k ratios. The ESS and the value of S depend critically on the accuracy with which the two contestants can estimate their relative roles from V/k ratios. S is given by:

$$S=-\frac{\bar{V}}{\bar{k}}\ln\left[1-\frac{\phi_l(1-\phi_w)+(k_w k_l)}{k_w(1-\phi_w)(k_l\phi_l)}\right]$$

If accuracy is high, S will be small, the loser will select a very low persistence time, and the battle will be short. If accuracy is low (random choice, $\phi = 0.5$), S

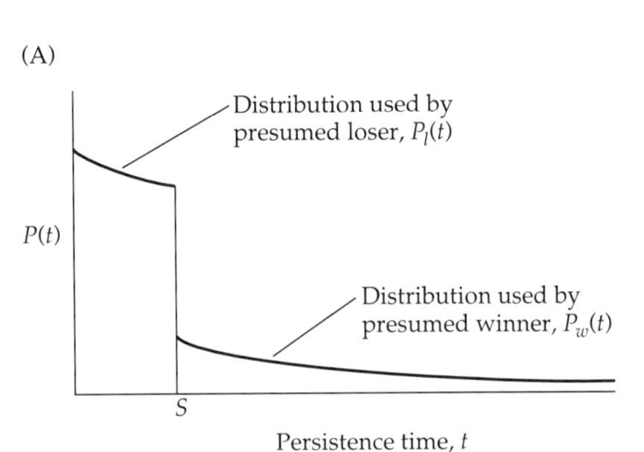

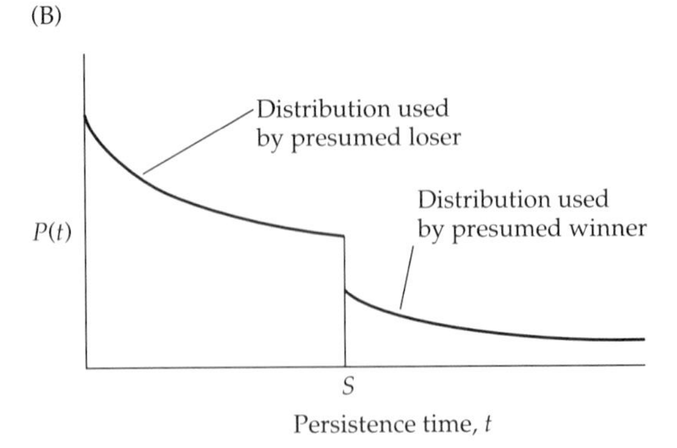

Figure 21.5 ESS for the asymmetric war of attrition. Graphs show the probability of choosing a particular persistence time, $P(t)$, as a function of persistence time t. Here t could also refer to fighting intensity or any other measure of contest investment. The contestant that estimates it is likely to lose (indicated by the subscript l) chooses a persistence time from the distribution on the left, and the contestant that estimates it is likely to win (subscript w) chooses a persistence time from the distribution on the right. S is the persistence time that delineates the presumed loser's and winner's distributions. (A) When role assessment is relatively accurate, S is small and contests are short. (B) When role assessment is relatively inaccurate, S is large and contests are both longer on average and more variable in duration. (After Hammerstein 1981.)

will be larger, and even the presumptive loser will benefit by sometimes choosing high persistence times. Since assessment is unlikely to be perfect, errors will sometimes be made in role estimation, and both contestants may occasionally decide they would lose or both may decide they would win. If both contestants choose the winning role, the contest will be very long, and if both contestants choose the losing role, the contest may be quite short. Thus the variance in contest duration increases when assessment is difficult (i.e., when the contestants have very similar V/k ratios). Finally, S also depends on the absolute value of V and k. Contest duration will therefore increase when the value of the resource is high and decrease when the cost of fighting is large. It makes intuitive sense that animals would be more willing to persist longer for more valuable resources.

Contest Characteristics for War-of-Attrition Games

As suggested above, the asymmetric war-of-attrition game makes some very specific predictions about the nature and duration of contests. (1) The type of behavior used during the battle must be a costly mutual interaction so that the point at which $V = k$ can establish the maximum investment a contestant is willing to expend in a fight. The cost could be strictly energetic or involve a risk of injury, or both. Examples include ritualized fighting such as head-butting, mouth-wrestling, grappling, and chasing. (2) As the difference in body size between opponents decreases, role estimation will be more uncertain, and both the average duration of contests and the variance in fight duration should increase. (3) As the value of the resource increases, the contest duration should increase. Laboratory experiments with staged contests in which either the size difference between opponents or the value of the contested resource was systematically varied provide good support for these last two predictions (Figures 21.6B and 21.7B).

Since both V and k can differ for the two contestants, simultaneous manipulations of the two parameters show that the outcome is truly dependent on the V/k ratio. Austad (1983) was able to vary the perceived value of a female resource to one male bowl and doily spider while holding the other male's estimate constant. The value of a female to a male in this species depends on how long he has been copulating with her; if copulation time has been short, the female's value is high because he has not yet inseminated her, and if copulation time has been long, her value is low because many of her eggs have already been fertilized. The intruder is introduced to the copulating pair after different amounts of time. He doesn't know the female's true value and can only assume that 50% of her eggs have already been fertilized; the original copulating male knows the female's value with certainty. The relative cost of fighting can also be controlled by varying the relative size difference between the resident and intruder males. Larger size predicted the winner in most cases. However, if a smaller resident had not been with the female very long, his V was higher than the intruder's V and the resident persisted longer, often winning.

Another study that initially seemed to fit the conditions of the war-of-attrition well is Marden and Waage's (1990) study of territorial damselflies.

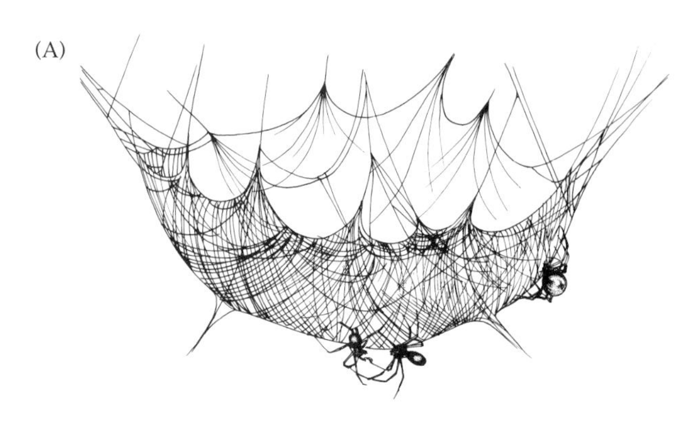

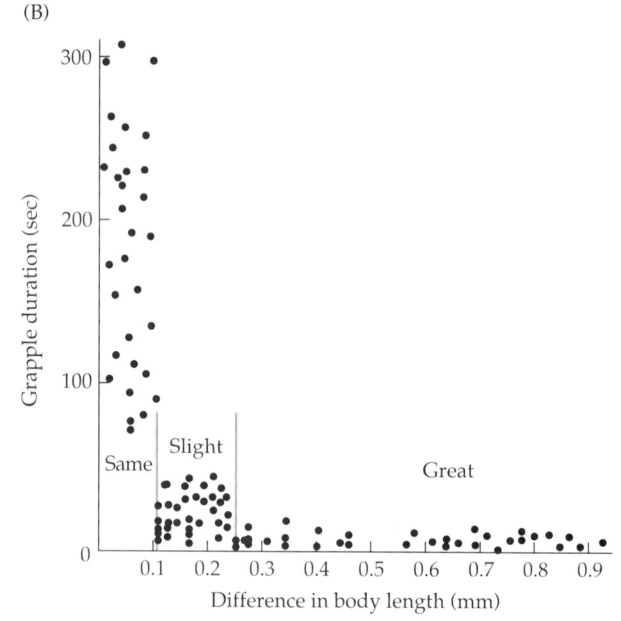

Figure 21.6 Contest duration versus difference in body size in male bowl and doily spiders. (A) The web of the bowl and doily spider (*Frontinella pyramitela*) consists of an upper tangle that knocks down flying insects into the bowl-shaped structure below. The female remains on the underside of the bowl and reaches through the web to grab the prey. Males interested in mating must enter on the underside of the web or else be treated as prey. When two males encounter each other on the web, they spar briefly with forelegs, then grapple for varying lengths of time. One male eventually retreats by dropping off the web. Here, two males grapple while the female stands aside. (B) Grapple duration as a function of the difference in body length. Duration is both longer and more variable when contestants are more similar in size. (A drawn from photos courtesy of Steve Austad; B from Austad 1983.)

Male damselflies defend areas of pond containing a single clump of vegetation used by females for oviposition. If the vegetation clumps of two adjacent males are slowly pulled together so that both think they own it, an escalated fight ensues. The male with the greater fat reserve tends to win the fight. Territorial males gradually lose condition since they don't feed and eventually lose their territories to younger males. If males effectively make a "sealed bid" by persisting until their fat reserve reaches a lower physical limit, then post-contest losers should all have a similar and low remaining fat supply compared to winners. But Marden and Rollins (1994) found high levels of variance in fat for both winners and losers. Fat reserve appears to determine the rate at which short-term energy mobilization declines during a fight. The longer the contest, the more likely was the fatter male to win the contest. This suggests that males assess each other's fat reserves by observing flying speed

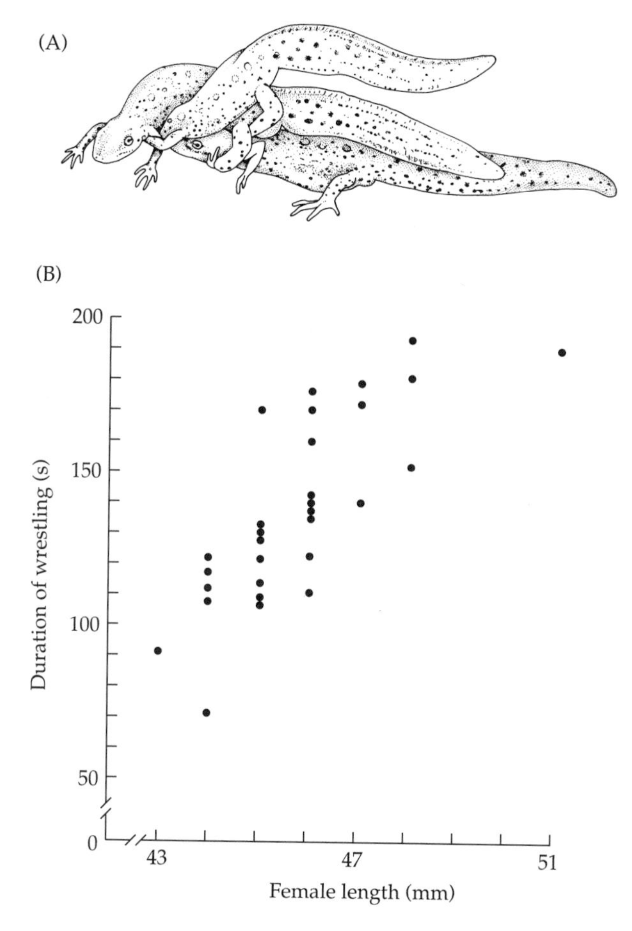

Figure 21.7 Fight duration as a function of female size in newts. (A) Two male newts (*Notophthalmus viridescens*) are wrestling over access to the gravid female below. (B) Contest duration increases as the size of the female increases. Males fight longer for larger females, who are more valuable because they lay more eggs. (A after Verrell 1986; B from Verrell 1986.)

and stamina during the course of the contest. Interestingly, older experienced males perform better than their low fat reserves predict. One possible explanation for this is their tendency to drag the chase into another neighbor's territory and involve this individual in the fight; this action interrupts the opponent's ability to assess the older male's decline in stamina.

Signals and Assessment of Roles

The war-of-attrition game makes some implicit predictions about the signals and sources of information used for the initial assessment of roles (Parker 1974; Maynard Smith and Parker 1976; Caryl 1979; Hauser and Nelson 1991). Assessment of relative fighting ability and resource value must be imperfect to reach an ESS. If role estimation is perfect, then $S = 0$, the lower V/k ratio player retreats immediately, and the entire game collapses. Furthermore, the probabilistic distributions of persistence times emerge as outcomes of the ESS. If one individual were to specify at the beginning of the conflict the amount of time it was prepared to persist, then the opponent could set its time one second longer and win the contest. In fact, it is specifically the war-of-attrition model that challenged the ethologist's notion that displays evolved to convey intentions and suggested that true intentions should be concealed (see historical review on pages 651–653). Therefore, contestants would not be expected to perform any displays during the assessment phase or at any time during the battle that revealed their intentions, aggressive motivation level, or willingness to persist. Direct evaluation of strength from tactile interactions also would not be expected in a war-of-attrition game. This is the reason why the same behavior is performed at the same intensity throughout the battle.

The assessment period before the battle is presumed to be a brief affair in which primarily visual cues are used to assess the opponent. Contestants use body size and any visual manifestations of condition, age, and health to judge the opponent's fighting ability. Inexpensive signals that reveal static information about body size and condition such as color badges or low-frequency vocalizations described earlier could also be exchanged. The accuracy of assessment, of course, depends on how well correlated the variations in the cues are to the probability of winning a fight and on the ability of contestants to measure these cues. If ability to inflict injury is the main determinant of winning, then evaluation of body size and strength cues may provide sufficient information, but if stamina is the criterion, cheap prefight information may be inadequate. Displays that reveal one's stamina, such as countersinging, calling rate, or vigorous movement display, could be used if they were inexpensive and reliable (but reliable signals are usually costly). Similarly, evaluation of resource value must be based on some combination of prior knowledge and any visual cues available to the contestants. If two males are contesting a female and they both can estimate her value, then assessment will be more accurate. If only one individual knows the value of the resource, it should not signal this information to its opponent, and the opponent will have to make a guess about relative values.

The Sequential-Assessment Game

The war-of-attrition model captures some aspects of real animal contests but still contains a number of unrealistic assumptions. The assumption of the sealed bid decision on persistence time at the beginning of the contest seems to be biologically unlikely. Consistent evidence in other studies for poor initial assessment abilities and strong negative correlations between weight asymmetry and contest duration suggest that animals do not decide on their persistence time at the outset, but continue to gather more information during escalated phases of the contest. Such observations led to the development of an iterated, dynamic game model called the **sequential-assessment game** (Enquist and Leimar 1983, 1987, 1990; Leimar and Enquist 1984). In this model, animals begin a contest with maximum uncertainty about their opponent, and gradually acquire information with repeated rounds of interaction; this process enables them to update their estimates of each other's relative fighting abilities. This decision-making process is essentially the same as the general communication model described in Part II of this book.

Fights consist of the repetition of one type of interaction or behavior that reveals some information about fighting ability to each opponent. As we learned in Chapter 13, the first few repetitions of the signal provide a great deal of information, but subsequent repetitions provide diminishing amounts of new information as the receiver's estimate gets closer to the truth. At each point in time, contestants have a current estimate of their probability of winning and an uncertainty term (standard deviation) for that estimate. As the contest proceeds, the estimate becomes more accurate and the uncertainty decreases in a manner similar to statistical sampling. In quantitative terms, the current estimate x at step n for contestants A and B (x_n^A and x_n^B, respectively) is the running average of the true difference in relative fighting ability plus the assessment error at each step:

$$x_n^A = \frac{1}{n}\sum_{i=1}^{n} \theta_{\text{AB}} + z_i^A \quad \text{and} \quad x_n^B = \frac{1}{n}\sum_{i=1}^{n} \theta_{\text{BA}} + z_i^B$$

where θ is the true relative fighting ability of each contestant against the other, measured as ln (c_A/c_B) or the ratio of the cost that each contestant can inflict on the other ($\theta_{AB} = -\theta_{BA}$). The variable z is random observer error, drawn from a distribution with a mean of zero, so its value at each step can be positive or negative. The standard deviation of the estimate is $\text{SD}_x = \sigma/\sqrt{n}$. At the beginning of the contest, each individual has an initial or a priori estimate of its chances of winning but the standard deviation is large. As the contest progresses, the effects of the error begin to average out to zero, and the current estimate approaches the true relative fighting ability. The model predicts that once one animal is fairly certain it will lose, it will end the contest by retreating or giving up. The ESS for such a dynamic game is an evolutionarily stable policy called the **giving-up line**. When one contestant's current estimate crosses this line, it is sufficiently certain of its lower fighting ability that it makes the decision to quit (Figure 21.8).

The smaller the difference in fighting ability, and the higher the error of assessment, the longer the fight will be on average. Contests between near equals are also more variable in duration than those between unequal competitors and the (slightly) poorer contestant may sometimes win. This is a much more reasonable model than the war of attrition, because it assumes that the outcome of a contest is determined by the relative fighting ability of the contestants, and it allows them to control the cost (duration) of the fight as it proceeds and to decide whether to continue or quit based on information gained during the fight. However, it yields the same predictions as the asymmetric war of attrition, namely that contest length is short if the individuals are very different in size or fighting ability, and very variable but generally longer when they are similar.

Austad reanalyzed his fight data on the bowl and doily spider with a sequential assessment model (Leimar et al. 1991). As in the previous analysis,

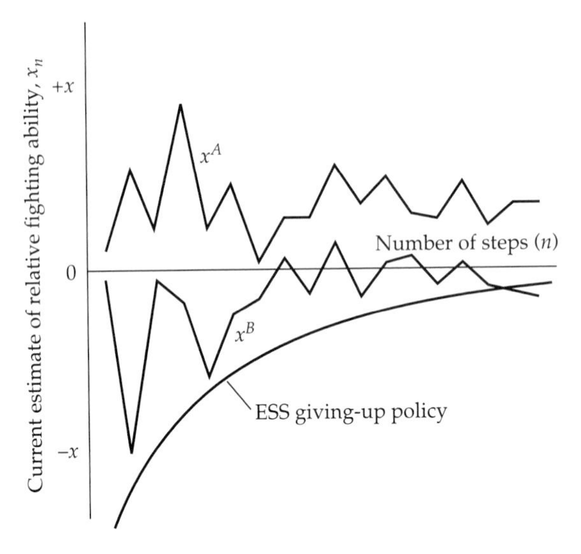

Figure 21.8 The sequential assessment game. The horizontal axis represents time or the number of repetitions of the aggressive behavior. The vertical axis represents the current estimate at step n of the probability of winning or losing for the two contestants A and B, x_n^A and x_n^B, respectively. The ESS is a policy of giving up when a certain minimum estimate has been reached, shown by the curvelike line. At the beginning of the contest illustrated here, the opponents assume that they have a 50% chance of winning (or use a priori estimates), but after each repetition of the behavior, they gain some information and update their estimate of true relative fighting abilities (θ_{AB} and θ_{AB}). The jagged lines show the trajectories of the updated estimates of probabilities of winning for each contestant. They vary more at the beginning of the contest because of estimation errors. When one individuals estimate crosses the giving-up line (in this case B's), it is sufficiently certain it will lose that it quits the contest. If the contestants are very different in fighting ability, they will learn this quickly, and one trajectory will cross the ESS line after a few rounds. If they are similar, it will take longer to determine the winner, and chance effects will sometimes result in the slightly poorer contestant winning. The vertical position and shape of the ESS policy curve depends on the level of error in assessment and the value of winning. As both assessment error and the value of winning increase, the policy curve will drop down, and contests will become longer in duration. (After Enquist and Leimar 1983.)

fights were found to be longer when the two males were more similar in size and when the female was perceived as having a higher value. Because the sequential assessment model specifies the behavioral mechanism of decision making, it also generates some highly specific predictions that do not emerge from the war-of-attrition model. A narrower range of variation in contest duration is predicted for fights between two simultaneous intruders that value the female similarly, compared to owner-intruder contests, and this was confirmed by the data. Furthermore, the model predicts the probability that the larger individual wins based on the size asymmetry (Figure 21.9). The observed probability of winning is only slightly less accurate than this prediction, indicating that when contestants are allowed to interact and acquire information during the contest, they can assess their relative fighting abilities better. It was con-

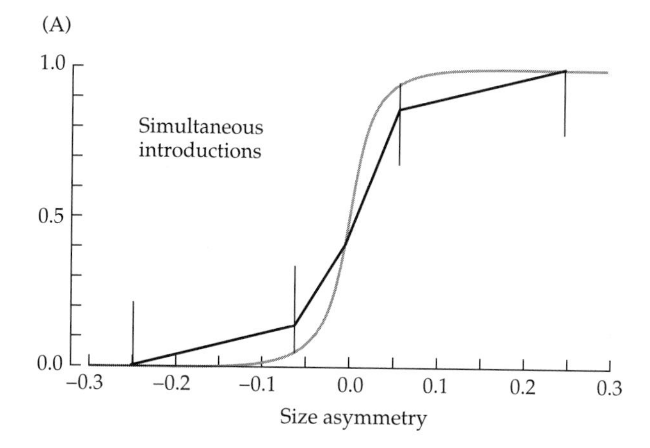

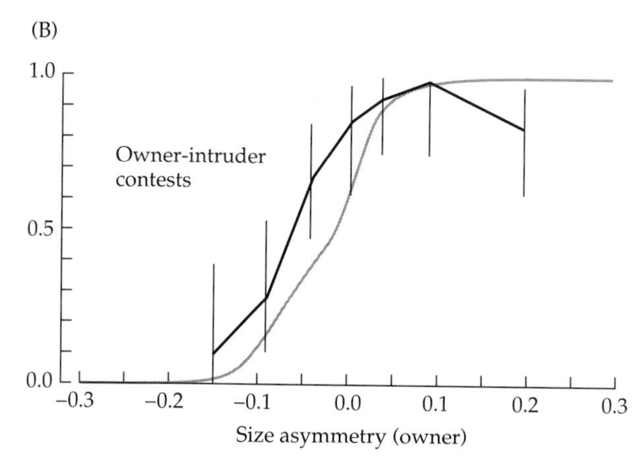

Figure 21.9 Predicted and observed relationship between size asymmetry and probability of winning in the bowl and doily spider. The dashed line shows the quantitative prediction from the sequential assessment game. The solid line with error bars shows the observed probability of winning. Size asymmetry is computed as the natural log of the ratio of the larger to the smaller contestant. (A) Simultaneously introduced males. There is a good fit between observed and predicted probabilities. (B) Owner-intruder contests. Given their size, owners have a higher probability of winning than expected when a female is present. (After Leimar et al. 1991.)

cluded that the sequential-assessment model is both a more powerful and more biologically appropriate game model for understanding conflict behavior.

The sequential assessment game avoids the difficult issue of whether displays signal intentions. It assumes that animals don't know their relative fighting ability at the beginning of the contest and use subsequent interactions to figure this out. A single type of display or action is repeated, not because players are concealing their motivations, but because repetition reveals more accurate information about fighting ability. As in the war of attrition, the behavior needs to be moderately costly so it will reveal this information. Aggressive motivational level is not assessed during the game, but is set in advance by the value of winning and the certainty level of losing that each player will accept before making the decision to quit.

FIXED-SEQUENCE CONTESTS

In a few species whose fighting behavior has been studied in detail, contests progress through a predictable series of stages and each stage is characterized by a different type of behavior. Each subsequent stage appears to represent an escalation over the prior stage, and different information is obtained by the contestants. For example, in red deer contests (Clutton-Brock and Albon 1979) territorial males first engage in a roaring contest (see Figure 20.1). The frequency of the roar immediately distinguishes young males (higher frequency) from mature males (lower frequency), but among mature males there are no differences in frequency. Instead, the *rate* of roaring is evaluated for a period of time. Roaring uses the same thoracic muscles employed in fighting and is a reasonably good measure of a male's condition and his probability of winning a fight. If roaring rate is similar, they approach and perform a parallel walk (a broadside display) that may provide visual information on body height and antler size. If this doesn't resolve the contest, then they engage in a pushing contest with locked antlers.

A second example can be found in jumping spider contests (Forster 1982; Wells 1988), with five stages of escalation associated with different behaviors. Stage 1 consists of either the hunch display, in which the spiders approach each other, arch their first three pairs of legs forward, and raise their abdomens together; or the erected legs display, in which both spiders repeatedly erect and lower leg III. In stage 2 they perform the stave display, shown in Figure 21.10, which involves erecting leg III on one side, tilting the abdomen toward the opponent on that same side, and flicking the erected leg against the similarly erected leg of the opponent. Stage 3 involves a head-butt behavior with head-to-head pushing. Stage 4 is a cheliceral-lock in which the spiders interlock mandibles and try to lift the opponent; they often form a "tent" with their bodies almost vertical. Stage 5 is the final escalation level, involving grappling and

(A)

(B)

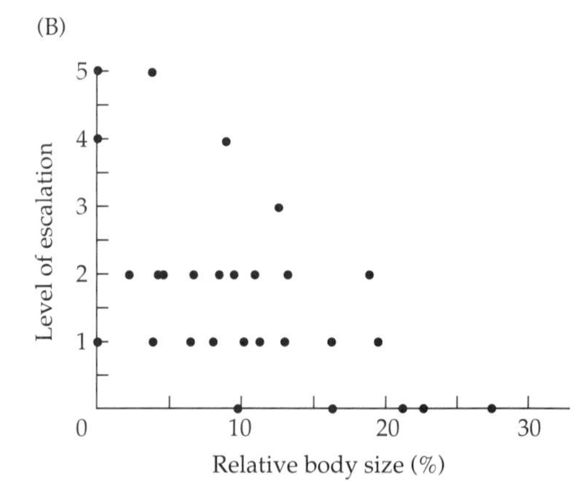

Figure 21.10 Stages in jumping spider contests. Contests in *Euophrys parvula* consist of distinct stages, always beginning with stage 1 and escalating to the next stage until one individual retreats. Each stage is associated with a different type of display or behavior, and each subsequent stage is associated with increasing risk of injury. All of the behaviors are mutual and performed by both spiders, i.e., when one escalates the other follows. (A) The stave display of stage 2. (B) The stage of escalation reached increases when the contestants are more similar in size. (A courtesy of Lyn Forster; B from Wells 1988.)

biting; it frequently leads to leg injuries. Each subsequent stage is associated with a greater risk of injury. In contrast to the examples in the prior section, the size asymmetry of contestants and the presence of a female resource affect the level of escalation reached in the contest but not the duration of the contest.

Enquist et al. (1990) developed another version of the **sequential assessment game with several behavioral options** to model contests that proceed in stages. They argue that opponents can obtain different types of information about relative fighting ability from different displays. In fish, for example, lateral display may provide only partial information on size, tail beating may improve the size estimation, and mouth-wrestling or head-butting may provide information on strength. Opponents initially choose the behavior that provides the most information at the lowest cost. This behavior is repeated and, as in the model with one behavior, each contestant's assessment of its chances of winning improves with time. If there is still uncertainty and the giving-up line has not been crossed, the next phase of the contest begins. They use (or add) a second behavior that has a higher cost but provides new information. The contestants escalate to several phases if neither individual believes it will lose, and end in a final fight stage (Figure 21.11). Besides predicting that more closely matched contestants will proceed through more stages, the model makes several specific predictions: the behavioral sequence should be independent of relative fighting ability, and the duration (number of acts) of completed phases should be independent of relative fighting ability.

These specific predictions were tested in a cichlid, *Nannacara anomala* (Enquist et al. 1990). Contests begin with broadside display and escalate to

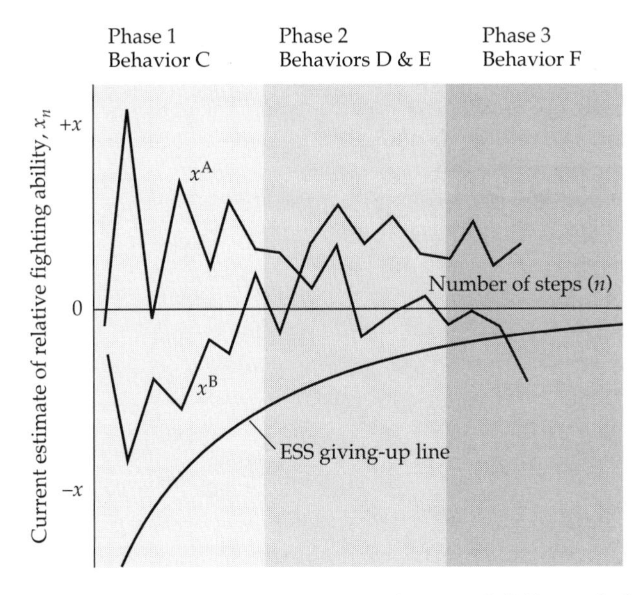

Figure 21.11 The sequential assessment game with several different behaviors. During the first stage, only behavior C is used; in the second stage, behaviors D and E are used; and in the final stage F (fighting) occurs. Axes are the same as in Figure 21.8. In this example, individual A gives up first. (After Enquist et al. 1990.)

tail beating, biting, mouth-wrestling, and finally to circling with mutual biting (Figure 21.12). Tail beating stops once mouth wrestling and biting begin. The predictions were supported in that the larger fish nearly always won

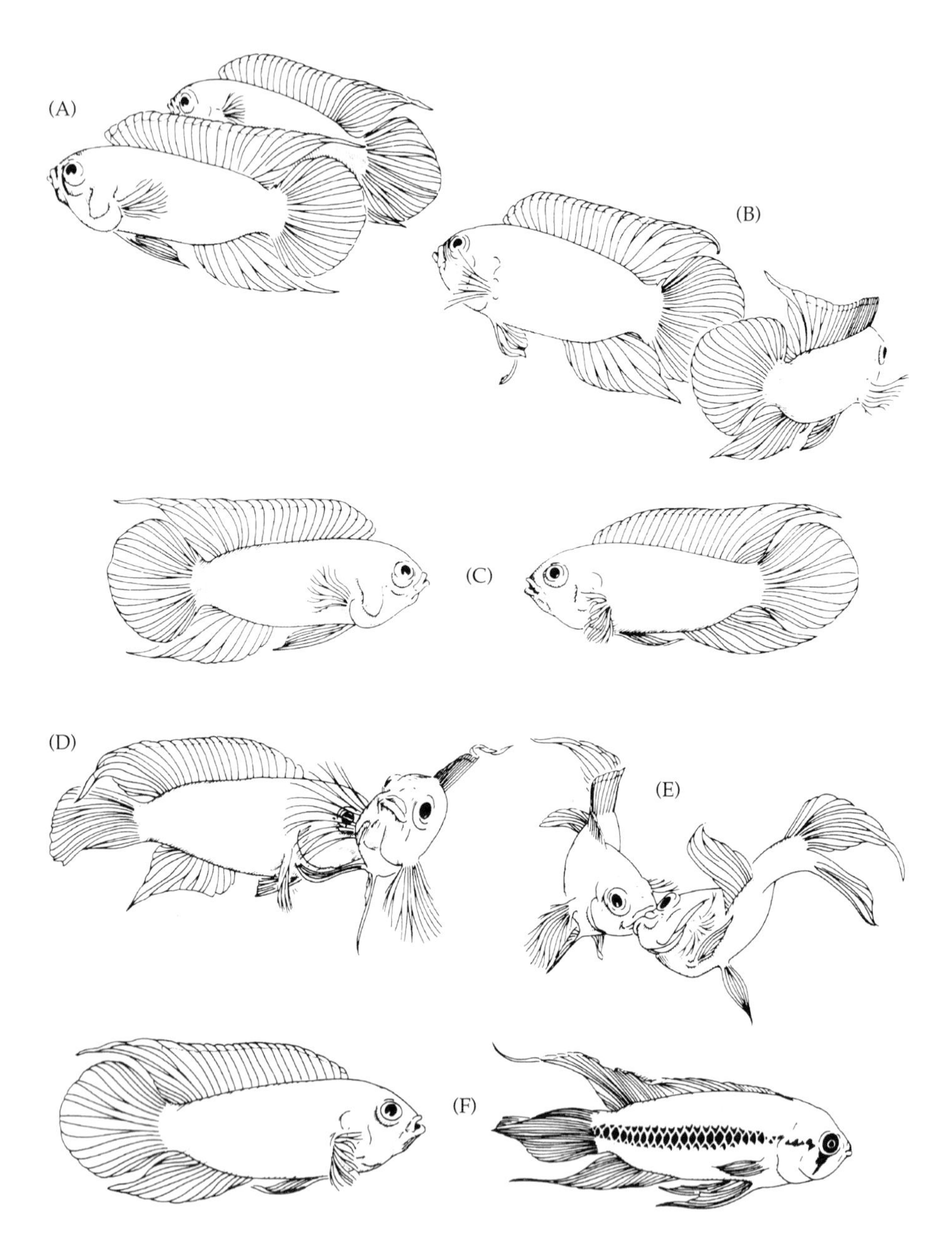

Figure 21.12 Fighting sequence in the cichlid fish *Nannacara anomala*. (A) Broadside display, (B) tail beating, (C) frontal orientation, (D) biting, (E) mouthwrestling, and (F) termination, in which the loser on the right adopts midline darkening and lowered fins. (From Krebs and Davies 1995, based on Jakobsson et al. 1979.)

and the duration and number of acts for completed phases were not correlated with weight asymmetry. However, the smaller fish performed significantly more tail beats in early phases, and in the final minute of a fight the winner bit twice as often. A similar test of the model was conducted on the convict cichlid (*Cichlosoma nigrofasciatum*) with similar results (Koops and Grant 1992).

Enquist and colleagues (Enquist and Jakobsson 1986; Enquist et al. 1987) assert that gradual escalation during the fight is a consequence of the fact that assessment of relative body size and weight in fish with purely visual cues is rather imprecise. They measured the precision of visual assessment in a controlled laboratory experiment. Two individual fish were allowed to view each other through a clear partition for half an hour, then the partition was removed and the fish permitted to interact. They noted whether a fish gave up immediately, gave up after a brief interaction, or did not give up, and compared this to known probabilities of winning as a function of the size asymmetry. If the smaller fish is 90% or less of the larger fish (i.e., they are 10% or more different in size) the smaller fish always loses, yet a fish must be 40 to 50% smaller than its opponent before it will give up after visual assessment only. Enquist argues that the discrepancy may arise because the fish have poor knowledge of their own size. However, visual assessment is by no means useless and can serve to reduce overall contest duration (Enquist et al. 1990; Englund and Olsson 1990; Koops and Grant 1992; Keeley and Grant 1993; Morris et al. 1995).

In another cichlid, *Oreochromis mosambicus*, studied by Turner and Huntingford (1986), a similar set of acts are performed but they are mixed randomly (not performed in any type of sequence or stages). Contest intensity, but not duration, is inversely related to relative size difference. For two of the behaviors, circling and tail beating, the investigators could predict the winner at the beginning of the contest. For a third behavior, ram-to-charge ratio, there was no predictive ability at the beginning of the contest, but this behavior became increasingly higher for the eventual winner as the fight progressed. The larger fish usually won. These observations and similar ones mentioned above for *Nannacara* do not fit Enquist and Leimar's model well and suggest that other types of processes are occurring during contests.

VARIABLE-SEQUENCE CONTESTS

Many animal species possess a large number of displays used in agonistic contexts and, like *Oreochromis* described above, different displays are used in mixed order rather than repeated in a consistent sequence. Crickets (*Acheta domesticus*), for example, use 13 different behaviors during contests; four are clearly displays with no physical contact, five are fighting tactics involving light physical contact, and four are tactics involving hard physical contact (Table 21.1) (Hack 1997a,b). Little blue penguins (*Eudyptula minor*) provide another example of a rich agonistic repertoire (Waas 1990a). They have 22 different displays that can be categorized into six risk levels according to the proximity the display generates between the sender and its opponent (see Table

Table 21.1 The agonistic display repertoire of the cricket *Acheta domesticus*

Intensity level	Energetic cost	Context or function
Displays without contact		
Cerci raise	0.02*	Defensive act by loser or subordinate
Prebout stridulation	0.02	Owner signal
During stridulation	0.02	Mild aggressive threat
Postbout stridulation	0.02	Postfight victory signal by winner
Mandible flare	0.08	Aggressive threat, intention to bite
Shake	0.49	Postfight victory signal by winner
Tactics with light contact		
Head butt	0.37	Ritualized fighting tactic
Foreleg punch	0.37*	Ritualized fighting tactic
Mandible spar	0.45	Mutual ritualized fighting tactic
Antennae lash	0.68	Dominance maintenance
Stridulation lash	0.70	Dominance maintenance
Tactics with hard contact		
Rear kick	0.37	Defensive act by loser or female guarder
Head charge	0.61	Overt attack, often by burrow intruder
Mandible lunge	0.61	Bite
Wrestling	0.83	Mutual mandible lock with head twisting

Note: Contests between males consist of a series of shorter interaction bouts interspersed with periods of avoidance. Within each bout, lower-cost behaviors are used at the beginning and change to higher-cost ones, but not all behaviors are used in a given bout. During one of the first few bouts, the interaction escalates from displaying to contact tactics to wrestling. Once the rivals have wrestled, the winner signals victory and subsequent interactions consist mainly of dominance/subordinance maintenance. Larger males tend to win, but prior experience and burrow ownership are more significant determinants of contest outcome. Behaviors are listed in order of increasing cost within each behavior category. Energetic cost is measured in ml O_2/gm · sec; values indicated by * are estimates.
Source: Hack 1997a,b.

21.2). The web-building spider *Agelenopsis* uses 33 different behaviors during contests (Reichert 1978). These can be divided into four categories that represent increasing levels of escalation and risk (Figure 21.13). Although some of the signals in each of these sets probably reveal aspects of fighting ability, there are obviously many more displays than are needed for this function alone.

These three species, cricket, penguin, and spider, are representative of the range of agonistic repertoire sizes seen in typical variable-sequence contests species, and we shall use them to illustrate existing ideas about this style of conflict resolution. Contests in these species and others like them share a number of characteristics in common that distinguish them from species with se-

Figure 21.13 Conflict behavior repertoire of female funnel-web spiders, *Agelenopsis aptera*. The large intruder on the right performs the spread leg display while the smaller resident performs a submissive wave legs display. Individuals can perform 33 different behaviors during such interactions, which can be grouped into four levels of risk: (1) *locating behaviors* ($n = 6$) such as turn, spread legs, palpate legs, and search while at a distance from the opponent, (2) *signals* ($n = 14$) such as drum pedipalps, stilt, crouch, put abdomen up, wave abdomen, wave legs, pluck web, put rear up, and flex, performed at closer distances, (3) *threats* ($n = 3$) including jump, chase, and lunge, and (4) *contact behaviors* ($n = 4$) including touch, shove, bite, and tumble. Four additional multiple-function acts include approach, circle web, lay silk, and retreat. Locating behaviors allow the contestants to assess relative body size by the amplitude of web vibrations. If the spiders are more than 10% different in weight, the locating behavior leads directly to threat and/or contact, and the smaller spider immediately retreats. If spiders are less than 10% different, extensive signaling occurs; multiple bouts of signaling, threat, attack, and retreat ensue, and contest duration increases. Displays performed at the end of a contest by winners include only bite web, circle web, and manipulate prey. (After Reichert 1978, 1984.)

quential assessment. Most of the displays and acts are performed unilaterally, rather than mutually. Thus, instead of both opponents performing a broadside display, countercalling, tail beating, or circling each other, opponents in the cricket, spider, and penguin species each perform different displays and acts at any point in time. Another difference is in the sequence and repetition of displays. Rather than performing a single type of display repeatedly and then escalating to another type of repeated display, acts tend not to be repeated. Contestant A begins the contest with an initiation act, contestant B responds with a different behavior, A then responds to this act, and so on as the chain of response and reaction continues. Contests can escalate and deescalate rapidly. In many cases, longer contests consist of a series of short bouts of approach and retreat with each opponent performing two to six different acts; the bouts may be repeated several times before a final resolution is achieved.

Three hypotheses, outlined in the following sections, have been invoked to explain the large number of threat displays in such species. They all propose that displays function to transmit additional information besides fighting ability, such as aggressive motivation, intentions, bluff, or specific contextual information. The hypotheses are not exclusive, so two or all three could operate in a single agonistic communication system.

Each Display Serves a Different Function

The classical hypothesis is that each display is adapted to one of a limited number of situations (Tinbergen 1959). For example, displays indicating a high probability of subsequent attack may be used against more threatening opponents, whereas milder displays are used against less threatening opponents. Alternatively, different displays may be used toward different types of opponents, such as adult versus young, or male versus female, or neighbor versus strange intruder. The form of displays might also be adapted for transmission to opponents at different distances or arriving from different positions, e.g., vocal displays might be more effectively transmitted to distant aerial opponents, whereas silent postural displays are used against nearby intruders on the ground. Finally, different displays may be used in conflicts over different types of resources such as females, food, and nest sites.

There is ample evidence that some threat displays are given in slightly different agonistic contexts. In the three focal species, agonistic repertoires are clearly divisible into broad functional categories. All three possess fighting tactics, contact threats, and signals. Signals can also be broken down into further categories. In the cricket, nearly every display or act can be ascribed to a distinctive function, and there are never more than two behaviors with what appears to be the same function. These functions include: (1) offensive threat display, (2) aggressive approach or contact tactic, (3) fighting tactic, (4) defensive tactic, (5) dominance-maintaining display, (6) submissive or quit display, (7) victory display by winner, and (8) owner-advertisement signal. Many other species with agonistic repertoires of from six to 15 acts exhibit one or two behaviors for each of these functions. In the penguin and spider display set, different displays are clearly associated with different distances from the opponent. For the spider these distance categories are partly a consequence of perceptual constraints. Web spiders cannot see well beyond a few centimeters, so the initial interaction of distant opponents involves transmission of tactile web vibrations that permit reasonably good assessment of relative body size. Once the opponents approach each other more closely, visual signals can be used for further information transfer. For the penguin, displays given at different distances from the opponent are associated with different probabilities of being pecked and therefore indicate the degree of risk an opponent is prepared to take. In all three species, winners give specific signals at the end of the fight that have been called victory displays. There has been no general discussion or specific study addressing the adaptive significance of this interesting type of signal. Finally, many studies have shown that levels of escalation and frequencies of displays change in different contest contexts such as presence or absence of a female or burrow, known neighbors versus strangers, arrival position of opponent, and other factors. Therefore, much of the variety in agonistic displays can be ascribed to different functions.

Degradation of Threat Effectiveness

A second hypothesis, promoted by Andersson (1980), is that threat displays lose their effectiveness over evolutionary time, requiring the evolution of new threats. The argument is as follows. Most threats evolve from intention movements. As they become ritualized, they become decoupled from the follow-through action. The frequency of bluff increases, and gradually the displays become less accurate at predicting what the sender will do next. Receivers are then selected to ignore these signals and look for more reliable cues of what the opponent will do next. New threat signals evolve, and useless displays eventually drop out of the repertoire. A species' repertoire at any point in time represents a mixture of predictive and nonpredictive behaviors.

Direct evidence for changes in display effectiveness over time is very difficult to obtain and does not yet exist. Paton and Caryl (1986) demonstrated that the correlation between displays and responses and the effectiveness of displays in predicting future acts by the displayer varied among different populations of great skuas (*Catharacta skua*), but it is difficult to know whether such differences are the consequence of differential selective pressures and display decay rates in different populations or merely a consequence of sampling. The penguin study clearly demonstrates that displays differ in their effectiveness at repelling opponents and that the more effective displays are used more frequently (Waas 1990a). However, we cannot assume that less effective threats are being ignored and going extinct; they may still perform an important function.

The theoretical possibility of changes in sender and receiver strategies over time has been explored in two recent simulation models. One is the neural net model of Krakauer and Johnstone (1995) on the evolution of exploitation and honesty discussed on page 673. Recall that both senders and receivers evolve in this simulation. Although honest communication is generally maintained as long as cheaters suffer higher costs, senders occasionally get ahead of receivers because they have hit on a highly effective but lower-cost display. This model provides support for the notion that new displays are continually arising, but not for the same reasons that Andersson postulated. In this case, displays are initially dishonest and become honest as receivers catch on to the trick and evolve modified networks for perceiving and discriminating against cheaters. As a possible example of such a process, Reichert (1978) proposed the evolution of **protean displays**, analogous to the surprise tactics some prey animals use to startle predators into dropping them, as a possible explanation for some of the signals in the spider signal set. Along the same line, displacement and redirected acts could be interpreted as sender strategies to mislead or catch receivers off guard rather than as outcomes of motivational conflict proposed in Chapter 16.

The second model, also a generational neural network simulation, compares the evolution of signal form in two contexts: conflict between sender and receiver, and conflict between several senders for the attention of a receiver (Arak and Enquist 1995). The first type of conflict is what we observe in agonistic interactions between rivals, and the second type is characteristic of mate-attraction signals. Only the receiver system is modeled as a neural net. Receiver biases act as a selective force on signals, causing them to become more costly and conspicuous as the intensity of conflict increases. Changes in the signal arise from random mutations and only spread when they correspond to biases in the receiver system. In sender-receiver conflict, receivers attempt to resist (ignore) exaggeration by senders; in sender-sender conflict, the degree of competition increases when more senders are present. The resulting patterns of signal evolution are qualitatively different for these two situations. In the competitive sender context, the display rapidly evolves to an exaggerated, costly signal and remains the same. In the sender-receiver conflict context, the cost of the display initially increases and may remain the same for long periods, but occasionally there is a shift to a lower cost but more effective display that overcomes receiver resistance. Like the previous model, it suggests that new displays can invade when there is a severe conflict between sender and receiver.

Displays Transmit Graded Information about Aggressive Intentions

The third hypothesis proposes that the choice of different behavioral options is an honest indicator of the actor's intentions or aggressive motivation; the actor reveals what it is likely to do next by its choice of action. The idea stems from the model by Enquist (1985; et al. 1985) described on pages 655–658. Different displays and tactics are assumed to differ in their effectiveness, measured as the probability of winning a contest if the particular behavior is used. As long as the cost of each behavior increases with increasing effectiveness, bluffing or indicating a higher motivation level than one really has will not pay, and honesty will be maintained. For example, a threat display derived from an attack intention movement is likely to possess a tactical element that renders the display highly effective in deterring opponents. If the most effective display also places the sender closer to the opponent where it is at greater risk of being attacked and injured, bluffers will pay a very high cost. The choice of more or less risky behaviors not only can reveal a combatant's valuation of the resource, but can in theory also reveal differences in fighting ability. To rigorously test this hypothesis, data on both the costs and consequences of each behavior are required.

A second relevant model on signals of intention has recently been developed as a dynamic two-stage extensive-form game by Kim (1995). Opponents engage in a round of signaling and then play the hawk-dove game (with $V < D$). Initially there are only two possible signals, a costly and a cheap one, and the strategies include different rules for responding to the opponent's signal with the same or different signal and for linking one's signal to a subsequent action in the hawk-dove game. Both players prefer to give the cheap signal. In this case, they have both paid the same signaling cost and therefore play the symmetric hawk-dove game, which means each will play a mixed strategy of hawk and dove in some proportion. There is no signaling of intentions here. This solution is a locally stable Nash equilibrium but it is not a global ESS because it can be invaded by a strategy in which the second individual to signal gives the opposite type of signal. Now they play the asymmetric hawk-dove game that results in an ESS for the policy: costly-signal sender plays hawk and cheap-signal sender plays dove. This is a strategy of signaling intentions that is globally stable because both contestants have higher net fitness than if they gave the same signal. Kim then goes on to show that as the number of different possible signals increases, the probability that the two opponents will have to fight a battle decreases. This outcome occurs because the likelihood that they will both give the same signal by chance becomes increasingly smaller. This elegant analysis not only offers an explanation for why opponents tend to give different signals in response to each other, but also why the number of distinctive threat signals tends to increase.

What is the evidence for signaling of intentions? Both before and during the heated debate over signal honesty described earlier, a host of investigators attempted to determine whether aggressive displays predicted what the actor would do next (Stokes 1962; Dunham 1966; Dingle 1969; Rubenstein and Hazlett 1974; Steinberg and Conant 1974; Rand and Rand 1976; Andersson 1976; Sinclair 1977; Caryl 1979; Rubenstein 1981; Hazlett 1980; Nelson 1984; Piersma and Veen 1988). A popular method in the 1970s was to construct contingency tables of the transition probabilities between pairs of sequential sender acts as well as between pairs of sender acts and receiver responses (see Boxes 15.1 and 15.2). All studies found moderate correlations between an actor's prior and subsequent acts, indicating some level of predictability in displays. However, most found even stronger correlations between the sender's act and the receiver's response. Furthermore, in dissecting the ability of specific displays to predict the actor's next act, it was generally concluded that escape or retreat behavior was more strongly associated with certain displays than was attack behavior (Caryl 1979; Paton and Caryl 1986). The war of attrition model, which rests on the assumption that contestants don't reveal their willingness to persist at the beginning of the contest, generated the notion that opponents should not provide information about the level to which they will escalate or what they will do next (Maynard Smith 1979). Several models countered this point of view by identifying specific conditions under which honest information of intentions would be favored, such as the presence of individual recognition and knowledge of rivals (van Rhijn 1980; van Rhijn and Vodegel 1980), careful scrutiny of both signals and unritualized intention movements to detect bluffers (Moynihan 1982; Bossema and Burgler 1980), and bargaining over divisible resources (Maynard Smith 1982b). The theoretical work of Enquist and Leimar played a major role in shifting the tide back to a belief in the validity of honest signaling of intentions. Their sequential assessment game, with it assumptions of information gathering and decision-making during the contest, provided an alternative to the war of attrition. Enquist's (1985) choice-of-behavior model then formally demonstrated that multiple acts could evolve to reveal honest information about intentions as long as the effectiveness of each act is associated with increased costs. This model has received recent empirical support from the studies on fulmars (Enquist et al. 1985), goldfinches (Popp 1987), elephants (Poole 1989), and penguins (Waas 1991a,b).

Waas (1990a,b; 1991a,b) has contributed significantly to the intention signal controversy by showing that it is important to control for the receiver's response to a signal when analyzing the relationship between signal and sender's next act. If a sender gives an effective threat that causes the receiver to retreat, the necessity of following through with an attack is removed. The threat therefore transmits the message, "I will attack *if you don't retreat*." Furthermore, an effective threat also bears the cost of a retaliatory strike by a highly motivated receiver, which may then cause the sender to retreat (see for example the goldfinch data given in Table 20.2). These receiver responses confound the analysis of sender intentions and explain the weak correlations between threat and subsequent sender attack reported by prior researchers. Waas statistically controlled for receiver responses and found strong positive associations between the cost of a given display and both its effectiveness in causing receiver retreat and the probability of subsequent sender attack.

Waas also provides an explanation for why so many different displays are needed by little blue penguins to signal aggressive motivation. Not only are the acts in the six major risk levels he identified associated with different probabilities of attack and effectiveness, but the different displays within a risk level also reveal fine differences in risk and effectiveness. For example, the three offensive stationary displays of risk level 3, direct look, directed flipper spread, and point, tend to be given at distances from the opponent of greater than 3, 2, and 1 meters, respectively (Table 21.2). The direct look actually causes opponents to approach and escalate, the directed flipper spread tends to halt any further approach by the opponent, and the point display causes opponents to deescalate slightly to the next-lowest risk level. Similarly, the addition of a vocalization to a threat display increases the likelihood that the opponent will deescalate compared to the silent version of the same display. Lastly, cave-dwelling populations of little blue penguins use almost twice the number of displays as burrow-dwelling populations. Cave-dwellers nest colonially, with nests in close proximity and no physical barriers between them, whereas burrow-dwellers nest in crevices isolated from other pairs. The frequency of conflict is three times higher in cave populations compared to burrow populations, but the level of escalation, fight duration, and severity of injury is much higher in burrow populations. Although the constraint of movement and approach within burrows partly explains the different forms of display in the two populations, Waas argues that the high frequency of agonistic interaction and high potential cost of fighting in the cave populations has favored the evolution of many more displays to resolve conflicts at lower levels of escalation. Burrow nesting sites may also be more limited and valuable than cave nesting sites. Note the particularly high diversity of contact behaviors in cave populations; cave penguins also tend to use a more ritualized and less injurious method of fighting than burrow penguins (Figure 21.14). These results concur with the general conclusion of the hawk-dove game, namely that high costs of injury favor the use of displays

Table 21.2 The agonistic display repertoire of cave-dwelling populations of little blue penguin *Eudyptula minor*

Category	Risk level	Distance	Movement
Defensive, distance-increasing			
Low walk	1	< 1	Away
Submissive hunch	1	< 1	Away
Defensive, stationary			
Face away	2	< 1	Stay
Indirect look	2	< 1	Stay
Offensive, stationary			
Direct look	3	> 3	Stay
Directed flipper spread	3	2–3	Stay
Point	3	1–2	Stay
Bowed flipper spread[a]	3	1–3+	Stay
Offensive, distance-reducing			
Zig-zag approach	4	2	Toward
Flipper spread approach	4	1	Toward
Contact			
Bill to back	5	0	Toward
Breast butt	5	0	Toward
Bill to bill	5	0	Toward
Bill slap	5	0	Toward
Bill lock/twist	5	0	Toward
Overt aggression			
Attack	6	0	—
Bite	6	0	—
Fight	6	0	—

Source: Waas 1990a,b.

Note: Risk level is an estimate of the probability that the actor will be pecked by its opponent; displays are listed in order of increasing risk. Distance indicates the typical physical distance in meters between the sender and opponent at which each display is given. Movement shows the direction the sender moves with respect to the opponent during the execution of the display.
[a]The bowed flipper spread is a victory signal by the winner.

to resolve conflicts. Injury cost is not only high in predatory species with prey capture weapons; nonpredatory species sometimes also evolve nasty weapons for use in conspecfic fights (Figure 21.15).

Why Are There So Many Threat Displays?

The models of Enquist and Kim clearly show that the signaling of motivational information requires many different threat displays. The alternation of different signals and responses between two rivals, along with observa-

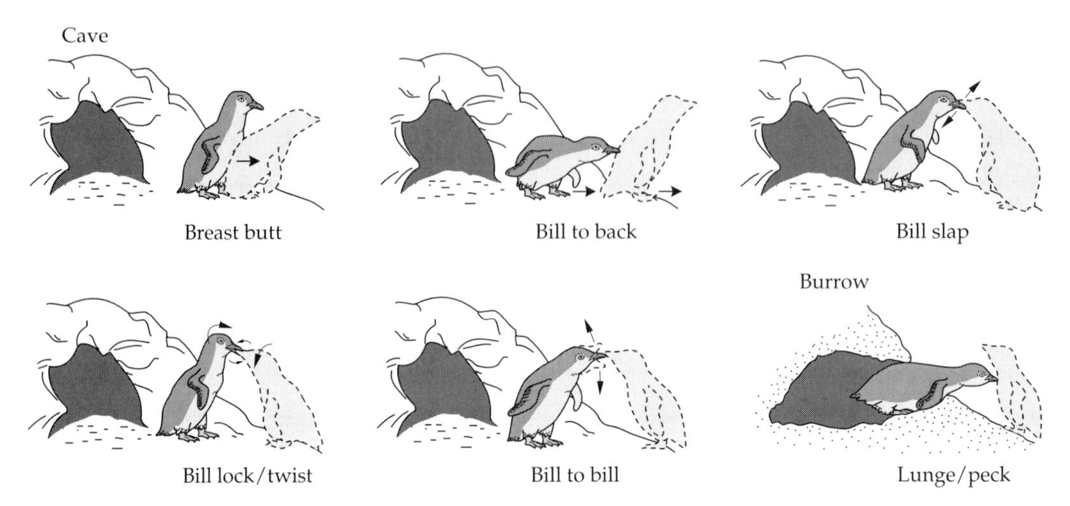

Figure 21.14 Contact behaviors used by cave- and burrow-dwelling little blue penguins during conflicts. Breeding pairs in cave-dwelling populations of the penguin (*Eudyptula minor*) nest colonially in large caves, whereas pairs in burrow-dwelling populations nest in relatively isolated crevices. Cave-dwellers fight much more frequently than burrow-dwellers and possess five different contact signals shown here, compared to a single equivalent display (lunge/peck) in burrow-dwellers. Opponents are indicated with dashed outlines. (After Waas 1990b.)

tional evidence of predictive signaling, provides strong support for the motivational signaling hypothesis. However, this does not mean that the other two hypotheses, different functions and signal erosion, are not important; in fact, all three processes probably work together. A graded motivational system requires a set of signals ranging from fearful submissive signals to aggressive threats to offensive attack tactics to deal with situations in which the opponent is either superior or inferior (or more or less motivated) compared to the sender. For a majority of species, six to 15 signals appear sufficient to achieve this range. As we saw in the cricket, it is possible to ascribe a nearly unique function to each signal in the agonistic repertoire. In the penguin, the burrow-dwellers also fall into this repertoire size range and exhibit little redundancy within functional categories based on distance to and movement toward or away from the opponent. Cave-dwellers, on the other hand, appear to have more redundancy in several functional categories, especially in approach and contact tactics and submissive signals. As Kim's model suggests, they may need more signals to reduce the frequency of escalated fights, but such signals may have invaded because they were more effective than others. The spider seems to have many more signals than it needs; even after subtracting out the victory and submissive signals, there are 10 with an apparently similar function. Bluffing, sender exploitation of receiver biases, and discounting of signals may have played a more important role in this species where loss of a web means significant loss of future reproduction.

Figure 21.15 Weapons for intraspecific contests. (A) Male stag beetles (Lucanidae) use their horns in male-male contests. (B) Male gladiator frogs (*Hyla rosenbergi*) have sharp, curved scythe-like spines on their thumbs. (C) Pythons (*Python molurus*) fight by coiling their bodies together, jostling for position, hissing, and secreting musk from the cloaca. They also raise their sharp pelvic spines and attempt to force these into the delicate skin between their opponent's scales. (D) The primary function of horns in antelope and deer (here, sable antelope *Hyppotragus equinus*) is intraspecific fighting. Unlike the previous examples, however, the elaborate structures are not designed for injuring opponents. Instead, they facilitate frontal pushing matches and contests of strength. (A–C after Huntingford and Turner 1987; D after Fraser 1968.)

Although Enquist et al. (1985) suggest that choice of behavior could be used to signal fighting ability as well as aggressive motivation, it seems that fighting ability is better assessed with the repetition of one or a few displays as modeled in the sequential assessment game. What determines whether a species signals mainly fighting ability or motivational level during contests? The answer may be quite simple. If the major source of variation between contestants in a population is fighting ability (k), then signals that reveal fighting ability should evolve and contests will resemble the sequential assessment game. This would tend to occur in species with large variations in body size or in body condition. If the major source of variation between contestants is resource valuation (V), then signals that reveal aggressive motivational level should evolve, and contests will escalate and deescalate quickly and involve rapid exchanges of nonrepeated signals and responses. This is more likely to occur in species with little body-size variation, in territorial systems with ownership asymmetries, or where previous experience affects contest outcome.

SUMMARY

1. Conflict occurs when two rivals compete for the same resource. **Conflict of interest** is high because each contestant wants the other to retreat and concede the resource, but both would prefer to settle the conflict without a fight. Resolution involves reciprocal exchange of information about **relative fighting ability** and **relative resource value**. The behaviors observed during contests comprise a mixture of displays and tactical acts. Senders will be tempted to bluff, exaggerate, and manipulate, and receivers will be selected to evaluate information from all sources skeptically and devalue any exaggeration. Several game-theoretical models have been developed to describe the dynamics of animal contests.

2. The **hawk-dove game** can be applied to contests with two distinct phases: display and fight. If fighting is very costly, some displaying will be favored. A third strategy, assessor, can easily invade this game and become an ESS even when fighting is not very expensive. Assessors use information in the display to evaluate their fighting ability relative to the opponent. Assessment requires the evolution of a graded signal whose magnitude is well correlated with fighting ability, either because signal form is physically constrained and unbluffable, or because cheaters suffer higher retaliation costs.

3. The **asymmetric war-of-attrition game** can be applied to species that perform the same behavior repeatedly for variable lengths of time. Each player makes a quick evaluation of its likelihood of winning based on an estimate of its relative **V/k ratio**, where V is the value placed on the resource and k is the contest cost accrual rate (lower for better fighters). Each player then **chooses a persistence time** from a high or low distribution depending on whether it figures it will win or lose. The battle commences and ends when the first player reaches its **giving-up time**. The model assumes that initial assessment of fighting ability is imperfect and that players never reveal their bid times at the beginning of the contest.

Specific predictions about the duration of contests as a function of the size asymmetry of contestants and the value of the resource are well supported by empirical data.

4. The **sequential assessment game** is an alternative model to the war of attrition for repeated-display contests. It assumes that contestants have poor knowledge of each other's fighting ability at the beginning of the battle but **acquire this information during the contest** by updating their estimates at each repeated round of interaction. When one individual becomes reasonably sure it is likely to lose, it quits the contest. The model generates the same predictions about contest duration as the war of attrition, but conforms better to our understanding of the decision-making process of communication. A second version of the model assumes that if a single repeated behavior is unable to reveal fighting ability asymmetries with sufficient certainty, the contestants shift to a second phase with a different, more costly behavior that reveals new information.

5. Contests in many species don't fit any of the above models in that many displays are used, behaviors are not repeated, chain reactions of stimulus and response occur between the contestants, and contests escalate and de-escalate rapidly. Three hypotheses are proposed to explain the large number of displays. (1) Displays serve slightly different functions and/or are designed for different contexts. (2) Many displays are bluffs, arising either from sender manipulation of receiver biases or ritualization and devaluation of formerly indicative intention acts. (3) Displays encode different aggressive motivation levels or intentions.

6. Good evidence exists that some displays serve different functions such as submission, offensive threat, defensive threat, dominance maintenance, victory, and ownership. Evidence that different displays are associated with different levels of fear or aggressive motivation also exists, and the effectiveness of the display in deterring opponents is correlated with the risk or cost of performing the display that guarantees its honesty. There is limited evidence of overt bluffing. Ineffective threat displays seem to be a part of motivational-level display sets that encode submission or low probability of attack. Devaluation of intention movements and receiver exploitation cannot be entirely ruled out, however.

7. Where opponents differ most in fighting ability (body size or condition), contests will tend to involve assessment of fighting ability, and one or a few graded displays will be used to reveal this asymmetry. Where opponents differ most in resource valuation (territorial systems, prior experience), contests will tend to involve intentional signals and multiple display sets that reveal aggressive motivational level via choice of high- versus low-risk acts.

8. Both game theory and observations show that animals attempt to resolve conflicts at the lowest possible energetic cost or risk of injury, escalating

only if low-cost displays cannot reveal contestant asymmetries. Fighting costs are generally high in predatory species with prey capture weapons, but even nonpredatory species will evolve nasty weapons for use in intraspecific contests.

FURTHER READING

Two recent books review aggressive behavior in animals: Archer (1988) stresses the hormonal determinants of aggression but also provides a good review of the game-theoretical models; Huntingford and Turner (1987) give a good feeling for fighting tactics in a variety of species and describe the ecological and evolutionary context of conflict.

Chapter 22

Territorial Signaling Games

TERRITORIAL ADVERTISEMENT SIGNALS transmit the information that a particular area is owned and that the owner is present. But there is much more to territorial signaling than just declaring one's ownership of an area and pointing out its boundaries. Territory owners control a valuable resource that others may want to exploit or take over. In order for a territorial signal to effectively keep intruders out of the territory, it must warn intruders that an attack is likely if the intruder persists in trespassing. A territorial signal is therefore a type of long-distance threat signal. Threats are effective only if they are followed by some increased probability of escalated aggressive behavior. Furthermore, an efficient territory owner must also be an astute receiver of the territorial signals of others. The owner needs to determine whether another individual it detects is truly invading or just declaring its own territory and whether or not to approach and chase this individual away. Conflicts over territorial ownership and territorial boundaries must be resolved with some combination of signaling and aggressive acts, as discussed in the last chapter. But resolution of territorial conflicts involves one more element besides the assessment of fighting ability and payoffs—there is an ownership asymmetry. This chapter will examine the game context of territorial conflict. The game-theoretical approach not only clarifies the role of territorial signals in assessing ownership asymmetries, but also explains the elaboration of signal form that arises from competitive territorial interactions.

TYPES OF TERRITORIES AND INTRUDERS

The costs and benefits of territorial defense depend on the size and resources of the territory and on the number and intentions of intruders. We shall begin the discussion of territorial signaling by defining terms and describing the players of the game.

Definition of Territoriality

A territory is a fixed area from which intruders are excluded by some combination of advertisement, threat, and attack (Kaufmann 1983). The benefit of territorial defense is usually exclusive access to the resources on the territory such as food, mates, and nest sites. The costs of defense include time lost; the energetic costs of signaling, patrolling, and chasing; the risk of predation while exposed; and the cost of injury if an intruder does not back down but instead fights. Territories in different species vary enormously in size depending on the type of resource defended. Three main types of territories are distinguished. **Breeding territories** are relatively small and contain only a nesting or mating site. This type of territory is characteristic of colonially nesting species that cluster their nests at limited safe sites and of lekking and chorusing species in which males aggregate to attract females. Colonial, lekking, and chorusing species forage away from the breeding territory, and foraging areas are not defended. **Feeding territories**, on the other hand, are larger because they must contain sufficient food to support the territory owner. Such territories are usually defended during nonbreeding periods by single individuals and when food is sufficiently scarce that exclusive access is worth the cost of defense. **All-purpose territories** contain both a breeding site and a food supply. This type of territory is usually larger than a feeding territory because it must contain enough food to support the growing offspring and often one or more mates. For all three types of territories, adjacent territories are usually contiguous and each owner may have two to six immediate neighbors with which it shares common boundaries. Networks of contiguous territories are called **neighborhoods** (Figure 22.1).

Two Types of Intruders

Territory owners may be challenged by two types of intruders: neighbors and floaters (McGregor 1993). **Neighbors** are, of course, also territorial owners, so they have demonstrated their fighting ability, possess the resources necessary to survive and/or breed, and have other neighbors with whom they must contend. Although the initial interactions with a new neighbor are likely to be intensely aggressive, over a period of time an owner develops a long-lasting

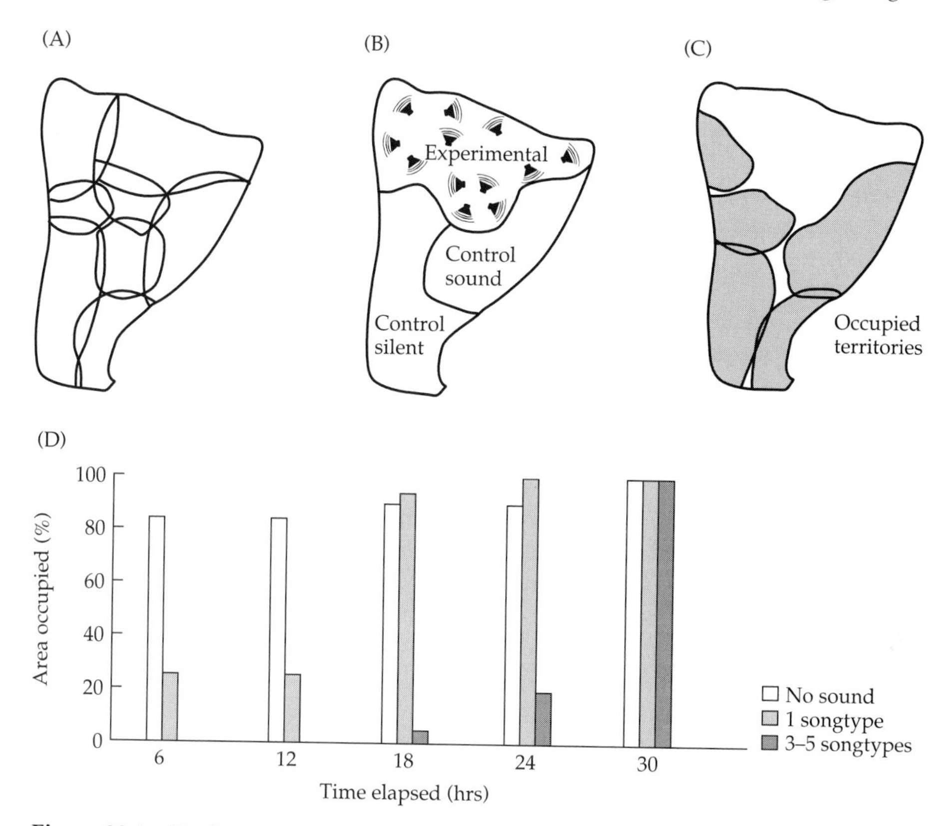

Figure 22.1 Territory neighborhoods and the effect of auditory signals. (A) A neighborhood of eight male great tit (*Parus major*) territories in a patch of woodland habitat. Each male is mated to a single female. There is very little overlap in the boundaries of adjacent territories, and essentially no unclaimed space. (B) All eight pairs were removed from their territories and a speaker-replacement experiment conducted. In the experimental area, great tit songs were broadcast from an array of speakers; sounds from a tin whistle were similarly broadcast in the control sound area, and the control silent area contained no speakers. (C) Twelve hours later, the two control areas were entirely occupied by new territorial birds but only one had entered the experimental area, showing that the territory advertisement songs are effective in deterring intruders. The new birds were either floaters or individuals that had owned territories in nearby less preferred habitat. Eventually, the entire experimental area was occupied because intruders realized that the auditory threats were not backed up by threats from live birds. (D) Great tit males possess a repertoire of three to five songtypes. In a similar speaker-replacement experiment, the time to occupancy was compared for speakers broadcasting a repertoire of songtypes versus speakers broadcasting only one songtype. The bar charts show the percent of area occupied after different periods of elapsed daylight hours since the start of the experiment. A repertoire of songtypes was more effective in deterring settlers than a single songtype. (After Krebs 1977b; Krebs et al. 1978.)

stable association with its neighbors, and interactions between them are characterized by repeated, usually low-level encounters. The owner needs to signal its presence to its neighbors from time to time and monitors the presence of the neighbors by their signals. An escalated response is then only needed when the neighbor actually trespasses; when the neighbor is on its own territory, such a response would be unnecessary and costly. Most of the time neighbors maintain a truce between themselves because both are likely to be of similarly high competitive ability and both possess a similarly valued resource and investment. It is therefore crucial that an owner be able to both recognize each neighbor as an individual and to judge the neighbor's location. Over a longer term, owners must monitor any changes in the presence and breeding status of the neighbor, as well as changes in the habitat that may require shifts in territory boundaries.

Floaters, on the other hand, are nonowners who may be searching for a vacated territory. They typically skulk through a neighborhood, occasionally emitting the territorial advertisement signal to test the owner's presence and condition (Smith 1978; Arcese 1987). An owner's interactions with floaters are characterized by nonrepeated encounters requiring immediate escalation. Figure 22.2 illustrates the dramatic nature of interactions between a territorial owner and an intruder. In contrast to neighbors, owners and floaters are likely to differ in several ways. Since owners tend to hold on to their territories for long periods of time, sometimes for the major portion of their adult lives, floaters tend to be younger than owners. There may also be a large difference in fighting ability between an owner and a floater. If territories are frequently contested and stronger individuals generally win contests, then the average owner has a higher resource-holding potential than the average floater. Finally, owners and floaters may place a different value on the territory. The owner usually values the territory it has worked hard to defend and in which mating and breeding may be underway much more than does the floater, who can always look elsewhere for a vacant territory. The value of the territory to the intruder will increase if alternative territories are very scarce. A more precise definition of territory value is given below.

Owners must be able to distinguish neighbors from floaters because the responses to the two types of intruders are usually quite different. Mistaking a floater for a neighbor may permit the floater to take over the territory, and mistaking a neighbor for a floater might result in a protracted fight with a formidable opponent. Selection on receivers is thus very strong in territorial signaling, and receiver strategies and responses exert a major role in shaping the characteristics of the signal (McGregor 1991).

Quantifying the Value of a Territory

It is important that we define the value of a territory carefully because it is a critical variable in the game models that follow. Territory value must be measured in terms of expected future reproductive success, i.e., in the units of number of offspring. In the context of fights over territorial ownership, the value of a territory is the number of offspring expected if a given fight is won minus the number of offspring expected if the fight is lost (Grafen 1987). This value is likely to differ for the two contestants depending on their roles in the contest as well as the type of territory. For example, in early breeding season

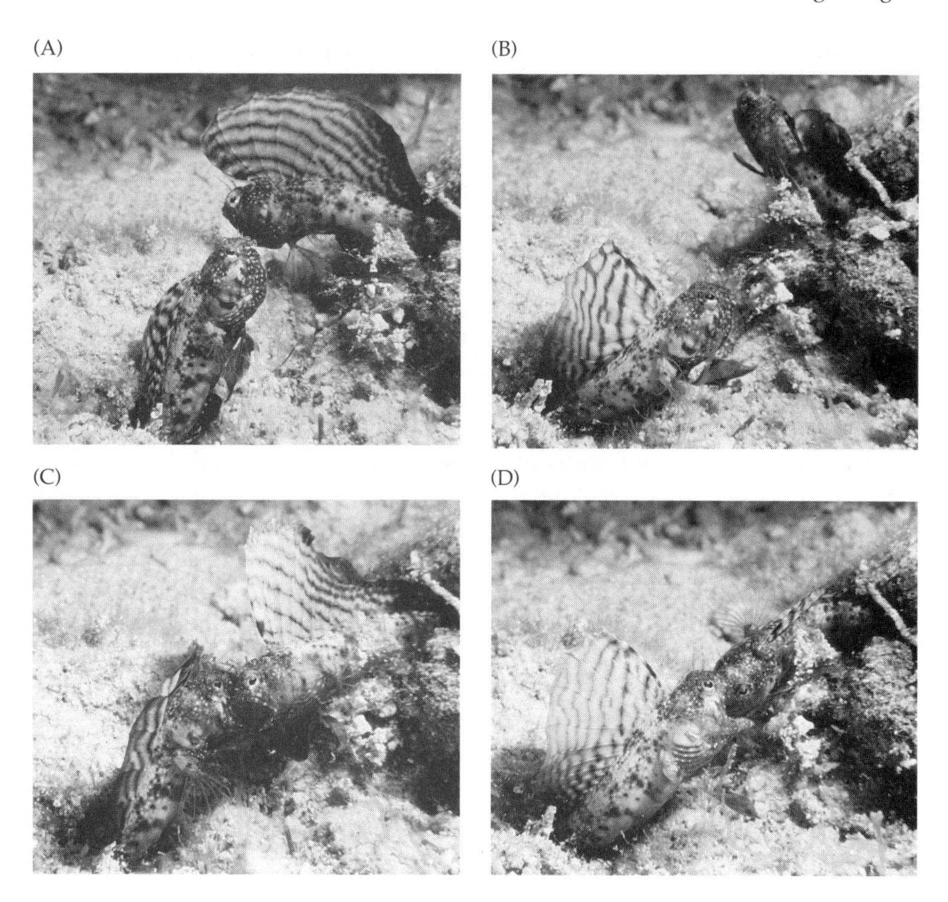

Figure 22.2 Escalation in owner-intruder encounters. In sailfin blennies (*Emblemaria pandionis*), males defend a burrow and guard eggs laid there by females. This series of photos shows an encounter between an owner (left) and an intruder (right); who has wedged himself into a small crevice for support. (A) Intruder approaches and gives raised fin threat display. Both males are relatively dark. (B) Owner gives raised fin plus open-mouth display while intruder turns away. (C) Intruder escalates and both males fight with fins raised. Both males are becoming paler in color. (D) Intruder gives up and lowers his fin. (Photographs courtesy of Ken Clifton.)

contests over all-purpose territories, owners and floaters may both place a moderately high value on the territory, but for different reasons. The owner stands to forfeit all of its reproductive and boundary negotiating investment to date if it loses to a floater, so the fitness decrement of losing is high. The floater currently has no such investment but faces further searching costs, so it has little to lose by defeat and much to gain by winning. As the breeding season progresses, the owner's valuation of the territory increases, while the floater's opportunity to reproduce decreases. In neighbor-neighbor interactions, on the other hand, both contestants have similar investments and the benefits of acquiring another territory may be small given the costs of defending a double territory. The value of winning such a fight against a neighbor may therefore be low, and the costs of a battle potentially high. As we shall see, these economic considerations play a large role in determining whether a contestant will fight, display, or retreat.

WHY DO OWNERS USUALLY WIN?

Naturalists have long noted that owners almost always win in conflicts against intruders. If there were not some type of bias favoring the owner, territories would change hands frequently, and stable territorial neighborhoods would not exist. The game-theoretical treatment of animal conflict resolution described in the prior chapter showed how the assessment of role asymmetries between the contestants can resolve conflicts without resorting to fatal fighting. In the case of conflicts over short-term rewards such as a mating opportunity or a food item, opponents use threat displays and other aggressive acts to assess each other's relative fighting abilities and aggressive motivation levels. In the case of territorial conflicts there are three possible asymmetries that could be used (Maynard Smith and Parker 1976). Asymmetry in fighting ability is, of course, still an important determinant of which contestant would win an escalated fight. Secondly, asymmetry in resource valuation is more likely to occur in territorial conflicts, especially in owner-floater encounters where prior knowledge and investment in the territory differ greatly. Conflict resolution based on these two asymmetries requires that contestants assess each other's relative costs and benefits of winning, so they are called **payoff-relevant asymmetries**. Thirdly, the owner-intruder asymmetry itself could be used to settle the conflict if a convention were adopted to respect ownership. This is a **payoff-irrelevant asymmetry**, but one that is easily assessed. Which asymmetry is used clearly affects the type of information encoded in the territorial signal and therefore determines signal form. In this section we shall review the game models and empirical studies that have sought to determine the relative importance of each of these three asymmetries in settling owner-intruder contests. (At this stage, a distinction will not be made between neighboring versus floating intruders.) We will first examine the likelihood that a game convention based only on the owner-intruder role asymmetry can settle territorial contests, then evaluate the role of payoff asymmetries.

The Bourgeois Game

In all natural territorial conflicts, one individual is always an owner and the other is an intruder, and this difference in ownership roles is perfectly known for both opponents as discussed earlier. If conflicts are settled only on the basis of this owner-intruder asymmetry, then the strategy or decision rule used by both the intruder and the owner must be to escalate if one is the owner and to retreat if one is the intruder. This strategy has been named **bourgeois** (Maynard Smith 1979). Ownership is always respected with this conditional strategy, and any other asymmetries in fighting ability or territory valuation are completely ignored. Bourgeois is therefore a payoff-irrelevant or uncorrelated asymmetry because it is based only on the ownership role and is uncorrelated with fighting ability and territory value. One way to ask whether this strategy can be evolutionarily stable is to include it as a third strategy in the symmetric hawk-dove game. The three discrete strategies are therefore: hawk (always escalate immediately), dove (display, retreat against escalators, and flip a coin or share with another displayer), and bourgeois (play hawk if owner, play dove if intruder). Figure 22.3 shows the payoff matrix for this game. Bourgeois is a pure ESS when the cost of losing a fight, D, is greater than the benefit of winning, V. Hawk is the ESS whenever $V > D$. This classic territory game thus suggests that an arbitrary convention of respecting ownership can in fact invade a population, but only in the situation in which the value of the territory is less than the cost of fighting for it.

The form that signals must take to settle territorial conflicts with a bourgeois strategy is consistent with the design rules outlined in Chapter 18 (pages 592–593): signals must advertise the territory location and identify the owner. Olfactory signals are ideal for this function. The territory boundaries can be outlined with a series of deposited marks containing individual identity signatures, the entire territory can be diffused with additional marks, and an intruder that detects a match between the substrate marks and an encountered individual can identify the owner (Gosling 1982). Laboratory studies of staged contests in neutral arenas have demonstrated a clear effect of odor matches on the behavior of contestants consistent with the bourgeois strategy. In rabbits, an ani-

	Hawk	Dove	Bourgeois
Hawk	$\frac{1}{2}(V-D)$	V	$\frac{1}{2}V + \frac{1}{4}(V-D)$
Dove	0	$\frac{1}{2}V$	$\frac{1}{4}V$
Bourgeois	$\frac{1}{4}(V-D)$	$\frac{3}{4}V$	$\frac{1}{2}V$

ESS Conditions
1. Hawk is a pure ESS when $V > D$.
2. Bourgeois is a pure ESS when $V < D$.

Figure 22.3 The game matrix and ESS conditions for the hawk-dove-bourgeois game. The three discrete strategies are hawk (always escalate), dove (display), and bourgeois (escalate if owner, display if intruder). V is the increase in fitness gained by the winner and D is the cost paid by the loser. When bourgeois meets either hawk or dove, we assume it is owner half the time (and therefore plays hawk), and intruder half the time (and therefore plays dove). Its payoffs are the average of the two cells above it in the matrix. Similarly, when bourgeois meets bourgeois, on half of the occasions it is the owner and wins, while on half of the occasions it is the intruder and retreats, i.e., it never pays the cost of losing in contests against other bourgeois individuals. (After Maynard Smith 1982a.)

mal is more likely to win the contest if its own territorial smell is wafted into the arena (Mykytowycz et al. 1976). In mice, males are less likely to fight if their unknown opponent's odor is present in the substrate (Gosling and McKay 1990). Visual bourgeois signals in open habitats are most likely to be arbitrary color patches, structures, and/or postures that identify the position and presence of the owner at all times. Visual movement displays and auditory signals are less likely to qualify as bourgeois signals because they are more expensive to produce continuously and can encode payoff-relevant information.

The classic experimental study by Davies (1978) of the speckled wood butterfly (*Pararge aegeria*) provided the strongest evidence for an arbitrary ownership convention. This northern European species overwinters as an adult and breeds very early in the spring. Males defend sunspots hitting the forest floor to increase their conspicuousness to females. One value of the sunspots is thermoregulation; both sexes are attracted to them to increase body temperature and maintain high activity levels. If an intruding male enters an occupied sunspot, the two males perform a short spiral flight and the intruder leaves. Davies removed territory owners for a short time, allowing a replacement male to take over the territory. When the original owner was subsequently released on the territory, he engaged in a short spiral flight then left, i.e., the new owner always won. Thus ownership for only a few minutes appeared to determine the outcome of the contest. Davies argued that the ownership convention was favored because sunspots were abundant and territory value therefore low.

Other investigators questioned Davies's result, arguing that fighting ability probably plays a role in this system. A study of the same butterfly species in Sweden failed to demonstrate a bourgeois convention (Wickman and Wiklund 1983). Temporarily absent or removed owners engaged in a significantly longer spiral flight upon return to their territory but they always got their territory back. Davies (1979) suggested that the difference in escalation rules in the two populations could be due to a difference in territory value as predicted by the game. Davies's study was conducted during a particularly warm British spring where sunspots were very ephemeral and of low thermal value, compared to the Swedish population. Even in Sweden, spiral flight duration decreased during the season as the weather warmed. An alternative explanation for Davies's result is that body temperature in these butterflies is an important determinant of fighting ability. Butterflies that were removed from their sunspots and held for a few minutes could have become cold and sluggish, whereas the temporary owners were warm and agile and quickly outmaneuvered the original owners (Austad et al. 1979). Thus ownership could confer improved fighting ability on territory owners. Similar removal studies on territorial birds show that territory owners almost always get their territories back if held for a short period of time (a few days), but as holding time increases and residence time of the temporary owner increases, original owners are less likely to reclaim their territories (Krebs 1982; Beletsky and Orians 1987). This type of evidence suggests that ownership increases the value of the territory and an animal's willingness to escalate a fight. If either fighting ability or territory value increase with territory ownership, then owners are increasingly likely to win against most intruders, and the superficial resemblance to an arbitrary bourgeois convention is really caused by a correlated payoff or fighting-ability asymmetry.

Theoretical considerations also mitigate against the likelihood of a payoff-irrelevant ownership convention maintaining itself in a natural population. Grafen (1987) outlines several convincing arguments for why bourgeois may not be an ESS. The condition under which bourgeois is stable, $V < D$, requires that contest costs be high relative to benefits. As bourgeois begins to invade such a population, however, both V and D change in a frequency-dependent manner that prevents the condition for bourgeois from being maintained. This happens because the benefits of winning and the costs of losing depend on the fraction of individuals in the population adopting each alternative strategy. (That is, territory games should be modeled as scrambles rather than contests.) In a population that does not respect ownership, territories change ownership frequently, but as the fraction of ownership-respecting bourgeois individuals increases, owners retain their territories for longer periods of time and vacant territories become rarer. The value of winning is therefore higher in a population that respects ownership. Furthermore, as the territorial opportunities for nonowners become scarcer, the cost of losing an escalated fight against an owner becomes lower. This favors what Grafen calls a **desperado strategy** in which intruders have nothing to lose and much to gain by ignoring the convention and playing hawk. Both the reduction in D and the increase in V will result in $V > D$ and the conditions for bourgeois are no longer met. Grafen concludes that owners usually win because they are in fact better fighters, and that territory turnover occurs occasionally when a superior intruder challenges a tired or older owner in a long, escalated fight (Smith and Arcese 1989; Rosenberg and Enquist 1991).

Games with Assessment of Payoffs

Territory games that include assessment of relative fighting ability and territory valuation have had much greater success in explaining what takes place in nature. Empirical studies clearly show that body size plays an important role, but that ownership confers some type of additional advantage on owners. For example, in encounters between small owners and large intruders, owners still usually win as long as the size differential is not too large (reviewed in Enquist and Leimar 1987 and Archer 1988; several examples can be found in Rosenberg and Enquist 1991; Englund and Otte 1991; Jackson and Cooper 1991; Olsson 1992). Contest duration and intensity are often quite high in such encounters, suggesting that ownership partially compensates for small size. The three basic conflict-resolution models we have discussed in prior chapters—the hawk-dove, war-of-attrition, and sequential assessment games—can each be expanded to evaluate the importance of fighting ability and value asymmetries that are correlated with ownership. Although they all reach the same basic conclusion—that payoff-relevant asymmetries can explain why owners usually win—each is based on somewhat different assumptions and generates slightly different predictions that can be tested with field observations and experiments.

The symmetric hawk-dove game can be converted into an asymmetric game with a size (i.e., fighting-ability) asymmetry as well as an owner-intruder asymmetry (Hammerstein 1981). There are two subgames here—larger owner versus smaller intruder and smaller owner versus larger intruder—but the subgame of greatest interest is the second one in which the two asymmetries are opposed. The asymmetry that is most important in determining the escalation strategy of the two contestants changes as the parameter values in the model change. Figure 22.4 shows the game matrix and the zones for the optimal strategies as a function of the relative size difference and the ratio of contest cost to territory value (D/V). Both players should play hawk when the benefit V is much greater than the cost D. The size asymmetry dictates the optimal strategy when the size difference is large and D is only slightly larger or smaller than V. In this zone, the larger individual should escalate and the smaller should display, a strategy we called assessor in Chapter 21. When the value of the territory is less than the cost of defense and the size differential is small, an ambiguous zone results in which the conventions of complementary strategies are the ESS. These may consist of either bourgeois (owner escalates and intruder displays) or anti-bourgeois (owner displays and intruder escalates). The first of these sets is the **common sense solution** and the second is called a **paradoxical solution**.

The hawk-dove game can also be modified to examine the effect of an ownership asymmetry that causes a territory value asymmetry. The result is that differences in the value of a territory to owners versus intruders have lit-

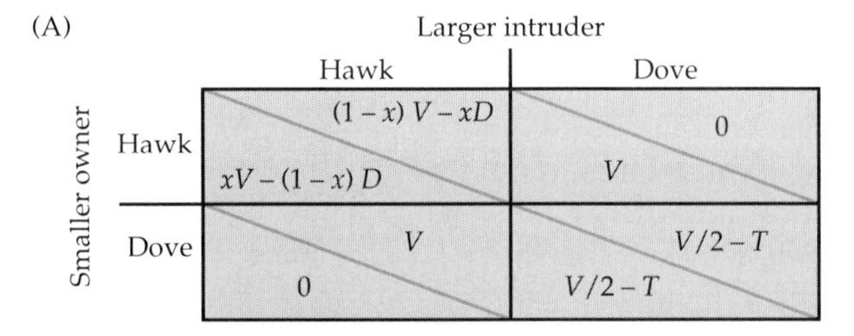

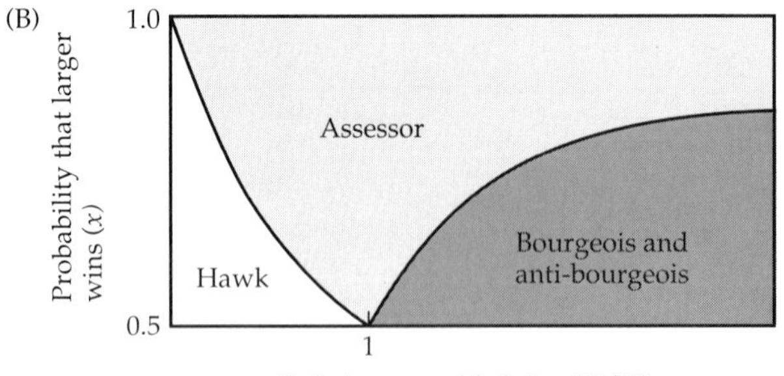

Figure 22.4 The asymmetric hawk-dove game with a size asymmetry and an ownership asymmetry. (A) Payoff matrix for the subgame in which the owner is smaller than the intruder. V and D are the same as in the bourgeois game of Figure 22.3, T is the time cost of a lengthy mutual display when two doves meet, and x is the probability that the larger individual would win an escalated fight. (B) The ESS depends on x and the relative cost of fighting expressed as the ratio of D/V. The zones for each type of ESS are illustrated here. Hawk is a pure ESS when $V > D$ and the size asymmetry is small. Respect for the size asymmetry, a strategy called assessor, is the ESS when the size asymmetry is large and $V \approx D$. Bourgeois and anti-bourgeois are the ESS when $V < D$ and the size asymmetry is small. This game is similar to the one depicted in Figure 19.9. (After Hammerstein 1981.)

tle effect on the ESSs. A final relevant modification of the hawk-dove game explores the situation in which ownership causes a reduction in the risk of injury or costs of fighting, i.e., ownership improves fighting ability. This game results in a larger ambiguous region where bourgeois can be the ESS. Although these modifications of the hawk-dove game capture some of the realities we see in natural contests, such as an advantage of ownership to slightly smaller owners and the likelihood of a takeover when the intruder is very large, the model fails in a number of respects. It predicts that differences in the value of the territory are unlikely to affect contestant strategies, it predicts paradoxical conventions, and it predicts a payoff-irrelevant bourgeois strategy under the difficult-to-maintain condition of $V < D$. The hawk-dove framework suffers because it does not allow contestants to vary the cost of escalation along a continuum and it assumes perfect knowledge of the size-based probability of winning.

In the asymmetric war of attrition (pages 640–641 and Figure 21.5), two contestants assess their relative V/k ratios and select a high or low contest persistence time based on whether they decide they are likely to win or lose, respectively. Different V's and k's for the two opponents are built into the model, so it can easily be adapted to explore the effects of payoff-relevant asymmetries. However, it cannot easily determine the relative importance of valuation asymmetries versus fighting-ability asymmetries because an increase in V has the same effect as a decrease in k. The model also cannot explore the conditions for a payoff-irrelevant ownership convention. In field and laboratory experiments undertaken to determine how animals respond to variations in V and k, most researchers concluded that the owner's advantage arises because the owner is more willing to escalate quickly or persist longer. The interpretation has been that the owner possesses a higher valuation of the territory, not a greater intrinsic fighting ability (Krebs 1982; Austad 1983; Yasukawa and Bick 1983; Beletsky and Orians 1987). For example, Krebs removed territorial pairs of great tits (*Parus major*) and held them for variable lengths of time after replacement owners had taken over the territories. If owners always won upon their release, a fighting-ability asymmetry would be supported, and if temporary owners always won, an arbitrary bourgeois convention would be supported, but neither of these possibilities occurred. Instead, the outcome depended on the length of time the temporary owner had resided on the territory. Original owners released after a short time delay nearly always got their territories back with a short fight, owners released after a rather long time nearly always lost their territories with short fights, and owners released after an intermediate time fought long battles and had a mixed chance of winning. Krebs argues that this result is consistent with an asymmetric war-of- attrition model in which territory value changes with time for the two contestants (Figure 22.5). He speculates that the value of several days of territory ownership is the resolution of stable boundaries with neighbors.

The sequential assessment game provides the most complete framework for evaluating both payoff-relevant and -irrelevant asymmetries in territorial contexts (Leimar and Enquist 1984; Enquist and Leimar 1987). In this model, relative fighting ability is initially uncertain and assessed during the contest,

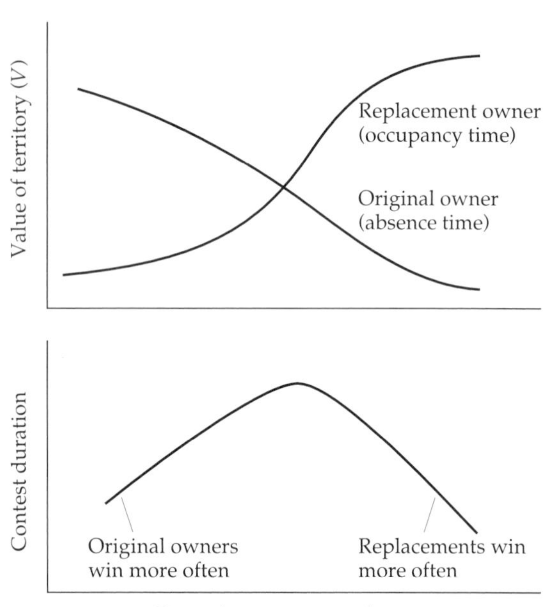

Figure 22.5 Prediction of a peak in contest duration and escalation between residents and replacements at intermediate replacement times. The horizontal axis indicates how long the original owner was absent from its territory and the replacement owner has occupied it. Upper graph: The value of the territory for the original resident decreases with absence time, while the value of the territory for the replacement increases with occupancy time; k is assumed to change little with time. Lower graph: The asymmetric war of attrition predicts longer contest durations when the V/k ratios for the contestants are similar compared to when they are different. (From Krebs 1982, © Springer–Verlag.)

whereas the value of the territory is fixed at the beginning of the contest. The relative importance of fighting ability versus territory value to the ESS outcome can thus be evaluated. In the version of the game presented in the prior chapter (Figure 21.8), fighting ability was asymmetric, but resource value was the same for both contestants. The result was an ESS policy or set of giving-up lines that, when crossed by either contestant, ends the contest. When the value of the territory is higher for one contestant (the owner), the result is an equilibrium pair of ESS line sets, one for the owner and one for the intruder (Figure 22.6). The higher the territory value, the lower the giving-up line and the longer the owner is willing to persist in a fight. This game has the interesting property that the persistence time of the intruder decreases as the persistence time of the owner increases. The first three predictions of this model listed in the figure are the same as the predictions of the war of attrition. In addition, because the sequential assessment game evaluates contest duration more precisely, it generates one more specific prediction—that fights won by the intruder are longer than those won by the owner. Comparisons between the ESS conditions with a fighting-ability asymmetry (territory value held constant) versus a territory-value asymmetry (fighting ability held constant) revealed that territory value had a stronger effect on the decision strategies. Finally, with this model it is possible to evaluate the payoff-irrelevant asymmetry of ownership role. Parameter values leading to a bourgeois convention can be found, and such a convention can make the ESS solution much more stable when the fighting-ability asymmetry is very small. Interestingly, the conditions for bourgeois with this game model do not require that the value of the territory be less than the cost of fighting, i.e., V can be greater than D.

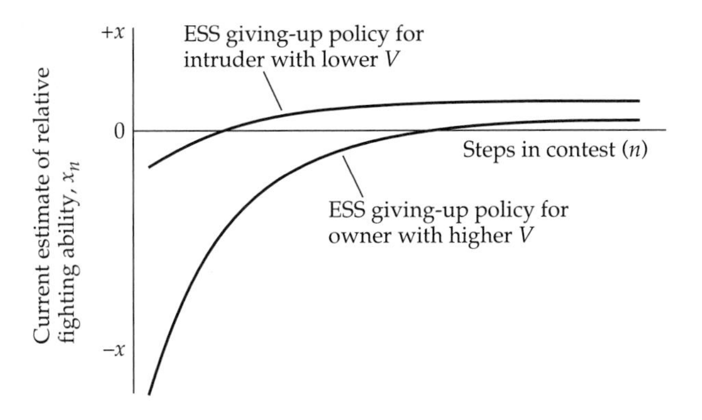

Predictions

1. When contestants have similar fighting ability, owners win because they have a higher V and a lower giving-up policy line.
2. Contest duration is longer when the intruder has the higher fighting ability because the intruder's giving-up line is lowered.
3. Contests are longer if both contestants consider themselves owners.
4. Contests won by intruders are longer than fights won by owners.

Figure 22.6 Sequential assessment game with an owner-intruder territory value asymmetry and a fighting-ability asymmetry. The vertical axis shows the current estimate of fighting ability for the two contestants as a function of the number of steps in the contest (shown on the horizontal axis). The ESS is a pair of giving-up policy lines. Whenever a contestant crosses its ESS line it quits the fight. Therefore, the lower the line, the longer it will take that contestant to quit and the more likely it is to win. (After Leimar and Enquist 1984.)

Paradoxical solutions also exist, but their domain of attraction is much smaller than for bourgeois, so they are not likely to occur.

Several specific tests of the sequential assessment model have now been made, but because the predictions are so similar to those of the war of attrition, it is difficult to distinguish them. The fact that virtually all studies designed to examine the relative importance of value asymmetries versus fighting-ability asymmetries find value asymmetries to be most important is in itself good support for this model. An excellent example is the analysis of long fights between male elephant seals (Haley 1994). Neither fighting ability (size asymmetry) nor ownership (beach tenure prior to the fight) alone can predict the outcome of a fight, but there is a significant interaction between these, with fights won by either small residents or large intruders. Longer-term residents who win fights obtain a higher subsequent reproductive success than winning intruders. Haley concludes that residency increases a male's motivation to fight longer and against heavier males. A second recent study compared contests for case ownership in caddisfly larvae in a dense and a sparse population (Englund and Olsson 1990). Owners usually win in both populations, but ownership is more important in the sparse population, whereas weight asymmetry is more important in the dense population. The ownership asymmetry may be used in the sparse population because size assessment is more costly or because cases are of overall lower value, compared to the dense population. Finally, several studies have found that takeover contests are longer than owner-won contests (Reichert 1978; Smith and Arcese 1989; Leimar et al. 1991).

The general conclusion from both the game models and empirical studies is that territorial ownership increases an owner's motivation to fight to defend its territory. The fact that it takes a certain amount of time to develop this aggressive motivational advantage implies that it is territory value, more than fighting ability, that increases with territorial ownership. Factors responsible for the increased valuation of a territory include: (1) acquisition of a mate and investment in reproduction and nesting on all-purpose territories, (2) increased probability of additional matings at a proven mating site (de Vos 1983), (3) increased knowledge of food and hiding places on the territory (Metzgar 1967; Davies and Houston 1981), and (4) negotiation of territory boundaries with neighbors (Eason and Hannon 1994). Fighting-ability asymmetries still play some role, especially when intruders are significantly larger than owners. Fighting ability also seems to be important in organisms that do not feed while defending territories and therefore lose condition with time (Balmford et al. 1992; Marden and Rollins 1994). The game-theoretical analyses therefore tell us that territorial signals should be designed primarily to reveal the owner's aggressive motivation and secondarily to reveal its fighting ability. Motivation level is best encoded with a series of displays reflecting different cost or risk levels and permitting rapid escalation and deescalation. The rapid signal or approach response that owners exhibit upon detecting an intruder has undoubtedly been selectively favored because it reveals the owner's confidence and motivation. Fighting ability is best encoded with signals that are either constrained to correlate with body size such as call frequency or energetically expensive to reveal current condition such as long, loud vocalizations, rapidly repeated displays, and large conspicuous weapons. An interesting example of the independent coding of both types of information has recently been described for cricket frog (*Acris crepitans*) males defending calling sites (Wagner 1989, 1992). The dominant frequency of the initial threat call is tightly negatively correlated with body size. If the intruder responds by calling and is larger than the owner, the owner calls again with a drop in frequency. The degree of frequency drop is correlated with the owner's likelihood of attacking. Thus the call initially reveals body size, but subsequent repetitions signal honest information about intentions. In principle, lowering the call frequency should be more energetically expensive for males but this cost has not yet been established.

ASSESSMENT OF NEIGHBORS AND FLOATERS

Up to this point we have not distinguished between floaters and neighbors as opponents of a territorial owner, but in fact this distinction makes a great deal of difference. Encounters between two neighbors are frequent because of their spatial proximity; both opponents value their territories similarly and both know that escalated fights between them will be costly. Encounters between owners and floaters are not likely to be repeated; there is a greater asymmetry in territory value and the opponents have no prior knowledge of each other's fighting ability. In this section we shall look at the signal characteristics and assessment strategies that territorial players use to evaluate different opponents and then show how this information affects the territory defense game and fighting and signaling strategies.

Neighbor Recognition

The need to recognize neighbors and distinguish them quickly from strangers is the key factor that places receivers in the position of exerting strong selection on territorial signal form. Not only does the owner, as a receiver, want the neighbor to produce an individually distinctive signal, but the sender also benefits from doing so. Imagine a territory owner who produces a highly variable or nondistinctive signal. Its neighbors will believe it is a floater and immediately attack if it approaches their common boundary. Most territorial signals studied to date show evidence of permitting individual discrimination, but this requires a sufficiently complex signal with parameter space for individual variation as well as a brain that can remember these variants. Ample evidence now exists that animals do recognize their neighbors and distinguish them from strangers (Stoddard 1996). We described the features used to encode individual distinctiveness in territorial signals in Chapter 18 and Figure 18.8.

The strongest evidence that individual variants are established for the primary purpose of distinguishing oneself from one's territorial neighbors comes from a few species in which animals alter their signal characteristics when they move to new territories or acquire new neighbors. Randall (1995) shows how the foot-drumming patterns of territorial kangaroo rats that shift territory location change in temporal patterning so as to make them maximally distinctive in the new neighborhood (Figure 22.7). Temporal pattern is the only element that individuals can vary, since there is no frequency or note length or shape variation possible in this auditory percussive signal. The signal consists of a series of footrolls in which the number of drums in each roll and the total number of rolls in the series varies in an individually consistent manner. Patterns of individuals differed significantly within immediate neighborhoods but overlapped extensively between non-neighbors. Rats that moved were significantly more likely to change their signature than rats that did not move.

Evidence that owners discriminate between neighbors and floaters has been obtained from a wide variety of territorial birds, mammals, reptiles, amphibians, fish, and insects by using both observational and experimental techniques (Temeles 1994). The design of the classic neighbor-stranger discrimination experiment is shown in Figure 22.8. A neighbor's territorial signal is broadcast to a focal animal from a position near the boundary with that neighbor, and the focal animal's aggressive response is compared to its response to the playback of a stranger's signal from the same position. A reduced aggressive response to the neighbor clearly demonstrates that the focal individual recognizes the signal of its neighbor and has habituated to it, in contrast to the stranger signal. If that same neighbor's signal is broadcast

Individual

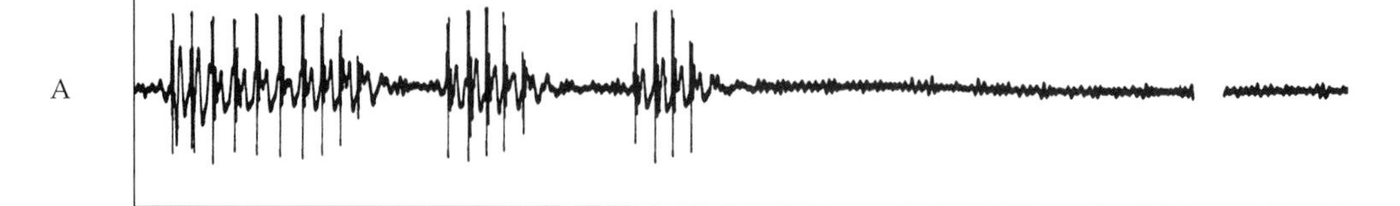

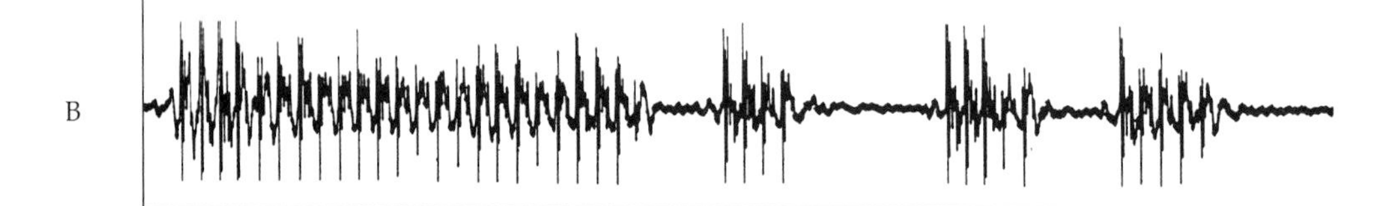

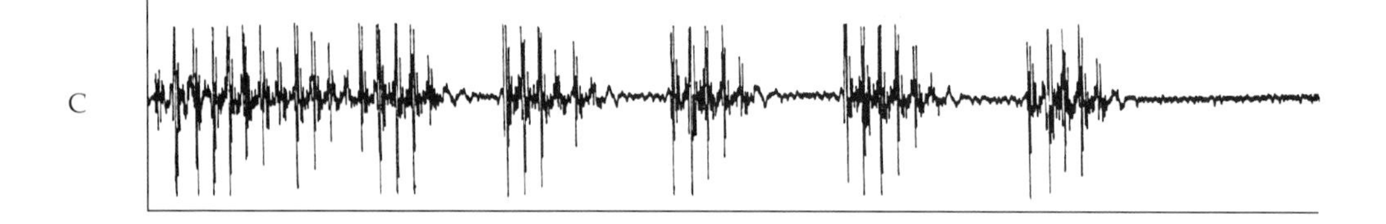

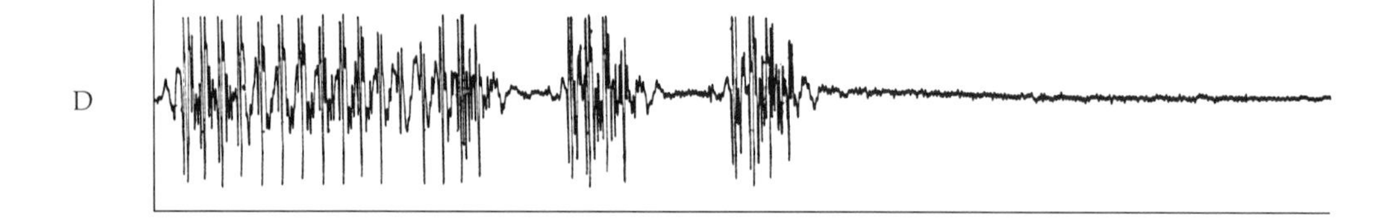

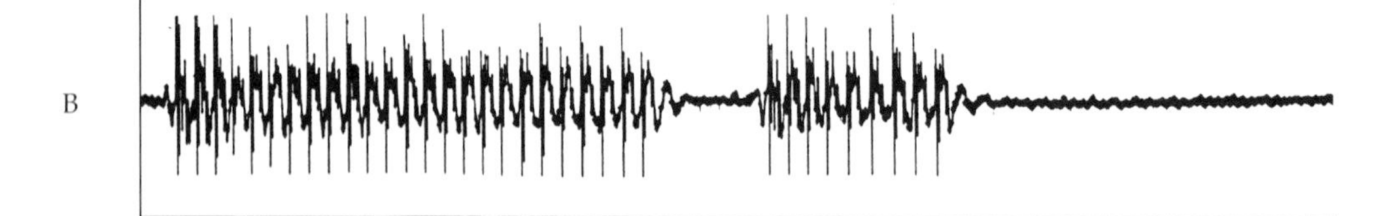

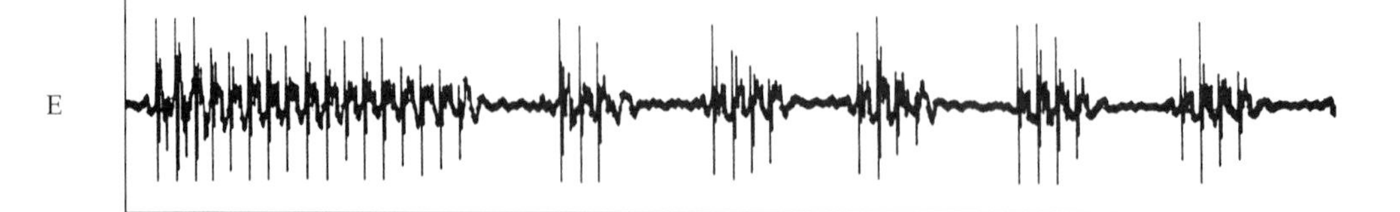

◀ **Figure 22.7 Change in individual signature patterns associated with the acquisition of new neighbors.** Male and female kangaroo rats, *Dipodomys spectabilis*, defend individual territories with a percussive hindfoot drumming signal. The temporal patterning of rolls and gaps is unique to each individual and remains stable for long periods of time unless the animal moves. The top three waveforms show the initial signature of focal animal B (a female) and two of her immediate neighbors A and C (also females). Rat B subsequently moved to a new territory about 40 meters away. The bottom three waveforms show her subsequent signature pattern and that of her two new neighbors, D (a female and likely daughter of A) and E (a male). (After Randall 1989, 1995; original waveforms courtesy of Jan Randall.)

from the opposite territorial boundary, however, the focal animal responds as aggressively as it would to a stranger's signal. Thus the familiar signal must come from the correct location in order to produce the differential aggressive response. The owner's rule of thumb is therefore to recognize the neighbor's signal from the correct location and treat all unfamiliar and inappropriately positioned signals as arising from invading floaters.

Distance Estimation

Territory owners with an ability to determine whether a neighbor is signaling from within or outside of their territorial boundaries gain two benefits. One is the reduction in energy expended to approach and investigate the exact location of the sender. Another is the minimization of the risk of injury from an escalated encounter with a neighbor who is occupying and advertising its own territory. The term **range** is used to refer to the process of estimating the distance between the receiver and the sender (Morton 1982). Ranging is only an issue for auditory territorial signals. In the case of visual signals, the exact

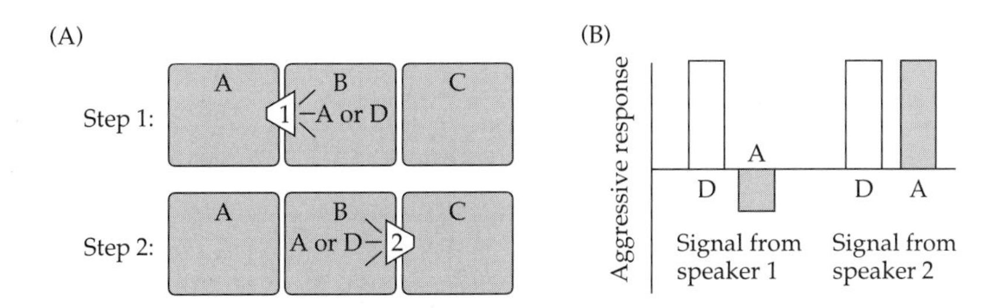

Figure 22.8 Neighbor-stranger discrimination experiments. (A) Experimental design: Animal B is the focal territory owner, with neighbors A and C on the left and right; animal D is a floater. Speaker 1 is placed on the boundary with neighbor A and speaker 2 on the boundary with neighbor C. The response of B to the territorial signal of A from speaker 1 is compared to the response of B to the signal of bird D from speaker 1. The signals of A and D are then played from speaker 2. (B) The graph shows the typical level of aggressive response to these four trials. The response is lower to the A's signal when played from A's boundary than to the signal of D from the same location, and equally strong to the signal of A and D when played from an inappropriate boundary. (After McGregor 1993.)

location of the sender is always known once the signal is detected, and in the case of deposited olfactory markers the sender and the signal are separated, so there is no information available on the location of the sender inherent in the mark.

There are two possible cues that auditory receivers could use for ranging: amplitude and degradation. Amplitude declines with the inverse of the squared distance from the sender, and degradation or distortion increases with distance. It has been argued that degradation is the more reliable cue for ranging because the sender can behaviorally alter the amplitude of its signal with its orientation and source output level. As we learned in Chapter 2, degradation is caused by scattering, interference and echoes from the medium, the ground, and objects in the environment such as tree trunks and vegetation. Degradation changes auditory signals in two ways: (1) frequency-dependent attenuation causes the higher amplitude components of the signal to drop out, and (2) reverberations cause notes with sharp onset and offset such as trills to blur (Figure 22.9). If a receiver knows the undegraded characteristics of the signal, it can compare this to a received degraded version and estimate the distance.

Playback experiments with birds show that territorial owners have the ability to range based on degradation information alone. For example, the Carolina wren (*Thryothorus ludovicianus*) responds to playback of relatively undegraded songs by silently approaching the loudspeaker, whereas it responds to playback of degraded songs at the same amplitude by singing at a distance (Morton 1986). This is similar to the behavior of an owner when he is played a song from a position just inside his territorial boundary versus well outside his boundary. In other species such as Kentucky warblers (*Oporonis formosus*) that always approach strange intruders, birds accurately approach the speaker when the song is undegraded and fly past the speaker when the song is degraded (Wiley and Godard 1996). Furthermore, ranging can only be accomplished when the focal bird is familiar with the song; a differential response does not occur between degraded and undegraded versions of an unfamiliar song (McGregor et al. 1983; McGregor and Falls 1984). Finally, Naguib (1995) has shown that birds can independently use either frequency-dependent attenuation or reverberation information to accurately range the sender's location.

Differences between Contests with Neighbors versus Floaters

During a typical neighbor-stranger discrimination playback experiment, as described above, an owner responds less aggressively to the neighbor. Not only does this tell us that the owner recognizes its neighbor, but also that it treats the neighbor differently. In the case of an auditory signal, playback from the territory boundary should be perceived as occurring from well within the neighbor's territory since the recording itself was partially degraded by the distance between the microphone and the sender. The reduced response of the owner is therefore easily explicable as adaptive habituation to the neighbor's advertisement signal, in contrast to a stranger's signal. However, even when the playback speaker is placed well within the owner's territory, simulating a

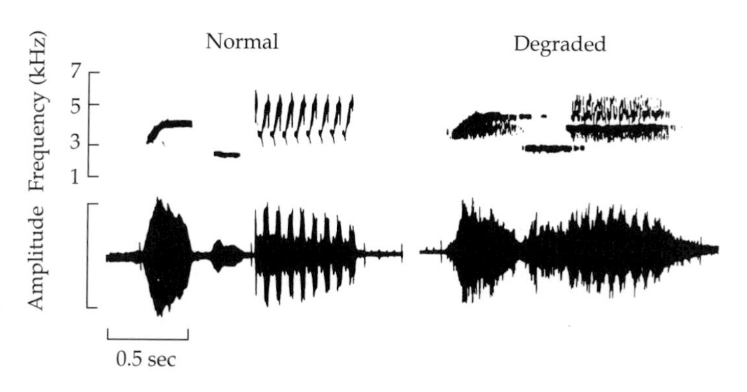

Figure 22.9 Effects of degradation on a complex vocal signal. Spectrogram (top) and waveform (bottom) of the song of the rufous-sided towhee (*Pipilo erythrophthalmus*) recorded at a very short distance from the bird (left) and at a long distance from the bird (right). (After Richards 1981.)

clear intrusion by the neighbor, the owner's response is still significantly less aggressive to the neighbor's signal than to a stranger's signal. The reduced aggressiveness to intrusions by known neighbors has been called the **dear enemy phenomenon**. The term arose from the observation that two neighbors are formidable foes in an escalated fight, but once they have defined their common boundary they appear to develop a truce and refrain from aggression. The individual signature information in the territorial signal thus enables owners to play a different game with neighbors versus strangers.

Is the owner less aggressive toward a neighbor compared to a floater because the neighbor is more intimidating or because a floater is perceived as a greater threat? Not surprisingly, game models were used to investigate this question. Ydenberg et al. (1988) invoked the war-of-attrition model, proposing that the degree of familiarity between the two contestants affected their accuracy of role assessment. Recall from Chapter 21 (pages 686–687) that when two contestants can ascertain their relative V/k ratios fairly accurately, the cutoff value S between the ranges of "lose" and "win" bid distributions decreases. Average contest duration is therefore short. When two contestants are uncertain about their roles, S increases and average contest duration increases. Thus two neighbors who are familiar with each other and presumably know their roles engage in shorter contests compared to an owner-floater encounter where role uncertainty is greater. Getty (1989) criticized some details of this argument and suggested that the sequential assessment game was a better model. Here, fights between owners and floaters are longer because the two contestants need more time to learn about each other's fighting abilities compared to two known neighbors. These familiarity arguments ignore potential differences in the costs and benefits of defense against different intruders. Temeles (1994) proposed that degree of threat from invasion could explain the difference in behavior. He reviewed all studies of neighbor-stranger discrimination and found that the dear enemy effect occurred primarily in species with all-purpose territories, but not in species with feeding territories or very small breeding territories. In the latter cases, invasion by neighbors is potentially as damaging to fitness as invasion by strangers so both should be vigorously attacked, whereas with all-purpose territories neighbor invasion is less damaging than stranger invasion.

Detailed observations of the behaviors performed during interactions with neighbors versus strangers show that owners are not only less likely to attack a neighbor but also more likely to engage in display with them; the reverse is true for interactions with strangers. In songbirds, for example, neighbors tend to engage in a countersinging bout from a distance rather than approach each other (Catchpole and Slater 1995). In rock ptarmigan (*Lagopus mutus*) studied by Brodsky and Montgomerie (1987), the male territorial signal is a long aerial flight display with vocalization that gives the owner a chance to get off the ground and survey his territory and allows neighbors to see and hear him. When a neighbor invades a territory it enters with an aerial flight, the owner immediately approaches and the two resolve the conflict quickly with ritualized visual displays; the invading neighbor then retreats to his own territory. Floaters, on the other hand, always sneak into territories and, when the floater is detected, the owner immediately approaches and escalates to fighting and chasing (Figure 22.10). These encounters are much longer, perhaps because the floater does not know the boundaries of the owner's territory.

Ptarmigan, songbirds, and other all-purpose territory owners appear to use this rule: If the opponent is a neighbor, play bourgeois, and if a stranger, play hawk. However, the extensive displaying to neighbors may occasionally be used to challenge them, test their motivation and condition, and verify the exact position of the boundary. Choice or vigor of display may be used to transmit this information, as argued by Enquist (Figure 20.2). An alternative explanation is that two neighbors have a common mutual interest in maintaining the status quo of their territorial boundary and keeping contest costs low. Several recent game models have shown that when interactants have similar interests in the outcome, reliable and low-cost signals can be stable (Johnstone and Grafen 1992a; Maynard Smith 1994). Bargaining and other economic models that end in subdivision of the disputed resource also predict resolution with reduced aggression and low-cost signals or **cheap talk** (Maynard Smith 1982a; Kim 1995; Farrell 1995). Such models may explain why interactions between two neighbors often end in a draw rather than a win by one contestant.

EFFECTS OF COMPETITION ON TERRITORIAL SIGNALS

Competition for access to valuable and limited territories has led to the elaboration of territorial signaling behavior beyond that which is necessary for owner and territory boundary identification. Just as competitive arms races have led to the evolution of status badges, weapons, and other short-range signals for conflict resolution (Chapter 21), competition among territory owners has led to increased signal complexity and strategies of deceit, bluff, manipulation, and eavesdropping. Oscine songbirds, in particular, have raised the complexity of auditory territorial signals to an art form as a result of competition among males for territories and mates (Craig and Jenkins 1982). Similar strate-

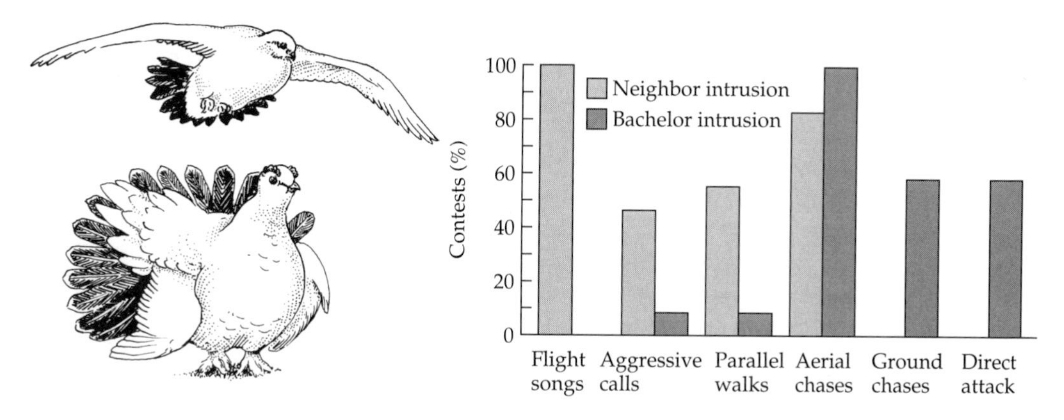

Figure 22.10 The frequency of behaviors performed in territorial conflicts in rock ptarmigan (*Lagopus mutus*). Black bars show the frequency of different behaviors used in interactions with intruding neighbors, and white bars show behaviors used in interactions with intruding bachelor floaters. Displays are clearly more frequent in neighbor interactions, whereas aggressive acts characterize bachelor interactions. (From Brodsky and Montgomerie 1987, © Springer–Verlag.)

gies are also found in other organisms and modalities. This section describes some of these strategies and provides examples of the signals involved.

Countersinging, Song Matching, and Population Dialects

The omni-directional property of loud auditory signals is ideal for the general advertisement and associated mate-attraction function of territorial signals. However, during escalated contests with a particular individual such as an invading floater or a new neighbor, owners need to direct their signal to a specific receiver. As outlined briefly in Chapter 18, the primary mechanism for directing an auditory signal to a particular signaling rival is a rapid answer to the rival's signal with the same type of signal. This sets up a bout of countercalling or countersinging between the two opponents in which they alternate calls on a one-to-one basis. An optimally designed threat signal for countercalling interactions should therefore consist of repeated, relatively short duration calls interspersed with significant silent gaps for listening to responses. Competitive countercalling has been described in orthopterans, birds, and mammals (Greenfield and Minckley 1992; Bremond and Aubin 1992; Clutton-Brock and Albon 1979). It provides an accurate mechanism for assessing relative stamina and aggressive motivation level. As a countercalling bout escalates with an increase in call rate or call duration, failure of one individual to keep up the alternating rhythm determines the loser (Greenfield 1994).

Oscine songbirds have carried this type of interaction one step further. Their ability to learn the fine features of their species-specific song (discussed on pages 475–480) gives them an opportunity to precisely match the details of their rival's or neighbor's song. In species with one songtype per male, the ontogeny of song learning clearly indicates that the function of song learning is the matching of neighbor song (Baptista 1985; Slater 1989). Recently fledged male birds float among the territories of territorial adults and memorize their songs. Early in the next spring they attempt to set up and defend a territory and emit several different songtypes. However, they gradually drop from their repertoire all but the one songtype that is most similar to that of their neighbors; this songtype is reinforced during countersinging interactions (Nelson 1992; DeWolfe et al. 1989). No further learning takes place after territory establishment. The consequence of this dispersal and copying system is a pattern of population-specific song dialects. Two classic and well-studied examples of dialects are the white-crowned sparrow (*Zonotrichia leucophrys*) and the indigo bunting (*Passerina cyanea*). Dialect regions in the sparrow are relatively large and stable over time because copying is extremely precise (Marler and Tamura 1964; Baptista 1975; DeWolfe and Baptista 1995) (Figure 22.11). In the bunting, dialect neighborhoods are much smaller, consisting of from 2 to 12 birds, and the half-life of a songtype is only 3.8 years because birds modify the songtype slightly and sometimes fail to match a neighbor (Payne 1996).

Considerable effort has been expended attempting to understand why song sharing is important in neighbor interactions. In the white-crowned sparrow, playback experiments of home dialect versus foreign dialect songs resulted in a lower response to the foreign dialect, indicating that the foreign dialect is a less threatening stimulus (Tomback et al. 1983). Some males with territories straddling the border between two dialect areas are bilingual and use only the matching songtype when interacting with neighbors on each side (Baptista 1975). Matched countersinging therefore appears to be more effective in dealing with neighbors. In the bunting, 80% of first-year males matched the songtype of a dominant adult neighbor, while the remaining 20% did not; matchers had significantly higher survival and fledging success than

Figure 22.11 West coast song dialects of the white-crowned sparrow (*Zonotrichia leucophrys*). The northern (*Z. l. pugetensis*) and southern (*Z. l. nuttalli*) subspecies are morphologically distinctive and exhibit consistent differences in song syntax and note types. Dialect areas for *pugetensis* are very large and boundary zones are characterized by a mosaic of birds singing either one dialect or the other. Dialect areas for *nuttalli* are small, between 2 to 5 km in diameter. Only a few different dialects are illustrated here. Within *nuttalli* dialect areas there are also subtle subdialect areas. Boundaries between dialects are narrow and characterized by a low level of pure dialect interdigitation, bilingual birds singing both dialects, and birds singing blended intermediate songs. The difference in dialect area size between the two subspecies is believed to be a consequence of the fact that *pugetensis* birds migrate to southern California and Mexico for the winter, whereas *nuttalli* birds are nonmigratory. Errors in migratory homing mean that *pugetensis* males become established on territories some distance away from the site where they learned their song in the months after hatching. In sedentary populations, on the other hand, males typically acquire territories a few hundred meters from their hatching site. Short dispersal distances and reinforcement of the learned song by territorial neighbors results in a finer dialect scale, as well as the occasional bilingual and blended strategies observed. Dialects: 1 = Vancouver; 2 = Seattle; 3 = Raymond; 4 = Astoria; 5 = Newport; 6 = Port Orford; 7 = Crescent City; 8 = Eureka; 9 = Mackerricher State Park; 10 = Inverness; 11 = Bear Valley Trail; 12 = Bolinas; 10–12 = Point Reyes dialects; 13 = Fort Baker; 14 = San Francisco; 15 = Lake Merced; 16 = Pacifica; 17 = Pacific Grove; 18 = Point Lobos; 19 = Lompoc. (After Baptista 1975; DeWolfe and Baptista 1995; spectrograms courtesy of Luis Baptista and Andrea Jesse.)

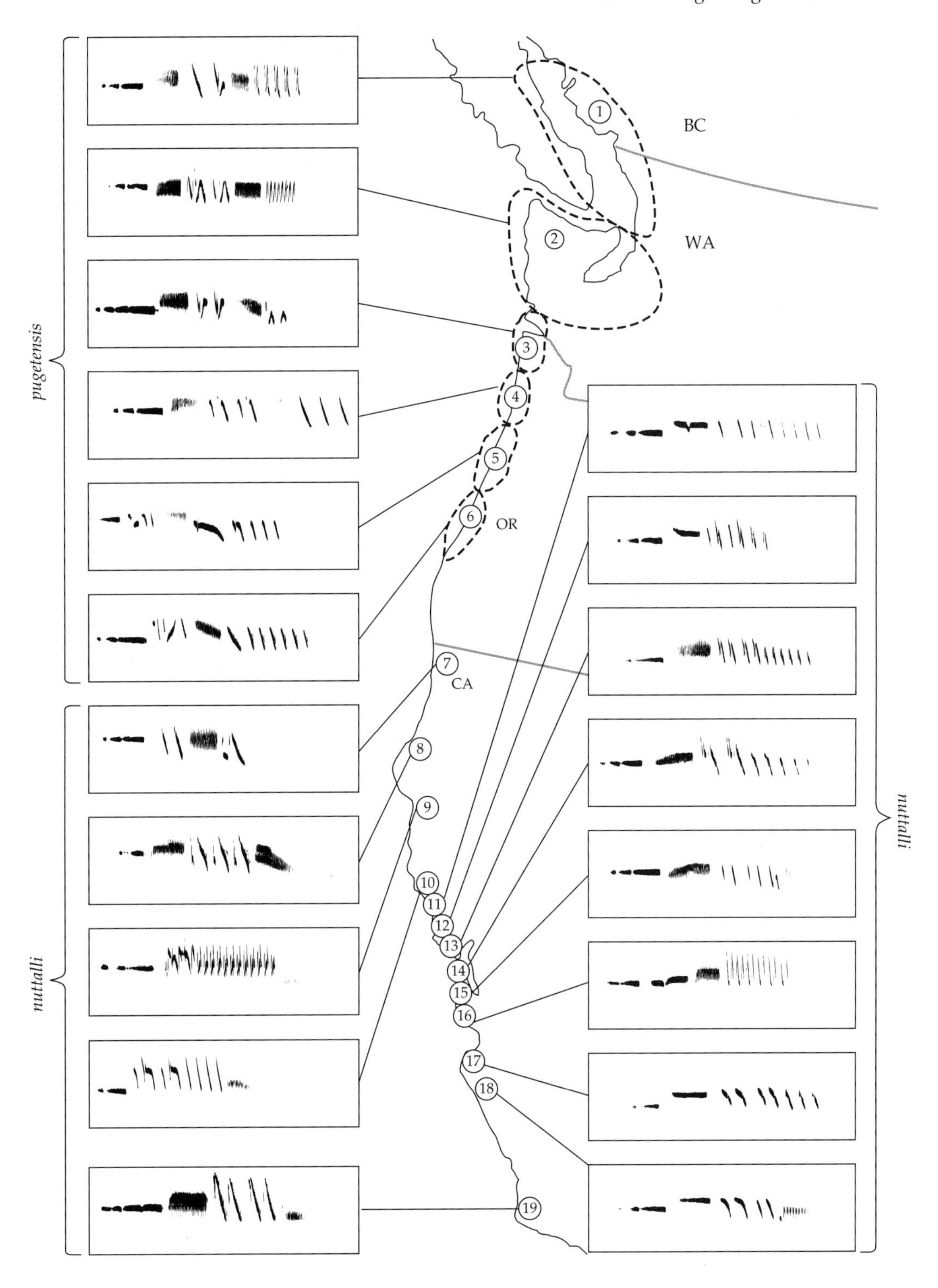
pugetensis
nuttalli
nuttalli
BC
WA
OR
CA
1
2
3
4
5
6
7
8
9
10
11
12
13
14
15
16
17
18
19

nonmatchers (Payne 1982, 1983; Payne et al. 1988). However, Payne was unable to determine the reason for this finding. One hypothesis, called **deceptive mimicry**, is that young floater males that match are able to deceive other males (or females) into believing they are experienced residents (Rohwer 1982; McGregor and Krebs 1984). If a floater can pass itself off as an owner, or perhaps imitate the prior owner of a vacated territory, then residents may hesitate to attack immediately and the floater may gain a foothold. But young buntings always copied the adult male with which they countersang the most, suggesting that the refinement of matching develops during the territory establishment process. Some experiments have shown that neural and behavioral responses of a bird change as a stimulus song becomes more similar to its own song (Margoliash and Konishi 1985; Falls et al. 1982; McArthur 1986). Further study on the psychological effects of matching is needed.

Songtype Repertoires

A majority of oscine songbirds possess multiple songtype repertoires (Kroodsma 1982). Repertoire size, song length, duty cycle, and sequential patterning of songtypes differ greatly among species as shown in Table 22.1 (Slater 1981; Read and Weary 1992). About 15% of songbirds have extremely large or infinite repertoires. They tend to sing almost continuously (high duty cycle), maximizing the variety of sounds they produce and often improvising and mimicking other species. Males of these species also tend to stop or reduce singing once they have mated. Examples include mockingbirds (*Mimus polyglottos*), nightingales (*Luscinia megarhynchos*), and brown thrashers (*Toxostoma rufum*). The function of this singing strategy appears to be primarily mate attraction. About a quarter of songbird species possess a single songtype as described above. The remainder possess small to moderate repertoires of short, discrete songtypes that are usually copied from other males and delivered at intervals spaced with silent gaps (Figure 22.12). For the most part these songtypes are believed to be functionally equivalent, but a few species such as *Dendroica* warblers and European blackbirds (*Turdus merula*) have distinctive songtypes that are used in specific contexts such as mate attraction, territory advertisement, and short-distance threat (Spector 1992, Dabelsteen and Pedersen 1990). Two songtype delivery strategies are observed: switching to a new songtype after every song, called **mixed-mode** (or **immediate variety**) **singing** (e.g., ABCDBADCAB), and repeating a given songtype several times before switching, called **bout** (or **eventual variety**) **singing** (e.g., AAAAAABBBCCCCCCDDDDBBBBB). These species appear to be using their songtype repertoire in male-male territorial interactions, and males routinely engage in countersinging bouts throughout the breeding season. In several species, individuals with larger repertoires were found to maintain ownership longer and possess higher-quality territories (Hiebert et al. 1989; McGregor et al. 1981). Speaker replacement studies such as the one described in Figure 22.1D also show that a large repertoire is a more effective keep-out signal than a small one. There is little doubt that competitive interactions are responsible for this complex signaling behavior, but the mechanism remains elusive. Females also tend to prefer males with larger reper-

Table 22.1 A summary of the mean values of song variables for songbird species employing different types of singing strategies

Song variable	Singing Strategies			
	One songtype	Bout	Mixed	Infinite[a]
Size of songtype repertoire	1.0	8.1	20.2	∞
No. of syllable types per song	3.8	4.6	7.0	3–∞
Song duration (sec)	6.1	1.8	2.0	0.3–∞
Intersong interval (sec)	6.6	5.7	5.2	0–10.5
Duty cycle (%)	31	27	24	22–100
Song delivery rate (per min)	7.2	12.3	12.4	–∞–130.4
No. of species	24	34	28	15

Source: Based on data from Read and Weary 1992.
[a]Minimum and maximum values are given for infinite repertoire species.

toires in many of these birds (Catchpole and Slater 1995; see also Chapter 23). Several hypotheses for the function of such repertoires have been proposed.

The **anti-habituation hypothesis** (Hartshorne 1956) proposes that birds that sing at high rates must use a greater diversity of songtypes to reduce habituation in listeners. In a similar vein, the **anti-exhaustion hypothesis** (Lambrechts and Dhondt 1988) suggests that birds singing at high rates use a greater diversity of songtypes to avoid tiring the same muscles involved in sound production. These hypotheses beg the question of why song delivery rates differ among species. Although both ideas are supported by the fact that song delivery rate and continuity (duty cycle) of singing are positively correlated with repertoire size, this correlation is only apparent when infinite and single songtype species are included (Read and Weary 1992). Differential selection on song rate and repertoire size in mate attraction versus territory defense contexts may be responsible for this correlation (Catchpole and Slater 1995).

The **Beau Geste hypothesis** (Krebs 1977a) acquired its name from the story of the French legionnaire who successfully defended a fort alone by imitating many different men's voices and deceiving the enemy into believing there were many more troops in the fort. In the territorial bird version, owners deceive potential floaters into believing there are many more territorial owners in a neighborhood, discouraging them from attempting to acquire a territory there. Tests of the model indicate that song delivery patterns are not consistent with the hypothesis and that floaters are not deceived in the predicted manner (McGregor 1993).

The **ranging hypothesis** involves an element of spite (Morton 1982). The suggestion here is that birds acquire new songtypes in order to make it more difficult for their neighbors to judge their distance and position accurately. Neighbors therefore must interrupt their foraging to approach and investigate the singer, activities that prevent the neighbors from feeding or breeding and that lower their success relative to the singer. This process could lead to an arms race of increasing repertoire size if birds must sing a songtype in order to range it accurately. As discussed above on page 728, birds do need to be familiar with

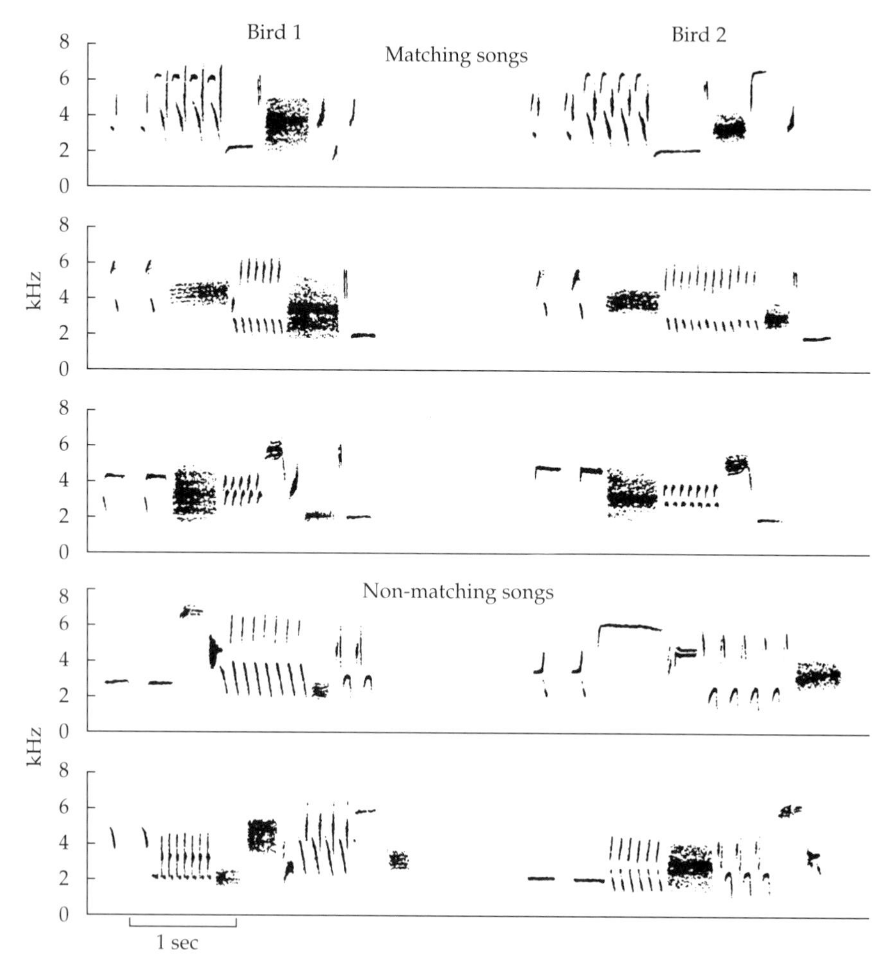

Figure 22.12 Partial songtype repertoires of two neighboring song sparrow males. In Washington state song sparrows, *Melospiza melodia*, each male has between 8 and 12 songtypes in its repertoire and shares on average 30% of these with a given neighbor. The repertoire becomes crystallized during a bird's first year, and no new songtypes are added after this point. The two males depicted here share the first three songtypes. Song sparrows are bout singers and can therefore vary the rate of singing and the rate of switching to a new songtype to indicate aggressive motivation. Two countersinging birds will often switch songtypes in synchrony. A bird frequently responds to the song of its neighbor by singing a song it shares with that neighbor, but usually avoids singing the matching songtype. If a stranger appears and sings a songtype that the owner has in its repertoire, however, the owner will match it immediately. Males with larger repertoires maintain ownership of their territories for longer periods and therefore have higher lifetime reproductive success. (Spectrograms courtesy of M. D. Beecher.)

the undegraded version of a songtype before they can use degradation to evaluate distance to the singer, but they do not have to sing it themselves. New songtypes do not arise frequently enough for this hypothesis to function.

Finally, the **escalation-level hypothesis** suggests that variations in songtype delivery pattern encode different levels of contest escalation (Krebs et al.

1981; Kramer and Lemon 1983; McGregor et al. 1992; Nielsen and Vehrencamp 1995; Beecher et al. 1996). The idea is that choice of songtype relative to the songtype being sung by the rival transmits information on motivation to escalate or deescalate a countersinging interaction. If songtypes are copied from several tutor males in a local population, such that a given individual shares some but not all songtypes with a given neighbor, then territorial males can opt to songtype match, repertoire match (i.e., sing another song that the neighbor has in its repertoire; Figure 22.12), or fail to match. These rules could be used to signal changing levels of aggressive engagement. Bout singing provides two additional features—switching rate, which may be a general indicator of aggressive motivation, and synchronous switching, which can be used to direct a song threat to a specific rival. Searcy et al. (1995) have clearly shown that a playback switch to a new songtype stimulates a renewed aggressive approach in song sparrows. For switching events to be most effective, repertoire size should be small so that songs are maximally different. Note that bout singers do have smaller repertoires and fewer note types per song than mixed-mode singers (Table 22.1). Mixed-mode singers can vary songtype order and short-term songtype diversity (Kroodsma 1977). Each species appears to settle on a different set of rules, with mixed singing, bout singing, and single songtype singing representing the endpoints of different evolutionary games between territorial senders and receivers. Since different songtypes probably do not entail differential energetic production costs, the varied song delivery patterns must represent conventions in which honesty is maintained via incidental or punishment costs. The benefit of such strategies is most likely a reduction in the frequency, duration, and risks of physical chases and escalated fights.

Eavesdropping

So far we have considered the signaling interactions between territory owners and intruders as strictly pairwise dyadic interactions and contests. However, any signaling animal is within receiving range of a number of other potential receivers and senders. Long-range territorial signals are particularly likely to be detected by other animals. Individuals not directly involved in a signaling interaction can nevertheless gather information from it. This is an example of eavesdropping (McGregor 1993). As discussed in Chapter 12, eavesdropping can occur between different species, for example, between predators and prey, where predators cue in on the location of prey via their territorial or mate-attraction signals. Intraspecific eavesdropping can also occur. In the context of a territorial neighborhood, eavesdroppers can obtain several types of information. They can learn about the presence of intruders by paying attention to the territorial signals of their neighbors. In addition, eavesdroppers may be able to make an initial assessment of the relative fighting ability of another animal (neighbor or intruder) on the basis of that individual's signaling interaction with a known neighbor. This provides a low-risk means of evaluating a potential future rival. A receiver might also take advantage of his neighbor's interaction with an intruder to trespass onto the neighbor's territory. There is some evidence that birds do pay attention to the engagements of their neighbors. For instance, during playback experiments on the fan-tailed warbler (*Cisticola juncidis*), neighbors decrease their own vocalization rate and immediately increase it when the playback has stopped (Gray 1992, cited in McGregor 1993). In cardinals (*Paroaria gularis*), neighbor interactions clearly function as an early warning system against intruders. Territory holders detected an intruder immediately in most cases if the intruder had just been expelled from a neighbor's territory, but owners frequently failed to detect intruders that were undetected by their neighbors or arrived from a direction with no adjacent neighbors (Eason and Stamps 1993). Not surprisingly, intruding floaters avoid entering a territory adjacent to one in which they have had a vocal interaction with the owner, and instead skip to a nonadjacent territory.

Olfactory territorial marks can also be used for eavesdropping to a certain extent since the marks leave a record of an individual's presence (Figure 22.13). House mice and many canids including the domestic dog leave olfactory marks to advertise their presence and maintain their status in a community (Hurst 1989, 1990; Macdonald 1980). Carefully controlled experiments in mice have shown that a subordinate male who fails to mark on communal posts is considered to have dispersed, and if the dominant male then encounters him on the territory, he is treated more aggressively than subordinate males that continue to mark (Hurst et al. 1993). Floaters that move among several territories obtain most of their information on owners via eavesdropping, and by not signaling can hide their own presence. **Countermarking** is another competitive strategy in olfactory territorial systems. Territory owners cover over any foreign urine marks they find in their areas (Macdonald 1980). In areas where two territories overlap, the adjacent owners both increase their marking density (Charles-Dominique 1977; Stoddart 1980).

Territorial Cheaters

When territories are defended by males as a means of obtaining access to reproductive females, interactions among the males in a population may be extremely competitive. If good territories are very limited or a few males are able to control a large fraction of the females, only the males with the highest resource-holding potential will breed, and lesser males will be excluded. The greater the discrepancy in fitness between territorial and nonterritorial males, the stronger is the selective pressure on nonterritorial males to adopt alternative mating strategies. On pages 642–644 we discussed a scramble competition game between displayers that defend territories and roamers that intercept females in roving groups. Roaming is basically a strategy of cheating on the territorial system. Recall that the ESS for this game is a cutoff value of some male trait such as body size, above which a male should display and below which a male should roam. There are many examples of such secondary strategies in which smaller and/or younger males usurp the efforts of territorial males. Examples include fish that streak into a nesting male's territory and spawn just after a female has laid her eggs in the nest, noncalling frogs that intercept females attracted to calling males, and sneaky non-harem-

Figure 22.13 Olfactory territory marking in mammals. (A) Like mice, rabbits (*Oryctolagus cuniculus*) live in mixed sex communities. Dominant males mark any protruding surface in their home ranges with a chin gland secretion. (B) Canids such as jackals and the red fox (*Vulpes vulpes*) shown here mark their territories with urine, single feces, and feces piles. Single feces are left in visually conspicuous raised sites throughout the territory, whereas piles are mainly deposited around the border. The density of piles is highest in perimeter zones most frequently contested with neighbors. (C) Thomson's gazelle males (*Gazella thomsoni*) mark the tips of grass stems and twigs with the thick black secretion of their preorbital gland. (D) The brown hyena (*Hyaena brunnea*) marks blades of grass in its territory with an everted anal gland. The blade is drawn through a crevice in the gland and coated with a white secretion, a behavior called pasting. (A after Bell 1985; B and D after Macdonald 1985b; C after Gosling 1985.)

holding deer (Arak 1984). In most cases, males adopt these strategies after discovering that they are unable to compete for a territory and the sneaky behavior is thus conditional.

Sneaky males are usually more successful if they are either inconspicuous, lack the adult male secondary sexual characteristics used in display, or mimic the appearance of females. Young males of many species delay the development of secondary sexual traits to avert the aggression of dominant adult males and therefore may succeed at these sneaky strategies because they look like females. In other species the development of male traits is dependent on environmental conditions, and small individuals engage in secondary strategies to make the best of a bad situation. In a few species, however, a genetically determined polymorphism determines the appearance and behavioral strategy adopted by males. These represent examples of evolved, sneaky anti-territorial signals. The ruff, a lek-breeding shorebird species, is one of the best-described examples (van Rhijn 1983; Lank and Smith 1987; Lank et al. 1995). Territorial males have a dark-colored ruff and head tuft and aggressively fight other dark males for territorial ownership, whereas satellite males have light-colored plumage and nonaggressively occupy a territory along with a territorial male (Figure 22.14). Another example is the bluegill sunfish, where males are either large, territorial, and parental or small, female-mimic, sneaky spawners (Gross 1982). Similarly, the blue throat patch of the tree lizard illustrated in Figure 21.3B is associated with territorial behavior, whereas orange-throated males are non-territorial. Coloration is permanent, not age-based, and arises from complex hormonal mechanisms during development (reviewed in Moore et al., in press).

In conclusion, competition for territories often leads to signaling arms races with complex rules for engagement, display, and escalation to avoid direct attack. The direction and final level that these games take vary enormously among species and result in elaborate and deceptive signaling in some cases.

Figure 22.14 Genetically determined polymorphism in the ruff (*Philomachus pugnax*). The photo shows light and dark ruffed males on a lek. A dark territorial male in the typical forward aggressive posture is shown in the foreground and a light satellite male in the upright submissive posture is standing behind him; a female can be seen on the left. Dark males attempt to defend a small mating territory on a traditional lek arena or at temporary leks adjacent to female traffic areas, whereas light males never defend territories. Both types of males will follow and display to foraging females, but most matings occur on the lek. Unlike dark males, light males can follow females when they move onto a lek and are tolerated by the territorial males. Satellites increase the attractiveness of territorial males to females, and they sometimes succeed in copulating with a female when the territorial male is occupied with another female or a neighboring territoriai male. Among dark males, a few individuals are extremely successful and the remainder are largely unsuccessful. Most individual satellites, on the other hand, achieve a few matings. The average copulation success of light males equals the average success of dark males, thus maintaining the polymorphism in a balanced ratio of about 16% to 84%, respectively. The inheritance pattern of male morphs is consistent with a single locus two-allele autosomal trait with the light allele dominant to the white allele. (After Lank and Smith 1987; Lank et al. 1995; photo courtesy of Ola Jennersten.)

SUMMARY

1. Territories are defended areas containing food, breeding sites, or both. Owners must contend with two types of intruders: **neighbors** and **floaters**. Interactions between two neighbors are usually characterized by repeated, low-level, ritualized encounters, whereas interactions between an owner and a floater are aggressive, escalated chases and attacks.

2. Owners generally win contests against intruders. One possible explanation for this bias is a fighting strategy based on the ownership asymmetry that is always present and easy to assess: escalate if one is the owner and retreat if the intruder. This rule of thumb, called **bourgeois**, is a conventional, **payoff-irrelevant** strategy that respects only ownership and ignores asymmetries in fighting ability and territory value. Game-theoretical models show that the conditions required for bourgeois to be a stable ESS are extreme and unlikely to be met in nature. Field studies claiming to have demonstrated this rule are better explained with alternative models.

3. An alternative explanation for the ownership bias is some type of **payoff-relevant** asymmetry favoring owners over intruders, such as greater fighting ability or higher territory valuation. An owner has obtained its territory because it is a good fighter, and a period of ownership then confers additional advantages on the owner by increasing self-confidence, increasing knowledge of the area, and increasing the value of the territory if breeding is under way. Asymmetric game theory models that incorporate assessment of fighting ability and territory-value asymmetries predict that owners will win in most, but not all, circumstances and are well supported by field studies. Territory value, in particular, is higher for owners, and territorial signal form and delivery pattern are designed to reveal the owner-sender's aggressive motivation level.

4. Territory owners must not only signal their presence and ownership status, they also must **perceive and assess** other individual's territorial signals to evaluate the status and position of potential intruders, the seriousness of an intrusion, and the need for a strong defensive response. Owners benefit if they can recognize their immediate neighbors and avoid engaging them in unnecessary escalated contests. For the same reason, territorial owners also benefit if their neighbors can recognize them and are therefore selected to incorporate a distinctive individual signature into their signal. With long-distance auditory signals, owners can use degradation information to determine the distance to a sender and therefore whether it is encroaching on the territory.

5. The **identification information** in territorial signals enables owners to play a different game with neighbors as compared to strangers. Neighbors are known and intimidating adversaries, escalated contest costs are high, and they are unlikely to take over an owner's territory because they already have one, so contests are brief and characterized by ritualized display. A floater is an unknown adversary with the intent of taking over the whole territory, so contests are long and aggressive. The conflict-resolu-

tion game models predict certain aspects of these two types of encounters, but field work suggests that the relative threat and potential losses to neighbors versus floaters best explains the different responses.

6. Competition for territories can lead to several forms of exaggerated and deceitful territorial signaling. Song matching of neighbors in songbirds that learn the final rendition of their song may be the consequence of **deceptive mimicry** in which young floater males attempt to deceive established resident males of their true status. Songtype repertoires in other songbird species could function to maintain a high threat stimulus level, deceive listeners, or regulate different levels of contest escalation. Neighborhood **communication networks** permit both eavesdropping to gain information on the position and abilities of other individuals in the population and cooperative alliances among neighbors. Finally, **territorial sneaks** obtain some of the benefits of territorial behavior by adopting alternative signaling strategies that are cryptic, mimic females, or deflect aggression from territorial males.

FURTHER READING

The vast majority of work on territorial signaling has been carried out on birds. McGregor (1991, 1993) and Catchpole and Slater (1995) have provided recent and thoughtful reviews on this subject, and the two volumes edited by Kroodsma and Miller (1992, 1996) contain encyclopedic information on birdsong. Territory games have been developed and reviewed by Leimar, Enquist, Maynard Smith, Hammerstein, and Parker. Arak (1984) reviews sneaky breeding strategies, and Stoddart (1980) provides much useful information on olfactory signaling systems.

Chapter 23

Mating Games and Signaling

IN PART II OF THIS BOOK, WE CHARACTERIZED THE COMMUNICATION system involved in the meeting and mating of males and females as a mutualistic endeavor for achieving successful reproduction. Both sexes were presumed to want accurate signals to identify the species and sex of potential mates and to exchange honest information on receptivity so that fertilization could proceed efficiently. It is certainly true that each sex is dependent on the other for successful reproduction and both benefit from mating. However, the interactions between males and females during periods of mating are bristling with conflict. This conflict is the result of fundamental differences in male and female strategies that arose very early in the evolution of eukaryotic diploid organisms with haploid gametes. Sexual strategies determine the basic social systems of all higher organisms and most aspects of the signaling behavior involved in mating. In all sexual species, reproduction occurs in a series of stages from attracting and finding members of the opposite sex to deciding whether an encountered individual is an appropriate conspecific mate, evaluating whether the individual is the most qualified mate one could find, attaining successful fertilization, and promoting offspring survival. The conflict at each of these stages can be characterized by different game-theoretical models that specify the ways in which the conflict can be resolved and the nature of the signals involved in mediating the dispute. In this chapter we shall first identify the differences in male and female strategies that set the scene for conflict, then examine each of the sequential stages of reproduction. For each stage we shall briefly examine the relevant game models, review the ecological and social factors that favor certain solutions over others, and finally describe how the outcome determines the signals used at that stage.

SEXUAL STRATEGIES

The two sexes in animals are defined on the basis of the size and mobility of their gametes. **Females** are individuals that produce large, relatively immobile gametes, called **ova**, that are provisioned with a store of nutrients the fertilized zygote will use during its early development. **Males** are individuals that produce small, highly mobile, and often flagellated gametes, called **sperm**, that contribute only DNA to the zygote. These differences in size and mobility, collectively called **anisogamy**, simultaneously facilitate the meeting and fusion of conspecific gametes and set the stage for conflict between the sexes.

Anisogamy is presumed to have arisen from an initial state of **isogamy** in which all gametes are similar in size and have the ability to fuse with any other conspecific gamete. This state appears to be highly unstable. Some models of the evolution of anisogamy (e.g., Parker et al. 1972) suggest that evolutionary forces favoring larger gametes that survived better led to a secondary strategy of small, parasitic gametes. Other models propose that the need to recognize conspecifics leads to a chemical signaling system that favors differentiation into separate senders and receivers (Hoekstra 1982, 1987). A third scenario focuses on the evolution of a small, mobile, and fast gamete that searches for larger, less mobile gametes (Hoekstra et al. 1984). In each case, parental individuals that produce gametes of one of the extreme types fare better in conjunction with the opposing type than parental individuals with the ancestral intermediate type. Two mating types is the only stable solution.

Anisogamy has several important consequences. Because sperm are considerably smaller, males can afford to produce large numbers of gametes and they have the potential to fertilize many females. Males will therefore compete intensely among themselves to gain access to as many females as possible (Parker 1984b). Male-male competition leads to a variety of fitness-maximizing strategies. In immobile, externally fertilizing species males have little option but to maximize sperm number and mobility, and their gamete production costs are likely to be very high. If the organisms are mobile, however, several different **male mating strategies** become available. Males can move rapidly in search of females or position themselves in good locations for encountering as many females as possible. If males can sequester themselves alone with a spawning female or insert their sperm directly into her body, they will have a higher probability of fertilizing all of a female's ova without competition from other males' sperm. In some cases, males can engage in postmating activities that serve to increase the survival of zygotes such as provisioning or guarding them. Male mating strategies are listed in Table 23.1. As we shall see, these strategies have a major impact on the type of long-distance mate-attraction signal used and the form of courtship signals. Females, on the other hand, do not usually benefit from maximizing the number of mates or competing with other females for males. Instead, they increase their fitness by adopting strategies that maximize their food intake for the provisioning of their gametes and zygotes. In short, females are limited by food, whereas males are limited by access to females.

Once strategies such as internal fertilization, copulation, male territoriality, and female guarding have evolved and males are released from the extreme sperm competition inherent in the group-spawning external fertilization context, males can afford to reduce their overall expenditure on sperm. This serves to further exacerbate the discrepancy between male and female strategies. Males require very little time and energy to replenish their sperm supply after a mating, whereas females need a longer time, depending on the size of their eggs and clutches. During this renewal time, animals are unreceptive. The greater the duration of the female's reproductive cycle relative to the male's, the more the ratio of receptive males to receptive females is skewed in favor of males. The ratio of receptive males to receptive females is called the

Table 23.1 Male mating strategies in mobile animals

Mating strategy	Description
Female defense	Males defend one or more females directly.
Long-term association	Monogamous bond with one female, or permanent harem with a few females when females associate in small cohesive groups.
Dominance hierarchy	Animals live in large cohesive mixed-sex groups, dominant males have priority of access to females and form short-term consortship bonds. Females asynchronously receptive.
Scramble competition	Females generally solitary or loosely clumped in colonies; males roam from female to female, associating briefly for mating only. Females synchronously receptive.
Coercion	Males physically force females to copulate.
Resource defense	Males defend a resource females require such as food or breeding sites. Males mate with receptive females that enter their territory. Females solitary or in unstable groups.
Self-advertisement	Males position themselves in locations visited by many females and display to attract them. Brief association for mating only. Males may be clustered on leks or solitary. Females highly mobile and nonterritorial.

Source: Vehrencamp and Bradbury 1984.

Note: The strategy that evolves in a given species depends on the spatial and temporal distribution of receptive females, which in turn depends on the distribution of food and other resources.

operational sex ratio (Emlen and Oring 1977). When it is strongly skewed toward males, competition among males for receptive females is very high. Males are thus even further limited in their access to females.

Sexual selection is the evolutionary process that arises from competition among members of one sex for access to the other (limiting) sex. It typically leads to the evolution of specific traits in the competing sex that improve their mating success. Although sexual selection is a subset of the process of natural selection, Darwin made a point of distinguishing sexual selection because the traits that evolve via this process often appear to have a negative effect on the survival of their bearers. Competition for mates can take two basic forms that affect the types of traits that evolve. **Intrasexual selection** includes overt competition among members of the competitive sex to control or monopolize the limiting sex. This leads to selection for traits such as weapons, large body size, strength, aggressiveness, speed, and endurance. **Intersexual selection** arises when the limiting sex can exercise a choice of mates and leads to the elaboration of structures, displays, vocalizations, and odors in the competitive sex that attract the limiting sex.

The mate-attraction and courtship signaling systems that evolve in different species are strongly affected by the operational sex ratio, the male mating strategy, and the relative importance of intra- and intersexual selection. These themes will reoccur frequently throughout the discussion of mating signals. In this chapter we shall first examine the stage of finding and attracting mates and the conflict over which sex should perform the long-distance mate-attraction signal. Given that a male and female have encountered each other, we then specify the degree to which they are likely to agree on whether a mating should take place and show how this conflict affects the kinds of behaviors and displays performed during courtship. Next we examine the process of mate choice and ask how prospective mates can obtain information about the potential partner's suitability. Finally, we look at the small but very interesting set of species in which the usual sex roles are reversed.

SEARCHING FOR AND ATTRACTING MATES

The first step leading to mating is of course locating a conspecific member of the opposite sex. For some species this is not an issue because males and females live together in permanent mixed-sex groups, so this discussion will focus on solitary and sexually segregated species. Searching for receptive members of the opposite sex not only entails a loss of time, but it also exacts an energetic locomotory cost and an increased risk of mortality due to exposure and predation. Intuitively, it seems that if one sex exerted a high searching effort, the other sex could profit by countering this pattern with low mobility. Each sex would therefore benefit if the other one did the searching, so a clear conflict exists. The mobility game described below clarifies the dynamics of this conflict and attempts to determine the conditions under which the male or the female is most likely to win. When we then examine the pattern of sexual biases in searching within different animal groups, we find a strong linkage with the sex that performs the long-distance attraction signal. The sex that wins the mobility game, i.e., sits and waits for mates, is likely to be the mate-attraction sender.

Searching Games

The mobility game (Hammerstein and Parker 1987) is a continuous asymmetric scramble among all of the adult members of a breeding population. It assumes an equal ratio of males to females. An adult goes through a series of several reproductive cycles consisting of three phases: (1) a mate searching period, (2) a period involving gamete production, courtship, copulation, and parental care, and (3) a recovery period of reproductive inactivity during which the animal pays for the expenses of its movement during the searching period. The order of these phases is unimportant for the model but can be visualized as follows:

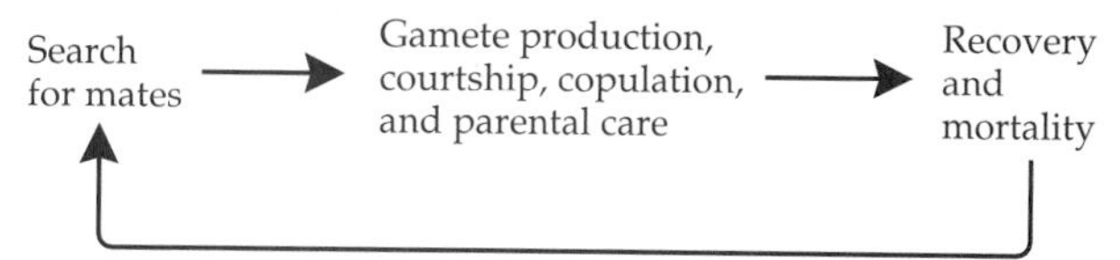

The total cycle length for each sex is thus the sum of the duration of these three periods. To maximize fitness, each sex seeks the strategy that minimizes its cycle time, given the strategy of the opposite sex.

The males and females in the population are envisioned as moving about on a grid or checkerboard. When a male and female meet on a square, they mate and go through the gamete and recovery phases, during which time they are temporarily removed from the board. Only searching-phase individuals are therefore actively playing on the board, and the sex ratio of searching males and females, i.e., the operational sex ratio, will be skewed in favor of the sex with the shortest gamete plus recovery time. Time progresses in small steps. The critical strategic parameter in this game is the level of mobility or velocity, v, defined as the probability of moving to another square in a given time step of the searching phase. Each sex has a continuous range of mobility options from 0 to 1 as alternative strategies.

The game matrix and solution are shown in Figure 23.1. There are two ESSs: either the male searches and the female waits, called the roaming male

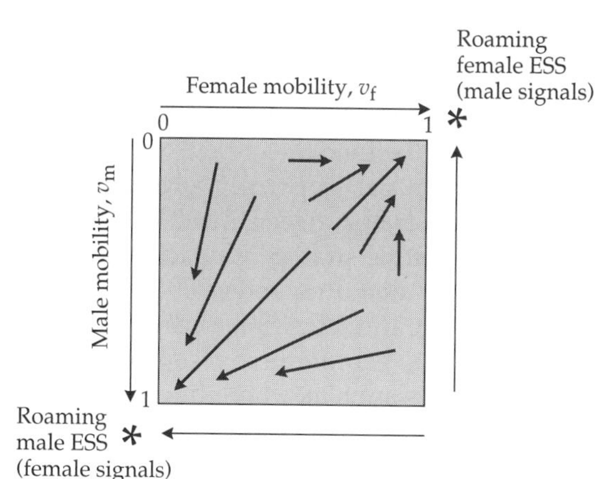

Figure 23.1 The mobility game. This is a continuous-strategy, asymmetric scramble game. Each sex can vary its mobility from 1 (the animal moves during every time interval until it finds a mate) to 0 (the animal sits and waits until a mate finds it). The male's velocity options are given by v_m and the females by v_f. The outcome is a double ESS denoted by the asterisks for either maximum male mobility and stationary females (the roaming male ESS) or for maximum female mobility and stationary males (the roaming female ESS). The domain of attraction is stronger for the roaming male ESS. (After Hamerstein and Parker 1987.)

ESS, or the female searches and the male waits, called the roaming female ESS. Solutions in which both sexes move some of the time are unstable. All other things being equal, the male is more likely to lose this game and get stuck with the searching. The domain of attraction is much stronger for the roaming male ESS compared to the roaming female ESS. This is a result of the greater gametic investment of females relative to males, which produces an operational sex ratio favoring males. Males that move more are much more likely to encounter females and shorten their search phase in competition with other males. On the other hand, if a population begins with a large number of searching females and waiting males, lazy females will do worse than continually moving females, and the roaming female ESS will be locally stable. The valuable lessons from this game are: (1) when one sex is selected to search, the other sex's best reply is to wait, and (2) either sex may lose the game. However, the game does not provide many insights into the conditions under which males or females lose.

Searching and Signaling Patterns

A broad survey of major animal groups reveals that, as predicted by the model, males are more likely to be the searching sex (Table 23.2). However, there are a substantial number of examples in which females search, notably most birds, a few mammals, frogs and toads, fish with paternal care, and some insects. Thus the fact that at the gamete level, sperm are always more mobile than eggs, does not necessarily mean that at the organism level this same relationship will hold. The roaming male strategy may nevertheless be the default strategy that arises from the primordial difference between males and females and remains in place unless specific factors operate to favor roaming females. Table 23.2 also reveals that when one sex searches, the other sex usually performs the long-distance attraction signal. Several selective factors have been proposed that shift the searching and signaling strategies toward one sex or the other.

Avoidance of inbreeding is a frequently invoked explanation for the differential movement patterns of the two sexes. If there is inbreeding depression, one sex will be selected to move away from the natal area while the other takes advantage of the increased benefits or reduced costs of staying close to home. The pressure to avoid inbreeding is stronger for females. This is a consequence of the greater gametic investment of females, which makes inbreeding a more costly error for them. The presence of inbreeding depression is therefore likely to favor the roaming female strategy rather than the stronger roaming male strategy predicted by the model.

Another important determinant of the natal dispersal pattern and the mate searching strategy is the male mating strategy (see Table 23.1) (Greenwood 1980). Table 23.2 also gives the male mating strategy for each taxon and shows the strong linkage between it and the signaling and searching strategies. In resource-defense systems, males are by definition tied to their defended territories and females move around to exploit these resources. Males that provide any type of resource to females such as a breeding site, food, nuptial gift, or parental care are likely to attract even more females to them-

Table 23.2 Patterns of searching sex, long-distance advertising sex, and mating system in a variety of mobile animals

Taxon	Searching sex	Signaling sex and modality	Mating system
Insects			
Crickets, katydids, grasshoppers, cicadas	F	M/Aud	Burrow defense, nuptial gift or self-advertisement
Bark beetle, carrion beetle, boll weevil	F	M/ Olf	Oviposition site defense or self-advertisement
Hawaiian *Drosophila,* fireflies, hill-topping and swarming species, dragonflies	F	M/Vis	Self-advertisement or oviposition site defense
Some bees, other nectivores, parasitoids, dung beetles	F	none	Defense of food sites, males grab females
Most moths	M	F/Olf	Female defense
Many Hymenoptera, Diptera, some butterflies, scarab beetle	M	none	Males wait at female emergence site
Many butterflies, solitary bees	M	F/Vis	Female defense, male patrolling
Fish			
Reptiles and Amphibians			
Urodeles	M	F/Vis	Female defense
Anurans	F	M/Aud	Oviposition site defense or self-advertisement
Lizards	M	F/Vis	Resource defense
Geckos	F	M/Aud	Resource defense
Snakes	M	F/Olf	Female defense
Birds			
Most songbirds, many nonpasserines	F	M/Aud + Vis	Resource defense or self-advertisement
Ducks, geese	M	M/Vis	Female defense
Some corvids, quail	M	F/Aud	Female defense
Mammals			
Most primates	M	F/Olf + Vis	Female defense
Chimpanzee, gorilla	F	M/Aud	Resource defense
Most rodents	M	F/Olf	Female defense
Pika, house mouse, white-lined bat	F	M/Aud + Olf	Resource defense
Horses	M	F/Olf.	Female defense
Most ungulates	F	M/Vis + Olf	Resource defense

Source: Greenwood 1980; Thornhill and Alcock 1983; Macdonald 1984.
Note: Aud = auditory; Olf = olfactory; Vis = visual.

selves and the resources they control if they advertise their presence, and thus are frequently selected to produce a long-range attraction signal. Male signals are often auditory to maximize signal range and hence tend to be expensive. In female defense systems on the other hand, males guard females directly.

Males will therefore fight to gain access to females, young males will be ejected from their natal territories by dominant males, and males will be selected to seek out receptive females, all of which favor male mobility. Expensive long-distance mate-attraction signals are not necessary, but females produce short-distance visual or olfactory signals to advertise their receptivity.

A final factor that may operate to tip the mobility game toward one sex or the other is direct selection on one sex to perform the long-distance mate-attraction signal. As we learned earlier in one of the scenarios for the evolution of anisogamy, selection for a gamete chemical-recognition system favors one gamete type that releases the pheromone and another that detects and is attracted to the pheromone. The linkage between mobility in one sex and attraction signals in the other is very tight, so selection on one of these is likely to impact the other one. If one sex produces a chemical, sound, or image that is detectable by the other sex, the receiver of the signal will become the searching sex. It is therefore not surprising that in the majority of species in which the male roams, the female produces a chemical mate-attraction signal. Such pheromones are easy to evolve from the hormonal and cellular products that reproductively active females generate. Again, this may simply be the default system that arises from the primordial sexual differences. As mentioned on pages 585–586, females may also produce visual or auditory signals to attract males, but these are usually short-distance, inexpensive signals for species recognition and reproductive state identification.

Similarly, systems in which the males signal and females search could have evolved as a result of selection on the male to produce a very long-ranging signal to attract mates. Repeated and vigorous auditory or visual signals are given by lekking birds, chorusing frogs and fireflies, cicadas, and other self-advertisement species. In male-attraction systems such as these, the female may have technically lost the mobility game and pays the cost of searching, but the stationary male is also paying a large cost of signaling. Males can afford to produce more costly mate-attraction signals than females because their gametic investment is lower. However, the primary reason for expensive male signals is strong sexual selection arising from male-male competition for females. An arms race situation then selects for increasing intensity and duration of male signals as males vie for the attention of the limiting opposite sex. Females usually do not need to compete among themselves for access to preferred males; even when they are the signaling sex, the male-biased operational sex ratio and competitive mate-searching efforts by the males reduce the need for elaborate signals. In a later section we take a look at the few role-reversed species in which females are the more competitive sex.

MATING DECISIONS

Once a male and female have met, they must decide whether or not to mate. Although both can only receive the fitness benefits of reproduction by ultimately mating, the success of a mating with the currently encountered partner depends on their genetic compatibility. If they are members of different species or ecotypes or if one individual is genetically inferior, then the number and/or quality of the offspring they would produce will be reduced compared to a mating with a more compatible individual. Male and female interest in proceeding with such a mating is likely to differ. The cost of a heterospecific mating is considerably lower for males because of their smaller gametic investment, and they may even benefit if the hybrid offspring are partially viable. For females, mating-decision errors translate into a huge fitness loss because of their larger investment, and they are therefore much more concerned about correct species identification. This conflict results in clear sexual differences in mating behavior and the assessment of the potential partner's qualifications.

The extent of agreement and disagreement over mating is well illustrated in the **hybrid mating game** model developed by Parker (1979). The game is a discrete asymmetric scramble in which males and females encounter each other and must opt to either mate or withdraw and search for another mate. They assess the decrement in viability of the offspring they would produce, d, which ranges from 0 to 1. Low values of d correspond to genetic quality variation within a local conspecific population, intermediate values correspond to the reduced offspring viability often found in hybrid or intergradation zones between two ecotypes or subspecies, and values close to 1 correspond to incompatible heterospecifics. Offspring viability relative to the ideal conspecific mate is therefore $(1 - d)$. Figure 23.2 shows how the ESS decision thresholds for the two sexes change as a function of the decrement in offspring viability d and the ratio of male to female gametic investment (g_m/g_f). For each sex, mating should be rejected in the region above its threshold curve and accepted in the region below the curve. The male's decision threshold curve lies above that of the female's. Thus both male and female should agree to withdraw when viability is poor and male gametic investment is relatively high. When viability is high (d is low) and g_m/g_f is high, both sexes should agree to mate. In the region of intermediate viability and low male investment there is a conflict between the two sexes. In this region the male is selected to mate, whereas the female is selected to withdraw.

The lower the male gametic investment relative to that of the female, the wider is the range of d values over which the two sexes disagree. The reason for this is as follows. The female becomes more choosy as her investment relative to the male's becomes larger, not only because the cost of a reproductive mistake increases but also because the strongly male-biased operational sex ratio enhances her probability of finding a better mate. At the same time, the male becomes less choosy and more eager to mate because his reproductive cost is low and his probability of finding another receptive female is low. The consequence of this conflict is that females are behaviorally coy upon first encounter with a male and take more time to assess and discriminate among alternate males. Males, on the other hand, are generally less discriminating and are selected to engage in persuasive or forceful behaviors to obtain the mating.

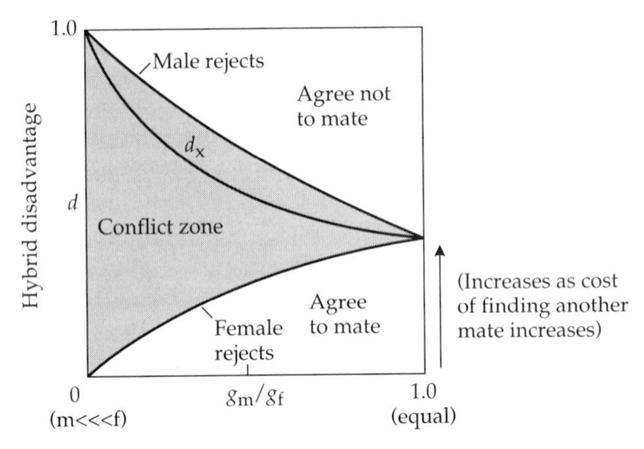

Figure 23.2 The hybrid mating game. This graph is the outcome of a discrete asymmetric scramble game between the male and female of two different ecotypes or species. Each sex has two strategies, to accept the mating or to reject the mating. If they mate, the offspring suffer a hybrid disadvantage of d ($0 < d < 1$) relative to the fitness of offspring of a conspecific pair. If the mating is accepted, the offspring viability is $(1 - d)$ and the only cost is the time spent renewing gametes for the next round of reproduction. If the mating is rejected, the individual pays the cost of searching for a conspecific plus the cost of gamete renewal but obtains an offspring viability of 1. The operational sex ratio, which is determined by the relative gamete investment levels g_m for the male and g_f for the female, affects the searching time. The graph shows the threshold values of rejection by the male and female as a function of the gamete investment ratio g_m/g_f and the hybrid disadvantage d. When d is small and male gametic investment is relatively high, both sexes agree to mate. When d is large and male gametic investment is high, both agree to withdraw. But when d is intermediate and male investment is low, the sexes disagree and males are more eager to mate than females. Within the stippled conflict zone, selection is stronger on females to reject than on males to accept for the region above the line d_x, and below the line, selection is stronger on males to accept than on females to reject. (After Parker 1979.)

Which sex will win this game? Selection is generally stronger on the male to achieve a mating than it is on the female to hold out for a more compatible mate. The d_x line in the conflict zone of Figure 23.2 separates the region in which selection is stronger on the female to resist than on the male to persist from the region in which selection is stronger on the male to persist. On this basis alone, one might expect the male to win more frequently and hybrids to be common. However, the outcome of this conflict is more likely to depend on which sex can physically control copulation, sperm precedence, and fertilization, factors that are determined by a species' body form, genitalia, and reproductive biology. Females in most species must cooperate with males to achieve a successful insemination and therefore usually have ultimate control over mating decisions. If males can physically dominate females, then females may not be able to exercise much choice of mates and forceful courtship behaviors will be favored in males. As a last resort, females will evolve mechanisms to select sperm or prevent fertilization of their eggs by clearly inferior heterospecific males (Howard and Gregory 1993). This will have the evolutionary consequence of increasing *d* in the model to 1 so that even males will develop mechanisms for avoiding heterosexual females.

The mating-decision conflict between males and females may also result in different pressures on males and females to make their mate-attraction signals distinctly species-specific. When females perform the long-distance attraction signal, they have evolutionary control over the species specificity of their signal. Females that produce a signal that attracts only conspecific males have a selective advantage over females that attract nonconspecifics as well. It therefore comes as no surprise that one the strongest examples of sexual **character displacement** in the overlap region between two congeneric species involves a female signal (Figure 23.3). When males perform the long-distance signal, the outcome depends on their gametic costs and hybrid viability. If gametic costs are low, males could potentially benefit by failing to produce a distinguishable signal and attracting both conspecific females and females of a partially compatible congeneric species. If male gametic costs are high or hybrid viability is zero, they will be under strong selective pressure to produce a species-specific signal to avoid mismating, and male and female inter-

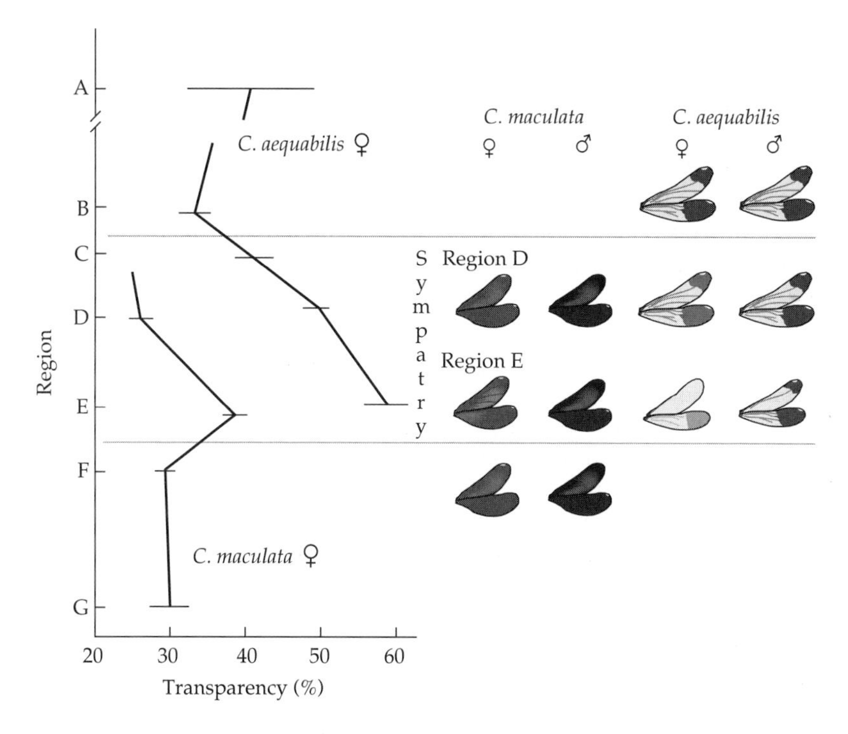

Figure 23.3 An example of character displacement in the visual signal of female damselflies (*Calopteryx aequabilis* and *C. maculata*). Wing coloration (transparency of dark area) of females is similar and dark for the two species in areas of allopatry (A, B and F, G). In areas of sympatry (C, D and E), *C. aequabilis* females exhibit a significant shift toward lighter and more transparent wings. Note that the male *C. aequabilis* do not change their wing pattern in a parallel way. (After Waage 1979.)

ests will largely concur. The species recognition aspect of mate-attraction signals is an issue we shall take up in the next chapter once we have covered the mechanisms of mate choice within a species.

COURTSHIP PERSISTENCE

The hybrid mating decision game described above can be generalized to mating decisions within species where the viability of offspring varies depending on the quality of the mate. Males in general are more eager to mate because they can benefit greatly from each additional female they meet and fertilize. Females do not benefit from additional matings in the same way, but they may benefit from resisting the current male and holding out for a better-quality mate. In this section we will attempt to resolve which sex wins in the conflict zone depicted in Figure 23.2. The critical issue is whether females are able to exercise mate choice or not. When the female wins, males will have to persuade females to mate, and lengthy courtship with elaborate male signals should evolve. When the male wins, courtship is expected to be brief or forceful.

Courtship Persistence Games

Most existing **courtship persistence games** treat the interaction between male and female over intraspecific mating decisions as a contest. This means that those aspects of a contestant's payoff that are dependent on the frequency of individuals in the population playing different strategies have not been incorporated. Figure 23.4 shows a simple discrete strategy game (Parker 1979). The female's two alternative strategies are to reject the mating or to passively allow the mating. If the mating occurs, she loses an amount of fitness relative to what she could achieve with a higher-quality male, and she therefore adopts the strategy that minimizes her net loss. The male's alternative strategies are to attempt to persist in mating or to withdraw if the female rejects him. The male benefits from the mating and attempts to maximize his net gain. If both sexes play the more aggressive strategy, they pay the costs of escalation, and there is a certain probability that a mating will occur anyway. Several outcomes are possible. The passive female/nonpersistent male solution is unstable because both rejection and persistence can invade. However, as the frequency of the nonpassive strategy increases in one sex, the strength of selection on the nonpassive strategy in the other sex begins to lessen and eventually reverses direction. Three ESSs are therefore possible, depending on the relative costs to the two sexes. (1) If female escalation costs are high, the male persists and the female is passive; the consequence would be a lack of male courtship, and females would not be able to exercise any mate choice. (2) If male escalation costs are high, the female rejects and the male is nonpersistent; here, females would have a choice of mates, and males would court persuasively until females reject them. (3) When both have low costs, the female rejects and the male is persistent; in this case, courtship could be a prolonged and perhaps violent affair with a highly variable outcome.

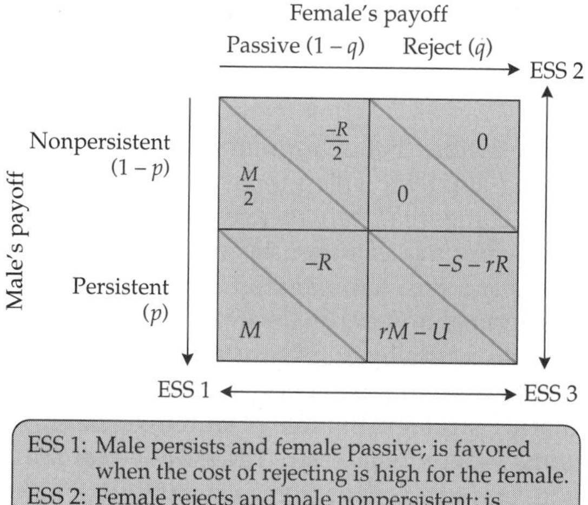

Figure 23.4 Parker's discrete courtship persistence game. This is a discrete asymmetric contest game. The female's two alternative strategies are to reject the mating or to passively allow the mating, and the male's alternative strategies are to persist in mating or to withdraw if the female rejects him. The value of the mating to the male is M. If the female allows the mating, she loses R units of fitness. If the female rejects the mating but the male persists, she pays a cost S and he pays a cost U, and there is a certain probability r that the mating will occur anyway. When males are nonpersistent and females are passive, the pair is assumed to mate with a 50% probability so the payoff is the average of the mating and nonmating alternatives. (After Parker 1979.)

Other game models that we have described previously for intrasexual conflict have also been applied to male-female contests (Clutton-Brock and Parker 1995b). A **sexual arms race** can be used to model the ability of males to physically force females to mate. In arms races, contest costs do not depend on decisions made during each game but rather on the armament level chosen by each contestant at the beginning of its life. A stable ESS can only be achieved if there is some variation in the arms level attained by a given investment strategy due to growth or random environmental variation. Since males generally place a higher value on winning and can afford to invest more in armaments, they will win unless the costs of forcing females are high. High costs will be incurred when females are the larger or more aggressive sex, when females can easily evade males, and when males would have to leave their territories or waste time in long chases to follow a female.

The **sexual war-of-attrition** and **sexual sequential assessment games** can be used to model the variable duration of individual courtship contests as a function of the perceived costs and benefits to each opponent. Females are assumed to have complete control over copulation, and contest duration depends on the quality of the male. If the male is high quality, the value of resisting for the female is low, she will chose a short persistence time, and the courtship will be brief and successful. If the male is poor quality, the value of resisting for the female is high, so she will chose a long persistence time. The male would suffer a high time cost if he attempted to outlast her, so he should give up after a short courtship in search of more willing females. The longest courtship durations may occur with intermediate-quality males where the

V/k ratios for the two contestants are very similar; in these contests the sex that wins will vary (Clutton-Brock and Parker 1995b).

Which Sex Controls Courtship?

A useful way to evaluate some of the factors that determine whether courtship will be characterized by elaborate and lengthy display, forceful control maneuvers, or a brief noncontested act is to compare closely related species that differ in courtship behavior. Table 23.3 provides a brief summary of such contrasts in a few taxa. Species in the center column are characterized by passive female behavior, and species in the right-hand column are characterized by greater female control and either elaborate male courtship display or extended contests.

A very important factor affecting courtship behavior is the mobility game outcome and the related male mating strategy. In male resource-defense and self-advertisement mating systems where females are the mobile searching sex, females have a choice of mates and can benefit from searching for a high-quality male. Males would pay large costs if they left their territories to chase or restrain females. Courtship in such species consists of male displays designed to persuade females to mate, such as colorful plumage, elaborate dances, circling around the female, and in some cases soft sounds and gentle tactile signals; the male is never aggressive toward the female. Examples include most birds, ungulates with resource defense, lekking and chorusing species, fish with paternal care, and a few insects. Females have clearly won the courtship persistence game in these species. In female defense systems, males are the mobile sex and they attempt to control one or more females on a long-term basis or consort briefly with a series of fertile females. Because males must engage in extensive contests with other males to gain access to females, they are often larger and stronger than females as a result of intrasexual selection. This places them in a better position to control copulation, and as the sexual arms race predicts, males often win. In permanent harem systems, females have little option for choosing mates, so they typically do not resist at all and passively accept the harem male. They signal their receptivity to the male, who copulates without any significant display. In scramble competition systems with serial consortship, females have a choice of mates but must succeed in rejecting unwanted suitors if they are to exercise their preferences. Maximum conflict can occur in this situation. Female responses may include noncooperation, aggression, or fleeing. In several diverse species ranging from lizards to insects to marine mammals, females can avoid forced copulations by flipping over on their backs. Females of lizard species employing this tactic develop a bright red or yellow ventral color patch after their receptive period to enhance this rejection signal (Olsson 1995).

Horses and their relatives demonstrate another effect of the mating system on courtship behavior (Ewer 1968; Rubenstein 1986). All equids exhibit female defense in the general sense, but as shown in Table 23.1 there are several substrategies within this category. Plains and mountain zebra and wild horses (*Equus burchelli*, *zebra*, and *przewalskii*, respectively) exhibit a permanent single-male harem system and their courtship is characterized by fe-

Table 23.3 Courtship patterns in taxa exhibiting contrasts in male versus female control

Taxon	Greater male control	Greater female control
Fiddler crabs (*Uca*) Live in large communities on muddy or sandy shores of rivers, estuaries and bays. Both sexes dig and defend individual burrows. Males have a single enlarged claw that they wave in a species-specific pattern to attract females and repel neighboring males. (Crane 1975)	In Indo-Pacific species, male moves the claw in a lethargic vertical display. He displays from his own territory, but approaches a receptive female on her territory, seizes her, taps or strokes her carapace with his walking legs, and attempts copulation. Mating occurs on the surface near the female's burrow.	In many American species, the male waves his claw in a vigorous vertical or horizontal arc while females search. Display intensifies when females approach, curtseys and bobs may be added. He then enters his burrow; if the female is willing, she follows him. Mating occurs in the male's burrow.
Fruit flies (*Drosophila*) Aggregate on decaying fruit and fungi, where mating takes place and eggs laid. Males roam in search of females, tapping any fly-like insect with the forelegs to identify conspecific females by taste. (Bastock 1967)	In subgenus *Pholadoris*, males forgo all display elements. After tapping, the male crouches behind a female with his abdomen curled under and suddenly attempts intromission. If the intromission is successful, he then mounts.	In subgenus *Sophophora*, courtship is elaborate with 3 phases: circling around the female, wing vibration, and genital licking. Male may attempt to mount, but female can easily escape by twisting, kicking, or moving away. She signals acceptance by spreading her wings and genitalia.
Even-toed ungulates (*Artiodactyla*) Large-bodied ruminating mammals that feed on grass (grazers) or leaves (browsers). Highly social, herd size depends on resource distribution. Resource defense a common male strategy. (Franklin 1983; Ewer 1968)	In camelids (camels, guanaco, and vicuna) copulation occurs with the female lying on her chest. Males pursue and drive an estrous female, then force her to adopt the prone mating posture by biting at her legs or by neck wrestling and pushing her to the ground, the same aggressive behaviors used in male-male combat.	In antelope and other bovids, females must be enticed to stand still for copulation. Males employ a restrained courtship strategy that reduces the female's readiness to flee or fight such as prancing visual displays, resting the head on the female's back, and tapping her rear legs before attempting to mount.
Horses, asses, zebra (*Equus*) Large-bodied non-ruminating mammals. Highly social, living in herds of various sizes. Different female defense strategies most common. (Ewer 1968; Rubenstein 1986)	In horses, mountain and plains zebra adapted to richer habitats, females form small stable herds defended by a single male. Male performs visual prancing display before estrous females. Females have little choice of mates and passively accept the harem male.	In wild asses and Grevy's zebra adapted to arid environments, patchy food prevents stable herd formation. Sexes live in separate unstable aggregations that merge occasionally for reproduction. Males pursue, bite and kick estrous females until they stand still; females fight back.
Primates Males and females live together in stable single-male or multi-male groups. Potential mates are therefore well known in advance of mating. Courtship is not elaborate; female signals receptivity with olfactory cues. (Dixson 1983; Dunbar 1988)	In single-male species (langurs, guenons, collobus monkeys, gorilla), females have no choice of mate and are passive. Rump swellings are uncommon; males initiate courtship in response to female olfactory signal. Hamadryas baboons are an extreme example of aggressive male control.	In multi-male species (baboons, macaques, chimpanzee), females have several potential mates and advertise estrous with a conspicuous red rump swelling. Males establish dominance hierarchy for access to estrous females, but females may solicit copulation from other more preferred males.

male passivity and little display, as in other mammals with this system. In contrast, the wild asses and Grevy's zebra (*E. africanus*, *hemionus*, and *grevyi*) exhibit a scramble competition form of mating. The sexes spend most of the year in separate herds of unstable membership, potential mates are not known to each other, no bonds can be established, and males have no established dominance hierarchy among themselves. When opposite-sex herds meet and females are in estrous, males rush among the females to mate. Since females also have a choice of mates, they are selected to reject nonpreferred males. Mating is rather violent in these species, with males biting and kicking females to force them to stand still and females fighting back.

A second type of factor controlling courtship behavior is the method of fertilization and constraints on the location of egg deposition. These affect the ability of males to force females to copulate. The fiddler crabs and even-toed ungulates of Table 23.3 exhibit contrasts that appear to be related to such constraints. In fiddler crab species with copulation in the male's burrow, males must entice females to approach them with elaborate display, whereas in species that mate on the surface, males approach females but do not display (Crane 1975). In the camelids, the induced ovulation strategy of females may require that copulation take place with the female lying, so males have a forceful strategy available to them that is not possible in other ungulates that copulate while standing (Franklin 1983). If males have a means of grabbing females, such as in frogs and toads, ducks, some pinnipeds, and many insects, females may not be able to exercise choice of mates. Some insect males encounter females at their eclosion site and copulate with them as they emerge, giving them no choice of mates (e.g., fig wasps and digger bees; Thornhill and Alcock 1983). Forced copulation is also common in turtles (M. Andersson 1994). The ultimate example of male dominance occurs in a few opistobranch sea slug species and the bedbugs (Cimicidae) in which males inject their sperm into the body cavity of the female with a sharp, piercing penis (Eberhard 1985). On the other hand, if females must lay eggs in a specific location or nest, or if synchrony of gamete release is critical, then males cannot force females but must display and entice until they are willing to mate.

Another factor that affects courtship is the possession of predatory weapons or armor. Females tend to win in such species because the injury cost to males of controlling them is high or the tactical ability to control is poor. Examples include birds of prey, lion, praying mantis, and porcupine. Females also win the arms race when they are larger and/or more aggressive than males. Some examples of male and female arms race winners are shown in Figure 23.5. In many species, it is difficult to declare a winner because females accept the first conspecific male that finds them. Conflict is probably minimal because both parties benefit from rapid conspecific mating. This is the expected ESS when both sexes are solitary and nonconspecific mating produces inviable offspring.

HONESTY OF MATE QUALITY INFORMATION

If the limiting sex has won the courtship persistence game and has access to several members of the opposite sex, it has the luxury of mate choice. Be-

Figure 23.5 Examples of arms-race winners. (A) Males win in the elephant seal *Mirounga angustirostris*. The two- to threefold weight advantage of males makes it difficult for females to resist copulation attempts. (B) Males win in the hamadryas baboon *Papio hamadryas*. Males acquire a harem by "kidnapping" young females and forcing them to remain close to him. Here a male directs a threat yawn toward another male in defense of his females. Nevertheless, female preferences do modulate male behavior. (C) Females win in most spiders. The small male *Araenus pallidus* on the left plucks the mating thread in a characteristic manner as he approaches the receptive female on the right. (D) Female spotted hyenas (*Crocuta crocuta*) are socially dominant over males. Their enlarged clitoris not only looks like the male genitalia but it may make it more difficult for males to copulate by force. (A courtesy of Burney LeBoeuf; B from Kummer 1971, 1995, photo courtesy of Hans Kummer; C courtesy of M. Grasshoff; D from East et al. 1993, photo courtesy of H. Hofer and M. East.)

cause the limiting sex has the greater gametic investment, it is more likely to be concerned with mate quality and to be coy and choosy during courtship. Mate quality means different things in different species but could include size, physiological condition, foraging ability, parenting ability, agility, intelligence, dominance, age, viability, parasite resistance, or genetic complementarity. The choosy sex will use some combination of prior knowledge, direct assessment, and information from the attraction signal of the competing sex to make this determination. Since there is a potential for coded quality information in the signal, the problem of signal honesty arises, and the games discussed in Chapter 20 become highly relevant. In this section we examine how the mechanisms of mate choice determine the honesty of mate-attraction and courtship signals.

Mechanisms of Mate Choice

There are three basic models for the evolution of conspicuous mate-attraction signals: the Fisherian runaway model, good genes models, and direct benefit models. Females are assumed to be the choosy sex and males the competitive advertising sex (reversed role situations can be examined in a similar way). In all cases, there is simultaneous evolution of a female preference for a specific male trait and evolution of the male trait. The models differ in their assumptions, their mechanisms of operation, and the honesty of the resulting attractive traits.

The **Fisherian runaway selection model** (Fisher 1930) operates best in polygynous species. Initially, there must be some genetic variability associated with phenotypic variability in a male trait such as a bright color spot and some genetic variability in the female's tendency to prefer males with the color spot. Males with the preferred trait obtain more mates, and because their mates tend to be those females with the preference, the genes for the female preference and the genes for the male trait become linked in their offspring. Females with the preference benefit because their sons possess the preferred trait and are more successful themselves, i.e., females with the preference have more grandchildren. Given the right conditions, a runaway evolutionary process may then ensue with the male trait becoming larger and the female preference for it becoming stronger. The male trait eventually becomes so large that it imposes a cost on males, and the runaway process stops when the cost of the male trait just balances its mating advantage (Lande 1981; Kirkpatrick 1982). Traits that evolve via this process are called **arbitrary traits** because they can take any type of conspicuous form and provide no information to the female about the male except his attractiveness to females.

The Fisherian process requires a rather special set of circumstances to get it going, but several different situations could bring about the initial trait and the preference for it. One possibility is sensory exploitation that we discussed on pages 527–534. If females have preexisting sensory biases, feature detectors for prey items or food plants, or hidden preferences arising from sensory recognition mechanisms, males with traits that mimic these stimuli are more likely to attract the attention of females. Alternatively, the male trait could initially be favored by natural selection, and a female preference for this trait could subsequently evolve. As it undergoes the runaway process, however, the male trait is likely to become exaggerated to the point that it loses its natural selection advantage. A third possibility is that females evolve preferences for certain male traits because they distinguish conspecific males from a closely related species and serve to prevent interspecific hybridization (Lande 1982; Liou and Price 1994).

Unlike the Fisherian model, **good genes models** propose that costly and conspicuous male traits become the targets of female choice because such traits indicate some aspect of male quality. A female benefits from mating with a preferred male by producing both male and female offspring with higher survivorship or viability. The resulting male traits are therefore called **indicator traits**, and the costs to males are presumed to be necessary costs (page 588). Genetic models of this process are based on the coevolution of three characters, the male indicator trait, intrinsic male viability, and female preference for the trait. Several different models have been proposed (Zahavi 1975; Maynard Smith 1976b, 1985; Andersson 1986; Kirkpatrick 1986; Pomiankowski 1987, 1988; Tomlinson 1988; Hasson 1989b, 1990; Iwasa et al. 1991). In the **classic handicap model**, males that acquire a trait such as a heavy set of antlers pay a survivorship cost. Low-quality males cannot support the cost of this handicap and die, whereas higher-quality males survive. The trait acts like a quality filter, so females that select males with the trait are assured of higher-quality mates on average. Quantitative versions of this model show that the trait and the preference for it cannot invade because males without the trait and its cost have higher net fitness. In the **condition-dependent indicator model**, males can vary their expression of a preferred trait such as call rate or tail length to optimize their mating success and survival given their quality. Males sustain only the cost of the phenotypic magnitude of their trait. High-quality males can afford to expend more on trait expression than low-quality males, so trait magnitude is a good index of male quality. Display rate, tail length, and other energetically costly signals are possible examples. In the **revealing indicator model**, all males attempt to develop the trait to the same magnitude and pay the same cost, but the condition of the trait is lower in poor-quality males. Call frequency, feather condition, and other traits that are physically or physiologically constrained would be possible examples. Box 23.1 graphically illustrates the distinction between condition-dependent and revealing traits in a game-theoretical format. With these latter two models, the trait can evolve more readily because genes for the male trait, male viability, and female preference become associated in both sons and daughters. Such models work even when there is a cost of choice to females and when the mating system is monogamous (monogamy reduces the operation of the mating advantage and runaway process of the Fisherian model). Indicator traits arising from both models are honest signals because low-quality males would have to pay relatively more than high-quality males to produce a very attractive signal; cheating is therefore disadvantageous.

Both the Fisherian and good genes models described above assume that the fitness consequences of the male trait remain heritable despite the strong directional selection. Fitness heritability normally erodes under these circumstances, and the advantage of choosing good or attractive males is therefore expected to decline. Theoreticians have argued that special mechanisms must operate to maintain or reintroduce heritable variations in the trait such as biased mutation pressure, selection for variability, or temporal and spatial variation in environmental selection pressures (Hamilton and Zuk 1982; Pomiankowski et al. 1991; Iwasa et al. 1991). The **direct benefits model** is another type of indicator model that eliminates the requirement of heritable fitness. Males vary in a nonheritable phenotypic trait that indicates the potential for direct benefits to the female and/or her young. Examples might be the high song rate of a bird that indicates the amount of resources on his territory or a pheromone in a male moth that reveals the amount of egg-protecting alkaloids in his seminal fluids. These models produce the same qualitative results as the condition-dependent genetic quality models based on heritable fitness (Heywood 1989;

Box 23.1 Graphical Game Theoretical Analyses of Condition-Dependent and Revealing Traits

THESE TWO GAMES (Figures A and B) are asymmetric, continuous-strategy scrambles among males that differ in phenotypic quality, in which subscript 1 indicates the highest quality and subscript 3 the lowest quality. The strategic parameter is a male's investment level in some costly display trait, x, which increases from left to right along the horizontal axis. Females exhibit a relative preference for higher-quality males and use the trait to assess male quality. Male fecundity, F, is a function of their mating success and increases with increasing trait value. Male survival, S, decreases as trait value increases. Each graph shows the ESS solution, with fecundity at the ESS indicated by black dots and survivorship at the ESS indicated by gray dots. The differences among high-, medium- and low-quality males in terms of their trait values, fecundity, and survivorship at the ESS are summarized in the small box to the right of each graph. These models are not genetic coevolutionary models between males and females; female preferences do not evolve.

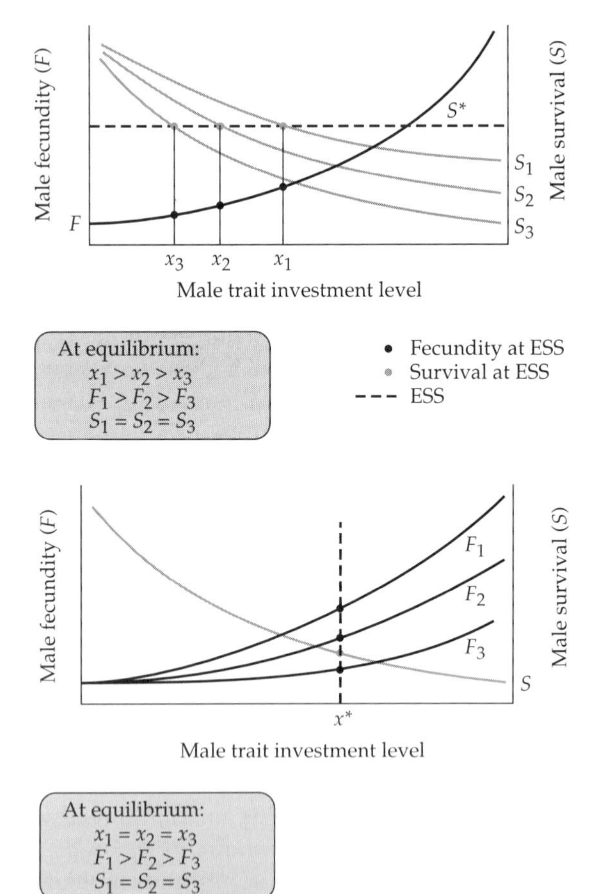

Figure A Condition-dependent trait. For the condition-dependent trait, lower-quality males suffer a more severe survivorship cost for a given trait investment level. Their survivorship thus declines more precipitously as trait investment increases. The ESS, computed with differential calculus, is for males to adjust their investment in the trait such that all males have the same survivorship, S^*. Any male that deviates from the ESS has lower fitness. This means that lower-quality males invest less and have smaller phenotypic traits. Females therefore use the size of the trait as an indicator of male quality and higher-quality males enjoy higher reproductive success.

Figure B Revealing trait. For the revealing trait, the expression or appearance of the trait is inferior for lower-quality males, even though they have invested the same amount of energy to produce the trait as higher-quality males. Trait value therefore increases not only as a function of investment level but also as a function of male quality. Thus there is a series of female preference curves for males of different quality classes. At the ESS, all males have the same survivorship and trait investment level, x^*, but higher-quality males are recognized and preferred by females. (After Parker 1979.)

Hoelzer 1989; Price et al. 1993). As discussed on pages 658–660, Grafen's (1990a) game-theoretical model of male courtship signal honesty also falls into this category, since females improve their fecundity when they can accurately choose high-quality males and males adopt an informative and honest strategy in which trait expression is correlated with their phenotypic quality. Again, the crucial condition that guarantees honesty is that males pay a very large cost if they display at a level that is too high for their quality.

These different models of intersexual selection are not exclusive; they may operate simultaneously or in an evolutionary sequence and the distinctions between them begin to blur. For example, an indicator trait can act as the initial stage of the Fisherian process. An arbitrary Fisherian trait could become an indicator trait as its cost increases if it begins to reveal some aspect of quality that benefits females or their offspring. An amplifier trait could become a condition-dependent indicator if low-quality males elected to hide their poor quality by failing to display the trait. Finally, a Fisherian mating advantage is likely to occur in any type of good genes indicator model, and both processes operating together facilitate the evolution of the trait and the preference. Since it is difficult to reconstruct the history of the evolution of a given male trait, our task in the following sections will be to review examples of male traits shown to be preferred by females in a variety of species and categorize their present-day fit to one of the three basic models: (1) arbitrary Fisherian traits that provide no quality information or direct benefit to the choosy female but are favored because of mating advantages achieved by sons, (2) indicator traits that are correlated with heritable aspects of male quality and provide only good genes to the female, and (3) indicator traits that are correlated with direct benefits females receive from males. It is very difficult to show that a trait is purely Fisherian because both direct and indirect benefits must be disproven. Indicator traits (2 and 3) are easier to demonstrate, and here we can often determine the factors that maintain signal honesty, i.e., identify the linkages between signal form and information about quality or benefit.

Anatomical Traits

This category includes visually transmitted physical traits such as color patches, feather plumes, elongated tails and fins, and other nonweapon body structures collectively called **ornaments**. If any Fisherian traits are to be found, they will mostly likely belong to this category of signal. Some possible examples of Fisherian arbitrary traits are illustrated in Figure 23.6. Many ornaments appear to be examples of good genes indicators, direct benefit indicators, or combinations of the three trait types.

As mentioned in earlier chapters, carotenoid-based red, orange, and yellow color patches are difficult for animals to produce because the pigments cannot be synthesized; carotenoids are obtained from dietary sources. Where red items are rare in the diet, the size or intensity of red color patches may indicate a male's foraging skill or nutritional condition. In Figure 20.5 we saw that female house finches (*Carpodacus mexicanus*) obtain direct benefits in the form of better paternal care for nestlings by selecting males with larger and

Figure 23.6 Some possible arbitrary Fisherian traits. (A) Female preferences for elaborate calls in the tungara frog *Physalaemus pustulosus*. The diagrams depict stylized waveforms of the calls and artificially generated call variants presented to female frogs in choice experiments. See Figure 16.16 for spectrogram versions of these calls. Females recognize and are attracted to the simple whine of the male's call (top row), but they show even stronger preferences for a whine with one to five chucks by conspecific males (second row), for the whine plus a prefix from *P. pustulatus* (third row), and for a double whine (fourth row) that is characteristic of *P. coloradorum* but not *P. pustulosus*. (B) Female zebra finches (*Poephilia guttata*) prefer males with red color bands, a behavior that is linked with female preference for males with redder bills (Burley 1988). Females also prefer males with an artificial white plume glued to their heads, a novel trait that clearly has no direct or indirect benefit. (C) The penis of some rodents, such as the South American *Agouti paca*, is covered with spines. It seems unlikely that the structure provides direct information about male quality. Eberhard (1985) notes that the movement of this penis in the vagina cannot go unnoticed by the female, and suggests that it produces a tactile signal that is preferred by females. (D) Males of all bird-of-paradise species exhibit elaborate feather development, but each genus has taken off with a different and seemingly arbitrary pattern: *Astrapia* has super-elongated tails, *Paradisea* has long filamentous body feathers, *Pteridophora* has specialized head plumes, and *Cicinnurus* has curly tail feathers and erectable nape and chest shields. (A after Ryan and Rand 1993; B photo by Kerry Klayman, courtesy of Nancy Burley; C after Cooper and Forshaw 1977; D after Hooper 1962.)

redder plumage patches. Male redness is also heritable, so females obtain the Fisherian benefit of more attractive sons as well. In addition, brighter males have a higher migratory return rate (Hill 1991), so evidence for good genes benefits may eventually be documented in this species. Another constraint on the expression of carotenoid color patches is parasite load. Carotenoid levels are associated with immunological competency in many vertebrates (Anderson and Theron 1989; Bendich 1989; Leslie and Dubey 1992; Lozano 1994). Sick males often exhibit a reduction in red coloration compared to healthy males, and males that overinvest in color-patch development may compromise their immune system function. Only genetically resistant males can afford to invest in intense color production and remain viable. In the stickleback *Gasterosteus aculeatus*, female preference for red males helps them avoid leaving their eggs with parasitized males, a direct benefit (Milinski and Bakker 1990). In the guppy, redness is also correlated with male health and condition, but since there is no male parental care, direct benefits can be ruled out. Evidence for higher growth rate and survival in the offspring of redder males supports a good genes model (Reynolds and Gross 1992). Both fish species are polygynous, so the Fisherian process may also be operating.

In contrast, noncarotenoid color patches such as white, black, and structural blue, green, and red are not energetically difficult to produce and may be good candidates for arbitrary Fisherian signals. Examples include the white patches on the tail of the great snipe *Gallinago media* (Höglund et al. 1990), the blue spots of the guppy *Poecilia* (Kodric-Brown 1989), and the structural blue and white patches of forest manakins *Corapipio* (Théry and Vehrencamp 1995). These species are either promiscuous or lekking, so females receive no direct benefits and all display in dark environments where the flashing or iridescent colors are highly stimulating to females. There is no evidence yet that the size or intensity of the color patches is correlated with male condition, parasite load, or viability. However, variation in noncarotenoid color signals is frequently correlated with age and dominance. If having a dominant mate benefits females directly in the form of protection from harassment by other males during copulation (Clutton-Brock et al. 1992) or better quality resources in a resource defense system, then a direct benefits model is supported. Furthermore, badges of dominance could also reveal good genes. As discussed on pages 662–664, badges require both receiver-dependent (incidental) punishment costs and receiver-independent (necessary) costs of general aggressiveness to maintain honesty. The development of noncarotenoid color patches, as well as most other sexually dimorphic anatomical traits, is under the control of testosterone. High testosterone levels increase aggressiveness but reduce immunocompetency. Badge size may therefore indicate genetically based viability and resistance to disease (Folstad and Karter 1992). A possible example of such a trait is the the black breast stripe of the great tit *Parus major*, a monogamous bird with shared parental care (see Figure 21.3A). Females prefer males with wider stripes. Wide-stripe males are not only better parents (i.e., direct benefits), but also more dominant against other males and produce sons that have wider chest stripes and higher survival in cross-fostered nests (Norris 1990a,b, 1993).

Even though the effects are restricted to male offspring, females appear to be obtaining good genes benefits as well as attractive sons.

Wattles and similar bare patches of skin that reveal the red color of blood may also be honest revealing indicators of a male's health, parasite load, vigor, and dominance. The best evidence comes from the jungle fowl *Gallus gallus* in which comb size and color brightness appear to be heritable traits; males with larger and brighter combs produce heavier offspring (Zuk et al. 1990; Johnson et al. 1993). Since there is no direct male parental care, females may be obtaining good genes benefits.

Tails are another common focus of sexual selection. Experimental studies have verified a female preference for longer tails in four bird species with a range of mating systems from monogamy to resource defense polygyny and leks (M. Andersson 1994). In the polygynous (no male parental care) widowbirds *Euplectes jacksoni* and *E. progne* and the whydah *Vidua regia*, there is indirect evidence that tail length is correlated with male condition. Although this suggests a good genes effect, we still need to know whether good condition is heritable and passed on to offspring. Even more suggestive, female peahens *Pavo cristatus* mating with males possessing more eye spots on their tails produce offspring with higher growth rates (Petrie 1994). Since larger tails require energy to produce and impair locomotion, their size may be a condition-dependent indicator. The condition of long tail feathers may also be a revealing indicator of fighting ability or parasite load. Interestingly, in the Australian *Cisticola exilis*, males moult to a shorter tail before the breeding season (M. Andersson 1994). The functional significance of this counterintuitive signal awaits discovery.

The symmetry of tail feathers and other bilateral ornaments may also be a sensitive index of male quality. The degree of symmetry is argued to be indicative of a harmonious genome, since random deviations from symmetry reflect how accurately the genome can maintain developmental homeostasis in the face of environmental variation and stress. This concept is called **fluctuating asymmetry** (Møller et al. 1993). Møller's (1990a, 1992) extensive study of the long forked tail of barn swallows (*Hirundo rustica*) provides some of the

Figure 23.7 Evidence for good genes indicator traits. Four types of evidence must be collected to make a convincing case for a good genes signal trait. These have all been demonstrated for tail length in barn swallows *Hirundo rustica* (A). (B) Females must show a mate choice preference for the trait. Male barn swallows with experimentally lengthened tails pair up faster than males with experimentally shortened tails. (C) The trait must be costly. Barn swallows are aerial foragers; artificial elongation of the tail increases drag (Figure 17.4) and reduces foraging efficiency (prey size) and survival. (D) Trait magnitude must positively reflect some other component of fitness or survival. Males with longer tails possess fewer blood-sucking mite parasites than shorter-tailed males, indicating that they are more parasite-resistant. Manipulation of parasite loads affects tail growth, i.e., reducing parasites leads to longer tails. (E) The fitness component has a genetic basis and is inherited by the male's offspring. The nestlings of males with longer tails (and fewer parasites) possess fewer parasites themselves, even when raised in foster nests. (A by Hugh Clark/Frank Lane Picture Agency; B–E after Møller 1989, 1990b, et al. 1995, © Macmillan Magazines.)

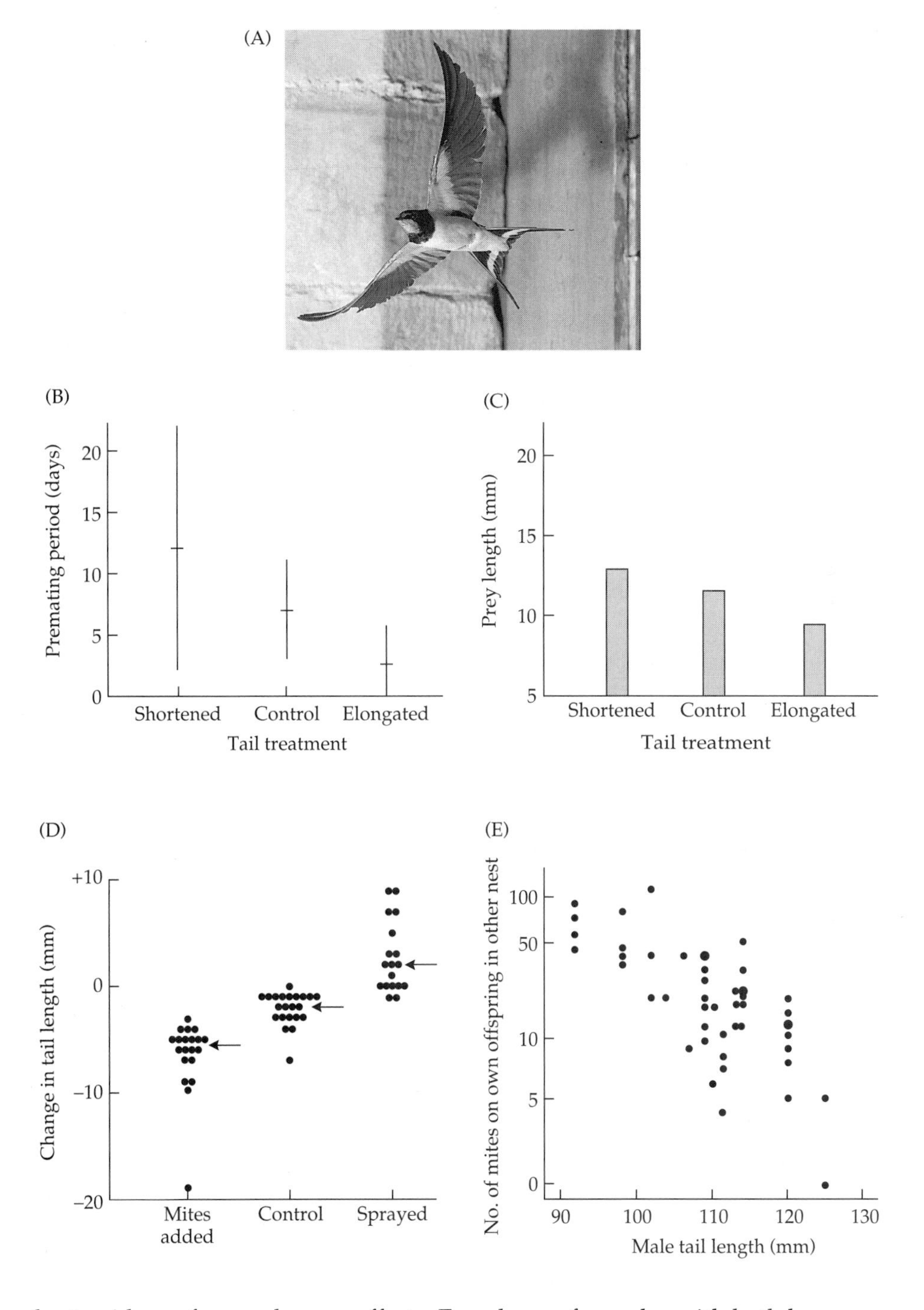

best evidence for good genes effects. Females prefer males with both longer and more symmetrical tails and benefit by producing more parasite-resistant offspring (Figure 23.7). The peacock is another bird in which females prefer more symmetrical as well as longer tails (Petrie et al. 1991). A growing body of examples in a variety of species, including humans, show that symmetry is a preferred trait (Watson and Thornhill 1994).

A final sexually selected anatomical trait that has been recently examined for the relative roles of Fisherian, direct, and indirect selective pressures is the elongated eye stalks of some fly species (Burkhardt and de la Motte 1985, 1988; Wilkinson and Reillo 1994). Eye stalks have evolved independently several times in Diptera. In *Cyrtodiopsis* species from Malaysia (Figure 23.8), the flies forage solitarily during the day but aggregate in the evening on root hairs under banks along streams. Males defend a root hair against other males with head-to-head confrontations that are won by the longer-eyed individual. Females prefer to roost and mate with longer-eyed males. Eye stalk length is heritable and positively correlated with male body length and fighting ability. This correlation initially seemed to be an example of female choice for a trait that determined male dominance. Selection experiments verified that the female preference was genetically correlated with the magnitude of the male trait, in support of all of the coevolutionary models. Females do not receive any direct benefits from mating with longer-eyed males such as nutrients from a spermatophore, improved fecundity, or improved survival. There are no indirect immunity or survival benefits for offspring, and the larval development time of the offspring of longer-eyed males is actually longer, a distinct fitness disadvantage. Recent evidence

(A)

(B)

Figure 23.8 Sexual dimorphism in stalk-eyed flies. (A) A male stalk-eyed fly *Cyrtodiopsis dalmanni* defending a root hair. The male is third from top, the other short eye-stalked individuals are all females. (B) Two males confronting each other. They use eye span to assess relative body size. Males fight with their forelegs, not with their heads or eyes. (Photos courtesy of Jerry Wilkinson.)

suggests that longer-eyed males possess a sex-ratio distorter gene that produces more sons at the expense of daughters. These sons are invariably longer-eyed. The distorter gene surely decreases the fitness of genes on the female's sex chromosomes, but the higher reproductive success of these sons must enhance the spread of the female's autosomal genes. That females prefer such males suggests that choice is an autosomal trait, but this remains to be investigated. Genetic exploitation of one sex by the other has been found in a number of taxa (Rice 1996). This example points out the complex evolutionary games that can operate concurrently in sexually selected systems.

Visual Displays

Visual movement displays are likely to be energetically costly to perform on a repeated basis. In nonterritorial, nonpaternal care species such as lekking birds and mammals in which direct benefits to females can be ruled out, display vigor may be an honest condition-dependent indicator of endurance or physiological fitness. If low-quality males pay a greater cost for a given level of display, the conditions are set for honest signaling. Sage grouse (*Centrocercus urophasianus*) appear to fit this model well (Figure 23.9). In *Drosophila* fruitflies, males vigorously court females with visual, auditory, and tactile displays and females frequently reject males. When females are allowed to choose a mate among several males, their offspring survive better than when they are randomly assigned a mate (Taylor et al. 1987). However, the precise source of the information females are obtaining from male display is not yet clear. Female bicolored damselfish (*Stegastes partitus*) favor parental males with high courtship display rates and thereby gain direct benefits (see Figure 20.4). Vigorous display is indicative of high energy reserve and such males are more likely to guard rather than eat their eggs (Knapp and Kovach 1991).

Males of some insect and avian species offer prey items as nuptial gifts during courtship (Thornhill and Alcock 1983). Females may base their choice of mates on the size, quality, or provisioning rate of such items. Egg-producing females clearly benefit from receiving food, and if provisioning rate is correlated with offspring feeding rate in paternal care species, the behavior is clearly an indicator with direct benefits. Furthermore, if the cost to the male increases with gift magnitude, it could be a indicator of heritable fitness. In the marsh hawk (*Circus cyaneus*), courtship provisioning is a good indicator of a male's nestling provisioning ability, but males can use this signal deceptively to attract females into disadvantageous polygynous matings (R. Simmons 1988).

External Structures

Bowerbirds and a few other species construct elaborate bowers to attract females (Borgia 1985a,b; Diamond 1991). The bowers are decorated with colored flowers, fruit, shells, and butterfly wings. Those species with the most elaborate bowers are themselves the most drab in body coloration, suggesting a

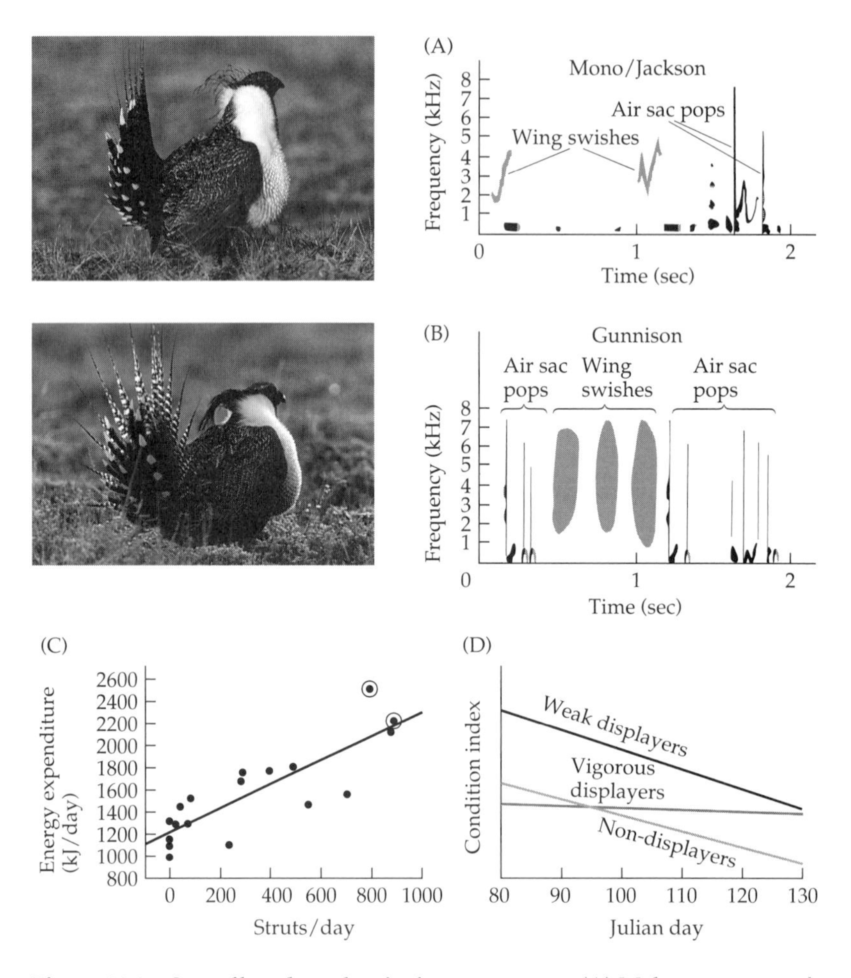

Figure 23.9 Sexually selected traits in sage grouse. (A) Male sage grouse from the main contiguous Mono-to-Jackson populations possess several sexually dimorphic plumage and behavioral traits, including the elongated tail, head plumes, esophageal sacs for production of the pop sounds, and white chest patch with stiff feathers for wing swish sounds. (B) Males of the isolated Gunnison sage grouse have longer head plumes, more white tail spots, an exaggerated tail quiver during the strut display, and give more pops during the display. These traits evolve quickly in response to female choice, as predicted for arbitrary Fisherian traits. (C) Energetic expenditure increases with increasing time spent displaying. Females prefer males that perform the strut display at high rates. Successful males (double circles) have twice the daily energetic expenditure as nondisplaying males. (D) Vigorous displayers are very lean but do not lose weight during the season, whereas less vigorous displayers and nondisplayers lose more weight. The vigor of a male's display is positively correlated with the distance he travels from the lek to forage. Display vigor may therefore be an indicator of endurance and foraging skill. (After Vehrencamp et al. 1989; Young et al. 1994; photos courtesy of Marc Dantzker.)

gradual evolutionary shift from body ornamentation to bower ornamentation (Figure 23.10). Once this shift has occurred, a new element is added to intrasexual competition—males can steal ornaments from each other (Pruett-Jones and Pruett-Jones 1994). Bower quality is therefore an indicator of male age, experience, and dominance. Bowers may be examples of arbitrary traits arising from sensory bias and then subjected to the Fisherian runaway process. It is not known whether females benefit only from the success of their sons or whether they produce more viable offspring in general from their choice.

Auditory Signals

The auditory signals of male anurans, birds, insects, mammals, and fish are obvious targets of female choice. In all of these groups, there are examples in which females prefer males with greater calling rates, sound intensity, and/or

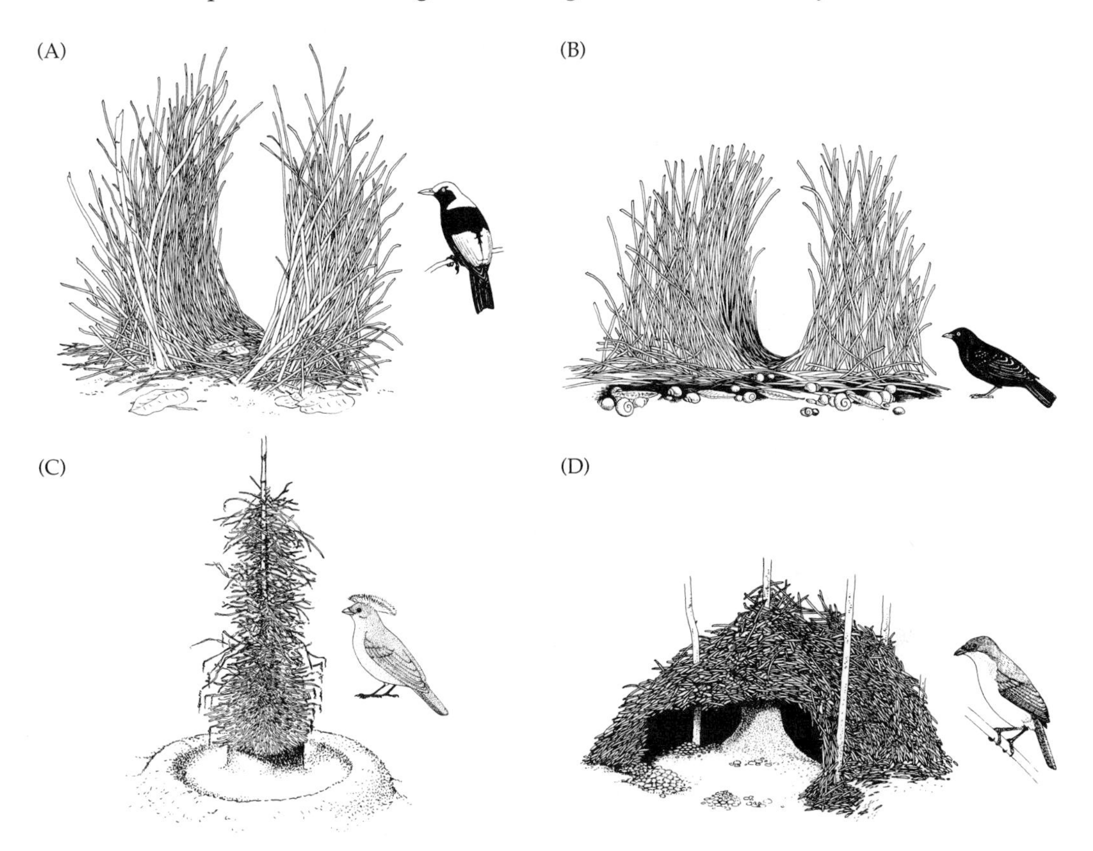

Figure 23.10 Bowerbirds. During the course of evolution in bowerbirds, the focus of female choice has been shifted from male plumage ornamentation to bower elaboration. Females are attracted to the bower, and copulation takes place there. (A) The Australian regent bowerbird (*Sericulus chrysocephalus*) sports brilliant yellow and black plumage but builds a very simple undecorated bower. (B) The satin bowerbird (*Ptilonorhynchus violaceus*) has shiny black plumage and a bright blue iris; the bower is still structurally simple but it is decorated with colorful petals, shells, and fruit, especially blue objects. (C) MacGregor's bowerbird (*Amblyornis macgregoriae*) builds a maypole bower containing a central sapling piled with twigs and surrounded by a circular raised court; colorful objects are hung on the tips of the twigs. The bird is uniform brown with a large golden crown. (D) The most elaborate bower is the covered maypole structure of the very plain brown bowerbird (*Amblyornis inornatus*), which is decorated with butterfly wings, beetle elytra, fruit, leaves, shells, flowers, and fruit. (After Cooper and Forshaw 1977; Borgia 1986.)

call duration (M. Andersson 1994). Ryan and Keddy-Hector (1992) point out that these features increase the stimulation value of the signal, so the preference may have arisen from the sensory bias of female receivers and the traits could be arbitrary. However, these call characteristics are also energetically expensive, and if sustained calling capacity is correlated with overall physiological fitness, the signals may be condition-dependent indicators. In fact, females seem to prefer precisely those call characteristics that are the most energetically expensive. For example, female gray tree frogs (*Hyla versicolor*) prefer long duration calls repeated at a slow rate over a short-call/high-rate stimulus with the same acoustic "on" time (Gerhardt et al. 1996). Field and laboratory studies show that production of long–duration calls is more costly energetically than calling at high rates (Wells and Taigen 1986; Wells et al. 1995). In a related species, *Hyla gratiosa*, males with higher chorus attendance not only had a higher mating rate, but also maintained their body condition better during the breeding season (Murphy 1994). Such evidence for differential costs suggests that display vigor is an honest indicator of male quality. A few studies have shown that costly call characteristics are correlated with age, size, dominance, or parasite load and provide females with good genes benefits such as faster growing, more viable, or parasite-resistant young (Houtman 1992; L. Simmons 1987). In some species, call rate is a good index of direct benefits females can expect to receive, such as large spermatophores or resource-rich territories (Gwynne 1982; Reid 1987; Alatalo et al. 1990).

Song features such as frequency, mimicry quality, note variability, or repertoire size are not correlated with energetic costs and are more likely to operate as revealing indicators or Fisherian traits. Female anurans use the negative correlation between male body size and call frequency to select ideally sized mates. Large body size in several species results in higher fertilization rates, so females appear to obtain direct benefits from choosing males with low-frequency calls. Figure 23.11 illustrates a fascinating case study of a frog species in which females use call frequency to select optimally sized males. There is some evidence in *Hyla crucifer* that offspring sired by larger males grow faster and/or survive better than offspring sired by smaller males (Woodward et al. 1988). In songbirds, females of many species have been shown to prefer males with larger repertoires (Figure 23.12A) (Searcy and Yasukawa 1996). The most commonly documented correlate of repertoire size is male age. This arises in some species as a consequence of continued song learning throughout life but in other cases is due to attrition of smaller-repertoire individuals. Older males in many of these studies were found to possess higher-quality territories. Thus females obtain direct benefits from the male's territory, but it is not always clear whether females use repertoire size as a revealing indicator of age and territory quality or whether they evaluate territory quality directly and as a consequence tend to mate with older males. A recent study on the great reed warbler (*Acrocephalus arundinaceus*) has demonstrated that in addition to direct benefits, females obtain good genes benefits by selecting males with larger repertoires (Hasselquist et al. 1996). Not only fledgling success, but also offspring survival and recruitment rate the follow-

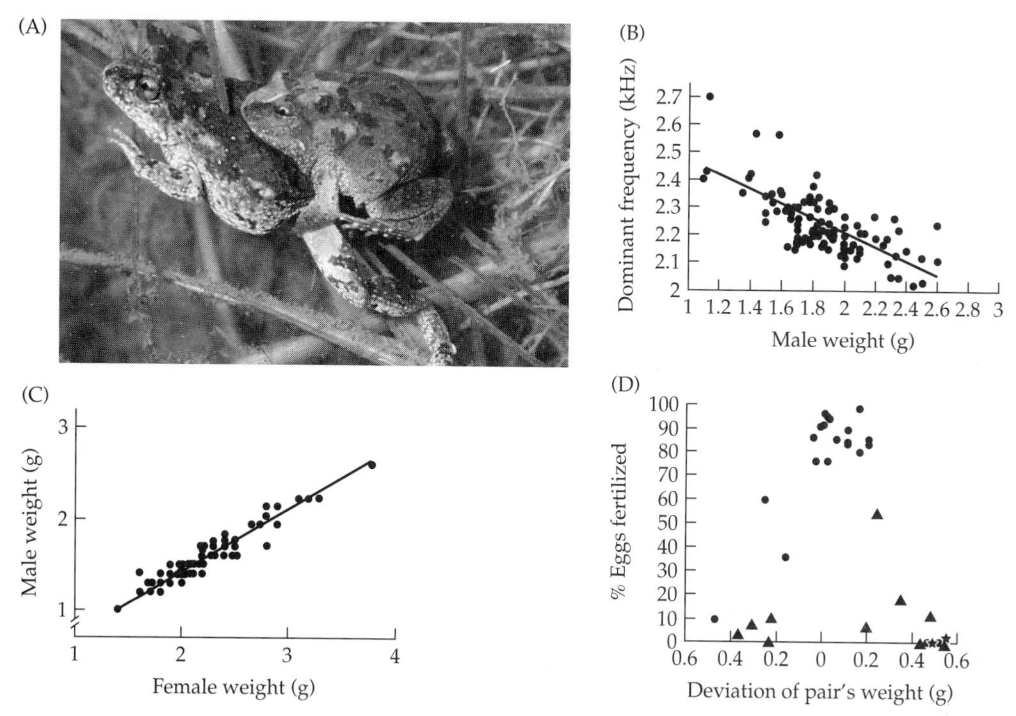

Figure 23.11 Benefits of female preference for call frequency. In the Australian frog *Uperoleria rugosa*, females use call frequency to select males that are 70% of their own weight, an optimal size relationship that maximizes their direct fitness. (A) A female laying eggs with a smaller male amplexed on her back. Each egg is fertilized and attached to vegetation singly. (B) Plot of the strong negative relationship between body weight and dominant frequency of a male's call. (C) The very tight relationship between female weight and male weight of amplexing pairs collected in the field. Males average 70% of female weight. (D) Plot of the deviation of the male's weight from the female's preferred weight versus percent of eggs fertilized in pairs forced to amplex. When the male is too small, he is unable to fertilize all of the female's eggs and females may bring a halt to egg-laying if the male is very small. When the male is larger than ideal, females may lay no eggs and in some cases are drowned. • = female laid whole clutch; ▲= female laid few or no eggs; ★ = female drowned. (After Robertson 1986a,b, 1990; photo courtesy of Jeremy Robertson.)

ing spring, increase for males with larger repertoires (Figure 23.12B). Females even sought the genetic benefits without the direct benefits. All documented cases of extra pair fertilizations were attributed to neighboring males with larger repertoires than the female's social mate. Repertoire size does increase with age in this species, so the presumption is that males who have demonstrated their ability to survive produce offspring that also survive well.

Olfactory Signals

Females in some mammals and insects use odors to select mates. In most cases, the preferred odor is correlated with male dominance. As with visual badges of dominance, honesty may be maintained by intrasexual interactions and social costs. However, in some species females preferred the odor of

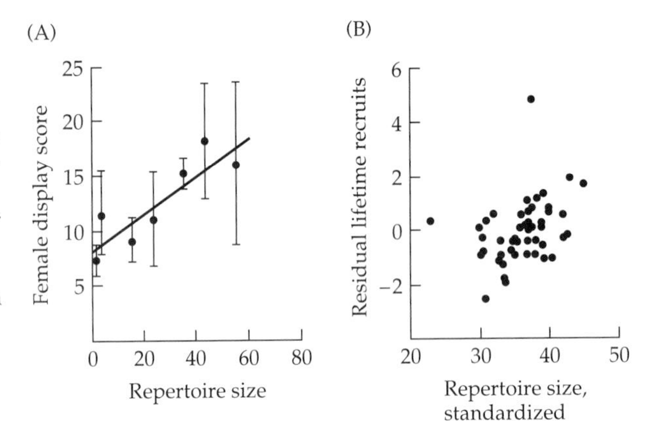

Figure 23.12 Preference for large repertoire size in birds. (A) In laboratory experiments on hormonally primed female sedge warblers (*Acrocephalus schoenobaenus*), birds gave a stronger receptive display upon hearing playback of a larger repertoire of songtypes. In the field, males with larger song repertoires pair at an earlier date than males with smaller repertoires. (B) Positive relationship between repertoire size of male great reed warblers (*Acrocephalus arundinaceus*) and the number of their offspring recruited into the breeding population the next spring. Offspring sired by males with nonmate females on neighboring territories are included. (A from Catchpole 1980, Catchpole et al. 1984; B from Hasselquist et al. 1996; A and B © Macmillan Magazines.)

males that would later become dominant, so they may be detecting some underlying quality in males (Huck et al. 1981; Moore 1988). The male moth *Utethesia ornatrix* provides direct benefits to females in the form of protective alkaloids in the seminal fluid that females use to help defend their eggs. The attraction pheromone of the male moth is derived from the same alkaloids and provides females with an honest signal of the quality of his nuptial gift (Dussourd et al. 1988, 1991).

Multiple Traits

Males of some species have several extravagant sexually selected traits, whereas in other species there is only one dimorphic trait. Several hypotheses have been suggested for the evolution of multiple traits (Møller and Pomiankowski 1993). (1) The multiple message hypothesis argues that each trait signals different properties of the quality of a male such as foraging ability, resistance to parasites, and strength. (2) The redundant signal hypothesis is based on the assumption that each signal only provides a partial measure of overall quality, but that all traits are correlated with quality and must be evaluated together to obtain an accurate estimation. (3) The unreliable signal hypothesis suggests that many ornaments of species with multiple sex traits do not currently signal male quality because each has in turn undergone Fisherian runaway selection.

Expanded versions of both the Fisherian and the indicator models have examined the conditions under which multiple traits can be stable (Pomiankowski and Iwasa 1993; Iwasa and Pomiankowski 1994; Schulter and Price 1993). Two Fisherian traits can coexist even if females must expend more effort to evaluate them. Females will give greater weight to the trait that provides the greater Fisherian benefit (i.e., more attractive male offspring). Two indicator traits can only coexist when the additional evaluation cost to female receivers is negligible. Otherwise, multiple traits are unstable and revert to a single locally stable trait. Once a preference for a single indicator trait has evolved, another cannot invade because a single, costly (to the male) and therefore reliable, trait is better than two less-costly and less-reliable traits. A Fisherian and an indicator trait can coexist, even if female evaluation cost is high, but females will place more weight on the indicator trait. Additional Fisherian traits could invade, but not another indicator trait. Finally, in a system with many Fisherian traits, an indicator trait may not be able to invade if female evaluation cost is high.

The evidence, primarily from bird feather ornaments, generally supports the third hypothesis (Table 23.4). Multiple ornaments are found primarily in polygynous and lekking species in which female evaluation cost is low due to clustering of the males and male trait costs are low because males don't perform parental care. Traits in these species are not condition-dependent, as judged by the lack of any correlation between the size of the trait and its fluctuating asymmetry. Therefore, multiple traits tend to be Fisherian traits as the models suggest, and the Fisherian process is facilitated by the polygynous mating system. Single-ornament species, on the other hand, tend to be monogamous. The trait is almost always condition-dependent, as judged by the presence of a strong negative correlation between the size of the trait and its fluctuating asymmetry. Thus the traits are likely to be indicators. It has been suggested that the cost of evaluating multiple indicator traits is high because males are spread out, so only a single trait is stable as predicted.

The sage grouse is an excellent example of a lekking species with multiple traits (Figure 23.9). Males have three feather ornaments: head plumes, an elongated tail with white spots, and stiff white chest feathers, in addition to inflatable esophageal sacs. These traits do not vary much among males and are not currently associated with female mating preferences, so they probably arose as Fisherian traits. Mate preference is based instead on the energetic effort males put into their strut display. This effort affects the loudness of the two sac pops,

Table 23.4 Comparison of sexually dimorphic feather ornaments in birds with one versus several ornaments

	Single ornament	Multiple ornaments
No. of monogamous families	9	1
No. of polygynous and lekking families	5	9
Coefficient of variation in ornament size among males	8.49	12.36
Ornament asymmetry relative to body size	0.031	0.032
Correlation coefficient between ornament asymmetry and ornament size	–0.26	0.01

Source: Møller and Pomiankowski 1993.

Note: Single ornament species tend to be monogamous, and males with larger ornaments have significantly lower bilateral asymmetry.

the frequency of the intervening whistle, the rate of strutting, and the total time spent strutting on the lek each day. These associated components of male vigor are a costly (to the male) but reliable indicator to the female of male condition and foraging behavior. In the Gunnison sage grouse, all three of the plumage traits are conspicuously different (longer, fuller, head plumes; more white on the tail; reduced chest feather roughness), the strut display consists of nine pops and no elaboration of the whistle, and a vigorous terminal tail shake has been added (Young et al. 1994). It appears likely that females in this population are using the number of pops and tail shaking as indicators of vigor, and the head plumes and white tail spots are used as amplifiers.

In summary, there is good evidence for indicator traits that provide direct benefits to females and growing evidence for genetic viability advantages to offspring (often only sons) (Johnstone 1995). Furthermore, there are many examples of female preference in which no direct benefit is obtained, but it is not always clear whether this is the result of a good genes effect or a Fisherian process. Published studies tend to report only positive results, so there is no clear example of a preferred trait that lacks both direct benefits and good genes effects. Many types of signals appear to have the expected characteristics of Fisherian traits, but few studies have examined the effects of mate choice on the attractiveness of sons.

POST-COPULATION SIGNALS

Conspicuous signals that occur during or immediately after copulation have been described in a few species. Such signals are usually vocal and may be given by the male, female, or both sexes. In most cases it has proved difficult to determine the function of the signal and the intended receiver. Copulatory vocalizations are unlikely to be an incidental effect of mating, because the displays are highly structured and individually distinctive (Hauser 1993b). Several adaptive hypotheses have been proposed.

One hypothesis for female copulation signals is synchronization of orgasm with the male (Hamilton and Arrowood 1978). This might explain the occurrence of female vocalizations during copulation in gibbons and humans, both monogamous species in which there is typically no other possible intended receiver in the vicinity of the copulating pair besides the male. All other examples, however, involve socially mating species where the intended receiver could be external to the pair. Female signals could be intended for other females and serve to increase their dominance status in the group. Male baboons provide subsequent support to females with whom they have mated, and vocal copulation signals by subordinate females could advertise the likelihood of a supporter (O'Connell and Cowlishaw 1994). Female signals could also serve to recruit additional male mates and incite competition among them as a way of obtaining dominant fathers for their offspring (Cox and LeBoeuf 1977; Montgomerie and Thornhill 1989). In elephant seals, female calls attract the dominant harem male and result in interrupted copulation attempts by subordinate males. In baboons and chimpanzees, however, females are more likely to call after a copulation with an adult, relatively high-ranking male; the significance of this behavior is not yet clear (Hauser 1989; O'Connell and Cowlishaw 1994; Hanson 1996; Henzi 1996). The female hammerheaded bat (*Hypsignathus monstrosus*) gives a vocalization that is similar to the sound she makes when grabbed by a human (Bradbury 1977). However, in this lek-breeding species, females might benefit by advertising their choice of mate to other females.

Male copulation signals could transmit information about the male's mating success to other females. Hauser (1993b) suggests that male post-copulation calls in rhesus monkeys are honest indicators of male quality. Males that give calls are more likely to be attacked by other males, but they obtain more matings subsequently compared to males that don't call. Alternatively, the intended recipient of male copulation signals could be other males. Post-copulatory male rats repeatedly emit long ultrasonic whistles (22 kHz) similar to vocalizations given in alarm and defensive threat contexts. During this post-ejaculatory period the male appears lethargic and inactive, but he will aggressively attack another male that tries to approach the mated female. The signal seems to indicate mate-guarding intentions (Barfield and Thomas 1986). Male ducks perform a brief visual and/or vocal display immediately following copulation. It is most elaborate in the mallard-like species in which males engage in social courtship and forced mating with extra-pair females (Johnsgard 1965). The function and intended receiver, however, are unclear.

SEX-ROLE REVERSAL

No discussion of courtship would be complete without mentioning the handful of species that exhibit sex-role reversal. These species are the exceptions that prove the rule about the basic gametic investment differences responsible for male-female conflict. Sex-role reversal occurs when male gametic investment is high and approaches or exceeds that of females. Male gametic investment increases in one of two different ways: (1) males assume parental care duty, or (2) sperm production costs increase because a significant nutritional supplement for the zygotes is added. If these male strategies cause the male gametic investment period to be longer than the female gametic investment period, then the operational sex ratio of receptive adults is biased in favor of females. This bias can lead to the cascade of consequences we have been discussing, but with the sexes reversed. Female reproduction is limited by access to males, so they become the more competitive sex. Females may evolve secondary sexual characteristics designed to fight off rival females and attract male mates. Females become the persuasive, persistent sex, while males become the discriminating, choosy, and rejecting sex because the costs of a mistake are greater for them. In short, one or more aspects of the usual sexual roles are reversed (Ridley 1978; Gwynne 1991).

The question of which sex should perform the parental care has been modeled as a 2 × 2 discrete asymmetric contest by Maynard Smith (1977). In this game, each sex has two options: to guard the offspring or to desert them. Deserters are free to search for additional mates, and female deserters can also lay more eggs. There are four main ESSs that not only establish the options played but also the total parental investment in offspring and the mating system: (1) both sexes guard (yielding monogamy), (2) both sexes desert (promiscuity), (3) the female guards while the male deserts (polygyny), and (4) the male guards while the female deserts. In this last case, the mating system depends on whether males can simultaneously guard the eggs of one versus several females (Vehrencamp and Bradbury 1984). If several female's eggs can be guarded, the operational sex ratio is still biased toward males, and males will compete for females. Males are likely to be the courting and more ornamented sex, and only the parental care role is reversed. This results in a **polygynandrous** mating system (males are simultaneously polygynous, females are sequentially polyandrous). If only one female's clutch can be guarded and the period of parental care is longer than the time it takes the female to produce a new clutch, then the operational sex ratio will be biased toward females, and courtship roles as well as parental care roles will be reversed. This results in a **polyandrous** mating system.

Examples of species with male parental care and normal courtship roles include most fish with uniparental care such as sticklebacks and many cichlids, ratite birds such as rheas and ostriches, and Belostomatid waterbugs (M. Andersson 1994). Males defend a nesting site to which they attract several gravid females by using bright colors, visual displays, auditory signals, or in the case of the waterbugs, low-frequency water-surface waves produced with a push-up display (Kraus 1989a). Females are the mobile sex, and they are highly selective of males that demonstrate good parental care potential. Males guard the eggs alone while females move on to produce another clutch for a different male. Examples of species with completely reversed sexual roles include most seahorses and pipefish, cardinal fish and other male mouthbrooders, dendrobatid frogs, and several birds such as the spotted sandpiper, jacana, and phalarope (Figure 23.13). Males can only care for the eggs of one female and they show a strong preference for larger or more fecund females. Females are usually more brightly colored and aggressive. In the avian species, females defend large resource territories within which several males reside and females lay a series of clutches for each male. These birds can only incubate four eggs at a time. Pipefish and seahorse males carry the eggs on their ventral side or in a pouch, dendrobatid frogs carry tadpoles on their backs, and the cardinal fish mouth-broods the eggs. These carrying strategies clearly limit the number of eggs or offspring a parent can care for and create a constraint on males that shifts the operational sex ratio toward females.

High sperm production costs are found in a few orthopteran species (Gwynne 1990). Males transfer sperm in a large, nutritious spermatophore to the female during copulation. The female eats the packet and can produce larger clutches and eggs as a result. The degree to which the operational sex ratio is biased toward males or females depends on the availability of food for

Figure 23.13 Sex-role reversed species. (A) A male dendrobatid frog, *Colostethus trinitatus*, carrying tadpoles. Females are aggressively territorial and display their conspicuous yellow throats from rocky perches. Males turn black when approaching and courting females with vocalizations, but noncourting and parental males are cryptic brown. (B) The female waterbug (*Abedus indentatus*) lays her eggs on the back of the male after copulating with him; he carries the eggs until hatching. Males are more stationary than females and produce the long-distance mate-attraction signal, a body-pumping display that sends ripples across the water's surface. (C) A breeding pair of Northern phalaropes (*Phalaropus lobatus*). Although not very apparent in this black and white photo, the female (left) is larger and has a deeper red neck and more contrasting white patches. Phalaropes are completely role-reversed; the female actively courts one or a series of males and the male performs all incubation and parental care. (D) The male mormon cricket (*Anabrus simplex*) produces a large spermatophore that increases offspring survival. This photo shows a female eating a spermatophore. When food is scarce, males are the limiting sex and females compete for access to them. (A from Wells 1980, photo courtesy of K. D. Wells; B from Smith 1979, Kraus 1989a, photo courtesy of Bob Smith; C courtesy of John Reynolds; D from Gwynne 1981, photo courtesy of Darryl Gwynne.)

males (Gwynne and Simmons 1990). When food is abundant, the spermatophore can be produced relatively quickly and males compete for females. When food abundance is low, spermatophore production time is slow and females are also particularly hungry. In this context, females compete and fight for males, and males reject small females in favor of large ones. In a similar vein, the direction of mate competition in the male-brooding fish and waterbugs changes as a function of water temperature. When water temperature is high, the time required for males to brood is reduced and the operational sex ratio is biased in favor of males, whereas when water temperature is low, the sex ratio favors females (Kraus 1989b; Ahnesjö 1995; Kvernemo 1996).

In summary, complete role reversal is very rare because the conditions required for male gametic investment to completely counterbalance the basic anisogamy asymmetry are difficult to attain (Elgar 1996).

SUMMARY

1. Conflict between males and females arises from **anisogamy,** the fundamental gamete size and mobility differences that define the sexes. Male gametes (**sperm**) are small, mobile, inexpensive, and numerous, whereas female gametes (**ova**) are large, immobile, expensive, and rare. Since males have the potential to fertilize many females, their fitness is limited by their access to females and they evolve **mating strategies** to maximize this access. Females are limited by the resources they need to produce eggs and by the quality of their mates.
2. **Sexual selection** is a form of natural selection that favors traits in the more competitive sex that increase their mating success. **Intrasexual** selection arises from aggressive interactions among members of the competitive sex to control the opposite sex, and **intersexual** selection arises from mate choice preferences for attractive traits in the competitive sex. The greater the disparity in the gametic investment of the two sexes, the more strongly skewed the **operational sex ratio** becomes, and the stronger the forces of sexual selection.
3. In order for the sexes to meet for mating, they must search for and attract each other. The **mobility game** examines the conflict over which sex should search. The ESS is for one sex to be maximally mobile and the other to be stationary, but both the male search and female search strategies are locally stable so either sex can win. The stationary sex is likely to produce a long-distance mate-attraction signal. The male mating strategy is an important determinant of the mobility/signaling pattern. Females search and males signal in resource defense and self-advertisement mating systems, whereas males search and females signal in female defense systems.
4. Since females invest much more in each reproductive episode, they are more concerned about making an accurate assessment of the species and quality of a potential mate because mistakes are costly. Males are playing a numbers game and tend to mate indiscriminately with as many females as they can encounter. The **hybrid mating game** establishes the zone of conflict in a meeting between the sexes in which the male prefers to mate but the female prefers not to mate. If females have control over mate

choice or if they are the senders, strong species specificity will be favored in the attraction signals.

5. **Courtship** is characterized by one of three patterns: (a) female passivity and little or no male courtship display, (b) choosy females and elaborate, persuasive male courtship display, or (c) forceful male control over females and resistance by females. The game context of courtship is a contest of persistence between male and female and can be modeled as a sexual war of attrition or sexual arms race. If the female wins, courtship consists of elaborate display; if the male wins, courtship is forceful or brief. Factors that affect which sex will win include the outcome of the mobility game and the male mating strategy, constraints imposed by mode of fertilization and egg deposition site, and adaptations such as body size, weapons, and armor.

6. When females are free to make mate-choice decisions, they attempt to select preferred males who can directly or indirectly improve their fitness. Strong female preferences lead to the elaboration of male traits that attract females. Three models of mate choice have been proposed that lead to different trait characteristics. The **Fisherian runaway model** leads to the evolution of arbitrary male traits that provide no information about intrinsic male quality but enable females to produce sons that are in turn chosen by females. **Good genes indicator models** propose that females select high-quality mates that enhance the viability of their offspring. **Direct benefits indicator models** lead to traits in males that are correlated with more immediate benefits females can obtain. There is good evidence for the direct benefits model in species with male investment. Firm evidence for distinguishing Fisherian and good genes models is lacking, but some cases strongly suggest a good genes effect.

7. Species with **multiple sexually selected traits** are typically polygynous or lekking species, and their traits appear to be mostly Fisherian. Species with a single trait tend to be monogamous, and variation in the size of the trait is an indicator of aspects of male condition, performance, or genetic quality.

8. **Copulation calls** and visual displays are found in many primates, ducks, and a few other species. Females may give these signals to incite competition among males in polygynous societies or to advertise the subsequent support they may receive from their consorts to other females. Males may give these signals to advertise their copulation success to other females.

9. **Sex-role reversal** results from the evolution of male parental investment, but not all paternal care species show complete reversal of courtship roles. If males can care for the offspring of several females simultaneously, the operational sex ratio is still skewed in favor of males, males are the competitive sex, and they will perform the aggressive or persuasive courtship behavior. If males can only care for the offspring of one female, then males become a limiting resource for females. Females will compete for males and develop aggressive behaviors, ornaments, and mate-attraction displays.

FURTHER READING

The February 1996 issue of *TREE* contains several thought-provoking reviews of the evolution of sex. M. Andersson (1994) gives a comprehensive review of the theory and consequences of sexual selection. Parker (1979, 1982, 1983, 1984b, and other co-authored papers) is the driving force behind all of the sexual conflict game research.

Chapter 24

Social Integration

THERE ARE A VARIETY OF SOCIAL CONTEXTS in which two or more individuals must integrate their activities to achieve a common goal. Examples include the coordination between a male and female attempting to mate and care for offspring, communication between parents and dependent offspring to assure that offspring needs are met, and integration of the members of group-living species to maintain group cohesion, coordinate movement, and organize communal activities. Synchronization is a key function of signals in such interactions. Senders are usually in the position of requiring some type of aid or cooperation. Integrative signals should transmit honest information about the sender's state, needs, or what it intends to do next, but the risk of exaggeration by senders is frequently present. Receivers are usually in the position of donors of aid. They must not only evaluate the sender's true need, but they must also be certain that the sender is the correct recipient. Aid-giving and cooperative behaviors can only evolve when the donor obtains some type of direct or indirect fitness benefit for its actions. This means that the aid must be given to an offspring, a relative, or a reciprocating partner. **Recognition** is therefore a second key component of integration signals. Here again there is a risk of deceit by cheaters who hide or disguise their true identity to receive the benefits of aid. Receivers use both signals and cues to evaluate the identity and honesty of senders. In this chapter we shall first examine the general properties of recognition and then describe the specific recognition mechanisms and signals used to synchronize activities in three social contexts: male-female coordination, parent-offspring interaction, and group cohesion. Although most of the signals described in this chapter are honest because they serve mutualistic interests, conflicts of interest still arise, senders may exaggerate or manipulate, and receivers may be subjected to incidental costs.

GENERAL PROPERTIES OF RECOGNITION

The Process of Recognition

Recognition is the discrimination and identification of a target individual or group among a field of similar nontarget individuals or groups. It implies that a receiver has an innate or learned memory of characteristics unique to the target and has found a good match to that standard (Beecher 1990). The process of recognition follows the same basic steps that we identified for the evolution of a signal in Part II (Figure 16.1): (1) The sender provides **information** about its identity with some combination of signature signals and cues. (2) The receiver **perceives** the signal against background noise. (3) The receiver **compares** the information received to a model of the target individual's signature stored in memory. (4) The receiver **decides** whether the sender is the target or not. This decision is based on the prior odds that the sender is the target gleaned from the cues, the evidence provided by the signals, and the costs and benefits of correct and erroneous assessment. (5) The receiver takes some **action** toward the sender, such as to attack if the sender is perceived as nontarget or to feed if perceived as target. Recognition will not be observed if any one of these steps fails, i.e., if there is inadequate signature information, inability to perceive signature differences, a decision not to discriminate, or lack of an appropriate or distinctive bias in behavior toward targets.

There are three important points to bear in mind throughout this chapter. The first point is that the complexity of the discrimination task varies greatly depending on the number of classes that must be distinguished. The number of classes depends in part on the identification level required by the social context, i.e., species, populations, colonies, foraging groups, kin groups, families, age classes, status classes, or individuals (see Chapter 18). The greater the number of classes that must be distinguished, the more difficult the perceptual task, the greater the memory requirements, and the more complex the signal needs to be to encode different variants. Animals have evolved several different sets of signaling, perception, and decision strategies, called recognition mechanisms, to deal with these different levels of complexity. Describing the use of these different mechanisms is one objective of this chapter. The second point is that recognition will never be perfect. As we learned in Part II, it does not pay the sender to encode perfect information, and it does not pay the receiver to extract all information. What matters is that the sender benefits more from transmitting some information than it costs to signal, and that the receiver benefits more from paying attention to the signal than from ignoring it. The third point is that the sender and receiver may not always agree on how much information should be provided. The potential for inaccurate assessment by receivers opens up the possibility of deception on the part of senders. Recognition must therefore be viewed as a game of sender signal distinctiveness and receiver discrimination accuracy.

Recognition Games

The game context of recognition can be viewed in several ways. In many respects it is analogous to the Sir Philip Sidney begging game discussed in Chapter 20. The players in this game are a sender who begs for aid either honestly, dishonestly, or not at all and a receiver who provides aid either all of the time, only in response to signals, or never. In the continuous version of the game, sender need, sender begging level, and receiver aid are all continuous variables. The solution at best is a double ESS for either honest signaling and aid giving or no signaling and no aid giving. Recall that the cost to the sender need not be very high, but the benefit to senders must increase with increasing need and an indirect benefit must accrue to the receiver. Converting this to a recognition game, the sender produces a signal that varies in its distinctiveness from the masses of possible senders and the receiver varies in its ability to distinguish the sender. The receiver would only respond to the sender if it made a positive identification so the probability of the sender (and receiver) benefiting increases as the signal becomes more distinctive. As above, the ESS is the positive coevolution of signal distinctiveness and receiver discriminability.

Signal detection theory (Chapter 14) could also be used as the basis for a recognition game (but such a model has not yet been developed). Here, the sender's strategic option would be to vary the distance d' between its signal character distribution and the distribution of all others in the population. The receiver would vary its cutpoint for acceptance or rejection, i.e., its recognition threshold. The game should be a scramble, so the payoffs for both players depend on the fraction of discriminators and distinctive senders in the population, as well as the various benefits of hits, misses, false alarms, and correct rejections. Envision a mother bat or penguin returning to the communal crèche with a load of food. All of the young are begging at the parent in the hopes of being fed, and she must find her own offspring among them. In a population of undistinguished offspring and indiscriminate parents, offspring would randomly receive food from all parents with young in the crèche. A parent that can distinguish its own young will clearly invade. Offspring that make distinctive signals obtain more food from their true parent but less food from other parents. Selection for receiver discrimination will therefore be stronger than selection for sender distinctiveness. In this particular example of parents and offspring, parents with both good discrimination ability and distinctive offspring will prevail, so these components will become linked and coevolve. As the population of senders from which target individuals must be recognized increases, signal complexity must increase to encode distinguishable variants, and the discrimination ability of receivers must also increase.

Recognition Mechanisms

The mechanisms animals use for recognition fall into four categories: spatial location, familiarity, phenotype matching, and allele matching (Waldman 1987; Hepper 1991a). These mechanisms are based upon different strategies for signal or cue production, perception, and decision-making. They vary in their accuracy, vulnerability to cheaters, and usefulness for recognition at different social levels. Specific examples will be provided in later sections of the chapter.

Spatial location is the simplest recognition mechanism. If the probability is high that only target individuals or groups will be encountered at a particular place, then the prior-odds ratio for target identification at this site is high, and location is a good cue for recognition. There is no selection on recipients to give signature signals, and the action rule is to vary agonistic, cooperative, or aid-giving behavior as a function of distance from this site. For example, an animal's nest usually contains only its own offspring, so directing all parental care to this site ensures that offspring receive the parent's attention. We have already seen (pages 727–728) how location is used for neighbor recognition in territorial systems. Location can be used as a mechanism for nondescendent kin recognition if one sex tends to be philopatric. Location can even facilitate species recognition in animals such as swarming insects that form mating aggregations at specific times and sites. Using spatial location as a cue for recognition is a very simple rule of thumb that requires a small amount of learning and spatial memory but is easily subverted by cheaters. Both conspecific and nonconspecific parasites can insert themselves or their offspring into the target location and receive the benefits of aid-giving behavior bestowed there. The best evidence for the operation of this mechanism is immediate appropriate behavior directed toward nontarget individuals placed into the target site.

Familiarity is a recognition mechanism that requires prior experience with target individuals or groups, followed by the learning and memorization of their specific characteristics. An important feature of the familiarity mechanism is that the classes among which distinctions are made are arbitrary (Grafen 1990c). As the number of classes increases, the prior odds for correctly assigning a stimulus to a target class become smaller, so there is strong selection for signature signals. Individual-level recognition is always based on learned familiarization and entails the largest number of classes and the greatest need for complex signature signals. In Figure 18.8 we illustrated some of the mechanisms for achieving individual distinctiveness in territorial signals. Recognition signals use similar mechanisms: hue variations and contrasting patterns for visual signals, chemical mixes for olfactory signals, and frequency, temporal patterning, and note shape for auditory signals. Figure 24.1

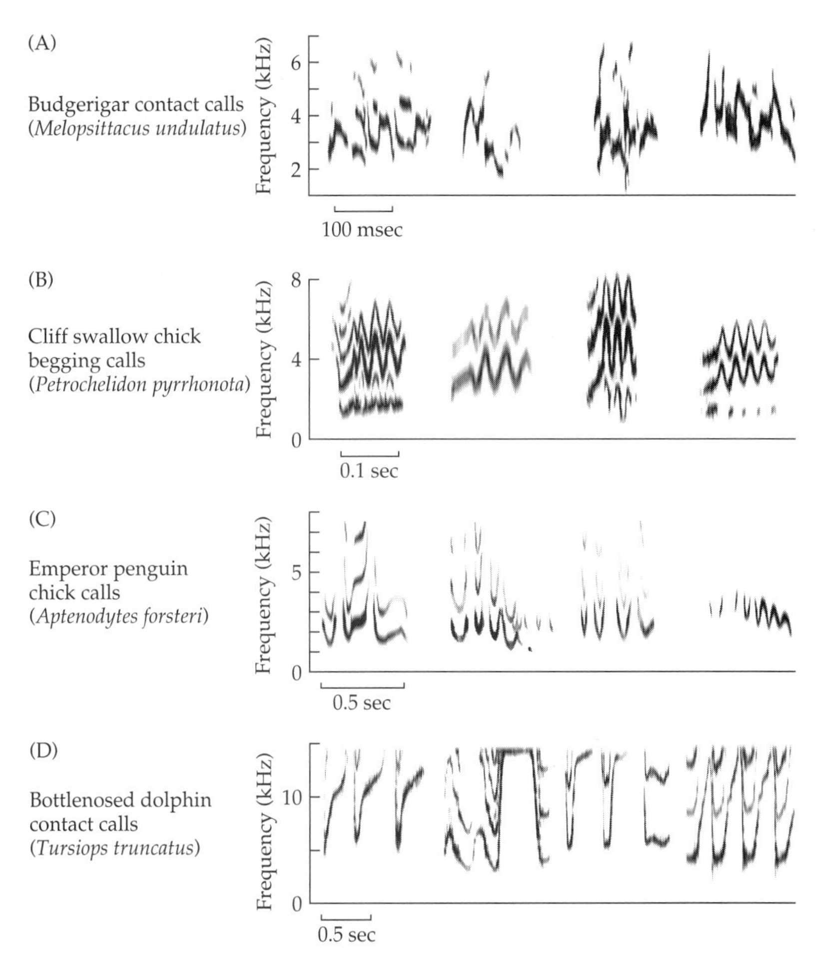

Figure 24.1 Design mechanisms for vocal signature signals. The examples for each species show the call of four different individuals. The two mechanisms commonly used to permit the encoding of individual distinctiveness are harmonic structure and strong frequency modulation. See also Figures 15.11 and 24.10. (After Ali et al. 1993; Medvin et al. 1992; Jouventin 1982; Sayigh et al. 1990.)

illustrates a remarkable convergence in the design of vocal contact signals that are short and frequently repeated to maintain individual identity.

Familiarity can also be used to identify kin, but target relatives must be unambiguously associated with receivers during the initial learning period. Group and colony recognition can be achieved with familiarity if all group members possess or acquire a common badge, such as the odor of a communally used food source, that distinguishes them from other groups Finally, familiarity may be involved in population and species recognition in birds and perhaps other vertebrates, where exposure to the songs and appearance of parents during a critical early learning period causes young to prefer similar individuals as mates later in life. Familiarity is more difficult to subvert than spatial location, but examples still exist. Memory saturation in individual recognition systems may cause receivers to sometimes misidentify target individuals, as occasional nontarget individuals by chance may have fairly similar characteristics to targets. If nontarget individuals are inserted into a family before learning has been completed, offspring and sibling recognition can be subverted. In colonial insects, parasites can sometimes gradually acquire the colony smell by interacting with peripheral colony members and may eventually be accepted into the main colony. To demonstrate that receivers are basing recognition on signature cues, it is important to remove any locational cues.

Phenotype matching is a mechanism that functions to identify genetic similarity. In contrast to the familiarity mechanism, recognition via phenotype matching is the ability to assign stimuli to classes of relatedness relative to the receiver (Grafen 1990c). When phenotype similarity is well correlated with genetic similarity, receivers can distinguish kin from nonkin in the absence of any prior experience with the target individual or group. The receiver must compare the phenotype of an encountered individual with a visual, olfactory, or auditory template acquired from a referent. The referent can be a known familiar relative or oneself. If the referent is the family, then the receiver acquires a template based on average genetically based family characteristics and the recognition system will do little more than permit the distinction between own versus foreign family members. If the referent is oneself, then the receiver may be able to distinguish fairly fine levels of relatedness such as half versus full siblings. To demonstrate phenotype matching clearly, it is important to control for locational and familiarity cues by showing that unfamiliar siblings are distinguished from unfamiliar nonrelatives or that full versus half siblings within a litter are distinguished. Phenotype matching is relatively difficult to subvert because similarity is based on a set of polygenic characters that are unlikely to be matched by an unrelated individual.

Allele matching is a very precise chemical recognition mechanism for identifying close kin. Specific alleles in senders encode both the phenotypic cues indicative of genetic similarity and the ability to recognize these cues in receivers without having any previous experience with them. In other words, the template is not learned, but innate and inexorably linked to the phenotypic cues. A single hypervariable locus with many rare alleles is usually involved. Similar hypervariable loci are used in plants to distinguish outcrossed from own pollen and in the vertebrate immune system to distinguish foreign from self cells. A high rate of unique allele creation is favored specifically because of the advantages of recognition. When used as a kin-recognition mechanism, a rule of thumb such as accepting all individuals with at least one matching allele at this locus guarantees a high degree of relatedness at all other loci. The mechanism risks occasionally excluding an otherwise closely related individual that shares no alleles at the recognition locus. Although biochemically sophisticated, this system requires no complex memory or perception and is not subvertible.

These mechanisms are not exclusive. Several different mechanisms could operate simultaneously in the same species to achieve different levels of recognition. For example, honeybee guards may use colony-specific odor cues to distinguish incoming colony foragers from noncolony foragers and a self-referent-based phenotype matching mechanism to distinguish paternal lineages within the colony for making decisions about brood care and colony fissioning (Arnold et al. 1996). A single mechanism can be used to simultaneously discriminate different levels. For example, animals can learn by familiarity which individuals are kin if there is a reliable period of kin association. Any recognition system that is based on genetically determined phenotypic cues will also permit discrimination of levels of genetic similarity because relatives are always more similar than nonrelatives. Thus a phenotype-matching species recognition mechanism will also permit some discrimination of close and distant kin (Grafen 1990), and an individual recognition system based on perception of genetically determined signature signals will also encode family membership (Beecher 1982; Medvin et al. 1992). It is therefore important to consider alternative hypotheses for recognition and to determine the primary evolutionary function of an observed discrimination ability.

MALE-FEMALE INTEGRATION

In Chapter 23 we focused on the conflicting interests of males and females over mating decisions. In this section we shall now examine some of the more integrative aspects of male-female interactions. Given the chapter themes of recognition and synchronization, we shall first examine the mechanisms of species recognition and outline the controversy surrounding the roles of species recognition and sexual selection in speciation. Then we shall consider the interactive signaling and receiving strategies used by males and females to coordinate reproduction and look at the communication aspects of long-term pair-bond maintenance in monogamous species.

Species Recognition

From the point of view of an ideal sender, species distinctiveness is achieved by making the mate-attraction signal sufficiently different in form from that of closely related sympatric species. We examined the specific features senders can vary to ensure species distinctiveness in Chapter 18. For auditory signals the features that are commonly varied for species specificity include frequency, note shape, temporal pattern, and syntax; for visual signals the varied signals include color, shape, spatial pattern, and movement pattern; and for olfactory signals the varied signals include one or a small number of highly specific organic compounds. From the point of view of an ideal receiver, a species recognition mechanism can be hard-wired into receivers because only one signal variant needs to be recognized by all members of the species. The most common method of species recognition is therefore a peripheral sensory filter or a highly tuned receptive organ that is sensitive only to the species-specific mate-attraction signal. Examples include the chemoreceptors on the antennae of male insects that respond only to conspecific female-produced contact or airborne pheromones (Linn and Roelofs 1995), the frequency-tuned external ears of anurans (Capranica 1965), and the color-sensitive eyes of some butterflies, fish, and birds (Bernard and Remmington 1991; Barlow 1992). When the reception mechanism is hard-wired into receivers, receiver tuning and signal characteristics coevolve so that the transmission of species information is as error-free as possible.

Other recognition mechanisms can also play a role in species recognition. Early learning of parental and familial characteristics affects subsequent mating preferences in many vertebrate species with parental care of offspring. In ground-nesting avian species such as ducks, geese, and galliforms with precocial young that leave the nest shortly after hatching, the chicks are programmed to follow after a large moving object and learn its characteristics. Normally the object would be the chick's parent and the learning process brings about parental recognition via familiarity. Upon reaching sexual maturity they exhibit a mating preference for individuals with similar characteristics. This type of rapid focused learning during a short critical period of development, which is irreversible with subsequent experience, is called **imprinting** (Hess 1959; see page 481–482). Birds and mammals that have been imprinted on foster species or humans often fail to recognize true conspecifics as mates (Immelmann 1972). Familiarity is too risky a mechanism for species recognition in most cases, especially for females, and hard-wired mechanisms or template-driven learning is more commonly observed (Gould and Marler 1987). Phenotype matching can be used as a species recognition mechanism with considerably less risk of heterospecific mating error. In some frog and toad tadpoles, individuals exhibit a slight tendency to school with close relatives and even prefer unfamiliar paternal half-sibs over unrelated individuals, demonstrating that they have the ability to distinguish by genetic similarity. Although initially touted as an example of kin recognition, the kin bias is more likely to be a byproduct of an olfactory phenotype-matching species recognition mechanism or a habitat recognition mechanism (Grafen 1990c; Pfennig 1990; Blaustein et al. 1993). Phenotype matching can also be used to recognize and avoid inbreeding with close relatives (Bateson 1978a; Simmons 1989; Barnard and Aldhous 1991; Waldman and Rice 1992). Finally, location and timing contribute toward mate attraction in many insects that swarm or form mating aggregations, for example, hilltopping butterflies that aggregate at the peaks of local hills, ants that swarm after the onset of the season's first heavy rain, and lekking flies that aggregate in species-specific sites (Thornhill and Alcock 1983).

There is considerable controversy surrounding the precise evolutionary processes that select for species-specific signals and receivers and surrounding the process of speciation itself (recently reviewed by Rice and Hostert 1993; M. Andersson 1994). These models are relevant to our discussion because they imply different means of avoiding species deception. The classic **geographic isolation speciation model** emphasizes the importance of geographic barriers that can bisect a species into two independently evolving populations (Mayr 1963). The two populations diverge gradually as a consequence of genetic drift and adaptation to the local environment. Reproductive isolation develops as a result of premating isolation mechanisms (such as differential timing and location of breeding or positive assortative mating) and postmating isolating mechanisms (such as zygote or hybrid inviability). If the geographic barrier should break down before reproductive isolation is complete, a process called **reinforcement** takes place in which the low fitness of hybrid offspring selects for positive assortative mating and completes the speciation process. Support for this model comes from field observations of **character displacement** (divergence of mate-attraction characters, as in Figure 23.3) and positive assortative mating between two overlapping populations in the zone of sympatry but not between members of the allopatric populations (Coyne and Orr 1989). According to this model, signal honesty is maintained by the cost of producing low-fitness hybrid offspring.

A second speciation process, the **ecological divergence model**, does not require a physical barrier but relies on a cline or disjunction in environmental conditions within the species' range (Maynard Smith 1966; Endler 1977; Felsenstein 1981). The different ecological selective regimes cause a divergence in ecological adaptation even in the presence of gene flow. Reproductive isolation develops gradually because genes for mate attraction and recognition are linked to or pleiotropically affected by the ecologically adapted genes. No reinforcement process need be invoked. This model is supported by genetic models as well as by laboratory selection experiments that find that two lines separately selected for viability in different ecological regimes are often incompatible when subsequently allowed to interact. Signal honesty is maintained by the cost to receivers of failing to recognize ecologically well-adapted mates.

A third view proposes that **sexual selection** is an important force driving the evolution of new species (West-Eberhard 1983; Lande 1981; Carson 1995). As we learned in Chapter 23, preferences for either arbitrary or quality-indicating traits can lead to rapid coevolution of both the trait and the preference for it. The mate-attraction systems of geographically isolated populations under sexual selection are likely to diverge in different directions without the necessity of either ecological adaptation or the challenge of a potentially hybridizing sympatric species, and thus they can result in premating isolation and speciation. Furthermore, sexual selection can speed up the speciation process when there is an ecological cline and increase the chances of reinforcement in a hybridization context (Lande 1982; Liou and Price 1994). Sexual competition, exaggeration, and female preferences drive mate-attraction signals to greater extremes than required for species identification, but ironically this maintains species honesty. Several good examples of interpopulational divergence in both male signal characteristics and female receiver tuning support the sexual selection view (Butlin 1995; Ryan and Wilczynski 1988; Bakker 1993). The observation that females sometimes prefer heterospecific males with exaggerated versions of their own male's traits can only be explained with sexual selection (Ryan 1990). Mounting evidence suggests that innovative adaptations in either the sender or the receiver that permit greater mate discrimination are accompanied by rapid speciation. Ryan (1986b) showed that in frogs, an increase in the complexity of the inner ear (amphibian papilla) is associated with more speciose families. He argued that the mating calls in such taxa can diverge over a wider frequency range, making speciation more likely than in taxa with simpler inner ears. Finally, Raikow (1986) showed that more speciose passerine bird taxa possess more advanced syringes. An advanced organ for sound production could facilitate more rapid song divergence among species, but the cause and effect relationship here is controversial.

Coordination of Reproduction

Males and females of most species possess very different reproductive plumbing. As we saw in Chapter 23, males are typically in a continuously receptive state with stored sperm available at all times, whereas females are not receptive until they have developed a clutch of eggs or are prepared to ovulate or gestate internal embryos. Female reproduction is physiologically a more complex process, regulated by the serial release of steroid and pituitary hormones. In most species, the female requires a species-specific stimulus from a male at some point in her reproductive cycle. Examples include the induction of ovulation with vigorous copulation in some mammals, the enhanced stimulation of oviduct development in many birds with highly versatile or vigorous male song, and the tranquilizing effect of the male pheromone in pigs, salamanders, and butterflies. These are mechanisms by which females can guarantee a conspecific and possibly also a high-quality mate. However, there is a fine line between male stimulation that benefits the female and manipulation of the female that benefits the male. Finally, males also require a species-specific signal from the female that stimulates them to mount and release sperm at the correct stage of the female's cycle. Communication of receptive stages by both parties is therefore critical.

Figure 24.2 shows a number of examples of visual spawning and copulation synchronization signals. These represent the final triggering signals and may have been preceded by other courtship and stimulatory signals. Most mating synchronization signals are visual or tactile, because olfactory signals are too slow and auditory signals too conspicuous. Synchronous gamete release is extremely important for both sexes in fish and other aquatic organisms with pelagic eggs released directly into the water. In externally fertilizing substrate spawners, males must release sperm immediately after the female releases her eggs. In many birds and mammals, males give females a signal of their intent to mount, and the female then signals acceptance by assuming a mating posture.

Classic studies of doves, rats, and other species illustrate the mutually stimulating interactions between male and female signals, internal hormones, and external stimuli in coordinating the entire reproductive cycle (Lehrman 1965). Hormones can affect behavior in three ways: (1) by influencing the development of special structures involved in the performance of a behavior,

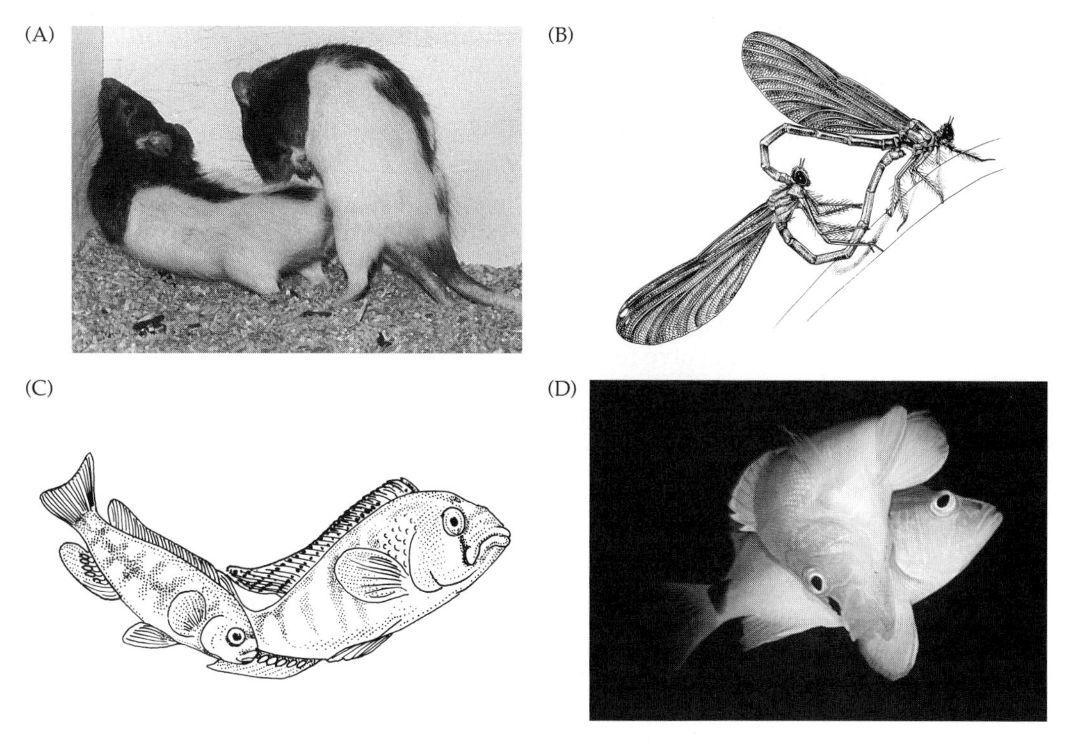

Figure 24.2 Copulation synchronization signals. (A) Lordosis in the female rat signals the male to mount. The arched-back posture tilts the female's pelvis forward and makes copulation possible. A male cannot achieve intromission if the female does not adopt this position. (B) A copulating pair of damselflies, *Calopteryx maculata*. The female (left) must twist her abdomen around so that her genitalia are over the male's penis. The sperm transferring structure is located on the underside of his abdomen. The male grasps the female by the head with graspers on the tip of his abdomen. (C) Fertilization in the mouth-brooding fish *Haplochromis* involves visual and tactile signals. The female first lays the eggs on the substrate, then proceeds to pick them up in her mouth. The spots on the anal fin of the male mimic the size and color of eggs and when the female attempts to pick them up the male releases his sperm. (D) Spawning synchronization in the hermaphroditic hamlet, *Hypoplectrus unicolor*, is achieved with auditory signals. The individual playing the male role produces a series of pulses at about 500 Hz prior to the spawn. The individual playing the female role assumes a head-down position and gives a brief sound at the moment of egg release. (E) Courtship in the roadrunner, *Geococcyx californianus*, involves extensive tail signals in this long-tailed bird. The male first grabs a prey item, approaches the female, waves his raised tail back and forth, and at intervals bows his head and fans the tail. The female signals her acceptance of mating by orienting her backside to the male and raising her tail up. She then crouches for mounting. (A courtesy of R. Barfield; B from Alcock 1993; C after McFarland 1981; D from Lobel 1992, photo courtesy of Phillip Lobel; E after Whitson 1977.)

such as male claspers, display structures, oviduct enlargement, or lactation gland development; (2) by influencing the peripheral nervous system that controls sensory input to the brain; and (3) by triggering behavioral mechanisms in the brain directly. In the presence of appropriate external stimuli, olfactory, auditory and visual displays by one sex affect hormone production in the other sex and modify this individual's behavior, which in turn influences the hormonal state and behaviors of the first sex. Figure 24.3 summarizes these interactions in the ring dove *Streptopelia risoria*, a monogamous species in which both sexes brood the eggs and young and feed them by regurgitating a milky substance from a specialized crop sac.

Signals that evolve for mating synchronization can also be exploited by males eager to mate a reluctant female or by third-party competitive males. Let us look at a second example of courtship in another group of organisms, salamanders and newts, where these additional elements have been examined. Most species breed in ponds and rivers, although adults of many species breathe air and live in adjacent moist terrestrial sites. Fertilization is internal but males have no penis. Instead, the male deposits a spermatophore packet on the substrate that the female picks up with her cloaca and stores. This creates a great challenge for the male, since the female must stay with him after he has produced a spermatophore and align her body correctly to pick it up. Courtship therefore tends to be long and energetically costly for males, and they risk the intervention of another male who sneaks his spermatophore into position at the last moment (Halliday 1977). Males use a combination of displays, courtship pheromones, and physical restraint to persuade females to cooperate. Pheromone delivery patterns vary and appear quite forceful in some species (Figure 24.4). Experiments with some of these species suggest that the pheromone stimulates or primes the female into reproduction condition faster, whereas in other cases it seems to operate more like a tranquilizer (Verrell 1988; Houck and Reagan 1990).

Many other examples of manipulation and competitive interruption exist. In mice and other mammals, the normal estrous cycle can be manipulated by the olfactory signals of other individuals. The presence of other females can prolong or suppress cycling, the presence of a male can speed the maturation of a young female, and the introduction of a new male can cause a pregnant female to abort and become estrous again (Nelson 1995). Males of some butterflies and moths produce a pheromone that they deposit on the antennae of females that causes them to alight and remain still for mating (Thornhill and Alcock 1983). In externally fertilizing fish with male egg guarding, a male must entice females to his nest and fertilize the eggs as soon as they are laid; outside males can easily streak in and fertilize some of the eggs. In garter snakes where emergent females are surrounded by a mass of males attempting to mate, some males exude the female attractiveness pheromone, which confuses normal males and gives the impostors a competitive edge in obtaining the mating (Mason et al. 1989). The queens of most eusocial hymen-

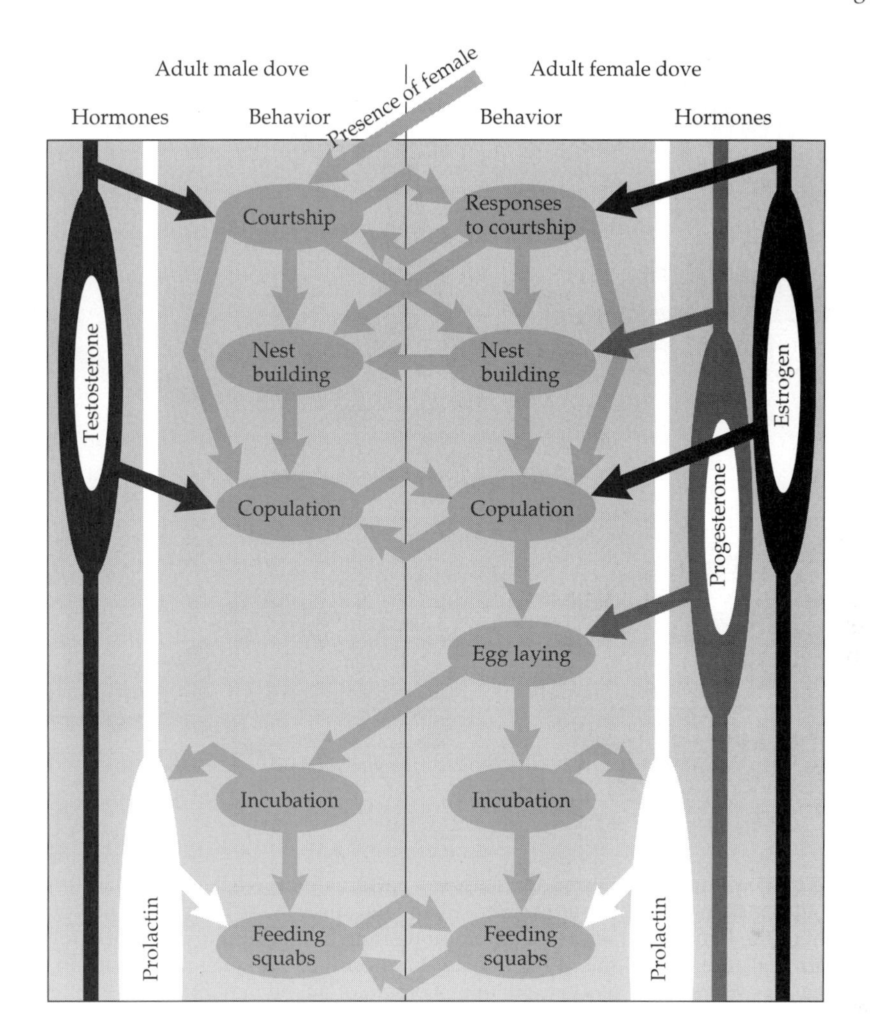

Figure 24.3 Reciprocal stimulation of reproductive behaviors in the ring dove. When a male and a female are first placed together, the sight of the female causes the male to increase his testosterone production, and he begins to perform the bow-coo audiovisual display. In response to the courting male, the female begins to coo, and her own vocalization stimulates estrogen production. Estrogen stimulates the development of the female's oviduct. If a nest site and nesting material are present, the reciprocal courtship behavior causes both sexes to begin nest building. After about a week, estrogen declines and progesterone increases in the female, the pair copulates, and two eggs are laid. The sight of the eggs in the nest induces incubation behavior and suppresses courtship, testosterone, and progesterone. Incubation is maintained with the onset of prolactin, which also causes the development of the crop sac and brood patch. When the young hatch, the parents initially feed the young by regurgitating crop milk and then bring them grain. When the young are a few weeks old, prolactin decreases, testosterone in the male and estrogen in the female increases, and the cycle is repeated. (After Nelson 1995.)

Figure 24.4 Courtship pheromone delivery patterns in newts and salamanders. (A) In the smooth newt *Triturus carnifex*, an aquatic species, the male (right) displays in front of the female and wafts the pheromone from a huge abdominal gland toward her with his large crested tail. If the female is reluctant or not quite receptive and the male has to interrupt the sequence to take a breath of air, he may find his place taken by another male. (B) In *Plethodon jordani*, the male (rear) attempts to rub the secretions of his mental gland directly on to the nares of the female. (C) In the mountain dusky salamander *Desmognathus ochrophaeus*, the male (right) rasps the female's back with his teeth and rubs the pheromone from his mental gland directly into her circulatory system. (D) A fourth strategy in which males dispense with the pheromone altogether is found in *Euproctus montanus*. A male forcefully grabs the female by biting her tail and wrapping his tail around her pelvic region. In this position, the male's cloaca is pressed against the female's cloaca and transfer of simplified spermatophores takes place. (A after Halliday 1977; B after Arnold 1976; C after Houck and Reagan 1990; D after Brizzi et al. 1995; all photos courtesy of Stevan J. Arnold.)

opteran species suppress the ovarian development of female workers with chemical signals to prevent them from producing eggs. Most of these examples involve olfactory signals. Pheromones have a greater potential for manipulating a potential mate or competitor to the benefit of the sender and the detriment of the receiver because they can modify the receiver's internal hormonal state.

Pair-Bond Maintenance

Monogamy with long-term pair bonds is common in birds and found occasionally in some mammals, fish, and insects. The function of the extended bond is usually to promote the sharing of parental care, but monogamy can also be favored when the competitive sex can only defend one individual of the limiting sex (Wittenberger and Tilson 1980). Mated animals, of course, learn to recognize their mate as an individual. In addition, special signals between mated individuals appear to maintain the pair bond and discourage infidelity. Some examples of mutual visual and tactile displays are shown in Figures 24.5 and 24.6, and examples of auditory displays in Figure 24.7.

In monogamous primates such as gibbons, marmosets, and tamarins it is clearly the highly aggressive behavior of the female that maintains monogamy (Kleiman 1977). Females are very active in territory defense. They use auditory signals in the case of gibbons and olfactory signals in the case of marmosets and tamarins. Each sex primarily defends the territory against intruders of the

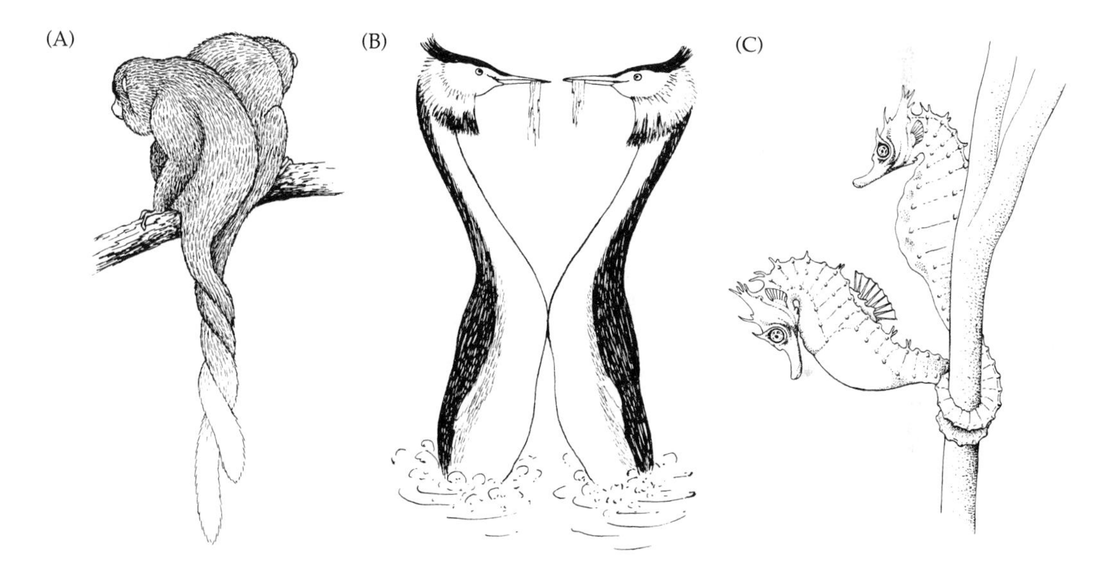

Figure 24.5 Mutual tactile and visual displays between mated individuals. (A) Tail twining in dusky titi monkeys (*Callicebus moloch*). (B) The penguin dance of the male and female great crested grebe *Podiceps cristatus*. Each has previously dived for a bill full of weeds (nesting material). They rise up out of the water, and breast to breast, they sway back and forth together. This is one component of a longer series of displays used in courting and greeting. (C) Greeting ceremony in the seahorse *Hippocampus whitei*. Each morning the female approaches the male, they both lighten in color from a brown-gray to pale yellow, and they perform a mutual greeting ceremony that lasts about six minutes. Side by side they both grasp a plant shoot, and with the male on the outside they circle around the shoot in a maypole dance illustrated here. They then release the shoot and swim slowly parallel to each other with the male grasping the female's tail. They swim back to the shoot, repeat the maypole dance, and alternate with the slow parallel swim until one of them, often the female, darkens and swims away. If the male has released a brood, the mutual display is prolonged, and egg deposition and fertilization follow. (A after Moynihan 1966; B after Storer 1969; C after Vincent and Sadler 1995; Vincent 1994.)

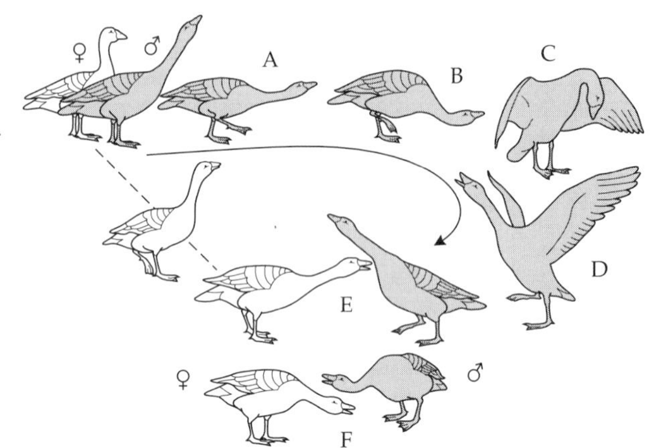

Figure 24.6 Sequential steps of the triumph ceremony in the graylag goose. The male is shown in gray, the female in white. The male approaches and aggressively challenges another male (on right, not shown). He then returns to his female with outstretched wings. She approaches him, and they both adopt aggressive postures resembling those the male used in his aggressive drive, except that they orient obliquely past each other. (After Fischer 1965.)

same sex, so each member of a pair enforces fidelity in the other. In the presence of the mate each member will also take an aggressive stance against members of the opposite sex, as if demonstrating their commitment to the relationship. However, in the absence of the mate they may approach and attempt to court such an intruder. Mutual grooming, contact resting (Figure 24.5A), and duetting are some of the mutual displays found in these animals. Similar relationships have also been described for birds (Zann 1977).

In birds, mutual displays in species with long-term pair bonds include duetting and coordinated visual and vocal displays. Most mutual displayers are also sexually monomorphic. Ducks and geese perform what is called a triumph ceremony, in which the mated female threatens a third party or pair, her mate aggressively approaches the third party and then returns to the mate's side, and both perform a mutual visual and vocal display (Figure 24.6). It suggests that aggression toward outside individuals strengthens the pair bond. Visual displays are more commonly found in ground-nesting birds inhabiting open habitats such as marine colonial species, divers, grebes, ducks, and geese (Figure 24.5B). Duetters, on the other hand, tend to be territorial species in more densely vegetated habitats such as songbirds, owls, woodpeckers, and parrots (Figure 24.7) (Malacarne et al. 1991). Duetting not only serves to keep the members of a pair in contact with each other, but may also enable them to keep their mate under surveillance and reduce extra-pair copulations (Farabaugh 1982; Levin 1996).

Monogamous fish exhibit several different patterns of integration (Turner 1993). In biparental neotropical cichlid fish, pairs remain in close contact on their small territory and adopt a color pattern that advertises their paired, territorial status. However, there is clear division of labor, with males primarily defending the territory boundary while females guard the brood (Barlow 1974). In the male-brooding seahorse *Hippocampus whitei*, the time required to brood is equal to the time needed by the female to develop a clutch of eggs, so the operational sex ratio is equal and there is no courtship role re-

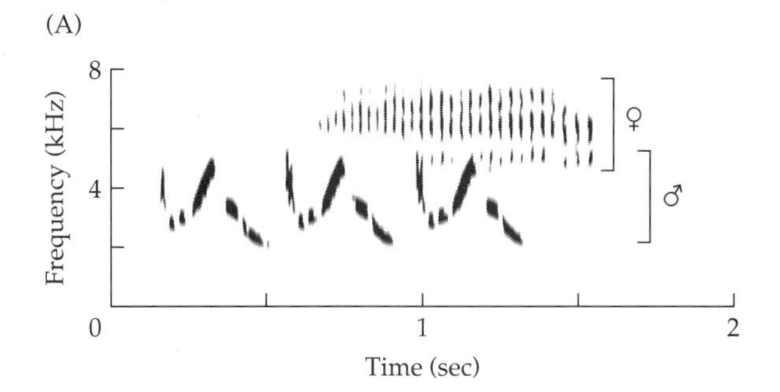

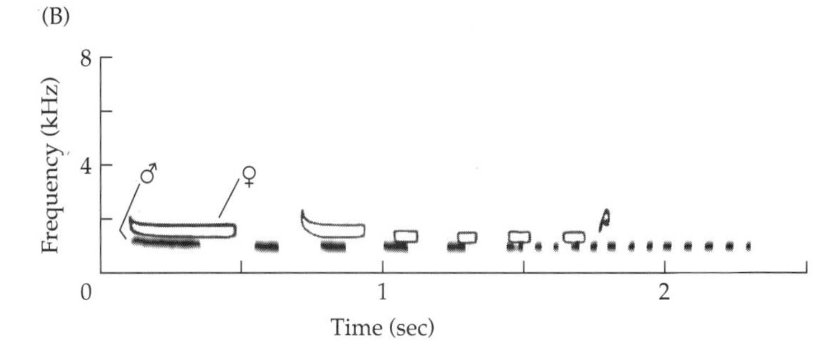

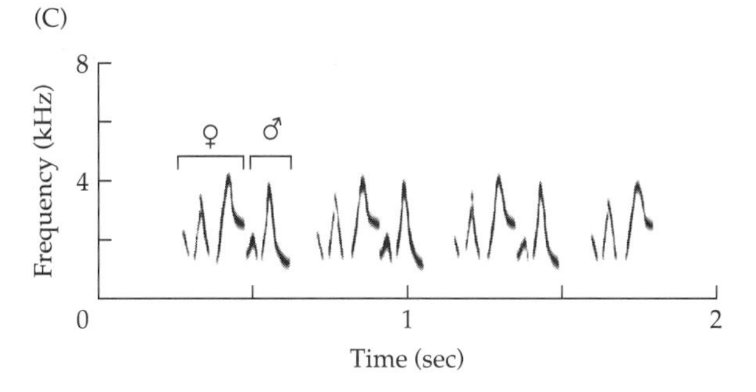

Figure 24.7 Duets in three species of wrens. (A) Carolina wren *Thryothorus ludovicianus*. The male sings a songtype from his repertoire of about 30 songs while the female churrs. (B) The rufous-and-white wren *T. rufalbus*. Songs of the male (black) and female (white) have the same general structure and overlap each other. (C) Buff-breasted wren *T. leucotis*. Males and females have consistently different syllable types, but the female sings her notes immediately after the male's without overlapping. (From Farabaugh 1982, © Academic Press.)

versal. Females have a larger home range that includes the smaller territory of a male but the male and female spend the day apart. Each morning they engage in a greeting ceremony (Figure 24.5C) that probably allows the female to determine whether the male has hatched his brood. If so, she immediately gives him another clutch (Vincent and Sadler 1995; Vincent 1995). The hermaphroditic harlequin bass *Serranus tigrinus* produces pelagic eggs and performs no parental care, but the close association between two individuals on a defended territory promotes a daily double spawn with each acting as male and female; this close association also prevents male philandering (Pressley 1981).

PARENT-OFFSPRING INTEGRATION

Parental investment is defined as time and energy devoted to offspring that increases the offspring's survival while decreasing the parent's ability to survive and invest in future offspring (Trivers 1974). Gametic investment, such as the provisioning of eggs by females, is just one component of parental investment. Another component, zygotic investment, may be contributed by the male, the female, or both parents and includes aspects of parental care such as incubating, guarding, and feeding of young. It is important that these costly behaviors be directed toward true offspring, so accurate recognition mechanisms are strongly favored. In this section we shall examine the mechanisms used by parents and offspring to recognize each other, and we look at the communication signals employed to regulate the amount of parental effort given to offspring.

Parent-Offspring Recognition

The optimal mechanism used by parents to recognize their offspring depends on three factors: the number of offspring per brood, the mobility of the offspring, and the number of nearby nonoffspring young. Location and familiarity are the most commonly used mechanisms. Offspring may also be selected to recognize their parents. The following examples show how these factors affect recognition.

For species with immobile young placed in a nest and well separated from adjacent broods, such as in territorial birds and mammals, the nest site is the focus of parental recognition of offspring. Neither signature signals by offspring nor offspring recognition of parents are required. Although location is an optimal recognition cue in this context, it leaves the parent vulnerable to brood parasites that dump nontarget offspring into the nest. Avian host species that are heavily parasitized by cuckoos and cowbirds eventually evolve the ability to learn to recognize their own eggs and eject foreign eggs (Rothstein 1990; Davies and Brooke 1989a,b). Nestling recognition, however, is both less effective and more difficult to evolve than egg recognition (Lotem 1993), resulting in the occasional spector of a small warbler parent feeding a huge conspicuously non-warbler-like cuckoo nestling. The evolution of egg or nestling ejection behavior also involves the risk of erroneously rejecting one's own offspring (Marchetti 1994). Recognition of the eggs and nestlings of *conspecific* brood parasites is an even more difficult problem and rarely evolves in birds because the probability of parasitism is relatively low, discrimination is difficult, the cost of raising one more (nontarget) offspring is low, and the cost of ejecting one's own young is high (Petrie and Møller 1991).

In colonial species such as seabirds, swallows, and ground squirrels that place their nests or burrows in close proximity, the total number of nontarget offspring in the vicinity is high, and the low prior odds favor an individual signature system (Beecher 1991). Initially, the young are immobile and restricted to the nest site, so parents use nest location as a recognition cue. As the young become mobile, they too use nest location as a parental recognition mechanism. The period of attachment to the nest site provides the reliable kin association required for learned familiarization between parents and offspring. Offspring may learn to recognize their parents first if soliciting to nonparental adults results in injury or cannibalism (Holley 1984). Shortly before fledging, the parents learn to recognize their young as individuals and there is strong selection for complex signature signals. The recognition challenge is greatest in species that place their offspring in communal crèches; multiple-modality signature signals and cues are the rule here. The mere existence of crèches implies an effective offspring recognition mechanism. Nursery colonies in the Mexican free-tailed bat may contain thousands of pups (McCracken 1984, 1993; McCracken and Gustin 1987, 1991). A mother remembers the approximate location of the site where she last left her pup and thereby reduces the number of individuals she must examine by a large fraction. She then uses olfactory and auditory cues to distinguish her own pup. Pups will beg from every passing female. Since 17% of mother-baby pairings were found to be unrelated, the decision rule results in a substantial number of misses. In penguins, terns, gulls, and swallows, visual and vocal signature signals are important (Davies and Carrick 1962; Jouventin 1982; Buckley and Buckley 1972; Beer 1979; Stoddard and Beecher 1983; Shugart 1990; Seddon and van Heezik 1993). Parents require the young to vocalize before they will feed them, which prevents offspring deception by silence. If accosted by many begging young, parents may lead them on a feeding chase away from the crèche, which increases the cost to cheaters (Thompson and Emlen 1968; Spurr 1977).

In species whose young are mobile shortly after birth or hatching, recognition must be learned during the first few hours with a rapid imprinting process. Either the parent must learn the characteristics of the offspring or the offspring must learn to recognize the parents, or both. In ungulates such as goats, mothers quickly learn the characteristic scent of their single offspring (Halpin 1991). The mother may subsequently label its offspring with her own olfactory cues but this is not required for recognition (Gubernick 1981; Romeyer et al. 1993). Later the mother learns the voice and appearance of her offspring for long-distance recognition. Calves also learn the voice, appearance, and smell of their mothers. In ground-nesting precocial birds, the initial burden is placed on the chicks to recognize their parents during the imprinting process (Cowan and Evans 1974). Since there are many chicks and parents only guard, but do not feed them, parental effort per chick is low and there is less pressure on parents to individually learn their offspring. Similarly in fish that guard mobile fry, the fry initially learn to recognize their parents, and parents eventually recognize their offspring using olfactory cues (McKaye and McKaye 1977).

When brood size is very large, individual recognition of each offspring is not feasible and a family-level recognition mechanism is favored. The desert isopod *Hemilepistus reaumuri* is an excellent example of a learned family-specific olfactory badge (Linsenmaier 1987). Monogamous pairs raise broods of 50 to 100 in an underground burrow that is vigorously defended against invasion by nonfamily members. If a family member is temporarily removed and rubbed against the body of a nonfamily member, it will subsequently be rejected by its own family. The family odor is a meld of organic compounds produced by each family member. The chemicals are exuded on the cuticular exoskeleton, wiped onto the tunnel walls as the animals move around in the borrow, and redeposited onto the backs of all other members. This system relies upon genetically based individual differences in pheromone production to generate differences among families. The two (presumably unrelated) parents are different from each other. Since their offspring represent a genetic mix of parental genes affecting odor, the family badge must blend all members' odors. This is an example of phenotype matching in which the referent is an average of all family members.

Parent-Offspring Conflict

Parents are the providers of food, aid, and protection to their offspring, and offspring fitness and survival increase as a result. Although both parties share an interest in the successful rearing of the young, parents and offspring are likely to disagree over the exact length and extent of parental care. Parents are selected to maximize their lifetime production of young, and given that several broods may be raised in a lifetime, an excessive investment in one brood or offspring may reduce the parent's ability to invest in future offspring. Parents should thus allocate an equal share of resources to all offspring. Individual offspring, on the other hand, are selected to obtain a disproportionate share of parental care for themselves. However, offspring selfishness is not unbounded, but limited by the fact that siblings represent an important component of an individual offspring's inclusive fitness (see page 365).

Using the formula for kin selection developed by Hamilton (1964), we can express the extent of conflict between parents and offspring quantitatively. Parental inclusive fitness is maximized when the benefit B of increased aid to an offspring is equal to or greater than the cost C of lost future reproductive success, or $B/C \geq 1$. An offspring's fitness is maximized when this B/C ratio is equal to or greater than r, the coefficient of relatedness between the offspring and its siblings. Thus when $1 > B/C > r$, there is a conflict between the two parties in which the offspring prefers more investment than is optimal for the parent (Trivers 1974). The coefficient of relatedness is 0.5 for full siblings and 0.25 for half siblings, so the degree of conflict increases when relatedness decreases. Figure 24.8 gives a graphical representation of the zone of conflict.

How is this conflict resolved? Trivers argued that parents would be selected to monitor offspring needs and therefore be subject to exaggeration of true need by the offspring, and Zahavi (1977b) suggested that offspring might blackmail parents by threatening to scream loudly and attract predators, both of which would lead to greater investment than is optimal for the parent.

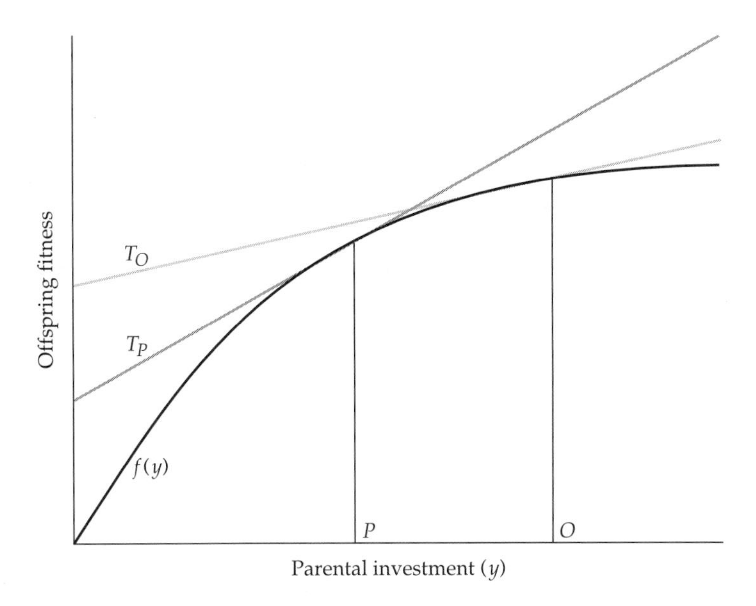

Figure 24.8 The parent-offspring conflict game. Offspring fitness is a function $f(y)$ of the resources, y, it receives from the parent. The benefit of investment decelerates at higher levels of investment. The parent suffers a linear reduction in future fitness as it invests more in current offspring. The optimum investment from the parent's point of view, P, occurs when the marginal benefits of feeding the current young equals the marginal costs of lost future reproductive success. This occurs at the point where a line with slope T_P is just tangent to the offspring fitness curve. The optimal investment from the offspring's point of view, O, occurs where the line with slope T_O is tangent to the offspring fitness curve. The parental line slope is always greater than the offspring line slope, and the difference between P and O is a measure of the extent of parent-offspring conflict. (From Godfray 1995a.)

Alexander (1974) argued that the parent would always win because it was in the position of control. Several game-theoretical models seeking a quantitative resolution of the conflict and using very different approaches and assumptions found ESS solutions intermediate between the parental and offspring optima (summarized in Godfray 1995a). A model by Godfray (1991, 1995b) that focuses on the signaling aspects of the interaction found the ESS to be the parental optimum. This model assumes that begging intensity can vary as a function of offspring need and that parents monitor the signal and respond with an appropriate level of provisioning. The benefit of more food must be lower for an offspring in good condition compared to an offspring in poor condition, i.e., the decelerating-benefits curve of Figure 24.8. Cheating by the offspring is prevented by two costs: the cost of signaling itself, and the loss in inclusive fitness. Thus this model, like the general begging model discussed in Chapter 20, predicts that honesty will prevail only when the signal is costly and when benefits increase with need.

Offspring Signals and Parental Responses

Evidence for and against parent-offspring conflict over investment duration and provisioning levels has been sought primarily in birds and mammals. Trivers (1974) cites examples of weaning conflict in mammals in which mothers refuse nursing attempts by juveniles. Bateson (1994), however, argues that such evidence is scarce, and predicts that optimal weaning times for mothers and offspring will be found to coincide. This coincidence would occur if the offspring benefit versus parental investment curve of Figure 24.8 rises sharply and approaches an asymptote so quickly so that parental and offspring optima are similar. An example of this idea is that avian parents and offspring may both favor maximum investment to meet offspring needs so that the duration of the vulnerable nestling period is minimized (Nilsson and Svensson 1993). Alternatively, weaning conflict may not be apparent in mammals because the mother controls milk production at her optimum level and offspring accept what is offered, as predicted by Godfray. As the juvenile grows, its nutritional needs increase and solid food is more abundant than the mother's declining milk supply.

Conflict between parents and offspring is easier to document in provisioning systems where offspring actively beg for food. The signaling behavior quantifies offspring demand, and the feeding rate reflects parental responsiveness and investment. Several studies have shown that nestling birds increase the rate, duration, or intensity of begging calls when hungry, and parents respond by increasing their rate of food delivery (reviewed by Mondloch 1995). When pigeon squabs are young, parents are maximally sensitive to offspring calls, but as the squabs age, parents exert greater control over feeding rates. A recent study of begging and provisioning in a tropical jay provides clear evidence of postfledging conflict, with begging rates peaking 11 days later than the peak of parental provisioning (Langen 1996b). In a study of infant solicitation calls in vervet monkeys, Hauser (1993a) found an increase in call rate between 8 to 10 weeks of age associated with a decrease in the probability of a maternal response. Recently fledged or weaned offspring may be genuinely hungry, but mothers appear to be discounting excessively high infant begging rates once offspring reach this stage.

Hauser's study implies that infants are exaggerating their need. How honest are begging signals? When the hunger level of one nestling is experimentally manipulated by temporarily removing it from the nest or feeding it, its begging behavior subsequently increases (Redondo and Castro 1992a; Weary 1995). In some cases such a nestling will obtain an increased proportion of the feedings, but this advantage may be diminished by two factors: (1) the other nestlings in the nest increase their begging intensity in response to the increased begging of the hungry nestling (Smith and Montgomerie 1991), and (2) asynchronous hatching results in a large differential in competitive ability among siblings and overwhelms the effects of hunger (Bengtsson and Ryden 1983; Magrath 1990). In some asynchronously hatching species, parents actively ignore or discount the begging behavior of nestlings and use cues to discern the nestlings in greatest need. The budgerigar (*Melopsittacus undulatus*) female, for example, avoids feeding the loudest nestlings, turns each offspring onto its back, and seeks out the smaller, younger, and quieter individuals. Males, on the other hand, respond to the average begging level in the nest by adjusting feeding rates and always feed the loudest, largest, closest nestling (Stamps et al. 1985). Similar sex-differential parental behavior has also been described for the great tit (*Parus major*) (Bengtsson and Ryden 1981, 1983). The female strategy appears to counterbalance the negative effects of disparate nestling ages.

Recent evidence indicates that avian begging calls are not very expensive energetically (McCarty 1996). Therefore, the critical cost that maintains honesty may be predator attraction (Redondo and Castro 1992b; Haskell 1994). This type of cost could not maintain honesty in a Grafen-like model with differential sender costs (pages 658–660) because all other nestlings in a nest pay for the extravagant signaling of any one nestmate. But the begging models described on pages 665–666 only require a general cost of signaling, and honesty is maintained by differential benefits instead. Thus begging behavior meets the conditions for honest signaling.

GROUP INTEGRATION

Group living serves a variety of adaptive functions: warning and defense against predators, location of food, territory defense, migration, and cooperative reproduction, among others. The type of cooperative endeavor affects the size and stability of groups, the composition of groups, and the signals group members need to coordinate activities. In this section we shall examine the mechanisms used to recognize group members and describe some of the signals used to maintain group cohesion. Signals used to inform group members of environmental information such as the presence of predators or the location of food will be discussed in the next chapter.

Group Recognition

The mechanisms used for group recognition depend on the size and stability of groups, whether or not group members are tied to a fixed site such as a nest or territory, and the importance of recognizing specific individuals or classes within groups. If natal dispersal is restricted, most of the recognition mechanisms outlined earlier in this chapter can be adapted to distinguishing kin. Many forms of cooperative and altruistic behaviors can evolve only when restricted to close kin so the recognition mechanism plays an important role in determining the kinds of interactions that take place among group members.

Individual recognition based on familiarity with group members is a common mechanism used to distinguish group members from nonmembers. Groups must be relatively small and stable in composition for this mechanism to operate. Individual recognition facilitates the development of stable dominance hierarchies within groups because each individual can remember the outcome of prior encounters with other members (Barnard and Burk 1979). In house mice (*Mus domesticus*), group members are treated with aggression if they fail to maintain their olfactory marks (Hurst 1989; Hurst et al. 1993). Furthermore, individual recognition and long-term association are essential for maintaining coalitions and reciprocal alliances (Trivers 1971). We know that individual recognition occurs in such groups because animals that encounter a nongroup member immediately exhibit aggressive behavior. Some experimental studies have examined the key modalities and cues used for recognition: olfactory in most mammals, visual and vocal in primates and birds, and visual in fish. In species with female recruitment into natal groups, individual familiarity can also be linked with kinship. Cheney and Seyfarth (1982, 1990) have used clever playback experiments with vocalizations from known individuals to reveal the complex perceptual world of vervet monkeys. Within groups, animals distinguish group members as both individuals and as members of matrilineal kinships. Across troops, monkeys associate vocalizations of particular individuals with particular troops. In species with a fission-fusion spatial organization such as horses, elephants, dolphins, and parrots, the smallest cohesive units sometimes merge to form larger temporary associations. For elephants, one can describe concentric circles of association levels that include increasingly larger numbers of individuals of lower degrees of relatedness that associate for progressively shorter periods of time (Moss 1988). Killer whales (*Orcinus orca*) and *Amazona* parrots develop vocal dialects that reveal group membership at different social levels (Ford 1991; Wright 1996).

When group membership is unstable, a group label is meaningless and groups are more appropriately called aggregations. Examples include some wintering and migratory flocks of birds and herds of antelopes. Individual recognition is difficult to maintain because of the temporary nature of associations. However, if there is competition for resources within such aggregations and individuals vary in fighting ability, then an ability to recognize relative dominance status is favored (Rohwer 1982). Visually based dominance badges evolve under these circumstances. On pages 662–664 we noted that the conditions required for the evolution of honest status badges include frequent testing of true fighting ability by individuals with similar badges and a significant cost of cheating. Honest badges reduce the overall level of fighting for all individuals in the association because they settle contests quickly without escalation. In a similar vein, age badges reduce aggression of older individuals against younger individuals.

When group members are obligatorily associated with a communal nest or roost site and some unique feature of this site such as its odor is imparted to the occupants, a true group label exists. Group members can be unambiguously distinguished from nonmembers without prior knowledge of each individual. This familiarity mechanism can function equally well for groups of related or unrelated individuals. Social insects such as bees and some ants and termites that forage cooperatively on plant material (nectar and leaves) commonly use olfactory cues derived from their diet to recognize nestmates (Jaisson 1991; Pfennig and Sherman 1995). Diet-based cues vary depending on the food item currently being exploited by the colony, so neighboring colonies will differ, but group members must constantly update their olfactory template. Predatory solitarily foraging wasps commonly base nestmate recognition on odors derived from the nest material. Environmentally derived cues may play a more important role in colonies founded by multiple queens where genetic cues are too variable within colonies (Breed and Bennett 1987). Given the great importance in social insects of cooperating with close relatives and excluding nest parasites, foreigners, and nonkin, many species supplement environmental cues or exclusively use genetically based signals for colony recognition (Michener and Smith 1987). The recognition mechanism then becomes a case of phenotype matching for genetic similarity, based on an average family referent. Like the desert isopod example of family recognition mentioned earlier (page 802), a learned chemical mix based on all colony members is used to distinguish nestmates from others. When genetic cues are used, however, close relatives from nearby colonies are sometimes treated as nestmates (Greenberg 1979; Provost 1991).

If groups are composed of a mixture of individuals with different degrees of relatedness and it is important to cooperate only with the closest kin or avoid cannibalism of kin, then an accurate kin-recognition mechanism is required. Phenotype matching with oneself as the referent is one way this recognition can be achieved. This mechanism may occur in species with large, multiple-paternity broods where discriminating between full and half sibs is important (Sherman 1991). In bees and ground squirrels that have the potential to discriminate at this level, the difference in behavioral bias toward full and half sibs is very small (Page et al. 1989; Holmes and Sherman 1982). Allele matching is a second mechanism for true kin recognition that is favored when a very high precision of kin recognition is required, when group-specific environmental variation does not occur, and/or when there is no opportunity to learn one's own or one's immediate family phenotype. Colonial invertebrates that fuse to form differentiated reproductive structures will join only with very close relatives sharing at least one allele at the recognition locus (Grosberg and Quinn 1986).

Appeasement Signals

Animals living in stable cooperative groups use a variety of appeasement signals to reduce aggression and maintain friendly relationships among group members. A useful way to categorize these signals is to separate them along two dichotomous axes as shown in Figure 24.9. One axis reflects the duration of the consequences of the signal and can be dichotomized into signals that result in immediate resolution versus those that have longer-lasting effects. The second axis reflects the symmetry of the interaction. Symmetrical interactions involve individuals who are approximately equal in status and the signals are usually mutual, whereas asymmetrical interactions involve signals given unilaterally by one individual.

Asymmetric unilateral displays given by either the dominant or subordinate individual serve to reduce aggression in potential or imminent conflict situations. Most of the signals in this category are **submissive displays** given by subordinates to reduce or inhibit attack by dominants. In Chapter 16 we

Figure 24.9 Categorization of appeasement signals. Appeasement signals can be categorized along two dichotomous axes: the duration of the consequences of the signal (immediate resolution or long-term maintenance) and the symmetry of the partners (asymmetric or symmetric). The figure shows where a variety of signals would be placed with this schema.

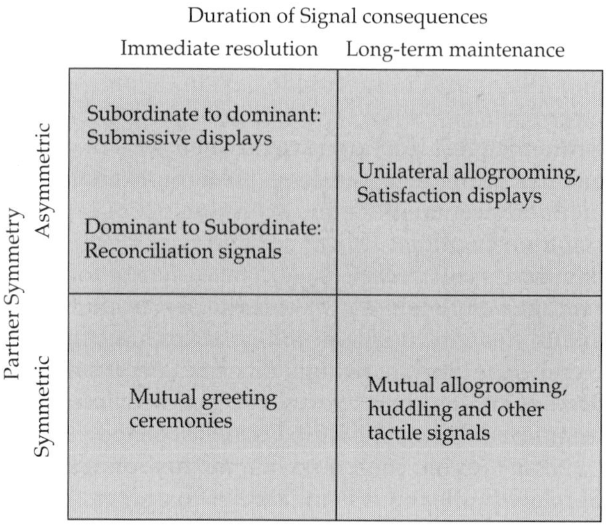

Partner Symmetry	Duration of Signal consequences: Immediate resolution	Long-term maintenance
Asymmetric	Subordinate to dominant: Submissive displays Dominant to Subordinate: Reconciliation signals	Unilateral allogrooming, Satisfaction displays
Symmetric	Mutual greeting ceremonies	Mutual allogrooming, huddling and other tactile signals

pointed out that the nonaggressive meaning of such displays often arises from the fact that their form is the opposite of the species' threat displays. Examples include hiding weapons, looking or turning away, exposing a vulnerable part of the body, and uttering high-frequency sounds. Another source of submissive displays is the arousal of conflicting motivations in the attacker with signals that stimulate other, nonaggressive drives. For example, threatened subordinates of both sexes may give the female sexual invitation to deflect aggression (Figure 16.9), or subordinates may adopt a juvenile behavior such as begging for food to solicit parental care behavior. Signals given by dominants to subordinates that transmit nonaggressive intent are called **reconciliation signals**. In baboons, an approaching dominant giving the conciliatory grunt vocalization is less likely to supplant and more likely to be friendly toward a subordinate. The grunt is also used to reconcile two opponents following a fight (Cheney et al. 1995). Other reconciliation signals include kiss (Figure 24.10A), embrace, hold-out-hand, touch, lip-smacking, and redirected threats (de Waal 1989, 1996; Petit and Thierry 1994).

Asymmetric maintenance signals are unilateral signals between unequal individuals that serve to cement bonds, maintain friendly relationships, and establish alliances for future support. The primary example of such a behavior is **allogrooming**, the preening or grooming of another individual (Figure 24.10B). Allogrooming is common in social animals living in stable groups such as primates, horses, and avian cooperative breeders and is conspicuously absent in most ungulates and flocking aggregations. The grooming tends to be directed to those parts of the body that the self-groomer cannot reach such as the head, neck, and back, so it clearly serves a hygienic function. In primates in particular, allogrooming has been shown to result in significant future benefits (Dunbar 1991). Males in multimale troops groom certain females, often receive grooming from them in return, and may later become preferred mating partners of these females. Subordinate individuals

Figure 24.10 Examples of appeasement signals. (A) Kissing as a reassurance gesture between two female chimpanzees *Pan troglodytes*. (B) Allogrooming in olive baboons, *Papio anubis*. An adult female is grooming another female with an infant. (C) Greeting ceremony of elephants *Loxodonta africana*. The female in the center has probably rumbled, drawing in distant family members from left and right and a male from the rear. (D) Social huddling in cuis *Galea musteloides*. (A courtesy of Frans de Waal; B courtesy of Jim Moore/Anthro-Photos; C courtesy of Cynthia Moss; D courtesy of E. Schwarz-Weig and N. Sachser.)

frequently groom dominants and are subsequently more likely to obtain the dominant's support in conflicts (Hemelrijk and Anneke 1991; Gust 1995). Dominants in some species use allogrooming as a reconciliation gesture toward subordinates (O'Brien 1993). Horses may rest their head on the back of another individual or allogroom the neck and withers area. Gestures such as these are known to reduce tension and heart rate in recipients (Feh and Mazieres 1993). Lip-smacking and other motions used during allogrooming are used to both initiate and solicit allogrooming sessions in primates. Birds solicit allogrooming by bending the head down or to the side and fluffing the head and neck feathers to expose the areas of preferred grooming.

Satisfaction signals occur in the context of consummatory activity such as feeding, copulation, and cuddling and appear to transmit the message of pleasure or contentment. Such signals are rare among animals, and are probably best categorized as asymmetric maintenance signals. The most familiar example is the purring of cats and raccoons. Purring is performed by both kittens and mother during nursing as an indication of contentment, but it is also given by adults to other adults during symmetric mutual greeting and grooming exchanges. It is a graded signal, with the loudness and roughness of the sound correlated with the degree of contentment. Wolves and dogs perform a contentment facial expression with relaxed mouth and droopy eyelids that has been classified as a satisfaction signal (Peters 1980). Captive beavers give an exhaled breath sound when being fed. When the communal nesting groove-billed ani rests during the day in physical contact with its mate or roosts at night in a tightly huddled linear array with the entire group, it gives a very soft warble vocalization that, like purring, seems to indicate satisfaction or desire to huddle.

Social animals engage in a variety of **affiliative behaviors** that are usually performed mutually and serve to maintain peaceful relationships. Examples include nose-touching, sniffing, nuzzling, and licking the face or genitals observed in a wide variety of mammals (Peters 1980). Huddling is observed in many social birds and mammals (Figure 24.10D), and while it frequently serves a thermoregulatory function, it is also used in other contexts for social bonding (Schwarz-Weig and Sachser 1996). Other forms of physical contact for bonding purposes include tail twining, hand-holding, body rubbing, and hugging. Soft, continuous, and variable vocalizations serve a similar function in beavers. Mutual or reciprocal grooming is another common affiliative behavior. In impala, one of the few antelopes living in stable groups, two individuals will engage in alternating reciprocal allogrooming without regard to dominance, sex, or kinship (Hart and Hart 1992). Finally, another function of play behavior may be to reduce aggressive tension (Caro 1988).

The final category of group cohesion signals includes **greeting ceremonies** that serve to reunite group members that have been separated. These ceremonies are most conspicuous in social mammals and almost always involve mutual tactile displays and elements of appeasement and affiliation. Chimpanzees hug and embrace upon meeting, behaviors derived from infant clasping behavior. A gentle head butt or rub to the chest or flank area of the recipient follows meeting and recognition in cats, horses, and dolphins (Figure 17.9A). Canids perform mouth licking and gentle jaw wrestling upon meeting in conjunction with a variety of excited whines, squeaks, and barks (Figure 12.7). These behaviors are derived from the begging solicitation of juveniles and subsequent regurgitation of food to the young. Greetings in the spotted hyena involve sexual elements: two individuals stand head to tail, lift the rear leg, and lick the partner's exposed genitalia. Dolphins also include touching the genitals in their greeting ceremonies. Probably the most spectacular ceremony to behold is the meeting of two closely related female elephant groups (Figure 24.10C). They detect each other from a distance, probably with olfactory and low-frequency auditory signals, then commence a full running charge toward each other with loud trumpeting and ear flapping. Upon contact, the groups mingle and engage in excited rubbing and trunk-touching of the mouth and head. In humans, greeting typically involves a recognition eyebrow flash, an appeasement grin, a hand wave to show no weapon is present, and some type of tactile signal such as an embrace or hand shake.

Coordination of Group Movement

For the majority of group-living species, the advantages of grouping can only accrue when group members remain in close proximity. When groups are not tied to a specific location, such as a nesting colony, but move over large areas as a unit, mechanisms for maintaining group cohesion are required. The signals used to coordinate group movement vary quite a bit depending on the function of grouping, the precision required, and the mode of locomotion. Coordination of group movement can be important in both stable and unstable groups.

For migratory birds such as ducks, geese, and waders, the critical task is to take off synchronously with an optimally sized group of conspecifics. Migrating in groups increases the accuracy of direction-finding, and flying in a linear or V-formation provides an aerodynamic energetic benefit for all but the lead bird (Badgerow and Hainsworth 1981). Flocks must be of optimal size and composed of same-sized conspecifics to maximize this benefit. Since many birds are traveling in the same direction at the same time and landing in the same meadows and marshes to feed between flights, migratory flocks form anew in each instance with different flock members. Most birds fly at night, so sunset serves as a general timing cue. Birds prepared to fly begin to vocalize and/or give visual preflight signals (Figure 18.12G). The rate of vocalization increases as more birds join in, and a group may take off. They typically circle around in an unorganized bunch formation, and will land again if the group is too small. Still vocalizing, the birds in a successful take-off gradually find their places in the V-formation. If the group is larger than the optimal size, it will split in two at this point. Vocalizing probably plays some as yet unknown role in deciding which bird will lead and in which direction the flock will fly, because vocalizing stops once the formation has been achieved (Piersma et al. 1990).

Foraging flocks of birds and mammals feeding on fine-grained food resources face the difficulty of continually staying together while individuals hunt for food items. Group members are usually somewhat spaced out and hidden by vegetation. Communication signals are needed to steer the direction of travel toward known food patches and prevent stragglers from getting lost. In a wide variety of such species, three specific types of vocalizations are found—a soft, frequently repeated contact call with individual signature, a separation call, and a movement initiation call (Figure 24.11). Where group members frequently stray large distances apart, loud contact calls are given (Mitani and Nishida 1993; East and Hofer 1991, Boinski 1991).

Fish schools exemplify the most precise level of movement coordination in groups of unstable composition. Grouping improves homing and orientation accuracy in migratory schools and increases foraging efficiency up to a

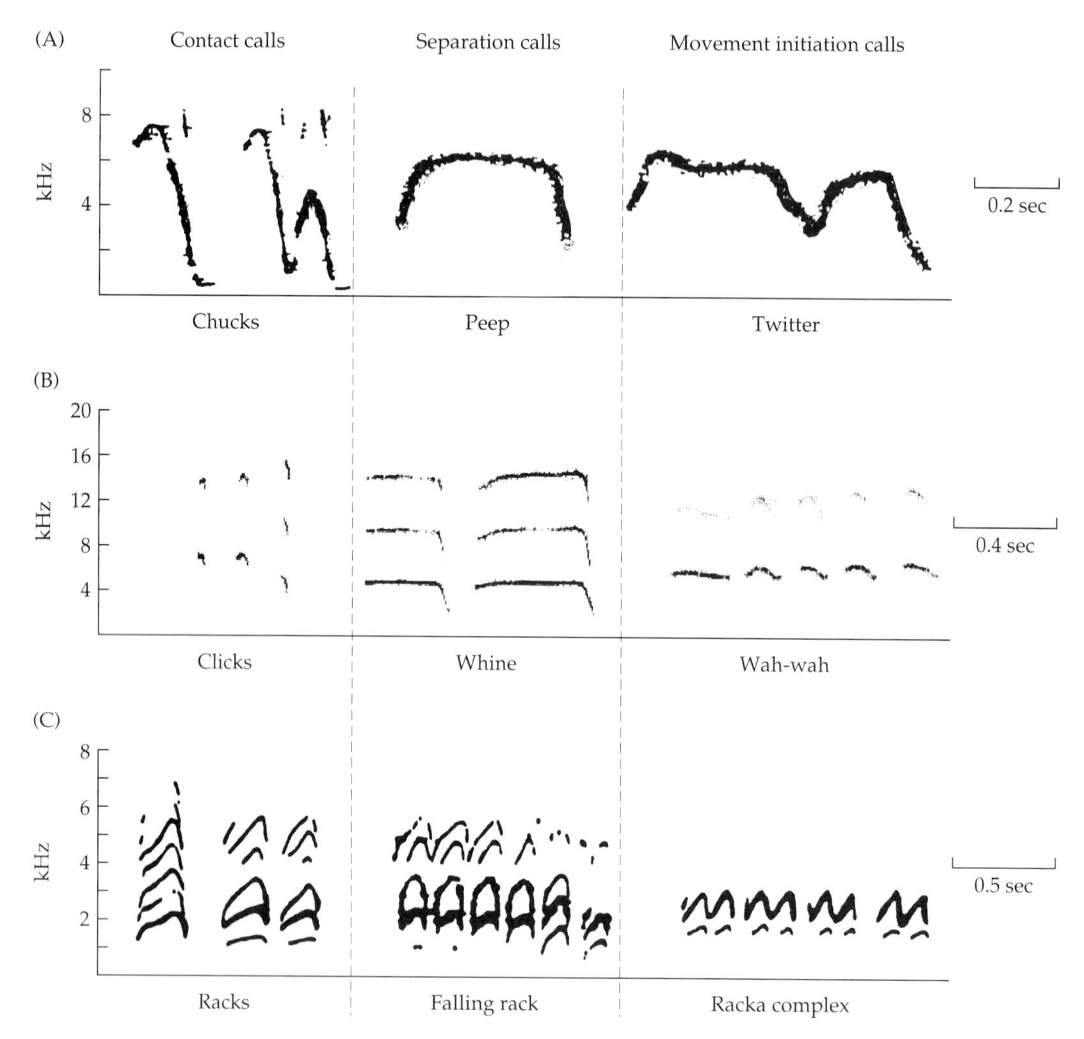

Figure 24.11 Parallel examples of foraging flock cohesion calls in three different species. (A) Squirrel monkey *Saimiri oerstedi*. (B) Golden lion tamarin *Leontopithecus rosalia*. (C) Pinyon jay *Gymnorhinus cyanocephalus*. Contact calls are soft, short, frequently repeated and usually contain individual signature information. Separation calls by individuals who have become separated from the group are louder and longer. Movement initiation calls are the loudest of the three types and are structured to be easily locatable. (A from Boinski 1996; B from Boinski et al. 1994; C from Marzluff and Balda 1992.)

certain size by decreasing vigilance and improving sampling efficiency and location of food patches (Pitcher and Parrish 1993). However, the primary selective pressure for precision schooling is predator detection, confusion, and evasion (Pitcher and Wyche 1983). Predator confusion is a passive effect of the extraordinary uniformity of the school, with all individuals regularly spaced and swimming at exactly the same speed parallel to each other. Individuals that deviate in any way are quickly picked out. This favors highly sensitive spatial perception, cryptic silvery body coloration or longitudinal body stripes, and size-assortative schools. Evasive maneuvers, illustrated in Figure 24.12, can result in the condensation, explosion, splitting, or reversal of the school. The signals and cues employed by the fish to coordinate these different moves are not known. Research has instead focused on the reception of spatial information by schoolmates. Schooling fish use both their visual and lateral line sensory systems to maintain uniform spacing and parallel orientation (Partridge and Pitcher 1980). Vision is responsible for the attraction to conspecifics and the maintenance of position and angle between fish; when vision is blocked, a fish tends to swim too far away from schoolmates. The lat-

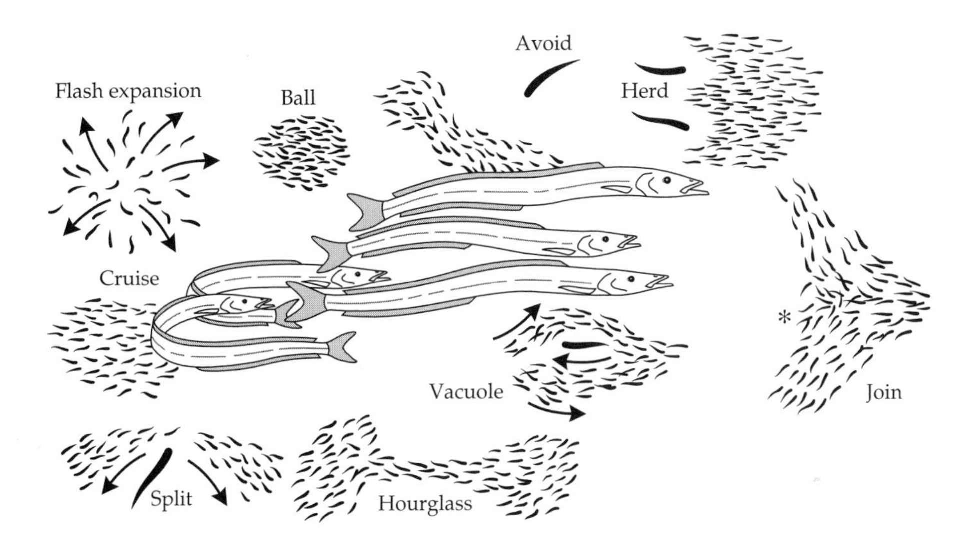

Figure 24.12 Patterns of fish school movements. Schematic diagram of the repertoire of antipredator tactics in schools of sand-eels (*Ammodytes* sp.) under threat from hunting mackerel. Tactics are drawn as seen from above, and a close-up side view of the eels is provided for illustrative purposes only. *Cruise* is the normal pattern for an undisturbed school. *Ball* is a fear response observed when fish are placed in an unfamiliar tank. *Flash expansion* occurs when a predator compresses a school against a wall or barrier. The most common maneuver to a slowly approaching mackerel is to change the angle of travel, as in *Avoid*. When several mackerel swim behind the school, the eels at the back of the school crowd into those in front of them to maintain a minimum distance between them and the predator as shown in *Herd*. *Vacuole* is the response to a single mackerel when the predator is swimming inside the sand-eel school. *Split* occurs when the school is compressed between two groups of mackerel. One way to regroup after a split is via *Hourglass*, with sand-eels streaming at high velocity across a pseudopodium like bridge two or three fish wide. When two schools *Join* at an oblique angle, the fish at the front of each school align themselves along the vector of the two original tracks, but the fish as the back continue to travel on their original course part way into the other school creating a confusion zone designated by *. (From Pitcher and Wyche 1983, © Kluwer Academic Publishers.)

eral line system is responsible for repulsion between fish, maintenance of nearest neighbor distance, and monitoring of neighbors' swimming speed and direction of travel; when this system is blocked, the fish swims too close to its schoolmates. There is some redundancy between the two systems. Precision schooling thus appears to be the consequence of cues and a few simple response rules to neighboring fish by individual school members.

Worker Organization in Social Insects

The colonies of social insects—ants, bees, wasps, and termites—represent the pinnacle of group organization. The number of individuals in a colony can reach into the thousands. The tasks that must be completed include constructing a complex nest structure, laying eggs, caring for the brood, cleaning the nest, foraging for food, and protecting the colony against predators and parasites. How are all these activities organized and integrated? How do individuals know what to do from moment to moment? The queen controls reproduction and egg laying, but information about specific colony needs is not disseminated from a central source. It is worker-controlled, and each worker is an autonomous unit that can perform a range of chores and does whatever it perceives needs doing. To be sure, there are specific signals to coordinate many tasks. Social ant repertoires contain about 12 different signals, mainly olfactory and tactile, that regulate recognition, foraging, food exchange, alarm, nest building, and brood care (Wilson 1971). Depending on the species, there may be several castes that are specialized to perform certain tasks. Ants tend to have more morphological castes, for example, majors with huge mandibles for colony defense, submajors with extra-long legs for carrying prey, medium workers acting as jacks-of-all-trades, and minims that are small and care for the brood. In bees, on the other hand, all workers are the same size and each can perform any of the required tasks, but the tendency to perform certain tasks changes with age.

The key to understanding colony organization is identifying rules under which workers operate and the cues they use to make momentary decisions. A worker spends a substantial amount of time patrolling the nest and gathering information. If it discovers a brood cell that is only partially completed, it can begin construction where the former left off because the construction rules are fixed. If the nest temperature is too high, bees begin to fan their wings and additional bees will join in until the temperature has been reduced. In army ants that specialize on capture of large arthropod prey, individuals form teams to carry prey back to the nest organized by a larger porter ant. Additional workers will join the team only as long as the prey item is moving below the standard retrieval speed. Army ants also possess systematic rules for orienting the direction of subsequent days' raids from the central nest such that the same area is not exploited twice. Forager bees use the delay time after entering the colony with food before a storage bee finds it and unloads the food as an indicator of colony nutritional status. When this delay time is short, it means the colony is hungry and the forager increases its foraging effort. Similar subtle cues, described more fully in the next chapter, also affect the direction of foraging and the location of food. The social insects have been likened to a superorganism or a human brain, in which the collective intelligence of the mass action of many simple units can find flexible solutions to complex problems (Franks 1989; Seeley 1989).

SUMMARY

1. Social integration requires both **recognition** of key target individuals such as mates, offspring, and group members and **honest signals** to synchronize activities.
2. The **process of recognition** involves transmission of information about individual identity via signals and cues from a potential target, perception of the information by a Bayesian receiver, a decision on the degree of match between the information and an innate or learned template, and actions taken by the receiver toward the target. The interaction between the receiver and the target sender can be viewed as a game of receiver discrimination accuracy versus sender distinctiveness. Animals use four different recognition mechanisms: **spatial location**, **familiarity**, **phenotype matching**, and **allele matching**.
3. Both the production and reception of species-specific signals are usually hard-wired, but location, familiarity, and phenotype matching may also be used in **species recognition**. Selection for species-specific signals may be driven by the cost of low-fitness hybrid offspring, the adaptation of a population to specific environmental conditions, or sexual selection for attractive, high-quality mates.
4. Males and females both benefit from giving each other honest signals to synchronize spawning and copulation. **Synchronization** of reproductive cycles is mediated by reciprocal stimulation of hormonal sequences in the partner. Olfactory signals in particular can be used by males to persuade or manipulate reluctant females into mating.
5. Most species with parental care recognize their offspring by using the cue of the nest site or by familiarizing themselves with the young as individuals during a close association period after birth. Where dependent offspring typically mingle with the offspring of other parents, strong selection pressure favors an individually distinctive **signature signaling system**. When brood size is extremely large and associated with a fixed site, phenotype matching may be used for offspring recognition.
6. Parents and offspring disagree to a certain extent over the amount of **parental investment** to be provided, with offspring preferring more than the parent is optimally selected to give. Game models predict either that they reach a compromise between the two optima, or that the parental optimum is the ESS. There is little evidence for weaning conflict in mammals, but in birds and mammals that solicit provisioning with begging calls, parents discount the calls of older juveniles.
7. Mechanisms for **group recognition** range from individual identification of each group member to recognition of status and age classes only. When group members are obligatorily tied to a site such as a nest, an olfactory

group label to which all members become familiar can be used to distinguish groups. If groups are formed via natal recruitment and consist only of kin, then genetic similarity can be used for group recognition. When there is no opportunity to learn kin characteristics at a common site, allele matching is used to recognize kin.

8. Signals for maintaining harmonious relationships among group members can be categorized along two dichotomous axes: duration of the benefits of the signal, and symmetry of the sender and receiver. Immediate-benefit asymmetric partner signals include **submissive** signals by subordinates and **reconciliation** signals by dominants. **Allogrooming** of dominants by subordinates to build future alliances is an example of long-term benefits between asymmetric partners. Long-term benefit symmetric partner interactions involve physical contact, play, and mutual allogrooming. Finally, **mutual greeting ceremonies** fall into the immediate-benefit symmetric partner category.

9. Animals that range widely in cohesive groups use a combination of signals and cues to **coordinate group movement**. Foraging monkey troops use contact, separation, and movement initiation calls to remain together. Migratory flocks of birds combine temporal and locational cues with vocalizations to sort out into optimally sized V-formation groups. Fish school coordination is based almost entirely on visual and mechanical cues from neighbors.

10. **Social insects** coordinate the complex activities of cooperative brood care, foraging, colony defense, and nest construction and maintenance with surprisingly few signals. Most individual workers in a colony are jacks-of-all-trades who patrol the colony, gather information on the tasks currently needed, and step in to continue any uncompleted job. Individuals use simple rules of thumb to make decisions about what to do next.

FURTHER READING

Hepper (1991b) and Fletcher and Michener (1987) are relatively recent multiauthored books on the popular subject of kin recognition. Otte and Endler (1989) and Lambert and Spencer (1995) are both multiauthored books on the topic of speciation. Godfray (1995a) provides an outstanding review of parent-offspring conflict models. The best compendium of integrative signals is found in Peter's (1980) review of mammalian communication signals. Many of the original articles cited in the sections on male-female integration signals, parent-offspring integration signals, and group integration signals include taxon-specific reviews of the relevant literature.

Chapter 25

Environmental Signals

THE PRIOR CHAPTERS DEAL WITH SIGNALS that provide a receiver with information about the sender: focal questions include the sender's aggressive intentions, territorial status, suitability as a mate, current actions, identity, or group membership. We now turn to signals in which the sender provides the receiver with **referential information** about objects external to itself. Such objects could include conspecifics, accessible resources, or predators. In this chapter, we shall limit our attention to resource and predator referents (under the general rubric of **environmental information**), because they pose the more interesting theoretical problems. Some environmental signals are shared by sympatric species at the same trophic level. For example, forest guenon monkeys (genus *Cercopithecus*) often form multispecies groups; the alarm calls of these species are very similar structurally, and all members of a mixed troop respond to any species' alarms (Gautier and Gautier 1977). Savannah guenons respond to the alarm calls of sympatric starlings (Cheney and Seyfarth 1985). In contrast, some signals are exchanged between species of different trophic levels. Although the most common exchange is between species in adjacent levels (e.g, prey to their predators), communication can also occur between nonadjacent trophic levels (e.g., prey to the predators of their predators). Environmental signals exchanged by animals within the same trophic level are likely to be cooperative in the sense that the sender is providing useful information to the receiver. They are most likely when conflicts of interest between sender and receiver are low, and honesty is then favored by both parties. Signals exchanged between trophic levels occur when conflicts of interest may be very high; they usually entail some form of persuasion, and guarantees of honesty may be required before receivers attend to them. The dynamics and mechanisms of within- versus between-trophic-level environmental signaling are thus quite different. We take up each kind of exchange in turn.

WITHIN-TROPHIC-LEVEL ENVIRONMENTAL SIGNALING

Are There Any Truly Cooperative Environmental Signals?

The existence of environmental signaling between conspecifics or species within the same trophic level has been met with some skepticism. One argument is that most supposedly environmental signals show variations in signal form, amplitude, or emission rate that are clearly related to the motivational state of the sender. This has led some workers to argue that the primary function of these signals is to communicate the sender's motivational intentions; because motivations are likely to be modulated by ambient social and environmental contexts, the provision of any environmental information is then accidental (Smith 1965, 1977, 1981). A ground squirrel that sights a hawk emits a call and flees into its burrow. Although this incidentally warns other nearby squirrels that a hawk is present, environmental signal skeptics would argue that this is not the major function for which the call has evolved. Instead, it is just another signal indicating sender motivation. Most current workers take a more intermediate position and acknowledge that both environmental and motivational information may be present in such signals (Marler et al. 1992). We humans typically include motivational overtones in our declarations about the proximity of some danger or the availability of a preferred food. That animals also show such a mix is not surprising, and should not be used to argue that the provision of environmental information cannot be the primary reason why a given signal evolved. There are in fact numerous examples of environmental signals that are not modulated by urgency or motivation, or where receivers routinely ignore the motivational information. In addition, many studies have shown that senders will not produce signals in the absence of receivers that can benefit from the environmental content of the signals. This would not be the case were the environmental exchange an incidental outcome of motivational communication.

A more serious concern is why a sender should bother to provide environmental information. Environmental signaling within a trophic level (with the exception of distress calls) is usually the reverse of the Sir Philip Sidney game. In the latter, a sender begs a receiver to aid the sender at an immediate cost to the receiver. We have seen that a receiver will only attend to such signals when the receiver obtains a sufficiently large indirect or deferred benefit from helping the sender; this could be a result of close kinship, long-term reciprocity, or a mutual endeavor beneficial to both parties. In environmental signaling, the sender pays a direct and immediate cost to the benefit of the receiver; it should not do so unless it gains a sufficiently large deferred, indirect, or mutual benefit as a consequence. Whereas we can invoke the same kinds of indirect and deferred benefits for environmental communication that we invoked for begging, the two games are not simple opposites. The sender in a begging game pays the costs of signaling, but is likely to be compensated immediately by the response of the receiver. A sender providing environmental information must pay signaling costs, and responses of receivers may provide no direct sender benefits; they may even lead to additional sender costs. For example, advertising a food find takes time and energy (necessary costs), and will only attract competitors that reduce the amount the sender gets to eat (incidental costs). Environmental signaling thus requires particularly large indirect or deferred benefits to senders to justify the payment of both types of sender cost.

As we saw in Part II, the amount of information increases with the number and equal likelihood of alternative messages, and the necessary costs of signaling tend to increase with the amount of information. Costs may be particularly high for environmental signals. This is because the number of alternative messages for environmental signals is usually greater than for equivalent signals about sender motivational states and intentions. Environmental signals can provide information about which possible object is present, what state it is in, and where it is located. Each of these categories of information usually has a large number of reasonably probable answers. For example, most animals eat a variety of foods and are at risk from a variety of predators. The provision of a separate environmental signal for each will surely impose high necessary costs on both parties. Communication of object state also allows for a large number of alternative messages: foods vary in quality, abundnace, and accessibility, and predators may be sleeping, sitting alert, or on the prowl.

Perhaps the most costly type of environmental information is the provision of referent location. A begging sender need not indicate its location to a receiver because the two are usually in close proximity. However, the referent for an environmental signal may be far from appropriate receivers. Communicating its location (azimuth, altitude, and distance) may thus be of great value. There are basically three ways in which the location of a referent may be communicated between animals: (1) the discoverer broadcasts a signal from the referent's location and receivers follow the signal to the site; (2) the discoverer returns to suitable receivers, communicates the fact of discovery, and leads interested receivers back to the referent location; or (3) the discoverer returns to suitable receivers and provides directional information that allows them to find the referent on their own. Broadcast signals are the cheapest to produce, but they allow unsuitable eavesdroppers to benefit from the advertisement. As Trivers (1971) noted, such cheaters will undermine any deferred benefits of reciprocity, and they may reduce nepotistic benefits as well. The alternative mechanisms allow selection of suitable receivers and thus higher indirect benefits, but are much more costly to the sender.

As a result of the multiple costs inflicted on senders by environmental signals, and the exacerbation of these costs if the sender must be selective about receiver identity, the transmission of environmental information is often uneconomical. True environmental signals are likely to be uncommon in animals, and most of those seen will be simple. For example, there may be one food call for all contexts and one predator alarm call for all possible predators, with the provision of no information about object location. Only where conflicts of interest are unusually low (for example in the close kin associations of social insects) and the ecological benefits particularly high should we expect multiple signals for different objects and the provision of locational information to specific parties. As we shall see, this situation is generally what is found, and the few exceptions prove the rule either because they occur in contexts where sender costs are unusually low, or because there are hidden direct benefits to senders that offset the direct costs of signaling; high indirect or deferred benefits are not necessary in these cases.

Resource-Recruitment Signals

Resource-recruitment signals are more easily studied than predator alarm signals because there is rarely any uncertainty about whether the call is aimed at a receiver in the same trophic level or not—few predators benefit from providing signals to their prey. In contrast, it is not always clear whether a signal given in the presence of a predator is intended for other individuals in the same trophic level (a true alarm signal) or instead directed at the predator to discourage it from attacking. Vertebrate parents often provide food-recruitment signals to their young. We have discussed this type of signal in the prior chapter and because it raises no new theoretical issues, do not discuss it further here. However, recruitment of adults to a newly located resource by other adults faces all the problems we outlined in the prior section. Does this kind of signaling occur often, and is it ever very sophisticated?

Several bird species produce signals at food finds that attract other adults. The ways in which these are used are instructive. Domestic roosters produce a specific call when they encounter a local patch of food (Marler et al. 1986a,b). The more preferred the food, the higher the calling rate. Hens respond by approaching the calling rooster. The higher calling rate for preferred foods results in higher rates of hen approach. Roosters are more likely to give this call when females are present, and are less likely to give it when other adult males are nearby. They may also give it falsely when no food is present (Gyger and Marler 1988). The data strongly suggest that roosters announce food finds to females to attract them and thereby increase chances of mating. Any direct costs to roosters of making the call and sharing the food are thus made up by direct benefits via mating. Indirect or deferred benefits are not necessary to explain this behavior.

North American ravens (*Corvus corax*) feed on carrion during the winter. Local populations are divided into territorial mated pairs and vagrant unmated individuals that are largely juveniles (Heinrich 1988). Vagrants produce a distinct call when a new carcass is discovered and this attracts other nearby ravens; territorial pairs do not advertise new finds and vigorously chase off other birds in the vicinity. If enough vagrants can be recruited to a carcass, they are able to displace the resident territorial pair and utilize the carcass. It is not clear whether the calling of vagrants has recruitment as its main function. In captive flocks, only dominant birds produce this call and it may be primarily a declaration of status during competition around a carcass, with the incidental consequence that other birds use it as a cue to locate new food finds (Heinrich and Marzluff 1991). In either case, there are direct benefits to the caller and again no need to invoke indirect or deferred compensations.

House sparrows (*Passer domesticus*) that discover a new food source may produce a chirrup call that attracts other nearby foragers (Elgar 1986a,b). Food that is easily divisible among several foragers is advertised, whereas the same total amount of food in a single indivisible clump is not. The farther the feeder from a safe perch and the closer the feeder to a potential threat (in this case a human observer), the higher the chirrup rate by an initial forager. Elgar's interpretation is that solitary discoverers of shareable food seek to attract other birds to help reduce predation risks while feeding. This again suggests that there are direct benefits to senders that compensate for calling costs.

Cliff swallows (*Hirundo pyrrhonota*) nest colonially and forage in flocks on aerial insects. Brown et al. (1991) have shown that these birds will produce a specialized call when a rich swarm of insects is located on the foraging grounds (Figure 25.1). The call is most likely to be given on days when foraging conditions are poor (roughly 25% of the time). Because the swarms of insects are large and ephemeral, they are not worth defending and no single bird could use them up before they disappear. There is thus little secondary cost to a sender of notifying others of the swarm. On the contrary, because the swarm locations move, reciprocal calling by different members of the flock facilitates the tracking of the current position of the swarm. The low sender costs and the short-term direct benefits thus make this call easy to justify from an economic point of view.

Although a large number of bird species nest or roost communally and show following of successful foragers by unsuccessful ones to resources (*sensu* Ward and Zahavi 1973), few if any of these species perform food recruitment signals back at the communal roosts. For example, black vultures (*Coragyps atratus*) share large communal night roosts, successful foragers are followed on successive days, and foraging groups often consist of kin (Parker et al. 1995; Rabenold 1987). One might think this an ideal context in which successful foragers could perform invitations to follow or more symbolic indications of the location of food finds. Despite this, neither type of signal has

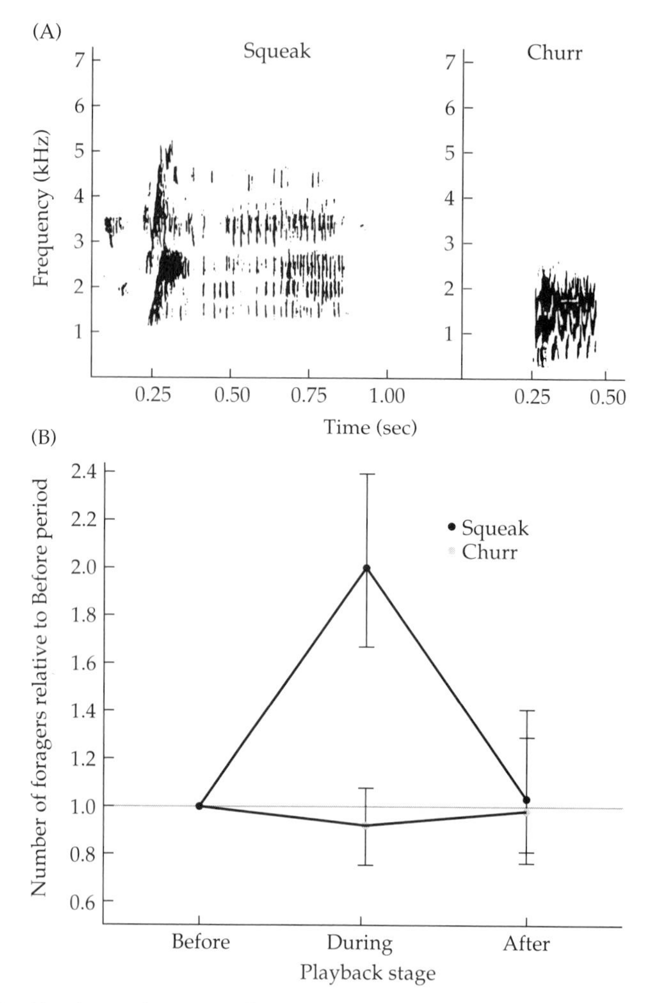

Figure 25.1 Food recruitment calls in cliff swallows (*Hirundo pyrrhonota*). (A) Brown et al. (1991) hypothesized that the squeak call (left spectrogram) is used by cliff swallows to recruit colony members to a newly discovered swarm of aerial insects. In contrast, the churr call (right spectrogram) is used in a wide variety of contexts and thus is not a likely food-recruitment signal. They played back both types of calls in a large field in which some foraging birds were already present and then compared the number of foragers before, during, and after the playback to see whether playbacks attracted more birds. (B) The average number of foragers during 16 playbacks of squeak calls was double that before the playbacks or after; playbacks of churrs had no significant effect on numbers of swallows foraging over the field. These results were statistically significant ($p < 0.004$ or less), indicating that the calls can recruit new foragers to a site. (After Brown et al. 1991.)

been observed. Male ospreys (*Pandion haliaetus*) will perform a sky dance display when returning to a nesting colony with preferred fish species, and colony mates will then fly off in the direction from which the displaying male arrived (Greene 1987). However, such displays are performed only rarely (on no more than 8% of preferred fish captures), and are more likely directed to female mates (who do not fish) than to adjacent nesting pairs (Bretagnolle and Thibault 1993).

A variety of social primates produce at least one food call while at a find. The small tamarins (*Saguinus*) and marmosets (*Callithrix*) of South America give one call upon discovering food and a second distinct call while consuming it (Benz 1993; Caine et al. 1995; Cleveland and Snowdon 1982; Elowson et al. 1991; Pola and Snowdon 1975). Call rates usually increase with food preferences and quantity found. In at least one species, an animal out of sight of other group members is more likely to produce a discovery call than one with others nearby. Animals do not share food, but compete for access at a given location. This suggests that (like house sparrows) the calls are used to ensure sufficient group sizes for predator surveillance during feeding.

Spider monkeys (*Ateles geoffroyi*) may produce a whinny when they discover a new tree in fruit (Chapman and Lefebvre 1990). The likelihood and rate of whinny production increase the larger and more divisible the patch of fruit, and the higher the current average density of available food. They are also given more by dominant individuals than by subordinates. The calls are thus used when a sender has least to lose by recruiting conspecifics.

Several species of macaque monkeys (*Macaca*) produce food calls. Toque macaques (*M. sinica*) give a whee vocalization when they discover a rich food source; other nearby members of the group immediately move to the location of the caller (Dittus 1984). Only very large food patches or ones highly preferred are so advertised. Rhesus macaques (*M. mulatta*), like tamarins, produce calls from one set (two alternative vocalizations) upon discovering a new food source and calls from another set (three distinct types) when consuming it (Hauser and Marler 1993a,b). Discovery calls are associated with staple foods, but may also be given in nonfood contexts; consumption calls are used primarily when rare or preferred items are found and are uniquely associated with food advertisement. Consumption calls are twice as likely as discovery calls to attract recruits to the new food site. Females produce food calls more than males, and females are more likely to produce these calls the more close kin they have nearby. In provisioning experiments, discoverers of new food sources that failed to call and were subsequently found eating were punished by other members of the troop. In this species, a combination of direct (avoidance of punishment) and indirect (kin selection) payoffs seems to be relevant to the evolution of food-recruitment signals.

Like many of the other social primates, chimpanzees (*Pan troglodytes*) have a discovery call (pant hoot) and a consumption call (rough grunt) (Marler and Tenaza 1977). The pant hoot is also used in nonfood contexts and in some populations is one of the most common calls heard. Males tend to pant hoot more than females. Wild chimps are more likely to produce this call when a particularly rich or preferred food item is found (Ghiglieri 1984; Wrangham 1977). Captive chimps only produce pant hoots when sufficiently large amounts of preferred foods are provided (Hauser and Wrangham 1987). Hauser et al. (1993) provisioned captive chimps with small divisible, large divisible, and large indivisible amounts of melon. Pant hoots were only given to large amounts of melon, whereas rough grunts were produced for any amount of this preferred food. The total number of calls was highest for divisible items when the amount was held constant. Chimpanzees do share food, and complex reciprocity associations are known (de Waal 1989). Unlike most other mammals, females are the dispersing sex in chimpanzees, causing local groups of males to be genetically related. A combination of kinship and reciprocity may explain why pant hoots are mostly produced by male chimps.

Finally, naked mole-rats (*Heterocephalus glaber*) form subterranean colonies in which genetic relatedness of members is so high that most of the colony members give up their own reproduction to aid that of a close female relative (the colony "queen"). These animals thus have a social system more like that of social insects than of most other vertebrates. Workers in captive colonies have been shown to return to the main nest area with newly found food and vocalize to attract the attention of other workers. The latter then appear able to follow scent trails of the advertiser back to the new food source (Judd and Sherman 1996). As we shall see below, these food-recruitment behaviors of naked mole-rats are very similar to those of some ants. Again, very high levels of kinship appear to compensate for the costs of new food site advertisement in this species.

Food-recruitment signals in most nonhuman vertebrates are thus relatively simple. Many species give none at all, and those that do distinguish between at most two types of food, characterize quality only crudely, and usually indicate food location by broadcasting a localizable call from the site. In contrast, many social insects have developed resource recruitment to a fine art. Since all of the social insects show high levels of genetic relatedness, indirect benefits may be sufficient to compensate for the costs of exchanging large amounts of information. All have fixed nest sites to which workers return bearing samples of the discovered food item. This indicates what has been found. Most also have a mechanism for indicating the quality of the source, although this may be no more sophisticated than the vertebrate techniques. Where social insects excel, however, is in their ability to indicate the locations of food sources to nest mates and incite them to seek these out. The only vertebrate that comes close to the social insects in this regard is the naked mole-rat, but even it cannot compare with the sophisticated resource-recruitment signals of honeybees.

Let us first consider the ants. Foragers often wander far from their nests. When a worker finds a new source of food, it returns to the nest using visual landmarks and recruits other workers to the new site. Workers that live in small colonies often use **tandem running** to lead fellow workers to new food finds (Hölldobler and Wilson 1990). The returning forager performs a dance or other alerting display at the nest, and then runs back to the food find with a recruit following close behind. In many species, the leading ant releases puffs of pheromone from its posterior to facilitate following by the recruit. It may have also laid a pheromonal trail that both leader and follower can follow. Although many vertebrates roost communally and successful foragers may be followed back to food sites, in very few cases do the leaders display signals to ensure following by roost mates. Such signals are the rule among ants.

In contrast to tandem running with a single recruit, other ant species show **group leading** in which a returning forager lays a trail by using sting gland secretions or hindgut fluids or both. It then incites a number of potential recruits and uses its trail to lead a group of workers back to the food site. In ants with large colonies, the pheromonal trail laid by a returning forager often contains components that both stimulate other workers to follow the trail and enable them to do so. The returning worker need not lead workers to the site. The result is **mass recruitment** of many workers. Recruits at the new site attempt to obtain some of the food; if they are successful, they return and reinforce the initial trail with their own secretions. As more successful foragers return, more pheromones are deposited and the active space of the trail becomes greater. This then attracts even more workers. Once the advertised site becomes sufficiently crowded, new arrivals are unable to obtain food and they return without adding trail marks. The number of workers recruited to any site is thus self-regulating (Deneubourg et al. 1983; Strickland et al. 1995). Once at the general foraging site, some ant species may also produce air- or substrate-transmitted sounds to gather workers around particularly large or struggling food items (Markl and Hölldobler 1978).

Many ant nests establish a branching system of trails. Major trunk routes radiate out from the nest and are often used for weeks at a time. In one species (*Pheidole militicida*), 68% of the ants using a given trunk route had used the same route the day before (Hölldobler and Möglich 1980). They thus are familiar with a route. Trunks divide into branches that are used for days without changing. Branches in turn divide into twig routes that change daily. Because of the persistence of familiar trunk and branch routes, the task of following food trails is greatly simplified. Visual landmarks can be used for much of the route, and careful following of pheromone trails is necessary only on the terminal twig sections. One cost of such routes is that they are often conspicuous, allowing competing species to follow them to the food sources. Even if the interloper species only discovers a pheromonal trail, the continuous nature of this type of mark makes it relatively easy to follow.

How accurately do ants recruit nestmates to a food source? Recruits following a pheromonal trail make two kinds of errors: they may stop too soon or overshoot the target (longitudinal errors), or they may veer from the trail (lateral errors). Experiments with various species indicate that workers on average travel only 80 to 90% of the true trail length; most stop too soon (Hölldobler and Wilson 1990; Wilson 1962). Lateral error varies with the study. Fire ants (*Solenopsis saevissima*) show fixed lateral errors to either side of the trail regardless of trail length; most workers would thus be found within a 2 degree sector at 5 cm from the nest and a 1 degree sector at 10 cm (Wilson 1962). In the ant *Formica fusca*, lateral errors were much greater, with most recruits being found within a 50 to 55 degree sector at 2, 12, and 22 cm from the nest (Möglich and Hölldobler 1975). As a result of both kinds of errors, only 18%

of workers leaving the nest in a group-leading species (*Tetramorium impurum*) reached an advertised target 10 cm from the nest (Deneubourg et al. 1983). The average recruit traversed only 17% of the trail before it wandered off track. In contrast, 73% of a mass-recruitment species (*Tapinoma erraticum*) setting out on the trail arrived at the target, and the average worker traversed 68% of the trail before it deviated significantly or turned back. Ant species clearly vary in the accuracy of their trails. However, the fact that ant colonies can have thousands to millions of workers means that even a low-percentage recruitment rate can result in the arrival of large numbers of workers at any new food site in a short time.

In addition to recruitment to food, ants also advertise and lay trails to new nest sites (Hölldobler and Wilson 1990). Some small colony species make only simple nests that they frequently abandon and rebuild at new locations. Scouts search for suitable new nest sites and perform advertisement dances, trail laying, and either tandem running, group leading, or mass recruitment to lead other workers to their selection. The same posterior glands used for food trails are also enlisted for nest-site advertisement. In ponerine ants, some species forage solitarily and never lay food trails, yet use sting gland secretions for recruiting other workers to new nest sites. Related species use the same organs and similar pheromones to lay trails for mass recruitment to patches of aggregated prey such as termite colonies (Hölldobler 1984). There is thus considerable flexibility in the functions of pheromonal trails once they evolve in an ant taxon.

Termites also use pheromonal trails to recruit workers from the nest to suitable food sites (Figure 25.2). Most species have very large colonies with millions of workers and utilize mass-recruitment methods. In contrast, most colonial wasps and bumblebees do not provide information to nestmates about the locations of food sources. A few species return to nests and perform agitated displays, but the only information provided is the identification of the resource via its odor.

There are two groups of social bees that actively recruit workers to new resource sites. The stingless bees are widespread in the tropics. Workers of the genus *Trigona* return to the nest from a food find leaving small dabs of head gland pheromones on objects every 2 to 3 m along the route (Kerr et al. 1981; Lindauer 1961). Then the forager performs an agitated dance at the hive, accompanied by specific sounds to alert potential recruits, provides samples to indicate the type of food, and finally leads as many as 50 foragers along the scent trail to the food find. Note that in contrast to the relatively continuous pheromone trails of ants, stingless bee trails are like a dotted line. A recruit following the trail must be able to extrapolate from the last few pheromone dots to know where the next one is likely to be. It is thus essential that a new recruit be given an initial sense of the trail's direction by a bee familiar with the target site. No accuracy figures are available for *Trigona* trail following; however, one might expect this interrupted form of trail to be lost more easily than the continuous trail of ants. One advantage of interrupted trails is that interlopers encountering a dot of pheromone will have a more difficult time

Figure 25.2 Foraging termites following an odor trail. Tropical termites like these *Nasutitermes* use pheromones to mark trails to food sources in ways very similar to those of ants. A gland on the ventral side of the abdomen produces pheromones and leaves a trail when pressed against the substrate as the termite moves. In some species only certain subsets of the worker castes have working glands and thus the responsibility of trail routing. All termites eat cellulose that is broken down into useful nutrients by symbiotic protozoans in their gut, or fungi cultivated in their nests. Some species never leave single masses of wood except to found new colonies: their pheromone trails function to redirect traffic within existing tunnels as needs change. Many tropical species, such as the one shown here, actively forage for leaves, wood, and seeds outside their nests at night. Pheromone trails are a critical element of these forays. The oddly-shaped individuals in this genus are soldiers with special head glands and delivery tubes for squirting sticky materials on attacking ants. (Photo courtesy of Carl Rettenmeyer.)

finding and following the trail than is the case for the more continuous trails of ants and termites.

Another common genus of stingless bees, *Melipona*, does not lay scent trails to food sites but dances and produces specific sounds at the nest to elicit foraging by fellow workers. Esch (1967) has shown that the duration of the pulsed sounds produced increases with the distance to the target. When enough potential recruits are attracted, dancers emerge from the hive and use a zig-zag flight to indicate the direction to the food. Recruits then fly off towards the food source. Field studies on a Panamanian species showed that recruits may also be provided with information about the height of the source above the ground (Nieh and Roubik 1995). How this is encoded into forager dances or zig-zag flights is unknown; scent trails are definitely not the mechanism.

The most sophisticated signals for food recruitment known occur in honeybees (genus *Apis*). Most of the seven known species occur in tropical southeast Asia. All species are highly social and build comb nests using wax they secrete from their own bodies. Foragers collect nectar as an energy source and pollen for protein. They also collect propolis resins from various plants that they use as glue to attach combs to trees and seal holes in nest sites. Water and new nest sites are additional resources sought by scout workers. The dwarf honeybee (*Apis florea*) is a small tropical species that builds a single uncovered comb on a thin tree branch. The top of this nest extends over the branch. Returning foragers or scouts land on the horizontal top of the nest and perform a **waggle dance** to other workers (Dyer 1985). This involves a figure-8 circuit with a wagging of the dancer's abdomen while traversing the portion between the two circles (Figure 25.3).

The waggle dance provides coded information about the location of a resource (von Frisch 1967). The angle of the dance relative to the position of the sun tells recruits which azimuth they should fly (also relative to the sun) to find the advertised resource (Figure 25.4). The dancers advertise the direction a recruit ought to fly were there no obstructions or deviations (hence the source of the term "beeline"). If there are deviations, the dancer extrapolates the direct angle of its return flight even though it did not fly directly along that line. The duration of a complete circuit of the dance indicates the distance to the resource: the longer the circuit, the farther the advertised site (Figure 25.5). When foragers must fly in a headwind or up a hill to get to the resource, the distance advertised by the dance increases. The height of the site above the ground is not provided. The dancer then identifies the type of resource advertised (pollen, nectar, water, or propolis) by providing samples to attendant recruits. Dances advertising new nest sites are only performed when a swarm of bees has left the parental hive and is temporarily bivouacked awaiting agreement on a new site. The quality of the source is indicated by the

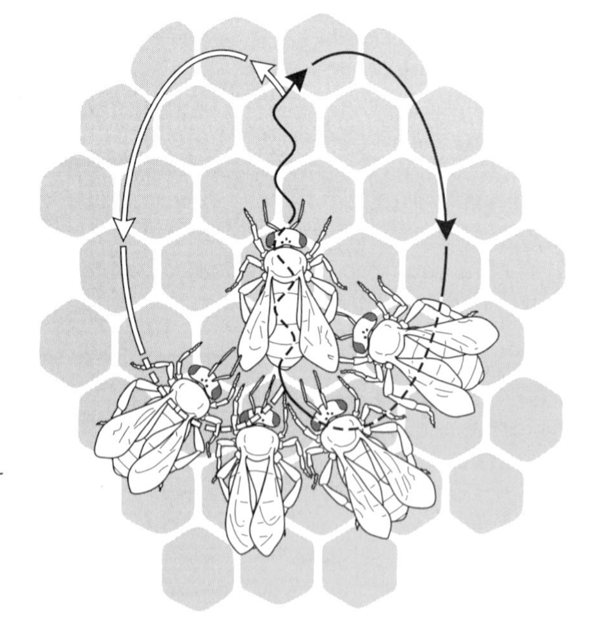

Figure 25.3 Waggle dance of honeybees (*Apis* spp.). Returning worker repeatedly dances in figure-8 pattern. During a run along the center of the two loops, it maintains a constant angle on the comb and waggles its body from side to side at a constant rate. The angle and the rate of waggling provide recruits with information about the location of the advertised resource. (See Figures 25.4 and 25.5.)

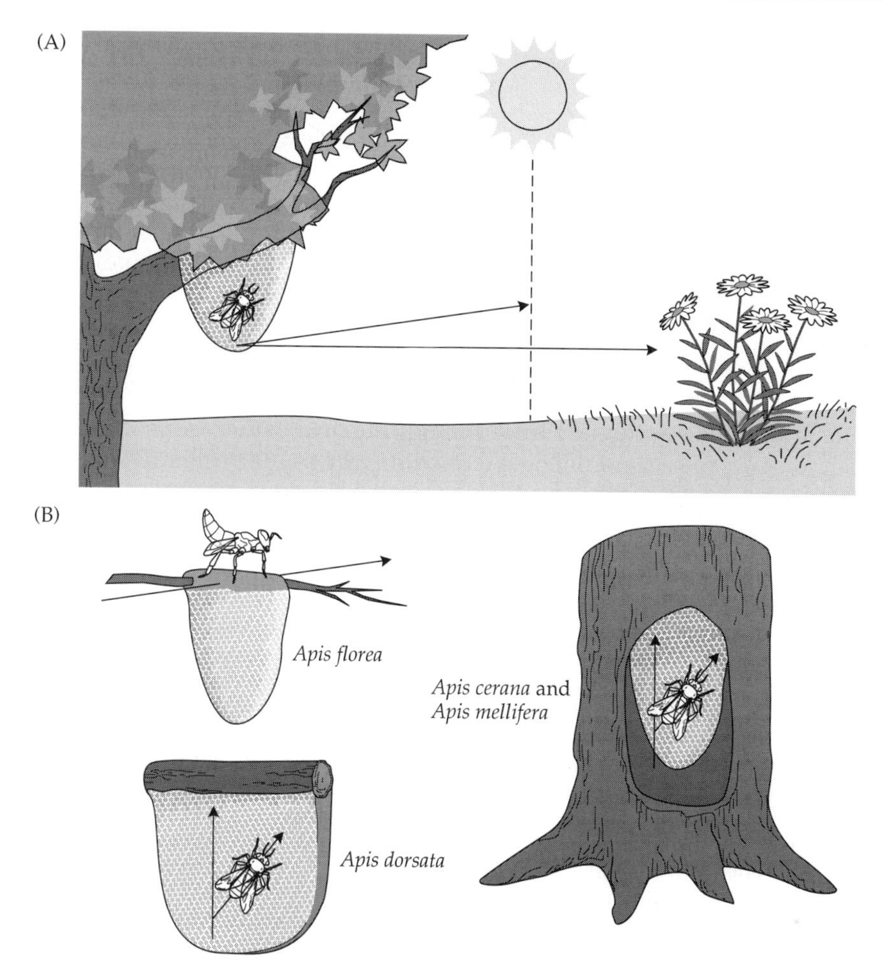

Figure 25.4 Communication of azimuth angle in the honeybee waggle dance. (A) A returning forager determines the angle between the line connecting the hive to the sun and the line connecting the hive to the food source. In this case, food is 40° to the right of the line to sun. (B) Different tropical honeybee species use somewhat different codes to communicate resource angle. Dwarf honeybees (*Apis florea*) suspend their nests from small branches. A waxy horizontal platform is built over the branch on which foragers dance. Because the hive is open to the elements, the dancer and recruits all know where the sun is located. The dancer thus adjusts its dance to the correct angle relative to the sun (in this case 40° to right). If the sun has moved significantly before the forager can dance, it adjusts the angle to what it should now be to find the resource. Giant honeybees (*A. dorsata*) also have exposed combs, but these are built below large branches or rocks; there is no horizontal surface on the hive available for dancing. Foragers thus dance on the vertical surface of the comb and use the convention that up (with respect to gravity) means fly towards the sun, and down means fly away from the sun. A giant honeybee advertising the resource shown above would thus dance at an angle 40° to the right of vertically upwards. This use of gravity to represent the direction to the sun has allowed other species such as *A. cerana* and *A. mellifera* to nest inside hollow trees where the sun can no longer be seen. This type of site provides greater protection for the colonies and the ability to withstand more extreme weather conditions. Both species have exploited this ability by invading temperate climates.

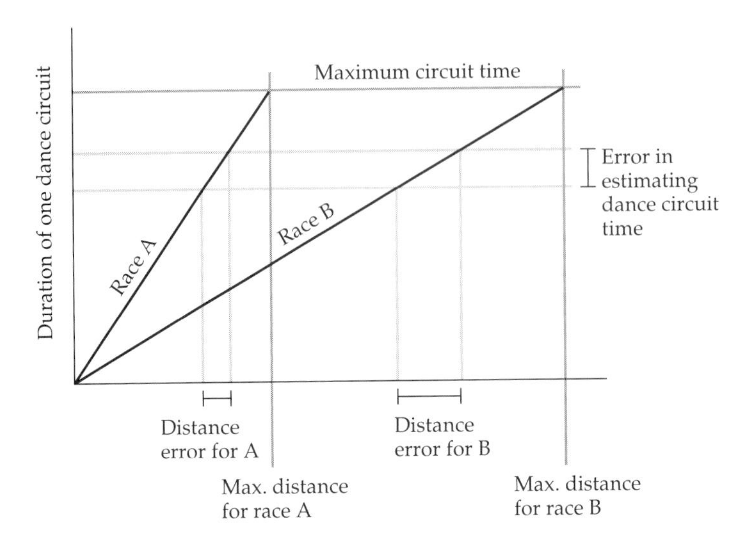

Figure 25.5 Communication of distance to resource in the honeybee waggle dance. A dancer bee indicates the distance to an advertised resource by adjusting the number of waggles per circuit of its dance. Longer distances are encoded by more waggles and thus slower circuit times. The rate at which the dance-circuit durations increase with the distance to the advertised site varies among different honeybee species and races. Two races of the domestic honeybee are represented here. Race A shows a steep slope relating dance-circuit time to target distance, whereas race B shows a shallower slope. If both races have the same maximal circuit time that they can monitor, this limit is reached sooner for race A than for race B. Thus race A has a smaller maximum distance that it can advertise than does race B. Assuming both races have the same amount of dancer error in the performance of a given circuit time, this translates into a smaller error in distance as perceived by a recruit of race A than for a recruit of race B. Thus steeper slopes lead to greater accuracy in the communication of distance information, but they are limited to a shorter range of advertisable distances.

length of time that the dancer persists in its advertisement. Potential recruits attend to a number of dances and then fly off to find the resource. They are not led, as in some stingless bees.

Note that the use of the sun as a reference point introduces some difficult problems. The most obvious is that the sun moves over time. Honeybees have internal clocks that allow them to extrapolate sun positions. Thus if the resource were 10 degrees to the left of the sun when the worker discovered it, the dancer corrects for the passage of time during the return flight and perhaps does its waggle run at the currently correct angle of 12 degrees to the left of the sun. Workers need to be exposed to normal sun movements to learn the patterns and make correct temporal adjustments. What do dancer and recruit bees do when the sun is obscured by clouds or nearby landmarks? As we saw in Box 7.1, ultraviolet light (UV) from the sun is scattered and polarized as it passes through the atmosphere. The eyes of honeybees (and many other Hymenoptera) are sensitive to polarized UV and can use the pattern of polarized light vectors in any patch of blue sky to estimate the position of the sun. When the sky is completely covered by clouds, the bees can still locate the sun. Because most UV is scattered and reaches the bees by indirect routes away from the direct route of longer wavelengths, any appropriately sized patch of bright cloud with little UV is taken by the bees to be the sun. Finally, dancing dwarf honeybees can combine the known positions of nearby trees and their learned knowledge of daily sun patterns to estimate the sun's position when it is not visible (Dyer 1985).

The giant honeybee (*Apis dorsata*) is also a tropical species that builds uncovered nests. However, the combs are suspended beneath very large branches or rocks, making it difficult for the bees to dance on a horizontal nest top. Instead, these bees dance on top of a sheet of nestmates covering the vertical surface of the comb. The vertical orientation of the dance makes it impossible to dance relative to the sun's position. The bees have thus evolved a symbolic convention in which dancing vertically upward represents a flight directly at the sun, and vertically downward a resource away from the sun (Dyer 1991; Lindauer 1961). Any other angle between a resource and the sun is converted to one relative to gravity by using the same convention.

The final step in honeybee dance evolution occurs in *Apis cerana* and *Apis mellifera*. Both species build their nests inside hollow trees or caves. While this provides additional protection for the nests, recruits are no longer able to see the dances of returning foragers. In these species, the visibly conspicuous dance postures of open-nesting species are absent, and in their place are dances that emphasize lateral movements and allow for tactile sampling. These are accompanied by special sounds and near-field air movements (Figure 25.6; Esch 1961; Michelsen et al. 1986; Towne 1985). Although the open-nest species also produce sounds during dancing, it is in the cavity-nesting species that sounds (near or far field) appear critical for the exchange of resource location information (Dyer 1991). There is also evidence that it is only those recruits who are oriented along the line of the waggle run (and are thus aimed in the right direction) that leave and find the resource (Judd 1995). The sounds may be the mechanism that allows recruits to know whether they are or are not aligned with the dancer during this part of the dance. In addition to the waggle dance, returning foragers of *Apis mellifera* perform a simpler circling advertisement for resources within a short distance of the hive called a **round dance**. This dance conveys no utilized angular or distance information, but simply incites workers to leave the hive looking for a nearby resource of the type provided by the dancer. Dances for slightly greater distances are intermediate in form between round and waggle dances. These are called **transition dances**.

Cavity nesting appears to be the critical adaptation that allowed *Apis mellifera* to move into temperate western Asia, Africa, and Europe, and *Apis cerana* to invade similar habitats in eastern Asia. The former species has spread so widely that it now consists of at least 20 different races. Although all of these races perform round and waggle dances to advertise resource finds, they have adjusted the dance codes to meet their own environmental require-

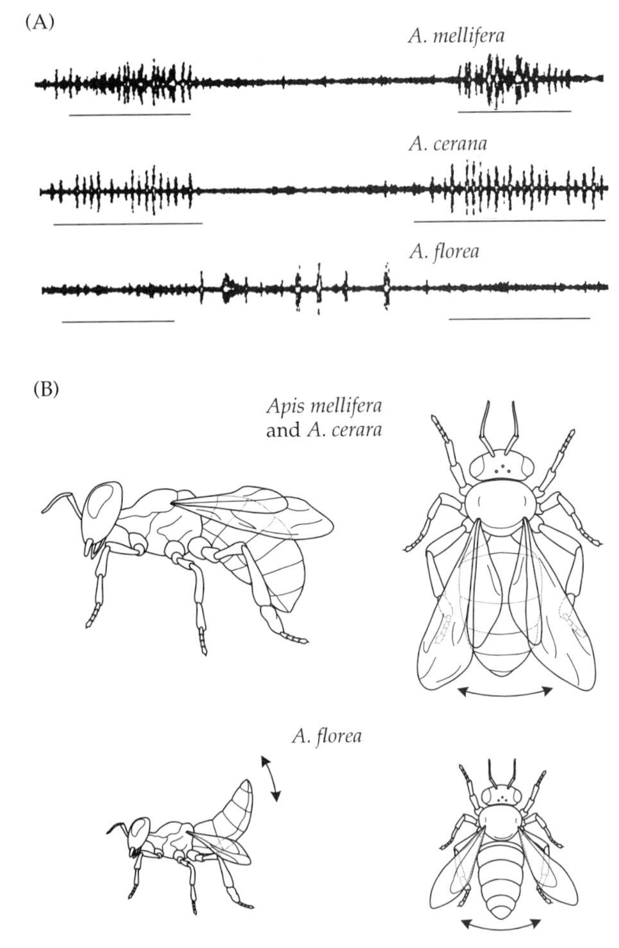

Figure 25.6 Differences in cues of waggle dances used by open-nesting versus close-cavity-nesting honeybees. (A) Time domain waveforms of sounds produced by two cavity nesters (*Apis mellifera* and *A. cerana*), and one open-nesting species, *A. florea*. Bars under plots indicate periods when a dancer was performing the waggle part of the circuit. The two cavity nesters always produce distinct sounds while waggling, whereas the open nester is relatively silent. In this example, the dancer produced some sounds between waggle runs. These sounds are irregularly produced if at all. (B) Differences in waggle dance postures of the same three species. *A. florea* dancers elevate their abdomens above the mass of attendant recruits. This position plus vertical oscillations of the abdomen during waggling make the circuit duration and angle conspicuously visible to all. In contrast, *A. mellifera* and *A. cerana* dancers keep their abdomens close to the substrate and show only a horizontal oscillation during waggling. This is presumably because it is acoustic, not visual cues, that cavity-nester recruits use to extract dance information. (From Towne 1985, © Springer–Verlag.)

ments. For example, the distances from the hive at which round dances are replaced by transition and eventually waggle dances vary with the race of bee from as little as 3 m to over 80 m (Lindauer 1961). Presumably, this distance is adjusted within each race depending upon the relative amount of time that foragers can feed close to home instead of at long distances from the hive. Similarly, the code relating the duration of a waggle circuit to the distance of the site advertised varies with race and with species of *Apis*. There may be an adaptive logic to these variations. For example, the steeper the relationship between waggle duration and target distance, the more accurately a recruit will perceive the advertised target's distance (Figure 25.5). High accuracy may come with a cost, however. If there is a maximum circuit duration that a recruit can measure, this limit will be hit at a much shorter distance from the hive if the circuit duration-target distance relationship is steep than if it is shallow (Dyer 1991; von Frisch 1967). Thus races or species that regularly forage at long distances from the hive may have to adopt circuit duration-target distance relationships that are shallow and sacrifice accuracy in favor of range. Is it the case that all races and species have the same maximum circuit duration? This appears to be roughly true for the various races of *Apis mellifera* and thus may explain distance-code variation within this species. It is clearly not true for different species of *Apis*. *A. florea* and *A. dorsata* show maximum dance durations of about 30 sec whereas *A. cerana* and *A. mellifera* circuits rarely exceed 10 sec (Dyer 1991). As we have noted, the former two open-nesting species use vision to interpret dances, whereas the latter cavity-nesting forms rely more on tactile and acoustical cues. It is not unlikely that visual assessment allows for measurement of longer circuits than acoustical or tactile means. If one therefore divides the different species according to nesting habitat, within each group a tradeoff between accuracy and maximum distance does appear to hold (Figure 25.7).

How accurate is the honeybee waggle dance communication? Most recruits monitor 6 to 7 successive dances and then use the average angle and distance to seek the resource. Domestic honeybees start looking for the target with a 2 to 10% error in expected distance and a 9 to 12 degree error in angle (Winston 1987). Although this distance error is somewhat smaller than for ants, the angular error is about the same. Some authors have argued that the angular error of honeybee dances can be smaller (e.g., when the advertised resource is a possible nest site), but because most flower patches are large, the dancers purposely add angular error to spread recruits out over new food locations (Towne and Gould 1988). On average, about half of the recruits to a

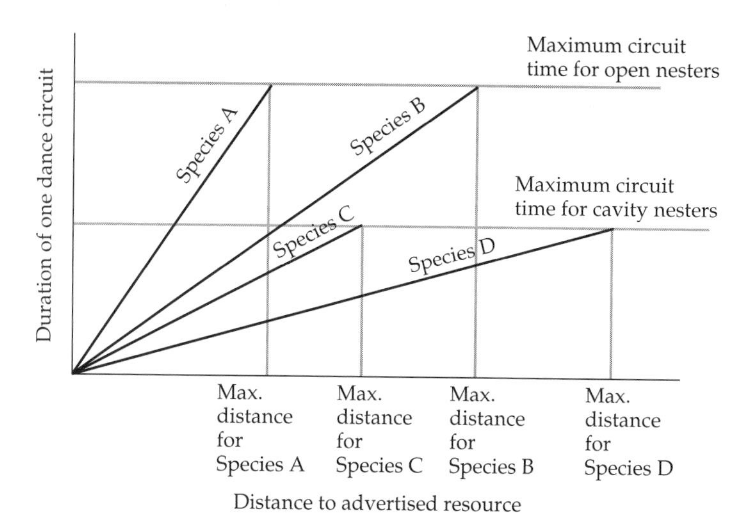

Figure 25.7 Coding differences for distance in the waggle dance of different honeybee species. It has been suggested that there is a different maximal circuit time for hole-nester honeybees and open-nesting species that may be a result of the reliance on different dance cues (see Figure 25.6). If so, then the tradeoff between maximum range and accuracy shown in Figure 25.5 no longer holds for all species taken together. However, for species sharing the same circuit duration limit (e.g., species A and B or species C and D in Figure 25.5), the tradeoff should be retained.

dancer make it to the advertised target. Maximal foraging ranges vary with the species. *Apis cerana* remain within 2 km of their hive, *A. mellifera* usually forage within 8 km of their hive, and both *A. florea* and *A. dorsata* may feed up to 12 km away (Dyer and Seeley 1991; Visscher and Seeley 1982). It is amazing that these small insects can forage so far from their hives and still be able to indicate the locations of food finds to nestmates with such accuracy. This accuracy of communication plus large hive sizes result in a very efficient operation: a typical hive of 20,000 to 40,000 honeybees can collect up to 5 kg of nectar daily (Winston 1987).

We are thus left with some marked differences between social insects and vertebrates in the kinds of resource recruitment signals they produce. Whereas ants and two groups of bees return to their nests to incite, lead, and in some cases direct recruits to food finds, vertebrates (with the exception of humans and the naked mole-rat) at most produce broadcast signals from the resource. One could argue that the small size of social insects precludes them from using the simpler and cheaper mechanism of producing loud, long-range signals from a food find. Why should a large vertebrate perform a more costly signal when a cheaper one will do as well? This does not explain the absence of explicit recruitment signals at the colonial roosts of birds and mammals whose foraging ranges exceed the maximum ranges of any broadcast signal they could produce. Signals given at the roost would surely be useful, but they are conspicuously absent. One answer is that colony-mates are more closely related in bees and ants than in most vertebrates. This allows indirect benefits to compensate enough for the high costs of exchanging sophisticated resource signals in the social Hymenoptera, but not so in equivalent vertebrates. The only deviant nonhuman vertebrate is the naked mole-rat, which is unusual in the degree of genetic relatedness within colonies. It is thus the exception that proves the rule.

Alarm Signals

As noted earlier, a signal emitted in the presence of a predator could be an alarm intended to warn others or it could be designed to deter the predator from attack. Our focus in this section is the former. However, many published studies do not explicitly consider both alternatives, and this should be kept in mind when considering specific examples below. It is possible that both functions are operating in some systems.

Alarm signals can be classified according to the urgency or level of predator threat (Klump and Shalter 1984). When predators are detected at some distance, a sender has time to move to a secure location before producing a **low-risk warning**; this may allow it to include some information about the type and location of the predator in the signal. An initial warning may be followed by **on-guard** signals, indicating continued alert until the risk has passed. These are usually variants of the low-risk warning that are emitted at intervals. A predator that is not an immediate threat because it is immobile, engaged in nonthreatening activities, or relatively small may elicit **inspection** or **mobbing** signals that are given as prey aggregate around the predator or harass it, respectively. A predator discovered at close proximity or in the process of attacking may elicit a **high-risk warning**. Such signals are usually short in duration and lacking in detailed information about the situation. Where these are different in form from low-risk warnings, receivers may respond differently to them. Just prior to and during a predator attack, a victim may produce **distress** signals. These are usually loud and highly localizable. As discussed in Chapter 18, each of these contexts favors somewhat different design features in the associated signals.

The production of at least two different types of warning call is common in birds and mammals. Most ground squirrel species give a short whistle in high-risk situations, and longer trills or chatters in low-risk contexts (Macedonia and Evans 1993; Sherman 1985). Because raptor attacks are often swift and dangerous, these are most often associated with the high-risk warning; the slower approach of terrestrial predators such as coyotes, weasels, or badgers more often elicits low-risk calls. However, the strength of this association varies with species of squirrel. In some, distant raptors may elicit trills, while the sudden appearance of a terrestrial predator close to callers leads to whistles. This variation reinforces the notion that it is the urgency of the alarm, not the type of predator per se, that is communicated in most ground squirrels.

Ring-tailed lemurs (*Lemur catta*) and domestic roosters also produce separate warning calls for aerial and terrestrial predators; in both cases, however, neither proximity of the predator nor urgency of risk causes a switch between call types (Evans et al. 1993; Periera and Macedonia 1991). Many corvids such as jays, crows, and nutcrackers mimic their predator's calls; whether these imitations are used as predator-specific warning signals is still unclear (Hope 1958; Löhrl 1980; Slagsvold 1984, 1985). Vervet monkeys (*Cercopithecus aethiops*) produce different warning calls for each of four predator classes: uncommon mammalian predators, leopards, snakes, and eagles (see pages 479–480). The matching of call type to predator class is very accurate and unaffected by level of risk (Cheney and Seyfarth 1990). In all of these taxa, the type of predator, not urgency, seems to be the relevant information communicated.

Differences in the number of warning signals in a repertoire and their application appears best predicted by the number of alternative responses available to receivers (Macedonia and Evans 1993). Ground squirrels exhibit one of two alternative responses to warning signals. High-risk calls cause them to run for their burrows or a nearby bush, whereas they find the nearest high point, sit up to watch, and scold the predator after low-risk calls. Vervet monkeys have different escape behaviors for each of the four classes of predators they encounter. These behaviors remain predator-specific regardless of the urgency of the threat, and so do the associated alarm calls. Playbacks of each type of warning call nearly always elicit the appropriate response (Cheny and Seyfarth 1990).

This tight linkage between suitable response and call type might suggest that no referential information is really transferred in warning signals. However, a clever experiment by Seyfarth and Cheney (1990) shows that this supposition need not be the case. Vervets respond similarly to their own eagle warning call and the generalized raptor warning of sympatric starlings. Repeated playbacks of either of the warning calls to the monkeys over a short period result in habituation and reduced responsiveness. By habituating monkeys to their own eagle call, and then playing starling raptor calls, vervet leopard calls, or vervet snake calls, one can determine whether the monkeys treat their own eagle calls and starling raptor calls as having the same referent or not. The results show clearly that they do, indicating that some referential information is transferred in these calls.

There has been much debate over sender economics for warning signals. In a number of ground squirrel species, it has been shown that emission of warning signals can be very risky for senders. In Belding's ground squirrels (*Spermophilus beldingi*), nearly half of those killed by terrestrial predators had just given warning calls, and callers were much more at risk than noncallers (Sherman 1977, 1980). Guards in dwarf mongoose (*Helogale parvula*) groups also experience significantly higher risks than other group members (Rasa 1989a,b). As argued earlier in this chapter, these costs must be compensated in order to justify the production of alarm signals.

Table 25.1 lists seven hypothesized benefits that might compensate for the costs of true warning signals in animals; signals directed at predators are covered on pages 841–847. Discriminating between these potential benefits in actual examples is not always easy. However, different types of signals should be

Table 25.1 Compensatory benefits to senders of emitting warning signals

Hypothesized benefit	Authors
Manipulate fellow prey: The sender flushes group members that are more likely to flee toward the predator than is the sender, who knows where the predator is located.	Charnov and Krebs 1974; Brown 1982
Synchronize response: Sender spots predator and wants to flee or freeze, but would be conspicuous if it alone did so. It thus produces a signal that causes all group members to flee synchronously, or all group members to freeze and hide. This spreads the risk.	Dawkins 1989; Sherman 1985
Coordinate flight: Sender's risk during flight is minimized the more group members remain near to it during flight. Signals thus keep group together while fleeing.	Hirth and McCullough 1977; Magurran and Higham 1988
Protect mate: Where survival of a mate is important to its own fitness, the sender may benefit by producing a warning signal.	Witkin and Ficken 1979; Morton and Shalter 1977
Reduce future attack: By making it less likely predator will catch any prey in the sender's group, the predator is more likely to hunt elsewhere or hunt alternative prey. This reduces the future risk to sender.	Trivers 1971
Maintain optimal group size: If survival or reproduction is maximal at the current group size, reducing chances of loosing group members may benefit the sender in the future.	Smith 1986a; Rasa 1989a,b
Obtain kin-selection benefit of signal: If sufficiently close kin are nearby, sender may benefit indirectly by preventing their demise.	Maynard Smith 1965; Hasson 1989a

associated with different types of benefits. If senders were **manipulating** fellow group members to the former's benefit, we might expect senders to flee in a direction away from the predator, whereas other group members would scatter in random directions. Such senders would not give warning signals when alone. The **synchronization** of responses to alarms can be beneficial in several ways. For example, an animal that spots a predator and flees alone may be sufficiently conspicuous that it suffers higher risk than those that do not flee. It would be better to emit a warning signal so that all group members flee together. Such a signal can be loud and easily localizable at no cost. Where stealth is the best response, the animal spotting the predator should produce a signal that causes all group members to freeze or sneak into cover. This type of signal should be as soft and unlocalizable as possible to prevent the predator from detecting the hidden group. Signals used to **coordinate flight** should be given continuously as the group flees from the predator; a single alarm call is unlikely to be involved in such coordination. **Defense of a mate** is most likely when long-term cooperative efforts favor stable pair bonds (Figure 25.8); even here, it is not always obvious whether warning signals are aimed at mates or nearby offspring. Where pair bonds are seasonal, mate protection is an unlikely compensation for warning signals except during the breeding period. **Reduction of future predator attacks** is more likely for resident than for migratory species; unfortunately this is not a unique prediction, as resident species may be more likely to have kin in the vicinity than do migrants. Were **maintenance of optimal group size** the reason for warning signals, one would expect group members to call only when group size was near or below optimal values. Since most animal groups are larger than optimal (Giraldeau and Gillis 1985; Pulliam and Caraco 1984; Sibly 1983), many occurrences of warning signals are not easily explained by this scenario. Finally, the benefits of warning signals may be indirect. If kin of sufficiently close relatedness are sufficiently near to the sender, costs of warning signals may be made up by **nepotistic benefits**.

Sherman's (1980, 1985) studies of low- and high-risk warnings in Belding's ground squirrels consider a number of these alternatives. Lower-risk warnings, which are still quite risky for senders to produce, are only given when sufficiently related kin are nearby. Note that these need not be descendant kin (e.g., offspring or grand-offspring) to justify emitting warning calls. This strongly suggests that the indirect benefits of emitting such calls are the major compensation for increased risks to senders. Nepotism cannot explain the production of higher-risk signals (usually given for raptors), as the rates of calling are independent of the number and dispersion of kin. Because senders and receivers do not flee in different directions with respect to the location of the approaching predator, manipulation does not seem a likely factor here (Figure 25.9). Flight is not coordinated spatially but scattered, and because the calls tend to be given only once, flight coordination can be excluded. Mate protection can be ruled out, since mating status had no relationship to calling. Sherman concluded that the most likely benefit was temporal synchronization of flight (as opposed to spatial coordination). In support of this

Figure 25.8 Warning signals of mammals with long-term pair bonds. A number of mammals exhibit long-term pair bonds in which protection of a current mate may provide direct benefits to compensate for the costs of alarm signals. (A) Patagonian cavy or mara pairs (*Dolichotis patagonum*). This large South American rodent lives in arid areas. Pairs mate for life and are intensely monogamous even though several pairs may share a single den to house young. The male spends much of its time guarding the female and is the most likely member of a pair to give an alarm signal. (B) Beavers (*Castor canadensis*) form long-term pair bonds, and both mates share in the construction of dams and lodges. Alarms consist of slapping the water's surface with the flat tail and are usually given by adults. (C) Elephant shrews (e.g., *Elephantulus rufescens* shown here) live in Africa. Most species form monogamous pairs that share the task of meticulously clearing a network of escape trails within their territories. Alarm signals include drumming of the feet on the ground. (A courtesy of Gerard Dubost).

interpretation, he noted that high-risk alarms are given by species that have clustered-burrow distributions (thus making synchronized flight possible), but not in species with widely scattered burrows and thus widely scattered individuals.

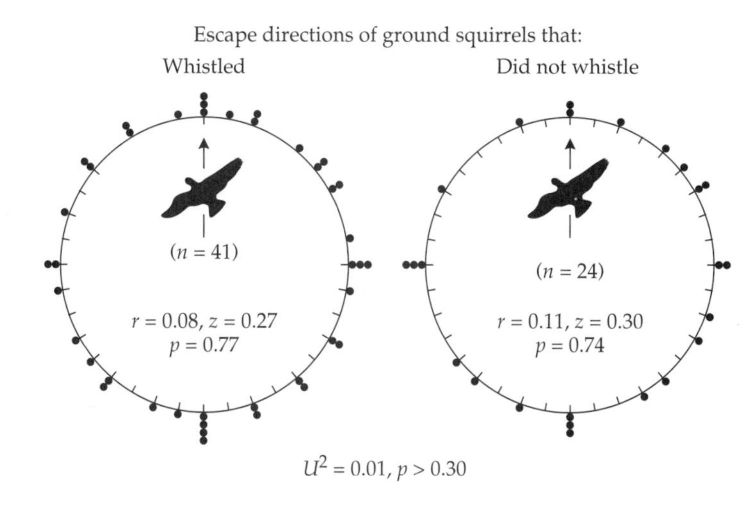

Figure 25.9 Direction of flight of senders and receivers after high-risk warning in Belding's ground squirrels (*Spermophilus beldingi*). Neither senders nor receivers show any directional pattern of flight relative to the location of the predator after emission of a high-risk warning. There is thus no indication that compensatory benefits of emitting high risk alarms is due to manipulation of fellow prey. (From Sherman 1985, © Springer–Verlag.)

Most chickadees and tits (*Parus* spp.) exhibit high-risk warning signals that have low intensities, high frequencies, and gradual onsets and offsets (Ficken 1990; Marler 1955). These warnings are often given in wintering flocks when birds are not paired and groups contain few, if any, kin. The response to such a call is for nearby receivers to sneak into the nearest cover and remain motionless. Klump et al. (1986) have shown that sparrowhawks, the most common predator of the great tit, are 30 dB less sensitive than great tits at the dominant frequency in the high-risk warning. The high frequencies in the call attenuate rapidly and, given ambient noise, cannot be heard by the hawks unless they are within about 10 m of the callers (see also pages 603–605). Since the tits never emit the call unless the hawk is at least 10 m distant, the sender experiences only minor risk by emitting it (Klump and Shalter 1984). This signal appears to be a stealth synchronization call.

Tits often form mixed associations with other species in winter. Many of these appear to have convergent high-risk warning calls (Marler 1955) or at least to respond to the tit's calls. Downy woodpeckers (*Dendrocopus pubescens*) may join such mixed groups, but they appear to give alarm calls only when a conspecific of the opposite sex is present. Unlike the benefits justifying calling in the tits, the woodpeckers appear to invoke mate or prospective mate defense (Sullivan 1985). Thus even within the same flocks, the compensatory benefits for emitting warning signals varies with the participant.

The costs and benefits of inspecting and/or mobbing predators have received considerable attention (Bildstein 1982; Curio 1978; Flasskamp 1994; Gehlbach and Leverett 1995; Kobayashi 1994; Pettifor 1990; Shedd 1982; Sordahl 1990; Stone and Trost 1991). Both direct and indirect benefits have been suggested. Most of these studies focus on the propriety of inspection and mobbing per se and not on the economics of signals produced to coordinate such actions. In fact, most birds and mammals that inspect or mob also have special calls that recruit conspecifics around a predator as opposed to inciting flight or static surveillance. Such calls are usually repeated, loud, and easily localizable. Ants and bees have separate pheromones to warn nestmates, attract them to the vicinity of the sender, and incite attack on a predator or intruder (Wilson 1971). This sophistication is again not surprising, given the close relatedness in these species. The functions of inspection and mobbing calls in vertebrates are less clear. Some cases are surely intertrophic level signals. Where inspection or mobbing is favored for within-trophic-level reasons, it seems likely that senders would benefit directly by attracting conspecifics before harassing a predator. Synchronization of action and large numbers of mobbing individuals surely reduce the risks to any one individual and this would favor appropriate signals. Mobbing calls may also be used in some species to teach naive individuals that a given organism is a dangerous entity (Curio et al. 1978). These calls may increase the efficacy of shared vigilance in the future or better prepare offspring for survival on their own. Finally, mobbing calls may selectively recruit kin who have a vested interest in protecting related offspring. The benefits of mobbing calls thus overlap largely with those adduced for warning signals.

Distress signals, given in the final stages of predator attack, are fundamentally different from the other environmental signals we have discussed this far. In species where conspecifics will come to the aid of a threatened individual, whether kin or nonkin, distress signals will be a form of begging and thus follow Sir Philip Sidney game models (pages 665–666). As we have seen, such models require strong nepotistic or reciprocity links between sender and receiver before it pays a receiver to respond. These conditions are sometimes met. For example, parents of many species will respond to distress calls of their own offspring. Ants trapped in cave-ins will stridulate and their closely related nestmates respond by digging them out (Markl 1965). It is also possible that receivers can benefit directly by disrupting a predator attack if the predator responds by leaving the area or switching to another prey type (Trivers 1971). Even though such conditions are sometimes met, there are many species in which nearby conspecifics get few if any benefits from responding to distress calls and thus ignore them. Despite this lack of benefit, distress calls are still emitted. We take up this paradox in a later section on distress signals.

A final issue concerning alarm signals is, Who gives them? It is unlikely that all group members experience the same costs and benefits of sending such signals. In many ground squirrels, adult females usually have kin nearby, whereas adult males do not. It is the females who give low-risk alarm calls; males rarely do so (Sherman 1980). An exception occurs in prairie dogs, in which both sexes have kin in the vicinity. In this species, both sexes produce alarm calls as a function of the numbers and relatedness of nearby conspecifics (Hoogland 1983). In other vertebrate groups, subordinate and nonbreeding individuals do the guarding and alarm signaling. For example, capybaras (the world's largest rodents, *Hydrochoerus hydrochaeris*) live in mixed sex groups of 10 or more individuals (Figure 25.10). The dominant male in each group does all the mating, but subordinate males do most of the guarding (Yber and Herrera 1994). The dominant male keeps subordinate

Figure 25.10 Alarm calling in capybaras (*Hydrochoerus hydrochaeris*). Capybaras are large neotropical rodents that live in groups consisting of a dominant male, females, and subordinate (nonbreeding males). The latter are tolerated in groups only if they remain at the periphery and assume the majority of guarding duties. One individual in this wild group has been fitted with an individual collar to facilitate field studies. (Photo courtesy of David W. Macdonald.)

males at the periphery of the group where the subordinates are most exposed to approaching predators. Furthermore, it appears that subordinate males are tolerated in the group only because they relieve the dominant male of the guarding task. The costs of alarm signaling to subordinate males are thus made up by the increased chances of surviving in groups and of moving up to dominant male status. Dwarf mongooses (*Helogale parvula*) also live in mixed sex groups; again, subordinate males do all the guard work (Rasa 1989a). Foraging groups of mongooses are harassed by raptors an average 1.5 times per hour, causing them to spend up to 18% of their potential foraging time in hiding. Despite this, the efficient guarding of subordinate males thwarts all but 7% of raptor attacks. The cost is that subordinate males are much more at risk than other group members. Because most animals in these groups are kin, and both survival and fecundity of mating animals depend on maintenance of a sufficient group size, subordinate males may obtain high enough indirect benefits to compensate for the survival risks.

BETWEEN-TROPHIC-LEVEL ENVIRONMENTAL SIGNALING

Many organisms produce signals to members of species at a different trophic level. Such signals can be divided into those that cause a receiver to approach the sender and those that encourage the receiver to withdraw. The recruitment of pollinators to flowers and seed dispersers to ripe fruit relies on plant signals such as visual patterns, bright or contrasting colors, or volatile chemicals. Where such communication requires persuasion on the part of the sender, that is, where pollinators or seed dispersers have a choice of plants to approach and plants differ in suitability, we might expect the signals to include relevant handicaps (Hasson 1994). Repellent signals are usually given by prey being approached or attacked by a predator. Predators should attend to them only if they provide accurate information about the likely profitability of continued attack. Hence such signals should be largely honest. The honesty may be enjoined by the context, or it may be enforced by incorporation of some handicap on the sender. In the next section, we discuss five different contexts in which prey signal to predators; the relevant economics are somewhat different in each case.

Notification of Predator Detection

Predators that stalk their prey rely heavily on surprise. If prey detect a stalking predator and signal this fact to it, the predator may give up its current hunt since it is now likely to be unsuccessful (Bildstein 1983; Caro 1995). Clearly, a predator will attend to such signals only if they are largely honest. Prey that perform the signal repeatedly, whether a predator is stalking or not, will soon be ignored. A number of vertebrates have conspicuous white patterns on their posterior ends that are flashed, flicked, or exaggerated by bouncing gaits as they flee a detected predator. Examples includes swamp hens (Woodland et al. 1980), a variety of ungulates (Figure 25.11), rabbits, and some rodents. Functions initially proposed for these signals included appeasement of dominants (Guthrie 1971), coordination of flight (Hirth and McCullough 1977), and incitation of predators to attack earlier than is optimal for the predator (Smythe 1970). The observations that these signals are given by solitary as well as social species, that their posterior location makes them more visible to predators than to fleeing conspecifics, that the probability of signaling is often independent of the number of nearby kin, and that they are usually given only when a predator is still at some distance have persuaded most workers that these are signals to the predator that it has been spotted (Caro 1995). By carefully comparing predictions from six alternative functions, Caro (1994) concluded that snorting (an acoustical signal produced by many African savannah antelopes) and prancing (a high-stepping gait exhibited by topi, wildebeest, hartebeest, and waterbuck) are both notifications to a predator that it has been spotted. Snorting in North American deer, however, is more likely a warning to conspecifics since it is most often given when kin are nearby (Hirth and McCullough 1977). Tail wagging by lizards (Hasson et al. 1989), alarm duets by pairs of klipspringers (Tilson and Norton 1981), foot stamping by kangaroo rats (Randall and Stevens 1987), and loud alarms by a variety of birds (Klump and Shalter 1984) all appear to be signals to notify a predator that it has been detected.

Notification of Condition

Where predators have already begun an attack, prey in good enough condition to escape pursuit may benefit by signaling that fact to the predator. Prey that fail to signal will be at a disadvantage; this provides a strong incentive to cheat and produce the signal regardless of condition. As discussed on page

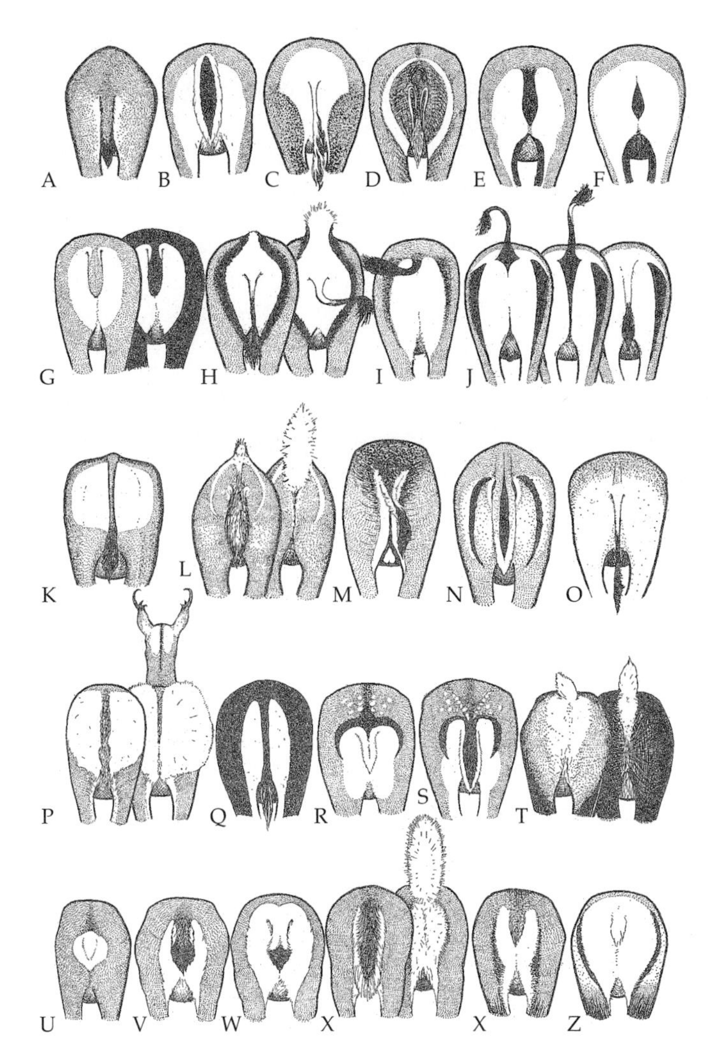

Figure 25.11 Conspicuous rump patches in selected ungulates. Dimorphic pairs (G and J) are female (left) and male (right), or winter and summer (T); other combinations show tail or rump in relaxed versus alert postures. Species illustrated are: (A) Uganda kob (*Adenota kob*); (B) Defassa waterbuck (*Kobus defassa*); (C) Bontebok (*Damaliscus pygargus*); (D) Common waterbuck (*Kobus ellipsiprymus*); (E) Stone sheep (*Ovis dallistoni*); (F) Bighorn sheep (*Ovis canadensis*); (G) Blackbuck (*Antilope cervicapra*); (H) Springbok (*Antidorcas marsupialis*); (I) Thomson's gazelle (*Gazella thomsoni*); (J) Grant's gazelle (*Gazella granti*); (K) Banteng (*Bossondaicus*); (L) Greater kudu (*Tragelaphus strepsiceros*); (M) Gerenuk (*Litocranius walleri*); (N) Impala (*Aepyceros malampus*); (O) Coke's hartebeest (*Alcephalus cokei*); (P) Pronghorn (*Antilocapra americana*); (Q) Sable (*Hippotragus niger*); (R) Sika deer (*Cervus nippon*); (S) Fallow deer (*Dama dama*); (*T*) Caribou (*Rangifer tarandus*); (*U*) Roe deer (*Capreolus capreolus*); (*V*) Black-tailed deer (*Odocoileus hemionus virginianus*); (*W*) Mule deer (*Odocoileus hemionus*); (*X*) White-tailed deer (*Odocoileus virginianus*); (*Y*) Red deer (*Cervus elaphus scoticus*); (Z) Wapiti (*Cervus elaphus canadensis*). (From Guthrie 1971, © Brill, Leiden.)

668, predators should rely on such signals only if they entail a handicap to enforce honesty (Vega-Redondo and Hasson 1993). This appears to be the case for the jumping gait called stotting used by a number of African antelopes (Caro 1994; Fitzgibbon and Fanshawe 1988). Thomson's gazelles (*Gazella thomsoni*) rarely stot for stalking predators like cheetahs, but do stot for coursing predators like wild dogs (Figure 25.12). Cheetah attacks are quick and there is little time for targeted prey to respond before the predator is upon them. Wild dogs however often circle and chase their prey for considerable distances before capturing them. Once wild dogs begin a chase, some gazelles stot and some do not, and those that do, stot at different rates. The stotting itself neither hinders nor helps the gazelle to escape once it is chased by a dog. However, it appears to be difficult for a gazelle in poor condition to stot at high rates. Wild dogs preferentially focus their chases on gazelles with lower

(A)

(B)

Figure 25.12 Notification-of-condition signals in ungulates. Signals indicating condition to a predator must incorporate a handicap to ensure honesty. (A) Stotting in Thomson's gazelles (*Gazella thomsoni*) and (B) pronking in springbok (*Antidorcas marsupialis*) are locomotory patterns that are difficult to do rapidly and for sustained periods. Wild dogs selectively chase gazelles that stot at low rates. (A courtesy of Hans Kruuk; B courtesy of Barrie Wilkins.)

stotting rates. On those occasions where dogs chase gazelles with different stotting rates, it is those with the higher rates that are more likely to outrun the dogs and escape. The evidence is thus persuasive that stotting is an honest indicator to the dogs of the likelihood that a gazelle will be able to outrun them, and that the dogs rely on this information in selecting which gazelles to chase. Leaping, which is seen in gazelles, topi, wildebeest, and impala with coursing predators also appears to be an honest indicator of prey condition (Caro 1994).

Predator Inspection and Mobbing

Many species of animals emit calls to recruit other prey around a predator. **Predator inspection** ensues when the assembled prey conspicuously monitor the predator's subsequent activities; in some species, inspection is followed by mobbing. In addition to alarm functions, predator inspection and mobbing are likely to discourage a predator from continuation of a stalk or pursuit and to encourage it to move to another site where fewer prey know of its location (Curio 1978; Ishihara 1987; Pettifor 1990). Thomson's gazelles that spot a stalking predator may snort, stand alert, and stare at the predator, or even run toward it a short distance. This behavior alerts other nearby gazelles, who repeat the responses until the entire herd is watching the predator. This behavior is costly to the gazelles both as lost foraging time and as increased risk of being chased. However, Fitzgibbon (1994) has shown that cheetahs move over twice as far between rests after being inspected by gazelles than when not inspected;

the bigger the inspecting group, the farther the cheetah moves away. Cheetahs also walked faster and for longer when approached by a wary herd of gazelles. These observations suggest that any short-term costs of predator inspection are more than compensated by encouraging the predator to move on.

Aposematic Signals

Animals that are toxic, armed with spines, or otherwise unpalatable often exhibit conspicuous colors or patterns. Skunks, many butterflies, and nudibranchs (marine slugs) are well-known examples (Figure 25.13). Such **aposematic signals** are clearly aimed at discouraging predators from attacking the sender. Most are highly conspicuous against the relevant background. This imposes a cost on aposematic prey as a predator is more likely to detect the sender. However, conspicuous and distinctive signals may be more easily remembered by predators, and the subsequent avoidance of the prey may more than make up for their enhanced detectability (Guilford 1990).

There are game-theoretical problems associated with aposematic signals. If prey are insufficiently unpalatable to disrupt a predator attack before they

(A)

(B)

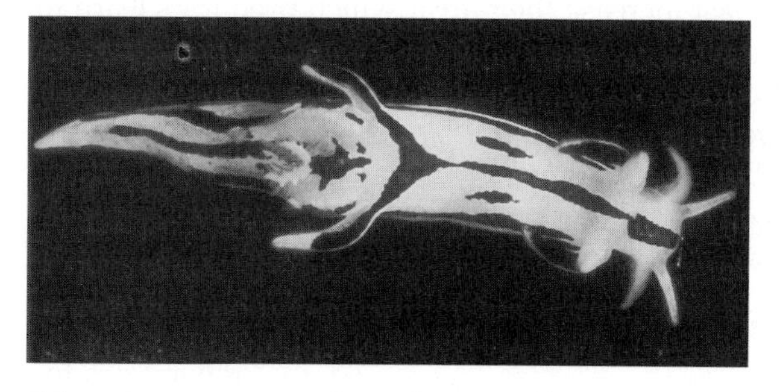

(C)

Figure 25.13 Aposematic coloration in nudibranchs. Nudibranchs are marine slugs that are unpalatable to fish and crustacean predators because of toxins or stinging cells they acquire from the foods they eat (Faulkner 1988, 1992). Most nudibranchs sport highly conspicuous patterns and colors. This aposematic coloration helps predators learn to avoid nudibranchs. (A) *Kentrodoris funebris*, a nudibranch that eats sponges. (B) *Trapania velox*, a nudibranch that eats bryozoans. (C) *Dirona albolineata*, a nudibranch that eats a wide variety of other marine invertebrates. (Photos courtesy of James Lance.)

are killed (i.e., they take a long time to make the predator sick), then the costs of aposematic signals are borne by those who are attacked, and the benefits enjoyed by the survivors. As stated, the trait cannot become established. Such a system might evolve if individuals within the range of a particular predator are close kin; then the costs to any one individual could be made up by indirect benefits (Harvey and Greenwood 1978). Unpalatable aposematic species of insects tend to be more socially gregarious than cryptic unpalatable ones, and aposematic species will often lay eggs in local clusters. Costs are also reduced on an individual basis if a given aposematic pattern is very common. This presumably explains the evolution of **Müllerian mimicry** in which different unpalatable species in a community evolve highly convergent warning coloration. However, lower costs also facilitate cheating by **Batesian mimics**. These are species that are palatable but that copy the aposematic pattern of sympatric unpalatable forms. Batesian mimics tend to be rare relative to the unpalatable species whose signals they imitate.

Aposematic signals generate a game similar to that for badges of status. In both cases, receivers use a signal to avoid having to determine a hidden trait directly. Possible prey strategies involve different combinations of signal conspicuousness and unpalatability. Honest signalers use conspicuous signals if they are unpalatable, and cryptic ones if not. Cheats use conspicuous signals but are palatable, and sneaks are unpalatable but remain cryptic. Predator strategies include variations in the rigor with which warning colors are respected or ignored, and in the speed with which new associations between aposematic pattern and palatability are learned. Predators sometimes ignore warning coloration (when they are young and naive), and cheating prey can invade an initially honest system. As the number of cheaters increases, reduced discrimination by predators will be favored. The payoffs of any predator or prey strategy will depend on the frequencies of alternative strategies. The relevant game is thus an asymmetric scramble (Guilford and Dawkins 1993). Leimar et al. (1986) have examined one version of this game. They find that ESS combinations are achieveable if (1) more conspicuous signals are inherently avoided or they at least increase the rate at which predators learn to avoid prey; and (2) increasing prey unpalatability increases the chances that an attacked prey will survive, or increases the rate at which predators learn to avoid this species. The degrees of unpalatability and signal conspicuousness at the ESS depend sensitively on the patterns of learning of the predator. The observations that predators do sample aposematic prey occasionally and that Batesian mimics are invariably rare both support the notion that such signals are at the relevant ESS.

Distress Signals

Why do animals that are solitary or have no nearby kin produce distress signals when attacked by a predator? A variety of organisms release pheromones from damaged tissues when attacked (Hews 1988; Pfeiffer 1977; Sabelis and de Jong 1988; Smith 1992; Wilson and Lefcort 1993). For which receiver are these signals intended? There is a growing consensus that many distress signals function to attract predators other than the one attacking the distressed organism (Högstedt 1983; Mathis et al. 1995; Perrone 1980). The simplest case is attraction of predators of the same trophic level as that currently attacking the prey. If these additional predators distract or interfere with the first, the prey may have a chance to escape. Pheromones released from the skins of damaged fish and tadpoles attract other predators, and both hawks and mammalian predators can be attracted to the distress calls of prey being harassed by another predator. Were this the original function of distress signals, then flight or defense responses of conspecific prey to distress signals would be a secondary adaptation. A more complex function of distress calls is the attraction of predators of the attacking predator. Some plants respond to herbivore attack by releasing chemicals that attract the herbivore's parasites or predators (Sabelis and de Jong 1988). Responding predators would prefer that a plant signal only if there are enough attacking herbivores to make it worth the trip; the plant, however, would like to attract predators regardless of herbivore numbers. This creates a conflict of interest that Godfray (1995c) has modeled as an asymmetric continuous scramble. He finds that honest signaling can be an ESS. However, this is only so if the costs of producing the chemical signal are sufficiently high to make cheating by the plants unprofitable. Again, honesty is enforced by a handicap.

CONSTRAINTS ON ENVIRONMENTAL SIGNAL SOPHISTICATION

Our review has shown that environmental signals between animals at the same trophic level are usually simple or even absent when they would otherwise seem appropriate. We interpreted this pattern as a consequence of the relevant game economics; it rarely pays for a sender to communicate complex environmental information if the major benefits accrue only to other individuals. The few exceptions to the pattern are colonies of naked mole-rats or honeybees in which genetic relatedness is very high, and thus senders can expect substantial indirect benefits. Yet even the remarkable honeybee system conveys only a fraction of the information that could be transferred and relies on a number of approximate rules of thumb. An alternative interpretation of the modest evolution of environmental signals is that the machinery required to encode and extract complex information about external object identity, state, and location is beyond the reach of nonhuman animals even were there sufficient sender benefits to compensate for the costs. Put simply: Is the modest state of most animal environmental signals a matter of insufficient benefits to offset costs, or is it instead the result of physical and physiological limits? We take up this issue in the final chapter.

SUMMARY

1. **Environmental signals** between members of the same trophic level provide information about resources or predators and are essentially cooper-

ative. Members of different trophic levels usually have conflicts of interest, and signals exchanged between them typically involve persuasion and honesty guarantees.

2. Environmental signals could provide information about the identity, state, and location of external objects. Because the possible alternatives for each of these parameters are numerous, the amount of environmental information that could be transmitted is very large and the potential sender costs high. The costs are exacerbated if receiver responses reduce the sender's immediate fitness, and if the sender must be selective about which receivers detect the signals. Sender costs must be compensated by sufficient indirect or deferred benefits to warrant emitting environmental signals. Because they often are not, many animals do not send environmental signals or, if they do, they transmit a very reduced set of possible messages. Exceptions occur where senders and receivers are close kin or where direct benefits to senders are sufficient to offset costs.

3. **Resource-recruitment signals** in nonhuman vertebrates occur sporadically and typically provide only minimal information about resource type and quality. Location is usually provided by broadcasting a signal from the site. Many species that might seem likely candidates for such signals (e.g., colonial roosting birds) do not use them. Those that do either limit resource advertisement to rich patches at which competition costs are minimal (large primates) or else derive some direct sender benefit that offsets the signal costs. Examples include mate attraction (roosters), displacement of territorial owners (ravens), shared predator surveillance (sparrows and tamarins), and tracking of moving prey (swallows). Indirect benefits apparently play a role in some primates and naked mole-rats.

4. A number of social insects use elaborate resource-recruitment signals. Ants and bees incite recruits to seek out a resource by performing some form of agitated dance at the nest and by providing samples of the resource found. Advertised resources may include food, nesting materials, water, and new nest sites. The quality of the resource or its salience relative to current colony needs is indicated by the duration and vigor of the sender's displays. Ants, termites, and stingless bees indicate location by laying odor trails to the resource; some species will also lead nestmates back to the site. Forager honeybees perform a symbolic dance at the hive that indicates the direction and distance that recruits should fly to locate the resource. Direction to the resource relative to the sun is indicated by the direction of the dance relative to gravity. Internal clocks in both sender and receiver correct for the sun's movement during and after dances. Distance is indicated by the duration of each circuit of the dance. In ants and honeybees, communication is remarkably accurate. Angular and distance errors average about 10 to 15%, and from 50 to 70% of all recruits may arrive at the advertised resource. Even higher accuracy may be possible, but may be avoided to ensure scatter of recruits over a large patch of resource. The sender benefit of such sophisticated and costly signals is surely nepotism due to the very high kinship within social insect colonies.

5. Within-trophic-level **alarm signals** can be given at each of four stages of predator-prey interaction: **low-risk warning signals** may be given on first sighting a predator, **inpection** or **mobbing signals** given if the predator is roosting or inactive, **high-risk warnings** given if an attack is imminent, and **distress signals** may be produced once attack has commenced.

6. Most social species have at least one signal for each of the four stages of alarm. Insects tend to use pheromones, whereas terrestrial vertebrates use sounds. The degree to which low- versus high-risk warnings are predator-specific depends on whether the propriety of alternative receiver responses is determined by predator type (in primates, ring-tailed lemurs, chickens) or urgency of the situation (in ground squirrels). Predator inspection, mobbing, and distress signals tend to be high intensity, localizable, and predator-independent. Potential sender benefits of alarm signals include startling of fellow prey toward the predator, synchronization of flight or stealth, retention of groupings during flight, protection of a mate, discouragement of future predation, maintenance of optimal group size, or protection of kin. Benefits appear to vary widely with situation and may even vary within a taxon: ground squirrel low-risk warnings are nepotistic, whereas high-risk warnings in the same species appear to synchronize flight.

7. Intertrophic level signaling can be either attractive or repellent. The latter type of signal can take any of five forms. **Detection-notification signals** indicate to a stalking predator that it has been spotted and might as well give up on the signaling individual. Unless predator attacks are very common, it does not pay prey to give this signal except when predators are detected. It is thus essentially honest. Examples include snorting of antelope and tail flashes of various ungulates. **Condition-notification signals** demonstrate the ability of a prey animal to escape pursuit. They are thus persuasive, and predators should only attend to those that incorporate uncheatable handicaps. Stotting in antelopes is an example. Only animals in good enough condition to outrun a pursuit can perform it rapidly and for a long period. **Predator inspection** and **mobbing signals** advertise the location of a stalking predator to all prey in an area and recruit them into the inspection aggregation. Such behavior usually causes the predator to move to another area. **Aposematic signals** make prey more conspicuous and memorable so that they will be recognized as unpalatable or toxic and avoided. Since unpalatability is usually learned, the costs of the signals are borne by the attacked animals and the benefits enjoyed by the survivors. Initial evolution of such signals requires nonfatal attacks or close kinship among those sharing the signal. **Distress signals** may be calls for assistance in social species. However, many solitary species also give such calls. Their proposed function in these cases is to attract competitor predators or predators of the predator that then disrupt the attack on the prey and allow it to escape.

FURTHER READING

The issue of whether environmental signals are referential or motivational is reviewed in Marler et al. (1992). Summaries of the economic problems associated with environmental signal evolution can be found in Trivers (1971), Wittenberger (1981), and Harvey and Greenwood (1978). Brief reviews of resource recruitment signals in vertebrates can be found in Elowson et al. (1991) and Hauser and Marler (1993a). The classic treatments of bee recruitment by Lindauer (1961) and von Frisch (1967) remain captivating reading; more recent accounts can be found in Seeley (1985) and Winston (1987). Hölldobler and Wilson (1990) review all types of environmental signals by ants. Good reviews of alarm signals in vertebrates include Klump and Shalter (1984, birds) and Macedonia and Evans (1993, primarily mammals). Interested readers may want to consult the detailed case studies on Belding's ground squirrels (Sherman 1985) and vervet monkeys (Cheney and Seyfarth 1990). Functions of mobbing are reviewed by Curio (1978). Caro (1986, 1995) examines ungulate predator notification signals, and Högstedt (1983) presents the case for distress calls as predator attractor signals. The evolution of aposematic signals is reviewed by Guilford (1990) and Endler (1991a).

Chapter 26

Autocommunication

THROUGHOUT THIS BOOK, we have invoked a very broad definition of communication. The disadvantage is that the boundaries between signal exchange in the ethological sense and processes such as direct and cue assessment or intertrophic interactions may remain blurred for some readers. The advantage is that examination of these related processes provides useful and even necessary perspectives on signal evolution. With the latter goal in mind, we here treat echolocation (often called biosonar) and electrolocation under the rubric of "autocommunication." An echolocator emits pulses of sound and uses the returning echoes to extract information about nearby objects; an electrolocator accomplishes the same task using electric fields. These processes can be called autocommunication because the same individual both sends and receives any given signal.

Sensory physiologists might object that neither activity is truly communication, but instead refined forms of direct or cue assessment. However, many of the same forces shape the evolution of traditional and autocommunication signals. Senders in both cases carefully adjust signal design to control for propagation effects, and in both cases, receivers use the structure of perceived signals to extract information they need to know. Because the focus of autocommunication is the provision of environmental data (e.g., what is the shape, size, and type of a nearby object, what is it doing, and where is it?), the relevant signals must convey a large amount of information. We saw in the prior chapter that such signals are expensive and thus rarely exchanged unless sender and receiver have very similar interests. In this case, sender and receiver are the same individual and there is no conflict of interest. Thus although the necessary costs of effective autocommunication should be near the upper limit seen for signals, there is the greatest incentive to pay them.

An examination of autocommunication systems might thus answer the question posed at the end of Chapter 25. Is the relative lack of sophistication in the environmental signals of most species a result of insufficient communality of interests (a game-theoretical answer), or is it simply too difficult and expensive to encode and extract such information (a neuroethological answer)? If it were the former, then we might expect autocommunication systems to be highly sophisticated because conflicts of interest are absent; if the latter, then autocommunication should not exhibit signal systems any more sophisticated than those used to exchange environmental information between individuals. In either case, the amounts of information provided in autocommunication ought to be as high or higher than for any other signaling exchange. This result would arise both from the absence of conflict of interest and from the opportunity for autocommunicators to fine-tune the efficiency of signal design by directly examining their own responses as the receiver. It is more difficult to maximize efficiency when sender and receiver are different individuals. The result is that autocommunication systems ought to provide benchmarks for information provision against which any other communication sytem can be compared. We can also compare their performance to that of an **ideal receiver**. As we saw on pages 442–444, an ideal receiver extracts as much information from a signal as is physically possible; no animal or device can do better. Whether autocommunicators fall closer to ideal receivers or to receivers handling signals from other animals may tell us much about the forces shaping signal evolution.

In the following pages, we shall focus on echolocation because it is the process that has been best studied. It also poses the most challenging problems because the determination of the location of an external object is much more difficult when the signal receptors are centralized in a few locations (e.g., ears) than when they are distributed over the body of the animal (e.g., electroreceptors). As we saw in the last chapter, object location is a very difficult type of environmental information to encode and extract from signals.

DESIGN CONSIDERATIONS FOR ECHOLOCATION

Echolocation is a substitute for vision when light is not available or is poorly transmitted. We thus find it used in nocturnal species and those that live underwater. It has evolved many different times in vertebrates. Simple forms of echolocation are seen in cave-dwelling swiftlets and oilbirds (Figure 26.1) and in small nocturnal mammals such as shrews, rats, and tenrecs. More sophisticated echolocation systems are found in many bats and toothed whales. The latter groups use echoes to locate and capture food as well as to avoid obstacles during rapid locomotion. They thus seek the same kinds of information as provided by most environmental signals: detection of an object at a distance, and subsequent estimation of its location (range and angular position), identity (shape, texture, and composition), relative velocity, and trajectory. Echolocators can improve the accuracy of this information by careful design of both outgoing signals and receiver processing mechanisms. As we shall show below, there are tradeoffs between the measurement accuracy of any one property and that of other properties. Depending on ecological contexts, different echolocators emphasize different target properties and thus use different types of echolocation signals and different types of receiver mechanisms. To understand this diversity, we need to examine how each property can be measured using echolocation.

Detecting Targets at a Distance

The first task faced by an echolocator is detection of a target. Detection requires that outgoing signals be (1) loud enough to produce echoes at sufficient intensities relative to ambient noise, and (2) emitted a large enough fraction of the time to ensure that the target is intercepted by the sound beam. Echo sig-

Figure 26.1 Adult oilbird (*Steatornis caripensis*) on nest in Trinidad cave. Oilbirds feed on fatty palm and laurel fruits at night. They build their nests in caves using regurgitated fruit pulp as a base. To navigate in the darkness of the cave and while foraging, oilbirds emit bursts of clicks with dominant frequencies between 6 and 10 kHz and durations of 1 to 1.5 msec. They use the echoes from these clicks to detect nearby obstacles. Given the large wavelengths of the sounds (relative to those emitted by echolocating bats and cetaceans), oilbirds probably only detect fairly large objects with these echolocation signals. (Photo courtesy of Louise H. Emmons.)

nal-to-noise ratios can be enhanced by using a single pure frequency because senders need not spread their sound-generating energy over several components, and receivers can specialize in detection of that frequency. Determining which frequency is optimal is a complicated matter. For small and distant targets, the ratio between the intensities of outgoing sounds and returning echoes decreases with the fourth power of the distance to the target. This relationship occurs because both the echolocator and the target act as point sources of sound, and for each, intensity falls off with the square of the distance between them (pages 31–32). By itself, this favors selection of that frequency that the sender can emit at the highest intensity. However, other factors come into play. Maximal echo formation occurs when incident wavelengths are the same size or smaller than the target (pages 32–33). This fact sets an upper limit on allowable wavelengths, and consequently a lower limit on useful sound frequencies. For bats chasing insects of 1 cm or less, the necessary frequencies will be 34 kHz or higher; for porpoises hunting 10 cm fish, frequencies must be 15 kHz or higher. The cost of using high frequencies, especially in air, is severe medium absorption during propagation (Griffin 1971). This cost favors using as low a frequency as possible. The optimal frequency for detection is thus a compromise between maximization of source intensity and echo formation, and minimization of absorption losses (Figure 26.2). This optimum will vary with body size of the echolocator, size of critical targets, and the type of medium.

The fraction of time that sound is emitted is called the **duty cycle** (page 574). It can be maximized by increasing signal durations, signal emission rates, or both. No echolocator has a 100% duty cycle. One reason is that most echolocators avoid temporal overlap between the emission of a pulse and the return of echoes from the prior pulse; if they did not, the much louder outgoing pulse would mask the returning echo. The degree to which the pulse-to-echo separation constrains duty cycle depends upon the minimal and maximal ranges at which an echolocator tracks targets. When targets are very close to the echolocator, pulses cannot be very long, or echoes will begin to return before the pulse emission is complete. For example, a bat cannot track insect prey closer than 17 cm unless it uses pulses with durations less than 1 msec. Porpoises, which must contend with a speed of sound 4.4 times that of air, must limit pulse durations to 23% of those of bats at the same minimal tracking distances. Maximal ranges of detection set an upper limit on the pulse repetition rate of echolocators. If pulses are emitted at too high a rate, echoes from distant targets will not have time to return to the sender before the next pulse is emitted. There is thus an inverse relationship between maximal allowable repetition rates and maximal target range. The combined constraints on pulse duration (via minimum range) and pulse repetition rate (via maximum range) together set an upper limit on duty cycle and thus the efficacy of detection. Minimum and maximum tracking ranges are in turn set by the ecological contexts in which echolocation is undertaken. These differ for different species and thus we expect pulse durations, pulse emission rates, duty cycles, and detection abilities all to vary with habitat (Neuweiler 1984).

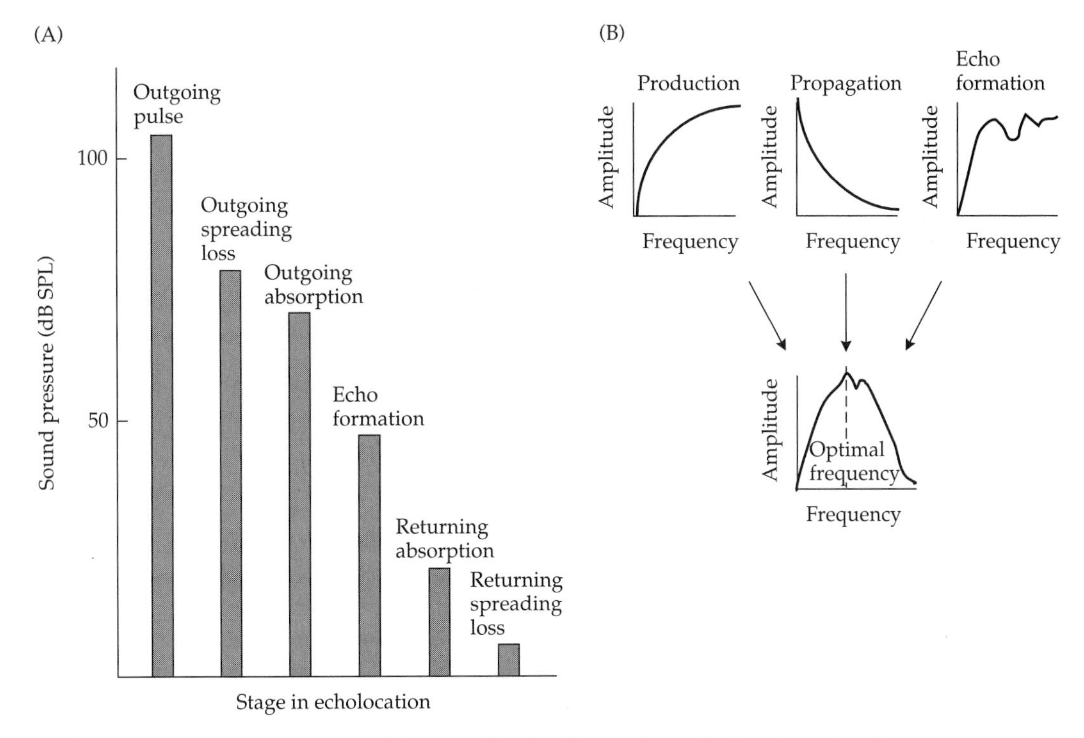

Figure 26.2 Factors affecting relative amplitude of outgoing echolocation signals and returned echoes. (A) Sources of signal attenuation between pulse emission and echo reception. The graph assumes an output signal amplitude of about 110 dB near the bat source, a 60 kHz pulse frequency, and a 19 mm spherical target at 3 m. At each stage, the remaining signal intensity in dB SPL is indicated. Note that the final echo is only 10 dB SPL. If a higher frequency were used, atmospheric attenuation would be greater and the echo increasingly would be close to the typical mammalian threshold of 0 dB SPL. (B) Typical relationships between signal amplitude and frequency at three successive stages in echolocation process (production, propagation, and echo formation) and when all factors are combined (bottom graph). The frequency at the peak of the bottom graph is the optimal one. (A after Laurence and Simmons 1982.)

Determining Target Distance

An echolocator has two sources of information about target range: echo amplitude and the delay between emission of a sound pulse and return of the echo. Because of spreading losses and medium attenuation, the farther the target, the less intense the echo. An echolocator familiar with these processes could thus use echo amplitude as an index of range. However, echo amplitude will also be affected by the size and shape of the target relative to the wavelengths of the sound in the emitted pulse. The larger the target relative to the wavelengths, the more intense the echo. If wavelengths and targets are of similar sizes, diffraction and interference may cause echo intensity to vary in complex ways depending upon signal wavelength, target size, and target shape (see pages 33–36). This confounding of target properties and distance makes echo amplitude a poor index of range. On the other hand, if an echolocator knows the speed of sound in the medium, c, and can measure the time delay, t, between outgoing pulse and returning echo, then target range R is simply $R = ct/2$. The factor of 2 in this expression corrects for the fact that the pulse must travel the distance between sender and target twice.

Most echolocators appear to use this delay between pulse emission and echo detection to measure target range. How accurately might they do this? To maximize detection, we might expect the echolocator to use a single pure frequency. Consider a 5 msec pulse of a single frequency emitted by a bat at a distant insect flying with a wingbeat frequency of 50 Hz. Suppose that the insect is far enough away that a detectable echo is formed only when the insect's wings are perpendicular to the sound beam; this will occur at most once during the 5 msec that sound hits the insect and for only a fraction of that time. The resulting echo will therefore be only a piece of the outgoing pulse; which piece will be a matter of chance. Unless the bat can be sure which piece is returned, it cannot time precisely how long that piece took to travel to the target and back. The longer the outgoing pulse, the larger the number of alternative pieces that could have been returned, and thus the greater the uncertainty. In fact, it can be shown that the minimal average error in range estimation, ΔR, of a small and distant target by an ideal receiver responding to a constant frequency echo is

$$\Delta R \approx \frac{cT}{4\pi d'}$$

where T is a weighted measure of pulse duration called **rms duration**, and d' is the sensitivity parameter from signal detection theory (Au 1993; Bird 1974; Kingsley and Quegan 1992; Woodward 1964). For most realistic echolocation pulses, T is approximately equal to total pulse duration. As we saw on pages 442–444, $d' = \sqrt{2S/N}$ where S is the energy of the echo and N is the energy per Hz of ambient noise. A 5 msec constant frequency pulse when $d' = 5$ will thus allow range measurements with a minimal error of 2.7 cm in air and 12 cm in water. No receiver can do better with such a signal.

An echolocator could improve its range estimation if it had some way to label the parts of the outgoing pulse and remember at what point in the pulse each part was emitted. Many bats produce a pulse that begins at a high frequency and is frequency-modulated smoothly to a low value by the end of emission. Thus each part is labeled by frequency. If only a single piece of outgoing pulse is returned as an echo, the accuracy of any delay time measure would depend on the accuracy with which the frequency of that piece could be determined. For a single piece of given duration (Δt), the uncertainty principle for sound sets a limit on how accurately the returned frequency can be measured (Δf; see pages 63–68). Real echolocators would do worse than this because of the presence of ambient noise. However, if several pieces are returned, the effects of random noise can be reduced by averaging the delay times estimated from each piece.

If more than one piece of a labeled waveform is returned, this additional information facilitates another method to estimate range. This method compares the echo with copies of the outgoing signal that have been stored in the brain but delayed by different times. When the true delay is used, the waveforms of the two signals will show the greatest overall similarity at corresponding time points. If noise is present, it will cause the waves of some components to peak earlier than they should, and others to peak later than they should. Trying to align the two signals using any one component is likely to lead to error; obtaining the best average alignment for all components is most likely to average out the effects of noise and give the true delay time. How well this works depends upon the type of correlator used by the echolocator. As outlined in Box 26.1, no receiver designed by man or beast can do better than a **coherent cross-correlator**. Such a receiver uses frequency, amplitude, and both absolute and relative phase information to compare outgoing pulses and echoes and find the best alignment. Because it is the phase information that is responsible for the smallest differences in alignment, and because the assessment of phase is easier the higher the average frequency of the signal, it can be shown that a coherent correlator will have a minimal error of range measurement given a single echo of

$$\Delta R = \frac{c}{4\pi\beta_{\text{rms}}d'}$$

where β_{rms} is a weighted average of all frequencies in the signal (called the **rms bandwidth**), and other terms are as defined above (Menne and Hackbarth 1986). This formula is only appropriate when the relative velocity between bat and target is zero, and the signal-to-noise ratio as measured by $d' >> 1$; at lower signal-to-noise ratios, the error becomes worse very rapidly. For a bat sweeping from 50 kHz down to 20 kHz in each pulse, the rms bandwidth will be about 30 kHz. The minimum range error when $d' = 5$ would then be 0.2 mm. This is a 135-fold improvement over the constant frequency pulse.

What if the echolocator's ears and brain cannot measure the absolute phase of the signals, or this phase changes as the sound is reflected by the target? The tagging of pulse components with different frequencies should still allow for more accurate alignments of delayed pulse and echo than is true for a constant-frequency pulse. Such a receiver is called a **semi-coherent cross-correlator**, and it can be shown to have a minimal accuracy (again for $d' >> 1$) of

$$\Delta R = \frac{c}{4\pi\beta_{\text{crms}}d'}$$

where β_{crms} (the **centralized rms bandwidth**) is a weighted average of the variation in frequency during the pulse around the value of β_{rms} (Menne and Hackbarth 1986). This measure is more similar to our use of bandwidth in other contexts; it is always less than β_{rms}. (Note that the equation for a constant frequency pulse given earlier is a special case of this equation because centralized bandwidth for a short pulse of constant frequency depends

Box 26.1 Cross-Correlations and Ideal Receivers

THE CORRELATION BETWEEN ANY TWO DATA SETS with the same number of samples is computed by multiplying the corresponding values from each of the two sets, adding up the products, and normalizing the sum with respect to the average variation in the two sets. Two signal waveforms can be correlated by multiplying their amplitudes at each of a number of successive times, summing the products, and making appropriate normalizations. This computation is called the **cross-correlation** between the signals, and it yields a single real number as a result. If we cross-correlate a signal with a copy of itself and begin the sampling at the start of each signal, we expect to obtain as perfect a correlation as is possible with a finite set of samples. If we then perform a correlation between the signal and copies delayed by increasing lags, we shall find a generally decreasing level of correlation. In part, this is because some of the early samples in the original signal and the same number of late samples in the copy will be multiplied by zero (since the corresponding signal has no amplitude at those times). As the lag increases, the number of samples multiplied by zero increases, and thus the correlation sum decreases. By the time the lag equals or exceeds the duration of the original signal, T, the correlation sum is zero. One can do the same thing by comparing a signal to copies that begin at times increasingly prior to its own onset. Clearly the same thing will happen; the earlier the copy begins, the lower the correlation value. A plot of the correlation value versus the relative time (lag) at which the copy begins is called a **cross-correlation function**. (Cross-correlations between signals and strict copies of themselves are also called **autocorrelations**. Since echoes are not strict copies of the outgoing pulses, we shall refer to all computations here as cross-correlations.)

In practice, the cross-correlation function does not decline monotonically with lag. To see why, consider a signal consisting of some constant frequency, f. The period of this signal will be $t = 1/f$. A correlation between the signal and an undelayed copy of itself will show the highest value. If we then delay the copy by $t/2$ seconds, the signal and its copy will be exactly out of phase and the resulting cross-correlation will be negative. Increasing the delay further to t seconds will bring the signal and its copy back into phase again. However, the cross-correlation value is now less than for the undelayed copy because some end samples of each signal are multiplied by zero. As the delay is increased further, the cross-correlation rises and falls but with diminishing amplitude as more samples are multiplied by zero. A graph of the cross-correlation value for two signals versus different possible delays appears in Figure A.

The structure of the cross-correlation function reflects various characteristics of the original signal. The width of peaks in this function, v, is inversely related to a weighted average of the frequencies in the signal (the **rms bandwidth**); distances between peaks, z, are equal to the reciprocal of the fundamental frequency if the original signal has harmonics present or equal to v if not. If there is no added noise, then the amplitude of the peaks in this function is maximal for a lag of 0 (i.e., for perfect alignment of signal and its copy), and decreases for larger lags in either direction. We can speak of the **envelope** of the function as the line connecting successive positive peaks in the correlation plot. The width of this envelope depends upon the **centralized bandwidth** of the original signal—a weighted average of the amount of variation around the mean frequency in the signal. A convenient measure of envelope width is W, which is the interval on each side of a lag of 0 within which all peaks are greater than

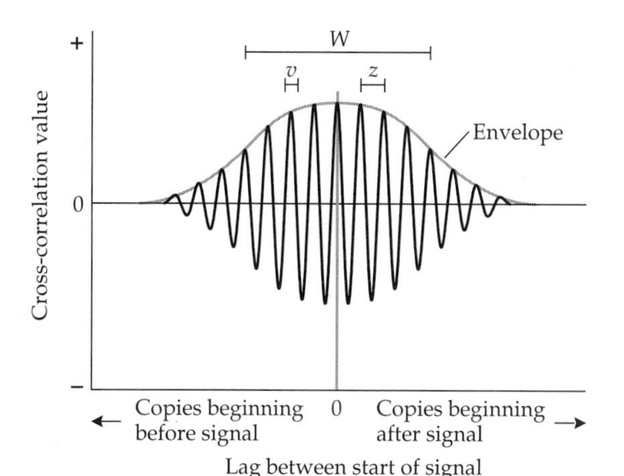

Figure A Typical autocorrelation function between a signal and a copy of itself at varying lags. Lags can be negative (copy begins before signal) or positive (copy begins after signal).

than half the value at the maximum. It can be shown that the value of W is equal to the reciprocal of the signal's centralized bandwidth. If signal bandwidth is increased, W is decreased, and this means fewer peaks around 0 lag will have large amplitudes.

For an echolocator, the task is to identify the most likely delay between an outgoing pulse and its returned echo. One way to do this is to cross-correlate the echo with images of the outgoing pulse resurrected after each of several possible delays. In the absence of ambient noise, the image that generates the highest cross-correlation indicates the most likely delay. The example above indicates that this determination may not always be easy. If the amplitude of the peak (when the lag between the image and the echo is zero) is not that different from the amplitudes of adjacent peaks, then random noise might cause the receiver to err by selecting the lag under one of those peaks instead of the correct one. The more closely packed the peaks, the wider they are, and the larger their amplitudes relative to the maximum, the more likely the chance of receiver error. Error can thus be reduced by using high average frequencies (reducing peak width), large bandwidths (reducing envelope width and thus the amplitudes of lateral peaks), and including harmonics (moving peaks laterally so that their amplitudes will drop due to the envelope). Given a particular signal, no receiver can do better than the error specified by the cross-correlation function of that signal.

A **coherent cross-correlator receiver** is one that can assess the component amplitudes, absolute phases, and frequencies of both outgoing signal and perceived echo. The probability it will identify any given lag as the actual delay time is directly related to the relative amplitudes of adjacent peaks in the cross-correlation function, and thus it makes maximal use of the available information. It represents the most accurate type of receiver possible. A **semi-coherent receiver** is unable to keep track of the absolute phases of the signal and echo. Instead of seeing the individual peaks of the cross-correlation function, it sees only the function envelope. The probability that any given delay is selected by a semi-coherent receiver is thus directly related to the height of the

Box 26.1 (continued)

envelope above that delay. As noted above, the width of this envelope at the half-amplitude points is inversely related to centralized bandwidth; thus signals with wider bandwidths should provide better range accuracy even for a semi-coherent receiver. For low enough signal-to-noise ratios, coherent and semi-coherent receivers do equally badly at measuring ranges; for higher values, both improve, but semi-coherent receivers do worse by a fixed amount for any given signal (Menne and Hackbarth 1986).

Other types of receivers are possible. A spectrogram correlator compares the spectrogram of the outgoing pulse and that of the returning echo for alternative delay periods. It requires a 5 dB higher signal-to-noise ratio to produce the same performance as that of a coherent correlator in a detection task. This type of receiver will also show an accuracy that increases with signal bandwidth, but the amount of improvement will be less than for coherent correlators (Altes 1980, 1988).

mostly upon the reciprocal of pulse duration; see pages 63–68.) For a bat sweeping from 50–20 kHz in each pulse, a typical value of β_{crms} might be 3–4 kHz. This value would produce a minimal error, for an ideal semi-coherent receiver, of 1.6 mm. This result clearly shows that echolocation pulses with many frequencies are preferable to constant frequency ones for measuring target range even if absolute phase cannot be measured. In addition to frequency modulation, echolocators can increase bandwidth by including harmonics in their pulses, or by using short enough pulses that bandwidth has to increase as predicted by the uncertainty principle.

A number of other types of receivers have been postulated for various echolocators. None perform better than coherent cross-correlators with a single echo, but some can achieve remarkable accuracies if the signal-to-noise ratios are high enough and the results from several successive pulses can be averaged. One currently popular model envisions receivers that generate spectrograms from echoes and then compare these to spectrograms of outgoing calls (the **spectrogram correlator** of Altes 1980, 1988). The accuracy of this method is also limited by the uncertainty principle, and it generally requires a 5 dB higher signal-to-noise ratio to achieve an accuracy equivalent to a coherent cross-correlator. It also requires constraints on the range of possible echo intensities. However, it seems to fit well with the kinds of neural structures that have been discovered in the ears and brains of both echolocating and nonecholocating mammals (Neuweiler 1990; Simmons 1993; Simmons et al. 1990a,b).

The accuracy of range measurement will be reduced if the echolocator and its target are moving relative to each other. This is because Doppler shifts (pages 21–22)will change all emitted frequencies in proportion to this relative velocity, and each part of the emitted signal originally tagged with one frequency will return at a slightly different one. If the bat is approaching the target, Doppler shifts will cause echoes to have shorter durations than outgoing pulses. Shortening of the echo duration relative to the outgoing pulse also alters the apparent delay and phase. This type of error can be minimized, although not eliminated, by careful control of the frequency modulation pattern. The optimal solution for an FM pulse is to increase the period of successively emitted waves as a linear function of time. This increase will result in a hyperbolic decrease in emitted frequency during the course of the pulse (Figure 26.3). Any other pattern of frequency modulation results in higher errors due to Doppler shifts (Altes and Titlebaum 1970). A signal that optimally resists the effects of target or echolocator velocity is said to be **Doppler-tolerant**.

Determining Target Angle

Echolocators have two tools available to measure target angle: (1) focusing of the outgoing sound field into a narrow forward beam, and (2) directional sensitivity of their ears. Outgoing sound beams can be made wide or narrow depending upon the shape and structure of the emission site (Au 1993; Hartley and Suthers 1987, 1989, 1990). A narrow beam has the advantage that the echolocator can scan a small area and receive echoes only from that zone; it has the disadvantage that a moving insect is more likely to escape detection by moving out of the narrow sound beam. The shape of an emitted sound beam is unlikely to be simple and will depend upon the component frequencies of the emitted sound. High frequencies are likely to generate narrower

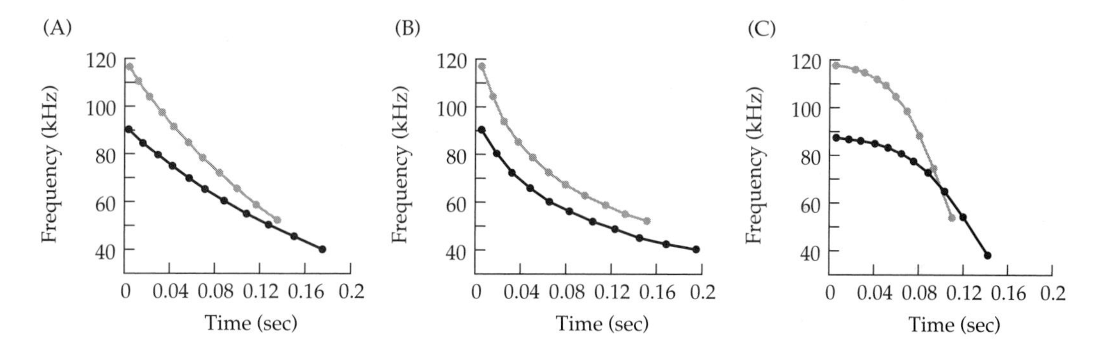

Figure 26.3 Effects of Doppler shifts on different types of frequency modulated echolocation pulses. In each case, a nonzero relative velocity of echolocator towards the target causes all frequencies in the returning echo to be raised and the echo durations to be shortened relative to the outgoing pulses. To make differences visible, an unrealistic 30% shift was used for this graph; 1 to 3% shifts would be more realistic values. Dark points plot the frequency modulation of the outgoing pulse; open circles show the frequency composition of the returning echo. (A) Linear modulation of frequency in an outgoing pulse results in a different slope and thus a different rate of modulation in the echo. This modulation causes large increases in errors in range measurement. (B) Linear modulation of the period results in a hyperbolic decrease in frequency of both the outgoing pulse and the returning echo. Because the two modulation patterns are parallel, the receiver experiences minimum additional error in range measurements. (C) Convex modulation of an outgoing pulse results in echo with a very different modulation pattern and thus a large increase in error of range measurements.

beams than low frequency ones. Thus the beam width for an FM pulse will vary during the scan, and different targets may be detected by different parts of the echo. High frequencies and long duration pulses are also likely to generate "lobes" in the sound field; different angular locations would then be hit by different intensities of outgoing pulse. This variation can confound the benefits of ear directionality outlined below.

Echolocators rely on the same properties of ears to find a sound source as are used by nonecholocators (pages 145–149). For example, differences in the time of arrival of an echo at the two ears could provide information on target angle. Pulses that have rapid onsets and short durations would make such timing measurements easier, and repeated pulses of short duration would allow for several measurements to be averaged. Pinnae or other directional sound-gathering organs can also compare relative amplitudes of arriving sounds to estimate the target location. Such organs are present in most echolocators. Finally, we saw in Chapter 6 that auditory source localization is enhanced when there are many frequencies in a sound. This is because the nature of any interference between sounds following different paths from target to eardrum varies with wavelength. Having a sound with many frequencies allows a receiver to compare the outgoing and returning power spectra and infer the location of the target from any differences. The technique is complicated by any spatial pattern in the outgoing sound beam. As long as different locations experience different intensities of outgoing sound, it will be unclear whether echo intensities at any given frequency are due to different source intensities or different locations. An effective echolocator might get around this problem by learning all of the eccentricities of its emission beam and ear patterns, or it could evolve adaptations so that ear and beam eccentricities tended to cancel out. Actually, the latter option seems to be common in animals (Figure 26.4).

The accuracy of target angle measurement will thus be maximized by using the same kinds of outgoing pulses that are optimal for range measurement. These are short in duration, avoid overlap between echo and outgoing pulse, and have wide bandwidths. Optimal receiver requirements are also similar for the two tasks. These include accurate measurement of very short time delays, small amplitude differences, and component frequencies. Although the two tasks rely on similar signals and receiver properties, they are unlikely to confound each other since binaural comparisons can be used to identify target angle, whereas the delay recorded at the nearer ear can be used for distance assessment.

Determining Target Properties

Echolocators may benefit from being able to identify or at least categorize different targets from their echoes. At the minimum, they may want to identify which echoes come from inanimate background objects (called **clutter** in the radar and sonar literature) and which from prey, or they might need to choose between two detected prey. How might they do this? Echo amplitude can be used to infer target size relative to incident sound wavelengths as

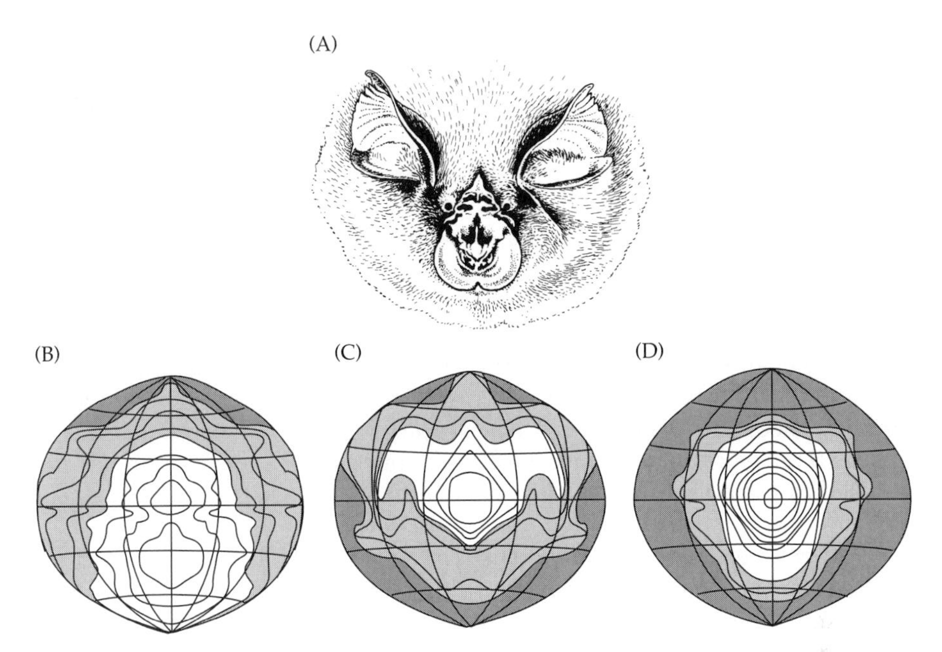

Figure 26.4 Directionality of echolocation in horseshoe bats (*Rhinolophus*). (A) View of the head showing structured ears and a complex nose-leaf around nostrils that is used to beam outgoing echolocation pulses. (B) Polar plot of the sound field generated around the central axis of the bat. The center of the plot corresponds to a point directly ahead of the bat. Darker regions represent lower intensity sound with contour lines separated by 3 dB. (C) Polar plot of ear sensitivity using same coordinate system as in B. (D) Directionality of entire echolocation system obtained by combining beam and hearing patterns. Note the much more focused and symmetrical pattern around the main forward axis than is true for either beam or hearing alone. Eccentricities in two patterns tend to cancel out. (After Schnitzler and Henson 1980.)

long as the echolocator has an independent measure of target range. As we have seen, echolocators use the delay between pulse emission and echo detection for range estimation, leaving echo amplitude as a measure of target size. Target shape and texture can be best assessed by emitting a wide-bandwidth pulse. Diffraction by targets of a size similar to incident wavelengths (pages 32–33) will cause some frequencies to be back-scattered more than others; this will change the power spectrum of the echo relative to the outgoing pulse. Larger targets with multiple reflection points will generate many mini-echoes. These arrive at the receiver with phase relationships dependent upon the wavelengths of the sound. If they arrive sufficiently separated in time, the ear of the echolocator will perceive them as separate events in the time domain; if they arrive too close together, then their echoes overlap temporally and they will be perceived as changes in the frequency domain. By comparing outgoing and returning power spectra, an echolocator may thus be able to classify the shape and surface texture of the target. Classification may not be an easy task, however. Because many of the patterns of interference depend on different path lengths, compound echo spectra are likely to change with shifts in the relative positions of the target and echolocator. If either target or echolocator is moving, successive echoes from the same target may show quite different spectra.

In air, the acoustic impedance of most targets is so much higher than that of the medium that most of the incident pulse energy is reflected by the target's surface. Bats are thus unlikely to be able to discover much about the internal composition of their targets by examining echoes. In water, most targets have acoustic impedances similar enough to the medium that incident sound energy is absorbed by the target. The absorption may cause the target to resonate, and the interactions between these resonant oscillations and the incident sound waves can have a dramatic effect on the power spectra of returned echoes (see pages 33–34). The shape of the power spectra of aquatic echoes thus depends on the composition of the targets. This information can be used by porpoises or whales to discriminate between different classes of targets.

As with determination of target range and angular location, wide-bandwidth signals are needed to provide maximal information about target properties. Accurate assessment of target size will depend on the ability of the receiver to measure echo amplitudes after independently correcting for target range. As we have seen, the latter requires wide bandwidths. The characterization of target shape and texture and that of target angle relative to the echolocator both rely on analyses of echo spectra. There is thus a risk that angular location and target shape may be confounded. Binaural comparisons of the arrival times of the leading edge of an echo (the part least likely to consist of multiple reflections) for target angle, and spectral comparisons of the echoes averaged over both ears for target shape might resolve this conflict. Again, wide-bandwidth pulses are necessary to provide sufficient spectral comparisons to accomplish any of these tasks.

Determining Target Velocity and Trajectory

Because echolocators and/or their targets are usually moving, their signals suffer Doppler shifts (see pages 21–22). In principle, these changes could be used to estimate the relative velocities and directions of movement of detected prey. However, this would require fairly accurate determination of the small differences between outgoing pulse and returning echo frequencies. For a bat approaching a prey item at a relative velocity of 2 m/sec, all components in returned echoes will have frequencies 1% higher than those emitted, and durations 1% shorter than the original pulses. An 80 kHz pure tone pulse would thus return an echo of 81 kHz. Changes for a porpoise pursuing an equivalent fish would be only 23% of those experienced by a bat.

Given the uncertainty principle, accurate determination of frequency is best accomplished if a long sample of the returning echo is provided. The more waves of each component that are available for estimation, the more accurate the resulting value. A short-duration and frequency-modulated pulse would thus be an ineffective way to measure velocity because each frequency component would be present for only a few waves. A better signal would be a long-duration pulse with a single constant frequency (abbreviated **long-CF** in the literature). An ideal receiver that looked for maximal correlation between echoes and various Doppler-shifted copies of an outgoing long-CF pulse would have a minimal average error in velocity estimation, ΔV, given that target range is already known, of

$$\Delta V = \frac{c}{4\pi f_o T d'}$$

where f_o is the frequency of the outgoing pulse, T is the rms pulse duration as defined earlier in this discussion, and c and d' are as defined earlier (Pye 1983; Woodward 1964). This equation fits our intuitive notions since the product $f_o \cdot T$ is the total number of cycles of the frequency f_o in the outgoing pulse. The higher the constant frequency and the longer the pulse, the greater the number of cycles available to measure echo frequencies, and thus the more accurate that measurement. This statement simply reflects again the uncertainty principle. Note that this is the minimal velocity error that can be obtained by even an ideal receiver given a pulse of frequency f_o and rms duration T; whether living echolocators do this well remains to be discussed.

If long-duration pulses provide accurate measurement of velocity, and wide-bandwidth pulses provide accurate range measurement, why not produce long-duration wide-bandwidth pulses and obtain high accuracy for all parameters? Note that the uncertainty principle for sound does not prohibit this result. It simply says that the product of bandwidth and pulse duration must be greater than a constant; it says nothing about how large that product can be. Unfortunately, long durations give accurate velocity measures only when the target range is known; high bandwidths give accurate range measures only when the relative velocity of echolocator and target is known. When both velocity and range have to be estimated from the same pulse, there will be a set of combinations of possible velocity and range that cannot be discriminated from each other or from the true values even by an ideal receiver. Velocity and range errors will then be confounded. The number of indiscriminable combinations is set only by the value of d'. It is unrelated to the bandwidth or pulse duration (Burdic 1968; Woodward 1964). If signal bandwidth is increased, but d' is held fixed, this reduces the number of possible range values that will be confused with each other and thus reduces ΔR. However, because the total number of indiscriminable combinations is fixed, it follows that the number of indiscriminable velocity values must increase, and this will result in greater values of ΔV. The confounding of velocity and range errors thus creates an **uncertainty principle for sonar and radar** that states: For any given signal-to-noise ratio, reduction of error for either velocity or range below a certain value results in an increased error in the other.

These relations generate many of the tradeoffs faced by echolocating animals. Signals optimal for range measurement are the worst ones to use for velocity measures. Similarly, signals optimal for measurement of velocity are poor choices for assessment of target angle and identity. The only exception is the discrimination between moving and nonmoving targets. Optimal signals for velocity measurement are, however, also optimal for detection; these two tasks can thus be undertaken by the same signal with no conflicts. In the next sections, we examine the degree to which the various tradeoffs have shaped the types of signals produced by different echolocator species. Not surprisingly, ecological context and mode of hunting will be seen to play big roles in shaping signal form.

ECHOLOCATION IN BATS

With more than 800 species throughout the world, the bats constitute the second largest order of mammals. There are two suborders of bats. The Megachiroptera inhabit tropical Africa and Asia, feed largely on fruit, nectar, or flowers, and rely on sophisticated nocturnal eyes for flight and foraging. One genus, *Rousettus*, lives in caves and uses a crude form of echolocation to avoid obstacles when in complete darkness. Echolocation pulses are generated by clicking the tongue against the side of the mouth. The Microchiroptera live on every continent except Antarctica, and all use echolocation for obstacle avoidance and foraging. Their echolocation sounds are produced in the larynx. Although about 70% of this group feeds on insects and other arthropods, tropical forms are ecologically diverse and include species that capture frogs, lizards, and birds in dense vegetation, three vampires that take blood from other vertebrates, several species that fish, and a variety of species that eat fruit or collect nectar from specialized "bat-flowers." What types of signals do these diverse species use for echolocation?

Open-Site Foragers

Bats foraging in more open sites invariably feed on flying insects. They must detect tiny targets at a distance and intercept them; they usually do not need to worry about echoes from clutter. Consider the small European pipistrelle bat, *Pipistrellus pipistrellus* (Kalko 1995). Although this species sometimes forages close to vegetation, it then flies parallel to the foliage and thus minimally detects nearby clutter. Each attack sequence can be divided into three phases: search, approach, and terminal "buzz" (Figure 26.5). During the search phase, these bats emit moderate length (8 msec) pulses of nearly constant frequency (around 55 kHz) or constant-frequency portions preceded by a gentle FM sweep starting around 80 kHz. Bandwidths and pulse repetition rates are both low. Duty cycles average 5 to 9%. The long-duration, nearly constant-frequency portion allows for detection of targets at maximal range. The intermittent small bandwidth FM probably provides ranging information on nearby clutter. As soon as the bat detects an insect (usually at a distance of less than 2 m), it reduces and eventually eliminates the constant-frequency part of the pulse and doubles the bandwidth of the remaining FM portion. Typical pulses sweep from 100 kHz down to 55 kHz. With further approach, pulse rates in-

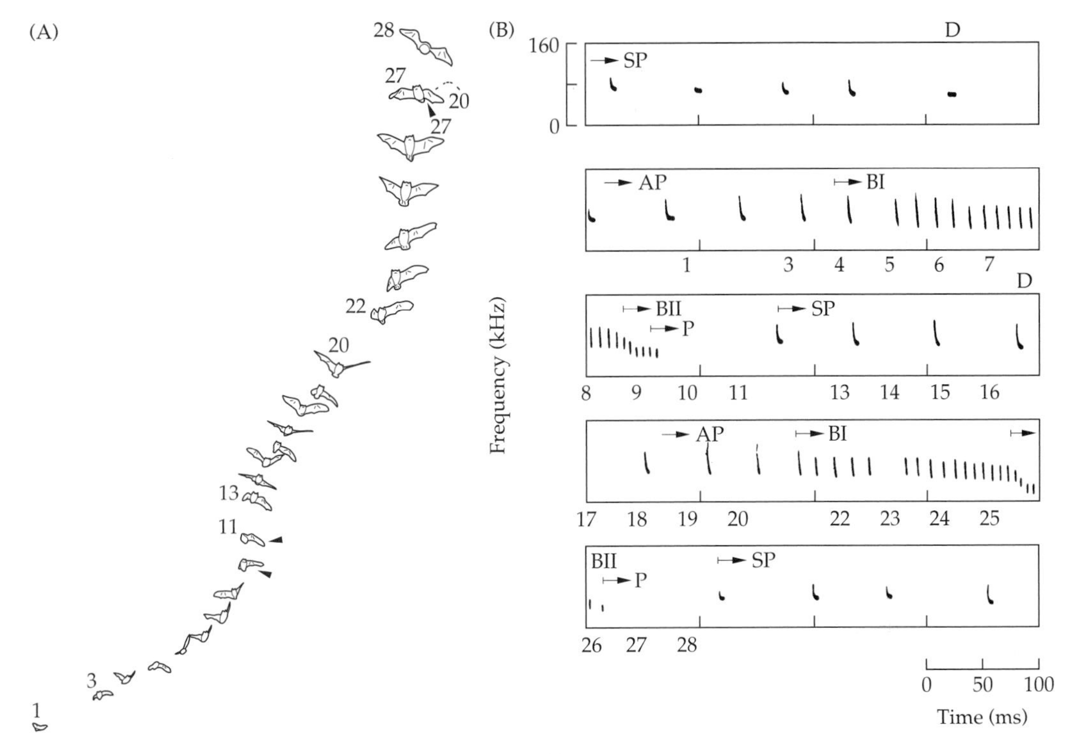

Figure 26.5 Typical hunt of aerial insects by European pipistrelle bat *Pipistrellus pipistrellus*. (A) Tracings from high-speed camera record of the bat's trajectory during two successive insect captures. Numbers indicate successive flashes of camera. Prey were captured at flashes 10–11, and 27. (B) Spectrograms of echolocation pulses emitted during the same sequence. Flash numbers are indicated below the time line of spectrograms. SP = search phase, D = detection of prey, AP = approach phase, BI = first part of terminal buzz, BII = last part of terminal buzz, P = prey capture. (From Kalko 1995.)

crease steadily, pulse durations drop by half, and duty cycles rise to 10 to 14%. These very short and broad bandwidth signals are optimal for determining target range, angular location, and target shape or texture. The terminal buzz is characterized by even shorter duration pulses (0.7 to 1 msec), very high emission rates, and duty cycles of 16 to 22% (thus producing a "buzz"). Bandwidths are initially high (with both the fundamental and a second harmonic present), but these and the mean frequency drop until the final five to eight pulses have only a tiny FM sweep. The short durations of these last pulses may provide sufficient bandwidth for final ranging and angle information without requiring modulation over a large range. This species, like many others, carefully adjusts pulse durations to avoid pulse-echo overlap (Figure 26.6). Because it would be impossible to make pulses short enough to avoid overlap when very close, the bats stop echolocating when 4 to 10 cm from the target. Shortly thereafter, the bat scoops up the insect with either its wing or

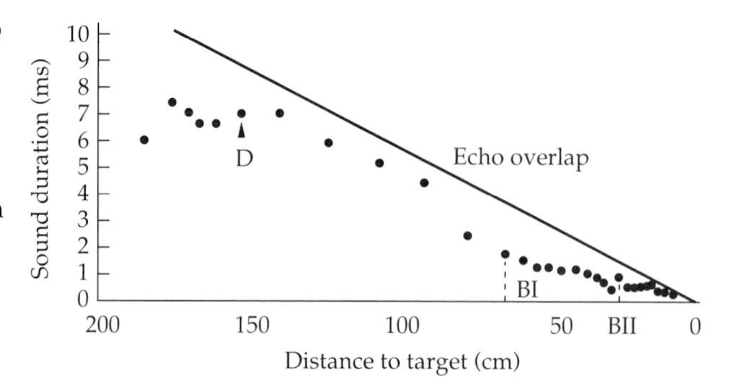

Figure 26.6 Avoidance of pulse-echo overlap by foraging pipistrelle bat as it attacks an insect. The plot shows the maximum pulse duration required at decreasing target distances to avoid pulse-echo overlap (diagonal line) and the actual pulse durations used at each target distance by the bat shown in Figure 26.5. (From Kalko 1995.)

the membrane between its legs and bends its head forward to bite the prey. The entire sequence from detection to capture takes 0.5 to 1.0 sec and is repeated roughly every 4 sec if there are sufficient insects nearby. Capture efficiencies range from 30 to 40% for capture of moths and 60 to 70% for capture of mosquitoes or caddis flies (Kalko 1995).

To maximize detection at sufficient distances, open-area foragers emit sounds through their mouths and thus generate wide sound beams (of from 60 to 90°). They also use extremely intense pulses; typical amplitudes are 100 to 120 dB SPL (Griffin 1958). Because echolocation pulses are generated by the larynx in Microchiropteran bats, their intensities are a direct function of the air pressure that the lungs can produce in the trachea below the glottis (Suthers 1988). This pressure cannot be higher than blood pressure or it would collapse the blood vessels in the lungs. In fact, the measured air pressures in echolocating open-country bats are just below this level. Sounds emitted are thus as intense as is physiologically possible. One might think that such intense sound production at rapid rates would be very costly energetically. Microchiropteran bats have reduced this problem by coupling the buildup of air pressure, and the attendant release by sound emission, to flight by using the same muscles. All species emit pulses only during exhalation, and exhalation always occurs on the upstroke of a wingbeat (Jones 1994; Speakman et al. 1989; Suthers 1988). Long-duration pulses are thus emitted singly per wingbeat, whereas short duration pulses are emitted in a staccato burst within the exhalation phase of each wingbeat-respiration cycle. Thus sound production adds little additional cost to that already required to beat the wings. Once bats have detected prey and begin their approach, they reduce emitted pulse amplitudes to keep echo amplitudes as constant as possible (Hartley 1992; Kick and Simmons 1984). For certain types of receivers, this reduced amplitude enhances ranging accuracy.

The distance at first detection is also maximized in these species by careful choice of pulse frequency. A good match between typical prey size and wavelength requires frequencies of 35 kHz and higher. The advantages of matching target size and wavelength are diminished by increasing atmospheric attenuation as frequency is increased. Spreading and absorption losses alone will reduce the intensity of a 30 kHz pulse by 80 dB when the target is

7 m away; the target need only be 4 m away to cause the same decrement in a 120 kHz signal. The result is that species using pulses such as those of the pipistrelle in Figure 26.5 can detect items as small as 0.2 mm but only at a maximum range of 1.5 to 2 m. Open-area foragers of larger body size tend to use lower frequencies (Waters et al. 1995). This relationship could be a result of a focus on larger prey (thus allowing lower frequencies for wavelength-prey size matches), but it could also arise because larger bats fly faster or are less agile and thus require greater distances of detection (Barclay and Brigham 1991). The frequencies used by open-country foragers during the search phase will constrain the maximum pulse durations that can be used; the higher the frequency, the shorter the range of detection, and thus the shorter the pulse duration must be to avoid pulse-echo overlap. This relationship is in fact what is found (Figure 26.7).

A number of studies have examined how well bats using short-duration FM pulses can estimate range, angular location, target properties, and velocities of targets. Like the pipistrelle, the large brown bat, *Eptesicus fuscus*, also forages in uncluttered contexts. Typical pulses contain two to three harmonics, with the fundamental sweeping hyberbolically from 55 kHz down to 25 kHz in about 1.5 msec; we have seen that this type of FM sweep is maximally Doppler-tolerant. Representative values of β_{rms} and β_{crms} are 55 kHz and 14

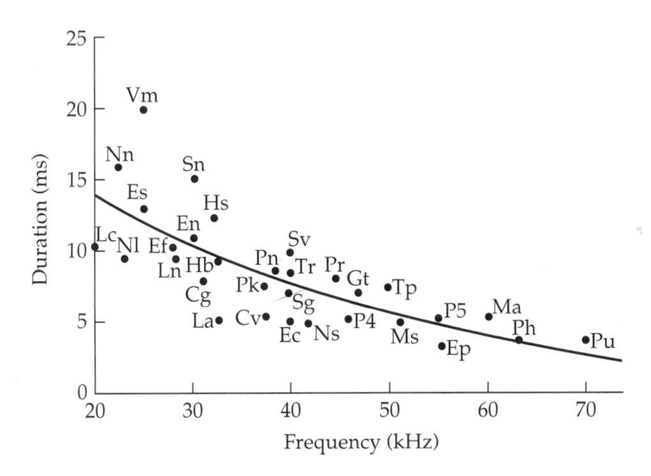

Figure 26.7 Maximum pulse duration versus average frequency in search pulses of various vespertilionid bats. The line shows result of the regression of pulse duration on the logarithm of frequency. Code to species plotted: Cg, *Chalinolobus gouldii;* Cv, *Chalinolobus variegatus;* Ec, *Eptesicus capensis;* Ef, *Eptesicus fuscus;* En, *Eptesicus nilssonii;* Ep, *Eptesicus pumilus;* Es, *Eptesicus serotinus;* Gt, *Glisochropus tylopus;* Hs, *Hypsugo savii;* Hb, *Hesperoptenus blandfordi;* La, *Laephotis angolensis;* Lc, *Lasiurus cinereus;* Ln, *Lasionycteris noctivagans;* Ma, *Miniopterus australis;* Ms, *Miniopterus schreibersii;* Nl, *Nyctalus leiserli;* Nn, *Nyctalus noctula;* Ns, *Nycticeinops schleffeni;* P4, *Pipistrellus sp1;* P5, *Pipistrellus sp2;* Ph, *Pipistrellus hesperus;* Pk, *Pipistrellus kuhli;* Pn, *Pipistrellus nathusii;* Pr, *Pipistrellus ruepelli;* Pu, *Pipistrellus nanus;* Sg, *Scotorepens greyii;* Sn, *Scotophilus nigrita;* Sv, *Scotophilus viridis;* Tp, *Tylonycteris pachypus;* Tr, *Tylonycteris robustula;* Vm, *Vespertilio murinus.* (From Waters and Jones 1995, © Springer-Verlag.)

kHz respectively. In laboratory experiments where head movements by the bat are minimized, other sources of echoes are carefully eliminated, and d' was estimated to be about 63, the bats exhibited a minimal ΔR of 0.007 mm (Simmons et al. 1990b). This ΔR is within experimental error of that predicted for a coherent correlator (0.008 mm) but much better than the semi-coherent-correlator prediction (0.031 mm). Although two targets of 1 to 4 mm and 6 mm or more separation are very accurately discriminated, the bats show a significantly higher error rate for separations of from 4 to 6 mm. This range of high-error distances corresponds to the first lateral peak in the cross-correlation function of their signals (see Box 26.1). Simmons et al. (1990b) count this as additional evidence that the bats operate like coherent cross-correlators and thus can monitor signal phases. Additional evidence suggesting accurate phase measurement is obtained by reversing the bats' echoes in time before they hear them so that the reversed echoes have the same power spectrum as a normal echo, but have quite different phases. Echo reversal greatly reduces the accuracy of *Eptesicus* range measurements, supporting the notion that the bats are able to measure phase (Masters and Jacobs 1989).

These results reflect the latest evidence in a continuing debate about the kinds of receivers used by FM bats to measure target range (Altes 1980; Menne and Hackbarth 1986; Pollak 1993; Schnitzler and Henson 1980; Simmons 1993; Simmons and Grinnell 1988; Simmons et al. 1990a,b). Although the kinds of neural structures necessary to support either cross-correlator or spectrogram correlator receivers have been found in echolocating bats (see review by Suga 1988), the degree to which phase information can be reconstructed from the combined responses of neuronal ensembles remains contentious. For our purposes, it is sufficient to note that the accuracy of range measurement by FM bats is phenomenal and close to that of ideal receivers. How they manage to do this remains a major challenge to neuroethologists. From an ecological point of view, it is not clear that any bat requires a range accuracy of less than 1 cm to capture its prey. Natural head movements between pulse emission and receipt of echo may be on the order of millimeters, and field values of ΔR can be no smaller than these changes. Where fine temporal accuracy may be important is in the binaural determination of target angle, and/or in the characterization of target shape. Here, very small differences in arrival times make big differences in the information extracted. Simmons et al. (1990a) have even argued that bats convert all frequency domain information, such as that extracted from differences in echo and pulse power spectra, into a spatial map. Differences in spectra become translated into differences in distance between the bat and each reflecting part of the target. In this view, the determination of target range, target angle, and target properties are all seen as part of the same task.

Gleaners

Bats that select food items close to surfaces face quite different tasks from those preying on flying insects in open spaces. The biggest challenge is the discrimination of a prey echo from those of the abundant and nearby clutter. This background clutter puts a very high premium on the ability to identify target angle and target properties; both tasks are favored by the use of very short pulses with large bandwidths. As an example, the notch-eared bat, *Myotis emarginatus*, is sympatric with the common pipistrelle bat, but often gleans insect prey from foliage or inside barns. The sequence of FM pulses used in a gleaning attack by the notch-eared bat is similar to that shown in Figure 26.5 except that pulse bandwidths are larger (e.g., the fundamental sweeps from 124 kHz down to 45 kHz) and pulse durations much shorter (all pulses are 1 msec or less) (Schumm et al. 1991). In many gleaners, bandwidths are further increased by the presence of from two to four harmonics with significant energies. High-intensity pulses and broad sound beams during gleaning would generate far too many clutter echoes. Gleaners are thus known as "whispering bats" because they use much lower-intensity signals (Griffin 1958). Many species emit the signals from their nostrils and have special nose-leaves to produce a narrow beam of sound for scanning the foliage (Figure 26.8).

The ability of a gleaner to discriminate between moving and static targets requires no greater skill than the perception of change in target angle over several successive pulses. Many echolocating bats, and especially gleaner species, appear to have angular accuracies of 1.5 to 2° (Neuweiler 1989; Simmons and Grinnell 1988); at a range of 2 m, a prey would have to move at least 5 to 7 cm to be detectable by its position change. Smaller movements by prey may not be detectable as differences in angle, but they may be perceived by the bat as changes in the power spectrum of the echo since spectra will change if the prey moves any part of its body (Neuweiler 1990). If the bat is not moving significantly, static targets would generate echoes with stable power spectra, whereas targets with moving body parts would produce echoes with fluctuating power spectra. A hovering gleaner would then be able to discriminate between echoes of slightly moving prey and those of inanimate background clutter.

A more difficult task is the discrimination between echoes from completely static prey and those from other inanimate targets. Bats would here need to compare target shape, size, or texture. Discrimination might occur in either the time or the frequency domains. The different reflecting portions of a single target will create echoes that arrive at either ear at slightly different times. These echoes might be perceived as separate events or, alternatively, as overlapping ones creating a characteristic power spectrum (Beuter 1980). Given the acute range and spectral accuracies of FM bats (Simmons et al. 1990a,b), one might expect the ability to classify target shapes to show similar acuity. In fact, target classification is not as easy as it sounds. One problem is that the power spectrum of the echo from an irregular target will change depending upon the orientation of the target in the sound beam. A bat would thus have to learn the power spectra for all likely orientations to recognize a given target in every circumstance. There may, however, be certain power-spectrum traits for a given target that occur regardless of orientation. These would allow some degree of target classification, but not at the accuracy predicted by range measurements. This type of target discrimination is in fact

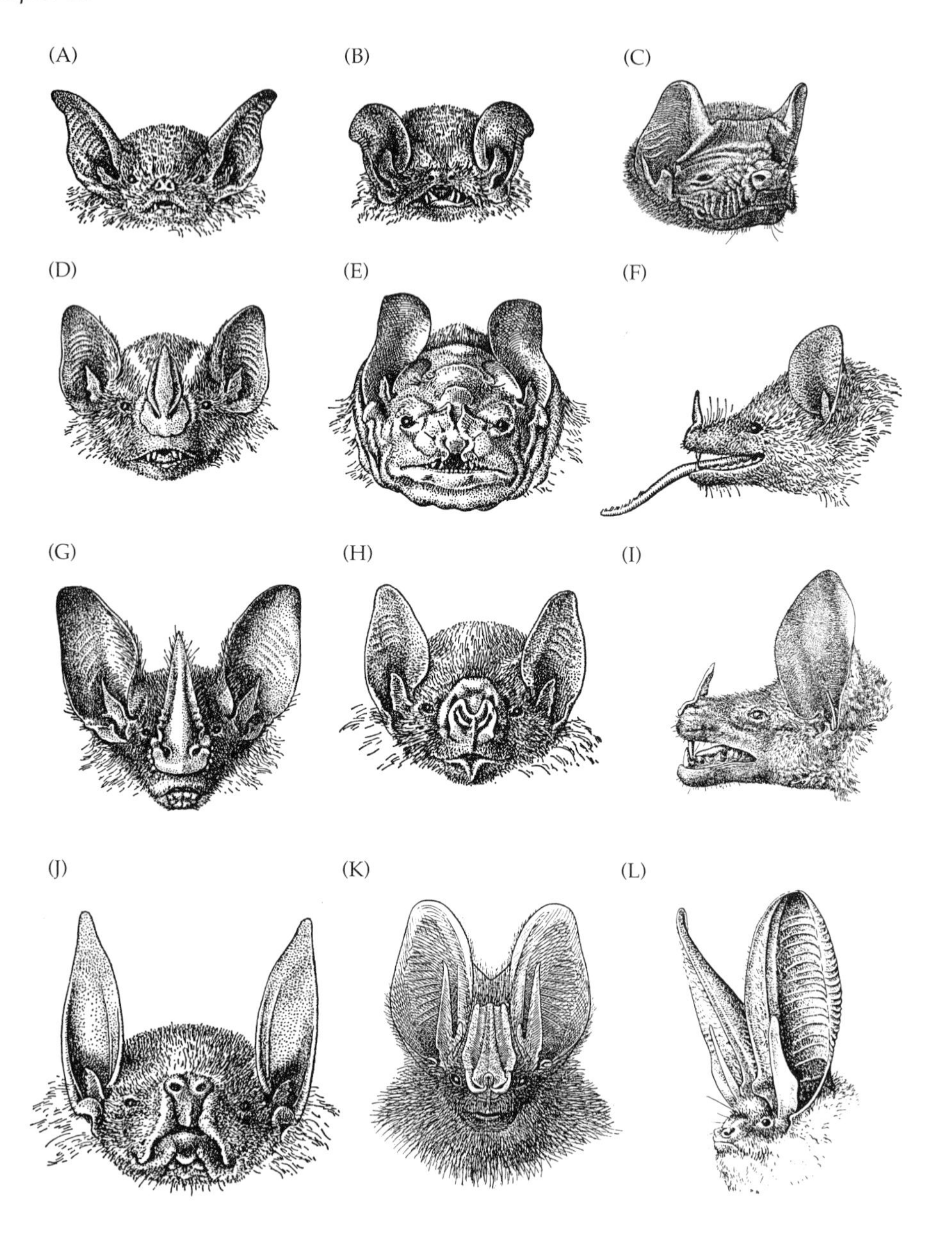

what is found. The large gleaner, *Megaderma lyra*, can discriminate between targets at different distances down to a 0.2 mm separation; its ability to discriminate between two sandpaper surfaces using echolocation is limited to differences in grain size of 2 mm or greater (Schmidt 1988). There is thus a 10-fold difference between the ranging acuity and the target texture discrimination. The difficulty of target classification may be one reason that many predatory gleaners, including *Megaderma*, rely heavily on cues generated by their

◀ **Figure 26.8 Heads of selected bats, showing special ear, nose, or mouth adaptations for echolocation.** The top row shows heads of three species of insectivorous bats that fly above vegetation: (A) *Saccopteryx bilineata*, (B) *Lasiurus borealis*, and (C) *Tadarida condylura*. All three emit echolocation pulses from the mouth. The second row shows the Neotropical frugivores *Uroderma bilobatum* (D) and *Centurio senex* (E), and the nectar-feeding *Glossophaga longirostris* (F). All have nose-leaves for emitting narrow beams of pulses through the nostrils. The third row includes *Mimon crenulatum* (G), a predator on insects, spiders, and lizards; *Desmodus rotundus* (H), a true vampire taking blood from large mammals; and *Vampyrum spectrum* (I), the world's largest echolocating bat and a predator of sleeping doves and parrots. All emit sounds through the nostrils and have very large ears for finding prey in cluttered environments. The bottom row features the fishing bat *Noctilio leporinus* (J), the large African *Lavia frons* (K) that hawks large insects from a perch, and a European gleaner of insects from foliage, *Plecotus auritus* (L). (A–B, D–J from Goodwin and Greenhall 1961, courtesy of the American Museum of Natural History; C and K from Rosevear 1965; L from Brosset 1966.)

prey to discriminate between living and nonliving targets. Sounds produced by prey are used by a wide variety of gleaner hunters (Neuweiler 1989), and the large carnivorous *Vampyrum spectrum*, that favors sleeping birds, probably utilizes odor cues (Vehrencamp et al. 1977).

Hawkers and Fishers

A number of bat species hawk like birds; they hang in a suitable location, scanning for potential prey, and then fly off in a rapid sally to capture detected prey. We might expect this strategy to require estimation not only of the location of prey, but of their trajectories and velocities. A small number of bat species have also become fishers. Because sound in air is largely reflected at the surface of water, a fishing bat must rely on disturbances to the water's surface caused by movements of nearby fish; its sonar cannot penetrate the water. Such surface disturbances will be fleeting, and there is again a premium on predicting where the fish will be by the time the bat arrives and can make an attack.

Hawking is common in the bat families Rhinolophidae and Hipposideridae. Rather than hunt from a static position, some species patrol very closely to vegetation and make quick sallies after moving insects. This type of foraging is also seen in the neotropical moustache bat, *Pteronotus parnellii* (family Mormoopidae). The typical search signals emitted by hawking bats contain a long-duration and constant-frequency (long-CF) segment, with short FM sweeps on one or both ends. The frequency of this long-CF portion is specific to each individual animal and varies only within a very small range for any given species (Neuweiler et al. 1987; Vater et al. 1985). Consider the greater European horseshoe bat, *Rhinolophus ferrum-equinum*, which is a typical hawker. The major harmonic in its search pulses begins with a 2 msec FM sweep upward from 70 to around 83 kHz, a 50 msec long-CF portion around 83 kHz (the exact value differing among individuals), and a 2 msec downward FM sweep to about 64 kHz (Griffin and Simmons 1974; Jones and Rayner 1989). This component is the second harmonic; the fundamental and the third harmonic may also be present but are usually much weaker in intensity. Typical duty cycles average 50 to 60%, which is many times higher than for FM open-site bats. When prey are detected, the bat leaves its perch and the pursuit is characterized by an increase in pulse emission rate, a decrease in pulse duration down to 3 to 5 msec, an increase in duty cycles to a maximum of 75%, and an accentuation of the bandwidth of the FM portions of the pulses (Figure 26.9).

The long-CF portions of the search signals in these bats should enhance their detection abilities over those of species using alternative pulse types. In fact, the smallest objects that a *Rhinolophus* can detect at 2 m are one-fourth the mimimal size of those detectable by FM bats emitting pulses of similar intensity (Schnitzler and Henson 1980). The high duty cycles also increase the chances of prey detection. On the other hand, the high frequencies used by most of these bats must result in major atmospheric absorption losses and thus lower detection ranges. The advantages of using a long-duration constant frequency thus seem compromised by the concurrent choice of such high frequencies. Why use such high frequencies?

The answer is surely related to the utility of long-CF pulses in the detection and measurement of target motion. For a static hawker like *R. ferrum-equinum*, immobile targets will return echoes with a long-CF frequency identical to that of outgoing pulses, whereas moving insect prey will return Doppler-shifted echoes. The differences in echo frequencies of moving prey

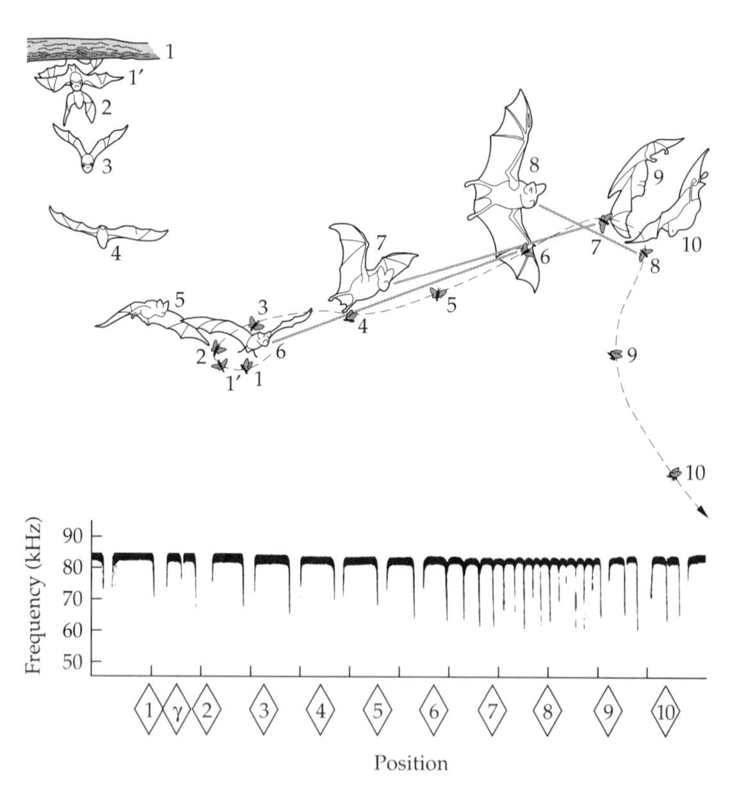

Figure 26.9 Hawking of a moth by a greater horseshoe bat (*Rhinolophus ferrum-equinum*). (A) Trajectory of the bat and moth. (B) Spectrograms of emitted pulses. Underlying numbers refer to numbered stages in location of bat. Note the long CF pulses with small initial and terminal FM portions. (From Trappe and Schnitzler 1982, © Springer-Verlag.)

and static objects will depend upon the direction in which the insect is moving, its velocity, and the frequency in the outgoing pulse. The first two factors usually limit Doppler shifts to small values. Differences between moving and static target echoes will thus be detectable only if outgoing frequencies are sufficiently high. Even for a long-CF frequency of 83 kHz in the outgoing pulse (as in *R. ferrum-equinum*), an insect flying directly away from the bat at 1 m/sec will produce echoes differing in frequency from the output pulse by only 484 Hz. The frequency resolution of the echolocator must be very accurate to detect this type of change. In fact, the cochlea of long-CF bats is expanded around each individual's favored long-CF frequency. Some authors have even referred to this portion of the cochlea as an "acoustic fovea." In *Rhinolophus rouxi*, the preferred long-CF frequency is about 75 kHz, and nearly 25% of the cochlear basilar membrane is devoted to detection of frequencies between 72 and 77 kHz (Neuweiler 1989, 1990). The combination of pulse structures optimized for velocity detection with unique cochlear and brain specializations allow bats like rhinolophids to detect echo deviations from the outgoing long-CF frequency as low as 10 Hz. The higher the frequency in the outgoing pulse, the lower the minimal velocity that is measurable using a 10 Hz minimal frequency resolution. For *R. ferrum-equinum*, an 83 kHz outgoing pulse and a 10 Hz frequency error will permit discrimination between static targets and those moving as slowly as 2 cm/sec. Most insect prey will be flying this fast or faster. Note that this mechanism of discriminating between prey and inedible objects requires that prey be moving; as predicted, most of these long-CF bats ignore static prey (Neuweiler 1989; Schnitzler and Henson 1980).

The discriminability of flying insects from static background is enhanced by the beating of their wings. Wingbeating will produce both amplitude and frequency modulations in any echo with a duration long enough to capture several wingbeats. These echo modulations are called "glints." A bat pulse of 55 msec duration could pick up at least one glint in the echo if the insect's wingbeat rate were 18 Hz or higher. Typical bat insect prey (and wingbeat rates) include lacewings and geometrid moths (25 Hz), caddis flies and nongeometrid moths (30 to 50 Hz), crane flies and beetles (50 to 100 Hz), and most flies (>100 Hz) (Kober and Schnitzler 1990). Long-CF bats should thus be able to use the presence of glints in echoes as a criterion for discrimination between flying insects and static clutter. Where wingbeat rates are fast enough that several cycles occur per echolocation pulse, bats might even be able to measure the rates of amplitude and/or frequency modulation, compare these to known wingbeat frequencies of different prey types, and thus discriminate between more versus less profitable prey before bothering to attack. Several studies have shown that rhinolophids can do precisely this. With training, captive bats can learn to discriminate between insect wingbeat rates that differ by as little as 3 to 5 Hz; some insects with the same wingbeat rates can even be distinguished because of different modulation patterns (von der Emde 1988; Roverud et al. 1991).

Once a *Rhinolophus ferrum-equinum* begins to fly, and at any time when other species such as *R. hipposideros* patrol foliage, static as well as moving targets will return Doppler-shifted echoes. If the bat moves at a high enough velocity, the frequencies returned may lie outside the acoustic fovea and resolution will be decreased. For this reason, flying rhinolophids, hipposiderids, and the mormoopid *Pteronotus* compensate by reducing all frequencies in the outgoing pulse just enough that the long-CF portions of echoes from static objects return at that individual's favored frequency (Schnitzler and Henson 1980). Such Doppler compensation allows long-CF bats to ignore all echoes at their favored frequencies, and instead concentrate on those with slightly higher or lower frequencies. Several species are exceptional among bats because they allow returning echoes and outgoing pulses to overlap temporally. This overlap may be one mechanism by which Doppler compensation is achieved. Note that only the long-CF portions are allowed to overlap. Pulse durations are carefully adjusted to prevent overlap between corresponding FM portions of pulse and echo. Thus the ability of these bats to use the FM portion for assessment of target range and angle is preserved. The FM portions of long-CF bats do not show the Doppler-tolerant hyperbolic pattern. However, because Doppler compensation shifts all frequencies, including the FM portions, by the amount necessary to cancel out the effects of bat velocity, Doppler tolerance may not be necessary.

Although long-CF species could use a single echo to measure target velocity and then predict a prey's position at the point of anticipated interception, there is no evidence that any insectivorous bats do so (Masters 1988; Neuweiler 1990). Instead, they sequentially alter their own trajectories as the location of the prey item changes. The one long-CF species that does appear able to predict the future location of its prey is the large fishing bat, *Noctilio leporinus* (Campbell and Suthers 1988). A foraging individual usually has only a few brief ripples in the water's surface to indicate a fish's presence. It uses these to estimate the fish's velocity and direction of motion and drags the water's surface with its large hooked claws at the point where it predicts an intersection with the fish's trajectory (Figure 26.10). Although this bat has long-CF signals, it apparently extrapolates target velocity using at least a few successive echoes instead of assessing Doppler shifts from a single echo (Wenstrup and Suthers 1984). In the field, only 0.5 to 2% of drags are successful (Schnitzler et al. 1994). This low value surely reflects random drags in the absence of indicator ripples. However, the very fact that the bats drag at random supports the notion that trajectory interception from ripples is a difficult and not always profitable task.

Tradeoffs and Accuracy in Bat Echolocation

Bats relying primarily on short-duration FM pulses have better range resolution than long-CF bats. Minimum range error (allowing head movements) is about 6 mm for *Eptesicus fuscus* (FM) and 12.5 mm for *Rhinolophus ferrumequinum* (long-CF). In contrast, target velocity can be measured by a long-CF bat such as *Rhinolophus* down to 0.02 m/sec, whereas *Eptesicus* can do no better than several m/sec. Measurement of insect wingbeat frequencies is another expression of velocity measurement. *Rhinolophus* can detect differences in cyclic movement down to 2.7 Hz, whereas *Eptesicus* discrimination is limited

Figure 26.10 Fishing bat (*Noctilio leporinus*) dragging water to catch fish. Estimating the right location hinges on the accurate measurement of direction and velocity of the fish by using echoes from ripples. (Photo courtesy of Merlin D. Tuttle, Bat Conservation International.)

to 15.8 Hz (Roverud et al. 1991). The fact that both species show a tradeoff between range and velocity accuracy suggests that resolution in one measure is sufficiently small that the other must be large to satisfy the sonar uncertainty principle. We have seen that FM bats tend to use optimally Doppler-tolerant signals, and when head movements are minimal, they can achieve range resolutions close to that of the best ideal receivers. Long-CF bats do not appear to achieve ideal-receiver accuracy in velocity measurement. Typical echoes used by *R. ferrum-equinum* imply d' values less than 1 given the observed velocity errors, the high emitted frequencies, the long pulse durations, and an assumption of ideal reception. It is more likely that d' values are in the range of from 40 to 60, and that the bats are not operating ideally. Since they apparently do not use velocity to plot prey trajectories, but instead use velocity to discriminate between static and moving prey or to select between alternative prey species, the nonideal velocity measures may be sufficient.

ECHOLOCATION IN CETACEANS

Like the bats, the order Cetacea is divided into two suborders that differ in their use of echolocation. As far as is known, most toothed whales and porpoises echolocate, whereas none of the baleen whales appear to do so. This is not surprising since toothed whales are active predators of moving prey, whereas baleen whales are filter-feeders.

The aquatic environment of toothed cetaceans requires a different set of optimal echolocation pulse designs. Because wavelengths for any frequency are 4.4 times longer in water than for air, good echo formation and extraction of target property information require higher frequencies than in air for the same sized targets. Being larger animals, however, cetaceans might be expected to attend to larger targets, and thus not need such small wavelengths. In the end, these factors balance out, and cetaceans largely use frequency ranges similar to those of bats. Social offshore species favor echolocation pulses at 20 to 80 kHz; those that are solitary and live inshore or in fresh water tend to use frequencies of 100 to 150 kHz (Ketten 1992). Each species favors its own range of frequencies, although some may expand the range when ambient noise is high (Au 1993).

The 4.4-times higher speed of sound in water requires very short pulse durations to prevent pulse-echo overlap. Typical observed values are 0.05 to 0.1 msec. This is much shorter than necessary to avoid overlap except when very close to targets. As a result, most cetaceans do not have to reduce pulse duration as they near targets. In fact, even when close to targets, duty cycles rarely exceed 1%. One advantage of very short pulse durations is that bandwidths will be correspondingly high due to the uncertainty principle. Even at the high frequencies used, most pulses consist of at most five to eight complete cycles, and this results in rms bandwidths of 16 to 28 kHz. Cetacean pulses thus require no frequency modulations to create large bandwidths. It is even possible that the short durations required to avoid overlap in water make modulation physiologically impossible. Reducing pulse durations even further would then be the only way to ensure sufficient bandwidth for accurate range, angle, and target property assessment. Although sound pressure levels for cetaceans seem high (over 200 dB for porpoises), when corrected for the higher acoustic impedance in water (see pages 23–24) actual energy intensities are only slightly higher than for bats (Au 1993). Absorption losses during propagation are much lower in water than in air. As a result of loud pulses and low absorption, a porpoise can thus detect a 2.5 cm sphere at 73 m, whereas a typical bat cannot detect such a target at less than 2 to 4 m.

There is continuing debate about how cetaceans create their echolocation pulses (Au 1993; Ketten 1992; Morris 1986; Norris and Harvey 1974). The most widely accepted mechanism involves moving air between sacs in the head and through thin membranes that vibrate (Figure 26.11). The skull is shaped like a parabola and the combination of skull and air sacs beams the sounds forward. A oil-filled sac called the melon lies in the forehead and provides a gradual impedance match (e.g., an "acoustic lens") between animal and medium. Outgoing pulses are beamed within a 6 to 10° cone ahead of the animal. Echoes are thought to be captured by the lower jaw and transmitted through special fat-filled channels to the inner ears. Target angle can be measured to within 1 to 4°.

The ability of dolphins to detect faint echoes in noise is only 6 to 8 dB worse than that predicted for ideal receivers (Au 1993). At reasonable values of d', they, like bats, have minimal range errors of about 1 cm; because of the higher speed of sound in water, this means porpoises achieve better temporal resolution. This resolution by porpoises is more likely due to broader bandwidths in the pulses than to better receiver performance. Both groups can recognize complex targets whose multiple facets differ in range by only several

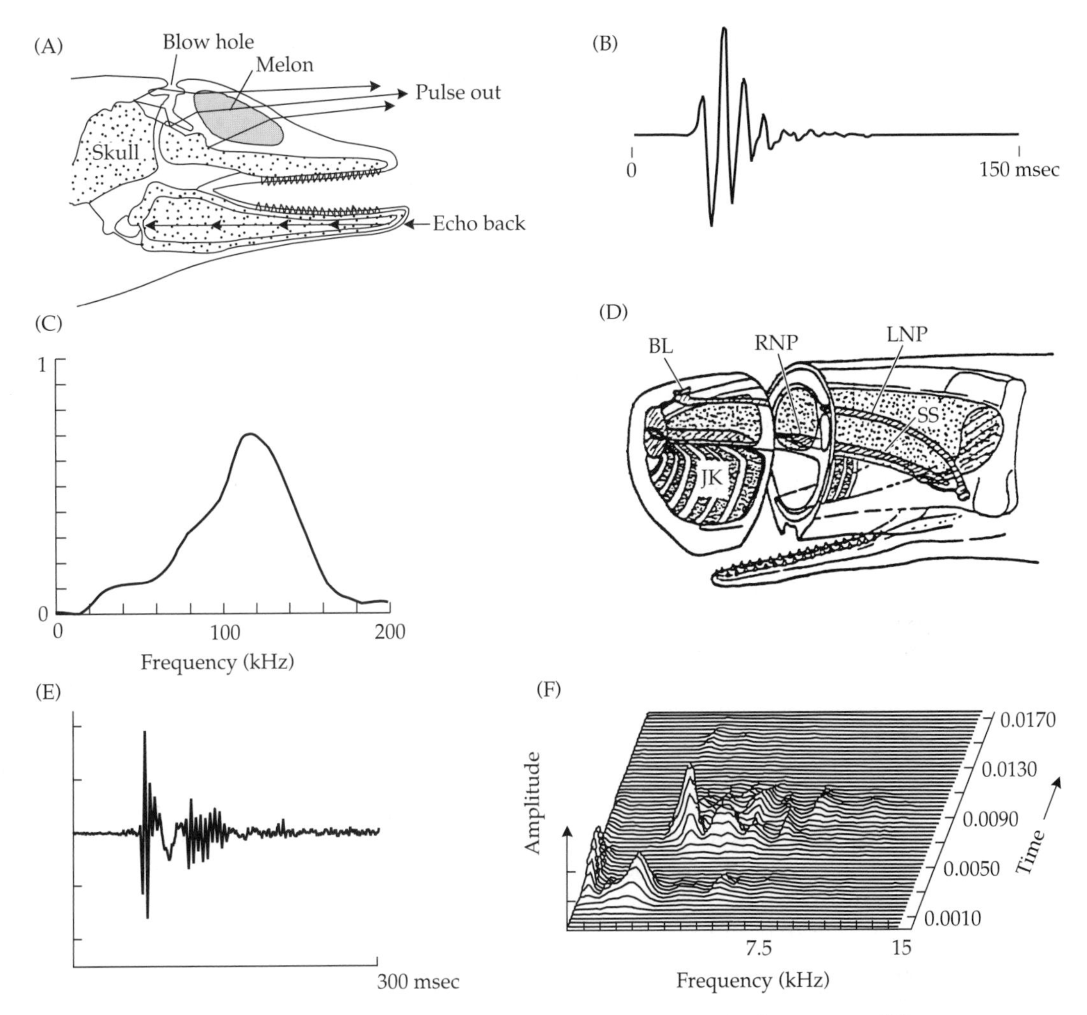

Figure 26.11 Echolocation pulse production and reception in toothed cetaceans. (A) Longitudinal section of the head of a porpoise. Blowhole in the neck region opens into a nasal passage. En route to the lungs, the nasal passage has a number of sacs branching off into the head. The pulses seem to be produced by the movement of air across membranes between air sacs and along the nasal passage. The air sacs and skull reflect the sound forward. The melon matches the internal to external impedances and helps focus the sound beam. Echoes are retrieved by the lower jaw, which is filled with fatty material conducting sounds directly to inner ear. Typical waveform (B) and power spectrum (C) of a single click of the porpoise, *Tursiops truncatus*. (D) Head of a sperm whale (*Physeter macrocephalus*). The blowhole (BL) is much more anterior than in porpoises. The forehead contains a massive oil-filled spermaceti organ. The left nasal passage (LNP) runs posterior to the lungs; the right passage (RNP) opens through fibrous lips into a sac near the front of the animal, which in turn opens into another tube connecting this anterior sac with a second just behind the spermaceti organ. It is thought that trapped air is moved through the fibrous lips and into the sacs to produce pulses. Several pulses are generated in quick succession, and the resulting sounds reverberate in the spermaceti organ before being emitted. (E) Typical-pulse waveform and (F) "waterfall display" of power spectrum of a sperm whale (A–C after Au 1993; D–F from Goold and Jones 1995.)

millimeters. Porpoises, unlike bats, should experience changes in echo power spectra as a function of target composition. Au (1993) reviews the many experiments he and colleagues have performed, showing how acutely cetaceans can extract such information. So far, all echolocating cetaceans use broad bandwidth signals like those of FM bats; none uses long-CF signals to measure target velocities or to discriminate between static and moving targets through Doppler shifts.

LIMITS ON ENVIRONMENTAL COMMUNICATION

It should be clear from this review that echolocating bats and cetaceans have evolved signals and decoding procedures that provide themselves with far more information than animals ever provide for each other. For some tasks, both taxa perform close to the limits set by ideal receivers; further sophistication is physically impossible. This suggests an answer to the question posed at the beginning of this chapter. Communicating animals could exchange much more information with each other than they do; that they do not is most likely a result of nonidentical interests. It could be argued, however, that few senders ever have access to as detailed information about the environment as bats or dolphins can extract from echoes; if senders don't have the information to pass on, then it is not surprising they do not attempt to communicate it. Similarity of interests would then be irrelevant. As a reply, we need only note that echolocating bats and dolphins also happen to be among the most social of all mammals. Is there evidence that individuals of either group ever exchange the highly detailed information they acquire through echolocation? The answer to date is no. Although some species in both groups show remarkable examples of cooperative behaviors (e.g., Connor and Norris 1982; Wilkinson 1984), there is currently no persuasive evidence for sophisticated exchanges of environmental information beyond that seen in nonecholocating species. If anything, bats personalize their own emitted pulses to make eavesdropping by other individuals more difficult (Masters et al. 1995; Neuweiler 1989; Obrist 1995). This confirms the notion that conflict of interests is a major constraint on the sophistication of communication systems.

There is, of course, one mammal that is a conspicuous exception to all we have argued in this and the prior chapter. Humans (*Homo sapiens*) routinely exchange highly sophisticated information about the environment or about any topic. Whether speaking or writing, we all do this at a level quite beyond that of any animal system we have discussed in this book and to individuals with whom we have few common interests. While the same selective forces have surely shaped animal and human communication systems, it is striking how rarely any organism evolves this level of sophistication. Still, social insects do pretty well, and it may turn out that parrots, corvids, elephants, higher primates, or cetaceans engage in much more sophisticated exchanges than we currently acknowledge (Griffin 1984, 1992). It remains the case that any such examples, including humans, are rare exceptions to a general pattern in which animals provide only as much information as is optimal given the physical constraints, basic economic tradeoffs, and game-theoretical considerations that have been the respective foci of each part of this book. If this economic approach has any merit, it should also be able to explain the exceptions. Perhaps this text will encourage the completion of that task.

SUMMARY

1. **Autocommunication** (e.g., echolocation or electrolocation) should set an upper bound on the amount of environmental information that signals can convey. If observed amounts are greater than that exchanged between individuals, then the low levels seen in the latter case are likely due to conflicts of interest (game-theoretical explanation); if they are about the same, then the low levels are likely due to physical or neural limits to the encoding and decoding of environmental information (neuroethological explanation).
2. An echolocator emits pulses of sound and listens for an echo to avoid obstacles or to locate prey. **Detection** of small or distant targets is maximized by emitting intense pulses, selecting a single constant frequency that maximizes echo formation while minimizing atmospheric absorption, and maximizing **duty cycle** (the fraction of time that the sound is emitted). Because echolocators usually avoid overlap between outgoing pulses and returning echoes, **pulse durations** are constrained by the proximity of nearest targets, and the **pulse emission rates** are constrained by the distance to farthest targets. The product of pulse duration and pulse emission rate sets a limit on duty cycle.
3. Echolocators use time delay between the outgoing pulse and the echo return to measure target range. **Range accuracy** increases the wider the **bandwidth** of the signal and the higher the **signal-to-noise ratio** of the echo. Ideal receivers using **coherent correlation** attend to signal phase and are the most accurate physically possible; receivers ignoring phase will do worse. If the echolocator is moving relative to the target, Doppler shifts will reduce range accuracy. **Doppler-tolerant signals** are those that minimize ranging errors due to Doppler shifts.
4. Echolocators can assess the **angular position** of a target and its **shape and texture** by comparing the power spectra of outgoing pulses and returned echoes. Echolocators in water can also use power spectrum differences to determine the composition of targets. Measurements of angular position and target properties are most accurate with short-duration wide-bandwidth signals.
5. Echolocators can use Doppler shifts in echoes to measure the **velocity** of targets relative to the echolocator. Accuracy is maximized by using long-duration, constant-frequency pulses. Such signals also allow echolocators to discriminate between static and moving targets, and if a target is flying, to determine target wingbeat rate.
6. Because the pulses that most accurately measure target range, target angle, and target properties (short duration, wide bandwidth) are exactly opposite to those that maximize detection and most accurately measure

velocity (long duration, single constant frequency), echolocators must either specialize in one or the other task, or instead shift pulse types at different stages of approach.

7. **Microchiropteran bats** all use echolocation to navigate and find food at night. Typical frequencies range from 30 to 120 kHz. Bats in open sites use long-duration constant frequencies to detect prey, and then shift to short-duration pulses (down to 0.5 msec) with frequency modulation (FM) to increase bandwidth for approach and attack. Gleaners attempting to discriminate static prey from background clutter use low-intensity FM pulses with harmonics to improve bandwidths and allow for spectrum comparisons. Hawkers use long-duration, constant-frequency pulses to detect the movement of flying insects against static clutter, and may measure wingbeat rates to select between alternative prey. FM bats show levels of range accuracy close to those of coherent ideal receivers; long-duration, constant-frequency bats show remarkable levels of detection, and although velocity measures are worse than ideal receiver predictions, they are still extremely accurate (e.g., ± 0.02 m/sec).
8. **Toothed whales** and **porpoises** use echolocation to navigate and find prey at ranges greater than is possible with aquatic vision. All use short-duration pulses (0.05 to 0.1 msec) and each uses some band of frequencies between 20 and 150 kHz. Pulses are too short for frequency modulation, but the short durations alone create bandwidths as high or higher than those seen in bats. Cetaceans show detection sensitivities and ranging accuracies commensurate with bats; unlike bats, they are able to exploit information about target compositions provided in echo spectra.
9. The fact that echolocators can evolve signal and receiver designs that provide amounts of information close to maximal levels allowable, whereas no animals (except humans) exchange environmental information with equivalent sophistication, argues that it is conflicts of interest, and not design constraints, that limit environmental exchanges between individuals. The exceptional exchanges of humans remain an exciting challenge for future work.

FURTHER READING

An older but lucid treatment of the tradeoffs in echolocator design can be found in Pye (1983). Readers interested in where the equations came from can consult the classic treatments by Woodward (1964) and Burdic (1968), or the more recent texts by Bird (1974) or Kingsley and Quegan (1992). Good synopses of alternative bat echolocation strategies include Griffin (1958), Simmons and Grinnell (1988), and Neuweiler (1989), and more details on the neurobiology of bat echolocation are available in Neuweiler (1990). The most current review of cetacean echolocation is Au (1993). The symposium volume edited by Nachtigall and Moore (1988) is a mammoth collection of exciting studies on echolocation in both taxa.

Literature Cited

The numbers in brackets following each reference denote the pages on which the reference is cited.

Ache, B. W. 1988. Integration of chemosensory information in aquatic invertebrates. In *Sensory Biology of Aquatic Animals* (Ed. by J. Atema, R. R. Fay, A. N. Popper and W. N. Tavolga), pp. 387–402. New York: Springer-Verlag. [306, 313]

Adams, E. S. and R. L. Caldwell. 1990. Deceptive communication in asymmetric fights of the stomatopod crustacean *Gonadodactylus bredini*. *Anim. Behav.* 39:706–717. [668, 669]

Agosta, W. C. 1992. *Chemical Communication:: The Language of Pheromones*. New York: W. H. Freeman. [317, 547, 558]

Ahnesjö, I. 1995. Temperature affects male and female potential reproductive rate differentially in the sex-role reversed pipefish, *Sygnathus typhle*. *Behav. Ecol.* 6:229–233. [780]

Aho, A. C. 1997. The visual acuity of the frog (*Rana pipiens*). *J. Comp. Physiol. A* 180:19–24. [269]

Alatalo, R. V., A. Carlson and A. Lundberg. 1988. The search cost in mate choice of the pied flycatcher. *Anim. Behav.* 36:289–291. [561]

Alatalo, R. V., D. Glynn and A. Lundberg. 1990. Singing rate and female attraction in the pied flycatcher: An experiment. *Anim. Behav.* 39:601–603. [772]

Albers, V. M. 1960. *Underwater Acoustics Handbook*. Lancaster, PA: Pennsylvania State Univ. Press. [127, 128, 139]

Alberts, A. C. 1989. Ultraviolet visual sensitivity in desert iguanas: Implications for pheromone detection. *Anim. Behav.* 38:129–137. [288]

Alberts, A. C. 1992. Constraints on the design of chemical communication systems in terrestrial vertebrates. *Am. Nat.* 139S:S62-S89. [304, 576, 594, 607, 609]

Albone, E. S. 1984. *Mammalian Semiochemistry*. New York: John Wiley. [283, 318]

Albone, E. S., P. E. Gosden and G. C. Ware. 1977. Bacteria as a source of chemical signals in mammals. In *Chemical Signals in Vertebrates* (Ed. by D. Müller-Schwarze and M. M. Mozell), pp. 35–44. New York: Plenum Press. [286, 524]

Alcock, J. 1993. *Anim. Behav.* 5th Edition. Sunderland, MA: Sinauer Associates. [10]

Alexander, J. and W. H. Stimson. 1988. Sex hormones and the course of parasitic infection. *Parasitol. Today* 4:189–193. [664]

Alexander, R. D. 1961. Aggressiveness, territoriality, and sexual behavior in field crickets (Orthoptera: Gryllidae). *Behaviour* 17:130–223. [609, 610]

Alexander, R. D. 1962. Evolutionary change in cricket acoustical communication. *Evolution* 16:443–467. [515]

Alexander, R. D. 1968. Arthropods. In *Animal Communication* (Ed. by T. A. Sebeok), pp. 167–216. Bloomington, IN: Indiana Univ. Press. [580, 609, 610]

Alexander, R. D. 1974. The evolution of social behaviour. *Annu. Rev. Ecol. Syst.* 5:325–383. [803]

Ali, M. A. 1984. *Photoreception and Vision in Invertebrates*. New York: Plenum Press. [278]

Ali, M. A. and M. A. Klyne. 1985. *Vision in Vertebrates*. New York: Plenum Press. [278, 243]

Ali, N. J., S. Farabaugh and R. Dooling. 1993. Recognition of contact calls by the budgerigar (*Melopsittacus undulatus*). *Bull. Psychon. Soc.* 31:468–470. [787]

Allan, S. E. and R. A. Suthers. 1994. Lateralization and motor stereotypy of song production in the brown-headed cowbird. *J. Neurobiol.* 25:1154–1166. [103, 105, 112]

Altes, R. A. 1980. Models for echolocation. In *Animal Sonar Systems* (Ed. by R. G. Busnel and J. F. Fish), pp. 625–671. New York: Plenum Press. [860, 870]

Altes, R. A. 1988. Some theoretical concepts for echolocation. In *Animal Sonar* (Ed. by P. E. Nachtigall and P. W. B. Moore), pp. 725–752. New York: Plenum Press. [860]

Altes, R. A. and E. L. Titlebaum. 1970. Bat signals as Doppler-tolerant waveforms. *J. Acoust. Soc. Am.* 48:1014–1020. [861]

Altmann, S. A. 1965. Sociobiology of rhesus monkeys. II. Stochastics of social communication. *J. Theor. Biol.* 8:490–522. [411]

Altmann, S. A. 1967. The structure of primate social communication. In *Social Communication among Primates* (Ed. by S. A. Altmann), pp. 325–362. Chicago: Chicago Univ. Press. [572]

Ames, P. L. 1971. The morphology of the syrinx in passerine birds. *Bull. Peabody Mus. Nat. Hist.* 37:1–194. [112]

Anderson, R. and A. J. Theron. 1989. Physiological potential of ascorbate, beta-carotene, and alpha-tocopherol individually and in combination in the prevention of tissue damage, carcinogenesis, and immune dysfunction mediated by phagocyte-derived reactive oxidants. *World Rev. Nutr. Diet* 62:27–58. [765]

Andersson, M. 1976. Social behaviour and communication in the great skua. *Behaviour* 58:40–77. [658, 703]

Andersson, M. 1980. Why are there so many threat displays? *J. Theor. Biol.* 86:773–781. [473, 672, 674, 699]

Andersson, M. 1986. Evolution of condition-dependent sex ornaments and mating preferences: Sexual selection based on viability differences. *Evolution* 40:804–816. [761]

Literature Cited

Andersson, M. 1994. *Sexual Selection*. Princeton, NJ: Princeton Univ. Press. [549, 570, 615, 660, 758, 766, 772, 778, 782, 790]

Andersson, S. 1989. Sexual selection and cues for female choice in leks of Jackson's widowbird *Euplectes jacksoni*. *Behav. Ecol. Sociobiol.* 25:403–410. [561, 590]

Andersson, S. 1994. Costs of sexual advertising in the lekking Jackson's widowbird. *Condor* 96:1–10. [549]

Andrew, R. J. 1956. Intention movements of flight in certain passerines. *Behaviour* 10:179–204. [506]

Andrew, R. J. 1957. The aggressive and courtship behaviour of certain emberizines. *Behaviour* 10:255–308. [506]

Arak, A. and M. Enquist. 1993. Hidden preferences and the evolution of signals. *Philos. Trans. R. Soc. Lond. B Biol. Sci.* 340:207–213. [529, 676]

Arak, A. and M. Enquist. 1995. Conflict, receiver bias and the evolution of signal form. *Proc. R. Soc. London B* 349:337–344. [702]

Arak, K. 1984. Sneaky breeders. In *Producers and Scroungers: Strategies of Exploitation and Parasitism* (Ed. by C. J. Barnard), pp. 154–194. London: Croom Helm. [738, 742]

Arcese, P. 1987. Age, intrusion pressure and defense against floaters by territorial male song sparrows. *Anim. Behav.* 35:772–784. [714]

Archer, J. 1988. *The Behavioural Biology of Aggression*. Cambridge: Cambridge Univ. Press. [681, 710, 719]

Archer, S. N. and J. N. Lythgoe. 1990. The visual basis for cone polymorphism in the guppy, *Poecilia reticulata*. *Vision Res.* 30:225–233. [260]

Armstrong, E. A. 1950. The nature and function of displacement activities. *Symp. Soc. Exp. Biol.* 4:361–387. [508]

Arnold, G. 1996. Kin recognition in honeybees. *Nature* 379:498. [788]

Arnold, S. J. 1972. Sexual behavior, sexual interference, and sexual defense in the salamanders *Ambystoma maculatum*, *Ambystoma tigrinum* and *Plethodon jordani*. *Z. Tierpsych.*, 42: 247–300. [796]

Atema, J. 1986. Review of sexual selection and chemical communication in the lobster, *Homarus americanus*. *Can. J. Fish. Aquat. Sci.* 43:2283–2290. [288]

Atema, J. 1988. Distribution of chemical stimuli. In *Sensory Biology of Aquatic Animals* (Ed. by J. Atema, F. F. Fay, A. N. Popper and W. N. Tavolga), pp. 29–56. New York: Springer-Verlag. [314]

Atema, J., R. R. Fay, A. N. Popper and W. L. Tavolga. 1988. *Sensory Biology of Aquatic Animals*. Berlin: Springer-Verlag. [317]

Attneave, F. 1959. *Applications of Information Theory to Psychology*. New York: Holt. [411, 418]

Au, W. W. L. 1993. *The Sonar of Dolphins*. New York: Springer-Verlag. [856, 861, 878, 879, 880, 882]

August, P. V. and J. G. T. Anderson. 1987. Mammal sounds and motivation-structural rules: A test of the hypothesis. *J. Mammal.* 68:1–9. [522]

Austad, S. 1983. A game theoretical interpretation of male combat in the bowl and doily spider *Frontinella pyramitela*. *Anim. Behav.* 31:59–73. [687, 688, 720]

Austad, S. N., W. T. Jones and P. M. Waser. 1979. Territorial defense in speckled wood butterflies: Why does the resident always win? *Anim. Behav.* 27:690–691. [718]

Autrum, H. 1979. *Comparative Physiology and Evolution of Vision in Invertebrates. A: Invertebrate Photoreceptors*. Berlin: Springer-Verlag. [278]

Axelrod, R. and W. D. Hamilton. 1981. The evolution of cooperation. *Science* 211:1390–1396. [634]

Aylor, D. 1971. Noise reduction by vegetation and ground. *J. Acoust. Soc. Am.* 51:197–205. [139]

Backwell, P. R. Y. and N. I. Passmore. 1996. Time constraints and multiple choice criteria in the sampling behaviour and mate choice of the fiddler crab, *Uca annulipes*. *Behav. Ecol. Sociobiol.* 38:407–416. [560]

Badgerow, J. P. and F. R. Hainsworth. 1981. Energy savings through formation flight? A re-examination of the Vee formation. *J. Theor. Biol.* 93:41–52. [810]

Baerends, G. P. 1975. An evaluation of the conflict hypothesis as an explanatory principle for the evolution of displays. In *Function and Evolution in Behaviour*. (Ed. by G. P. Baerends, C. Beer and A. Manning), pp. 187–227. Oxford: Clarendon Press. [508]

Bagnara, J. T. and M. E. Hadley. 1973. *Chromatophores and Color Change*. Englewood Cliffs, NJ: Prentice-Hall. [214, 217, 238]

Bailey, W. J. 1990. The anatomy of the tettigoniid hearing system. In *The Tettigoniidae: Biology, Systematics, and Evolution* (Ed. by W. J. Bailey and D. C. F. Rentz), pp. 217–245. Bathhurst, New South Wales, Australia: Crawford House Press. [173]

Bailey, W. J. 1991. *Acoustic Behavior of Insects: An Evolutionary Perspective*. London: Chapman and Hall. [111, 173]

Baker, M. C. 1983. The behavioral response of female Nuttall's white-crowned sparrows to male song of natal and alien dialects. *Behav. Ecol. Sociobiol.* 12:309–315. [482]

Baker, M. C. and K. J. Spitler-Nabors. 1981. Early experience determines song dialect responsiveness of female sparrows. *Science* 214:819–821. [482]

Baker, M. C., K. J. Spitler-Nabors and D. C. Bradley. 1982. The response of female mountain white-crowned sparrows to songs from their natal dialect and an alien dialect. *Behav. Ecol. Sociobiol.* 10:175–179. [482]

Baker, M. C., K. J. Spitler-Nabors, A. D. Thompson and M. A. Cunningham. 1987. Reproductive behaviour of female white-crowned sparrows: Effect of dialects and synthetic hybrid songs. *Anim. Behav.* 35:1766–1774. [482]

Baker, R. R. and G. A. Parker. 1979. The evolution of bird colouration. *Phil. Trans. Roy. Soc. Lond. B* 287:63–130. [610, 611]

Bakker, T. C. M. 1993. Positive genetic correlation between female preference and preferred male ornament in sticklebacks. *Nature* 363:255–257. [791]

Bakker, T. C. M. and M. Milinski. 1991. Sequential mate choice and the previous male effect in sticklebacks. *Behav. Ecol. Sociobiol.* 29:205–210. [448]

Balaban, E. 1988. Cultural and genetic variation in swamp sparrows (*Melospiza georgiana*) II. Behavioural salience of geographic song variants. *Behaviour* 105:292–322. [482]

Balmford, A., A. L. R. Thomas and I. L. Jones. 1993. Aerodynamics and the evolution of long tails in birds. *Nature* 361:628–631. [553]

Balmford, A., I. L. Jones and A. L. R. Thomas. 1994. How to compensate for costly sexually selected tails: The origin of sexually dimorphic wings in long-tailed birds. *Evolution* 48:1062–1070. [553]

Balmford, A., S. Albon and S. Blakeman. 1992. Correlates of male mating success and female choice in a lek-breeding antelope. *Behav. Ecol.* 3:112–123. [551, 724]

Baptista, L. F. 1975. Song dialects and demes in sedentary populations of the White-crowned Sparrow (*Zonotrichia leucophrus nuttalli*). *Univ. Calif. Publ. Zool.* 105:1–52. [732]

Baptista, L. F. 1978. Territorial, courtship and duet songs of the Cuban grassquit (*Tiaris canora*). *J. Ornithol.* 119:91–101. [598]

Baptista, L. F. 1985. The functional significance of song sharing in the White-crowned Sparrow. *Can. J. Zool.* 63:1741–1752. [731]

Baptista, L. F. 1996. Nature and its nurturing in avian vocal development. In *Ecology and Evolution of Acoustic Communication in Birds* (Ed. by D. E. Kroodsma and E. H. Miller), pp. 39–60. Ithaca, NY: Comstock Publishing Associates.

Baptista, L. F. and S. L. L. Gaunt. 1994. Advances in studies of avian sound communication. *Condor* 96:817–830. [478, 482, 496]

Baptista, L. F. and M. L. Morton. 1982. Song dialects and mate selection in montane white-crowned sparrows. *Auk* 99:537–547. [482]

Baptista, L. F. and M. L. Morton. 1988. Song learning in montane white-crowned sparrows: From whom and when? *Anim. Behav.* 36:1753–1764. [482]

Barclay, R. M. R. and R. M. Brigham. 1991. Prey detection, dietary niche breadth, and body size in bats: Why are insectivorous bats so small? *Am. Nat.* 137:693–703. [869]

Barfield, R. J. and D. A. Thomas. 1986. The role of ultrasonic vocalizations in the regulation of reproduction in rats. In *Reproduction: A Behavioral and Neuroendocrine Perspective* (Ed. by B. R. Komisaruk, H. I. Siegel, M.-F. Cheng and F. H. H.), pp. 33–43. New York: New York Academy of Sciences. [777]

Barlow, G. W. 1974. Contrasts in social behavior between Central American cichlid fishes and coral-reef surgeon fishes. *Am. Zool.* 14:9–34. [284, 798]

Barlow, G. W. 1977. Modal action patterns. In *How Animals Communicate* (Ed. by T. A. Sebeok), pp. 98–134. Bloomington, IN: Indiana Univ. Press. [222, 460]

Barlow, G. W. 1992. Is mating different in monogamous species? The midas cichlid fish as a case study. *Am. Zool.* 32:91–99. [790]

Barnard, C. J. and P. Aldhous. 1991. Kinship, kin discrimination and mate choice. In *Kin Recognition.* (Ed. by P. G. Hepper), pp. 125–147. Cambridge: Cambridge Univ. Press. [790]

Barnard, C. J. and T. E. Burk. 1979. Dominance hierarchies and the evolution of 'individual recognition.' *J.Theor. Biol.* 81:65–73. [805]

Barrow, G. M. 1962. *Molecular Spectroscopy*. New York: McGraw-Hill. [179, 205]

Basolo, A. L. 1990. Female preference predates the evolution of the sword in swordtail fish. *Science* 250:808–810. [504, 673]

Bass, A. H. 1986. Electric organs revisted: Evolution of a vertebrate communication and orientation organ. In *Electroreception* (Ed. by T. H. Bullock and W. Heiligenberg), pp. 13–70. New York: John Wiley. [330, 331, 350]

Bastian, J. 1986. Electrolocation: Behavior, anatomy, and physiology. In *Electroreception* (Ed. by T. H. Bullock and W. Heiligenberg), pp. 577–612. New York: John Wiley. [335, 348, 350]

Bastock, M. 1967. *Courtship: An Ethological Study*. Chicago: Aldine. [preface, 615, 757]

Bateson, P. P. G. 1978. Early experience and sexual preferences. In *Biological Determinants of Sexual Behaviour.* (Ed. by J. B. Hutchison), pp. 29–53. London: [790]

Bateson, P. P. G. 1978. Sexual imprinting and optimal outbreeding. *Nature* 273:659–660. [482]

Bateson, P. 1983. Optimal outbreeding. In *Mate Choice* (Ed. by P. P. G. Bateson), pp. 257–277. Cambridge: Cambridge Univ. Press. [482]

Bateson, P. P. G. 1991. Are there principles of behavioural development? In *The Development and Integration of Behaviour* (Ed. by P. P. G. Bateson), pp. 19–40. Cambridge: Cambridge Univ. Press. [475]

Bateson, P. P. G. 1994. The dynamics of parent-offspring relationships in mammals. *TREE* 9:399–403. [804]

Baylis, J.R. 1974. The behavior and ecology of *Herotilapia multispinosa* (Teleostei, Cichlidae). *Z. Tierpsych.* 34:115–146. [597]

Becker, P. H. 1982. The coding of species-specific characteristics in bird sounds. In *Acoustic Communication in Birds* (Ed. by D. E. &. M. Kroodsma E H), pp. 213–252. New York: Academic Press. [474, 579, 580]

Beecher, M. D. 1982. Signature systems and kin recognition. *Am. Zool.* 22:477–490. [788]

Beecher, M. D. 1988. Spectrographic analysis of bird vocalizations: Implications of the uncertainty principle. *Bioacoustics* 1:187–208. [74]

Beecher, M. D. 1989. Signalling systems for individual recognition: An information theory approach. *Anim. Behav.* 38:248–261. [442, 450]

Beecher, M. D. 1990. The evolution of parent-offspring recognition in swallows. In *Contemporary Issues in Comparative Psychology* (Ed. by D. Dewsbury), pp. 360–380. Sunderland, MA: Sinauer Associates. [450, 451, 454, 784]

Beecher, M. D. 1991. Successes and failures of parent-offspring recognition in animals. In *Kin Recognition* (Ed. by P. G. Hepper), pp. 94–124. Cambridge: Cambridge Univ. Press. [449, 482, 450, 801]

Beecher, M. D. 1996. Birdsong learning in the laboratory and field. In *Ecology and Evolution of Acoustic Communication in Birds.* (Ed. by D. E. Kroodsma and Miller), pp. 61–78. Ithaca, NY: Cornell Univ. Press. [559]

Beecher, M. D., P. K. Stoddard, S. E. Campbell and C. L. Horning. 1996. Repertoire matching between neighboring song sparrows. *Anim. Behav.* 51:917–923. [737]

Beer, C. G. 1979. Vocal communication between Laughing Gull parents and chicks. *Behaviour* 70:118–146. [801]

Beletsky, L. D. and G. H. Orians. 1987. Territoriality among male red-winged blackbirds. II. Removal experiments and site dominance. *Behav. Ecol. Sociobiol.* 20:339–350. [718, 720]

Bell, C. C., J. C. Bradbury and C. J. Russell. 1976. The electric organ of a mormyrid as a current and voltage source. *J. Comp. Physiol.* 110:65–88. [348, 350]

Bell, D. J. 1985. The rabbits and hares: Order Lagomorpha. In *Social Odours in Mammals* (Ed. by R. E. Brown and D. W. Macdonald), pp. 507–530. Oxford: Clarendon Press. [739]

Bell, P. D. 1980. Transmission of vibrations along plant stems: Implications for insect communication. *J. N.Y. Entom. Soc.* 88:210–216. [115, 139]

Bell, W. J. and R. T. Carde. 1984. *Chemical Ecology of Insects.* Sunderland, MA: Sinauer Associates, Inc. [318]

Bell, W. J. and R. E. Gorton. 1978. Informational analysis of agonistic behavior and dominance hierarchy formation in a cockroach. *Behaviour* 67:217–235. [411]

Bendich, A. 1989. Carotenoids and the immune response. *J. Nutr.* 119:112–115. [765]

Bengtsson, H. and O. Ryden. 1981. Development of parent-young interaction in asynchronously hatched broods of altricial birds. *Z. Tierpsych.* 56:255–272. [805]

Bengtsson, H. and O. Ryden. 1983. Parental feeding rate in relation to begging behavior in asynchronously hatched broods of the great tit *Parus major*. *Behav. Ecol. Sociobiol.* 12:243 - 251. [666, 804, 805]

Bennet-Clark, H. C. 1970. The mechanism and efficiency of sound production in mole crickets. *J. Exp. Biol.* 52:619–652. [111, 550]

Bennet-Clark, H. C. 1971. Acoustics of insect song. *Nature* 234:255–259. [111]

Bennet-Clark, H. C. 1975. Sound production in insects. *Sci. Prog. Oxford.* 62:263–283. [111]

Bennet-Clark, H. C. 1989. The tuned singing burrow of mole crickets. *J. Exp. Biol.* 128:383–409. [550]

Bennett, A. T. D. and I. C. Cuthill. 1994. Ultraviolet vision in birds: What is its function? *Vision Res.* 34:1471–1478. [260]

Bennett, M. V. L. 1970. Comparative physiology: Electric organs. *Annu. Rev. Physiol.* 32:471–528. [330, 350]

Bensch, S. and D. Hasselquist. 1992. Evidence for active female choice in a polygynous warbler. *Anim. Behav.* 44:301–312. [561]

Benz, J. J. 1993. Food-elicited vocalizations in golden lion tamarins: Design features for representational communication. *Anim. Behav.* 45:443–455. [823]

Beranek, L. L. 1986. *Acoustics.* New York: American Inst. of Physics/Acoustical Society of America. [40]

Beranek, L. L. 1988. *Acoustical Measurements (Revised Edition).* New York: American Inst. of Physics/Acoustical Society of America. [40]

Berglund, A. 1993. Risky sex: Male pipefishes mate at random in the presence of a predator. *Anim. Behav.* 46:169–175. [560]

Bernard, G. D. and C. L. Remington. 1991. Color vision in *Lycaena* butterflies: Spectral tuning of receptor arrays in relation to behavioral ecology. *Proc. Natl. Acad. Sci, USA* 88:2783–2787. [260, 790]

Bertram, B. 1970. The vocal behaviour of the Indian hill mynah, *Gracula religiosa*. *Anim. Behav. Monog.* 3:79–192. [479]

Beuter, K. J. 1980. A new concept of echo evaluation in the auditory system of bats. In *Animal Sonar Systems* (Ed. by R. G. Busnel and J. F. Fish), pp. 747–761. New York: Plenum Press. [871]

Bialek, W., F. Rieke, R. R. de Ruyter van Steveninck and D. Warland. 1991. Reading a neural code. *Science* 252: 1854–1857. [416]

Bildstein, K. L. 1982. Responses of northern harriers to mobbing passerines. *J. Field Ornithol.* 53:7–14. [839]

Bildstein, K. L. 1983. Why white-tailed deer flag their tails. *Am. Nat.* 121:709–715. [842]

Birch, M. C. 1974. *Pheromones*. Amsterdam: North Holland Publishing. [318]

Birch, M. C. and K. F. Haynes. 1982. *Insect Pheromones*. London: Edward Arnold Ltd. [318]

Bird, G. J. A. 1974. *Radar Precision and Resolution*. London: Pentech Press. [882]

Black-Cleworth, P. 1970. The role of electrical discharges in the non-reproductive social behavior of *Gymnotus carapo* (Gymnotidae, Pisces). *Anim. Behav. Monogr.* 3:1–77. [350]

Blair, W. F. 1963. Acoustic behaviour of Amphibia. In *Acoustic Behaviour of Animals* (Ed. by R.-G. Busnel), pp. 694–708. Amsterdam: Elsevier Publishing Company. [112]

Blaustein, A., T. Yoskikawa, K. Asoh and S. Walls. 1993. Ontogenetic shifts in tadpole kin recognition: Loss of signal and perception. *Anim. Behav.* 46:525–538. [790]

Blomquist, G. J. and J. W. Dillwith. 1983. Pheromones: Biochemistry and physiology. In *Endocrinology of Insects* (Ed. by D. Laufer), pp. 527–542. New York: Alan R. Liss. [523, 535]

Boesch, C. and H. Boesch. 1983. Optimisation of nut-cracking with naturqal hammers by wild chimpanzees. *Behaviour* 83:265–286. [666]

Boesch, C. and H. Boesch. 1989. Hunting behavior of wild chimpanzees in the Tai National Park. *Am. J. Phys. Anthrop.* 78:547–573. [666]

Boinski, S. 1991. The coordination of spatial position: A field study of the vocal behaviour of adult female squirrel monkeys. *Anim. Behav.* 41:89–102. [811]

Boinski, S. 1996. Vocal coordination of troop movement in squirrel monkeys (*Samiri oerstedi* and *S. sciureus*) and white-faced capuchins (*Cebus capucinus*). In *Adaptive Radiations of Neotropical Primates.* (Ed. by M. Norconk, A. L. Rosenberger and P. A. Garber), pp. 251–269. New York: Plenum Press. [812]

Boinski, S., E. Moraes, D. G. Kleiman, J. M. Dietz and A. J. Baker. 1994. Intra-group vocal behaviour in wild golden lion tamarins, *Leontopithecus rosalia*: Honest communication of individual activity. *Behaviour* 130:53–75. [812]

Bolhuis, J. J. 1991. Mechanisms of avian imprinting: A review. *Biol. Rev.* 66:303–345. [475]

Bolwig, N. 1959. A study of the behavior of the chacma baboon, *Papio ursinus*. *Behaviour* 14:136–163. [514]

Bond, A. B. 1989. Toward a resolution of the paradox of aggressive displays: I. Optimal deceit in the communication of fighting ability. *Ethology* 81:29–46. [668]

Borgia, G. 1985a. Bower descruction and sexual competition in the satin bowerbird (*Ptilonorhynchus violaceus*). *Behav. Ecol. Sociobiol.* 18:91–100. [769]

Borgia, G. 1985b. Bower quality, number of decorations and mating success of male satin bowerbirds (*Ptilonorhynchus violaceus*): An experimental analysis. *Anim. Behav.* 33:266–271. [769]

Borgia, G. 1986. Sexual selection in bowerbirds. *Sci. Am.* 254:92–100. [771]

Bossema, I. and R. R. Burgler. 1980. Communication during monocular and binocular looking in European jays. *Behaviour* 74:274–283. [703]

Bossert, W. H. 1968. Temporal patterning in olfactory communication. *J. Theor. Biol.* 18:157–170. [280]

Bossert, W. H. and E. O. Wilson. 1963. The analysis of olfactory communication among animals. *J. Theor. Biol.* 5:443–469. [294, 299, 302]

Bowmaker, J. K. 1991a. The evolution of vertebrate visual pigments and photoreceptors. In *Evolution of the Eye and Visual System* (Ed. by J. R. Cronly-Dillon and R. L. Gregory), pp. 63–81. Boca Raton, FL: CRC Press. [262]

Bowmaker, J. K. 1991b. Visual pigments, oil droplets and photopigments. In *Perception of Colour* (Ed. by P. Gouras), pp. 108–127. Boca Raton, FL: CRC Press. [259]

Boyd, R. and P. J. Richerson. 1985. *Culture and the Evolutionary Process*. Chicago: Univ. of Chicago Press. [475, 477, 496]

Boyd, R. and P. J. Richerson. 1988. An evolutionary model of social learning: The effects of spatial and temporal variation. In *Social Learning: Psychological and Biological Perspectives* (Ed. by T. R. Zentall and B. G. Galef), pp. 29–48. Hillsdale, NJ: Lawrence Erlbaum. [475, 477, 496]

Brackenbury, J. H. 1977. Physiological energetics of cock crow. *Nature (Lond.)* 270:433–435. [101, 112]

Brackenbury, J. H. 1978. Respiratory mechanics of sound production in chickens and geese. *J. Exp. Biol.* 72:229–250. [101, 112]

Brackenbury, J. H. 1979a. Aeroacoustics of the vocal organ in birds. *J. Theor. Biol.* 81:341–349. [101, 112]

Brackenbury, J. H. 1979b. Power capabilities of the avian sound-producing system. *J. Exp. Biol.* 78:163–166. [101, 112, 550]

Brackenbury, J. H. 1980. Control of sound production in the syrinx of the fowl *Gallus gallus*. *J. Exp. Biol.* 85:239–251. [101, 112]

Brackenbury, J. H. 1982. The structural basis of voice production and its relationship to sound characteristics. In *Acoustic Communication in Birds* (Ed. by D. Kroodsma, E. H. Miller and H. Ouellet), pp. 53–73. New York: Academic Press. [88, 101, 112]

Brackenbury, J. H. 1989. Functions of the syrinx and the control of sound production. In *Form and Function in Birds* (Ed. by A. S. King and J. McLelland), pp. 193–220. New York: Academic Press. [101, 112]

Bradbury, J. W. 1972. The silent symphony: Tuning in on the bat. In *The Marvels of Anim. Behav.* (Ed. by T. B. Allen), pp. 112–126. Washington, D C: National Geographic Society. [288]

Bradbury, J. W. 1977. Lek mating behavior in the hammer-headed bat. *Z. Tierpsych.* 45:225–255. [551, 589, 777]

Bradbury, T. N. and F. D. Fincham. 1991. The analysis of sequence in social interaction. In *Personality, Social Skills, and Psychopathology: An Individual Differences Approach* (Ed. by D. G. Gilbert and J. J. Connolly), pp. 257–289. New York: Plenum Press. [471]

Bradshaw, J. W. S., R. Baker and P. E. Howse. 1979. Multicomponent alarm pheromones in the mandibular glands of major workers of the African weaver ant, *Oecophylla longinoda*. *Physiol. Entomol.* 4:15–25. [607]

Breed, M. and B. Bennett. 1987. Kin recognition in highly eusocial insects. In *Kin Recognition in Animals* (Ed. by D. J. C. Fletcher and C. D. Michener), pp. 2243–285. Chichester: John Wiley & Sons. [807]

Breed, M., K. R. Williams and J. H. Fewell. 1988. Comb wax mediates the acquisition of nestmate recognition cues in honey bees. *Proc. Natl. Acad. Sci.* 85:8766–8769. [478]

Breipohl, A. M. 1970. *Probabilistic Systems Analysis*. New York: John Wiley. [420]

Bremond, J.-C. and T. Aubin. 1992. Cadence d'emission du chant territorial du troglodyte (*Troglodytes*). *C. R. Acad. Sci. Paris Ser. III.* 314:37–42. [731]

Brenowitz, E. A. 1982a. The active space of red-winged blackbird song. *J. Comp. Physiol. A.* 147:511–522. [126, 594]

Brenowitz, E. A. 1982b. Long-range communication of species identity by song in the red-winged blackbird. *Behav. Ecol. Sociobiol.* 10:29–38. [126]

Brenowitz, E. A. 1986. Environmental influences on acoustic and electric animal communication. *Brain Behav. Evol.* 28:32–42. [139, 350]

Bretagnolle, V. 1996. Acoustic communication in a group of nonpasserine birds, the petrels. In *Ecology and Evolution of Acoustic Communication in Birds* (Ed. by D. E. Kroodsma and E. H. Miller), pp. 160–177. Ithaca, NY: Comstock Publishing Associates. [484]

Bretagnolle, V. and J.-C. Thibault. 1993. Communicative behavior in breeding ospreys (*Pandion haliaetus*): Descriptions and relationship of signals to life history. *Auk* 110:736–751. [823]

Bristowe, W. S. 1958. *The World of Spiders.* New York: Collins. [591]

Brizzi, R., C. Calloni, G. Delfino and G. Tanteri. 1995. Notes on the male cloacal anatomy and reproductive biology of *Euproctus montanus* (Amphibia: Salamadridae). *Herpetologica* 51:8–18. [796]

Brodsky, L. M. and R. D. Montgomerie. 1987. Asymmetrical contests in defense of rock ptarmigan territories. *Behav. Ecol. Sociobiol.* 21:267–272. [730, 731]

Brooks, D. R. and D. A. McLennan. 1991. *Phylogeny, Ecology and Behavior: A Research Program in Comparative Biology.* Chicago: Univ. of Chicago Press. [502]

Brosset, A. 1966. *La Biologie des Chiroptères.* Paris: Masson et Cie. [288, 873]

Brower, L. P., J. V. Z. Brower and F. P. Cranston. 1965. Courtship behavior of the queen butterfly, *Danaus gilippus* berenice (Cramer). *Zoologica* 50:1–39. [288]

Brown, C. H. 1982. Ventriloquial and locatable vocalizations in birds. *Z. Tierpsych.* 59:338–350. [836]

Brown, C. R., M. B. Brown and M. L. Shaffer. 1991. Food-sharing signals among socially foraging cliff swallows. *Anim. Behav.* 42:551–564. [821, 822]

Brown, E. D., S. M. Farabaugh and C. J. Veltman. 1988. Song sharing in a group-living bird, the Australian magpie, *Gymorhina tibicen*, Part I: Vocal sharing within and among social groups. *Behav.* 104:1–28. [597]

Brown, J. L. 1975. *The Evolution of Behavior.* New York: W. W. Norton. [503]

Brown, J. L. and A. Eklund. 1994. Kin recognition and the major histocompatibility complex: An integrative review. *Am. Nat.* 143:435–461. [485]

Brown, R. E. and D. W. Macdonald. 1985. *Social Odours in Mammals.* Oxford: Clarendon Press. [285, 317, 318]

Brown, S. D., R. J. Dooling and K. O'Grady. 1988. Perceptual organization of acoustic stimuli by budgerigars (*Melopsittacus undulatus*). III. Contact calls. *J. Comp. Psychol.* 102:236–247. [489]

Bruch, R. C., D. L. Kalinoski and M. R. Kare. 1988. Biochemistry of vertebrate olfaction and taste. *Annu. Rev. Nutr.* 8:21–42. [305, 306, 318]

Bucher, T. L., M. J. Ryan and G. Bartholomew. 1982. Oxygen consumption during resting, calling and nest building in the frog *Physalaemus pustulosus. Physiol. Zool.* 55:10–22. [549]

Buck, L. B. 1996. Information coding in the vertbrate olfactory system. *Annu. Rev. Neurosci.* 19:517–544. [305, 306, 318]

Buckley, P. A. and F. G. Buckley. 1972. Individual egg and chick recognition by adult royal terns (*Sterna maxima maxima*). *Anim. Behav.* 20:457–462. [801]

Buechner, H. K. and R. Schloeth. 1965. Ceremonial mating behavior in Uganda kob (*Adenota kob thomasi* Neumann). *Z. Tierpsych.* 22:209–225. [568]

Bullock, T. H. and W. Heiligenberg. 1986. *Electroreception.* New York: John Wiley. [337, 338, 350]

Burdic, W. S. 1968. *Radar Signal Analysis.* Englewood Cliffs, NJ: Prentice-Hall. [865, 882]

Burghardt, G. M. 1970. Defining "communication". In *Communication by Chemical Signals.* Vol 1 of *Advances in Chemoreception.* (Ed. by J. W. Johnston, D. G. Moulton and A. Turk), pp. 5–18. New York: Appleton-Century-Crofts. [4]

Burghardt, G. M. 1977. Ontogeny of communication. In *How Animals Communicate* (Ed. by T. A. Sebeok), pp. 71–97. Bloomington, IN: Indiana Univ. Press. [478]

Burk, T. 1988. Acoustic signals, arms races and the costs of honest signalling. *Fla. Entomol.* 71:400–409. [549]

Burkhardt, D. 1982. Birds, berries and UV. *Naturwiss.* 69:153–157. [262]

Burkhardt, D. and I. de la Motte. 1985. Selective pressures, variability, and sexual dimorphism in stalk-eyed flies (Diopsidae). *Naturwiss.* 72: 204–206. [768]

Burkhardt, D and I. de la Motte. 1988. Big 'antlers' are favored: Female choice in stalk-eyed flies (Diptera, Insecta), field collected harems and laboratory experiments. *J. Comp. Physiol. A* 162:649–652. [768]

Burley, N. 1988. Wild zebra finches have band-colour preferences. *Anim. Behav.* 36:1235–1237. [764]

Burtt, E. H. J. 1979. *The Behavioral Significance of Color.* New York: Garland Press. [209, 238]

Burtt, E. H. J. 1986. An analysis of physical, physiological and optical aspects of avian coloration with emphasis on wood warblers. *Ornithol. Monogr.* 38:1–126. [610]

Butcher, G. S. and S. Rohwer. 1988. The evolution of conspicuous and distinctive coloration for communication in birds. *Curr. Ornithol.* 6:51–108. [610, 611]

Butlin, R. 1995. Genetic variation in mating signals and responses. In *Speciation and the Recognition Concept.* (Ed. by D. M. Lambert and H. G. Spencer), pp. 327–366. Baltimore: Johns Hopkins Univ. Press. [791]

Butlin, R. K. and M. G. Richie. 1989. Genetic coupling in mate recognition systems: What is the evidence? *Biol. J. Linnean Soc.* 37:237–246. [478, 500]

Byrne, R. and A. Whiten. 1988. Tactical deception in primates. *Behav. Brain Sci.* 11:233–273. [668]

Cade, W. 1975. Acoustically orienting parasitoids: Fly phonotaxis to cricket song. *Science* 190:1312–1313. [546]

Cade, W. H. 1979. The evolution of alternative male reproductive strategies in field crickets. In *Sexual Selection and Reproductive Competition in Insects.* (Ed. by M. Blum and N. A. Blum), pp. 343–379. London: Academic Press. [546]

Caine, N. G., R. L. Addington and T. L. Windfelder. 1995. Factors affecting the rates of food calls given by red-bellied tamarins. *Anim. Behav.* 50:53–60. [823]

Caldwell, M. C., D. K. Caldwell and P. L. Tyack. 1990. Review of the signature-whistle hypothesis for the Atlantic bottlenose dolphin. In *The Bottlenose Dolphin* (Ed. by S. Leatherwood and R. R. Reeves), pp. 191–234. New York: Academic Press. [489]

Camhi, J. M. 1984. *Neuroethology.* Sunderland, MA: Sinauer Associates. [193, 221, 222]

Campbell, K. A. and R. A. Suthers. 1988. Predictive tracking of horizontally moving targets by the fishing bat, *Noctilio leporinus.* In *Animal Sonar* (Ed. by P. E. Nachtigall and P. W. B. Moore), pp. 501–506. New York: Plenum Press. [876]

Canady, R. A., D. E. Kroodsma and F. Nottebohm. 1984. Population differences in complexity of a learned skill are correlated with the brain space involved. *Proc. Natl. Acad. Sci. USA* 81:6232–6234. [558]

Capranica, R. R. 1965. *The Evoked Vocal Response of the Bullfrog: A Study of Communication by Sound.* Cambridge, MA: MIT Press. [112, 155, 790]

Capranica, R. R. 1992. The untuning of the tuning curve: Is it time? *Sem. Neurosci.* 4:401–408. [155]

Capranica, R. R. and A. J. M. Moffat. 1983. Neurobehavioral correlates of sound communication in anurans. In *Advances in Vertebrate Neuroethology* (Ed. by J. P. Ewert, R. R. Capranica and D. J. Ingle), pp. 701–730. New York: Plenum Press. [88]

Caprio, J. 1988. Peripheral filters and chemoreceptor cells in fishes. In *Sensory Biology of Aquatic Animals* (Ed. by J. Atema, R. R. Fay, A. N. Popper and W. N. Tavolga), pp. 313–338. New York: Springer-Verlag. [305, 310]

Carlin, N. F. and B. Hölldobler. 1986. The kin recognition system of carpenter ants (*Campanotus spp.*). I. Hierarchical cues in small colonies. *Behav. Ecol. Sociobiol.* 19:123–134. [478]

Carlin, N. F. and B. Hölldobler. 1987. The kin recognition system of carpenter ants (*Campanotus spp.*). II. Larger colonies. *Behav. Ecol. Sociobiol.* 20:209–217. [478]

Carlin, N. F. and B. Hölldobler. 1988. Influence of virgin queens on kin recognition in the carpenter ant *Camponotus floridanus* (Hymenoptera: Formicidae). *Insect. Soc.* 35:191–197. [478]

Caro, T. M. 1986. The functions of stotting: A review of the hypotheses. *Anim. Behav.* 34:649–662. [850]

Caro, T. 1988. Adaptive significance of play: Are we getting closer? *TREE* 3:50–54. [810]

Caro, T. M. 1994. Ungulate predator behaviour-preliminary and comparative data from African bovids. *Behaviour* 128: 189–228. [842, 843, 844]

Caro, T. M. 1995. Pursuit-deterrence revisited. *Trends Ecol. Evol.* 10:500–503. [842, 850]

Carpenter, C. C. 1978. Ritualistic social displays in lizards. In *Behavior and Neurology of Lizards* (Ed. by N. Greenberg and P. D. Maclean), pp. 253–267. Washington, D C: National Institute of Mental Health. [509]

Carr, W. E. S. 1988. The molecular nature of chemical stimuli in the aquatic environment. In *Sensory Biology of Aquatic Animals* (Ed. by J. Atema, R. R. Fay, A. N. Popper and W. N. Tavolga), pp. 3–28. New York: Springer-Verlag. [283]

Carson, H. L. 1995. Fitness and the sexual environment. In *Speciation and the Recognition Concept.* (Ed. by D. M. Lambert and H. G. Spencer), pp. 103–122. Baltimore: Johns Hopkins Univ. Press. [791]

Caruthers, J. W. 1977. *Fundamentals of Marine Acoustics.* Amsterdam: Elsevier Publishing. [29, 40, 131, 139]

Caryl, P. G. 1979. Communication by agonistic displays: What can games theory contribute to ethology? *Behaviour* 68:136–169. [653, 658, 689, 703]

Catchpole, C. K. 1980. Sexual selection and the evolution of complex songs among warblers of the genus *Acrocephalus. Behaviour* 74:149–166. [774]

Catchpole, C. K. 1982. The evolution of bird sounds in relation to mating and spacing behavior. In *Acoustic Communication in Birds* (Ed. by D. E. Kroodsma and E. H. Miller), pp. 297–319. New York: Academic Press. [598]

Catchpole, C. K. 1983. Variation in the song of the great reed warbler *Acrocephalus arundinaceus* in relation to mate attraction and territorial defence. *Anim. Behav.* 31:1217–1225. [598]

Catchpole, C. K. 1987. Bird song, sexual selection and female choice. *TREE* 2:94–97. [598]

Catchpole, C. K. and P. J. B. Slater. 1995. *Bird Song: Biological Themes and Variations.* Cambridge: Cambridge Univ. Press. [473, 478, 487, 496, 615, 730, 735, 742]

Catchpole, C. K., J. Dittami and B. Leisler. 1984. Differential responses to male song repertoires in female songbirds implanted with oestradiol. *Nature* 3122:563–564. [774]

Chamberlain, D. R., W. B. Gross, G. W. Cornwell and H. S. Mosby. 1968. Syringeal anatomy of the common crow. *Auk* 89:244–252. [101, 112]

Chance, M. R. A. 1962. An interpretation of some agonistic postures: The role of "cut-off" acts and postures. *Symp. Zool. Soc. London* 8:71–89. [512]

Chaplin, C.G. and P. Scott. 1972. *Fishwatcher's Guide to West Atlantic Coral Reefs.* Newton Square, PA: Harrowood Books. [356]

Chapman, C. A. and L. Lefebvre. 1990. Manipulating foraging group size-spider monkey food calls at fruiting trees. *Anim. Behav* 39:891–896. [823]

Chappuis, C. 1971. Un exemple de l'influence du milieu sur les missions vocales des oiseaux: l'evolution des chants en fort quatoriale. *Terre et la Vie* 118:183–202. [139]

Charles-Dominique, P. 1977. Urine marking and territoriality in *Galago alleni* (Waterhouse, 1837 – Lorisoidae, Primates): A field study by radio-telemetry. *Z. Tierpsych.* 43:113–138. [286, 738]

Charman, W. N. 1991. The vertebrate dioptric apparatus. In *Evolution of the Eye and Visual System* (Ed. by J. R. Cronly-Dillon and R. L. Gregory), pp. 82–117. Boca Raton, FL: CRC Press. [264]

Charnov, E. L. and J. R. Krebs. 1974. The evolution of alarm calls: Altruism and manipulation. *Am. Nat.* 109: 107–112. [836]

Cheney, D. L. and R. M. Seyfarth. 1982. Recognition of individuals within and between groups of free-ranging vervet monkeys. *Am. Zool.* 22:519–529. [806]

Cheney, D. L. and R. M. Seyfarth. 1985. Social and non-social knowledge in vervet monkeys. *Philos. Trans. R. Soc. Lond. B Biol. Sci.* 308:187–201. [470, 817]

Cheney, D. L. and R. M. Seyfarth. 1990. *How Monkeys See the World.* Chicago: Univ. of Chicago Press. [479, 488, 806, 835, 850]

Cheney, D., R. Seyfarth and J. Silk. 1995. The role of grunts in reconciling opponents and facilitating interactions among adult female baboons. *Anim. Behav.* 50:249–257. [808]

Cherry, C. 1966. *On Human Communication.* 2nd Edition. Cambridge, MA: MIT Press. [479, 488]

Chessell, C. I. 1977. The propagation of noise along a finite impedance boundary. *J. Acoust. Soc. Am.* 62:825–834. [139]

Chilton, G., M. R. Lein and L. F. Baptista. 1990. Mate choice by female white-crowned sparrows in a mixed dialect population. *Behav. Ecol. Sociobiol.* 27:223–227. [482]

Christy, J. H. 1995. Mimicry, mate choice, and the sensory trap hypothesis. *Am. Nat.* 146:171–181. [533]

Chvla, M., J. Doskocil, J. H. Mook and V. Pokorny. 1974. The genus *Lipara* Meigen (Diptera, Choropidae), systematics, morphology, behaviour, and ecology. *Tijdschr. Entomol.* 117:125. [134]

Clapperton, B. K., E. O. Minot and D. R. Crump. 1988. An olfactory recognition system in the ferret *Mustella furo* L. (Carnivora: Mustelidae). *Anim. Behav.* 36:541–553. [596]

Clark, C., P. Marler and K. Beeman. 1987. Quantitative analysis of animal vocal phonology: An application to swamp sparrow song. *Ethology* 76:101–115. [74, 464]

Clay, C. S. and H. Medwin. 1977. *Acoustical Oceanography: Principles and Applications.* New York: John Wiley. [139]

Clayton, N. S. 1989. Song, sex, and sensitive phases in the behavioural development of birds. *Trends Ecol. Evol.* 4:82–84. [482]

Cleveland, J. and C. T. Snowdon. 1982. The complex vocal repertoire of the adult cotton-top tamarin (*Saguinus oedipus oedipus*). *Z. Tierpsych.* 58:231–270. [520, 522, 823]

Cloney, R. and E. Florey. 1968. Ultrastructure of celphalopod chromatophore organs. *Z. Zellforsch.* 89:250–280. [218]

Clutton-Brock, T. H. and S. D. Albon. 1979. The roaring of red deer and the evolution of honest advertisement. *Behaviour* 69:145–170. [653, 692, 731]

Clutton-Brock, T. H. and G. A. Parker. 1995a. Punishment in animal societies. *Nature* 373:209–216. [655]

Clutton-Brock, T. H. and G. A. Parker. 1995b. Sexual coercion in animal societies. *Anim. Behav.* 49:1345–1365. [755, 756]

Clutton-Brock, T. H., O. Price and A. McColl. 1992. Mate retention, harassment and the evoltuion of ungulate leks. *Behav. Ecol.* 3:234–242. [765]

Coemans, M. A. J. M., J. J. vos Hzn and J. F. W. Nuboer. 1994. The relation between celestial colour gradients and the position of the sun, with regard to the sun compass. *Vision Res.* 34:1461–1470. [195, 260, 564]

Connor, D. 1984. The role of an acoustic display in territorial maintenance in the pika. *Can. J. Zool.* 62:1906–1909. [598]

Connor, D. 1985. The function of the pika short call in individual recognition. *Z. Tierpsych.* 67:131–143. [596, 598]

Connor, R. C. and K. S. Norris. 1982. Are dolphins and whales reciprocal altruists? *Am. Nat.* 119:358–374. [880]

Cook, N. D. 1995. Artefact or network evolution? *Nature* 374:313. [531]

Coombs, C. H., R. M. Dawes and A. Tversky. 1970. *Mathematical Psychology: An Elementary Introduction*. Englewood Cliffs, NJ: Prentice Hall. [454]

Cooper, W. E. and L. J. Vitt. 1986. Tracking of female conspecific odor trails by male broad-headed skinks (*Eumeces laticeps*). *Ethology* 71:242–248. [315]

Cooper, W. T. and J. M. Forshaw. 1977. *The Birds of Paradise and Bowerbirds*. Sydney: Collins. [515, 764, 771]

Cott, H. B. 1940. *Adaptive Coloration in Animals*. London: Methuen. [209, 232, 235, 236, 238]

Cowan, P. J. and R. M. Evans. 1974. Calls of different individual hens and the parental control of feeding behaviour in young *Gallus gallus*. *J. Exp. Zool.* 188:353–360. [801]

Cox, C. R. and B. LeBoeuf. 1977. Female incitation of male competition: A mechanism in sexual selection. *Am. Nat.* 111:317–335. [776]

Coyne, J. A. and H. A. Orr. 1989. Patterns of speciation in *Drosophila*. *Evolution* 43:362–381. [791]

Craig, J. L. and P. F. Jenkins. 1982. The evolution of complexity in broadcast song of passerines. *J. Theor. Biol.* 95:415–422. [730]

Crane, J. 1975. *Fiddler Crabs of the World. Ocypodidae: Genus Uca*. Princeton, NJ: Princeton Univ. Press. [757, 758]

Crawford, F. S. J. 1965. *Waves*. New York: Education Development Center. [205]

Crawford, F. S. 1968. *Waves*. New York: McGraw-Hill Book Company. [40]

Crawford, J. D. 1992. Individual and sex specificity in the electric organ discharges of breeding mormyrid fish (*Pollimyrus isidori*). *J. Exp. Biol.* 164:79–102. [474]

Crawford, V. P. 1989. Learning and mixed-strategy equilibria in evolutionary games. *J. Theor. Biol.* 140:537–550. [630]

Crawford, V. P. 1990. Nash equilibria and evolutionary stability in large-population and finite-population playing the field models. *J. Theor. Biol.* 145:83–94. [630]

Crawford, V. P. 1995. Adaptive dynamics in coordination games. *Econometrica* 63:103–143. [630]

Crespi, B. J. 1986. Size assessment and alternative fighting tactics in *Elaphrothrips tuberculatus*. *Anim. Behav.* 34:1324–1335. [683, 684]

Crocroft, R. B. and M. J. Ryan. 1995. Patterns of advertisement call evolution in toads and chorus frogs. *Anim. Behav.* 49:283–303. [556]

Cronin, T. W. and N. J. Marshall. 1989. A retina with at least ten spectral types of photoreceptors in a mantis shrimp. *Nature* 339:137–140. [261]

Cronly-Dillon, J. R. 1991. Origin of invertebrate and vertebrate eyes. In *Evolution of the Eye and Visual System* (Ed. by J. R. Cronly-Dillon and R. L. Gregory), pp. 15–51. Boca Raton, FL: CRC Press. [242, 244, 248]

Cronly-Dillon, R. L. and J. R. Gregory. 1991. *Evolution of the Eye and Visual System*. Boca Raton, FL: CRC Press. [278]

Crook, J. H. 1964. The evolution of social organization and visual communication in the weaver birds (Ploceinae). *Behaviour Suppl.* 10:1–178. [555]

Cullen, J. M. 1966. Reduction of ambiguity through ritualization. *Phil. Trans. Roy. Soc. Lond. B* 251:363–374. [651]

Curio, E. 1978. The adaptive significance of avian mobbing. I. Telenomic hypotheses and predictions. *Z. Tierpsych.* 48:175–183. [839, 844, 850]

Curio, E., U. Ernst and W. Vieth. 1978. The adaptive significance of avian mobbing. II. Cultural transmission of enemy recognition in blackbirds: Effectiveness and some constraints. *Z. Tierpsych.* 48:184–202. [840]

Daanje, A. 1950. On locomotory movements in birds and the intention movements derived from them. *Behaviour* 3:48–98. [501]

Dabelsteen, T. and P. K. McGregor. 1996. Dynamic acoustic communication and interactive playback. In *Ecology and Evolution of Acoustic Communication in Birds* (Ed. by D. E. Kroodsma and E. H. Miller), pp. 398–408. Ithaca, NY: Comstock/Cornell Univ. Press. [471]

Dabelsteen, T. and S. B. Pedersen. 1990. Song and information about aggressive responses of blackbirds, *Turdus merula*: Evidence from interactive playback experiments with territory owners. *Anim. Behav.* 40:1158–1168. [599, 734]

Daigle, G. A. 1979. Effects of atmospheric turbulence on the interference of sound waves above a finite impedance boundary. *J. Acoust. Soc. Am.* 65:45–49. [139]

Dale, S. and T. Slagsvold. 1994. Polygyny and deception in the pied flycatcher: Can females determine male mating status? *Anim. Behav.* 41:865–873. [448]

Darwin, C. 1872. *The Expression of the Emotions in Man and the Animals*. London: John Murray. [504]

David, C. T. 1986. Mechanisms of directional flight in wind. In *Mechanisms of Insect Olfaction* (Ed. by T. L. Payne, M. C. Birch and C. E. J. Kennedy), pp. 49–67. Oxford: Clarendon. [314, 318]

David, C. T., J. S. Kennedy and A. R. Ludlow. 1983. Finding of a sex pheromone source by gypsy moths released in the field. *Nature* 303:804–806. [314]

Davies, N. B. 1978. Territorial defence in the speckled wood butterfly (*Pararge aegeria*): The resident always wins. *Anim. Behav.* 26:138–147. [718]

Davies, N. B. 1979. Game theory and territorial behaviour in speckled wood butterflies. *Anim. Behav.* 27:691–962. [718]

Davies, N. B. and M. L. Brooke. 1989a. An experimental study of co-evolution between the cuckoo, *Cuculus canorus*, and its hosts. I. Host egg discrimination. *J. Anim. Ecol.* 58:207–224. [800]

Davies, N. B. and M. L. Brooke. 1989b. An experimental study of co-evolution between the cuckoo, *Cuculus canorus*, and its hosts. II. Host egg markings, chick discrimination and general discussion. *J. Anim. Ecol.* 58:225–236. [800]

Davies, N. B. and T. R. Halliday. 1978. Deep croaks and fighting assessment in toads *Bufo bufo*. *Nature* 274:683–685. [653, 681, 682]

Davies, N. B. and A. I. Houston. 1981. Owners and satellites: The economics of territory defense in the pied wagtail, *Motacilla alba*. *J. Anim. Ecol.* 52:621–634. [724]

Davies, S. and R. Carrick. 1962. On the ability of crested terns to recognize their own chicks. *Aust. J. Zool.* 10:171–177. [801]

Davis, E. A. and C. D. Hopkins. 1988. Behavioral analysis of electric signal localization in the electric fish, *Gymnotus carapo* (Gymnotiforms). *Anim. Behav.* 36:1658–1671. [350]

Davis, J. M. 1975. Socially induced flight reactions in pigeons. *Anim. Behav.* 23:597–601. [506]

Davis, M. S. 1987. Acoustically mediated neighbor recognition in the North American bullfrog, *Rana catesbeiana*. *Behav. Ecol. Sociobiol.* 21:185–190. [488]

Dawkins, M. S. 1993. Are there general principles of signal design? *Philos. Trans. R. Soc. Lond. B Biol. Sci.* 340:251–255. [538, 539, 575, 615, 651, 655, 668, 671, 676]

Dawkins, M. S. and T. Guilford. 1991. The corruption of honest signalling. *Anim. Behav.* 41:865–873. [651, 670, 676]

Dawkins, M. S. and T. Guilford. 1994. Design of an intention signal in the bluehead wrasse (*Thalassoma bifasciatum*). *Proc. R. Soc. Lond. B* 257:123–128. [576, 590]

Dawkins, R. 1989. *The Selfish Gene*. 3rd Edition. Oxford: Oxford Univ. Press. [630, 863]

Dawkins, R. and J. R. Krebs. 1978. Animal signals: Information or manipulation? In *Behavioural Ecology*. 1st Edition. (Ed. by J. R. Krebs and N. B. Davies), pp. 282–309. Oxford: Blackwell Scientific Publications. [652, 672, 673]

de Ruyter van Steveninck, R. R. and S. B. Laughlin. 1996. The rate of information transfer at graded-potential synapses. *Nature* 379:642–645. [416]

de Vos, G. J. 1983. Social behaviour of black grouse: An observational and experimental field study. *Ardea* 71:1–103. [724]

de Vries, H., W. J. Netto and P. L. H. Hanegraaf. 1993. Matman: A program for analysis of sociometric matrices and behavioural transition matrices. *Behaviour* 125:157–175. [466]

de Waal, F. B. M. 1986. Deception in the natural communication of chimpanzees. In *Deception: Perspectives on Human and Non-Human Deceit* (Ed. by R. W. Mitchell and N. S. Thompson), pp. 221–244. Albany, NY: State Univ. of New York Press. [668]

de Waal, F. B. M. 1989. Food sharing and reciprocal obligations among chimpanzees. *J. Hum. Evol.* 18:433–459. [824]

Demski, L. S. 1992. Chromatophore systems in teleosts and cephalopods: A levels oriented analysis of convergent systems. *Brain Behav. Evol.* 40:141–156. [9, 218, 238]

Demski, L. S. and J. G. Dulka. 1986. Thalamic stimulation evokes sex color changes and gamete release in a vertebrate hermaphrodite. *Experimentia* 42:1285–1287. [9, 218]

Deneubourg, J. L., J. M. Pasteels and J. C. Verhaeghe. 1983. Probabilistic behaviour in ants: A strategy of errors. *J. Theor. Biol. 105:259–271. [825, 826]*

Derby, C. D. and J. Atema. 1988. Chemoreceptor cells in aquatic invertebrates: Peripheral mechanisms of chemical signal processing in decapod crustaceans. In *Sensory Biology of Aquatic Animals* (Ed. by J. Atema, R. R. Fay, A. N. Popper and W. N. Tavolga), pp. 365–386. New York: Springer-Verlag. [307, 565]

Deutsch, J. C. and P. Weeks. 1992. Uganda kob prefer high-visibility leks and territories. *Behav. Ecol.* 3:223–233. [560]

DeVoogd, T. J., J. R. Krebs, S. Healy and A. Purvis. 1993. Relations between song repertoire size and the volume of brain nuclei related to song: Comparative evolution analyses amongst oscine birds. *Proc. R. Soc. Lond. B Biol. Sci.* 254:75–82. [493, 558, 559, 570]

deWaal, F. B. M. 1989. *Peacemaking among Primates*. Cambridge, MA: Harvard Univ. Press. [808]

deWaal, F. B. M. 1996. *Good Natured: The Origins of Right and Wrong in Humans and Other Animals*. Cambridge, MA: Harvard Univ. Press. [808]

DeWolfe, B. B., L. F. Baptista and L. Petrinovich. 1989. Song development and territory establishment in Nuttall's White-crowned Sparrows. *Condor* 91:397–407. [732]

Diamond, J. M. 1991. Borrowed sexual ornaments. *Nature* 349:105. [769]

Dingle, H. 1969. A statistical and information analysis of aggressive communication in the mantis shrimp *Gonodactylus bredini* Manning. *Anim. Behav.* 17:561–575. [703]

Dingle, H. 1972. Aggressive behavior in stomatopods and the use of information theory in the analysis of animal communication. In *Behavior of Marine Animals: Current Perspectives in Research. I. Invertebrates* (Ed. by H. E. Winn and B. L. Olla), pp. 126–155. New York: Plenum Press. [411, 418]

Ditchburn, R. W. 1963. *Light*. Glasgow: Blackie & Sons. [205]

Dittus, W. P. J. 1984. Toque macaque food calls: Semantic communication concerning food distribution in the environment. *Anim. Behav.* 32:470–477. [823]

Dixson, A. F. 1983. Observations on the evolution and behavioural significance of 'sexual skin' in female primates. *Adv. Study Behav.* 13:63–106. [757]

Dobler, S., A. Stumpner and K-G. Heller. 1994. Sex-specific spectral tuning for the partner's song in the duetting bush cricket *Ancistrura nigrovittata* (Orthoptera: Phaneropteridae). *J. Comp. Phys. A* 175:303–310. [590]

Dombrovsky, Y. and N. Perrin. 1994. On adaptive search and optimal stopping in sequential mate choice. *Am. Nat.* 144:355–361. [448]

Donato, R. J. 1976. Propagation of a spherical wave near a plane boundary with complex impedance. *J. Acoust. Soc. Am.* 60:34–39. [139]

Dooling, R. J., S. D. Brown, T. J. Park, S. D. Soli and K. Okanoya. 1987a. Perceptual organization of acoustic stimuli by budgerigars (*Melopsittacus undulatus*). I. Pure tones. *J. Comp. Psychol.* 101:139–149. [489]

Doupe, A. J. 1993. A neural circuit specialized for vocal learning. *Curr. Opin. Neurobiol.* 3:104–111. [529]

Doupe, A. J. and M. Konishi. 1991. Song-selective auditory circuits in the vocal control system of the zebra finch. *Proc. Natl. Acad. Sci.* 88:11339–11343. [529]

Dowling, J. E. 1987. *The Retina: An Approchable Part of the Brain*. Cambridge, MA: Belknap Press of Harvard Univ. Press. [248]

Downhower, J. F. and D. B. Lank. 1994. Effects of previous experience on mate choice by female mottled sculpins. *Anim. Behav.* 47:369–372. [448]

Dubrovin, N. N. and R. D. Zhantiev. 1970. Acoustic signals of katydids (Orthoptera, Tettigoniiidae). *Zool. J.* 49:1001–1014. [134, 173]

Dumortier, B. 1963. Insect sound production. In *Acoustic Behaviour of Animals* (Ed. by Busnel), pp. 277–345. Amsterdam: Elsevier. [516, 535]

Dumortier, B. 1963. Morphology of sound emission apparatus in Arthropoda. In *Acoustic Behaviour of Animals* (Ed. by R.-G. Busnel), pp. 277–345. Amsterdam: Elsevier Publishing. [80]

Dunbar, R. I. M. 1988. *Primate Social Systems*. Ithaca, NY: Cornell Univ. Press. [757]

Dunbar, R. I. M. 1991. Functional significance of social grooming in primates. *Folia Primatol.* 57:121–131. [808]

Dunham, D. W. 1966. Agonistic behaviour in captive rose-breasted grosbeaks, *Pheucticus ludovicianus* (L). *Behaviour* 27:160–173. [703]

Dusenbery, D. B. 1992. *Sensory Ecology*. New York: W.H. Freeman. [2, 11, 228, 230, 244, 278, 294, 300, 301, 317, 318]

Dussourd, D. E., C. A. Harvis, J. Meinwald and T. Eisner. 1988. Biparental defensive endowment of eggs with acquired plant alkaloid in the moth *Utethesia ornatrix*. *Proc. Nat. Acad. Sci. USA* 85:5992–5996. [774]

Dussourd, D. E., C. A. Harvis, J. Meinwald and T. Eisner. 1991. Pheromonal advertisementof a nuptial guft by a male moth (*Utethesia ornatrix*). *Proc. Natl. Acad. Sci. USA* 88:9224–9227. [774]

Duvall, D., D. Müller-Schwarze and R. M. Silverstein. 1986. *Chemical Signals in Vertebrates*. New York: Plenum Press. [318]

Dyer, F. C. 1985. Nocturnal orientation by the Asian honey bee *Apis florea*. *Anim. Behav.* 33:769–774. [828, 831]

Dyer, F. C. 1991. Comparative studies of dance communication: Analysis of phylogeny and function. In *Diversity in the Genus Apis* (Ed. by D. R. Smith), pp. 177–198. Boulder: Westview Press. [831, 832, 833]

Dyer, F. C. and T. D. Seeley. 1991. Dance dialects and foraging range in three Asian honey bee species. *Behav. Ecol. Sociobiol.* 28:227–233. [834]

Eakin, R. M. 1965. Evolution of photoreceptors. *Cold Spring Harbor Symp. Quant. Biol.* 30:363–370. [242]

Eason, P. and S. J. Hannon. 1994. New birds on the block: New neighbors increase defensive costs for territorial male willow ptarmigan. *Behav. Ecol. Sociobiol.* 34:419–426. [724]

Eason, P. K. and J. A. Stamps. 1993. An early warning system for detecting intruders in a territorial animal. *Anim. Behav.* 46:1105–1109. [738]

East, M. L. and H. Hofer. 1991. Loud calling in a female-dominated mammalian society: I. Structure and composition of whooping bouts of spotted hyaenas, *Crocuta crocuta*. *Anim. Behav.* 42:637–647, 650–669. [810]

East, M. L. and H. Hofer. 1991. Loud calling in a female dominated mammalian society. II: Behavioural contexts and functions of whooping of spotted hyaenas, *Crocuta crocuta*. *Anim. Behav.* 42:651–669. [597]

East, M. L., H. Hofer and W. Wickler. 1993. The erect 'penis' is a flag of submission in a female-dominated society: Greetings in Serengeti spotted hyenas. *Behav. Ecol. Sociobiol.* 33:355–370. [759]

Eberhard, W. G. 1977. Aggressive chemical mimicry by a bolas spider. *Science* 198:1173–1175. [561]

Eberhard, W. G. 1985. *Sexual Selection and Animal Genitalia.* Cambaridge, MA: Harvard Univ. Press. [758, 764]

Eberhardt, L. S. 1994. Oxygen consumption during singing by male Carolina wrens (*Thryothorus ludovicianus*). *Auk* 111:124–130. [549, 550]

Ebling, F. J. 1977. Hormonal control of mammalian skin glands. In *Chemical Signals in Vertebrates* (Ed. by D. Müller-Schwarze and M. M. Mozell), pp. 17–33. New York: Plenum Press. [283]

Egan, J. P. 1975. *Signal Detection Theory and ROC Analysis.* New York: Academic Press. [454]

Ehret, G., J. Tautz, B. Schmitz and P. M. Narins. 1990. Hearing through the lungs: Lung-eardrum transmission of sound in the frog *Eleutherodactylus coqui. Naturwiss.* 77:192–4. [173]

Eimas, P. D., P. Siqueland, P. Jusczyk and J. Vigorito. 1971. Speech perception in infants. *Science* 171:303–306. [470]

Elgar, M. A. 1986a. The establishment of foraging flocks in house sparrows: risk of predation and daily temperature. *Behav. Ecol. Sociobiol.* 19:433–438. [821]

Elgar, M. A. 1986b. House sparrows establish foraging flocks by giving chirrup calls if the resources are divisible. *Anim. Behav.* 34:169–174. [821]

Elgar, M. 1996. Role-reversed risky copulation. *TREE* 11:189–190. [780]

Elkington, J. S. and R. T. Carde. 1984. Odor dispersion. In *Chemical Ecology of Insects* (Ed. by J. W. Bell and R. T. Carde), pp. 73–91. Sunderland, MA: Sinauer Associates. [286]

Elowson, A. M., P. L. Tannenbaum and C. T. Snowdon. 1991. Food-associated calls correlate with food preferences in cotton-top tamarins. *Anim. Behav.* 42:931–937. [823, 850]

Elsner, N. 1983. Insect stridulation and its neurophysiological basis. In *Bioacoustics: A Comparative Approach* (Ed. by B. Lewis), pp. 69–92. London: Academic Press. [111]

Embleton, T. F. W., N. Olson, J. E. Piercy and D. Rollin. 1974. Fluctuations in the propagation of sound near the ground. *J. Acoust. Soc. Am.* 55:485(A). [139]

Embleton, T. F. W., J. E. Piercy and N. Olson. 1976. Outdoor sound propagation over ground of finite impedance. *J. Acoust. Soc. Am.* 59:267–277. [139]

Emlen, S. T. and L. W. Oring. 1977. Ecology, sexual selection, and the evolution of animal mating systems. *Science* 197:125–223. [746]

Endler, J. A. 1977. *Geographic Variation, Speciation, and Clines.* Princeton: Princeton Univ. Press. [791]

Endler, J. A. 1980. Natural selection on color patterns in *Poecilia reticulata. Evolution* 34:76–91. [661]

Endler, J. A. 1983. Natural and sexual selection on color patterns in Poeciliid fishes. *Env. Biol. Fishes* 9:173–190. [485, 544, 661]

Endler, J. A. 1987. Predation, light intensity, and courtship behaviour in *Poecilia reticulata. Anim. Behav.* 35:1376–1385. [544]

Endler, J. A. 1990. On the measurement and classification of colour in studies of animal colour patterns. *Biol. J. Linnean Soc.* 41:315–352. [199, 205, 223, 225, 230, 233, 238]

Endler, J. A. 1991a. Interactions between predators and prey. In *Behavioural Ecology: An Evolutionary Approach.* 3rd Edition. (Ed. by J. R. Krebs and N. B. Davies), pp. 169–196. Oxford: Blackwell Scientific Publications. [850]

Endler, J. A. 1991b. Variation in the apearance of guppy color patterns to guppies and their predators under different visual conditions. *Vision Res.* 31:587–608. [544, 545, 570]

Endler, J. A. 1992. Signals, signal conditions, and the direction of evolution. *Am. Nat.* 139:S125-S153. [233, 527, 535, 573]

Endler, J. A. 1993a. Some general comments on the evolution and design of animal communication systems. *Phil. Trans. R. Soc. Lond. B* 340:215–225. [573, 615]

Endler, J. A. 1993b. The color of light in forests and its implications. *Ecological Monographs* 63:1–27. [199, 226, 238]

Englund, G. and T. I. Olsson. 1990. Fighting and assessment in the net-spinning caddis larva *Arctopsyche ladogensis*: A test of the sequential assessment game. *Anim. Behav.* 39:55–62. [697, 723]

Englund, G. and C. Otto. 1991. Effects of ownership status, weight asymmetry and case fit on the outcome of case contests in two populations of *Agrypnia pagetana* (Trichoptera: Phryganeidae) larvae. *Behav. Ecol. Sociobiol.* 29:113–120. [719]

Enquist, M. 1985. Communication during aggressive interactions with particular reference to variation in choice of behaviour. *Anim. Behav.* 33:1152–1611. [355, 658, 669, 702, 704]

Enquist, M. and A. Arak. 1993. Selection of exaggerated male traits by female aesthetic senses. *Nature* 361:446–448. [528, 529, 673, 676]

Enquist, M. and A. Arak. 1994. Symmetry, beauty, and evolution. *Nature* 372:169–172. [529, 676]

Enquist, M. and S. Jakobsson. 1986. Decision making and assessment in the fighting behaviour of *Nannacara anomala* (Cichlidae, Pices). *Ethology* 72:143–153. [697]

Enquist, M. and O. Leimar. 1983. Evolution of fighting behaviour: Decision rules and assessment of relative strength. *J. Theor. Biol.* 102:387–410. [690, 692]

Enquist, M. and O. Leimar. 1987. Evolution of fighting behaviour: The effect of variation in resource value. *J. Theor. Biol.* 127:187–205. [690, 719, 721]

Enquist, M. and O. Leimar. 1990. The evolution of fatal fighting. *Anim. Behav.* 39:1–9. [690]

Enquist, M., E. Plane and J. Röed. 1985. Aggressive communication in fulmars (*Fulmarus glacialis*) competing for food. *Anim. Behav.* 33:1007–1020. [655, 656, 657, 658, 702, 704, 708]

Enquist, M., T. Ljungberg and A. Zandor. 1987. Visual assessment of fighting ability in the cichlid fish *Nannacara anomala. Anim. Behav.* 35:1262–1263. [697]

Enquist, M., O. Leimar, T. Ljungberg, Y. Mallner and N. Sgerdahl. 1990. A test of the sequential assessment game: Fighting in the cichlid fish *Nannacara anomala. Anim. Behav.* 40:1–14. [695, 697]

Erickson, R. P. 1978. Common properties of sensory systems. In *Handbook of Behavioral Neurobiology, Vol. 1, Sensory Integration* (Ed. by R. B. Masterton), pp. 73–90. New York: Plenum Press. [306]

Erlinge, S. 1968. Territoriality of the otter, *Lutra lutra* L. *Oikos* 19:81–98. [522]

Esch, H. 1961. Über die Schallerzeugung beim Werbetanz der Honigbiene. *Z. Verg. Physiol.* 45:1–11. [831]

Esch, H. 1967. Die Bedeutung der Lauterzeugung für die Verständigung der stachellosen Bienen. *Z. Verg. Physiol.* 56:408–411. [827]

Estes, R. D. 1972. The role of the vomeronasal organ in mammalian reproduction. *Mammalia* 36:315–341. [310]

Evans, C. S., L. Evans and P. Marler. 1993. On the meaning of alarm calls: Functional reference in an avian vocal system. *Anim. Behav.* 46:23–38. [835]

Evans, L. B., H. E. Bass and L. C. Sutherland. 1971. Atmospheric absorption of sound: Theoretical predictions. *J. Acoust. Soc. Am.* 51:1565–1575. [139]

Evans, M. 1991. The size of adornments of male scarlet-tufted malachite sunbirds varies with environmental conditions as predicted by handicap theories. *Anim. Behav.* 42:797–804. [664]

Evans, M. R. and B. J. Hatchwell. 1991. An experimental study of male adornment in the scarlet-tufted malachite sunbird. I. The role of pectoral tufts in territorial defense. *Behav. Ecol. Sociobiol.* 29:413–419. [664]

Evans, M. R. and A. L. R. Thomas. 1992. The aerodynamic and mechanical effects of elongated tails in the scarlet-tufted malachite sunbird: Measuring the cost of a handicap. *Anim. Behav.* 43:337–348. [522]

Everitt, B. S. 1977. *The Analysis of Contigency Tables.* London: Chapman and Hall. [464]

Everitt, B. S. and G. Dunn. 1992. *Applied Multivariate Data Analysis.* New York: Oxford Univ. Press. [466, 467]

Ewer, R. F. 1968. *Ethology of Mammals*. New York: Plenum Press. [506, 615, 756, 757]

Ewert, J. P. 1974. The neural basis of visually guided behavior. *Sci. Am.* 230:1394–1397. [528]

Ewing, A. W. 1989. *Arthropod Bioacoustics: Neurobiology and Behavior*. Ithaca, NY: Comstock/Cornell Univ. Press. [478]

Falls, J. B. 1982. Individual recognition by sounds in birds. In *Acoustic Communication in Birds* (Ed. by D. E. Kroodsma and E. H. Miller), pp. 237–278. New York: Academic Press. [474, 479, 566]

Falls, J. B., J. R. Krebs and P. K. McGregor. 1982. Song matching in the great tit *(Parus major)*: The effect of similarity and familiarity. *Anim. Behav.* 30:977–1009. [734]

Farabaugh, S. M. 1982. The ecological and social significance of duetting. In *Acoustic Communication in Birds* (Ed. by D.E. Kroodsma E. H. Miller), pp. 85–124. New York: Academic Press. [586, 798, 799]

Farabaugh, S. M. and R. J. Dooling. 1996. Acoustic communication in parrots: Laboratory and field studies of budgerigars, *Melopsittacus undulatus*. In *Ecology and Evolution of Acoustic Communication in Birds* (Ed. by D. E. Kroodsma and E. H. Miller), pp. 97–117. Ithaca, NY: Comstock/Cornell Univ. Press. [484]

Farabaugh, S. M., A. Linzenbold and R. J. Dooling. 1992. Vocal plasticity in budgerigars (*Melopsittacus undulatus*): Evidence for social factors in the learning of contact calls. *J. Comp. Psychol.* 108:81–92. [484]

Farrell, J. 1995. Talk is cheap. *Am. Econ. Rev.* 85:186–190. [730]

Faulkner, D. J. 1988. Feeding deterrents in molluscs. In *Biomedical Importance of Marine Organisms* (Ed. by D. G. Fautin), pp. 29–36. San Francisco: California Academy of Sciences. [845]

Faulkner, D. J. 1992. Chemical defenses of marine molluscs. In *Ecological Roles of Marine Natural Products* (Ed. by V. J. Paul), pp. 119–163. Ithaca, NY: Comstock Publishing Associates. [845]

Faulkner, E. A. 1969. *Introduction to the Theory of Linear Systems*. London: Chapman and Hall, Ltd. [74]

Fay, R. R. 1988. Peripheral adaptations for spatial hearing in fish. In *Sensory Biology of Aquatic Animals* (Ed. by J. Atema, R. R. Fay, A. N. Popper and W. N. Tavolga), pp. 711–731. New York: Springer-Verlag. [173]

Feh, C. and J. de Mazieres. 1993. Grooming at preferred site reduces heart rate in horses. *Anim. Behav.* 46:1191–1194. [809]

Felsenstein, J. 1981. Skepticism towards Santa Rosalia, or why are there so few kinds of animals? *Evolution* 35:124–138. [791]

Feynman, R. P. 1964. *The Feynman Lectures on Physics (II): Mainly Electromagnetism and Matter*. Reading, MA: Addison-Wesley. [350]

Feynman, R. P., R. B. Leighton and M. Sands. 1964. *The Feynman Lectures on Physics*. Reading, MA.: Addison-Wesley. [205]

Feynman, R. P., R. B. Leighton and M. Sands. 1989. *The Feynman Lectures on Physics*. Redwood City, CA: Addison-Wesley. [40]

Ficken, M. S. 1990. Acoustic characteristics of alarm calls associated with predation risk in chickadees. *Anim. Behav.* 39:400–401. [839]

Ficken, M., R. W. Ficken and S. R. Witkin. 1978. Vocal repertoire of the black-capped chickadee. *Auk* 95:34–48. [484]

Finger, E., D. Burkhardt and J. Dyck. 1992. Avian plumage colors: Origin of UV reflection in a black parrot. *Naturwiss.* 79:187–188. [216]

Finger, T. E. 1988. Organization of chemosensory systems within the brains of bony fishes. In *Sensory Biology of Aquatic Animals* (Ed. by J. Atema, R. R. Fay, A. N. Popper and W. N. Tavolga), pp. 339–364. New York: Springer-Verlag. [311]

Finlay, B. L. and R. B. Darlington. 1995. Linked regularities in the development and evolution of mammalian brains. *Science* 268:1578–1584. [565, 570]

Fischer, H. I. 1965. Das Triumphgeschrei der Graugans (*Anser anser*). *Z. Tierpsych.* 22:247–304. [798]

Fisher, R. A. 1930. *The Genetical Theory of Natural Selection*. Oxford: Clarendon Press. [760]

Fite, K. V. and S. Rosenfeld-Wessels. 1975. A comparative study of deep avian foveas. *Brain Behav. Evol.* 12:97–115. [270]

Fitzgibbon, C. D. 1994. The costs and benefits of predator inspection behaviour in Thomson's gazelles. *Behav. Ecol. Sociobiol.* 34:139–148. [844]

Fitzgibbon, C. D. and J. H. Fanshawe. 1988. Stotting in Thomson's gazelles: An honest signal of condition. *Behav. Ecol. Sociobiol.* 23:69–74. [668, 843]

Flasskamp, A. 1994. The adaptive significance of avian mobbing. 5. An experimental test of the move-on hypothesis. *Ethology* 96:322–333. [839]

Fleishman, L. J. 1985. Cryptic movement in the vine snake *Oxybelis aeneus*. *Copeia* 1985:242–245. [235]

Fleishman, L. J. 1992. The influence of the sensory system and the environment on motion paterns in the visual displays of anoline lizards and other vertebrates. *Am. Nat.* 139:S36-S61. [235, 532]

Fletcher, D. J. C. and C. D. Michener. 1987. *Kin Recognition in Animals*. Chichester: John Wiley & Sons. [816]

Fletcher, N. H. 1988. Bird song: A quantitative acoustic model. *J. Theor. Biol.* 135:455–481. [112]

Fletcher, N. H. 1989. Acoustics of bird song: Some unresolved problems. *Comments Theor. Biol.* 1:237–251. [112]

Fletcher, N. H. 1992. *Acoustic Systems in Biology*. Oxford: Oxford Univ. Press. [40, 101, 111]

Flood, P. 1985. Sources of significant smells: The skin and other organs. In *Social Odours in Mammals* (Ed. by R. E. Brown and D. W. Macdonald), pp. 19–36. Oxford: Clarendon Press. [283, 284]

Foelix, R. F. 1982. *Biology of Spiders*. Cambridge, MA: Harvard Univ. Press. [506, 507]

Folstad, I. and A. J. Karter. 1992. Parasites , bright males, and the immuno-competence handicap. *Am. Nat.* 139:603–622. [664, 765]

Ford, J. K. B. 1991. Vocal traditions among resident killer whales (*Orcinus orca*) in coastal waters off British Columbia. *Can. J. Zool.* 67:727–745. [806]

Ford, N. B. and J. R. Low. 1984. Sex pheromone source location by garter snakes: A mechanism for detection of direction in nonvolatile trails. *J. Chem. Ecol.* 10:1193–1199. [315]

Forsgren, E. 1992. Predation risk affects mate choice in gobiid fish. *Am. Nat.* 140:1041–1049. [560]

Forster, L. 1982. Visual communication in jumping spiders (Salticidae). In *Spider Communication* (Ed. by P. N. Witt and J. S. Rovner), pp. 161–212. Princeton, NJ: Princeton Univ. Press. [274, 694]

Foster, K. W. and R. D. Smyth. 1980. Light antennas in phototactic algae. *Microbiological Reviews* 44:572–630. [240]

Fowler, S. V. and J. H. Lawton. 1985. rapidly induced defenses and talking trees: The devil's advocate position. *Am. Nat.* 126:181–195. [524]

Fox, D. L. 1976. *Animal Biochromes and Structural Colours*. Berkeley, CA: Univ. of California Press. [213, 215, 216, 230]

Fox, D. L. 1979. *Biochromy: Natural Coloration of Living Things*. Berkeley: Univ. of California Press. [213, 215, 230]

Fraenkel, G. S. and D. L. Gunn. 1961. *The Orientation of Animals*. New York: Dover. [313]

Franklin, W. L. 1983. Contrasting socioecologies of south America's wild camelids: The vicuña and guanaco. In *Advances in the Study of Mammalian Behavior* (Ed. by J. F. Eisenberg and D. G. Kleiman), pp. 573–629. Pittsburgh: American Society of Mammalogists. [757, 758]

Franks, N. R. 1989. Army ants: A collective intelligence. *Am. Sci.* 77:138–145. [815]

Fraser, A. F. 1968. *Reproductive Behavior in Ungulates*. New York: Academic Press. [707]

Friedman, D. 1991. Evolutionary games in economics. *Econometrica* 59:637–666. [630]

Friedman, M. A. and C. D. Hopkins. 1996. Tracking individual mormyrid electric fish in the field using electric organ discharge waveforms. *Anim. Behav.* 51:391–407. [474, 489, 582]

Fritzsch, B., M. J. Ryan, W. Wilczynski, T. E. Hetherington and W. Walkowiak. 1988. *The Evolution of the Amphibian Auditory System.* New York: John Wiley. [173]

Frumhoff, P. C. and H. K. Reeve. 1994. Using phylogenies to test hypotheses of adaptation: A critique of some current proposals. *Evolution* 48:172–180. [504]

Fugle, G. N. and S. I. Rothstein. 1987. Experiments on the control of deceptive signals of status in white-crowned sparrows. *Auk* 104:188–197. [664]

Fugle, G. N., S. I. Rothstein, C. W. Osenberg and M. A. McGinley. 1984. Signals of status in wintering white-crowned sparrows, *Zonotrichia luecophrys gambelli. Anim. Behav.* 32:86–93. [664]

Fullard, J. H. 1994. Auditory changes in Noctuid moths endemic to a bat-free habitat. *J. Evol. Biol.* 7:435–445. [533]

Fullard, J. H. and J. E. Yack. 1993. The evolutionary biology of insect hearing. *TREE* 8:248–252. [533]

Galef, B. G. 1988. Imitation in animals: history, definition, and interpretation of data from the psychological laboratory. In *Social Learning: Psychological and Biological Perspectives* (Ed. by T. R. Zentall and B. G. Galef), pp. 3–28. Hillsdale, NJ: Lawrence Erlbaum. [475, 477, 478]

Galef, B. G. 1995. Why behaviour patterns that animals learn socially are locally adaptive. *Anim. Behav.* 49:1325–1334. [475, 477, 478]

Gamboa, G. J., H. K. Reeve and D. W. Pfennig. 1986. The evolution and ontogeny of nestmate recognition in social wasps. *Annu. Rev. Entomol.* 31:431–454. [478]

Gamboa, G. J., T. A. Grudzien, K. E. Espelie and E. A. Bura. 1996. Kin recognition pheromones in social wasps: Combining chemical and behavioural evidence. *Anim. Behav.* 51:625–629. [478]

Gans, C. 1973. Sound production in the Salientia: Mechanism and evolution of the emitter. *Am. Zool.* 13:1179–1194. [112, 556]

Gans, C. 1989. On phylogenetic constraints. *Acta Morphol. Neerlando* 27:133–138. [543]

Gardner, R. and M. R. Morris. 1989. The evolution of bluffing in animal contests: An ESS approach. *J. Theor. Biol.* 137:238–243. [668]

Gaunt, A. S. 1987. Phonation. In Bird Respiration (Ed. by T. J. Seller), pp. 71–94. Boca Raton, FL: CRC. [101, 112]

Gaunt, A. S. and S. L. L. Gaunt. 1977. Mechanics of the syrinx in Gallus gallus. II. Electromyographic studies of ad libitum vocalizations. *J. Morph.* 153:1–20. [101, 112]

Gaunt, A. S. and S. L. L. Gaunt. 1980. Phonation of the ring dove: The basic mechanism. *Am. Zool.* 20:757. [112]

Gaunt, A. S. and S. L. L. Gaunt. 1985. Syringeal structures and avian phonation. In *Current Ornithology* (Ed. by R. F. Johnston), pp. 213–245. New York: Plenum Press. [88, 101, 112]

Gaunt, A. S. and M. K. Wells. 1973. Models of syringeal mechanisms. *Am. Zool.* 13:1227–1247. [109, 112]

Gaunt, A. S., R. C. Stein and S. L. L. Gaunt. 1973. Pressure and air flow during distress calls of the starling, Sturnus vulgaris (Aves: Passeriformes). *J. Exp. Zool.* 183:241–262. [112]

Gaunt, A. S., S. L. L. Gaunt and D. H. Hector. 1976. Mechanics of the syrinx in Gallus gallus. I. A comparison of pressure events in chickens to those in oscines. *Condor* 78:208–223. [101, 112]

Gaunt, A. S., S. L. L. Gaunt and R. M. Casey. 1982. Syringeal mechanics reassessed: Evidence from *Streptopelia. Auk* 99:474–494. [101, 103, 112]

Gaunt, A. S., S. L. L. Gaunt, H. D. Prange and J. S. Wasser. 1987. The effects of tracheal coiling on the vocalizations of cranes (Aves; Gruidae). *J. Comp. Physiol. A* 161:43–58. [109, 112]

Gautier, J.-P. 1978. Répertoire sonore de *Cercopithecus cephus. Z. Tierpsych.* 46:113–169. [607]

Gautier, J.-P. and A. Gautier. 1977. Communication in old world monkeys. In *How Animals Communicate* (Ed. by T. A. Sebeok), pp. 890–964. Bloomington: Indiana Univ. Press. [817]

Gautier-Hion, A., J.-M. Duplantier, L. Emmons, F. Feer, P. Heckestweiler, A. Moungazi, R. Quris and C. Sourd. 1985. Fruit characters as a basis of fruit choice and seed dispersal in a tropical forest vertebrate community. *Oecologia* 65:324–337. [262]

Gehlbach, F. R. and J. S. Leverett. 1995. Mobbing of eastern screech owls: Predatory cues, risk to mobbers and degree of threat. *Condor* 97:831–834. [839]

Geist, V. 1966. The evolution of horn-like organs. *Behav.* 27:175–213. [682]

Geldard, F. A. 1977. Tactile communication. In *How Animals Communicate* (Ed. by T. A. Sebeok), pp. 211–232. Bloomington, IN: Indiana Univ. Press. [567]

Gerhardt, H. C. 1974. Vocalizations of some hybrid treefrogs: Acoustic and behavioral analyses. *Behaviour* 49:130–151. [500]

Gerhardt, H. C. 1978. Temperature coupling in the in the vocal communication system of the gray treefrog *Hyla versicolor. Science* 199:922–994. [481]

Gerhardt, H. C. 1982. Sound pattern recognition in some North American tree frogs (Anura: Hylidae). *Am. Zool.* 22:581–595. [481]

Gerhardt, H. C. 1991. Female mate choice in treefrogs: Static and dynamic acoustic criteria. *Anim. Behav.* 42:615–635. [473, 576, 582, 585]

Gerhardt, H. C. 1994. The evolution of vocalization in frogs and toads. *Annu. Rev. Ecol. Syst.* 25:293–324. [473, 478, 485, 533]

Gerhardt, H. C., M. L. Dyson and S. D. Tanner. 1996. Dynamic properties of the advertisement calls of gray tree frogs: Patterns of variability and female choice. *Behav. Ecol.* 7:7–18. [772]

Getty, T. 1989. Are dear enemies in a war of attrition? *Anim. Behav.* 37:337–339. [729]

Getty, T. 1995. Search, discrimination and selection: Mate choice by pied flycatchers. *Am. Nat.* 145:146–154. [448, 454]

Getty, T. 1996. Mate selection by repeated inspection: More on pied flycatchers. *Anim. Behav.* 51:739–745. [448, 454]

Getty, T., A. C. Kamil and P. G. Real. 1987. Signal detection theory and foraging for cryptic and mimetic prey. In *Foraging Behavior* (Ed. by A. C. Kamil, J. R. Krebs and H. R. Pulliam), pp. 525–548. New York: Plenum Press. [454]

Getz, W. M. 1991. The honey bee as a model kin recognition system. In *Kin Recognition* (Ed. by P. G. Hepper), pp. 358–412. Cambridge: Cambridge Univ. Press. [478]

Ghiglieri, M. P. 1984. *The Chimpanzees of Kibale Forest.* New York: Columbia Univ. Press. [823]

Gibbons, B. 1986. The intimate sense of smell. *Nat. Geogr.* 170:324–361. [315]

Gibson, R. M. 1996. Female choice in sage grouse: The roles of attraction and active comparison. *Behav. Ecol. Sociobiol.* 39:55–60. [561, 590]

Gibson, R. M. and G. C. Bachman. 1992. The costs of female choice in a lekking bird. *Behav. Ecol.* 3:300–309. [560]

Gibson, R. M. and J. W. Bradbury. 1985. Sexual selection in lekking sage grouse: Phenotypic correlates of male mating success. *Behav. Ecol. Sociobiol.* 18:117–123. [462]

Gibson, R. M., J. W. Bradbury and S. L. Vehrencamp. 1991. Mate choice in lekking sage grouse revisited: The roles of vocal display, female site fidelity, and copying. *Behav. Ecol.* 2:165–180. [485]

Giraldeau, L. A. and D. Gillis. 1985. Optimal group size can be stable: A reply to Sibly. *Anim. Behav.* 33:666–667. [837]

Godfray, H. C. J. 1991. Signalling of need by offspring to their parents. *Nature, London* 352:328–330. [665, 666, 667, 669, 803]

Godfray, H. C. J. 1995a. Evolutionary theory of parent-offspring conflict. *Nature* 376:133–138. [803, 816]

Godfray, H. C. J. 1995b. Signaling of need between parents and young: Parent-offspring conflict and sibling rivalry. *Am. Nat.* 146:1–24. [666, 667, 803]

Godfray, H. C. J. 1995c. Communication between the first and third trophic levels: An analysis using biological signalling theory. *Oikos* 72:367–374. [847]

Gogala, M., A. Cokl, K. Draslar and A. Blazevic. 1974. Substrate-borne sound communication in Cydnidae (Heteroptera). *J. Comp. Physiol.* 94:25–31. [116, 134, 139]

Goldsmith, T. H. 1990. Optimization, constraint and history in the evolution of the eyes. *Quart. Rev. Biol.* 65:281–322. [261, 262, 563, 570]

Goldsmith, T. H. 1994. Ultraviolet receptors and color vision: Evolutionary implications and a dissonance of paradigms. *Vision Res.* 34:1479–1487. [259]

Goller, F. and R. A. Suthers. 1995. Implication for lateralization of bird song from unilateral gating of bilateral motor patterns. *Nature* 373:63–66. [103, 112]

Goller, F. and R. A. Suthers. 1996a. Role of syringeal muscles in controlling the phonology of bird song. *J. Neurophysiol.* 76:287–300. [101, 103, 105, 112]

Goller, F. and R. A. Suthers. 1996b. Role of syringeal muscles in gating airflow and sound production in singing brown thrashers. *J. Neurobiol.* 75:867–876. [101, 103, 105, 112]

Goodenough, J., B. McGuire and R. A. Wallace. 1993. *Perspectives on Animal Behavior.* New York: John Wiley & Sons. [11, 587]

Goodwin, G.G. and A. M, Greenhall. 1961. A review of the bats of Trinidad and Tobago. *Bull. Am. Mus. Nat. Hist.* 122:187–302. [168, 873]

Goold, J. C. and S. E. Jones. 1995. Time and frequency domain characteristics of sperm whale clicks. *J. Acoust. Soc. Am.* 98:1279–1291. [879]

Gorman, M. L. and M. G. L. Mills. 1984. Scent marking strategies in hyaenas (Mammalia). *J. Zool. (London)* 202:535–547. [552]

Gosling, L. M. 1982. A reassessment of the function of scent marking in territories. *Z. Tierpsych.* 60:89–118. [717]

Gosling, L. M. 1985. The even-toed ungulates: Order Artiodactyla. In *Social Odours in Mammals* (Ed. by R. E. Brown and D. W. Macdonald), pp. 550–618. Oxford: Clarendon Press. [285, 288, 739]

Gosling, L. M. and H. V. McKay. 1990. Competitor assessment by scent-matching: An experimental test. *Behav. Ecol. Sociobiol.* 26:415–420. [718]

Gottlander, K. 1987. Parental feeding behavior and sibling competition in the pied flycatcher *Ficedula hypoleuca. Ornis Scand.* 18:269–276. [666]

Gottman, J. M. 1981. *Time Series Analysis: A Comprehensive Introduction for Social Scientists.* Cambridge: Cambridge Univ. Press. [471]

Gould, J. L. 1982. *Ethology: The Mechanisms and Evolution of Behavior.* New York: W. W. Norton. [253, 528, 581]

Gould, J. L. and P. Marler. 1987. Learning by instinct. *Sci. Am.* 256:62–73. [790]

Gouras, P. 1991a. Precortical physiology of colour vision. In *The Perception of Colour* (Ed. by P. Gouras), pp. 163–178. Boca Raton, FL: CRC Press. [251]

Gouras, P. 1991b. *Perception of Colour.* Boca Raton,FL: CRC Press. [278]

Gouzoules, H. and S. Gouzoules. 1989. Design features and developmental modification of pigtail macaque, *Macaca nemestrina,* agonistic screams. *Anim. Behav.* 37:383–401. [480]

Grafen, A. 1987. The logic of divisively asymmetric contests: Respect for ownership and the desperado effect. *Anim. Behav.* 35:462–467. [714, 719]

Grafen, A. 1990a. Biological signals as handicaps. *J. Theor. Biol.* 144:517–546. [355, 658, 660, 669, 676, 763]

Grafen, A. 1990b. Sexual selection unhandicapped by the Fisher process. *J. Theor. Biol.* 144:473–516. [658, 660, 669, 676]

Grafen, A. 1990c. Do animals really recognize kin? *Anim. Behav.* 39:42–54. [786, 788, 790]

Grafen, A. and R. A. Johnstone. 1993. Why we need ESS signalling theory. *Phil. Trans. Roy. Soc. London, B* 340:245–250. [671]

Green, R. F. 1980. Bayesian birds: A simple example of Oaten's stochastic model of optimal foraging. *Theor. Pop. Biol.* 18:244–256. [411, 418]

Green, S. and P. M. Marler. 1979. The analysis of animal communication. In *Handbook of Behavioral Neurobiology: Vol. 3. Social Behavior and Communication* (Ed. by P. Marler and J. G. Vandebergh), pp. 73–158. New York: Plenum Press. [4, 457, 467, 469, 496]

Greenberg, L. 1979. Genetic component of bee odor in kin recognition. *Science* 206:1095–1097. [807]

Greene, E. 1987. Individuals in an osprey colony discriminate between high and low quality information. *Nature* 329:239–241. [822]

Greenewalt, C. H. 1968. *Bird Song: Acoustics and Physiology.* Washington, D.C.: Smithsonian Institution Press. [74, 88, 103, 108, 109, 112]

Greenfield, M. D. 1994. Cooperation and conflict in the evolution of signal interactions. *Annu. Rev. Ecol. Syst.* 25:97–126. [731]

Greenfield, M. D. and R. L. Minckley. 1992. Acoustic dueling in tarbush grasshoppers. *Ethol.* 95:309–326. [731]

Greenwood, P. J. 1980. Mating systems, philopatry and dispersal in birds and mammals. *Anim. Behav.* 28:1140–1162. [748, 749]

Griffin, D. R. 1958. *Listening in the Dark.* New Haven, CT: Yale Univ. Press. [preface, 868, 870, 882]

Griffin, D. R. 1971. The importance of atmospheric attenuation for the echolocation of bats (Chiroptera). *Anim. Behav.* 22:672–678. [115, 139, 854]

Griffin, D. R. 1984. *Animal Thinking.* Cambridge, MA: Harvard Univ. Press. [880]

Griffin, D. R. 1992. *Animal Minds.* Chicago: Chicago Univ. Press. [880]

Griffin, D. R. and J. A. Simmons. 1974. Echolocation of insects by horseshoe bats. *Nature* 250:731–732. [873]

Grosberg, R. K. and J. F. Quinn. 1986. The genetic control and consequences of kin recognition by the larvae of a colonial marine invertebrate. *Nature* 322:456–459. [807]

Groschupf, K. 1985. Changes in five-striped sparrow song in intra- and intersexual contexts. *Wilson Bull.* 97:102–105. [598]

Gross, M. R. 1982. Sneakers, satellites and parentals: Polymorphic mating strategies in North American sunfishes. *Z. Tierpsych.* 60:1–26. [740]

Gross, W. B. 1954. Voice production by the chicken. Poult. Sci. 43:1005–1008. [101, 112]

Gross, W. B. 1964. Devoicing the chicken. *Poult. Sci.* 43:1143–1144. [112]

Grossman, C. J. 1985. Interactions between the gonadal steroids and the immune system. *Science* 227:838–840. [664]

Gubernick, D. J. 1981. Mechanisms of maternal 'labeling' in goats. *Anim. Behav.* 29:305–306. [801]

Guenther, R. D. 1990. *Modern Optics.* New York: John Wiley. [193, 205]

Guilford, T. 1990. The evolution of aposematism. In *Insect Defenses: Adaptive Mechanisms and Strategies of Prey and Predators* (Ed. by D. L. Evans and J. O. Schmidt), pp. 23–61. State Univ. of New York Press. [845, 850]

Guilford, T. and M. S. Dawkins. 1991. Receiver psychology and the evolution of animal signals. *Anim. Behav.* 42:1–15. [573]

Guilford, T. and M. S. Dawkins. 1993. Are warning colors handicaps? *Evolution* 47:400–416. [846]

Guilford, T. and M. S. Dawkins. 1995. What are conventional signals? *Anim. Behav.* 49:1689–1695. [655, 676]

Gust, D. 1995. Moving up the dominance hierarchy in young sooty mangabeys. *Anim. Behav.* 50:15–21. [809]

Guthrie, R. D. 1971. A new theory of mammalian rump patch evolution. *Behaviour* 38:132–145. [842, 843]

Gwynne, D. T. 1981. Sexual difference theory: Mormon crickets show role reversal in mate choice. *Science* 213:779–780. [779]

Gwynne, D. T. 1982. Mate selection by female katydids (Orthoptera: Tettigoniidae, *Conocephalus nigropleurum*). *Anim. Behav.* 30:734–738. [772]

Gwynne, D. T. 1990. Testing parental investment and the control of sexual selection in katydids: The operational sex ratio. *Am. Nat.* 136:474–484. [778]

Gwynne, D. T. 1991. Sexual competition among females: What causes courtship role reversal? *TREE* 6:118–121. [777]

Gwynne, T. and L. W. Simmons. 1990. Experimental reversal of courtship roles in an insect. *Nature* 346:171–174. [778]

Gyger, M. and P. Marler. 1988. Food calling in the domestic fowl, Gallus gallus: The role of external referents and deception. *Anim. Behav.* 36:358–365. [820]

Gyger, M. P. Marler and R. Pickert. 1987. Semantics of an avian alarm system: The male domestic fowl, *Gallus domesticus*. *Behaviour* 102:15–40. [607]

Haccou, P. and E. Meelis. 1992. *Statistical Analysis of Behavioural Data*. Oxford: Oxford Univ. Press. [471]

Hack, M. 1997a. The energetic costs of fighting in the house cricket, *Acheta domesticus* L. *Behav. Ecol.* 8:28–36. [697, 698]

Hack, M. 1997b. The determinants of fighting success and tests for assessment in the contests of male crickets, *Acheta domesticus* (L). *Anim. Behav.* 53:733–747. [602, 697, 698]

Hagedorn, M. 1986. The ecology, courtship, and mating of gymnotiform electric fish. In *Electroreception* (Ed. by T. W. Bullock and W. Heiligenberg), pp. 497–525. New York: John Wiley. [350]

Hagedorn, M. 1988. Ecology and behaviour of a pulse-type electric fish, *Hypopomus occidentalis* (Gymnotiformes, Hypopomidae) in a fresh-water stream in Panama. *Copeia* 1988:324–335. [350]

Hagedorn, M. and W. Heiligenberg. 1985. Court and spark: Electric signals in the courtship and mating of gymnotid fish. *Anim. Behav.* 33:254–265. [346, 350]

Hagedorn, M. and R. Zelick. 1989. Relative dominance among males is expressed in the electric organ discharge characteristics of a weakly electric fish. *Anim. Behav.* 38:520–525. [350]

Haigh, J. 1975. Game theory and evolution. *Adv. Appl. Prob.* 7:8–11. [628, 647]

Hailman, J. P. 1977a. *Optical Signals: Animal Communication and Light*. Bloomington, IN: Indiana Univ. Press. [233, 418, 535]

Hailman, J. P. 1977b. Communication by reflected light. In *How Animals Communicate* (Ed. by T. A. Sebeok), pp. 184–210. Bloomington: Indiana Univ. Press. [504, 515, 535]

Hailman, J. P. 1979. Environmental light and conspicuous colors. In *The Behavioral Significance of Color* (Ed. by E. H. J. Burtt), pp. 289–354. New York: Garland STPM Press. [209, 232, 233, 234, 236, 238]

Hailman, J. P. and M. S. Ficken. 1996. Comparative analysis of vocal repertoires, with reference to chickadees. In *Ecology and Evolution of Acoustic Communication in Birds* (Ed. by D. E. Kroodsma and E. H. Miller), pp. 136–159. Ithaca, NY: Comstock/Cornell Univ. Press. [472]

Hailman, J. P., M. S. Ficken and R. W. Ficken. 1987. Constraints on the structure of combinatorial "chick-a-dee" calls. *Ethology* 75:62–80. [461]

Haley, M. P. 1994. Resource-holding power asymmetries, the prior residence effect, and reproductive payoffs in male northern elephant seal fights. *Behav. Ecol. Sociobiol.* 34:427–434. [723]

Hall, S. E. and D. E. Mitchell. 1991. Grating acuity of cats measured with detection and discrimination tasks. *Behav. Brain Res.* 44:1–9.[269]

Hall-Craggs, J. 1979. Sound spectrographic analysis: Suggestions for facilitating auditory imagery. *Condor* 81:185–192. [74]

Halliday, T. R. 1977. The courtship of European newts: An evolutionary perspective. In *Reproductive Biology of Amphibians* (Ed. by D. H. Taylor and S. I. Guttman), pp. 185–232. New York: Plenum Press. [794, 796]

Halliday, T. R. and P. J. B. Slater. 1983. *Communication*. New York: W.H. Freeman and Company. [11]

Halpern, M. 1987. The organization and function of the vomeronasal system. *Annu. Rev. Neurosci.* 10:325–362. [312, 318]

Halpin, Z. T. 1991. Kin recognition cues of vertebrates. In *Kin Recognition*. (Ed. by P. G. Hepper), pp. 220–259. Cambridge: Cambridge Univ. Press. [801]

Hamilton, W. D. 1964. The genetical evolution of social behaviour. *J.Theor. Biol.* 7:1–52. [802]

Hamilton, W. D. 1971. Geometry for the selfish herd. *J. Theor. Biol.* 31:295–311. [632]

Hamilton, W. D. and M. Zuk. 1982. Heritable true fitness and bright birds: A role for parasites? *Science* 218:384–387. [761]

Hamilton, W. J. III. 1973. *Life's Color Code*. New York: McGraw-Hill. [610]

Hamilton, W. J. and P. C. Arrowood. 1978. Copulatory vocalizations of chacma baboons (*Papio ursinus*), gibbons (*Hylobates hoolock*), and humans. *Science* 200:1405–1409. [776]

Hammerstein, P. 1981. The role of asymmetries in animal contests. *Anim. Behav.* 29: 193–205. [639, 686, 720]

Hammerstein, P. and G. A. Parker. 1982. The asymmetric war of attrition. *J. Theor. Biol.* 96:647–682. [685]

Hammerstein, P. and G. A. Parker. 1987. Sexual selection: Games between the sexes. In *Sexual Selection: Testing the Alternatives*. (Ed. by J. W. Bradbury and M. Andersson), pp. 119–142. New York: John Wiley. [747]

Hanlon, R. T. and J. B. Messenger. 1996. *Cephalopod Behaviour*. Cambridge: Cambridge Univ. Press. [612]

Hanlon, R. T., M. J. Smale and W. H. H. Sauer. 1994. An ethogram of body patterning behavior in the squid *Loligo vulgaris reynaudii* on spawning grounds in South Africa. *Biol. Bull.* 187:363–372. [612]

Hansen, A. J. 1986. Fighting behavior in bald eagles-a test of game theory. *Ecology* 67:787–797. [658]

Hansen, A. J. and S. A. Rohwer. 1986. Coverable badges and resource defence in birds. *Anim. Behav.* 34:69–76. [597, 664]

Hanson, J. H. 1996. Rhesus macaque copulation calls: Re-evaluating the honest signal hypothesis. *Primates* 37:145–154. [777]

Harkness, L. 1977. Chameleons use accomodation cues to judge distance. *Nature* 267:346–349. [273]

Harnad, S. 1987. *Categorial Perception*. Cambridge: Cambridge Univ. Press. [476, 496]

Harper, A. B. 1986. The evolution of begging: sibling competition and parent-offspring conflict. *Am. Nat.* 128:99–114. [666]

Harper, D. G. C. 1991. Communication. In *Behavioural Ecology: An Evolutionary Approach*. 3rd Edition. (Ed. by J. R. Krebs and N. B. Davies), pp. 374–397. Oxford: Blackwell Scientific Publications. [675]

Harrington, F. H. 1989. Chorus howling by wolves: Acoustic structure, pack size and the beau geste effect. *Bioacoustics* 2:117–136. [597]

Harris, C. L., W. B. Gross and A. Robeson. 1968. Vocal acoustics of the chicken. *Poult. Sci.* 47:107–112. [112]

Harris, G. G. 1964. Considerations on the physics of sound production by fishes. In *Marine Bio-Acoustics* (Ed. by W. N. Tavolga), pp. 233–247. New York: Pergamon Press. [112]

Harris, V. E. and J. W. Todd. 1980. Male-mediated aggregation of male, female, and 5th-instar southern green stink bugs and concomitant attraction of a tachinid parasite, *Trichopoda pennipes*. *Entomol. Exp. Applic.* 27:117–126. [8]

Hart, B. J. 1983. Flehmen behavior and vomeronasal organ function. In *Chemical Signals in Vertebrates* (Ed. by D. Müller-Schwarze and R. M. Silverstein), pp. 87–104. New York: Plenum Press. [318]

Hart, B. and L. Hart. 1992. Reciprocal allogrooming in impala, *Aepyceros melampus*. *Anim. Behav.* 44:1073–1083. [810]

Hartigan, J. A. 1975. *Clustering Algorithms*. New York: John Wiley. [465]

Hartley, D. J. 1992. Stabilization of perceived echo amplitudes in echolocating bats. 1. Echo detection and automatic gain control in the big brown bat, *Eptesicus fuscus*, and the fishing bat, *Noctilio leporinus*. *J. Acoust. Soc. Am.* 91:1120–1132. [868]

Hartley, D. J. and R. A. Suthers. 1987. The sound emission pattern and the acoustical role of the noseleaf in the echolocating bat, *Carollia perspicillata*. *J. Acoust. Soc. Am.* 82:1892–1900. [112, 861]

Hartley, D. J. and R. A. Suthers. 1989. The emission pattern of the echolocating bat, *Eptesicus fuscus*. *J. Acoust. Soc. Am.* 85:1348–1351. [861]

Hartley, D. J. and R. A. Suthers. 1990. Sonar pulse radiation and filtering in the mustached bat, *Pteronotus parnellii rubiginosus*. *J. Acoust. Soc. Am.* 87:2756–2772. [112, 861]

Hartley, R. S. 1990. Expiratory muscle activity during song production in the canary. *Respir. Physiol.* 81:177–188. [100, 105, 112]

Hartley, R. S. and R. A. Suthers. 1989. Airflow and pressure during canary song: Direct evidence for mini-breaths. *J. Comp. Physiol.* A 165:15–26. [105]

Hartley, R. S. and R. A. Suthers. 1990. Lateralization of syringeal function during song production in the canary. *J. Neurobiol.* 21:1236–1248. [103]

Hartshorne, C. 1956. The monotomy threshold in singing birds. *Auk* 83:176–192. [735]

Harvey, P. H. and P. J. Greenwood. 1978. Anti-predator defence strategies: Some evolutionary problems. In *Behavioural Ecology: An Evolutionary Approach*. 1st Edition. (Ed. by J. R. Krebs and N. B. Davies), pp. 129–151. Oxford: Blackwell Scientific Publications. [846, 850]

Harvey, P. H. and J. R. Krebs. 1990. Comparing brains. *Science* 249:140–145. [565]

Haskell, D. 1994. Experimental evidence that nestling begging behavior incurs a cost due to nest predation. *Proc. Roy. Soc. Lond. B* 257:161–164. [666, 667, 805]

Haskell, P. T. 1974. Sound production. In The Physiology of Insecta (Ed. by M. Rockstein), pp. 354–410. New York: Academic Press. [111]

Hasselquist, D., S. Bensch and T. von Schantz. 1996. Correlation between male song repertoire, extra-pair paternity and offspring survival in the great reed warbler. *Nature* 381:229–232. [772, 774]

Hasson, O. 1989a. The effect of uncertainty on the relationship between the frequency of warning signals and prey density. *Theor. Pop. Biol.* 36:241–250. [836]

Hasson, O. 1989. Amplifiers and the handicap principle in sexual selection: A different emphasis. *Proc. R. Soc. Lond. B Biol. Sci.* 235:383–406. [355, 669, 761]

Hasson, O. 1990. The role of amplifiers in sexual selection: An integration of the amplifying and the Fisherian mechanisms. *Evol. Ecol.* 4:277–289. [666, 669, 761]

Hasson, O. 1991. Sexual displays as amplifiers: Practical examples with an emphasis on feather decorations. *Behav. Ecol.* 2:189–197. [355]

Hasson, O. 1994. Cheating signals. *J. Theor. Biol.* 167:223–238. [355, 650, 675, 842]

Hasson, O., R. Hibbard and G. Ceballos. 1989. The pursuit-deterrent function of tail-wagging in the zebra-tailed lizard (*Callisaurus draconoides*). *Can. J. Zool.* 67:1203–1209. [842]

Hasson, O., D. Cohen and A. Shmida. 1992. Providing or hiding information: On the evolution of amplifiers and attenuators of perceived quality differences. *Acta Biotheor.* 40:269–283. [357, 669]

Hausberger, M., J. M. Black and J.-P. Richard. 1991. Bill opening and sound spectrum in barnacle goose loud calls: Individuals with "wide mouths" have higher pitched voices. *Anim. Behav.* 42:319–322. [109]

Hauser, M. D. 1989. Do chimpanzee copulatory calls incite male-male competition. *Anim. Behav.* 39:596–597. [777]

Hauser, M. D. 1991. Sources of acoustic variation in rhesus macaque (*Macaca mulatta*) vocalizations. *Ethology* 89:29–46. [467]

Hauser, M. 1993a. Do vervet monkey infants cry wolf? *Anim. Behav.* 45:1242–1244. [804]

Hauser, M. D. 1993b. Rhesus monkey copulation calls: Honest signals for female choice? *Proc. R. Soc. Lond. B Biol. Sci.* 254:93–96. [776, 777]

Hauser, M. D. 1996. *The Evolution of Communication*. Cambridge, MA: MIT Press. [4, 11, 355]

Hauser, M. D. and P. Marler. 1993a. Food-associated calls in rhesus macaques (Macaca mulatta). 1. Socioecological factors. *Behavioral Ecology* 4:194–205. [823, 850]

Hauser, M. D. and P. Marler. 1993b. Food-associated calls in rhesus macaques (*Macaca mulatta*). 2. Costs and benefits of call production and suppression. *Behavioral Ecology* 4:206–212. [823]

Hauser, M. D. and D. A. Nelson. 1991. "Intentional" signalling in animal communication. *TREE* 6:186–189. [689]

Hauser, M. D. and R. W. Wrangham. 1987. Manipulation of food calls in captive chimpanzees. *Folia Primatologia* 48:207–210. [823]

Hauser, M. D., P. Teixidor, L. Field and R. Flaherty. 1993. Food-elicited calls in chimpanzees-effects of food quantity and divisibility. *Anim. Behav.* 45:817–819. [823]

Hawkins, A. D. and A. A. Myrberg. 1983. Hearing and sound communication under water. In *Bioacoustics: A Comparative Approach* (Ed. by B. Lewis), pp. 347–405. London: Academic Press. [132, 139, 173]

Hazlett, B. 1980. Patterns of information flow in the hermit acrab *Calcinus tibicen*. *Anim. Behav.* 28:1024–1032. [703]

Hazlett, B. A. and G. F. Estabrook. 1974. Examination of agonsitic behavior by character analysis. I. The spider crab *Microphrys bicornutus*. *Behaviour* 48:131–144. [411]

Hedrick, A. V. and L. M. Dill. 1993. Mate choice by female crickets in influenced by predation risk. *Anim. Behav.* 46:193–196. [560]

Heiligenberg, W. 1975. Theoretical and experimental approaches to spatial aspects of electrolocation. *J. Comp. Physiol.* 103:247–272.

Heiligenberg, W. 1977. *Principles of Electrolocation and Jamming Avoidance in Electric Fish*. Berlin: Springer-Verlag. [335, 336, 348, 350]

Heiligenberg, W. and C. D. Hopkins. 1976. Electrolocation and electrical communication by gymnotid fishes from coastal Surinam. *Nat. Geog. Res. Rep.* 1976:461–474. [350]

Heinrich, B. 1988. Winter foraging at carcasses by three sympatric corvids, with emphasis on recruitment by the raven, Corvus corax. *Behav. Ecol. Sociobiol.* 23:141–156. [821]

Heinrich, B. and J. M. Marzluff. 1991. Do common ravens yell because they want to attract others? *Behav. Ecol. Sociobiol.* 28:13–21. [821]

Heller, K.-G. 1992. Risk shift between males and females in the pair-forming behavior of bush crickets. *Naturwiss.* 79:89–91. [560, 590]

Hemelrijk, C. and E. Anneke. 1991. Reciprocity and interchange of grooming and 'support' in captive chimpanzees. *Anim. Behav.* 41:923–935. [809]

Henderson, B. A. 1975. Role of the chick's begging behavior in the regulation of parental feeding behavior of *Larus glaucescens*. *Condor* 77:488–492. [666]

Henzi, S. P. 1996. Copulation calls and paternity in chacma baboons. *Anim. Behav.* 51:233–234. [777]

Hepper, P. G. 1991a. Recognizing kin: Ontogeny and classification. In *Kin Recognition*. (Ed. by P. G. Hepper), pp. 259–288. Cambridge: Cambridge Univ. Press. [786]

Hepper, P. G. 1991b. *Kin Recognition*. Cambridge: Cambridge Univ. Press. [786, 816]

Herrnstein, R. J. 1991. Levels of categorization. In *Signal and Sense* (Ed. by G. M. Edelman, W. E. Gall and W. M. Cowan), pp. 385–412. Somerset, NJ: Wiley-Liss. [496]

Hess, E. H. 1959. Imprinting, an effect of early experience. *Science* 130:133–141. [790]

Hews, D. K. 1988. Alarm response in larval western toads, *Bufo boreas*: Release of larval chemicals by a natural predator and its effect on predator capture efficiency. *Anim. Behav.* 36:125–133. [846]

Heyes, C. M. 1993. Imitation, culture, and cognition. *Anim. Behav.* 46:99–110. [475]

Heyes, C. M. 1994. Social learning in animals: Categories and mechanisms. *Biol. Rev.* 69:207–231. [475, 477]

Heywood, J. S. 1989. Sexual selection by the handicap mechanism. *Evolution* 43:1387–397. [763]

Hiebert, S. M., P. K. Stoddard and P. Arcese. 1989. Repertoire size, territory acquisition and reproductive success in the song sparrow. *Anim. Behav.* 37:266–273. [734]

Hildebrand, J. G. and R. A. Montague. 1986. Functional organization of olfactory pathways in the central nervous system of *Manduca sexta*. In *Mechanisms in Insect Olfaction* (Ed. by T. L. Payne, M. C. Birch and C. E. J. Kennedy), pp. 279–286. Oxford: Clarendon Press. [306]

Hill, G. E. 1990. Female house finches prefer colourful mates: Sexual selection for a condition-dependent trait. *Anim. Behav.* 40:563–572. [662]

Hill, G. E. 1991. Plumage color is a sexually selected indicator of male quality. *Nature* 350:337–339. [661, 662, 765]

Hill, G. E. 1994. Trait elaboration via adaptive mate choice: Sexual conflict in the evolution of signals of male quality. *Ethol. Ecol. Evol.* 6:351–370. [661]

Hill, G. E. 1996. Redness as a measure of the production cost of ornamental coloration. *Ethol. Ecol. Evol.* 8:157–175. [549]

Hinde, R. A. 1982. *Ethology: Its Nature and Relations with Other Sciences*. Oxford: Oxford Univ. Press. [509, 683]

Hines, W. G. S. 1980. Three characterizations of population strategy stability. *J. Appl. Prob.* 17:333–340. [628, 647]

Hines, W. G. S. 1987. Evolutionary stable strategies: A review of basic theory. *Theor. Pop. Biol.* 31:195–272. [628, 647]

Hirth, D. H. and D. R. McCullough. 1977. Evolution of alarm signals in ungulates with special reference to white-tailed deer. *Am. Nat.* 111:31–42. [836, 842]

Hjorth, I. 1970. Reproductive behaviour in Tetraonidae, with special reference to males. *Viltrevy* 7: 183–596. [462]

Hockett, C. F. 1963. The problem of universals in language. In *Universals of Language* (Ed. by J. H. Greenberg), pp. 1–29. Cambridge, MA: M.I.T. Press. [572]

Hockett, C. F. and S. A. Altman. 1968. A note on design features. In *Animal Communication* (Ed. by T. A. Sebeok), pp. 61–72. Bloomington, IN: Indiana Univ. Press. [572, 615]

Hoekstra, R. F. 1982. On the asymmetry of sex: Evolution of mating types in isogamous populations. *J. Theor. Biol.* 98:427–451. [744]

Hoekstra, R. F. 1987. The evolution of sexes. In *The Evolution of Sex and Its Consequences*. (Ed. by S. C. Stearns), pp. 59–92. Basel: Birkhäuser. [744]

Hoekstra, R. F., R. F. Janz and A. J. Schilstra. 1984. Evolution of gamete motility differences. I. Relations between swimming speed and pheromonal attraction. *J. Theor. Biol.* 107:57–70. [744]

Hoelzer, G. A. 1989. The good parent process of sexual selection. *Anim. Behav.* 38:1067–1078. [763]

Hofbauer, J. and K. Sigmund. 1988. *The Theory of Evolution and Dynamical Systems*. Cambridge: Cambridge Univ. Press. [647]

Höglund, J., M. Eriksson and L. E. Lindell. 1990. Females of the lek-breeding great snipe, *Gallinago media*, prefer males with white tails. *Anim. Behav.* 40:23–32. [765]

Höglund, J., J. A. Kålås and P. Fiske. 1992. The costs of secondary sexual characters in the lekking Great Snipe (*Gallinago media*). [549]

Hogstad, O. 1987. Is it expensive to be dominant? *Auk* 104:333–336. [664]

Högstedt, G. 1983. Adaptation unto death: Function of fear screams. *Am. Nat.* 121:562–570. [847, 850]

Hölldobler, B. 1984. Evolution of insect communication. In *Insect Communication* (Ed. by T. Lewis), pp. 349–377. London: Academic Press. [826]

Hölldobler, B. and M. Möglich. 1980. The foraging system of Pheidole militicida (Hymenoptera: Formicidae). *Insectes Sociaux* 27:237–264. [825]

Hölldobler, B. and E. O. Wilson. 1990. *The Ants*. Cambridge, MA: Belknap Press of Harvard Univ. Press. [315, 525, 607, 615, 824, 825, 826, 850]

Holley, A. J. F. 1984. Adoption, parent-chick recognition and maladaptation in the herring gull. *Z. Tierpsych.* 64:9–14. [801]

Holloway, G. J., P. W. Dejong and M. Ottenheim. 1993. The genetics and cost of chemical defense in the two-spot ladybird (*Adalia bipunctata* L.). *Evolution* 47:1229–1239. [551]

Holmes, W. G. and P. W. Sherman. 1982. The ontogeny of kin recognition in two species of ground squirrels. *Am. Zool.* 22:491–517.

Hoogland, J. L. 1983. Nepotism and alarm calling in the black-tailed prairie dog (*Cynomys ludovicianus*). *Anim. Behav.* 31:472–479. [840]

Hooper, E. T. 1962. The glans penis in *Proechimys* and other caviomorph rodents. *Occ. Pap. Mus. Zool. Univ. Mich.* 623: 1–18. [764]

Hope, S. 1980. Call form in relation to function in the Stellar's jay. *Am. Nat.* 116:788–820. [834]

Hopkins, C. D. 1972. Sex differences in electric signalling in an electric fish. *Science* 176:1035–1037. [350]

Hopkins, C. D. 1973. Lightning as background noise for communication among electric fish. *Nature* 242:268–270. [337, 350]

Hopkins, C. D. 1974a. Electric communication: Functions in the social behavior of *Eigenmannia virescens*. *Behaviour* 50:270–305. [350]

Hopkins, C. D. 1974b. Electric communication in the reproductive behavior of *Sternopygus macrurus* (Gymnotoidei). *Z. Tierpsych.* 35:518–535. [350]

Hopkins, C. D. 1974c. Electric communication in fish. *Am. Sci.* 62:426–437. [350]

Hopkins, C. D. 1977. Electric communication. In *How Animals Communicate* (Ed. by T. A. Sebeok), pp. 263–289. Bloomington, IN: Indiana Univ. Press. [350]

Hopkins, C. D. 1983. Neuroethology of species recognition in electroreception. In *Advances in Vertebrate Neuroethology* (Ed. by J. P. Ewert, R. R. Capranica and D. Ingle), pp. 871–881. New York: Plenum Press. [582]

Hopkins, C. D. 1983a. Neuroethology of species recognition in electroreception. In *Advances in Vertebrate Neuroethology* (Ed. by J. P. Ewert, R. R. Capranica and D. Ingle), pp. 871–881. New York: Plenum Press. [350]

Hopkins, C. D. 1983b. Functions and mechanisms in electroreception. In *Fish Neurobiology. I. Brain Stem and Sense Organs* (Ed. by R. G. Northcutt and R. E. Davis), pp. 215–259. Ann Arbor, MI: Univ. of Michigan Press. [342, 350]

Hopkins, C. D. 1986. Behavior of Mormyridae. In *Electroreception* (Ed. by T. H. Bullock and W. Heiligenberg), pp. 527–576. New York: John Wiley. [350]

Hopkins, C. D. 1988a. Social communication in the aquatic environment. In *Sensory Biology of Aquatic Animals* (Ed. by J. Atema, R. R. Fay, A. N. Popper and W. N. Tavolga), pp. 233–268. New York: Springer-Verlag. [247, 350]

Hopkins, C. D. 1988b. Neuroethology of electric communication. *Annu. Rev. Neurosci.* 11:497–535. [345, 350]

Hopkins, C. D. and A. H. Bass. 1981. Temporal coding of species recognition signals in electric fish. *Science* 212:85–87. [342, 582]

Hopkins, C. D. and W. Heiligenberg. 1978. Evolutionary designs for electric signals and electroreceptors in gymnotid fishes of Surinam. *Behav. Ecol. Sociobiol.* 3:113–134. [350]

Hopkins, C., M. Rosetto and A. Lutjen. 1974. A continuous sound spectrum analyzer for animal sounds. *Z. Tierpsych.* 34:313–320. [74]

Horn, A. G. and J. B. Falls. 1996. Categorization and the design of signals: The case of song repertoires. In *Ecology and Evolution of Acoustic Communication in Birds* (Ed. by D. E. Kroodsma and E. H. Miller), pp. 121–135. Ithaca, NY: Comstock/Cornell Univ. Press. [467]

Horowitz, B. R. 1981. Theoretical considerations of the retinal receptor as a waveguide. In *Vertebrate Photoreceptor Optics.* (Ed. by J. M. Enoch and F. L Tobey, Jr.), pp. 219–300. Berlin: Springer-Verlag. [191]

Horridge, G. A. 1975. *The Compound Eye and Vision in Insects.* Oxford: Clarendon Press. [278]

Horridge, G. A. 1977. The compound eye of insects. *Sci. Am.* 237:108–120. [271]

Houck, L. D. and N. L. Reagan. 1990. Male courtship pheromones increase female receptivity in a plethodontid salamander. *Anim. Behav.* 39:729–734. [794, 796]

Houston, A. I. and J. M. McNamara. 1987. Singing to attract a mate: A stochastic dynamic game. *J. Theor. Biol.* 129:57–68. [623]

Houston, A. I. and B. H. Sumida. 1987. Learning rules, matching, and frequency dependence. *J. Theor. Biol.* 126:289–308. [411]

Houston, A. I., A. Kacelnick and J. M. McNamara. 1982. Some learning rules for acquiring information. In *Functional Ontogeny* (Ed. by D. J. MacFarland), pp. 148–191. London: Pitman. [410, 418]

Houtman, A. M. 1992. Female zebra finches choose extra-pair copulations with genetically attractive mates. *Proc. R. Soc. Lond. B Biol. Sci.* 249:3–6. [772]

Howard, D. J. and P. G. Gregory. 1993. Post-insemination signalling systems and reinforcement. *Philos. Trans. R. Soc. Lond. B Biol. Sci.* 340:231–236. [752]

Hoy, R. R. 1992. The evolution of hearing in insects as an adaptation to predation from bats. In *The Evolutionary Biology of Hearing* (Ed. by D. B. Webster, R. R. Fay and A. N. Popper), pp. 115–129. New York: Springer Verlag. [158, 515]

Hoy, R. R., J. Hahn and R. C. Paul. 1977. Hybrid cricket auditory behavior: Evidence for genetic coupling in animal communication. *Science* 195:82–83. [500]

Hsu, H. P. 1970. *Fourier Analysis.* New York: Simon and Schuster. [74]

Huck, U. W., E. M. Banks and S. C. Wang. 1981. Olfactory discrimination of social status in the brown lemming. *Behav. Neural Biol.* 33:364–371. [774]

Hueter, R. E. 1991. Adaptations for spatial vision in sharks. *J. Exp. Zool. Suppl.* 5:130–141. [269]

Hughes, A. 1971. Topographical relationships between the anatomy and physiology of the rabbit visual system. *Docum. Ophthal. (Den Haag)* 30:33–159. [274]

Hughes, A. 1976. A supplement to the cat schematic eye. *Vision Res.* 16:149–154. [274]

Hughes, A. 1977. The topography of vision in mammals of contrasting lifestyle: Comparative optics and retinal organization. In *Handbook of Sensory Physiology* (Ed. by F. Crescitelli), pp. 613–756. Berlin: Springer Verlag. [269]

Hund, A. 1942. *Frequency Modulation.* New York: McGraw-Hill. [74]

Hunsaker, D. 1962. Ethological isolating mechanisms in the *Sceloporus torquatus* group of lizards. *Evolution* 16:62–74. [581]

Hunt, D. M., A. J. Williams, J. K. Bowmaker and J. D. Mollon. 1993. Structure and evolution of the polymorphic photopigment gene of the marmoset. *Vision Res.* 33:147–154. [259]

Huntingford, F. and A. Turner. 1987. *Animal Conflict.* London: Chapman and Hall. [707, 710]

Hurd, P. L., C. A. Wachtmeister and M. Enquist. 1995. Darwin's principle of antithesis revisited: A role for perceptual biases in the evolution of intraspecific signals. *Proc. R. Soc. Lond. B Biol. Sci.* 259:201–205. [506]

Hurst, J. L. 1989. The complex network of olfactory communication in populations of wild house mice *Mus domesticus* Rutty: Urine markings and investigation within family groups. *Anim. Behav.* 37:705–725. [738, 806]

Hurst, J. L. 1990. Urine marking in populations of wild house mice (*Mus domesticus* Rutty). I. Communication between males. *Anim. Behav.* 40:209–222. [738]

Hurst, J. L., J. Fang and C. J. Barnard. 1993. The role of substrate odours in maintaining social tolerance between male house mice, *Mus musculus domesticus. Anim. Behav.* 45:997–1006. [738, 806]

Hurvich, L. M. 1981. *Color Vision.* Sunderland, MA: Sinauer Associates. [278]

Hussell, K. J. T. 1988. Supply and demand in tree swallow broods: A model of parent-offspring food provisioning interactions in birds. *Am. Nat.* 131:175–202. [666]

Hutchings, M. and B. Lewis. 1983. Insect sound and vibration receptors. In *Bioacoustics: A Comparative Approach* (Ed. by B. Lewis), pp. 181–205. London: Academic Press. [173]

Hyatt, G. W. and M. Salmon. 1979. Comparative statistical and information analysis of combat in the fiddler crabs, *Uca pugilator* and *U. pugnax. Behaviour* 68:1–23. [411]

Hyder, D. E. and C. Y. Oseto. 1989. Structure of the stridulatory apparatus and analysis of the sound produced by *Smicronyx fulvus* and *Smicronyx sordidus* (Coleoptera, Curculionidae, Erirrhininae, Smicronychini). *J. Morph.* 201:69–84. [111]

Ichikawa, T. 1976. Mutual communication by substrate vibrations in the mating behavior of planthoppers (Homoptera, Delphacidae). *Appl. Entomol. Zool.* 11:8–21. [111, 116, 134, 139]

Imafuku, M. and H. Ikeda. 1990. Sound production in the land hermit crab *Coenobita purpureus Stimpson*, 1858 (Decapoda, Coenobitidae). *Crustaceana* 58:168–174. [111]

Immelmann, K. 1972. Sexual and other long-term aspects of imprinting in birds and other species. *Adv. Study Behav.* 4:147–174. [790]

Irwin, R. E. 1988. The evolutionary importance of behavioral development: The ontogeny and phylogeny of bird song. *Anim. Behav.* 36:814–825. [487, 502]

Ishihara, M. 1987. Effect of mobbing toward predators by the damselfish *Pomacentrus coelestis* (Pisces: Pomacentridae). *J. Ethology* 5:43–52. [844]

Iwasa, Y. and A. Pomiankowski. 1991. The evolution of costly mate preferences. 2. The 'handicap' principle. *Evolution* 45:1431–1442. [660]

Iwasa, Y. and A. Pomiankowski. 1994. The evolution of mate preferences for multiple sexual ornaments. *Evolution* 48:852–867. [774]

Jackobsson, S., T. Radesäter and T. Järvi. 1979. On the fighting behaviour of *Nannocara anomala* (Pisces: Cichlidae) males. *Z. Tierpsych.* 49:210–220. [696]

Jackson, R. R. and K. J. Cooper. 1991. The influence of body size and prior residency on the outcome of male-male interactions of *Marpissa marina*, a New Zealand jumping spider (Araneae, Salticidae). *Ethol. Ecol. Evol.* 3:79–82. [719]

Jacobs, G. H. 1981. *Comparative Color Vision.* New York: Academic Press. [254, 256, 259, 278]

Jacobs, G. H. 1993. The distribution and nature of colour vision among the mammals. *Biol. Rev.* 68:413–471. [254, 257, 259, 278]

Jacobs, G. H. and J. Nietz. 1985. Color vision in squirrel monkeys: Sex-related differences siggest the mode of inheritance. *Vision Res.* 25:141–143. [257]

Jacobs, G. H., J. Neitz and M. Neitz. 1993. Genetic basis of polymorphism in the color vision of platyrrhine monkeys. *Vision Res.* 33:269–2274. [259]

Jacobson, M. 1974. Insect pheromones. In *The Physiology of the Insecta* (Ed. by M. Rockstein), pp. 229–276. New York: Academic Press. [586]

Jaisson, P. 1991. Kinship and fellowship in ants and social wasps. In *Kin Recognition* (Ed. by D. G. Hepper), pp. 60–93. Cambridge: Cambridge Univ. Press. [806]

Janetos, A. C. 1980. Strategies of female mate choice: A theoretical analysis. *Behav. Ecol. Sociobiol.* 7:107–112. [448]

Janetos, A. C. and B. J. Cole. 1981. Imperfectly optimal animals. *Behav. Ecol. Sociobiol.* 9:203–209. [448]

Järvi, T. and M. Bakken. 1984. The function of variation of the breast stripe of the great tit (*Parus major*). *Anim. Behav.* 32:590–596. [664]

Jerison, H. J. 1985. Animal intelligence and encephalization. *Philos. Trans. R. Soc. Lond. B Biol. Sci.*308:21–35. [565]

Jerlov, N. G. 1968. *Optical Oceanography*. London: Elsevier Publishing. [238]

Johnsgard, P. A. 1965. *Handbook of Waterfowl Behavior*. Ithaca, NY: Cornell Univ. Press. [490]

Johnson, C., R. Thornhill, D. Ligon and M. Zuk. 1993. The direction of mothers' and daughters' preferences and the heritability of male ornaments in red jungle fowl (*Gallus gallus*). *Behav. Ecol.* 4:254–259. [766]

Johnston, R. E. 1977. Sex pheromones in golden hamsters. In *Chemical Signals in Vertebrates* (Ed. by D. Müller-Schwarze and M. M. Mozell), pp. 45–59. New York: Plenum Press. [282]

Johnston, T. D. 1982. The selective costs and benefits of learning: An evolutionary analysis. *Adv. Study Behav.* 12:65–106. [475]

Johnston, T. D. and M. T. Turvey. 1980. An ecological metatheory for theories of learning. In *The Psychology of Learning and Motivation: Advances in Research and Theory* (Ed. by G. H. Bower), pp. 147–205. New York: Academic Press. [475]

Johnstone, R. 1994. Honest signalling, perceptual error, and the evolution of "all or nothing" displays. *Proc. R. Soc. Lond. B Biol. Sci.* 256:169–175. [671, 672]

Johnstone, R. A. 1995. Sexual selection, honest advertisement and the handicap principle: Reviewing the evidence. *Biol. Rev.* 70:1–65. [776]

Johnstone, R. 1997. The evolution of animal signals. In *Behavioural Ecology: An Evolutionary Approach.* 4th Edition. (Ed. by J. R. Krebs and N. B. Davies), pp. 155–178. Oxford: Blackwell Science. [675]

Johnstone, R. A. and A. Grafen. 1992a. The continuous Sir Philip Sidney game: A simple model of biological signalling. *J. Theor. Biol.* 156:215–234. [656, 669, 730]

Johnstone, R. A. and A. Grafen. 1992b. Error-prone signalling. *Proc. R. Soc. Lond. B Biol. Sci.* 248:229–233. [671]

Johnstone, R. A. and A. Grafen. 1993. Dishonesty and the handicap principle. *Anim. Behav.* 46:759–764. [674]

Johnstone, R. A. and K. Norris. 1993. Badges of status and the cost of aggression. *Behav. Ecol. Sociobiol.* 32:127–134. [661, 663, 664, 669]

Jolly, A. 1966. *Lemur Behavior.* Chicago: Chicago Univ. Press. [600]

Jones, G. 1994. Scaling of wingbeat and echolocation pulse emission rates in bats: Why are aerial insectivorous bats so small? *Funct. Ecol.* 8:450–457. [868]

Jones, G. and J. M. V. Rayner. 1989. Foraging behavior and echolocation of wild horseshoe bats *Rhinolophus ferrumequinum* and *Rhinolophus hipposideros* (Chiroptera, Rhinolophidae). *Behav. Ecol. Sociobiol.* 25:183–191. [873]

Jouventin, P. 1982. *Visual and Voal Signals in Penguins, their Evolution and Adaptive Characteristics.* Berlin: Verlag Paul Parey. [787, 801]

Judd, T. 1995. The waggle dance of the honeybee: Which bees following a dancer successfully acquire the information? *J. Insect Behav.* 8:343–354. [831]

Judd, T. M. and P. W. Sherman. 1996. Naked mole-rats recruit colony mates to food sources. *Anim. Behav.* 52:957–969. [824]

Jürgens, U. 1979. Vocalization as an emotional indicator: A neuroethological study in the squirrel monkey. *Behaviour* 69:88–117. [522]

Kafka, W. A. 1987. Peripheral coding by graded overlapping reaction spectra? In *Olfaction and Taste IX* (Ed. by S. D. Roper and J. Atema), pp. 391–395. New York: Ann. N. Y. Acad. Sci. [307]

Kaissling, K. E. 1971. Insect olfaction. In *Handbook of Sensory Physiology, Vol. IV, Chemical Senses 1, Olfaction* (Ed. by L. M. Beidler). Berlin: Springer-Verlag. [308, 309, 318]

Kaissling, K. E. 1979. Recognition of pheromones by moths, especially in saturnids and *Bombyx mori*. In *Chemical Ecology: Odour Communication in Animals* (Ed. by F. J. Ritter), pp. 43–56. Amsterdam: Elsevier. [308]

Kalko, E. K. V. 1995. Insect pursuit, prey capture and echolocation in pipistrelle bats (Microchiroptera). *Anim. Behav.* 50:861–880. [866, 867, 868]

Kalmijn, A. J. 1988a. Detection of weak electric fields. In *Sensory Biology of Aquatic Animals* (Ed. by J. Atema, R. R. Fay, A. N. Popper and W. N. Tavolga), pp. 151–186. New York: Springer-Verlag. [337, 341, 345, 350]

Kalmijn, A. J. 1988b. Hydrodynamic and acoustic field detection. In *Sensory Biology of Aquatic Animals* (Ed. by J. Atema, R. R. Fay, A. N. Popper and W. N. Tavolga), pp. 83–130. New York: Springer-Verlag. [173]

Kalmring, K. and N. Elsner. 1985. *Acoustic and Vibrational Communication in Insects.* Heidelberg: Springer-Verlag. [173]

Kalmring, K., W. O. C. Kaiser and R. Kühne. 1985. Coprocessing of vibratory and auditory information in the CNS of different tettigoniids and locusts. In *Acoustic and Vibrational Communication in Insects* (Ed. by K. Kalmring and N. Elsner), pp. 193–202. Berlin: Paul Parey. [156]

Kauer, J. S. 1991. Contributions of topography and parallel processing to odour coding in the vertebrate olfactory pathway. *TINS* 14:79–85. [306]

Kaufmann, J. H. 1983. On the definitions and functions of dominance and territoriality. *Biol. Rev.* 58:1–20. [712]

Kavanagh, M. W. and D. Young. 1989. Bilateral symmetry of sound production in the mole cricket, *Gryllotalpa australis. J. Comp. Physiol. A* 166:43–49. [111]

Keeley, E. R. and J. W. Grant. 1993. Visual information, resource value and sequential assessment in convict cichlid (*Cichlasoma nigrofasciatum*) contests. *Behav. Ecol.* 4:345–349. [697]

Keidel, W. D. and W. D. Neff. 1974. *Auditory System. Vol. 5/1. Handbook of Sensory Physiology.* Berlin: Springer-Verlag. [173]

Kelemen, G. 1963. Comparative anatomy and performance of the vocal organ in vertebrates. In *Acoustic Behaviour of Animals* (Ed. by R.-G. Busnel), pp. 489–521. Amsterdam: Elsevier Publishing Company. [112]

Kennedy, J. S. 1986. Some current issues in orientation to odour sources. In *Mechanisms of Insect Olfaction* (Ed. by T. L. Payne, M. C. Birch and C. E. J. Kennedy), pp. 11–26. Oxford: Clarendon Press. [313, 318]

Kerr, W. E., M. Blum and H. M. Fales. 1981. Communication of food source between workers of *Trigona (Trigona) spinipes. Rev. Brasil Biol.* 41:619–623. [826]

Kessel, E. L. 1955. Mating activities of balloon flies. *Syst. Zool.* 4:97–104. [507]

Ketten, D. R. 1992. The marine mammal ear: Specializations for aquatic audition and echolocation. In *The Evolutionary Biology of Hearing* (Ed. by D. B. Webster, R. R. Fay and A. N. Popper), pp. 717–750. New York: Springer Verlag. [878]

Kick, S. A. and J. A. Simmons. 1984. Automatic gain control in the bat's sonar receiver and the neuroethology of echolocation. *J. Neurosci.* 4:2725–2737. [868]

Kiley, M. 1972. The vocalizations of ungulates, their causation and function. *Z. Tierpsych.* 31:171–222. [112]

Kim, Y.-G. 1995. Status signaling games in animal contests. *J. Theor. Biol.* 176:221–231. [702, 730]

King, A. P. and M. J. West. 1983. Epigenesis of cowbird song: A joint endeavor of males and females. *Nature* 305:704–706. [484]

Kingdon, J. 1988. What are face patterns and do they contribute to reproductive isolation in guenons? In *A Primate Radiation:*

Evolutionary Biology of the African Guenons (Ed. by A. Gautier-Hion, F. Bourlière and J.-P. Gautier), pp. 227–245. Cambridge: Cambridge Univ. Press. [582]

Kingsley, S. and S. Quegan. 1992. *Understanding Radar Systems.* London: McGraw-Hill. [856, 882]

Kinsler, L. E. and A. R. Frey. 1962. *Fundamentals of Acoustics.* 2nd Edition. New York: John Wiley and Sons. [40]

Kirk, J. T. O. 1983. *Light and Photosynthesis in Aquatic Ecosystems.* Cambridge: Cambridge Univ. Press. [205, 230, 238]

Kirk, J. T. O. 1994. *Light and Photosynthesis in Aquatic Ecosystems.* 2nd Edition. Cambridge: Cambridge Univ. Press. [198, 205, 238]

Kirkpatrick, M. 1982. Sexual selection and the evolution of female choice. *Evolution* 36:1–12. [760]

Kirkpatrick, M. 1986. Sexual selection and cycling parasites: A simulation study of Hamilton's hypothesis. *J. Theor. Biol.* 119:263–271. [761]

Kirschfeld, K. 1976. The resolution of lens and compound eyes. In *Neural Principles in Vision* (Ed. by F. Zettler and R. Weiler), pp. 354–370. Berlin: Springer-Verlag. [563]

Kivett, V. K. 1978. Integumentary glands of Columbian ground squirrels (*Spermophilus columbianus*): Sciuridae. *Can. J. Zool.* 56:374–381. [284]

Kleiman, D. 1977. Monogamy in mammals. *Q. Rev. Biol.* 52:39–69 [797]

Klump, G. M. and E. Curio. 1983. Why don't spectra of songbirds' vocalizations correspond with the sensitivity maxima of their absolute threshold curves? *Versh. Dtsch. Zool. Ges.* 1983:182. [127, 128, 139]

Klump, G. M. and M. D. Shalter. 1984. Acoustic behavior of birds and mammals in the predator context. I. Factors affecting the structure of alarm signals. II. The functional significance and evolution of alarm signals. *Z. Tierpsych.* 66:189–226. [127, 128, 139, 575, 603, 834, 839, 842, 850]

Klump, G. M., E. Kretzschmar and E. Curio. 1986. The hearing of an avian predator and its avian prey. *Behav. Ecol. Sociobiol.* 18:317–323. [839]

Knapp, R. A. and J. T. Kovach. 1991. Courtship as an honest indicator of male parental quality in the bicolour damselfish, *Stegastes partitus. Behav. Ecol.* 2:295–300. [660, 661, 769]

Knudsen, E. I. 1975. Spatial aspects of the electric fields generated by weakly electric fish. *J. Comp. Physiol.* 99:103–118. [335, 350]

Kobayashi, T. 1994. The biological function of snake mobbing by Siberian chipmunks. 1. Does it function as a signal to other conspecifics? *J. Ethol.* 12:89–95. [839]

Kober, R. and H.-U. Schnitzler. 1990. Information in sonar echoes of fluttering insects available for echolocating bats. *J. Acoust. Soc. Am.* 87:882–896. [875]

Kodric-Brown, A. 1989. Dietary carotenoids and male mating success in the guppy: An environmental component to female choice. *Behav. Ecol. Sociobiol.* 25:393–401. [485, 661, 765]

Konishi, M. 1993. Neuroethology of sound localization in the owl. *J. Comp. Physiol. A* 173:3–7. [529]

Konishi, M. and F. Nottebohm. 1969. Experimental studies in the ontogeny of avian vocalizations. In *Bird Vocalizations* (Ed. by R. A. Hinde), pp. 29–48. Cambridge: Cambridge Univ. Press. [479]

Koops, M. A. and J. W. A. Grant. 1992. Weight asymmetry and sequential assessment in convict cichlid contests. *Can. J. Zool.* 71:475–479. [697]

Krakauer, D. C. and R. A. Johnstone. 1995. The evolution of exploitation and honesty in animal communication: A model using artificial neural networks. *Philos. Trans. R. Soc. Lond. B Biol. Sci.*348:355–361. [673, 676, 701]

Kramer, H. G. and R. E. Lemon. 1983. Dynamics of territorial singing between neighboring song sparrows. *Behavior* 85:198–223. [737]

Kraus, W. F. 1989a. Surface wave communication during courtship in the giant waterbug, *Abedus indentatus* (Hemiptera, Belostomatidae). *J. Kansas Ent. Soc.* 62:316–328. [778, 779]

Kraus, W. F. 1989b. Is male back space limiting: An investigation into the reproductive demobraphy of the giant water bug, *Abedus indentatus* (Heteroptera, Belostomatidae). *J. Insect Behav.* 623–648. [780]

Krebs, J. R. 1977a. The significance of song repertoires: The Beau Geste hypothesis. *Anim. Behav.* 25:475–478. [735]

Krebs, J. R. 1977b. Song and territory in the great tit *Parus major.* In *Evolutionary Ecology* (Ed. by B. Stonehouse and C. Perrins), pp. 47–62. London: Macmillan [713]

Krebs, J. R. 1982. Territorial defense in the great tit (*Parus major*): Do residents always win? *Behav. Ecol. Sociobiol.* 11:185–194. [718, 720, 722]

Krebs, J. R. and N. B. Davies. 1993. *An Introduction to Behavioral Ecology.* 3d Edition. Oxford: Blackwell Publishing. [10, 136]

Krebs, J. R. and N. B. Davies. 1995. *An Introduction to Behavioural Ecology.* 4th Edition. London: Blackwell Scientific Publications. [615, 696]

Krebs, J. R. and R. Dawkins. 1984. Animal signals: Mind-reading and manipulation. In *Behavioural Ecology: An Evolutionary Approach* (Ed. by J. R. Krebs and N. B. Davies), pp. 380–402. Oxford: Blackwell Scientific Publications. [3, 12, 652]

Krebs, J. R., R. Ashcroft and M. Webber. 1978. Song repertoires and territory defence in the great tit. *Nature* 271:539–542. [713]

Krebs, J. R., R. Ashcroft and K. van Orsdol. 1981. Song matching in the great tit, *Parus major. Anim. Behav.* 29:918–923. [736]

Krebs, J. R., N. S. Clayton, S. D. Healy, D. A. Cristol, S. N. Patel and A. R. Jolliffe. 1996. The ecology of the avian brain: Food-storing memory and the hippocampus. *Ibis* 138:34–46. [566]

Kroodsma, D. E. 1977. Correlates of song organization among North American wrens. *Am. Nat.* 111:995–1008. [737]

Kroodsma, D. E. 1982. Learning and the ontogeny of sound signals in birds. In *Acoustic Communication in Birds* (Ed. by D. E. Kroodsma and E. H. Miller), pp. 1–23. New York: Academic Press. [478, 734]

Kroodsma, D. E. 1988. Song types and their use: Developmental flexibility of the male blue-winged warbler. *Ethology* 79:235–247. [487]

Kroodsma, D. E. and R. A. Canady. 1985. Differences in repertoire size, singing behavior, and associated neuroanatomy among Marsh Wren populations have a genetic basis. *Auk* 102:439–446. [558]

Kroodsma, D. E. and E. H. Miller. 1982. *Acoustic Communication in Birds.* New York: Academic Press. [615, 742]

Kroodsma, D. E. and E. H. Miller. 1996. *Ecology and Evolution of Acoustic Communication in Birds.* Ithaca, NY: Cornell Univ. Press. [615, 742]

Kroodsma, D. E., M. E. Vielliard and F. G. Stiles. 1996. Study of bird sounds in the neotropics: Urgency and opportunity. In *Ecology and Evolution of Acoustic Communication in Birds* (Ed. by D. E. Kroodsma and E. H. Miller), pp. 269–281. Ithaca, NY: Comstock Publishing Associates. [478]

Kruuk, H. 1972. *The Spotted Hyaena.* Chicago: Chicago Univ. Press. [552]

Kruuk, H. 1978. Social organization and territorial behaviour of the European badger *Meles Meles. J. Zool.* 185:205–212. [595]

Kummer, H. 1971. *Primate Societies: Group Techniques of Ecological Adaptation.* Chicago: Aldine Atherton, Inc. [759]

Kummer, H. 1995. *In Quest of the Sacred Baboon.* Princeton, NJ: Princeton Univ. Press. [759]

Kvernemo, C. 1996. Temperature affects operational sex ratio and intensity of male-male competition: An experimental study of sand gobies. *Behav. Ecol.* 7:208–212. [780]

Ladich, F. 1989. Sound production by the river bullhead, Cottus gobio L. (Cottidae, Teleostei). *J. Fish Biol.* 35:531–538. [112]

Ladich, F. and H. Kratochvil. 1989. Sound production by the marmoreal goby Proterorhinus marmoratus (Pallas) (Gobiidae, Teleostei). *Allgem. Zool. Physiol. Tiere* 93:501–504. [112]

Laland, K. N., P. J. Richerson and R. Boyd. 1993. Animal social learning: Towards a new theoretical approach. In *Perspectives in Ethology* (Ed. by P. P. G. Bateson, P. H. Klopfer and N. S. Thompson), pp. 249–277. New York: Pergamon Press. [475]

Lambert, D. M. and H. Spencer. 1995. *Speciation and the Recognition Concept.* Baltimore: Johns Hopkins Univ. Press. [816]

Lambrechts, M. and A. A. Dhondt. 1988. The anti-exhaustion hypothesis: A new hypothesis to explain song performance and song switching in the great tit. *Anim. Behav.* 36:327–334. [735]

Land, M. F. 1980. Compound eyes: Old and new mechanisms. *Nature* 287:681–686. [248]

Land, M. F. 1981. Optics and Vision in Invertebrates. In *Comparative Physiology and Evolution of Vision in Invertebrates. B: Invertebrate Visual Centers and Behavior I* (Ed. by H. Autrum). Berlin: Springer-Verlag. [243, 244, 245, 265, 267, 268, 272, 275, 278]

Lande, R. 1981. Models of speciation by sexual selection on polygenic traits. *Proc. Natl. Acad. Sci. USA* 78:3721–3725. [760, 791]

Lande, R. 1982. Rapid origin of sexual isolation and character divergence in a cline. *Evolution* 36:213–223. [760, 791]

Langen, T. 1996a. The mating system of the white-throated magpie jay *Calocitta formosa* and Greenwood's hypothesis for sex-biased dispersal. *Ibis* 138:506–513. [666]

Langen, T. A. 1996b. Skill acquisition and the timing of natal dispersal in the white-throated magpie jay, *Calocitta formosa. Anim. Behav.* 51:575–588. [804]

Lank, D. L. and C. M. Smith. 1987. Conditional lekking in ruff (*Philomachus pugnax*). *Behav. Ecol. Sociobiol.* 20:137–146. [740]

Lank, D. L., C. M. Smith, O. Hanotte, T. Burke and F. Cooke. 1995. Genetic polymorhism for alternative mating behaviour in lekking male ruff *Philomachus pugnax. Nature* 378:59–62. [740]

Laska, M. and R. Hudson. 1995. Ability of female squirrel monkeys (*Saimiri sciureus*) to discriminate between conspecific urine odors. *Ethology* 99:39–52. [489]

Laughlin, S. B. 1990. Coding efficiency and visual processing. In *Vision: Coding and Efficiency.* (Ed. by C. Blakemore), pp. 25–31. Cambridge: Cambridge Univ. Press. [416, 418]

Laughlin, S. B. 1994. Matching coding, circuits, cells and molecules to signals—General principles of retinal design in the fly's eye. *Progr. Retinal Eye Res.* 13: 165–196. [416, 418]

Laurence, B. D. and J. A. Simmons. 1982. Measurements of atmospheric attenuation of ultrasonic frequencies and the significance for echolocation by bats. *J. Acoust. Soc. Am.* 71:585–590. [855]

Laverack, M. S. 1988. The diversity of chemoreceptors. In *Sensory Biology of Aquatic Animals* (Ed. by J. Atema, R. R. Fay, A. N. Popper and W. N. Tavolga), pp. 287–312. New York: Springer-Verlag. [306]

Le Grand, Y. 1970. *An Introduction to Photobiology.* New York: American Elsevier Publishing. [205]

Lee, J. 1989. Bioluminescence. In *The Science of Photobiology* (Ed. by K. C. Smith), pp. 391–417. New York: Plenum Press. [221]

Legendre, L. and P. Legendre. 1983. *Numerical Ecology.* Amsterdam: Elsevier Publishing. [464]

Legendre, P. and A. Vaudor. 1991. *The R Package: Multidimensional Analysis and Spatial Analysis.* Montreal: Département de Sciences Biologiques, Université de Montréal. [466]

Leger, D. W. and D. H. Owings. 1978. Responses to alarm calls by California ground squirrels: Effects of call structure and maternal status. *Behav. Ecol. Sociobiol.* 3:177–186. [607]

Lehrman, D. S. 1965. Interaction between internal and external environments in the regulation of the reproductive cycle of the ring dove. In *Sex and Behavior* (Ed. by F. A. Beach), pp. 355–380. New York: John Wiley. [792]

Leibovic, K. N. 1990. *The Science of Vision.* New York: Springer-Verlag. [278]

Leimar, O. and M. Enquist. 1984. Effects of asymmetries in owner-intruder conflicts. *J. Theor. Biol.* 111:475–491. [690, 721, 723]

Leimar, O., M. Enquist and B. Sillén-Tullberg. 1986. Evolutionary stability of aposematic coloration and prey unprofitability: A theoretical analysis. *Am. Nat.* 128:469–490. [673, 846]

Leimar, O., S. Austad and M. Enquist. 1991. A test of the sequential assessment game: Fighting in the bowl and doily spider *Frontinella pyramitela. Evolution* 45:862–874. [691, 693, 724]

Lemon, R. E. 1973. Nervous control of the syrinx in white-throated sparrows (*Zonotrichia albicollis*). *J. Zool.* 171:229–262. [103]

Lemon, R. E., C. W. Dobson and P. G. Clifton. 1993. Songs of American redstarts (*Setophaga ruticilla*): Sequencing rules and their relationships to repertoire size. *Ethology* 93:198–210. [487]

Lemon, R. E., S. Monette and D. Roff. 1987. Song repertoires of American warblers (Parulinae): Honest advertising or assessment? *Ethology* 74:265–284. [360, 487]

Lenington, S. 1994. Of mice, men and the MHC. *Trends in Ecology and Evolution* 9:455–456. [483]

Leslie, C. A. and D. P. Dubey. 1992. Carotene and natural killer cell activity. *Fed. Proc.* 41:331. [765]

Leston, D. and J. W. S. Pringle. 1963. Acoustical behaviour of hemiptera. In *Acoustic Behaviour of Animals* (Ed. by R.-G. Busnel), pp. 391–411. Amsterdam: Elsevier. [517]

Levin, R. N. 1996. Song behaviour and reproductive strategies in a duetting wren, *Thryothorus nigricapillus*: II. Playback experiments. *Anim. Behav.* 5222:1107–1117. [798]

Levins, R. 1968. *Evolution in Changing Environments.* Princeton, NJ: Princeton Univ. Press. [371, 385]

Lewicki, M. S. and M. Konishi. 1995. Mechanisms underlying the sensitivity of songbird forebrain neurons to temporal order. *Proc. Natl. Acad. Sci. USA* 922:5582–5586. [529]

Lewis, B. 1983. Directional cues for auditory localization. In *Bioacoustics: A Comparative Approach* (Ed. by B. Lewis), pp. 233–257. London: Academic Press. [173]

Lewis, D. B. and D. M. Gower. 1980. *Biology of Communication.* New York: Wiley. [11]

Lindauer, M. 1961. *Communication Among Social Bees.* Cambridge, MA: Harvard Univ. Press. [826, 831, 832, 850]

Lindemann, W. 1955. Uber die Jugendentwicklung beim Luchs (*Lynx lynx* Kerr.) und bei der Wildkatze (*Felis s. silvestrus* Schreb.). *Behaviour* 8:1–45. [568]

Linn, C. E. J. and W. L. Roelofs. 1995. Pheromone communication in moths and its role in the speciation process. In *Speciation and the Recognition Concept* (Ed. by D. M. Lambert and H. G. Spencer), pp. 263–300. Baltimore: Johns Hopkins Univ. Press. [790]

Linsenmaier, K. E. 1987. Kin recognition in subsocial arthropods, in particular in the desert isopod *Hemilepistus reaumuri.* In *Kin Recognition in Animals* (Ed. by D. J. C. Fletcher and C. D. Michener), pp. 121–208. New York: John Wiley & Sons. [802]

Liou, L. W. and T. P. Price. 1994. Speciation by reinforcement of premating isolation. *Evolution* 48:1451–1459. [760, 791]

Lloyd, J. E. 1966. Studies on the flash communication system in *Photinus* fireflies. *Univ. Mich. Museum Zool. Misc. Publ.* 130:1–95. [590]

Lloyd, J. E. 1977. Bioluminescence and communication. In *How Animals Communicate* (Ed. by T. A. Sebeok), pp. 164—183. Bloomington, IN: Indiana Univ. Press. [581]

Lloyd, J. E. 1983. Bioluminescence and communication in insects. *Annu. Rev. Entomol.* 28:131–160. [221]

Lloyd, J. E. 1986. Firefly communication and deception: "Oh, what a tangled web." In *Perspectives on Human and Nonhuman Deceit* (Ed. by R. W. Mitchell and N. S. Thompson), pp. 113–128. Albany, NY: State Univ. of New York Press. [382]

Lobel, P. S. 1992. Sounds produced by spawning fishes. *Environ. Biol. Fishes* 33:351–358. [793]

Lockner, F. R. and D. E. Murrish. 1975. Interclavicular air sac pressures and vocalization in mallard ducks Anas platyrhynchos. *Comp. Biochem. Physiol.* 52A:183–187. [112]

Lockner, F. R. and O. M. Youngren. 1976. Functional syringeal anatomy of the mallard. I. In situ electro-myograms during ESB elicited calling. *Auk* 93:324–342. [112]

Löfstedt, C., N. J. Vickers, W. L. Roelofs and T. C. Baker. 1989. Diet related courtship success in the oriental fruit moth, *Grapholita molesta* (Tortricidae). *Oikos* 55:402–408. [485]

Löhrl, H. 1958. Das Verhalten des Kleibers. *Z. Tierpsych.* 15:191–252. [835]

Lorenz, K. 1941. Vergleichende Bewegungsstudien an Anatinen. *J. Ornithol.* 89:19–29. [512]

Lorenz, K. 1953. Die Entwicklung der vergleichenden Verhaltensforschung in den letzten 12 Jahren. *Zool. Anz. Suppl.* 16:36–58. [510]

Lorenz, K. 1971. Comparative studies of the motor patterns of Anatinae. In *Studies in Animal and Human Behavior* (Ed. by R. Martin), pp. 14–114. Cambridge, MA: Harvard Univ. Press. [490]

Losey, G. S. 1978. Information theory and communication. In *Quantitative Ethology* (Ed. by P. W. Colgan), pp. 43–78. New York: John Wiley. [411, 418]

Lotem, A. 1993. Learning to recognize nestlings is maladaptive for cuckoo *Cuculus canorus* hosts. *Nature* 362:743–745. [800]

Lozano, G. A. 1994. Carotenoids, parasites, and sexual selection. *Oikos* 70:309–311. [765]

Lundgren, B. and N. Højerslev. 1971. Daylight measurements in the Sargasso Sea. *Københavns Univ. Inst. Fysisk Oceanog. Rep.* 14:1–33. [198]

Luttbeg, B. 1996. A comparative Bayes tactic for mate assessment and choice. *Behav. Ecol.* 7: 451–460. [448]

Lythgoe, J. N. 1979. *The Ecology of Vision.* Oxford: Clarendon Press. [191, 198, 224, 230, 234, 238, 257, 262, 278]

Lythgoe, J.N. 1988. Light and vision in the aquatic environment. In: *Sensory Biology of Aquatic Animals* (Ed. by J. Atema, R. R. Fay and A. N. Popper), pp. 57–82. Heidelberg: Springer-Verlag. [221, 223, 230]

Macdonald, D. 1984. *The Encyclopedia of Mammals.* Oxford: Equinox (Oxford) Ltd. [602, 615, 749]

Macdonald, D. W. 1980. Patterns of scent marking with urine and feces among carnivore communities. *Symp. Zool. Soc. Lond.* 45:107–139. [738]

Macdonald, D. W. 1985a. The rodents IV: Suborder Hystricomorpha. In *Social Odours in Mammals* (Ed. by R. E. Brown and D. W. Macdonald), pp. 480–506. Oxford: Clarendon Press. [288]

Macdonald, D. W. 1985b. The carnivores: order Carnivora. In *Social Odours in Mammals* (Ed. by R. E. Brown and D. W. Macdonald), pp. 619–722. Oxford: Clarendon Press. [288, 739]

Macedonia, J. M. and C. S. Evans. 1993. Variation among mammalian alarm call systems and the problem of meaning in animal signals. *Ethology* 93:177–197. [835, 850]

Macmillan, N. A. and C. Creelman. 1991. *Detection Theory: A User's Guide.* Cambridge: Cambridge Univ. Press. [437, 438, 454]

MacNally, R. C. and D. Young. 1981. Song energetics of the bladder cicada. *J. Exp. Biol.* 90:185–196. [459, 550]

Maddison, W. P. and D. R. Maddison. 1992. *MacClade: Analysis of Phylogeny and Character Evolution. Version 3.* Sunderland, MA: Sinauer Associates. [502]

Madsen, T. and J. Loman. 1987. On the role of colour display in the social and spatial organization of male rainbow lizards (*Agama agama*). *Amphib. Reptil.* 8:365–372. [682]

Maeda, T. and N. Masataka. 1987. Locale-specific vocal behaviour of the tamarin (*Saguinus l. labiatus*). *Ethology* 75:25–30. [474]

Magnhagen, C. 1991. Predation risk as a cost of reproduction. *TREE* 6:183–186. [544, 560, 570]

Magrath, R. D. 1990. Hatching asynchrony in altricial birds. *Biol. Rev.* 65:587–622. [804]

Magurran, A. E. and A. Higham. 1988. Information transfer across fish shoals under predator threat. *Ehtology* 93:177–197. [836]

Maier, E. J. 1994. Ultraviolet vision in a Passeriform bird: From receptor spectral sensitivity to overall spectral sensitivity in *Leiothrix lutea. Vision Res.* 34:1415–1418. [259]

Mairy, F. 1976. Tentative models of sound-figure and coo production in the ring dove (*Streptopelia risoria*). *Biophon.* 4:2–5. [112]

Malacarne, G., M. Cucco and S. Camanni. 1991. Coordinated visual displays and vocal duetting in different ecological situations among Western palearctic non-passerine birds. *Ethol. Ecol. Evol.* 3:207–219. [798]

Mammen, D. L. and S. Nowicki. 1981. Individual differences and within-flock convergence in chickadee calls. *Behav. Ecol. Sociobiol.* 9:179–186. [479, 484]

Mangel, M. 1990. A dynamic habitat selection game. *Math. Bios.* 100:241–248. [623]

Mangel, M. and C. W. Clark. 1988. *Dynamic Modeling in Behavioral Ecology.* Princeton, NJ: Princeton Univ. Press. [375, 385, 647]

Manning, A. and M. S. Dawkins. 1992. *An Introduction to Animal Behaviour.* 4th Edition. Cambridge: Cambridge Univ. Press. [11]

Mantel, N. 1967. The detection of disease clustering and a generalized regression approach. *Cancer Res.* 27:209–220. [466]

Mappes, J., R. V. Alatalo, J. Kotiaho and S. Parri. 1996. Viability costs of condition-dependent sexual male display in a drumming wolf spider. *Proc. R. Soc. Lond. B Biol. Sci.* 263:785–789. [549]

Marchetti, K. 1993. Dark habitats and bright birds illustrate the role of the environment in species divergence. *Nature* 362:149–152. [230, 231, 597, 664]

Marchetti, K. 1994. Costs to host defense and the persistence of parastic cuckoos. *Proc. Natl. Acad. Sci. USA* 248:41–45. [800]

Marden, J. H. and R. A. Rollins. 1994. Assessment of energy reserves by damselflies engaged in aerial contests for mating territories. *Anim. Behav.* 48:1023–1030. [688, 724]

Marden, J. H. and J. K. Waage. 1990. Escalated damselfly territorial contests are energetic wars of attrition. *Anim. Behav.* 39:954–959. [687]

Margoliash, D. and M. Konishi. 1985. Auditory representation of autogenous song in the song system of white-crowned sparrows. *Proc. Nat. Acad. Sci.* 82:5997–6000. [734]

Markl, H. 1965. Stridulation in leaf-cutting ants. *Science* 149:1392–1393. [840]

Markl, H. 1968. Die Verstndigung durch Stridulationssignale bei Blattschneiderameisen. II. Erzeugung und Eigenshaften der Signale. *Z. Vergl. Physiol.* 60:103–150. [115, 134, 139]

Markl, H. 1983. Vibrational communication. In *Neuroethology and Behavioral Physiology* (Ed. by F. Huber and H. Markl), pp. 332–353. Berlin: Springer. [111, 139, 567]

Markl, H. 1985. Manipulation, modulation, information, cognition: Some of the riddles of communication. In *Experimental Behavioral Ecology and Sociobiology (Fortsch. Zool, Vol. 31)* (Ed. by B. Hölldobler and M. Lindauer), pp. 164–194. Stuttgart: Gustav Fisher Verlag. [2]

Markl, H. and B. Hölldobler. 1978. Recruitment and food-retrieving behavior in Novomessor (Formicidae, Hymenoptera). II. Vibration signals. *Behav. Ecol. Sociobiol.* 4:183–216. [115, 139, 825]

Marler, C. A. and M. C. Moore. 1988. Evolutionary costs of aggression revealed by testosterone manipulations in free-living male lizards. *Behav. Ecol. Sociobiol.* 23:21–26. [664]

Marler, C. A. and M. C. Moore. 1989. Time and energy costs of aggression in testosterone-implanted free-living male mountain spiny lizards (*Sceloporous jarrovi*). *Phys. Zool.* 62:1334–1350. [664]

Marler, P. 1955. Characteristics of some animal calls. *Nature* 176:6–8. [603, 839]
Marler, P. 1959. Developments in the study of animal communication. In *Darwin's Biological Work* (Ed. by P. R. Bell), pp. 150–206. New York: Cambridge Univ. Press. [575, 603, 607]
Marler, P. 1961. The logical analysis of animal communication. *J. Theor. Biol.* 1:295–317. [573, 6175
Marler, P. 1969. Tonal quality of bird sounds. In *Bird Vocalizations: Their Relations to Current Problems in Biology and Psychology* (Ed. by R. A. Hinde), pp. 5–18. Cambridge: Cambridge Univ. Press. [74, 103]
Marler, P. 1970. A comparative approach to vocal learning: Song development in white-crowned sparrows. *J. Comp. Physiol. Psychol.* 71 (Suppl):1–25. [479]
Marler, P. 1977. The evolution of communication. In *How Animals Communicate* (Ed. by T. A. Sebeok), pp. 45–70. Bloomington, IN: Indiana Univ. Press. [2, 496]
Marler, P. 1984. Song learning: Innate species differences in the learning process. In *The Biology of Learning* (Ed. by P. Marler and H. S. Terrace), pp. 289–309. Berlin: Dahlem Konferenzen. Springer-Verlag. [475, 478]
Marler, P. and W. J. Hamilton. 1967. *Mechanisms of Anim. Behav.*. New York: John Wiley and Sons. [preface]
Marler, P. and S. Peters. 1982. Developmental overproduction and selective attrition: New processes in the epigenesis of bird song. *Dev. Psychobiol.* 15:369–378. [479]
Marler, P. and S. Peters. 1988. The role of song phonology and syntax in vocal learning preferences in the song sparrow, *Melospiza melodia. Ethology* 77:125–149. [461, 479]
Marler, P. and R. Pickert. 1984. Species-universal microstructure in a learned birdsong: The swamp sparrow (*Melopspiza georgiana*). *Anim. Behav.* 32:673–689. [479]
Marler, P. and M. Tamura. 1964. Culturally transmitted patterns of vocal behavior in sparrows. *Science* 146:1483–1486. [732]
Marler, P. and R. Tenaza. 1977. Signalling behaviour of apes with special reference to vocalization. In *How Animals Communicate* (Ed. by T. Sebeok), pp. 965–1033. Bloomington, IN: Indiana Univ. Press. [823]
Marler, P., Dufty, A. and Pickert, R. 1986a. Vocal communication in the domestic chicken: I. Does a sender communicate information about the quality of a food referent to a receiver? *Anim. Behav.* 34:188–193. [820]
Marler, P., Dufty, A. and Pickert, R. 1986b. Vocal communication in the domestic chicken: II. Is a sender sensitive to the presence and nature of a receiver? *Anim. Behav.* 34:194–198. [820]
Marler, P., C. S. Evans and M. D. Hauser. 1992. Animal signals: Motivational, referential, or both? In *Nonverbal Vocal Communication* (Ed. by H. Papoušek, U. Jürgens and M. Papoušek), pp. 66–86. Cambridge: Cambridge Univ. Press. [496, 818, 850]
Marten, K. and P. Marler. 1977. Sound transmission and its significance for animal vocalizations. I. Temperate habitats. *Behav. Ecol. Sociobiol.* 2:271–290. [117, 139]
Marten, K., D. B. Quine and P. Marler. 1977. Sound transmission and its significance for animal vocalizations. II. Tropical habitats. *Behav. Ecol. Sociobiol.* 2:291–302. [117, 139]
Martin, A. A. 1972. Evolution of vocalizations in the genus *Bufo*. In *Evolution in the genus Bufo* (Ed. by W. F. Blair), pp. 279–309. Austin, TX: Univ. of Texas Press. [556]
Martin, A. C. 1994. A brief history of the "feature detector". *Cerebral Cortex* 4:1–7. [528]
Martin, W. F. 1971. Mechanics of sound production in toads of the genus *Bufo*: Passive elements. *J. Exp. Zool.* 176:273–294. [88, 90, 98, 112]
Martin, W. F. 1972. Evolution of vocalization in the toad genus *Bufo*. In *Evolution in the Genus Bufo* (Ed. by W. F. Blair), pp. 279–309. Austin: Univ. of Texas Press. [99, 112]
Martin, W. F. and C. Gans. 1972. Muscular control of the vocal tract during release signalling in the toad *Bufo valliceps*. *J. Morph.* 137:1–28. [112]
Martins, E. P. 1993a. A comparative study of the evolution of *Sceloporus* push-up displays. *Am. Nat.* 142:994–1018. [504, 585]
Martins, E. P. 1993b. Contextual use of the push-up display by the sagebrush lizard, *Sceloporus graciosus*. *Anim. Behav.* 45:25–36. [464]
Martins, E. P. 1996. *Phylogenies and the Comparative Method in Anim. Behav.*. New York: Oxford Univ. Press. [554]
Marzluff, J. M. and R. P. Balda. 1992. *The Pinyon Jay: Behavioral Ecology of a Colonial and Cooperative Corvid*. London: T. & A. D. Poyser. [520, 812]
Mason, R. T., H. M. Fales, T. H. Jones, L. K. Pannell, J. W. Chinn and D. Crews. 1989. Sex pheromones in snakes. *Science* 245:290–293. [284, 794]
Masters, W. M. 1988. Prey interception: Predictive and nonpredictive strategies. In *Animal Sonar* (Ed. by P. E. Nachtigall and P. W. B. Moore), pp. 467–470. New York: Plenum Press. [876]
Masters, W. M., K. A. S. Raver and K. A. Kazial. 1995. Sonar signals of big brown bats, *Eptesicus fuscus*, contain information about individual identity, age and family affiliation. *Anim. Behav.* 50:1243–1260. [489, 880]
Masters, W. W. and S. C. Jacobs. 1989. Target detection and range resolution by the big brown bat (*Eptesicus fuscus*) using normal and time-reversed model echoes. *J. Comp. Physiol. A* 166:65–73. [870]
Mathis, A., D. P. Chivers and R. J. F. Smith. 1995. Chemical alarm signals: Predator deterrents or predator attractants? *Am. Nat.* 145:994–1005. [847]
Maynard Smith, J. 1965. The evolution of alarm calls. *Am. Nat.* 99:59–63. [836]
Maynard Smith, J. 1966. Sympatric speciation. *Am. Nat.* 100:637–650. [791]
Maynard Smith, J. 1976a. Evolution and the theory of games. *Am. Sci.* 64:41–45. [679]
Maynard Smith, J. 1976b. Sexual selection and the handicap principle. *J. Theor. Biol.* 57:239–242. [653, 761]
Maynard Smith, J. 1977. Parental investment: A prospective analysis. *Anim. Behav.* 25:1–9. [778]
Maynard Smith, J. 1979. Game theory and the evolution of behaviour. *Proc. R. Soc. Lond. B Biol. Sci.* 205:475–488. [679, 703, 716]
Maynard Smith, J. 1982a. *Evolution and the Theory of Games*. Cambridge: Cambridge Univ. Press. [628, 647, 680, 717, 730]
Maynard Smith, J. 1982b. Do animals convey information about their intentions? *J. Theor. Biol.* 97:1–5. [703]
Maynard Smith, J. 1985. Sexual selection, handicaps and true fitness. *J. Theor. Biol.* 115:1–8. [761]
Maynard Smith, J. 1991. Honest signalling: The Sir Philip Sidney game. *Anim. Behav.* 42:1034–1035. [656, 669]
Maynard Smith, J. 1994. Must reliable signals always be costly? *Anim. Behav.* 47:1115–1120. [665, 730]
Maynard Smith, J. and D. G. C. Harper. 1988. The evolution of aggression: Can selection generate variability? *Philos. Trans. R. Soc. Lond. B Biol. Sci.* 319:557–570. [663, 666]
Maynard Smith, J. and G. A. Parker. 1976. The logic of asymmetric contests. *Anim. Behav.* 24:159–175. [680, 689, 719]
Mayr, E. 1963. *Animal Species and Evolution*. Cambridge, MA: Harvard Univ. Press. [791]
McArthur, P. D. 1986. Similarity of playback songs to self song as a determinant of response strength in song sparrows (*Melospiza melodia*). *Anim. Behav.* 34:199–207. [734]
McCarty, J. P. 1996. The energetic cost of begging in nestling passerines. *Auk* 113:178–188. [667, 805]
McComb, K., C. Packer and A. Pusey. 1994. Roaring and numerical assessment in contests between groups of female lions, *Panthera leo*. *Anim. Behav.* 47:379–387. [597]
McCowan, B. and D. Reiss. 1995. Quantitative comparisons of whistle repertoires from captive adult bottlenose dolphins (Delphinidae, *Tursiops truncatus*): A re-evaluation of the signature whistle hypothesis. *Ethology* 100:194–209. [489]

McCracken, G. F. 1984. Communal nursing in Mexican free-tailed bat maternity colonies. *Science* 223:1090–1091. [801]
McCracken, G. and M. Gustin. 1991. Nursing behavior in Mexican free-tailed bat maternity colonies. *Ethology* 89:305–321. [801]
McCracken, G. F. 1993. Locational memory and female-pup reunions in Mexican free-tailed bat maternity colonies. *Anim. Behav.* 45:811–813. [801]
McCracken, G. F. and M. F. Gustin. 1987. Batmom's daily nightmare. *Natural History* 96:66–73. [801]
McDonald, D. 1989. Cooperation under sexual selection: Age-graded changes in a lekking bird. *Am. Nat.* 134:709–730. [552]
McFarland, D. 1981. *The Oxford Companion to Animal Behaviour*. Oxford: Oxford Univ. Press. [507, 510, 512, 515, 535, 793]
McFarland, D. 1985. *Animal Behavior*. Menlo Park, CA: Benjamin Cummings. [506, 509, 518, 535]
McGillvray, W. B. and H. Levenson. 1986. Distribution of food within broods of barn swallows. *Wilson Bull.* 98:286–291. [666]
McGregor, P. K. 1989. Pro-active memory interference in neighbour recognition by a songbird. In *Acta XIX Congressus Internationalis Ornithologici* (Ed. by H. Ouellet), pp. 356–374. Ottawa: Univ. of Ottawa Press. [566]
McGregor, P. K. 1991. The singer and the song: On the receiving end of bird song. *Biol. Rev.* 66:57–81. [714, 742]
McGregor, P. K. 1993. Signalling in territorial systems: A context for individual identification, ranging and eavesdropping. *Philos. Trans. R. Soc. Lond. B Biol. Sci.* 340:237–244. [712, 727, 735, 737, 738, 742]
McGregor, P. K. and M. I. Avery. 1986. The unsung songs of great tits (*Parus major*): Learning neighbor's songs for discrimination. *Behav. Ecol. Sociobiol.* 18:311–316. [566]
McGregor, P. K. and J. B. Falls. 1984. The response of western meadowlarks (*Sturnella neglecta*) to the playback of undegraded and degraded songs. *Can. J. Zool.* 62:2215–2218.
McGregor, P. K. and J. R. Krebs. 1984. Song learning and deceptive mimicry. *Anim. Behav.* 32:280–287. [728]
McGregor, P. K. and G. W. M. Westby. 1992. Discrimination of individually charcteristic electric organ discharges by a weakly electric fish. *Anim. Behav.* 43:977–986. [474, 596]
McGregor, P. K., J. R. Krebs and C. M. Perrins. 1981. Song repertoires and lifetime reproductive success in the great tit (*Parus major*). *Am. Nat.* 118:149–159. [734]
McGregor, P. K., J. R. Krebs and L. M. Ratcliffe. 1983. The reaction of great tits (*Parus major*) to playback of degraded and undegraded songs: The effect of familiarity with the stimulus song type. *Auk* 100:898–906. [728]
McGregor, P. K., T. Dabelsteen, M. Shepherd and S. B. Pedersen. 1992. The signal value of matched singing in Great Tits: Evidence from interactive playback experiments. *Anim. Behav.* 43:987–998. [737]
McKaye, K. and N. M. KcKaye. 1977. Communal care and kidnapping of young by parental cichlids. *Evolution* 31:674–681. [801]
McKinney, F. 1965. The spring behavior of wild Steller eiders. *Condor* 67:273–290. [545]
McKitrick, M. C. 1993. Phylogenetic constraint in evolutionary theory. *Annu. Rev. Ecol. Syst.* 24:307–330. [543]
McNamara, J. M. and A. I. Houston. 1980. The application of statistical decision theory to animal behaviour. *J. Theor. Biol.* 85:673–690. [418]
McNamara, J. M. and A. I. Houston. 1985. Optimal foraging and learning. *J. Theor. Biol.* 117:231–249. [418]
McNamara, J. M. and A. I. Houston. 1987. Memory and the efficient use of information. *J. Theor. Biol.* 125:385–395. [418]
McShane, L. J., J. A. Estes, M. L. Riedman and M. M. Staedler. 1995. Repertoire, structure, and individual variation of vocalizations in the sea otter. *J. Mammal.* 76:414–427. [520]
Medvin, M. B., P. K. Stoddard and M. D. Beecher. 1992. Signals for parent-offspring recognition: Strong sib-sib call similarity in cliff swallows but not in barn swallows. *Ethology* 90:17–28. [787, 788]
Medvin, M. B., P. K. Stoddard and M. D. Beecher. 1993. Signals for parent-offspring recognition: A comparative analysis of the begging calls of cliff swallows and barn swallows. *Anim. Behav.* 45:841–850. [450]
Meisenheimer, J. 1921. *Geschlecht und Geschlecter im Tierreich*. Jena: G Fischer. [507]
Menne, D. and H. Hackbarth. 1986. Accuracy of distance measurement in the bat *Eptesicus fuscus*: Theoretical aspects and computer simulations. *J. Acoust. Soc. Am.* 79:386–397. [857, 860, 870]
Menzel, R. 1975. Color receptors in insects. In *The Compound Eye and Vision of Insects* (Ed. by G. A. Horridge), pp. 121–153. Oxford: Clarendon Press. [278]
Menzel, R. 1981. Spectral Sensitivity and Colour Vision in Invertebrates. In *Comparative Physiology and Evolution of Vision in Invertebrates. A: Invertebrate Photoreceptors* (Ed. by H. Autrum), pp. 503–580. Berlin: Sringer-Verlag. [260, 278]
Menzel, R. and W. Backhaus. 1989. Color vision honey bees: Phenomena and physiological mechanisms. In *Facets of Vision* (Ed. by D. G. Stavenga and R. C. Hardie), pp. 281–297. Berlin: Springer-Verlag. [252]
Metzgar, L. A. 1967. An experimental comparison of screech own predation on resident and transient white-footed mice. *J. Mammal.* 48:387–390. [724]
Meyer, A., J. M. Morrissey and M. Schartl. 1994. Recurrent origin of a sexually selected trait in *Xiphophorus* fishes inferred from a molecular phylogeny. *Nature* 368:539–542. [504]
Meyer, J. H. 1982. Behavioral responses of weakly electric fish to complex impedances. *J. Comp. Physiol.* 145:459–470 [350]
Meyerreicks, A. J. 1960. Comparative breeding behavior of four species of North American herons. *Nuttal. Ornithol. Club Publ.* 2:1–158. [504, 507]
Michael, R. P. and R. W. Bonsall. 1977. Chemical signals and primate behavior. In *Chemical Signals in Vertebrates* (Ed. by D. Müller-Schwarze and M. M. Mozell), pp. 251–271. New York: Plenum Press. [285]
Michael, R. P., E. B. Keverne and R. W. Bonsall. 1971. Pheromones: Isolation of male sex attractants from a female primate. *Science* 172:964–966. [285]
Michelsen, A. 1966. On the evolution of tactile stimulatory actions in long-horned beetles (Cerambycidae, Coleoptera). *Z. Tierpsych.* 23:257–266. [516]
Michelsen, A. 1978. Sound reception in different environments. In *Perspectives in Sensory Ecology* (Ed. by B. A. Ali), pp. 345–373. New York: Plenum Press. [139]
Michelsen, A. 1979. Insect ears as mechanical systems. *Am. Sci.* 67:696–706. [153, 159, 173]
Michelsen, A. 1983. Biophysical basis of sound communication. In *Bioacoustics: A Comparative Approach* (Ed. by B. Lewis), pp. 3–38. London: Academic Press. [111, 153, 173]
Michelsen, A. 1992. Hearing and sound communication in small animals: Evolutionary adaptations to the laws of physics. In *The Evolutionary Biology of Hearing* (Ed. by D. B. Webster, R. R. Fay and A. N. Popper), pp. 61–77. New York: Springer-Verlag. [111]
Michelsen, A. and O. Larsen. 1985. Hearing and sound. In *Comprehensive Insect Physiology, Biochemistry, and Pharmacology* (Ed. by G. A. Kerkut and L. I. Gilbert), pp. 496–556. Oxford: Pergamon Press. [153, 173]
Michelsen, A. and H. Nocke. 1974. Biophysical aspects of sound communication in insects. *Adv. Insect Physiol.* 10:247–296. [87, 111, 134, 150, 153, 159, 173]
Michelsen, A., F. Fink, M. Gogala and D. Traue. 1982. Plants as transmission channels for insect vibrational songs. *Behav. Ecol. Sociobiol.* 11:269–281. [112, 115, 122, 135, 139]
Michelsen, A., W. H. Kirchner and M. Lindauer. 1986. Sound and vibration signals in the dance language of the honeybee, *Apis mellifera*. *Behav. Ecol. Sociobiol.* 18:207–212. [112, 139, 666, 831]

Michener, C. D. and B. H. Smith. 1987. Kin recognition in primitively eusocial insects. In *Kin Recognition in Animals* (Ed. by D. J. C. Fletcher and C. D. Michener), pp. 209–242. Chichester: John Wiley & Sons. [807]

Michod, R. E. and O. Hasson. 1990. On the evolution of reliable indicators of fitness. *Am. Nat.* 135:788–808. [669]

Milinksi, M. and T. C. M. Bakker. 1990. Female sticklebacks use male coloration in mate choice and hence avoid parasitized males. *Nature* 344:330–333. [661, 765]

Milinski, M. and T. Bakker. 1992. Costs influence sequential mate choice in sticklebacks, *Gasterosteus aculeatus. Proc. R. Soc. Lond. B Biol. Sci.* 2250:229–233. [448, 560]

Miller, D. B. 1977. Two-voice phenomenon in birds: Further evidence. *Auk* 94:567–572. [112]

Miskimen, M. 1951. Sound production in passerine birds. *Auk* 68:493–504. [112]

Mitani, J. C. and T. Nishida. 1993. Contexts and social correlates of long-distance calling by male chimpanzees. *Anim. Behav.* 45:735–746. [811]

Möglich, M. and B. Hölldobler. 1975. Communication and orientation during foraging and emigration in the ant *Formica fusca. J. Comp. Physiol.* 101:275–288. [825]

Møller, A. P. 1987a. Social control of deception among status signalling house sparrows *Passer domesticus. Behav. Ecol. Sociobiol.* 20:307–311. [663, 664]

Møller, A. P. 1987b. Variation in badge size in male house sparrows *Passer domesticus*: Evidence for status signalling. *Anim. Behav.* 35:1637–1644. [663, 664]

Møller, A. P. 1988a. Badge size in the house sparrow *Passer domesticus*: Effects of intra- and intersexual selection. *Behav. Ecol. Sociobiol.* 22:373–378. [664]

Møller, A. P. 1988b. False alarm calls as a means of resource usurpation in the great tit *Parus major. Ethology* 79:25–30. [561, 668]

Møller, A. P. 1989. Natural and sexual selection on a plumage signal of status and on morphology in house sparrows, *Passer domesticus. J. Evol. Biol.* 2:125–140. [664]

Møller, A. P. 1989. Viability costs of male tail ornaments in a swallow. *Nature* 339:132–135. [552, 570, 766]

Møller, A. P. 1990a. Fluctuating asymmetry in male sexual ornaments may reliably reveal male quality. *Anim. Behav.* 40:1185–1187. [766]

Møller, A. P. 1990b. Effects of a haematophagoous mite on the barn swallow (*Hirundo rustica*): A test of the Hamilton and Zuk hypothesis. *Evolution* 44: 711–784 . [766]

Møller, A. P. 1992. Female swallow preference for symmetrical male sexual ornaments. *Nature* 357: 238–240. [766]

Møller, A. P. 1996. The cost of secondary sexual characters and the evolution of cost-reducing traits. *Ibis* 138:112–119. [552, 553, 570]

Møller, A. P. and F. de Lope. 1994. Differential costs of a secondary sexual character: An experimental test of the handicap principle. *Evolution* 48:1676–1683. [552]

Møller, A. P. and A. Pomiankowski. 1993. Why have birds got multiple sexual ornaments? *Behav. Ecol. Sociobiol.* 3: 167–176. [774, 775]

Møller, A. P., P. H. Harvey, S. Nee, A. F. Read, I. C. Cuthill, J. P. Swaddle and M. S. Witter. 1993. Fluctuating asymmetry. *Nature* 363: 217–218. [766]

Møller, A. P., F. de Lope and J. M. Lopez Caballero. 1995. Foraging costs of a tail ornament: Experimental evidence from two populations of barn swallows *Hirundo rustica* with different degrees of sexual size dimorphism. *Behav. Ecol. Sociobiol.* 37: 289–296. [552, 766]

Mollon, J. D. 1991. Uses and evolutionary origins of primate colour vision. In *Evolution of the Eye and Visual System* (Ed. by J. R. Cronly-Dillon and R. L. Gregory), pp. 306–319. Boca Raton, FL: CRC Press. [262]

Mondloch, C. J. 1995. Chick hunger and begging affect parental allocation of feeding in pigeons. *Anim. Behav.* 49:601–613. [666, 804]

Montgomerie, R. and R. Thornhill. 1989. Fertility advertisement in birds: A means of inciting male-male competition? *Ethology* 81:209–220. [776]

Moore, A. J. 1988. Female preferences, male social status and sexual selection in *Nauphoeta cinerea. Anim. Behav.* 36:303–305. [485, 774]

Moore, B. P. 1968. Studies on the chemical composition and function of the cephalic gland secretion in Australian termites. *J. Insect Physiol.*, 14: 33–39.

Moore, M. C., D. K. Hews and R. Knapp. (In Press). Hormonal control and evolution of alternative male phenotypes: Generalizations of models of sexual differentiation. *Am. Zool.* [740]

Moran, D. T. 1987. Evolutionary patterns in sensory receptors. In *Olfaction and Taste IX* (Ed. by S. D. Roper and J. Atema), pp. 1–8. New York: Ann. N. Y. Acad. Sci. [305]

Morris, D. 1956. The feather postures of birds and the problem of the origin of social signals. *Behaviour* 9:75–113. [514]

Morris, D. 1957. "Typical intensity" and its relation to the problem of ritualisation. *Behaviour* 11:1–12. [519]

Morris, D. 1959. The comparative ethology of grassfinches (*Erythrurae*) and mannikins (*Amadinae*). *Proc. Zool. Soc. London* 131:389–439. [509, 511]

Morris, M. R., L. Bass and M. J. Ryan. 1995. Assessment and individual recognition of opponents in the pygmy swordtails *Ziphophorus nigrensis* and *X. multilineatus. Behav. Ecol. Sociobiol.* 37:303–310. [697]

Morris, R. J. 1986. The acoustic faculty of dolphins. In *Research on Dolphins* (Ed. by M. M. Bryden and R. Harrison), pp. 369–399. Oxford: Clarendon Press. [878]

Morse, D. and E. Meighen. 1986. Pheromone biosynthesis and role of functional groups in pheromone specificity. *J. Chem. Ecol.* 12:335–351. [282, 526, 558, 583]

Morse, D. H. 1970. Territorial and courtship songs of birds. *Nature* 226:26–35. [598]

Morse, P. M. and K. U. Ingard. 1968. *Theoretical Acoustics.* New York: McGraw-Hill. [40]

Morton, E. S. 1975. Ecological sources of selection on avian sounds. *Am. Nat.* 109:17–34. [126, 127, 128]

Morton, E. S. 1977. On the occurrence and significance of motivation-structural rules in some bird and mammal sounds. *Am. Nat.* 111:855–869. [519, 520, 535, 652]

Morton, E. S. 1982. Grading, discreteness, redundancy and motivational-structural rules. In *Evolution and Ecology of Acoustic Communication in Birds.* (Ed. by D. E. Kroodsma and E. H. Miller), pp. 183–212. New York: Academic Press. [520, 727, 735]

Morton, E. S. 1986. Predictions from the ranging hypothesis for the evolution of long distance signals in birds. *Behaviour* 99:65–86. [728]

Morton, E. S. and M. D. Shalter. 1977. Vocal response to predators in pair-bonded Carolina wrens. *Condor* 79:222–227. [836]

Moss, C. J. 1988. *Elephant Memories.* Boston: Houghton Mifflin. [806]

Moynihan, M. 1955. Some aspects of reproductive behaviour in the black-headed gull (*Larus ridibundus ridibundus* L.) and related species. *Behaviour Suppl.* 4:1–201. [504]

Moynihan, M. 1966. Communication in *Callicebus. J. Zool. London* 150:77–127. [797]

Moynihan, M. 1970. The control, suppression, decay, disappearance and replacement of displays. *J. Theor. Biol.* 29:85–112. [490, 651]

Moynihan, M. 1982. Why is lying about intentions rare during some kinds of contests. *J. Theor. Biol.* 97:7–12. [703]

Moynihan, M. 1985. *Communication and Noncommunication by Cephalopods.* Bloomington, IN: Indiana Univ. Press. [218]

Moynihan, M. and A. F. Rodaniche. 1982. The behavior and natural history of the Caribbean reef squid *Sepioteuthis sepioidae. Fortschr. Verhaltensforsch.* 25:1–150. [612]

Müller-Schwarze, D. and M. M. Mozell. 1977. *Chemical Signals in Vertebrates.* New York: Plenum Press. [318]

Mundinger, P. C. 1970. Vocal imitation and individual recognition of finch calls. *Science* 168:480–482. [479, 484]

Mundinger, P. C. 1979. Call learning in the Carduelinae: Ethological and systematic considerations. *System. Zool.* 28:270–283. [479, 484]

Munsell, A. H. 1975. *Munsell Book of Color.* Baltimore: Munsell Color Co. [210]

Murlis, J. and C. D. Jones. 1981. Fine-scale structure of odour plumes in relation to insect orientation to distant pheromone and other attractant sources. *Physiol. Entomol.* 6:71–86. [302, 303]

Murphy, C. G. 1994. Chorus tenure of male barking treefrogs, *Hyla gratiosa. Anim. Behav.* 48:763–777. [772]

Mykytowycz, R., E. R. Kesterman, S. Gambale and M. L. Dudzinski. 1976. A comparison of the effectiveness of the odors of rabbits, *Oryctolagus cuniculus,* in enhancing territorial confidence. *J. Chem. Ecol.* 2:13–24. [718]

Myrberg, A. A. 1978. Ocean noise and the behaviour of marine animals: Relationships and implications. In *Effects of Noise on Wildlife* (Ed. by J. L. Fletcher and R.-G. Busnel), pp. 169–208. New York: Academic Press. [139]

Myrberg, A. A. 1980. Fish bioacoustics: Its relevance to the 'not so silent world.' *Env. Biol. Fish.* 5:297–304. [139]

Myrberg, A. A. 1981. Sound communication and interception in fishes. In *Hearing and Sound Communication in Fishes* (Ed. by W. N. Tavolga, A. N. Popper and R. R. Fay), pp. 395–426. New York: Springer-Verlag. [112, 128, 139, 173]

Myrberg, A. A. and R. J. Riggio. 1985. Acoustically mediated individual recognition by a coral reef fish (*Pomacentrus partitus*). *Anim. Behav.* 33:411–416. [488]

Myrberg, A. A., S. J. Ha and M. J. Shamblott. 1993. The sounds of bicolor damselfish (*Pomacentrus partitus*)- predictors of body size and a spectral basis for individual recognition and assessment. *J. Acoust. Soc. Am.* 94:3067–3070. [489]

Nachtigall, P. E. and P. W. B. Moore. 1988. *Animal Sonar.* New York: Plenum Press. [882]

Naguib, M. 1995. Auditory distance assessment of singing conspecifics in Carolina wrens: The role of reverberation and frequency-dependent attenuation. *Anim. Behav.* 50: 1297–1307. [728]

Nassau, K. 1983. *The Physics and Chemistry of Color.* New York: John Wiley. [205, 211, 215, 216, 221, 238, 549]

Negus, V. E. 1949. *The Comparative Anatomy and Physiology of the Larynx.* New York: Hafner Publishing. [112]

Nelson, D. A. 1984. Communication of intentions in agonistic contexts by the pigeon guillemot, *Cepphus columba. Behaviour* 88:145–189. [522,658, 703]

Nelson, D. A. 1992. Song overproduction and selective attrition lead to song sharing in the Field Sparrow (*Spizella pusilla*). *Behav. Ecol. Sociobiol.* 30:415–424. [732]

Nelson, D. A. and P. Marler. 1989. Categorical perception of a natural stimulus continuum: Birdsong. *Science* 244:976–978. [470]

Nelson, R. J. 1995. *An Introduction to Behavioral Endocrinology.* Sunderland, MA: Sinauer Associates. [794, 795]

Neumeyer, C. 1991. Evolution of color vision. In *Vision and Visual Dysfunction* (Ed. by J. R. Cronly-Dillon and R. L. Gregory), pp. 282–305. London: Macmillan, Houndsmills. [256, 259, 262]

Neumeyer, C. 1992. Tetrachromatic color vision in goldfish: Evidence from color mixture experiments. *J. Comp. Physiol. A* 171:639–649. [259]

Neumeyer, C. and K. Arnold. 1989. Tetrachromatic color vision in the goldfish becomes trichromatic under white adaptation light of moderate intensity. *Vision Res.* 29: 1719–1727. [261]

Neuweiler, G. 1984. Foraging, echolocation, and audition in bats. *Naturwiss.* 71:446–455. [854]

Neuweiler, G. 1989. Foraging ecology and audition in echolocating bats. *Trends Ecol. Evol.* 4:160–166. [871, 873, 875, 880, 882]

Neuweiler, G. 1990. Auditory adaptations for prey capture in echolocating bats. *Physiol. Rev.* 70:615–641. [860, 871, 875, 876, 882]

Neuweiler, G., W. Metzner, U. Heilmann, R. Rübsamen, M. Eckrich and H. H. Costa. 1987. Foraging behaviour and echolocation in the rufous horseshoe bat of Sri Lanka. *Behav. Ecol. Sociobiol.* 20:53–67. [873]

Newman, E. A. and P. H. Hartline. 1982. The infrared "vision" of snakes. *Sci. Am.* 246:116–127. [181]

Nieh, J. C. and D. W. Roubik. 1995. A stingless bee (*Melipona panamica*) indicates food location without using a scent trail. *Behav. Ecol. Sociobiol.* 37:63–70. [827]

Nielsen, B. M. B. and S. L. Vehrencamp. 1995. Responses of song sparrows to song-type matching via interactive playback. *Behav. Ecol. Sociobiol.* 37:109–117. [599, 737]

Nilsson, J. A. and M. Svensson. 1993. Fledging in altricial birds: Parental manipulation or sibling competition? *Anim. Behav.* 46:379–386. [804]

Norberg, A. 1994. Swallow tail streamer is a mechanical device for self-deflection of tail leading edge, enhancing aerodynamic efficiency and flight manoeuverability. *Proc. R. Soc. Lond. B Biol. Sci.* 257:227–233. [553]

Norris, K. 1990a. Female choice and the evolution of the conspicuous plumage colouration of monogamous male great tits. *Behav. Ecol. Sociobiol.* 26:129–138. [664, 765]

Norris, K. 1990b. Female choice and the quality of parental care in the great tit *Parus major. Behav. Ecol. Sociobiol.* 27:275–281. [664, 765]

Norris, K. 1993. Heritable variation in a plumage indicator of viability in male great tits *Parus major. Nature* 362:537–539. [765]

Norris, K. S. and G. W. Harvey. 1974. Sound transmission in the porpoise head. *J. Acoust. Soc. Am.* 56:659–664. [112, 878]

Northmore, D. M. P. and C. A. Dvorak. 1979. Contrast sensitivity and acuity in the goldfish. *Vision Res.* 19:255–261. [269]

Nottebohm, F. 1971. Neural lateralization of vocal control in a passerine bird. I. Song. *J. Exp. Zool.* 177:229–262. [103, 106, 112]

Nottebohm, F. 1972. Neural lateralization of vocal control in a passerine bird. II. Subsong, calls, and a theory of learning. *J. Exp. Zool.* 179:35–49. [103, 112]

Nottebohm, F. and M. E. Nottebohm. 1976. Left hypoglossal dominance in the control of canary and white-crowned sparrow song. *J. Comp. Physiol.* 108:171–192. [103, 112]

Nowicki, S. 1987. Vocal tract resonances in oscine bird sound production: Evidence from birdsongs in a helium atmosphere. *Nature (Lond.)* 325:53–55. [88, 109, 112]

Nowicki, S. 1989a. Peripheral lateralization of bird-song reanalyzed: Comparison of multiple techniques for unilateral disablement of syringeal function in sparrows. In *Neural Mechanisms of Behavior* (Ed. by J. Erber, R. Menzel, H.-J. Pflueger and D. Todt), pp. 121. New York: Georg Thieme & Verlag. [112]

Nowicki, S. 1989b.Vocal plasticity in captive black-capped chickadees: The acoustic basis and rate of call convergence. *Anim. Behav.* 37:64–73. [479, 484]

Nowicki, S. and R. Capranica. 1986. Bilateral syringeal interaction in vocal production of an oscine bird sound. *Science* 231: 1297–1299. [105, 112]

Nowicki, S. and P. Marler. 1988. How do birds sing? *Music Percep.* 5:391–426. [112]

Nowicki, S. and D. A. Nelson. 1990. Defining natural categories in acoustic signals: Comparison of three methods applied to 'chick-a-dee' call notes. *Ethology* 86:89–101. [465]

Nowicki, S., P. Marler, A. Maynard and S. Peters. 1992. Is the tonal quality of birdsong learned? *Ethology* 90:225–235. [479]

O'Brien, T. G. 1993. Allogrooming behaviour among adult female wedge-capped capuchin monkeys. *Anim. Behav.* 46:499–510. [809]

Obrist, M. K. 1995. Flexible bat echolocation: The influence of individual, habitat and conspecifics on sonar signal design. *Behav. Ecol. Sociobiol.* 36:207–219. [489, 880]

O'Connell, R. J. and A. J. Grant. 1987. Electrophysiological responses of olfactory receptor neurons to stimulation with mixtures of individual pheromone components. In *Olfaction and Taste IX* (Ed. by S. D. Roper and J. Atema), pp. 79–85. New York: Ann. N. Y. Acad. Sci. [308]

O'Connell, S. M. and G. Cowlishaw. 1994. Infanticide avoidance, sperm competition and mate choice: The function of copulation calls in female baboons. *Anim. Behav.* 48:687–694. [776, 777]

Offutt, G. C. 1974. Structures for the detection of acoustic stimuli in the Altantic codfish, *Gadus morhua*. *J. Acoust. Soc. Am.* 56:665–671. [173]

Okubo, A. 1980. *Diffusion and Ecological Problems: Mathematical Models*. Berlin: Springer-Verlag. [298]

Olson, D. J., A. C. Kamil, R. P. Balda and P. J. Nims. 1995. Performance of four seed-caching corvid species in operant tests of nonspatial and spatial memory. *J. Comp. Psychol.* 109:173–181. [566]

Olsson, M. 1992. Contest success in relation to size and residency in the sand lizards, *Lacerta agilis*. *Anim. Behav.* 44:386–388. [719]

Olsson, M. 1994. Nuptial coloration in the sand lizard, *Lacerta agilis*: An intra-sexually selected cue to fighting ability. *Anim. Behav.* 48:607–613. [682, 683]

Olsson, M. 1995. Forced copulation and costly female resistance behavior in the Lake Eyre dragon, *Ctenophorus maculosus*. *Herpetologica*, 51: 19–24. [756

Orians, G. H. and G. M. Christman. 1968. A comparative study of the behavior of red-winged, tricolored, and yellow-headed blackbirds. *Univ. Calif. Publ. Zool.* 84:1–81. [591]

Otte, D. 1974. Effects and functions in the evolution of signaling systems. *Annu. Rev. Ecol. Syst.* 5:385–417. [4, 499, 535, 573, 615]

Otte, D. and J. A. Endler. 1989. *Speciation and its Consequences*. Sunderland, MA: Sinauer Associates. [816]

Owens, I. P. F. and P. M. Bennett. 1994. Mortality costs of parental care and sexual dimorphism in birds. *Proc. R. Soc. Lond. B Biol. Sci.* 257:1–8. [549]

Owens, I. P. F. and I. R. Hartley. 1991. 'Trojan sparrows': Evolutionary consequences of dishonest invasion for the badges of status model. *Am. Nat.* 138:1187–1205. [663, 669]

Page, R. E., G. E. Robinson and M. K. Fondrk. 1989. Genetic specialists, kin recognition and nepotism in honey-bee colonies. *Nature* 338:576–579. [807]

Pak, M. A. and S. J. Cleveland. 1991. The high spatial frequency limits and resolving power of the visual system of the pigeon. *EEG-EMC Z. Elektronenzeph. Elektronmyo. verwand. Gebiete* 22:194–199. [269]

Parker, G. A. 1974. Assessment strategy and the evolution of fighting behaviour. *J. Theor. Biol.* 47:223–243. [689]

Parker, G. A. 1979. Sexual selection and sexual conflict. In *Sexual Selection and Reproductive Competition in Insects*. (Ed. by M. S. Blum and N. A. Blum), pp. 123–166. New York: Academic Press. [751, 752, 754, 755, 762, 782]

Parker, G. A. 1982. Phenotype limited evolutionarily stable strategies. In *Current Problems in Sociobiology* (Ed. by B. R. Bertram, T. H. Clutton-Brock, R. I. M. Dunbar, D. I. Rubenstein and R. Wrangham), pp. 173–201. Cambridge: Cambridge Univ. Press. [644, 782]

Parker, G. A. 1983. Mate quality and mating decisions. In *Mate Choice* (Ed. by P. Bateson), pp. 141–166. Cambridge: Cambridge Univ. Press. [448, 782]

Parker, G. A. 1984a. Evolutionarily stable strategies. In *Behavioural Ecology: An Evolutionary Approach*. 2nd Edition. (Ed. by J. R. Krebs and N. B. Davies), pp. 30–61. Oxford: Blackwell Scientific Publications. [640, 647]

Parker, G. A. 1984b. Sperm competition and the evolution of animal mating strategies. In *Sperm Competition and the Evolution of Animal Mating Systems*. (Ed. by R. L. Smith), pp. 1–60. New York: Academic Press. [744, 782]

Parker, G. A. and D. I. Rubenstein. 1981. Role assessment, reserve strategy, and acquisition of information in asymmetric animal conflicts. *Anim. Behav.* 29:221–240. [685]

Parker, G. A., R. R. Baker and V. G. F. Smith. 1972. The origin and evolution of gamete dimorphism and the male-female phenemenon. *J. Theor. Biol.* 36:529–533. [744]

Parker, G. H. 1948. *Animal Color Changes and their Neurohumours*. Cambridge: Cambridge Univ. Press. [217, 238]

Parker, P. G., T. A. Waite and M. D. Decker. 1995. Kinship and association in communally roosting black vultures. *Anim. Behav.* 49:395–401. [821]

Parsons, J. and L. F. Baptista. 1980. Crown color and dominance in the white-crowned sparrow. *Auk* 97:807–815. [664]

Partridge, B. L. and T. J. Pitcher. 1980. The sensory basis of fish schools: Relative roles of lateral line and vision. *J. Comp. Physiol.* 135:315–325. [813]

Paton, D. and P. G. Caryl. 1986. Communication by agonistic displays. I. Variation in information content between samples. *Behaviour* 98:213–239. [653, 658, 701, 703]

Paul, C. R. and S. A. Sasar. 1987. *Introduction to Electromagnetic Fields*. New York: McGraw-Hill. [350]

Payne, R. and D. Webb. 1971. Orientation by means of long range acoustic signalling in baleen whales. *Annu. N.Y. Acad. Sci.* 188:110–141. [126, 139]

Payne, R. B. 1982. Ecological consequences of song matching: Breeding success and intraspecific song mimicry in indigo buntings. *Ecology* 63:401–411. [734]

Payne, R. B. 1983. The social context of song mimicry: Song-matching dialects in indigo buntings. *Anim. Behav.* 31:788–805. [734]

Payne, R. B. 1996. Song traditions in indigo buntings: Origin, improvisation, dispersal and extinction in cultural evolution. In *Ecology and Evolution of Acoustic Communication in Birds* (Ed. by D. E. Kroodsma and E. H. Miller), pp. 198–220. Ithaca, NY: Cornell Univ. Press. [732]

Payne, T. L., M. C. Birch and C. E. J. Kennedy. 1986. *Mechanisms of Insect Olfaction*. Oxford: Clarendon Press. [318]

Payne, R. B., L. L. Payne and S. M. Doehlert. 1988. Biological and cultural success of song memes in indigo buntings. *Ecology* 69:104–117. [734]

Pereira, M. E. and J. M. Macedonia. 1991. Ringtailed lemur antipredator calls denote predator class, not response urgency. *Anim. Behav.* 41:543–544. [835]

Perrone, M. 1980. Factors affecting the incidence of distress calls in passerines. *Wilson Bull.* 92:404–408. [847]

Perry, G. C., R. L. S. Patterson, H. J. H. Macfie and G. C. Stimson. 1980. Pig courtship behaviour: Pheromonal property of androstene steroids in male submaxillary secretion. *Anim. Prod.* 31:191–199. [285, 524]

Peters, R. 1980. *Mammalian Communication: A Behavioral Analysis of Meaning*. Monterey, CA: Brooks/Cole. [522, 615, 810, 816]

Petit, O. and B. Thierry. 1994. Aggressive and peaceful interventions in conflicts in Tonkean macaques. *Anim. Behav.* 48:1427–1436. [808]

Petrie, M. 1988. Intraspecific variation in structures that display competitive ability: Large animals invest relatively more. *Anim. Behav.* 36:1174–1179. [664]

Petrie, M. 1994. Improved growth and survival of offspring of peacocks with more elaborate trains. *Nature* 371:598–599. [766]

Petrie, M. and A. P. Møller. 1991. Laying eggs in others' nests: Intraspecific brood parasitism in birds. *TREE* 6:315–320. [800]

Petrie, M., T. Halliday and C. Sanders. 1991. Peahens prefer peacocks with elaborate trains. *Anim. Behav.* 41:323–331. [768]

Petrinovich, L. 1990. Avian song development: Methodological and conceptual issues. In *Contemporary Issues in Comparative*

Psychology (Ed. by D. A. Dewsbury), pp. 340–359. Sunderland, MA: Sinauer Associates. [475]

Petrovskaya, E. D. 1969. On frequency selectivity of cercal trichoid receptors in the cricket *Gryllus domesticus*. *Z. Evol. Biokhim. Fiziol.* 5:337–338. [142]

Pettifor, R. A. 1990. The effects of avian mobbing on a potential predator, the European festrel, Falco tinnunculus. *Anim. Behav.* 39:821–827. [839, 844]

Pettigrew, J. D. 1991. Evolution of binocular vision. In *Evolution of the Eye and Visual System* (Ed. by J. R. Cronly-Dillon and R. L. Gregory), pp. 271–283. Boca Raton, FL: CRC Press. [273]

Pettigrew, J. D. and S. P. Collin. 1995. Terrestrial optics in an aquatic eye: The sandlance, *Limnichthyes fasciatus* (Creediidae, Teleostei). *J. Comp. Physiol. A* 177:397–408. [273]

Pettigrew, J. D., B. Dreher, C. S. Hopkins, M. J. McCall and M. Brown. 1988. Peak density and distribution of ganglion cells in the retinae of microchiropteran bats: Implications for visual acuity. *Brain Behav. Evol.* 32:39–56. [269]

Pfeiffer, W. 1977. The distribution of fright reaction and alarm substance cells in fishes. *Copeia* 1977:653–665. [846]

Pfennig, D. W. 1990. "Kin recognition" among spadefoot toad tadpoles: A side-effect of habitat selection? *Anim. Behav.* 44:785–798. [790]

Pfennig, D. W. and P. K. Sherman. 1995. Kin recognition. *Sci. Am.* 272:98–103. [806]

Pidgeon, R. 1981. Calls of the galah *Cacatua roseicapilla* and some comparisons with four other species of Australian parrot. *Emu* 81:158–168. [478]

Pierce, J. R. 1980. *An Introduction to Information Theory*. New York: Dover Publications. [418, 496]

Piercy, J. E., T. F. W. Embleton and L. C. Sutherland. 1977. Review of noise propagation in the atmosphere. *J. Acoust. Soc. Am.* 61:1402–1418. [139]

Piersma, T. and J. Veen. 1988. An analysis of the communication function of attack calls in little gulls. *Anim. Behav.* 36:773–779. [703]

Piersma, T., L. Zwarts and J. H. Bruggemann. 1990. Behavioural aspects of the departure of waders before long-distance flights: Flocking, vocalizations, flight paths and diurnal timing. *Ardea* 78:157–184. [810]

Pilleri, G. 1983. *The Sonar System of the Dolphins*. Endeavor, New Series 7:59–64. [112]

Pimentel-Souza, F. and N. Fernandes-Souza. 1987. Electric organ discharge rhythms and social interactions in a weakly electric fish, *Rhamphichthys rostratus*, (Rhamphichthydae, Gymnotiformes) in an aquarium. *Exp. Biol.* 46:169–176. [350]

Pitcher, T. J. and J. K. Parrish. 1993. Functions of shoaling behaviour in teleosts. In *Behaviour of Teleost Fishes* (Ed. by T. J. Pitcher), pp. 363–440. London: Chapman & Hall. [812]

Pitcher, T. J. and C. J. Wyche. 1983. Predator avoidance behaviour of sand-eel schools: Why schools seldom split? In *Predators and Prey in Fishes* (Ed. by D. L. G. Noakes, B. G. Lindquist, G. S. Helfman and J. A. Ward), pp. 193–204. The Hague: Junk. [812, 813]

Pockington, R. and L. M. Dill. 1995. Predation on females or males: Who pays for bright male traits? *Anim. Behav.* 49:1122–1124. [560]

Podos, J., S. Peters, T. Rudnicky, P. Marler and S. Nowicki. 1992. The organization of song repertoires in song sparrows: Themes and variations. *Ethology* 90:89–106. [465]

Pola, Y. V. and C. T. Snowdon. 1975. The vocalizations of pygmy marmosets (*Cebuella pygmaea*). *Anim. Behav.* 23:826–842. [823]

Pollak, G. D. 1993. Some comments on the proposed perception of phase and nanosecond time disparities by echolocating bats. *J. Comp. Physiol.* A 172:523–531. [870]

Pomiankowski, A. 1987. Sexual selection: The handicap principle does worksometimes. *Proc. R. Soc. Lond. B Biol. Sci.* 231:123–145. [761]

Pomiankowski, A. 1988. The evolution of female mate preferences for male genetic quality. *Oxford Surv. Evol. Biol.* 5:136–184. [761]

Pomiankowski, A. and Y. Iwasa. 1993. Evolution of multiple sexual ornaments by Fisher's process of sexual selection. *Proc. R. Soc. Lond. B Biol. Sci.* 2253:173–181. [774]

Pomiankowski, A., Y. Iwasa and S. Nee. 1991. The evolution of costly mate preference I. Fisher and biased mutation. *Evolution* 45:1422–1430. [761]

Poole, J. H. 1989. Announcing intent: The aggressive state of musth in African elephants. *Anim. Behav.* 37:140–152. [704]

Poole, J., K. B. Payne, W. R. Jr. Lanagabauer and C. J. Moss. 1988. The social context of some very low-frequency calls of African elephants. *Behav. Ecol. Sociobiol.* 22:385–392. [586]

Popp, J. W. 1987. Risk and effectiveness in the use of agonistic displays by American goldfinches. *Behaviour* 103:141–156. [656, 658, 704]

Popper, A. N., P. H. Rogers, W. M. Saidel and M. Cox. 1988. Role of the fish ear in sound processing. In *Sensory Biology of Aquatic Animals* (Ed. by J. Atema, R. R. Fay, A. N. Popper and W. N. Tavolga), pp. 687–710. New York: Springer-Verlag. [173]

Pressley, P. H. 1981. Pair formation and joint territoriality in a simultaneous hermaphrodite: The coral feef fish *Serranus tigrinus*. *Z. Tierpsych.* 56:33–46. [800]

Prestwich, G. D. 1985. Communication in insects. II. Molecular communication of insects. *Q. Rev. Biol.* 60:437–456. [607]

Prestwich, K. N., K. E. Brugger and M. Topping. 1989. Energy and communication in three speices of hylid frogs: Power input, power output and efficiency. *J. Exp. Biol.* 144:53–80. [550]

Price, T. D., D. Schulter and N. E. Heckman. 1993. Sexual selection when the female directly benefits. *Biol. J. Linn. Soc.* 48:187–211. [763]

Pringle, J. W. W. 1954. A physiological analysis of cicada song. *J. Exp. Biol.* 31:525–560. [112]

Proctor, H. C. 1991. Courtship in the water mite *Neumania papillator*: Males capitalize on female adaptations for predation. *Anim. Behav.* 42:589–598. [533]

Proctor, H. C. 1992. Sensory exploitation and the evolution of male mating behaviour: A cladistic test using water mites (Acari: Parasitengona). *Anim. Behav.* 44:745–752. [533]

Promislow, D. E. L. 1992. Costs of sexual selection in natural populations of mammals. *Proc. R. Soc. Lond. B Biol. Sci.* 247:203–210. [549]

Promislow, D. E. L., R. Montgomerie and T. E. Martin. 1992. Mortality costs of sexual dimorphism in birds. *Proc. R. Soc. Lond. B Biol. Sci.* 250:143–150. [549]

Provost, E. 1991. Non-nestmate kin recognition in the ant *Leptothorax lichtensteini*: Evidence that genetic factors regulate colony recognition. *Behav. Genet.* 21:151–167. [807]

Pruett-Jones, S. G. and M. A. Pruett-Jones. 1990. Sexual selection through female choice in Lawes parotia, a lek-mating bird of paradise. *Evolution* 44:486–501. [134]

Pruett-Jones, S. and M. Pruett-Jones. 1994. Sexual competition and courtship disruptionsWhy do male bowerbirds destroy each other's bowers? *Anim. Behav.* 47:607–602. [770]

Pulliam, H. R. and T. Caraco. 1984. Living in groups: Is there an optimal group size? In *Behavioural Ecology: An Evolutionary Approach*. 2nd Edition. (Ed. by J. R. Krebs and N. B. Davies), pp. 122–147. Oxford: Blackwell Scientific Publications. [837]

Pumphrey, R. F. 1948. The theory of the fovea. *J. Exp. Biol.* 25:299–312. [270]

Pye, J. D. 1983. Echolocation and countermeasures. In *Bioacoustics* (Ed. by B. Lewis), pp. 407–429. London: Academic Press. [865, 882]

Quastler, H. 1958. A primer on information theory. In *Symposium on Information Theory in Biology* (Ed. by H. P. Yockey and R. L. Platzman), pp. 3–49. New York: Pergamon Press. [411, 418]

Quay, W. B. 1977. Structure and function of skin glands. In *Chemical Signals in Vertebrates* (Ed. by D. Müller-Schwarze and M. M. Mozell), pp. 10–16. New York: Plenum Press. [283, 317]

Rabbitt, R. D. 1990. A hierarchy of examples illustrating the acoustic coupling of the eardrum. *J.Acoust. Soc. Am.* 87:2566–2582. [173]

Rabenold, P. P. 1987. Recruitment to food in black vultures: Evidence for following from communal roosts. *Anim. Behav.* 35:1775–1785. [821]

Raiffa, H. 1968. *Decision Analysis: Introductory Lectures on Choices under Uncertainty*. Reading, MA: Addison-Wesley. [420, 454]

Raikow, R. J. 1986. Why are there so many kinds of passerine birds? *Syst. Zool.* 35:255–259. [556, 792]

Rand, W. M. and A. S. Rand. 1976. Agonistic behavior in nesting iguanas: A stochastic analysis of despotic settlement dominated by the minimization of energy. *Z. Tierpsych.* 40:279–299. [411, 703]

Randall, J. A. 1989. Individual footdrumming signatures in banner-tailed kangaroo rats *Dipodomys spectabilis*. *Anim. Behav.* 38:620–630. [727]

Randall, J. A. 1995. Modification of footdrumming signatures by kangaroo rats: Changing territories and gaining new neighbours. *Anim. Behav.* 49:1227–1237. [725, 727]

Randall, J. A. and C. M. Stevens. 1987. Footdrumming and other anti-predator responses in the bannertail kangaroo rat (*Dipodomys spectabilis*). *Behav. Ecol. Sociobiol.* 20:187–194. [842]

Rasa, O. A. E. 1989a. Behavioural parameters of vigilance in the dwarf mongoose: Social acquisition of a sex-biased role. *Behaviour* 110:125–145. [836, 841]

Rasa, O. A. E. 1989b. The costs and effectiveness of vigilance behaviour in the dwarf mongoose: Implications for fitness and optimal group size. *Ethology, Ecology and Evolution* 1:265–282. [836]

Ratcliffe, L. and K. Otter. 1996. Sex differences in song recognition. In *Ecology and Evolution of Acoustic Communication in Birds* (Ed. by D. E. Kroodsma and E. H. Miller), pp. 339–355. Ithaca, NY: Comstock Publishing Associates. [482]

Read, A. F. and D. M. Weary. 1992. The evolution of bird song: Comparative analyses. *Phil. Trans. R. Soc. Lond. B* 165–187. [609, 734, 735]

Real, L. 1990. Search theory and mate choice. I. Models of single sex discrimination. *Am. Nat.* 136:376–405. [448]

Real, L. 1991. Search theory and mate choice. II. Mutual interaction, assortative mating, and equilibrium variation in male and female fitness. *Am. Nat.* 138:901–917. [448]

Redondo, T. and E. Castro. 1992a. Signalling of nutritional need by magpie nestlings. *Ethology* 92:193–204. [666, 804]

Redondo, T. and F. Castro. 1992b. The increase in risk of predation with begging activity in broods of magpies *Pica pica*. *Ibis* 134:180–187. [666, 667, 805]

Reeve, H. K. 1989. The evolution of conspecific acceptance thresholds. *Am. Nat.* 133:407–435. [448, 454]

Regnier, F. E. and M. Goodwin. 1977. On the chemical and environmental modulation of pheromone release from vertebrate scent marks. In *Chemical Signals in Vertebrates* (Ed. by D. Müller-Schwarze and M. M. Mozell), pp. 115–134. New York: Plenum Press. [304]

Reichenbach, A. and S. R. Robinson. 1995. Phylogenetic constraints on retinal organization and development. *Prog. Ret. Eye Res.* 15:139–172. [252, 563, 570]

Reichert, S. E. 1978. Games spiders play: Behavioral variability in territorial disputes. *Behav. Ecol. Sociobiol.* 3:135–162. [698, 699, 701, 724]

Reichert, S. E. 1984. Games spiders play. III. Cues underlying contest-associated changes in agonistic behaviour. *Anim. Behav.* 32:1–15. [699]

Reid, M. L. 1987. Costliness and reliability in the singing vigour of Ipswich sparrows. *Anim. Behav.* 35:1735–1743. [772]

Reidenberg, J. S. and J. T. Laitman. 1988. Existence of vocal folds in the larynx of Odontoceti (toothed whales). *Anat. Rec.* 221:884–891. [112]

Reiss, D. and B. McCowan. 1993. Spontaneous vocal mimicry and production by bottlenose dolphins (*Tursiops truncatus*): Evidence for vocal learning. *J. Comp. Psychol.* 107:301–312. [478]

Reitz, J. R., F. J. Milford and R. W. Christy. 1980. *Foundations of Electromagnetic Theory*. Reading, MA: Addison-Wesley. [205]

Rendall, D., P. S. Rodman and R. E. Emond. 1996. Vocal recognition of individuals and kin in free-ranging rhesus monkeys. *Anim. Behav.* 51:1007–1015. [489]

Reynolds, J. D. and I. M. Côté. 1995. Direct selection on mate choice: Female redlip blennies pay more for better mates. *Behav. Ecol.* 6:175–181. [559]

Reynolds, J. D. and M. R. Gross. 1992. Female mate preference enhances offspring growth and reproduction in a fish, *Poecilia reticulata*. *Proc. R. Soc. Lond. B Biol. Sci.* 250:57–62. [765]

Rice, W R. 1996. Sexually antagonistic male adaptation triggered by experimental arrest of female evolution. *Nature* 361: 232–234. [769]

Rice, W. R. and E. E. Hostert. 1993. Laboratory experiments on speciation: What have we learned in 40 years? *Evolution* 47:1637–1653. [790]

Richards, D. G. 1981. Alerting and message components in songs of rufous-sided towhees. *Behavior* 76:223–249. [729]

Richards, D. G. and R. H. Wiley. 1980. Reverberations and amplitude fluctuations in the propagation of sound in a forest: Implications for animal communication. *Am. Nat.* 115:381–399. [130, 132, 135, 139]

Ricklefs, R. E. 1990. *Ecology*. 3rd Edition. New York: W.H. Freeman. [385]

Ridley, M. 1978. Paternal care. *Anim. Behav.* 26:904–932. [777]

Rigden, J. S. 1985. *Physics and the Sound of Music*. 2nd Edition. New York: John Wiley and Sons. [40]

Robert, D., R. N. Miles and R. R. Hoy. 1996. Directional hearing by mechanical coupling in the parasitoid fly *Ormia ochracea*. *J. Comp. Physiol. A* 179:29–44. [565]

Roberts, J., A. Kacelnik and M. C. Hunter. 1979. A model of sound interference in relation to acoustic communication. *Anim. Behav.* 27:1271–1273. [139]

Roberts, O. F. T. 1923. The theoretical scattering of smoke in a turbulent atmosphere. *Proc. Royal Soc. Lond. A* 104:640–654. [299]

Robertson, B. C. 1996. Vocal mate recognition in a monogamous, flock-forming bird, the silvereye, *Zosterops lateralis*. *Anim. Behav.* 51:303–311. [484]

Robertson, J. G. M. 1986a. Male territoriality, fighting and assessment of fighting ability in the Australian frog *Uperoleria rugosa*. *Anim. Behav.* 34:763–772. [681, 773]

Robertson, J. G. M. 1986b. Female choice, male strategies and the role of vocalizations in the Australian frog *Uperoleia rugosa*. *Anim. Behav.* 34: 773–784. [773]

Robertson, J. G. M. 1990. Female choice increases fertilization success in the Australian frog *Uperoleia laevigata*. *Anim. Behav.* 39:639–645. [681, 773]

Robinson, S. R. 1980. Antipredator behaviour and predator recognition in Belding's ground squirrel. *Anim. Behav.* 28:840–852. [607]

Rodeick, R. W. 1988. The primate retina. In *Comparative Primate Biology* (Ed. by H. D. Steklis and J. Erwin), pp. 203–278. New York: Alan R. Liss. [248]

Rogers, P. H. and M. Cox. 1988. Underwater sound as a biological stimulus. In *Sensory Biology of Aquatic Animals* (Ed. by J. Atema, R. R. Fay, A. N. Popper and W. N. Tavolga), pp. 131–149. New York: Springer-Verlag. [127, 139]

Röhmer, H. 1992. Ecological constraints for the evolution of hearing and sound communication in insects. In *The Evolutionary Biology of Hearing* (Ed. by D. B. Webster, R. R. Fay and A. N. Popper), pp. 79–93. New York: Springer-Verlag. [157]

Rohwer, S. A. 1975. The social significance of avian winter plumage variability. *Evolution* 29:593–610. [662]
Rohwer, S. A. 1977. Status signalling in Harris sparrows: Some experiments in deception. *Behaviour* 61:107–129. [664]
Rohwer, S. 1982. The evolution of reliable and unreliable badges of fighting ability. *Am. Zool.* 22:531–546. [662, 734, 806]
Rohwer, S. A. 1985. Dyed birds achieve higher social status than controls in Harris' sparrows. *Anim. Behav.* 33:1325–1331. [664]
Rohwer, S. A. and P. W. Ewald. 1981. The cost of dominance and advantage of subordination in a badge signalling system. *Evolution* 35:441–454. [664, 665]
Rohwer, S. A. and F. C. Rohwer. 1978. Status signalling in Harris' sparrows: Experimental deceptions achieved. *Anim. Behav.* 26:1012–1022. [664]
Romeyer, A., R. H. Porter, F. Levy, R. Nowak, P. Orgeur and P. Poindron. 1993. Maternal labelling is not necessary for the establishment of discrimination between kids by recently parturient goats. *Anim. Behav.* 46:705–712. [801]
Rood, J. P. 1972. Ecological and behavioural comparisons of three genera of Atrintine cavies. *Anim. Behav. Mono.* 5:1–83. [591]
Roper, S. D. and J. Atema. 1987. *Olfaction and Taste IX.* New York: Ann. N. Y. Acad. Sci. [318]
Rosenberg, R. H. and M. Enquist. 1991. Contest behaviour in Weidemeyer's admiral butterfly *Limenitis weidemeyerii* (Nymphalidae): The effect of size and residency. *Anim. Behav.* 42:805–811. [719]
Rosevear, D.R. 1965. *The Bats of West Africa.* London: Trustees of the British Museum (Natural History). [873]
Rosie, A. M. 1973. *Information and Communication Theory.* London: Van Nostrand Reinhold. [418, 496]
Røskaft, E. and S. A. Rohwer. 1987. An experimental study of the function of the epaulets and the black body colour of male red-winged blackbirds. *Anim. Behav.* 35:1070–1077. [664]
Røskaft, E., T. Jarvi, M. Bakken, C. Bech and R. E. Reinertsen. 1986. The relationship between social status and resting metabolic rate in great tits (*Parus major*) and pied flycatchers (*Ficedula hypoleuca*). *Anim. Behav.* 34:838–842. [664]
Rossing, T. D. 1990. *The Science of Sound.* 2nd Edition. Reading, MA: Addison-Wesley Publishing. [40, 111]
Rossotti, H. 1983. *Colour: Why the World Isn't Grey.* Princeton, NJ: Princeton Univ. Press. [205]
Rothstein, S. I. 1990. A model system for coevolution: Avian brood parasitism. *Annu. Rev. Ecol. Syst.* 21:481–508. [800]
Roverud, R. C., V. Nitsche and G. Neuweiler. 1991. Discrimination of wingbeat motion by bats, correlated with echolocation sound pattern. *J. Comp. Physiol.* A 168:259–263. [875, 877]
Rowland, W. J. 1979. The use of color in intraspecific communication. In *The Behavioral Significance of Color* (Ed. by E. H. J. Burtt), pp. 381–421. New York: Garland STPM Press. [209]
Rowley, I. and G. Chapman. 1986. Cross-fostering, imprinting, and learning in two sympatric species of cockatoo. *Behaviour* 96:1–16. [478, 479]
Rowley, I. and G. Chapman. 1991. The breeding biology, food, social organization, demography and conservation of the Major Mitchell or pink cockatoo, *Cacatua leadbeateri*, on the margin of the western Australian wheatbelt. *Aust. J. Zool.* 39:211–261. [478, 479]
Rubenstein, D. I. 1981. Combat and communication in the Everglades pygmy sunfish. *Anim. Behav.* 29:249–258. [703]
Rubenstein, D. I. 1986. Ecology and sociality in horses and zebras. In *Ecological Aspects of Social Evolution: Birds and Mammals* (Ed. by D. I. Rubenstein and R. W. Wrangham), pp. 282–302. Princeton: Princeton Univ. Press. [756, 757]
Rubenstein, D. I. and M. A. Hack. 1992. Horse signals: The sounds and scents of fury. *Evol. Ecol.* 6:254–260. [469]
Rubenstein, D. I. and B.A. Hazlett. 1974. Examination of the agonistic behaviour of the crayfish, *Oronectes virilis*, by character analysis. *Behaviour* 50:193–216. [411, 703, 793]
Rubner, J. and K. Schulten. 1990. Development of feature detectors by self-organization. *Biol. Cybern.* 62:193–199. [531]
Rüppel, W. 1933. Physiologie und Akustik der Vögelstimme. *J. Ornith.* 74:433–542. [112]
Rupprecht, R. 1968. Das Trommeln von Plecopteren. *Z. Vergl. Physiol.* 59:38–71. [135]
Ryan, M. 1988. Energy, calling, and selection. *Am. Zool.* 28:885–898. [90, 112]
Ryan, M. 1990. Sexual selection, sensory systems, and sensory exploitation. *Oxford Surveys in Evolutionary Biology* 5:156–195. [673]
Ryan, M. and A. S. Rand. 1990. The sensory bias of sexual selection for complex calls in the tungara frog. *Evolution* 44:305–314. [527]
Ryan, M. A. and G. H. Walter. 1992. Sound communication in *Nezara viridula* (L.) (Heteroptera: Pentatomidae): Further evidence that signal transmission is substrate-borne. *Experientia* 48:1112–1115. [8]
Ryan, M. J. 1985a. *The Tungara Frog.* Chicago: Univ. of Chicago Press. [557]
Ryan, M. J. 1985b. Energetic efficiency of vocalization by the frog *Physalaemus pustulosus*. *J. Exp. Biol.* 116:47–52. [90, 112, 550]
Ryan, M. J. 1986a. Factors influencing the evolution of acoustic communication: Biological constraints. *Brain Behav. Evol.* 28:70–82. [550, 570]
Ryan, M. J. 1986b Neuroanatomy influences speciation rates among anurans. *Proc. Natl. Acad. Sci. USA* 83:1379–1382. [792]
Ryan, M. J. 1988. Energy, calling and selection. *Am. Nat.* 28:885–898. [570]
Ryan, M. J. 1990. Signals, species, and sexual selection. *Am. Sci.* 78:46–52. [791]
Ryan, M. J. and E. A. Brenowitz. 1985. The role of body size, phylogeny, and ambient noise in the evolution of bird song. *Am. Nat.* 126:87–100. [112, 126, 127. 128, 139, 557]
Ryan, M. R. and A. Keddy-Hector. 1992. Directional patterns of female mate choice and the role of sensory biases. *Am. Nat.* 139:S4-S35. [529, 532, 772]
Ryan, M. R. and A. S. Rand. 1993. Sexual selection and signal evolution: The ghost of biases past. *Philos. Trans. R. Soc. Lond. B* 340:187–195. [527, 532, 533, 535, 764]
Ryan, M. J. and A. S. Rand. 1995. Female responses to ancestral advertisement calls in tungara frogs. *Science* 269:390–392. [504]
Ryan, M. J. and W. Wilczynski. 1988. Coevolution of sender and receiver: Effect on local mate preference in cricket frogs. *Science* 240:1786–1788. [791]
Ryan, M. J., M. D. Tuttle and A. S. Rand. 1982. Bat predation and sexual advertisement in a neotropical frog. *Am. Nat.* 119:136–139. [383, 546]
Ryan, M. J., J. Fox, W. Wilczyski and A. S. Rand. 1990. Sexual selection for sensory exploitation in the frog *Physalaemus pustulosus*. *Nature* 343:66–67. [530, 673]

Sabelis, M. W. and M. C. M. de Jong. 1988. Should all plants recruit bodyguards? Conditions for a polymorphic ESS of synomone production in plants. *Oikos* 53:247–252. [846, 847]
Saibil, H. 1990. Cell and molecular biology of photoreceptors. *Sem. Neurosci.* 2:15–23. [240, 241]
Salmon, M. 1967. Waving display and sound production by Florida fiddler crabs (genus *Uca*). *Anim. Behav.* 15:449–459. [581]
Salmon, M. and S. P. Atsaides. 1968. Visual and acoustical signalling during courtship by fiddler crabs (genus: *Uca*). *Am. Zool.* 8:623–639. [515]
Salvador, A., J. P. Veiga, J. Martin, P. Lopez, M. Abelenda and M. Puerta. 1996. The cost of producing a sexual signal: Testosterone increases the susceptibility of male lizards to ectoparasitic infestation. *Behav. Ecol.* 7:145–150. [549]
Sayigh, L. S., P. L. Tyack, R. S. Wells and M. D. Scott. 1990. Signature whistles of free-ranging bottlenose dolphins,

Tursiops truncatus: Stability and mother-offspring comparisons. *Behav. Ecol. Sociobiol.* 26:247–260. [478, 787]

Scala, G., M. Corona and G. V. Pelagalli. 1990. The structure of the syrinx in the duck (Anas platyrhinchos). *Anatomia, Histologia, Embryologia* 19:135–142. [112]

Scheich, H., G. Langner, C. Tidemann, R. Coles and A. Guppy. 1986. Electroreception and electrolocation in platypus. *Nature* 319:401–402. [337, 350]

Schellart, N. A. M. and A. N. Popper. 1992. Functional aspects of the evolution of the auditory system of actinopterygian fish. In *The Evolutionary Biology of Hearing* (Ed. by D. B. Webster, R. R. Fay and A. N. Popper), pp. 295–322. New York: Springer-Verlag. [163]

Schenkel, R. 1956. Zur Deutung der Balzleistungen einiger Phasianiden und Tetraoniden. *Ornithol. Beobacht.* 53:182–201. [502, 503]

Scherer, K. R. 1985. Vocal affect signaling: A comparative approach. *Adv. Study Behav.* 15:189–244. [522]

Scherrer, J. A. and G. S. Wilkinson. 1993. Evening bat isolation calls provide evidence for heritable signatures. *Anim. Behav.* 46:847–860. [486, 493]

Schiller, P. H. 1995. The ON and OFF channels of the mammalian visual system. *Prog. Ret. Eye Res.* 15:173–195. [251]

Schiller, P. H., N. K. Logothetis and E. R. Charles. 1990. Functions of the colour-opponent and broad-band channels of the visual system. *Nature* 343:68–70. [261]

Schmidt, R. S. 1965. Larynx control and call production in frogs. *Copeia* 1965:143–147. [112]

Schmidt, S. 1988. Evidence for a spectral basis of texture perception in bat sonar. *Nature* 331:617–619. [872]

Schneider, D. 1964. Insect Antennae. *Annu. Rev. Ent.* 9:103–122. [308, 318]

Schneider, H. 1988. Peripheral and central mechanisms of vocalization. In *The Evolution of the Amphibian Auditory System* (Ed. by B. Fritzsch, M. Ryan, W. Wilczynski, T. E. Herrington and W. Walkowiak), pp. 537–558. New York: John Wiley. [112]

Schnell, G. D., D. J. Watt and M. Douglas. 1985. Statistical comparison of proximity matrices: Applications in animal behaviour. *Anim. Behav.* 33:239–253. [466]

Schnitzler, H.-U. and O. W. Henson. 1980. Performance of airborne animal sonar systems. I. Microchiroptera. In Animal Sonar Systems (Ed. by R. G. Busnel and J. F. Fish), pp. 109–181. New York: Plenum Press. [863, 870, 874, 875, 876]

Schnitzler, H. U., E. K. V. Kalko, I. Kaipf and A. D. Grinnell. 1994. Fishing and echolocation behavior of the greater bulldog bat, *Noctilio leporinus*, in the field. *Behav. Ecol. Sociobiol.* 35:327–345. [876]

Schulter, D. and T. Price. 1993. Honesty, perception and population divergence in sexually selected traits. *Proc. R. Soc. Lond. B Biol. Sci.* 253:117–122. [774]

Schumm, A., D. Krull and G. Neuweiler. 1991. Echolocation in the notch-eared bat, *Myotis emarginatus*. *Behav. Ecol. Sociobiol.* 28:255–261. [871]

Schutz, F. 1965. Sexuelle Prägung bei Anatiden. *Z. Tierpsych.* 22:50–103. [482]

Schwarz-Weig, E. and N. Sachser. 1996. Social behaviour, mating system and testes size in Cuis (*Galea musteloides*). *Z. Saugetierk.* 61:25–38. [810]

Searcy, W. A. 1979. Morphological correlates of dominance in captive male red-winged blackbirds. *Condor* 81:417–420. [664]

Searcy, W. A. and P. Marler. 1981. A test for responsiveness to song structure and programming in female sparrows. *Science* 213:926–928. [482]

Searcy, W. A. and P. Marler. 1984. Interspecific differences in the response of female birds to song repertoires. *Z. Tierpsych.* 66:128–142. [482]

Searcy, W. A. and P. Marler. 1987. Response of sparrows to songs of deaf and isolation-reared males: Further evidence of innate auditory templates. *Dev. Psychobiol.* 20:509–519. [482]

Searcy, W. A. and K. Yasukawa. 1996. Song and female choice. In *Ecology and Evolution of Acoustic Communication in Birds* (Ed. by D. E. Kroodsma and E. H. Miller), pp. 454–473. Ithaca, NY: Cornell Univ. Press. [772]

Searcy, W. A., J. Podos, S. Peters and S. Nowicki. 1995. Discrimination of song types and variants in song sparrows. *Anim. Behav.* 49:1212–1226. [737]

Sebeok, T. A. 1968. *Animal Communication: Techniques of Study and Results of Research.* Bloomington, IN: Indiana Univ. Press. [11, 615]

Sebeok, T. A. 1977. *How Animals Communicate.* Bloomington, IN: Indiana Univ. Press. [11, 112, 615]

Seddon, P. J. and Y. van Heezik. 1993. Parent-offspring recognition in the jackass penguin. *J. Field Ornithol.* 64:27–31. [801]

Seeley, T. 1989. The honey bee colony as a superorganism. *Am. Sci.* 77:546–553. [3, 355]

Seeley, T. D. 1985. *Ecology: A Study of Adaptation in Social Life.* Princeton, NJ: Princeton Univ. Press. [815, 850]

Sekulic, R. 1982. The function of howling in red howler monkeys (*Alouatta seniculus*). *Behaviour* 81:38–54. [597]

Seller, T. J. 1979. Unilateral nervous control of the syrinx in Java sparrows (Padda oryzivora). J. Comp. Physiol. 129:281–288. [103]

Sevenster, P. 1961. A causal analysis of a displacement activity (fanning) in *Gasterosteus aculeatus*. *Behaviour Suppl.* 9:1–170. [511]

Seyfarth, R. M. and D. L. Cheney. 1990. The assessment by vervet monkeys of their own and another species' alarm calls. *Anim. Behav.* 40:754–764. [835]

Shannon, C. E. and W. Weaver. 1949. *The Mathematical Theory of Communication*, (Illini Books Edition, 1963). Urbana, IL: Univ. of Illinois Press. [395, 418]

Shedd, D. H. 1982. Seasonal variation and function of mobbing and related antipredator behaviors of the American robin (Turdus migratorius). *Auk* 99:342–346. [839]

Sherman, P. W. 1977. Nepotism and the evolution of alarm calls. *Science* 197:1246–1253. [836]

Sherman, P. W. 1980. The limits of ground squirrel nepotism. In *Sociobiology: Beyond Nature/Nurture* (Ed. by G. W. Barlow and J. Silverberg), pp. 505–544. Boulder, CO: Westview. [836, 837, 840]

Sherman, P. W. 1985. Alarm calls of Belding's ground squirrels to aerial predators: Nepotism or self-preservation? *Behav. Ecol. Sociobiol.* 17:313–323. [835, 836, 837, 839, 850]

Sherman, P. W. 1991. Multiple mating and kin recognition by self-inspection. *Ethol. Sociobiol.* 12:377–386. [807]

Shettleworth, S. J. 1993. Varieties of learning and memory in animals. *J. Exp. Psych., Anim. Behav. Proc.* 19:5–14. [478]

Shichi, H. 1983. *Biochemistry of Vision.* New York: Academic Press. [240]

Shugart, G. W. 1990. A cue-isolation experiment to determine if Caspian tern parents learn their offspring's down color. *Ethology* 84:155–161. [801]

Sibly, R. M. 1983. Optimal group size is unstable. *Anim. Behav.* 31:947–948. [837]

Sieber, O. J. 1984. Vocal communication in raccoons (*Procyon lotor*). *Behaviour* 90:80–113. [520]

Signoret, J. P. 1970. Reproductive behaviour of pigs. *Reprod. Fert. Suppl.* 11:105–117. [285, 524]

Simmons, J. A. 1993. Evidence for perception of fine echo delay and phase by the fm bat, *Eptesicus fuscus*. *J. Comp. Physiol. A* 172:533–547. [860, 870]

Simmons, J. A. and A. D. Grinnell. 1988. The performance of echolocation: Acoustic images perceived by echolocation bats. In *Animal Sonar Processes and Performances* (Ed. by P. E. Nachtigal and P. W. B. Moore), pp. 353–385. New York: Plenum Press. [870, 871, 882]

Simmons, J. A., C. F. Moss and M. Ferragamo. 1990a. Convergence of temporal and spectral information into acoustic images of complex sonar targets perceived by the echolocating bat, *Eptesicus fuscus*. *J. Comp. Physiol. A* 166:449–470. [860, 870, 871]

Simmons, J. A., M. Ferragamo, C. F. Moss, S. B. Stevenson and R. A. Altes. 1990b. Discrimination of jittered sonar echoes by the echolocating bat, *Eptesicus fuscus*: The shape of target images in echolocation. *J. Comp. Physiol. A* 167:589–616. [860, 870, 871]

Simmons, L. W. 1987. Heritability of a male character chosen by females of the field cricket *Gryllus bimaculatus*. *Behav. Ecol. Sociobiol.* 21:129–133. [772]

Simmons, L. 1989. Kin recognition and its influence on mating preferences of the field cricket, *Grillus bimaculatus* (deGaer). *Anim. Behav.* 38:68–77. [790]

Simmons, R. E. 1988. Food and the deceptive acquisiton of mates by polygynous male harriers. *Behav. Ecol. Sociobiol.* 23:83–92. [769]

Simon, H. 1971. *The Splendor of Iridescence. Structural Colors in the Animal World.* New York: Dodd and Mead. [215, 238]

Sinclair, M. E. 1977. Agonistic behaviour of the stone crab, *Menippe mercenaria* (Say). *Anim. Behav.* 25:193–207. [703]

Sinclair, S. 1985. *How Animals See : Other Visions of Our World.* New York: Facts on File Publications. [257]

Sissom, D. E. F., D. A. Rice and G. Peters. 1991. How cats purr. *J. Zool., Lond.* 223:67–78. [112]

Sivak, J. G. and D. B. Allen. 1975. An evaluation of the "ramp" retina of the horse eye. *Vision Res.* 15:1353–1356. [271]

Slagsvold, T. 1984. The mobbing behavior of the hooded crow, *Corvus corone cornix*: Antipredator defence or self-advertisement? *Fauna Norv. Ser C Cinclus* 7:127–131. [835]

Slagsvold, T. 1985. Mobbing behavior of the hooded crow, *Corvus corone cornix*, in relation to age, sex, size, season, temperature and kind of enemy. *Fauna Norv. Ser. C. Cinclus* 8:9–17. [835]

Slagsvold, T., J. T. Lifjeld, G. Stenmark and T. Breiehagen. 1988. On the cost of searching for a mate in female pied flycatchers *Ficedula hypoleuca*. *Anim. Behav.* 36:433–442. [561]

Slater, P. J. B. 1981. Chaffinch song repertoires: Observations, experiments and a discussion of their significance. *Z. Tierpsych.* 56:1–24. [598, 734]

Slater, P. J. B. 1989. Bird song learning: Causes and consequences. *Ethol. Ecol. Evol.* 1:19–46. [731]

Slobodkin, L. B. and A. Rapoport. 1974. An optimal strategy of evolution. *Quart. Rev. Biol.* 49:181–200. [475]

Smith, C. A. and T. Takasaka. 1971. Auditory receptor organs of reptiles, birds, and mammals. In *Contributions to Sensory Physiology* (Ed. by D. Neff), pp. 129–178. New York: Plenum Press. [173]

Smith, D. G. 1972. The role of the epaulettes in the red-winged blackbird (*Agelaius phoeniceus*) social system. *Behaviour* 41:251–268. [597]

Smith, F. J. F. 1986. The evolution of chemical alarm signals in fishes. In *Chemical Signals in Vertebrates* (Ed. by D. Duvall, D. Müller-Schwarze and R. M. Silverstein), pp. 99–116. New York: Plenum Press. [605]

Smith, H. G. and R. Montgomerie. 1991. Nestling American robins compete with siblings by begging. *Behav. Ecol. S ociobiol.* 9:307–312. [804]

Smith, J. N. M. and P. Arcese. 1989. How fit are floaters? Consequences of alternative territorial behaviors in a nonmigratory sparrow. *Am. Nat.* 133:830–845. [719, 724]

Smith, R. J. F. 1986a. Evolution of alarm signals: Role of benefits of retaining group members or territorial neighbors. *Am. Nat.* 128:604–610. [836]

Smith, R. J. F. 1986b. The evolution of chemical alarm signals in fishes. In *Chemical Signals in Vertebrates* (Ed. by D. Duvall, D. Müller-Schwarze and R. Silverstein), pp. 99–115. New York: Plenum Press. [525]

Smith, R. J. F. 1992. Alarm signals in fishes. Reviews in *Fish Biology and Fisheries* 2:33–63. [846]

Smith, R. L. 1979. Paternity assurance and altered roles in the mating behviour of a giant water bug, *Abedus herberti* (Heteroptera: Belastomatidae). *Anim. Behav.* 27:716–725. [779]

Smith, S. M. 1978. The "underworld" in a territorial sparrow: Adaptive strategy for floaters. *Am. Nat.* 11:571–582. [714]

Smith, W. J. 1965. Message, meaning, and context in ethology. *Am. Nat.* 99:405–409. [818]

Smith, W. J. 1968. Message-meaning analyses. In *Animal Communication* (Ed. by T. A. Sebeok), pp. 44–60. Bloomington, IN: Univ. of Indiana Press. [472, 476, 490]

Smith, W. J. 1977. *The Behavior of Communicating: An Ethological Approach.* Cambridge, MA: Harvard Univ. Press. [11, 385, 472, 473, 476, 490, 496, 651, 818]

Smith, W. J. 1981. Referents of animal communication. *Anim. Behav.* 29:1273–1275. [818]

Smith, W. J. 1986. Signaling behavior: Contribution of different repertoires. In *Dolphin Cognition and Behavior: A Comparative Approach* (Ed. by R. J. Schusterman, J. A. Thomas and F. G. Wood), pp. 315–330. Hillsdale, NJ: Erlbaum. [472, 490, 496]

Smith, W. J. and A. M. Smith. 1996. Information about behaviour provided by Louisiana waterthrush, *Seiurus motacilla* (Parulinae), songs. *Anim. Behav.* 51:785–799. [496]

Smithe, F. B. 1975. *Naturalist's Color Guide.* New York: American Museum of Natural History. [210]

Smythe, N. 1970. On the existence of "pursuit invitation" signals in mammals. *Am. Nat.* 104:491–484. [842]

Smythe, R. H. 1975. *Vision in the Animal World.* New York: St. Martin's Press. [274]

Sommer, L. 1989. *Analytical Absorption Spectrophotometry in the Visible and Ultraviolet: The Principles.* Amsterdam: Elsevier. [200, 202]

Sordahl, T. A. 1990. The risks of avian mobbing and distraction behavior: An anecdotal review. *Wilson Bull.* 102:349–352. [839]

Sorjonen, J. 1986. Factors affecting the structure of song and the singing behaviour of some northern European passerine birds. *Behaviour* 98:286–304. [579, 585]

Sotavalta, O. 1963. The flight sounds of insects. In *Acoustic Behaviour of Animals* (Ed. by R.-G. Busnel), pp. 374–411. Amsterdam: Elsevier Publ. Co. [518, 586]

Speakman, J. R., M. E. Anderson and P. A. Racey. 1989. The energy cost of echolocation in pipistrelle bats (*Pipistrellus pipistrellus*). *J. Comp. Physiol.* A 165:679–685. [868]

Spector, D. A. 1992. Wood-warbler song systems. A review of paruline singing behaviors. In *Current Ornithology* Vol. 9 (Ed. by D. M. Power), pp. 199–238. New York: Plenum Press. [734]

Spurr, E. B. 1977. Behavior of the Adelie penguin chick. *Condor* 77:272–280. [801]

Stacey, N. E., A. L. Kyle and N. R. Liley. 1986. Fish reproductive pheromones. In *Chemical Signals in Vertebrates* (Ed. by D. Duvall, D. Müller-Schwarze and R. Silverstein), pp. 117–134. New York: Plenum Press. [524]

Staddon, J. E. R. 1975. A note on the evolutionary significance of "supernormal stimuli". *Am. Nat.* 109:541–545. [673]

Staddon, J. E. R. 1983. *Adaptive Behavior and Learning.* New York: Cambridge Univ. Press. [475]

Staddon, J. E. R., L. W. McGeorge, R. A. Bruce and F. F. Klein. 1978. A simple method for the rapid analysis of animal sounds. *Z. Tierpsych.* 48:306–330. [74]

Stamps, J. A., A. Clark, P. Arrowood and B. Kus. 1985. Parent-offspring conflict in budgerigars. *Behaviour* 94:1–40. [805]

Stamps, J., A. Clark, P. Arrowood and B. Kus. 1989. Begging behavior in budgerigars. *Ethology* 81:177–192. [666]

Stavenga, D. G. and R. C. Hardie. 1989. *Facets of Vision.* Berlin: Springer-Verlag. [278]

Stebbins, R. C. 1966. *A Field Guide to Western Reptiles and Amphibians.* Boston: Houghton Mifflin. [285]

Steger, R. and R. L. Caldwell. 1983. Intraspecific deception by bluffing: A defense strategy of newly molted stomatopods (Arthropoda: Crustacea). *Science* 221:558–560. [668, 670]

Stein, R. C. 1968. Modulation in bird sounds. *Auk* 85:229–243. [103, 112]

Steinberg, J. B. and R. C. Conant. 1974. An informational analysis of the inter-male behaviour of the grasshopper (*Chortophaga viridifasciata*). *Anim. Behav.* 22:617–627. [411, 703]

Stephen, R. O. and W. J. Bailey. 1982. Bioacoustics of the ear of the bushcricket Hemisaga (Saginae). *J. Acoust. Soc. Am.* 72:13–25. [173]

Stephen, R. O. and H. C. Bennet-Clark. 1982. The anatomical and mechanical basis of stimulation and frequency analysis in the locust ear. *J. Exp. Biol.* 99:279–314. [159, 173]

Stephens, D. W. 1989. Variance and the value of information. *Am. Nat.* 134:128–140. [385, 454]

Stephens, D. 1991. Change, regularlity, and value in the evolution of animal learning. *Behav. Ecol.* 2:77–89. [475, 478, 496]

Stephens, D. W. 1993. Learning and behavioral ecology: Incomplete information and environmental unpredictability. In *Insect Learning: Ecological and Evolutionary Perspectives* (Ed. by D. R. Papaj and A. C. Lewis), pp. 195–218. New York: Chapman and Hall. [475, 496]

Stephens, D. W. and J. R. Krebs. 1986. *Foraging Theory*. Princeton, NJ: Princeton Univ. Press. [411, 418]

Stiebler, I. B. and P. M. Narins. 1990. Temperature-dependence of auditory nerve response properties in the frog. *Hearing Res.* 46:63–82. [481]

Stoddard, P. K. 1996. Vocal recognition of neighbors by territorial passerines. In *Ecology and Evolution of Acoustic Communication in Birds* (Ed. by D. E. Kroodsma and E. H. Miller), pp. 356–376. Ithaca, NY: Comstock Publishing Associates. [482, 566, 725]

Stoddard, P. K. and M. D. Beecher. 1993. Parental recognition of offspring in the cliff swallow. *Auk* 100:795–799. [801]

Stoddard, P. K., M. D. Beecher, C. L. Horning and S. E. Campbell. 1991. Recognition of individual neighbors by song in the song sparrow, a species with song repertoires. *Behav. Ecol. Sociobiol.* 29:211–215. [566]

Stoddard, P. K., M. Beecher, P. Loesche and S. Campbell. 1992. Memory does not constrain individual recognition in a bird with song repertoires. *Behaviour* 122:274–287. [566]

Stoddart, D. M. 1979. External nares and olfactory perception. *Experientia* 35:1456–1457. [313]

Stoddart, D. M. 1980. *The Ecology of Vertebrate Olfaction*. London: Chapman and Hall. [288, 310, 311, 318, 738, 742]

Stokes, A. W. 1962. Agonistic behavior among blue tits at a feeding station. *Behaviour* 19:118–138. [703]

Stone, E. and C. H. Trost. 1991. Predators, risks and context for mobbing and alarm calls in black-billed magpies. *Anim. Behav.* 41:633–638. [839]

Storer, R. W. 1969. The behavior of the horned grebe in spring. *Condor* 71:180–205. [797]

Strickland, T. R., N. F. Britton and N. R. Franks. 1995. Complex trails and simple algorithms in ant foraging. *Proc. R. Soc. Lond. B Biol. Sci.* 260:53–58. [825]

Studd, M. V. and R. J. Robertson. 1985. Evidence for reliable badges of status in yellow warblers (*Dendroica petechia*). *Anim. Behav.* 33:1102–1113. [664]

Suga, N. 1988. Parallel-hierarchical processing of biosonar information in the mustached bat. In *Animal Sonar* (Ed. by P. E. Nachtigall and P. W. B. Moore), pp. 149–159. New York: Plenum Press. [870]

Sullivan, K. 1985. Selective alarm calling by downy woodpeckers in mixed-species flocks. *Auk* 102:184–187. [839]

Sullivan, M. S. 1990. Assessing female choice for mates when the male's characters vary during the sampling period. *Anim. Behav.* 40:780–782. [560]

Sullivan, M. S. 1994. Mate choice as an information gathering process under time constraint: Implications for behaviour and signal design. *Anim. Behav.* 47:141–151. [560, 570]

Sullivan, S. L., K. J. Ressler and L. B. Buck. 1995. Spatial patterning and information coding in the olfactory system. *Cur. Opin. Genet. Dev.* 5:516–523. [307]

Suthers, R. A. 1988. The production of echolocation signals by bats and birds. In *Animal Sonar* (Ed. by P. E. Nachtigall and P. W. B. Moore), pp. 23–45. New York: Plenum Press. [88, 112, 868]

Suthers, R. A. 1990. Contributions to birdsong from the left and right sides of the intact syrinx. *Nature* (Lond.) 347:473–477. [103, 112]

Suthers, R. A. 1994. Variable asymmetry and resonance in the avian vocal tract: A structural basis for individually distinct vocalizations. *J. Comp. Physiol.* A 175:457–466. [109, 112]

Suthers, R. A., D. J. Hartley and J. J. Wenstrup. 1988. The acoustic role of tracheal chambers and nasal cavities in the production of sonar pulses by the horseshoe bat, *Rhinolophus hildebrandti*. *J. Comp. Physiol., A* 162:799–813. [112]

Suthers, R. A., F. Goller and R. S. Hartley. 1994. Motor dynamics of song production by mimic thrushes. *J. Neurobiol.* 25:917–936. [103, 105, 106, 109, 112]

Sutton, O. G. 1953. *Micrometerology*. New York: McGraw-Hill. [299, 317]

Taigen, T. L. and K. D. Wells. 1985. Energetics of vocalization by an anuran amphibian (*Hyla versicolor*). *J. Comp. Physiol. B* 155:163–170. [90, 549, 550]

Tandy, M. and R. Kieth. 1972. *Bufo* of Africa. In *Evolution in the genus Bufo* (Ed. by W. F. Blair), pp. 119–170. Austin, TX: Univ. of Texas Press. [556]

Tautz, J. and H. Markl. 1978. Caterpillars detect flying wasps by hairs sensitive to airborne vibration. *Behav. Ecol. Sociobiol.* 4:101–110. [142]

Taylor, C. E., A. D. Pereda and J. A. Ferrari. 1987. On the correlation between mating success and offspring quality in *Drosophila melanogaster*. *Am. Nat.* 129:721–729. [769]

Tembrock, G. 1963. Acoustic behavior of mammals. In *Acoustic Behaviour of Animals* (Ed. by R.-G. Busnel), pp. 751–786. Amsterdam: Elsevier Publishing Company. [112]

Temeles, E. J. 1994. The role of neighbours in territorial systems: When are they 'dear enemies'? *Anim. Behav.* 47:339–350. [725, 729]

Tennekes, H. and Lumley. 1990. *A First Course in Turbulence*. Cambridge, MA: MIT Press. [317]

Théry, M. 1992. The evolution of leks through female choice: Differential clustering and space utilization in six sympatric manakins. *Behav. Ecol. Sociobiol.* 30:227–237. [551]

Théry, M. and S. L. Vehrencamp. 1995. Light patterns as cues for mate choice in the lekking white-throated manakin (*Corapipo gutturalis*). *Auk* 112:133–145. [765]

Thielke, G. 1970. Die sozialen Funktionen der Vogelstimmen. *Vogelwarte* 25:204–229. [478]

Thiessen, D. D. 1977. Thermoenergetics and the evolution of pheromone communication. In *Progress in Psychobiology and Physiological Psychology* , *Vol. 7*. (Ed. by J. M. Spragauet and A. N. Epstein), pp. 91–191. New York: Academic Press. [551]

Thomas, J. W., F. M. Robinson and R. G. Marburger. 1965. Social behavior in a white-tailed deer herd containing hypogonadal males. *J. Mammal.* 56:314–327. [509]

Thomas, T. C. 1986. *Games, Theory, and Applications*. New York: Halstead Press. [647]

Thomasson, S. I. 1977. Sound propagation above a layer with a large refraction index. *J. Acoust. Soc. Am.* 61:659–674. [139]

Thompson, C. W. and M. C. Moore. 1991. Throat colour reliably signals status in male tree lizards, *Urosaurus ornatus*. *Anim. Behav.* 42:745–754. [682]

Thompson, D. H. and J. T. Emlen. 1968. Parent-chick individual recognition in the Adelie penguin. *Antartic J.* 3:132. [801]

Thompson, E., A. Palacios and F. J. Varel. 1992. Ways of coloring: Comparative color vision as a case study for cognitive science. *Behav. Brain Sci.* 15:1–74. [236]

Thompson, I. 1991. Considering the evolution of vertebrate neural retina. In *Evolution of the Eye and Visual System* (Ed. by J. R. Cronly-Dillon and R. L. Gregory), pp. 136–151. Boca Raton, FL: CRC Press. [248, 252]

Thompson, T. J., H. E. Winn and P. J. Perkins. 1979. Mysticete sounds. In *Behavior of Marine Animals: Current Perspectives in Research* (Ed. by H. E. Winn and B. L. Olla), pp. 403–431. New York: Plenum Press. [112, 126]

Thomson, A. L. 1964. *A New Dictionary of Birds*. London: Thomas Nelson and Sons Ltd. [515]

Thornhill, R. and J. Alcock. 1983. *The Evolution of Insect Mating Systems*. Cambridge, MA: Harvard Univ. Press. [749, 758, 769, 790, 794]

Tilson, R. L. and P. M. Norton. 1981. Alarm duetting and pursuit deterrence in an African antelope. *Am. Nat.* 118:455–462. [842]

Timney, B. and K. Keil. 1992. Visual acuity in the horse. *Vision Res.* 32:2289–2293. [269]

Tinbergen, N. 1948. Social releasers and the experimental method required for their study. *Wilson Bull.* 60:6–52. [529]

Tinbergen, N. 1951. *The Study of Instinct*. Oxford: Oxford Univ. Press. [511, 512]

Tinbergen, N. 1952a. "Derived" activities; their causation, biological significance, origin, and emancipation during evolution. *Q. Rev. Biol.* 27:1–32. [499, 501, 508]

Tinbergen, N. 1952b. The curious behavior of the stickleback. *Sci. Am.* 187:22–26. [587]

Tinbergen, N. 1959. Comparative studies of the behaviour of gulls (Laridae). *Behaviour* 15:1–70. [512, 700]

Tinbergen, N. and A. C. Perdeck. 1950. On the stimulus situation releasing the begging response in the newly-hatched herring gull chick (*Larus a. argentatus* Pont). *Behaviour* 11:1–38. [529]

Tolstoy, I. and C. S. Clay. 1966. *Ocean Acoustics*. New York: McGraw-Hill. [128, 139]

Tomback, D. F., D. B. Thompson and M. C. Baker. 1983. Dialect discrimination by white-crowned sparrows: Reactions to near and distant dialects. *Auk* 100:452–460. [732]

Tomlinson, I. P. M. 1988. Diploid models of the handicap principle. *Heredity* 60:283–293. [761]

Towne, W. F. 1985. Acoustic and visual cues in the dances of four honeybee species. *Behav. Ecol. Sociobiol.* 16:185–187. [831, 832]

Towne, W. F. and J. L. Gould. 1988. The spatial precision of the honeybees' dance communication. *J. Insect Behav.* 1:129–155. [833]

Towne, W. F. and W. H. Kirchner. 1989. Hearing in honey bees: Detection of air-particle oscillations. *Science* 244:686–688. [173]

Trappe, M. and H.-U. Schnitzler. 1982. Doppler-shift compensation in insect-catching horseshoe bats. *Naturwiss.* 69:193–194. [874]

Trivers, R. L. 1971. The evolution of reciprocal altruism. *Quart. Rev. Biol.* 46:35–57. [634, 806, 820, 836, 840, 850]

Trivers, R. L. 1974. Parent-offspring conflict. *Am. Zool.* 14:249–264. [666, 800, 802, 804]

Turner, G. F. 1993. Teleost mating behaviour. In *Behaviour of Teleost Fishes*. (Ed. by T. J. Pitcher), pp. 307–332. London: Chapman & Hall. [798]

Turner, G. F. and F. A. Huntingford. 1986. A problem for game theory analysis: Assessment and intention in male mouthbrooder contests. *Anim. Behav.* 34:961–970. [697]

Urick, R. J. 1967. *Principles of Underwater Sound for Engineers*. New York: McGraw-Hill. [139]

Valberg, A. and B.B. Lee. 1991. *From Pigments to Perception: Advances in Understanding Visual Processes*. NATO ASI Series A, Life Sciences. [278]

Valone, T. J. 1991. Bayesian and prescient assessment: Foraging with pre-harvest information. *Anim. Behav.* 41:569–577. [411, 418]

Valone, T. J. 1992. Information for patch assessment: A field investigation with black-chinned hummingbirds. *Behav. Ecol.* 3:211–222. [411, 418]

Valone, T. J. and J. S. Brown. 1989. Measuring patch assessment of desert granivores. *Ecology* 70:1800–1810. [418]

van Bergeijk, W. A. 1967. The evolution of vertebrate hearing. In *Contributions to Sensory Physiology* (Ed. by W. D. Neff), pp. 1–49. New York: Plenum Press. [173]

van der Pers, J. N. C. and C. Löfstedt. 1986. Signal-response relationship in sex pheromone communication. In *Mechanisms in Insect Olfaction* (Ed. by T. L. Payne, M. C. Birch and C. E. J. Kennedy), pp. 235–241. Oxford: Clarendon Press. [308]

van Heel, A. C. S. and C. H. F. Velzel. 1968. *What Is Light?* New York: McGraw-Hill. [205]

Van Hooff, J. A. R. A. M. 1967. The facial displays of the catarrhine monkeys and apes. In *Primate Ecology* (Ed. by D. Morris), pp. 7–68. Chicago: Aldine. [504]

van Rhijn, J. G. 1980. Communication by agonistic displays: A discussion. *Behaviour* 74:284–293. [703]

van Rhijn, J. G. 1983. On the maintenance and origin of alternative strategies in the ruff *Philomachus pugnax*. *Ibis* 125:482–498. [740]

van Rhijn, J. and R. Vodegel. 1980. Being honest about one's intentions: An evolutionary stable strategy for animal conflicts. *J. Theor. Biol.* 85:623–641. [703]

Vandenbergh, J. G. 1983. *Pheromones and Reproductive Behavior in Mammals*. New York: Academic Press. [318]

Vater, M., A. S. Feng and M. Betz. 1985. An HRP-study of the frequency-place map of the horseshoe bat cochela: Morphological correlates of the sharp tuning to a narrow frequency band. *J. Comp. Physiol.* 157:671–686. [873]

Vega-Redondo, F. and O. Hasson. 1993. A game-theoretic model of predator-prey signaling. *J. Theor. Biol.* 162:309–319. [668, 669, 843]

Vehrencamp, S. L. and J. W. Bradbury. 1984. Mating systems and ecology. In *Behavioural Ecology: An Evolutionary Approach*. (Ed. by J. R. Krebs and N. B. Davies), pp. 251–278. Oxford: Blackwell. [745, 778]

Vehrencamp, S. L., F. G. Stiles and J. W. Bradbury. 1977. Observations on the foraging behavior and avian prey of the neotropical carnivorous bat, *Vampyrum spectrum*. *J. Mammal.* 58:469–478. [873]

Vehrencamp, S. L., J. W. Bradbury and R. M. Gibson. 1989. The energetic cost of display in male sage grouse. *Anim. Behav.* 38:885–896. [549, 552, 770]

Verrell, P. A. 1986. Wrestling in the red-spotted newt (*Notophythalamus viridescens*): Resource value and contestant asymmetry determine contest duration and outcome. *Anim. Behav.* 34:398–402. [689]

Verrell, P. 1988. The chemistry of sexual persuasion. *New Scientist* 118:40–43. [794]

Vicario, D. S. 1991. Contributions of syringeal muscles to respiration and vocalization in the zebra finch. *J. Neurobiol.* 22:36–73. [105, 112]

Vincent, A. 1994. The improbable seahorse. *Nat. Geogr.*, 186 (4): 126–141. [797]

Vincent, A. C. J. 1995. A role for daily greetings in maintaining seahorse pair bonds. *Anim. Behav.* 49:258–260. [799]

Vincent, A. C. J. and L. M. Sadler. 1995. Faithful pairbonds in wild seahorses *Hippocampus whitei*. *Anim. Behav.* 50:1557–1569. [797]

Vinnikov, Y. A. 1975. The evolution of olfaction and taste. In *Olfaction and Taste* (Ed. by D. A. Denton and J. P. Coghlan), pp. 175–187. New York: Academic Press. [305]

Visscher, P.K. and T. D. Seeley. 1982. Foraging strategy of honeybee colonies in a temperate deciduous forest. *Ecology* 63:1790–1801. [834]

Vogt, R. G. and L. M. Riddiford. 1986. Pheromone reception: A kinetic equilibrium. In *Mechanisms in Insect Olfaction* (Ed. by T. L. Payne, M. C. Birch and C. E. J. Kennedy), pp. 201–208. Oxford: Clarendon Press. [305]

von Bekesy, G. 1964. Olfactory analogue to directional hearing. *J. Appl. Physiol.* 19:369–373. [313]

von Campenhausen, C. 1986. Photoreceptors, lightness constancy and color vision. *Naturwiss.* 73:674–675. [261]

von der Emde, G. 1988. Greater horseshoe bats learn to discriminate simulated echoes of insects fluttering with different wingbeat rates. In *Animal Sonar* (Ed. by P. E. Nachtigall and P. W. B. Moore), pp. 495–499. New York: Plenum Press. [875]

von Frisch, K. 1967. *The Dance Language and Orientation of Bees.* Cambridge, MA: Harvard Univ. Press. [488, 828, 832, 850]

Waage, J. K. 1979. Reproductive character displacement in *Calypteryx. Evol.* 33:104–116. [753]

Waas, J. R. 1990a. An analysis of communication during aggressive interactions of little blue penguins (*Eudyptula minor*). In *Penguin Biology* (Ed. by L. S. Davis and J. T. Darby), pp. 366–371. San Diego: Academic Press. [697, 700, 704, 705]

Waas, J. R. 1990b. Intraspecific variation in social repertoires: Evidence from cave-and burrow-dwelling little blue penguins. *Behaviour* 115:63–99. [704, 705, 706]

Waas, J. R. 1991a. Do little blue penquins signal their intentions during aggressive interactions with strangers? *Anim. Behav.* 41:375–382. [658, 704]

Waas, J. R. 1991b. The risks and benefits of signalling aggressive motivation: A study of cave-dwelling little blue penquins. *Behav. Ecol. Sociobiol.* 29:139–146. [554, 658, 704]

Wagner, W. E. Jr. 1989. Fighting, assessment, and frequency of alteration in Blanchard's cricket frog. *Behav. Ecol. Sociobiol.* 25:429–436. [681, 724]

Wagner, W. E. Jr. 1992. Deceptive or honest signallingof fighting ability? A test of alternative hypotheses for the function of changes in call dominant frequency by male cricket frogs. *Anim. Behav.* 44:449–462. [681, 724]

Wagner, W. E. Jr. 1996. Convergent song preferences between female field crickets and acoustically orienting parasitoid flies. *Behav. Ecol.* 7:279–285. [546, 547]

Waldman, B. 1987. Mechanisms of kin recognition. *J. Theor. Biol.* 128:159–185. [786]

Waldman, B. and J. E. Rice. 1992. Kin recognition and incest avoidance in toads. *Am. Zool.* 32:18–30. [790]

Walker, T. J. 1957. Specificity in the response of female tree crickets (Orthoptera, Gryllidae, Oecanthinae) to calling songs of the males. *Annu. Ent. Soc. Am.* 50:626–636. [481]

Walls, G. L. 1967. *The Vertebrate Eye and its Adaptive Radiation.* New York: Hafner. [261, 267, 270, 272, 274, 278]

Walther, F. R. 1958. Zum Kampf- und Paarungsverhalten einiger Antilopen. *Z. Tierpsych.* 15:340–380. [591]

Walther, F. R., E. C. Mulgall and G. A. Grau. 1983. *Gazelles and their Relatives: A Study in Territorial Behavior.* Park Ridge, NJ: Noyes Publications. [552]

Ward, D. 1989. The morphology of the syrinx in some southern African barbets (Capitonidae). *Ostrich* 60:44–45. [112]

Ward, P. and A. Zahavi. 1973. The importance of certain assemblages of birds as information-centres for food finding. *Ibis* 115:517–534. [821]

Waring, H. 1963. *Color Change Mechanisms of Cold-blooded Vertebrates.* New York: Academic Press. [217, 238]

Warner, R. 1984. Deferred reproduction as a response to sexual selection in a coral reef fish: A test of the life historical consequences. *Evolution* 38:148–162. [643]

Warner, R. R. and S. G. Hoffman. 1980. Local population size as a determinant of mating system and sexual competition in two tropical marine fishes (*Thalassoma* spp.). *Evolution* 34:508–518. [643]

Warner, R. W. 1971a. The syrinx in the family Columbidae. *J. Zool.* 166:385–390. [112]

Warner, R. W. 1971b. The structural basis of the organ of voice in the genera *Anas* and *Aythya* (Aves). *J. Zool.* 164:197–207. [112]

Warner, R. W. 1972. The anatomy of the syrinx in passerine birds. *J. Zool.* 168:381–393. [112]

Waser, P. M. and M. S. Waser. 1977. Experimental studies of primate vocalization: Specializations for long distance propagation. *Z. Tierpsych.* 43:239–263. [126, 127, 128, 139]

Waterman, T. H. 1981. Polarization Sensitivity. In *Comparative Physiology and Evolution of Vision in Invertbrates: B: Invertebrate Visual Centers and Behavior I* (Ed. by H. Autrum), pp. 281–470. Berlin: Springer-Verlag. [182, 198, 205, 243, 264, 564]

Waterman, T. H. 1984. Natural polarized light and vision. In *Photoreception and Vision in Invertebrates* (Ed. by M. A. Ali), pp. New York: Plenum Press. [264]

Waters, D. A. and G. Jones. 1995. Echolocation call structure and intensity in five species of insectivorous bats. *J. Exp. Biol.* 198:475–489. [869]

Watkins, W. A. 1967. The harmonic interval: Fact or artifact in spectral analysis of pulse trains. In *Marine Acoustics* (Ed. by W. N. Tavolga), pp. 15–43. New York: Pergamon Press. [74]

Watson, P. J. and R. Thornhill. 1994. Fluctuating asymmetry and sexual selection. *TREE* 9:21–25. [768]

Weary, D. M. 1995. Calling by domestic piglets: Reliable signals of need? *Anim. Behav.* 50:1047–4055. [804]

Webster, D. B., R. R. Fay and A. N. Popper. 1992. *The Evolutionary Biology of Hearing.* New York: Springer Verlag. [173]

Wedekind, C. 1992. Detailed information about parasites revealed by sexual ornamentation. *Proc. R. Soc. Lond. B Biol. Sci.* 247:169–174. [664]

Wehner, R. 1976. Polarized light navigation by insects. *Sci. Am.* 235:106–115. [197, 264]

Wehner, R. 1989. Neurobiology of polarization vision. *Trends Neurosci.* 12:353–359. [260, 264]

Weiner, J. 1992. Physiological limits to sustainable energy budgets in birds and mammals: Ecological implications. *TREE* 7:384–388. [552]

Weisskopf, V. F. 1968. How light interacts with matter. *Sci. Am.* 219:60–71. [184, 205]

Weldon, P. J. 1983. The evolution of alarm pheromones. In *Chemical Signals in Vertebrates* (Ed. by D. Müller-Schwarze and R. M. Silverstein), pp. 309–312. New York: Plenum Press. [605]

Wellington, W. G. 1974. A special light to steer by. *Nat. Hist.* 83:46–53. [197]

Wells, K. D. 1980. Social behavior and communication of a dendrobatid frog *(Colostethus trinitatis). Herpetologica,* 36: 189–199. [779]

Wells, K. D. and T. L. Taigen. 1984. Reproductive behavior and aerobic capacity of male American toads (*Bufo americanus*): Is behavior constrained by physiology? *Herpetologica* 40:292–298. [90]

Wells, K. D. and T. L. Taigen. 1986. The effect of social interactions on calling energetics in the Gray Treefrog (*Hyla versicolor*). *Behav. Ecol. Sociobiol.* 19:8–18. [90, 549, 772]

Wells, K. D. and T. L. Taigen. 1989. Calling energetics of a neotropical treefrog, *Hyla microcephala. Behav. Ecol. Sociobiol.* 25:13–22. [90, 570]

Wells, K. D., T. L. Taigen, S. W. Rusch and C. C. Robb. 1995. Seasonal and nightly variation in glycogen reserves of calling gray treefrogs (*Hyla versicolor*). *Herpetologica* 51: 359–368. [772]

Wells, M. S. 1988. Effects of body size and resource value on fighting behaviour in a jumping spider. *Anim. Behav.* 36:321–326. [694]

Wenstrup, J. J. and R. A. Suthers. 1984. Echolocation of moving targets by the fish catching bat, *Noctilio leporinus. J. Comp. Physiol. A* 155:75–89. [876]

West, M. J. and A. P. King. 1988. Female visual displays affect the development of male song in the cowbird. *Nature* 334:244–246. [484]

Westby, G. W. M. 1975. Comparative studies of the aggressive behaviour of two gymnotid electric fish (*Gymnotus carapo* and *Hypopomus artedi*). *Anim. Behav.* 23:192–213. [350]

Westby, G. W. M. 1988. The ecology, discharge diversity, and predatory behaviour of gymnotiform electric fish in the coastal streams of French Guiana. *Behav. Ecol. Sociobiol.* 22:341–354.

West-Eberhard, M. J. 1983. Sexual selection, social competition, and speciation. *Q. Rev. Biol.* 58:155–183. [791]

Westfall, J. A. 1982. *Visual Cells in Evolution.* New York: Raven Press. [242]

Westneat, M. W., J. H. Long, W. Hoese and S. Nowicki. 1993. Kinematics of birdsong: Functional correlation of cranial movements and acoustic features in sparrows. *J. Exp. Biol.* 182:147–171. [109]

Wheeler, J. W. 1977. Properties of compounds used as chemical signals. In *Chemical Signals in Vertebrates* (Ed. by D. Mller-Schwarze and M. M. Mozell), pp. 61–70. New York: Plenum Press. [281]

Whitfield, D. P. 1986. Plumage variability and territoriality in breeding turnstone *Arenaria interpres*: Status signalling or individual recognition? *Anim. Behav.* 34:1471–1482. [596]

Whitfield, D. P. 1987. Plumage, variability, status signalling and individual recognition in avian flocks. *Trends Ecol. Evol.* 2:13–18. [664]

Whitson, M. 1977. Courtship behavior of the greater roadrunner. *Living Bird* 14: 215–257. [218, 793]

Wickler, W. 1964. Das Problem der stammesgeschichtlichen Sackgassen. *Naturwiss. Med.* 1:6–29. [489]

Wickman, P. O. and C. Wiklund. 1983. Territorial defense and its seasonal decline in the speckled wood butterfly. *Anim. Behav.* 31:1206–1216. [718]

Wilcox, R. S. 1988. Surface wave reception in invertebrates and vertebrates. In *Sensory Biology of Aquatic Animals* (Ed. by J. Atema, R. R. Fay, A. N. Popper and W. N. Tavolga), pp. 643–663. New York: Springer-Verlag. [156, 157]

Wiley, R. H. 1973. The strut display of the male sage grouse: A "fixed" action pattern. *Behaviour* 47:129–152. [462]

Wiley, R. H. 1983. The evolution of communication: Information and manipulation. In *Communication* (Ed. by T. R. Halliday and P. J. B. Slater), pp. 156–189. New York: W.H. Freeman. [4, 135, 380, 385]

Wiley, R. H. 1991. Associations of song properties with habitats for territorial oscine birds of eastern North America. *Am. Nat.* 138:973–993. [135, 139, 585]

Wiley, R. H. 1994. Errors, exaggeration, and deception in animal communication. In *Behavioral Mechanisms in Evolutionary Ecology* (Ed. by L. Real), pp. 157–189. Chicago: Univ. of Chicago Press. [4, 448, 454, 651, 675]

Wiley, R. H. and R. Godard. 1996. Ranging of conspecific songs by Kentucky warblers and its implications for interactions of territorial males. *Behaviour* 133:81–102. [728]

Wiley, R. H. and D. G. Richards. 1978. Physical constraints on acoustic communication in the atmosphere: Implications for the evolution of animal vocalizations. *Behav. Ecol. Sociobiol.* 3:69–94. [139]

Wiley, R. H. and D. G. Richards. 1982. Adaptations for acoustic communication in birds: Sound transmission and signal detection. In *Acoustic Communication in Birds* (Ed. by D. Kroodsma, E. H. Miller and H. Ouellet), pp. 131–181. New York: Academic Press. [117, 130, 131, 133, 135, 139]

Wilkinson, G. S. 1984. Reciprocal food sharing in the vampire bat. *Nature* 308:181–184. [635, 666, 880]

Wilkinson, G. S. and P. R. Reillo. 1994. Female choice response to artificial selection on an exaggerated male trait in a stalk-eyed fly. *Proc. R. Soc. Lond. B Biol. Sci.* 255:1–6. [768]

Williams, D. M. 1983. Mate choice in the mallard. In *Mate Choice* (Ed. by P. P. G. Bateson), pp. 297–309. Cambridge: Cambridge Univ. Press. [482]

Williams, H., J. Cynx and F. Nottebohm. 1989. Timbre control in zebra finch (*Taeniopygia guttata*) song syllables. *J. Comp. Physiol.* 103:366–380. [109]

Williams, H., L. A. Crane, T. K. Hale, M. A. Esposito and F. Nottebohm. 1992. Right-side dominance for song control in the zebra finch. *J. Neurobiol.* 23:1006–1020. [103]

Williams, J. 1970. *Optical Properties of the Sea.* Annapolis, MD: United States Naval Institute. [185, 205, 238]

Wilson, D. J. and H. Lefcort. 1993. The effect of predator diet on the alarm response of red-legged frog, *Rana aurora*, tadpoles. *Anim. Behav.* 46:1017–1019. [846]

Wilson, E. O. 1962. Chemical communication among workers of the fire ant *Solenopsis saevissima* (Fr. Smith). *Anim. Behav.* 10:134–164. [825]

Wilson, E. O. 1963. Pheromones. *Sci. Am.* 208:100–111. [282, 317]

Wilson, E. O. 1970. Chemical communication within animal species. In *Chemical Ecology* (Ed. by E. Sondheimer and J. B. Simeone), pp. 133–155. New York: Academic Press. [281]

Wilson, E. O. 1971. *The Insect Societies.* Cambridge, MA: Harvard Univ. Press. [286, 288, 317, 525, 561, 615, 666, 814, 840]

Wilson, E. O. 1975. *Sociobiology.* Cambridge, MA: Belknap Press/Harvard Univ. Press. [4, 11, 418]

Wilson, E. O. and W. H. Bossert. 1963. Chemical communication among animals. *Rec. Prog. Horm. Res.* 19:673–716. [281, 317, 573, 576, 615]

Winn, H. E. and L. K. Winn. 1978. The song of the humpback whale, *Megaptera novengliae*, in the West Indies. *Mar. Biol.* 47:97–114. [139]

Winston, M. L. 1987. *The Biology of the Honey Bee.* Cambridge, MA: Harvard Univ. Press. [833, 834, 850]

Witkin, S. R. and M. S. Ficken. 1979. Chickadee alarm calls: Does mate investment pay dividends? *Anim. Behav.* 27:1275–1276. [836]

Wittenberger, J. F. 1981. *Animal Social Behavior.* Boston: Duxbury Press. [850]

Wittenberger, J. F. and R. L. Tilson. 1980. The evolution of monogamy: Hypotheses and evidence. *Annu. Rev. Ecol. Syst.* 11:197–232. [797]

Wolken, J. J. 1975. *Photoprocesses, Photoreceptors and Evolution.* New York: Academic Press. [244, 278]

Wolken, J. J. 1995. *Light Detectors, Photoreceptors, and Imaging Systems in Nature.* New York: Oxford Univ. Press. [205, 241, 253, 267, 272, 278]

Woodland, D. J., Z. Jaafar and M.-L. Knight. 1980. The "pursuit deterrent" function of alarm signals. *Am. Nat.* 115:748–753. [842]

Woodward, B. D., J. Travis and S. Mitchell. 1988. The effects of the mating system on progeny performance in *Hyla crucifer* (Anura: Hylidae). *Evolution* 42:784–794. [772]

Woodward, P. M. 1964. *Probability and Information Theory, with Applications to Radar.* 2nd Edition. New York: Pergamon Press. [856, 865, 882]

Wrangham, R. W. 1977. Feeding behavior of chimpanzees in Gombe National Park, Tanzania. In *Primate Ecology* (Ed. by T. Clutton-Brock), pp. 504–538. London: Academic Press. [823]

Wright, T. F. 1996. Regional dialects in the contact call of a parrot. *Proc. R. Soc. Lond. B Biol. Sci.* 263:867–872. [566, 806]

Wright, W. D. 1991. The measurement of colour. In *The Perception of Colour* (Ed. by P. Gouras), pp. 10–21. Boca Raton, FL: CRC Press. [210]

Wysocki, C. J., J. L. Wellington and G. K. Beauchamp. 1980. Access of urinary nonvolatiles to the mammalian vomeronasal organ. *Science* 207:781–783. [318]

Wyszecki, G. and W. S. Stiles. 1982. *Color Science. Concepts and Methods, Quantitative Data and Formulae.* 2nd Edition. New York: John Wiley. [209]

Yasukawa, K. 1989. Costs and benefits of a vocal signal: The nest-associated 'Chit' of the female red-winged blackbird, *Agelaius phoeniceus*. *Anim. Behav.* 38:866–874. [546, 548]

Yasukawa, K. and E. I. Bick. 1983. Dominance hierarchies in dark-eyed juncos (*Junco hymenalis*): a test of a game-theory model. *Anim. Behav.* 31:439–456. [720]

Yber, M. C. and E. A. Herrera. 1994. Vigilance, group size and social status in capybaras. *Anim. Behav.* 48:1301–1307. [840]

Ydenberg, R. C., L. A. Giraldeau and J. B. Falls. 1988. Neighbours, strangers and the asymmetric war of attrition. *Anim. Behav.* 36:343–347. [729]

Yokoi, M., K. Mori and S. Nakanishi. 1995. Refinement of odor molecule tuning by dendrodentritic synaptic inhibition in the olfactory bulb. *Proc. Natl. Acad. Sci. USA* 92:3371–3375. [529]

Yokoyama, S. and R. Yokoyama. 1996. Adaptive evolution of photoreceptors and visual pigments in vertebrates. *Annu. Rev. Ecol. Syst.* 27:543–568. [563]

Young, J. R., J. W. Hupp, J. W. Bradbury and C. E. Braun. 1994. Phenotypic divergence of secondary sexual traits among sage grouse, *Centrocercus urophasianus*, populations. *Anim. Behav.* 47:1353–1362. [462, 770, 776]

Zahavi, A. 1975. Mate selection: A selection for a handicap. *J. Theor. Biol.* 53:205–214. [658, 761]

Zahavi, A. 1977a. Reliability in communication systems and the evolution of altruism. In *Evolutionary Ecology* (Ed. by B. Stonehouse and C. M. Perrins), pp. 253–259. [652]

Zahavi, A. 1977b. The cost of honesty (further remarks on the handicap principle). *J. Theor. Biol.* 67:603–605. [802]

Zahavi, A. 1980. Ritualization and the evolution of movement signals. *Behaviour* 72:77–81. [461, 652, 672, 676]

Zahavi, A. 1987. The theory of signal selection and some of its implications. In *International Symposium of Biological Evolution* (Ed. by V. P. Delfino), pp. 305–327. Bari, Italy: Adriatica Editrice. [355, 356, 652, 672, 676]

Zahavi, A. 1993. The fallacy of conventional signalling. *Philos. Trans. R. Soc. Lond. B Biol. Sci.* 340:227–230. [652, 672, 676]

Zahavi, A. and A. Zahavi. 1997. *The Handicap Principle.* Oxford: Oxford Univ. Press. [652, 676]

Zakon, H. H. 1986. The electroreceptive periphery. In *Electroreception* (Ed. by T. H. Bullock and W. Heiligenberg), pp. 103–156. New York: John Wiley. [350]

Zakon, H. 1988. The electroreceptors: Diversity in structure and function. In *Sensory Biology of Aquatic Animals* (Ed. by J. Atema, R. R. Fay, A. N. Popper and W. N. Tavolga), pp. 813–850. New York: Springer-Verlag. [350]

Zann, R. 1977. Pair-bond and bonding behaviour in three species of grassfinches of the genus *Poephila* (Gould). *Emu* 77:97–106. [798]

Zar, J. H. 1984. *Biostatistical Analysis.* Englewood Cliffs, NJ: Prentice-Hall, Inc. [464]

Zimmerman, E. and C. Lerch. 1993. The complex acoustic design of an advertisement call in male mouse lemurs (*Microcebus murinus*, Prosimii, Primates) and sources of its variation. *Ethology* 93:211–224. [465, 467]

Zippelius, H. 1972. Die Karawanenbildung bie Feld- und Hausspitzmaus. *Z. Tierpsych.* 30:305–320. [482]

Zuk, M. 1990. Reproductive strategies and disease susceptibility: An evolutionary viewpoint. *Parasitol. Today* 6:231–233. [664]

Zuk, M., R. Thornhill, J. D. Ligon, K. Johnson, S. Austad, S. H. Ligon, N. W. Thornhill and C. Costin. 1990. The role of male ornaments and courtship behavior in female mate choice of red jungle fowl. *Am. Nat.* 136:459–473. [766]

Zuker, C. S. 1996. The biology of vision in *Drosophila*. *Proc. Nat. Acad. Sci. USA* 93:571–576. [243]

Index

About the Book

Editor: Andrew D. Sinauer
Project Editor: Nan Sinauer
Production Manager: Christopher Small
Electronic Book Production: Ed Schell, Michele Ruschhaupt, and Maggie Haddad
Illustration Program: Precision Graphics, Inc., Nancy Haver, and Abigail Rorer
Copy Editor: Roberta Lewis
Indexer: Robie Grant
Book Design: Janice Holabird
Cover Design: Nina Dudley
Book Manufacturer: Transcontinental Printing Inc.